Using AutoCAD® 2004:
Basics

Using AutoCAD® 2004: Basics

RALPH GRABOWSKI

autodesk Press

THOMSON
DELMAR LEARNING

Australia • Canada • Mexico • Singapore • Spain • United Kingdom • United States

Using AutoCAD® 2004: Basics
Ralph Grabowski

Vice President, Technology and Trades SBU:
Alar Elken

Editorial Director:
Sandy Clark

Senior Acquisitions Editor:
James DeVoe

Senior Development Editor:
John Fisher

Marketing Director:
Cindy Eichelman

Channel Manager:
Fair Huntoon

Marketing Coordinator:
Sarena Douglass

Production Director:
Mary Ellen Black

Production Manager:
Andrew Crouth

Production Editor:
Thomas Stover

Art/Design Coordinator:
Mary Beth Vought

Technology Project Specialist:
Kevin Smith

Editorial Assistant:
Mary Ellen Martino

Typesetting:
Ralph Grabowski

© COPYRIGHT 2004 by Delmar Learning, a division of Thomson Learning, Inc. Thomson Learning™ is a trademark used herein under license.

Printed in the United States
1 2 3 4 5 XXX 06 05 04 03

For more information contact Delmar Learning
Executive Woods
5 Maxwell Drive, PO Box 8007,
Clifton Park, NY 12065-8007
Or find us on the World Wide Web at
www.delmarlearning.com

ALL RIGHTS RESERVED. No part of this work covered by the copyright hereon may be reproduced in any form or by any means—graphic, electronic, or mechanical, including photocopying, recording, taping, Web distribution, or information storage and retrieval systems—without the written permission of the publisher.

For permission to use material from the text or product, contact us by
Tel.   (800) 730-2214
Fax   (800) 730-2215
www.thomsonrights.com

Library of Congress Cataloging-in-Publication Data:
Card Number: [Number]

ISBN: 1-4018-5060-X

NOTICE TO THE READER

Publisher does not warrant or guarantee any of the products described herein or perform any independent analysis in connection with any of the product information contained herein. Publisher does not assume, and expressly disclaims, any obligation to obtain and include information other than that provided to it by the manufacturer.

The reader is expressly warned to consider and adopt all safety precautions that might be indicated by the activities herein and to avoid all potential hazards. By following the instructions contained herein, the reader willingly assumes all risks in connection with such instructions.

The publisher makes no representation or warranties of any kind, including but not limited to, the warranties of fitness for particular purpose or merchantability, nor are any such representations implied with respect to the material set forth herein, and the publisher takes no responsibility with respect to such material. The publisher shall not be liable for any special, consequential, or exemplary damages resulting, in whole or part, from the readers' use of, or reliance upon, this material.

BRIEF CONTENTS

INTRODUCTION xli

INTRODUCTION TO AUTOCAD
  1 QUICK TOUR OF AUTOCAD 2004 1
  2 UNDERSTANDING CAD CONCEPTS 33
  3 SETTING UP DRAWINGS 77

DRAWING AND EDITING
  4 DRAWING WITH BASIC OBJECTS 117
  5 DRAWING WITH PRECISION 171
  6 DRAWING WITH EFFICIENCY 229
  7 CHANGING OBJECT PROPERTIES 311
  8 CORRECTING MISTAKES 351
  9 DIRECT EDITING OF OBJECTS 369
  10 CONSTRUCTING OBJECTS 395
  11 ADDITIONAL EDITING OPTIONS 435

TEXT AND DIMENSIONS
  12 PLACING AND EDITING TEXT 481
  13 PLACING DIMENSIONS 535
  14 EDITING DIMENSIONS 583
  15 GEOMETRIC DIMENSIONING & TOLERANCES 637

LAYOUTS AND PLOTTING
  16 LAYOUTS 661
  17 PLOTTING DRAWINGS 699
  18 REPORTING ON DRAWINGS 751

APPENDICES

- A COMPUTERS AND WINDOWS 773
- B AUTOCAD COMMANDS, ALIASES, AND KEYBOARD SHORTCUTS 789
- C AUTOCAD SYSTEM VARIABLES 807
- D AUTOCAD TOOLBARS AND MENUS 841
- E AUTOCAD FONTS 855
- F AUTOCAD LINETYPES 859
- G AUTOCAD HATCH PATTERNS 861
- H AUTOCAD LINEWEIGHTS 867
- I DESIGNCENTER SYMBOLS 869
- J AUTOCAD TEMPLATE DRAWINGS 877

INDEX 889

CONTENTS

INTRODUCTION xli

1 QUICK TOUR OF AUTOCAD 2004

STARTING AUTOCAD 2
    TUTORIAL: STARTING AUTOCAD 2
    DRAWING IN AUTOCAD 4
        Command Prompt Area 4
        UCS Icon 4
        Status Bar 4
        Drawing Area 4
        Toolbars 4
        Command Prompt Area 4
        Layout Tabs 4
        Crosshair Cursor 4
        Menu Bar 4
        Scrollbars 4
        Mouse 5
    USING MENUS 5
    TUTORIAL: SELECTING MENU ITEMS 5
    TOOLBARS 7
        Using Toolbar Buttons 8
    TUTORIAL: DRAWING WITH TOOLBAR BUTTONS 8
        Flyouts 9
    TUTORIAL: ACCESSING FLYOUTS 9
        Manipulating the Toolbar 10

EXITING DRAWINGS 11
    DISCARD THE DRAWING, AND EXIT AUTOCAD 11
        Save the Drawing, and Exit AutoCAD 12
        Save the Drawing, and Remain in AutoCAD 12
        Remain in AutoCAD, and Start a New Drawing 12
        Close the Drawing and Remain in AutoCAD 12

TUTORIAL 13
GETTING STARTED 13
Tutorial: Setting Up New Drawings 14
SNAP AND GRID 15
TUTORIAL: SETTING UP DRAWING AIDS 15
CREATING THE DRAWING 18
Drawing the Baseplate 18
Drawing the Shaft 19
Drawing Cylinders 20
Copying Cylinders 21
Creating Holes 22
VIEWING THE DRAWING IN 3D 23
Rounding the Corners 24
Removing Hidden Lines 25
PLOTTING THE DRAWING 26
EXERCISES 29
CHAPTER REVIEW 30

2 UNDERSTANDING CAD CONCEPTS
BEGINNING A DRAWING SESSION 34
AUTOCAD USER INTERFACE 34
Title Bar 35
Toolbars 35
Layer 35
Color 36
Linetype 36
Lineweight 36
Plot Styles 37
Text Styles 37
Dimension Styles 38
Drawing Area 38
UCS Icon 38
Layout Tabs 39
Command Prompt Area 39
Status Bar 40
X, Y, Z Coordinates 40
Mode Indicators 40
Tray Settings 40
SCREEN POINTING 41
SHOWING POINTS BY WINDOW CORNERS 41

The icon indicates the command or option is new to AutoCAD 2004

ENTERING COMMANDS 41
KEYBOARD 41
Keyboard Shortcuts 42
Function Keys 43
Control Keys 43
Repeating Commands 44
MOUSE BUTTONS 44
Right Mouse Button 45
The Mouse Wheel 46
TOOLBARS 46
TABLET MENUS 46
MENUS 47
Using Menus 47
DIALOG BOXES 48
Hyphen Prefix 48
System Variables 49
Alternate Commands 50
Using Dialog Boxes 50
Buttons in Dialog Boxes 51
Check Boxes 51
Radio Buttons 52
List Boxes 52
Scroll Bars 52
Drop Lists 52
Text Edit Boxes 53
Image Tiles 53

THE DRAFT-EDIT-PLOT CYCLE 53
DRAFT 54
Templates and Blocks 55
EDIT 56
PLOT 57
ePlots 58

COORDINATE SYSTEMS 59
AUTOCAD COORDINATES 59
Cartesian Coordinates 61
Polar Coordinates 62
Cylindrical Coordinates 63
Spherical Coordinates 64
Relative Coordinates 65
User-defined Coordinate Systems 66
ENTERING COORDINATES 66
Screen Picks 66
Keyboard Coordinates 67
Snaps, Etc. 67
Point Filters 67
Snap Spacing 68
Polar Snap and Ortho Mode 68

        Direct Distance Entry 68
        Tracking 68
        Object Snaps 69
        Object Tracking 69
    DISPLAYING COORDINATES 70
EXERCISES 71
CHAPTER REVIEW 73

3 SETTING UP DRAWINGS

FINDING THE COMMANDS 78

NEW AND QNEW 79
    TUTORIAL: STARTING NEW DRAWINGS 79
    STARTUP DIALOG BOX 79
        Start From Scratch 80
            English 80
            Metric 80
        Use a Template 80
        QNew 81
    TUTORIAL: STARTING NEW DRAWINGS WITH TEMPLATES 81
        Wizards 82
    ADVANCED WIZARD 83
    TUTORIAL: USING THE ADVANCED WIZARD 83

OPEN 86
    TUTORIAL: OPENING DRAWINGS 86
        Files List 88
        File Spec 88
            Making the Dialog Box Bigger 90
        Standard Folders Sidebar 90
        Toolbar 91
    FINDING FILES 93
        Named 93
        Types 93
        Look In 93
        Date Modified 93

MVSETUP 94
    TUTORIAL: SETTING UP NEW DRAWINGS 94

CHANGING THE SCREEN COLOR 96
    TUTORIAL: CHANGING COLORS 97

UNITS 97
    TUTORIAL: SETTING UNITS 97
    SETTING SCALE FACTORS 99

SAVE AND SAVEAS 101
   TUTORIAL: SAVING DRAWINGS BY OTHER NAMES 101
   SAVING DRAWINGS: ADDITIONAL METHODS 102
      My Network Places 102
   TUTORIAL: SAVING DRAWINGS ON OTHER COMPUTERS 102
      FTP 103
   TUTORIAL: SENDING DRAWINGS BY FTP 103
      Files of Type 104
   TUTORIAL: SAVING DRAWINGS IN OTHER FORMATS 104
      Options 105

QSAVE 107
   TUTORIAL: SAVING DRAWINGS 107
   AUTOMATIC BACKUPS 107
   TUTORIAL: MAKING BACKUP COPIES 107
   MAKING BACKUPS: ADDITIONAL METHODS 108
      SaveFilePath 108
      Backup Files 109

QUIT 110
   TUTORIAL: QUITTING AUTOCAD 110

EXERCISES 111

CHAPTER REVIEW 114

4 DRAWING WITH BASIC OBJECTS

FINDING THE COMMANDS 118

LINE 119
   TUTORIAL: DRAWING LINE SEGMENTS 119
   DRAWING LINE SEGMENTS: ADDITIONAL METHODS 121
      Close 121
      Undo 121
      Relative Coordinates and Angles 122
      Continue 122

RECTANG 122
   TUTORIAL: DRAWING RECTANGLES 122
   ADDITIONAL METHODS FOR DRAWING RECTANGLES 123
      Dimensions 123
      Width 124
      Elevation 125
      Thickness 125
      Fillet 126
      Chamfer 126

POLYGON 127

TUTORIAL: DRAWING POLYGONS 128
ADDITIONAL METHODS FOR DRAWING POLYGONS 128
    Edge 128
    Inscribed in Circle 129
    Circumscribed About Circle 130

CIRCLE 131

TUTORIAL: DRAWING CIRCLES 131
ADDITIONAL METHODS FOR DRAWING CIRCLES 132
    Diameter 132
    2P 133
    3P 133
    Ttr (tan tan radius) 134
    Tan, Tan, Tan 134

ARC 135

TUTORIAL: DRAWING ARCS 136
ADDITIONAL METHODS FOR DRAWING ARCS 136
START, CENTER, END & CENTER, START, END 137
START, CENTER, ANGLE & START, END, ANGLE & CENTER, START, ANGLE ... 138
START, CENTER, LENGTH & CENTER, START, LENGTH 138
START, END, RADIUS 139
START, END, DIRECTION 140
CONTINUE 140

DONUT 141

TUTORIAL: DRAWING DONUTS 141

PLINE 142

TUTORIAL: DRAWING POLYLINES 142
ADDITIONAL METHODS FOR DRAWING POLYLINES 143
    Undo 143
    Close 143
    Arc 143
    Width 143
    Halfwidth 144
    Length 145

ELLIPSE 145

TUTORIAL: DRAWING ELLIPSES 146
ADDITIONAL METHODS FOR DRAWING ELLIPSES 147
    Center 147
    Rotation 148
    Arc 148

POINT 150
- TUTORIAL: DRAWING POINTS 150
- ADDITIONAL METHODS FOR DRAWING POINTS 150
  - Multiple 150
  - DdPModes 151

CONSTRUCTING OTHER OBJECTS 152
- EQUILATERAL TRIANGLES 152
- ISOSCELES TRIANGLES 153
- SCALENE TRIANGLES 156
- RIGHT TRIANGLES 157

EXERCISES 158

CHAPTER REVIEW 168

5 DRAWING WITH PRECISION

FINDING THE COMMANDS 172

GRID 173
- TUTORIAL: TOGGLING THE GRID 173
- CONTROLLING THE GRID DISPLAY: ADDITIONAL METHODS 174
  - DSettings 174
  - Grid 175
- LIMITS 175

SNAP 176
- TUTORIAL: TOGGLING SNAP MODE 176
- CONTROLLING THE SNAP: ADDITIONAL METHODS 176
  - DSettings 177
  - Snap 178
- TUTORIAL: DRAWING WITH GRID AND SNAP MODES 179

ORTHO 180
- TUTORIAL: TOGGLING ORTHO MODE 180
- TUTORIAL: DRAWING IN ORTHO MODE 181

POLAR 181
- TUTORIAL: TOGGLING POLAR MODE 181
- CONTROLLING POLAR MODE: ADDITIONAL METHODS 182
  - DSettings 182
  - PolarAng 184
  - PolarDist 184
  - PolarMode 184

OSNAP 186
- APERTURE 187
- RUNNING AND TEMPORARY OSNAPS 187
- SUMMARY OF OBJECT SNAP MODES 187
  - APParent Intersection 188
  - CENter 188

    ENDpoint 188
    EXTension 189
    INSertion 189
    INTersection 189
    MIDpoint 190
    NEArest 190
    NODe 190
    PARallel 190
    PERpendicular 191
    QUAdrant 191
    TANgent 191
  ADDITIONAL OBJECT SNAP MODES 192
    From 192
    NONe 192
    TT 192
    QUIck 193
    OFF 193
  TUTORIAL: TOGGLING AND SELECTING OSNAP MODE 193
  CONTROLLING ORTHO MODE: ADDITIONAL METHODS 194
    OSnap (DSettings) 194
    -OSnap 195
  TUTORIAL: USING OBJECT SNAP MODES 196

OTRACK 196
  TUTORIAL: TOGGLING OTRACK MODE 197

RAY AND XLINE 197
  TUTORIAL: DRAWING CONSTRUCTION LINES 197
  DRAWING CONSTRUCTION LINES: ADDITIONAL METHODS 198
  TUTORIAL: DRAWING WITH CONSTRUCTION LINES 200

REDRAW AND REGEN 205
  TUTORIAL: USING THE REDRAW COMMAND 205
  TRANSPARENT REDRAWS 205
  REGENAUTO 205

ZOOM 206
  TUTORIAL: USING THE ZOOM COMMAND 206
  ZOOM COMMAND: ADDITIONAL METHODS 207
    Zoom All 208
    Zoom Center 208
    Zoom Dynamic 208
    Zoom Extents 210
    Zoom Previous 210
    Zoom Scale 210
    Zoom Window 210

	Zoom Real Time	211
	Transparent Zoom	211
PAN		**212**
	TUTORIAL: PANNING THE DRAWING	212
	TRANSPARENT PANNING	212
	SCROLL BARS	212
	-PAN	213
CLEANSCREENON		**214**
	TUTORIAL: TOGGLING THE DRAWING AREA	214
DSVIEWER		**215**
	TUTORIAL: OPENING THE AERIAL VIEW WINDOW	215
	AERIAL VIEW WINDOW: ADDITIONAL METHODS	216
	View Menu	216
	Options Menu	216
VIEW		**216**
	PREDEFINED VIEWS	216
	TUTORIAL: STORING AND RECALLING NAMED VIEWS	217
	TUTORIAL: RENAMING AND DELETING NAMED VIEWS	218
	TUTORIAL: STARTING AUTOCAD WITH A NAMED VIEW	219
VIEWRES		**220**
	TUTORIAL: CHANGING THE VALUE OF VIEWRES	221
EXERCISES		**222**
CHAPTER REVIEW		**227**

6 DRAWING WITH EFFICIENCY

FINDING THE COMMANDS		**230**
ADCENTER		**232**
	TUTORIAL: DISPLAYING THE DESIGNCENTER	232
	NAVIGATING THE DESIGNCENTER: ADDITIONAL METHODS	233
	Docking DesignCenter	233
	Floating DesignCenter	233
	Minimizing DesignCenter	234
	UNDERSTANDING DESIGNCENTER'S USER INTERFACE	234
	Tabs	235
	Folders	235
	Open Drawings	235
	Custom Content	235
	History	236
	DC Center	236
	Toolbar	236
	Load	236

Back	237
Forward	237
Up	237
Search	237
Favorites	237
Home	237
Tree View Toggle	237
Preview	237
Description	237
Views	238
DC Online	238
Stop	238
Reload	238
Palette	238
Copying Content	238
Adding Content	239
Working with Blocks and Xrefs	239
Create Tool Palette	240
Tree View	240
Description	240
Title Bar	240

BLOCK AND INSERT 241

BLOCKS ARE HARD TO EDIT	241
TUTORIAL: CREATING BLOCKS	242
Notes on Designing Blocks	244
CREATING BLOCKS: ADDITIONAL METHODS	244
-Block Command	245
INSERTING BLOCKS	246
TUTORIAL: PLACING BLOCKS	246
INSERTING BLOCKS: ADDITIONAL METHODS	247
Scale	248
Rotation Angle	249
Explode	249
DESIGNCENTER AND TOOLPALETTES	250
Inserting Blocks with DesignCenter	250
Inserting Blocks with Toolpalettes	251
CREATING AND INSERTING BLOCKS: ADDITIONAL METHODS	252
-Insert Command	252
Inserting Exploded Blocks	252
Redefining Blocks	252
Replacing Blocks with Drawings	252
Base	253
WBlock	253
MInsert	254
BlockIcon	255
REDEFINING BLOCKS	256

TOOLPALETTES 256
TUTORIAL: DISPLAYING AND USING THE TOOLPALETTE 256
NAVIGATING THE TOOLPALETTES: ADDITIONAL METHODS 257
Right-click Commands: Tabs 257
View Options Dialog Box 258
Right-click Commands: Icons 259
Tool Properties Dialog Box 259
Right-click Commands: Palettes 260
Transparency Dialog Box 260
Customize Dialog Box 261
Right-click Commands: Titlebar 262
Sharing Tool Palettes 262

BHATCH AND HATCHEDIT 263
BASIC TUTORIAL: PLACING HATCH PATTERNS 265
PLACING HATCH PATTERNS: ADDITIONAL METHODS 266
Select a Type of Hatch 266
Select a Pattern Name 266
Specify the Angle and Scale 266
Specifying User Defined Patterns 267
Selecting Gradient Fills 268
Selecting Other Hatch Parameters 269
Selecting the Hatch Boundary 269
Defining the Boundary by Selecting Objects 270
Defining the Boundary by Selecting an Area 270
Removing Islands 271
Selecting Advanced Hatch Options 272
Island Detection Style 272
Object Type 273
Boundary Set 273
Island Detection Method 274
Pattern Alignment 274
CREATING HATCHES: COMMAND LINE 274
TUTORIAL: CHANGING HATCH PATTERNS 275

BOUNDARY 276
TUTORIAL: CREATING BOUNDARIES 277
CREATING BOUNDARIES: COMMAND LINE 278
Boundary Set 278
Island Detection 278
Object Type 278

SELECT 278
TUTORIAL: SELECTING OBJECTS BY LOCATION 279
Select Options 280
Pick (Select Single Object) 280
ALL (Select All Objects) 281

- W (Select Within a Window) 281
- C (Select with a Crossing Window) 282
- BOX (Select with a Box) 283
- F (Select with a Fence) 283
- AU (Select with the Automatic Option) 284
- WP (Select with a Windowed Polygon) 284
- CP (Select with a Crossing Polygon) 284
- SI (Select a Single Object) 285
- M (Select Through the Multiple Option) 285
- Special Selection Modes 285
  - P (Select the Previous Selection Set) 285
  - L (Select the Last Object) 285
  - G (Select the Group) 285
- Changing Selected Items 285
  - U (Undo the Selected Option) 285
  - R (Remove Objects from the Selection Set) 286
  - A (Add Objects to the Selection Set) 286
- Cancelling the Selection Process 286
- Selecting Objects During Commands 286
- DIRECT SELECTION AND GRIPS EDITING 286
  - Grip Color and Size 287
  - Tutorial: Using Grips for Editing 287

QSELECT 288
- TUTORIAL: SELECTING OBJECTS BY PROPERTIES 288
- CONTROLLING THE SELECTION: ADDITIONAL METHODS 289
  - Pickbox 289
  - DdSelect 290
    - Pickbox Size 291
    - Noun/Verb Selection 291
    - Use Shift to Add to Selection 291
    - Press and Drag 291
    - Implied Windowing 291
    - Object Grouping 292
    - Associative Hatch 292
  - Filter 292

GROUP 294
- TUTORIAL: CREATING GROUPS 295
- WORKING WITH GROUPS: ADDITIONAL METHODS 296
  - Highlighting Groups 296
  - Changing Groups 296

DRAWORDER 297
- TUTORIAL: SPECIFYING THE DRAWORDER 297
- CHANGING DRAWORDER: ADDITIONAL METHODS 298

EXERCISES 299

CHAPTER REVIEW 308

7 CHANGING OBJECT PROPERTIES

FINDING THE COMMANDS 312

COLOR 313
- TUTORIAL: CHANGING COLORS 313
- WORKING WITH AUTOCAD COLORS 314
  - RGB 315
  - HSL 316
  - Color Books 316
- CHANGING COLORS: ADDITIONAL METHODS 317

LWEIGHT 319
- TUTORIAL: CHANGING LINEWEIGHT 320
- AUTOCAD'S DEFAULT LINEWEIGHTS 321
- CHANGING LINEWEIGHTS: ADDITIONAL METHODS 322

LINETYPE 324
- TUTORIAL: CHANGING LINETYPES 325
- CHANGING LINETYPES: ADDITIONAL METHODS 327
  - Global and Object Linetype Scale 328

LAYER 330
- TUTORIAL: CREATING LAYERS 330
  - Turning Layers On or Off 332
  - Freezing and Thawing Layers 332
  - Locking and Unlocking Layers 332
  - Setting the Layer Color 332
  - Setting the Layer Linetype 332
  - Setting the Layer Lineweight 333
  - Setting the Layer Plot Style 333
  - Setting the Layer Print Toggle 334
  - Show Details 334
  - Layout Modes 334
  - Sort by Columns 335
  - Show List 335
  - Deleteing Layers 336
- CONTROLLING LAYERS: ADDITIONAL METHODS 338
  - Changing the Current Layer 338
  - Changing the Layers' Status 338

PROPERTIES 340
- TUTORIAL: CHANGING ALL PROPERTIES 340
- CHANGING PROPERTIES: ADDITIONAL METHODS 344

MATCHPROP 344
- TUTORIAL: MATCHING PROPERTIES 345
- MATCHING PROPERTIES: ADDITIONAL OPTION 345

EXERCISES 347
CHAPTER REVIEW 349

8 CORRECTING MISTAKES

FINDING THE COMMANDS 352

U AND UNDO 353
TUTORIAL: UNDOING COMMANDS 353
UNDOING COMMANDS: ADDITIONAL METHODS 354
Undo 354
Auto 354
Control 354
BEgin/End 354
Mark/Back 355
Standard Toolbar 355
Undo, Previous Options and Cancel Buttons 355
Undo Option 355
Previous Option 355
Cancel Buttons 356

REDO AND MREDO 356
TUTORIAL: REDOING UNDOES 356
REDOING UNDOES: ADDITIONAL METHODS 356
MRedo 356
All 356
Last 356
Standard Toolbar 357

RENAME 357
TUTORIAL: RENAMING TABLES 357
RENAMING TABLES: ADDITIONAL METHODS 358
-Rename 358

PURGE 358
TUTORIAL: PURGING UNUSED OBJECTS 359
PURGING UNUSED OBJECTS: ADDITIONAL METHODS 361
-Purge 361
WBlock 361

RECOVER AND AUDIT 362
TUTORIAL: RECOVERING CORRUPT DRAWINGS 362
RECOVERING CORRUPT DRAWINGS: ADDITIONAL METHODS 363
Audit 363

FILE MANAGER 364

QUIT AND CLOSE 364
TUTORIAL: DISCARDING DRAWINGS 364

EXERCISES 365

CHAPTER REVIEW 367

9 DIRECT EDITING OF OBJECTS

FINDING THE SHORTCUT MENUS 370

SHORTCUT MENUS 371
    TITLE BAR 371
    TOOLBARS 372
    BLANK TOOLBAR AREA 372
    DRAWING AREA 373
    SHIFT+BLANK DRAWING AREA 373
    ACTIVE COMMANDS 374
    HOT GRIPS 375
    LAYOUT TABS 375
    COMMAND LINE 375
    DRAG BAR 376
    STATUS BAR 376
    BLANK STATUS BAR 376
    TUTORIAL: CONTROLLING RIGHT-CLICK BEHAVIOR 376
        Turn on Time-Sensitive Right-Click 377
        Default Mode 377
        Edit Mode 377
        Command Mode 378
    SHORTCUT MENUS: ADDITIONAL METHOD 378

GRIPS 378
    GRIP COLOR AND SIZE 380
    TUTORIAL: DIRECT OBJECT SELECTION 380

GRIP EDITING OPTIONS 381
    TUTORIAL: EDITING WITH GRIPS 381
    GRIP EDITING MODES 382
    GRIP EDITING OPTIONS 383
        Base point 383
        Copy 383
        Undo 383
        Reference 383
        eXit 383
    GRIP EDITING OPTIONS 383
        Stretch 383
        Move 385
        Rotate Mode 385
        Scale Mode 385
        Mirror Mode 385
    OTHER GRIPS COMMANDS 386

DIRECT DISTANCE ENTRY 387
    TUTORIAL: DIRECT DISTANCE ENTRY 387

TRACKING 388
    TUTORIAL: TRACKING 388

EXERCISES 390

CHAPTER REVIEW 394

10 CONSTRUCTING OBJECTS

FINDING THE COMMANDS ... 396

COPY ... 397
- TUTORIAL: MAKING A COPY ... 397
- MAKING COPIES: ADDITIONAL METHODS ... 398
  - Multiple ... 398
  - Displacement ... 398

MIRROR ... 399
- TUTORIAL: MAKING MIRRORED COPIES ... 399
- MAKING MIRRORED COPIES: ADDITIONAL METHODS ... 400
  - MirrText ... 400

OFFSET ... 401
- TUTORIAL: MAKING OFFSET COPIES ... 402
- MAKING OFFSET COPIES: ADDITIONAL METHODS ... 402
  - Through ... 403
  - OffsetDist ... 403
  - OffsetGapType ... 403

DIVIDE AND MEASURE ... 404
- TUTORIAL: DIVIDING OBJECTS ... 404
- TUTORIAL: "MEASURING" OBJECTS ... 405
- DIVIDING AND MEASURING: ADDITIONAL METHODS ... 406
  - Block ... 406
  - DdPType ... 406

ARRAY ... 407
- TUTORIAL: LINEAR AND RECTANGULAR ARRAYS ... 408
- TTUTORIAL: POLAR AND SEMICIRCULAR ARRAYS ... 410
- CONSTRUCTING ARRAYS: ADDITIONAL METHODS ... 411
  - -Array ... 411
  - 3DArray ... 412

FILLET AND CHAMFER ... 413
- TUTORIAL: FILLETING OBJECTS ... 413
- DIFFERENT FILLET RESULTS FOR DIFFERENT OBJECTS ... 414
  - Lines ... 414
  - Polylines ... 415
  - Arcs and Circles ... 415
- CONSTRUCTING FILLETS: ADDITIONAL METHODS ... 417
  - Polyline ... 417
  - Radius ... 417
  - Trim ... 417
  - mUltiple ... 417

CHAMFER POINTS	418
TUTORIAL: CHAMFERING OBJECTS	418
CONSTRUCTING CHAMFERS: ADDITIONAL METHODS	419
Polyline	419
Angle	419
Trim	420
Method	420
mUltiple	420

REVCLOUD

TUTORIAL: REDLINING OBJECTS	420
REDLINING OBJECTS: ADDITIONAL METHODS	421
Arc length	422
Object	422
RMLin	423
TUTORIAL: INSERTING REDLINES FROM VOLO VIEW	423

EXERCISES 425

CHAPTER REVIEW 434

11 ADDITIONAL EDITING OPTIONS

FINDING THE COMMANDS 436

ERASE 437

TUTORIAL: ERASING OBJECTS	437
RECOVERING ERASED OBJECTS	437
Oops	437
U	437

BREAK 438

TUTORIAL: BREAKING OBJECTS	438
BREAKING OBJECTS: ADDITIONAL METHODS	438
First	439

TRIM, EXTEND, AND LENGTHEN 440

TUTORIAL: SHORTENING OBJECTS	440
Trimming Closed Objects	441
Trimming Polylines	441
TRIMMING OBJECTS: ADDITIONAL METHODS	442
Project	442
Edge	442
Undo	442
Extend (shift-select)	442
EXTEND	443
TUTORIAL: EXTENDING OBJECTS	443
Extending Polylines	444

EXTEND OPTIONS	444
LENGTHEN	444
TUTORIAL: LENGTHENING OBJECTS	444
LENGTHENING OBJECTS: ADDITIONAL METHODS	445
Delta	445
Percent	446
Total	446
Dynamic	446
STRETCH	**447**
STRETCHING — THE RULES	447
TUTORIAL: STRETCHING OBJECTS	447
MOVE	**448**
TUTORIAL: MOVING OBJECTS	448
ROTATE	**449**
TUTORIAL: ROTATING OBJECTS	449
ROTATING OBJECTS: ADDITIONAL METHODS	450
Reference	450
Dragging	451
AngBase	451
AngDir	451
SCALE	**451**
TUTORIAL: CHANGING SIZE	451
CHANGING SIZE: ADDITIONAL METHODS	452
Reference	452
CHANGE	**453**
TUTORIAL: CHANGING OBJECTS	454
CHANGING OBJECTS: ADDITIONAL METHODS	454
Lines	454
Circles	455
Blocks	455
Text	455
Attribute Definitions	455
EXPLODE AND XPLODE	**456**
TUTORIAL: EXPLODING COMPOUND OBJECTS	456
Arcs	456
Circle	456
Blocks	456
Polylines	457
3D Polylines	457
Dimensions	457
Leaders	457
Multiline Text	457

Multilines	457
Polyface Meshes	457
3D Solids	458
XPLODE	458
TUTORIAL: CONTROLLED EXPLOSIONS	458
All	458
Color	458
Layer	458
LType	458
Inherit from Parent Block	458
Explode	458

PEDIT 459

TUTORIAL: EDITING POLYLINES	459
Close/Open	459
Join	460
Width	460
Edit vertex	460
Next/Previous	461
Break	461
Insert	462
Move	462
Regen	462
Straighten	462
Tangent	462
Width	462
Exit	462
Fit	462
Spline	463
Decurve	463
Ltype gen	463
Undo	464
EDITING POLYLINES: ADDITIONAL METHODS	464
Select	464
Mutiple	465
Jointype	465
SplFrame	465
SplineType	466
SplineSegs	467

EXERCISES 468
CHAPTER REVIEW 478

12 PLACING AND EDITING TEXT

FINDING THE COMMANDS 482

TEXT 484

TUTORIAL: PLACING TEXT 485
PLACING TEXT: ADDITIONAL METHODS 486
   Justify 487
      Align 488
      Fit 488
   Style 489
   Height 489
   Rotation Angle 490
   Control Codes 491
      ASCII Characters 491
   -Text 492

MTEXT 492

TUTORIAL: PLACING PARAGRAPH TEXT 493
   Text Formatting Toolbar 494
      Style 494
      Font 494
      Text Height 494
      Bold 494
      Italic 494
      Underline 495
      Undo 495
      Redo 495
      Stack 495
      Color 496
   Text Entry Area 496
      First Line Indent 496
      Paragraph Indent 496
      Tabs 496
      Indents and Tabs 497
      Set Mtext Width 497
   MText Shortcut Menu 498
      Indents and Tabs 498
      Justification 498
      Find and Replace 498
      Select All 499
      Change Case 499
      AutoCAPS 499
      Remove Formatting 499
      Combine Paragraphs 499
      Stack 499
      Properties 499
      Symbol 500
      Import Text 500

MText Control Codes	501
PLACING PARAGRAPH TEXT: ADDITIONAL METHODS	503
Height	503
Justify	503
Linespacing	503
Rotation	503
Style	504
Width	504
-MText	504
MTextEd	504
QLEADER AND LEADER	**505**
TUTORIAL: PLACING LEADERS	506
PLACING LEADERS: ADDITIONAL METHODS	507
Settings	507
Annotation Type	508
MText Options	508
Annotation Reuse	509
Leader Line	509
Number of Points	510
Arrowhead	510
Angle Constraints	510
Multiline Text Attachment	511
Leader	511
DDEDIT	**512**
TUTORIAL: EDITING TEXT	512
Single-line Text Editor	512
Multiline and Leader Text Editor	513
Attribute Text Editor	513
EDITING TEXT: ADDITIONAL METHODS	513
Undo	514
Properties	514
STYLE	**515**
TUTORIAL: DEFINING TEXT STYLES	515
USING TEXT STYLES	518
Styles Toolbar	518
DEFINING STYLES: ADDITIONAL METHODS	518
-Style	519
TextFill	519
TextQlty	519
FontAlt	519
SPELL	**520**
TUTORIAL: CHECKING SPELLING	520

FIND	521
TUTORIAL: FINDING TEXT	521
WIPEOUT	522
TUTORIAL: WIPING OUT OBJECTS	522
WIPING OUT OBJECTS: ADDITIONAL METHODS	522
Frames	523
Polyline	523
Undo	523
Close	523
QTEXT	523
TUTORIAL: QUICKENING TEXT	524
SCALETEXT AND **JUSTIFYTEXT**	525
TUTORIAL: RESIZING TEXT	525
TUTORIAL: REJUSTIFYING TEXT	526
COMPILE	526
TUTORIAL: COMPILING POSTSCRIPT FONTS	526
EXERCISES	528
CHAPTER REVIEW	532

13 PLACING DIMENSIONS

FINDING THE COMMANDS	536
INTRODUCTION TO DIMENSIONS	537
THE PARTS OF DIMENSIONS	537
Dimension Line	538
Extension Lines	538
Arrowheads	539
Dimension Text	539
Tolerances	540
Limits	540
Alternate Units	540
DIMENSIONING OBJECTS	541
Lines and Polyline Segments	541
Arcs, Circles, and Polyarcs	541
Wedges and Cylinders	542
Cones and Pyramids	542
Holes	543
Vertices	543
Single Points	543
Chamfers and Tapers	544
DIMENSION MODES	545
Dimension Variables and Styles	545

- Dimension Standards 545
- Object Snaps 545

DIMHORIZONTAL 546
- TUTORIAL: PLACING DIMENSIONS HORIZONTALLY 546
- HORIZONTAL DIMENSIONS: ADDITIONAL METHODS 546
  - Extension 547
  - MText 547
  - Text 548
  - Angle 548

DIMVERTICAL AND DIMROTATED 549
- TUTORIAL: PLACING DIMENSIONS VERTICALLY 549
- TUTORIAL: PLACING DIMENSIONS AT AN ANGLE 549

DIMLINEAR 550
- TUTORIAL: PLACING LINEAR DIMENSIONS 550
- LINEAR DIMENSIONS: ADDITIONAL METHODS 550
  - Horizontal and Vertical 551
  - Rotated 551

DIMALIGNED 552
- TUTORIAL: PLACING ALIGNED DIMENSIONS 552

DIMBASELINE AND DIMCONTINUE 553
- TUTORIAL: PLACING ADDITIONAL DIMENSIONS FROM A BASELINE 553
- TUTORIAL: CONTINUING DIMENSIONS 554
- CONTINUED DIMENSIONS: ADDITIONAL METHODS 555
  - Undo 555
  - Select 555

QDIM 555
- CHANGING DIMENSION TYPE 555
- TUTORIAL: QUICK CONTINUOUS DIMENSIONS 556
- QUICK DIMENSIONS: ADDITIONAL METHODS 556
  - Continuous 556
  - Staggered 557
  - Baseline 557
  - Ordinate 557
  - Radius 558
  - Diameter 558
  - datumPoint 558
  - Edit 558
  - seTtings 559

DIMRADIUS AND DIMDIAMETER 559
- TUTORIAL: PLACING RADIAL DIMENSIONS 560

DIMCENTER	561
TUTORIAL: PLACING CENTER MARKS	561

DIMANGULAR 562

TUTORIAL: PLACING ANGULAR DIMENSIONS	563
TUTORIA: DIMENSIONING	564

EXERCISES 567

CHAPTER REVIEW 580

13 EDITING DIMENSIONS

FINDING THE COMMANDS 584

GRIPS EDITING 585

TUTORIAL: EDITING DIMENSIONED OBJECTS	585
GRIPS EDITING OPTIONS	586
EDITING LEADERS WITH GRIPS	586
Arrowhead	586
Vertex	587
Endpoint	587
Text	587
GRIPS EDITING LINEAR AND ALIGNED DIMENSIONS	588
Dimension Line	589
Extension Lines	589
Text	589
GRIPS EDITING ANGULAR DIMENSIONS	590
Dimension Arc	590
Text	591
Extension Lines	591
Vertex	592
EDITING DIAMETER AND RADIUS DIMENSIONS	592
On Circumference	593
Center Mark	593
Text	594

DIMEDIT AND DIMTEDIT 594

TUTORIAL: OBLIQUING EXTENSION LINES	595
TUTORIAL: REPOSITIONING DIMENSION TEXT	596
TUTORIAL: EDITING DIMENSION TEXT	597

AIDIM AND AI_DIM 597

AIDIMPREC	598
TUTORIAL: CHANGING DISPLAY PRECISION	598
AIDIMTEXTMOVE	599
TUTORIAL: MOVING DIMENSION TEXT	599
Ai_Dim_TextAbove	600
Ai_Dim_TextCenter	600
Ai_Dim_TextHome	600

DIMENSION VARIABLES AND STYLES 601
CONTROLLING DIMVARS 601
SOURCES OF DIMSTYLES 601
Differences Between International Standards 602
Comparing Dimvars 602
Dimstyles from Other Sources 603
SUMMARY OF DIMENSION VARIABLES 604
General 604
Dimension Line 604
Extension and Offset 604
Suppression and Position 605
Color and Lineweight
Extension Lines 605
Color and Lineweight 605
Extension and Offset 605
Suppression and Position 605
Arrowheads 605
Size and Fit 605
Names of Blocks 606
Center Marks 606
Text 606
Color and Format 606
Units, Scale, and Precision 607
Justification 607
Alternate Text 608
Limits 608
Tolerance Text 608
Primary Tolerance 608
Alternate Tolerance 608
ALPHABETICAL LISTING OF DIMSTYLES 609

DIMSTYLE 614
TUTORIAL: WORKING WITH DIMENSION STYLES 614
Setting the Current Dimstyle 615
Renaming and Deleting Dimstyles 615
CREATING NEW DIMSTYLES 616
TUTORIAL: CREATING NEW DIMENSION STYLES 616
MODIFYING THE DIMSTYLE 616
LINES AND ARROWS 617
Dimension Lines 617
Extension Lines 618
Arrowheads 618
TUTORIAL: CREATING CUSTOM ARROWHEADS 618
Center Marks for Circles 620
TEXT 621
Text Appearance 621
Text Placement 621
Text Alignment 622

- FIT ... 622
  - Fit Options ... 622
  - Text Placement ... 623
  - Scale for Dimension Features ... 623
  - Fine Tuning ... 623
- PRIMARY UNITS ... 623
  - Linear Dimensions ... 623
  - Measurement Scale ... 624
  - Zero Suppression ... 624
  - Angular Dimensions ... 624
- ALTERNATE UNITS ... 625
  - Placement ... 625
- TOLERANCES ... 626
  - Tolerance Format ... 626
- MODIFYING AND OVERRIDING DIMSTYLES ... 627
- TUTORIAL: GLOBAL RETROACTIVE DIMSTYLE CHANGES ... 627
- TUTORIAL: GLOBAL TEMPORARY DIMSTYLE CHANGES ... 628

STYLE TOOLBAR AND DIMOVERRIDE ... 629
- TUTORIAL: LOCAL DIMSTYLE CHANGES ... 629
- TUTORIAL: LOCAL DIMVAR OVERRIDES ... 630
  - Clearing Local Overrides ... 630

EXERCISES ... 631

CHAPTER REVIEW ... 633

15 GEOMETRIC DIMENSIONING & TOLERANCES

FINDING THE COMMANDS ... 638

DIMORDINATE ... 639
- TUTORIAL: PLACING ORDINATE DIMENSIONS ... 640
- PLACING ORDINATES: ADDITIONAL METHODS ... 641
  - Xdatum ... 641
  - Ydatum ... 641
  - Mtext ... 641
  - Text ... 641
  - Angle ... 642
- DIMCONTINUE ... 642
- UCS ORIGIN AND 3POINT ... 643
- TUTORIAL: CHANGING THE ORDINATE BASE POINT ... 643
- TUTORIAL: ROTATING THE ORDINATE ... 644
- QDIM ... 646
- TUTORIAL: ORDINATE DIMENSIONS WITH QDIM ... 646
- EDITING ORDINATE DIMENSIONS ... 647
  - Grips Editing ... 647
    - Leader Line ... 647
    - Text ... 648
    - Datum ... 648
  - Editing Commands ... 649

TOLERANCE 649
GEOMETRIC DIMENSIONING AND TOLERANCES (GD&T) 649
GEOMETRIC CHARACTERISTIC SYMBOLS 650
Profile Symbols 650
Profile of a Line 650
Profile of a Surface 650
Orientation Symbols 650
Angularity 650
Perpendicularity 651
Parallelism 651
Location Symbols 651
True Position 651
Symmetry 651
Concentricity 651
Runout Symbols 651
Circular Runout 651
Total Runout 651
Form Symbols 651
Straightness 651
Flatness 651
Circularity 651
Cylindricity 651
ADDITIONAL SYMBOLS 651
Tolerance Symbols 651
No Symbol 651
Tolerance Diameter 651
Projected Tolerance Zone 652
Material Characteristics Symbols 652
Maximum Material Condition (MMC) 652
Least Material Condition (LMC) 652
Regardless of Feature Size (RFS) 652
SYMBOL USAGE 652
TUTORIAL: PLACING TOLERANCE SYMBOLS 652
QLEADER 654
TUTORIAL: PLACING TOLERANCES WITH QLEADER 654

EXERCISES 656

CHAPTER REVIEW 659

16 LAYOUTS

FINDING THE COMMANDS 662
TILEMODE 663
ABOUT LAYOUTS 663
Exploring Layouts 664
TUTORIAL: CREATING LAYOUTS 666
TUTORIAL: WORKING IN PAPER MODE 668
TUTORIAL: HIDING THE VIEWPORT BORDER 670
TUTORIAL: SWITCHING FROM PAPER TO MODEL MODE 671

TUTORIAL: SCALING MODELS IN VIEWPORTS	672
TUTORIAL: SELECTIVELY DISPLAY DETAILS IN VIEWPORTS	673
TUTORIAL: INSERTING TITLE BLOCKS	674
TUTORIAL: USING TILEMODE	676

LAYOUTWIZARD 676

TUTORIAL: MANAGING LAYOUTS 676

LAYOUT 681

TUTORIAL: MANAGING LAYOUTS 681

VIEWPORTS 682

TUTORIAL: CREATING RECTANGULAR VIEWPORTS 682

-VPORTS AND VPCLIP 684

TUTORIAL: CREATING POLYGONAL VIEWPORTS 684
TUTORIAL: CREATING VIEWPORTS FROM OBJECTS 685

PSLTSCALE 687

TUTORIAL: MAKING LINETYPE SCALES UNIFORM 688

'SPACETRANS 689

EXERCISES 691

CHAPTER REVIEW 695

17 PLOTTING DRAWINGS

FINDING THE COMMANDS 700

PAGESETUP 701

TUTORIAL: PREPARING FOR PLOTTING 701

PREVIEW 702

TUTORIAL: PREVIEWING PLOTS 702
PREVIEWING THE PLOT: ADDITIONAL METHODS 703

PLOT 703

TUTORIAL: PLOTTING DRAWINGS 704
- Step 1: Select a Plotter/Printer 705
- Step 2: Select a Plot Style 705
- Step 3: Select the Layout to Plot 705
- Step 4: Select the Media Size 706
- Step 5: Select the Orientation 706
- Step 6: Select the Plot Area 706
- Step 7: Select Plot Scale 707

Table of Contents

- Step 9: Preview the Plot ... 707
  - Full Preview ... 708
  - Partial Preview ... 708
- Step 10: Save the Settings ... 708
- Step 11: Plot the Drawing ... 709
- PLOTTING DRAWINGS: ADDITIONAL METHODS ... 710
  - Plot to File ... 710
  - AutoSpool ... 710
  - Plot Offset ... 711
  - Shaded Viewport ... 712
  - Plot Options ... 714
    - Plot Object Lineweights and Plot with Plot Styles ... 714
    - Plot Paperspace Last ... 714
    - Hide Paperspace Objects ... 714

PLOTSTAMP ... 714
- TUTORIAL: STAMPING PLOTS ... 714
- STAMPING PLOTS: ADVANCED OPTIONS ... 716

PUBLISH ... 717
- TUTORIAL: PUBLISHING DRAWING SETS ... 717

PLOTTERMANAGER ... 719
- TUTORIAL: CREATING PLOTTER CONFIGURATIONS ... 720
- ABOUT PRINTERS ... 724
  - Local and Network Printer Connections ... 724
  - Differences Between Local and Network Printers ... 725
  - Nontraditional Printers ... 725
  - System Printers ... 725
- IMPORTING OLDER CONFIGURATION FILES ... 726
- TUTORIAL: EDITING PLOTTER CONFIGURATIONS ... 726
- THE PLOTTER CONFIGURATION EDITOR OPTIONS ... 728
  - Media ... 728
    - Duplex Printing ... 728
    - Media Destination ... 729
    - Physical Pen Configuration ... 729
  - Graphics ... 729
  - Custom Properties ... 731
  - Initialization Strings ... 731
  - User-Defined Paper Sizes and Calibration ... 731
  - Save As ... 732

STYLESMANAGER ... 732
- CONVERTCTB AND CONVERTPSTYLES ... 733
  - Setting Default Plot Styles ... 734
  - Assigning Plot Styles ... 735
    - Objects ... 735
    - Layers ... 736
    - Layouts ... 736
- TUTORIAL: CREATING PLOT STYLES ... 737

TUTORIAL: EDITING PLOT STYLES 740
BATCHPLT 742
TUTORIAL: BATCH PLOTTING 742
EXERCISES 746
CHAPTER REVIEW 748

18 REPORTING ON DRAWINGS
FINDING THE COMMANDS 752

ID 753
TUTORIAL: REPORTING POINT COORDINATES 753
REPORTING POINT COORDINATES: ADDITIONAL METHODS 753
   Blipmode 754
   Marking Points 754

DIST 754
TUTORIAL: REPORTING DISTANCES AND ANGLES 754

AREA 756
TUTORIAL: REPORTING AREAS AND PERIMETERS 756
REPORTING AREA: ADDITIONAL METHODS 757
   Add 757
   Subtract 757
   Object 757

MASSPROP 758
TUTORIAL: REPORTING INFORMATION ABOUT REGIONS 758

LIST AND DBLIST 759
TUTORIAL: REPORTING INFORMATION ABOUT OBJECTS 759
TUTORIAL: REPORTING INFORMATION ABOUT ALL OBJECTS 760
PROPERTIES 760

DWGPROPS 761
TUTORIAL: REPORTING INFORMATION ABOUT OBJECTS 761

'TIME 763
TUTORIAL: REPORTING ON TIME 763
REPORTING ON TIME: ADDITIONAL METHODS 764
   Display 764
   ON, OFF, and Reset 764

STATUS 764

'SETVAR 766
TUTORIAL: LISTING SYSTEM VARIABLES 766

EXERCISES	768
CHAPTER REVIEW	772

Appendices

A COMPUTERS AND WINDOWS

HARDWARE OF AN AUTOCAD SYSTEM	773
COMPONENTS OF COMPUTERS	774
System Board	774
Central Processing Unit	774
Memory	775
Disk Drives	775
DISK CARE	776
Write-Protecting Data	776
PERIPHERAL HARDWARE	776
PLOTTERS	776
Inkjet Plotters	776
Laser Printers	777
Pen Plotters	777
MONITORS	777
INPUT DEVICES	777
Mouse	778
Digitizers	778
Scanner	778
DRIVES AND FILES	779
FILE NAMES AND EXTENSIONS	780
Wild-Card Characters	780
DISPLAYING FILES	780
Windows Explorer	780
File Manager	780
File-Related Dialog Boxes	781
DOS Session	781
WINDOWS EXPLORER	781
Viewing Files	781
Manipulating Files	782
Copying And Moving Files	783
AUTOCAD FILE DIALOG BOXES	784
AUTOCAD FILE EXTENSIONS	785
Drawing Files	785
AutoCAD Program Files	785

Support Files 785
Plotting Support Files 786
Import-Export Files 786
LISP and ObjectARX Programming Files 786
Miscellaneous Files 787
REMOVING TEMPORARY FILES 787

B AUTOCAD COMMANDS, ALIASES, AND KEYBOARD SHORTCUTS

AUTOCAD 2004 COMMANDS 789
COMMAND ALIASES 799
KEYBOARD SHORTCUTS 802
CONTROL KEYS 802
COMMAND LINE KEYSTROKES 803
FUNCTION KEYS 803
ALTERNATE KEYS 804
MOUSE AND DIGITIZER BUTTONS 804
MTEXT EDITOR SHORTCUT KEYS 805
MTEXT CONTROL CODES 806

C AUTOCAD SYSTEM VARIABLES

SYSTEM VARIABLES 807

D AUTOCAD TOOLBARS AND MENUS

AUTOCAD 2004 TOOLBARS 841
AUTOCAD 2004 MENUS 846

E AUTOCAD FONTS

FONTS 855

F AUTOCAD LINETYPES

LINETYPES 859

G AUTOCAD HATCH PATTERNS

ANSI PATTERNS 861
ISO PATTERNS 862
OTHER PATTERNS 863
OTHER PATTERNS 864
GRADIENT PATTERNS 865

H AUTOCAD LINEWEIGHTS
LINEWEIGHTS 867

I DESIGNCENTER SYMBOLS
AEC 870
MECHANICAL 872
ELECTRONICS 875

J AUTOCAD TEMPLATE DRAWINGS
GENERIC TEMPLATES 878
AMERICAN NATIONAL STANDARDS INSTITUTE TEMPLATES 879
CHINESE STANDARD TEMPLATES 881
GERMAN INDUSTRIAL STANDARDS TEMPLATES 883
INTERNATIONAL ORGANIZATION OF STANDARDS TEMPLATES ... 885
JAPANESE INDUSTRIAL STANDARDS TEMPLATES 887

INDEX 889

xl

INTRODUCTION

With more than 3,600,000 users around the world, AutoCAD offers engineers, architects, drafters, interior designers, and many others, a fast, accurate, extremely versatile drawing tool.

Now in its 12th edition, *Using AutoCAD 2004* makes using AutoCAD a snap by presenting easy-to-master, step-by-step tutorials covering all of AutoCAD's commands. *Using AutoCAD 2004* consists of two volumes: *Basics* gets you started with AutoCAD, teaching commands for basic 2D drafting; *Advanced* explores sophisticated 2D features, customizing, and 3D design.

BENEFITS OF COMPUTER-AIDED DRAFTING

CAD is more efficient and versatile than traditional drafting techniques. Some of the advantages are:

Accuracy. Computer-generated drawings are created and plotted to an accuracy of up to fourteen decimal places. The numerical entry of critical dimensions and tolerances is more reliable than traditional manual scaling.

Speed. The ability of the CAD operator to copy and array objects, and edit the work on the screen speeds up the drawing process. When the operator customizes the system for specific tasks, the speed of work increases even more markedly.

Neatness and Legibility. The ability of the plotter to produce exact and legible drawings is an obvious advantage over the traditional methods of hand-drawn work. CAD drawings are uniform: they contain lines of constant thickness, evenly-spaced hatch patterns, and print-quality lettering. CAD drawings are clean: they are free of smudges and other editing marks.

Consistency. Because the system is constant in its methodology, the problem of individual style is eliminated. A company can have a number of drafters working on the same project and produce a consistent set of graphics.

CAD APPLICATIONS

CAD is being applied to many industries. The flexibility of this software has a major impact on how tasks are performed. CAD applications include architecture, engineering, interior design, manufacturing, mapping, piping design, and entertainment. A brief discussion of each of these applications follows.

Architecture

Architects have found that CAD allows them to formulate designs in a shorter period of time than was possible by traditional techniques. The work is neater and more uniform. The designer can use 3D modeling capabilities to help the client better visualize the finished design. Changes can be made and resubmitted in a very short time.

Architects can assemble construction drawings using stored details, as illustrated at right.

The database of AutoCAD allow architects to extract information from the drawings, perform cost estimates, and prepare bills of materials.

Many third-party applications help architects customize AutoCAD for their discipline. Software is available for doing quick 3D conceptual designs, providing building details, generating automatic stairs, and more.

Revit

Autodesk promotes its Revit product as the future of architectural design software. You create the building in 3D, and then let Revit generate 2D section views. Worksets let team members work together on the same model, fully coordinating their work. Detailed graphic control and view-specific graphics make drawings show more or fewer details.

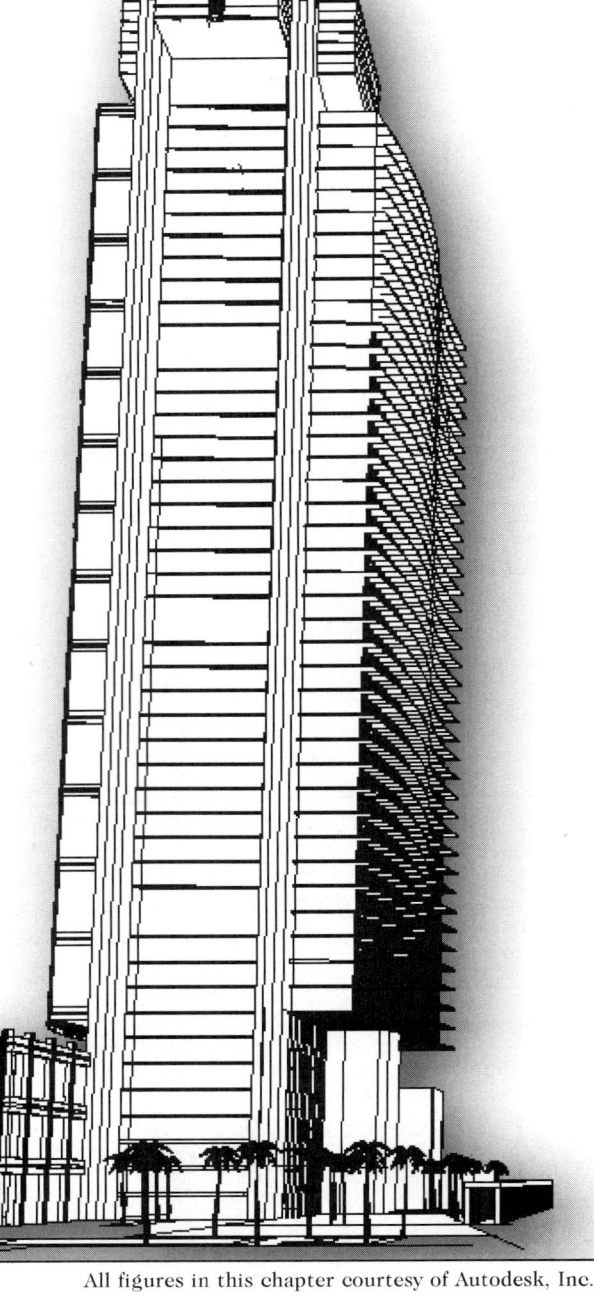

All figures in this chapter courtesy of Autodesk, Inc.

The software can be configured to your office standards for graphic style, data and layer export, and additional drafting and CAD standards. The software package includes thousands of building components, and can output its data to ODBC-compliant databases.

www.autodesk.com/revit

Architectural Desktop

Autodesk has an add-on called Architectural Desktop, which customizes AutoCAD for architectural design. The Content Browser provides access to catalogs, tool palettes, and design content in the form of blocks and multiview blocks. With direct manipulation you can modify designs directly, using either grip manipulation or in-place object editing.

Materials provide graphic and nongraphic detail to drawings for design, documentation, and visualization. You can attach materials to building model objects and their respective components for greater visual detail.

Scheduling allows you to track any object in drawings. Because schedules are linked to your data, they automatically update as the design changes. Schedule tables can be created with page breaks, objects tagged through xrefs, and schedules of external drawings created.

Third-party developers have add-on software for roadway design, digital terrain modeling, and more.

www.autodesk.com/architecturaldesktop

Architectural Studio

Another Autodesk product, called Architectural Studio, is useful for creating conceptual architectural designs. Sketching tools, such as pencils, pens, markers, and erasers, allow you to create free-form sketches. Hard-line drawing tools are CAD-like, allowing you to create straight lines, Layers can be dragged like transparent paper over any image, drawing, or 3D model and sketched directly.

www.autodesk.com/archstudio

Engineering

Engineers use AutoCAD for many different kinds of design; in addition, they use programs that interact with AutoCAD to perform calculations that would take more time using traditional techniques. Among the many engineering disciplines benefitting from AutoCAD are civil, structural, and mechanical.

This structural engineering drawing illustrates AutoCAD being used for specifying steel beams. The circled number refers to rows of columns.

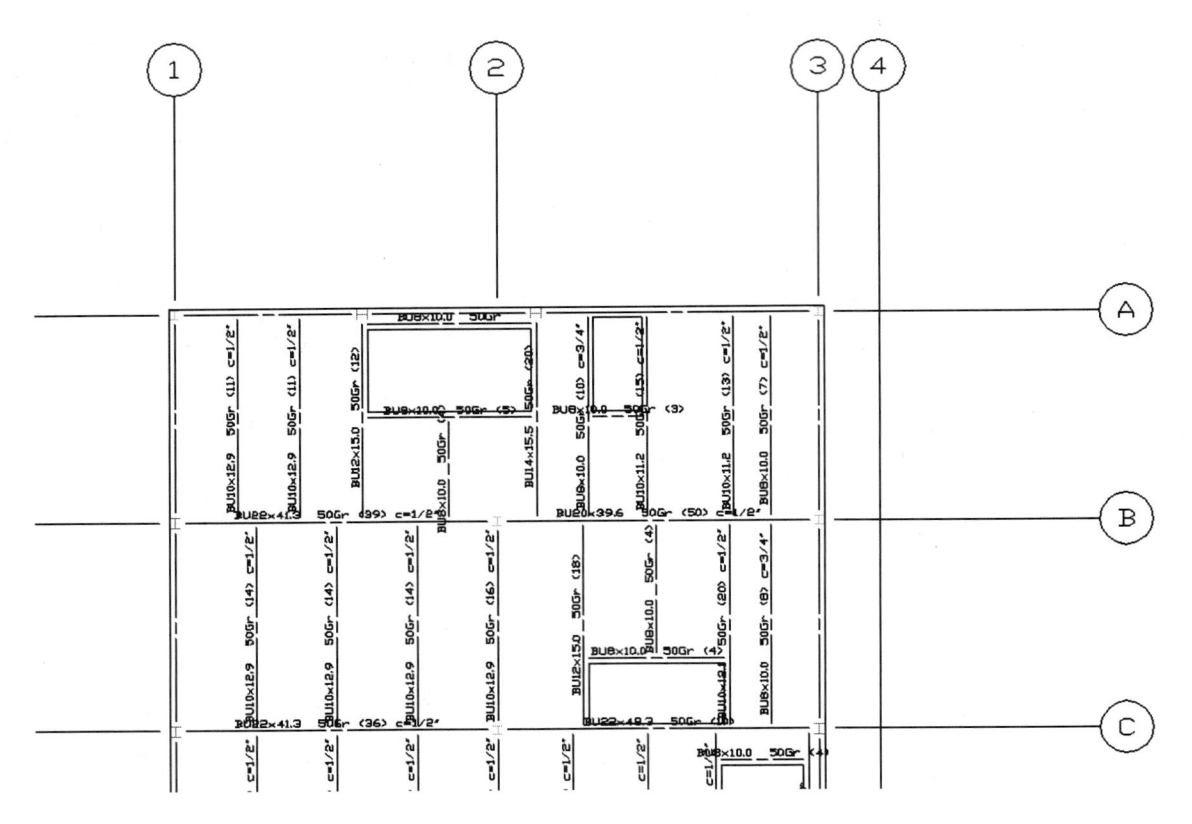

Building Systems

Autodesk's Building Systems software is meant for designing mechanical, electrical, and plumbing systems for buildings, such as offices. In this case, "mechanical" means heating, cooling, and fire protection. The Building Systems add-on helps you design splined flex ducts with editable grips; single-line, 2D, or 3D pipes for creating chilled water, hot water, steam, and other piping systems; rise/drop symbology for connecting ductwork; and fire protection content, such as sprinklers, deluge valves, and retard chambers.

The electrical design capabilities of Building Systems software include automatic calculation of service and feeder sizes, as well as panel schedules and single, two-, and three-pole intelligent circuit objects that work across all referenced files.

For plumbing, the software includes schematic tools for piping layout; built-in plumbing code systems for automatically calculating pipe sizing and flow requirements; and 3D plumbing content and 3D piping tools.

www.autodesk.com/buildingsystems

The AutoCAD drawing below illustrates the HVAC (heating, ventilation, air conditioning) ducts for carrying warm and cold air to individual offices.

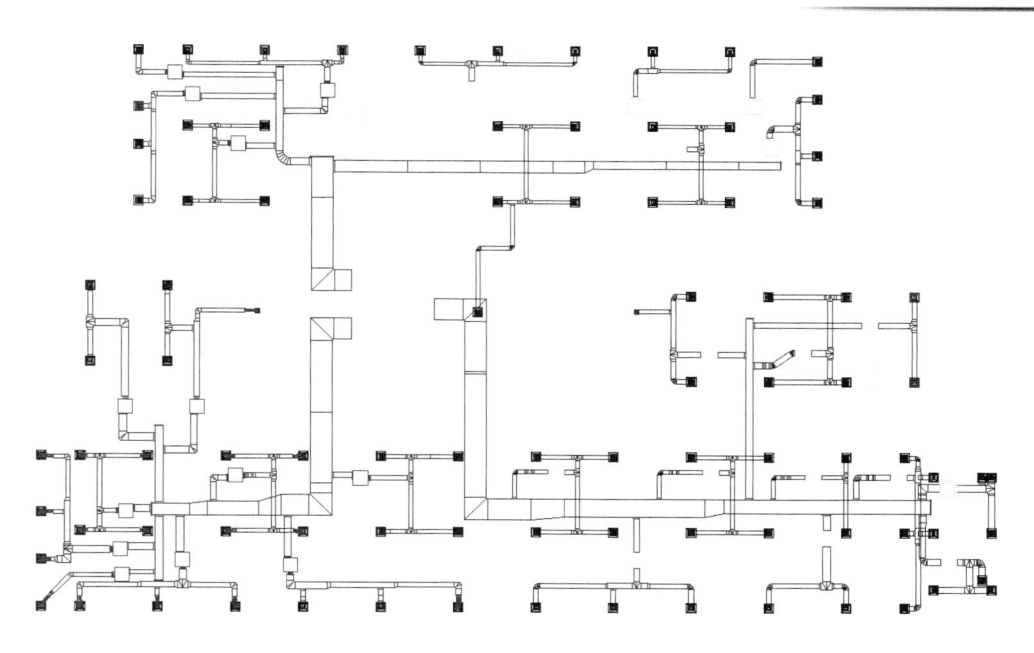

Civil Design

Autodesk's Civil Design software designs, analyzes, and drafts sanitary and storm drain systems. It allows you to evaluate grading scenarios, and work interactively among plan, profile, and sectional designs. You can apply design rules, transitional strings, and sectional criteria when designing all details of a road.

www.autodesk.com/civildesign

The civil engineering drawing below illustrates manholes and covers designed with AutoCAD.

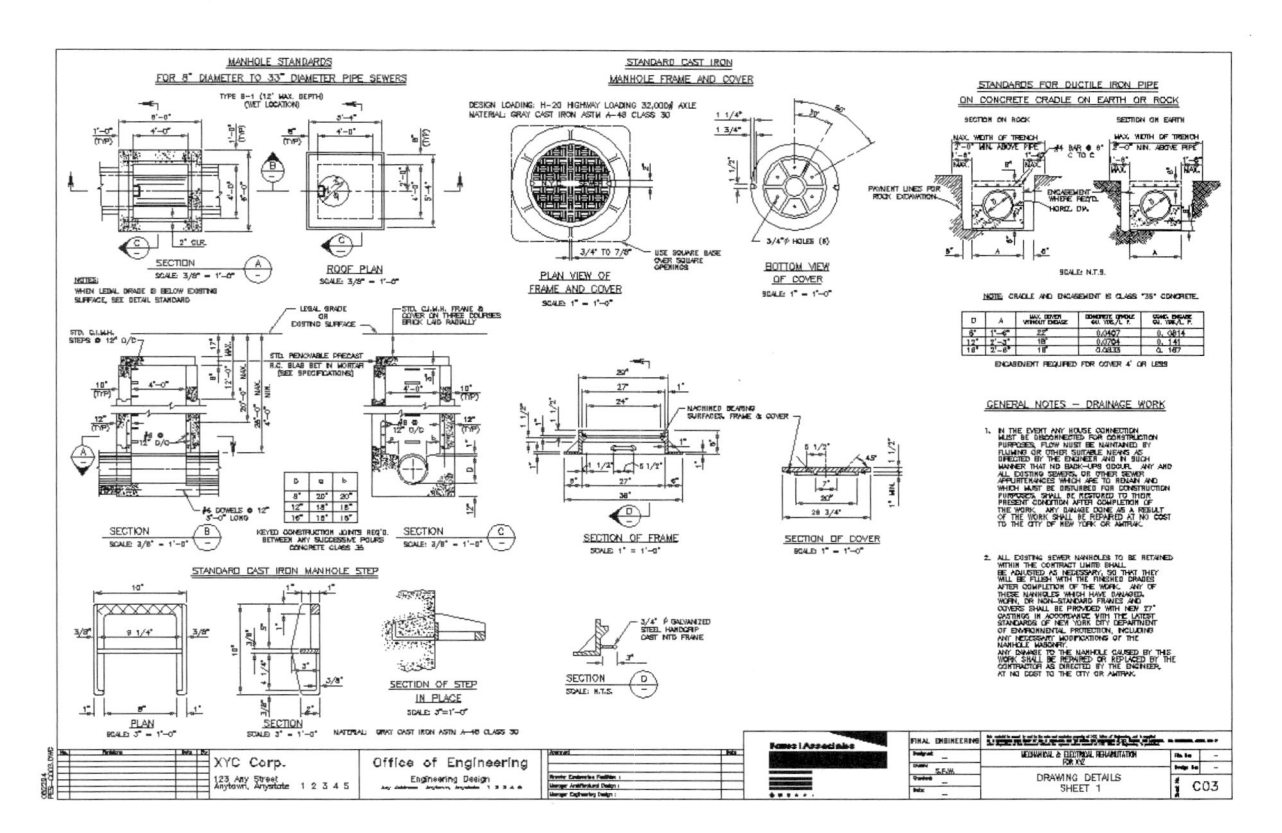

Land Desktop

Autodesk's Land Desktop performs COGO (coordinate geometry) tasks, as well as creates maps, models terrain, designs alignments, and and creates parcels for land planning. Other features include topographic analysis, real-world coordinate systems, volume totals, and roadway geometry.

The software allows you to create and label COGO points, define and edit parcels and roadway alignments, analyze and visualize terrain models, and calculate volumes and contours.

www.autodesk.com/landdesktop

The AutoCAD drawing below maps elevation contours for a development site.

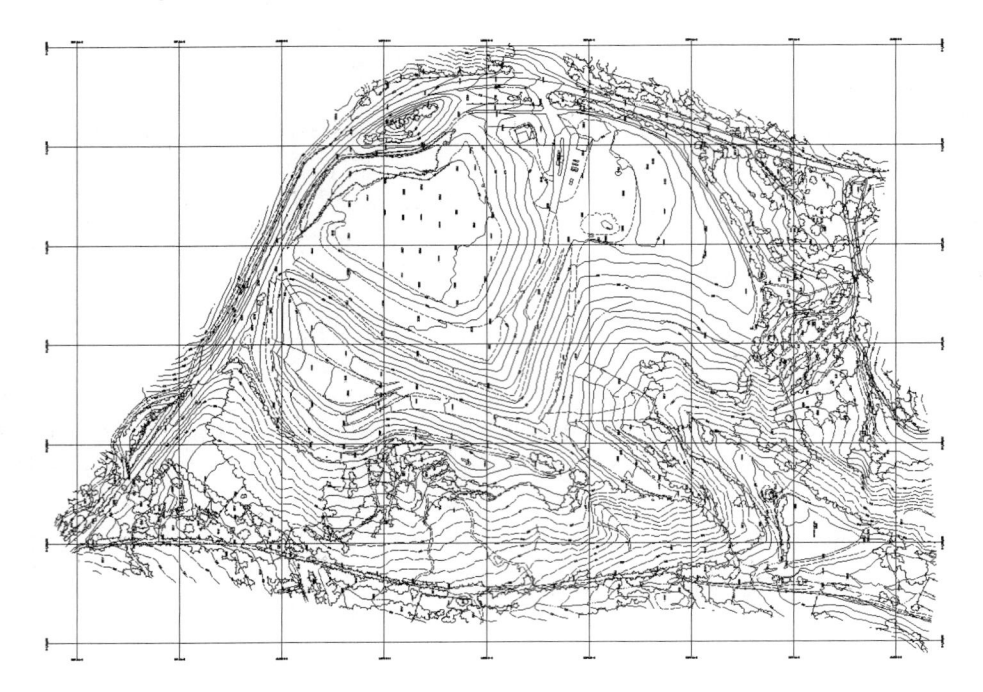

Autodesk does not have software specific to structural engineers, but design software is available from third-party developers. The girder drawing below was created by AutoCAD.

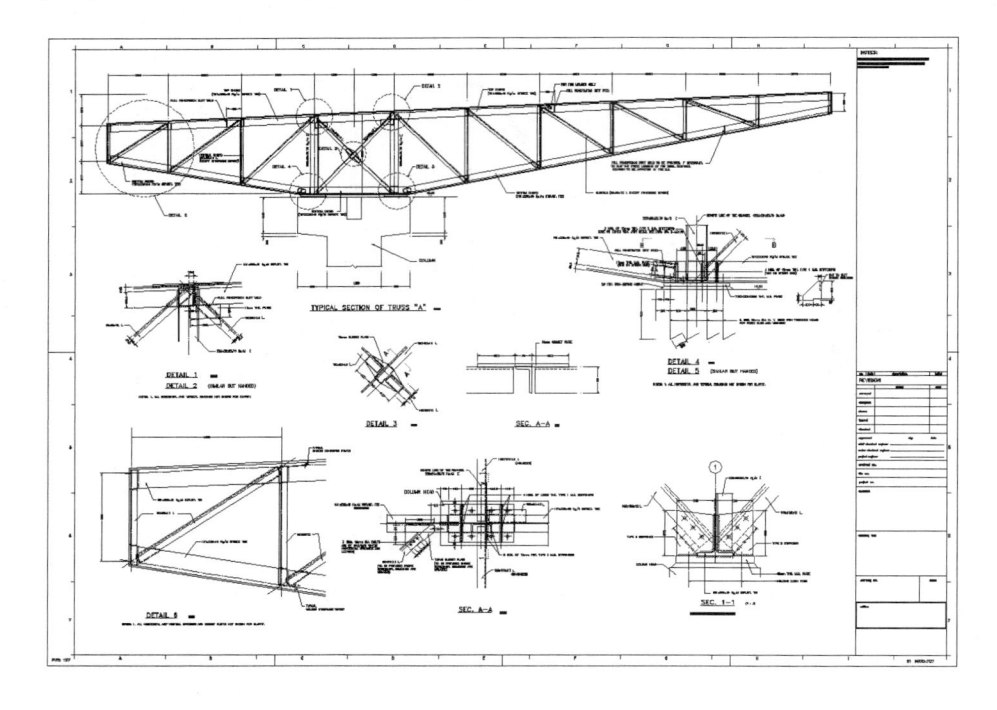

Interior Design

AutoCAD is a valuable tool for interior designers. The 3D capabilities can model interiors for clients. Floor plan layouts can be drawn and modified very quickly. Many third-party programs are available that make 3D layouts of areas such as kitchens and baths relatively simple.

VIZ

Autodesk's VIZ software creates high quality renderings of buildings and their interiors. Third-party developers have additional images and symbols for populating scenes.

VIZ has a "modeless" modeling environment with a unified workspace; surface finishes and lighting systems can be created, mapped, and manipulated on-the-fly. Global illumination rendering technology creates more accurate simulations of lighting effects — either natural or artificial. VIZ comes with a ready-to-use library of common lighting fixtures.

"DWG linking" lets VIZ interact with drawings created by AutoCAD, Architectural Desktop, Mechanical Desktop, and so on. For example, after designing a 3D floor plan in Architectural Desktop, you can link the file to VIZ to study materials and lighting. After modifying the floor plan in Architectural Desktop, the changes can be updated in a VIZ session, preserving materials and lighting defined in the linked model.

Substitution makes it easier to deal with complex details. For example, a chair represented as a simple 2D symbol in the production drawing is substituted in VIZ with a 3D equivalent that contains all the material information.

www.autodesk.com/viz

The lifelike rendering shown below was generated by VIZ from an AutoCAD drawing.

Facilities Management

After tenants move into the newly-constructed building, facilities managers keep track of how the building is used. When staff move to different offices, for example, they need to know whether the correct desks, chairs, cupboards, and telecommunications connections are available in their new office, or whether their existing furniture has to move with them. Keeping track of hundreds of offices is made easier using third-party facilities management (or "FM" for short) software dveloped for AutoCAD.

Shopping malls and airports use AutoCAD linked to a database to maximize the profit from leased stores. Owners of these complexes need to track the mix of stores, and the revenue they produce. Some stores don't want to be located near competitors; others want to be in prime locations, such as near the entrance or at a corner. Other data stored in the database includes agreements with cleaning and maintenance contractors, even details such as floor coverings and air conditioner capacities.

The AutoCAD drawing below uses colored hatch patterns to show clearly the different uses of this office building. The legend explains the patterns, and lists the total square footage of each use.

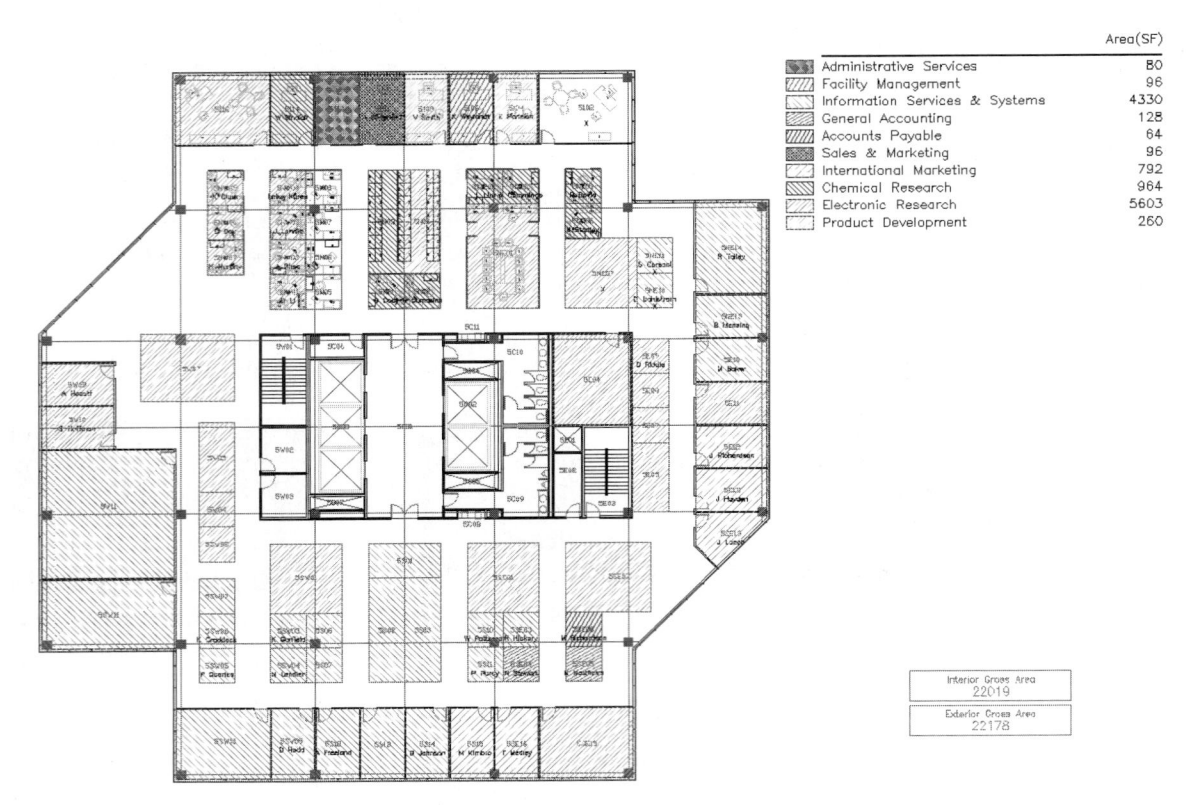

Manufacturing

Manufacturing uses for CAD are legion. One main advantage is integrating the program with a database for record keeping and tracking. Maintaining information in a central database simplifies much of the work required in manufacturing. Technical drawings used in manufacturing are constructed quickly and legibly.

The AutoCAD drawing below determines how to fit the components of a truck cab around the driver.

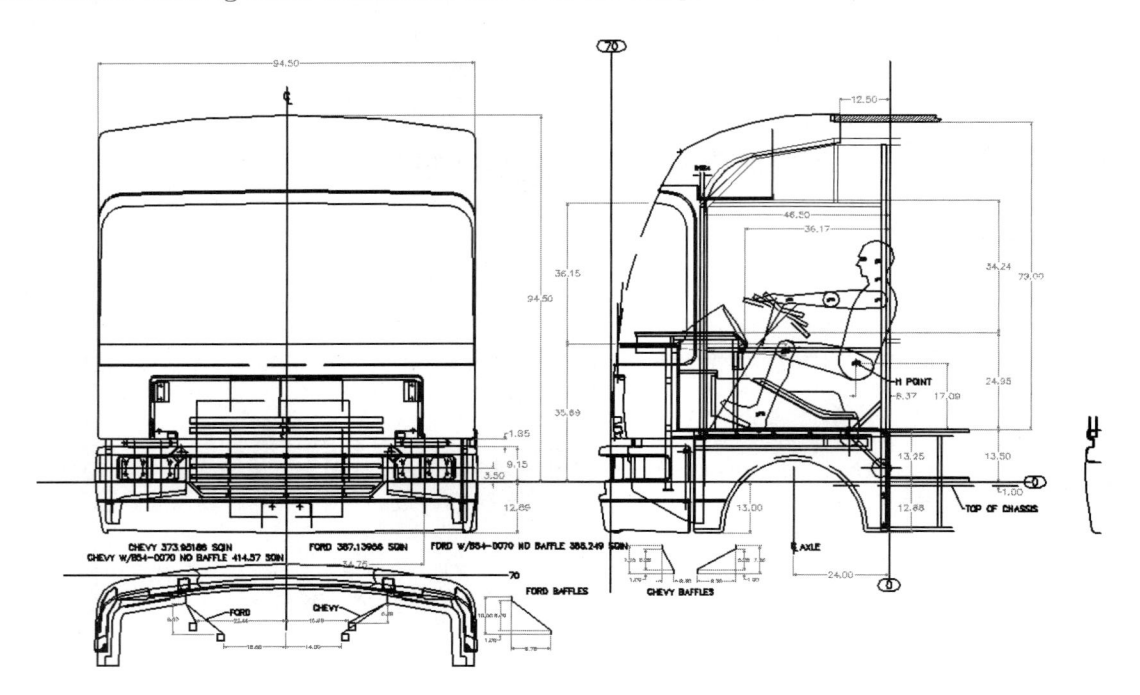

Autodesk has two programs for mechanical engineers. For 2D drafting, there is AutoCAD Mechanical, and for 3D design, there is Inventor.

AutoCAD Mechanical

AutoCAD Mechanical is an add-on program that runs on AutoCAD. It is meant for 2D drafting of mechanical parts. Its parts tracking feature and large pre-drawn content make it easy for mechanical engineers to create, manage, and reuse their drawing data.

www.autodesk.com/autocadmechanical

Inventor

The Inventor software uses a paradigm different from "regular" AutoCAD, because mechanical parts can be very complex. You begin drawing *parts* that make up *assemblies*. The assembly is the finished device, whether an MP3 music player or a plastic injection molding machine. The parts are, as the name suggests, the individual parts making up the assembly. Programs like Inventor are designed to handle tens of thousands of parts.

Once the assembly is created, Inventor can generate 2D plans, parts lists, and assembly instructions. Commands help place standard bolts, route wires and tubing, and determine whether the parts can withstand stresses from pressure and temperature.

www.autodesk.com/inventor

The mechanical drawing below illustrates an assembly plant.

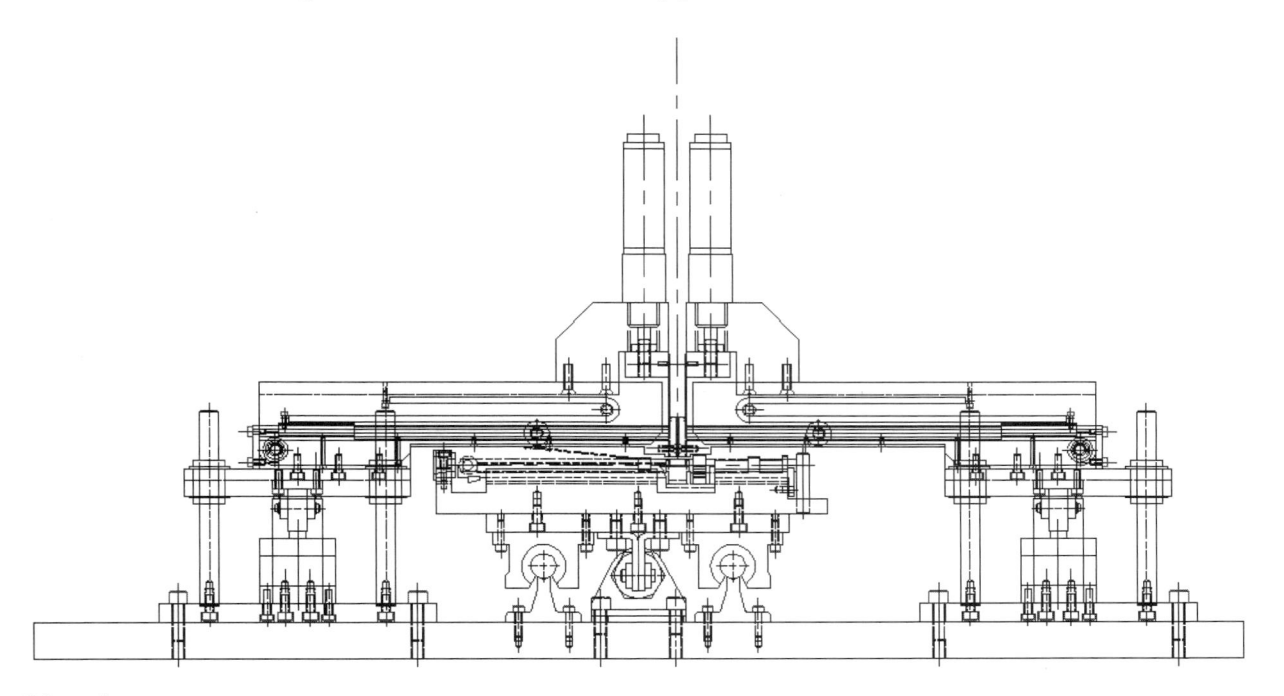

Mapping

Mapmakers use CAD to construct maps that store much data. For example, a municipal map can contain information about all buildings, transportation systems, and underground services (such as storm sewer and water pipes). By using search criterion, the CAD system returns a map showing, for example, all water pipes older than 20 years. This map then shows where municipal engineers should replace old piping.

GPS (global positioning system) generates the data used to create maps in CAD. GPS consists of 20 satellites that always circle the earth, broadcasting data about their location. A handheld GPS receiver fixes your location by analyzing data broadcast by at least three satellites.

Map

Autodesk has a program specifically for AutoCAD users called Map. It assign properties to objects according to classification, and then identifies, manipulates, and selects objects by their description (e.g., road, river) rather than by their simple primitives (lines and arcs). Topology functions define how nodes, links, and polygons connect to each other. This permits network tracing, shortest-path routing, polygon overlay, and polygon buffer generation.

Map has tools that help correct errors due to incorrect surveying, digitizing, or scanning. It connects to an external database, and can import data from other GIS products.

www.autodesk.com/map

The GIS drawing below uses layers and color to differentiate between geographic elements, such as homes, properties, streets, and underground services.

Entertainment

The entertainment industry is largely based on digital media, so tCAD graphics are a perfect match. The graphics you view on television and in the newspaper are often a form of CAD graphics. Movie and theater set designs are often done in CAD. Movie makers are turning to forms of electronic graphics for manipulations and additions to their filmed and animated work.

The AutoCAD drawing below shows details for a new live theatre.

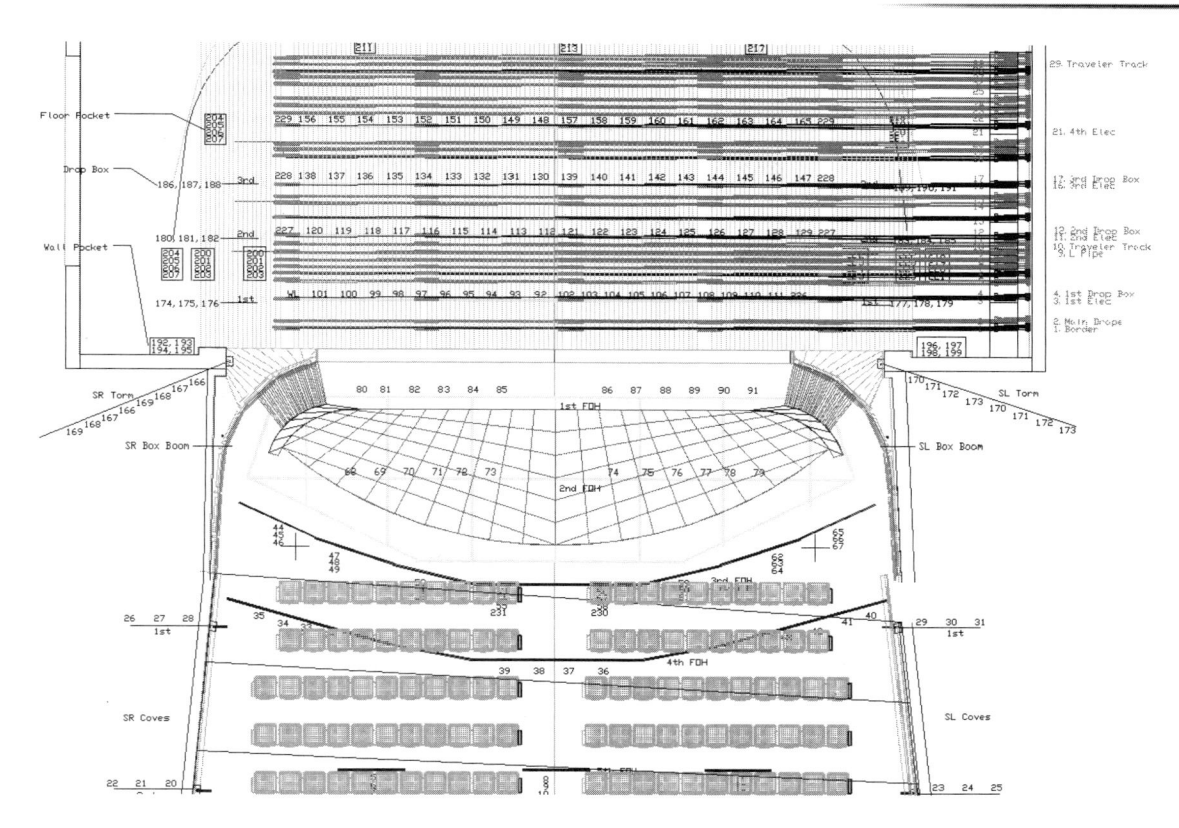

Yacht Design

And AutoCAD is even used to design yachts. The most difficult aspect is designing the hull, which requires a series of continuously changing curves, called splines.

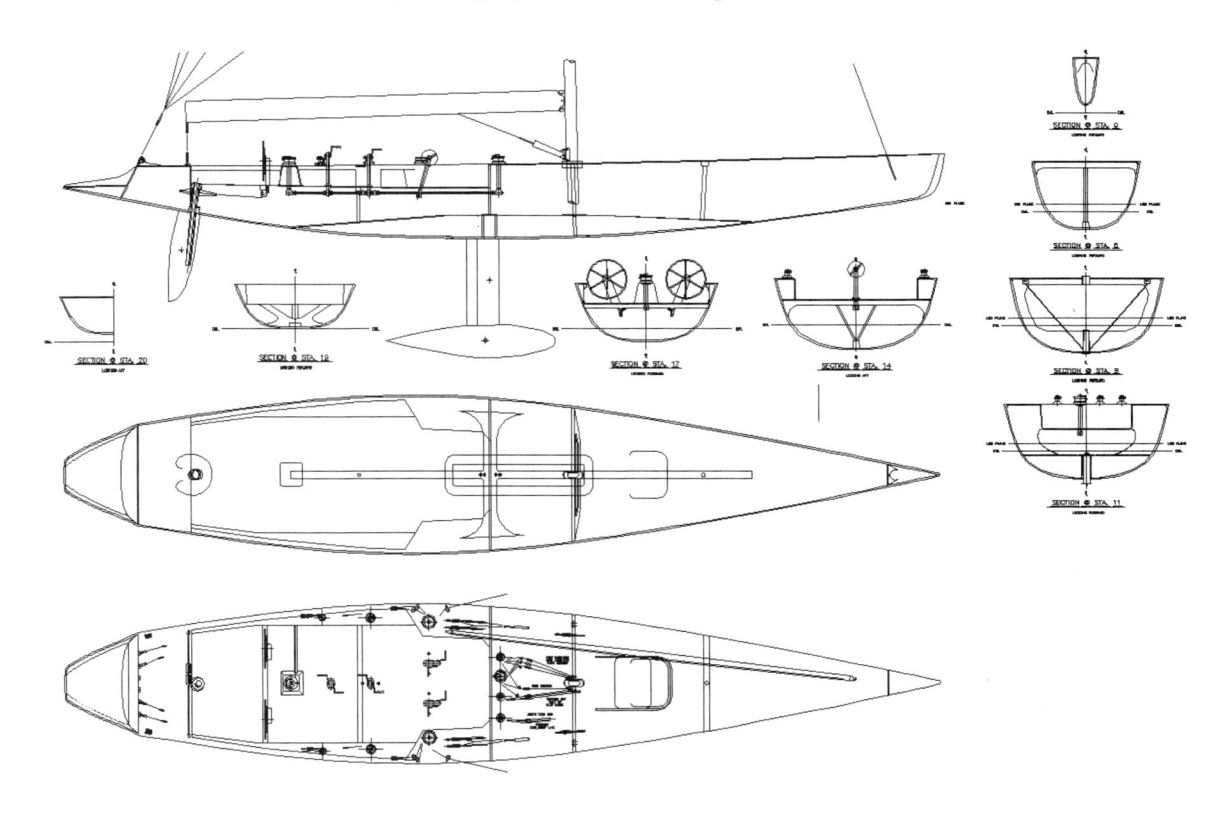

OTHER BENEFITS

Using a computer to draft and design also open a new world of information technology. Many CAD systems now incorporate access to email and the Web. There are Web sites specific to AutoCAD, as well as discussion groups and email newsletters.

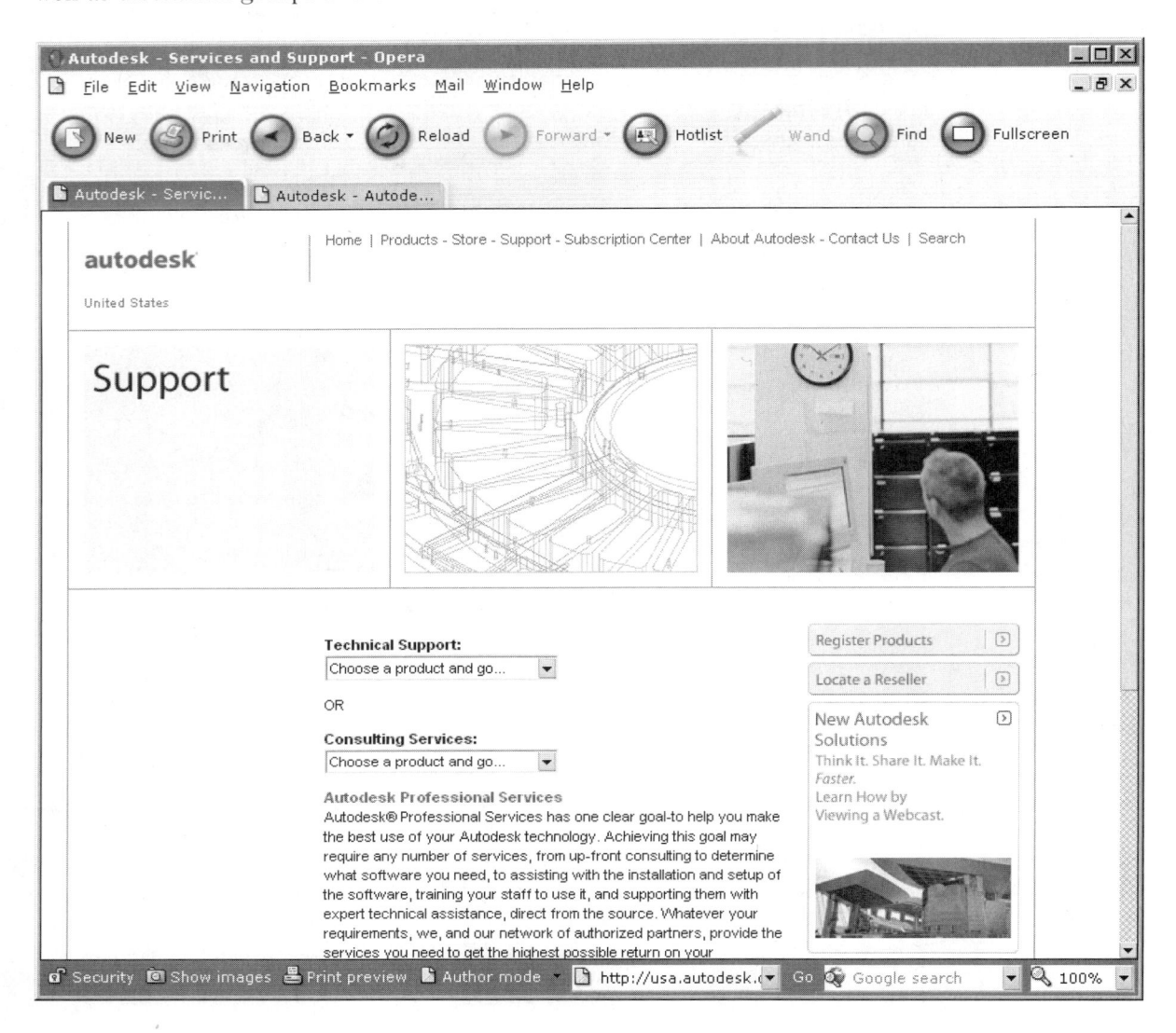

Autodesk maintains many kinds of information at its Web site, www.autodesk.com. You can learn about updates to AutoCAD, download useful utilities, read up tips on for using AutoCAD, and access large libraries of symbols.

AUTOCAD OVERVIEW

AutoCAD displays your drawings on the monitor or screen. The screen takes the place of paper in traditional drafting techniques. All the additions and changes to your work are shown on the screen as you perform them.

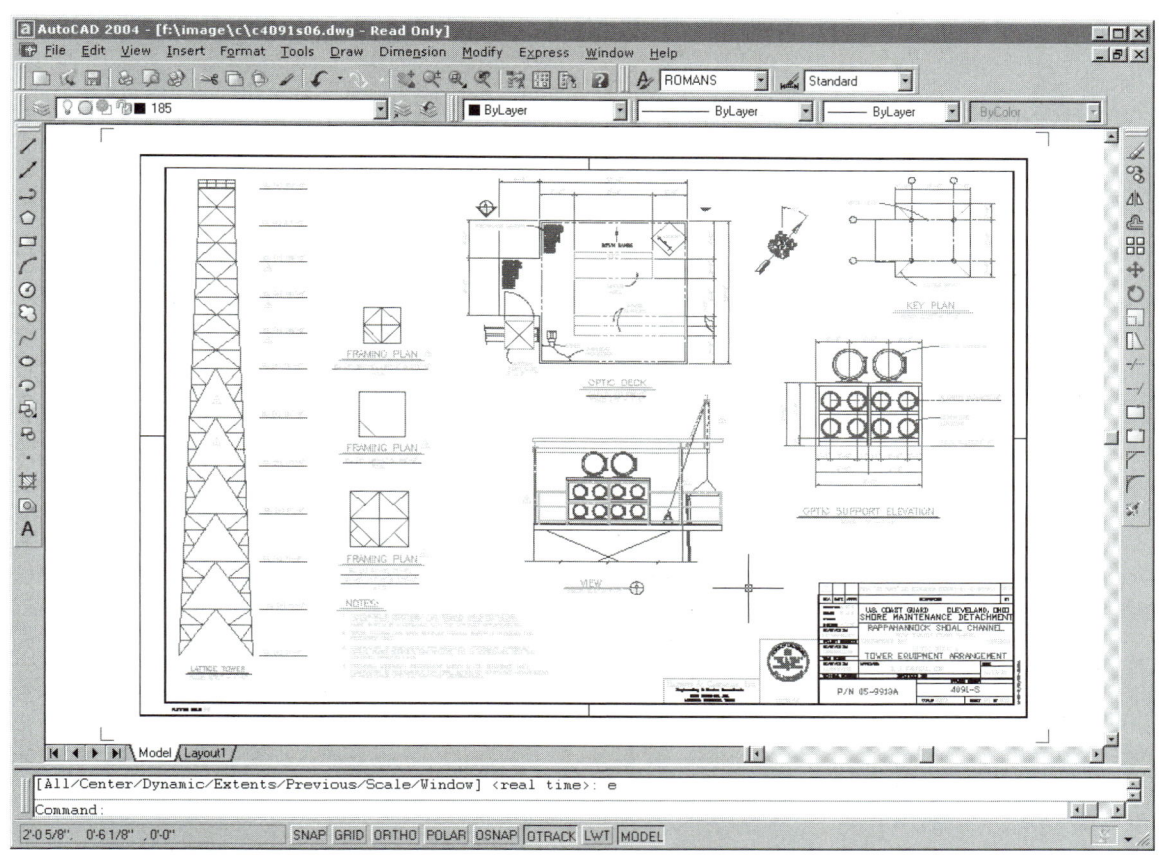

A drawing is made up of separate elements, consisting of lines, arcs, circles, strings of text, and others. These elements are usually called "objects" or sometimes "entities."

Objects are placed in the drawing by means of *commands*. You start commands by several methods: you can type the command's name at the keyboard; select the command from a menu; or choose the command from a toolbar button.

Some commands start when you select an object in the drawing. You are then asked to identify the parameters of the command. These parameters are called "options." After you identify all the information that AutoCAD requests, the new objects, or changes to objects, are shown on the screen.

TERMINOLOGY

This book contains terms and concepts that you need to understand to use AutoCAD properly. Some of the terms are explained briefly in this chapter. If you need additional help, refer to Appendix B "AutoCAD Command Summary" and Appendix H "Glossary." In addition, you may consult the index, which contains all the terms.

Coordinates

The Cartesian coordinate system is used in AutoCAD. The horizontal represents the x axis, while the vertical represents the y axis. Any point on the graph can be represented by an x and y value shown in the form of x,y. For example, 2,10 represents a point 2 units in the x direction and 10 units in the y direction.

The intersection of the x and y axis is the 0,0 point. This point, called the "origin," is normally the lower left corner of the screen. (You can, however, specify a different point for the lower left corner.)

AutoCAD also works with 3D (three-dimensional) drawings. The third axis is called the z axis, which normally points out of the screen at you.

Display

In this book, the term "display" refers to the part of the drawing that is currently visible on the monitor.

Drawing Files

Drawing files contain the information used to store the objects you draft. A drawing file automatically has a file extension of *.dwg* added to it.

Units

The distance between two points is described in *units*. The units for AutoCAD can be set to any of the following:

1. Scientific
2. Decimal (Metric)
3. Engineering
4. Architectural
5. Fractional
6. Surveying

Zooming and Panning

You often want to enlarge a portion of the drawing to see the work in greater detail, or to reduce it to see the entire drawing. The ZOOM command facilitates this. AutoCAD's zoom ratio is many trillions to one!

The PAN command allows you to move around the drawing while at the same zoom level. You can pan the view up, down, and side to side, for example, to see parts of the drawing that might be off the screen.

USING THIS BOOK

Each chapter in *Using AutoCAD 2004: Basics* builds on the previous chapters. The basic use for each command is given, followed by one or more tutorials that help you understand how the command works. This is followed by a comprehensive look at the command and its many variations.

The problems at the end of a chapter are specifically designed to use the commands covered in the book to that point. Review questions reinforce the concepts taught in the chapter. This method allows you to pace learning. Remember, not everyone grasps each command in the same amount of time.

CONVENTIONS

This book uses the following conventions.

Keys

Several references are made to keyboard keystrokes in this book, such as ENTER, CTRL, ALT, and function keys (F1, F2, and so on). Note that these keys might be found in different locations on different keyboards.

Control and Alternate Keys

Some commands are executed by holding down one key, while pressing a second key. The control key is labeled CTRL, and is always used in conjunction with another key or a mouse button.

To access menu commands from the keyboard, hold down the ALT key while pressing the underlined letter. The ALT key is also used in conjunction with other keys.

Flip Screen

The text and graphics screens can be alternately displayed by using the Flip Screen key. The **F2** key is used for this function.

Command Nomenclature

When a command sequence is shown, the following notations are used:

> Command: **line** *(Press* ENTER.*)*
>
> Specify first point: *(Pick a point.)*
>
> Specify next point or [Undo]: *(Pick point 1.)*

Boldface text designates user input. This is what you type at the keyboard. If the command and its options can be entered by other methods, this is described in the book. For example, commands can be selected from menus and toolbars, or entered as keyboard shortcuts.

(Press ENTER.*)* means you press the ENTER key; you do not type "enter."

(Pick a point.) means that you enter a point in the drawing to show AutoCAD where to place the object. To pick the point, you can either click on the point, or type the x, y, z coordinates.

(Pick point 1.) The book often shows you a point on a drawing. The points are designated, such as "point 1."

<Default> is how AutoCAD indicates *default* values. The numbers or text are enclosed in angle brackets. The default value is executed when you press ENTER.

[Undo] is how AutoCAD lists command *options*. The words are surrounded by square brackets. At least one letter is always capitalized. This means you enter the capital letter as the response to select the option. For example, if the option is **Undo**, simply entering **U** chooses the option.

Additional options are separated by the slash mark. When two options start with the same letter, then two letters are capitalized, and you must enter both.

NEW IN 2004 — WHAT'S NEW IN AUTOCAD 2004

Fully updated to AutoCAD 2004, this book includes new information:

- The DesignCenter, Properties window, and icons have a new user interface. Transparency lets you see the drawing through these windows.
- The new **CLEANSCREEN** command lets you see more of the drawing and less of the user interface.
- The new **QNEW** command starts new drawings based on *.dwt* template files.
- AutoCAD now supports 16.7 million colors (up from 255), gradient fill patterns, and colors defined by the PANTONE and RAL standards.
- The new **MREDO** command redoes more than one undo action; the new drop list shows all undo-able and redo-able actions.
- The new **REVCLOUD** command draws revision clouds.
- Tool Palettes provide access to commonly used hatch patterns, fill colors, and block (symbols).
- The new **WIPEOUT** command places blank polygons over busy drawing areas to help text and details stand out.

In volume 2 of *Using AutoCAD 2004: Advanced*, you learn about these new features in AutoCAD 2004:

- The new **TIFOUT**, **JPGOUT**, and **PNGOUT** commands save the current view as TIFF, JPEG, and PNG raster images.
- **securityoptions** keeps your drawing safe through passwords and encryption.
- The new **hlsettings** dialog box makes it easier to control the look of hidden-line removal views.
- The new **xopen** command opens externally-referenced drawings in their own window.
- And the new **3DORBITCTR** command enters the 3D view.

ONLINE COMPANION

The Online Companion™ is your link to AutoCAD on the Internet. We've compiled supporting resources with links to a variety of sites. You not only find out about training and education, industry sites, and the online community, but also about valuable archives compiled for AutoCAD users from various Web sites.

In addition, there is information of special interest to readers of *Using AutoCAD 2004*. This includes updates, information about the author, and a page where you can send your comments. You can find the Online Companion at www.autodeskpress.com/resources/olcs/index.asp. When you reach the Online Companion page, click on All AutoCAD 2004 Titles.

E.RESOURCE

e.Resource™ is an educational resource that creates a truly electronic classroom. It is a CD-ROM containing tools and instructional resources that enrich the classroom and make preparation time shorter. The elements of e.Resource link directly to the text and combine to provide a unified instructional system. With e.Resource you can spend your time teaching, not preparing to teach.

Features contained in e.Resource include: :

- Syllabus: Lesson plans created by chapter. You have the option of using these lesson plans with your own course information.
- Chapter Hints: Objectives and teaching hints that provide the basis for a lecture outline that helps you to present concepts and material.
- Answers to Review Questions: Solutions that enable you to grade and evaluate end-of-chapter tests.
- PowerPoint® Presentation: Slides that provide the basis for a lecture outline that helps you to present concepts and material. Key points and concepts can be graphically highlighted for student retention.
- Exam View Computerized Test Bank: Over 800 questions of varying levels of difficulty are provided in true/false and multiple-choice formats, so that you can assess student comprehension.
- AVI Files: AVI files, listed by topic, which allow you to view a quick video illustrating and explaining key concepts.
- DWG Files: A list of *.dwg* files that match many of the figures in the textbook. These files can be used to stylize the PowerPoint presentations.

WE WANT TO HEAR FROM YOU!

Many of the changes to the look and feel of this new edition came by way of requests from and reviews by users of our previous editions. We'd like to hear from you as well! If you have any questions or comments, please contact:

>The CADD Team
>
>c/o Autodesk Press
>
>5 Maxwell Drive
>
>Clifton Park NY 12065-8007

or visit our Web site at www.autodeskpress.com

ACKNOWLEDGMENTS

We would like to thank and acknowledge the professionals who reviewed the manuscript to help us publish this AutoCAD text.

Technical Editor: Bill Fane, British Columbia Institute of Technology, Burnaby BC, Canada

Copy Editor: Stephen Dunning, Douglas College, Coquitlam BC, Canada

ABOUT THE AUTHOR

Ralph Grabowski has been writing about AutoCAD since 1985, and is the author of over four dozen books on computer-aided design. He received his B.A.Sc. degree in Civil Engineering from the University of British Columbia.

Mr. Grabowski now publishes *upFront.eZine*, a weekly email newsletter about the business of CAD. He is the former Senior Editor of *CADalyst* magazine, the original magazine for AutoCAD users. He was also editor of *AutoCAD User*. His email address is grabowski@telus.net and his Web site is at www.upfrontezine.com.

CHAPTER 1

Quick Tour of AutoCAD 2004

Welcome to *Using AutoCAD 2004: Basics*!

As a new user, you need to experience the feel of AutoCAD. The concept and operation of graphic design software can be disorienting to first-time users. This quick-start chapter gets your feet wet, and introduces you to some of the features discussed in detail in other chapters. This chapter is specifically designed to acquaint you with the AutoCAD drawing program. Subsequent chapters cover the subject more thoroughly.

In this chapter, you learn these commands:

NEW starts a fresh drawing.

LINE draws straight line segments.

U undoes mistakes.

QUIT exits AutoCAD.

CLOSE and **CLOSEALL** close drawings.

SAVEAS gives drawings their name, and save them.

PLOT prints the drawing on paper.

Let's get started!

STARTING AUTOCAD

Before you start AutoCAD 2004 for the first time, your computer must have Microsoft® Windows™ 2000 or XP installed and running.

TUTORIAL: STARTING AUTOCAD

1. To start AutoCAD 2004, use one of these methods:
   - On the Windows desktop, double-click the icon labeled AutoCAD 2004.
     Double-click means to press the left mouse button twice quickly.

   - From the Windows taskbar, click the **Start** button, and then choose **Programs**. Choose **Autodesk**, and then **AutoCAD 2004**.

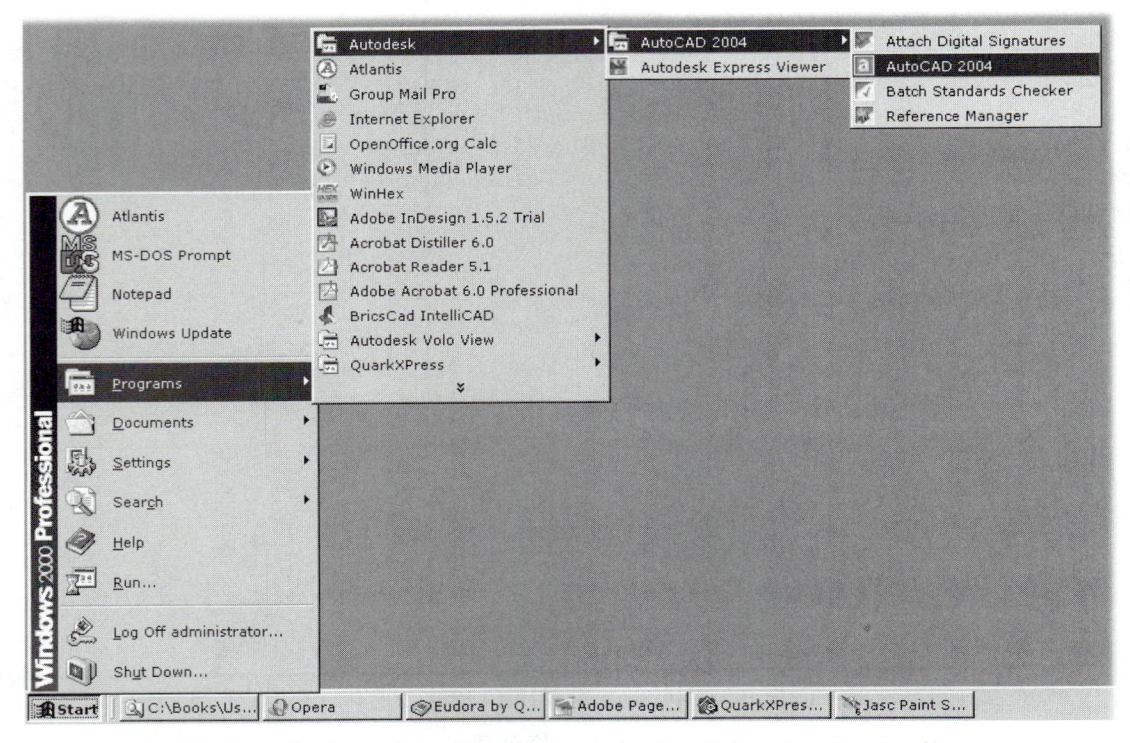

   - In the Windows Explorer (a.k.a File Manager), double-click a *.dwg* drawing file.
   - In the **Run** dialog box, enter *acad.exe*.

     Run: **acad.exe** *(Press ENTER.)*

2. In all cases, Windows opens the AutoCAD software.

   An opening screen is displayed, called the "splash screen," which shows the Stade de France stadium <www.stadefrance.com/presentation/vv.cfm>, designed by Macary, Zublena et Regembal, Costantini - Architectes, Paris.

   After the splash screen disappears, you see AutoCAD.

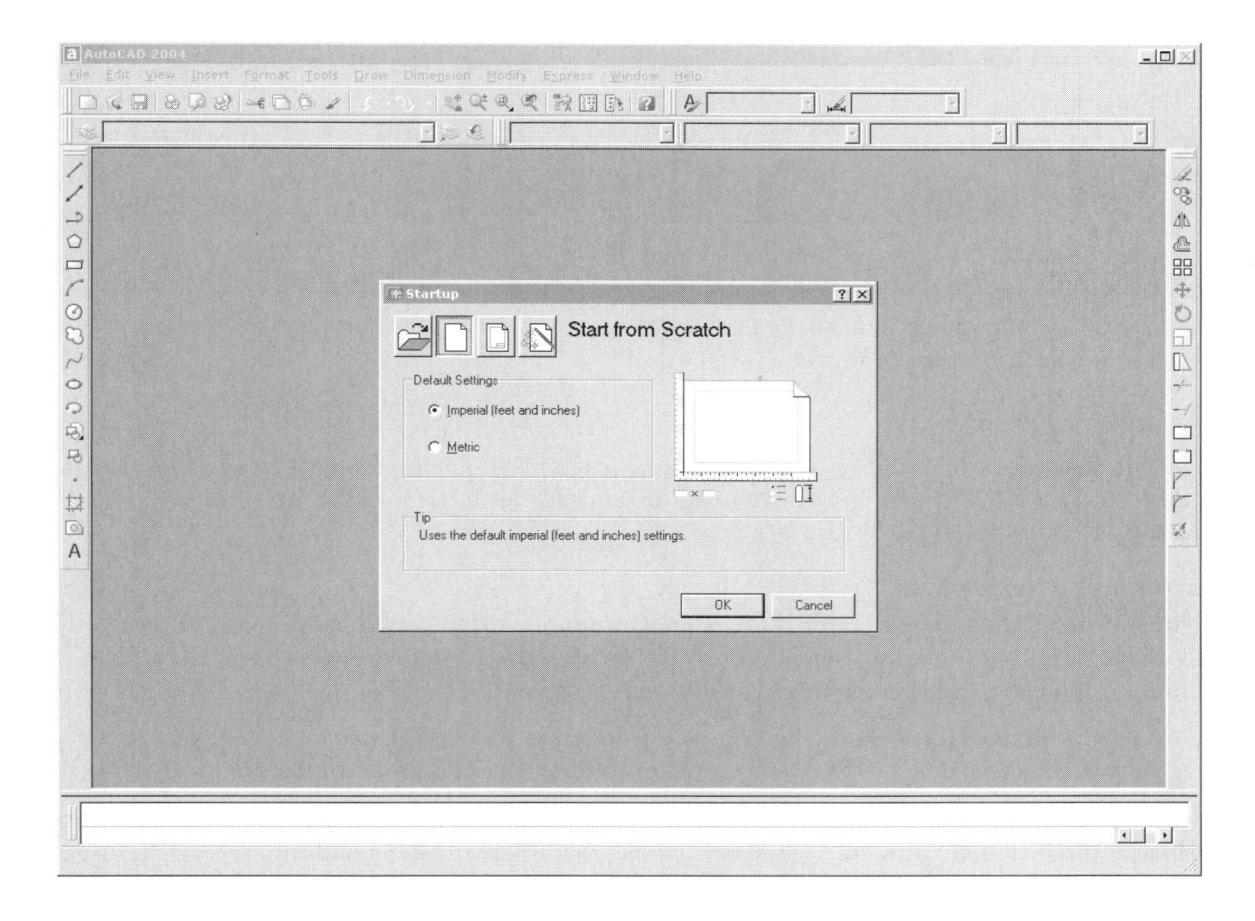

3. If you see the Startup dialog box, choose the **Start from Scratch** button, and then select **Imperial (feet and inches)**.

   If you do not see the Startup dialog box, carry on to the next page.

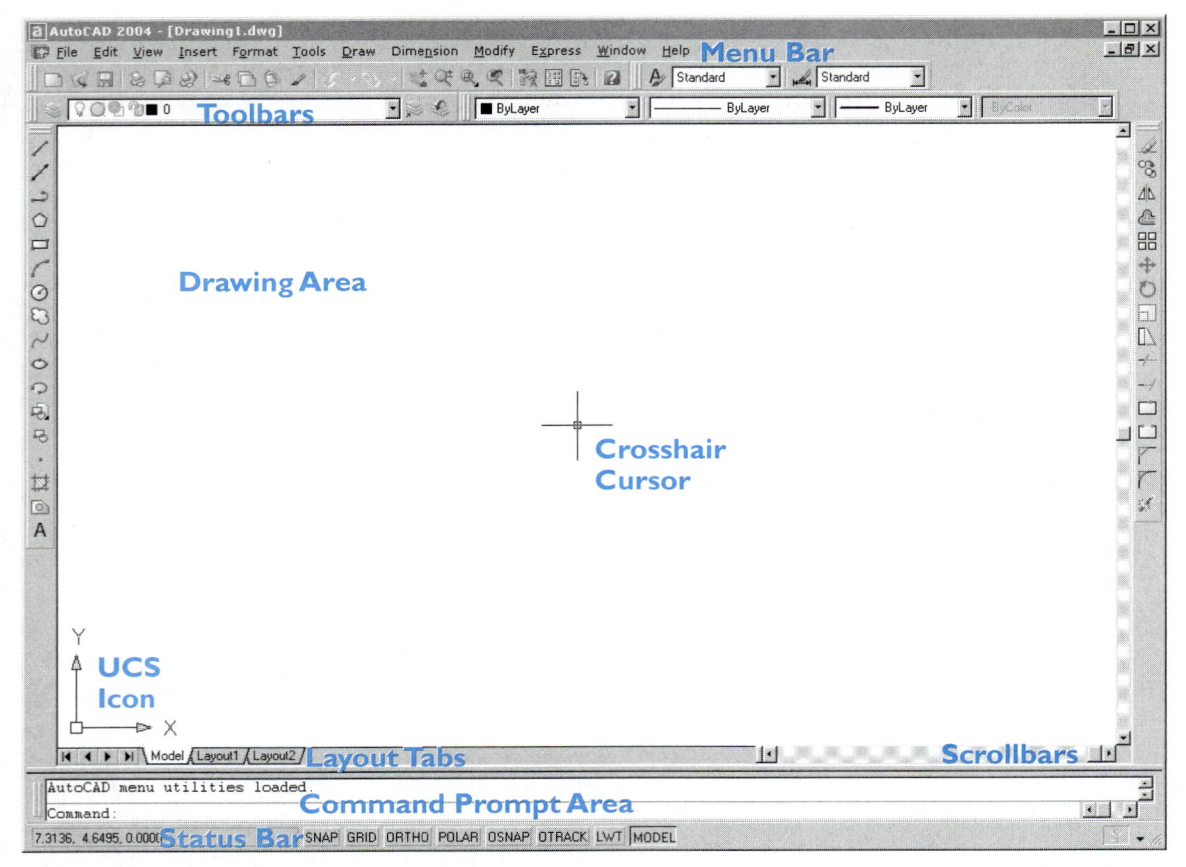

The AutoCAD window looks similar the figure above, except that the background color of the drawing area may be black, instead of white.

DRAWING IN AUTOCAD

AutoCAD draws or edits objects by using commands, which are often simple words such as LINE, CIRCLE, and ERASE. *Commands* describe the objects to be drawn — LINE and CIRCLE — or the operation to perform — ERASE and MOVE.

Command Prompt Area

At the bottom of the drawing editor is the *command prompt area*. This is where you type commands, and where AutoCAD responds. In addition to typing commands, you can select commands from the menu, the toolbars, and by clicking the right mouse button.

(AutoCAD has one other window that you see from time to time. It is called the "Text window," because it displays text only. It is a larger version of the command prompt area.)

You should see the word "Command:" on that line now. If you wish to type a command from the keyboard, the command line must be clear: the command line must display only:

Command:

...with nothing after it. If there is text following the colon (:), clear it by pressing the ESC key (short for "Escape") on the keyboard.

Mouse

In addition to typing commands, you can use the mouse, or sometimes a digitizing tablet or other pointing device to control AutoCAD. (From this point forward, we refer to the mouse only, being more common than the tablet and alternative input devices.)

1. Move the mouse, and notice how the crosshair cursor moves around the screen. The intersection of the crosshairs specifies points on drawings.
2. Move the crosshair cursor out of the drawing area. Notice that the cursor changes to an arrow. This lets you select commands from menus and toolbars.

USING MENUS

A common method for specifying commands is via a menu, as in almost all other Windows software. Menus contain commands that are selected with the mouse. To see how they work, select a command from the menu — choose the **Line** command from the **Draw** menu, as follows:

TUTORIAL: SELECTING MENU ITEMS

1. To select a command from the menu, move the crosshair cursor to the menu bar.
2. As you move the arrow across the words — **File**, **Edit**, **View**, and so on — on the menu bar, each is highlighted.

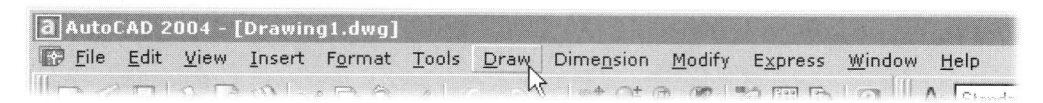

3. Select the **Draw**, menu: when **Draw** is highlighted by the cursor, click the left mouse button. Notice that a menu drops down into the drawing area.

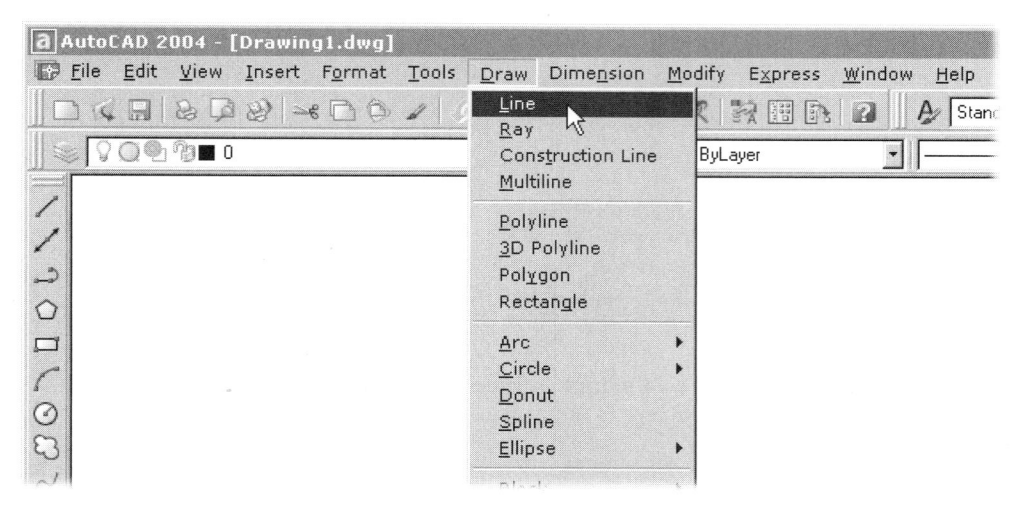

4. Move the pointer down the menu. past Line, Ray, and so on. As you do, notice that each word is highlighted.
   Click on **Line**.
5. Notice the words that appear on the command line at the bottom of the window:
   Command: _line Specify first point:

   This is called a *prompt*. On the command line, AutoCAD always tells you what it expects from you. Here, AutoCAD is asking you to specify the point from which the line starts (called the "first point").
6. Move the crosshair cursor into the drawing area, and then enter a *point* (click the left mouse button).

7. Continue to move the crosshair cursor around the screen. Notice how a line "sticks" to the intersection of the cursor crosshairs. This is called "rubber banding": the line stretches and follows your movement.
8. In the command prompt area, the prompt has changed:

    Specify next point or [Undo]:

    Move the cursor, and pick a point approximately at the point shown in figure below.
    (The **Undo** option "undraws" the previous line segment, which is useful when you make a mistake.)

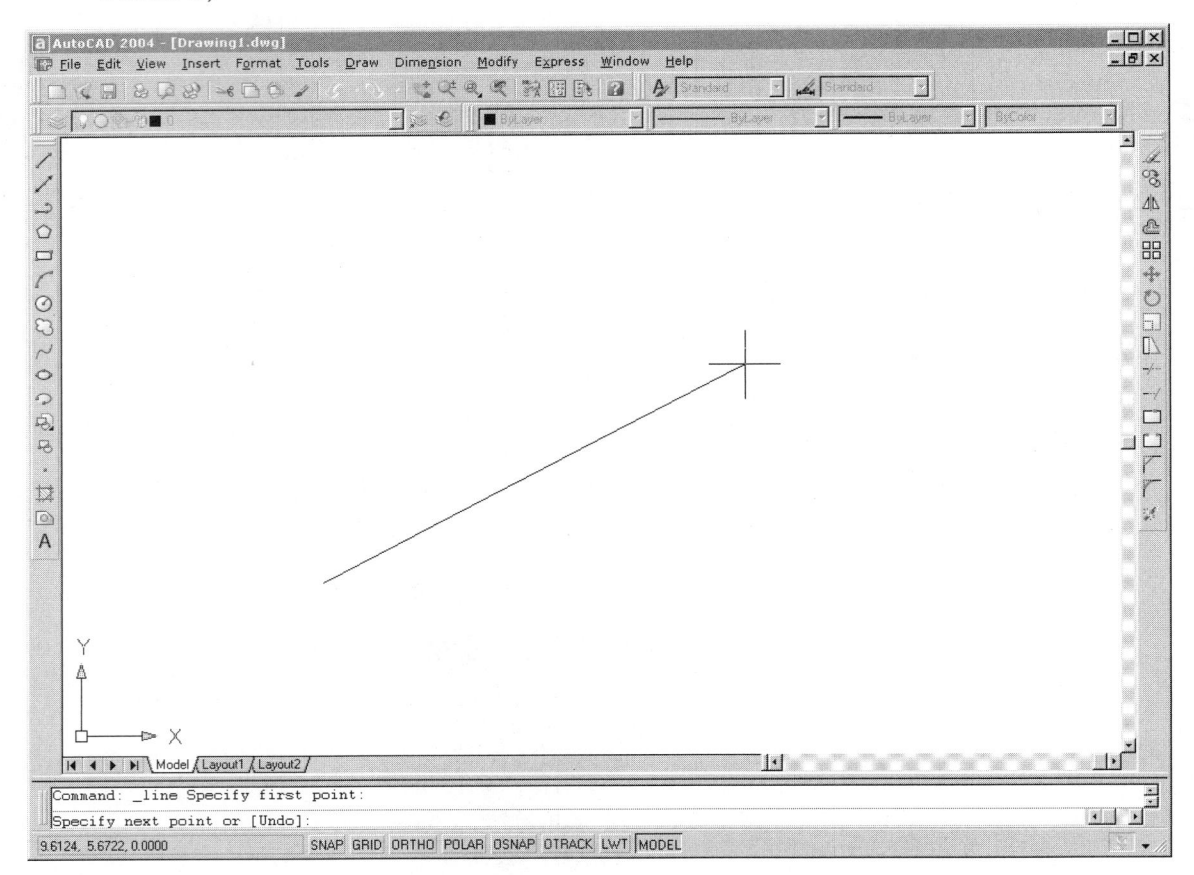

9. Continue to enter lines as shown in by the figure. As you do, the prompt changes to:

    Specify next point or [Close/Undo]:

    (The **Close** option draws a line segment from the end of the last segment to the start of the first segment.)

    The LINE command remains active, allowing you to draw as many line segments as you need without reselecting the command.

10. To end the command, click the right mouse button.

    From the shortcut menu, choose **Enter**, as illustrated by the figure below.

    Alternately, press ESC on the keyboard; for some commands, you need to press ESC twice before they end.

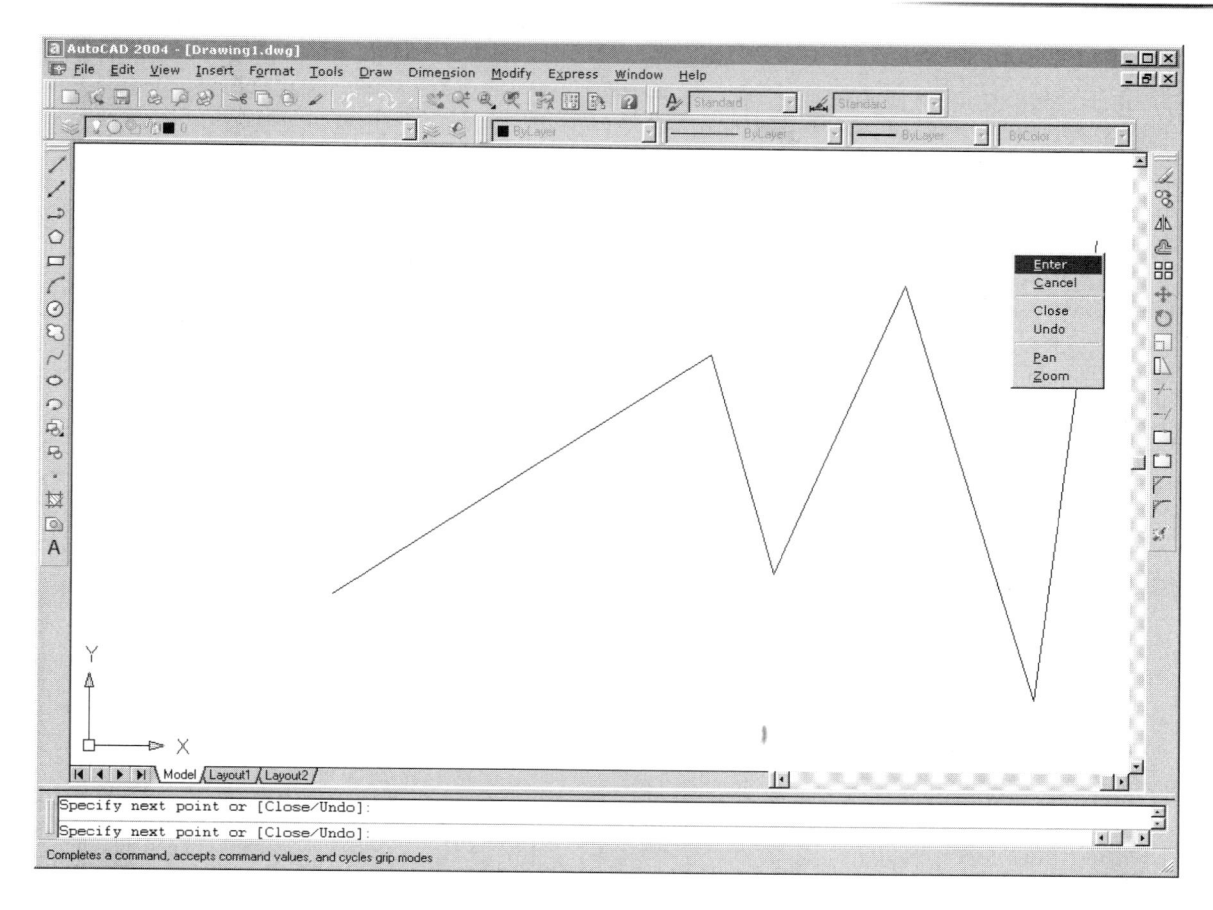

11. Enter the **U** command (short for "undo") to erase the lines you drew.
    Command: **u** *(Press ENTER.)*

    AutoCAD confirms:
    Everything has been undone.

TOOLBARS

AutoCAD places a strong emphasis on toolbars. Toolbars consist of buttons, and each button has a little picture called an "icon." AutoCAD has more than two dozen toolbars. In a default installation, AutoCAD normally displays the toolbars illustrated by the figure above. These toolbars access most of the commands you use everyday.

- **Standard** toolbar (below the menu bar) contains many of the same commands found in other Windows applications.

- **Styles** toolbar (to the right of the Standard toolbar) provides access to text and dimensions styles; new to AutoCAD 2004.

- **Layers** toolbar (below the Standard toolbar) controls layers in the drawing; new to AutoCAD 2004.

- **Properties** toolbar (to the right of the Layers toolbar) changes the properties of objects; renamed from earlier versions of AutoCAD, when it was known as Object Properties.

- **Draw** toolbar (vertical, at the left of the AutoCAD window) draws many different objects, such as lines, arcs, and hatch patterns.

- **Modify** toolbar (at the right of the AutoCAD window) edits objects to change their size, location, and other properties.

Using Toolbar Buttons

Every button on every toolbar activates a command. For example, the first button of the **Draw** toolbar is **Line**. It starts the LINE command. The icon for **Line** is the diagonal line with a dot at either end.

When you cannot remember the purpose of a button, move the cursor over the button and wait a second or two. AutoCAD then displays a tooltip, a one- or two-word description of the button's purpose.

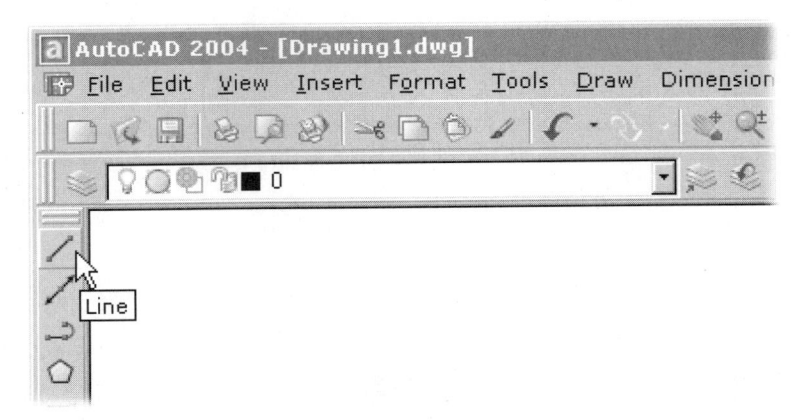

The status bar (at the bottom of the AutoCAD window) displays a one-sentence description of the command's purpose. In the case of the **Line** button, the status bar reads, "Creates straight line segments: LINE."

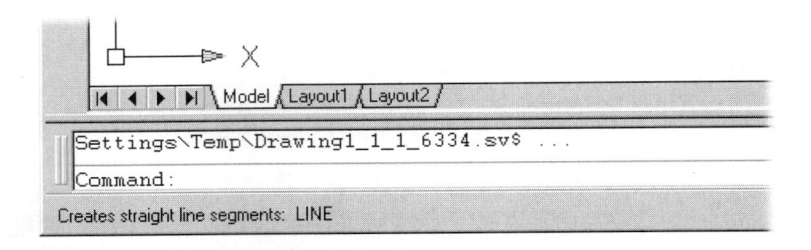

TUTORIAL: DRAWING WITH TOOLBAR BUTTONS

To draw a line with the icon button:
1. Position the cursor over the **Line** button.
2. Click the left mouse button.
   At the command line, you see the now-familiar prompt:
   Command: _line Specify first point:
3. Draw some lines, and then press **ESC** to exit the command.

Flyouts

Sometimes buttons represent a secondary toolbar. You recognize these buttons by the tiny black triangle in the corner of the icon. The triangle indicates the presence of a *flyout*, a group of buttons that "flies out" from a button.

TUTORIAL: ACCESSING FLYOUTS

1. To access flyouts:
    Position the cursor over any button with the tiny triangle, such as the **Zoom** button.

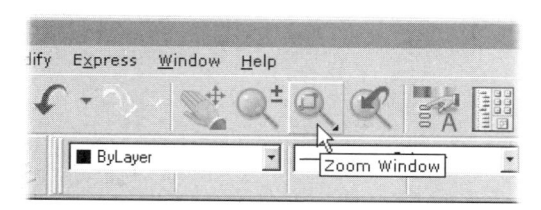

2. Hold down the left mouse button. Notice that AutoCAD displays a flyout toolbar.

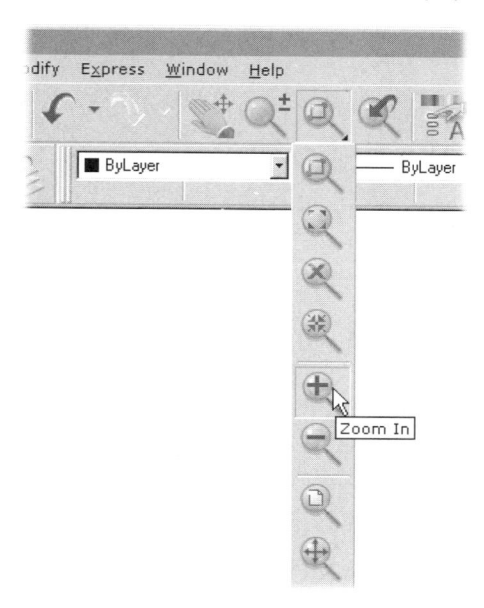

3. Without letting go of the mouse button, drag the cursor over the buttons on the flyout. As the cursor passes over a button, it changes slightly to give the illusion of being depressed.
4. When you reach the button you want, release the mouse button.
5. AutoCAD starts the command associated with the button.
    In addition, the button moves to the top of the toolbar where it is quick and easy to access.

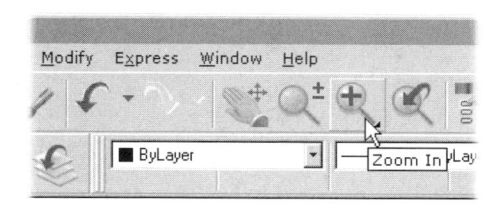

Manipulating the Toolbar

The toolbar has many controls hidden in it. You can move, resize, float, dock, and dismiss the toolbar. Here's how:

To move a docked toolbar, position the cursor over the two "bars" at the left end of horizontal toolbars (or the top end of vertical toolbars). Hold down the mouse button, and drag the toolbar away from the edge of the window. The toolbar now *floats* in the drawing window.

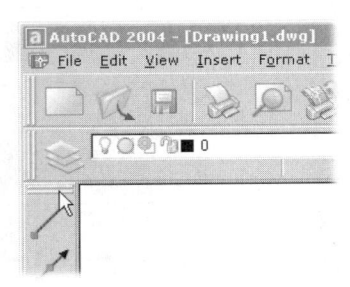

To move a floating toolbar, position the cursor over the title bar. Hold down the mouse button and drag the toolbar to another location.

To dock the toolbar means to move it to the side or top of the drawing area, like the two docked toolbars at the top of the window in the figure. To dock the toolbar, move it against one of the four sides of the drawing area.

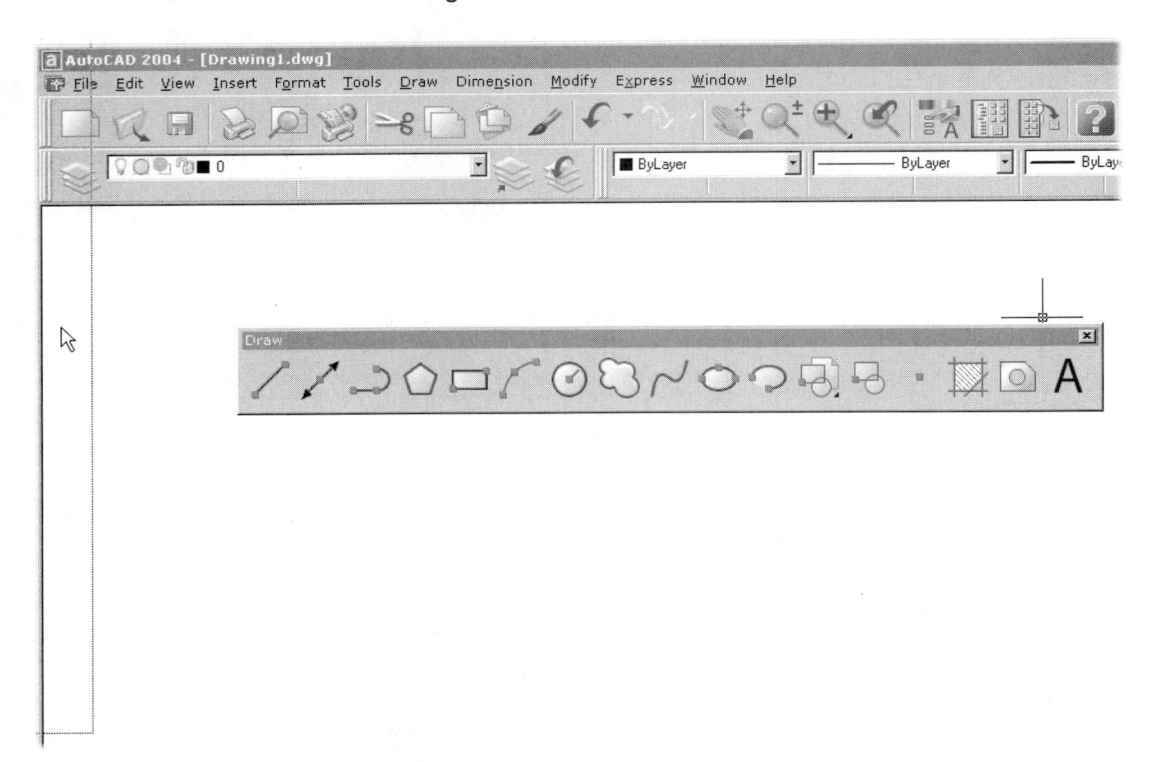

To resize (or stretch) the toolbar, position the cursor over one of the four edges of the toolbar. Hold down the left mouse button, and then drag the toolbox to achieve a new shape.

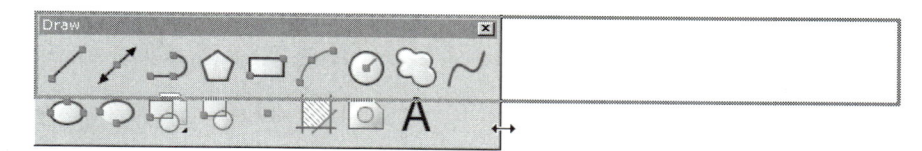

To **dismiss** (get rid of) the floating toolbar, click the tiny **x** in the upper right corner. The toolbar disappears.

To **get back** the toolbar, right-click any toolbar button. From the shortcut menu, select a toolbar name from the list.

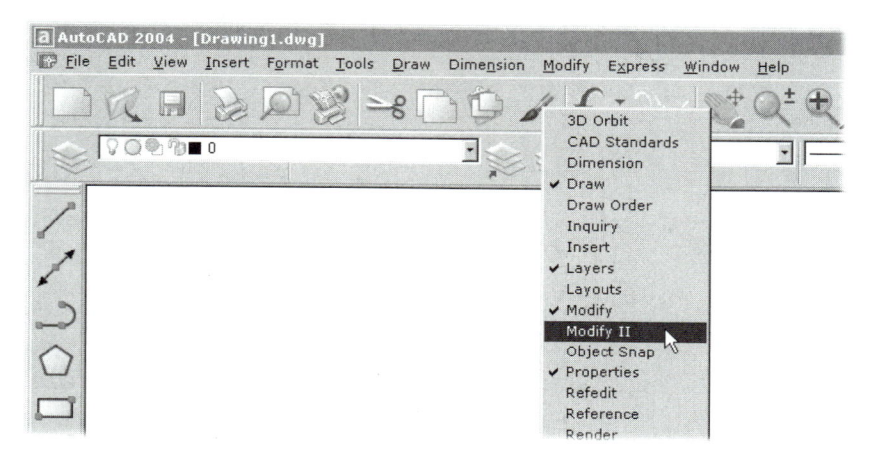

Alternatively, from the **View** menu, select **Toolbars**, and then the name of the toolbar.

Or, enter the **Customize** command, and then select the toolbar's name from the list displayed by the dialog box.

EXITING DRAWINGS

AutoCAD provides several different ways to close drawings and exit AutoCAD. The following sections outline the possibilities. Choose the option you want, and follow the instructions.

DISCARD THE DRAWING, AND EXIT AUTOCAD

When you don't want to keep your drawing, and wish to stop work, enter the **QUIT** or **EXIT** commands (you do this from the keyboard), or choose **Exit** from the **File** menu. AutoCAD displays a dialog box that asks you to confirm your choice.

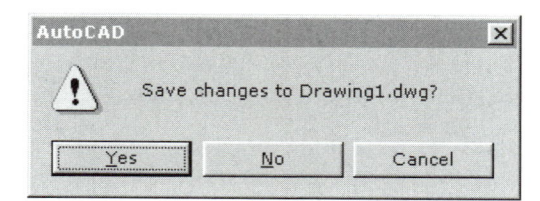

Move the cursor to choose the **No** button to exit without saving your work. AutoCAD does not record your work to disk, and exits to the Windows desktop.

Save the Drawing, and Exit AutoCAD

To save your work and exit AutoCAD, enter the same QUIT command (or select **Exit** from the **File** menu). This time choose **Yes** to save your work. Notice that AutoCAD displays the Save Drawing As dialog box.

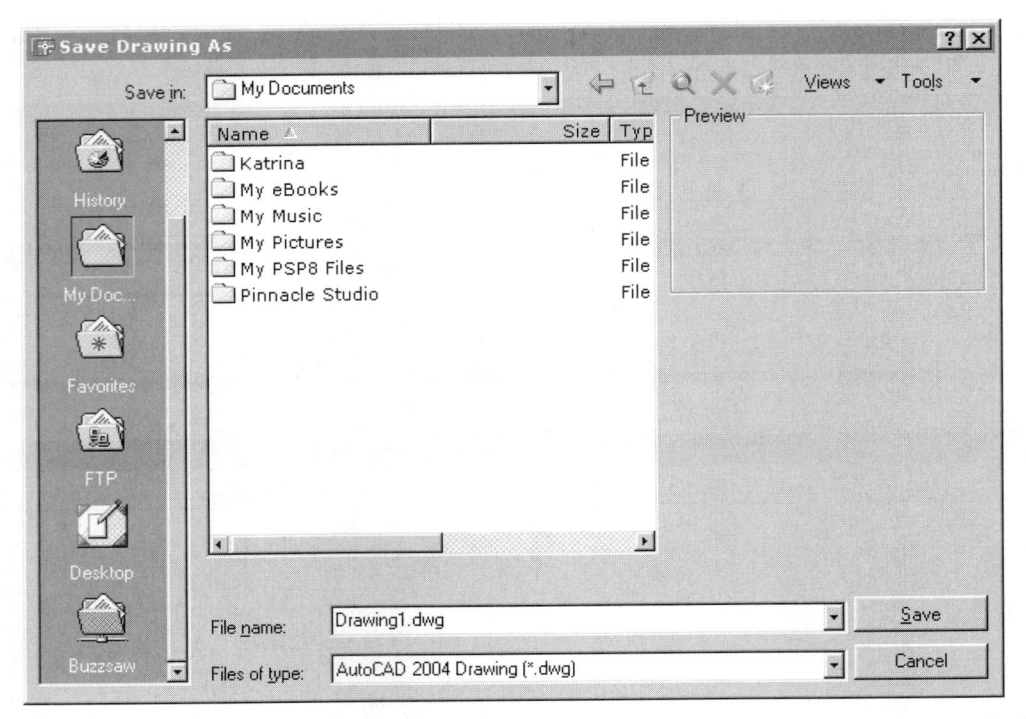

Enter a name for the drawing in the **File name** text box, and then choose **Save**. AutoCAD saves your work under the name, and exits to the Windows desktop.

Save the Drawing, and Remain in AutoCAD

If you want to keep your work *and* remain in AutoCAD, enter the QSAVE command (or select **File | Save**). If the drawing has not yet been named, the same Save Drawing As dialog box appears; name the file, and and then choose **Save**.

Remain in AutoCAD, and Start a New Drawing

If you would like to start a new AutoCAD drawing, use the NEW command. AutoCAD displays the Create New Drawing dialog box, which looks like the Startup dialog box shown earlier in this chapter.

Close the Drawing and Remain in AutoCAD

You can have more than one drawing open in AutoCAD. To close the current drawing, enter the CLOSE command (or from the **File** menu, select **Close**). If necessary, AutoCAD asks if you wish to save the drawing. AutoCAD then closes the drawing.

To close all drawings at once, enter the CLOSEALL command; from the **Window** menu, select **Close All**.

With the basics of navigating AutoCAD behind you, practice drawing in AutoCAD. In the following tutorial, you draw a three-dimensional object.

TUTORIAL

The figure below illustrates a rendering of the drawing you construct in this tutorial. It consists of a base with four holes, and a shaft with rounded edges.

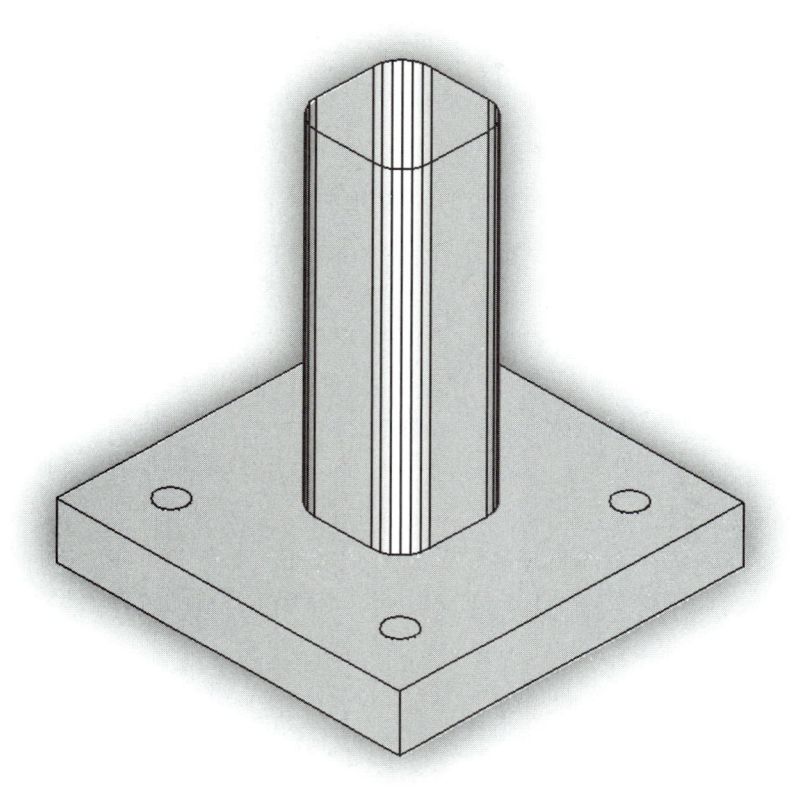

GETTING STARTED

The following tutorial describes the commands to complete the drawing. The commands you type appear on the command line. Your responses are shown by boldface in this tutorial. (You can type the items in either uppercase or lowercase.)

Items enclosed in parentheses (such as these) are instructions and are not typed. If ENTER is shown, press the **Enter** key on the keyboard. For example:

>Command: **box** *(Press* ENTER.*)*

If you make a mistake, just type **U**, and then press ENTER . (Or, you can click the **Undo** button on the toolbar, or press CTRL+Z.) This undoes the previous step. You may use it several times to undo each step in reverse order.

Before entering a command, the prompt on the command line must be "empty," with nothing following the colon:

>Command:

If not, press ESC. This cancels the current command, and makes AutoCAD ready for your command.

Tutorial: Setting Up New Drawings

1. To make sure AutoCAD displays the Create New Drawing dialog box, enter the following command:

    Command: **startup**

    Enter new value for STARTUP <0>: **1** *(Press* ENTER.*)*

2. Start a new drawing.
    From the **File** menu, choose **New**.

3. In the Create New Drawing dialog box, choose the **Use a Wizard** button.

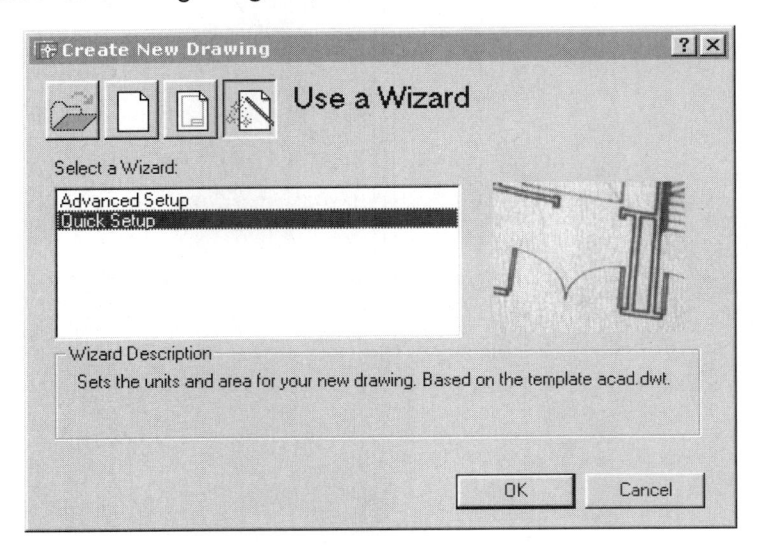

4. Select **Quick Setup**, and then choose the **OK** button.
    Notice that AutoCAD displays the Quick Setup dialog box.

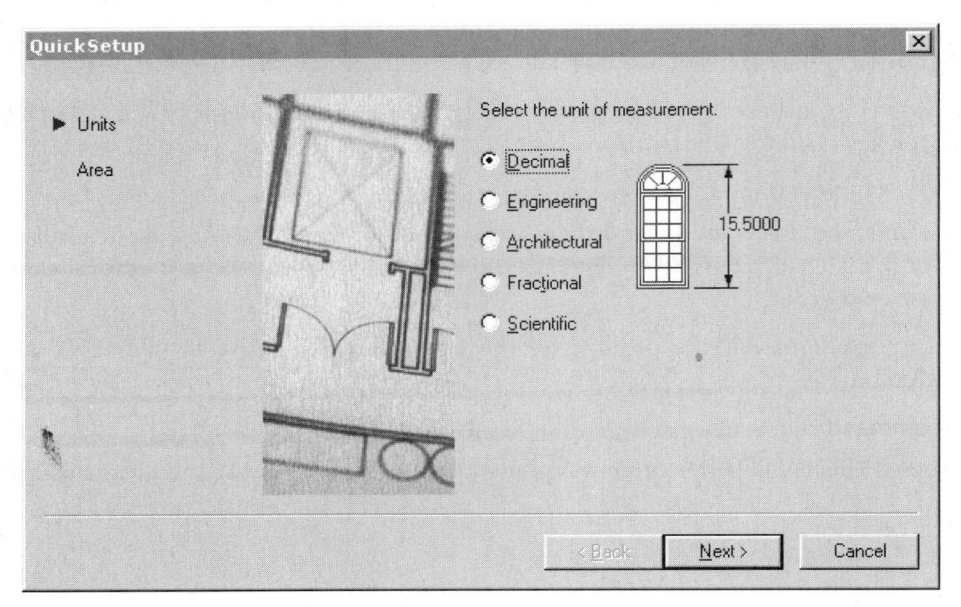

    Quick Setup presents a short series of setup dialog boxes that lead you through the steps of quickly setting up a new drawing with units and limits.

5. In the first dialog box, you specify the units.
   Accept the default of **Decimal** by choosing the **Next** button.
   Notice that AutoCAD displays the Area dialog box.

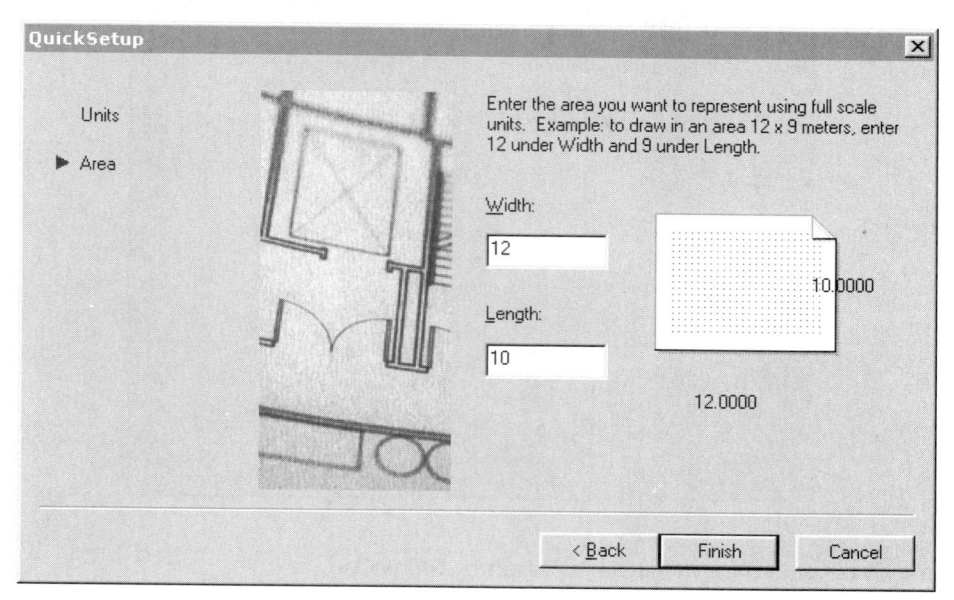

6. Here you specify the limits of the drawing: 12 units wide by 10 units tall (long).
   In the Area dialog box, define the area by clicking the **Length** text box.
   Clear the current value using the DEL or BACKSPACE key.
   Type **10**, and then press TAB key to see the effect in the preview image.
   Keep the **Width** at 12.

7. Choose **Finish**. AutoCAD dismisses the dialog box.

SNAP AND GRID

To help you draw accurately, turn on some drawing aids, such as *snap* and *grid*.

The grid is a visual guide that shows distances. Think of it as graph paper with dots, instead of lines. When the grid is set to 1 inch, for example, the drawing is covered with dots spaced one inch apart. The grid is displayed, but not printed.

Snap is like drawing resolution. When on, snap causes the cursor to move in increments. When the snap is set to 1 inch, for example, the cursor moves in one-inch increments. The grid and snap can have the same spacing, or a spacing different from each other.

TUTORIAL: SETTING UP DRAWING AIDS

1. From the **Tools** menu, choose **Drafting Settings**.
   Notice that AutoCAD displays the Drafting Settings dialog box.
   Select the **Snap and Grid** tab.

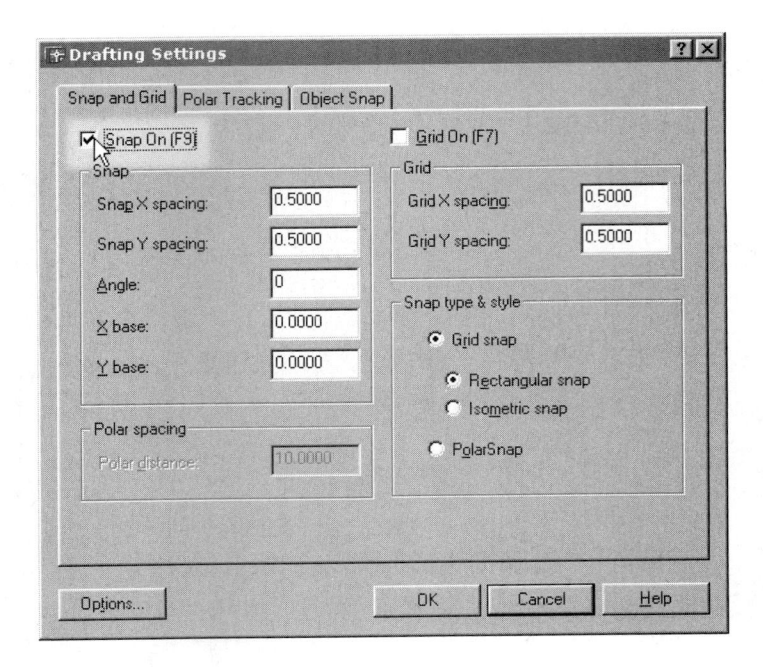

2. Turn on snap by clicking the check box next to **Snap On**.
    ☑ Check mark indicates the option is turned on.
    ☐ No check mark means option is turned off.
   The **F9** next to **Snap On** is a reminder that you can press the F9 function key to turn snap on and off. This is handy when you need to *toggle* (turn on and off) snap during commands.
   Alternatively, click the word **SNAP** on the status bar at the bottom of the AutoCAD window.

3. Change the **Snap X spacing** and **Snap Y spacing** to 1.

4. Do the same for the grid: turn it on, and set the grid spacing to 1.
   The settings in the dialog box should look like the highlighted area in the figure below.

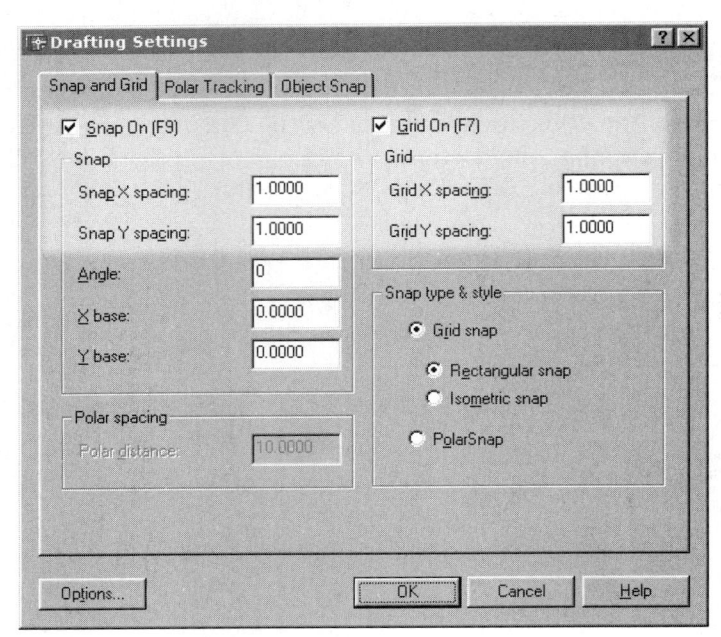

5. Choose **OK** to dismiss the dialog box. Notice that a grid of fine dots fills the drawing area.

   If the dots don't fill the drawing area, use the ZOOM command with the **All** option to see all of the drawing:

   Command: **zoom**

   Specify corner of window, enter a scale factor (nX or nXP), or

   [All/Center/Dynamic/Extents/Previous/Scale/Window] <real time>: **all**

6. Move the mouse to see the crosshairs cursor move in increments of one unit.

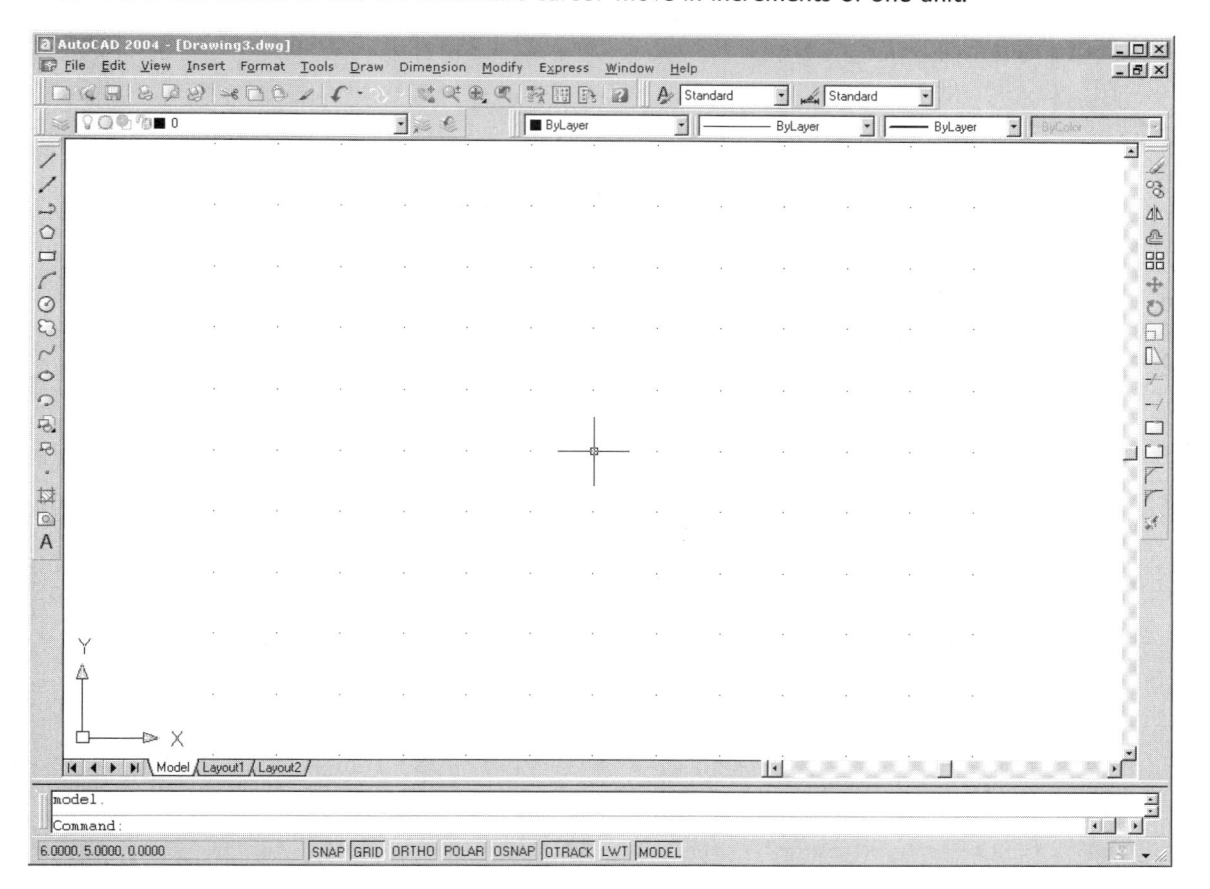

CREATING THE DRAWING

The drawing contains a baseplate, which you draw first. In later stages, you add the holes and the post.

In new drawings, you look down in plan (2D) view. Later, you change to a 3D view.

Drawing the Baseplate

The baseplate is drawn with the **BOX** command, which draws box shapes in three-dimensions. AutoCAD needs to know two things to draw a box: (1) the position of the two opposite corner, and (2) the height.

1. From the **Draw** menu bar, select **Solids**, and then **Box**.

    As an alternative, enter the **BOX** command at the keyboard:

    Command: **box** *(Press* ENTER.*)*

2. AutoCAD needs to know where to place the box.

    Specify corner of box or [CEnter] <0,0,0>: **3,2** *(Press* ENTER.*)*

    The numbers "3,2" specify the x,y coordinates of one corner.
    The opposite corner is at 9,8:

    Specify corner or [Cube/Length]: **9,8** *(Press* ENTER.*)*

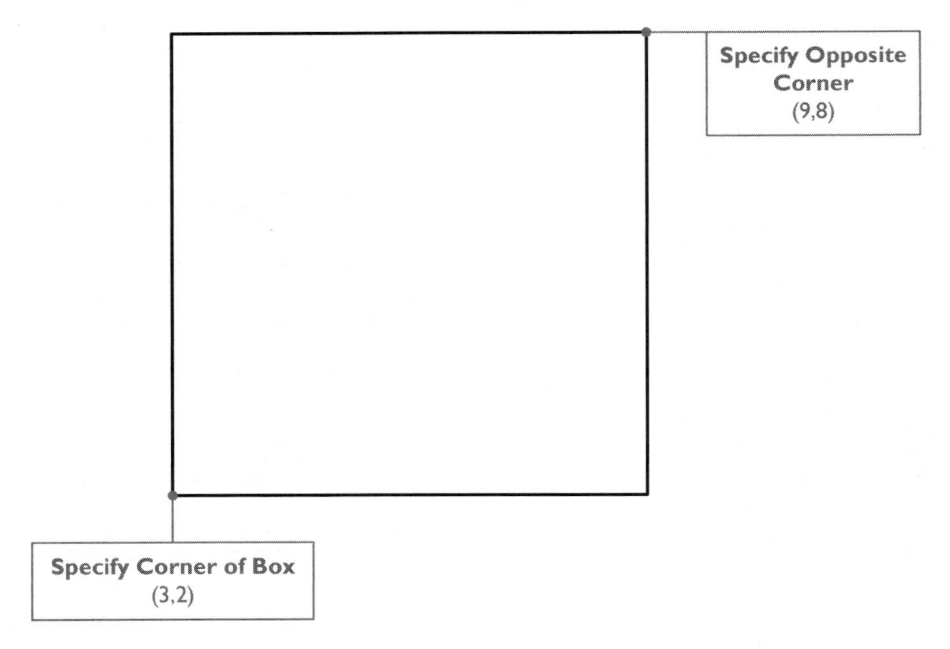

    Specify Opposite Corner (9,8)

    Specify Corner of Box (3,2)

3. The base is one unit tall:

    Specify height: **1** *(Press* ENTER.*)*

    Unlike the **LINE** command, the **BOX** command stops by itself; there is no need to press **ESC**.

4. Your drawing should look similar to the figure. If necessary, enter the **ZOOM** command, and select the **All** option:

    Command: **zoom** *(Press* ENTER.*)*

    Specify corner of window .. .<real time>: **all** *(Press* ENTER.*)*

Drawing the Shaft

Draw the shaft by repeating the BOX command.

1. Pressing the **SPACEBAR** repeats the previous command without requiring you to type its name a second time.

    Command: *(Press SPACEBAR.)*

2. This time, you specify the corners of the box using x,y,z coordinates, because the shaft sits on top of the one-unit tall base:

    BOX Specify corner of box or [CEnter] <0,0,0>: **5,4,1** *(Press ENTER.)*

    Specify corner or [Cube/Length]: **7,6,1** *(Press ENTER.)*

    Specify height: **6** *(Press ENTER.)*

Your drawing should look similar to the figure.

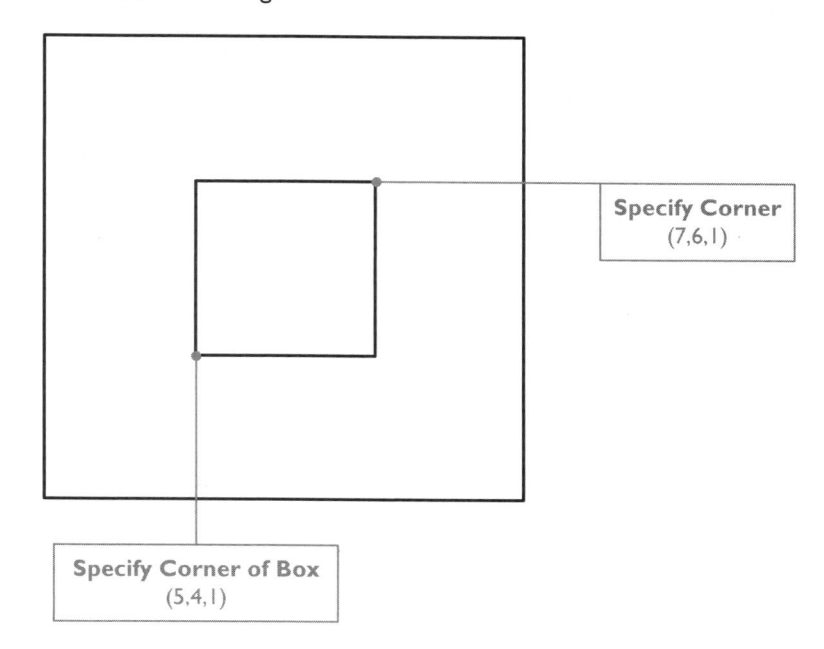

Specify Corner (7,6,1)

Specify Corner of Box (5,4,1)

Drawing Cylinders

Now draw the holes. Holes are created in two steps: draw cylinders, and then subtract the cylinders from the box.

To draw cylinders, AutoCAD needs three things: (1) the position of the cylinders in the drawing, based on their center points; (2) the radius or diameter of the cylinders; and (3) their height.

1. From the menu bar, select **Draw | Solids | Cylinder**.
    (The notation is shorthand for three steps: choose **Solids** from the **Draw** menu, and then choose **Cylinder**. The vertical bars, |, separate menu picks.)
    AutoCAD responds, as follows:
    Command: _cylinder Current wire frame density: ISOLINES=4

2. The cylinder is located at 4,3 in the drawing:
    Specify center point for base of cylinder or [Elliptical] <0,0,0>: **4,3** (Press ENTER.)

3. The cylinder has a radius of 0.25 units:
    Specify radius for base of cylinder or [Diameter]: **0.25** (Press ENTER.)

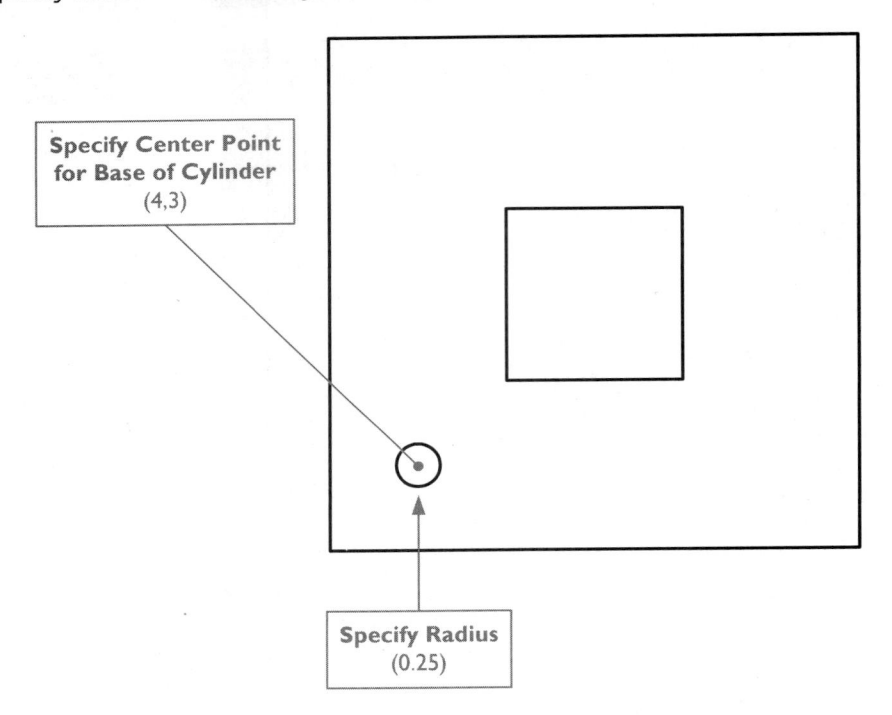

4. And the cylinder has the same height as the base, one I unit:
    Specify height of cylinder or [Center of other end]: **1** (Press ENTER.)

Copying Cylinders

To draw the other three cylinders, copy the first cylinder to the other locations on the baseplate.

1. The **COPY** command has a **Multiple** option that allows you to make multiple copies easily. Copy the last-drawn object using the **L** (short for "last") object selection option, as follows:

    Command: **copy** *(Press ENTER.)*

    Select objects: **L** *(Press ENTER.)*

    1 found Select objects: *(Press ENTER to end object selection.)*

2. Enter **m** to specify the **Multiple** option:

    Specify base point or displacement, or [Multiple]: **m** *(Press ENTER.)*

3. The *base point* is the point from where the copying takes place:

    Specify base point: **4,3** *(Press ENTER.)*

4. The *displacement* is the point at which the copy is placed:

    Specify second point of displacement or <use first point as displacement>: **8,3** *(Press ENTER.)*

    Specify second point of displacement ... : **8,7** *(Press ENTER.)*

    Specify second point of displacement ... : **4,7** *(Press ENTER.)*

    Specify second point of displacement ... :*(Press ESC to end the command.)*

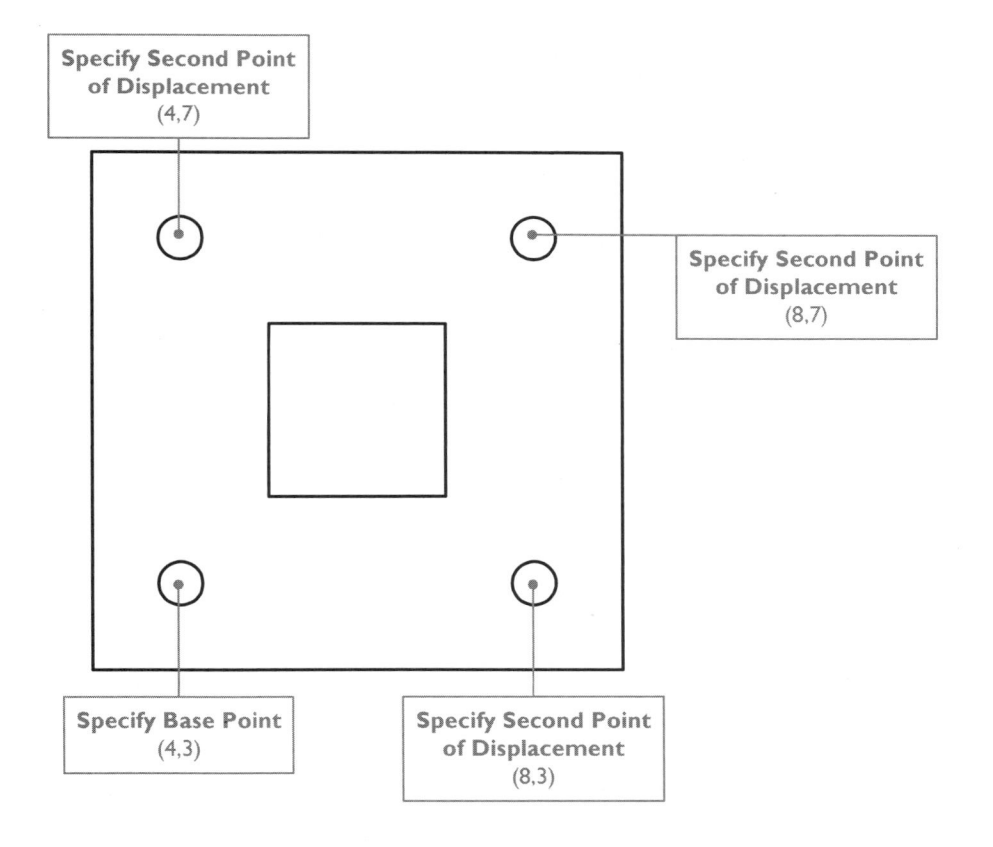

Creating Holes

You have drawn four cylinders that represent the holes, but they are "solid" cylinders. You need to remove the cylinders from the baseplate so that they become holes. To do so, use the **SUBTRACT** command.

This command needs to know two things: (1) the object(s) to subtract *from*; and (2) the objects to be subtracted.

1. From the **Modify** menu, select **Solids Editing**, and then choose **Subtract**.
    Or, enter **SUBTRACT** at the Command: prompt:
    Command: **subtract** *(Press ENTER.)*
2. Select the baseplace, from which the cylinders are subtracted:
    Select solids and regions to subtract from ..
    Select objects: *(Select the baseplate)*
3. And then select the four cylinders using **Fence** object selection:
    1 found Select objects: *(Press ENTER to end object selection.)*
    Select solids and regions to subtract...
    Select objects: **fence**
    First fence point: *(Place cursor at center of one circle, and then click.)*
    Specify endpoint of line or [Undo]: *(Click at center of second circle.)*
    Specify endpoint of line or [Undo]: *(Click at center of next circle.)*
    Specify endpoint of line or [Undo]: *(Click at center of last circle.)*
    Specify endpoint of line or [Undo]: *(Press ENTER to end fence selection.)*

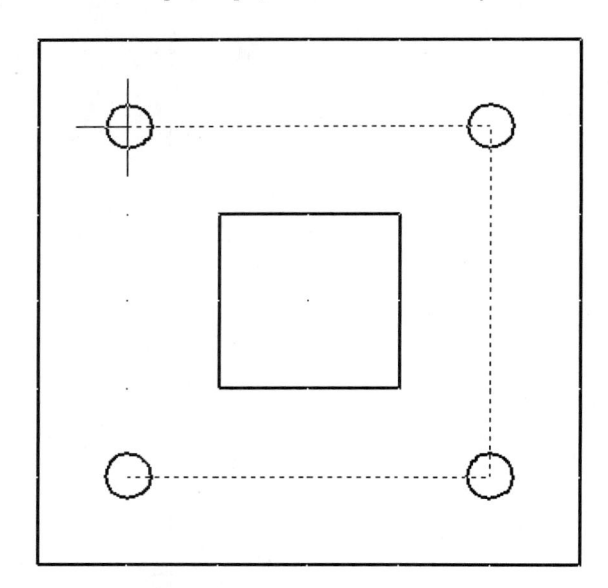

    4 found Select objects: *(Press ENTER to end object selection.)*

The baseplate doesn't look any different after the cylinders are subtracted — until you remove the hidden lines at the end of this tutorial.

VIEWING THE DRAWING IN 3D

Let's have some fun, and view the drawing in 3D.

1. From the **View** menu, select **3D Views**, and then choose **SW Isometric**.
   Notice that you see all of the drawing from a point in the "air," looking down at the model — called the "isometric viewpoint."

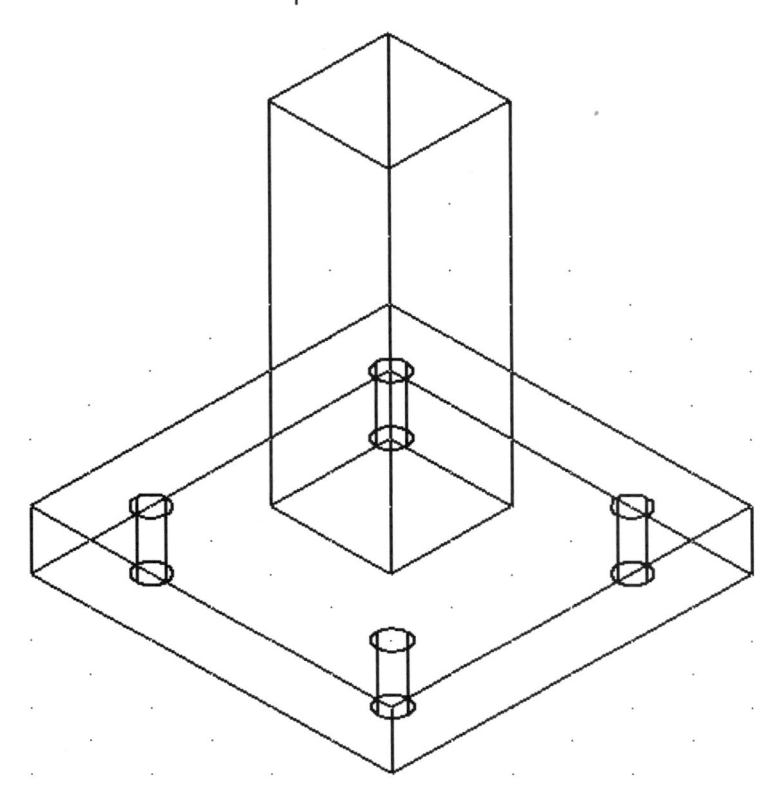

Rounding the Corners

The corners of the shaft are rounded with the FILLET command. This command needs to know two things: (1) the radius of the rounding, and (2) which edges to round.

1. From the **Modify** menu, select **Fillet**.

    Command: _fillet Current settings: Mode = TRIM, Radius = 0.5000

2. Select the shaft:

    Select first object or [Polyline/Radius/Trim]: *(Select one vertical edge of the shaft.)*

3. Specify a radius of 0.5 units:

    Enter radius <0.0>: **0.5** *(Press ENTER.)*

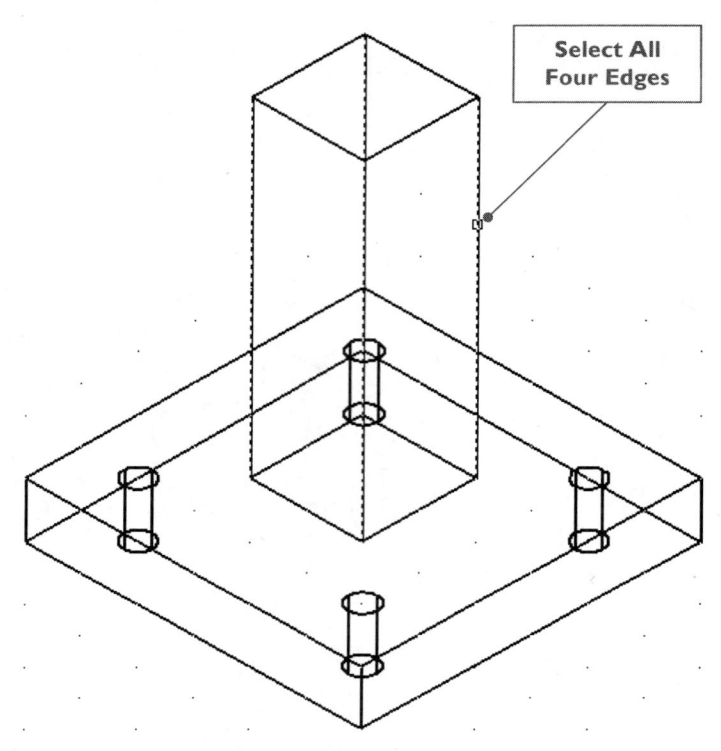

4. And now select each one of the four vertical edges. You won't see the fillets until you end the command.

    Select an edge or [Chain/Radius]: *(Select another vertical edge of the shaft.)*

    Select an edge or [Chain/Radius]: *(Select another vertical edge.)*

    Select an edge or [Chain/Radius]: *(Select last vertical edge)*

    Select an edge or [Chain/Radius]: *(Press ENTER to end edge selection.)*

    4 edge(s) selected for fillet.

    Notice that AutoCAD fillets the four corners of the shaft.

5. Save the drawing by the name "tutorial.dwg", as you learned earlier in this chapter.

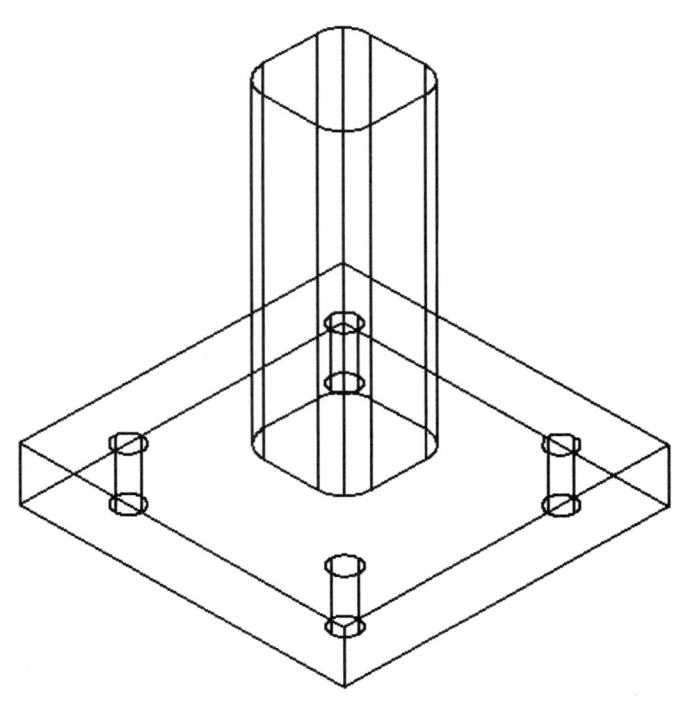

Removing Hidden Lines

You are seeing the 3D object in *wireframe* view, which shows you *all* the lines in the drawing. To make the drawing look more like a real object, AutoCAD can remove the lines that should be hidden from view.

1. Run the **HIDE** command:

    Command: **hide**

    Regenerating model.

    Notice that the model looks more realistic.

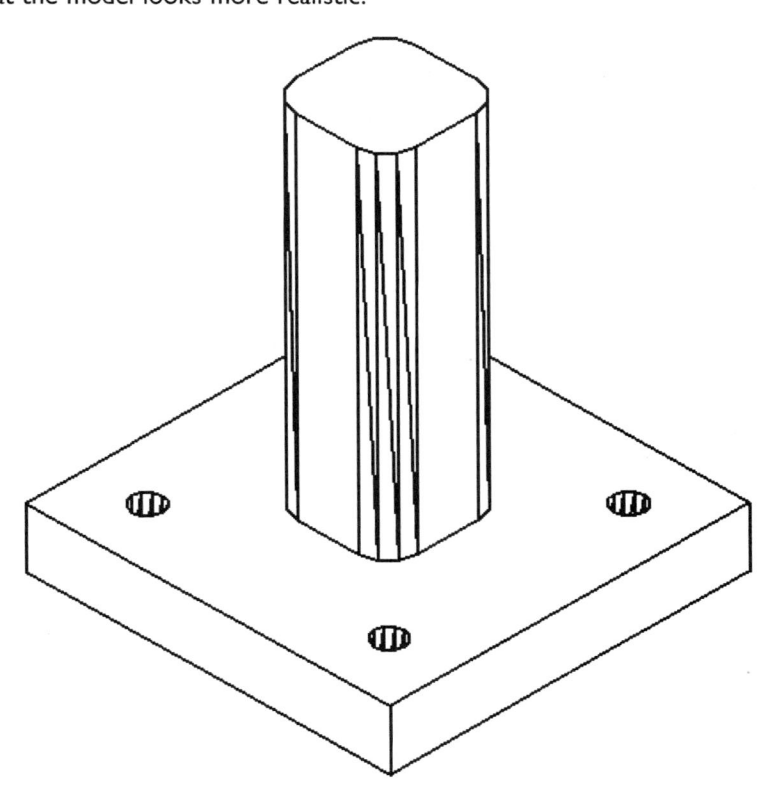

PLOTTING THE DRAWING

With the drawing complete, print out a copy to show your fine work to your instructor, family, and friends. This is done with the **PLOT** command.

Before printing, AutoCAD needs to know: (1) the printer on which the drawing will be plotted, (2) the view to plot, and (3) the scale. There are many other options, but those are the basic ones.

1. From the menu bar, select **File | Plot.**
   AutoCAD displays the Plot dialog box.

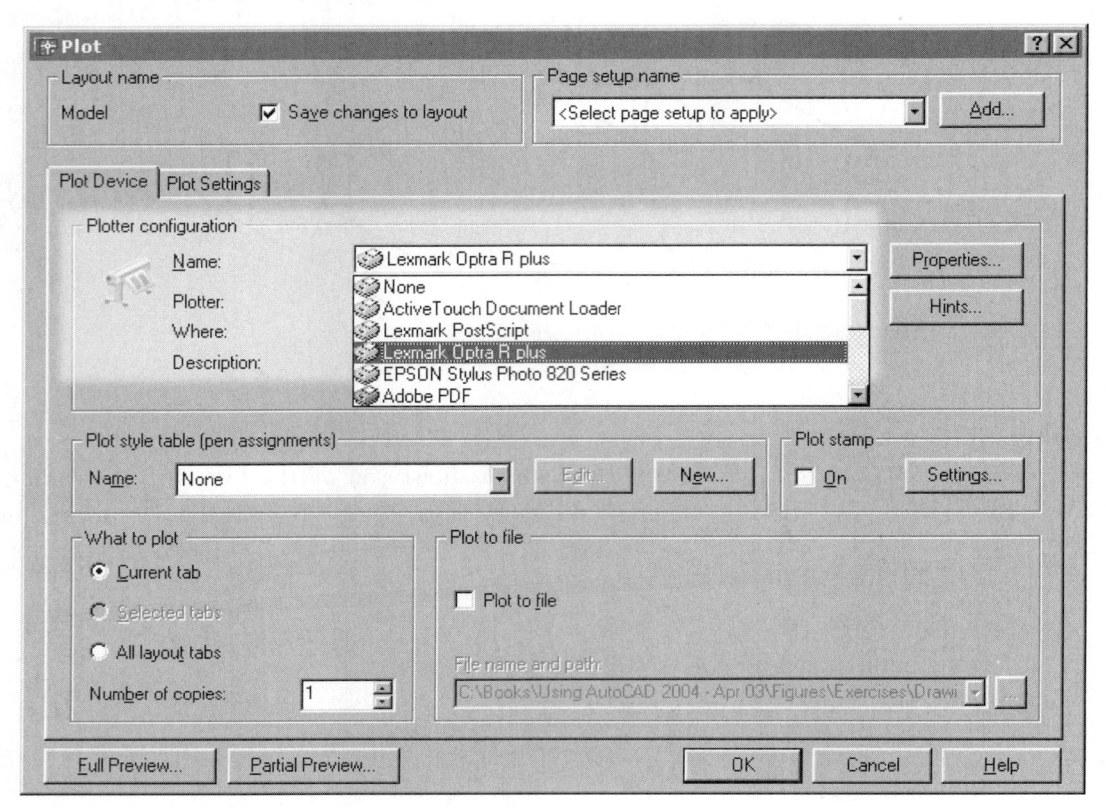

2. Select the **Plot Device** tab.
   In the Plotter Configuration area, select a printer from the **Name** drop list.
   You can afford to ignore all other settings in this tab.

3. Select the **Plot Settings** tab.
   In the Plot Area section, select **Extents**. This option ensures the entire drawing is plotted.

4. In the Plot Scale area, select **Scaled to Fit**. This ensures the drawing fits the paper, no matter what size the paper is.

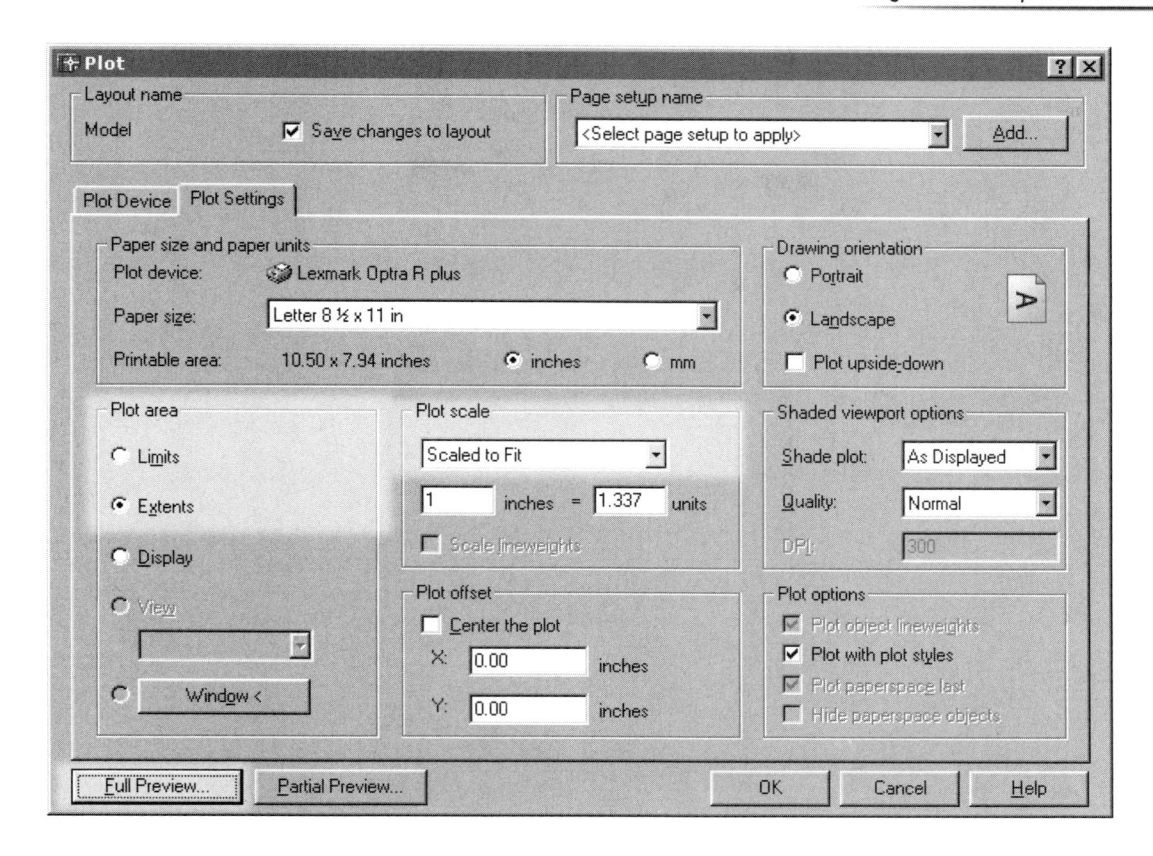

5. To ensure the plot works out correctly, and to save paper, click **Full Preview**. If AutoCAD displays the following dialog box, choose **Yes**.

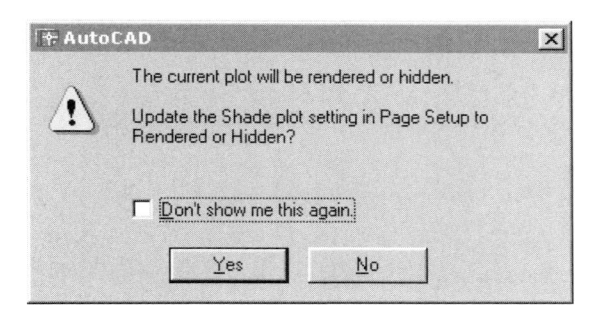

AutoCAD shows you what the plot will look like, as illustrated by the figure below.

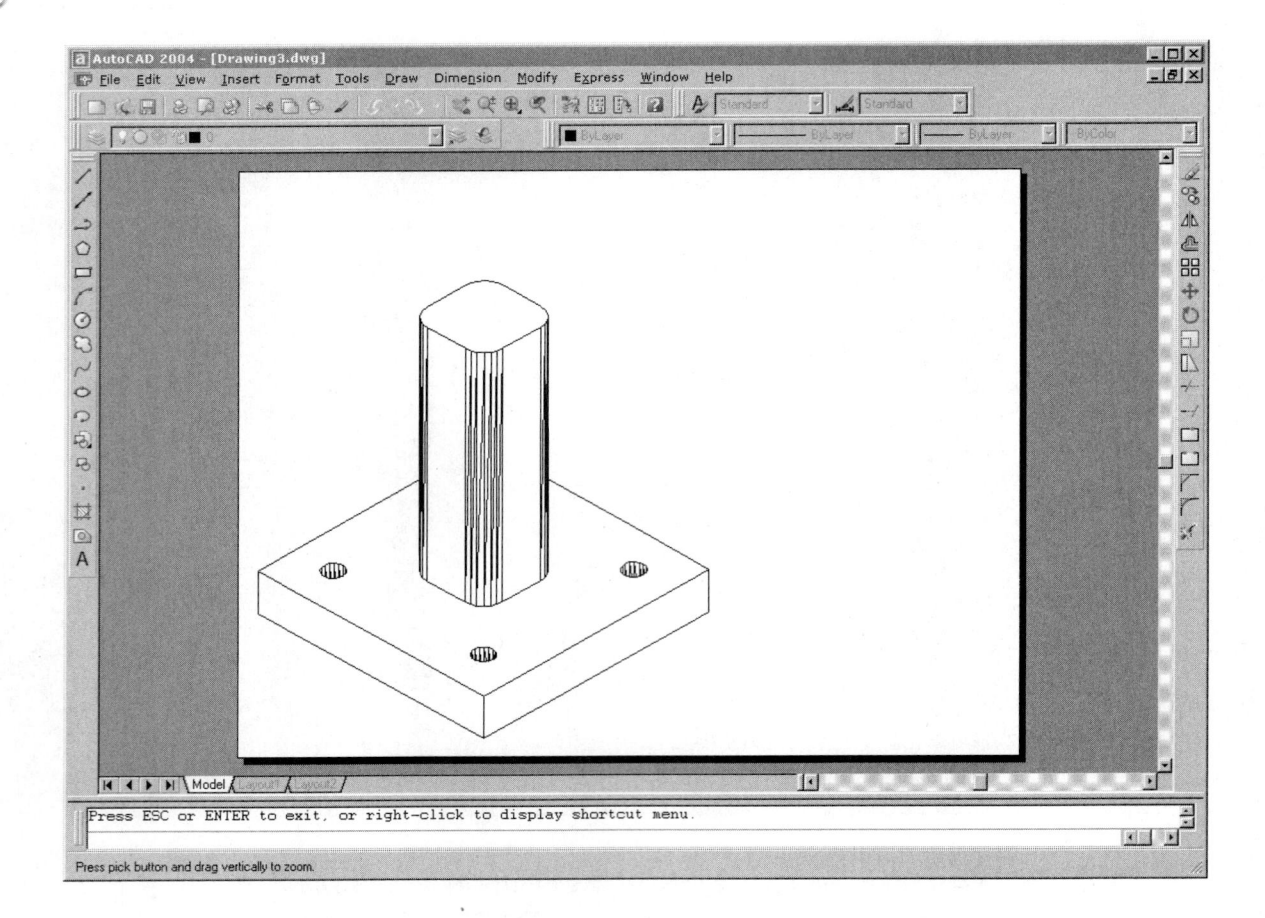

6. Press **ESC** to exit the print preview mode.
   Press ESC or ENTER to exit, or right-click to display shortcut menu. *(Press* **ESC**.*)*

7. Back in the Plot dialog box, choose **OK** to start the plot.
   AutoCAD reports:
   Effective plotting area: 7.94 wide by 6.35 high
   Plotting viewport 2.
   After a moment, the drawing should emerge from your printer.

8. To return to the *plan* (2D) view, enter the **PLAN** command, and then press **ENTER** twice.

9. Save the drawing with the **QSAVE** command, and then exit AutoCAD, as you learned earlier in this chapter.

So, that was a quick start to using AutoCAD. Welcome to the rest of *Using AutoCAD 2004: Basic*. Tthe remaining chapters take you step by step through the program. Before long you'll be using AutoCAD like a pro — enjoy!

EXERCISES

1. In this exercise, you "play around" with drawing in AutoCAD.
   Start AutoCAD. (If the Start New Drawing dialog box appears, click **Cancel**.)
   With the **LINE** command, draw several shapes:
   a. Rectangle.
   b. Triangle.
   c. Irregular polylgon (any shape you like).

2. Use the **U** command. What happens to the last object you drew?

3. Press function key **F9**, and then look at the status line.
   Does the button look depressed (pressed in)?

4. Repeat the **LINE** command, and again try drawing these shapes:
   a. Rectangle.
   b. Triangle.
   c. Irregular polylgon.
   Do you find it easier?

5. Draw a 3D rectangle with the **BOX** command with these parameters:

Corner	**2,3**
Other corner	**6,7**
Height	**3.0**

   Use the **View | 3D Views | SW Isometric** command to see the box in three dimensions.
   Do you find drawing three-dimensional objects easy?

6. Draw a 3D cylinder with the **CYLINDER** command with these parameters:

Center point	**4,5**
Radius	**0.75**
Height	**3.0**

   Is the cylinder drawn "inside" the box?

7. Use the **SUBTRACT** command to remove the cylinder from the box.
   Check that you selected objects in the correct order with the **HIDE** command.
   Which does your drawing show — a cylinder, or a box with a hole?

8. Print your drawing with the **PLOT** command.
   Does the plot look like the drawing on your computer screen?

CHAPTER REVIEW

1. What are AutoCAD drawings constructed with?
2. Name three areas from which commands are entered:
   a.
   b.
   c.
3. Describe the purpose of the 'Command:' line.
4. Must the command line must be "clear" before typing a new command from the keyboard?
5. What is the purpose of the **LINE** command?
6. How is the **U** command helpful?
7. Describe how to cancel commands.
8. What does the mouse control?
9. What is an *icon*?
10. Name two methods by which you can determine the function of a toolbar button:
    a.
    b.
11. What is a *tooltip*?
12. What is a *flyout*?
13. Can toolbars be moved around the AutoCAD window?
14. Which commands close a drawing without exiting AutoCAD?
15. Which commands exit AutoCAD?
16. Which command saves drawings?
17. Which command starts new drawings?
18. Describe how snap and grid are useful:
    Snap:
    Grid:
19. Is the grid plotted?
20. List three things AutoCAD needs to know, as a minimum, before plotting drawings:
    a.
    b.
    c.
21. Explain the advantage of the **PLOT** command's **Fit to Scale** option.
22. Why is the **Full Preview** option environmentally friendly?
23. How do you exit from print preview mode?
24. Which command converts a 3D view back to 2D view?

25. Identify the user interface elements illustrated below:
    a.
    b.
    c.
    d.
    e.

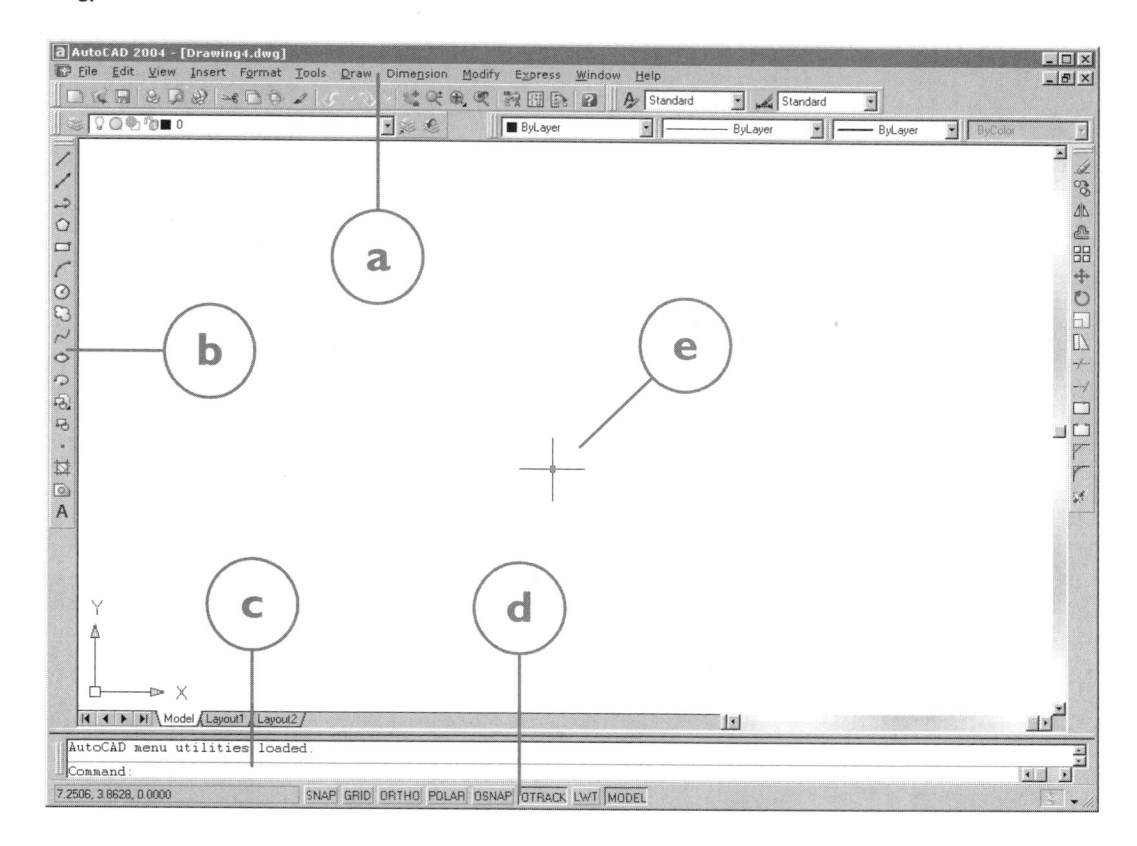

CHAPTER 2

Understanding CAD Concepts

CAD software contains concepts that distinguish it from other software programs. In part, the uniqueness stems from CAD being based on *vectors*, geometric objects that have length and direction. Other concepts are probably familiar to you, such as using dialog boxes and menus. This chapter covers the following areas:

- Touring the AutoCAD user interface.
- Understanding the information AutoCAD presents you.
- Entering commands.
- Picking points on the screen.
- Learning shortcut keystrokes and command aliases.
- Using mouse buttons and menus.
- Identifying the elements of dialog boxes.
- Understanding the draft-edit-plot cycle.
- Recognizing AutoCAD's coordinate systems.
- Entering coordinates from the keyboard and on the screen.

BEGINNING A DRAWING SESSION

Your drawing sessions begin by starting the AutoCAD program. As described in Chapter 1, "Quick Start in AutoCAD," double click the AutoCAD icon to start the program. (If there is no icon on the Windows desktop, click the Windows **Start** button, then choose **Programs**. Choose the **Autodesk** folder, choose the **AutoCAD 2004** folder, and then choose the **AutoCAD 2004** program.)

As an alternative, double-click .*dwg* files in Windows Explorer. (Avoid this method when another Autodesk application, such as Mechanical Desktop, is installed.)

When AutoCAD loads, you see either a blank drawings or the Startup dialog box — depending on how it is configured. (The Today window was removed from AutoCAD 2004.)

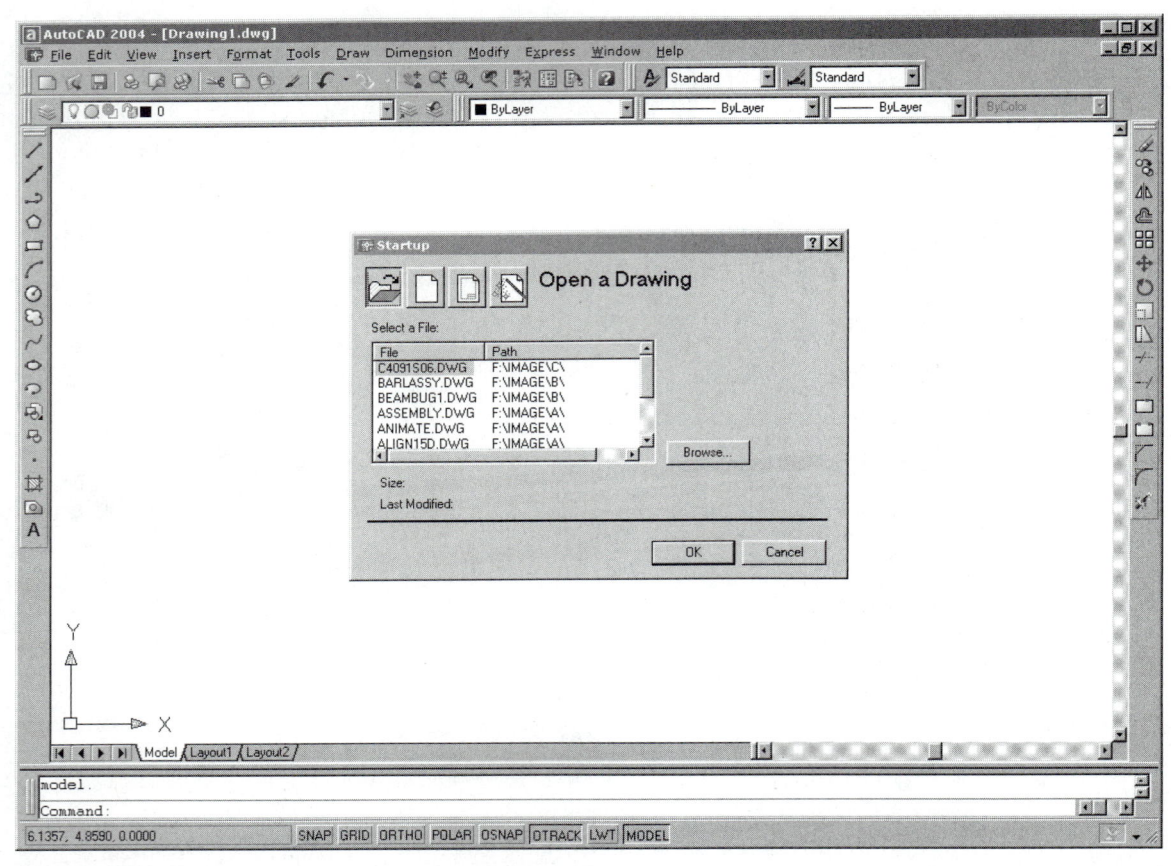

If the Startup dialog box appears, choose **Cancel** to dismiss it. You now see the *drawing area*, the window in which your drawing takes place.

AUTOCAD USER INTERFACE

Notice that AutoCAD displays information in several areas. Let's look at these areas, starting at the top and moving down the window.

Title Bar

At the very top of the AutoCAD window is the *title bar*. The title bar tells you the name of the software (AutoCAD 2004) and the name of the drawing, such as *drawing1.dwg*.

Toolbars

Just above the drawing area, AutoCAD displays the state of layers and objects. These toolbars list the names of layers, colors, linetypes, lineweights, plot styles, text styles, and dimension styles.

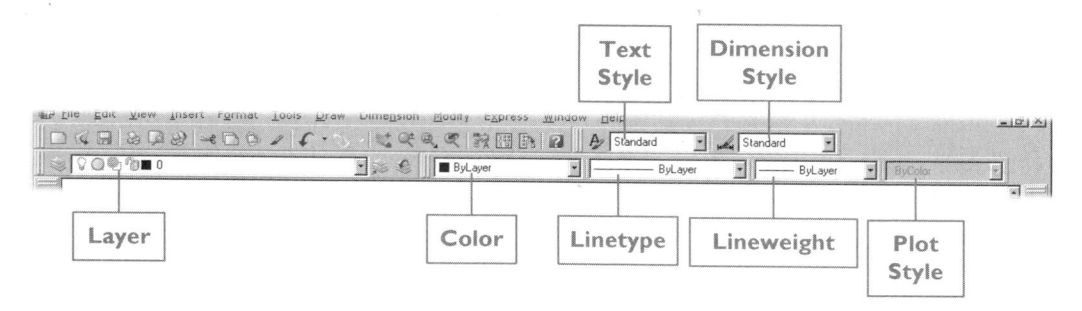

Layer

The layer property reports the status, color, and name of the current layer. Every new drawing contains at least one layer called "0" (zero). The icons prefixing the layer name indicate the layer status:

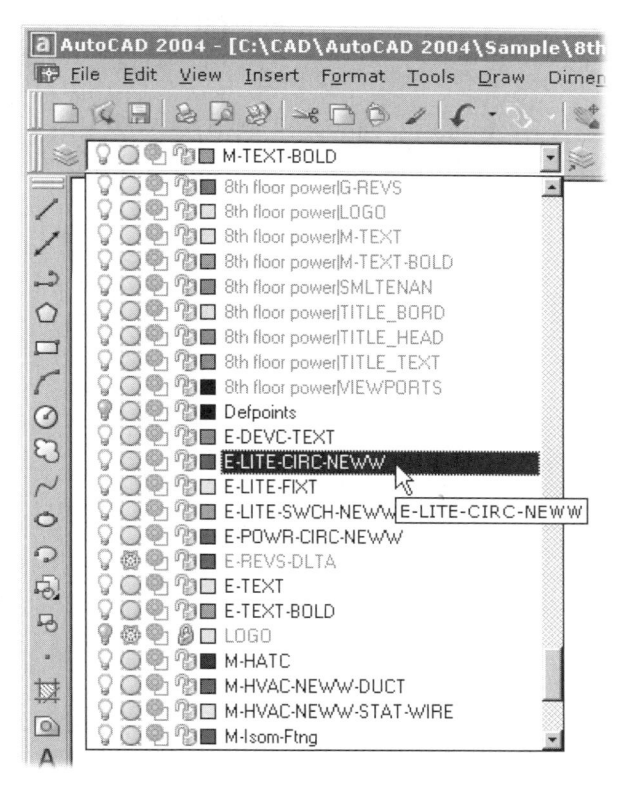

Lightbulb means the layer is on or off.

Shining sun means the layer is thawed.

Snowflake means the layer is frozen.

Sun with rectangle means the layer is thawed in the current viewport

Padlock means the layer is locked or unlocked.

The small square prefixing the layer name indicates the color of the layer.

Color

The color property lists the current working color. AutoCAD works with as many as 16.7 million colors. In most cases, the color is called "BYLAYER," which means that the layer setting dictates the color of objects. This, however, can overridden on an object-by-object basis.

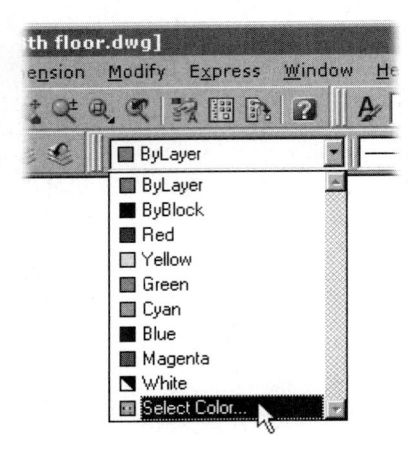

Linetype

The linetype property lists the look and name of the current *linetype* (a.k.a. line pattern). AutoCAD supports two styles of linetype: simple patterns made with dashes, dots, and gaps; and complex patterns with text and 2D shapes.

As with color, the linetype you see most commonly is BYLAYER, which means the layer setting determines the linetype displayed by objects. As with colors, this can be overridden for each object. You must load a linetype's definition before you can assign it.

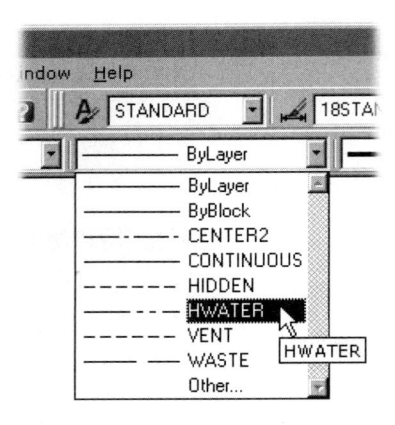

Lineweight

The lineweight property lists the *weight* (a.k.a width) of lines. As with colors and linetypes, most lines have a weight of BYLAYER, which means the layer decides the width of the line. This can be overridden with a specified width.

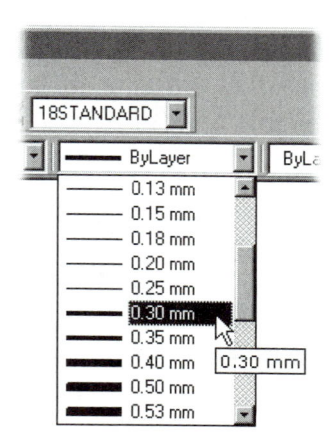

Plot Styles

The plot styles property lists the current *plot style*, if any, in the drawing. A plot style collects settings — such as lineweight, end caps, line fill, and screening — that affect the appearance of objects in two ways: (1) as they are drawn; and (2) as they are plotted. Plot styles are not available unless you explicitly turn them on.

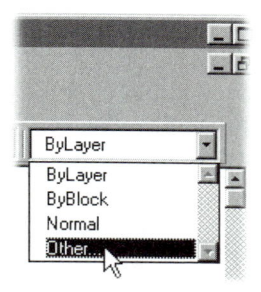

NEW IN 2004 The toolbar holding the text style and dimension style is new to AutoCAD 2004.

Text Styles

The text styles property lists the current *text style*; all new drawings have a text style named "Standard." Text styles define the font used for text placed in the drawing, as well as properties of the text, such as slant and height.

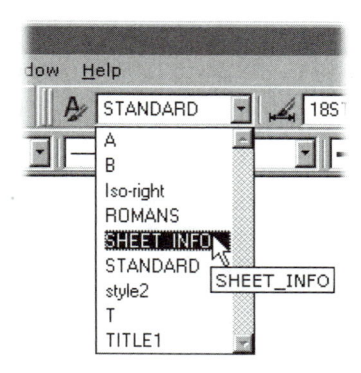

Dimension Styles

The dimension styles property lists the current dimension styles, which define the look of dimensions in the drawing, such as the type of arrowhead and the color of dimension lines. All new drawings have a dimension style called "Standard."

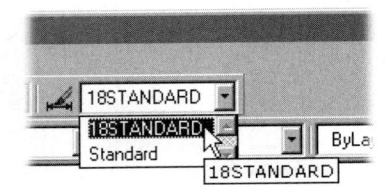

Drawing Area

Even the drawing area contains information. The *crosshair cursor* shows you where you are in the drawing.

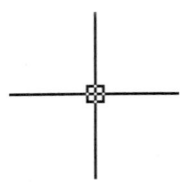

The small square at the center of the cursor indicates that you can *select* (or "pick") objects in the drawing. (Press the left mouse button to select objects.) The square is called the "pickbox." It's size can be made larger or smaller with a system variable named **PICKBOX**.

When you move the crosshair cursor out of the drawing area, it changes to an *arrow cursor*, which you are probably familiar with from other Windows software.

UCS Icon

In the lower-left corner is a double-ended arrow. This is called the *UCS icon*; UCS is short for "user coordinate system." The UCS icon is primarily used when you draw in three dimensions; it helps you orient yourself in three-dimensional space. Sometimes, the UCS icon will change to other shapes. You learn how to use the UCS icon in volume 2 of this book, *Using AutoCAD 2004: Advanced*.

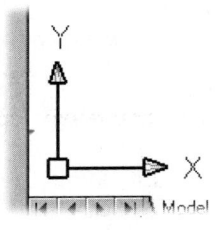

The UCS icon is not needed for 2D drafting; to turn off, enter the **UCSICON** command, as follows:

Command: **ucsicon** *(Press* ENTER.*)*
Enter an option [ON/OFF/All/Noorigin/ORigin/Properties] <ON>: **off** *(Press* ENTER.*)*

Layout Tabs

At the bottom of the drawing area are several tabs, labeled Model, Layout1, and so on. These tabs allow you to switch between model space and layout mode. You learn more about layout mode later in this book.

Command Prompt Area

Below the drawing area is the command prompt area. This is where you enter commands, and where AutoCAD responds with prompts and messages. If you become confused as to what AutoCAD expects of you, the prompt line provides the answer.

You can see the history of your commands by pressing function key **F2**. AutoCAD displays a text window. Press **F2** a second time to return to the drawing window.

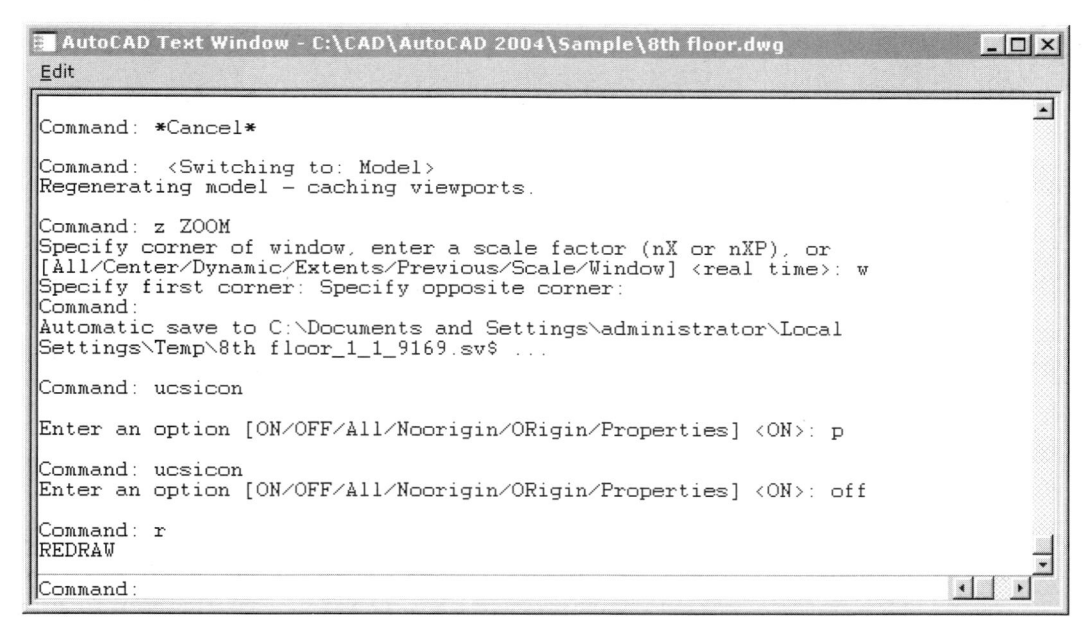

If you like, you can drag the command prompt area away from the AutoCAD window. With the mouse, drag the lower right corner (the square area between the scroll bars) away from the AutoCAD window. Release the mouse button.

Status Bar

At the bottom of the AutoCAD window is the status line. This line reports the x,y,z coordinates and status of certain drawing settings.

X, Y, Z Coordinates

At left of the status line are three numbers, which represent the X, Y, Z coordinates of the crosshairs cursor. The format of the numerical readout is controlled by the **UNITS** command, as you learn in Chapter 3, "Setting Up Drawings."

The coordinates constantly update as you move the cursor. You can change the display by pressing **CTRL+D** or **F4**. If the coordinates do not update as you move the cursor, then they are toggled off; click the coordinate display to turn them on.

Mode Indicators

The bottom of the screen, next to the coordinates, displays *mode* indicators. These modes are also called "toggles," because they are either on or off — just like a light toggle. When on, the mode's button looks as if it is pressed in. In the figure below, OTRACK and MODEL are on; the other modes are turned off.

The modes have the following meaning:

Mode	Meaning
SNAP	Cursor snap is on or off.
GRID	Grid display is on or off.
ORTHO	Orthogonal mode is on or off.
POLAR	Polar snap mode is on or off.
OSNAP	Object snap modes are on or off.
OTRACK	Object tracking mode is on or off.
LWT	Lineweights are displayed or not.
MODEL	Model space or paper space (layout mode) is current.

Tray Settings *(NEW IN 2004)*

At the right end of the status bar are three indicators new to AutoCAD 2004. This area is called the "tray."

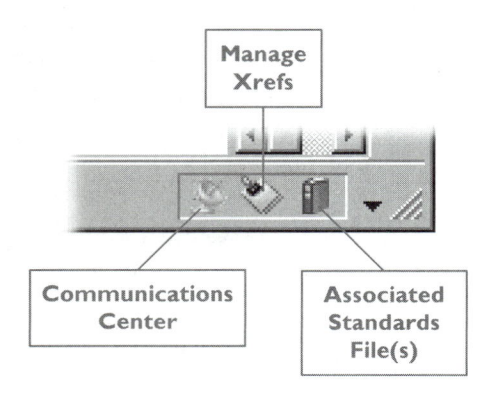

- Manage Xrefs
- Communications Center
- Associated Standards File(s)

The indicators are gray icons normally; when AutoCAD wants to communicate with you, the indicators become colored icons.

Communications Center notifies you of software updates from Autodesk through the Internet.

Manage Xrefs displays the Xref Manager dialog box; available only when an externally-referenced drawing is attached.

Associated Standards File(s) warns you when CAD standards have been breached; available only when the standards checker is active.

SCREEN POINTING

You enter points, distances, and angles in two ways: by "showing" AutoCAD the information on the screen; or by entering coordinates and angles at the keyboard.

Picking two points in the drawing could indicate, for example, a distance, an angle, or both — depending on the command and what it expects as input. With AutoCAD, you press the left mouse button to pick points in the drawing.

Showing Points by Window Corners

Some commands require input of both a horizontal and a vertical displacement. This is shown by a rectangle on the screen, called a "window." You pick a point in the drawing, and then move the cursor to form the rectangle. AutoCAD determines the x and y distances by measuring from the lower left corner to the upper right corner.

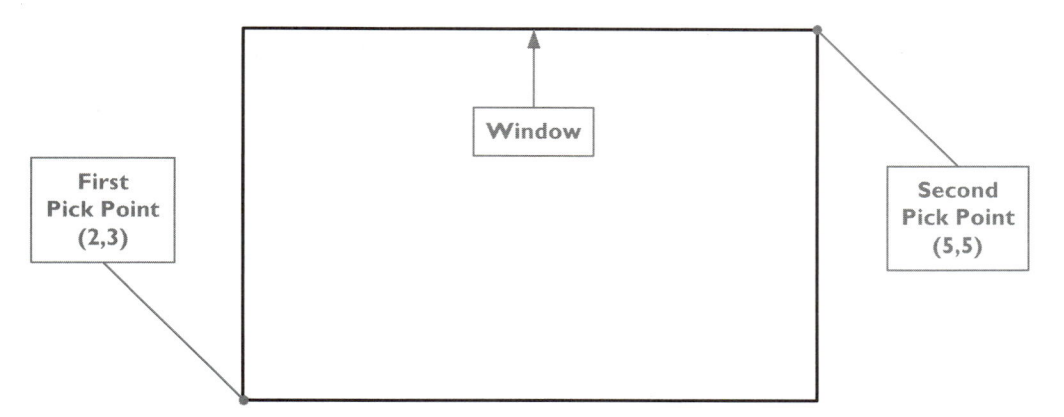

For example, the window in the figure above starts at (2,3) and rises to (5,5). To AutoCAD, this indicates a displacement of (3,2).

ENTERING COMMANDS

Commands can be entered in AutoCAD by several means. Let's look at each.

Keyboard

One method is through the keyboard — for historical reasons. The earliest versions of AutoCAD did not rely on a mouse, so everything could be done (had to be done) with the keyboard. Even though menus, toolbars, and right-click shortcut menus have replaced many commands, you still need to deal with the keyboard from time to time.

When you type commands at the keyboard, your input is displayed in the command prompt area.

To type a command from the keyboard, the command line must be "clear." That is, another command must not be in progress. If the command line is not clear, you can clear it by pressing ESC.

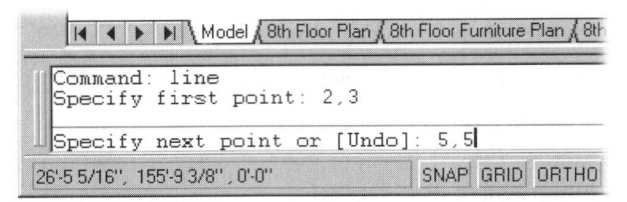

```
Command: line
Specify first point: 2,3
Specify next point or [Undo]: 5,5
```

AutoCAD doesn't care if you use uppercase or lowercase characters, or a combination of either. If you make a mistake, press the BACKSPACE key to correct (other editing keystrokes are listed below). After typing the command, you must press ENTER or the spacebar to activate the command.

Transparent commands are an exception. These commands can be entered during another command. Transparent commands are regular commands prefixed with the ' (apostrophe) character.

You edit the text on the command line using these keys:

Keystroke	Meaning
left arrow	Moves the cursor one character to the left.
right arrow	Moves the cursor to the right.
HOME	Moves the cursor to the beginning of the line of command text.
END	Moves the cursor to the end of the line.
DEL	Deletes the character to the right of the cursor.
BACKSPACE	Deletes the character to the left of the cursor.
INS	Switches between insert and typeover modes.
up arrow	Displays the previous line in the command history.
down arrow	Displays the next line in the command history.
PGUP	Displays the previous screen of command text.
PGDN	Displays the next screen of command text.
CTRL+V	Pastes text from the Clipboard into the command line.
ESC	Cancels the current command.

Press ESC at any time. If entered while an operation is in progress, it terminates the command.

Keyboard Shortcuts

In addition to accepting the full name of commands, AutoCAD allows you to type shortcuts. There are several types of shortcuts, such as aliases, function keys, and control keys.

An *alias* is an abbreviation of the full command name. For example, the alias of the LINE command is L. To execute the LINE command more quickly, simply press L, followed by ENTER. Examples of aliases are:

Alias	Command
a	Arc
adc	ADCenter
aa	Area
b	Block
-b	-Block
bh	BHatch
co	Copy
la	Layer
le	QLeader
z	Zoom

There are many other command aliases, which are defined in the *acad.pgp* file found in the \acad2004\support folder. The full list is provided in Appendix A.

Function Keys

Function keys appear on top row of your keyboard; they are all prefixed with the letter F. AutoCAD assigns the following meaning to function keys:

Function Key	Meaning
F1	Calls up the help window.
F2	Toggles between the graphics and text windows.
F3	Toggles object snap on and off.
F4	Toggles tablet mode on and off; you must first calibrate the tablet before toggling tablet mode.
CTRL+F4	Closes the current drawing.
ALT+F4	Exits AutoCAD.
F5	Switches to the next isometric plane when in iso mode; the planes are displayed in order of left, top, right, and then repeated.
F6	Toggles the screen coordinate display on and off.
F7	Toggles grid display on and off.
F8	Toggles ortho mode on and off.
ALT+F8	Displays the Macros dialog box.
F9	Toggles snap mode on and off.
F10	Toggles polar tracking on and off.
F11	Toggles object snap tracking on and off.
ALT+F11	Starts Visual Basic for Applications editor.
ALT+TAB	Switches to the next application.

Control Keys

You use a Ctrl-key shortcut by holding down the CTRL key, and then pressing another key. You may already be familiar with some, such as CTRL+C to copy to Clipboard and CTRL+Z for undo. AutoCAD displays its Ctrl-key shortcuts in the drop-down menus, next to the related command.

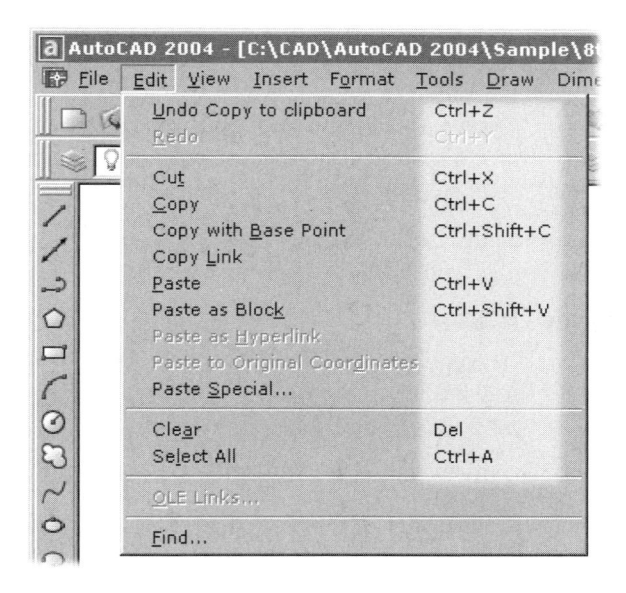

Ctrl-key	Meaning
CTRL+0	Toggles clean screen (new to AutoCAD 2004).
CTRL+1	Displays the Properties window.
CTRL+2	Opens the AutoCAD DesignCenter window.
CTRL+6	Launches dbConnect.
CTRL+A	Selects all unfrozen objects in the drawing.
CTRL+SHIFT+A	Toggles group selection mode (new to AutoCAD 2004).
CTRL+B	Turns snap mode on or off.
CTRL+C	Copies selected objects to the Clipboard.
CTRL+SHIFT+C	Copies selected objects with a base point to the Clipboard (new to AutoCAD 2004).
CTRL+D	Changes the coordinate display mode.
CTRL+E	Switches to the next isoplane.
CTRL+F	Toggles object snap on and off.
CTRL+G	Turns the grid on and off.
CTRL+H	Toggles pickstyle mode.
CTRL+K	Creates a hyperlink.
CTRL+L	Turns ortho mode on and off.
CTRL+N	Starts a new drawing.
CTRL+O	Opens a drawing.
CTRL+P	Prints the drawing.
CTRL+Q	Quits AutoCAD (new to AutoCAD 2004).
CTRL+R	Switches to the next viewport.
CTRL+S	Saves the drawing.
CTRL+SHIFT+S	Displays the Save Drawing As dialog box (new to AutoCAD 2004).
CTRL+T	Toggles tablet mode.
CTRL+V	Pastes from the Clipboard into the drawing or to the command prompt area.
CTRL+SHIFT+V	Pastes with an insertion point (new to AutoCAD 2004).
CTRL+X	Cuts selected objects to the Clipboard.
CTRL+Y	Performs the REDO command.
CTRL+Z	Performs the U command.
CTRL+TAB	Switches to the next drawing (window).

Repeating Commands

To repeat the previous command you entered in AutoCAD, press ENTER (or the spacebar) at the 'Command:' prompt. The last command entered repeats. As an alternative, right-click to display the cursor menu, and then select the **Repeat** option.

Some commands repeat automatically until you press ESC, while other commands do not. To force a command to repeat automatically, type **Multiple** before the command name, as in

> Command: **multiple circle**

This forces the CIRCLE command to repeat itself until you press ESC.

MOUSE BUTTONS

Earlier, I noted that you can pick points in the drawing with the cursor. You move the mouse, which controls the cursor. To pick a point, you press the left mouse button.

AutoCAD recognizes as many as 16 buttons on a mouse or digitizer puck. While most mouse devices are limited to two or three buttons, a digitizer puck commonly has four, 12 or 16 buttons.

It is possible to change the meaning of all the buttons except the first button (which is always the Select button). You learn more about button customization in *Using AutoCAD 2004: Advanced*.

By default, AutoCAD defines the first ten buttons found on an input device as follows:

Button #	Mouse Button	Meaning
...	Wheel	Zoom or pan.
1	Left	Select.
2	Right	Displays shortcut menus.
3	Center	Displays object snap menu when MBUTTONPAN = 0.
4		Cancel command.
5		Toggle snap mode.
6		Toggle orthographic mode.
7		Toggle grid display
8		Toggle coordinate display.
9		Switch isometric plane.
10		Toggle tablet mode.
SHIFT+1	SHIFT+Left	Toggle cycle mode.
SHIFT+2	SHIFT+Left	Display object snap shortcut menu.
CTRL+2	CTRL+Right	Display object snap shortcut menu.

Right Mouse Button

In many cases, pressing the right mouse button when inside the AutoCAD window displays a *context-sensitive shortcut menu*. "Context-sensitive" means that the content of the menu changes, depending on *where* you right-click or *which* command is in effect.

The shortcut menu (a.k.a cursor menu) appears at the cursor when you press the right-mouse button. The figure illustrates some of the shortcut menus that appear in different locations around AutoCAD.

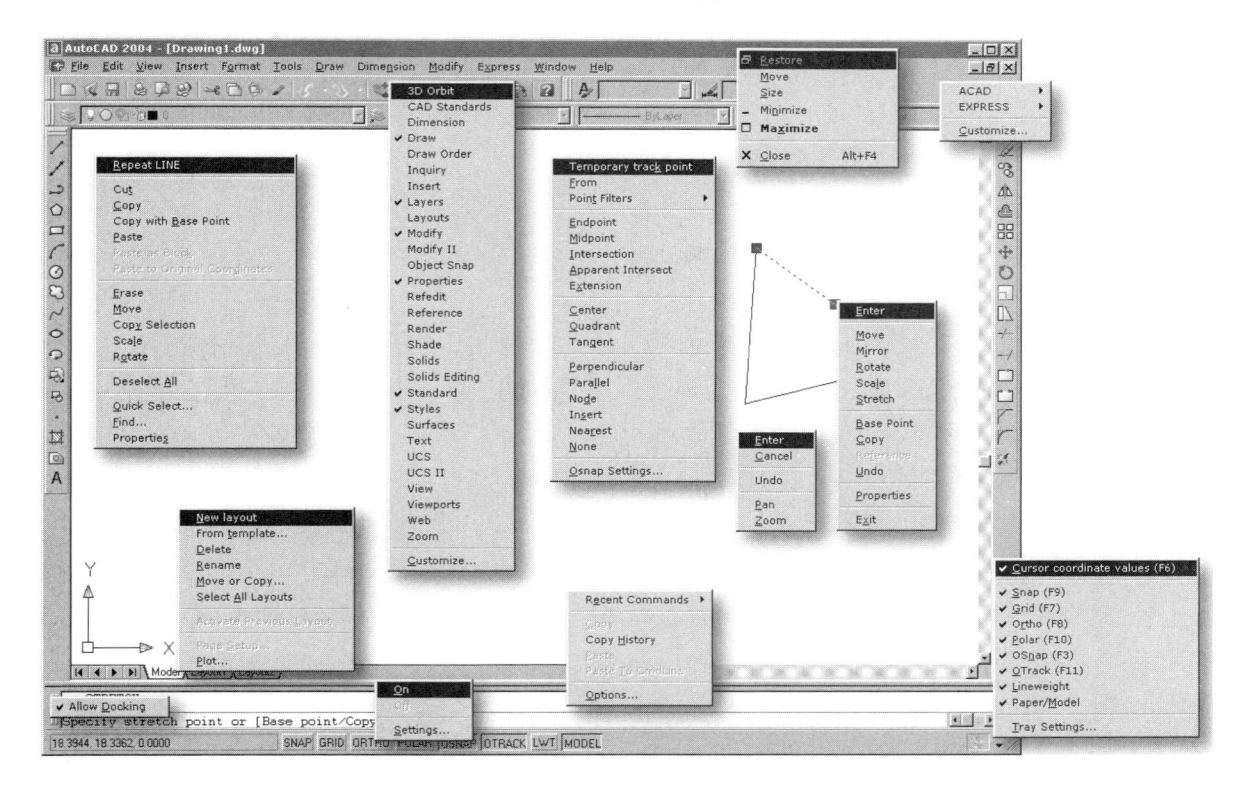

The Mouse Wheel

The wheel on a mouse can zoom and pan the drawing without invoking the ZOOM and PAN commands.

To zoom in or out, roll the wheel forward (zoom in) and backward (zoom out). To zoom the drawing to its extents, double-click the wheel button.

To pan about the drawing, hold down the wheel and drag the mouse.

In some cases, you might prefer that the wheel act as a button. You can do this by setting the value of the MBUTTONPAN system variable to 0. Now when you click the wheel, it acts like the middle button of a three-button mouse. Some mouse brands come with more than three buttons. This allows you to access additional AutoCAD functions.

TOOLBARS

AutoCAD provides *toolbars* that are close to conforming to the standard set by Microsoft for its Office line of software. All toolbars and icons can be changed, except for some portions of the Standard and Properties toolbars. To learn more about toolbar customization, read *Using AutoCAD 2004: Advanced*.

TABLET MENUS

The AutoCAD package includes support for *tablet menus*. A printed template is placed on a digitizing tablet. You can customize the tablet menu for a particular application. Items chosen from a tablet menu respond in the same manner as buttons chosen from the toolbar. Autodesk provides the *tablet.dwg* drawing file shown below in folder \autocad 2004\sample.

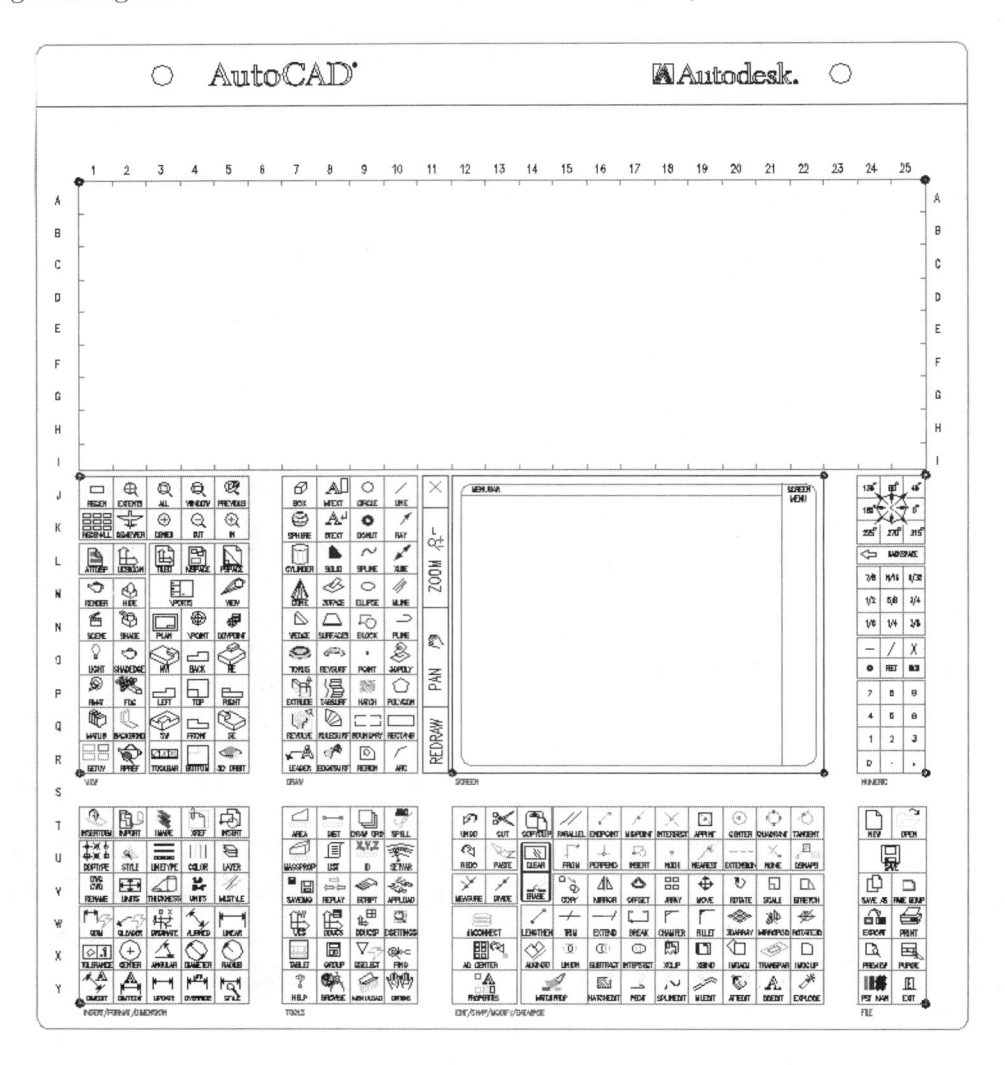

MENUS

The *menu bar* displayed under the title bar categorizes commands in groups. For example, all file-related commands are listed under **File**, while all drawing commands are listed under **Draw**.

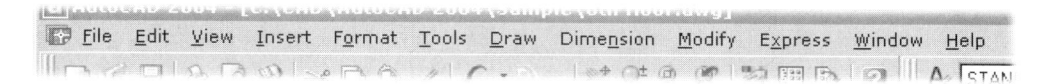

Using Menus

Clicking one of the items on the menu bar causes a menu to appear on the screen. You then choose a function from the menu. If you pull down a menu and do not want to select an item, you can exit the menu by clicking the menu item a second time.

To access items on the drop-down menu from the keyboard, press ALT, and then press the underlined letter. For example, to start the **Line** command, which is found in the **Draw** menu:

1. Press ALT.
2. Notice that the Draw menu has the D underlined. Press **d**.
3. Notice that the Line command has the L underlined. Press **l**.

The menu uses punctuation marks as shorthand to indicate special meaning. Look for these marks:

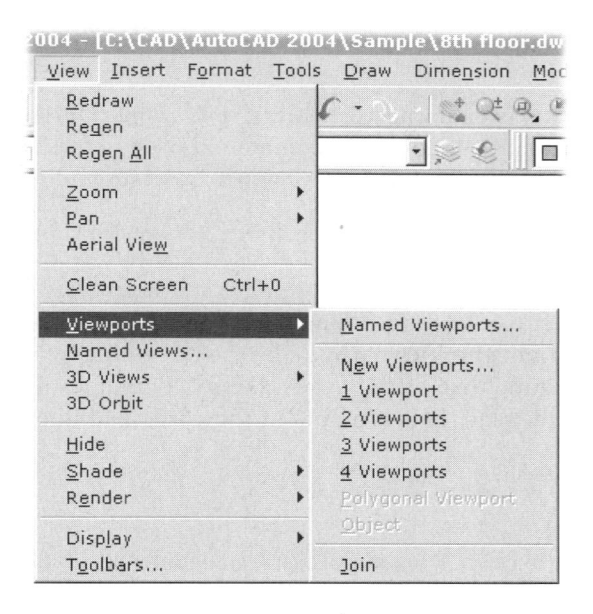

View — underlined letters on the menu bar are accessible from the keyboard. To access the **View** menu, hold down the ALT key, and then press **V**.

Tiled — underlined letters in menus are also accessible from the keyboard, but not by holding down the ALT key: just press the **D** key.

> — arrowhead indicates menu items that contain submenus. One menu item can have one, two, or more submenus. In the figure above, the **Viewports** menu item displays a submenu of nine items.

✓ — check mark indicates a menu item is turned on.

... — ellipses indicates a menu item that displays a dialog box; think of the ellipses as "more to come." In the figure, **New Viewport...** displays the Viewport dialog box.

Gray Text — menu items are not available at the current time. In the figure, **Polygonal Viewports** is not available in model space.

DIALOG BOXES

Many commands require you to specify option in dialog boxes. The benefit of dialog boxes is that they display all your options at the same time, unlike the command-line, which typically displays one option at a time. The dialog box and the command-line for creating arrays are compared below.

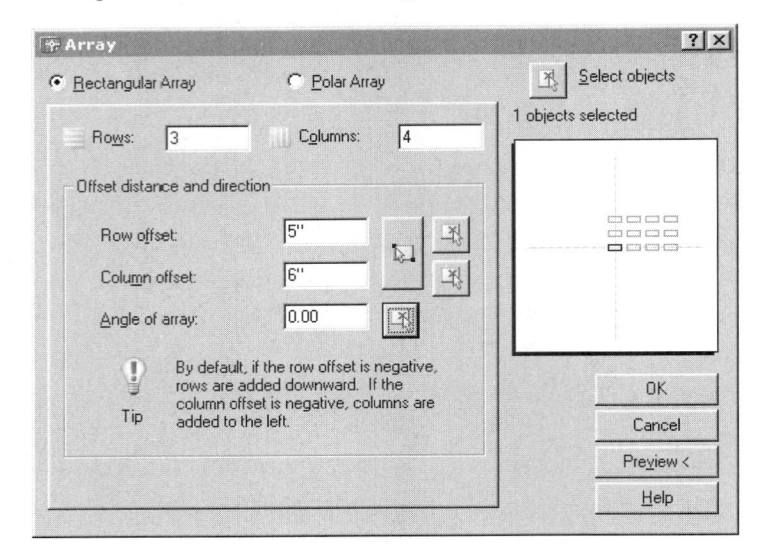

Command: **-array**

Select objects: *(Select one or more objects.)*

Select objects: *(Press* ENTER.*)*

Enter the type of array [Rectangular/Polar] <R>:
(Press ENTER.*)*

Enter the number of rows (---) <1>: **3**

Enter the number of columns (| | |) <1> **4**

Enter the distance between rows or specify unit cell (---): **5**

Specify the distance between columns (| | |): **6**

The drawback to dialog boxes is that they obscure the drawing. On the other hand, dialog boxes present options grouped logically; in contrast, command-line prompts guide you through the steps needed to complete the action, but lack the previews and hints found in dialog boxes.

A number of commands provide the option to use a dialog box or the command line. There are several reasons for this. One is historical: until Release 9, AutoCAD didn't use dialog boxes at all, and so all commands operated at the command prompt. Another reason is that you may prefer using the command line, because it can be faster than using dialog boxes. For example, I find it faster to use the **-VPORTS** command than the Viewports dialog box. Another reason is technical: script files and AutoLISP routines cannot control dialog boxes. (Many commands do not use dialog boxes, such as **LINE** and **COPY**; they only display their prompts at the command line.)

There are two ways to suppress a dialog boxes:

Hyphen Prefix

Prefixing commands with a hyphen (-) suppresses the dialog box, displaying the command line instead. For example, the **ARRAY** command normally displays the Array dialog box:

Command: **array**

(AutoCAD displays dialog box, shown above.)

Prefixing the command with a hyphen displays the command line options:

Command: **-array**

Select objects: *(Select one or more objects.)*

Select objects: *(Press* ENTER.*)*

(Command continues with the prompts listed above.)

Not all command have the hyphen option:

Hyphenated Commands			*Hyphenated Commands*	
ARRAY	-ARRAY		PAN	-PAN
ATTDEF	-ATTDEF		PARTIALOAD	-PARTIALOAD
ATTEDIT	-ATTEDIT		PLOT	-PLOT
ATTEXT	-ATTEXT		PLOTSTAMP	-PLOTSTAMP
BHATCH	-BHATCH		PLOTSTYLE	-PLOTSTYLE
BLOCK	-BLOCK		PSETUPIN	-PSETUPIN
BOUNDARY	-BOUNDARY		PUBLISH	-PUBLISH
COLOR	-COLOR		PURGE	-PURGE
DIMSTYLE	-DIMSTYLE		REFEDIT	-REFEDIT
GROUP	-GROUP		RENAME	-RENAME
HATCHEDIT	-HATCHEDIT		RMLIN	-RMLIN
IMAGE	-IMAGE		STYLE	-STYLE
IMAGEADJUST	-IMAGEADJUST		TOOLBAR	-TOOLBAR
INSERT	-INSERT		UNITS	-UNITS
LAYER	-LAYER		VBARUN	-VBARUN
LINETYPE	-LINETYPE		VIEW	-VIEW
MENULOAD	-MENULOAD		VPORTS	-VPORTS
MTEXT	-MTEXT		WBLOCK	-WBLOCK
OSNAP	-OSNAP		XBIND	-XBIND
			XREF	-XREF

There is one curious exception: **-PARTIALOPEN** is the true name, while **PARTIALOPEN** is an alias.

System Variables

Whether or not certain commands display dialog boxes is controlled by system variables. (*System variables* store the current state of AutoCAD and its many settings.)

When the **FILEDIA** system variable is set to 0, AutoCAD suppresses the display of file-related dialog boxes, such as with the **NEW**, **OPEN**, **SAVEAS**, and **VSLIDE** commands. Prompts are displayed at the command line:

> Command: **filedia**
> Enter new value for FILEDIA <1>: **0**
> Command: **open**
> Enter name of drawing to open: *(Enter path and name of .dwg drawing file.)*

To force the display of the dialog box, type the tilde (~) character when AutoCAD prompts for the file name:

> Command: **open**
> Enter name of drawing to open: ~
> *(AutoCAD displays the Select File dialog box.)*

When the **EXPERT** system variable is set to a value other than zero, it suppresses warning dialog boxes displayed by AutoCAD. Examples include "About to regen, proceed?" and "Really want to turn the current layer off?"

When the **ATTDIA** system variable is set to 0, the **INSERT** command displays prompts for attribute data at the command line; when set to 1, a dialog box is displayed instead.

Alternate Commands

In a very few cases, AutoCAD uses different command names for dialog boxes and command lines. Examples include:

Command Line	Dialog Box
FILEOPEN	OPEN
CHANGE	PROPERTIES
UCS	UCSMAN

Using Dialog Boxes

Move a dialog box by dragging its *title bar*. This means that you move the cursor over the title bar and, holding down the left mouse button, move the dialog box. Moving the dialog box is useful if you need to see objects under the box.

When a dialog box is displayed, the arrow cursor replaces the crosshair cursor. Use the cursor to choose options in the dialog box. Then choose the **OK** button to accept the changes and dismiss the dialog box.

Some dialog boxes have an **Apply** button that lets you apply changes you made in the dialog box, yet keep it open; others have a **Preview** button that temporarily dismisses the dialog box so that you can see the change to objects. To dismiss the dialog box without affecting the drawing, choose **Cancel**.

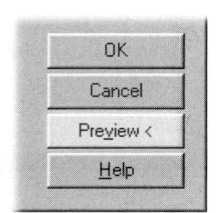

Almost all dialog boxes contain a **Help** button. Selecting the button displays a help box that explains the purpose and use of the dialog box. Many also have a **?** button in the upper-right corner: click it, and then click an item in the dialog box for a context-sensitive explanation.

Most dialog boxes are *modal*, which means you must dismiss them before you can continue working in AutoCAD. A few are *non-modal*, which means they continue to hover on your Windows desktop as you continue working in AutoCAD; these are called "windows." An example is the Properties window. To dismiss a non-modal dialog box, click the small **x** in the upper right corner.

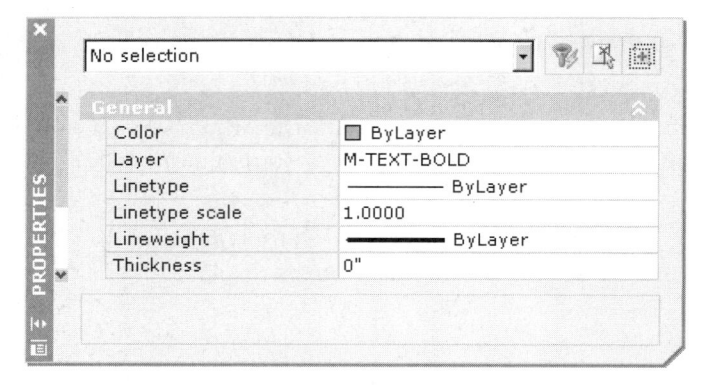

Other dialog boxes contain *tabs*, which allow a single dialog box to display a lot of information, separated into similar-looking boxes. For example, the **Tools | Options** command displays a dialog box with nine tabs!

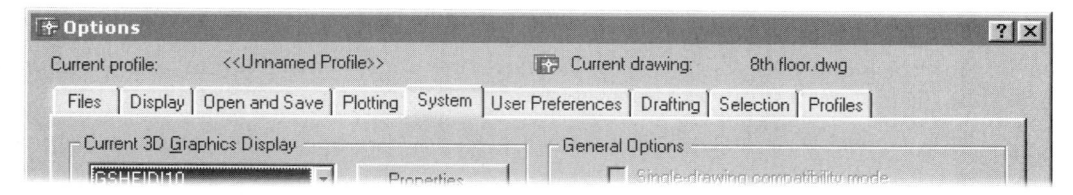

Some dialog boxes contain additional or sub-dialog boxes. If a sub-dialog box is displayed, you must choose **OK** or **Cancel** to return to the original dialog box. In the figure below, New Text Style is the sub dialog box displayed by the **New** button.

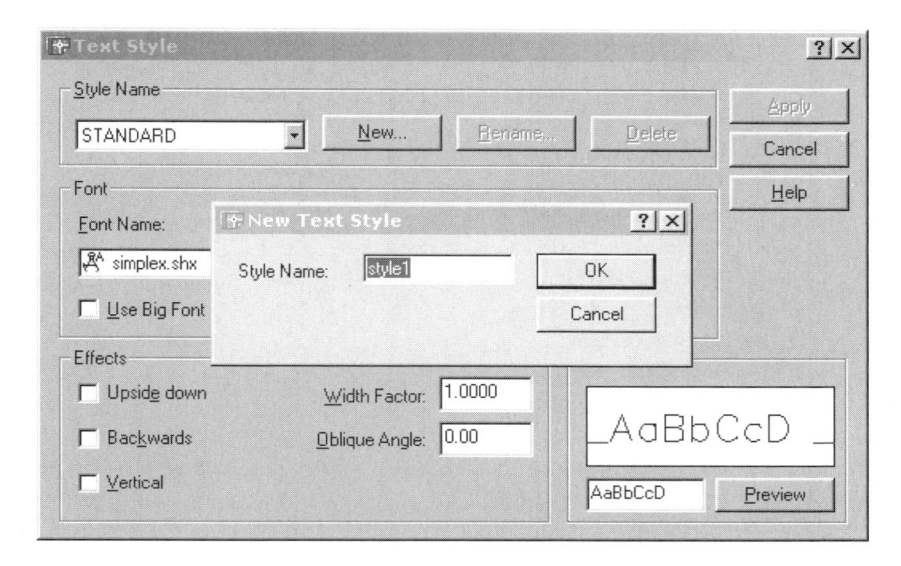

Buttons in Dialog Boxes

Buttons are used to select items. Select buttons with the cursor or by pressing keys on the keyboard.

- The button with the heavier border is the default. Pressing **ENTER** is like choosing the button with the cursor.

- Three periods (...) after a button's name opens a sub-dialog box. An example is the **New...** button in the Text Style dialog box illustrated above.

- Buttons with a single arrow pointing left (<) indicate action is required in the drawing, such as selecting an object. An example is the **Preview** button in the Array dialog box.

- Buttons with two arrow pointers (<<) expand the dialog box. Alternatively, the dialog box may have a **Show Details** button, such as in the Layer Properties Manager dialog box.

- If the text on the button is gray, the button's function is not available at this time, such as the **Delete** button in the Text Style dialog box, above.

The following is a summary of the buttons used in dialog boxes.

Check Boxes

Check boxes *toggle* options, which means the option is either on or off. A check mark indicates the option is selected or turned on; no check mark means the option is turned off.

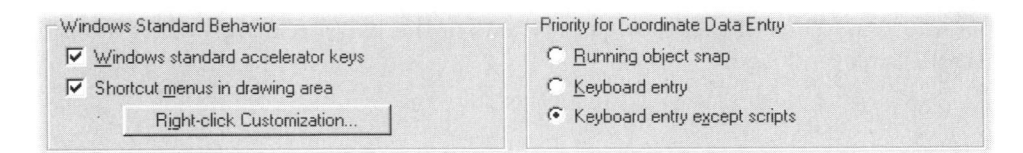

Radio Buttons

Two (or more) round radio buttons force you to select one option out of several. Only one option is active at a time. The black dot in the button indicates it is selected or turned on.

List Boxes

List boxes contain a list from which you choose one or more items. When the list is too long for the box, a scroll bar appears automatically, allowing you to scroll to choices not visible. To choose more than one item, hold down the **CTRL** key as you select items; not all list boxes allow this.

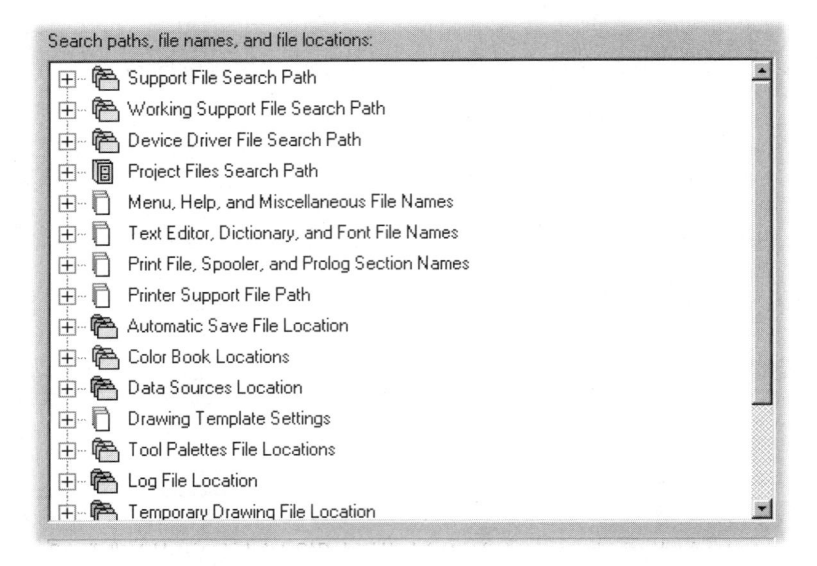

Scroll Bars

Some list boxes contain more entries than can be displayed at a time. Scroll bars exist on some list boxes to bring additional items or options into view by scrolling up or down. To display another entry below the current field of view, click on the down arrow of the scroll bar.

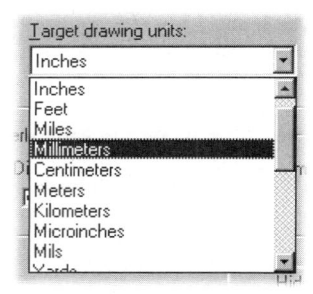

Drop Lists

Drop lists normally display a single item. It has an arrow button; when you click the button, a list "drops down" to show the entire list. An example is shown above.

You can also move through the scroll list by clicking on the bar and, while continuing to hold the mouse button down, sliding the bar up or down with the cursor. When you click again, the entries are redisplayed at the new location. Note that the position of the scroll bar is relative to the position of the displayed items. Thus, if there are many items in the list, a relatively small movement of the scroll bar scrolls by several items.

Text Edit Boxes

Text edit boxes contains a single line of text (or, in some cases, a paragraph of text) to be edited. To edit the text, click in the box. A cursor bar appears at the text. You move the cursor bar with the arrow keys on the keyboard.

Image Tiles

Image tiles are small windows that display graphical images. Select the image to select the option, such as the 3D surface objects illustrated below.

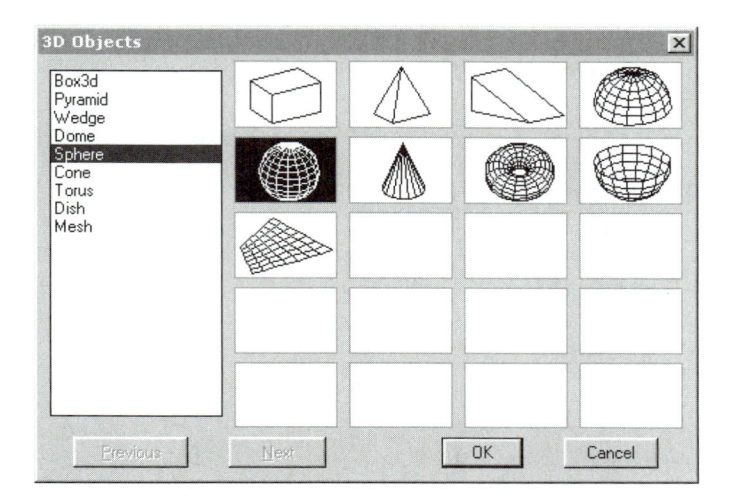

THE DRAFT-EDIT-PLOT CYCLE

Once you set up a drawing, working with AutoCAD involves a cycle consisting of draft, edit, and plot. You draft drawings, edit them, and then plot them. This book is arranged in a similar manner:

Cycle	Chapters
Setup	1: Quick Tour of AutoCAD 2004
	2: Understanding CAD Concepts
	3: Setting Up Drawings
Draft	4: Drawing with Basic Objects
	5: Drawing with Precision
	6: Drawing with Efficiency
	12: Placing and Editing Text
	13: Placing Dimensions
	14: Geometric Dimensioning and Tolerancing
Edit	7: Changing Object Properties
	8: Correcting Mistakes
	9: Direct Editing of Objects
	10: Constructing Objects

	11: Additional Editing Options
	12: Placing and Editing Text
	14: Editing Dimensions
Plot	16: Layouts
	17: Plotting Drawings

This book concentrates on two-dimensional drafting, editing, and plotting. Volume 2 of this book, *Using AutoCAD 2004: Advanced*, covers three-dimensional drafting, as well as customizing AutoCAD, and working with multiple drawings (layouts and xrefs).

DRAFT

Drafting consists of placing and constructing objects in the drawing. AutoCAD comes with a set of objects, from which everything else in the drawing must be made:

Objects	Drawn with Command(s)
Points	POINT
Lines	LINE, TRACE, PLINE
Parallel lines	MLINE
Construction lines	RAY, XLINE
Circles	CIRCLE, DONUT
Ellipses	ELLIPSE
Arcs	ARC, PLINE
Elliptical arcs	ELLIPSE
Regular Polygons	POLYGON
Rectangles	RECTANG
Splines	SPLINE
Irregular boundaries	BOUNDARY
Revision clouds	REVCLOUD
3D surfaces	AI_BOX, AI_DOME, AI_DISH, AI_PYRAMID, AI_CONE, CYLINDER, AI_SPHERE, AI_TORUS, AI_WEDGE, AI_MESH
3D solids	BOX, CONE, CYLINDER, SPHERE, TORUS, WEDGE
Text	TEXT, MTEXT
Dimensions	All commands starting with DIM, plus LEADER, TOLERANCE

When you cannot draw an object with one of AutoCAD's commands, then you need to *construct* it. For example, AutoCAD has no commands for drawing right-angle triangles and irregular polygons, so you construct them with the **LINE** or **POLYLINE** commands.

In addition, AutoCAD has a collection of commands that construct copies of objects:

Copy Method	Copied with Command(s)
Copy objects	COPY
Copy blocks	INSERT, XBIND, ADCENTER
Offset copies	OFFSET
Mirrored copies	MIRROR, MIRROR3D
Arrayed copies	ARRAY, 3DARRAY
Arrayed as block	MINSERT
Copy properties	MATCHPROP
Copy faces, edges	SOLIDEDIT
Copy along paths	DIVIDE, MEASURE

Templates and Blocks

Sometimes you start with a *template*, a drawing that contains settings and objects created previously. The most common templates consist of a drawing border, title block, and perhaps layers. Many firms also use templates for standard drawings of details.

One of Autodesk's cofounders, John Walker, said that you should never have to draw the same object twice in CAD. The best way to reuse drawing details is to turn them into *blocks*. Blocks can be used repeatedly in drawings, can be shared between drawings and offices, and can hold database information, called "attributes."

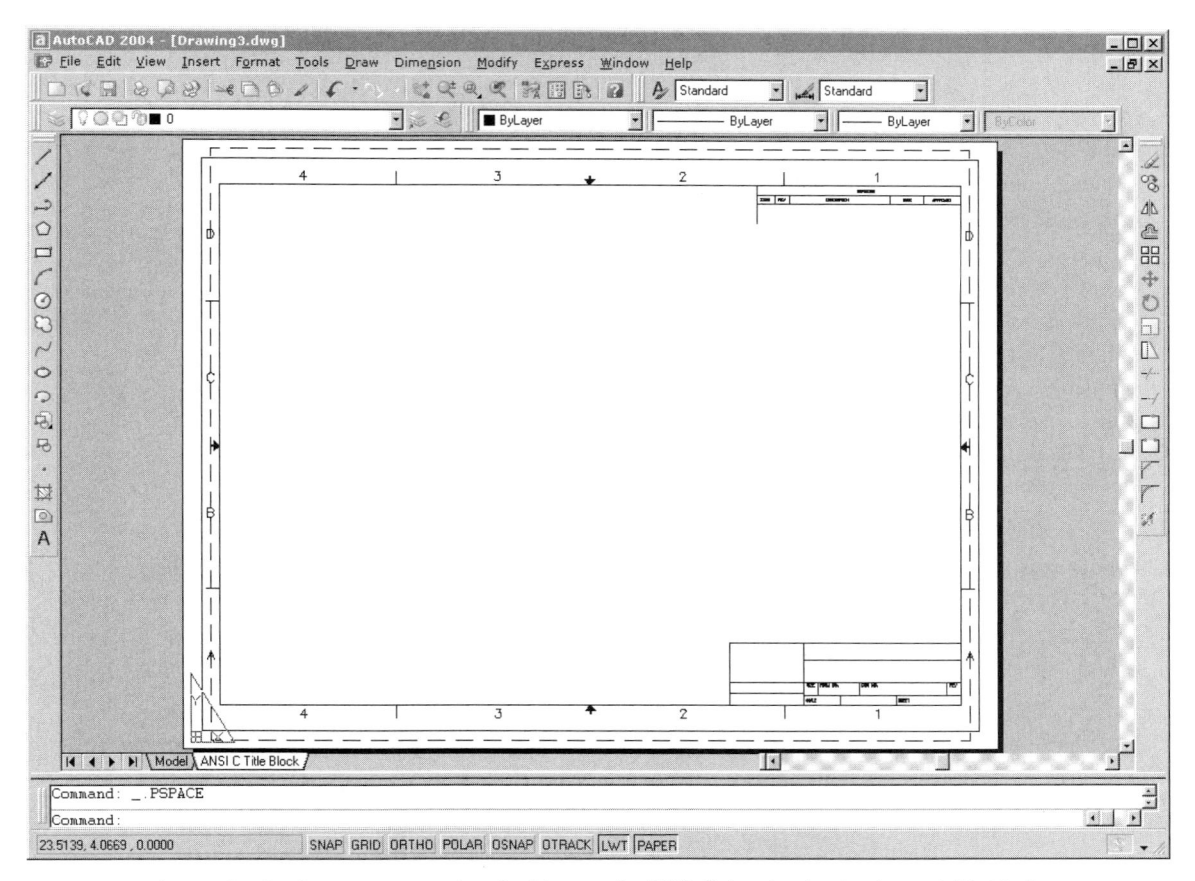

A new drawing begun as a template, in this case the ANSI C-size drawing border and title block.

EDIT

Editing consists of modifying objects in the drawing: correcting mistakes, moving objects, and so on.

Editing Method	Edited with Command(s)
Move objects	MOVE, STRETCH, ALIGN
Rotate objects	ROTATE, ROTATE3D
Resize objects	SCALE
Stretch objects	STRETCH, LENGTHEN, EXTEND
Mirror objects	MIRROR
Erase objects	ERASE
Trim objects	TRIM, BREAK
Fillet objects	FILLET
Chamfer objects	CHAMFER
Undo editing	U, UNDO, REDO, MREDO
Edit splines	SPLINEDIT
Edit multilines	MLEDIT
Edit text	DDEDIT
Change text size	SCALETEXT, STYLE
Text alignment	JUSTIFYTEXT, STYLE
Edit attributes	EATTEDIT
Edit polylines	PEDIT
Edit 3D solids	UNION, INTERSECT, SUBTRACT, SOLIDEDIT
Edit 3D surfaces	PEDIT
Reduce objects	EXPLODE, XPLODE
Change properties	CHANGE, PROPERTIES

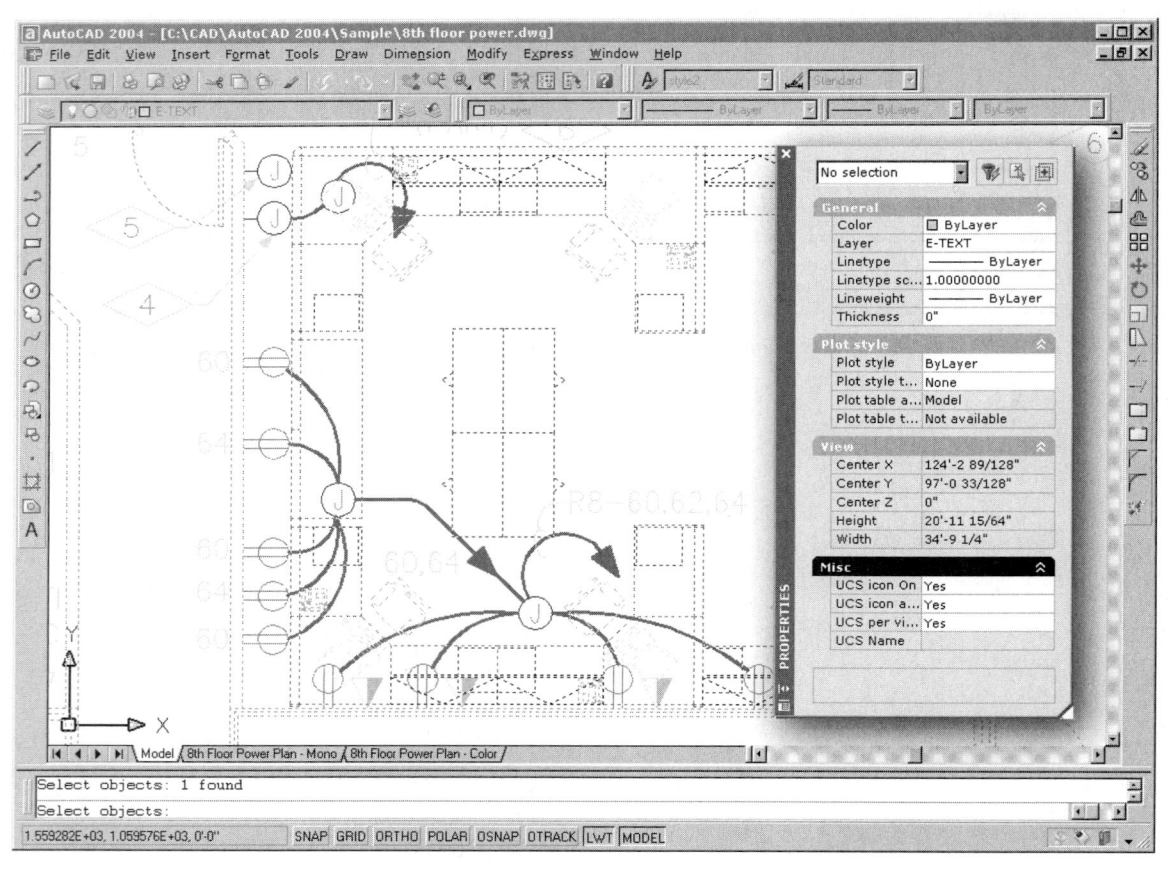

Editing properties of objects selected with the Properties window.

In addition to commands that edit, AutoCAD can edit objects directly. For example, double-clicking most objects brings up the Properties window, which lets you change their properties; double-clicking text brings up the appropriate text editor.

Single-clicking objects allows you to edit them though *grips* (or handles). With grips editing, you can stretch, move, copy, rotate, mirror, and resize objects.

In addition to editing objects, AutoCAD has a collection of commands that *construct* objects:

Construction Type	Constructed with Command(s)
Hatch patterns	BHATCH, HATCH, HATCHEDIT
Filled areas	SOLID, BHATCH
Blank areas	WIPEOUT
3D surface	REVSURF, TABSURF, RULESURF, EDGESURF
3D revolution	REVOLVE
3D extrusion	EXTRUDE
3D solids	INTERFERE, SLICE, SECTION

PLOT

When your drawings are complete, you print them on paper (a process called "plotting"), or send them through the Internet, called an "e-plot," short for "electronic plot." When plots are created before the drawing is complete, they are called "check plots."

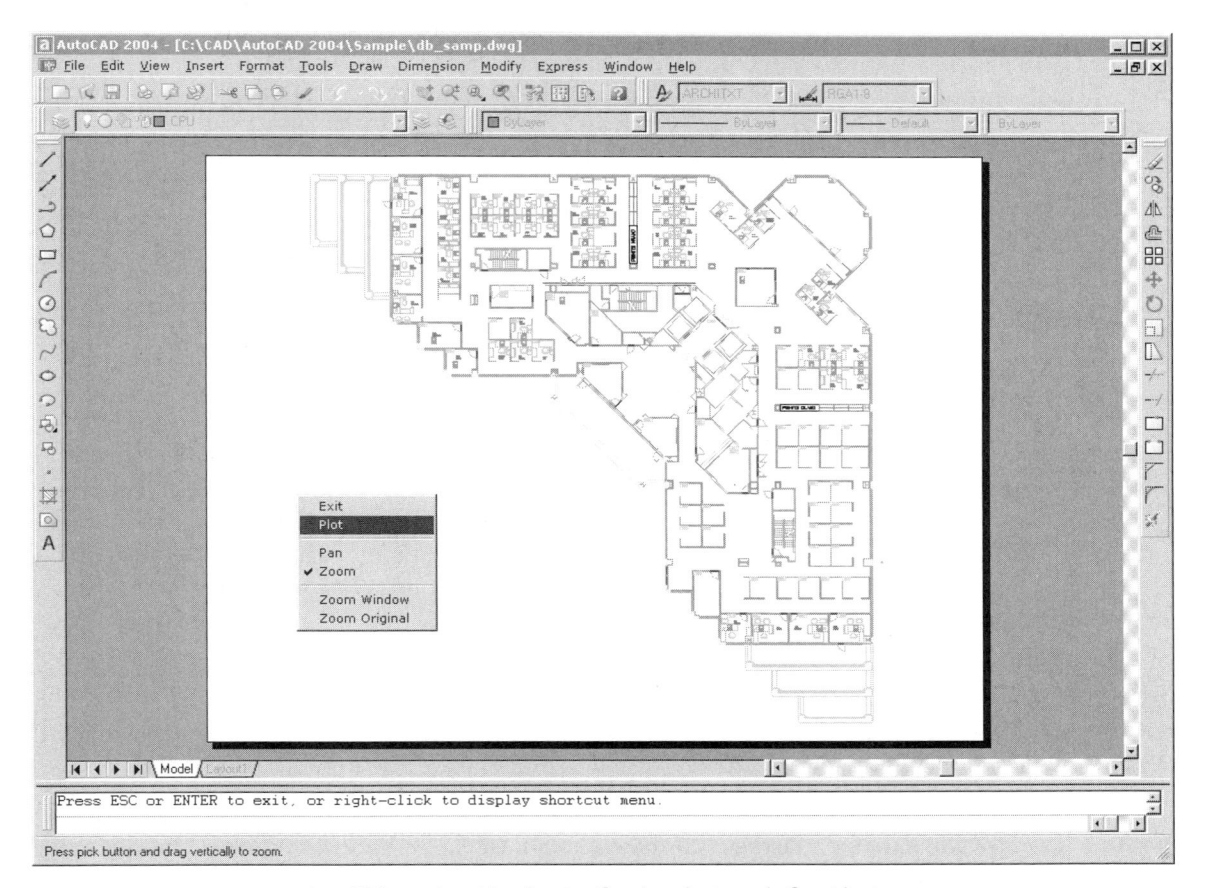

AutoCAD previews the drawing fitted to the page before plotting.

AutoCAD 2004 supports two kinds of plots: *color based* and *style based*. Color-based plotting is the older method, whereby the colors of objects determine the pens (or inkjets) with which they are

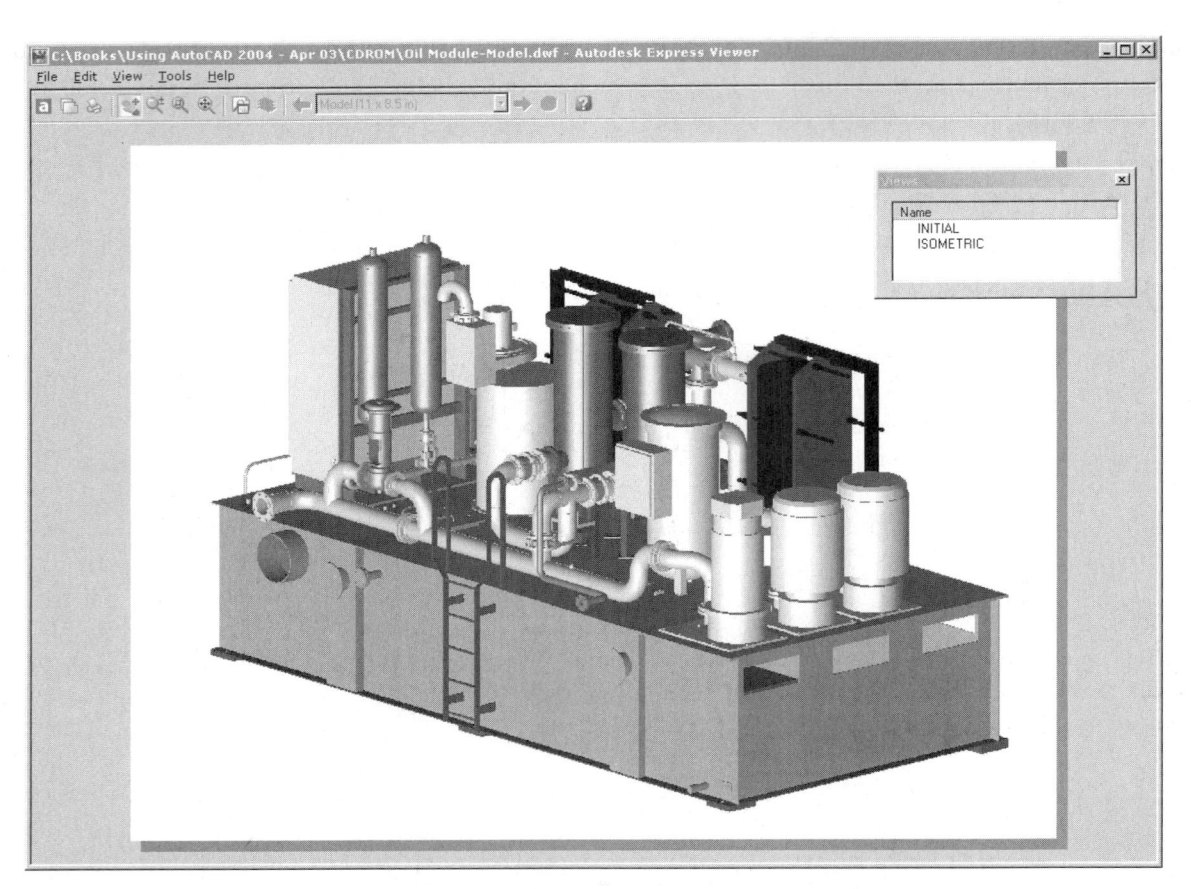

Express Viewer displays and prints .dwf e-plots.

plotted. Style-based plotting applies *styles* to layers and individual objects; these styles affect every aspect of how objects can be plotted: color of lines, width of lines, screen percentage, style of end caps, and even whether the objects are plotted at all.

AutoCAD's **PREVIEW** command allows you to see plots before they are committed to paper, while the **PLOTSTAMP** command allows you to label plots with the drawing file name, date and time of plot, and other information.

ePlots

In addition to plotting drawings on paper, you can also plot drawings to cyberspace. To send copies of drawings — but not the original *.dwg* drawing file — you save the drawing as a *.dwf* file (short for "design web format"). This is sometimes called an "electronic plot." The *.dwf* file looks just like the original, but cannot be edited. The "DWF ePlot" plotter configuration of the **PLOT** command creates *.dwf* eplots.

To view *.dwf* files on your computer and over the Internet, Autodesk provides the Express Viewer software with AutoCAD 2004, and as a free download from its Web site: www.autodesk.com/dwf. The software allows you to zoom in and out, select predefined views, and print drawings. The **PUBLISHTOWEB** command generates the Web page and images of the drawings.

To protect drawings, AutoCAD 2004 has implemented *password* protection and digital signatures. When you save drawings with passwords (through the **SECURITYOPTIONS** command), they cannot be opened unless the correct passwords are entered. *Be careful:* If you forget the password, the drawing is rendered useless. In this regard, it's best not to use passwords at all.

A different level of security is the *digital signature*. When drawings are "signed" digitally, a warning appears after they have been edited or altered. This alerts you that the drawings may not be originals; the system does not, however, tell you *what* has been changed.

COORDINATE SYSTEMS

CAD systems work with *coordinate* systems. Coordinates determine the location of objects in the drawing. (Technically, AutoCAD does not record lines and other objects, but records their endpoints and properties.) All 2D drafting is performed on the x,y-plane; for 3D drafting, you can redefine this plane to any orientation in space through UCSs.

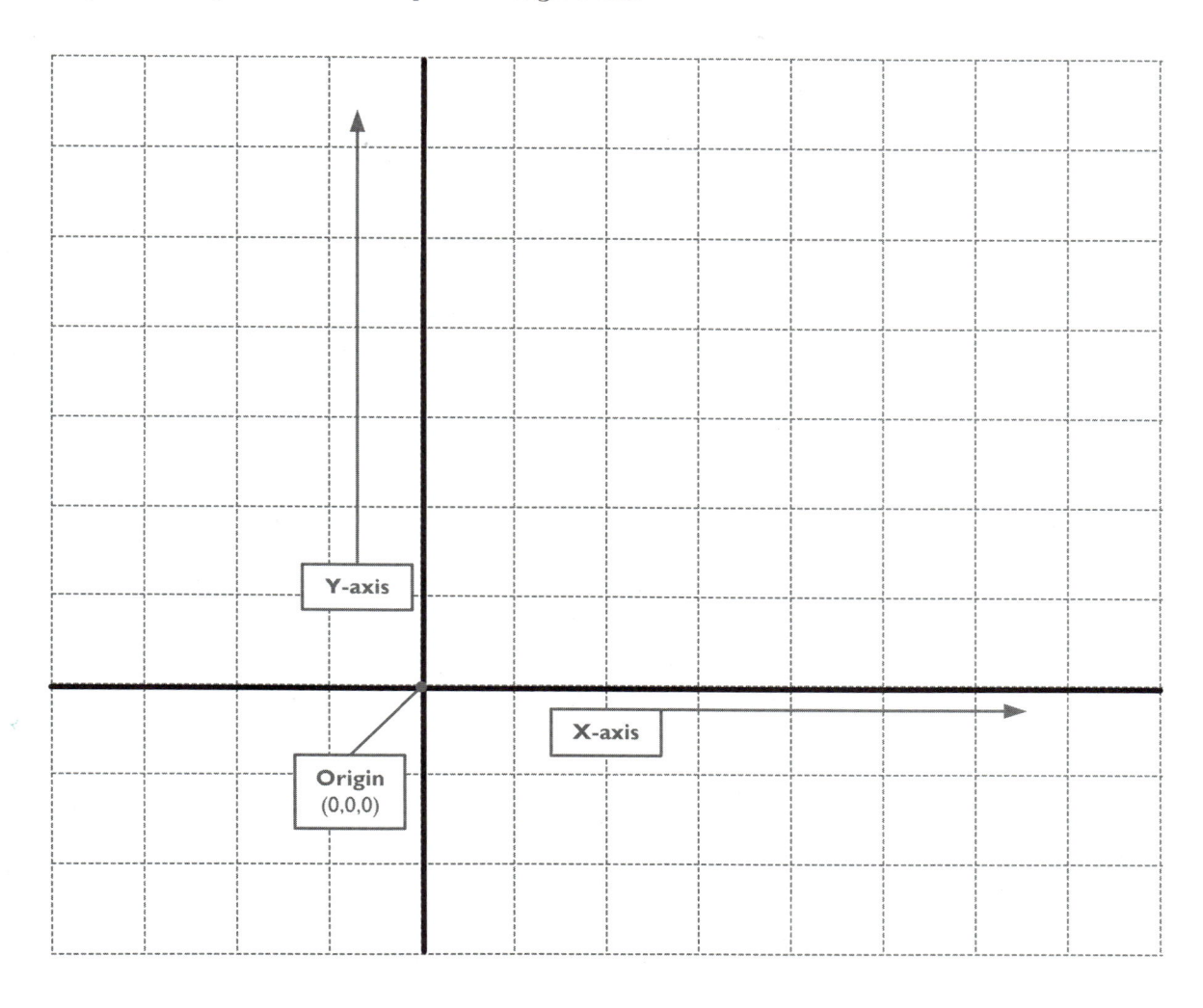

AUTOCAD COORDINATES

AutoCAD works with several coordinate systems, most of which measure distances relative to the *origin*, which is located at 0,0,0:

- **Cartesian** define 2D and 3D points by two or three distances along the x, y, and (sometimes) z axes.
- **Polar** define 2D points by a distance and an angle.
- **Cylindrical** define 3D points by two distances and an angle.
- **Spherical** define 3D points by a distance and two angles.
- **Relative** define 2D and 3D distances (and optionally angles) relative to the last point.
- **User-defined coordinate systems** (UCS) define an x,y-plane anywhere in 3D space.

You enter coordinates in formats of your choice: decimal (metric), engineering (decimal inches), architectural (Imperial units), fractional (inches only), and scientific notation (exponents). Accuracy ranges from 8 decimal places to zero (1/256th of an inch to whole inches).

Angles can be entered in degrees-minutes-seconds (360 degrees in a circle), radians (2*pi), grads (400), and in surveyor's units (N E).

In new drawings, angles are measured counterclockwise, with 0 degrees pointing from the origin to the East, or 3 o'clock. Using this convention, 90 degrees points North (12 o'clock), and 180 degrees points West (9 o'clock).

To measure an angle in the clockwise direction, use negative angles. Prefix the angle with a dash, as in -45 degrees. Because there are 360 degrees in a circle, -45 degrees is the same as 315 degrees.

In the same way, you specify negative distances by prefixing numbers with the dash. For example, -2,-3 means two units in the negative x-direction, and three units in the negative y-direction.

Recall that *absolute* coordinates are measured relative to the origin (0,0,0), while *relative* coordinates are measured relative to the last point. Similarly, absolute angles are measured from the x-axis in the counterclockwize directions, while relative angles are measured from the value stored in the **LASTANGLE** system variable.

If nothing else, AutoCAD is flexible in its measurement systems. You can make changes with the **UNITS** command so that angles are measured clockwise, and that 0 degrees points in any direction.

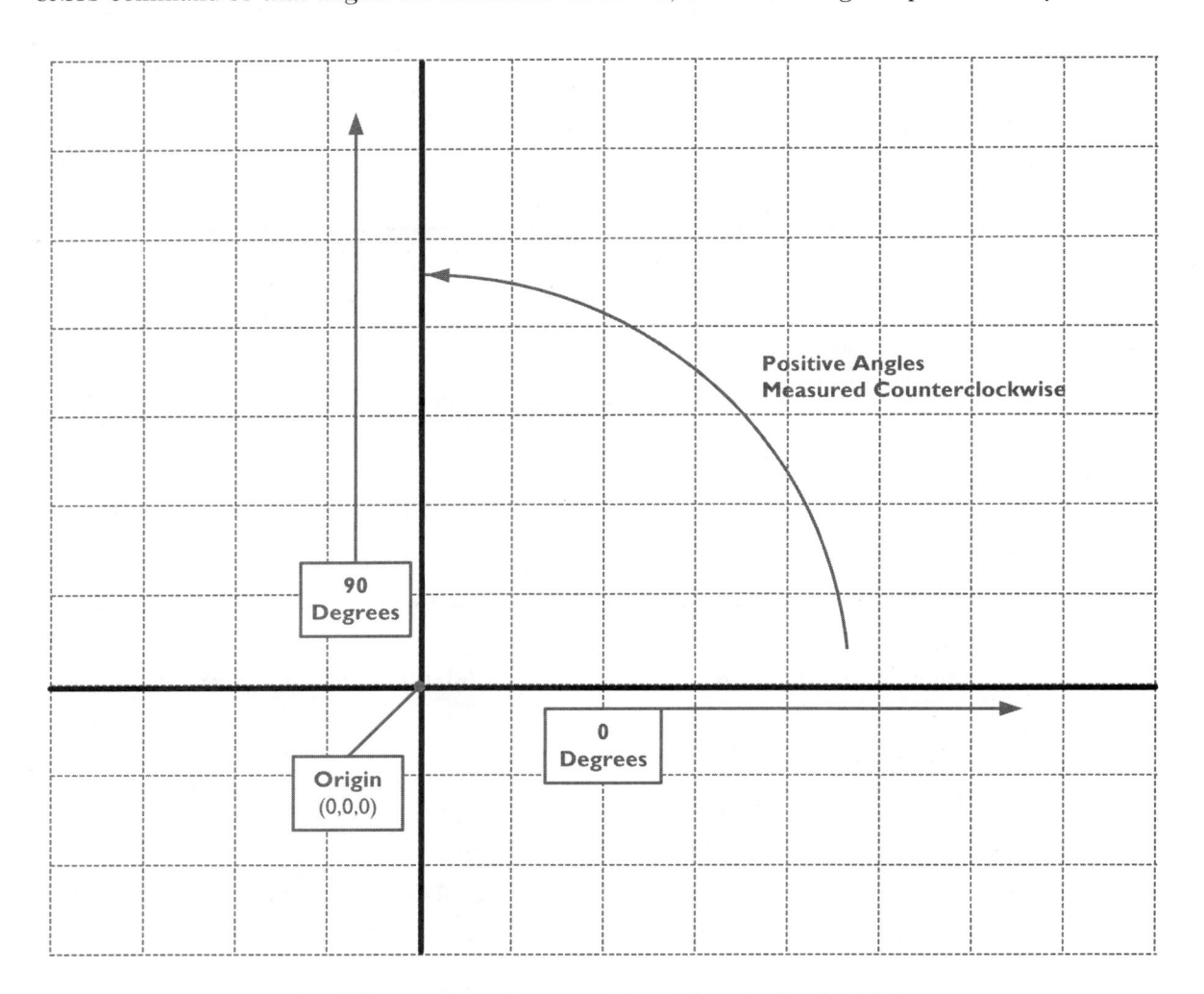

AutoCAD measures angles counterclockwise from the East, by default.

Cartesian Coordinates

Cartesian coordinates define 2D and 3D points by measuring two or three distances: each distance is along one of the three axes: x, y, and z, either positive or negative. In AutoCAD, the x-direction is horizontal, at 0 degrees; positive x is measured to the right. The y-direction is vertical along 90 degrees; positive y is measured upward. The z-direction comes out of the screen; positive z points at your face.

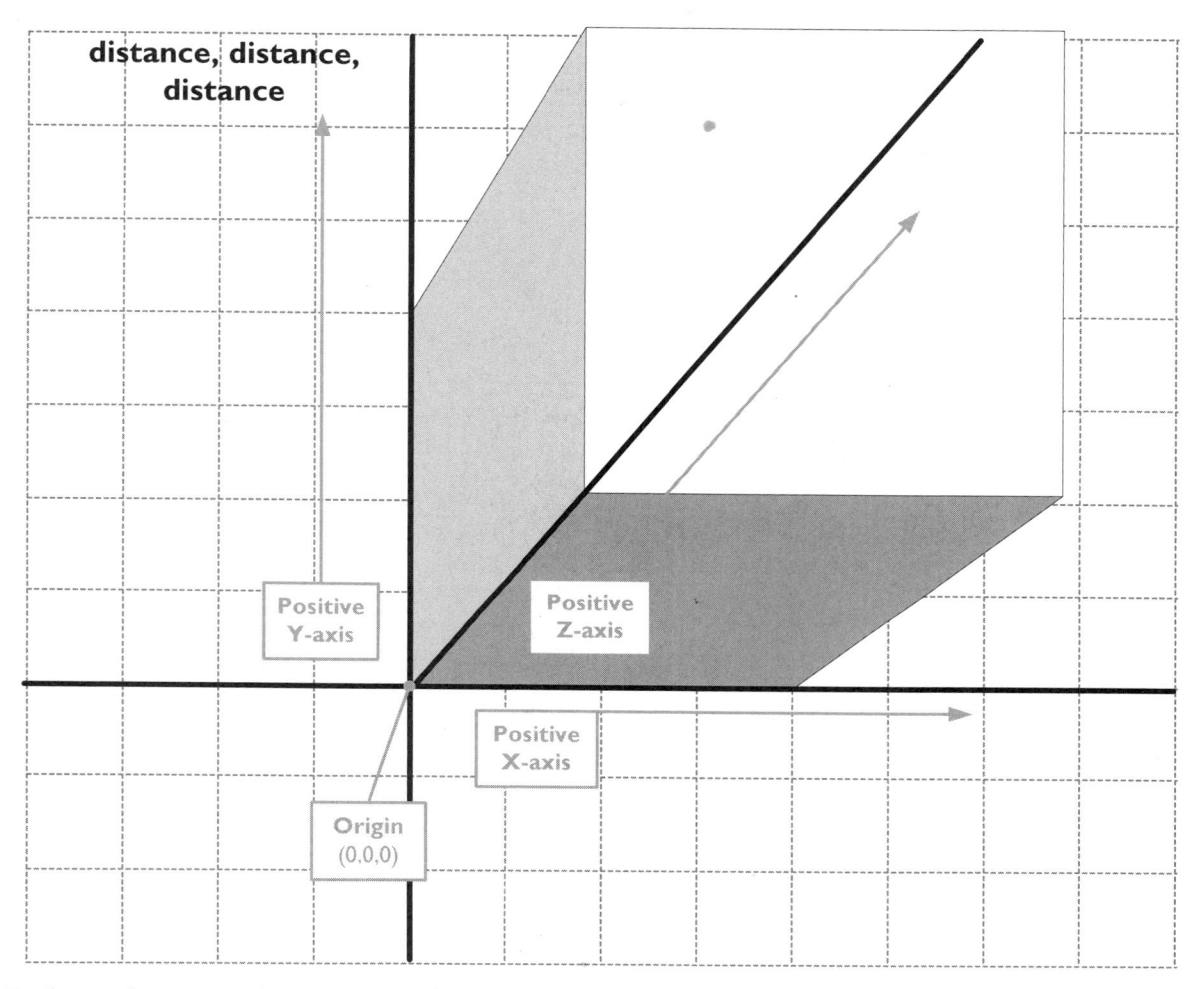

To draw a line using Cartesian coordinates with x,y (2D) and x,y,z (3D) coordinates:

> Command: **line**
> Specify first point: **1,2**
> Specify next point or [Undo]: **3,4,5**

Polar Coordinates

Polar coordinates define 2D points by measuring one distance and one angle in the x,y-plane. The distance is the straight-line distance measured from the origin to the point, while the angle is measured from 0 degrees. To indicate the angle to AutoCAD, you prefix the angle with an angle bracket (<), such as <45 for 45 degrees.

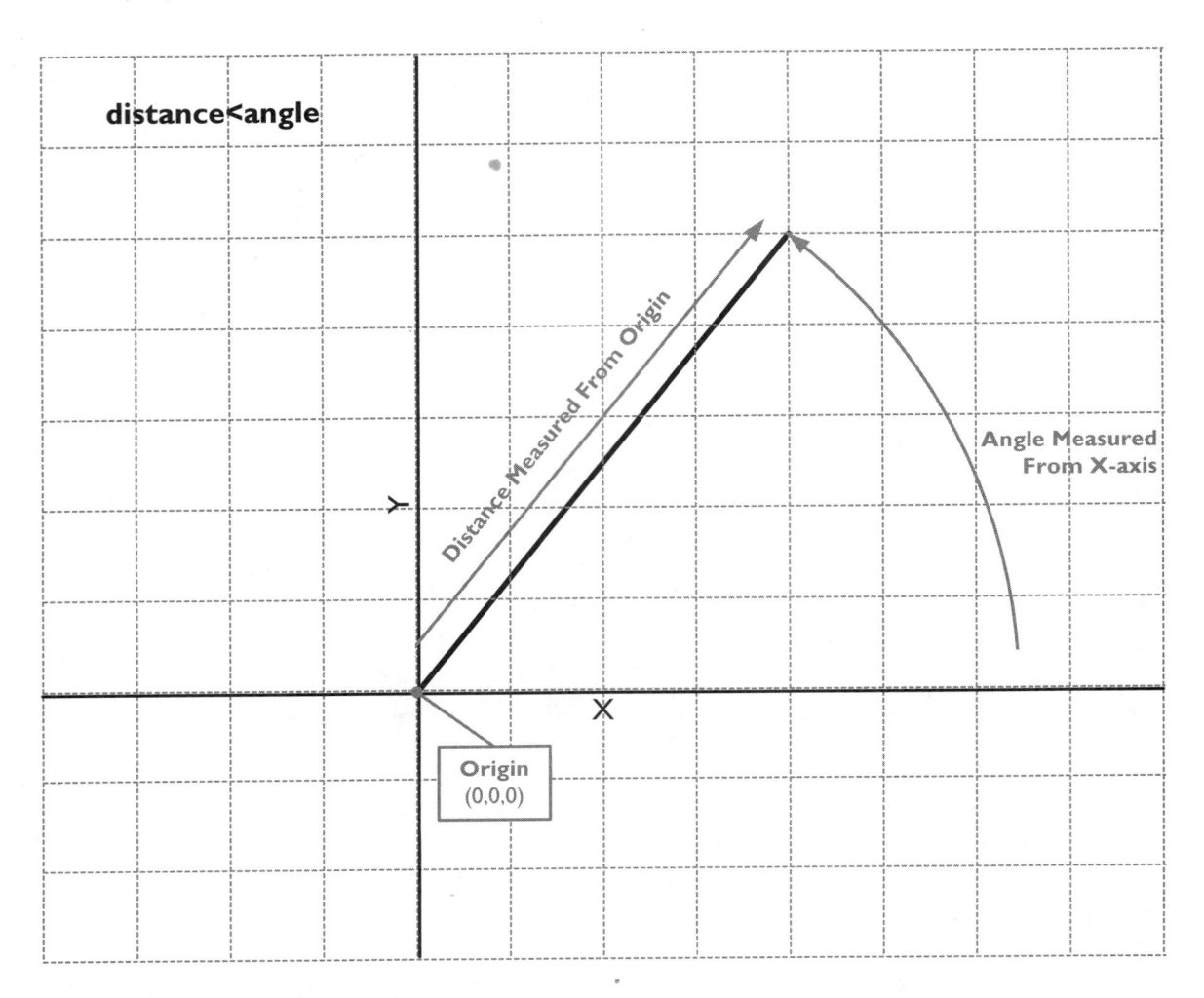

To draw a line using polar coordinates:

    Command: **line**
    Specify first point: **1<23**
    Specify next point or [Undo]: **4<55**

Cylindrical Coordinates

Cylindrical coordinates define 3D points by measuring two distances and one angle in 3D space. They are like polar coordinates, but with the z-distance added. The distances are measured from the origin to the point, while the angle is measured from 0 degrees.

The first distance lies in the x,y-plane; the second distance measures the distance along the z-axis. To indicate the measurement to AutoCAD, you use this notation: 1<23,4 — where 1 is the polar distance, <23 is angle from 0 degrees, and 4 is the height in the z direction.

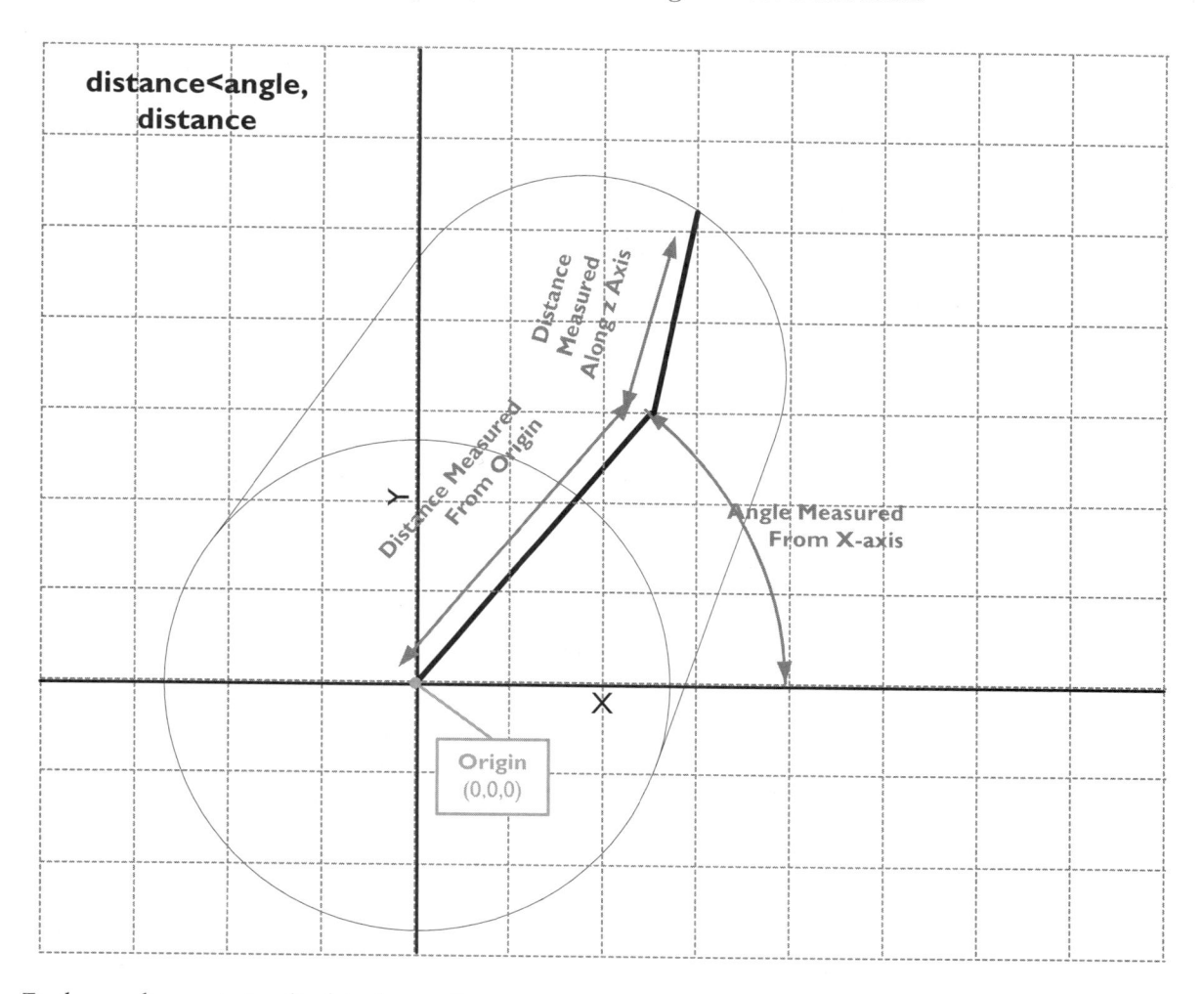

To draw a line using cylindrical coordinates:

    Command: **line**

    Specify first point: **1<23,4**

    Specify next point or [Undo]: **5<67,8**

Spherical Coordinates

Spherical coordinates define 3D points by measuring a distance and two angles from the origin. The first angle lies in the x,y-plane, while the second angle is up (or down) from the x,y-plane. The distance lies, of necessity, in the x,y,z-plane.

To indicate the angles to AutoCAD, you prefix each with angle brackets, such as <15<45 for 15 and 45 degrees.

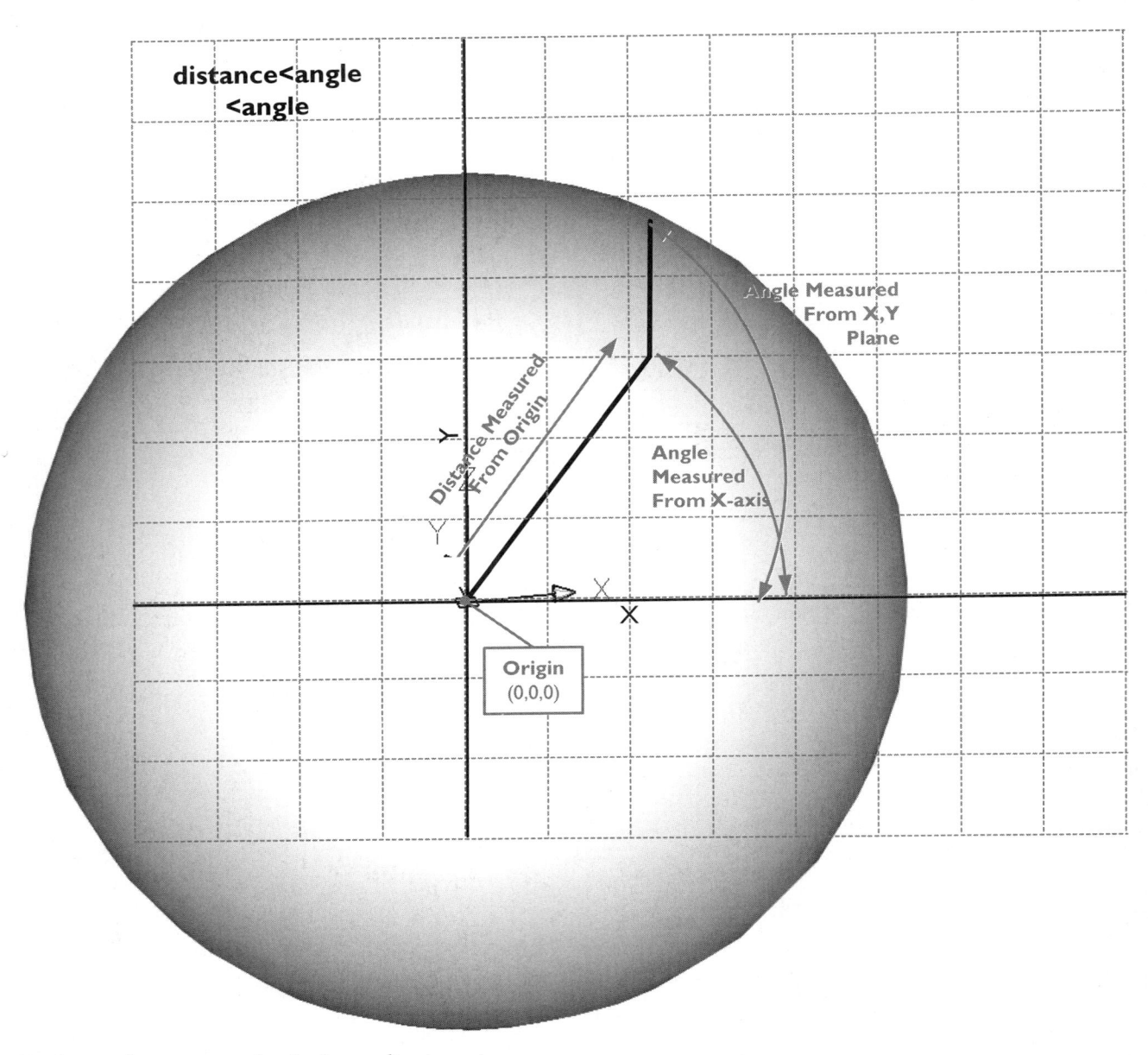

To draw a line using spherical coordinates:

    Command: **line**
    Specify first point: **1<23<45**
    Specify next point or [Undo]: **6<78<90**

Relative Coordinates

All coordinate systems can use *relative* coordinates, which measure the distance (and angle) from the last point. (The opposite of relative is *absolute*, which measures distances from the origin.) To indicate relative coordinates to AutoCAD, you prefix them with the at symbol (@), such as @2,3.

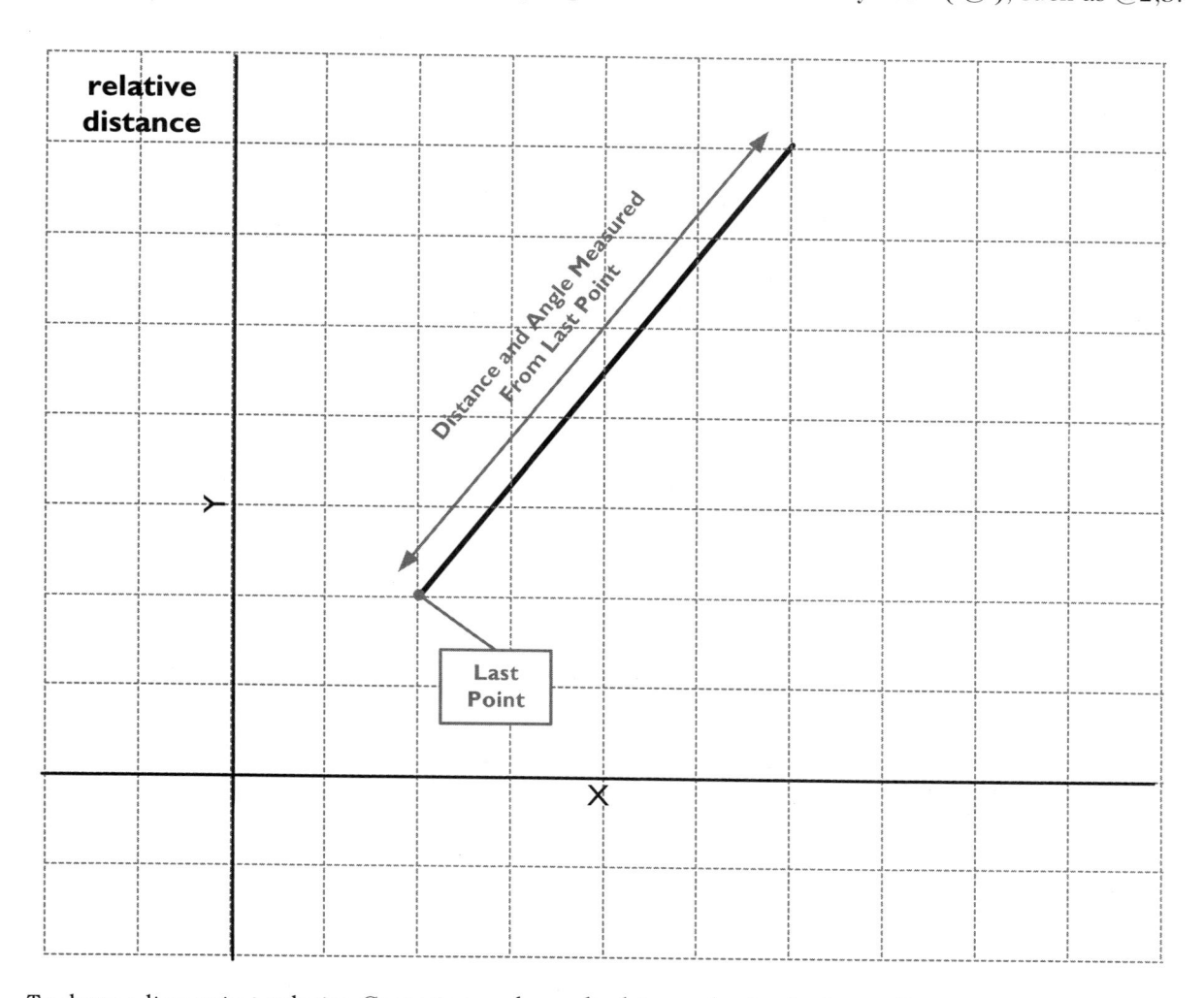

To draw a line using relative Cartesian, polar, cylindric, and spherical coordinates

> Command: **line**
> Specify first point: **@1,2**
> Specify next point or [Undo]: **@3,4,5**
> Specify next point or [Undo]: **@6<78**
> Specify next point or [Undo]: **@9<10<11**
> Specify next point or [Undo]: **@9<10<11**

Note that the first point can also be relative. It is measure from whatever last point was specified in the drawing with a previous drawing command. AutoCAD stores the coordinates of the last point in a system variable called **LASTPOINT**.

Similarly, relative angles are measured from the value stored in the **LASTANGLE** system variable.

User-defined Coordinate Systems

As I noted earlier, all drawing in AutoCAD takes place in the x,y-plane, also called the "working plane." How, then, do you draw in 3D space, which involves the z axis, along with the x,z and y,z planes? You can specify the z-coordinate, but not for all commands. In any case, specifying z coordinates becomes cumbersome when drafting details, say, on the side of a slanted roof.

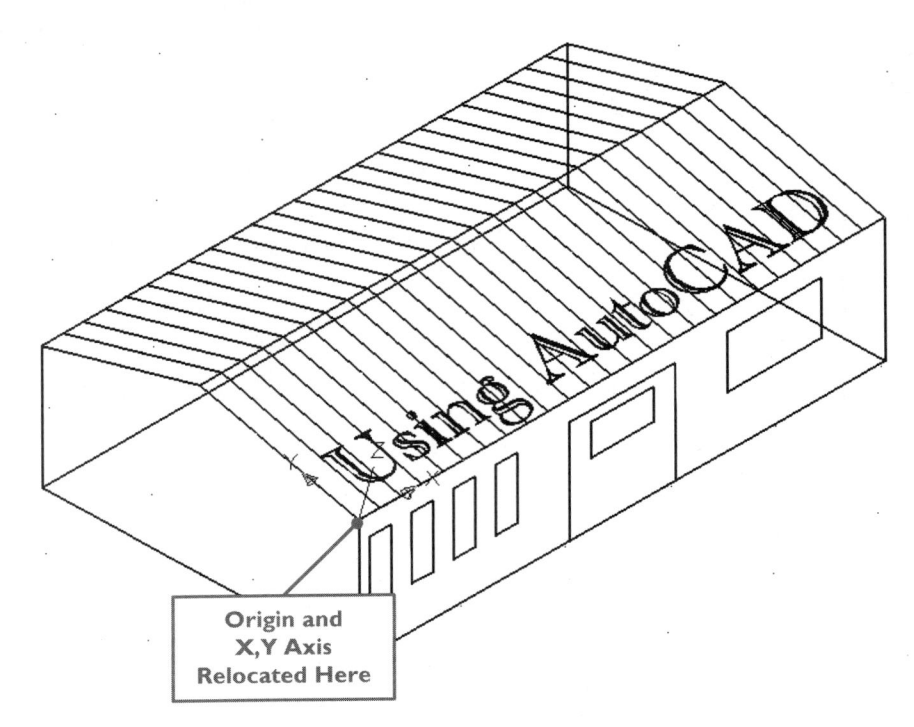

Origin and X,Y Axis Relocated Here

AutoCAD takes care of that through user-defined coordinate systems, or "UCS" for short. Through UCSs, you can define the working plane anywhere in 3D space. The UCS command lets you set the working plane relative to the current view, to three points picked in the drawing, and by several other methods. UCSs can be saved by name, so that you can return to previously-created working planes.

ENTERING COORDINATES

Much of working with AutoCAD involves specifying points on the screen. Typical are the prompts of the LINE command:

> Command: **line**
>
> Specify first point: *(Pick a point, or enter x,y,z coordinates.)*
>
> Specify next point or [Undo]: *(Pick another point, or enter more coordinates.)*

You can do this in three ways:

- Picking points on the screen.
- Entering coordinates at the keyboard.
- Using object snaps, tracking, direct distance entry, and point filters.

Screen Picks

When AutoCAD prompts you to specify a point, you can use the crosshair cursor to pick a point in the drawing. Move the mouse to move the cursor; press the left mouse button to pick the point. AutoCAD determines the x,y-coordinates of the pick point.

Screen picks are strictly 2D (two dimensional), even when working in 3D (three dimensional) drawings. "2D" means AutoCAD records the x and y coordinates only, and assumes the z-coordinate is 0. If you wish to specify a z-coordinate, you may do this with the **ELEVATION** command (which sets the distance along the z axis), or with the **.XY** *point filter*, which forces AutoCAD to prompt you for the z distance.

Keyboard Coordinates

You can enter coordinates at the keyboard in a variety of methods, as I describe earlier in this chapter. The formats permitted by AutoCAD are:

Absolute	Relative	Meaning
x,y	@x,y	2D Cartesian coordinates.
x,y,z	@x,y,z	3D Cartesian coordinates.
d<a	@d<a	2D polar coordinates.
d<a,z	@d<a,z	3D cylindrical coordinates.
d<a<r	@d<a<r	3D polar coordinates.

Where x, y, and z = distance along the x, y, and z axes, respectively.
   d = distance in x,y plane.
   a = angle from x axis (in x,y plane).
   r = angle from x,y plane (in z direction).

Recall that *absolute* coordinates are measured relative to the origin (0,0,0), while *relative* coordinates are measured relative to the last point. Similarly, absolute angles are measured from the x-axis in the counter-clockwise directions, while relative angles are measured from the value stored in the **LASTANGLE** system variable.

Snaps, Etc.

You may have a suspicion by now that entering coordinates at the keyboard can get tedious. There are alternatives and aids:

- **Point filters** combine screen picks with keyboard entry.
- **Snap spacing** restricts cursor movement to increments.
- **Polar snap** and **ortho mode** restrict cursor movement to angles.
- **Direct distance entry** shows angles, and types distances.
- **Tracking** shows distances.
- **Object snaps** snap the cursor to the nearest geometric feature.
- **Object tracking** shows geometric relationships to nearby objects.

You learn more about snap mode, polar snaps, objects snaps, and object tracking in Chapter 5, "Drawing with Precision."

Point Filters

Point filters combine screen picks with keyboard entry to specify, for example, the x coordinate with the keyboard and the y coordinate on the screen. AutoCAD recognizes five point filters:

Point Filter	AutoCAD Asks For...
.x	y and z coordinates.
.y	z and z coordinates.
.z	x and y coordinates.
.xy	z coordinate.
.xz	y coordinate.
.yz	x coordinate.

Here is an example of using point filters with the LINE command. Notice that you can enter coordinates at the keyboard or on the screen at any time:

Command: **line**
Specify first point: **.x**
of *(Pick a point; AutoCAD reads just the x-coordinate.)*
(need YZ): **2,3** *(AutoCAD reads these as the y and z coordinates.)*

Specify next point or [Undo]: **.y**
of **2** *(AutoCAD reads this as the y coordinate.)*
(need XZ): *(Pick a point; AutoCAD reads the x and z coordinates.)*

Snap Spacing

Snap mode restricts cursor movement. To restrict the cursor to moving in increments of 0.5, for example, set the snap spacing to 0.5 with the **SNAP** command. This allows you to pick points easily on the screen, accurate to the nearest 0.5 units. To take another example, say you measured the rooms in your house to the nearest one inch; when drawing the floor plan in AutoCAD, you would set the snap spacing to 1".

Snap is often used in conjunction with the grid; when the grid is set to the same spacing as the snap, then you can "see" the snap points.

Polar Snap and Ortho Mode

Polar snap snaps the cursor to angles, while ortho mode restricts the cursor to 90-degree increments. (Ortho is short for "orthographic.") You can specify the polar angle, although increments of 15 degrees are common. In addition, you can optionally specify polar distances, such as 1.

While drawing, AutoCAD displays a tooltip showing relative distance and angle from the last point.

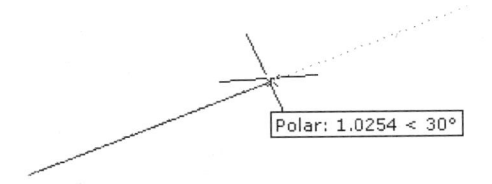

You turn on polar mode by clicking POLAR on the status bar; turn on ortho mode by clicking ORTHO.

Direct Distance Entry

Direct distance entry shows angles, and types distances. It allows you to draw and edit by showing relative distance: you move the cursor to indicate the angle, and then type a number to specify the distance. If you need a precise angle, then ensure that polar tracking or ortho mode are first turned on.

Specify first point: *(Move the cursor, and then enter a distance, such as 2.5.)*

Tracking

Tracking shows distances with the "pen up" during drawing and editing commands. You can think of it as the opposite of direct distance entry, which draws; tracking moves the cursor without drawing. During a prompt, such as "Specify first point:", enter **tk** to enter tracking mode. Move the cursor, in the direction to track, and then pick a point or enter a distance, such as **2.5**. You can continue tracking as far as you need.

>Specify first point: **tk**
>
>First tracking point: *(Move cursor, and then enter a distance or pick a point.)*
>
>Next point (Press ENTER to end tracking): *(Press ENTER.)*

Object Snaps

Object snaps snap the cursor to the nearest geometric feature, such as the endpoints of lines, the center points of circles and arcs, and the insertion points of text. To help you find these geometric features, AutoCAD displays icons and tooltips.

The figure below illustrates the line being drawn to the center of the circle. Because object snapping is turned on, AutoCAD displays the center icon (a circle), as well as the tooltip, "Center."

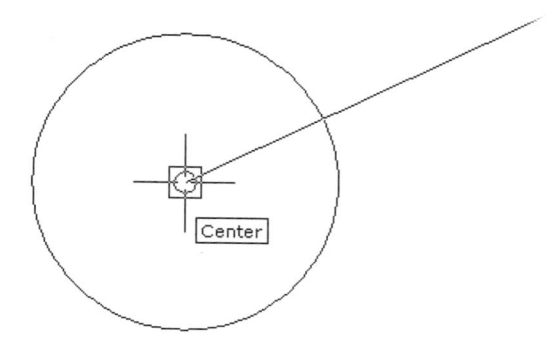

To use objects snaps, you can turn them on with the **OSNAP** command, or enter them at prompts, such as:

>Specify first point: **center**
>
>of *(Pick a circle or arc.)*

Object Tracking

Object tracking shows geometric relationships to nearby objects. When turned on with at least one object snap mode, object snap tracking shows *alignment paths* (dashed lines) based on object snap locations. The figure below illustrates the alignment path showing the quadrant object snap.

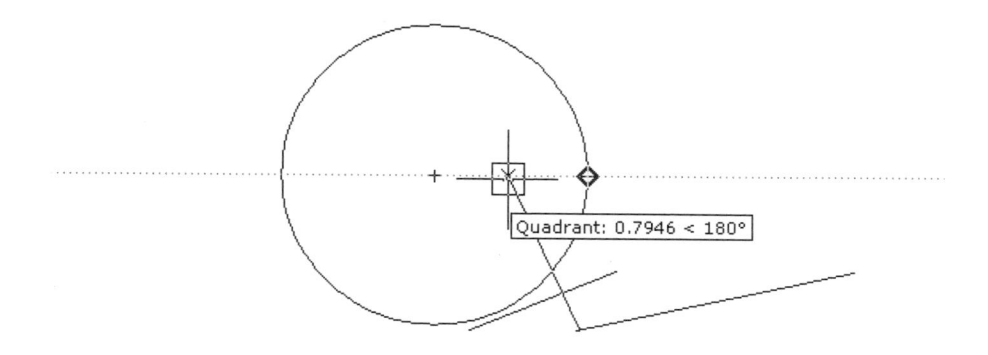

DISPLAYING COORDINATES

AutoCAD displays the cursor position by means of x,y,z coordinates displayed on the status line (look to the extreme left end), as illustrated below.

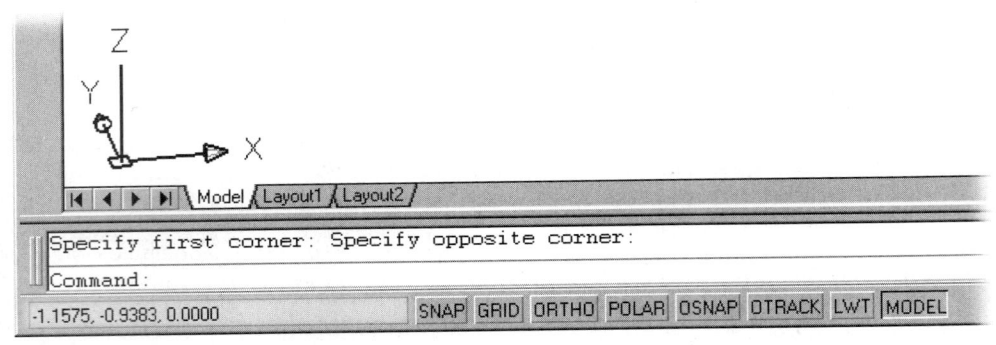

In most cases, the display shows absolute x,y,z coordinates; the z coordinate, however, always reads 0.0000 — unless changed by the ELEV command.

If you wish, you can turn off the display, although there is no point to doing that, at least not that I can think of. In addition, you can change the display to relative d<r coordinates. To change the coordinate display, press CTRL+D or function key F4 — or, simply click the coordinate display on the status line. In the figure below, notice that AutoCAD displays relative coordinates during line drawing:

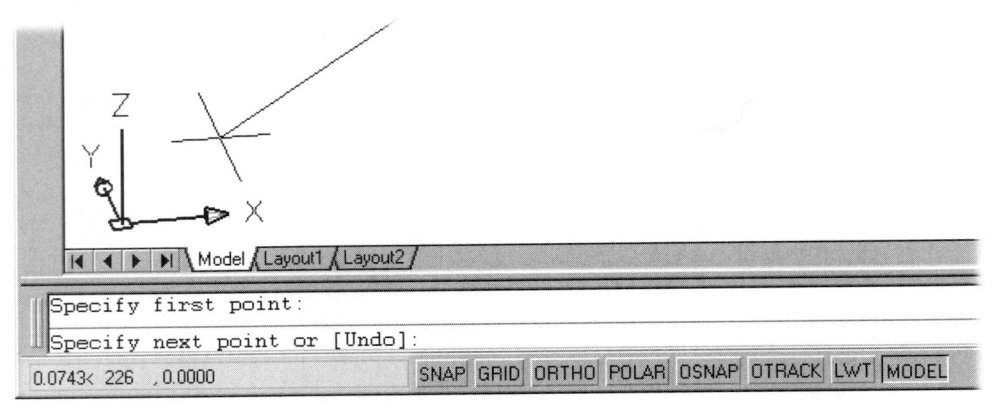

In addition to coordinates on the status line, AutoCAD also displays coordinate information through tooltips.

EXERCISES

1. In the following exercises, watch how the coordinates and cursor react to different modes.
   Move the cursor around with your mouse or digitizer. Do the coordinates at the lower left corner of the screen move?
   Press F6. Move the cursor again, and watch the coordinates. What is the difference?

2. Press F8. Check the status bar: is ortho mode turned on?
   Move the cursor. Do you notice a difference?

3. Press F9 to turn on snap mode.
   Move the cursor. Now is there a difference in cursor movement?

4. In this exercise, you practice using menus and keystrokes that affect commands.
   Move the crosshairs to the top of the drawing area and into the menu bar. Does the cursor change its shape?
   Move the arrow from left to right over the menu bar. What happens as the cursor moves over each word?
   Position the cursor over **Draw**, and then click. Does a menu drop down into the drawing area?
   Click on **Draw** a second time. What happens?

5. Choose the **Draw** menu again, and then choose **Line**.
   Draw some lines.
   To exit the command, right-click and select **Cancel** from the shortcut menu.

6. Press the ENTER key. Did the **Line** command repeat?
   Clear the command line with ESC.
   Press the spacebar. What happens?

7. Press F2. Do you see the Text window?
   Press F2 again to return to the drawing window.

8. Press F1. Do you see the Help window? Familiarize yourself with the Help window.

9. Use the LINE command to draw a line graph of insurance premiums for non-smoker men. Enter the coordinates listed by the table below:

X (Age)	Y (Rate $)
25	14
30	15
35	19
40	23
45	37
50	59
55	88
60	130
65	194

   Are these coordinates relative or absolute?
   (If you cannot see the drawing, use the ZOOM **Extents** command.)

10. Using the figure below, label each axis and the origin.

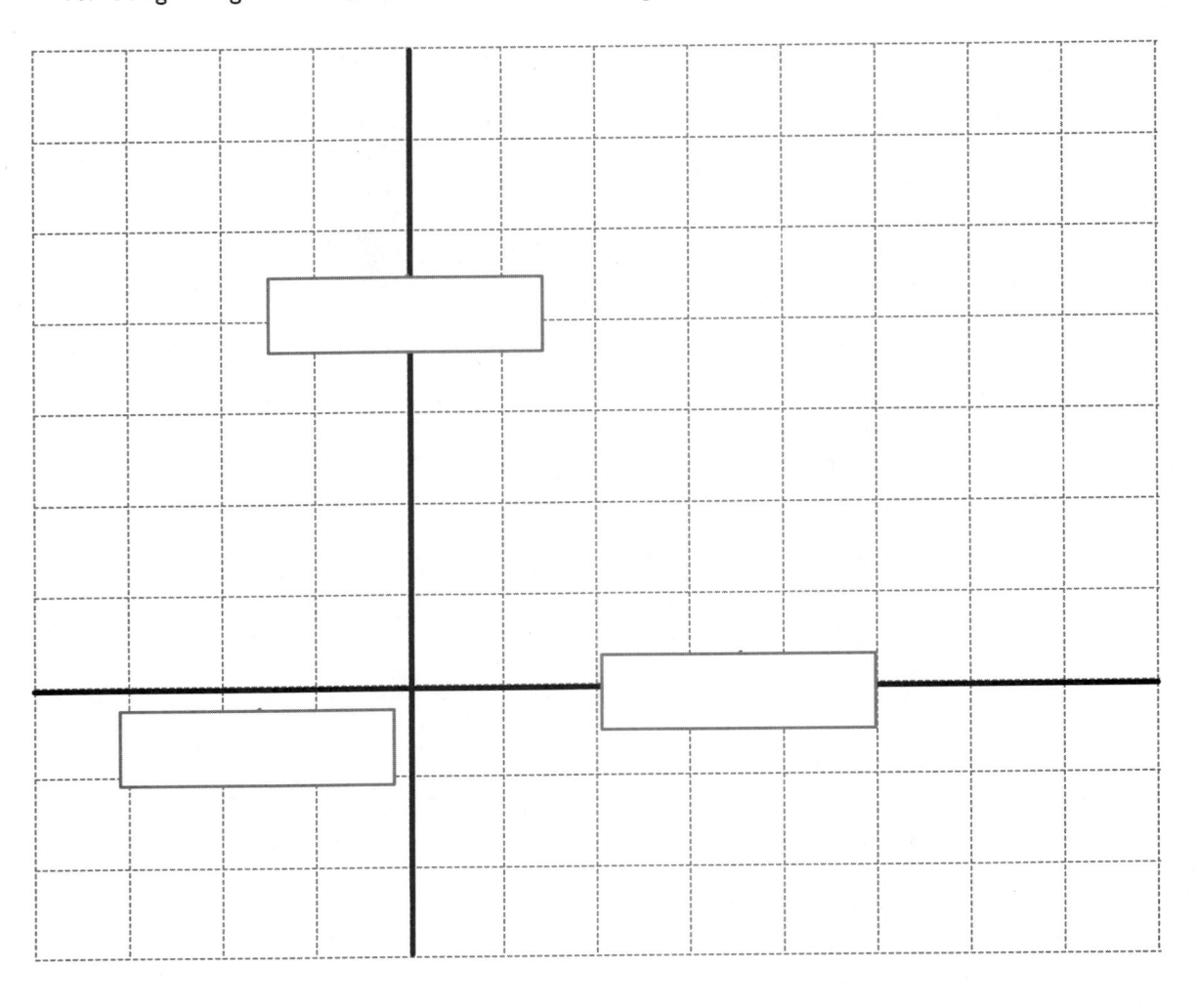

CHAPTER REVIEW

1. List the information found on AutoCAD's title bar.
2. What are benefits of using template drawings?
3. Where are the mode indicators located?
4. How does a menu indicate the presence of submenus?
   Of dialog boxes?
5. Must you exit AutoCAD to copy or delete files?
6. What happens when you press the **F2** key?
7. List the control and/or function key associated with the following actions:
   Turns ortho mode on and off.
   Selects all objects in the drawing.
   Allows you to toggle between graphic and text screens.
   Toggles tablet mode on and off.
   Cancels all the characters on the command line.
   Toggles to the next isometric plane in iso mode.
   Toggles the screen coordinate display modes.
   Toggles the grid on and off.
   Toggles snap on and off.
8. After a command ends, what happens when you press **ENTER**?
9. Name three ways to invoke the help facility:
   a.
   b.
   c.
10. What is the purpose of the prompt area?
11. Explain the meaning of the icons in the layer drop list:
    Lightbulb
    Shining sun
    Closed padlock
    Color square
    Gray layer name
12. What does the color BYLAYER mean?
13. What are linetypes?
14. Explain the purpose of the small square at the center of the crosshair cursor.
15. "UCS" is short for:
16. Can the UCS icon be turned off?
    If so, why?
17. Explain the meaning of these numbers on the status line:
    32'-11 3/8", 1'-2 1/2", 0'-0"
18. What does it mean when the following buttons on the status line look depressed?
    SNAP
    OSNAP
    MODEL
19. Where is the "tray" located?
20. Can you "show" AutoCAD an angle by picking points in the drawing?

21. How do you show AutoCAD x and y displacement simultaneously?
22. Describe how *transparent* commands work.
23. Can any command be transparent? <u>no</u>
24. What is an *alias*?
25. Write out the aliases for the following commands:
    ARC
    LINE
    BLOCK
    CIRCLE
    COPY
    ZOOM
26. Where are function keys located?
    How can you identify function keys?
27. Write out the function keys for the following actions:
    Toggle snap mode
    Toggle ortho mode
    Toggle grid display
    Exit AutoCAD
28. What does "toggle" mean?
29. Write out the Ctrl keys for the following actions:
    Display Properties window
    Toggle clean screen mode
    Copy objects to the Clipboard
    Save the drawing
    Switch to the next drawing window
30. Describe two methods to repeat commands.
31. Explain in two or three words the function of the mouse buttons:
    Left button
    Right button
    Scroll wheel
    SHIFT+Left button
32. What happens when you click an item on the menu bar a second time?
33. Describe two methods to access commands on menus.
34. Explain some pros and cons to using dialog boxes.
    Pro:
    Con:
35. Do all commands use dialog boxes?
36. Explain the difference between the **ARRAY** and the **-ARRAY** commands.
37. How do you force file-related commands to display their dialog box, when the function has been turned off?
38. Describe the purpose of the **Preview** button on dialog boxes.
39. What is a *non-modal* dialog box?
40. What is the purpose of *tabs* in dialog boxes?
41. Describe the difference between:
    Check boxes
    Radio buttons

42. What happens in dialog boxes when lists are too long?
43. How can you select more than one item from a list?
44. List the function of the commands:
    LINE
    CIRCLE
    ARC
    POLYGON
    TEXT
    Are these commands for drawing or editing?
45. List the function of the commands:
    COPY
    MIRROR
    INSERTs
    MATCHPROP
46. Give two reasons why you might need to edit drawings?
47. List the function of the commands:
    MOVE
    ERASE
    ROTATE
    DDEDIT
    PEDIT
    PROPERTIES
48. What happens when you double-click a line of text?
    Objects other than text?
49. Explain the difference between the two types of plotting:
    Color-based.
    Styles-based.
50. Describe the function of plot styles.
51. List one way that the **PREVIEW** command is useful.
52. What is an "e-plot"?
53. How would you email a drawing without sending the *.dwg* file itself?
54. What is *.dwf* short for?
    Name a benefit to *.dwf* files.
55. What happens if you forget the password to drawings?
56. Briefly explain the coordinate systems:
    Cartesian.
    Polar.
    Spherical.
    Cylindrical.
57. Describe the meaning of the following symbols:
    <
    @
    , (comma)
58. Name the coordinate system used by the following:
    25<45
    25,45,60
    25<45<60
    25<45,60

59. Write out relative coordinates for a distance of 5 units and an angle of 75 degrees.
60. Can coordinates be negative?
    If yes, when?
61. What are the coordinates of the *origin*?
62. Are relative coordinates measured relative to the origin?
    Are absolute coordinates measured relative to the origin?
63. In which direction are positive angle measured in AutoCAD, usually?
    Can the direction be changed?
64. In which direction is 0 degrees in AutoCAD, usually?
    Can the direction be changed?
65. Fill in the missing Cartesian coordinates to draw a line from the origin to 10,15.

    Command: **line**

    Specify first point:

    Specify next point:

66. Fill in the missing relative coordinates to draw a vertical line 10 units long from 20,15.

    Command: **line**

    Specify first point:

    Specify next point:

    What are the coordinates of the line's endpoint?
67. What happens when you press **ENTER** at the **LINE** command's "Specify first point" prompt?
68. What is the likely z coordinate when you pick a point in the drawing?
    When might the z coordinate change?
69. When might you use cylindrical and spherical coordinates?
70. When might you use UCSs?
71. Describe the distances and angles employed by spherical coordinates:
    Distance
    Angle 1
    Angle 2
    Write down an example of a spherical coordinate.
72. How are point filters helpful?
73. When you enter **.xy**, what does AutoCAD ask for?
74. Explain the purpose of:
    Snap spacing
    Object snaps
    Polar Snaps.
75. What is the difference between *direct distance entry* and *tracking*?
76. How do you enter tracking mode?
77. Can angles be negative?
    If so, when?

CHAPTER 3

Setting Up Drawings

After starting a new drawing, you must prepare it for use. After completing this chapter, you will have an understanding of these commands:

- **NEW** starts new drawings from scratch, or through the Startup dialog box.
- **QNEW** starts new drawings from templates (new to AutoCAD 2004).
- **Advanced Wizard** sets up units and angles.
- **OPEN** opens existing drawings, and performs file management functions.
- **MVSETUP** sets up units, and creates a scaled border.
- **OPTIONS** changes user interface colors, and specifies backup options.
- **UNITS** specifies unit and angle formats.
- **Scale factors** affect text, linetypes, hatch patterns, and dimensions.
- **SAVE** and **SAVEAS** save drawings by other names, in other formats, and to other computers.
- **QSAVE** saves drawings quickly.
- **QUIT** exits AutoCAD.

FINDING THE COMMANDS

On the **STANDARD** toolbar:

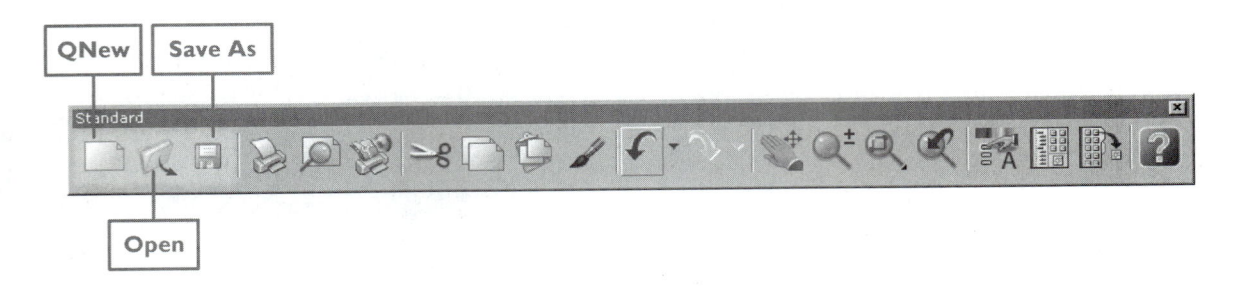

On the **FILE** menu:

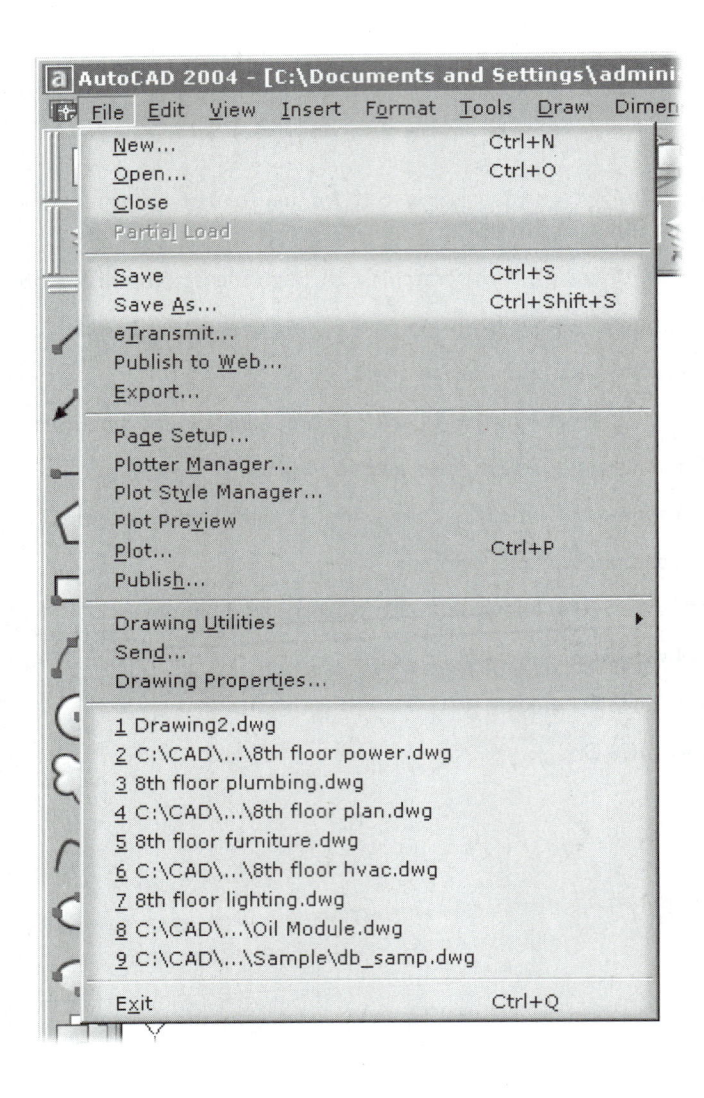

NEW AND QNEW

When you start AutoCAD, you either begin a new drawing or call up a previously-saved drawing. Let's look first at starting fresh, new drawings.

When AutoCAD first starts, it displays a new, blank drawing. At the same time, it may also display the Startup dialog box. After AutoCAD starts, repeat the **NEW** and **QNEW** commands when you need to start more new drawings. (You can have many drawings open in AutoCAD at the same time.)

TUTORIAL: STARTING NEW DRAWINGS

1. To open a new drawing, start the **NEW** command.
   - From the menu bar, choose **File**, and then **New**.
   - At the 'Command:' prompt, enter the **new** command.
   - Alternatively, enter the **CTRL+N** keyboard shortcut.

   Command: **new** (Press ENTER.)

2. In some cases, AutoCAD displays the Create New Drawing dialog box.
   In other cases, AutoCAD simply displays a new, blank drawing.

Note: Whether AutoCAD displays the Create New Drawing dialog box is determined by a setting in the Options dialog box. From the menu bar, select **Tools | Options**, and then choose the **System** tab. In the General Options section, the **Startup** option displays a list box with two options:

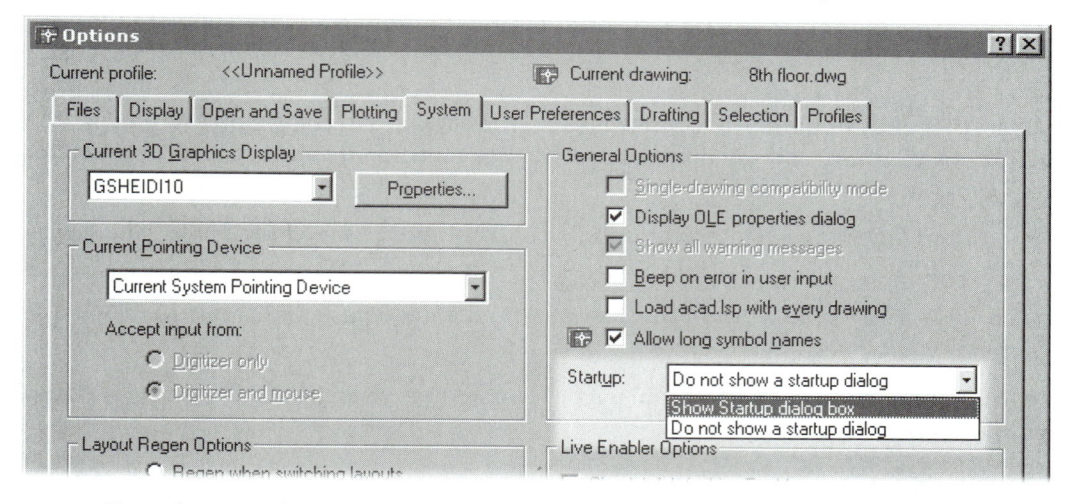

- **Show Startup dialog box** displays both dialog boxes, Startup and Create New Drawing.
- **Do not show a startup dialog box** suppresses the display of both dialog boxes.

STARTUP DIALOG BOX

The Startup and Create New Drawing dialog boxes are similar. The former is displayed when AutoCAD first starts; the latter when you enter the **NEW** command. The difference, other than the title, is that the Create New Drawing dialog box grays out the **Open a Drawing** button, a function handled by the **OPEN** command. The options are:

Start from Scratch starts new drawings in metric or English units.

Use a Template starts new drawings based on *.dwt* template files.

Use a Wizard starts new drawings based on two wizards, Quick and Advanced.

Open a Drawing opens existing drawings.

Start From Scratch

The Start from Scratch option is a fast way to start new drawings. The only option is English or Metric units.

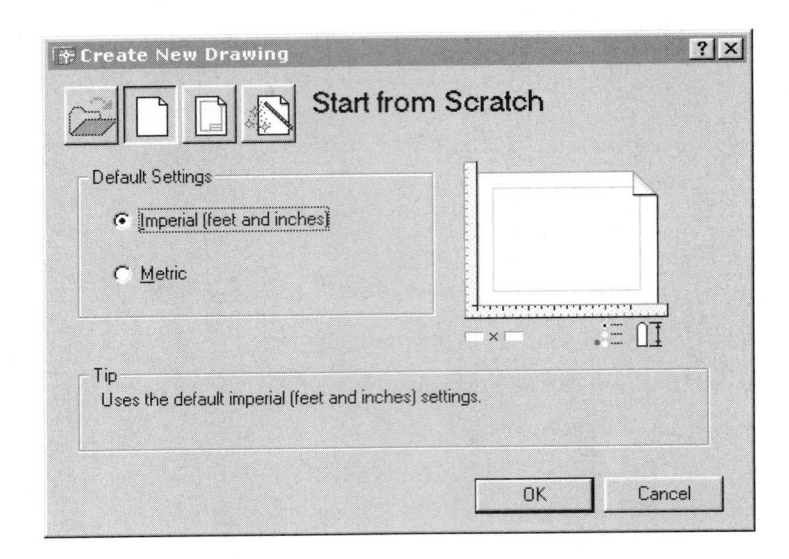

English

Select **English** for imperial units. New drawings use feet and inches for their measurement system. The drawing limits are set to 12" x 9", and dimension styles are set to inches. (You can change these settings later.) AutoCAD copies the settings stored in the *acad.dwt* template file.

Metric

Select **Metric** for metric units. New drawings use the metric system for measurement. Limits are set to 429 mm x 297 mm, and the dimension styles are metric. AutoCAD copies the settings stored in the *acadiso.dwt* template file.

In addition, AutoCAD stores your selection in the **MEASUREINIT** system variable, which has a value of 0 or 1, as follows:

0: (English) AutoCAD reads the hatch patterns and linetypes defined by the **ANSIHatch** and **ANSILinetype** registry settings.

1: (Metric) AutoCAD reads the hatch patterns and linetypes defined by the **ISOHatch** and **ISOLinetype** registry settings.

Use a Template

When you begin a new drawing, AutoCAD uses a standard list of settings, taken from *template* drawings (with the file extension of *.dwt*). Template drawings contain preset settings and, optionally, drawing elements that your new drawing can use as a template. The new drawing is identical to the template drawing. (Template files are equivalent to *prototype* files from earlier versions of AutoCAD.)

You can change any settings in the drawing, regardless of how they are set in the template drawing. In the dialog box, choose the **Use a Template** button. AutoCAD displays its list of template drawings, *.dwt* files stored in folder */autocad 2004/r16.0/env/template*. As you click on filenames, AutoCAD displays a small preview image. Click **Open** to open the template.

AutoCAD includes many template drawings for a variety of sizes and standards. Choose **Browse** to find other files.

Although template drawings have the *.dwt* extension, you are free to use *any* drawing as a template. In the Select a Template dialog box, click the **Files of type** drop list, and select *.dwg*.

You use this as a template to create a starting point for all your new drawings.

QNew

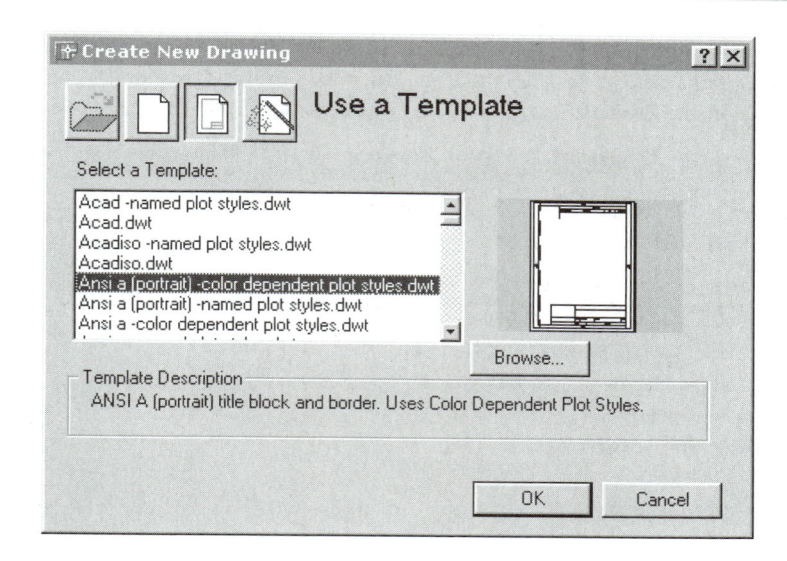

NEW IN 2004

The QNEW command starts a new drawing with a template. Depending on how AutoCAD is set up, the command either starts with a preselected *.dwt* file, or else prompts you to select a template.

TUTORIAL: STARTING NEW DRAWINGS WITH TEMPLATES

1. To open a new drawing based on a template, start the QNEW command.
   - From the **Standard** toolbar, choose **QNew**.
   - At the 'Command:' prompt, enter the **qnew** command.

      Command: **qnew** *(Press* ENTER.*)*

   In some cases, AutoCAD displays the Select Template dialog box.

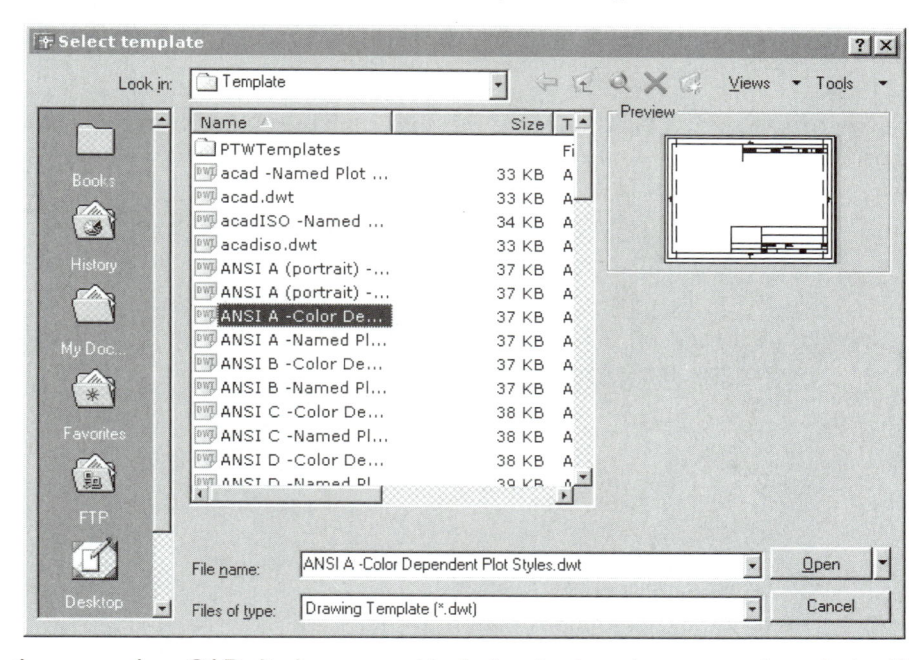

   In other cases, AutoCAD displays a new, blank drawing based on a preselected *.dwt* file.

2. Select a *.dwt* file, and then choose **Open**.

   Notice that AutoCAD opens the new drawing.

Note: Whether AutoCAD displays the Select Template dialog box is determined by a setting in the Files tab of the Options dialog box (from the menu bar, select **Tools | Options**). In the Drawing Template Settings section, the **Default Template File Name for QNEW** option has two options:

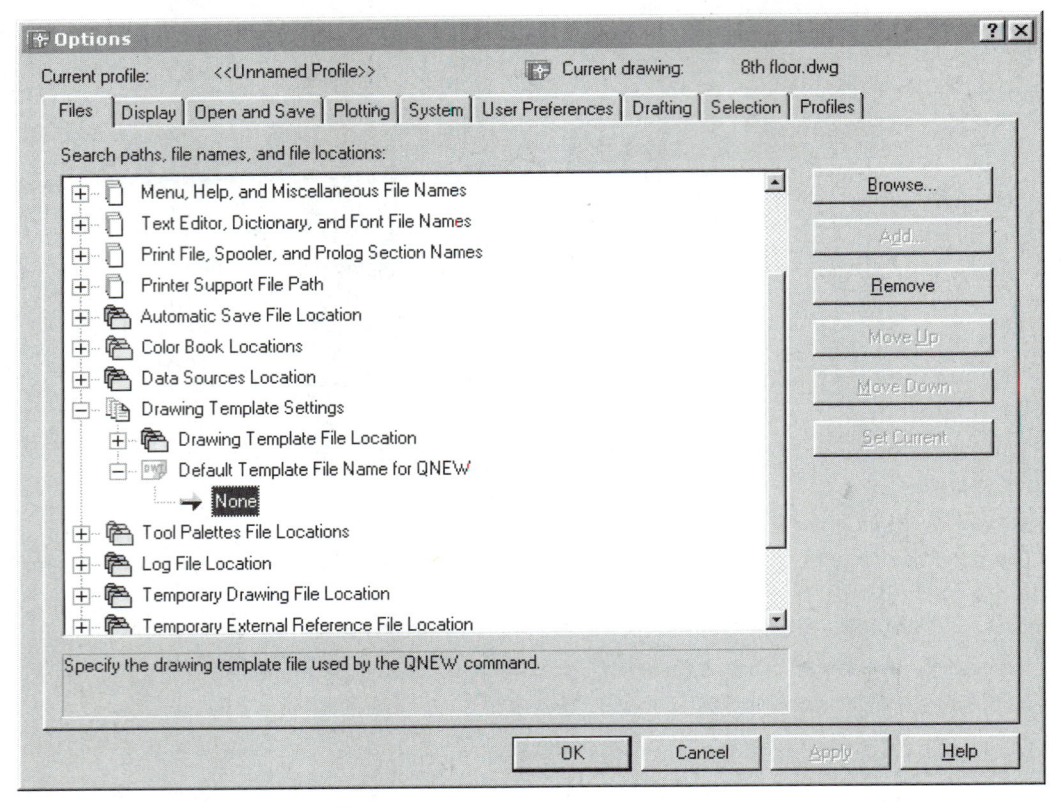

- **None** displays the Select Template dialog box.
- *filename.dwt* starts new drawings with the specified .dwt template file. Click the **Browse** button to select the file.

Wizards

AutoCAD includes two "wizards" that step you through the stages of setting up some of the many parameters that define a drawing: Quick and Advanced.

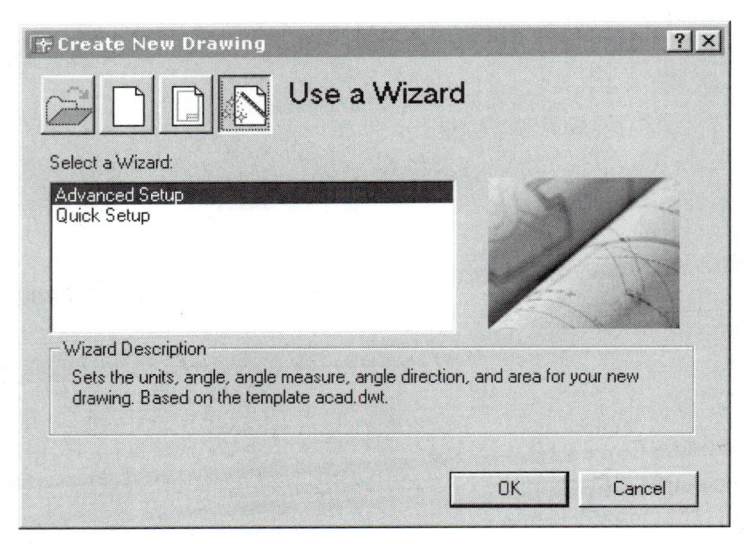

The Quick Wizard takes you through just two steps — units and area — in creating new drawings. You worked through the Quick Wizard in Chapter 1, "Quick Start in AutoCAD."

ADVANCED WIZARD

The Advanced Wizard takes you through five steps to set up units, angles, and area for new drawings.

TUTORIAL: USING THE ADVANCED WIZARD

1. To start a new drawing with the advanced wizard, choose the **Use a Wizard** button.
2. Select **Advanced Wizard**. Advanced Setup presents a series of dialog boxes that lead you through the steps of setting up new drawings. Curiously, no on-line help is available for this first encounter with AutoCAD.
3. The Units dialog box asks you to specify the units. Select a measurement unit.

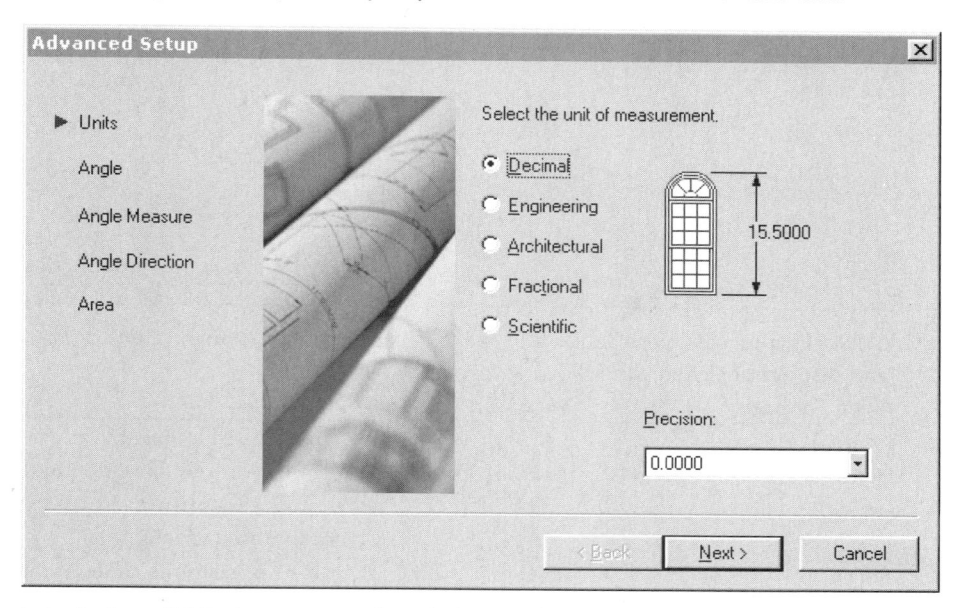

Technically, AutoCAD does not work with Imperial or metric units. Internally, the computer software uses scientific units. To help out human operators, however, AutoCAD accepts and displays units in the formats listed below, converting to and from scientific format on the fly.

AutoCAD Units	Example	Comment
Decimal	14.5	Also used for metric drawings.
Engineering	1'-2.5"	Feet and decimal inches.
Architectural	1'-2 1/2"	Feet and fractional inches.
Fractional	14 1/2	Fractional inches, no feet.
Scientific	1.45E+01	Exponent notation.

4. From the **Precision** drop list, select the number of decimal places (or fractional accuracy). This affect the display of units only; no matter which precision you select, AutoCAD calculates 16 decimal places of accuracy, and then displays the result rounded to the a maximum of eight places.
   - Decimal, engineering, and scientific units: 0 to 8 decimal places.
   - Architectural and fractional units: 1/1, 1/2, 1/4, 1/8, 1/16, 1/32, 1/64, 1/128, or 1/256.
5. Click **Next**.

6. The Angle dialog box asks you to specify the style of angle measurement. Select one.

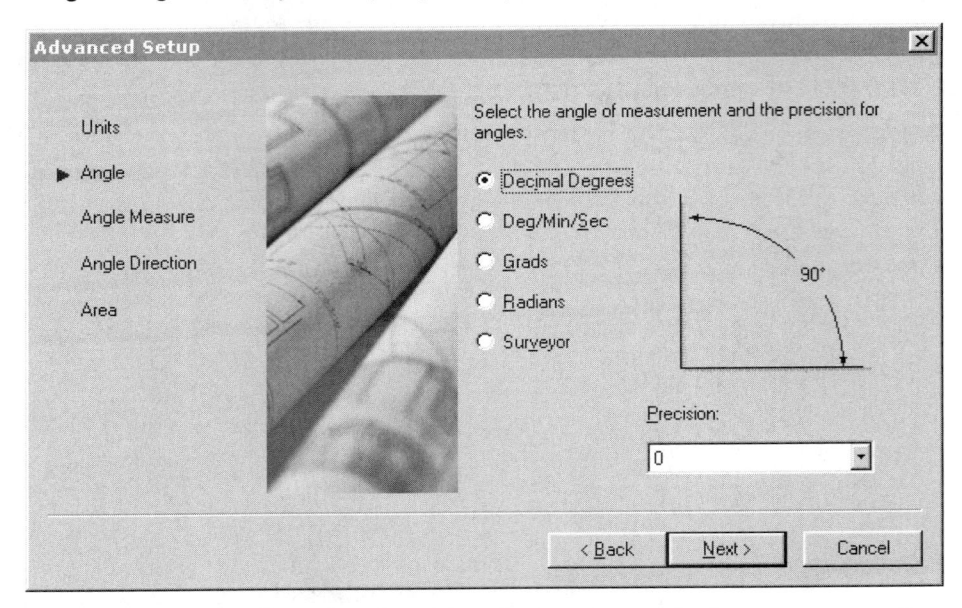

AutoCAD works internally with radians. To make it easier on the human operator, it accepts and displays angles in the following formats:

Angle Formats	Example	Comment
Decimal Degrees	45.00	360 degrees in a circle.
Degrees-minutes-seconds	45d00'00.00"	Fractions of seconds in decimals.
Grads	50.00g	400 grads in a circles.
Radians	0.79r	2pi radian in a circle.
Surveyor	N45d0'0.00"E	Fractions of seconds in decimals.

7. From the **Precision** drop list, select the number of decimal places (or fractional accuracy). As with units, this affects only the display of angles.

   - Decimal degrees, grads, and radians: 0 to 8 decimal places.

   - Deg/min/sec and surveyor angles: 45d, 45d00', 45d00'00", 45d00'00.0", 45d00'00.00", 45d00'00.000", 45d00'00.0000", or 45d00'00.00000".

8. Click **Next.**
9. The Angle Measure dialog box asks you to select a direction for 0 degrees.

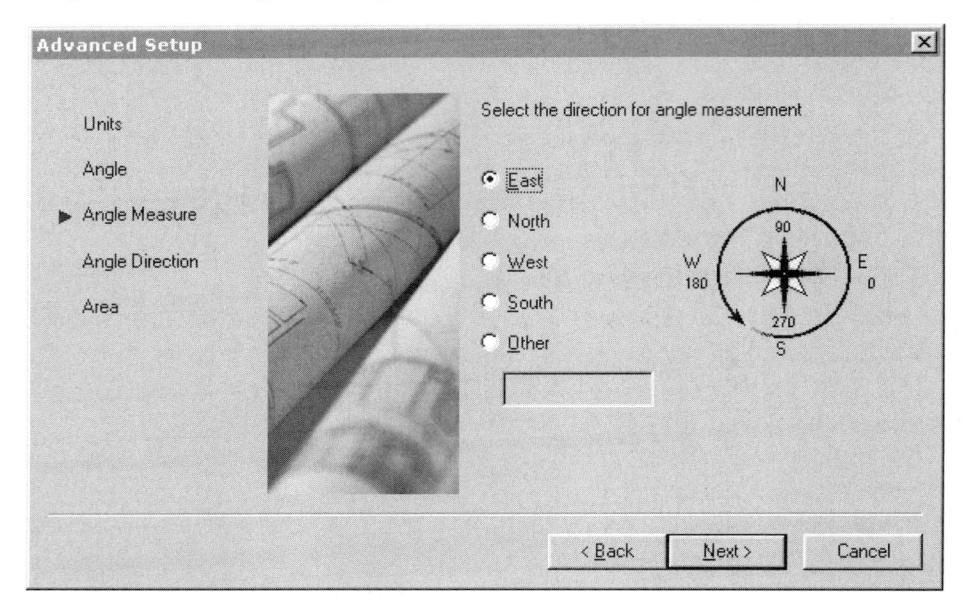

The default is East, which is the correct setting for most drawings.

If your drawings use a direction other than 0 degrees, select from **North**, **West**, or **South**, or **Other**.

In the **Other** text box, you can enter an angle between 0 and 360 degrees — even if you selected grads or radians in the previous dialog box.

10. Click **Next**.
11. The Angle Direction dialog box asks you the direction in which to measure positive angles — counterclockwise or clockwise. For most drawings, you keep the direction counterclockwise.

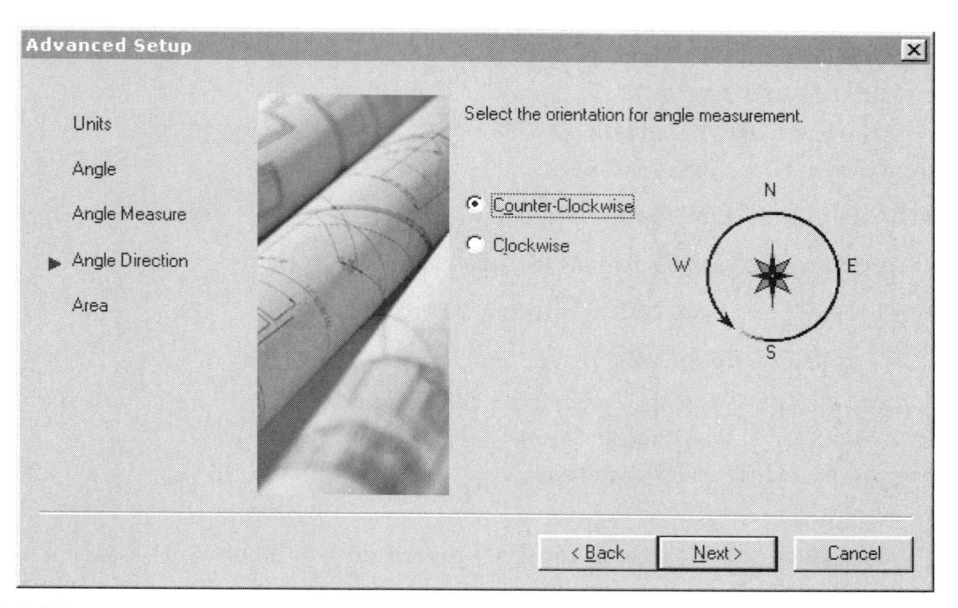

12. Click **Next**.
13. The Area dialog box asks you to specify the limits of the drawing. The default is 12 units wide (x direction) by 9 units tall (long, y direction).

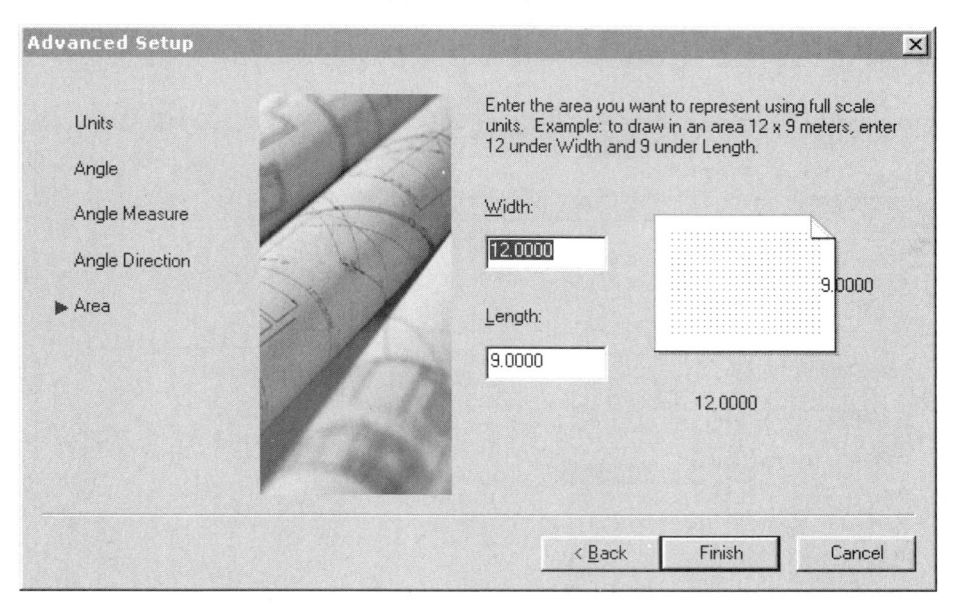

If you know how large the drawing will be, enter values in the **Width** and **Length** text boxes. Clear the current values using the DEL or BACKSPACE key.

AutoCAD uses these values to set the limits of the drawing, which affects the extent of the grid display (as shown in the preview image above) and the ZOOM All command.

Historically, AutoCAD prevented you from drawing outside the area defined by the limits; this didn't make much sense to users, and Autodesk soon dropped the restriction with version 2.17g in 1986.

14. Click **Finish**.
AutoCAD opens the new drawing with the settings you specified.

OPEN

More often than starting a new drawing, you open existing drawings for editing.

TUTORIAL: OPENING DRAWINGS

1. To open drawings, start the **OPEN** command.
   - From the **File** menu, choose **Open**.
   - From the **Standard** toolbar, choose **Open**.
   - At the 'Command:' prompt, enter the **open** command.
   - Alternatively, press the **CTRL+O** keyboard shortcut.

        Command: **open** *(Press* ENTER.*)*

   In some cases, AutoCAD displays the Select File dialog box.

2. Select a *.dwg* file, and then choose **Open**.
   Notice that AutoCAD opens the drawing.

Historically, the first several releases of AutoCAD presented a numbered *menu* of choices. To open a drawing, users pressed **2** (option 2 was called "Edit an EXISTING drawing"), and then **ENTER**. AutoCAD prompted for the name of the drawing, whose filename users had memorized, because then there was no selecting names from a list.

Release 9 introduced the basic dialog box, which allowed users to select drawing names from a list, as well as to select the drive and folder. AutoCAD 2000i greatly expanded the options for opening drawings. Among the capabilities of today's Select File dialog box are:

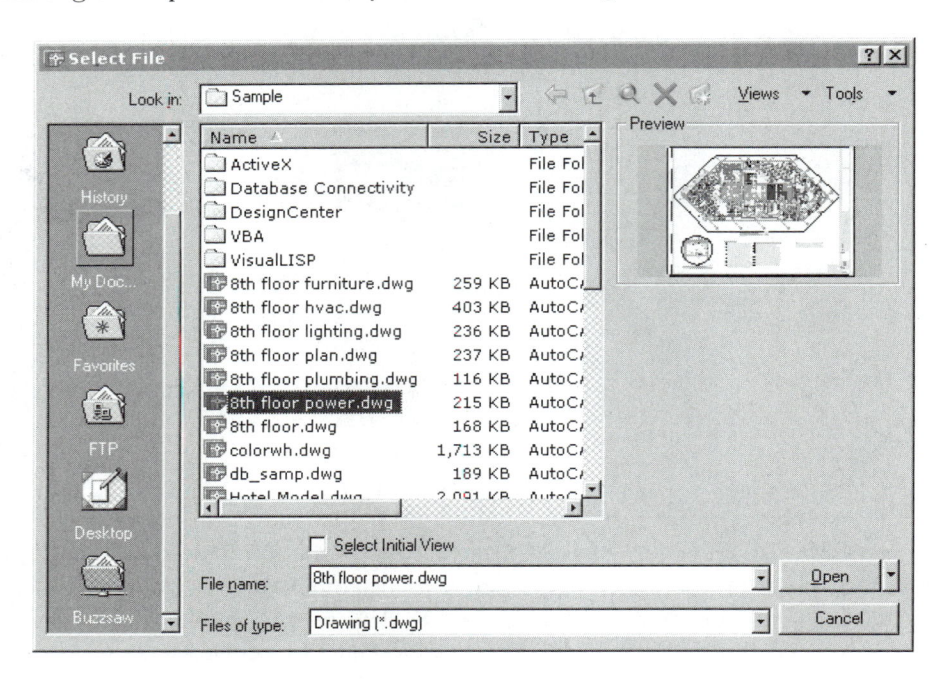

Setting Up Drawings **87**

- Open *.dwg* (drawing), *.dxf* (interchange), *.dws* (standards) and *.dwt* (template) files.
- Select one or more files to open at the same time.
- Preview drawings (via thumbnail representation) before opening.
- Sort files by name, date, type, and size.
- View drawings recently opened (History), and most often used (Favorites).
- Open drawings protected against change with read-only status.
- Load all or portions of drawings.
- Access drawings on your computer, any other computer located on your company network, or from any location accessible through the Internet (FTP).
- Search for drawings anywhere on your computer (Find).
- Create a list of frequently-accessed folders.
- Create new folders, and manipulate files and folders, including renaming and deleting files.
- Send drawings as email messages, or to removable drives.
- Search the Internet for information.
- Resize the Select Drawing dialog box.
- Access Autodesk's Buzzsaw.com Web site.

Note: The **OPEN** command is the most common method to open files in AutoCAD, but is not the only way to open files:

ACISIN opens *.sat* format solid models created by ACIS-compatible CAD programs. This command opens the model in a new drawing; it does not insert the model in the current drawing.
DXBIN opens *.,dxb* (drawing exchange binary) files created by CAD\camera.
DXFIN opens *.dxf* (drawing interchange format) files created by other CAD programs.
FILEOPEN opens drawings without a dialog box; useful for scripts and macros, but is limited to AutoCAD running in SDI mode (single drawing interface).
IMAGE inserts raster (bitmap) images into drawings.
INSERT inserts other drawings as blocks into the current drawing.
PARTIALOAD loads additional portions of partially-loaded drawings.
REPLAY displays images in *.bmp*, *.tif*, and *.tga* formats.
RMLIN inserts *.rml* (redline markup language) files created by Autodesk's Volo View software.
WMFIN inserts *.wmf* (Windows meta format) files.
XOPEN opens externally-referenced drawings in their own windows.
3DSIN opens *.3ds* models created by 3D Studio.

And the Clipboard pastes vector and raster graphics from most other programs.

The Select File dialog box consists of five primary sections. Going counterclockwise, they are:

Files List lists the names of files and folders.

Toolbar holds tools useful for manipulating files and folders.

Standard Folders provides shortcuts to folders and Web locations.

File Spec specifies the name and type of file to be opened.

Preview displays a thumbnail-size image of the drawing.

Files List

To open a drawing, select its name from the file list. Notice that its name also appears in the **File name** field. Click **Open**. (As a shortcut, double-click the filename. This is equivalent to selecting the filename, and then clicking the **Open** button.)

When you select a drawing in the Files List, notice that AutoCAD displays a preview image. The image is quite small, but provides sufficient detail to distinguish among most drawings. The preview image, also called a "thumbnail," is generated automatically by AutoCAD each time you save the drawing. Because the thumbnail shows the view when the drawing was last saved, the view may not include the entire drawing.

> **Note**: If you see no preview images, the **RASTERPREVIEW** system variable has been turned off (set to 0); turn it on by changing the value to 1:
>
> Command: **rasterpreview**
>
> Enter new value for RASTERPREVIEW <0>: **1** *(Press* ENTER.*)*
>
> The other possibility is that the drawing files were created by old versions of AutoCAD that didn't add preview images to files.

File Spec

By default, the dialog box displays the names of *.dwg* AutoCAD drawing files. In addition, you can open three other kinds of files:

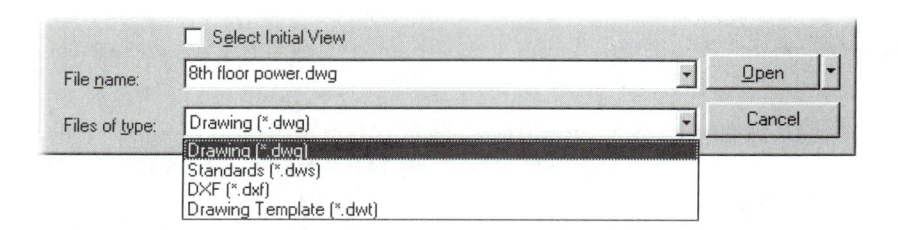

DXF (drawing interchange format) exchanges CAD drawings between incompatible programs.

DWT (drawing template) presets many settings for new drawings, such as dimension styles and drawing borders.

DWS (drawing standards) defines standard elements, such as linetypes and layers.

To select filetype different from *.dwg*, click the **Files of type** list. When you select "Drawing Template (*.DWT)", notice that the dialog box automatically switches to the *template* folder.

Whether you select *.dwg* or *.dxf*, there is more than one way to open the file. Click the down-arrow next to the **Open** button. Notice the menu that appears:

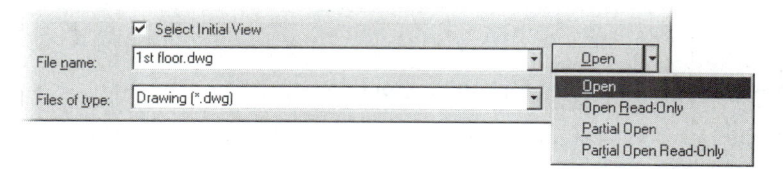

Open loads all of the drawing; you can edit the drawing, and save the changes.

Open Read-Only loads all of the drawing, and does not save editing changes.

Partial Open loads parts of the drawing, based on layer and view names.

Partial Open Read-Only loads parts of the drawing, and does not save editing changes.

When you select one of the two read-only options, AutoCAD displays the "Read Only" message on the title bar. In addition, if you attempt to save the drawing, a warning dialog box is displayed, as illustrated by the figure below.

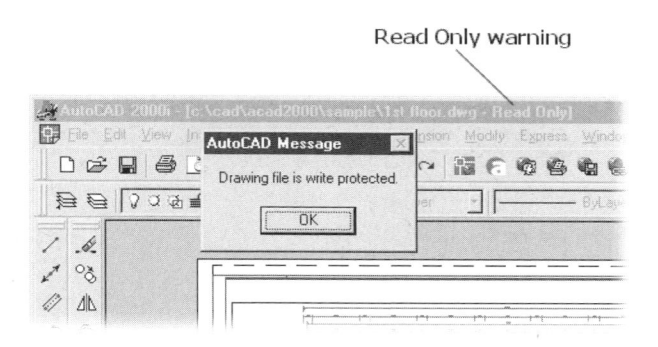

When you select one of the partial-open options, AutoCAD displays the Partial Open dialog box. Select a view from the lefthand column and the layers to be opened in the righthand column. When done, click **Open**.

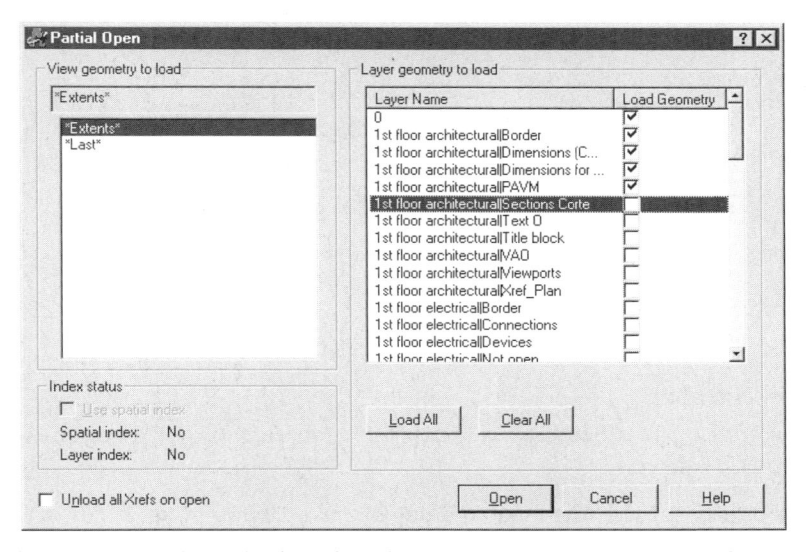

Partial opening of drawings is described in detail in *Using AutoCAD 2004: Advanced*.

The **Select Initial View** check box determines whether the drawing is opened with a saved view:

Off: When the drawing is opened, AutoCAD displays the same view as when the drawing was last saved (the default setting).

On: Before the drawing is opened, AutoCAD displays the Select Initial View dialog box, which lets you select a named view. If no views have been saved, then the only option is *Last View*.

Making the Dialog Box Bigger

A new feature is the ability to make dialog box larger. When you first see the dialog box, it is at its smallest. With the cursor, grab an edge of the dialog box, and stretch the dialog box larger, then smaller. Notice that this allows you to see more filenames at a time.

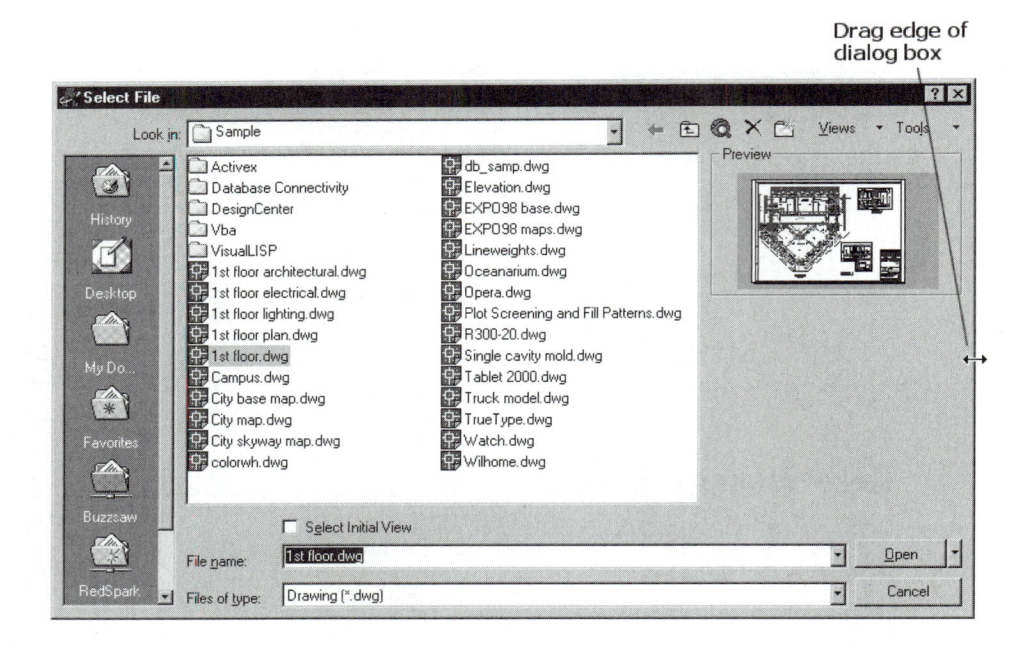

Drag edge of dialog box

Note: Press function key **F5** to update the Files List. You may need to do this when files appear to be missing or incorrectly named.

Standard Folders Sidebar

This vertical list of folders provides direct access to a location on your computer or the Internet. From top to bottom, these are:

History displays a list of the most-recently opened drawings.

My Documents displays files stored in your computer's \my documents folder.

Favorites displays files stored in the \windows\favorites folder; add files with **Tools | Add to Favorites**.

FTP browse FTP (file transfer protocol) sites on the Internet; add sites with **Tools | Add/Modify FTP Locations**.

Desktop displays the files and folders found on the Windows "desktop."

Buzzsaw connects you to Autodesk's www.buzzsaw.com Web site, after you install their ProjectPoint client software.

(The Red Spark item was removed from AutoCAD 2004; the Web site no longer exists.)

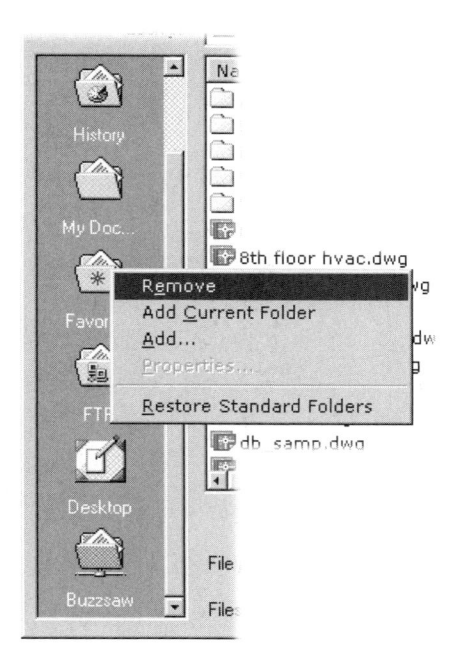

To add folders to the list, simply drag from the files list. Other commands are available by right-clicking the folder sidebar, as illustrated by the figure above.

Toolbar

The toolbar at the top of the Select File dialog box contains many commands.

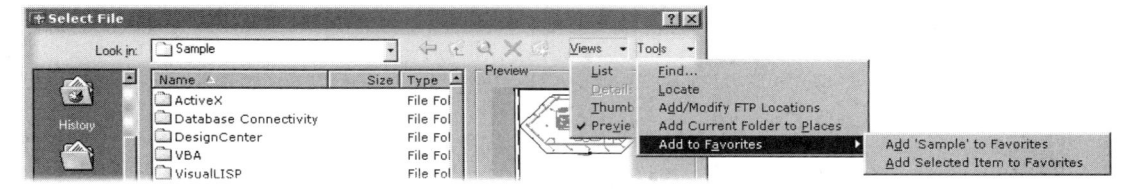

From left to right, these are:

    Look in selects the folder, drive, and network location from the drop list.

    Back moves back to the previous folder (keyboard shortcut is **ALT+1**).

    Up One Level moves up one level in the folder structure (**ALT+2**).

    Search the Web displays a simple Web browser (**ALT+3**).

    Delete erases selected files and folders (**DEL**); Windows asks if you are sure.

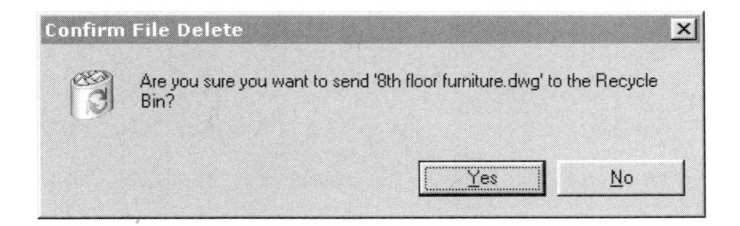

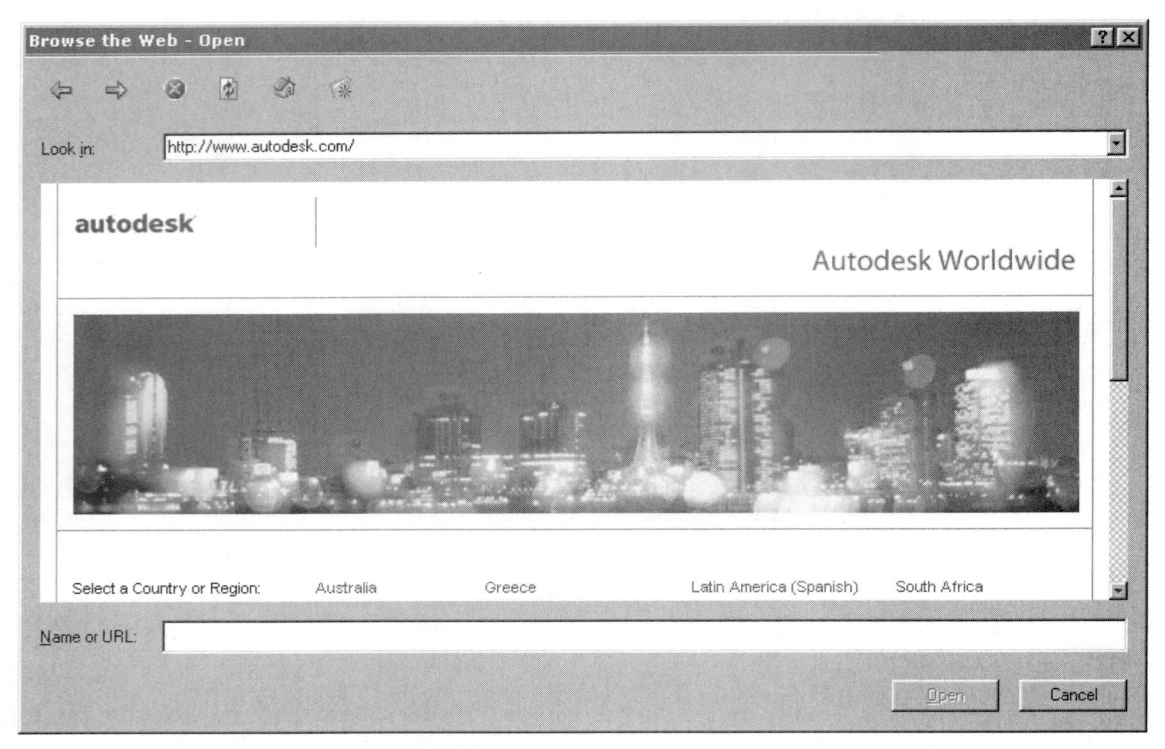

The basic Web browser built into AutoCAD.

Create New Folder creates new folders (**ALT+5**).

Views changes how files are listed:

List lists file names only.

Details lists file names, size, type, and date last modified.

Thumbnails displays files as thumbnail images.

Preview toggles the display of the preview image.

Tools provides additional commands, depending on the data displayed by the Files List:

Find displays the Windows Find dialog box that lets you search for files.

Locate displays files in folders listed by AutoCAD's search path.

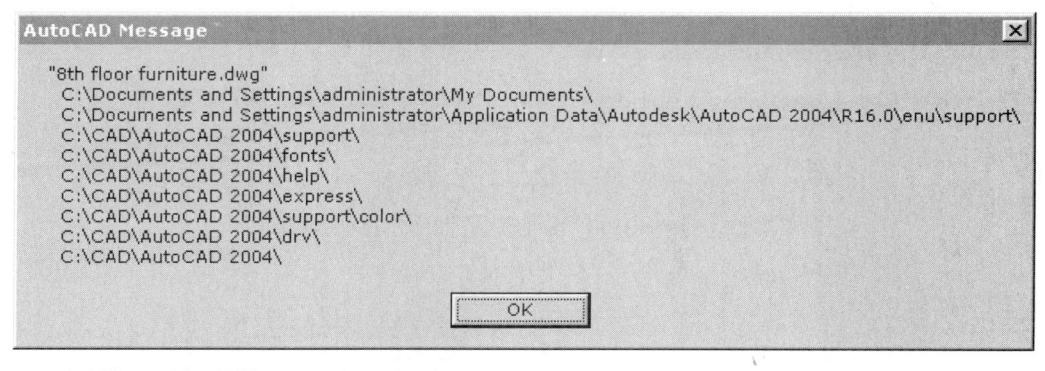

Add/Modify FTP Location displays the Add/Modify FTP Locations, which allows you to specify the URL, user name, and password for accessing FTP sites.

Add to Favorites adds selected files and folders to the Favorites list, which quickly accesses frequently-used drawings.

FINDING FILES

Sometimes you cannot remember the name of a drawing file or where it is located. To help you locate a drawing file, AutoCAD includes the **Find** option in the **Tools** item on the dialog box's toolbar. When you choose Find, AutoCAD displays the standard Windows dialog box for finding files.

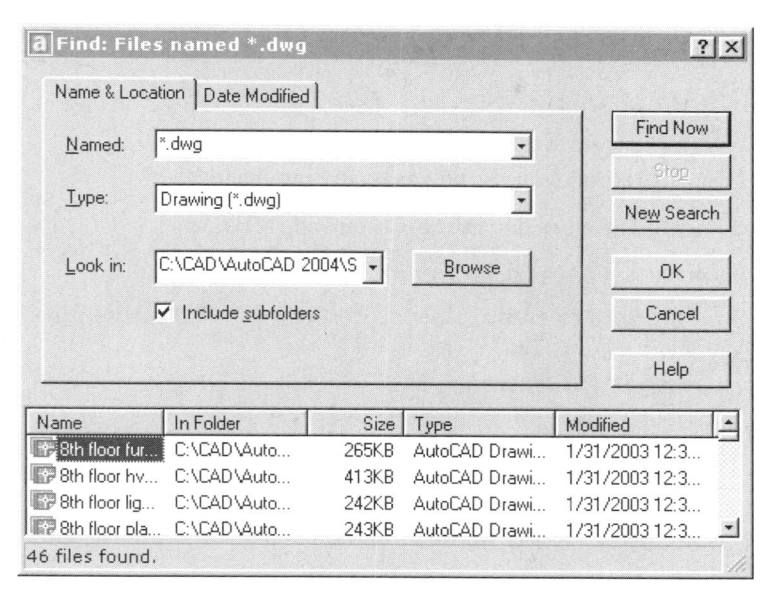

It is able to look at every folder on every drive in your computer. Drives include the hard drives, floppy drives, CD-ROM drives, and other storage devices. In addition, if your computer is connected to a network of other computers, the Find feature can search every folder of every networked drive your computer has permission to read.

Named

The **Named** text box specifies as much of the name as you can remember. Use wildcard characters to specify part of a name.

Recall that ? is a placeholder for any character, while * searches for all files that match the rest of the criteria you supply. For example, *door** searches for all files that start with **Door**, while *door????.dwg* searches for all AutoCAD *.dwg* drawing files that start with **Door**:

    door.dwg   door36.dwg   doorstop.dwg

Types

The **Type** drop list provides a filter for selecting *.dwg* drawing, interchange *.dxf*, template *.dwt*, and standards *.dws* files.

Look In

The **Look In** drop list narrows and broadens the search to specific drives, folders, and paths.

Date Modified

The **Date Modified** item narrows the search to specific dates. This is useful when you know roughly the date the drawing was created. For example, if the drawing was created last month, specify "Before this date", and enter this month.

After you enter the search parameters, choose **Find Now**. Windows spends a bit of time rummaging through your computer's drives, and produces a list of all matching files. At any time, choose **Stop** to bring the process to a premature halt.

MVSETUP

The Basic and Advanced Wizards are both pretty basic. They don't do many of the tasks required for setting up new drawings. AutoCAD "hides" a similar command, called **MVSETUP** (short for "model view setup"), that sets up drawings complete with a predrawn border. Optionally, it also sets the scale, multiple viewports, and aligns views in each viewport.

TUTORIAL: SETTING UP NEW DRAWINGS

1. To set up a new drawing, start the **MVSETUP** command:
   - At the 'Command:' prompt, enter the **mvsetup** command.
   - Alternatively, enter the **mvs** alias at the 'Command:' prompt.

      Command: **mvsetup** *(Press ENTER.)*

2. AutoCAD asks if you wish to enable paper space. Answer "N" (because paper space is dealt with in volume 2 of this book):

   Enable paper space? [No/Yes] <Y>: **n**

3. Specify the type of units:

   Enter units type [Scientific/Decimal/Engineering/Architectural/Metric]: *(Enter an option.)*

4. Depending on the units you specified, AutoCAD displays different scale factors.

   Enter the scale factor: *(Enter a number.)*

   AutoCAD expects you to enter the number in parentheses next to each scale factor. For example, enter 4 for "4 Times" decimal scale; enter 120 for 1"=10' engineering scale.

- Scientific and Decimal scale factors:

   (4.0)   4 TIMES
   (2.0)   2 TIMES
   (1.0)   FULL
   (0.5)   HALF
   (0.25) QUARTER

- Metric scale factors:

   (5000) 1:5000
   (2000) 1:2000
   (1000) 1:1000
   (500) 1:500
   (200) 1:200
   (100) 1:100
   (75)   1:75
   (50)   1:50
   (20)   1:20
   (10)   1:10
   (5)   1:5
   (1)   FULL

- Architectural scale factors:

   (480) 1/40"=1'
   (240) 1/20"=1'
   (192) 1/16"=1'
   (96) 1/8"=1'
   (48) 1/4"=1'
   (24) 1/2"=1'
   (16) 3/4"=1'
   (12) 1"=1'
   (4)   3"=1'
   (2)   6"=1'
   (1)   FULL

- Engineering scale factors:

   (120) 1"=10'
   (240) 1"=20'
   (360) 1"=30'
   (480) 1"=40'
   (600) 1"=50'
   (720) 1"=60'
   (960) 1"=80'
   (1200) 1"=100'

5. As a last step, AutoCAD prompts you to enter the width and height of the paper.

    Enter the paper width: *(Enter a value.)*

    Enter the paper height: *(Enter a value.)*

    Paper means the media the drawing is plotted on. Below are sizes for standard sheets.

Sheet	Size (inches)
ANSI Engineering:	
A	8.5" x 11"
B	11" x 17"
C	17" x 22"
D	22" x 34"
E	34" x 44"
US Architectural:	
A	9" x 12"
B	12" x 18"
C	18" x 24"
D	24" x 36"
E	36" x 48"
ISO/DIN Metric:	
A4	198mm x 278mm
A3	285mm x 408mm
A2	408mm x 582mm
A1	582mm x 819mm
A0	819mm x 1167mm

6. AutoCAD draws the outline of the sheet as a rectangle. The rectangle is multiplied by the scale factor. For example, an A-size sheet at scale factor 4 is drawn at 34" x 44".

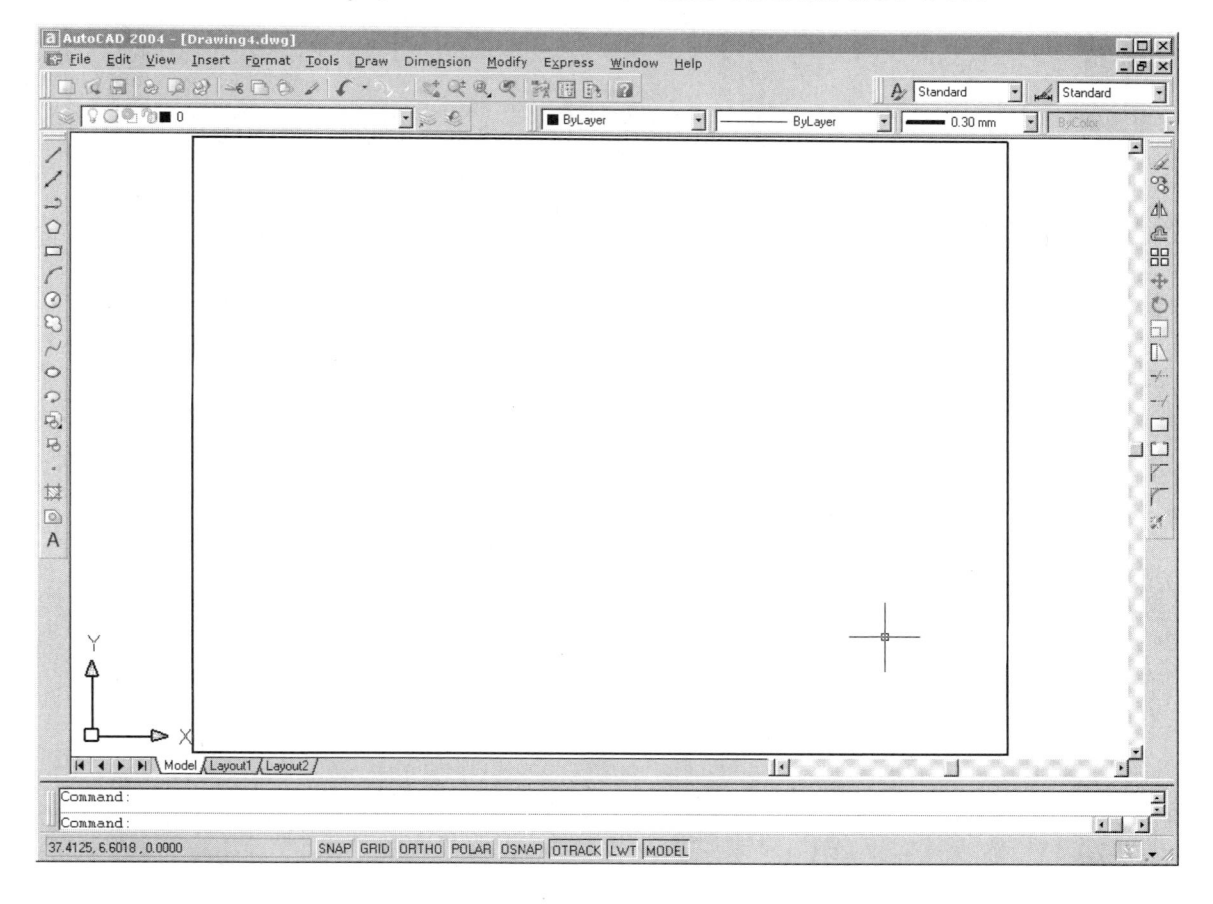

CHANGING THE SCREEN COLOR

AutoCAD installed out-of-the-box displays the drawing area in black. Historically, computer screens had a black background, displaying text and graphics in green, amber, or white — called "monochrome" displays. AutoCAD 2004 mimics history with the background.

Some drafters prefer the black background, because it makes lighter colors, such as yellow and green, stand out better. Others like black, because they find it creates less eye strain.

With the advent of color displays, the background color of choice became white, because it is the color paper documents and drawings are printed on.

The figure below illustrates AutoCAD with the black background. Notice that the cursor and grid dots show as white on the black background.

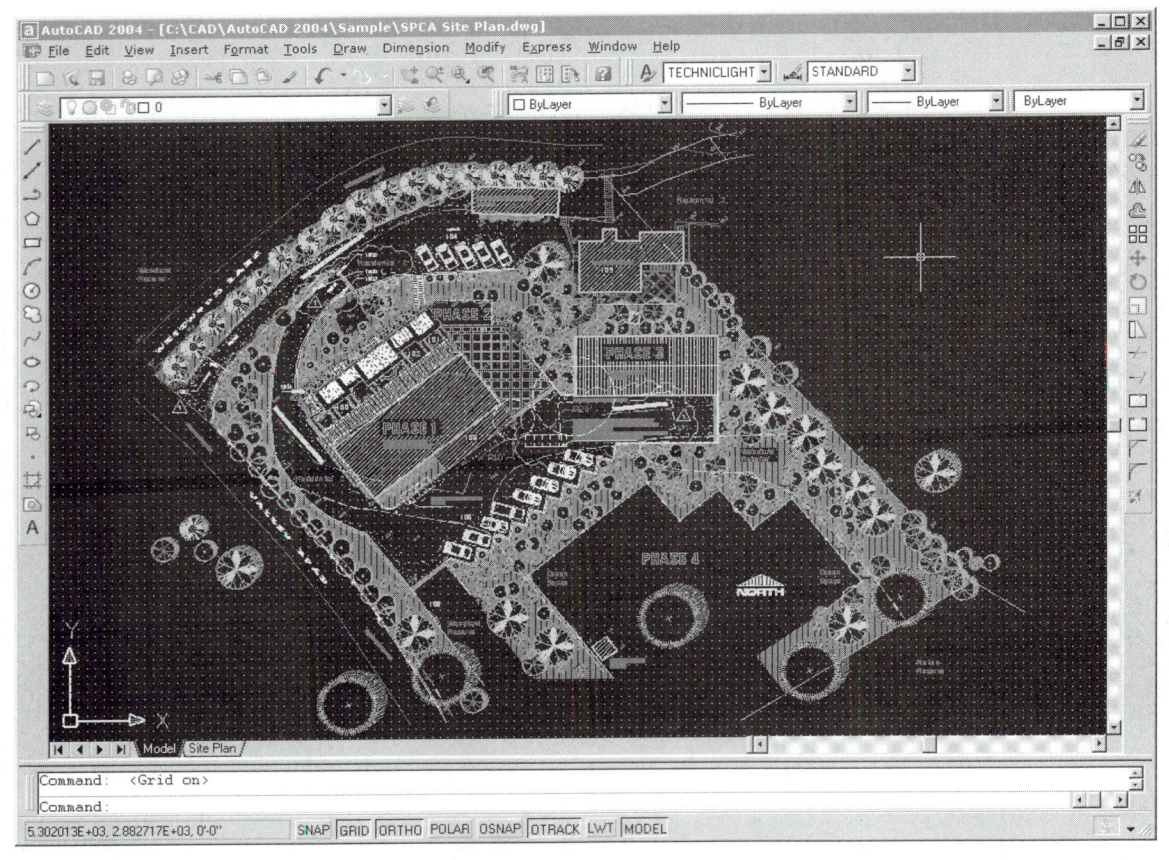

AutoCAD allows you to change the drawing background color, and some other parts of its user interface:

- **Model Tab Background** — background color of drawing area in model space.
- **Model Tab Pointer** — color of crosshair cursor and grid dots in model space.
- **Layout Tabs Background (Paper)** — background color of drawing area in all layouts.
- **Layout Tabs Pointer** — color of crosshair cursor and grid dots in all layouts.
- **AutoTracking Vector Color** — color of autotracking vectors (dotted lines displayed when object tracking and polar snap are turned on).
- **Command Line** — background color of the command prompt window.
- **Command Line Text** — color of text in the command prompt window.

TUTORIAL: CHANGING COLORS

1. To change the colors of the AutoCAD user interface, start the **OPTIONS** command:
   - From the **Tools** menu bar, choose **Options**, and then **Display**.
   - At the 'Command:' prompt, enter the **options** command.

   Command: **options** *(Press ENTER.)*

2. In the Display tab of the Options dialog box, click **Colors**.
   AutoCAD displays the Color Options dialog box.

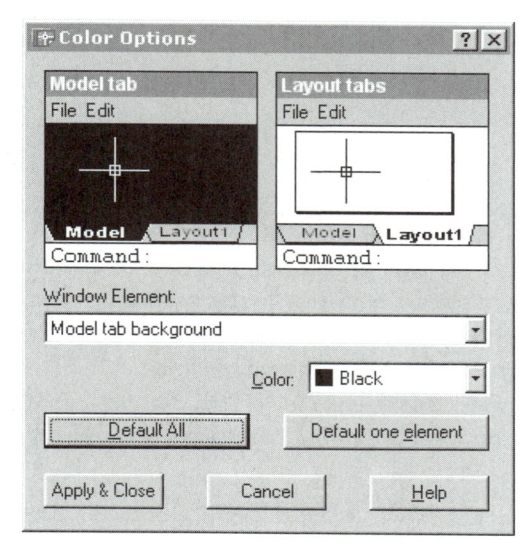

3. Select a user interface element to change. You can click on the graphical representation, or select an item from the **Window Element** drop list.
4. Assign a color to the element by selecting it from the **Color** drop list. If the basic eight colors are insufficient, click **Select Color**. AutoCAD displays the Select Color dialog box, which provides you with access to the full palette of 16.7 million colors.
5. If you mess up, click **Default one element** to change the element back to its default color. Or, choose **Default All** to change all elements back to their default colors.
6. Click **Apply & Close** to see the change in colors.
   Click **OK** to exit the Options dialog box.

UNITS

The **UNITS** command allows you to select the units of measurement, angles, and precision.

If you chose not to use the Advanced Wizard, you can set units with the Units dialog box.

TUTORIAL: SETTING UNITS

1. To set the units for drawings, start the **UNITS** command:
   - From the menu bar, choose **Format**, and then **Units**.
   - At the 'Command:' prompt, enter the **units** command.
   - Alternatively, enter the aliases **un** or **ddunits** (the old name for this command) at the 'Command:' prompt.

   Command: **units** *(Press ENTER.)*

   In all cases, AutoCAD displays the Units dialog box.

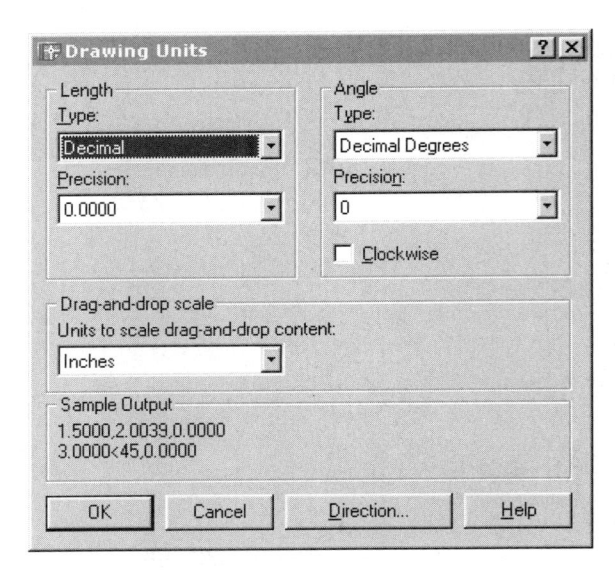

[Drawing Units dialog box]

2. In the **Length** area, select the type of unit and precision. To learn more about the units used by AutoCAD, see the discussion earlier in this chapter for the Advanced Wizard. Watch the **Sample Output** area for examples of the units you select.
3. In the **Angle** area, select the type and precision of angles.
4. **Drag-and-drop scale** specifies the units for blocks dragged into drawings from the DesignCenter, Tool Palettes, or from a Web site using i-drop. Available units are:

Unit	Comment
Inches	25.4mm
Feet	12 inches
Miles	5,280 feet
Millimeters	0.0393 inches
Centimeters	10 mm
Meters	100 cm
Kilometers	1000 m
Microinches	0.000001 inches
Mils	0.001 inches
Yards	3 feet
Angstroms	0.1 nanometers
Nanometers	10E-9 meters
Microns	10E-6 meters
Decimeters	0.1 meter
Decameters	10 meters
Hectometers	100 meters
Gigameters	10E9 meters
Astronomical Units	149.597E8 kilometers
Light Years	9.4605E9 kilometers
Parsecs	3.26 light years

Unitless inserts the block without scaling, so that it matches the units of the drawing.
5. To change the direction of angle measurement, click **Direction**.
AutoCAD displays the Direction Control dialog box.

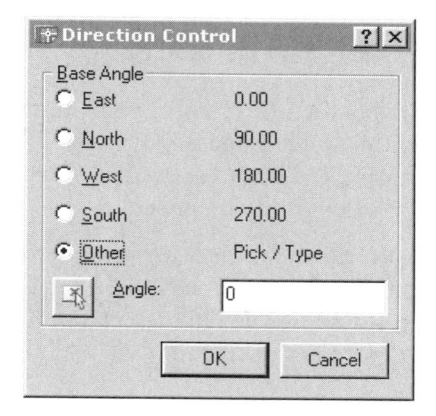

6. Select a direction for the angle.

    If you wish an angle that isn't one of the right angles listed, select **Other**, and then enter the angle in the **Angle** text box.

    Alternatively, choose the **Pick an Angle** button to pick the angle in the drawing:

    Pick angle: *(Pick a point in the drawing.)*

    Specify second point: *(Pick a second point.)*

7. Click **OK** to dismiss the Direction Control dialog box.

    Click **OK** to dismiss the Drawing Units dialog box.

8. Move the cursor, and watch the coordinate display on the status bar. It should match the units you selected. The figure below illustrates scientific units and grad angles:

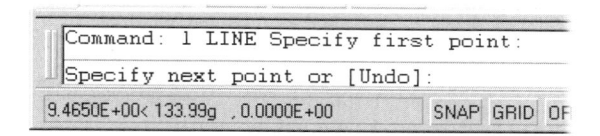

SETTING SCALE FACTORS

AutoCAD does not have a single command to set the scale factor for drawings — even though scale affects text, linetypes, hatch patterns, dimensions, and blocks inserted from other drawings.

One of the hardest concepts for the new CAD user to grasp is that of *scale*. Think of when you sketch a picture of your house on a piece of paper. You draw the house small enough to fit the paper (there isn't a piece of paper big enough for you to draw the house full size!).

A typical house is 50 feet long. To fit it on a 10"-wide sheet of paper, you sketch the house 60 times smaller. Where does the 60 come from? To fit 50' onto 10", the math works out like this: 50 feet x 12 inches/foot = 600 inches. Divide 600 inches by 10 inches = 60. The scale factor is 1:60. One way to think of scale factor is to realize, "One inch on my drawing is 60 inches on the house."

Here are additional examples of scale factors:

   A truck is 20 feet long. The paper is 10 inches wide. The scale factor works out to 1:24.

   The sailboat is 30 m long. The paper is 1m wide. The scale factor is 1:30. (It's much easier in metric, isn't it?)

   The gear is 1 inch in diameter. The paper is 8 inches wide. The scale factor is 8:1. (The scale factor numbers are "reversed" when small objects are drawn at a larger scale factor)

In CAD, scaling is done in reverse. Instead of drawing the house to scale, you create the CAD drawing full-size. That means the 50-foot long house is drawn 50 feet long; the one-inch gear is drawn one inch in diameter. That's much easier, isn't it?

(There is no limit to the size of drawing AutoCAD can handle; you could draw the entire known universe — full size — if you want.) Full size is shown as 1:1. There is no need to work out the scale factor — or is there?

Scale comes into the scene when it comes time to plot your drawing. That huge drawing, whether of your house or the solar system, must be made small enough to fit the paper it is plotted on. Thus, scale factor comes into play during the **PLOT** command, where you specify that AutoCAD should print the house drawing 60 times smaller to fit the paper.

There is, however, a catch. Some items in the drawing cannot be 60 times smaller. These items include text, hatch patterns, linetypes, and dimensions. These are called *scale-dependent*. If AutoCAD plots the text 60 times smaller, you would not be able to read it. The solution is to draw the text 60 times larger. When it is plotted, the text is the correct size.

You may be puzzled. How big should text be when plotted on a drawing? The standard in drafting is to draw text 1/8" tall for "normal" sizes. When you place text in your house drawing, specify a height of 7.5" tall (=1/8" x 60). That may seem much too high, but when plotted, it looks correct.

You must apply the scale factor to other scale-dependent objects, including dimensions, linetypes, and hatch patterns. Fortunately, you can use the same scale factor for all these; unfortunately, AutoCAD does not have one "scalefactor" command that applies to all, and there is not scale factor for text at all. Instead, scale factors for hatch patterns, linetypes, and dimensions are stored in these system variables:

System Variable	Scale Factor
HPSCALE	Affects hatch patterns.
LTSCALE	Affects linetype patterns.
DIMSCALE	Affects text height, arrowhead size, and leaders; does not affect dimension values.

To change the scale factor, enter the system variables listed above. For example, to set the hatch pattern scale to 60:

    Command: **hpscale**

    Enter new value for HPSCALE <1.0000>: **60**

Repeat for **LTSCALE** and **DIMSCALE**.

As mentioned, you cannot set a scale factor for text. Instead, specify the scaled height during the **TEXT** command:

    Command: **text**

    Current text style: "TECHNICLIGHT" Text height: 0'-2"

    Specify start point of text or [Justify/Style]: *(Pick a point.)*

    Specify height <0'-2">: **7.5**

Alternatively, use the **STYLE** command to create a text style with a fixed height of 7.5".

Note: **HPSCALE** and **DIMSCALE** are not retroactive to objects already in the drawing, but **LTSCALE** is.

SAVE AND SAVEAS

The **SAVE** and **SAVEAS** commands operate identically — both display the Save Drawing As dialog box.

The two commands save drawings by other names and in other formats. You may want to give drawings different names when you open them for editing, but don't want to save the changes to the original file. These commands also allow you to save the drawings in other folders, on other drives, and over networks on other computers.

Drawing files created by AutoCAD 2004 cannot be read by all earlier versions of AutoCAD until they are *translated*. In contrast, AutoCAD 2004 reads drawings created by earlier AutoCADs. For this reason, the **SAVEAS** command includes options to translate drawing to certain earlier formats.

TUTORIAL: SAVING DRAWINGS BY OTHER NAMES

1. To save drawings under different names, start the **SAVEAS** command:
   - From the menu bar, choose **File**, and then **Save As**.
   - At the 'Command:' prompt, enter the **saveas** or **save** command.
   - Alternatively, use the **CTRTL+SHIFT+S** keyboard shortcut.

   Command: **saveas** (*Press* ENTER.)

   In all cases, AutoCAD displays the Save Drawing As dialog box.

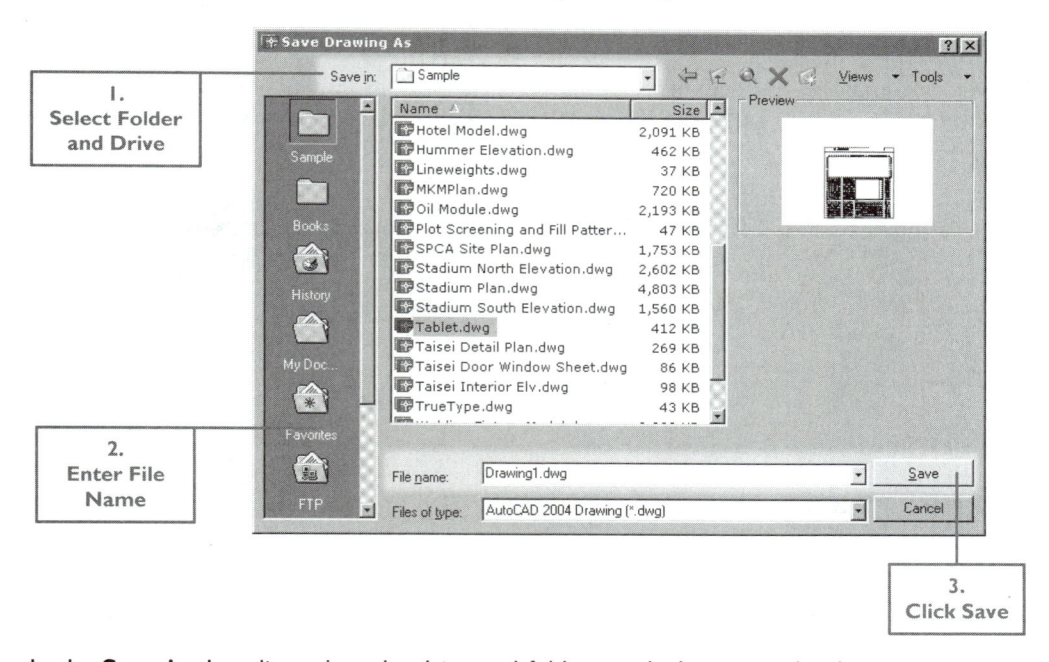

2. In the **Save in** drop list, select the drive and folder in which to save the drawing.
3. In the **File name** text box, enter a different name for the drawing. There is no need to include the *.dwg* extension; AutoCAD adds it automatically.
4. Click **Save**.

   When you enter the name of a drawing that already exists, AutoCAD displays the message box shown below. To replace the existing drawing with the new one, choose **Yes**. If not, choose **No**.

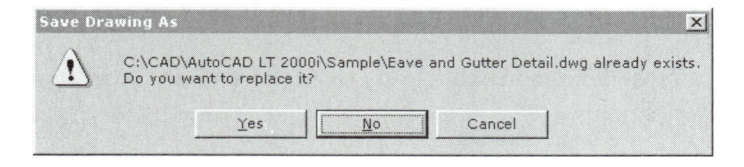

SAVING DRAWINGS: ADDITIONAL METHODS

The **SAVEAS** command allows you to save drawings on computers connected to networks, including the Internet. It also saves drawings in other formats.

- **My Network Places** option saves drawings on network-connected computers.
- **FTP** option saves drawings on the Internet.
- **Files of type** option saves drawings in other file formats.
- **Options** option specifies options for saving .dwg and .dxf files.

Let's look at each option.

My Network Places

The **My Network Places** item allows you to save drawings to other computers, provided two conditions are met: (1) the computers are connected to yours via a network; and (2) the other computers have given your access rights.

TUTORIAL: SAVING DRAWINGS ON OTHER COMPUTERS

1. In the Save Drawing As dialog box, click the **Save in** drop list.
2. From the list, select **My Network Places**. Windows displays the drives of computers connected to yours, as well as generic network groupings.
   - *Networked drives* are identified by names similar to "C on Downstairs", where **C** is the name of the drive, and **Downstairs** is the name of the computer.
   - *Network groupings* are identified by generic names, such as "Entire Network" and "Computers Near Me." If you regularly access another computer, click **Add Network Place** item to add it to the list.

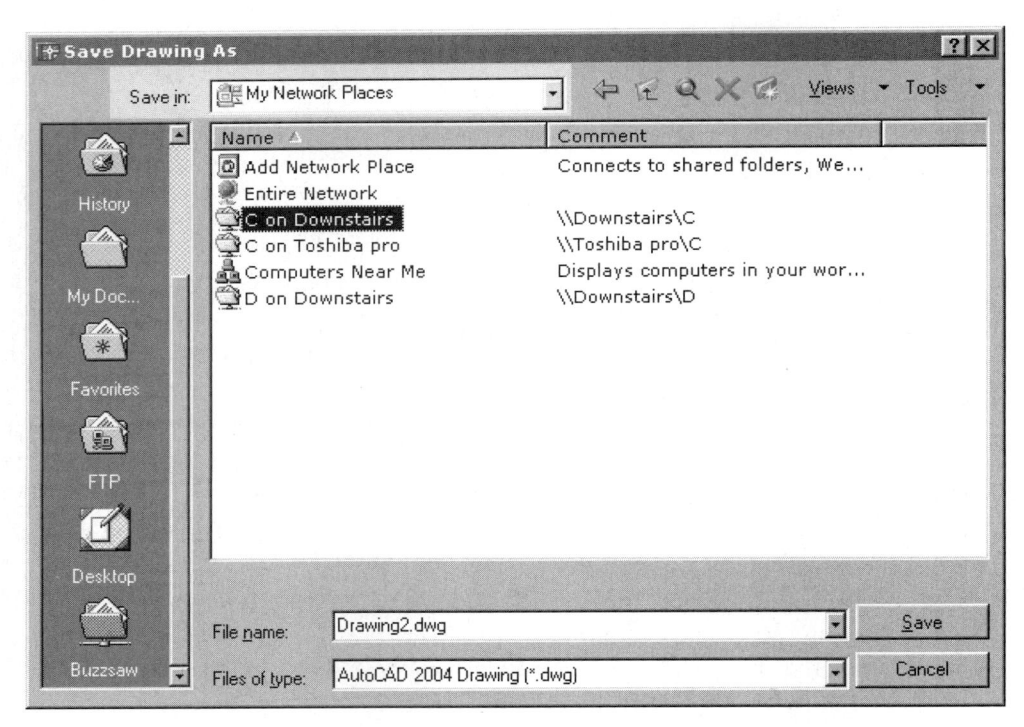

3. Select a network drive, and then choose the folder in which to save the drawing.
4. Click **Save**.

   Depending on the speed of the network, you may notice it takes longer to save drawings through the network.

FTP

FTP is a method of sending and receiving files over the Internet (short for "file transfer protocol"). You may be familiar with sending files through email at *attachments*. The drawback is that email was never designed to handle very large files, and some email providers limit attachments to 1.5MB or smaller.

FTP allows you to send files of any size, even entire CDs of files — 650MB or more. (FTP is also the method used to upload Web pages to Web sites.) The drawback is that FTP needs to be initially set up with a user name, password, and other data.

Clicking on **FTP** in the Places list results in an empty list. You need to first add a site.

TUTORIAL: SENDING DRAWINGS BY FTP

1. In the Save Drawing As dialog box, choose **Tools**, and then choose **Add/Modify FTP Locations**.

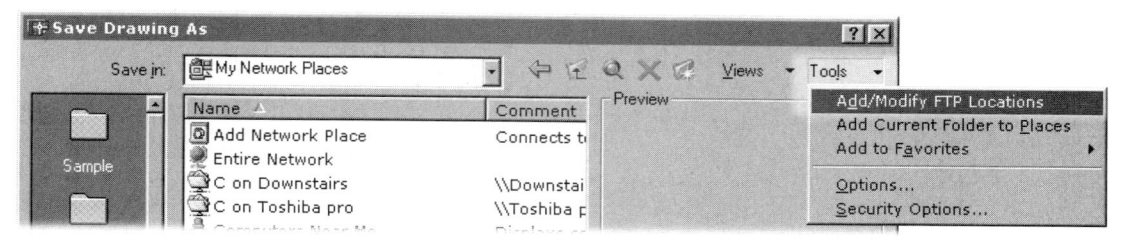

   AutoCAD displays the Add/Modify FTP Locations dialog box.

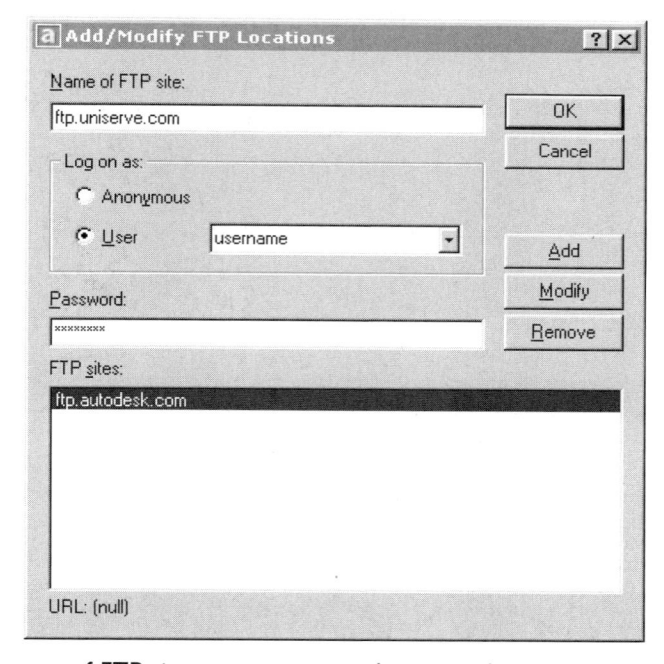

   There are two types of FTP sites: anonymous and password.

   - *Anonymous* sites allow anyone to access them (also known as "public FTP sites"). All you need is to provide your email address as the password; a username is not needed.

   - *Password* sites require that you have a username and a password, usually provided by the person running the FTP site.

2. Enter the details for the ftp site. For example, enter the following parameters to access Autodesk's public FTP site:

	Name of FTP site:	**ftp.autodesk.com**
	Logon as:	**Anonymous**
	Password:	**email@email.com**

3. Click **Add**.
4. Click **OK**.

   The FTP item has to be created. Notice that AutoCAD adds the FTP site to the list in the dialog box.

5. To save the drawing, double-click the FTP site.

   Notice that the dialog box displays a list of folders.

6. Select a folder, and then click **Save**.

   A dialog box shows the progress as AutoCAD transfers the file to the site. Expect it to be slower than saving to disk.

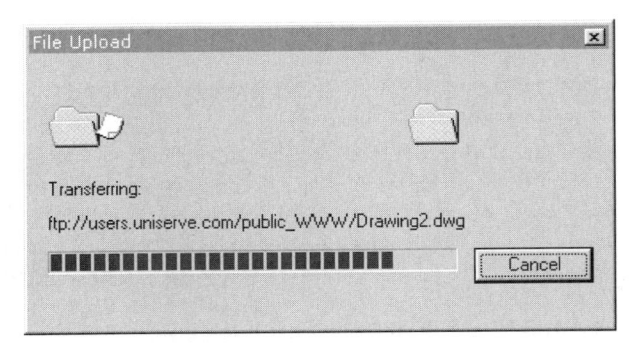

7. Some sites require a username and password before you can save files. If so, a dialog box appears requesting the information. Fill in the missing information, and click **OK**.

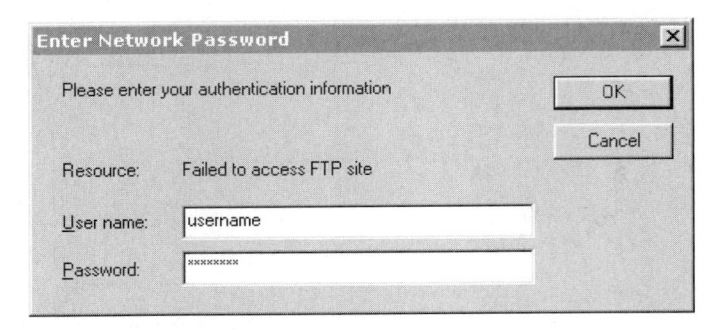

Files of Type

To save the drawing in an earlier release of AutoCAD, or in other AutoCAD formats, choose the **File of type** list box.

TUTORIAL: SAVING DRAWINGS IN OTHER FORMATS

1. To save drawings in other formats, start the **SAVEAS** command:

   Command: **saveas** *(Press ENTER.)*

   AutoCAD displays the Save Drawing As dialog box.

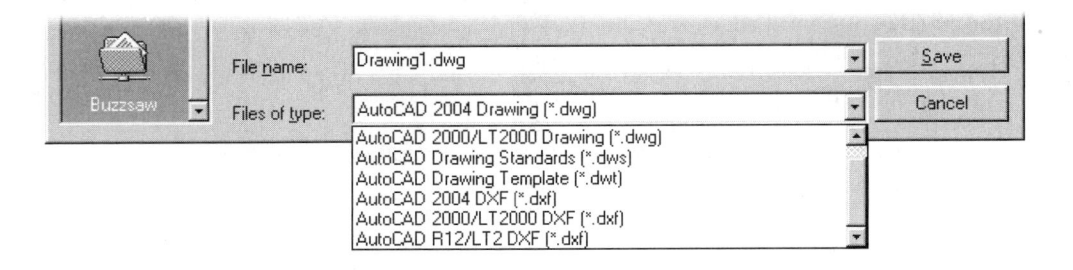

2. In the **Files of type** drop list, select a file format.
   - **AutoCAD 2000/LT 2000 Drawing (*.dwg)** saves drawings in 2000 format, which can also be read by AutoCAD 2000i, and 2002, as well as by AutoCAD LT 2000i and 2002.
   - **AutoCAD Drawing Standards (*.dws)** saves drawings as standards files.
   - **AutoCAD Drawing Template (*.dwt)** saves drawings as templates.
   - **AutoCAD 2004 DXF (*.dxf)** saves drawings in DXF format compatible with AutoCAD 2004.
   - **AutoCAD 2000/LT 2000 DXF (*.dxf)** saves drawings in DXF format compatible with AutoCAD 2000.
   - **AutoCAD R12/LT R2 DXF (*.dxf)** saves the drawing for use with AutoCAD Release 12. This saves the drawing in a format called DXF (short for "drawing interchange format"). To open the drawing in Release 12, Release 11, and AutoCAD LT Release 1 and 2, use the **DXFIN** command.
3. Click **Save**.

   Caution! Some objects may be lost in translation to older formats.

> **Note**: To save drawings in other formats — .wmf, .sat, .stl, .bmp, and .3ds — use the **EXPORT** command.
> To save drawings in .tif, .jpg, and .png formats, use the **TIFFOUT**, **JPGOUT**, and **PNGOUT** commands, respectively.
> To save drawings in PostScript (.eps) format, use the **PSOUT** command.
> To save drawings in .sld slide format, use the **MSLIDE** command.
> To save drawings in .dwf format, use the **ePlot** option of the **PLOT** command.

Options

The **Options** option allows you to specify how *.dwg* and *.dxf* files are saved.

1. In the Save Drawing As dialog box, choose **Tools**, and then **Options**.
   AutoCAD displays the Saveas Options dialog box.
2. The **DWG Options** tab has the following settings:

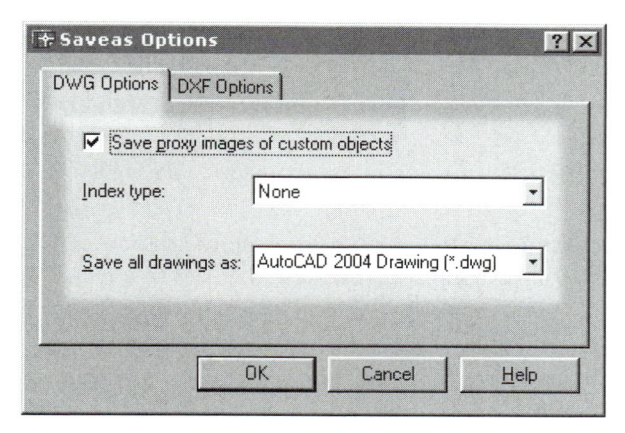

☑ **Save proxy images of custom objects** stores images of *custom objects* in the drawing file. Custom objects are created by ObjectARX programs, and are not understood by AutoCAD when the programs are not present. *Proxy images* are graphical representations of custom objects, which allow you to see them, but not edit them.
☐ When off, AutoCAD displays rectangles in place of custom objects. It usually makes sense to leave this option turned on.

Index type specifies whether AutoCAD creates indices when saving drawings. *Indices improve performance during demand loading, but may increase drawing save time.* The options are:

    None creates no indices.
    Layer loads on and thawed layers only.
    Spatial loads only drawing parts within clipped boundaries.
    Layer & Spatial combines both options above for optimal partial loading performance.

Save all drawing as determines the default format when drawings are saved with the **SAVEAS** and **QSAVE** commands. Keep this set to "AutoCAD 2004 Drawing (*.dwg)" to preserve all features unique to AutoCAD 2004. If, however, you regularly exchange drawings with clients who use AutoCAD 2002 or earlier, you may want to change the setting here.

3. The **DXF Options** tab has these settings:

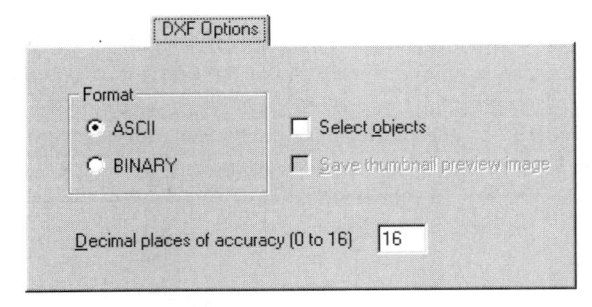

Format selects either ASCII or binary format.
ASCII formatted *.dxf* files are readable by a larger number of non-Autodesk software applications, and can be read by humans when opened in a word processor.
Binary formatted files are created and read faster than ASCII, and have smaller file sizes.

☑ **Select Objects** prompts you to select objects from the drawing.
☐ When off, AutoCAD outputs the entire drawing to the *.dxf* file.

☑ **Save Thumbnail Preview Image** includes a preview image in the *.dxf* file.

Decimal Places of Accuracy (0 to 16) creates more accurate *.dxf* files when set to a higher value, such as 16; creates smaller *.dxf* files when set to a lower value. CNC machines typically work to four decimal places, and get confused when files contain five decimal places.
This option is available only for ASCII *.dxf* files; binary files are always saved with 16 decimal places of accuracy.

QSAVE

The **QSAVE** command (short for "quick save") saves drawings as *.dwg* files on disc.

When you start a new drawing, AutoCAD gives it the generic name of *drawing1.dwg*. The first time you use **QSAVE**, AutoCAD displays the Save Drawing As dialog box, so that you can give the drawing a name. Subsequent uses of **QSAVE** quietly save the drawing without the dialog box.

TUTORIAL: SAVING DRAWINGS

1. To save drawings to disc, open the Options dialog box:
   - From the menu bar, choose **Options**, and then the **Open and Save** tab.
   - On the **Standard** toolbar, select the **Save** button.
   - At the 'Command:' prompt, enter the **options** command.
   - As an alternative, press the **CTRL+S** keyboard shortcut.

   Command: **qsave** *(Press ENTER.)*

2. AutoCAD saves the drawing.
   If the drawing's name is *drawing1.dwg*, then AutoCAD displays the Save Drawing As dialog box. See the **SAVE** command later in this chapter.

AUTOMATIC BACKUPS

Computers are notorious for *crashing*, where the software stops working for no apparent reason. Even with improvements to Windows 2000 and XP, AutoCAD can crash unexpectedly due to *bugs* (accidental errors in the programming code) and conflicts with the operating system. (Technically, the computer does not crash; it's the application software conflicting with the operating system.) When a crash occurs, you can lose the work you have done on your drawings.

As with any software program, it is wise to save your work periodically, even if AutoCAD automatically saves to a temporary file. (If you choose to discard your work with the **QUIT** command, only the part of the drawing changed since the last file save is discarded.)

To guard against this, AutoCAD includes two facilities to back up drawings: one is making backup copies of drawing files; the other is to create a second set of backup copies at set time intervals. Autodesk warns that it is possible to lose drawing data when the power fails during a save. I recommend your computers be plugged into uninterruptable power supplies that provide ten minutes of emergency power, giving you time to save drawings in case of a power outage.

TUTORIAL: MAKING BACKUP COPIES

1. To ensure your drawings are backed up, open the Options dialog box:
   - From the **Tools** menu, choose **Options**, and then the **Open and Save** tab.
   - At the 'Command:' prompt, enter the **options** command.

   Command: **options** *(Press ENTER.)*

   In either case, AutoCAD displays the Options dialog box.

2. Notice the items in the File Safety Precautions section:

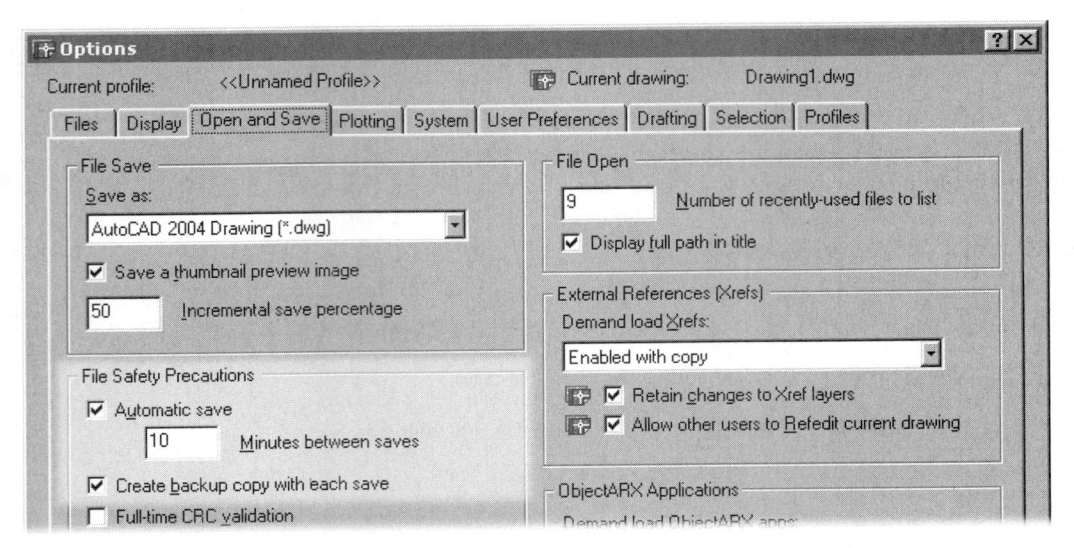

☑ **Automatic save** turns on the automatic backup facility. *This setting should always be turned on!* The only time you would consider turning it off is if your computer's hard drive is low on free space.

10 Minutes between saves means that AutoCAD automatically backs up the drawing every ten minutes. You can change this value to any number you like. A setting of 0 (zero) means that no backups are made. If your drawings are small, and your computer fast, then you may want to consider setting this value to five minutes or less. That way, at most, only five minutes' worth of work is lost.

☑ **Create backup copy with each save** creates a *.bak* backup copy each time you use QSAVE and CTRL+S. *This option also should always be turned on,* unless the hard drive is low on free space.

3. Click **OK** to save the settings, and exit the dialog box.

AutoCAD starts the **SAVETIME** timer when you first change the drawing, not when you first open the drawing. The timer resets each time you use the QSAVE, SAVE, and SAVEAS commands. This prevents unnecessary hard drive activity.

MAKING BACKUPS: ADDITIONAL METHODS

In addition to instructing AutoCAD to back up copies of your drawings, you can specify the folder in which backup files are stored.

- **SAVEFILEPATH** specifies the folder for storing backup copies.

Let's look at how it works.

SaveFilePath

AutoCAD normally saves backup files in the following folder:

C:\Documents and Settings*username*\Local Settings\Temp\

This is the same folder used by all other software on your computer. On my computer, for example, this folder holds 1,873 backup files from a variety of sources — word processors, Web browsers, paint programs, and AutoCAD. With all that clutter, you may prefer to store AutoCAD's backup files in a more suitable location, such as *\autocad 2004\dwgbackups*, so that the backup files are easier to access.

1. First, use Windows Explorer (or File Manager) to create the *dwgbackups* folder in *\autocad 2004*. (AutoCAD 2004 already has a folder called *\backup* for its own purposes.)
2. In AutoCAD, change the setting of the **SAVEFILEPATH** system variable, as follows:

    Command: **savefilepath**

    Enter new value for SAVEFILEPATH, or . for none <>: *(Enter a new path, such as* **"c:\autocad 2004\dwgbackups"**.*)*

    Backup files are now stored in the new folder.

When the path name includes spaces, as does the one shown above, you must surround it with quotation marks.

If the path name is not valid, AutoCAD complains, "Cannot set SAVEFILEPATH to that value. *Invalid*." Ensure the folders exist, and that you are typing the full path correctly (*full path* means *all* folder names, from the drive name down to the file folder).

Backup Files

The first time you save a drawing, AutoCAD adds the *.dwg* extension to the file name. With the second save, AutoCAD renames the file with the *.bak* file extension. Each time you use the **QSAVE** command, AutoCAD updates the backup file.

When AutoCAD crashes, it attempts to rename the *.bak* file as *.bk1*. (if *.bk1* exists, then AutoCAD renames it *.bk2*, and, if necessary, continues all the way through to *.bkz*). This prevents AutoCAD from replacing previous backup files.

Backup files are full drawing files, just with a different file extension. If necessary, you can open the backup files in AutoCAD, following these steps:

1. Use Windows Explorer to copy the *.bak* files to a different folder.
2. Rename the *.bak* files to *.dwg* files.
3. Open the renamed *.dwg* files in AutoCAD.

There are other files created by AutoCAD that may affect your drawings. When AutoCAD crashes, it sometimes leaves behind these files; if your computer's hard drive is getting low on free space, you can erase these leftover files — provided AutoCAD is not running at the time.

Drawings saved by the automatic backup process are given the extension *.ac$*.

When a drawing is in use, AutoCAD *locks* the drawing to prevent other AutoCAD users from editing it; the lock status is indicated by the presence of *.dwl* (drawing lock) files. These files are used by the **WHOHAS** command to determine which user is editing specific drawings.

The *acminidump.dmp* file is created by AutoCAD when it crashes, to help programmers determine the reason.

These files can be erased with Windows Explorer when AutoCAD is not running.

QUIT

Occasionally you create drawings that you do not wish to keep. To exit the current drawing and discard the changes, use the QUIT command.

TUTORIAL: QUITTING AUTOCAD

1. To exit AutoCAD, without saving changes to drawings, start the **QUIT** command:
   - From the menu bar, choose **File**, and then **Exit**.
   - At the 'Command:' prompt, enter the **quit** or **exit** commands.
   - Alternatively, use the **CTRTL+Q** or **ALT+F4** keyboard shortcuts.

   Command: **quit** *(Press ENTER.)*

2. If nothing has changed in the drawing since it was opened, AutoCAD exits.
   In all other cases, AutoCAD displays the strangely-named AutoCAD dialog box.

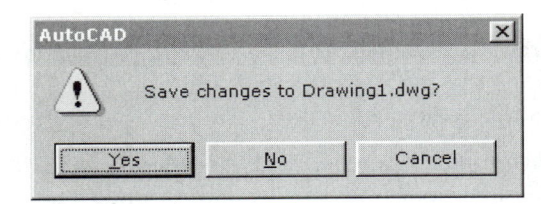

3. To save the drawing, choose **Yes**. AutoCAD displays the Save Drawing As dialog box. Enter a name, and then click **Save**.
   To not save the drawing, choose **No**. AutoCAD exits.
   To not save the drawing and return to AutoCAD, choose **Cancel**.

   Note: **CLOSEALL** closes all drawings. **QUIT** closes all drawings, and exits AutoCAD.

EXERCISES

1. In the following exercises, you start new drawings by a variety of methods.
   Start AutoCAD.
   Which do you see: a blank drawing, or the Startup dialog box?
   If you do not see the Startup dialog box, follow the instruction given in this chapter to turn it on.

2. Once the Startup dialog box is on the screen, click the **Use a Wizard** button.
   Select Advanced Setup, and then enter the settings for a engineering drawing with four decimal places precision.
   Write down a sample length expressed in the units you set: _____ .

   Select Degrees-minutes-seconds with 0 decimal places precision.
   Write down a sample angle expressed in the units you set: _____ .

   Select the 0 angle direction as North and clockwise.
   Draw a sketch below showing the direction of North and the direction of positive angle measurement:

   Specify a drawing size of 11" wide and 17" long.

3. On the Standard toolbar, click the New icon. What happens?
   If AutoCAD displays the Create New Drawing dialog box, select the Use a Template button.
   In either the Create New Drawing dialog box, or the Select Template dialog box, watch the preview window while clicking on the names of template drawings.
   When you see a template that looks interesting, open it.

4. In these exercises, you change the units of measurement.
   From the menu bar, select **Format | Units**.
   In the Length drop list of dialog box, select **Fractional Units**.
   Set the precision to 1/16.
   Click **OK** to exit the dialog box.
   Move the cursor, and watch the coordinates on the status line.
   What do they look like?

5. Press the **SPACEBAR** to return to the Units dialog box.
   In the **Angle** drop list, select **Surveyor's Units**.
   Change the precision to two decimal places.
   After exiting the dialog box, start the **LINE** command, and then click a point.
   Move the cursor, and watch the status line. What do the coordinates look like?
   Move the cursor to the coordinates, and click. Does the display change?
   Press **CTRL+D**, and then move the cursor. Is the coordinate display different?

6. In this exercise, you set the scale factor for hatch patterns to match the drawing.
   Start the **MVSETUP** command in a new drawing, with the following settings:

Units:	**Architectural**
Scale factor:	**240**
Paper width:	**22"**
Paper height:	**34"**

   Does the command draw a rectangle?
   Watching the coordinate display on the status line, move the cursor to the corners of the rectangle, and write down the x,y coordinates to the nearest feet and inches.

   Upper-left corner:

   _____ , _____

   Upper-right corner:

   _____ , _____

   Lower-left corner:

   _____ , _____

   Lower-right corner:

   _____ , _____

   Are the upper-right coordinates similar to the scale factor times the width and height dimensions?

   Change the **HPSCALE** system variable to **240**.
   Start the **BHATCH** command.
   In Hatch tab of the dialog box, look at the Scale setting. What do you see?
   Click the **Pick Points** button, and pick a point inside the border rectangle.
   Press **ENTER**, and then click **OK**.
   Does the hatch pattern appear to be sized correctly (neither too large or too small)?

7. Continuing from the previous exercise, practice using the save commands.
   Use the **QSAVE** command to save the drawing.
   Does AutoCAD ask for a name? If so, enter *mvsetup.dwg*.
   Repeat the **QSAVE** command. Does AutoCAD ask for a name?
   This time, use the **SAVE** command. Does AutoCAD ask for a name? If so, press **Cancel**.
   Finally, use the **SAVEAS** command. Does this work identically to the **SAVE** command?

8. If your computer is connected to others through a network, ask you instructor for the following information:

>Name of another computer.
>
>Name of the drive and folder in which to store drawings.

Use the **SAVEAS** command to save the drawing from exercise #6 using the network.

9. If your computer has a connection to the Internet, ask you instructor for the following information:

>Address of the school's FTP site.
>
>Username and password (or email address, if an anonymous FTP site).
>
>Name of a folder in which to store drawings.

Use the **SAVEAS** command to save the drawing from exercise #6 to the Web site using FTP transfer.

CHAPTER REVIEW

1. Is it necessary to add a *.dwg* extension to the drawing name when you save drawing files?
2. What is the danger of saving an AutoCAD 2004 drawing to an earlier version?
3. Which file types does the **OPEN** command open?
4. Write out the meaning of the following file extensions:
   DWG
   DXF
   DWF
   DWT
5. Does the Select Drawing File dialog box allow you to access drawings files located on the hard drive of your co-workers' computers and at sites located on the Internet?
6. Can the **OPEN** command access Web pages?
7. Is AutoCAD 2004 limited to working with just one drawing at a time?
8. Describe how to open more than one drawing in the Select Files dialog box.
9. How can you open a 3D model provided to you in SAT format?
10. Is AutoCAD able to open files other than DWG and DXF?
    If so, how?
11. Describe how to rename, move, and delete files with the **OPEN** command.
12. When can you erase a *.ac$ file?
13. AutoCAD starts up but does not display the Startup dialog box. Describe how to enable this dialog box.
14. Explain the difference between the **NEW** and **QNEW** commands.
15. What is the importance of the *acad.dwt* file?
16. Describe how template drawings are useful.
17. What is the purpose of *wizards*?
18. Write out the unit of measurement for each example shown below:
    1'-2 1/2"
    12.3456
    2'-2.2"
    6.54E+03
    43 3/4
19. Which unit of measurement is used for metric drawings?
20. Which unit of measurement does AutoCAD use internally?
21. How many decimal places can AutoCAD display?
    When would you not want to work with that many decimal places?
22. When you specify two decimal places, does this restrict the accuracy of AutoCAD's calculations?
23. Write out the angle of measurement for each example shown below:
    45.00
    45d00'00.00"
    50.00g
    0.79r
    N45d0'0.00"E
24. How many degrees are there in a circle?
    How many grads?

How many radian?
25. In which direction does AutoCAD think of 0 degrees, by default?
26. In which direction are positive angles measured, by default?
27. Can you change the direction of 0 degrees?
28. What does AutoCAD use *limits* for?
29. Does the Select File dialog box allow you to preview drawings?
30. How would you open a drawing but prevent changes from being saved?
31. What happens when you double-click a drawing name in the Select Files dialog box?
    What happens when you click a drawing name twice?
32. Explain the meaning of:
    Write-protected
    Read-only
    Read-write
33. Describe how to save a drawing that has been opened *read-only*.
34. What are .bak files?
    Why are they useful?
35. Describe how to open a .bak file.
36. How can you insure against losing drawing data from a power outage?
37. What happens when a drawing is "saved"?
38. Which command saves the drawing quickly?
    Can AutoCAD save the drawing on its own?
39. Describe the difference between the **QSAVE** and **SAVE** commands.
    And the difference between the **SAVE** and **SAVEAS** commands.
40. In the Select File dialog box, displayed by the **OPEN** command, the Folders sidebar provides quick access to:
    My Documents
    History
    FTP
    Desktop
    Buzzsaw
41. Can you create new folders with the Select Files dialog box?
42. Explain when you would use the Find tool?
43. What functions does the **MVSETUP** command perform?
44. What are the dimensions of a B-size sheet of ANSI Engineering paper?
45. Which metric sheet is larger — A4 or A0?
    Which imperial sheet is larger — A or E?
46. Can you change the background color of the drawing area?
    If so, why?
47. Describe the purpose of the **UNITS** command.
48. How does the **UNIT** command's **Unitless** setting help blocks being inserted into drawings?
49. Work out the scale factors for the drawing these objects on a page 10 inches wide:
    a. An automobile and trailer 12 feet long.
    b. A tool shed 10 feet wide.
    c. A telephone 9 inches long.
    Show all your calculations.

50. Work out the scale factors for the following objects drawn on a page 100 cm wide:
    a. A pen holder 7 cm in diameter.
    b. A stereo system 42 cm wide.
    c. A house 16m long.
    Show all your calculations.
51. Which system variable sets the scale factor for the following objects:
    Hatch patterns
    Linetypes
    Dimensions
    Text
52. Explain why text cannot be drawn full-size in drawings scaled at 1:50.
53. How tall should text be drawn in drawings with scale factors of:
    1:50
    1:1
    10:1
    1:2500
    Show all your calculations.
54. If the scale factor for a drawing is 1:50, what should the scale factor be for hatch patterns?
55. If the scale factor for linetypes is 10:10, what should the scale factor be for hatch patterns?
56. Under what conditions can drawings be saved onto others computers?
57. What happens when you use the **SAVE** command to save the drawing, when another drawing of the same name already exists?
58. Describe what happens when you quit AutoCAD without saving the drawing.
59. Explain a benefit and drawback to FTP.
    What is FTP short for?
    When might you use FTP?
60. What is the difference between *anonymous* and *private* FTP sites?
61. Describe the difference between the two types of *.dxf* file:
    ASCII
    Binary
38. Name the command that belongs to each keyboard shortcut:
    CTRL+S
    CTRL+SHIFT+S
    CTRL+O
    CTRL+N
    CTRL+Q

CHAPTER 4

Drawing with Basic Objects

AutoCAD drawing activities involve a primary set of drawing commands. These commands draw the objects basic to many drawings — lines, arcs, circles, and so on. Now that you know how to set up a new drawing, let's begin. In this chapter, you learn AutoCAD's basic drawing commands:

LINE draws line segments.

RECTANG constructs squares and rectangles.

POLYGON constructs regular polygons, from three to 1,024 sides.

CIRCLE constructs circles by several methods.

ARC constructs arcs by many methods.

DONUT draws thick and solid-filled circles.

PLINE draws connected lines, arcs, and curves.

ELLIPSE constructs the ellipses and the elliptical arcs.

POINT draws point objects.

FINDING THE COMMANDS

On the **DRAW** toolbar:

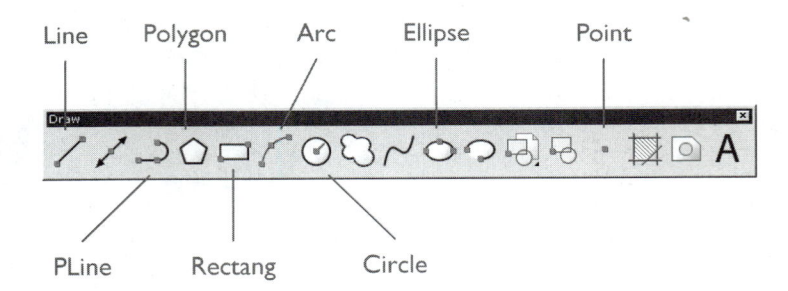

Line   Polygon   Arc   Ellipse   Point

PLine   Rectang   Circle

On the **DRAW** menu:

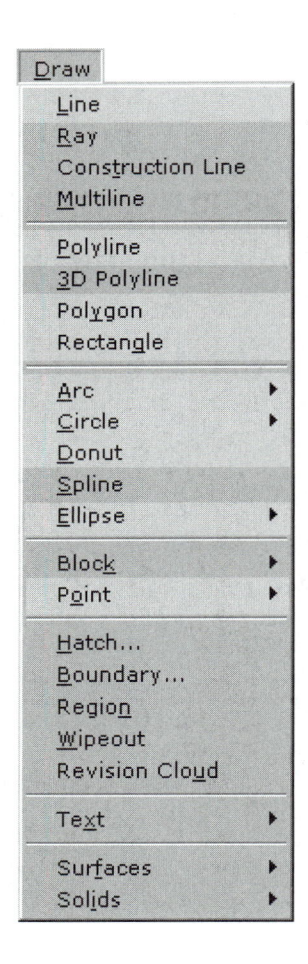

Draw
- Line
- Ray
- Construction Line
- Multiline
- Polyline
- 3D Polyline
- Polygon
- Rectangle
- Arc ▶
- Circle ▶
- Donut
- Spline
- Ellipse ▶
- Block ▶
- Point ▶
- Hatch...
- Boundary...
- Region
- Wipeout
- Revision Cloud
- Text ▶
- Surfaces ▶
- Solids ▶

LINE

Drawing lines in drawings is the most basic CAD operation. In AutoCAD, you can draw different types of lines and apply a variety of line options, but many lines begin with the **LINE** command. A single line is sometimes called a "segment." At each end, lines have *endpoints*.

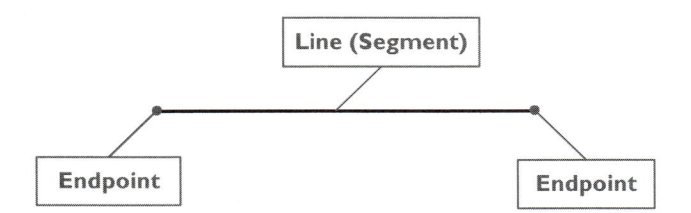

Let's first talk about "first points" and "next points." To draw a line, first you show AutoCAD the point from which to start drawing. AutoCAD calls this the "first point." Think of this as the place where you first put your pencil to the paper.

Next, determine the point at which the line segment should end. AutoCAD calls this is the "next point." The 'Specify next point:' prompt repeats, so that you can draw a connected series of lines without the inconvenience of restarting the **LINE** command for each segment.

AutoCAD repeats the 'Specify next point:' prompt until you press **ESC**, or press **ENTER** twice.

Note: AutoCAD has commands that draw different types of lines, some of which you encounter in this chapter:

- **MLINE** (multiline) draws as many as 16 parallel lines as a single object.
- **PLINE** (polyline) draws lines in the same manner as the **LINE** command, but a single polyline includes arcs, splines, and variable widths; the **3DPOLY** command draws three-dimensional polylines.
- **SKETCH** draws freehand lines.
- **SPLINE** draws splined curves.
- **XLINE** (construction line) and **RAY** draw infinite construction lines (xlines) and semi-infinite construction lines (rays).

TUTORIAL: DRAWING LINE SEGMENTS

1. To draw line segments, start the **LINE** command with one of these methods:
   - From the menu bar, choose **Draw**, then choose **Line**.
   - From the **Draw** toolbar, choose the **Line** button.
   - At the 'Command:' prompt, enter the **line** command.
   - Alternatively, enter the **l** alias at the 'Command:' prompt.

   Command: **line** *(Press* **ENTER**.*)*

2. Specify the starting point of the line segment:
   Specify first point: *(Pick point 1, or specify coordinates.)*
   - With the cursor, pick a point on the screen.
   - Or, at the keyboard, enter x,y coordinates.

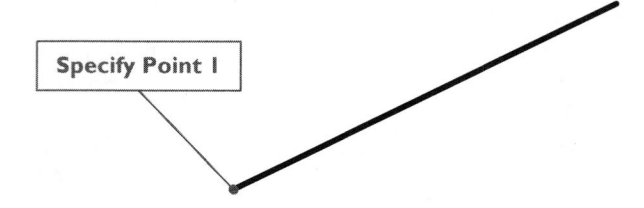

3. Specify the next point(s):
   Specify next point or [Undo]: *(Pick point 2, or specify coordinates.)*
   - Move the cursor, and then pick another point.
   - Or, enter another set of x,y coordinates.

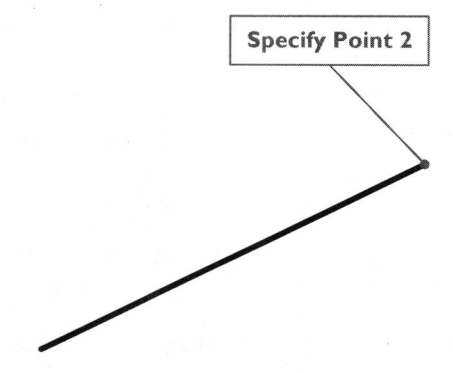

4. Press **ENTER** to end the command.
   Specify next point or [Undo/Close]: *(Press **ENTER** to exit the command.)*

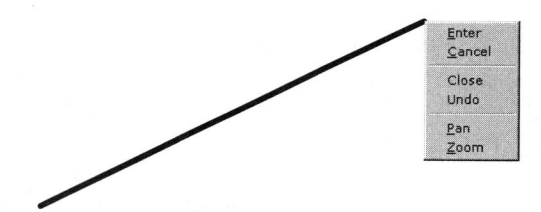

   Alternatives to pressing **ENTER** include:
   - Pressing **ESC**.
   - Right-clicking, and selecting **Enter** from the shortcut menu.

5. To repeat the **LINE** command, press **ENTER**.
   AutoCAD restarts the **LINE** command (and displays "LINE" at the 'Command:' prompt).
     Command: LINE Specify first point:

6. This time, draw several lines. Notice that each new line connects precisely at the endpoint of the previous line. This connection is called a "vertex."

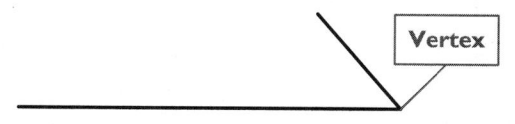

7. Press **ENTER** to exit the command. Notice that in AutoCAD, pressing **ENTER** ends (most) commands and it restarts commands.

DRAWING LINE SEGMENTS: ADDITIONAL METHODS

The LINE command contains these options that provide additional methods for drawing lines and ending the command:

- **Close** closes the polygon.
- **Undo** undraws the last segment.
- **@** and **<** draw with relative coordinates and specifies angles.
- **ENTER** continues from the last segment.

Some of the options are shown at the command prompt and on the right-click menu (below).

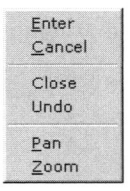

Let's look at each option.

Close

When you construct a polygon using the LINE command, the last segment connects to the first segment. Aligning the ends of segments can be tedious, so the **Close** option automatically does this for you. The **Close** option does not appear until you pick three points, because a minimum of two lines is needed before they can be closed with a third line to make the polygon.

To see how this works, consider drawing a right angle triangle. The illustration shows how the line at the final intersection is connected with the **Close** option.

    Command: **line**
    Specify first point: *(Pick point 1.)*
    Specify next point or [Undo]: *(Pick point 2.)*
    Specify next point or [Undo]: *(Pick point 3.)*
    Specify next point or [Close/Undo]: **c**

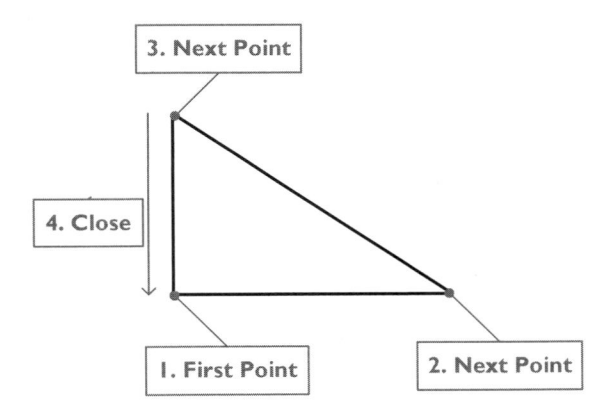

Undo

As you draw line segments with the LINE command, you may sometimes make a mistake. Instead of cancelling the command and starting over, use the **Undo** option to "undraw" the last segment. You may undo all the way back to the first segment.

Relative Coordinates and Angles

The @ and < symbols allow you to specify relative coordinates and angles. When you prefix x,y-coordinates with @, AutoCAD reads the numbers as distances, not coordinates. When you include the < symbol, AutoCAD reads the number following as an angle. Here are examples:

Command: **line**

Specify first point: @2,3   Draws the line 2 units up and 3 units left from that last point.

Specify next point or [Undo]: 10<45 .   Draws the line 10 units long at a 45-degree angle from the last point.

Continue

When you terminate the LINE command, you can begin a new line anywhere else in the drawing. If, however, you wish to go back and continue from the last endpoint, reconnect with the LINE command's **Continue** option.

This option is hidden by AutoCAD: the next time you start the LINE command, press ENTER at the "Specify first point:" prompt.

RECTANG

The RECTANG command draws rectangles and squares by a variety of methods, and in a variety of styles. It can draw them with thin or fat lines, tilted at an angle, and with rounded or cut-off corners. The primary parts of a rectangle are its length and width:

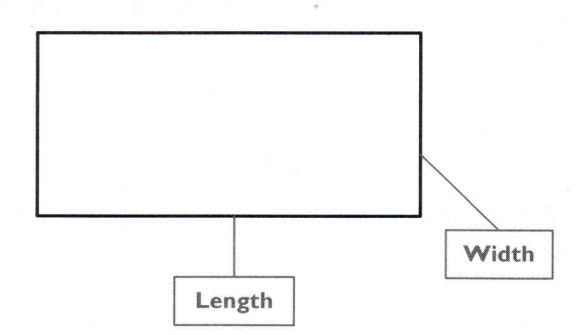

This command does not draw geometric variations on rectangles, such as parallelograms or rhomboids.

TUTORIAL: DRAWING RECTANGLES

To draw a basic rectangle, pick two points that define the opposite corners of the rectangle.

1. Start the **RECTANG** command by one of these methods:
   - From the menu bar, choose **Draw**, then choose **Rectangle**.
   - From the **Draw** toolbar choose the **Rectangle** button.
   - At the 'Command:' prompt, enter the **rectang** command.
   - Alternatively, enter the aliases **rec** or **rectangle** at the 'Command:' prompt.

   Command: **rectang** *(Press* ENTER.*)*

2. Pick a point for one corner of the rectangle, such as the lower-left corner:
   Specify first corner point or [Chamfer/Elevation/Fillet/Thickness/Width]: *(Pick point 1.)*

3. And pick a point for the opposite corner:

   Specify other corner point or [Dimensions]: *(Pick point 2.)*

   1. Specify First Point

   2. Specify Other Corner Point

   Notes: To draw a square, pick two points that create the square. As an alternative, you can use the **POLYGON** command to draw squares.

   The sides of the rectangle are parallel to the x and y axis of the current UCS (user-defined coordinate system).

ADDITIONAL METHODS FOR DRAWING RECTANGLES

The **RECTANG** command contains options that provide additional methods for drawing rectangles:

- **Dimensions** specifies the width and height of the rectangle.
- **Width** specifies the width of the four lines making up the rectangle.
- **Elevation** draws the rectangle a specific height above the x,y-plane.
- **Thickness** draws the rectangle with a thickness in the z-direction.
- **Fillet** rounds off the corners of the rectangle.
- **Chamfer** cuts off the corners of the rectangle.

The options are shown at the command prompt, and on the right-click menus shown below.

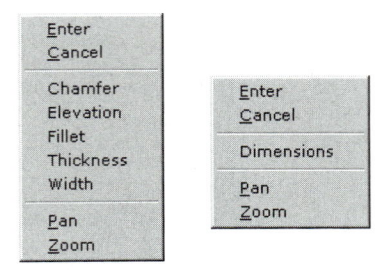

Let's look at each option.

Dimensions

The basic way to draw a rectangle is to specify the points of opposite corners. If you prefer, you can specify the size of the rectangle by the length and width to draw, for example, a 3" x 2" box.

1. Start the **RECTANG** command, pick a starting point, and then enter the **d** option:

   Command: **rectang**

   Specify first corner point or [Chamfer/Elevation/Fillet/Thickness/Width]: *(Pick point 1.)*

   Specify other corner point or [Dimensions]: **d**

2. Enter the *length*, which is in the x-direction (to the left):

   Specify length for rectangles <0.0000>: **3**

3. And enter the *width*, which is in the y-direction (up):

   Specify width for rectangles <0.0000>: **2**

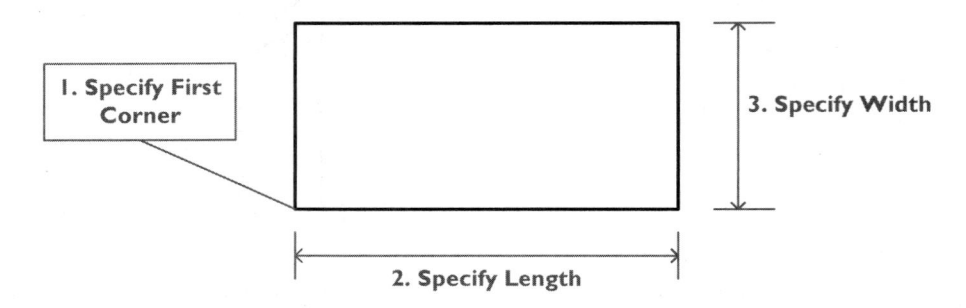

   [1. Specify First Corner] [2. Specify Length] [3. Specify Width]

4. At this point, AutoCAD asks you to specify the orientation of the rectangle. Depending on where you pick the "other corner point," you place the rectangle in one of four positions around the first corner point:

   Specify other corner point or [Dimensions]: *(Move the cursor to position the rectangle, and then pick point 4.)*

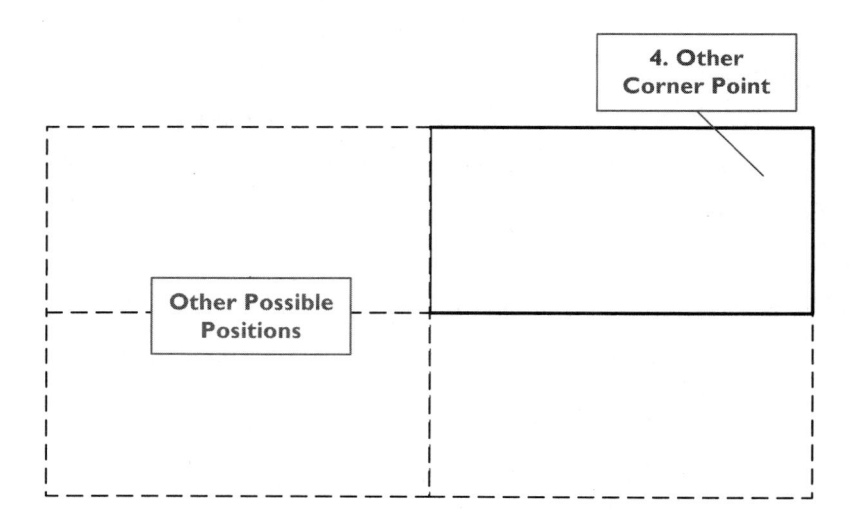

   [4. Other Corner Point] [Other Possible Positions]

Width

You can specify the width of the four lines making up the rectangle. (Technically, the "lines" are *polylines*, which are discussed later in this chapter.) Once you set the width, the same value is used by subsequent **RECTANG** commands, until you choose to change the width. To change the width, enter the **w** option, and then specify a width in units:

   Command: **rectang**

   Specify first corner point or [Chamfer/Elevation/Fillet/Thickness/Width]: **w**

   Specify line width for rectangles <0.0000>: **.1**

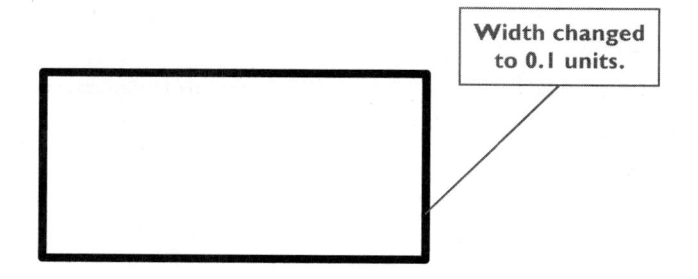

   [Width changed to 0.1 units.]

Elevation

You can specify the height of the rectangle above (or below) the x,y-plane — in the z-direction. Once you set the height, the same value is used by subsequent RECTANG commands until you change it. To change the elevation, enter the **e** option, and then specify an elevation in units:

Command: **rectang**

Specify first corner point or [Chamfer/Elevation/Fillet/Thickness/Width]: **e**

Specify the elevation for rectangles <0.0000>: **2**

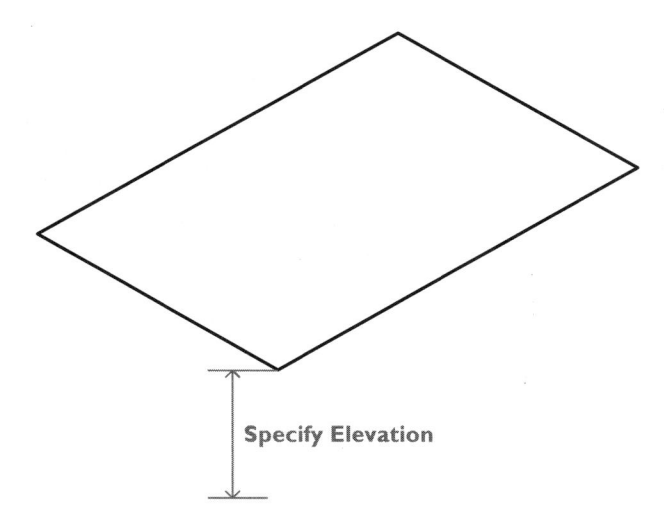

Specify Elevation

A positive value for elevation draws the rectangle above the x,y plane. A negative value pushes the rectangle below the x,y-plane.

Thickness

You can specify a thickness for the rectangle, which turns it into a 3D box. Once you set the thickness, the same value is used by subsequent RECTANG commands, until you change it. To change the thickness, enter the **t** option, and then specify a thickness in units:

Command: **rectang**

Specify first corner point or [Chamfer/Elevation/Fillet/Thickness/Width]: **t**

Specify thickness for rectangles <0.0000>: **2**

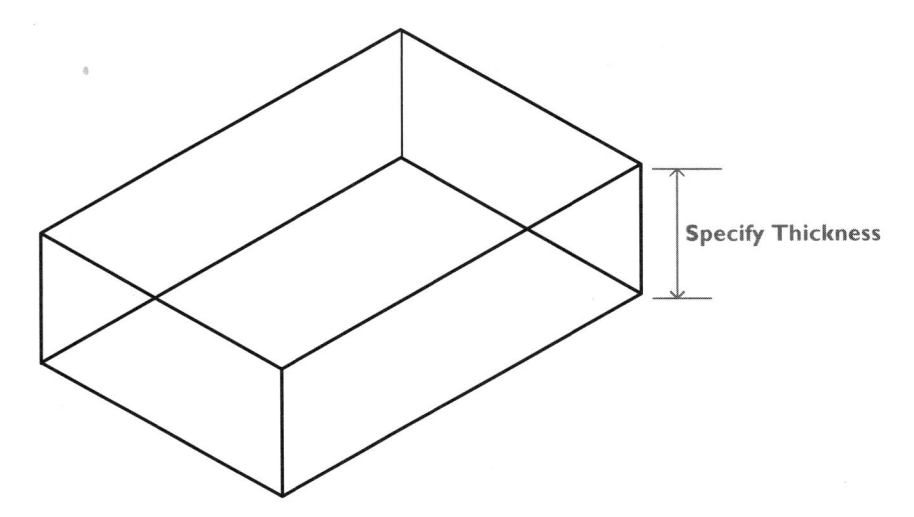

Specify Thickness

A positive value for thickness draws the rectangle upward from the base, while a negative value draws the rectangle downward.

Fillet

To round the corners of the rectangle, enter the **f** option:

>Command: **rectang**
>
>Current rectangle modes: Elevation=2.0000 Thickness=2.0000 Width=0.1000
>
>Specify first corner point or [Chamfer/Elevation/Fillet/Thickness/Width]: **f**
>
>Specify fillet radius for rectangles <0.0000>: **.25**

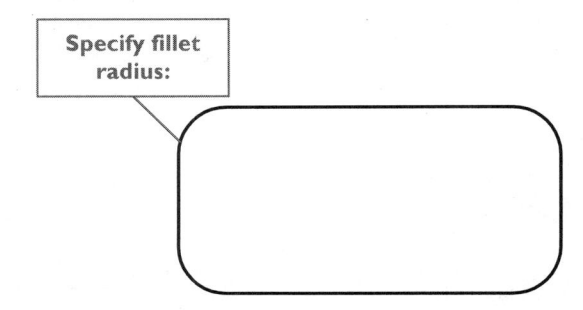

Specify fillet radius:

The single fillet radius applies to all four corners; you cannot specify a different radius for each corner. You can enter a negative radius, which produces this interesting effect:

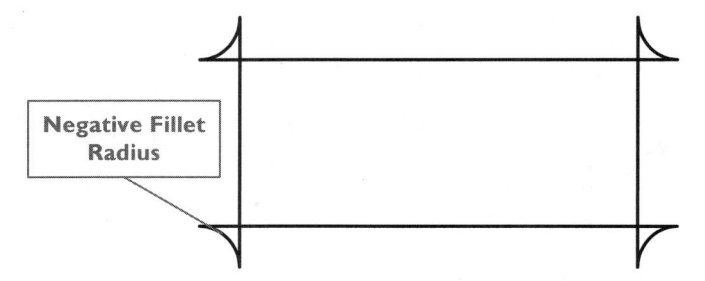

Negative Fillet Radius

Once you set the fillet, the same value is used until you change it.

Chamfer

To cut off (or "chamfer") the corners of the rectangle, enter the **c** option:

>Command: **rectang**
>
>Specify first corner point or [Chamfer/Elevation/Fillet/Thickness/Width]: **c**
>
>Specify first chamfer distance for rectangles <0.2500>: *(Press* ENTER.*)*
>
>Specify second chamfer distance for rectangles <0.2500>: *(Press* ENTER.*)*

The chamfer distance is measured from the corners of the rectangle. You can enter a different value for the first and second chamfer distances, as illustrated below. Once you set the chamfers, the same value is used by subsequent RECTANG commands, until you change it.

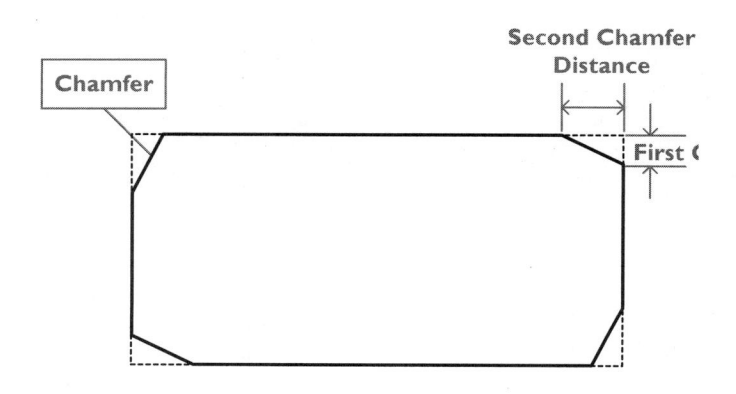

Chamfer

Second Chamfer Distance

First C

The pair of chamfer distances applies to all four corners; you cannot specify a different chamfer for each corner. You can enter negative and positive distances, which produces this pinwheel effect:

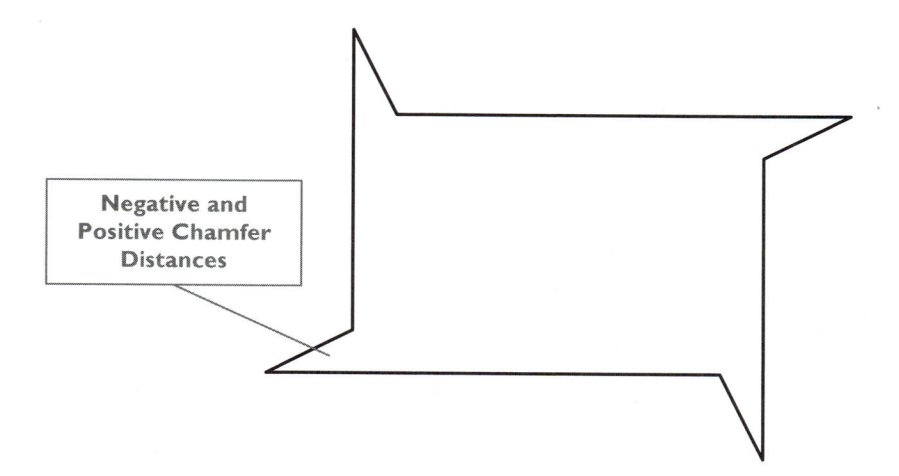

POLYGON

AutoCAD allows you to draw regular polygons, from three to 1,024 sides, using the **POLYGON** command. A *regular polygon* has all sides the same length; an *irregular polygon* has sides of different lengths. With this command, you can draw equilateral triangles, squares, pentagons, hexagons, octagons, and so on.

The polygons are constructed by any of three methods: on the basis of the length of one edge, or by fitting inside ("inscribed") or outside ("circumscribed") a circle. AutoCAD draws the polygon with a polyline, so you can change its width with the **PEDIT** command.

The primary parts of the polygon are its vertices, edges, radius of inscribed circle, and center point:

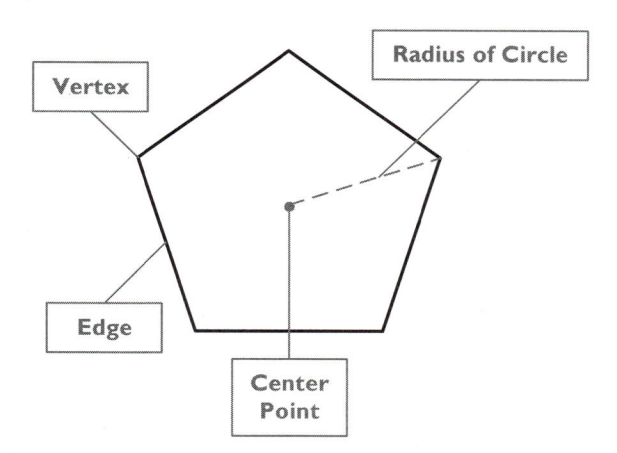

You can use either this command to draw "4-sided polygons," more commonly called squares, or you can use the **RECTANG** command. Neither command draws irregular polygons, such as acute triangles and parallelograms; you need to construct these using AutoCAD's other commands, as described at the end of this chapter.

TUTORIAL: DRAWING POLYGONS

1. To draw regular polygons, start the POLYGON command by one of these methods:
   - From the menu bar, choose **Draw**, then choose **Polygon**.
   - From the **Draw** toolbar, choose the **Polygon** button.
   - At the 'Command:' prompt, enter the **polygon** command.
   - Alternatively, enter the **pol** alias at the 'Command:' prompt.

   Command: **polygon** *(Press ENTER.)*

2. Specify the number of sides. For example, enter 3 for a triangle, 4 for a square, 5 for a pentagon, and so on.

   Enter number of sides <4>: *(Enter a value between 3 and 1024.)*

3. Pick a point for the center of the polygon:

   Specify center of polygon or [Edge]: *(Pick a point.)*

4. Decide if the polygon fits inside (is inscribed within) or outside (circumscribes) an imaginary circle;

   Enter an option [Inscribed in circle/Circumscribed about circle] <I>: *(Type I or C.)*

5. Specify the radius of the circle, which determines the size of the polygon:

   Specify radius of circle: *(Enter a radius, or pick a point.)*

ADDITIONAL METHODS FOR DRAWING POLYGONS

The POLYGON command contains options that provide these methods for drawing polygons:

- **Edge** defines the polygon by the length of one of its edges.
- **Inscribed in circle** fits the polygon inside an imaginary circle.
- **Circumscribed about circle** fits the imaginary circle inside the polygon.

The options are shown at the command prompt and on the right-click menus shown below.

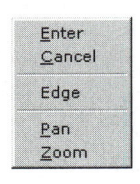

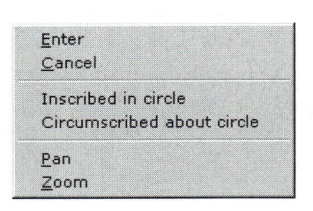

Let's look at each option.

Edge

Perhaps the easiest method is to construct polygons by specifying the length of one edge. Pick two points that define the length of the edge. Here is how to draw a pentagon with the **Edge** option.

1. Start the POLYGON command:

   Command: **polygon**

2. To draw a pentagon, enter **5** for the number of sides:

   Enter number of sides <4>: **5**

3. Select the **Edge** option by typing **e**:

   Specify center of polygon or [Edge]: **e**

4. Pick a point (or enter x,y coordinates) for the start of the edge:

   Specify first endpoint of edge: *(Pick a point.)*

5. And pick another point for the end of the edge:

   Specify second endpoint of edge: *(Pick another point.)*

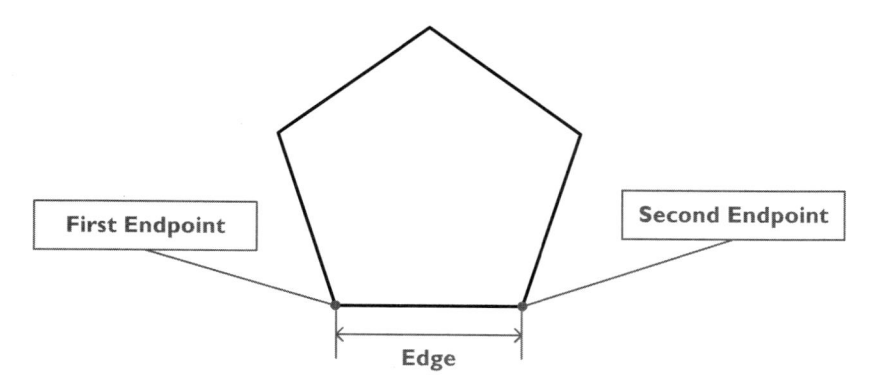

The angle of the two edge points determines the angle of the polygon. For example, to draw a diamond shape, pick the two points at a 45-degree angle:

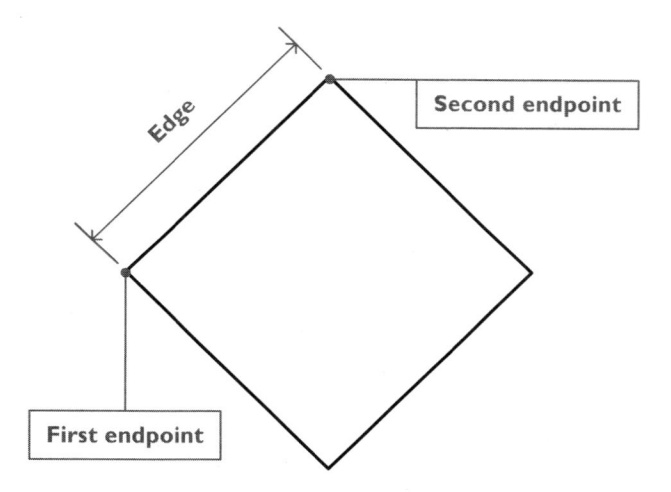

Inscribed in Circle

Inscribed polygons are constructed inside a circle of a specified radius (the circle itself is not drawn, but is imaginary). The vertices of the polygon fall on the circle. Here is an example of drawing a triangle inside a circle:

   Command: **polygon**

   Enter number of sides <4>: **3**

   Specify center of polygon or [Edge]: *(Pick a point.)*

   Enter an option [Inscribed in circle/Circumscribed about circle] <I>: **i**

   Specify radius of circle: *(Pick a point, or enter a radius.)*

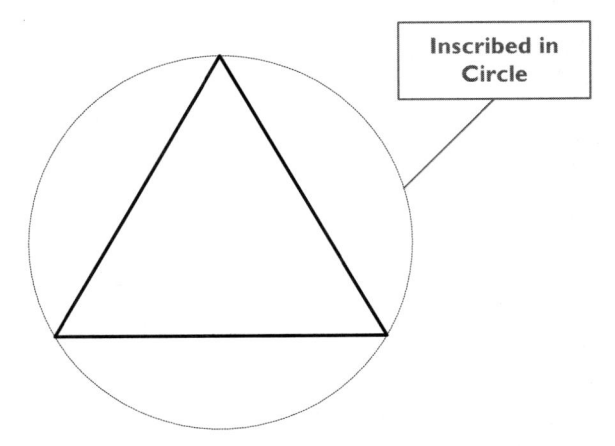

Just as the two points for the **Edge** option determine the rotation of the polygon, so too does the radius specification. As you move the cursor during the "Specify radius of circle" prompt, notice that one vertex moves with the cursor:

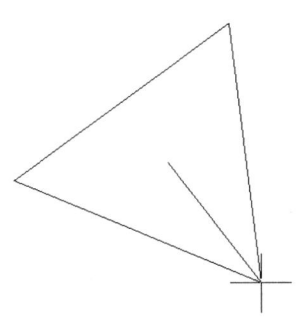

Circumscribed About Circle

Circumscribed polygons are constructed outside a circle of a specified radius. The midpoints of the polygon's edges are placed on circle's circumference. As with the inscribed option, the rotation of the polygon is determined at the "Specify radius of circle" prompt.

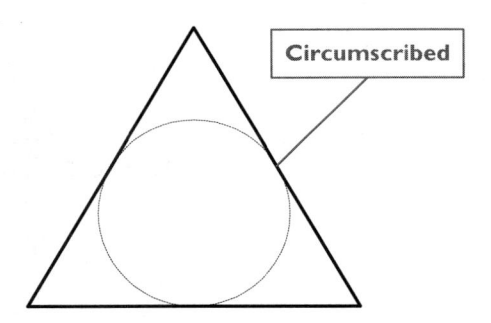

CIRCLE

AutoCAD draws circles by several methods. As you progress in your drafting, you will find that some methods work more easily than others. The important parts of the circle are the center point, the radius or diameter, and the circumference.

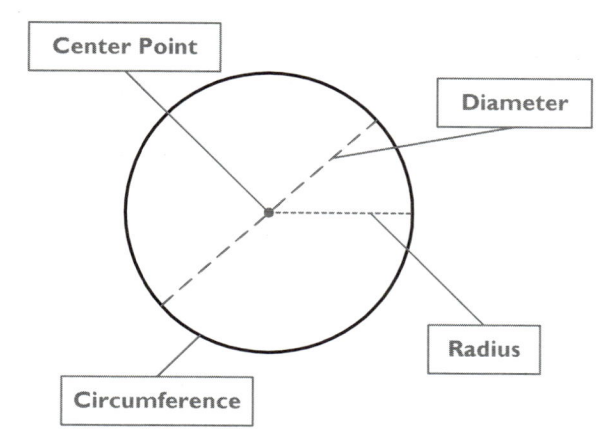

To draw isometric circles, use the **Isocircle** option of the ELLIPSE command, after turning on **Iso** mode with the SNAP command.

TUTORIAL: DRAWING CIRCLES

1. To draw circles, start the CIRCLE command by one of these methods:
   - From the menu bar, choose **Draw**, then choose **Circle**, and then choose **Center, Radius**.
   - From the **Draw** toolbar, choose the **Circle** button.
   - At the 'Command:' prompt, enter the **circle** command.
   - Alternatively, enter the **c** alias at the 'Command:' prompt.

   In all cases, you see the following prompts:
   Command: **circle** *(Press ENTER.)*

2. Pick a point for the center of the circle:
   Specify center point for circle or [3P/2P/Ttr (tan tan radius)]: *(Pick point 1, or enter x,y coordinates.)*

3. Enter the radius of the circle:
   Specify radius of circle or [Diameter]: *(Specify the radius, or pick point 2.)*

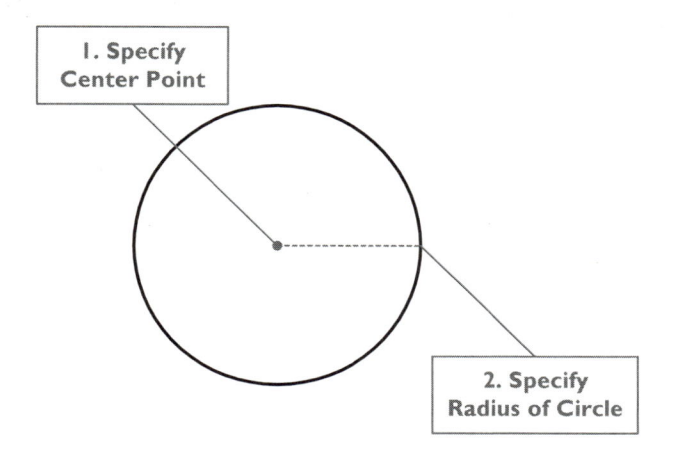

ADDITIONAL METHODS FOR DRAWING CIRCLES

The CIRCLE command has options that provide additional methods for drawing circles:

- **Diameter** draws circles based on a center point and diameter.
- **2P** draws circles based on two diameter points.
- **3P** draws circles based on three points along the circumference.
- **Ttr (tan tan radius)** draws circles touching two tangent points and a radius.
- **Tan, Tan, Tan** draws circles touching three tangent points.

The options are shown at the command prompt, on the **Draw | Circle** menu (left), and on the right-click menu (right).

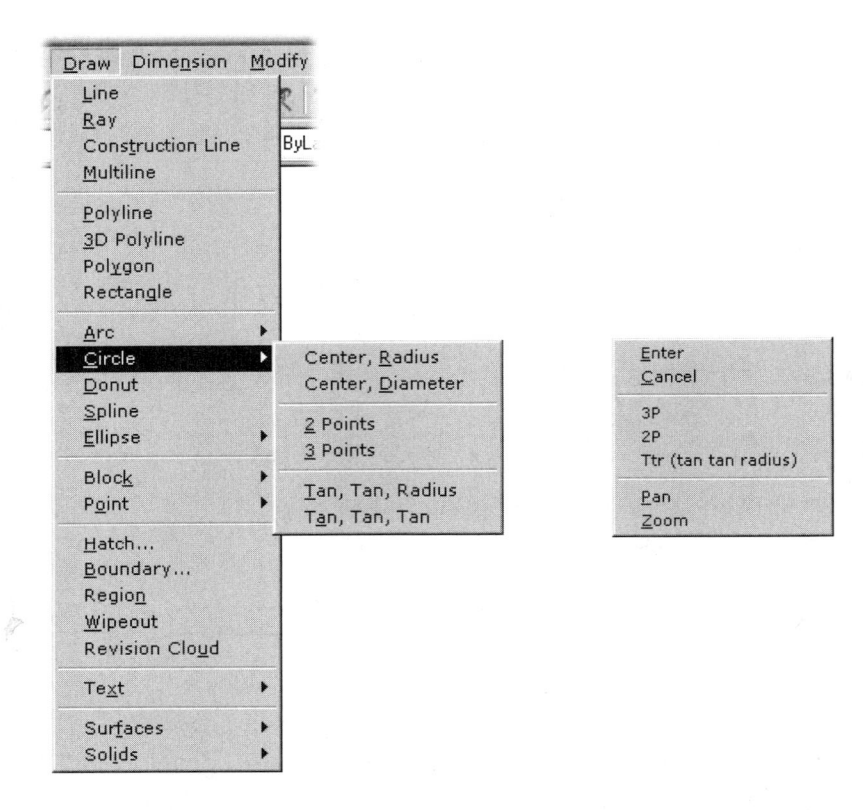

Let's look at each option.

Diameter

Instead of specifying the circle's center point and radius, you can specify the diameter. (The diameter is twice the radius.) From the menu bar, select **Draw | Circle | Center, Diameter.**

Command: **circle**

Specify center point for circle or [3P/2P/Ttr (tan tan radius)]: *(Pick point 1.)*

Specify radius of circle or [Diameter] <1.0>: **d**

Specify diameter of circle <2.0>: *(Pick point 2, or enter a value for the diameter.)*

Diagram: Circle with "1. Center Point" and "2. Diameter" labels

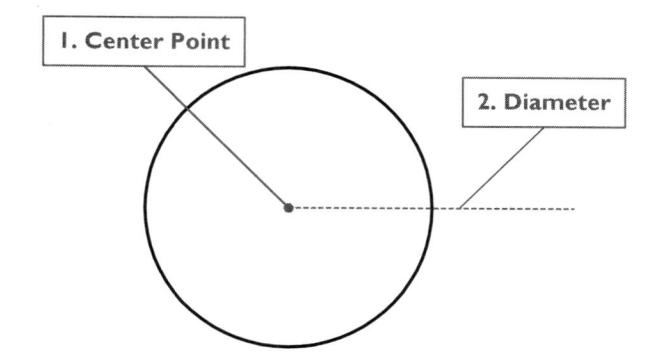

2P

Instead of specifying a center point and diameter, you can specify only the diameter by picking two points on the circumference. These two points define the diameter. From the menu bar, select **Draw | Circle | 2 Points.**

> Command: **circle**
> Specify center point for circle or [3P/2P/Ttr (tan tan radius)]: **2p**
> Specify first end point of circle's diameter: *(Pick point 1.)*
> Specify second end point of circle's diameter: *(Pick point 2.)*

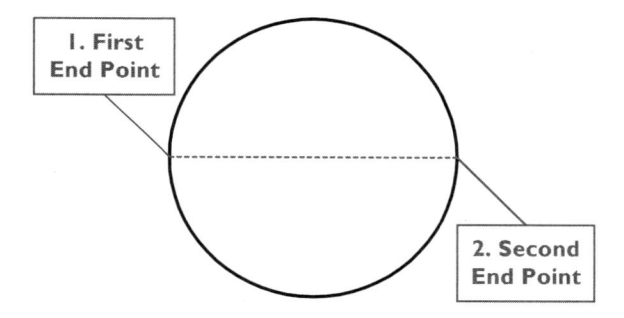

Diagram: Circle with "1. First End Point" and "2. Second End Point" labels

3P

As an alternative, you can pick any three points on the circumference, and AutoCAD constructs the circle. From the menu bar, select **Draw | Circle | 3 Points.**

> Command: **circle**
> Specify center point for circle or [3P/2P/Ttr (tan tan radius)]: **3p**
> Specify first point on circle: *(Pick point 1.)*
> Specify second point on circle: *(Pick point 2.)*
> Specify third point on circle: *(Pick point 3.)*

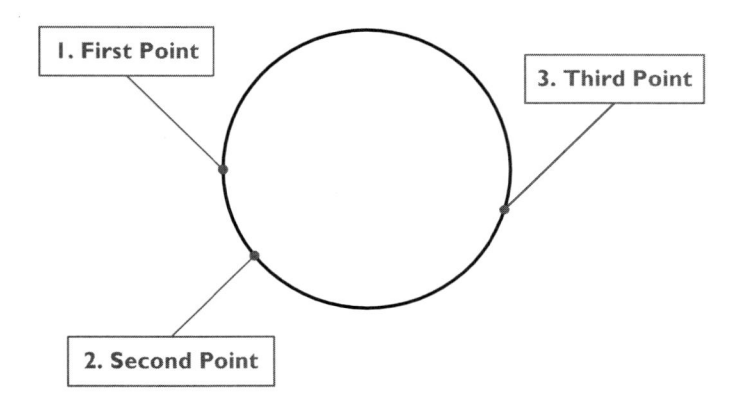

Diagram: Circle with "1. First Point", "2. Second Point", and "3. Third Point" labels

Ttr (tan tan radius)

Often, you need to draw a circle that's precisely tangent to other objects. The **Ttr** option automatically invokes **Tangent** object snap. (A *tangent* is the point at which the circle touches another object.) You can draw circles tangent to other circles, arcs, lines, polylines, and so on. From the menu bar, select **Draw | Circle | Tan Tan Radius**.

Command: **circle**

Specify center point for circle or [3P/2P/Ttr (tan tan radius)]: **t**

Specify point on object for first tangent of circle: *(Pick point 1.)*

Specify point on object for second tangent of circle: *(Pick point 2.)*

Specify radius of circle <0.9980>: *(Pick point 3, or enter a value for the radius.)*

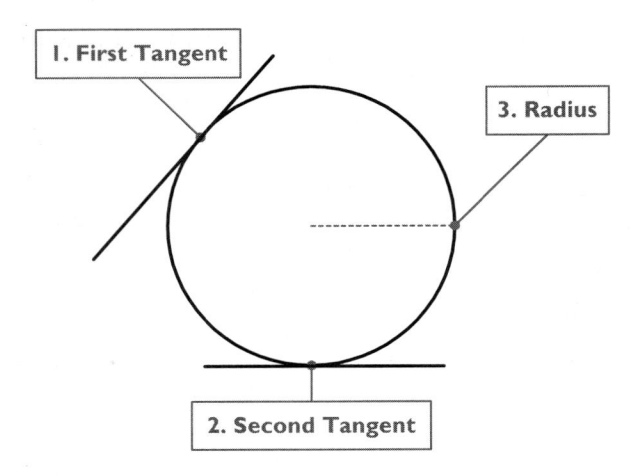

You may find it tricky placing the circle with tangents, because the radius determines where the tangents are placed.

Tan, Tan, Tan

AutoCAD has a sixth way of drawing circles that's not available from the CIRCLE command. Instead, you have to access it from the menu bar: select **Draw | Circle | Tan Tan Tan**. This "secret" option constructs circles tangent to three objects. After you select the command from the menu, notice that AutoCAD fills in several options for you, and automatically invokes **Tangent** object snap (displayed as "_tan"). All you do is pick three points, as follows:

Command: _circle

Specify center point for circle or [3P/2P/Ttr (tan tan radius)]: _3p

Specify first point on circle: _tan to *(Pick point 1.)*

Specify second point on circle: _tan to *(Pick point 2.)*

Specify third point on circle: _tan to *(Pick point 3.)*

Figure labels: 1. First Tangent, 2. Second Tangent, 3. Third Tangent

ARC

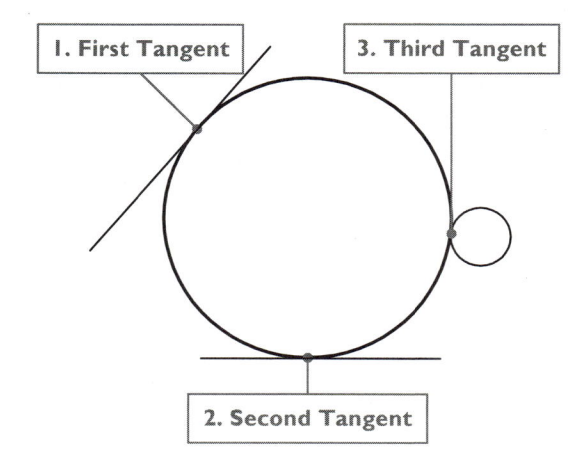

An arc is part of a circle. Like circles, arcs have a center point and a radius (or diameter). Unlike circles, however, arcs have starting and ending points. It can be difficult for beginners to remember that AutoCAD draws arcs in the counterclockwise direction, with a couple of exceptions: the three-point arc, and the start-end-direction arc.

If AutoCAD does not draw the arc as you expect, it may be because you are trying to draw it clockwise. Sometimes, you may find it easier to draw a circle, and then use the **BREAK** command to create an arc.

Some disciplines, such as railroad track design, define arcs by their *chords*, the straight distance between the start and end points. AutoCAD calls this distance the "length." The parts of an arc are shown below:

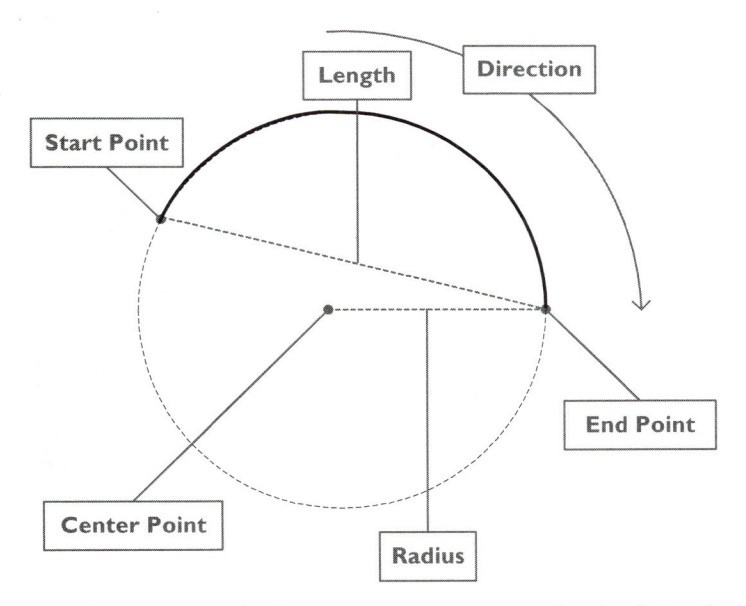

Figure labels: Start Point, Length, Direction, End Point, Radius, Center Point

AutoCAD constructs arcs by many methods — too many, you might think by the end of this section. This flexibility allows you to place arcs in many different situations.

TUTORIAL: DRAWING ARCS

1. To draw arcs, start the ARC command by one of these methods:
   - From the menu bar, choose **Draw**, then choose **Arc**, and then choose **Start, Center, End**.
   - From the **Draw** toolbar, choose the **Arc** button.
   - At the 'Command:' prompt, enter the **arc** command.
   - Alternatively, enter the **a** alias at the 'Command:' prompt.

   In all cases, you see the following prompts:
   Command: **arc** *(Press ENTER.)*

2. Pick the starting point of the arc:
   Specify start point of arc or [Center]: *(Pick point 1.)*

3. Pick a point that lies on the arc:
   Specify second point of arc or [Center/End]: *(Pick point 2.)*

4. Pick a point at the end of the arc:
   Specify end point of arc: *(Pick point 3.)*

   You can pick points in the drawing, or enter x,y coordinates. The arc drawn in this tutorial is called the "three-point arc." Unlike other arcs, this one is drawn in the direction determined by the first and second points.

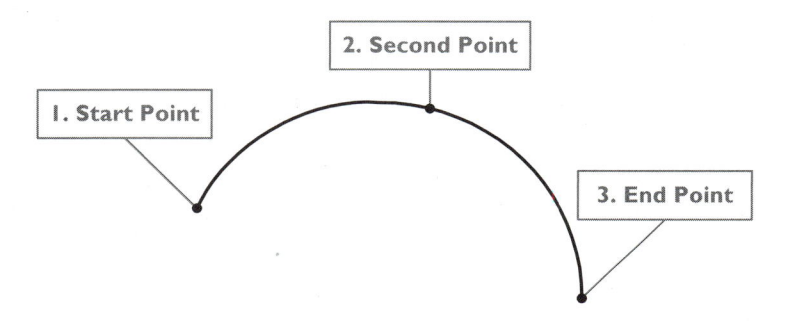

ADDITIONAL METHODS FOR DRAWING ARCS

The ARC command contains options that provide additional methods for drawing arcs. Several of the variations are similar, thus I have grouped them together:

- **Start, Center, End** constructs arcs from the start, center, and end points.
- **Center, Start, End** constructs arcs from the center, start, and end points.

- **Start, Center, Angle** constructs arcs from the start and center points, and the included angle.
- **Center, Start, Angle** constructs arcs from the center and start points, and the included angle.
- **Start, End, Angle** constructs arcs from the start and end points, and the included angle.

- **Start, Center, Length** constructs arcs from the start and center points, and the chord length.
- **Center, Start, Length** constructs arcs from the start, center, and end points.

- **Start, End, Radius** constructs arcs from the start and end points, and the radius.
- **Start, End, Direction** constructs arcs from the start and end points, and the direction.
- **Continue** continues the arc tangent to a line or another arc.

Drawing with Basic Objects 137

The options are shown at the command prompt, on the **Draw | Circle** menu (left), and on the right-click menus (center and right).

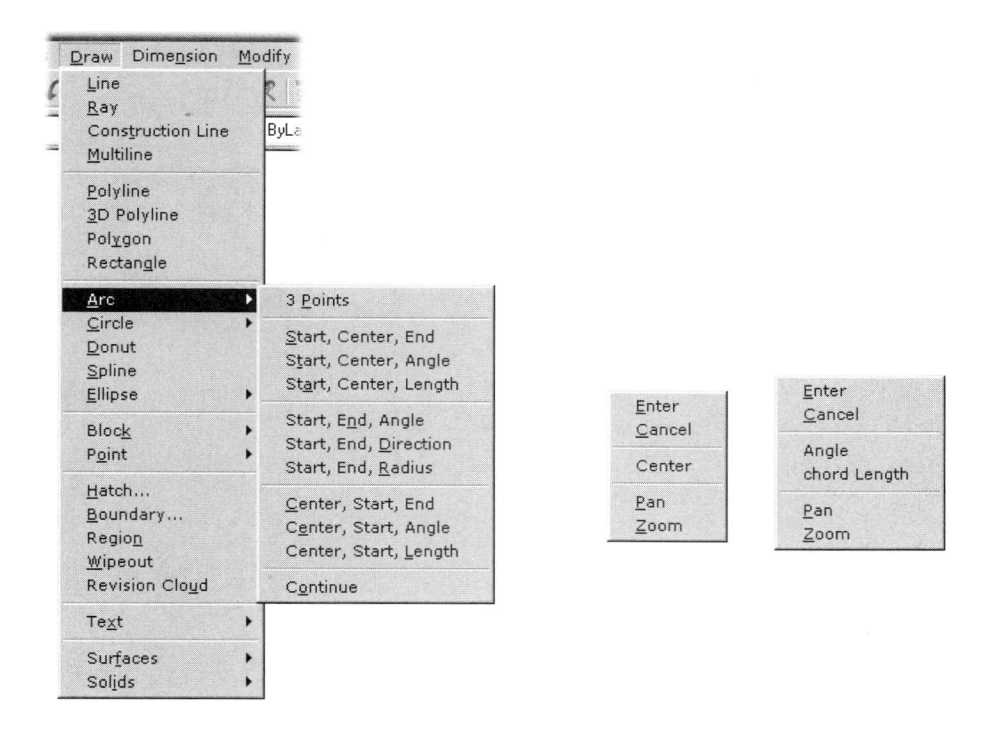

Let's look at each option.

START, CENTER, END & CENTER, START, END

The most common method of drawing arcs is to specify their start, center, and end points. AutoCAD constructs the arc in the counterclockwise direction, starting from the start and drawing to the end point. AutoCAD calculates the arc's radius as the distance from the start point to the center point.

If the arc is too big or looks wrong, it could be you started the arc at the point where it should end. From the menu bar, select **Draw | Arcs | Start, Center, End** or **Center, Start, End**.

Command: **arc**
Specify start point of arc or [Center]: *(Pick point 1, or enter x,y coordinates.)*
Specify second point of arc or [Center/End]: **c**
Specify center point of arc: *(Pick point 2, or enter x,y coordinates.)*
Specify end point of arc or [Angle/chord Length]: *(Pick point 3, or enter x,y coordinates.)*

After you pick the center point, notice that AutoCAD ghosts the end point as you move the cursor. The end point does not need to lie on the arc.

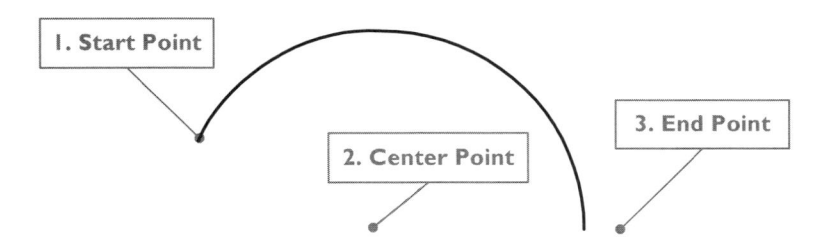

START, CENTER, ANGLE & START, END, ANGLE & CENTER, START, ANGLE

Arcs are sometimes specified by their *included angle*. This is the angle formed between the start and end points, with the angle's vertex at the center point.

Keep in mind that AutoCAD measures the arc in the counterclockwise direction, starting from the start to the end point. If you enter a negative angle, AutoCAD draws the arc clockwise from the start point.

From the menu bar, select **Draw | Arcs | Start, Center, Angle** or **Start, End, Angle** or **Center, Start, Angle**.

Command: **arc**
Specify start point of arc or [Center]: *(Pick point 1.)*
Specify second point of arc or [Center/End]: **c**
Specify center point of arc: *(Pick point 2.)*
Specify end point of arc or [Angle/chord Length]: **a**
Specify included angle: *(Enter an angle, 3.)*

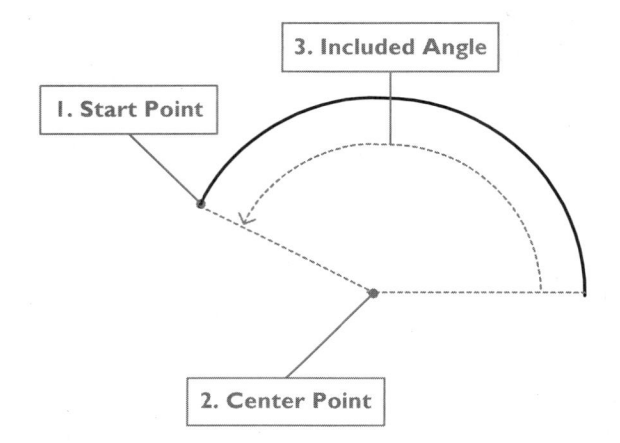

START, CENTER, LENGTH & CENTER, START, LENGTH

When you need to draw an angle with a specific chord, use the **Length** option. The length is the distance from the start point to the end point. From the menu bar, select **Draw | Arc | Start, Center, Length** or **Center, Start, Length**.

Command: **arc**
Specify start point of arc or [Center]: *(Pick point 1.)*
Specify second point of arc or [Center/End]: **c**
Specify center point of arc: *(Pick point 2.)*
Specify end point of arc or [Angle/chord Length]: **l**
Specify length of chord: *(Enter a length, 3.)*

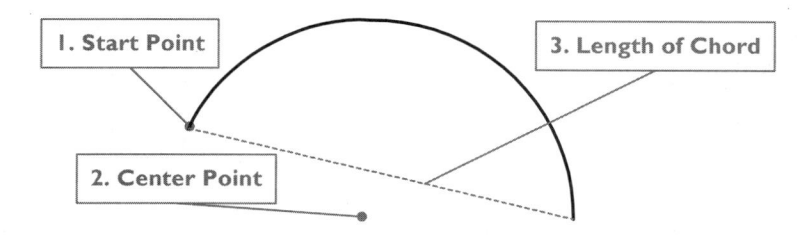

AutoCAD does not draw the arc if the chord length does not work with the start and center points you pick. When you see "*Invalid*" on the command prompt area, the chord is too long or too short. Note that the center point is usually *not* on the chord.

AutoCAD constructs a *major arc* or a *minor arc,* depending on the chord length. (Given that the chord divides a circle into two parts, the major arc is the larger arc, while the minor arc is the smaller one.) When the chord is positive, AutoCAD draws the minor arc; when negative, the major arc.

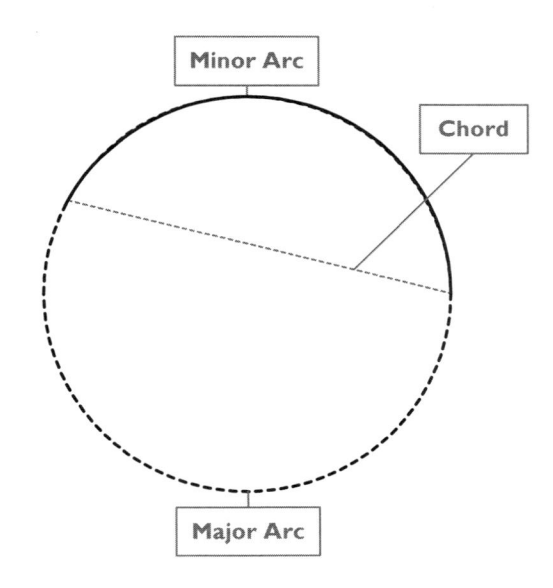

START, END, RADIUS

This option is most like drawing a circle: you specify two points and the radius. From the menu bar, select **Draw | Arc | Start, End, Radius.**

> Command: **arc**
> Specify start point of arc or [Center]: *(Pick point 1.)*
> Specify second point of arc or [Center/End]: **e**
> Specify end point of arc: *(Pick point 2.)*
> Specify center point of arc or [Angle/Direction/Radius]: **r**
> Specify radius of arc: *(Enter a positive or negative radius, or pick point 3.)*

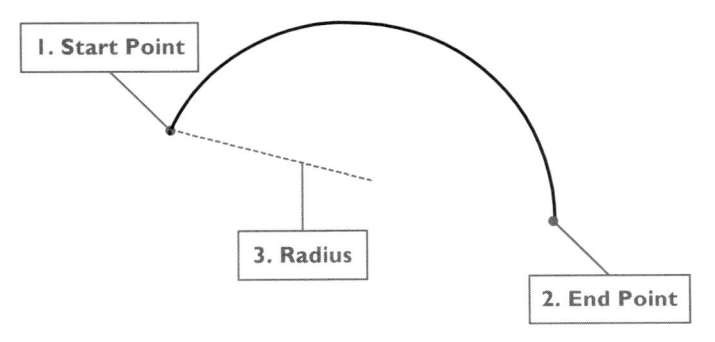

By default, AutoCAD draws the minor arc. If, however, you enter a negative value for the radius, AutoCAD draws the major arc.

START, END, DIRECTION

This option is perhaps the trickiest to understand. AutoCAD starts the arc tangent to the direction you specify. From the menu bar, select **Draw | Arc | Start, End, Direction**.

    Command: **arc**

    Specify start point of arc or [Center]: *(Pick point 1.)*

    Specify second point of arc or [Center/End]: **e**

    Specify end point of arc: *(Pick point 2.)*

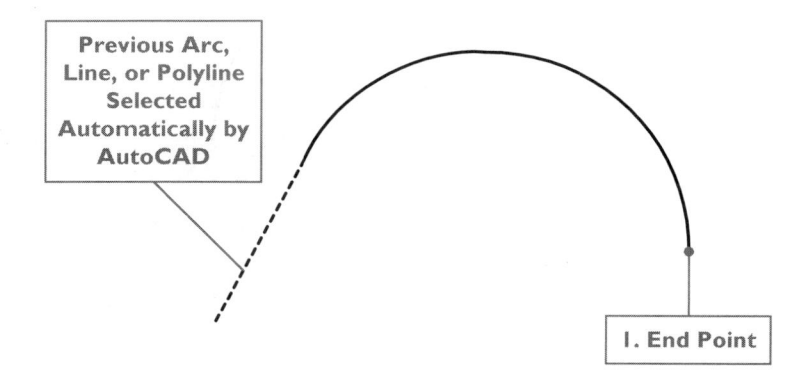

    Specify center point of arc or [Angle/Direction/Radius]: **d**

    Specify tangent direction for the start point of arc: *(Pick point 3.)*

At the "Specify tangent direction" prompt, move the cursor. Notice how AutoCAD ghosts the arc, depending in the cursor's location.

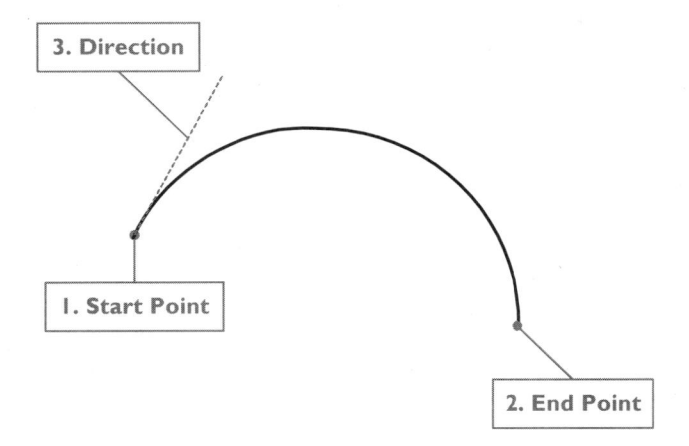

CONTINUE

Among the ARC command's options, one is "hidden." You can find it on the menu under **Draw | Arc | Continue**, but it is not visible at the 'Command:' prompt. The **Continue** option is designed to draw an arc tangent to the one last-drawn. Here's how it works:

    Command: **arc**

    Specify start point of arc or [Center]: *(Press* ENTER.*)*

Notice that AutoCAD automatically selects the end of the last line, polyline, or arc as the starting point for the new arc, which is drawn tangent to the last-drawn object. As you move the cursor, AutoCAD ghosts it in the drawing. There is just one option:

    Specify end point of arc: *(Pick point 1.)*

⊙ DONUT

The **DONUT** command draws thick and solid-filled circles. These kinds of circles are useful for PCB (printed circuit boards) designs. Donuts are drawn as polylines.

This is one of AutoCAD's few drawing commands that keeps on repeating itself until you press **ESC** or **ENTER**.

TUTORIAL: DRAWING DONUTS

1. To draw donuts, start the **DONUT** command by one of these methods:
   - From the menu bar, choose **Draw**, then choose **Donut**.
   - At the 'Command:' prompt, enter the **donut** command.
   - Alternatively, enter the **doughnut** alias at the 'Command:' prompt.

   In all cases, you see the following prompts:
   Command: **donut** *(Press ENTER.)*

2. Enter a value for the donut's hole, its inside diameter:
   Specify inside diameter of donut <0.5000>: *(Specify value, or pick two points, 1.)*

3. Enter a value for the outside of the donut:
   Specify outside diameter of donut <1.0000>: *(Specify value, or pick two points, 2.)*

4. Pick a point to place the donut.
   Specify center of donut or <exit>: *(Pick point 3.)*

5. Press **ENTER** to end the command.
   Specify center of donut or <exit>: *(Press ENTER to exit command.)*

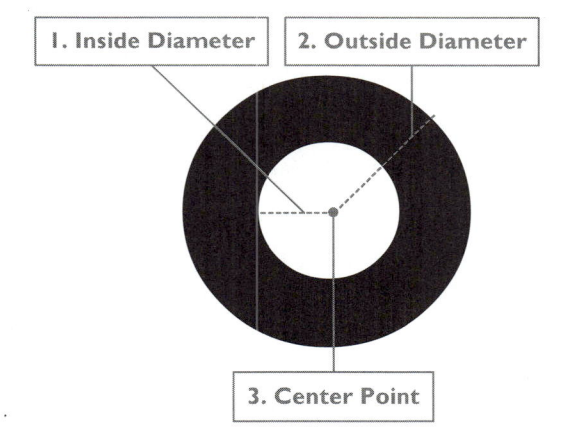

Note: When the inside diameter equals the outside diameter, AutoCAD draws a circle, albeit one made from a polyline. (It can be edited with the **PEdit** command.) When the inside diameter is zero, AutoCAD draws a solid-filled circle.

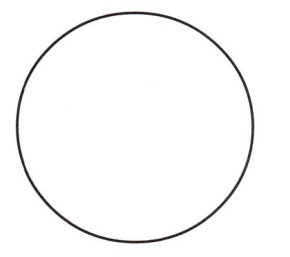

PLINE

We have referred to "polyline" in some commands earlier, specifically **POLYGON** and **DONUT**. The **PLINE** command draws the polyline, perhaps the most unique object created by AutoCAD.

A polyline is a single object that consists of connected lines and arcs. It can be curved, splined, and open or closed (like an irregular polygon). You can specify the width of each segment, and give each segment a tapered width. Until lineweights were recently introduced recently to AutoCAD, polylines were the only way to draw objects with width.

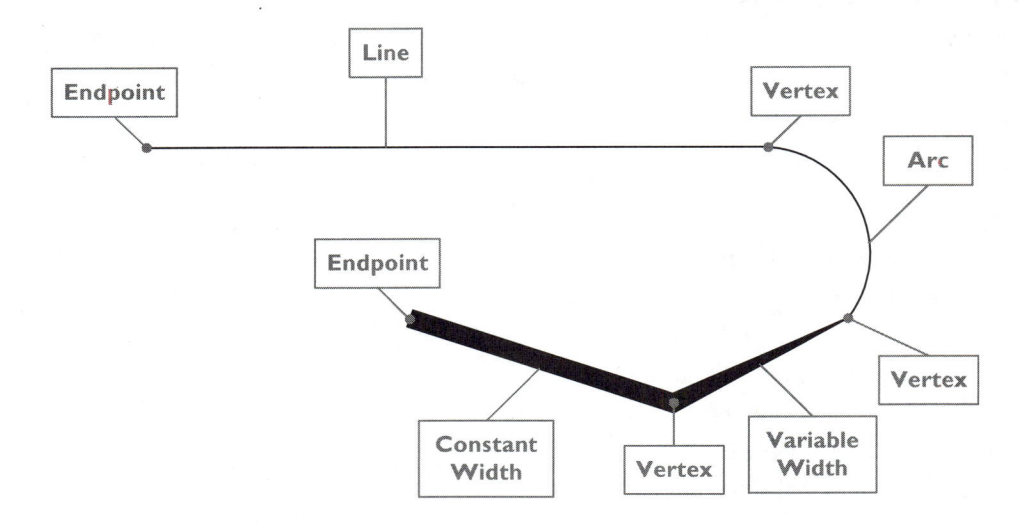

TUTORIAL: DRAWING POLYLINES

1. To draw polylines, start the **PLINE** command.
    - From the menu bar, choose **Draw**, then choose **Polyline**.
    - From the **Draw** toolbar, choose the **Polyline** button.
    - At the 'Command:' prompt, enter the **pline** command.
    - Alternatively, enter the **pl** alias at the 'Command:' prompt.

    Command: **pline** *(Press ENTER.)*

2. Pick a point from which to start drawing the polyline:

    Specify start point: *(Pick a point.)*

3. AutoCAD reminds you of the current linewidth, which is 0, unless you change it with the **Width** option:

    Current line-width is 0.0000

4. Pick the next point, or select an option:

    Specify next point or [Arc/Halfwidth/Length/Undo/Width]: *(Pick a point, or enter an option.)*

5. Continue picking points, and then press **ENTER** to exit the command:

    Specify next point or [Arc/Close/Halfwidth/Length/Undo/Width]: *(Press ENTER to exit the command.)*

ADDITIONAL METHODS FOR DRAWING POLYLINES

The PLINE command contains options that provide additional methods for drawing polylines:

- **Undo** removes the previous segment.
- **Close** draws a segment to the start point.
- **Arc** switches to arc-drawing mode.
- **Width** specifies the width of the polyline.
- **Halfwidth** specifies the halfwidth.
- **Length** draws a tangent segment of specific length.

The options are shown at the command prompt and on the right-click menus shown below.

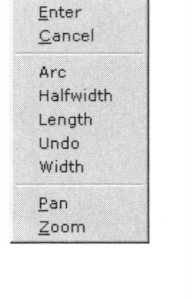

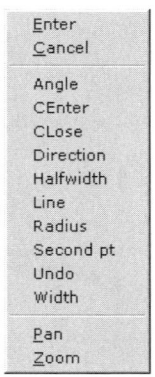

Let's look at each option.

Undo

As you draw polyline segments, you may sometimes make a mistake. Instead of cancelling the command and starting over, use the **Undo** option to "undraw" the last segment. You can undo all the way back to the first segment.

Close

When you construct a polygon using the PLINE command, the last segment connects to the first segment. Aligning the ends of segments be tedious, so the **Close** option automatically performs this for you. The **Close** option appears after you pick three points, because a minimum of two lines is needed before the polygon can be closed with a third line to make the polygon.

Arc

The **Arc** option switches the PLINE command to arc-drawing mode. This mode has options similar to those of the ARC command, with some additions:

> Specify next point or [Arc/Halfwidth/Length/Undo/Width]: **a**
>
> [Angle/CEnter/Direction/Halfwidth/Line/Radius/Second pt/Undo/Width]:

The **Width** and **Halfwidth** options specify the width of the arc segments, as described below.

The **Line** option switches out of arc-drawing mode and back to line drawing mode.

Width

The **Width** option allows you to specify the width of each segment. In addition, you can specify a different starting and ending width, which creates tapers. By default, the width starts 0 in new drawings, and then remembers the previous width setting.

1. Start the PLINE command, and pick a starting point:
   Command: **pline**

Specify start point: *(Pick a point.)*

2. Notice the current linewidth. Enter **w** to access the **Width** option:

   Current line-width is 0.1000

   Specify next point or [Arc/Halfwidth/Length/Undo/Width]: **w**

3. Enter a value for the "starting width." This is the width at the starting end of the polyline.

   Specify starting width <0.1000>: *(Press* ENTER *to accept the default, or enter another value.)*

4. Press ENTER if you want the ending width to be the same as the starting width. For a taper, enter a different value for the ending width:

   Specify ending width <0.1000>: **.25**

5. Pick the next point, or select another option:

   Specify next point or [Arc/Halfwidth/Length/Undo/Width]:

Ending Width = 0.25

Constant Width = 0.25

Starting Width = 0.1

Tapered Width

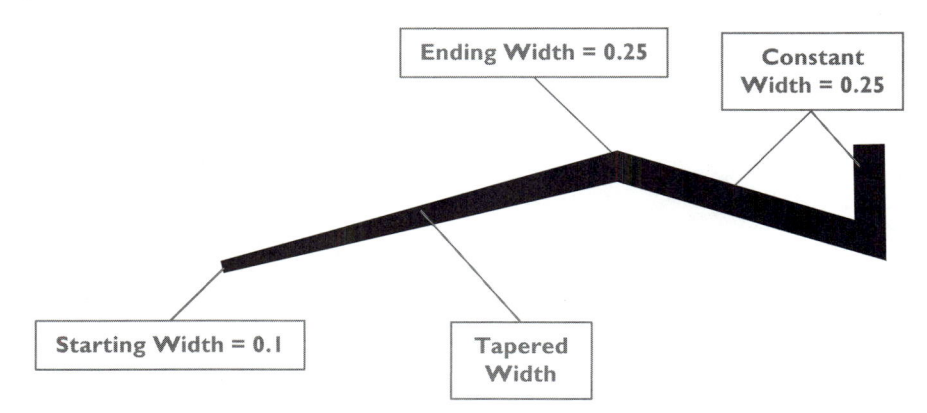

Notes: When you provide different values for the starting and ending widths, AutoCAD draws a taper but for that segment only! The next segment is drawn at the ending width of the previous segment.

When two polyline segments have a width other than 0, AutoCAD automatically bevels the vertex between them. To see the bevelled vertices, turn off fill mode with the **FILLMODE** system variable, followed by the **REGEN** command.

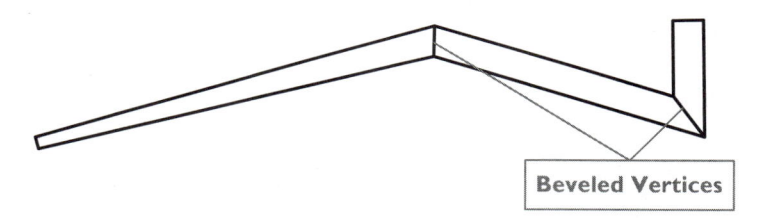

Beveled Vertices

To select a polyline of width other than 0, you must pick its edge. Picking the centerline of the polyline fails. For some reason, AutoCAD is unable to detect the center and fill of polylines.

Halfwidth

The **Halfwidth** option is identical to **Width**, except that you specify the width from the edge of the polyline to its centerline.

Specify next point or [Arc/Halfwidth/Length/Undo/Width]: **h**

Specify starting half-width <0.1250>: *(Enter a value, or press* ENTER *to keep the default.)*

Specify ending half-width <0.1250>: *(Press* ENTER *for constant width; enter a different value for tapered width.)*

Drawing with Basic Objects **145**

Halfwidth

"Centerline"

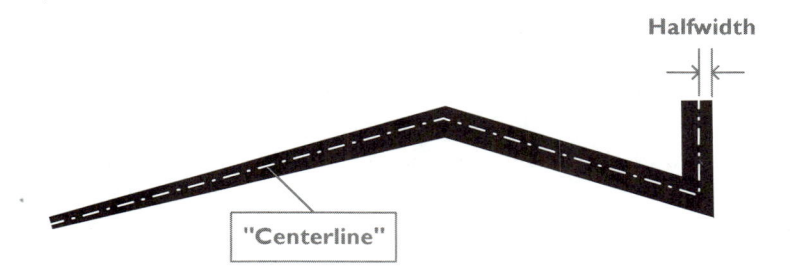

Length

The **Length** option draws the next segment at the same angle as the previous segment. You specify the length. This option is most useful when the previous segment is an arc, because the segment is drawn tangent to the arc's endpoint.

Specify next point or [Arc/Halfwidth/Length/Undo/Width]: **l**
Specify length of line: **3**

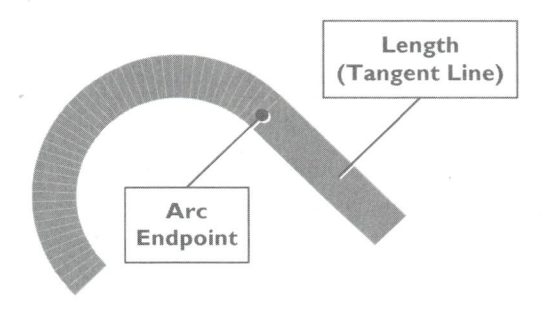

Length
(Tangent Line)

Arc
Endpoint

ELLIPSE

The **ELLIPSE** command draws ellipses and elliptical arcs. When isometric mode is turned on, this command adds the **Isocircle** option for drawing isometric circles.

The ellipse is an elongated circle drawn with two diameters called "axes." The *major axis* is the longer axis, the *minor axis*, the shorter.

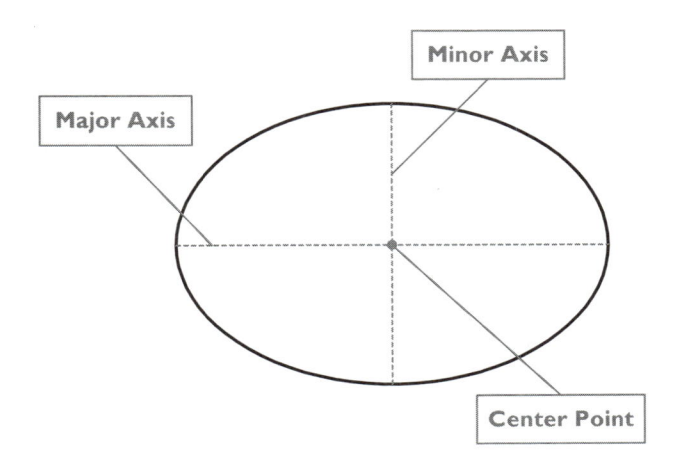

Minor Axis

Major Axis

Center Point

An ellipse is sometimes referred to by its *rotation*. A circle is an ellipse that has not been rotated: its rotation is zero degrees; the major and minor axes have the same length.

An ellipse is a circle that has been rotated, or viewed at an angle. For instance, a 40-degree ellipse is a circle that has been rotated by 40 degrees about the major axis. AutoCAD allows you to specify ellipse rotation between 0.0 (a circle) and 89.4 degrees (a very thin ellipse).

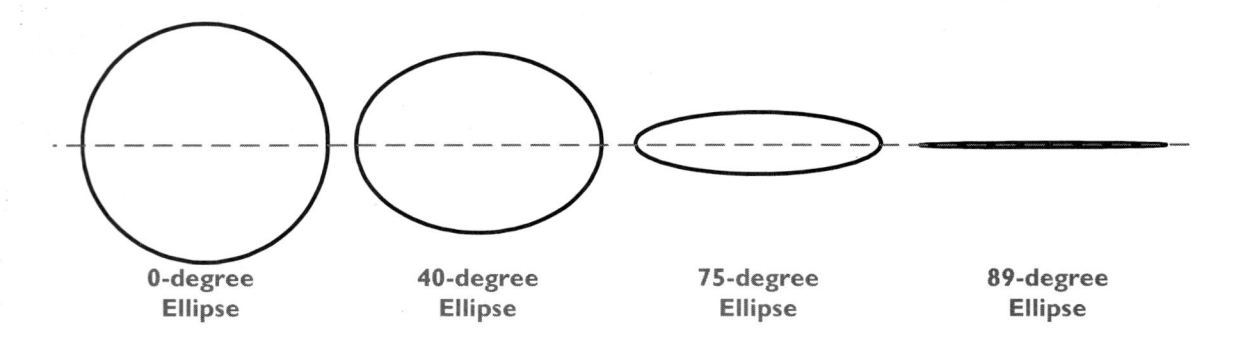

0-degree Ellipse   **40-degree Ellipse**   **75-degree Ellipse**   **89-degree Ellipse**

TUTORIAL: DRAWING ELLIPSES

1. To draw ellipses, start the **ELLIPSE** command.
   - From the menu bar, choose **Draw**, then choose **Ellipse**, and then choose **Center**.
   - From the **Draw** toolbar, choose the **Ellipse** button.
   - At the 'Command:' prompt, enter the **ellipse** command.
   - Alternatively, enter the **el** alias at the 'Command:' prompt.

   In all cases, you see the following prompts:
   Command: **ellipse** *(Press ENTER.)*

2. Pick a point to indicate an endpoint of one axis.
   Specify axis endpoint of ellipse or [Arc/Center]: *(Pick point 1.)*
3. Pick another point for the other end of the axis.
   Specify other endpoint of axis: *(Pick point 2.)*
4. Show the half-distance to the other axis.
   Specify distance to other axis or [Rotation]: *(Pick point 3.)*

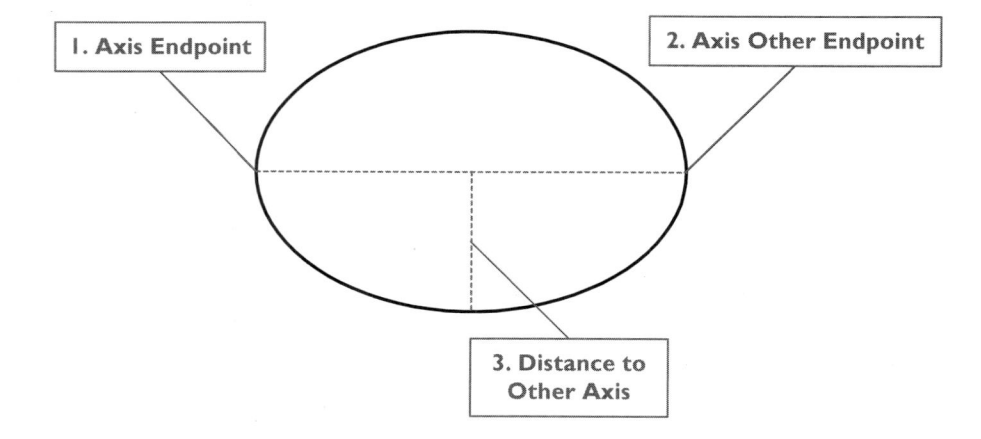

1. Axis Endpoint
2. Axis Other Endpoint
3. Distance to Other Axis

The "distance to other axis" does not need to be perpendicular to the first axis.

ADDITIONAL METHODS FOR DRAWING ELLIPSES

The **ELLIPSE** command contains options that provide additional methods for drawing ellipses:

- **Center** starts with the center point of the ellipse, followed by the axes.
- **Rotation** starts with the rotation angle about the major axis.
- **Arc** constructs an elliptical arc.

The options are shown at the command prompt, on the **Draw | Circle** menu (left), and on the right-click menus (center and right).

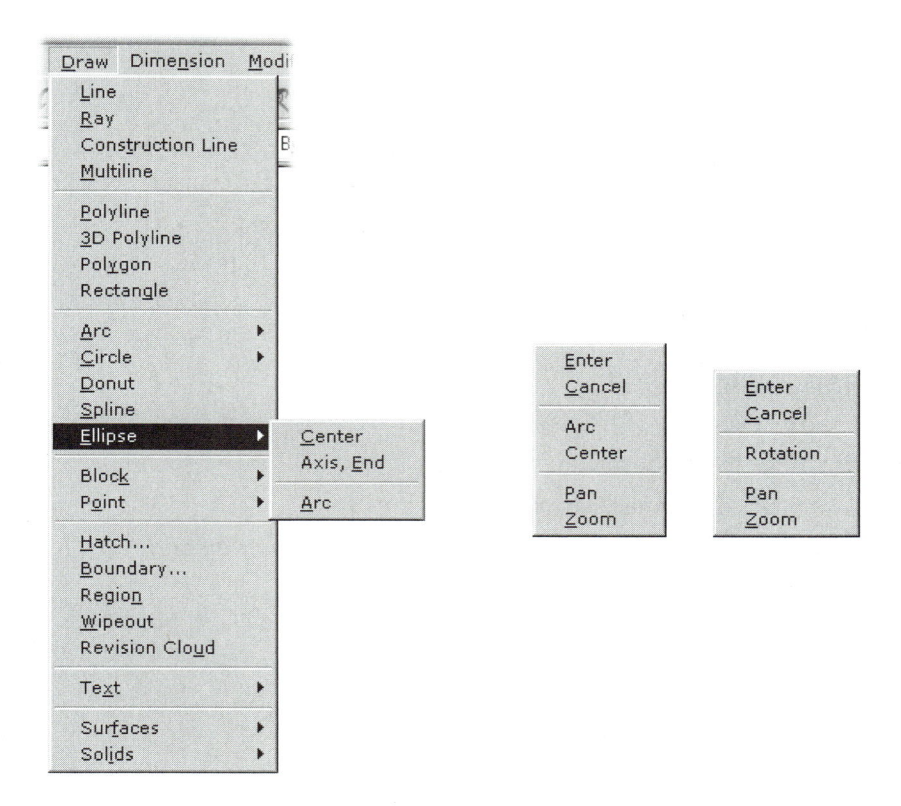

Center

The **Center** option allows you to pick the center of the ellipse, and then continues to let you specify distance from the center point to the end points of the axes.

Command: **ellipse**
Specify axis endpoint of ellipse or [Arc/Center]: **c**
Specify center of ellipse: *(Pick point 1.)*
Specify endpoint of axis: *(Pick point 2.)*
Specify distance to other axis or [Rotation]: *(Pick point 3.)*

As you move the cursor during the "Specify distance to other axis" prompt, AutoCAD ghosts in the ellipse. Note that AutoCAD doesn't care which axis you create first.

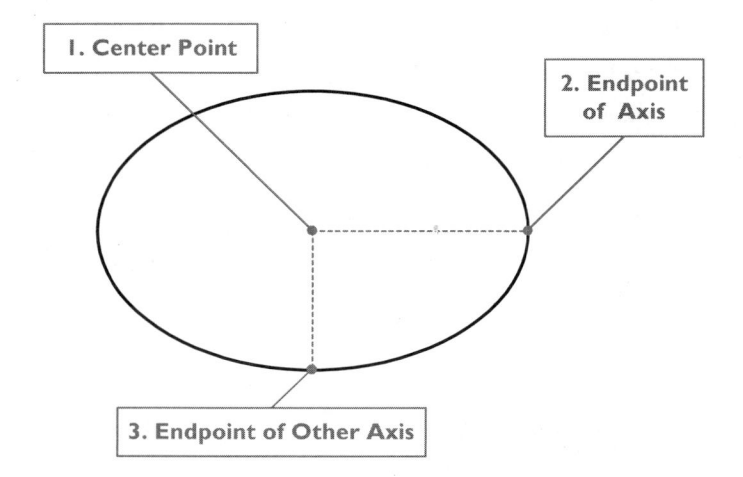

[Diagram: Ellipse with labels — 1. Center Point, 2. Endpoint of Axis, 3. Endpoint of Other Axis]

Rotation

The **Rotation** option rotates the ellipse about the major axis; it replaces the prompts for drawing the minor axis. (This option is also available when constructing ellipses with the **Center** option.) As you move the cursor during the "Specify rotation" prompt, AutoCAD ghosts the image of the ellipse.

    Command: **ellipse**

    Specify axis endpoint of ellipse or [Arc/Center]: *(Pick point 1.)*

    Specify other endpoint of axis: *(Pick point 2.)*

    Specify distance to other axis or [Rotation]: **r**

    Specify rotation around major axis: *(Pick point 3, or enter an angle.)*

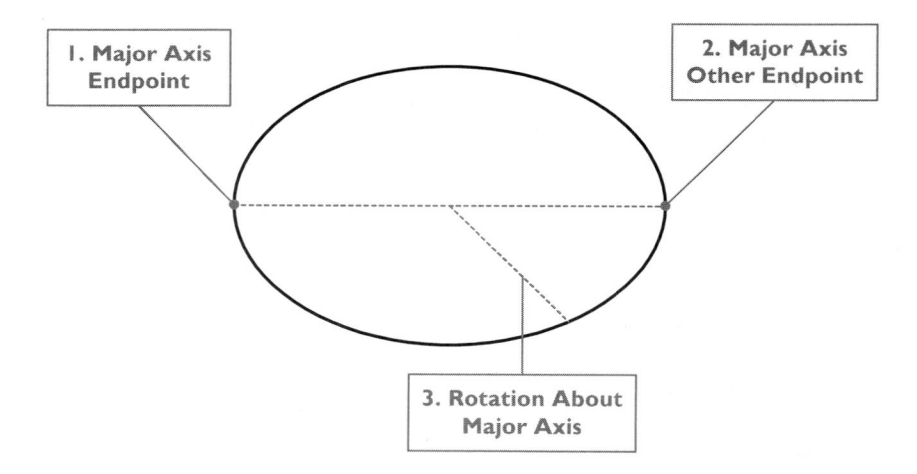

[Diagram: Ellipse with labels — 1. Major Axis Endpoint, 2. Major Axis Other Endpoint, 3. Rotation About Major Axis]

Arc

The **Arc** option draws elliptical arcs. The first few times you try drawing an elliptical arc, it might not turn out as desired. It takes some practice!

    Command: **ellipse**

    Specify axis endpoint of ellipse or [Arc/Center]: **a**

    Specify axis endpoint of elliptical arc or [Center]: *(Pick point 1.)*

    Specify other endpoint of axis: *(Pick point 2.)*

    Specify distance to other axis or [Rotation]: *(Pick point 3.)*

    Specify start angle or [Parameter]: *(Pick point 4.)*

    Specify end angle or [Parameter/Included angle]: *(Pick point 5.)*

As you move the cursor during the "Start angle" and "End angle" prompts, AutoCAD ghosts a preview of the elliptical arc. The figure below shows an arc that consists of the upper half of an ellipse.

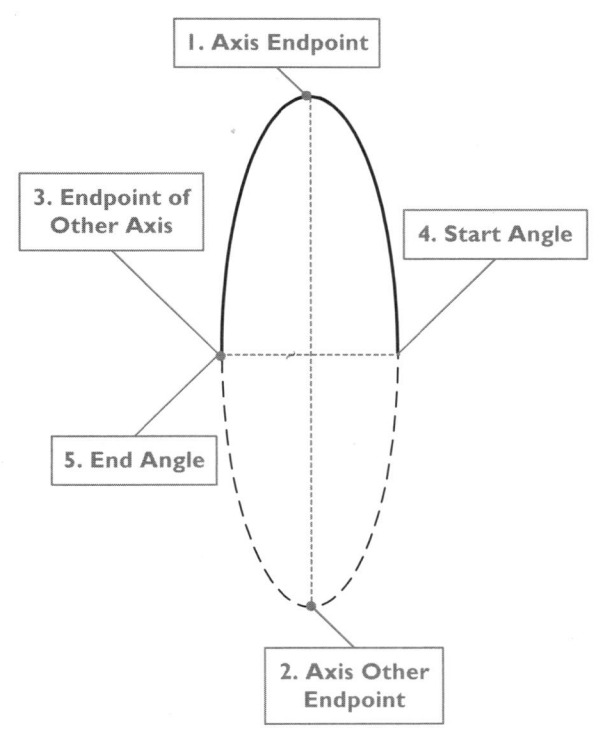

Figure: Ellipse with labels — 1. Axis Endpoint; 2. Axis Other Endpoint; 3. Endpoint of Other Axis; 4. Start Angle; 5. End Angle.

The **Arc** option has three methods of determining the arc: **Angle**, **Included Angle**, and **Parameter**.

The **Angle** option specifies the start angle and the end angle of the arc. The two angles are measured relative to the first axis endpoint. The *start angle* determines where the arc starts; the *end angle* where it ends. When you specify 0 degrees for the start angle, the arc starts at the first axis endpoint. AutoCAD draws the arc in the counterclockwise direction. If you run into trouble, it's probably because you are trying to draw the arc clockwise.

The **Included Angle** option defines the arc by an angle that starts from the start angle (described above). The angle is relative to the start angle. AutoCAD prompts:

> Specify end angle or [Parameter/Included angle]: **i**
> Specify included angle for arc <180>: *(Enter an angle, or pick a point.)*

The **Parameter** option uses this formula to construct the elliptical arc:

> p(u) = **c** + **a** × cos(u) + **b** × sin(u)

where:

> a is the major axis.
> b is the minor axis.
> c is the center point of the ellipse.

Does anyone use this formula in the real world? In any case, AutoCAD prompts you:

> Specify start angle or [Parameter]: **p**
> Specify start parameter or [Angle]: *(Pick a point.)*
> Specify end parameter or [Angle/Included angle]: *(Pick a point.)*

Notice how you can switch back and forth between the three arc definitions – Angle, Included Angle, and Parameter.

■ POINT

The **POINT** command constructs *point* objects. A point has no height or width: it is a dot. The point is the smallest object that output devices can produce: a single pixel on the screen, a single dot on the printed paper. Because points are so small, they can be hard to see.

Using the **PDMODE** and **PDSIZE** system variables, you can change the look and size of points. As illustrated below, points can be invisible, or combinations of lines, circles, and squares. The effect of the two system variables is retroactive, meaning changing their values changes the look and size of all previously-drawn points.

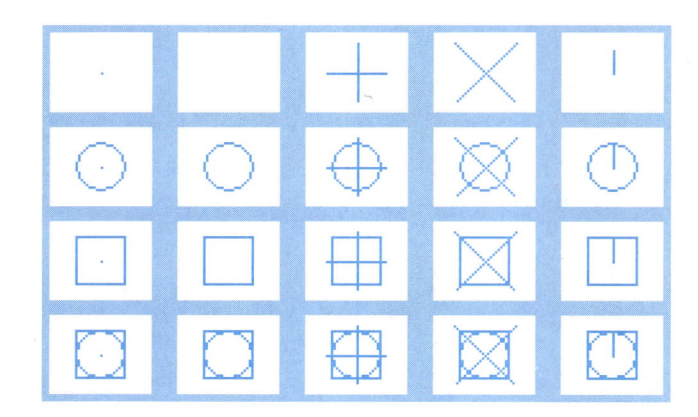

TUTORIAL: DRAWING POINTS

1. To draw points, start the **POINT** command.
   - From the menu bar, choose **Draw**, then choose **Points**, and then **Single Point**.
   - From the **Draw** toolbar, choose the **Point** button.
   - At the 'Command:' prompt, enter the **point** command.
   - Alternatively, enter the **po** alias at the 'Command:' prompt.

   In all cases, you see the following prompts:
   Command: **point** *(Press ENTER.)*

2. Notice that AutoCAD displays the value for the **PDMODE** and **PDSIZE** system variables.
   Current point modes: PDMODE=0 PDSIZE=0.0000

3. Pick a point to place the point.
   Specify a point: *(Pick a point.)*

ADDITIONAL METHODS FOR DRAWING POINTS

The **POINT** command does not contain any options, but there are related commands that provide additional methods for drawing points:

- **MULTIPLE** repeats the **POINT** command.
- **DDPMODES** changes the look of points.

Multiple

The **MULTIPLE** command repeats the **POINT** command, which is handy should you want quickly to place more than one point at a time. Indeed, the **MULTIPLE** command works with any command.

From the menu bar, select **Draw | Point | Multiple Point**. At the command prompt, enter:

> Command: **multiple**
> Enter command name to repeat: **point**
> Current point modes: PDMODE=0 PDSIZE=0.0000
> Specify a point: *(Pick a point.)*
> POINT Current point modes: PDMODE=0 PDSIZE=0.0000
> Specify a point: *(Press ESC or ENTER to exit repeating command.)*

The command repeats until you press ESC or ENTER.

DdPModes

The **DDPMODES** command displays a dialog box that allows you to select the style and size of point. There are several limitations: you can only chose from the styles listed in the dialog box, which excludes custom point styles; all points in the drawing take on the same style and size, which means different points cannot have different styles.

From the menu bar, select **Format | Point Style** to display the dialog box:

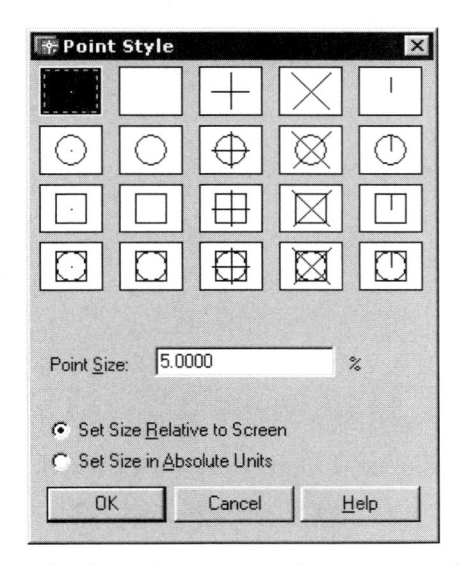

The upper half of the dialog box displays the point styles you can choose from. Note that the first style (the single dot) cannot be resized; the second style (the blank square) creates invisible points, which also cannot be sized.

The lower half changes the size of the point. The **Point Size** option sets the size as a percentage relative to the screen size or in units. The defaults are 5% and 5 units. I recommend you stick with the **Set Size Relative to Screen** (percentage) setting, because the **Absolute Units** setting can create points that are too large to see (when zoomed in) and too small to see (when zoomed out).

Click **OK** to accept the changes; AutoCAD automatically regenerates the drawing so that you can see the changes to the point display.

CONSTRUCTING OTHER OBJECTS

This chapter introduced you to AutoCAD's basic drawing commands. These allow you easily to draw lines, circles, rectangles, regular polygons, arcs, ellipses, elliptical arcs, points, and polylines. They don't, however, make it easy to draw other common 2D geometric shapes, such as isosceles and right triangles, spirals, parallelograms; and 3D objects, such as cubes and cylinders. You learn about 3D objects in volume 2 of this book, *Using AutoCAD: Advanced*.

To draw other kinds of 2D shapes, you need to *construct* them, using one or more commands available in AutoCAD. For instance, let's look at the four kinds of triangle:

- **Equilateral** all three sides the same length.
- **Isosceles** two sides the same length.
- **Scalene** all three sides at different lengths.
- **Right** one angle at 90 degrees.

As you may recall from your high school geometry class, triangles are drawn with a combination of lengths and angles, and sometimes require the use of sine, cosine, and tangent calculations. Although it's not obvious, AutoCAD provides all the tools.

EQUILATERAL TRIANGLES

All three sides of equilateral triangles have the same length. (Again, if you recall your high school geometry class, the three angles are also the same.) This kind of triangle is most easily drawn with the **POLYGON** command set to draw 3 sides. When you know the length of the sides, use the **Edge** option. (You have no choice over the angles: they are always 60 degrees.) For example, if the sides are 2.5 units long, using *direct distance entry* to specify the length, as follows:

1. Start the **POLYGON** command:
   Command: **polygon**

2. To draw triangles, specify 3 sides:
   Enter number of sides <4>: **3**

3. By using the **Edge** option, you specify the length of one side; AutoCAD draws the other two sides to match.
   Specify center of polygon or [Edge]: **e**
   Specify first endpoint of edge: *(Pick point 1.)*
   Specify second endpoint of edge: *(Move cursor in a direction, and then enter:)* **2.5**

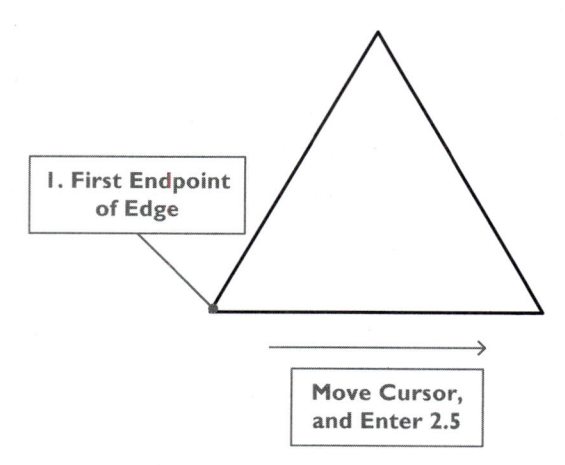

AutoCAD draws the equilateral triangle for you.

ISOSCELES TRIANGLES

Two sides of isosceles triangles have the same length (the two opposite angles are also the same). One method to this triangle is to use the **LINE**, **MIRROR**, and **TRIM** commands. For example, let's draw a triangle with a base of 3 units and isosceles angles of 63 degrees.

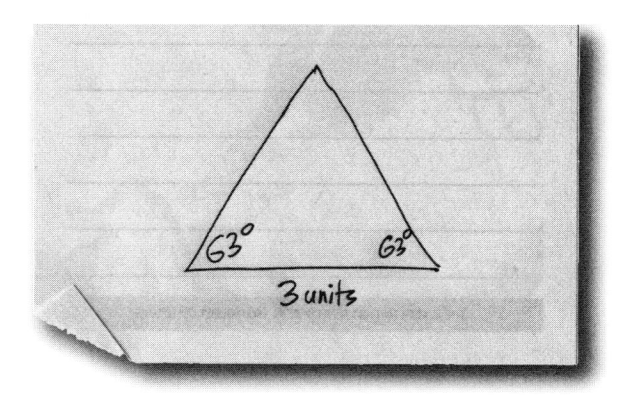

We don't know the length of the other two sides, so we draw them an arbitrary length, and then trim or extend to fit. For the best result, turn on polar mode.

1. Start the **LINE** command to draw the base and an angled leg:
   Command: **line**
   Specify first point: *(Pick point 1.)*

2. Use direct distance entry to draw the base 3 units long:
   Specify next point or [Undo]: *(Move cursor to the right, and then enter:)* **3**

3. Use relative coordinates to draw one leg 10 units long at 117 degrees. The 10 units is arbitrary, because we do not know its length; we edit the length later. The 117 degrees comes from 180 - 63 = 117.
   Specify next point or [Undo]: **@10<117**
   Specify next point or [Close/Undo]: *(Press* **ENTER** *to exit command.)*

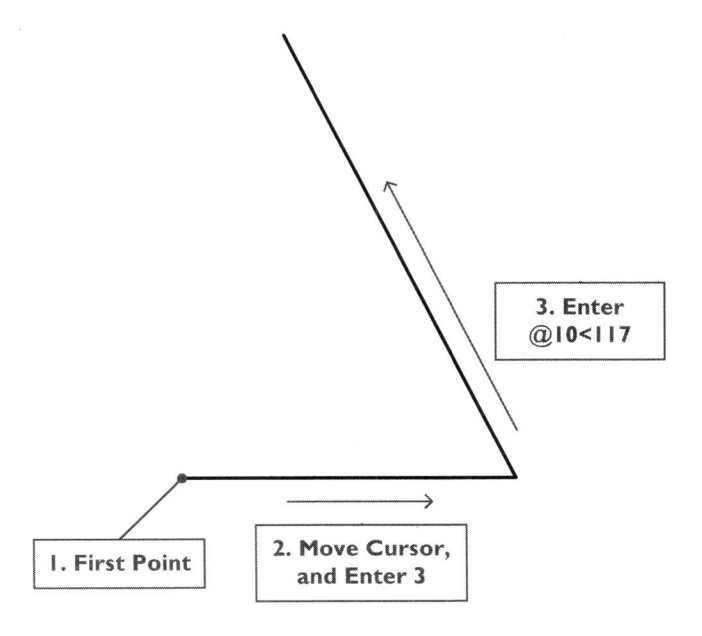

4. Start the MIRROR command to make a mirror copy of the angled leg:
   Command: **mirror**
5. Enter **L** to select the last-drawn object:
   Select objects: **L**
   1 found Select objects: *(Press ENTER to end object selection.)*
6. Use MIDpoint object snap to create an imaginary vertical mirroring line:
   Specify first point of mirror line: **mid**
   of *(Pick point 4.)*
   Specify second point of mirror line: *(Pick point 5.)*

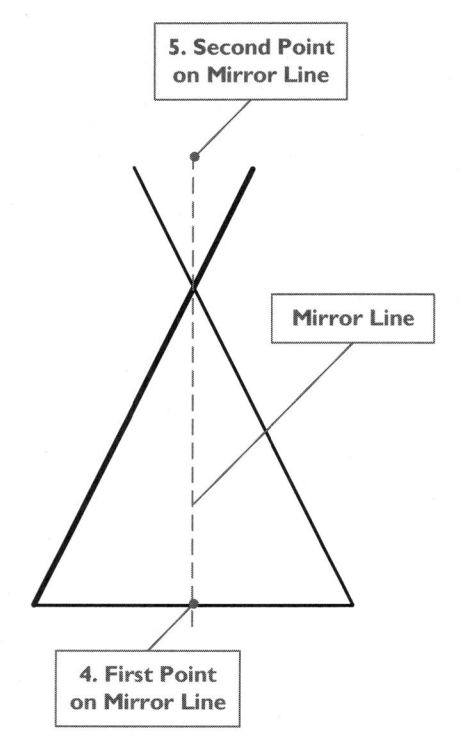

[5. Second Point on Mirror Line]

[Mirror Line]

[4. First Point on Mirror Line]

7. Keep the source object, and exit the command:
   Delete source objects? [Yes/No] <N>: *(Press ENTER to accept default, **No**, and to exit the command.)*
8. Use the TRIM command to trim back the two angled lines to their intersection point:
   Command: **trim**
   Current settings: Projection=UCS, Edge=None
   Select cutting edges ...
9. Select the two lines. They become each other's cutting edge:
   Select objects: *(Select one angled line, 6.)*
   1 found Select objects: *(Select the other angled line, 7.)*
   1 found, 2 total Select objects: *(Press ENTER to end object selection.)*

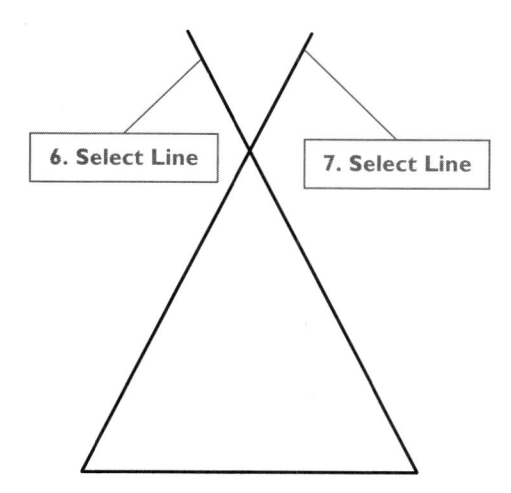

10. Select the two lines again to trim them. Be sure to pick the portion of the lines to be trimmed away:

> Select object to trim or shift-select to extend or [Project/Edge/Undo]: *(Select one angled line, 8.)*
>
> Select object to trim or shift-select to extend or [Project/Edge/Undo]: *(Select the other angled line, 9.)*

> Select object to trim or shift-select to extend or [Project/Edge/Undo]: *(Press ENTER to exit command.)*

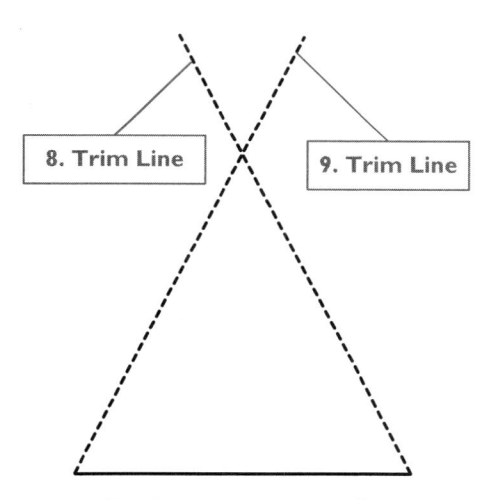

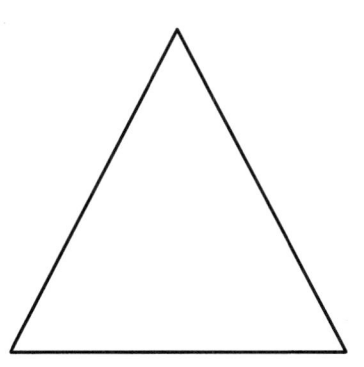

The isosceles triangle is complete.

SCALENE TRIANGLES

All three sides and corners of scalene triangles have different lengths and angles. This kind of triangle is drawn with the LINE command, using a combination of direct distance entry and polar coordinates, as required. Take, for example, a triangle with the following specifications:

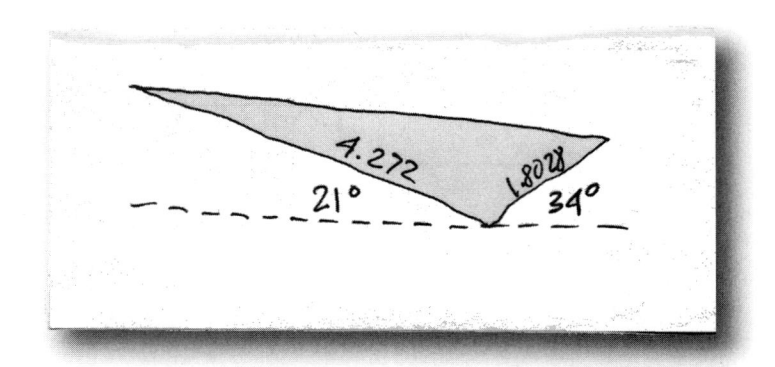

1. Start the LINE command:
   Command: **line**
   Specify first point: *(Pick point 1).*

2. Use relative coordinates to draw the short leg (at right):
   Specify next point or [Undo]: **@1.8028<34**

3. End the LINE command, and start over to draw the next left (at left):
   Specify next point or [Undo]: *(Press ENTER.)*
   Command: *(Press ENTER.)*
   LINE Specify first point: **endp**
   of *(Again, pick point 1).*

4. Use relative coordinates to draw the other leg. The angle of 159 degrees = 180 - 21=159.
   Specify next point or [Undo]: **@4.272<159**

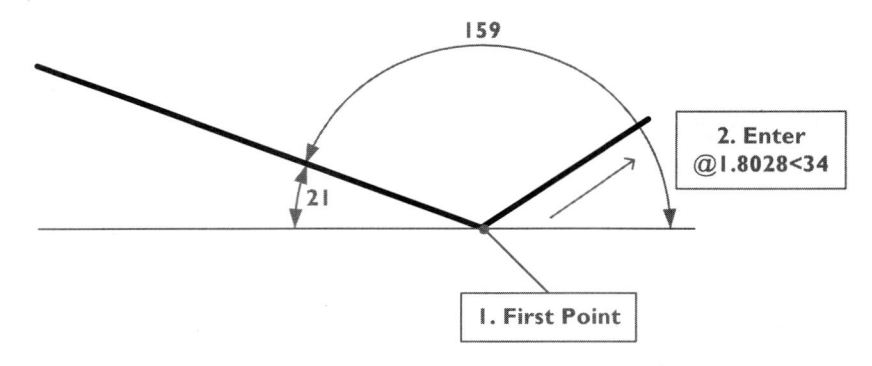

5. Use ENDpoint object snap to draw the third leg:
   Specify next point or [Undo]: **end**
   of *(Pick point 5.)*
   Specify next point or [Close/Undo]: *(Press ENTER.)*

Drawing with Basic Objects **157**

5. Use ENDpoint Object Snap to capture this point.

4. Enter @4.272<159

1. First Point

RIGHT TRIANGLES

One corner of right triangles is at 90 degrees; the two remaining corners have other angles, which can be the same (45 degrees) making them isosceles triangles, or different, making them scalene triangles. This kind of triangle is easily drawn with the **LINE** command. When you know the length of two sides, such as 3 and 2 units, follow these steps:

1. Turning on ortho mode helps you draw the 90-degree angle easily:

    (Press **F8** *to turn on ortho mode.)*

2. Start the **LINE** command:

    Command: **line**

    Specify first point: *(Pick point 1.)*

3. Notice that it is easier to draw this kind of triangle when you start at a corner away from the right-angle, and then draw toward the right-angle.

    Specify next point or [Undo]: *(Move cursor to the left, and then type:)* **3**

    Specify next point or [Undo]: *(Move cursor up, and then type:)* **2**

4. By using the **Close** option, you don't have to work out the length of angle of the third side.

    Specify next point or [Close/Undo]: **c**

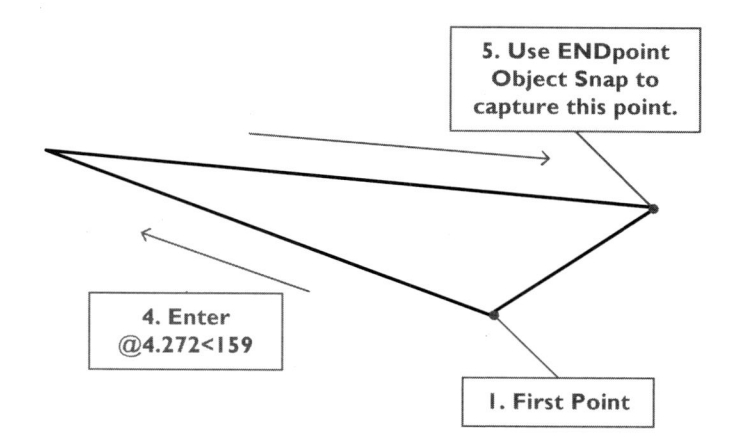

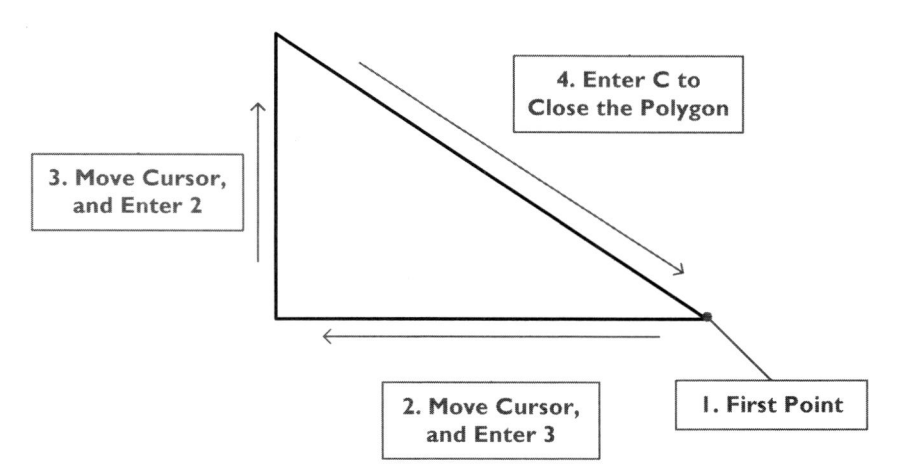

4. Enter C to Close the Polygon

3. Move Cursor, and Enter 2

2. Move Cursor, and Enter 3

1. First Point

The right triangle is complete.

EXERCISES

1. With the **LINE** command, draw a house shape:

    Command: **line**

    Specify first point: *(Enter point 1.)*

    Specify next point: *(Enter point 2.)*

    Specify next point: *(Enter point 3.)*

    Specify next point: *(Enter point 4.)*

    Specify next point: *(Enter point 5.)*

    Specify next point: *(Enter point 1.)*

    Notice that the line stretched behind the crosshairs. This is called "rubber banding."

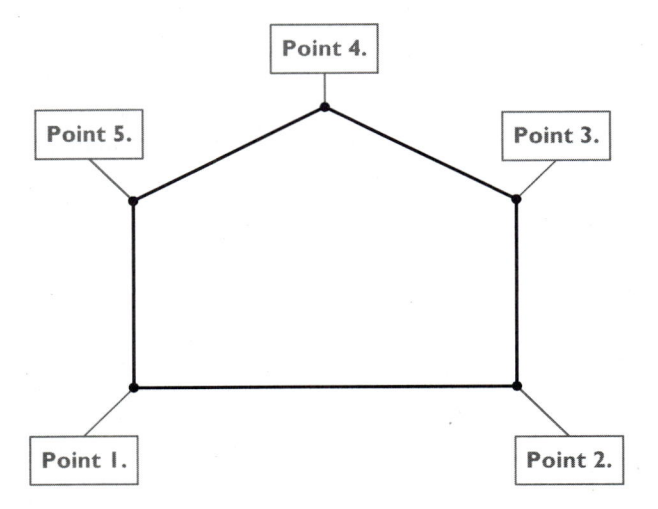

2. Using the figure below, fill in the missing absolute coordinates. Each side of the floorplan is dimensioned. Place the answers in the boxes provided. Assume the origin is at the lower left corner.

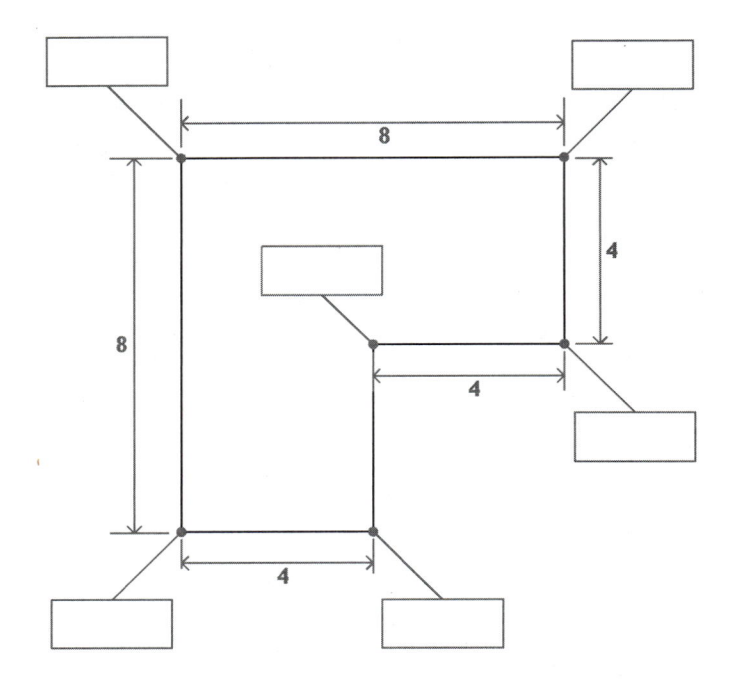

3. Using the LINE command, connect the following points designated by absolute coordinates.

   Point 1: **1,1**
   Point 2: **5,1**
   Point 3: **5,5**
   Point 4: **1,5**
   Point 5: **1,1**

   What shape did you draw?

4. Determine the length of each side of the floor plan in the figure below. (Calculate the lengths from the absolute coordinates given.)

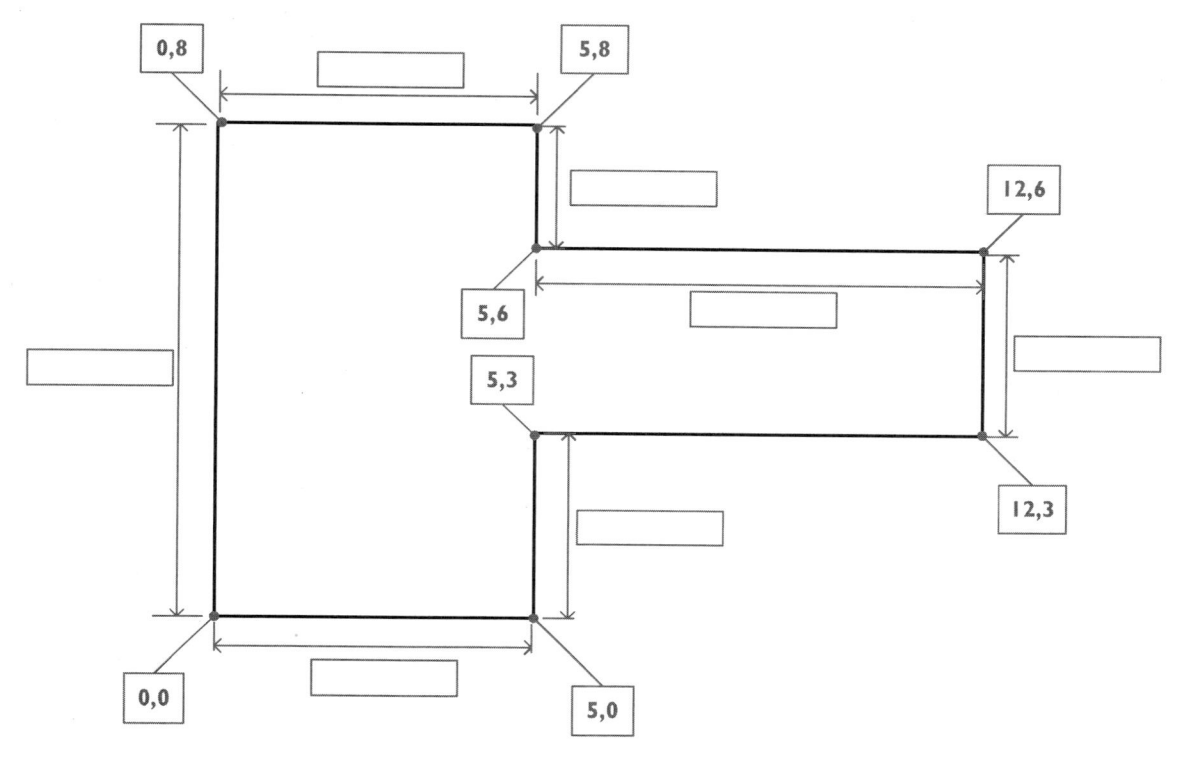

5. Use the following absolute and relative coordinates to draw an object.

   Point 1: **0,0**
   Point 2: **@3,0**
   Point 3: **@0,1**
   Point 4: **@–2,0**
   Point 5: **@0,2**
   Point 6: **@–1,0**
   Point 7: **0,0**

6. List the relative coordinates used to draw the following baseplate.

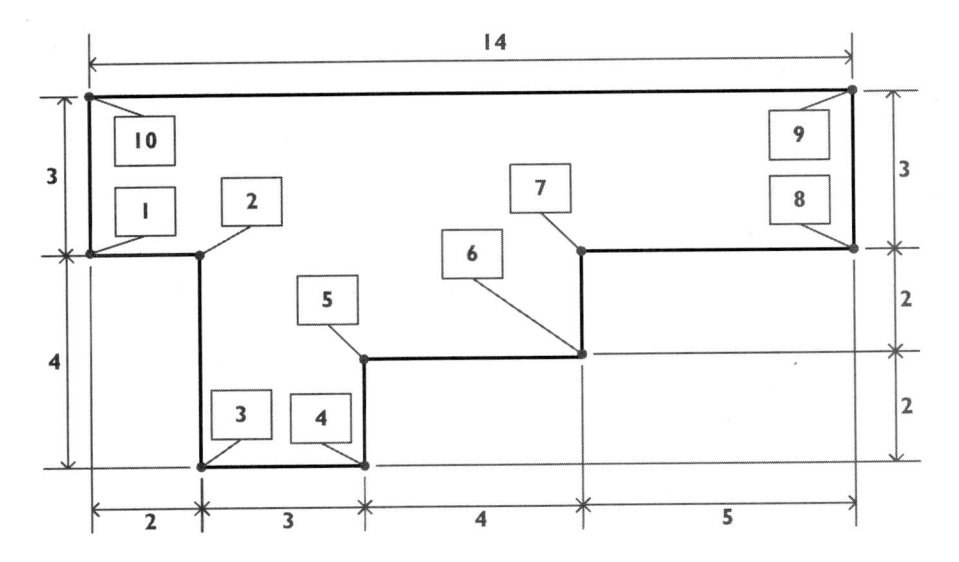

7. List the polar coordinates used to draw the baseplate above.

8. Write a list of the absolute coordinates used to construct the shim shown in the figure below. The origin is at the lower left corner.

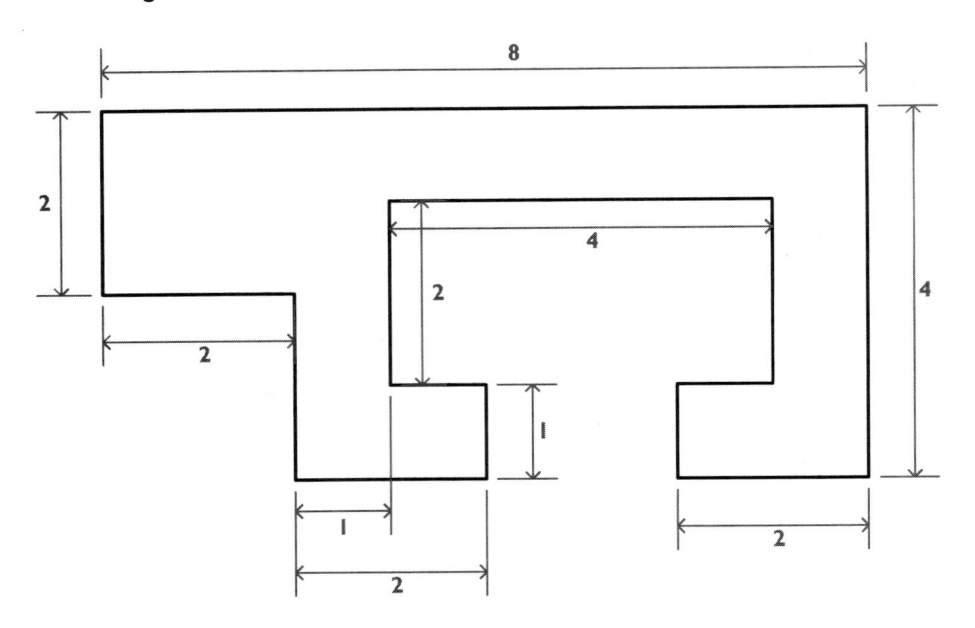

9. Use the following relative distances and angles to draw an object.
    Point 1: **0,0**
    Point 2: **@4<0**
    Point 3: **@4<120**
    Point 4: **@4<240**

10. Write a list of the relative coordinates used to construct the support plate shown in the figure below. The origin is at the lower left corner.

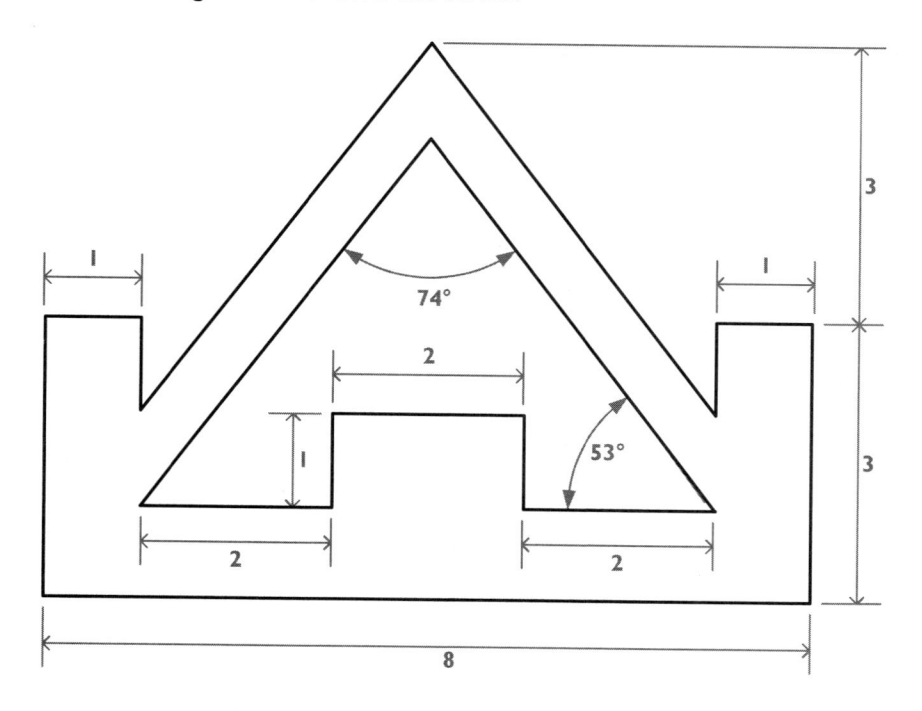

11. Write a list of the polar coordinates used to construct the corner shelf shown in the figure below.

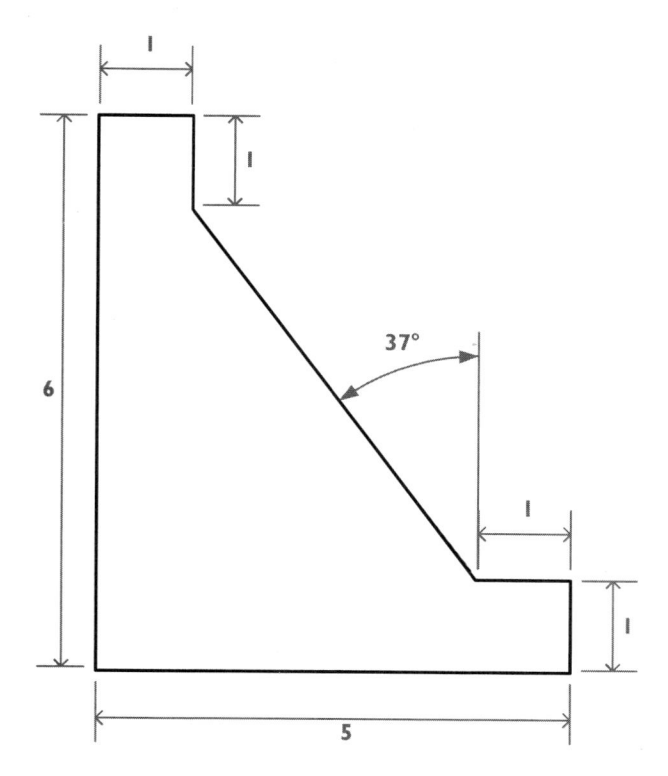

12. Set up a point type of 66, and then place it in the drawing:
    Command: **pdmode**
    New value for PDMODE <default>: **66**

    Recall that **PDMODE** affects all points in the drawing. Points placed previously are updated with the next command that causes a regeneration, such as **REGEN**. To place the point in the drawing, enter:
    Command: **point**
    Current point modes: PDMODE=66 PDSIZE=0.0000
    Specify a point: *(Pick a point.)*

13. Construct a circle with a *radius* of 5. From the menu bar, choose **Draw | Circle | Center, Radius**.
    Command: _circle
    Specify center point for circle or [3P/2P/Ttr (tan tan radius)]: *(Pick a point.)*
    Specify radius of circle or [Diameter]: **5**

14. Construct a circle using a center point and a *diameter* of 3.
    Command: **circle**
    Specify center point for circle or [3P/2P/Ttr (tan tan radius)]: *(Enter point 1.)*
    Specify radius of circle or [Diameter]: **d**
    Specify diameter of circle <5.0000>: **3**

15. With the **RECTANG** command, draw two rectangles. One rectangle represents a B-size drawing sheet (17" x 11"); the second represents the title block in the lower right corner (4" x 2").

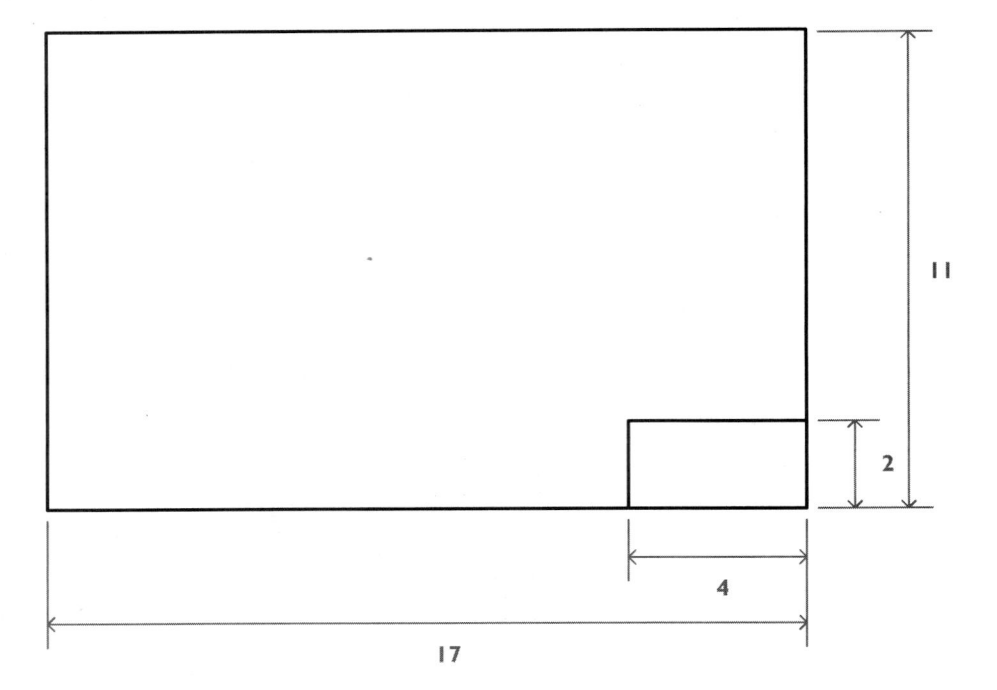

16. With the **RECTANGLE** command, draw the outline of a standard 24" x 36" speed limit sign; do not draw the text.

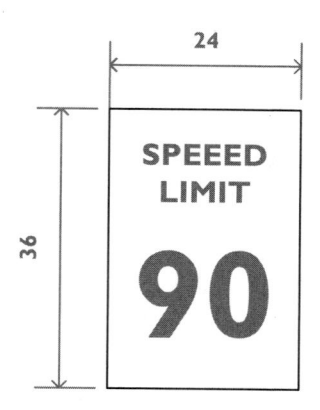

SPEEED LIMIT 90

24
36

17. Use the **POLYGON** command to draw the outline of a standard 30" warning sign; do not draw the text:

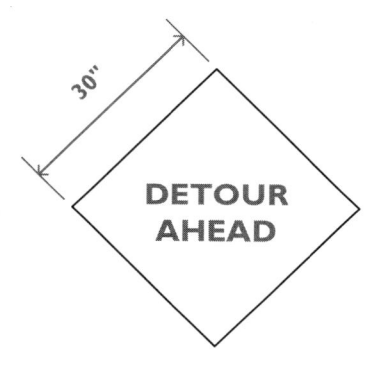

30"

DETOUR AHEAD

18. Use the **POLYGON** command to draw the outline of a standard 30" Stop sign; do not draw the text:

STOP

30

Drawing with Basic Objects 163

19. With the **POLYGON** command, draw the outline of a standard 36" Yield sign; do not draw the text:

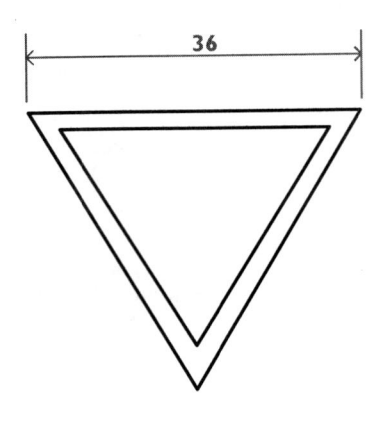

20. Draw donuts with the following diameters:
    a. ID = 1.0; OD = 1.5
    b. ID = OD = 2.5
    c. ID = 0.0; OD = 1.0

21. Use drawing commands learned in this chapter to draw the following base plates:
    a. Rectangular base plate with two holes:

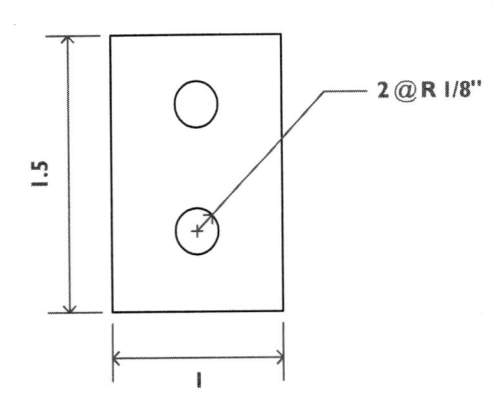

    b. Rectangular base plate with four holes:

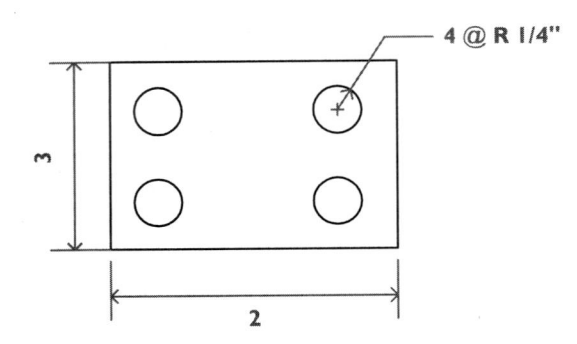

22. Use drawing commands learned in this chapter to draw the following base plates:
    a. Circular base plate with two square holes:

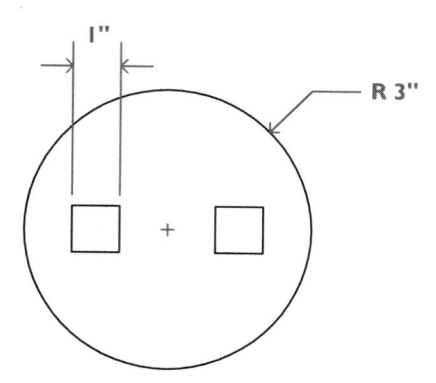

    b. Elliptical base plate with two square holes. The minor radius is 3", and the major radius is 4.5":

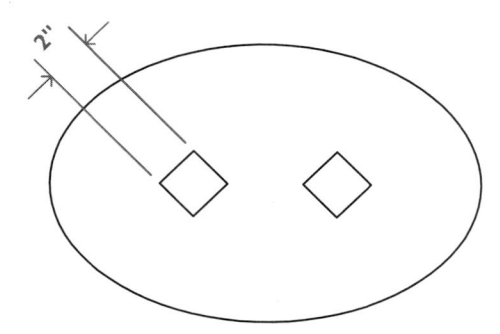

23. Use the **ELLIPSE** command to draw the face of an alien. Use the figure below to help you. Don't worry about the size.

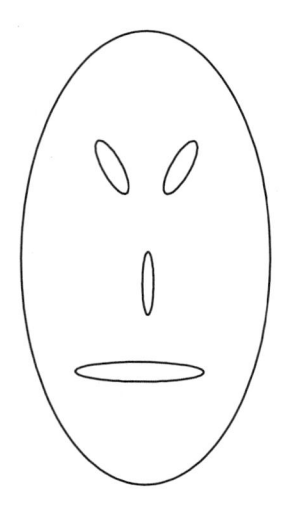

24. Draw a compact disc, which has an outer diameter of 4.75" and an inner diameter of 0.6". What are the equivalent measurements as radius?

25. Draw arcs with the following specifications:
    a. Start point   **7,5**
       Second point **9.5,3.5**
       End point   **6,2**

    b. Center point **4,4**
       Start point   **4,2**
       Angle   **270**

    c. Start point   **10.5,5.5**
       Center point **9,6**
       Length   **3**

26. Draw circles with the following specifications:
    a. Center point   **5,5**
       Radius   **3.5**

    b. Center point   **5,5**
       Diameter   **3.5**

    c. 2P: first point   **10,5**
       Second point   **3,2**

    d. 3P: first point   **4,7**
       Second point   **6,5**
       Third point   **5,2**

27. Draw the 4.5" × 2.75" electrical cover plate shown below. The screw holes are 1/8" in diameter. The light switch cover plate has an opening of 0.95" × 0.4".

Drawing with Basic Objects 167

28. With the **Line** and **Arc** options of the **PLINE** command, draw the outline of a standard credit card — 3.35" wide by 2.15" high, with a 0.1"-radius arc at each corner.

29. Draw the inserts for CD cases:
    a. The cover insert is 12cm x 12cm.
    b. The back cover insert is 15cm wide x 11.7 cm tall; the two spine strips are 6.4 mm wide.

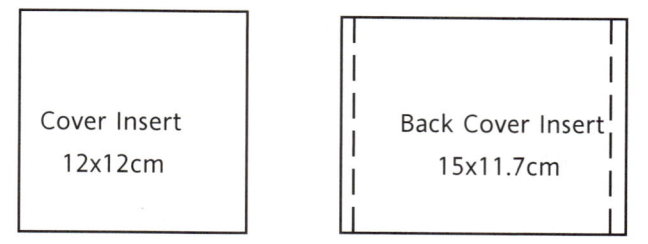

Cover Insert
12x12cm

Back Cover Insert
15x11.7cm

30. Draw the bicycle shown in the figure below.
    Use polylines for the frame, lines for the spokes, and donuts for the tires.

CHAPTER REVIEW

1. What is another name for a single line?
2. Is the **Close** option of the **Line** command more accurate at closing the polygon than attempting to lineup endpoints manually?
3. What does the @ symbol mean?
   The < symbol?
4. How do you continue drawing a line tangent to a previously drawn arc?
5. Name the best command for drawing the following triangles:
   Equilateral triangle.
   Scalene triangle.
6. What is AutoCAD's default method of constructing an arc?
7. What command can be used to change the look and size of points?
8. When a rectangle is drawn with the **RECTANG** command, can each corner have a different fillet radius?
9. The **POLYGON** command uses three methods to determine the size of polygons. Explain the meaning of the methods:
   Edge.
   Inscribed.
   Circumscribed.
10. **PDMODE** and **PDSIZE** are *retroactive*. What does this mean?
11. There are six methods of constructing circles. What four circle properties are used in various combinations to comprise these methods?
    a.
    b.
    c.
    d.
12. Describe the meaning of these options for drawing circles:
    2P
    3P
    Ttr
13. Describe a method of drawing arcs that does not involve the **ARC** command.
14. List three methods for ending the **LINE** command?
    a.
    b.
    c.
15. In which direction does AutoCAD draw arcs (by default).
16. What is the difference between a *minor* arc and a *major* arc?
17. What is the *chord* of an arc?
18. How do you continue drawing an arc tangent to a previously drawn arc or line?
19. What does the **ARC** command's **Direction** option specify?
20. What must you do to end the **DONUT** command?
21. What is the alias for the following commands:
    ARC
    LINE
    CIRCLE
    PLINE

22. Where on the menu bar do you find commands for drawing objects?
23. Name the three kinds of circles drawn by the **DONUT** command:
    a.
    b.
    c.
24. What do the **POLYLINE**, **DONUT**, and **POLYGON** commands have in common?
25. Can a polyline have varying width?
26. What is the purpose of the **PLINE** command's **Arc** option?
27. When selecting a wide polyline, where must you pick it?
28. Describe the purpose the **PLINE** command's **Length** option.
29. What is the difference between the *minor* axis and the *major* axis of an ellipse?
30. What is another name for an ellipse with a rotation of 0 degrees?
31. Which must you specify first, the minor axis or the major axis of an ellipse? *i*
32. Describe the purpose of the **MULTIPLE** command.

CHAPTER 5

Drawing with Precision

The advantage of CAD over hand drafting is precision. Instead of drawing with an accuracy to the nearest pencil (or pen) width, you draw with an accuracy to 14 decimal places in AutoCAD. In this chapter, you learn to use some of AutoCAD's commands for drawing with precision and seeing the drawing better. They are:

GRID displays a grid of dots.

SNAP specifies the cursor increment and additional options.

ORTHO constrains cursor movement to the horizontal and vertical.

POLAR helps you draw at specific angles.

OSNAP helps you accurately select geometric features.

OTRACK implements object snap tracking.

RAY and **XLINE** place construction lines in the drawing.

REDRAW and **REGEN** clean up and update the drawing.

ZOOM enlarges and reduces the view of the drawing.

PAN moves the view around.

CLEANSCREENON minimizes the user interface to maximize the drawing.

DSVIEWER provides a "bird's-eye view" of the drawing in an independent window.

VIEW stores and recalls views of the drawing by name.

VIEWRES controls the roundness of circular objects.

FINDING THE COMMANDS

On the status bar:

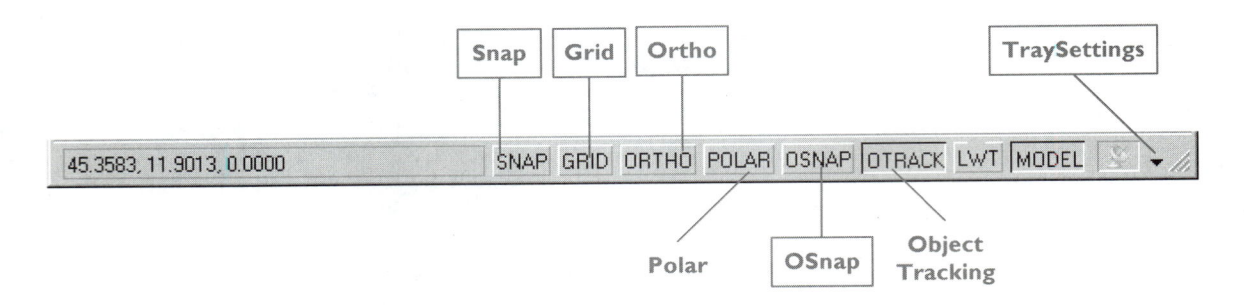

On the **VIEW** menu:

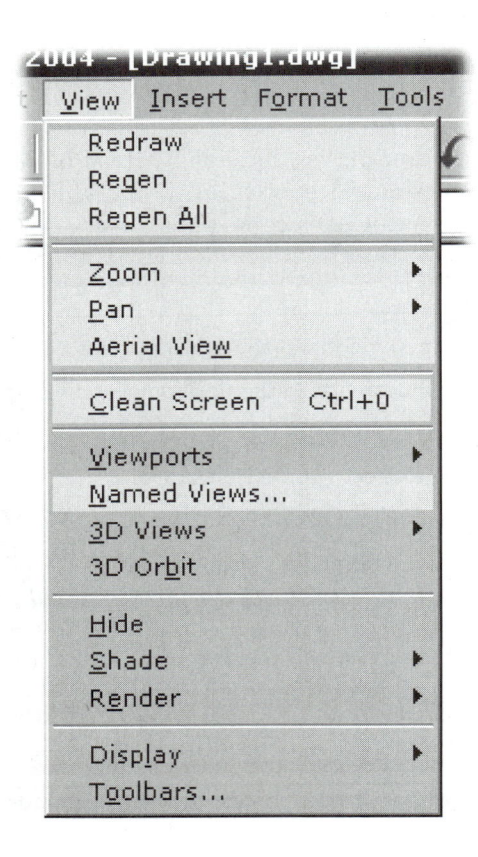

GRID

The **GRID** command *toggles* (turns on and off) the grid display of dots. Think of the grid as a sheet of graph paper. Instead of lines, however, AutoCAD displays dots — the intersections of the lines.

The purpose of the grid is to help you have an idea of the size the drawing, and of the distances within the drawing. Because it is purely a visual aid, the grid is not plotted.

You can change the spacing between dots, the angle of the grid, the style of grid (standard or isometric), and the extent of the grid.

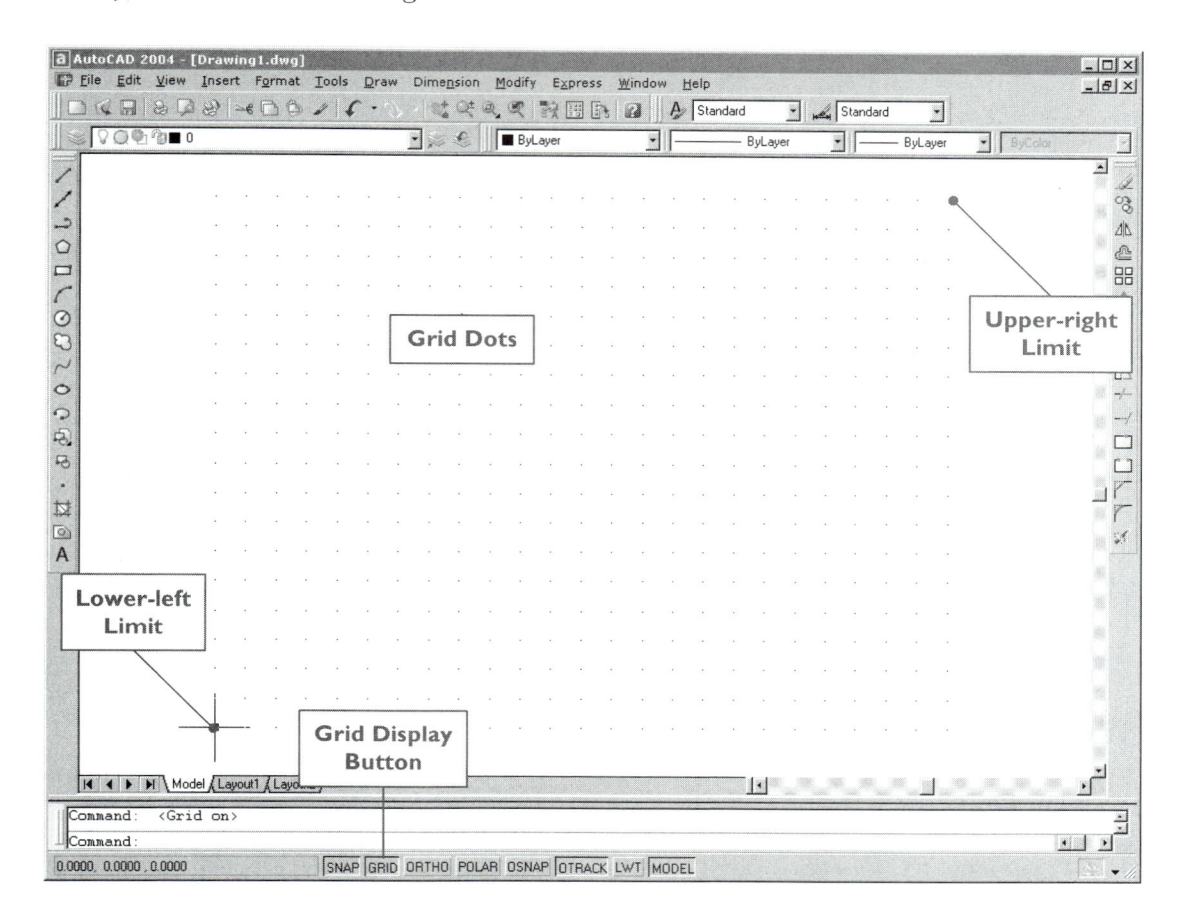

TUTORIAL: TOGGLING THE GRID

1. To turn on the grid, use the **GRID** command with one of these methods:
   - From the status bar, choose **GRID** (turned on when buttons looks depressed).
   - On the keyboard, press function key **F7** or **CTRL+G**.
   - At the 'Command:' prompt, enter the **grid** command.

     Command: **grid** (Press ENTER.)

2. Use the **ON** option to turn on the grid:
   Specify grid spacing(X) or [ON/OFF/Snap/Aspect] <0.5000>: **on**

   Notice that the grid is displayed.

3. To turn off the grid, repeat the command with the **OFF** option:
   Command: **grid** (Press ENTER.)

   Specify grid spacing(X) or [ON/OFF/Snap/Aspect] <0.5000>: **off**

CONTROLLING THE GRID DISPLAY: ADDITIONAL METHODS

When you start a new drawing in AutoCAD, the default grid spacing is set to 0.5 units. You can increase and descrease the spacing, make the spacing the same as the snap distance (described later in this chapter), and make the spacing different in the x- and y-direction. To effect these changes, you can use a dialog box, or enter the changes at the command-line using these commands:

- **DSETTINGS** displays a dialog box for changing grid settings.
- **GRID** specifies the grid spacing and aspect ratio via the command-line.
- **LIMITS** controls the extent of the grid.

Let's look at how each command affects the grid display.

DSettings

The DSETTINGS command (short for "drafting settings") displays a dialog box that controls many (but not all) grid display settings. The dialog box combines the options found in the GRID and SNAP commands; the settings missing from the dialog box are found in the LIMITS command.

1. To access the **Drafting Settings** dialog box:
   - From the menu bar, choose **Tools**, and then **Drafting Settings**.
   - At the 'Command:' prompt, enter the **dsettings** command.
   - Alternatively, enter the aliases **ds**, **se** (short for "settings") **ds**, or **ddrmodes** (the old name for this command) at the 'Command:' prompt.
   - The easiest way, I find, is to right-click **GRID** on the status line. From the short-cut menu, select **Settings**.

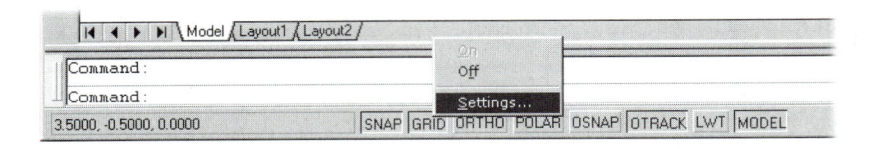

   In all cases, notice that AutoCAD displays the **Drafting Settings** dialog box, showing the **Snap and Grid** tab.

3. In this dialog box, you can make the following changes to the grid settings:
   - **Grid On (F7)** toggles the grid display. The (F7) notation reminds you that you can also press function key **F7** to turn on and off the grid.
   - **Grid X spacing** specifies the horizontal grid spacing. When you enter a value for the x-spacing, AutoCAD automatically changes the y-spacing to match.
   - **Grid Y spacing** specifies the vertical grid spacing.
   - **Grid Snap** selects between rectangular and isometric grids. The isometric grid is useful when drawing objects in isometric projection, and is covered in volume 2 of this book, *Using AutoCAD 2004: Advanced*.
   - **Angle** changes the angle of the grid, away from the horizontal. This setting also affects the angle of the snap, orthographic mode, and hatch patterns. You change the angle, for example, when you work on the wing of a building that extends at 45 degrees from the main hall.

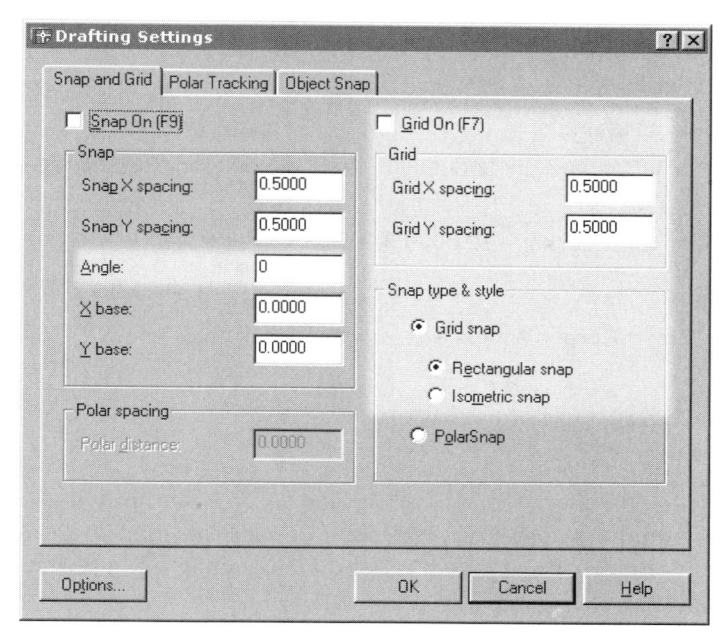

4. To accept the changes you make in the dialog box, click **OK**.

Grid

The GRID command contains options for changing the spacing between grid dots.

1. To set the spacing between grid dots, use the GRID command's default option. For example, to change the spacing to 10 units, enter 10, as follows:

    Command: **grid** *(Press ENTER.)*

    Specify grid spacing(X) or [ON/OFF/Snap/Aspect] <0.5000>: **1**

2. To make the grid spacing the same as the snap distance, use the **Snap** option:

    Command: **grid** *(Press ENTER.)*

    Specify grid spacing(X) or [ON/OFF/Snap/Aspect] <0.5000>: **s**

3. To set the grid spacing as a multiple of the snap distance, add an "x" to the GRID command's default option. For example, to make the grid spacing one-half of the snap spacing, enter 0.5x, as follows:

    Command: **grid** *(Press ENTER.)*

    Specify grid spacing(X) or [ON/OFF/Snap/Aspect] <0.5000>: **0.5x**

4. Some disciplines, such as highway design, use a different scale in the x-direction from the y-direction. To make the vertical (y) grid spacing ten times that of the horizontal (x) direction, use the **Aspect** option:

    Command: **grid** *(Press ENTER.)*

    Specify grid spacing(X) or [ON/OFF/Snap/Aspect] <0.5000>: **a**

    Specify the horizontal spacing(X) <0.0000>: **1**

    Specify the vertical spacing(Y) <0.0000>: **1**

Limits

The grid does not extend infinitely in all directions. The LIMITS command controls the extent of the grid. Typically, you want the grid to cover the area in which you are drafting. Also, typically, you keep the lower-left corner at 0,0 and change only the upper-right corner.

    Command: **limits**

    Reset Model space limits:

    Specify lower left corner or [ON/OFF] <0.0000,0.0000>: *(Press* ENTER, *or enter x,y coordinates.)*

    Specify upper right corner <12.0000,9.0000>: *(Enter x,y coordinates.)*

Note: The grid is set independently in each viewport and drawing.

SNAP

The SNAP command specifies the cursor increment and additional options. For example, if you are drafting a drawing to the nearest one-inch, it is useful to set the snap to 1". When turned on, snap restricts the cursor to the interval you specify; the cursor seems to jump from point to point. Snap is often used in conjunction with the grid.

As with the grid, you can change the snap spacing, have different spacing horizontally and vertically, rotate the snap angle, and select from a variety of snap styles — standard, isometric, and polar. Snap can be set individually for each viewport.

Unlike the grid, snap is not constrained by the limits: it extends throughout the entire drawing plane (the LIMITS command does not apply to snap).

TUTORIAL: TOGGLING SNAP MODE

1. To turn on snap, use the **SNAP** command by one of these methods:
   - From the status bar, click **SNAP** (turned on when buttons looks depressed).
   - On the keyboard, press function key **F9** or **CTRL+B**.
   - At the 'Command:' prompt, enter the **snap** command.
   - Alternatively, enter the **sn** alias at the 'Command:' prompt.

   Command: **snap** *(Press* ENTER.*)*

2. Use the **ON** option to turn on snap:

   Specify snap spacing or [ON/OFF/Aspect/Rotate/Style/Type] <0.5000>: **on**

   When you move the mouse, notice that the cursor jumps.

3. To turn off snap, repeat the command with the **OFF** option:

   Command: **grid** *(Press* ENTER.*)*

   Specify snap spacing or [ON/OFF/Aspect/Rotate/Style/Type] <0.5000>: **off**

   Notice that the cursor no longer jumps.

CONTROLLING THE SNAP: ADDITIONAL METHODS

When you start a new drawing in AutoCAD, the default snap spacing is set to 0.5 units — the same as the grid. You can increase and decrease the spacing, make the spacing the same as the grid distance, and specify different spacing in the x- and y-directions. To make these changes, you can use a dialog box, or enter the changes at the command-line, using these commands:

- **DSETTINGS** displays a dialog box for changing snap settings.
- **SNAP** specifies the snap spacing, aspect ratio, and modes via the command-line.

Let's look at how each command affects the snap.

DSettings

The DSETTINGS command displays a dialog box that controls all settings affecting the snap (the dialog box combines the SNAP and GRID commands).

1. To access the **Drafting Settings** dialog box:
   - From the menu bar, choose **Tools**, and then **Drafting Settings**.
   - At the 'Command:' prompt, enter the **dsettings** command.
   - Alternatively, enter the aliases **ds**, **se**, or **ddrmodes** at the 'Command:' prompt.
   - Or, right-click **SNAP** on the status line. From the short-cut menu, select **Settings**.

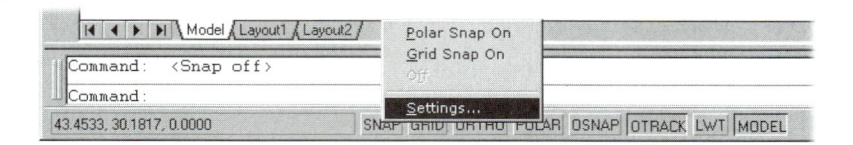

Note: The shortcut menu (shown above) has several settings that you can access without going through a dialog box:

   Polar Snap On turns on polar snap mode, where the cursor aligns with a distance and an angle.

   Grid Snap On makes the snap distance equal to the grid distance.

   Off turns off snap mode.

In all cases, notice that AutoCAD displays the **Drafting Settings** dialog box, showing the **Snap and Grid** tab.

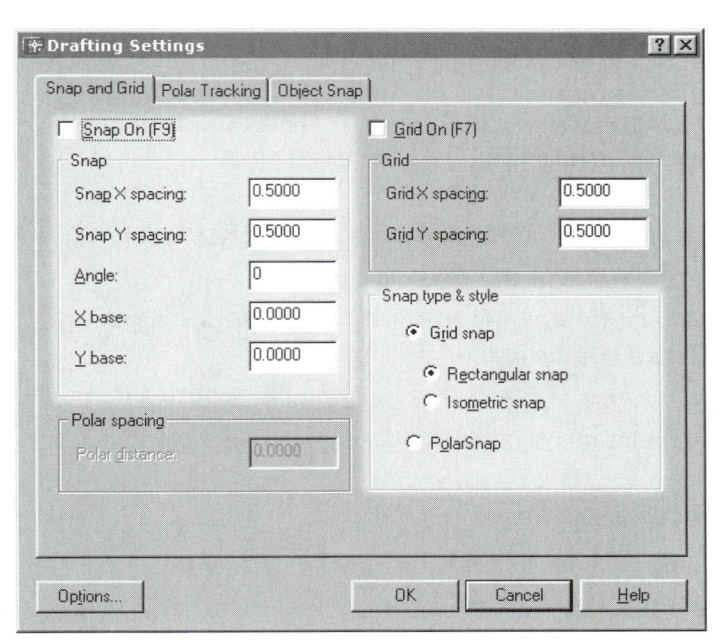

3. In this dialog box, you can make the following changes to the snap settings:
   - **Snap On (F9)** toggles the snap on and off. The (**F9**) is a reminder that you can press **F9** to turn on and off snap mode.
   - **Snap X spacing** specifies the horizontal snap spacing. When you enter a value for the x-spacing, AutoCAD automatically changes the y-spacing to match after you press **TAB**.

- **Snap Y spacing** specifies the vertical snap spacing.
- **Angle** changes the angle of snap, away from the horizontal. This setting also affects the angle of the gird, orthographic mode, and hatch patterns.
- **X Base** specifies the origin of the snap, more specifically the x-coordinate of the origin. Normally, the snap starts at 0,0, but you can change this to make the snap match any point in your drawing. This setting also affects the origin of hatch patterns.
- **Y Base** specifies the y-coordinate of the snap origin.
- **Regular Snap** turns on rectangular snap, the "normal" snap as opposed to isometric snap.
- **Isometric Snap** turns on isometric snap, which is useful for drawing isometric objects (see volume 2 of this book, *Using AutoCAD 2004: Advanced*).

4. To accept the changes you make in the dialog box, click **OK**.

Snap

The SNAP command has these options for changing the snap spacing:

1. To set the snap spacing, use the SNAP command's default option. For example, to change the spacing to 12 units (or inches), enter 12, as follows:

    Command: **snap** *(Press ENTER.)*

    Specify snap spacing or [ON/OFF/Aspect/Rotate/Style/Type] <0.5000>: **12**

2. To make the vertical (y) snap spacing twelve times that of the horizontal (x) direction, use the **Aspect** option:

    Command: **snap** *(Press ENTER.)*

    Specify snap spacing or [ON/OFF/Aspect/Rotate/Style/Type] <0.5000>: **a**

    Specify the horizontal spacing <0.5000>: **1**

    Specify the vertical spacing <0.5000>: **12**

3. To select the type of snap mode, use the **Style** and **Type** options. **Style** selects between regular and isometric modes, while **Type** selects between rectangular and polar modes. Here is how to enter polar mode:

    Command: **snap** *(Press ENTER.)*

    Specify snap spacing or [ON/OFF/Aspect/Rotate/Style/Type] <0.5000>: **t**

    Enter snap type [Polar/Grid] <Grid>: **p**

4. To rotate the snap by 45 degree, for example, use the **Rotate** option:

    Command: **snap** *(Press ENTER.)*

    Specify snap spacing or [ON/OFF/Aspect/Rotate/Style/Type] <0.5000>: **r**

    Specify base point <0.0000,0.0000>: *(Press ENTER, or enter x,y-coordinates for a new snap origin.)*

    Specify rotation angle <0>: **45**

Notice that the cursor rotates to the same angle, as does the grid, if it is turned on.

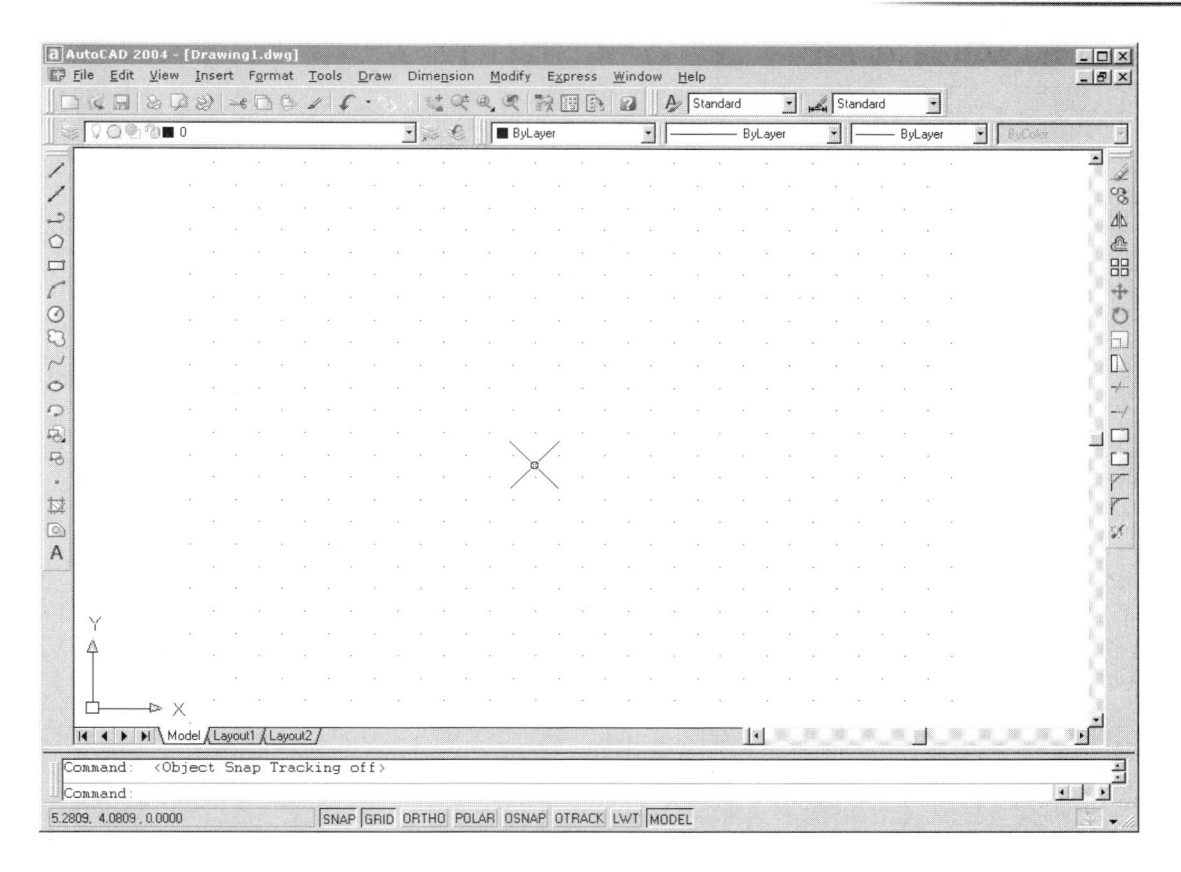

TUTORIAL: DRAWING WITH GRID AND SNAP MODES

Although the grid display can be independent of the snap spacing, the two are often used together. It is common to set the snap spacing to the drawing resolution (such as 1/8" or 1 cm) but to set the grid spacing to a larger value (such as 1' or 1 m) so that the dots don't clutter the screen.

In the first part of the tutorial, you turn on and off the grid display, and change its spacing.

1. On AutoCAD's status line, right-click **GRID**, and then choose **Settings** from the shortcut menu. Notice that the **Snap and Grid** tab of the **Drafing Settings** dialog box appears.
2. Select **Grid On** to turn on the grid display so that a check mark appears in the box. (The grid does not appear until after the dialog box closes.)
3. Set the **Grid X** and **Y Spacing** to 1.0.
4. Click **OK**. Notice the grid on the screen.
5. Press **F7**. This key toggles the grid off and on. Press **F7** again.
6. Return to the Snap and Grid dialog box. Set the **X Grid Spacing** to 1, and then the **Y Grid Spacing** to 0.5. Notice that the values for the the vertical and horizontal grid spacings are different. It is not necessary for the vertical and horizontal values to be the same.
7. Click **OK**, and notice the grid spacing has changed.

In the following steps, you draw a few lines with snap mode turned on.

8. Right-click the word **SNAP** on the status line, and then choose **Setting**.
9. In the **Snap and Grid** tab, choose **Snap On** to turn on snap mode.
   Change the **Snap X** and **Y Spacing** to 0.25.
   Click **OK**. You won't notice any difference in the display until you start drawing.

10. Start the **LINE** command, and then pick endpoints on the screen. Notice the crosshair cursor "snapping" to specific point on the screen.

Here, you make the grid spacing equal to the snap spacing.

11. Right-click **GRID** on the status line, and then choose **Settings**.
    Set the grid spacing to 0.25. The grid display has the same spacing as the snap.
    Click **OK** to exit the dialog box.
12. Use the **LINE** command to draw some lines. Notice that the crosshair cursor now lines up with the grid points.

Now you practice drawing rectangles with snap turned off and with it turned on.

13. Use **F9** to turn the snap mode off and on.
14. Draw two rectangles with the **LINE** command: one with snap mode on, and one with snap mode off. Notice how the endpoints are easier to line up with snap mode on.

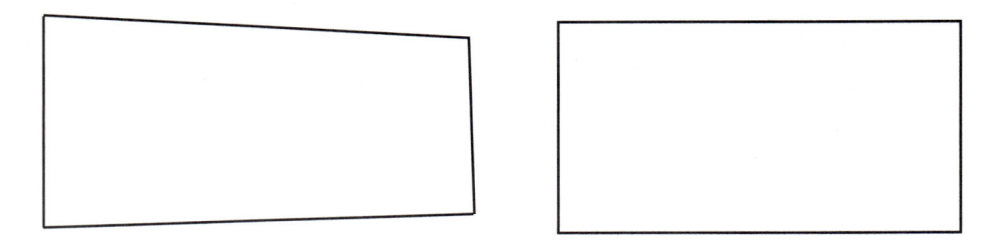

Finally, rotate the snap angle to see how it is easier for drawing at an angle.

15. Return to the **Snap and Grid** dialog box, and then change the snap angle to **45**.
    Click **OK** to exit the dialog box.
    Notice how the snap resolution and aspect ratio are maintained. The entire snap, grid, and crosshair cursor are rotated.
16. Draw a rectangle with the rotated snap grid and crosshairs. This is an excellent method of drawing objects that have many lines at the same angle.

ORTHO

When turned on, ortho mode (short for "orthographic") constrains cursor movement to the horizontal and vertical. This mode is very useful, because much drafting takes place at right angles — 0, 90, 180, and 270 degrees — in a manner similar to the T-squares and right-triangles used in manual drafting.

Ortho mode can be changed from the horizontal and vertical. Use the **SNAP** command's **Rotate** option to change the ortho angle, as described earlier.

TUTORIAL: TOGGLING ORTHO MODE

1. To turn on ortho mode, use the **ORTHO** command by one of these methods:
   - From the status bar, click **ORTHO** (turned on when button looks depressed).
   - On the keyboard, press function key **F8** or **CTRL+L**.
   - At the 'Command:' prompt, enter the **ortho** command.

     Command: **ortho** (Press ENTER.)

2. Use the **ON** option to turn on otho:

   Enter mode [ON/OFF] <ON>: **on**

   You don't notice any difference in the cursor movement until you use drawing and editing commands. Then the cusor moves only vertically or horizontally.

3. To turn off ortho mode, repeat the command with the **OFF** option:

   Command: **ortho** *(Press* ENTER.*)*

   Enter mode [ON/OFF] <ON>: **off**

Unlike the GRID and SNAP commands, the ORTHO command has no additional options. Ortho mode is, however, affected by these options of the SNAP command: **Rotate**, **Base**, and isometric **Style**.

Notes: While you can rotate the ortho angle, drawing still takes place at 90 degree increments. To draw at other angles, use polar tracking and PolarSnap. Ortho mode and polar tracking cannot, however, be both turned on at the same time. Turning on one turns off the other.

The ortho angle changes to 120 degrees when isometric mode is turned on. Ortho mode is ignored in 3D perspective views.

TUTORIAL: DRAWING IN ORTHO MODE

1. Click **ORTHO** on the status line to turn on ortho mode. You won't notice any difference in the display until you start drawing.
2. Press F8 to turn off ortho mode.
3. Draw a rectangle with the LINE command. Notice that is is hard to draw lines that are perfectly horizontal and vertical.
4. Press F8 to turn on ortho mode again.
5. Draw another rectangle with the LINE command. After picking the first endpoint, notice how the cursor moves precisely horizontal or vertically until you pick the next point.

POLAR

While ortho mode helps you draw at right angles, polar mode is more flexible, allowing you to draw at increments of specific angles, such as 15 degrees. If ortho mode is like manual drafting with the T-square with the right-triangle, then polar mode is like drafting the T-square with the 30-60-90 and 45-45-90 triangles.

Because ortho mode limits cursor movement to 90 degrees, but polar mode can be at any angle, AutoCAD turns off ortho mode when you turn on polar mode.

TUTORIAL: TOGGLING POLAR MODE

Curiously, there is no "POLAR" command, even though there are shortcut keystrokes to turn it on and off (as you later find out, polar mode is controlled by system variables and dialog boxes).

1. To turn on polar mode, use one of these methods:
   - From the status bar, click **POLAR** (turned on when button looks depressed).
   - On the keyboard, press function key **F10** or **CTRL+U**.
2. To initiate polar mode, start the line command, and then pick a point.

3. At the "Specify next point" prompt, move the cursor around. Notice that every so often, a *tooltip* appears, together with an *x-marker* and an *alignment path*.

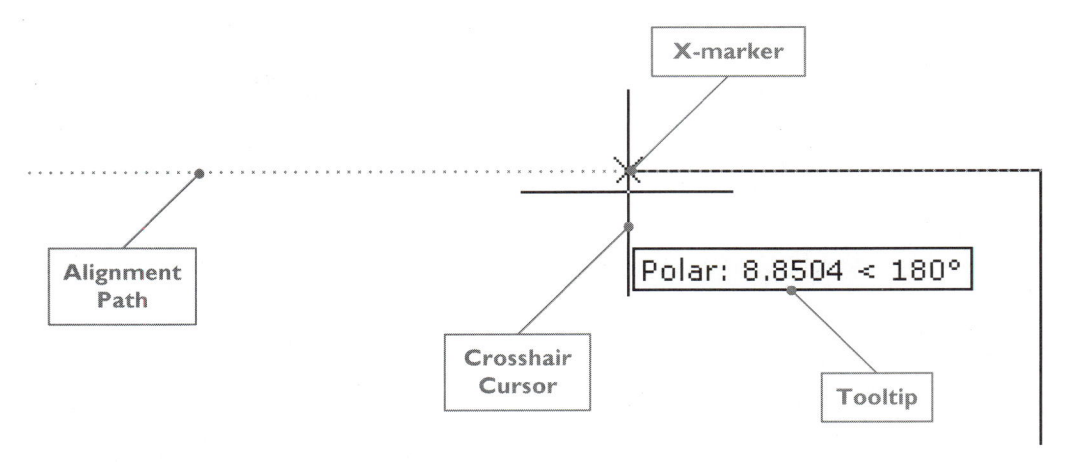

[Figure: Crosshair with X-marker, Alignment Path, Crosshair Cursor, and Tooltip showing "Polar: 8.8504 < 180°"]

Toolips are small, yellow, rectangles with explanitory text. In this case, the tooltip display the polar distance, such as

Polar: 4.7355 < 90°

This means that the line is 4.7355 units long, at an angle of 90 degrees.

X-marker emphasizes the end of your line, and the start of the alignment path. It also marks the point where the line would end if you were to click the mouse button.

Alignment paths are thin dotted lines that show where the line would be drawn, if you were to continue in that direction — sort of like a preview.

CONTROLLING POLAR MODE: ADDITIONAL METHODS

In new drawings, the default polar angle is set to 90 degrees — the same as ortho mode. You can add angles, and specify whether the angle measurement is absolute or relative to the last one. The changes are made through a dialog box or at the command-line using system variables.

- **DSETTINGS** command displays a dialog box for changing snap settings.
- **POLARANG** system variable specifies the increment for polar angles.
- **POLARDIST** system variable specifies the polar snap increment, but only when the **SNAPSTYL** system variable is set to 1.
- **POLARMODE** system variable specifies a number of variables for polar snap.

DSettings

The **DSETTINGS** command displays a dialog box that controls polar mode (because, as noted earlier, there is no "POLAR" command).

1. To access the **Drafting Settings** dialog box:
   - From the menu bar, choose **Tools**, and then **Drafting Settings**. Click the **Polar Tracking** tab.
   - At the 'Command:' prompt, enter the **dsettings** command, and then click **Polar Tracking** tab.
   - Alternatively, right-click **POLAR** on the status line. From the short-cut menu, select **Settings**.

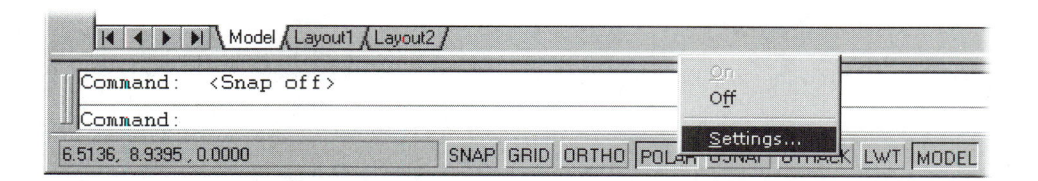

[Figure: Status bar screenshot showing Command prompt with "<Snap off>" and right-click menu with On/Off/Settings... options]

In all cases, AutoCAD displays the **Drafting Settings** dialog box with the **Polar Settings** tab.

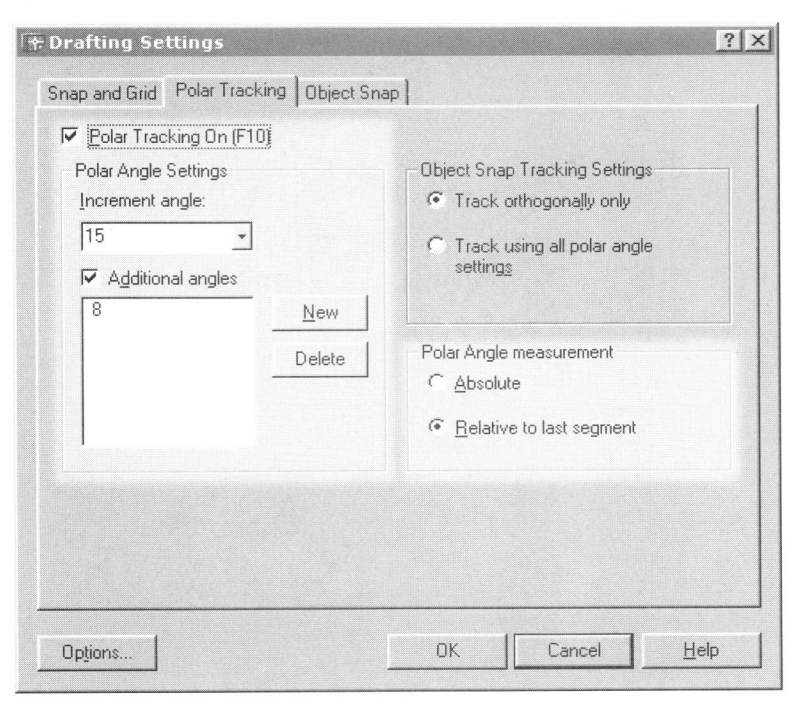

3. In this dialog box, you can make the following changes to the snap settings:
   - **Polar Tracking On (F10)** toggles the snap on and off. The (F10) reminds you that you can press **F10** to turn on and off polar mode.
   - **Increment Angle** specifies the polar tracking angle. AutoCAD presets a number of common angles, ranging from 5 to 90 dgrees.
   - **Additional Angles** allows you to add in any angle not provided by the list above. Adding an angle is a bit awkward: you need to click the **New** button, enter a value, and then press **ENTER**.
   - **Object Snap Tracking Settings** is an option described later in the chapter.
   - **Polar Angle Measurement** switches between absolute and relative angle measurements. The default, **Absolute**, means that the angle is measured from AutoCAD's 0-degree (usually the positive x-axis). **Relative** means the angle is measured relative to the last-drawn segment. The tooltip changes to read **Relative Polar**.

4. Not all polar settings are listed in this tab of the dialog box. Return to the **Snap and Grid** tab to uncover additional settings.

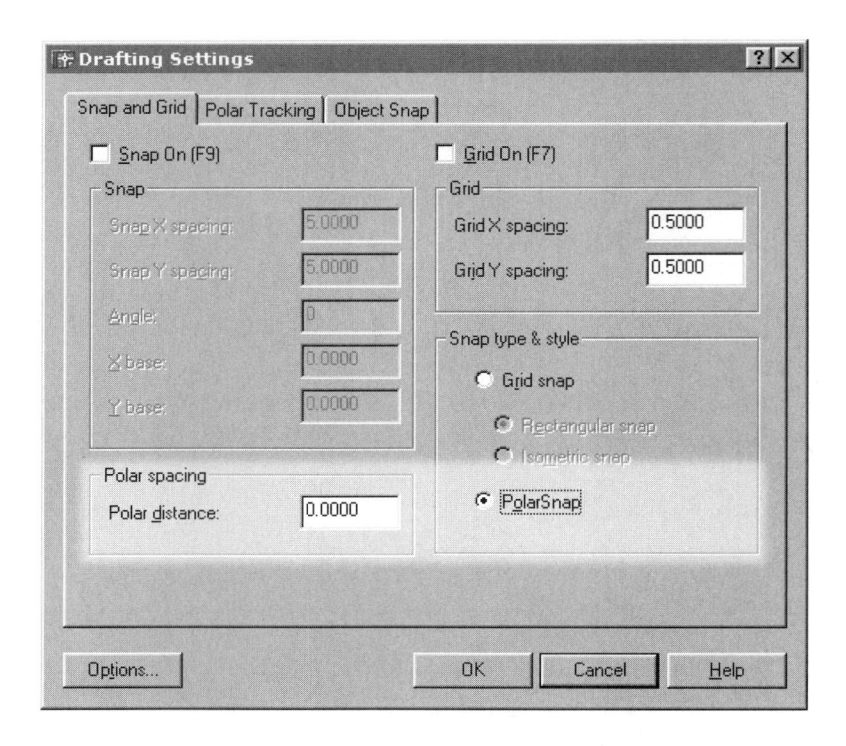

5. Under **Snap Type & Style**, select **PolarSnap**. This action causes all the **Snap** settings to gray out, which means they are unavailable. Instead, you now specify a distance under **Polar Spacing** , such as 2. This sets a snap spacing in the polar direction, i.e., in the direction of an angle specified in the previous tab.
6. Click **OK** to exit the dialog box.
7. With the **LINE** command, draw some lines, noticing the action of the cursor: you draw lines at specific angles, and the lines have lengths that are multiples of 2.

In general, you would use the **Drafting Settings** dialog box to control polar mode, but you may prefer the command line at times.

PolarAng

The **POLARANG** system variable specifies the increments for polar angles. (Even though you enter system variables at the 'Command:' prompt like commands, they are different, because they contain settings; system variables do not execute commands.)

> Command: **polarang**
>
> Enter new value for POLARANG <90>: *(Enter a new value, such as 15.)*

PolarDist

The **POLARDIST** system variable specifies the polar snap increment (distance), but only when the **SNAPSTYL** system variable is set to 1. Recall that polar measurements consist of a distance and an angle; this system variable specifies the distance.

PolarMode

The **POLARMODE** system variable specifies a number of variables for polar snap. It handles eight situations using a single number by means of *bitcodes*. Bitcodes are added up to create a single number. When all opions are turned on, for example, the value of **POLARMODE** is 15 (which comes from 1 + 2 + 4 + 8). Zero means no, off, or not.

PolarMode	Meaning
Measurement Mode:	
0	Absolute mode: Polar angle measurements are based on the current UCS.
1	Relative mode: Polar angles are measured from the last-drawn object.
Object Snap Tracking:	
0	Tracking uses orthogonal angles only.
2	Tracking uses polar angle settings.
Additional Polar Tracking Angles:	
0	Not used.
4	Used.
Acquire object snap tracking points:	
0	Acquire automatically.
8	Press SHIFT to acquire.

Some of the values used by this system variable are meant for otrack'ing, which is covered later in this chapter.

TUTORIAL: DRAWING WITH POLAR MODE

In the previous chapter, you saw that regular polygons can be drawn with the **POLYGON** command, but not irregular polygons. An excellent example is the 45-dgree isosceles triangle, which has two angles at 45 degrees and the third at 90 degrees. In this tutorial, you draw such a triangle with two sides 2.5 units long.

1. Before starting to draw, set up polar mode using the **DSETTINGS** command. In the **Drafting Settings** dialog box, make these changes:

   Polar Tracking tab:

Polar Tracking On	**On**
Increment Angle	**45 degrees**
Polar Angle Measurement	**Absolute**

   Snap and Grid tab:

Snap On	**On**
Snap Type & Style	**PolarSnap**
Polar Distance	**2.5**

   Click **OK** to exit the dialog box.

2. Start the **LINE** command, and then pick a point anywhere in the drawing:

   Command: **line**

   Specify first point: *(Pick point 1.)*

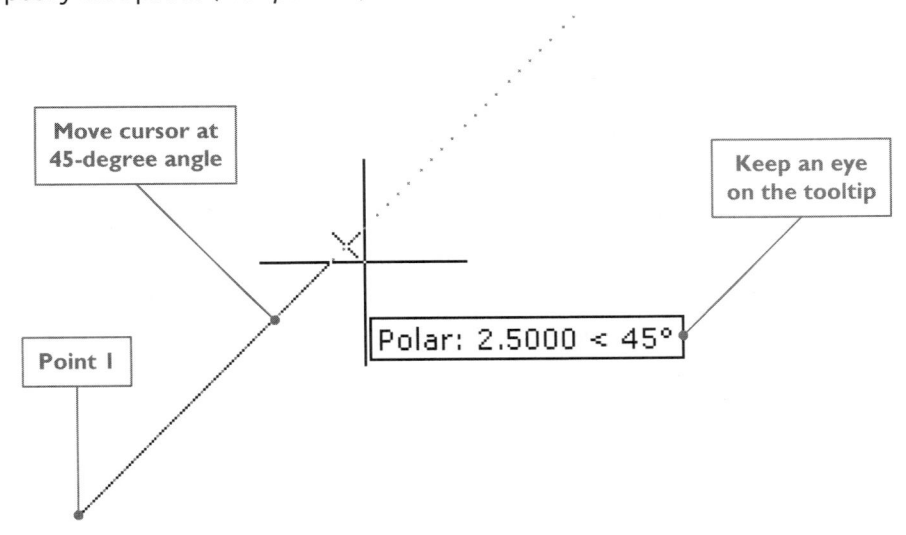

3. Move the cursor at a 45-degree angle to the upper-right. You may need to move the cursor around until the tooltip appears, reporting **Polar: 2.5000 < 45°**.

    Specify next point or [Undo]: *(Pick point 2.)*

4. Now move the cursor at a 45-degree angle to the lower-right. Click when the tooltip reports **Polar: 2.5000 < 315°**. (The angle of 315 comes from 45 + 270.)

    Specify next point or [Undo]: *(Pick point 3.)*

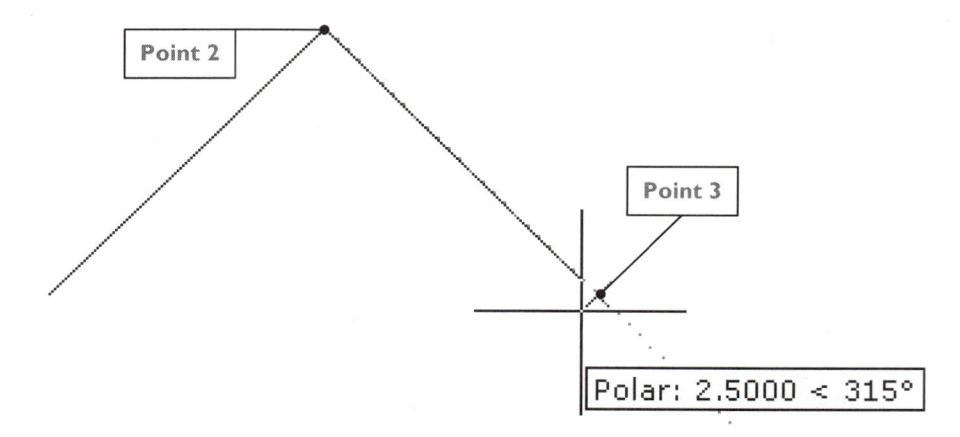

5. Close the triangle using the **Close** option:

    Specify next point or [Close/Undo]: **c**

    The isosceles triangle is complete. AutoCAD calculates the length of the third side.

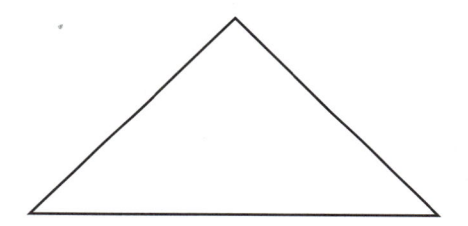

OSNAP

Object snap is probably *the most important tool* for drawing accurately in AutoCAD. You might be able to do without the other precise drawing aids described up to this point, but not object snap. As the name implies, object snap causes AutoCAD to snap to objects. More precisely, AutoCAD's cursor selects geometric features, such as the end of an arc, the center of a circle, or the intersection of two lines.

In all, AutoCAD has 13 object snaps, which are displayed in drawings as temporary icons (see figure at left). Object snaps are sometimes called "osnaps" for short.

In addition, AutoCAD has an entire toolbar devoted to object snaps:

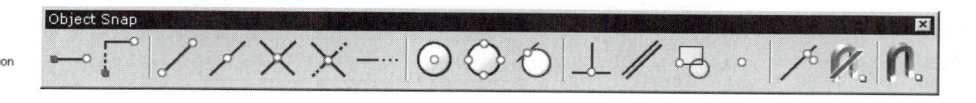

- □ Endpoint
- △ Midpoint
- ○ Center
- ⊗ Node
- ◇ Quadrant
- ✕ Intersection
- ⋯ Extension
- ⌂ Insertion
- ⊥ Perpendicular
- ○ Tangent
- ⊠ Nearest
- ⊠ Apparent intersection
- ∥ Parallel

APERTURE

The exciting aspect to object snap is that you don't have to be right on the geometric feature, just close enough — as in horseshoes, hand grenades, and dancing — to an object for AutoCAD to find the snap point. The close-enough distance is called the *aperture*, which is a square ten pixels in size. You can make the aperture as large as 50 pixels and and as small as 1 pixel with the **APERTURE** command.

The aperture appears as a square around the crosshair cursor. AutoCAD examines the objects that lie within the aperture for likely object snap connections.

RUNNING AND TEMPORARY OSNAPS

Object snaps can be used in two ways during drawing and editing commands: running and temporary. *Running* object snaps are set with the **OSNAP** and **-OSNAP** commands; in effect, the object snaps are always on, until you turn them off.

Temporary object snaps are in effect for the next object selection only. At a prompt, such as "Select first point:" or "Specify next point", enter one or more object snap modes (when entering more than one mode, separate them with commas). Here a line is drawn from the endpoint of one object to the midpoint or center point of another:

> Command: **line**
> Specify first point: **end**
> of *(Pick a point.)*
> Specify next point or [Undo]: **mid,cen**
> of *(Pick a point.)*
> Specify next point or [Undo]: *(Press ENTER to end command.)*

You can select object snap modes from a shortcut menu during commands. Hold down the **CTRL** key, and then press the right mouse button (or **SHIFT**+right click).

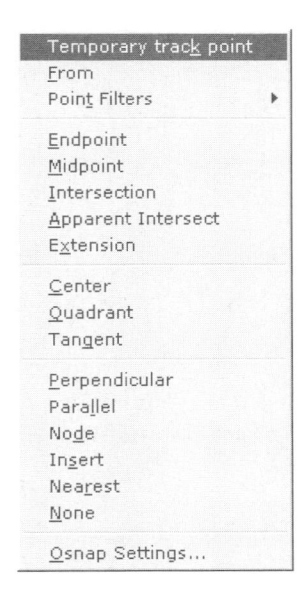

```
Temporary track point
From
Point Filters        ▶

Endpoint
Midpoint
Intersection
Apparent Intersect
Extension

Center
Quadrant
Tangent

Perpendicular
Parallel
Node
Insert
Nearest
None

Osnap Settings...
```

SUMMARY OF OBJECT SNAP MODES

In the review of object snap modes below, notice that the first three letters of each mode are capitalized. These three letters are the abbreviation for each mode. When you enter object snaps at the command line, you need enter only the three letters, not the entire name.

In alphabetical order, AutoCAD's object snap modes are:

APParent Intersection

The apparent intersection object snap (called "appint" for short) works differently, depending on whether you are working with a 2D or 3D drawing.

In 2D drawings, appint turns on *extended intersection* mode, so that AutoCAD snaps to the intersection of two objects that don't physically intersect. AutoCAD creates an imaginary extension to the two objects, and then determines if an intersection could occur.

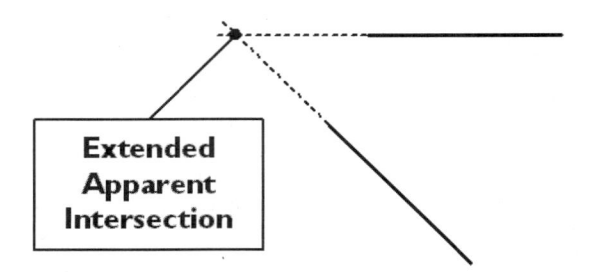

Extended Apparent Intersection

In 3D drawings, appint snaps to the point where two objects *appear* to intersect from your viewpoint; the objects do not actually need to intersect. The objects can be arcs, circles, ellipses, elliptical arcs, lines, multilines, polylines, rays, splines, and xlines.

CENter

The center object snap snaps to the centerpoint of arcs, circles, ellipses, and elliptical arcs.

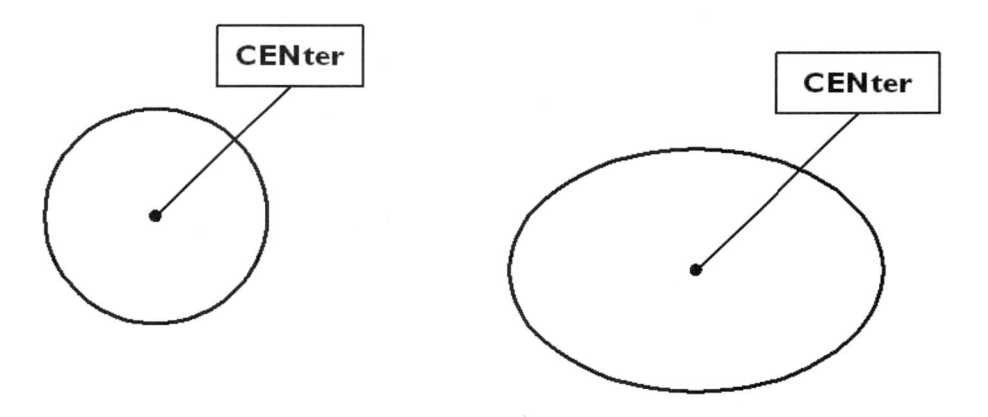

ENDpoint

The endpoint object snap snaps to the closest endpoint of arcs, elliptical arcs, lines, multilines, polyline segments, splines, regions, and rays. In addition, it snaps to the closest corner of traces, 2D solids, and 3D faces.

EXTension

The extension object snap displays an extension line when the cursor passes over the endpoints of objects. This allows you to draw objects that start and end on the extension line.

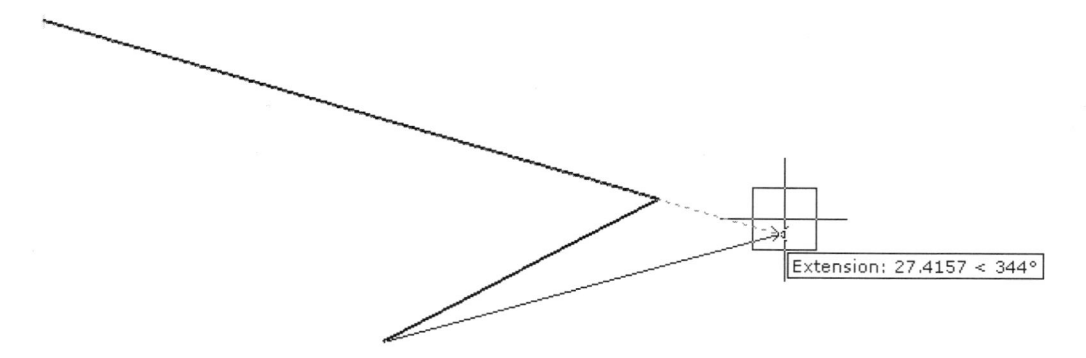

In the figure above, AutoCAD displays the entension line (shown dotted) from the endpoint of the arc. The tooltip reports the distance and angle. (*Extensions* are similar to, but a more limited form of object snap tracking, discussed later in this chapter.)

INSERTION

The insertion object snap snaps to the insertion point of attributes, blocks, shapes, and text.

INTersection

The intersection object snap snaps to the intersection of arcs, circles, ellipses, elliptical arcs, lines, multilines, polylines, rays, regions, splines, and xlines. Sometimes, more than one intersection is possible, such as when two circles intersect.

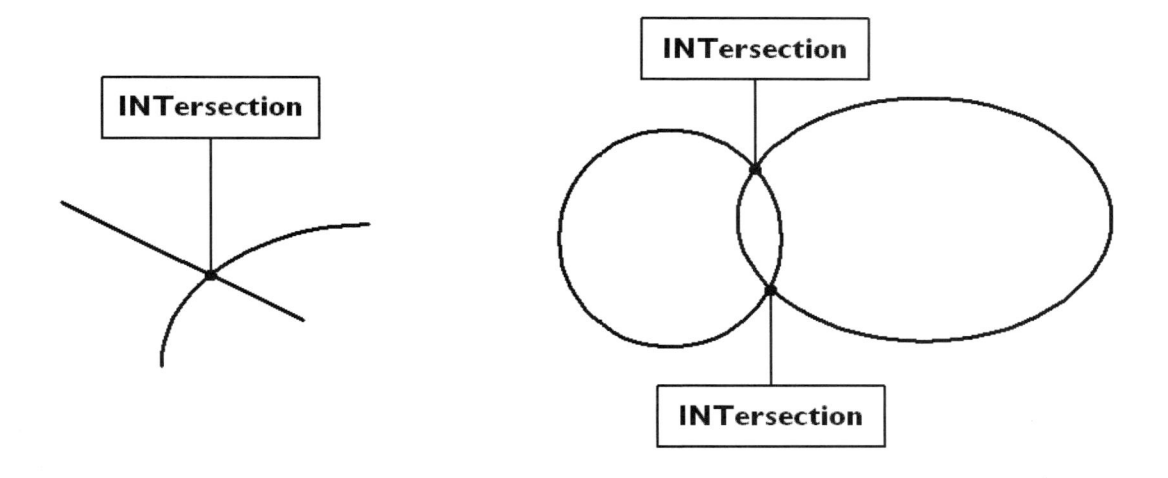

At the same time, extended intersection mode is turned on so that AutoCAD can snap to the intersection of two objects that don't physically intersect. An imaginary extension to the two objects is created to determine if an intersection could occur.

MIDpoint

The midpoint object snap snaps to the midpoint of arcs, ellipses, elliptical arcs, lines, multilines, polyline segments, regions, solids, splines, and xlines.

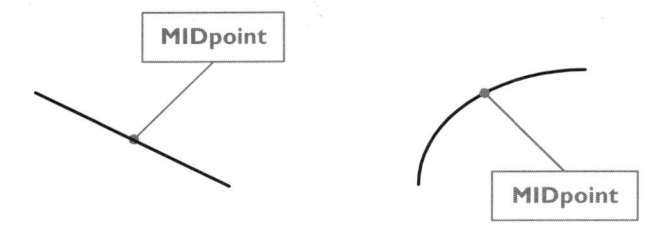

NEArest

The nearest object snap snaps to the nearest point on the nearest arc, circle, ellipse, elliptical arc, line, multiline, point, polyline, ray, spline, or xline.

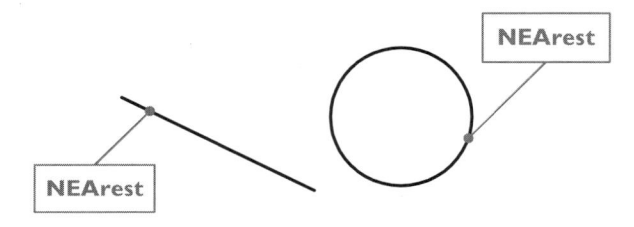

NODe

The node object snap snaps to points, dimension definition points, and dimension text origins.

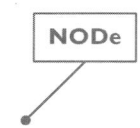

PARallel

The parallel object snap displays an alignment path parallel to a straight line segment, such as a line or polyline.

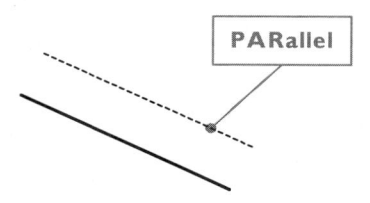

PERpendicular

The perpendicular object snap snaps to a point perpendicular to an arc, circle, ellipse, elliptical arc, line, multiline, polyline, ray, region, solid, spline, or xline.

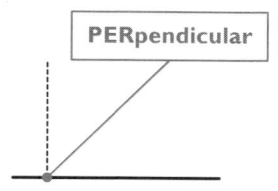

If necessary, AutoCAD turns on *deferred perpendicular* mode automatically when you need to pick more than one point to determine perpendicularity.

QUAdrant

The quadrant object snap snaps to the *quadrant points* of arcs, circles, ellipses, and elliptical arcs.

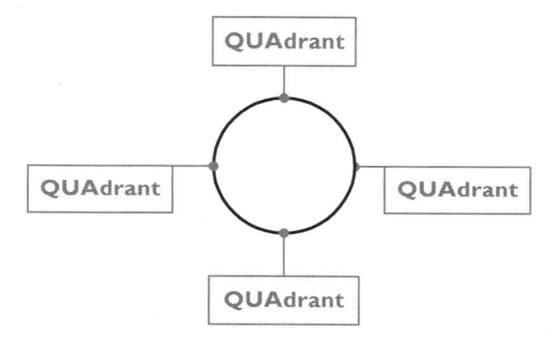

The quadrant points are located at the 0-, 90-, 180-, and 270-degree points of these curves. Thus, circles and ellipses always have four possible quadrant points, while arcs and elliptical arcs can have anywhere from one to four quandrant points.

TANgent

The tangent object snap snaps to the tangent points of arcs, circles, ellipses, elliptical arcs, and splines.

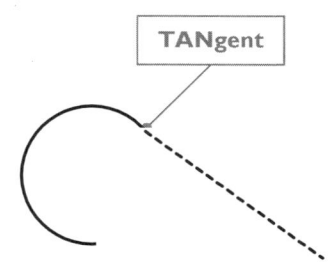

If necessary, AutoCAD turns on *deferred tangent* mode automatically when you draw more than one tangent, such as a line tangent to two arcs.

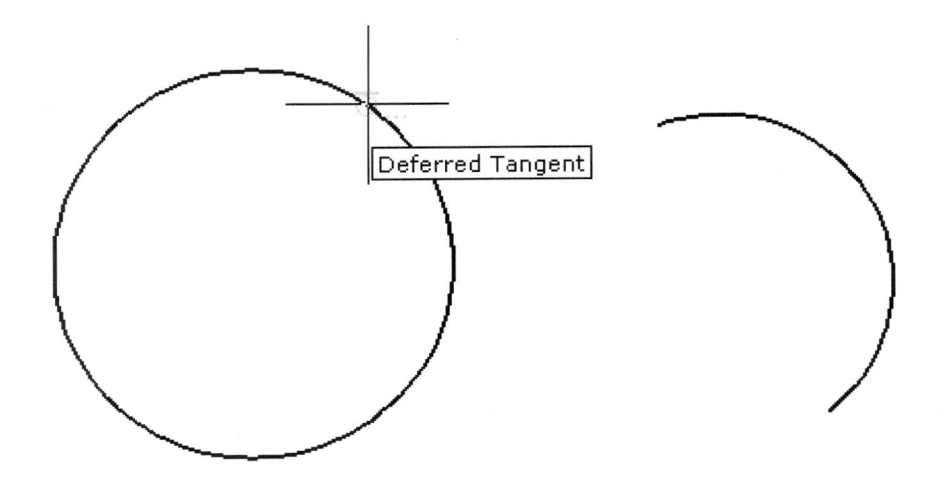

ADDITIONAL OBJECT SNAP MODES

There are several additional object snap modes that don't work with geometry. These find distances, or else turn off snap modes.

From

The from mode allows you to specify a temporary reference or base point, from which you can locate the next point.

> Command: **line**
>
> Specify first point: **from**
>
> Base point: *(Pick point 1.)* 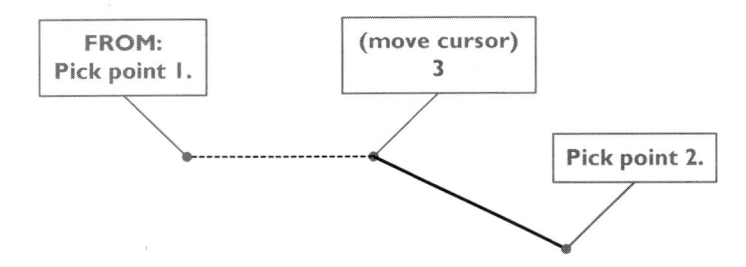
>
> <Offset>: *(Move the cursor to show the distance, or enter a distance:)* **3**
>
> Specify next point or [Undo]: *(Pick point 2.)*
>
> Specify next point or [Undo]: *(Press* **ENTER** *to end command.)*

NONe

The none mode turns off all object snap modes temporarily for the next pick point.

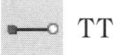

TT

The temporary tracking mode allows you to specify a *temporary tracking* point. AutoCAD places a + marker at the point, and then, as you move the crosshair cursor, it displays alignment paths relative to the + marker.

Command: **line**
Specify first point: **tt**
Specify temporary OTRACK point: *(Pick point 1.)*

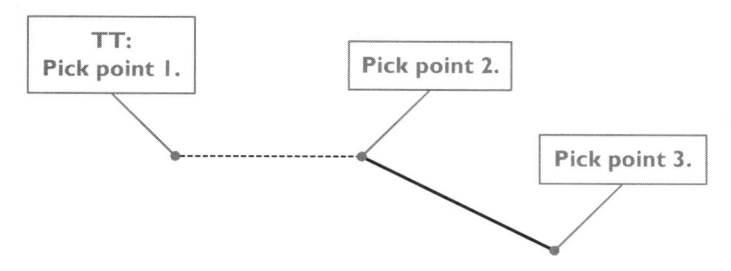

TT:
Pick point 1.

Pick point 2.

Pick point 3.

Specify first point: *(Pick point 2.)*
Specify next point or [Undo]: *(Pick point 3.)*
Specify next point or [Undo]: *(Press* ENTER *to end command.)*

QUIck

The quick object snap snaps to the first snap point AutoCAD finds. Quick mode works only when you have two or more other snap modes turned on. For example, when you have intersection and endpoint turned on, AutoCAD returns the first object snap point that is either an intersection or endpoint.

OFF

The off option turns off all object snap modes.

TUTORIAL: TOGGLING AND SELECTING OSNAP MODE

Object snap probably has the most complete set of command variations:

1. To toggle ortho mode, use one of these methods:
   * On the status bar, click **OSNAP** (turned on when button looks depressed).
   * At the keyboard, press function key **F3** or **CTRL+F**.

2. To toggle ortho mode *and* specify one or more object snaps, use one of these methods:
   * From the menu bar, select **Tools**, and then chose **Drafting Settings**. In the dialog box, choose the **Object Snap** tab, and then select object snap modes.
   * During another command, hold down the **CTRL** key, and then press the right mouse button to display a shortcut menu of object snap modes.
   * Similarly, you can use the the **Object Snap** toolbar during a command: from the toolbar, select a button corresponding to the object snap mode you wish to employ.
   * On the keyboard, enter the **-osnap** command, followed by one or more object snap mode names. (The hyphen prefix forces AutoCAD to display the command at the 'Command:' prompt; without the hyphen, the **OSNAP** command displays the dialog box.)
   * Alternatively, type the aliases **os** or **ddosnap** (the old command name) for the dialog box, and the **-os** alias for the command line.
   * And finally, the **OSMODE** system variable can be used to set object snap modes.

CONTROLLING ORTHO MODE: ADDITIONAL METHODS

In new drawings, AutoCAD remembers the object snap modes set previously. You change the modes through a dialog box, or at the command-line:

- **OSNAP** command displays a dialog box for changing osnap settings.
- **-OSNAP** command changes osnap settings at the command line.

OSnap (DSettings)

The OSNAP command displays a dialog box that allows you to select running object snap modes. (Strictly speaking, the OSNAP command is a shortcut for using the DSETTINGS command, followed by selecting the **Object Snap** tab of the **Drafting Settings** dialog box.)

1. To access the **Drafting Settings** dialog box:
   - From the menu bar, choose **Tools**, and then **Drafting Settings**. Click on the **Object Snap** tab.
   - At the 'Command:' prompt, enter the **osnap** command.
   - Alternatively, right-click **OSNAP** on the status line. From the short-cut menu, select **Settings**.

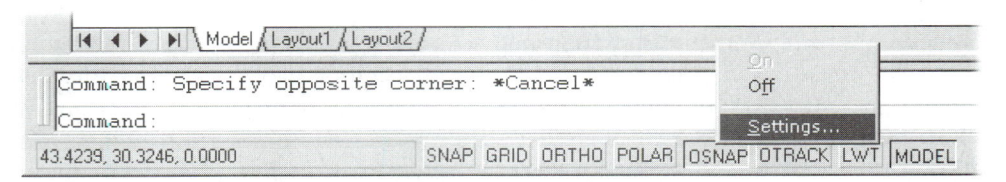

In all cases, AutoCAD displays the **Drafting Settings** dialog box with the **Object Snap** tab.

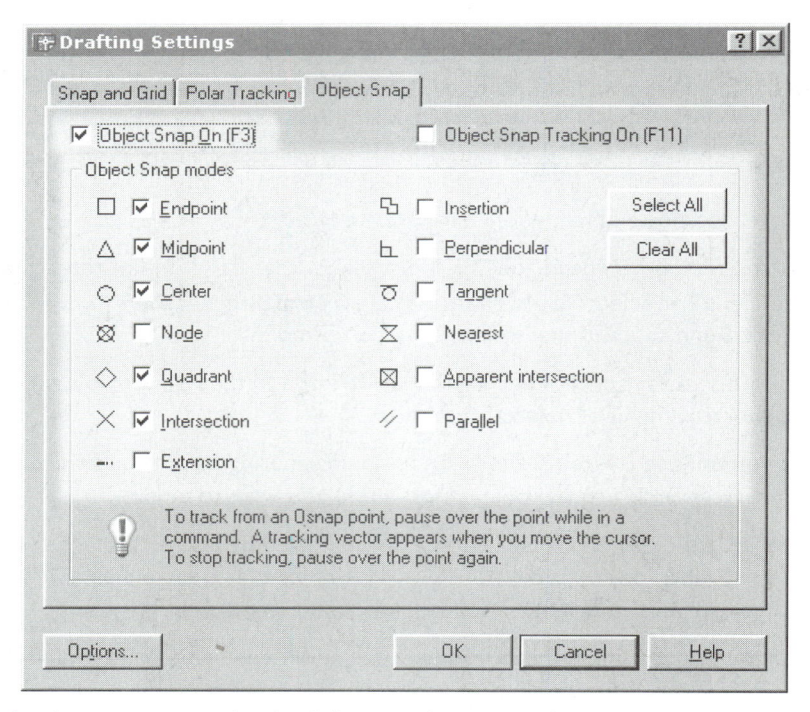

3. In this dialog box, you can make the following changes to the snap settings:
   - **Object Snap On (F3)** toggles the snap on and off. The (F3) reminds you that you can press F3 to turn on and off osnaps.
   - **Object Snap Modes** lists most of the modes available. Notice the icon next to each mode; this is the same icon AutoCAD displays in the drawing when it finds a geometric feature matching the selected mode(s).

4. Select one or more object snap modes. You can turn on all of osnap modes with the **Select All** button, and turn them all off with the **Clear All** button.
5. Not all object snap options are listed in this dialog box. Click the **Options** button to view more. AutoCAD displays the **Drafting** tab of the **Options** dialog box:

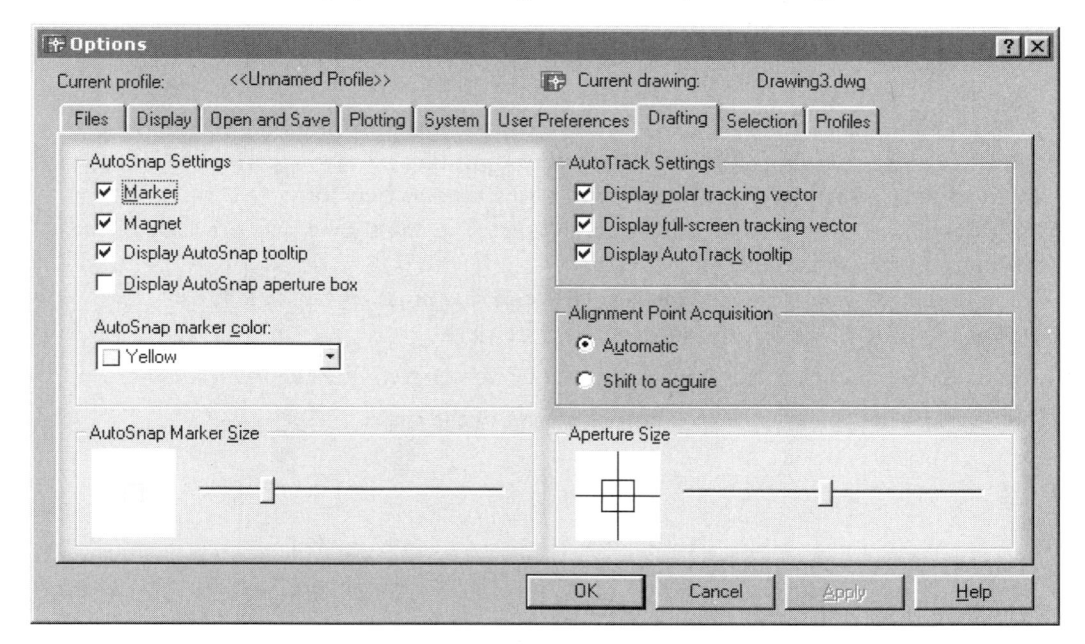

Here you select from the many options that affect how object snaps show up in the drawing. For example, you may want to change the color of the marker from the hard-to-see yellow to blue or another higher-visibility color. In summary, the options have the following meanings:

Marker toggles the display of the osnap icons.

Magnet toggles whether the cursor automatically moves to the osnap point.

Display AutoSnap Tooltip toggles whether the tooltip appears, which labels the object snap by name.

Display AutoSnap Aperture Box is a small square appearing at the center of the cursor to indicate that AutoSnap mode is on.

AutoSnap Marker Color allows you to select *any* color for the marker.

AutoSnap Marker Size allows you to make the marker (icon) larger and smaller.

Aperture Size allows you to make the object snap cursor larger and smaller.

6. Click **OK** to exit the dialog box. AutoCAD returns you to the Drafting Settings dialog box.

Click **OK** to exit the dialog box.

-OSnap

The -OSNAP command displays a prompt at the command line. It lists the current osnap modes, and then prompts you to enter additional modes. You can enter a single osnap mode, such as **Int**, or several modes separated by commas, as shown below:

> Command: **-osnap**
>
> Current osnap modes: End,Mid,Cen
>
> Enter list of object snap modes: **int,qua**

You need only type the first three letters of each mode name. For example, "int" means **INTersection** and "qua" means **QUAdrant**.

You cannot selectively turn off osnap modes. Instead, you turn off other modes by entering new modes. In the example above, entering **int** and **qua** turns off **end**, **mid**, and **cen**.

TUTORIAL: USING OBJECT SNAP MODES

1. Draw a circle and arc of any size.
2. From the **Draw** menu, choose the **Line** command. Notice that AutoCAD prompts you "Specify first point:"
3. Do not pick a point. Instead, hold down the CTRL key, and then press the right mouse button to access the shortcut menu. Choose **Endpoint** from the menu. Notice that the aperture square apperas at the intersection of the crosshair cursor.
4. Place the cursor over one end of the arc, and click. The line should snap precisely to the endpoint of the existing line.
5. Before placing the next point of the line, type **cen**, and press ENTER, as follows:

    Specify next point or [Undo]: **cen** *(Press ENTER.)*

    of *(Place the cursor over any part of the circle's circumference, and then click.)*

    Notice how the line snaps to the precise center of the circle.

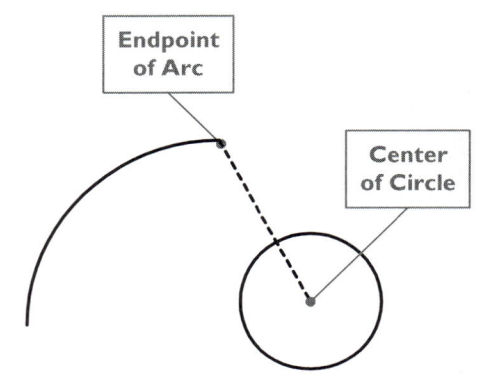

6. Continue the line, using the **Center** object snap on the arc. The line snaps to the centerpoint of the arc.
7. Try other object snap modes on the line, arc, and circle. For example, use the **Tangent** object snap to construct lines tangent with circles and arcs.

OTRACK

Otrack is short for "object snap tracking"; Autodesk also calls it AutoTrack™, complete with the trademark symbol. Otracking adds *alignment paths* to object snapping. Alignment paths are thin dotted lines that show the osnap relationships to other geometry. Think of otrack as "super extension osnap." At least one object snap mode must be turned on before otracking can work.

At any prompt that asks you to select objects, move the cursor around the drawing to see geometric relationships. In the figure below, AutoTrack has palced a marker (the **x**) at a point perpendicular to one segment and in-line with the endpoint of another segment.

You can use otracking together with temporary tracking. Enter **tt** for AutoTrack to show alignment paths relative to the temporary tracking point.

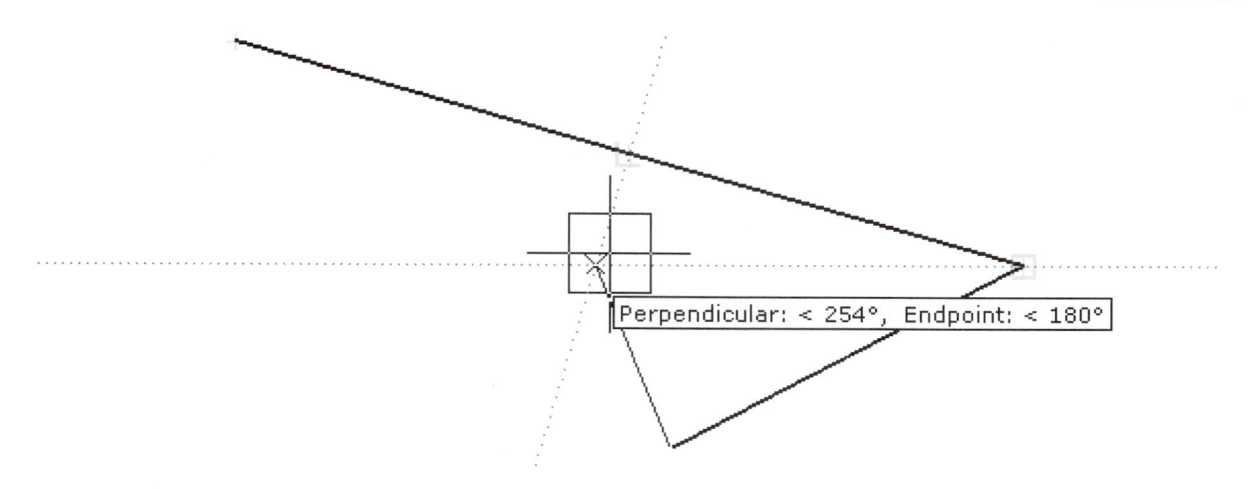

TUTORIAL: TOGGLING OTRACK MODE

Like polar mode, there is no "OTRACK" command, but there are shortcuts for turning it on and off. In addition, otrack mode is controlled by system variables and dialog boxes.

1. To turn on otrack mode, use one of these methods:
    - From the status bar, click **OTRACK** (turned on when button looks depressed).
    - On the keyboard, press function key **F11**.

RAY AND XLINE

The **RAY** and **XLINE** commands place *construction lines* in the drawing. Construction lines are useful for creating drawings, helping to reference things like centerlines and offset lines. The construction lines are, unfortunately, plotted by AutoCAD, so you should place them on frozen or no-plot layers.

RAY draws "semi-inifinite" lines, which have a start point, but no endpoint (it stretches to infinity). **XLINE** draws infinitely-long lines, which have no start or end points.

TUTORIAL: DRAWING CONSTRUCTION LINES

1. To draw construction lines, start the **RAY** or **XLINE** command with one of these methods:
    - From the menu bar, choose **Draw**, then choose **Ray** or **Xline**.
    - From the **Draw** toolbar, choose the **Construction Line** button.
    - At the 'Command:' prompt, enter the **ray** or **xline** commands.
    - Alternatively, enter the **xl** alias at the 'Command:' prompt.

        Command: **xline** *(Press* ENTER.*)*

2. Specify two points that lie along the construction line:
   Specify a point or [Hor/Ver/Ang/Bisect/Offset]: *(Pick point 1.)*
   Specify through point: *(Pick point 2.)*

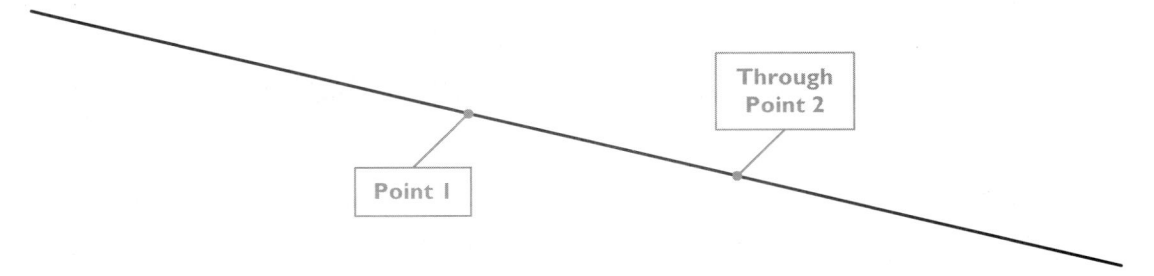

3. Because the **RAY** and **XLINE** command automatically repeat, press **ENTER** to exit them:
   Specify through point: *(Press ENTER to exit command.)*

The **RAY** command first asks for a starting point:

   Command: **ray**
   Specify start point: *(Pick point 1.)*
   Specify through point: *(Pick point 2.)*
   Specify through point: *(Press ENTER to exit command.)*

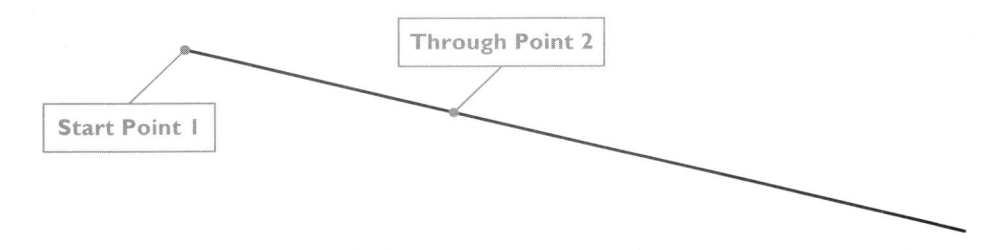

DRAWING CONSTRUCTION LINES: ADDITIONAL METHODS

The **XLINE** command contains these options that provide additional methods for drawing construction lines. The options are shown at the command prompt and on the right-click menu (below).

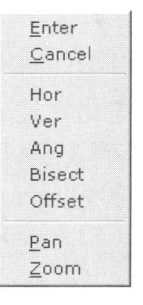

- **Hor** draws horizontal construction lines:

    Command: **xline**

    Specify a point or [Hor/Ver/Ang/Bisect/Offset]: **h**

    Specify through point: *(Pick a point.)*

    Specify through point: *(Press ENTER to exit command.)*

- **Ver** draws vertical construction lines:

    Command: **xline**

    Specify a point or [Hor/Ver/Ang/Bisect/Offset]: **v**

    Specify through point: *(Pick a point.)*

    Specify through point: *(Press ENTER to exit command.)*

- **Ang** draws construction lines at a specified angle:

    Command: **xline**

    Specify a point or [Hor/Ver/Ang/Bisect/Offset]: **a**

    Enter angle of xline (0) or [Reference]: *(Enter an angle, or type **r** for the reference option.)*

    Specify through point: *(Pick a point.)*

    Specify through point: *(Pick another point.)*

    Specify through point: *(Press ENTER to exit command.)*

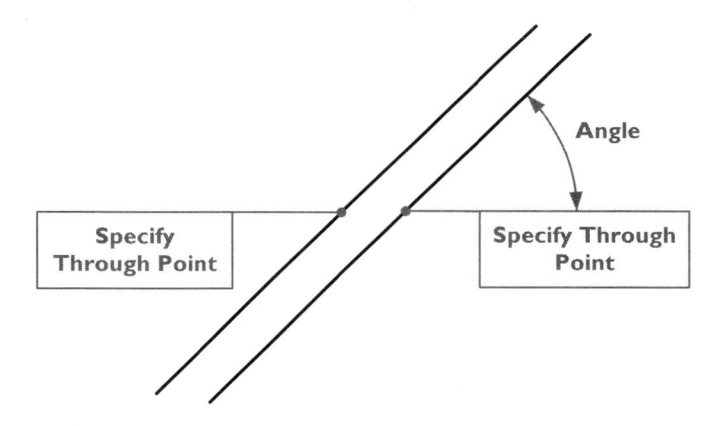

- **Bisect** draws construction lines that bisects an angle defined by a vertex and two endpoints:

    Command: **xline**

    Specify a point or [Hor/Ver/Ang/Bisect/Offset]: **b**

    Specify angle vertex point: *(Pick a point.)*

    Specify angle start point: *(Pick a point.)*

    Specify angle end point: *(Pick a point.)*

    Specify angle end point: *(Press ENTER to exit command.)*

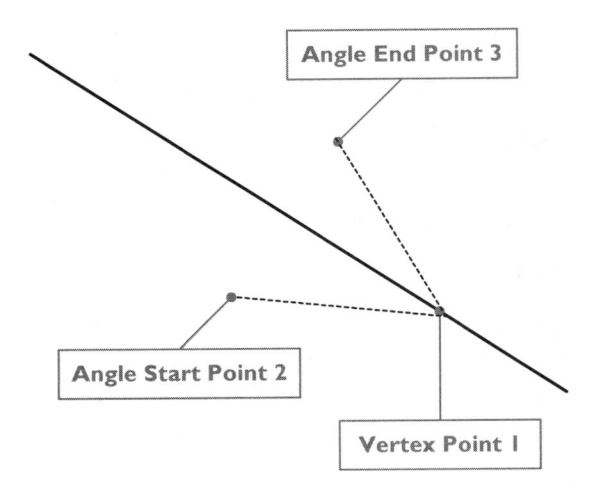

[Figure: Angle construction line with Vertex Point 1, Angle Start Point 2, and Angle End Point 3 labeled.]

- **Offset** draws construction lines parallel to existing lines and construction lines:

    Command: **xline**

    Specify a point or [Hor/Ver/Ang/Bisect/Offset]: **o**

    Specify offset distance or [Through] <Through>: *(Enter a distance, or type* **t** *for the through option.)*

    Select a line object: *(Pick a line.)*

    Specify side to offset: *(Pick a point.)*

    Select a line object: *(Press* ENTER *to exit command.)*

TUTORIAL: DRAWING WITH CONSTRUCTION LINES

In this tutorial, you draw the true surface of a rectangular inclined face. You use much of what you have learned so far in this chapter: grid, snap, ortho, object snaps, and construction lines.

[Figure: Hand-drawn sketch on graph paper showing TOP VIEW, AUXILLIARY VIEW, FRONT VIEW (with dimensions 0.5, 2.5, 0.5, 2.5), and SIDE VIEW (with dimension 1.0).]

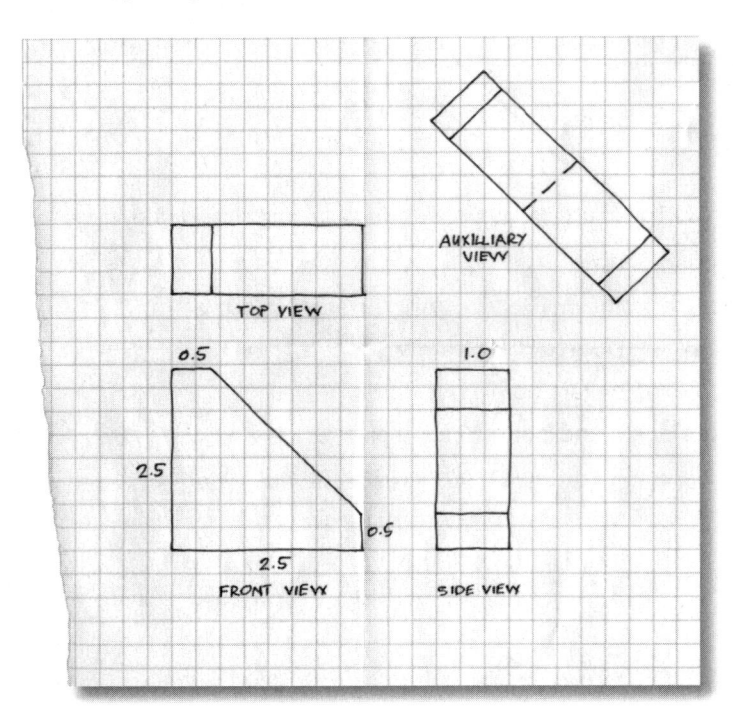

1. Start AutoCAD with a new drawing.

2. Change the following settings:
   Turn on **SNAP**; ensure the snap spacing is 0.5 units.
   Turn on **GRID**.
   Turn on **ORTHO**.
   Turn on **OSNAP**; ensure the object snap modes **Intersection** and **Perpendicular** are turned on.

3. With the LINE command (menu: **Draw | Line**), create the multiview drawing shown below.

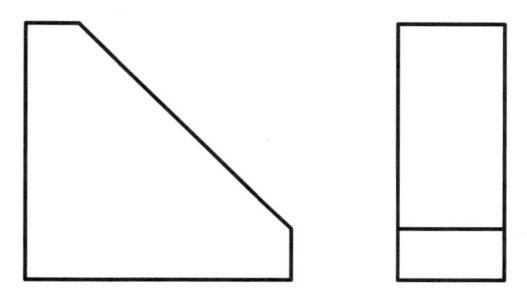

   Because the top view is a mirror image of the side view, consider using the MIRROR command (menu: **Modify | Mirror**) to create it. *Hint:* Turn off **ORTHO** mode, and use a 45-degree mirror line.

4. Use the SNAP command's **Rotate** option to rotate the snap and grid to align with the sloping edge of the front view:

   Command: **snap**
   Specify snap spacing or [ON/OFF/Aspect/Rotate/Style/Type] <0.5000>: **r**
   Specify base point <0.0000,0.0000>: *(Pick point 1.)*

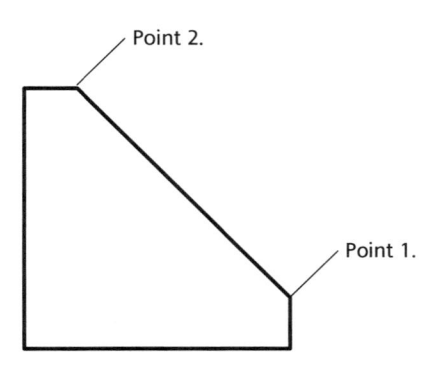

   Because you set the object snap modes in Step 2, you should have no difficulty snapping to the ends of the sloped edge.

Specify rotation angle <0>: *(Pick point 2.)*
Angle adjusted to 315

Notice that the grid changes to align itself with the sloped edge.

5. Turn off **SNAP**.
   With the **ray** command (menu: **Draw | Ray**), draw a series of parallel construction lines.
   Command: **ray**
   Specify start point: *(Pick point 1.)*
   Specify through point: *(Pick point 2.)*
   Specify through point: *(Press ESC.)*

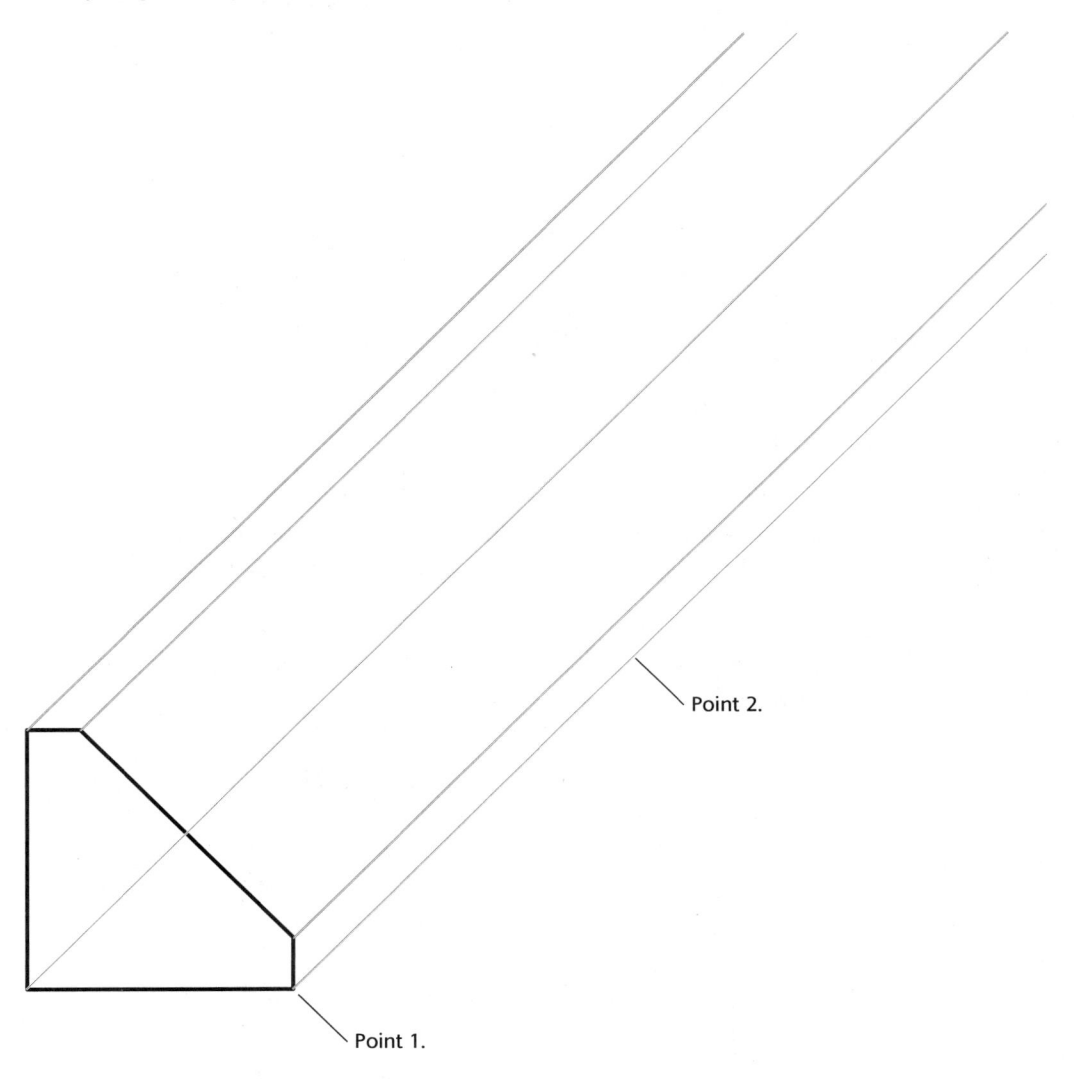

Repeat the command to draw four more construction lines.

Hint: Press ESC to end a command that repeats itself; press the spacebar to repeat the command.

6. Now draw the auxiliary view with the **LINE** command. Remember that the object is 1" wide.

   Hint: Use the **NEArest** object snap to start drawing from the construction line.

   Command: **line**

   Specify first point: **nea**

   to *(Pick point 1.)*

   Specify next point or [Undo]: *(Pick point 2.)*

   Specify next point or [Undo]: *(Move cursor toward point 3.)* **1**

   Specify next point or [Close/Undo]: *(Pick point 4.)*

   Specify next point or [Close/Undo]: **c**

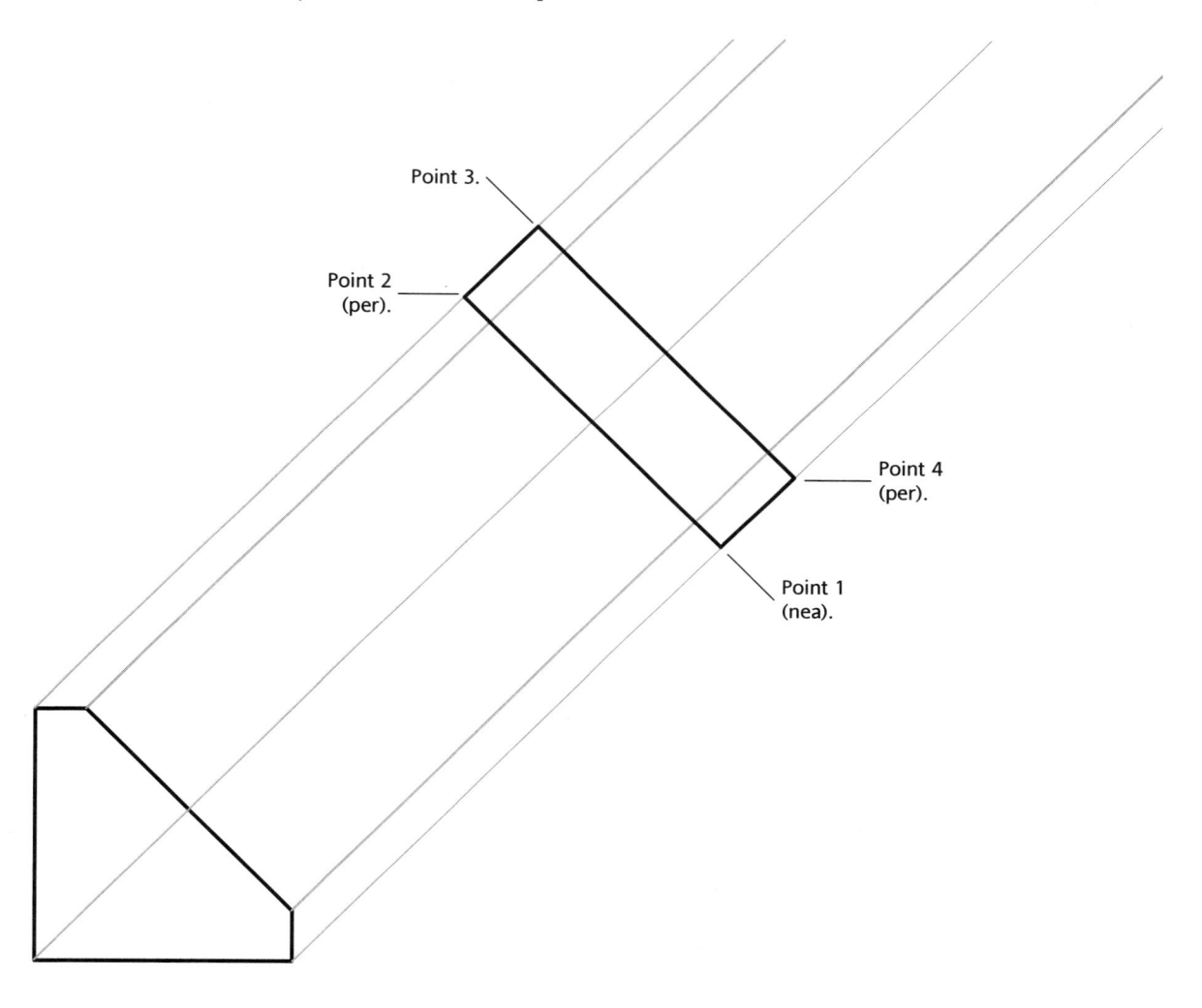

7. Draw the two edge lines, as follows:
   Command: **line**
   Specify first point: *(Pick point 1.)*
   Specify next point or [Undo]: *(Pick point 2.)*
   Specify next point or [Undo]: *(Press ESC.)*

   Command: *(Press ENTER to repeat the command.)*
   LINE Specify first point: *(Pick point 3.)*
   Specify next point or [Undo]: *(Pick point 4.)*
   Specify next point or [Undo]: *(Press ESC.)*

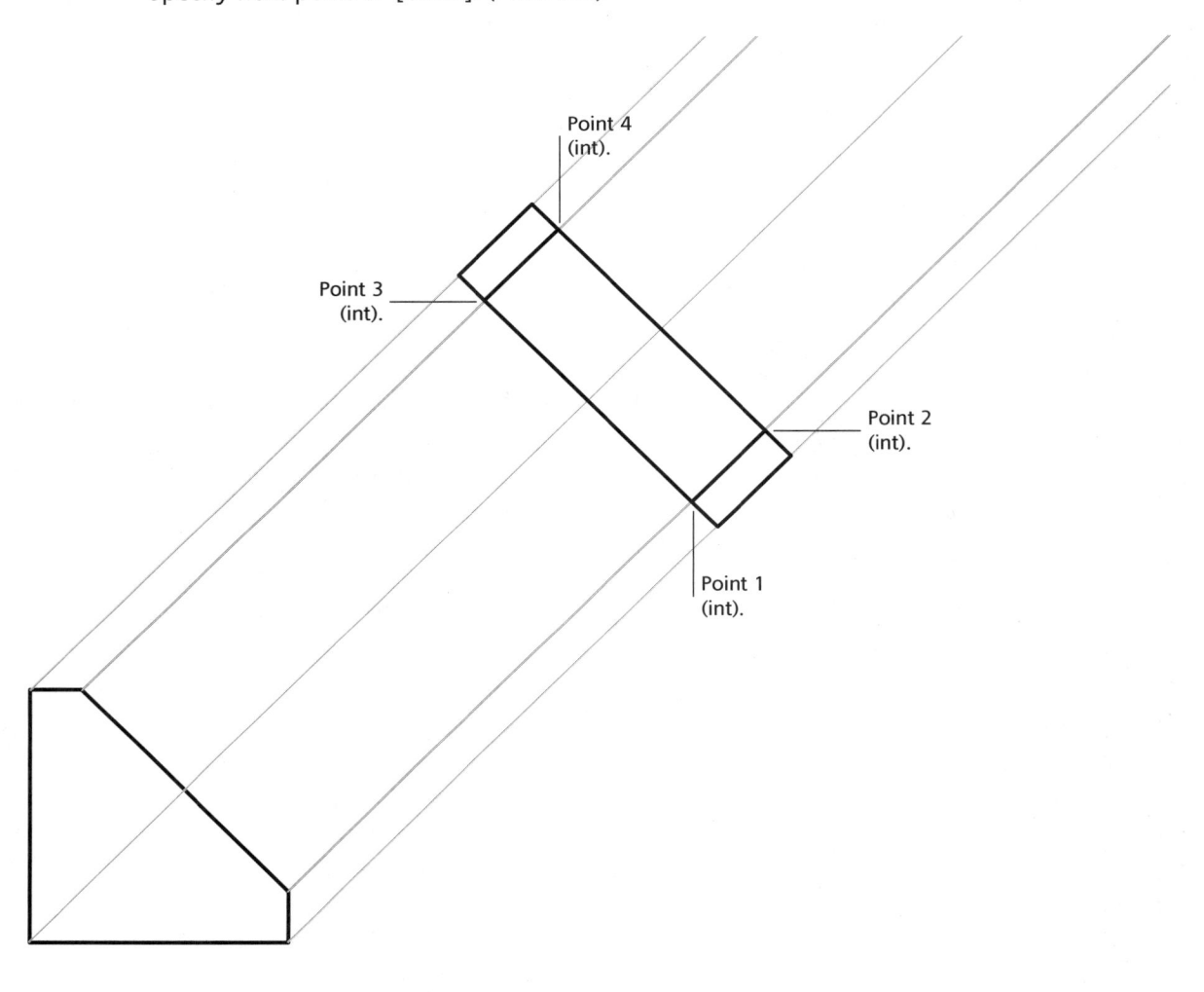

8. Save the drawing as *auxview5.dwg*.

REDRAW AND REGEN

The **REDRAW** and **REGEN** commands clean up and update the drawing. **REDRAW** cleans up the display by removing blipmarks (if any) and blanked out areas. **REGEN** is short for "regeneration," and updates the display by recalculating all the vectors making up the drawing. During a regeneration, AutoCAD recalculates line endpoints, hatched areas, and so on. A redraw is always faster than a regen.

If you have more than one viewport displayed, **REDRAW** and **REGEN** operate on the current viewport only. Use the **REDRAWALL** and **REGENALL** commands to clean up and update all viewports.

TUTORIAL: USING THE REDRAW COMMAND

1. To clean up the display, start the **REDRAW** command with one of these methods:
   - At the 'Command:' prompt, enter the **redraw** command.
   - Alternatively, enter the **r** alias at the 'Command:' prompt.

    Command: **redraw** (Press ENTER.)

    Notice that the screen flickers as AutoAD cleans up the drawing.

The associated commands are similar:

Command	View Menu	Alias
RedrawAll	Redraw	ra
Regen	Regen	re
RegenAll	Regen All	rea

The "REDRAW" command found on the menu actually executes the **REDRAWALL** command.

TRANSPARENT REDRAWS

A redraw can be executed while another command is active. This is called a *transparent* redraw. To perform this, enter an apostrophe (') before the command. For example, to perform a transparent redraw during the **LINE** command:

    Command: **line**
    Specify first point: (Pick a point.)
    Specify next point or [Undo]: **'redraw**
    Resuming LINE command.
    Specify next point or [Undo]: (Pick another point.)
    Specify next point or [Undo]: (Press ENTER.)

When the transparent redraw is completed, the previously-active command resumes. **REGEN** is not a transparent command.

REGENAUTO

Some commands cause an automatic regeneration under certain circumstances. Among these are **ZOOM**, **PAN**, and **VIEW Restore**. The regeneration occurs if the new display contains areas not within the currently-generated area. The **REGENAUTO** command controls automatic regeneration in AutoCAD.

Because each regeneration can take a long time on a slow computer or with a very complex drawing, AutoCAD provides a command to warn you before regenerations are performed.

    Command: **regenauto**
    Enter mode [ON/OFF] <ON>: **OFF**

When **REGENAUTO** is turned off, and a regeneration is required, AutoCAD displays a warning dialog box: "About to regen—proceed?" Choose **OK** to permit the drawing regeneration; choose **Cancel** to stop the regen.

As an alternative, press **ESC** to cancel the process. If the regen is interrupted in this way, some of the drawing might not be redisplayed. To redisplay the drawing, you must reissue the **REGEN** command.

ZOOM

The **ZOOM** command enlarges and reduces the view of the drawing. When zoomed out, you see more of the drawing; when zoomed in, you see less of the drawing, but you see more detail. Think of zoom as a magnifier and shrinker. Most drawings are too large and too detailed to work with on the computer monitor's small 15"—or even 20"—drawing screen. CAD operators routinely zoom in to small areas to work on details.

Let's consider this analogy: imagine that your drawing is the size of the wall in your room. The closer you walk to the wall, the more detail you see, but the less you see of the entire wall. When you move a great distance away from the wall, you see the entire wall but not much detail. Notice that the wall does not change size, just your viewing distance. The **ZOOM** command works in the same way: You enlarge and reduce the drawing size on the screen, but the drawing itself does not change size (or scale).

Because zooming is performed often, AutoCAD has many ways to zoom.

TUTORIAL: USING THE ZOOM COMMAND

1. To change the viewing size of the drawing, start the **ZOOM** command with one of these methods:
   - From the **View** menu bar, choose **Zoom**, and then one of its options.
   - On the **Zoom** toolbar, click one of the buttons.
   - Right-click anywhere in the drawing, and from the shortcut menu select **Zoom**.
   - At the 'Command:' prompt, enter the **zoom** command.
   - Alternatively, enter the **z** alias at the 'Command:' prompt.

    Command: **zoom** (*Press* ENTER.)

2. The command displays its many options:

    Specify corner of window, enter a scale factor (nX or nXP), or [All/Center/Dynamic/Extents/ Previous/Scale/Window] <real time>:

While the prompt indicates there are ten options, three more are hidden. Two are **Left** and **Vmax**, and the third is the mousewheel.

Most commands have just one default option, but Autodesk managed to endow this command with three defaults!

Specify corner of window — click a point on the screen, and AutoCAD goes into the **Window** option.

Enter a scale factor — type a number, and AutoCAD goes into the **Scale** option.

<real time> — press **ENTER**, and AutoCAD goes into real-time zoom mode.

ZOOM COMMAND: ADDITIONAL METHODS

The ZOOM command contains these options:

Zoom Option	Meaning
All	Shows everything in the drawing, or the limits of the drawing, whichever is larger.
Center	Zooms about a center point.
Dynamic	Displays a zoom/pan box for interactive zooming.
Extents	Shows everything in the drawing.
Previous	Shows the previous view, whether a zoom or pan.
Scale	Zooms by absolute and relative factors in model space and paper space.
Window	Zooms in to a rectangular area specified by two points.
real time	Zooms in real time with mouse movement.
Left	Zooms relative to a lower left point (hidden option).
Vmax	Zooms out to the maximum possible without invoking a regen (hidden option).

It may seem confusing to have so many options. I find I use only a few of the options frequently, specifically **Extents**, **Window**, and **Previous**. Other users find they don't need this command at all; instead, they use the mousewheel to zoom in and out of the drawing transparently. The options are shown at the command prompt, in the **View | Zoom** menu, and on the **Standard** and **Zoom** toolbars.

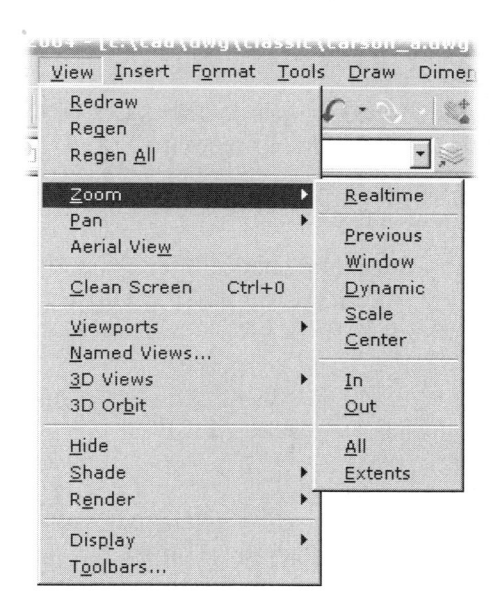

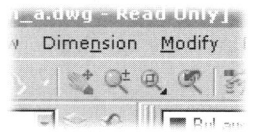

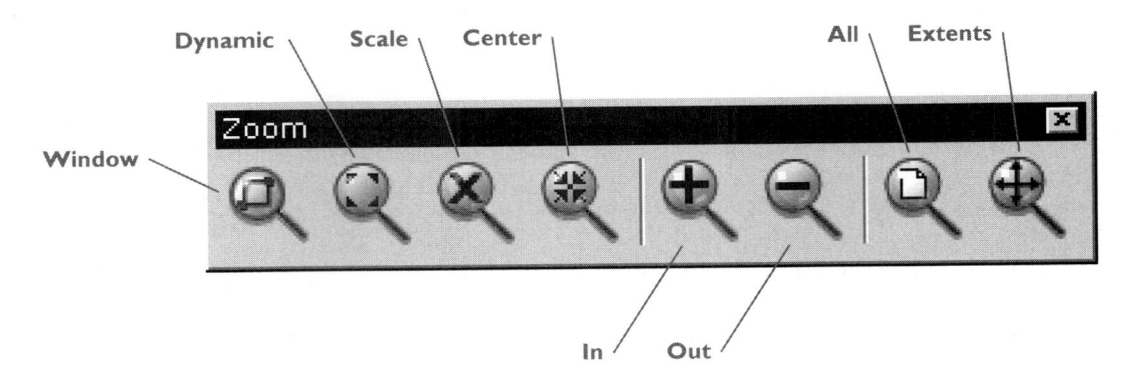

[Figure: Zoom toolbar with labels — Window, Dynamic, Scale, Center, In, Out, All, Extents]

Let's look at each option, in alphabetical order.

Zoom All

The **All** option displays the entire drawing on the screen. This typically displays the entire area of the limits extents of the drawing, including the area of the drawing outside the limits.

Note this difference between the ZOOM command's **Extents** and **All** options: **Extents** displays the extents of the drawing, while **All** displays either the drawing extents or the drawing limits, depending on which is larger.

> Command: **zoom**
> Specify corner of window, enter a scale factor (nX or nXP), or [All/Center/Dynamic/Extents/ Previous/Scale/Window] <real time>: **a**

Occasionally, ZOOM All has to regenerate the drawing twice. If this is necessary, AutoCAD displays the following message on the prompt line:

> ** Second regeneration caused by change in drawing extents.

When the limits are changed, the entire drawing area is not shown until a ZOOM All is performed.

Zoom Center

The ZOOM command's **Center** option allows you to choose a center point for the zoom. You then specify the magnification or height for the new views.

> Command: **zoom**
> Specify corner of window, enter a scale factor (nX or nXP), or [All/Center/Dynamic/Extents/ Previous/Scale/Window] <real time>: **c**
> Specify center point: *(Select a point.)*
> Enter magnification or height <5.0000>: *(Enter a number.)*

If an "x" follows the magnification value, such as 5x, the zoom factor is relative to the current display.

Zoom Dynamic

The **Dynamic** option allows *dynamic* zooming and panning. This option is an ancient predecessor to Aerial View and real-time zoom and pan, and is rarely used anymore.

> Command: **zoom**
> Specify corner of window, enter a scale factor (nX or nXP), or [All/Center/Dynamic/Extents/ Previous/Scale/Window] <real time>: **d**

When you specify **Dynamic**, you are presented with a new window displaying information about current and possible view selections. The window shown in the figure represents a typical display.

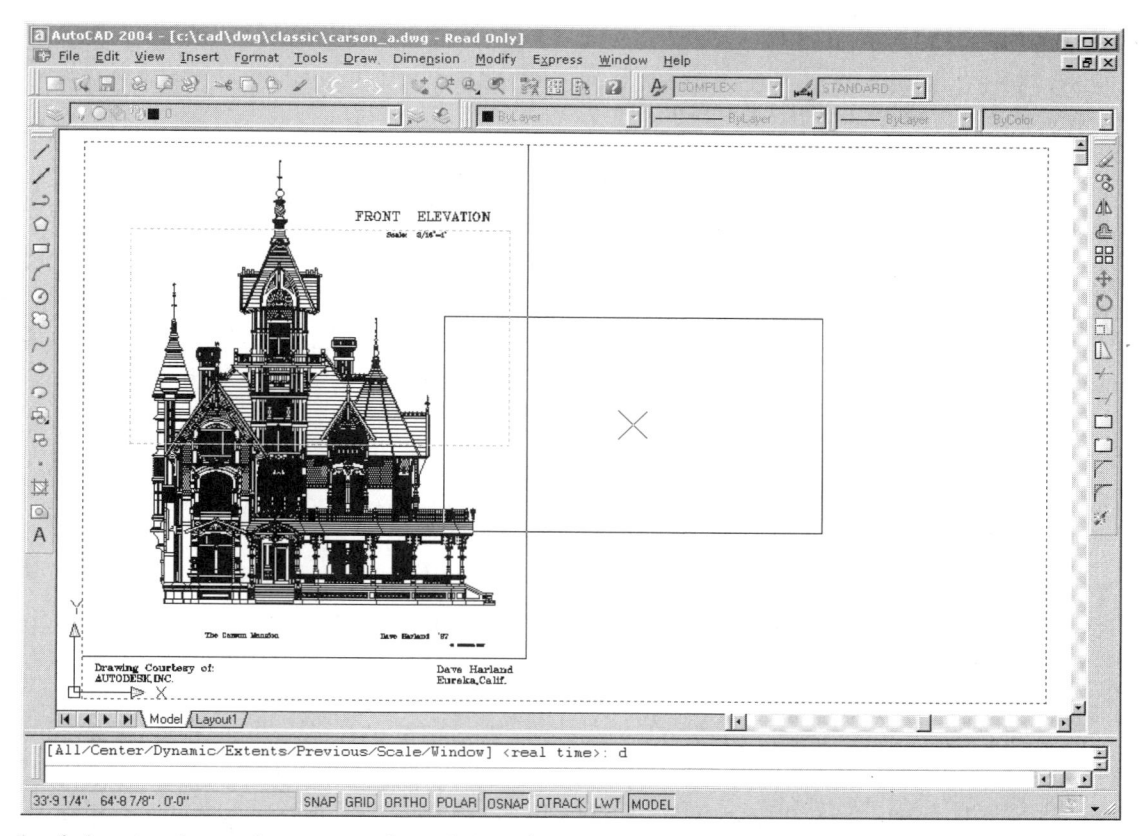

Each of the view boxes has a specific color and pattern:

> **Drawing Extents** is shown by the blue dotted rectangle. The drawing extents can be thought of as the "sheet of paper" on which the drawing resides.
>
> **Initial View Box** is shown by the green dotted rectangle. This is the view of the drawing when you began the **ZOOM Dynamic** command.
>
> **Desired View Box** is shown by the black solid-line rectangle. This defines the size and location of the desired view. The desired view box is initially the same as the current view box. Move this box to obtain the view you want. It operates in two modes: pan and zoom.
>
> **Pan Mode** moves the view box to another location. A large **X** is initially placed in the center of the box. This denotes panning mode. When the **X** is present, moving the cursor causes the box to move around the screen.
>
> **Zoom Mode** is denoted by the arrow. Moving the cursor right or left increases or decreases the size of the view box. The view box increases and decreases in proportion to your screen dimensions, resulting in a "what you see is what you get" definition of the zoomed area. This differs from the standard window zoom, which works from a stationary window corner and may show more of the screen, depending on the proportions of the defined zoom window. Pressing the pick button causes the **X** to change to an arrow at the right side of the box.

You can toggle between zoom and pan modes as many times as you wish to set the size and location of the view box. When you have windowed the desired area, press **ENTER** and the zoom is performed. The area defined by the zoom box is now the current screen view.

It is possible to use dynamic zoom without a mouse — something that harkens back to the days when the mouse was not so common. Use the arrow keys to move the view box. Press **ENTER** to toggle between pan mode and zoom mode.

Zoom Extents

The **Extents** option displays the drawing at its maximum size on the display screen. This results in the largest possible display, while showing the entire drawing.

> Command: **zoom**
>
> Specify corner of window, enter a scale factor (nX or nXP), or [All/Center/Dynamic/ Extents/ Previous/Scale/Window] <real time>: **e**

Zoom Previous

The **Previous** option allows you to return to the last zoom or pan you viewed. This option is useful if you need to move frequently between two areas. AutoCAD remembers the last ten view changes; simply use the "Z P" command several times in a row. Since the view coordinates are stored automatically, you do not need any special procedure to access them.

> Command: **zoom**
>
> Specify corner of window, enter a scale factor (nX or nXP), or [All/Center/Dynamic/ Extents/ Previous/Scale/Window] <real time>: **p**

Zoom Scale

The **Scale** option enlarges or reduces the entire drawing (original size) by a numerical factor. For example, entering **5** results in a zoom that increases the drawing to five times its normal size — AutoCAD zooms in. The zoom is centered on the screen's center point.

When an **x** follows the zoom factor, the zoom is computed relative to the current display. For example, entering **5x** makes the drawing five times larger than its current zoom factor.

Only positive values can be used for zoom scales. To zoom out (i.e., display the drawing smaller), use a decimal value smaller than 1. For example, **0.5** results in a view of the drawing that is one-half its normal size. (As of Release 14, **ZOOM All** and **Extents** leave a bit of room around the drawing, effectively doing a zoom 0.95x.)

> Command: **zoom**
>
> Specify corner of window, enter a scale factor (nX or nXP), or [All/Center/Dynamic/ Extents/ Previous/Scale/Window] <real time>: **s**
>
> Enter a scale factor (nX or nXP): **5**

It is not necessary to enter the "s" for the **Scale** option. Enter a number after starting the command. Then AutoCAD assumes you want the scaled zoom.

The **XP** option scales the model viewport relative to the paper space viewport. This is discussed in *Using AutoCAD 2004: Advanced*.

Zoom Window

The **Window** option uses a rectangular window to show the area, which you specify with two screen picks.

> Command: **zoom**
>
> Specify corner of window, enter a scale factor (nX or nXP), or [All/Center/Dynamic/ Extents/ Previous/Scale/Window] <real time>: **w**
>
> Specify first corner: *(Pick point 1.)*
>
> Specify other corner: *(Pick point 2.)*

A box is shown around the area to be zoomed.

It is not necessary to enter the "w" for the **Window** option. Simply pick two points after starting the command; AutoCAD assumes you want a windowed zoom.

Zoom Real Time

With Pentium computers and fast display drivers providing the horsepower, AutoCAD can zoom in real-time. *Real time* means that the zoom changes continuously as you move the mouse. To enter real-time zoom mode:

>Command: **zoom**
>
>Specify ... <real time>: *(Press ENTER.)*
>
>Press ESC or ENTER to exit, or right-click to display shortcut menu.

Once in real-time zoom mode, the cursor changes to a magnifying glass. To zoom, hold down the left mouse button, and then move the mouse up (toward the monitor) and down (away from the monitor). As you move the mouse up, the drawing becomes larger; as you move the mouse down, it becomes smaller.

While in real-time zoom mode, right-click the mouse to see the following shortcut menu:

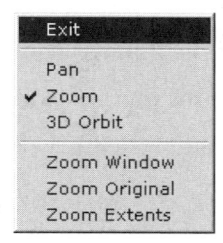

Option	Meaning
Exit	Exits the ZOOM command.
Pan	Switches to realtime pan mode (discussed later in this chapter).
3Dorbit	Allows real-time rotation of 3D objects.
Zoom Window	Displays a cursor with small rectangle. Select two points, which AutoCAD displays as the new zoomed view. Remains in real-time zoom.
Zoom Original	Displays the original view when you first entered real-time zoom.
Zoom Extents	Displays the drawing extents, and remains in real-time zoom.

To dismiss the shortcut menu without choosing one of its options, click anywhere outside the menu. When the drawing is the size you want, press **ENTER** or **ESC** to exit real-time zoom mode and the ZOOM command.

Transparent Zoom

A transparent zoom can be executed while another command is active. Enter 'ZOOM at almost any prompt (notice the apostrophe prefix). There are, however, restrictions in the use of transparent zooms:

Fast zoom mode must be turned on, which it is, by default (this option is set with the **VIEWRES** command).

You cannot perform a transparent zoom if a regeneration is required (zoom outside the generated area — see **ZOOM Dynamic**).

Transparent zooms cannot be performed when certain commands are in progress. These include the **VPOINT, PAN, VIEW,** and another **ZOOM** command.

The drop-down menus and toolbar buttons automatically use transparent zooms.

PAN

The PAN command moves around the view of the drawing.

Many times, you zoom into an area of the drawing to see more detail. You may want to "move" the screen a short distance, to continue working while remaining at the same zoom magnification. This sideways movement is called "panning."

A pan is similar to placing your eyes at a certain distance from a paper drawing, and then moving your head about the drawing. This would allow you to see all parts of the drawing at the same distance from your eyes.

TUTORIAL: PANNING THE DRAWING

1. To move the the view of the drawing, use the PAN command by one of these methods:
   - From the **View** menu, chose **Pan**, and then choose one of its options.
   - On the **Standard** toolbar, select the **Pan Realtime** button.
   - At the 'Command:' prompt, enter the **pan** command.
   - Alternatively, enter the **p** alias at the 'Command:' prompt.

   Command: **pan** *(Press ENTER.)*

2. Hold down the left mouse button; notice the hand cursor.
   Move the mouse; notice that the drawing moves with the mouse.
   Right-click to see the same shortcut menu as displayed during realtime zoom.

3. To exit pan mode, press **ENTER** or **ESC**.
   Press ESC or ENTER to exit, or right-click to display shortcut menu. *(Press ESC.)*

TRANSPARENT PANNING

A pan can be performed transparently while another command is in progress. To do this, enter 'PAN at any prompt, except those expecting you to enter text. The same restrictions apply as for transparent zooms.

SCROLL BARS

Like most other Windows applications, AutoCAD has a pair of scroll bars that allow you to pan the drawing horizontally and vertically (though not diagonally). To pan with a scroll bar, position the cursor over a scroll bar and click. AutoCAD pans the drawing transparently. There are three ways to use the scroll bars:

Click the arrow at either end of the scroll bar. This pans the drawing in one-tenth increments of the viewport size.

Click and drag on the scroll bar button. This pans the drawing interactively as you move the button. This is also the way to pan by a very small amount.

Click anywhere on the scroll bar, except on the arrows and button. AutoCAD pans the drawing by 80 percent of the view.

-PAN

The -PAN command is the command-line alternative to panning. It prompts you to pick a pair of points, much like the line command:

>Command: **-pan**
>
>Specify base point or displacement: *(Pick a point.)*

As you move the cursor, AutoCAD draws a dragline.

>Specify second point: *(Pick another point.)*

After the second pick point, AutoCAD pans by drawing by the distance indicated with the two pick points.

The panning commands are also available from the **View** menu:

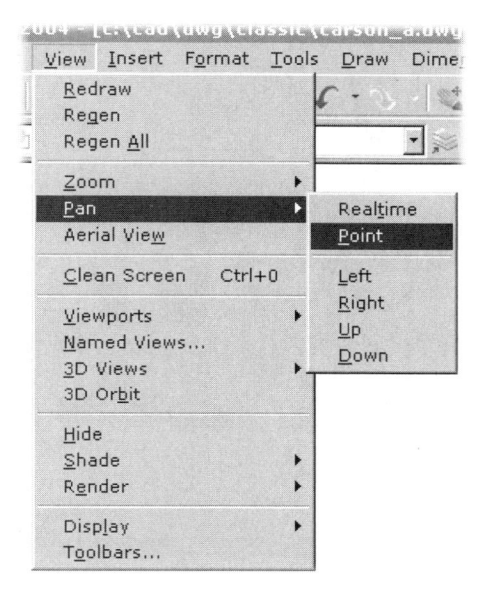

The **Realtime** menu pick executes the PAN command, while the **Point** menu pick executes the -PAN command. The **Left**, **Right**, **Up**, and **Down** menu picks pan the view by 25% in the indicated direction.

NEW IN 2004 — CLEANSCREENON

The **CLEANSCREENON** command minimizes the user interface to maximize the drawing. It turns off the title bar, toolbars, and window edges.

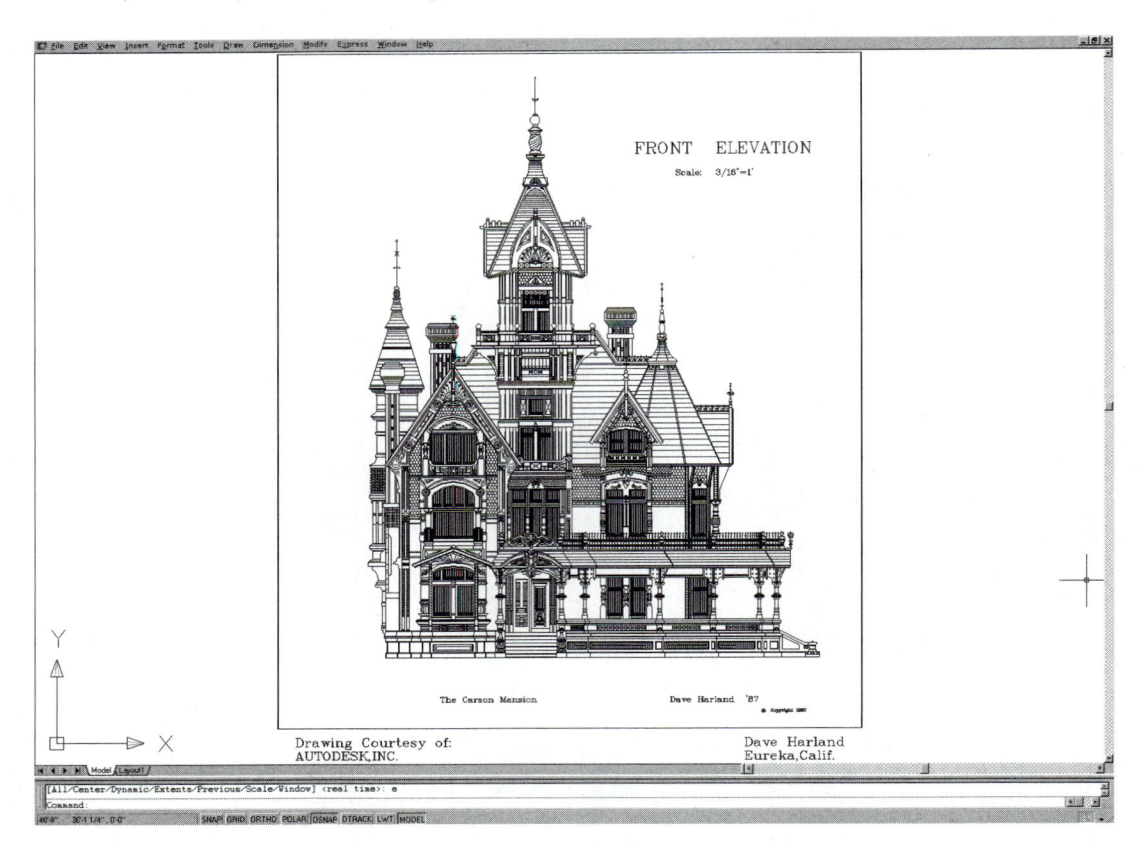

Note: You can make the drawing area even larger by turning off the scroll bars and layout tabs. This is done through the **Display** tab of the **Options** dialog box. In addition, you can drag the Command Prompt away from the bottom of the screen area to make it float, and then make it transparent.

TUTORIAL: TOGGLING THE DRAWING AREA

1. To make the drawing area as large as possible, start the **CLEANSCREENON** command with one of these methods:
   - From the menu bar, choose **View**, then choose **Clean Screen**.
   - On the keyboard, press **CTRL+0** (zero).
   - At the 'Command:' prompt, enter the **cleanscreenon** command.

   Command: **cleanscreenon** *(Press* ENTER.*)*

2. To return to the "normal" screen, use the **CLEANSCREENOFF** command.
   Command: **cleanscreenoff** *(Press* ENTER.*)*

You can press **CTRL+0** (zero) to toggle between maximized and regular views.

DSVIEWER

The **DSVIEWER** command provides a "bird's-eye view" of the drawing in an independent window called "Aerial View." This window is an an alternative to the realtime **ZOOM** and **PAN** commands and scroll bars. The Aerial View window lets you see the entire drawing at all times in an independent window. This is sometimes called the bird's-eye view.

After the drawing appears in the Aerial View window, zooming in is as simple as with the **ZOOM Window** command: pick two points. To zoom to another area, move the cursor. To pan, move the rectangle to the new location.

TUTORIAL: OPENING THE AERIAL VIEW WINDOW

1. To view the entire drawing in an independent window, start the **DSVIEWER** command with one of these methods:
   - From the menu bar, choose **View**, and then **Aerial View**.
   - At the 'Command:' prompt, enter the **dsviewer** command.
   - Alternatively, enter the **av** alias at the 'Command:' prompt.

   Command: **dsviewer** (Press ENTER.)

2. Notice that AutoCAD opens the Aerial View window.
   The Aerial View always shows the entire drawing. The heavy rectangle shows the current view in AutoCAD, while the light rectangle is the proposed view.

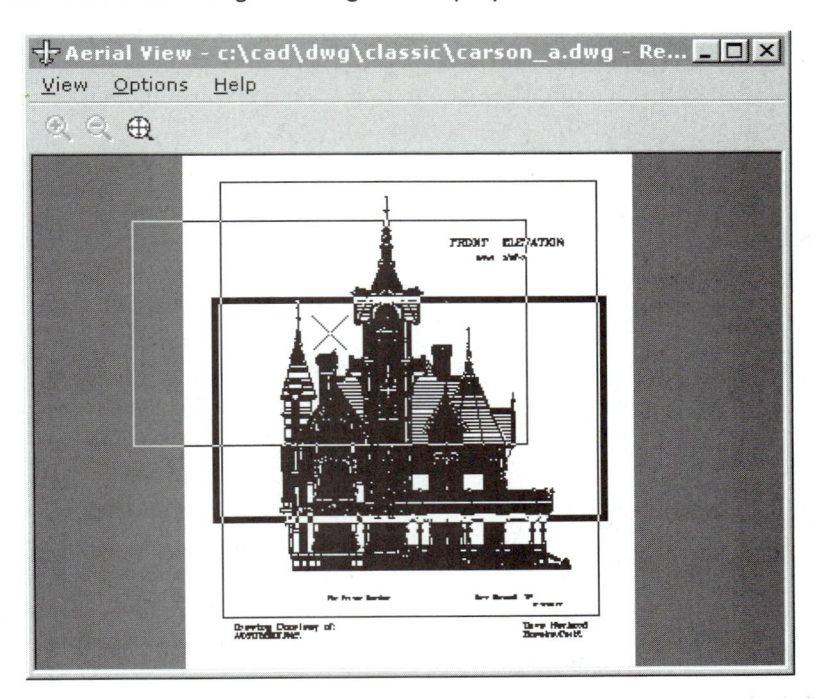

3. As wtih **ZOOM Dynamic,** you can switch between zoom and pan modes: left-click to switch. The **X** indicates pan mode, while the arrow indicates zoom mode.
   Moving the mouse in pan mode pans the view.
   Moving the mouse in zoom mode zooms in or out.
4. When you have the view you want, right-click in the Aerial View window.

AERIAL VIEW WINDOW: ADDITIONAL METHODS

The Aerial View window contains a menu bar and toolbar that provide options for viewing the drawing.

View Menu

The options in the **View** menu change the magnification of the Aerial View — not the drawing itself. The **Zoom In** option zooms in by a factor of 2, while the **Zoom Out** option zooms out by a factor of 2. The **Global** option displays the entire drawing of the current view in the Aerial View window.

Options Menu

The **Options** menu changes how the Aerial View window operates. The **Auto Viewport** option displays the model space view of the *current* viewport automatically. The **Dynamic Update** option updates the Aerial View window as you edit the drawing (when off, the image in the window is not updated until you click it). The **Realtime Zoom** option updates the drawing area in real time when you zoom using the Aerial View window.

VIEW

The **VIEW** command stores and recalls views of the drawing by name. This lets quickly move about a drawing. Think of it as combining the **ZOOM** and **PAN** commands, without the need to specify zoom ratios and pan directions.

You can save the current display as a view, or else window an area to define the view. Naturally, if you are saving the current view, you need to zoom and pan into that view first. Views can be stored relative to the current UCS (user-defined coordinate system), but this only makes sense when working with 3D drawings. See *Using AutoCAD 2004: Advanced*.

You take advantage of named views outside of the **VIEW** command in two areas. When AutoCAD starts, you can specify the name of a view to be displayed when the drawing opens. And, when AutoCAD plots, you can specify a named view with the **PLOT** command.

A drawback to named views is that they do not connect with layer names. That means you cannot access a view that automatically turn off certain layers for a more specific view.

You can change the name of views with the **RENAME** command, and delete views from the drawing with the **VIEW** command's "hidden" **Delete** option.

PREDEFINED VIEWS

In addition, AutoCAD includes predefined views — the six standard orthographic views and four standard isometric views. These are meant for viewing 3D drawings. The orthographic views show the top, bottom, left, right, front, and back of 3D drawings, while the isometric views show the southwest, northwest, northeast, and southeast views.

You can access these named views from the **View** menu and the **View** toolbar. From the menu bar, select **View | 3D Views**:

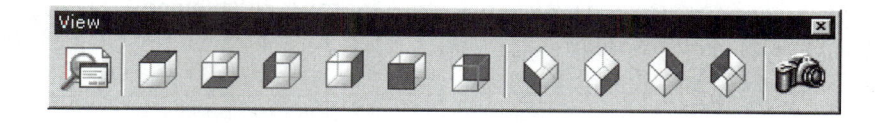

TUTORIAL: STORING AND RECALLING NAMED VIEWS

1. To create named views, start the **VIEW** command with one of these methods:
   - From the menu bar, choose **View**, and then **Named Views**.
   - From the **View** toolbar, choose the **Named Views** button.
   - At the 'Command:' prompt, enter the **view** command.
   - Alternatively, enter the aliases **v** or **ddview** (the old name for this command) at the 'Command:' prompt.

   Command: **view** *(Press ENTER.)*

   Notice that AutoCAD displays the **View** dialog box.

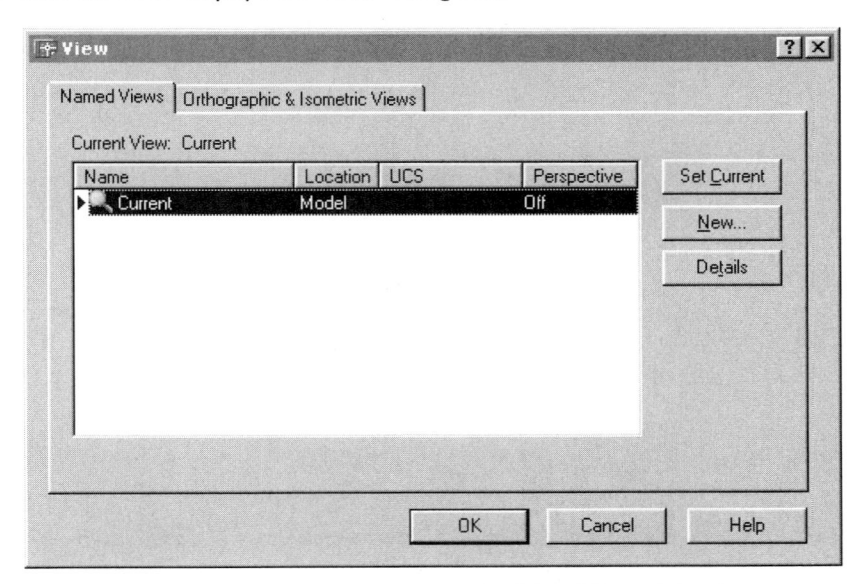

2. Click **New** to create a new named view. Notice the **New View** dialog box.

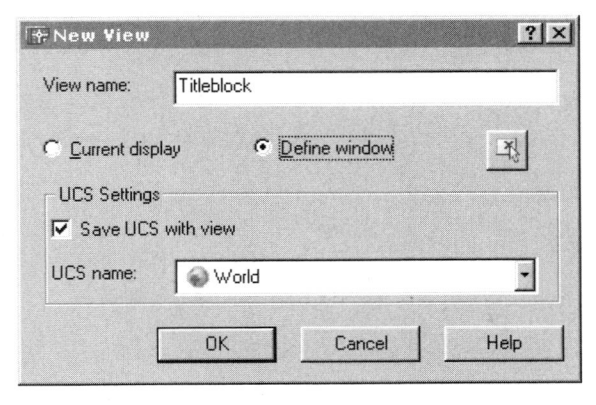

3. Enter a name for the view. Later, when it comes time to recall the view, you select it by name. That's why it makes sense to give the view a descriptive name, such as **Titleblock**.

4. Decide whether the view consists of:

   Current display. If you want the view to be the current display but the display isn't right, you need to get out of the dialog boxes (click **Cancel** twice), and then use the **ZOOM** and **PAN** commands to make the view right. Restart the **VIEW** command.

Define window. Click the **Define Window** button. Notice that AutoCAD dismisses the two dialog boxes temporarily, and prompts you to pick two points that define the rectangular window.

Specify first corner: *(Pick a point.)*

Specify opposite corner: *(Pick another point.)*

You can ignore the options relating to **UCS Settings**, because they are meaningful only when working with 3D drawings.

4. Click **OK**. AutoCAD adds the view name to the list.

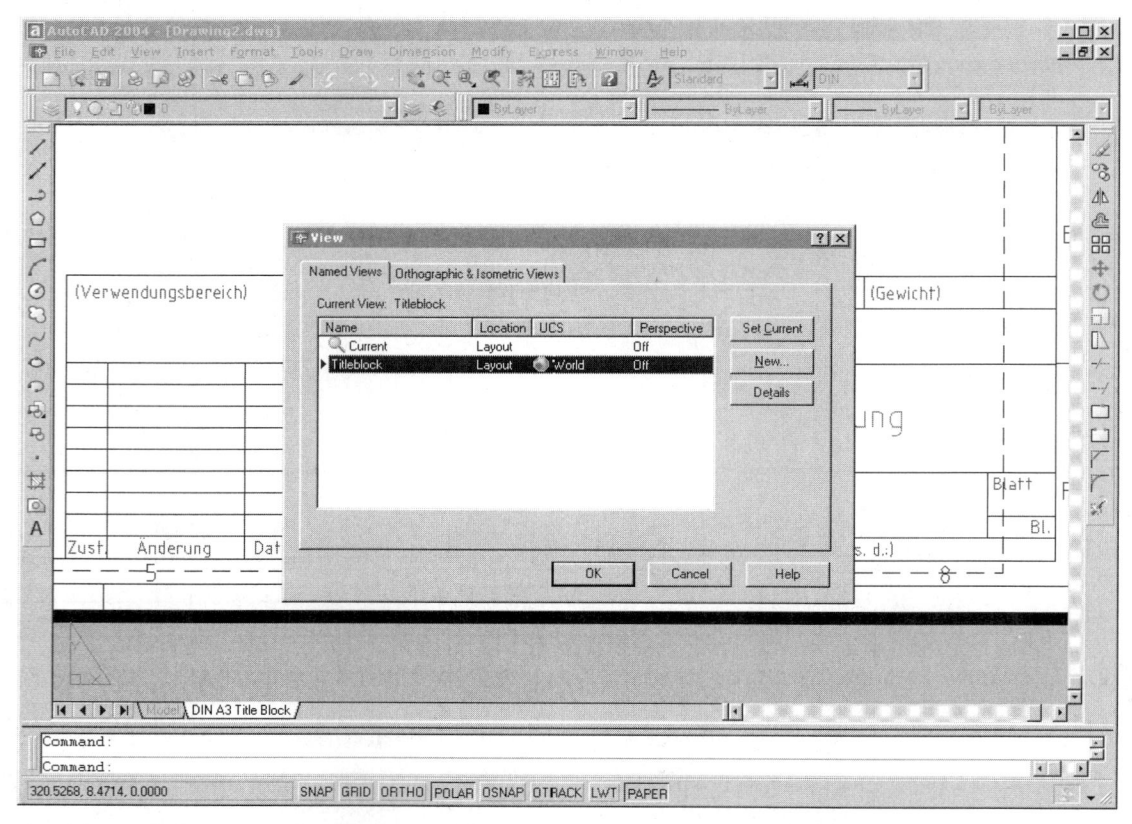

5. To recall the view, select its name (such as "Titleblock"), and then click **Set Current**.
   A small triangle points to the current view name.
6. Click **OK**, and notice how AutoCAD fills the screen with a closeup of the drawing's titleblock (or whatever you windowed).

TUTORIAL: RENAMING AND DELETING NAMED VIEWS

You can rename and delete views, but it's not entirely obvious at first glance, because the commands are "hidden" in a shortcut menu.

1. Open the View dialog box.

2. Select a named view, and right-click. The shortcut menu provides these options:

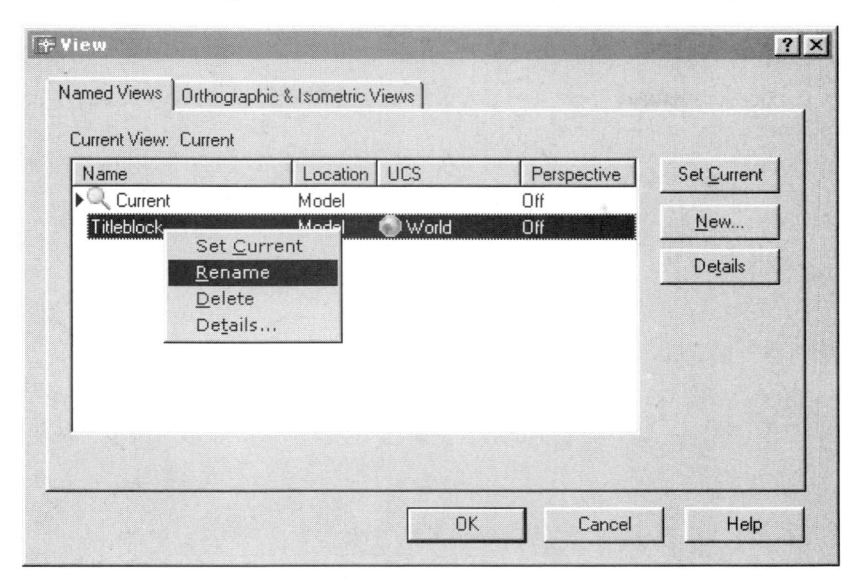

Set Current makes the named view current.

Rename changes the name of the view.

Delete removes the named view from the drawing. *Careful!* AutoCAD removes the view without warning.

Details displays a dialog box that descibes the saved view.

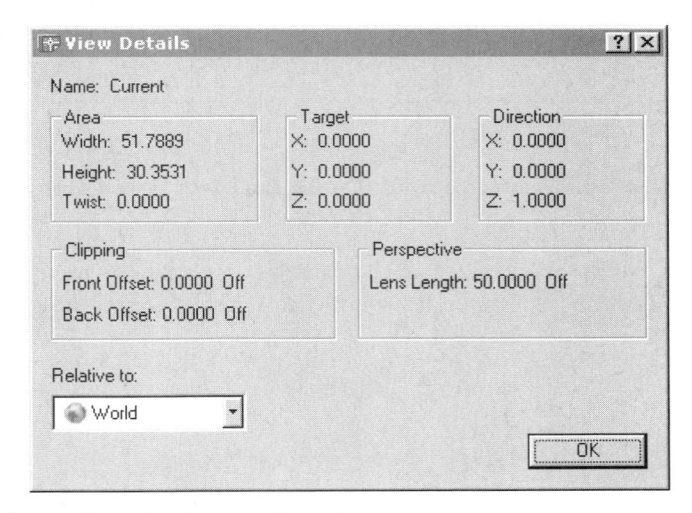

3. When done, click **OK** to dismiss the dialog box.

TUTORIAL: STARTING AUTOCAD WITH A NAMED VIEW

AutoCAD originally ran on another operating system called DOS (short for "disk-based operating system, also the building block upon which Windows was designed.) In those days, it was common to start a program with *command-line switches*, which instructed the program how to start. These switches are still available for AutoCAD, and the /v switch specifies the named view to display when AutoCAD opens a drawing. The drawing, naturally, needs to have the named view, otherwise AutoCAD ignores the /v switch, and instead shows the last saved view.

To use the /v switch, edit AutoCAD's command line, as follows:

1. On the Windows desktop, right-click the AutoCAD icon.
2. From the shortcut menu, select **Properties**.

3. In the **Properties** dialog box, select the **Shortcut** tab.
4. In the **Target** text box, add the boldface text to the command line:
   "C:\AutoCAD 2004\acad.exe" **"c:\autocad 2004\sample\file name.dwg" /v titleblock**
5. Click **OK** to close the dialog box, and double-click the AutoCAD icon to see if it loads the drawing with the named view.

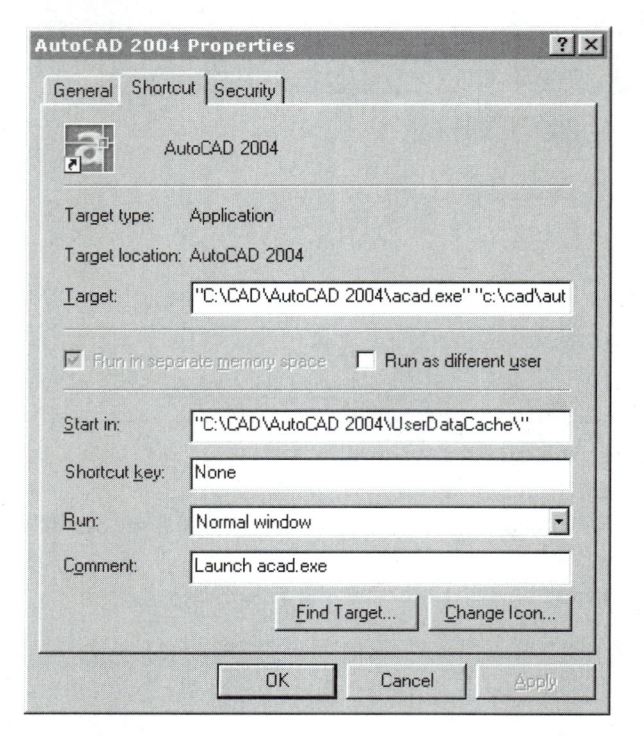

Notes: A space is required after the switch. Double quotes are needed when there are spaces in the filename.

You can have more than one shortcut icon on the Windows desktop. To make copies of icons, hold down the **CTRL** key, and then drag the icon. When you let go, Windows creates the copy, which you can then rename and edit.

VIEWRES

The VIEWRES command controls the roundness of circular objects. This is a technical command that is almost never needed anymore. It can be useful to know the purpose of the command, and hence the reason to use it when circles looks like octagons.

AutoCAD stores the objects you draw mathematically. When you draw a circle, for example, AutoCAD does not store a circle, but the circle's two primary parameters: its centerpoint and radius. If you drew the circle by specifying its centerpoint and diamter, AutoCAD internally converts the diameter to a radius. You can see this by using the LIST command on a circle: no matter how it was drawn, AutoCAD lists the circle's centerpoint and radius.

When it comes to diplaying the objects, AutoCAD displays them as vectors. *Vectors* are lines that have direction and length (Windows and the graphics board then convert the vector data into the raster image displayed by the screen). Everything is displayed by AutoCAD as vectors, even circles and arcs. They are composed of many very short vectors, so short that circular objects look round to you.

In most cases, you do not notice that circular objets are made of very short lines. Sometimes, however, circles and arcs look like octagons. This occurs when you zoomed into the drawing by a large amount. Regardless of the setting, AutoCAD never displays a circle with fewer than eight sides. On the other hand, AutoCAD does not display any more circle or arc segments than it calculates to be necessary for the current zoom. If the circle uses less than two screen pixels at the maximum zoom magnification that does not require a regeneration, the circle is displayed as a single pixel. Below, the circle at the right was regnerated with **VIEWRES** set to 1.

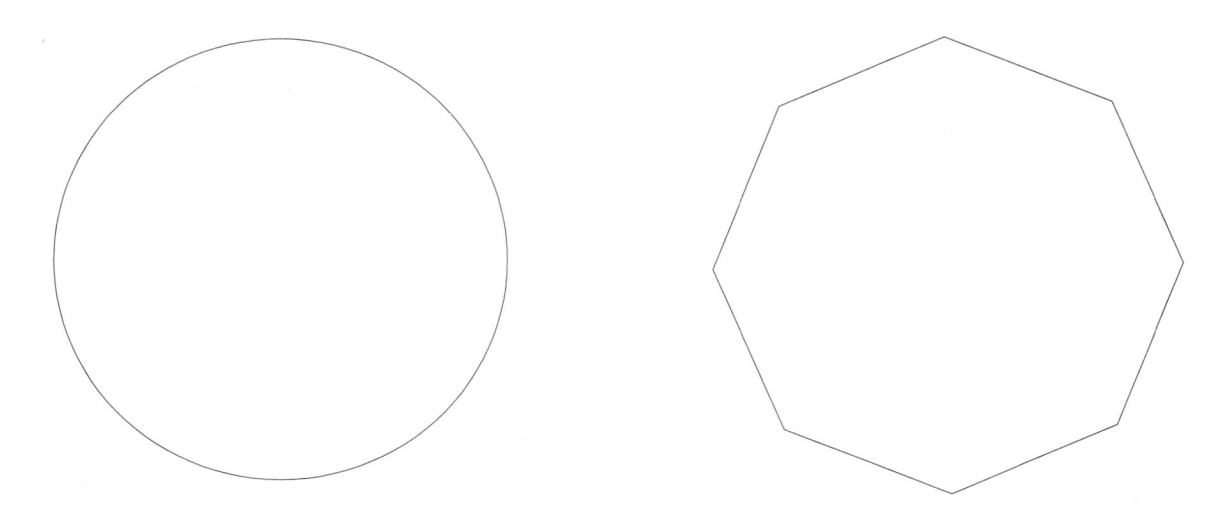

AutoCAD regenerates some zooms, some pans, and view restores, and when entering layout mode (paper space). The "Fast Zoom" mode allows AutoCAD simply to redisplay the screen wherever possible. This redisplay is performed at the faster redraw speed, instead of the slower regen speed. (Some extreme zooms still require a regeneration.)

TUTORIAL: CHANGING THE VALUE OF VIEWRES

1. Start the **VIEWRES** command, as follows:
    - At the 'Command:' prompt, enter the **viewres** command.

    Command: **viewres** *(Press* ENTER.*)*

2. Entering **n** at the first prompt causes all zooms, pans, and view restores to regenerate. For fastest speed, answer **y**:

    Do you want fast zooms? [Yes/No] <Y>: *(Enter* **y** *or* **n**.*)*

3. The default value for the circle zoom percent is 100. A value less than 100 diminishes the resolution of circles and arcs, but results in faster regeneration times.

    Enter circle zoom percent (1-20000) <1000>: **10000**

Values greater than 100 results in a larger number of vectors than usual being displayed for circles and arcs. For today's fast computers, Autodesk has increased **VIEWRES** to 1000, resulting in smooth circles and arcs at a zoom factor of 10. Increase the value to 10,000 for smooth circles at zoom factor 100.

EXERCISES

1. Draw the following baseplate from lines and circles.

   Set snap to 0.25 and grid to 1.0. Use the appropriate object snaps to assist your drafting. The dimensions need not be exact, but it may be helpful to know that the circles have a radius of 1.0.

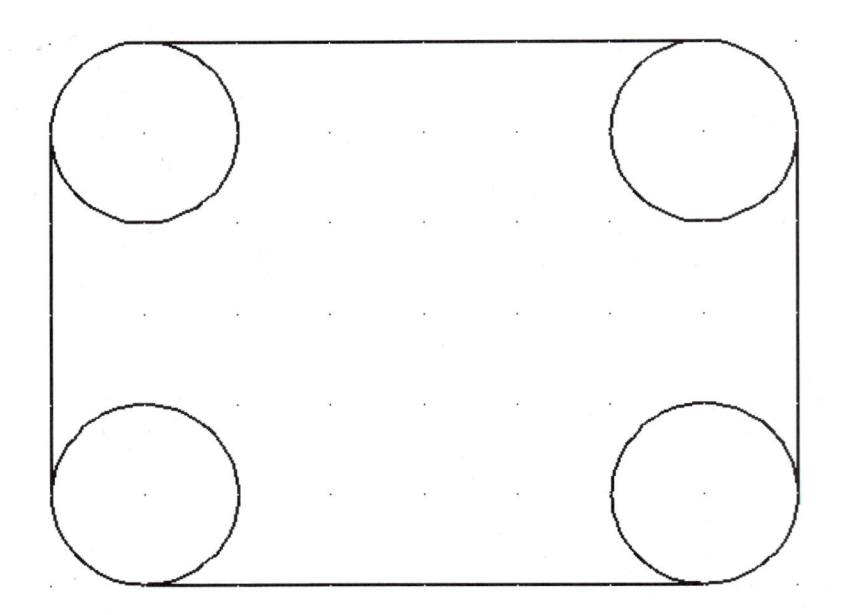

2. Draw the following floor using lines.

   Set snap and grid to 12.0. Set limits to 0,0 and 600,400. Use the ZOOM All command to see the entire drawing area. Turn on ortho mode. Use the appropriate object snaps to assist your drafting. The dimensions need not be exact, but it may be helpful to know that the longest wall is 500 units long.

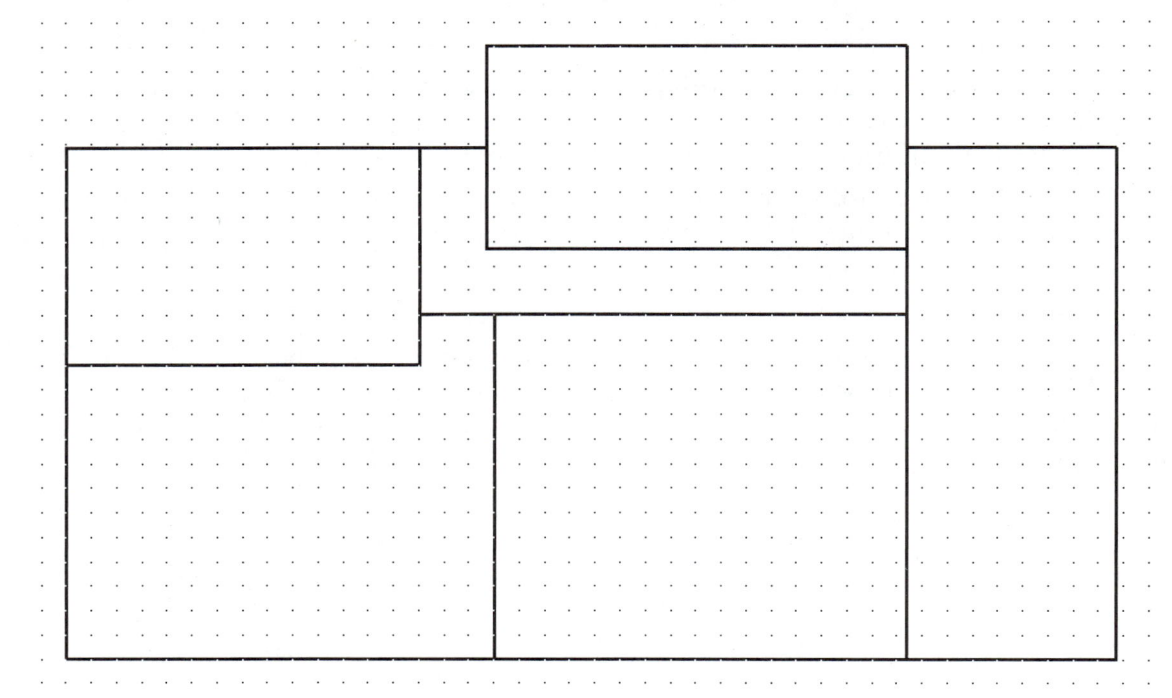

3. Draw the following L-shaped bracket using lines and arcs.
   Set the snap to 1.0 and the grid to 12.0.
   Set the limits to 0,0 and 216,144.
   Use the ZOOM All command to see the entire drawing area.
   Turn on ortho mode.
   Use the appropriate object snaps to assist your drafting. Remember that arcs are drawn counter-clockwise. The dimensions need not be exact, but it may be helpful to know that the longest edge is 10 units long, and that the metal is 1 unit thick.

4. Draw an isosolese triangle with lines.
   Its angles are 45-45-90, and the two sides are 5 units. Use polar tracking set to 45 degrees and polar snap set to 5 units.

5. Draw the profile of the electrical switch spring shown below.
   Use polylines, switching between line- and arc-drawing mode, as necessary.
   Set the pline width to 0.1 units. Snap and grid are 1.0 units.

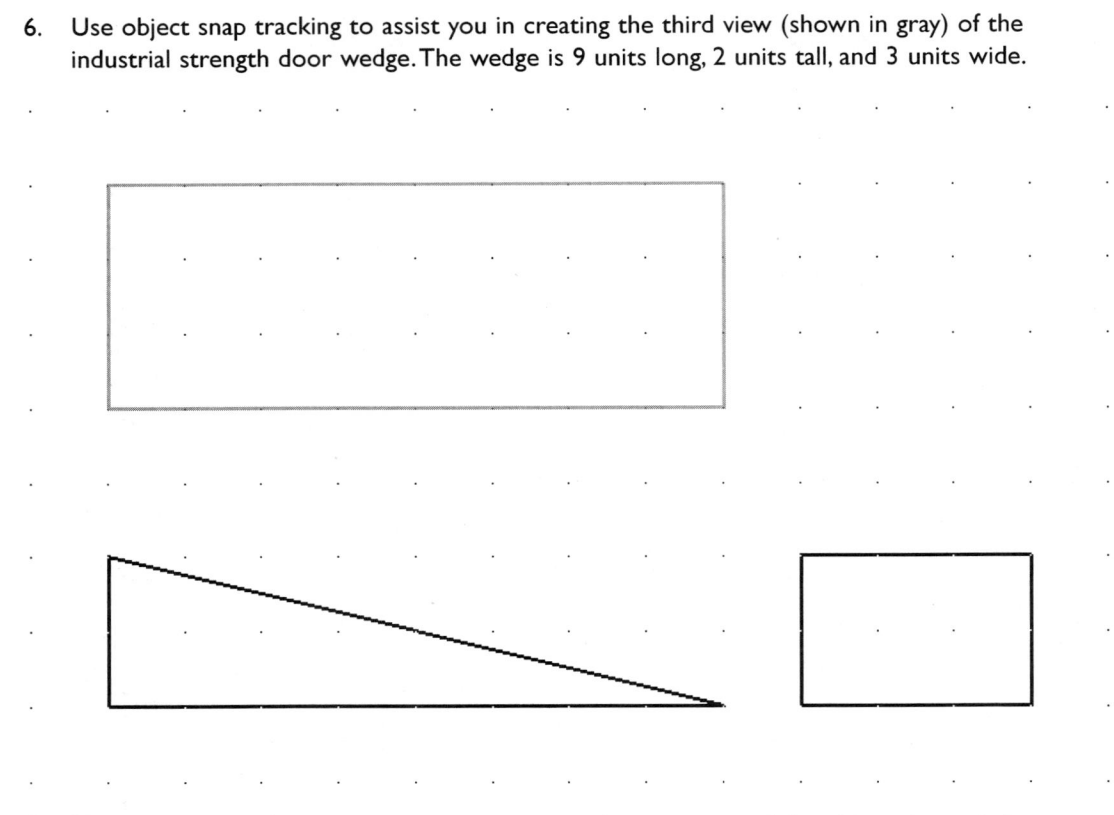

6. Use object snap tracking to assist you in creating the third view (shown in gray) of the industrial strength door wedge. The wedge is 9 units long, 2 units tall, and 3 units wide.

7. Use construction lines to assist you in creating the true view of the object shown below. The units are shown on the sketch.

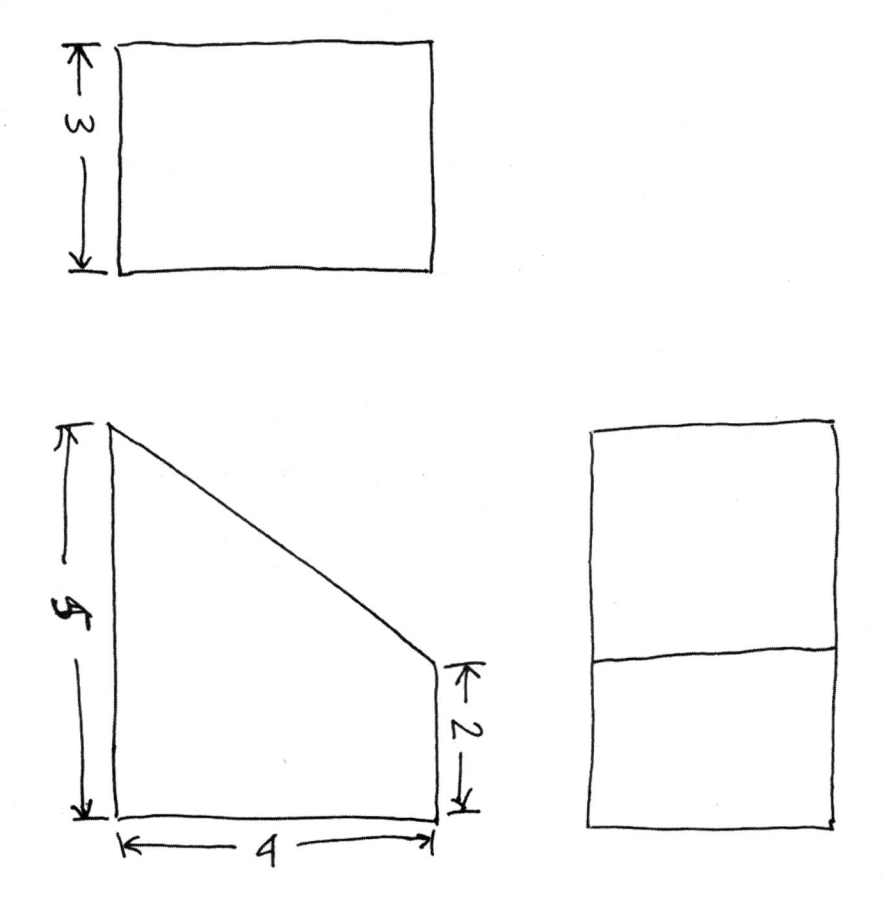

8. Draw the front, side, and top views of the bracket made of 1/4" sheetmetal. The units are shown on the sketch.

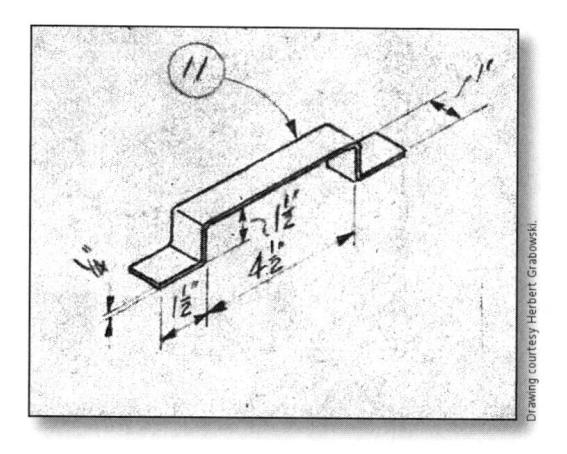

9. Using object snaps, place an arc tangent to the two circles.
   Then, draw the remaining lines, using the object snaps shown.
   The larger circle has a radius of 4 units; the smaller a radius of 2 units.

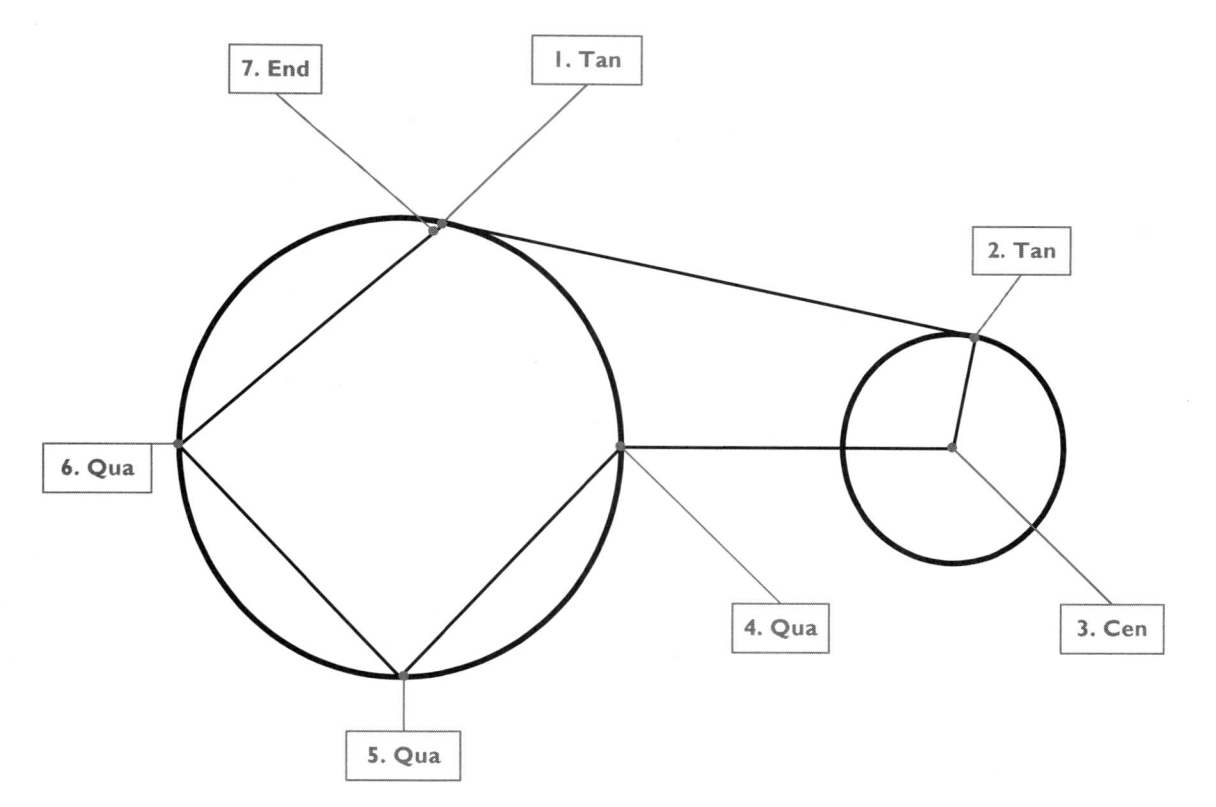

   1. Tan
   2. Tan
   3. Cen
   4. Qua
   5. Qua
   6. Qua
   7. End

10. Draw the 3.5" wide by 3.7" tall diskette.
    The figure below illustrates the diskette full size, 1:1; take your measurements directly from the photograph.

11. Draw the 3.9" wide by 1.75" tall crossbrace for ceiling lamps.
    Take your measurements directly from the photograph, which is shown below at full-size, 1:1 scale.

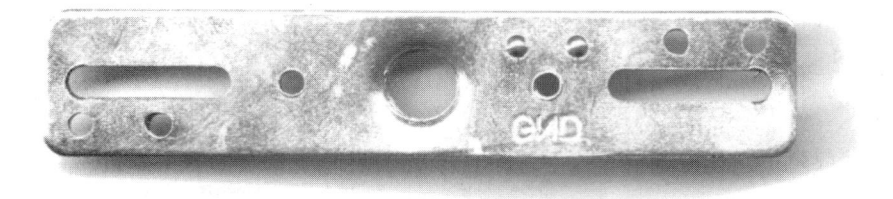

12. Draw the 2.2" wide by 1.0" tall doorjamb coverplate.
    Take your measurements directly from the full-size photograph, below.

CHAPTER REVIEW

1. What command allows you to enlarge and reduce the visual size of the drawing?
2. What is the purpose of the PAN command?
3. What is the **Realtime** option under the ZOOM command used for?
4. When a real-time zoom is in progress, can you override it? If so, how?
5. How do the PAN and ZOOM commands differ?
6. What does "toggle" mean?
7. What is a transparent command? How is this option invoked?
8. How do the REDRAW and REGEN command differ?
9. List four ways by which the grid can be toggled:
    a.
    b.
    c.
    d.
10. What is temporary tracking used for?
11. Name two ways to create construction lines:
    a.
    b.
12. What is the "aperture"?
13. What object snap mode would you choose to snap to a point object?
14. Can you snap to the midpoint of an arc?
15. When does AutoSnap come into effect?
16. Can different X and Y spacing be given to the snap? Can it coincide with the grid spacing?
17. What advantages do rays and xlines have over grids?
18. What mode provides great accuracy for creating true horizontal and vertical lines?
19. What purpose does APPint mode serve?
20. Which command controls the extent of the grid?
21. Although the grid is not a part of the drawing, can it be plotted?
22. Where does a ray start? End?
23. Name the function of the Aerial View.
24. In addition to the snap angle, the **Angle** option of the SNAP command also affects
    a.
    b.
    c.
    d.
25. Which mode does **F8** toggle?
26. What is the difference between ortho and polar modes?
27. What are tooltips?
28. When would you use the < symbol in AutoCAD?
29. What is an alignment path?
30. Describe the function of PolarSnap.
31. Provide the meaning of the following abbreviations:
    osnap
    otrack
    appint
    tt

32. Define the following object snap abbreviations:

    int

    cen

    qua

    end

33. Which part of the object geometry do the following object snaps snap to?

    ins

    nod

    per

    tan

34. When would you use the **From** object snap?
35. Describe how the EXTension object snap operates.
36. Can AutoCAD snap to the intersection of two circles?

    If so, in how many places could the object snap take place?
37. Why does AutoCAD sometimes defer snapping to a tangent?
38. Which command is toggled by the following function keys?

    F3

    F9

    F10

    F11

39. What is the purpose of AutoSnap's magnet?
40. How do the **REDRAW** and **REDRAWALL** commands differ?
41. When might you turn off **REGENAUTO**?
42. Name the command that matches the alias:

    z

    sn

    xl

    ds

43. Which command maximizes the drawing area?
44. What is the function of the **DSVIEWER** command?
45. Name four commands that take advantage of named views.

    a.

    b.

    c.

    d.

46. How do you delete a named view?
47. When do you need to increase the setting of the **VIEWRES** command?

CHAPTER 6

Drawing with Efficiency

In addition to precision, another advantage of CAD over hand drafting is *efficiency*. "You should never have to draw anything twice," said John Walker, one of the founders of Autodesk Inc. By drawing efficiently, you complete projects in less time — or more projects in the same time. In this chapter, you learn to use some of AutoCAD's commands for creating and reusing content, such as symbols (blocks), poches (hatch patterns), and groups of objects. The commands are:

ADCENTER (DesignCenter) provides access to content in other drawings.

ADCNAVIGATE sets the home path for DesignCenter

BLOCK and **-BLOCK** create reusable components called "blocks."

INSERT and **-INSERT** place blocks in drawings.

BASE changes the base point of drawings.

WBLOCK exports blocks and drawings as *.dwg* drawing files.

MINSERT inserts blocks as rectangular arrays.

BLOCKICON creates icons (preview images) of blocks.

TOOLPALETTES provide access to frequently-used content (new to AutoCAD 2004).

TOOLPALETTESCLOSE closes the Tool Palettes window (new to AutoCAD 2004).

BHATCH and **HATCH** place hatch patterns and colored fills.

HATCHEDIT edits associative hatch patterns and fills.

BOUNDARY and **-BOUNDARY** create single regions from disparate areas.

SELECT selects objects by their location.

QSELECT selects objects by their properties.

PICKBOX changes the size of the pick cursor.

DDSELECT provides control over selection options.

FILTER selects objects based on their properties and location.

GROUP creates selectable groups of objects.

DRAWORDER controls the order in which overlapping objects are displayed.

FINDING THE COMMANDS

On the **STANDARD** toolbar:

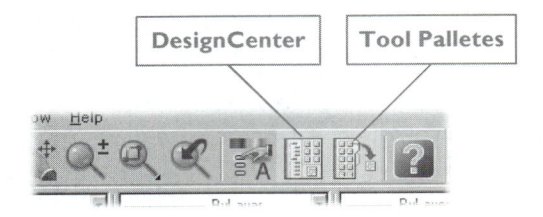

On the **DRAW** toolbar:

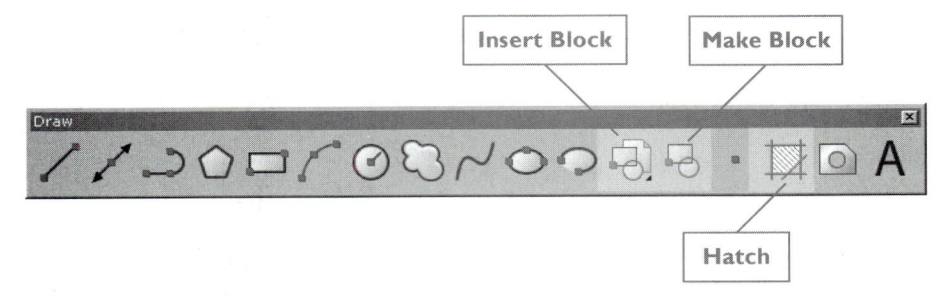

On the **MODIFY II** toolbar:

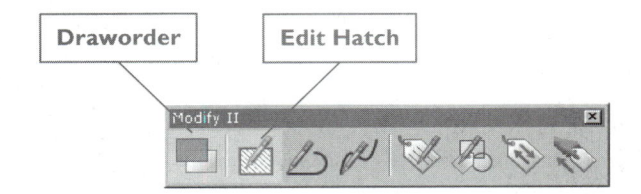

On the **INSERT** menu:

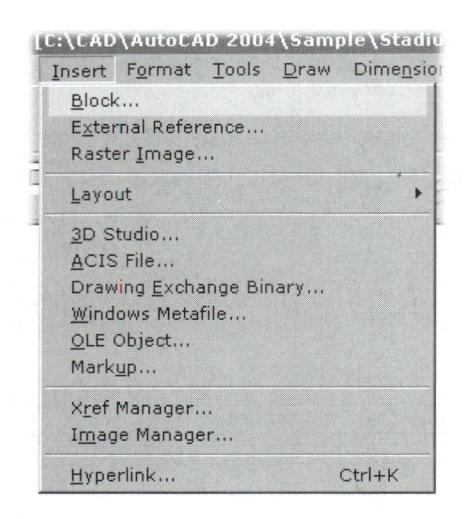

On the **DRAW** and **TOOLS** menus:

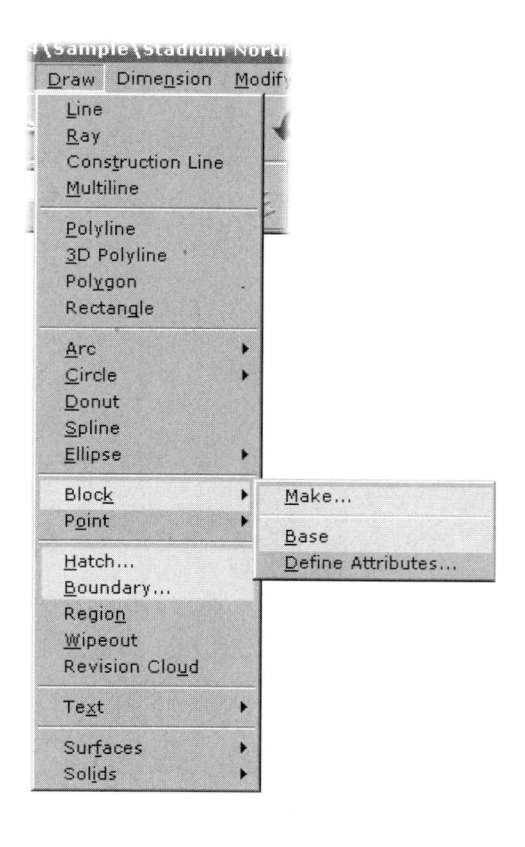

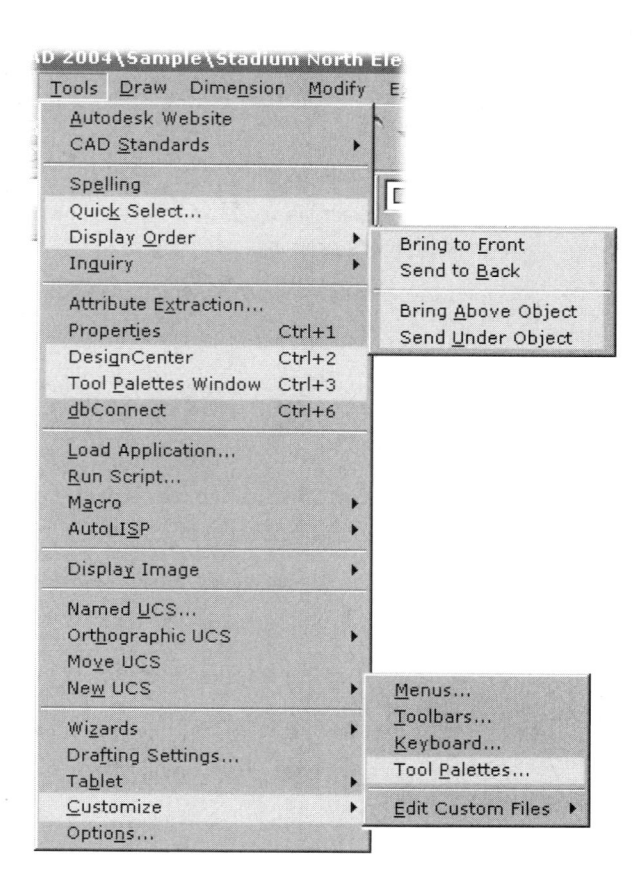

ADCENTER

The **ADCENTER** command (short for "AutoCAD Design Center") displays a window that provides access to content in other drawings. DesignCenter provides the following services:

- Views the content of folders and files on your computer and the network.
- Displays the layers, linetypes, text styles, blocks, dimension styles, external references, and layouts of every drawing.
- Copies content from one drawing to another.
- Accesses blocks (symbols) from Autodesk's Web site.
- Previews the content of drawing and raster files.

TUTORIAL: DISPLAYING THE DESIGNCENTER

1. To display the DesignCenter window, start the **ADCENTER** command by one of these methods:
   - From the menu bar, choose **Tools**, and then **DesignCenter**.
   - On the standard toolbar, pick the **DesignCenter** button.
   - On the keyboard, press **CTRL+2**.
   - At the 'Command:' prompt, enter the **adcenter** command.
   - Alternatively, enter the aliases **adc**, **dc**, **dcenter**, or **content** at the keyboard.

   Command: **adcenter** *(Press ENTER.)*

2. Notice the DesignCenter window.

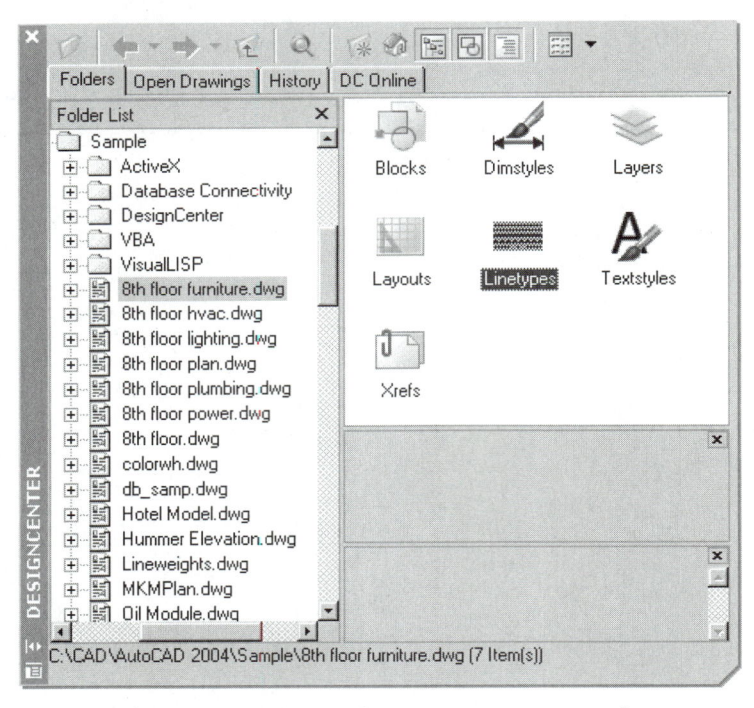

3. To turn off the DesignCenter window, use the **ADCCLOSE** command, or press **CTRL+2**. As an alternative, click the **x** in the window's upper corner.

4. The **ADCNAVIGATE** command specifies the initial path for DesignCenter:

   Command: **adcnavigate**

   Enter pathname <>: *(Enter a path, such as c:\autocad 2004\sample.)*

NAVIGATING THE DESIGNCENTER: ADDITIONAL METHODS

The DesignCenter is a collapsible window that can float independent of AutoCAD, be docked to the edge of the drawing area, or minimized to just its title bar.

Docking DesignCenter

To dock, drag DesignCenter by its title bar to an edge of the AutoCAD window. Notice that DesignCenter attaches itself to the edge. The drawback to docking is that the drawing area becomes significantly smaller.

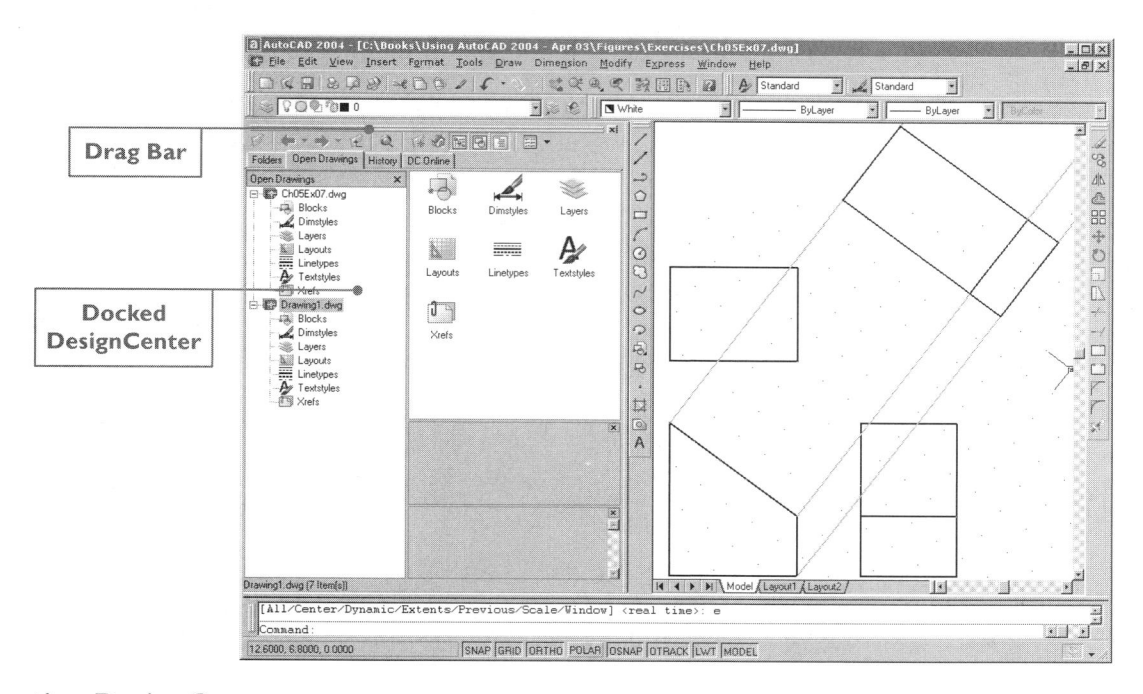

Floating DesignCenter

To float, drag the double-line (at the top of DesignCenter) away from the edge of the AutoCAD window. Notice that DesignCenter changes its shape.

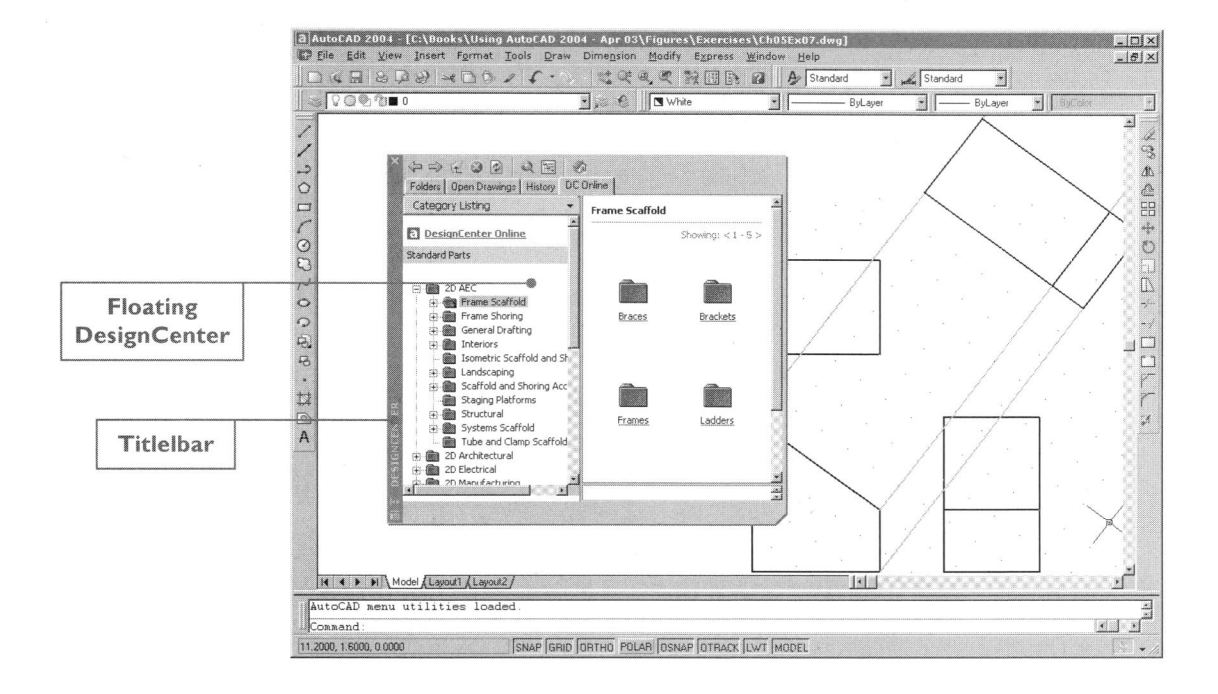

You can move DesignCenter anywhere on the screen, including outside of the AutoCAD window. If you have a two monitor system, consider dragging DesignCenter to the second monitor.

Minimizing DesignCenter

To minimize, click the double-arrow AutoHide button on DesignCenter's title bar. Notice that DesignCenter collapses to just its title bar. To restore, pass the cursor over DesignCenter, and it springs back to its original size. This method works only when DesignCenter is floating; when DesignCenter is docked, "minimize" it by dragging the boundary between it and the drawing area.

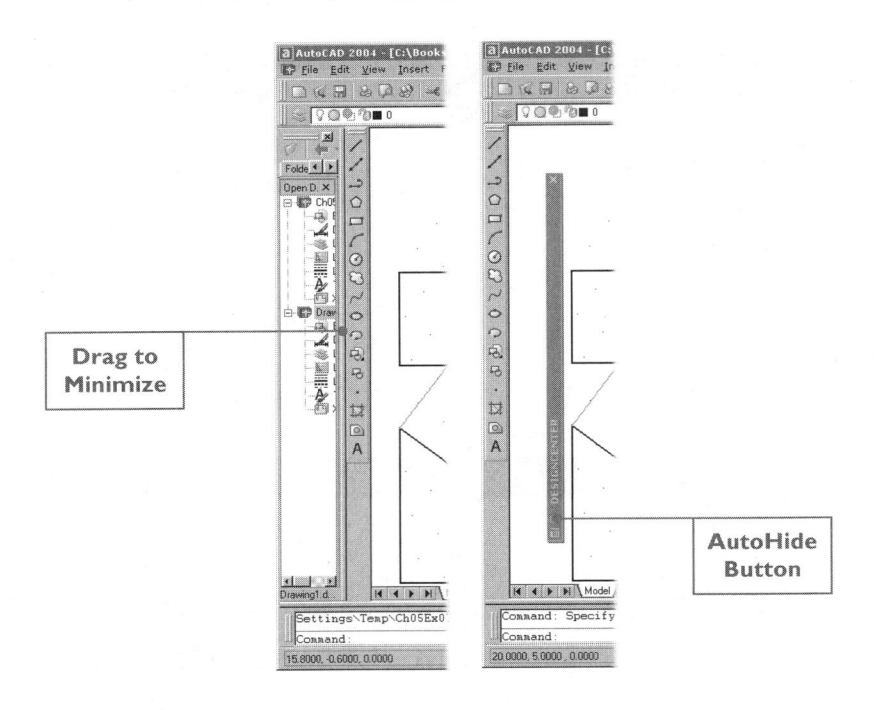

UNDERSTANDING DESIGNCENTER'S USER INTERFACE

The DesignCenter consists of many user interface elements, including toolbar, tabs, title bar, and several panes.

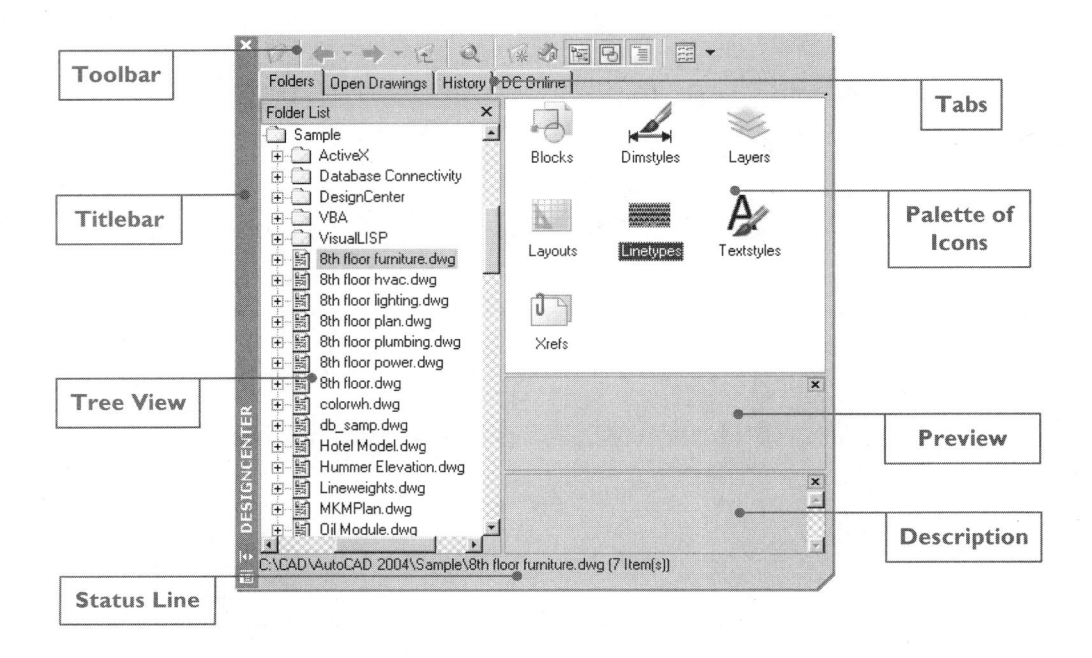

Tabs

DesignCenter uses tabs to display content in four different contexts.

Folders

The **Folders** tab displays the names of folders (subdirectories) and files on the drives of your computer and network — similarly to Windows Explorer. (In earlier release of AutoCAD, this tab was called "Desktop.")

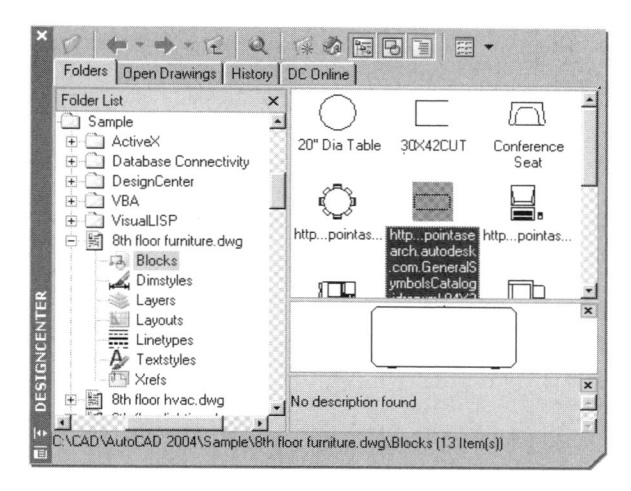

Open Drawings

The **Open Drawings** tab displays a list of the drawings currently open in AutoCAD. The named content of each drawing is also displayed: blocks, xrefs, layouts, layers, linetypes, dimension styles, and text styles.

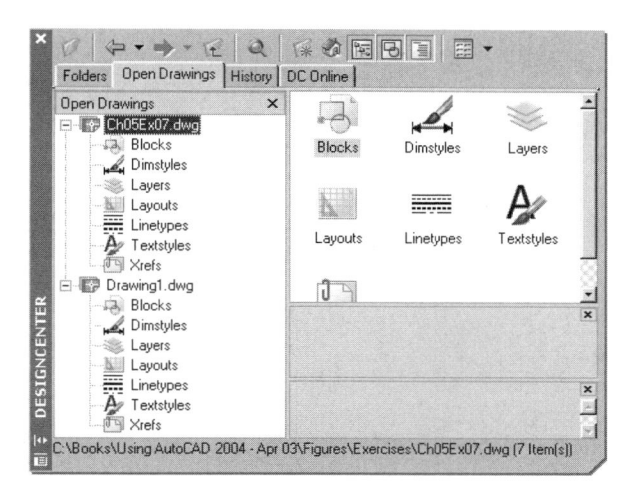

Custom Content

The **Custom Content** tab is present only if a drawing contains proxy data generated by registered (third-party) applications. Either the third-party application must be running or the application must be registered with DesignCenter.

History

The **History** tab displays the last 120 documents that the DesignCenter viewed. Double-click a file name to display its named content (if an AutoCAD drawing) or a thumbnail (if a raster image).

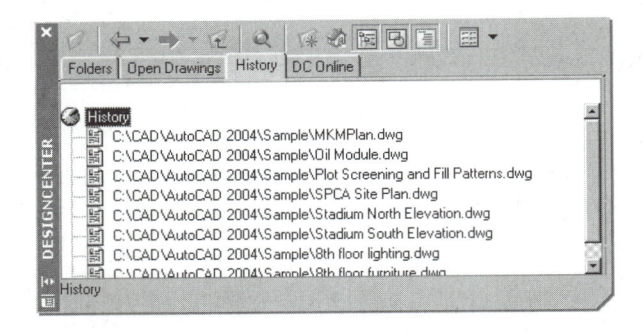

DC Center

NEW IN 2004

The **DC Center** tab displays content available for downloading from Autodesk's Web site. (This tab is new to AutoCAD 2004.)

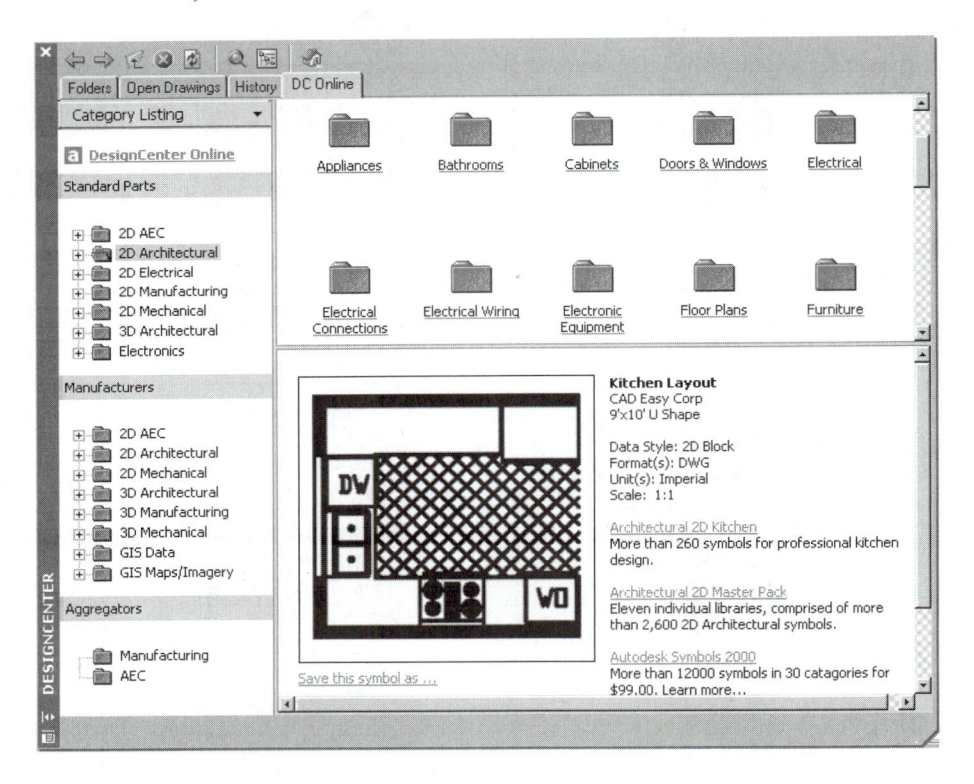

Toolbar

The toolbar changes, depending on which tab you select. For most tabs, the toolbar contains the following buttons:

Load

The **Load** button displays the Load dialog box for select content to load into DesignCenter. You can select content from files located on your computer, from other computers connected with a network, and from Web sites on the Internet.

Back

The **Back** button returns to the previous location in the history list. Click the arrow next to this button for the history list.

Forward

The **Forward** button lists the next location in the history list.

Up

The **Up** button moves up the tree view by one level.

Search

The **Search** button displays the Search dialog box, so that you can locate content by specific criteria.

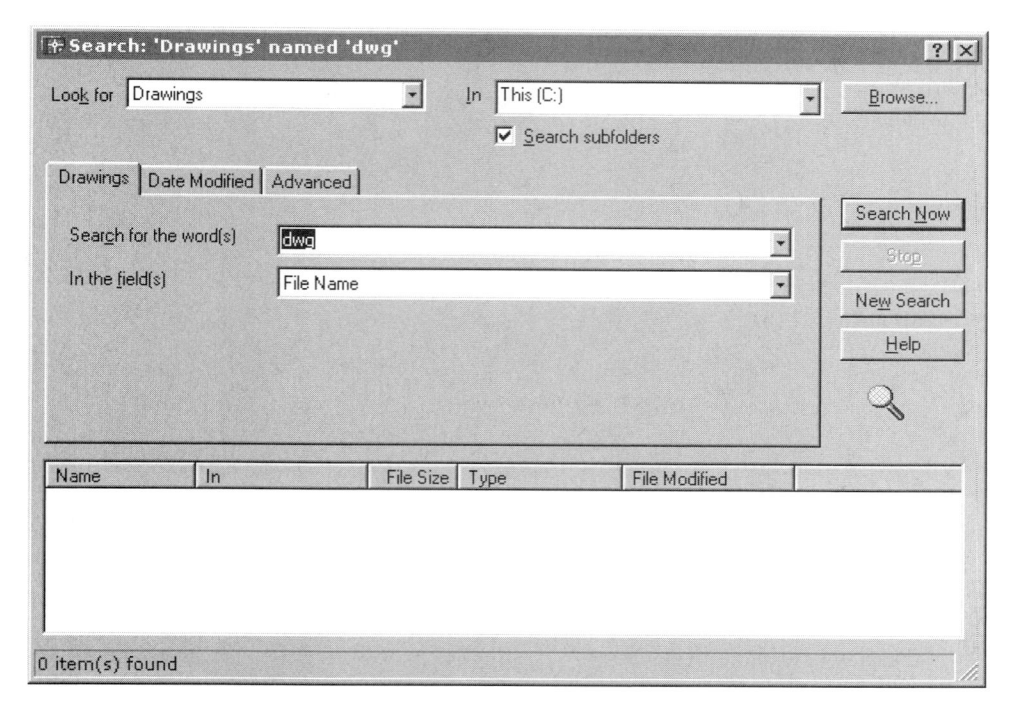

Favorites

The **Favorites** button displays the Windows' Favorites folder. Various software applications, including AutoCAD, automatically add files to the Favorites folder. AutoCAD, for example, provides several drawings with blocks that you may find useful for your own drawings. See Appendix I in this book.

Home

The **Home** button sends DesignCenter to the designated home folder, which is *AutoCAD 2004\Sample\DesignCenter* by default.

Tree View Toggle

The **Tree View Toggle** button shows and hides the tree view (the left-hand pane).

Preview

The **Preview** button shows and hides the preview pane.

Description

The **Description** button shows and hides the description pane.

Views

The **Views** button changes the display format of the content: large icons, small icons, list view (names only), and detail view (names and file information).

> **Note**: The "refresh" command is hidden in a shortcut menu. To refresh the display, right-click in the content area, and then select **Refresh** from the shortcut menu.

DC Online

The **DC Online** tab (short for "DesignCenter Online") displays these Web browser-like buttons on the toolbar:

Stop

The **Stop** button stops the current transfer of data.

Reload

The **Reload** button reloads the current page.

Palette

Right-click the palette (the area displaying the icons) to display a shortcut menu.

Copying Content

All palette cursor menus have one action in common: select **Copy** to copy the content from one drawing to another. For example, to copy a layer name from one drawing to another:

0. Navigate to a *.dwg* drawing file with DesignCenter, and then open the Layers folder.
1. Right-click the layer you want to copy.
2. From the shortcut menu, select **Copy**.

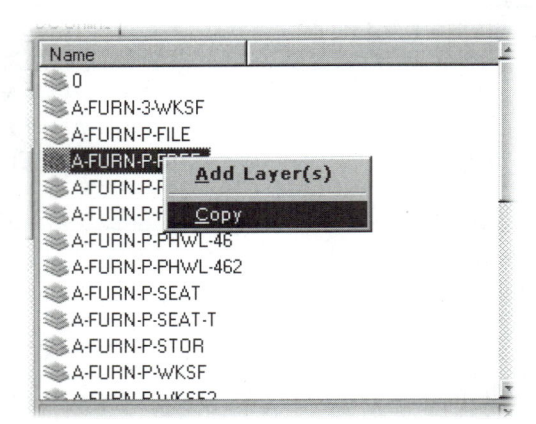

3. In the drawing to which you want the layer name copied, chose from the menu bar **Edit**, and then **Paste**. The layer and its properties are added to the drawing.\

As an alternative, drag content, such as a group of layers, from DesignCenter into the drawing:

4. Select the first layer name.

5. Hold down the **SHIFT** key, and then select another layer name. Notice that DesignCenter highlights all layer names between your two selections. (To select layers in a nonconsecutive order, hold down the **CTRL** key, and then pick them.)
6. Drag the group of highlighted layer names into the drawing. AutoCAD copies only the layer names and properties, not objects located on the layers. Sometimes AutoCAD warns:

> Duplicate names ignored.

This means that if a layer of that name already exists, AutoCAD does not duplicate it.

Adding Content

The second item in the shortcut varies, depending on the content; many have **Add**. This option adds the content to the current drawing.

It may seem to you that the **Copy** and **Add** commands are similar in purpose. The difference is that **Copy** copies the content to the Clipboard. From there you have to use the **Paste** command to get the data into the drawing. (This allows you to move content to another CAD system, such as AutoCAD LT.) In contrast, **Add** immediately adds the content to the current drawing.

For example, to add a linetype to the current drawing, right-click the linetype, and then select **Add** from the shortcut menu. As before, AutoCAD warns that duplicate names are ignored.

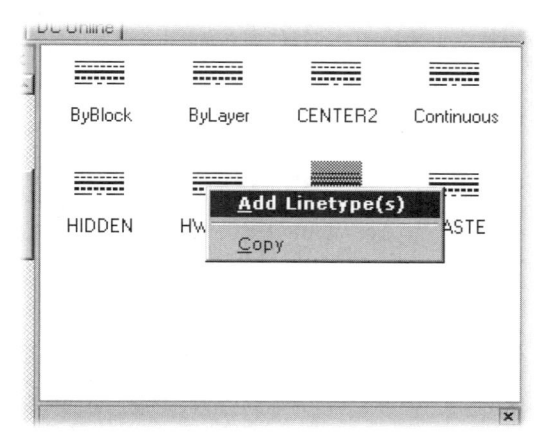

Working with Blocks and Xrefs

The shortcut menus for blocks and external references are different:

Blocks. The shortcut menu displays **Insert Block**, which displays the **Insert** dialog box. It is identical in function to AutoCAD's **Insert** dialog box described later in this chapter.

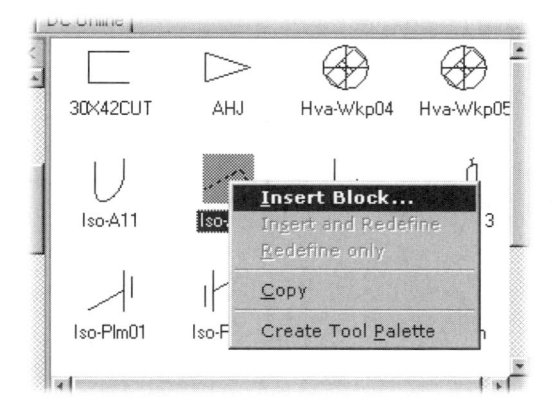

Insert and Redefine. When you try to add a layer or linetype that already exists in the drawing, AutoCAD ignores the command, and doesn't add it. Blocks are handled differently, however. When you add a block from DesignCenter to a drawing that contains a block of the same name, the **Redefine Only** and **Insert and Redefine** commands update blocks in the drawing with the definition you are adding.

External References. Selecting the **Attach Xref** item displays the **External Reference** dialog box, which also is identical to AutoCAD's **External Reference** dialog box. See *Using AutoCAD 2002: Advanced*.

Create Tool Palette

The **Create Tool Palette** option adds the selected object to the Tool palette window (discussed in detail next in this chapter).

Tree View

Right-click the tree view to display a shortcut menu that mimics the functions of some buttons on the toolbar: **Find** and **Favorites**.

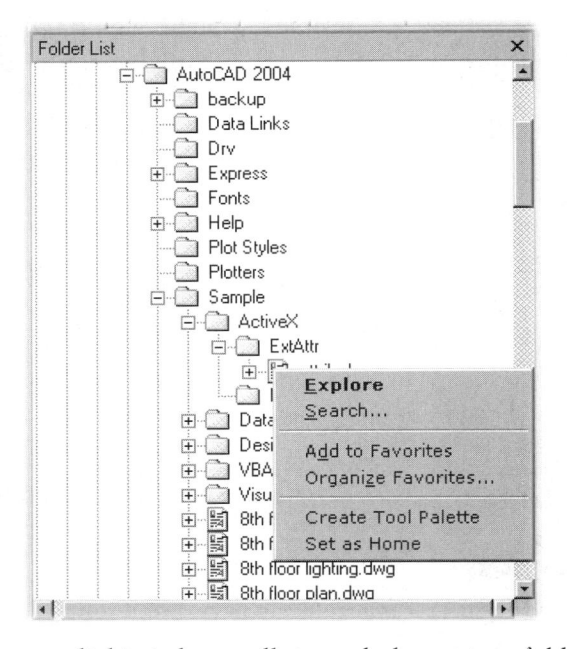

Explore has the same effect as clicking the small + symbol next to a folder or file name.

Search brings up the Search dialog box to hunt down content.

Add to Favorites adds the drawing or folder to the list of "favorites," which acts like a bookmark.

Organize Favorites displays the Windows folder showing the *Documents and Settings\ username\ Autodesk\ Favorites* folder.

Create Tool Palette adds *all* of the drawing's blocks to the **Tool Palette** window.

Set as Home is the same as using the ADCNAVIGATE command.

Description

Right-click the description area to display a cursor menu with commands suitable for copying the text to the Clipboard.

Title Bar

Double-click the title bar to dock DesignCenter on the left side of the AutoCAD window.

BLOCK AND INSERT

The **BLOCK** command creates reusable components called *blocks*, while the **INSERT** command places the blocks in drawings. Using blocks in drawings has several distinct advantages. Entire libraries of blocks can be used over and over again for repetitive details. AutoCAD includes 15 libraries of 300 blocks in the \AutoCAD 2004\Sample\DesignCenter folder (see Appendix I):

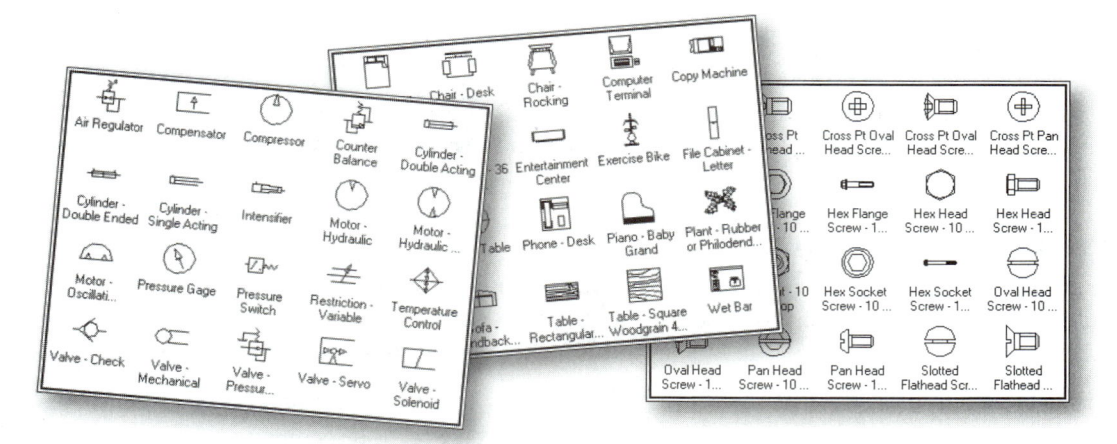

Analog Integrated Circuits
CMOS Integrated Circuits
Fasteners - Metric
Home - Space Planner
HVAC - Heating Ventilation Air Conditioning
Kitchens
Pipe Fittings
Welding

Basic Electronics
Electrical Power
Fasteners - US
House Designer
Hydraulic - Pneumatic
Landscaping
Plant Process

Blocks and *nested blocks* are excellent for building larger drawings from "pieces." (A nested block is a block placed within another block.) Several blocks require less space than several copies of the same objects, because AutoCAD stores only the information for each block *definition*.

Blocks can be used with *attributes*, which are text records that can be visible or invisible. (Attributes can only be attached to blocks.) The data from attributes can be exported from drawings into database and spreadsheet programs for further analysis. This is useful in facilities management, for example, where there are multiple occurrences of desks, chairs, and computers. Each of these is stored as a block, and has attributes, such as the person's name and telephone number, associated with it. Attributes are covered in *Using AutoCAD 2004: Advanced*.

BLOCKS ARE HARD TO EDIT

Blocks are groups of objects combined into a single object, and identified by a name. Blocks are considered a single object; you can move, scale, copy, and erase blocks as though they were single objects. Blocks are so "tight" that it is difficult to edit their individual members. There are two ways to edit the individual objects that make up a block:

- The **EXPLODE** command breaks the block into its individual parts. When done editing the parts, the **BLOCK** command recombines them.

- The **REFEDIT** command "checks out" the block, the **REFSET** command edits the block by adding and removing objects from it, and the **REFCLOSE** command adds the changes to the block definition.

If you need to work with groups of objects where members are easily edited, consider using the **GROUP** command, discussed later in this chapter.

> **Note**: It is more efficient to use blocks than to copy the same objects over and over. The **COPY** command makes a complete copy each time you use it; in contrast, AutoCAD creates a single definition of a block, and then points to it when you make "copies" with the **INSERT** command. This reduces the drawing's file size and improves the display time.

AutoCAD stores blocks in the drawing in which they were made. You can share them with other drawings using DesignCenter or the **WBLOCK** command, which exports blocks as *.dwg* drawing files on disk.

TUTORIAL: CREATING BLOCKS

There are three crucial pieces of information AutoCAD needs before it creates the block: (1) A name for the block, so that you can later identify it; (2) x,y,z coordinates for the insertion point, so that AutoCAD knows where to place the block; and (3) the objects to collect into a block.

With that in mind, let's see how to create blocks.

1. To create blocks, start the **BLOCK** command with one of these methods:
   - From the **Draw** menu, choose **Block**, and then choose **Make**.
   - On the **Draw** toolbar, pick the **Block** button.
   - At the 'Command:' prompt, enter the **block** command.
   - Alternatively, enter the aliases **b**, **bmod**, or **bmake** (the command's old name, short for "block make") at the keyboard.

   Command: **block** *(Press ENTER.)*

   Notice the **Block Definition** dialog box.

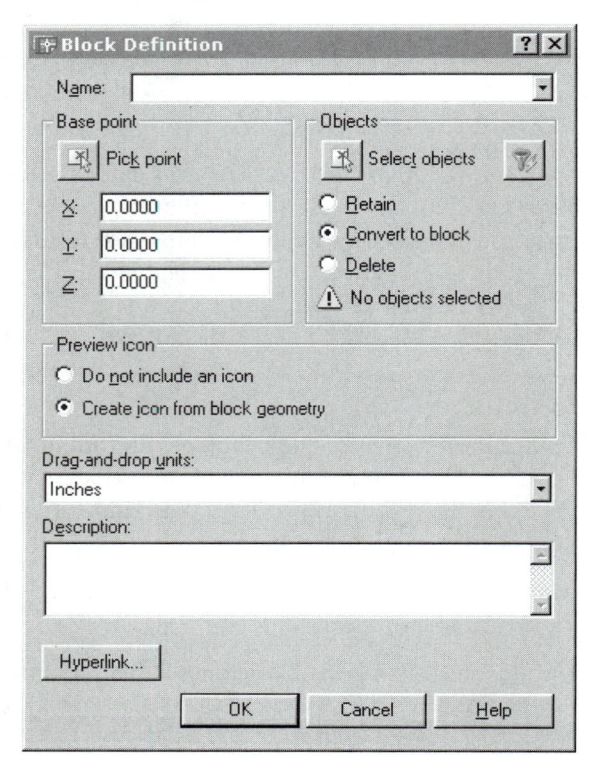

3. Give the block a name. In the **Name** field, you can enter anything from a single letter through to a word 255 characters long. If you try to give the block a name that already exists in the current drawing, AutoCAD warns:

> Blockname is already defined. Do you want to redefine it?

Most times you enter **No**, and then change the name.

Notes: Sometimes you deliberately want to redefine an existing block. Perhaps you made an error in creating the block or you left out attribute definitions. In this case, choose **Yes**.

If you need help seeing the names of blocks already defined in the drawing, click the down arrow in the **Name** field. AutoCAD displays the names of blocks in the current drawing. Block names that begin with * (asterisk), such as *X20, are blocks created by AutoCAD. These are called *anonymous blocks*.

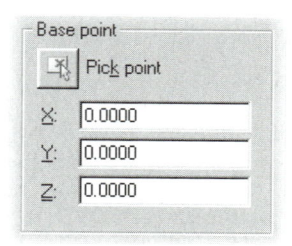

4. For the **Base Point, enter** coordinates in the **X, Y**, and **Z** fields, or else choose the **Pick Point** button to pick the insertion point in the drawing.

   Note: The base point defines the *insertion point*, which is where the block is later placed by the **INSERT** command.

   The base point is often the lower left corner of the block, but is sometimes located elsewhere, such as the center of the block. The figure below illustrates a sprinkler head; the logical location for the base point in the center of the head.

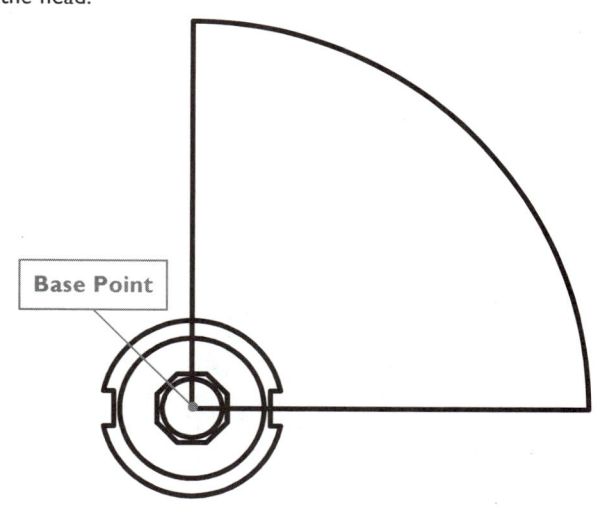

   Base Point

   When you choose the **Pick Point** button, the dialog box disappears temporarily, and AutoCAD prompts you,

   Specify insertion base point:

   I recommend using an object snap mode to make the pick accurate, such as ENDpoint for the end of a line or CENter for the center of a circle or arc.

   After you pick the insertion point, the dialog box reappears and AutoCAD fills in the **X, Y**, and **Z** fields. Normally, the **Z** value is 0, unless you have set the elevation or are selecting a point on a 3D object.

5. To select the objects that make up the block, choose the **Select Objects** button. The dialog box disappears and AutoCAD prompts you:

   Select objects: *(Select one or more objects.)*

   Select objects: *(Press ENTER to return to the dialog box.)*

   You can use any method of object selection, such as Window or Fence. After you finish selecting objects, press **ENTER** and the dialog box reappears. AutoCAD reports the number of objects you selected.

6. Click **OK**, and AutoCAD creates the block — even though it's hard to tell it happened. You access it later with the **INSERT** command.\

Notes on Designing Blocks

When creating blocks, it is best to make them the size you intend for use. For example, office desks are commonly 24" x 36", so it makes sense to draw blocks of office desks that size.

Blocks can be *nested*: you can place one block inside another.

Sometimes, however, you don't know at what size a block will be used. For example, the block of a tree symbol could be inserted at 2' or 3' or 5' or. . . . In these cases, draw the block at *unit size*. That means that the block fits inside a 1-unit square:

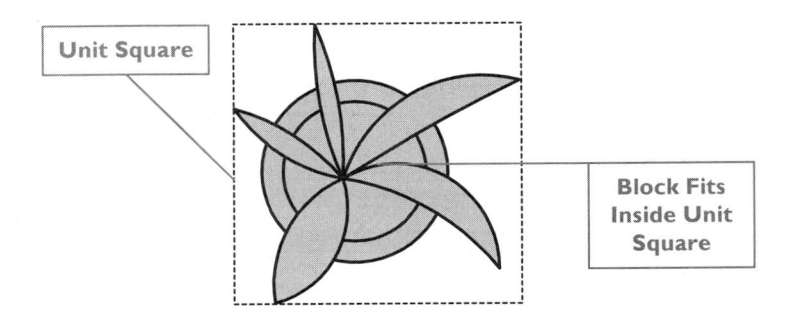

When it comes time to insert the block, the scale factor option sizes the block correctly.

CREATING BLOCKS: ADDITIONAL METHODS

The Block Definition dialog box contains a number of optional actions. The Objects section of the dialog box has these options:

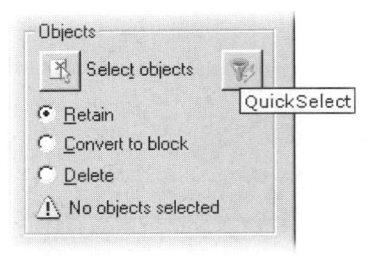

- **Retain** option keeps the objects making up the block in the drawing as individual objects, as well as stores the block definition.
- **Convert to Block** option erases the objects, and replaces them with the block. This is the default action.
- **Delete** option erases the objects, while storing the block definition in the drawing. This was formerly the "normal" behavior of the block command, which would freak out new AutoCAD users: "Wha' happened to my drawing!!??"

- **QSelect** button displays the Quick Select dialog box, which selects objects on the basis of common properties. See the **QSELECT** command elsewhere in this chapter.

In the lower half of the dialog box are these options:

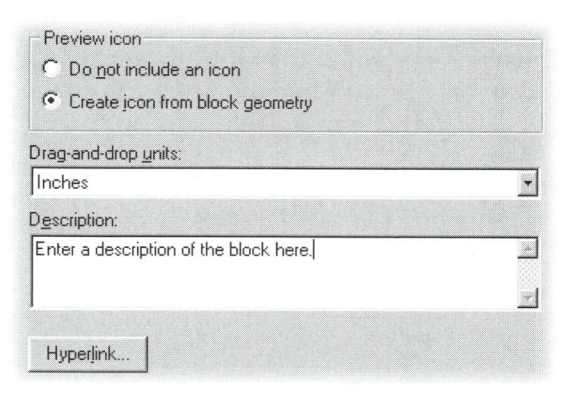

- **Preview Icon** determines whether an icon is included with the block definition. There isn't really any good reason to say "Do not include an icon," so include the icon.

- **Drag and Drop Units** specifies units for the block. Usually, AutoCAD makes one drawing unit equal the current unit. When you drag blocks from DesignCenter or *i-Drop*-enabled Web sites into drawings, AutoCAD automatically scales the block. (i-Drop is Autodesk technology that allows you to drag blocks from Web sites directly into drawings.) Select a drawing unit:

Inches	Nanometers	Feet	Microns
Miles	Decimeters	Millimeters	Decameters
Centimeters	Hectometers	Meters	Gigameters
Kilometers	Astronomical Units	Microinches	Light Years
Mils	Parsecs	Yards	Angstroms

- **Description** provides room for you to include a long (or short) description of the block. This is completely optional.

- **Hyperlink** displays the Insert Hyperlink dialog box so that you can attach a hyperlink to the block. (Hyperlinks are discussed in *Using AutoCAD 2004: Advanced*.)

-Block Command

AutoCAD also provides the **-BLOCK** command, which displays its prompts at the command line. This version of the command is meant for scripts and for power users who prefer to use the keyboard (like myself). The command displays the following prompts:

Command: **-block**

Enter block name or [?]: *(Enter a name, or type* **?** *to list the names of blocks already defined in the drawing.)*

Specify insertion base point: *(Pick a point, or enter x,y coordinates.)*

Select objects: *(Select the objects that make up the block.)*

Select objects: *(Press* ENTER *to end the command.)*

The objects disappear from the drawing, and AutoCAD creates the block definition. To bring them back, use the **OOPS** command:

Command: **oops**

INSERTING BLOCKS

After you define blocks with the **BLOCK** command, or copy the blocks from other drawings, you place them with AutoCAD's **INSERT** command. There are two pieces of information **INSERT** needs to place a block: (1) its name; and (2) its insertion point. In addition, you optionally specify: (3) the scale factors, normally 1.0; and (4) the rotation angle, usually 0 degrees.

The *insertion point* is the x,y-coordinate where the block is inserted. More specifically, the block is inserted at its base point, which was defined when the block was created.

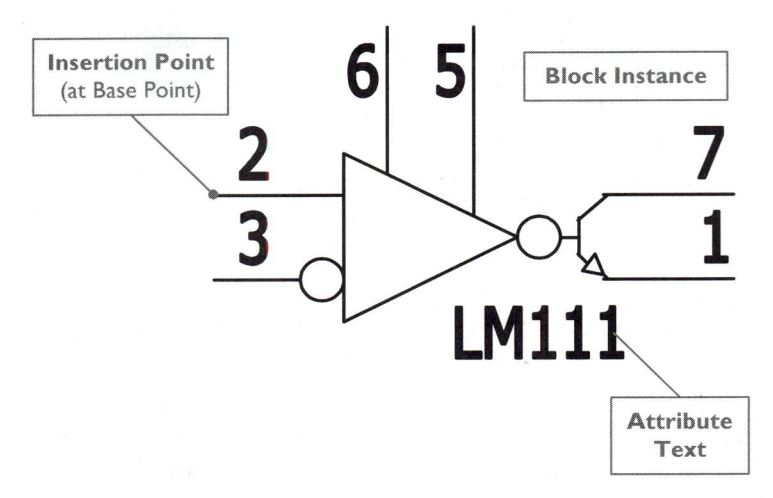

After blocks are inserted, use **INSertion** object snap to snap to their insertion points.

Note: When inserted, a block retains its original layer definitions. When an object is located on a layer named "PCBOARD," it will be on that layer when inserted. The exception is an object that was originally on layer 0. Objects created on layer 0 in the block are assigned the properties of the layer on which the block is inserted.

TUTORIAL: PLACING BLOCKS

1. To place blocks in the drawing, start the **INSERT** command with one of these methods:
   - From the **Insert** menu, choose **Block**.
   - On the **Draw** or **Insert** toolbars, pick the **Insert** button.
   - At the 'Command:' prompt, enter the **insert** command.
   - Alternatively, enter the aliases **i**, **ddinsert** (the command's old name), or **inserturl** at the keyboard.

   Command: **insert** (*Press* ENTER.)

2. Notice the **Insert** dialog box.

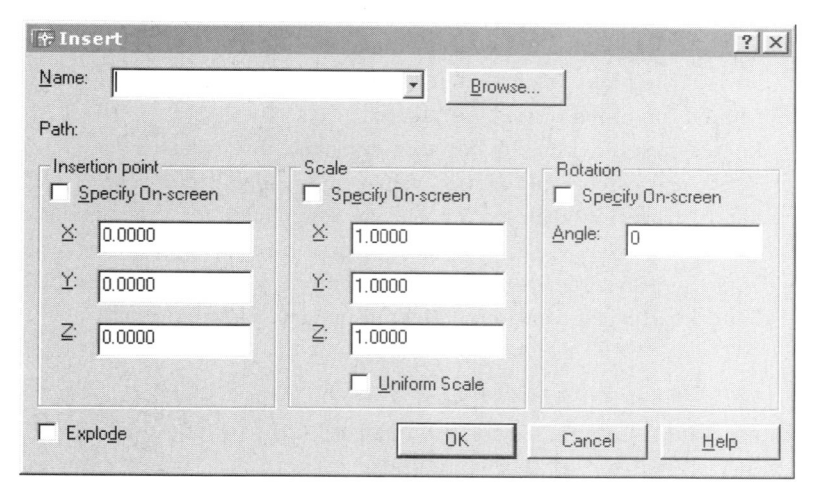

3. To select the block to insert, you must specify its name, which you can do in three ways:
   - In the **Name** field, type the name of the block, if you know it.
   - Click the arrow at the end of the **Name** field, and AutoCAD displays a list of all blocks defined in the drawing.

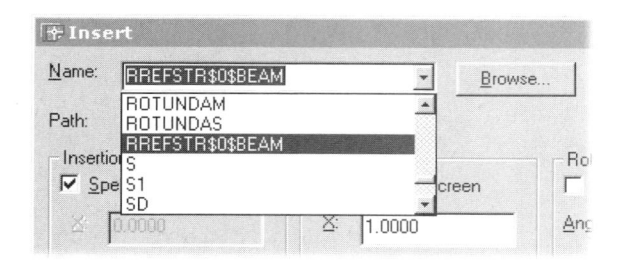

   - Click **Browse**, and from the file dialog box select *any* .dwg drawing file to insert as a block.

4. To specify the insertion point, you can enter the x, y, and z coordinates in the dialog box or in the drawing. Click the **Specify on Screen** option to determine where this takes place:
   - ☐ Specify the insertion point in the **X**, **Y**, and **Z** text entry boxes.
   - ☑ Specify the insertion point in the drawing. After clicking **OK** button, AutoCAD prompts you:

     Specify insertion point or [Scale/X/Y/Z/Rotate/PScale/PX/PY/PZ/PRotate]: *(Pick a point, enter an option, or type the x,y,z coordinates.)*

5. After you specify the insertion point, AutoCAD places the block in the drawing. Technically, this is called an "instance" of the block definition.

INSERTING BLOCKS: ADDITIONAL METHODS

The Insert dialog box contains several optional actions:
- **SCALE** changes the size of the block instance.
- **ROTATION** rotates the block instance.
- **EXPLODE** inserts the block with its original components, not as an instance.

In addition, the command-line lists a large number of options with puzzling abbreviations: **Scale**, **X**, **Y**, **Z**, **Rotate**, **PScale**, **PX**, **PY**, **PZ**, and **PRotate**. Let's look at each option.

Insert Option	Meaning
Scale	Specifies a single scale factor for the x, y, and z-directions.
X	Scales the block in the x-direction.
Y	Scales the block in the y-direction.
Z	Scales the block in the z-direction.
Rotate	Rotates the block in the counter-clockwise direction.
PScale	Sets the preview scale factor for the x, y, and z-directions.
PX	Sets the preview scale factor and insertion scale for the x-direction.
PY	Sets the preview scale factor and insertion scale for the y-direction.
PZ	Sets the preview scale factor and insertion scale for the z-direction.
PRotate	Sets the preview rotation angle and insertion scale factor.

Scale

AutoCAD normally uses a *unit* scale factor for the inserted block. ("Unit" means that the scale factor is 1.0 so that the block is inserted at its original size.) There are times, however, when you might want the block to be a different size. When the scale factor is more than 1.0, the block is larger; when the scale factor is less than 1.0, the block is smaller.

The three scale factors — X, Y, and Z — can be the same or different, negative or positive. When different from each other, the block is stretched; when negative, the block is mirrored, including its text; when positive, the block is inserted normally. The illustration below shows the effect of different scale factors on an inserted block:

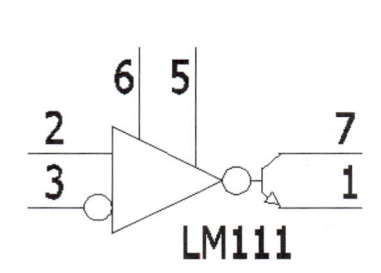

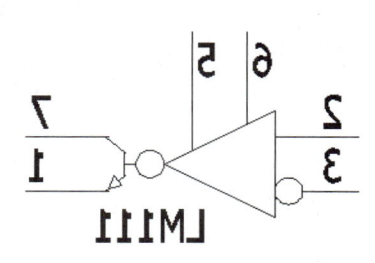

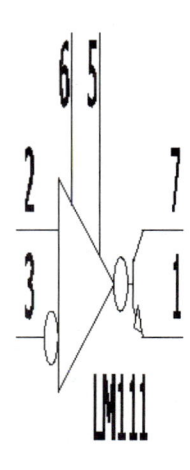

Block inserted with x and y scale factors = 1.0

Block inserted with x scale factor = -1.0

Block inserted with x scale factor = 0.5 and y scale factor = 2.0

The **Specify On-screen** option determines whether you specify the scale factor in the dialog box or in the drawing:

☐ Specify the scale factor in the **X**, **Y**, and **Z** text entry boxes.

☑ Specify the factor in the drawing. After you click the Insert dialog box's **OK** button, AutoCAD will prompt you:

> Enter X scale factor, specify opposite corner, or [Corner/XYZ] <1>: *(Press* ENTER *to accept the default factor, enter a different x-scale factor, pick a point, or specify an option.)*
>
> Enter Y scale factor <use X scale factor>: *(Press* ENTER *to make the y scale factor the same as x, or enter the y-scale factor.)*

AutoCAD presents a large number of options at the command line. Here's what they mean:

Scale Option	Meaning
X scale factor	Scales the block in the x-direction.
Opposite corner	Specifies a rectangle that defines the x and y scale factors; insertion point is the first corner.
Corner	Same effect as above option.
XYZ	Scales the block independently in the x, y, and z-directions.
Y scale factor	Scales the block in the y-direction.
Use X scale factor	Makes the y scale factor equal to the x scale factor.

The **Uniform Scale** option makes the y and z scale factors the same as the x factor.

Rotation Angle

You can have AutoCAD rotate the block about its insertion point relative to the original orientation of the block. The **Specify On-screen** option determines when the specification takes place:

☐ Specify the rotation angle in the **Angle** text entry box.

☑ Specify the angle in the drawing. After clicking **OK**, AutoCAD prompts you:

> Specify rotation angle <0>: *(Enter the rotation angle, or pick two points to indicate the angle.)*

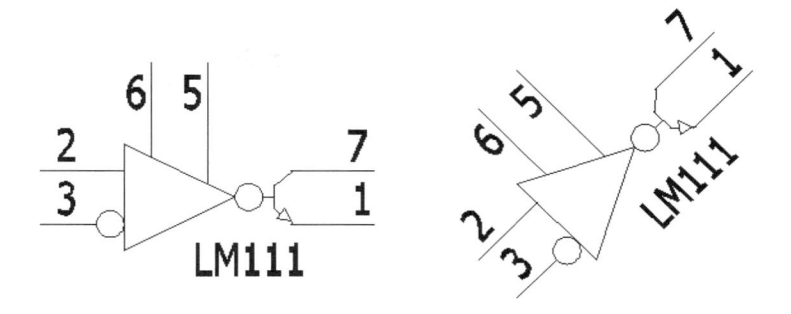

Block inserted with rotation angle = 0.   *Block inserted with rotation angle = 45 degrees.*

If you choose to indicate the angle, move the cursor. AutoCAD ghosts a rubberband cursor between the previously-set insertion point and the cursor. Move the cursor until the rubber-banded line shows the desired angle, and then click. The distance between the two points is irrelevant. The angle, shown by rubber-banded line between the insertion point and the angle point, determines the angle of insertion.

Explode

When you turn on the **Explode** option, AutoCAD inserts the block's constituent parts in the drawing — the lines, arcs, and text that make up the block definition. Note that AutoCAD limits you to specifying a single scale factor for exploded blocks, which is applied equally to X, Y, and Z.

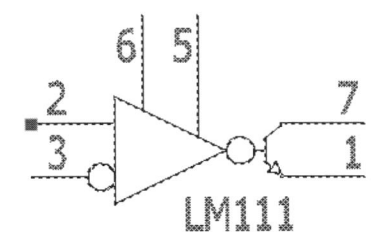

Block definition inserted as a block has a single grip.

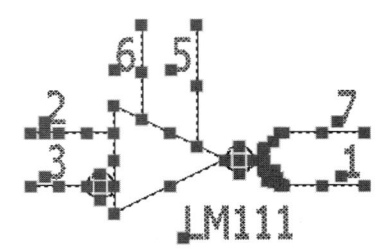

Block inserted exploded has grips for every constituent part.

DESIGNCENTER AND TOOLPALETTES

The drawback to the **INSERT** command is that it provides only a list of block names; you cannot preview them. Sometimes, it is easier to pick out a block by its looks, rather than by its name. Also, you may not be 100% sure of the block's location: which drawing or folder is it stored in? For these reasons, you may find it easier to use the DesignCenter or the Toolpalettes to insert blocks.

Inserting Blocks with DesignCenter

The advantage of the DesignCenter is that it locates and displays blocks stored in the current drawing, in other drawings on your computer, and in drawings located on the local network or the Internet.

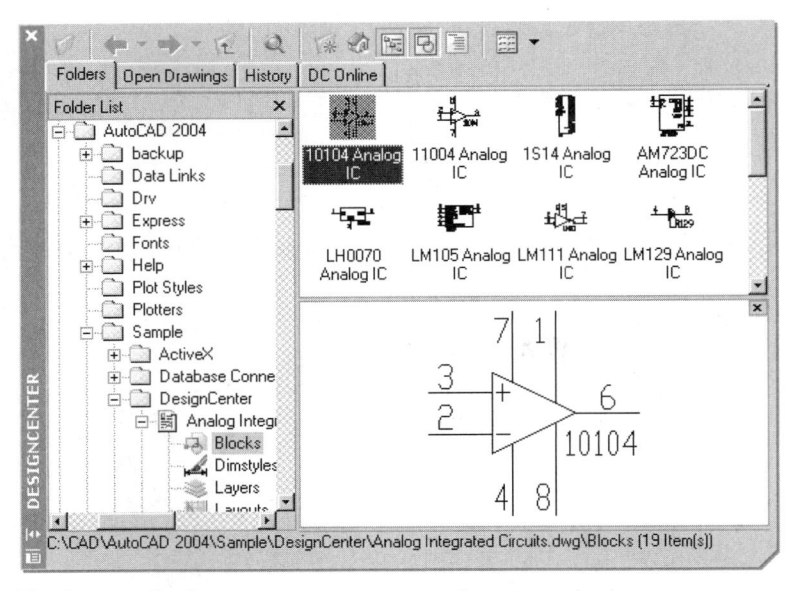

After finding the block, simply drag its icon into your drawing; there are no prompts to answer.

When you want control over placing the block, right-click the icon, and then select **Insert Block** from the shortcut menu; the familiar Insert dialog box is displayed.

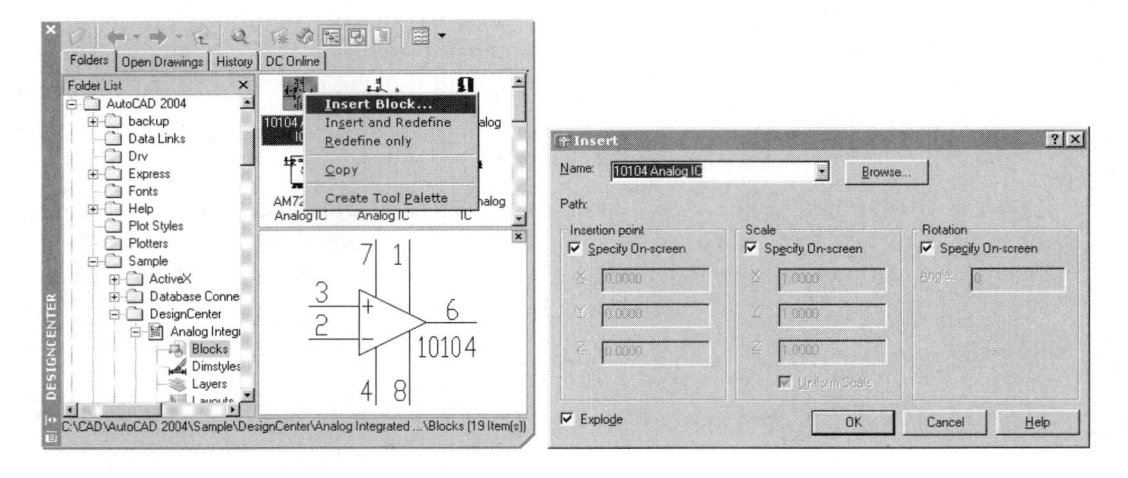

Inserting Blocks with Toolpalettes

The advantage of the toolpalettes is that it contains the blocks you commonly use. (Toolpalettes are discussed in detail later in this chapter.) It doesn't search for blocks, as does DesignCenter, but every drawing has access to the same tool palette. Simply drag a block's icon into the drawing; there are no prompts to answer.

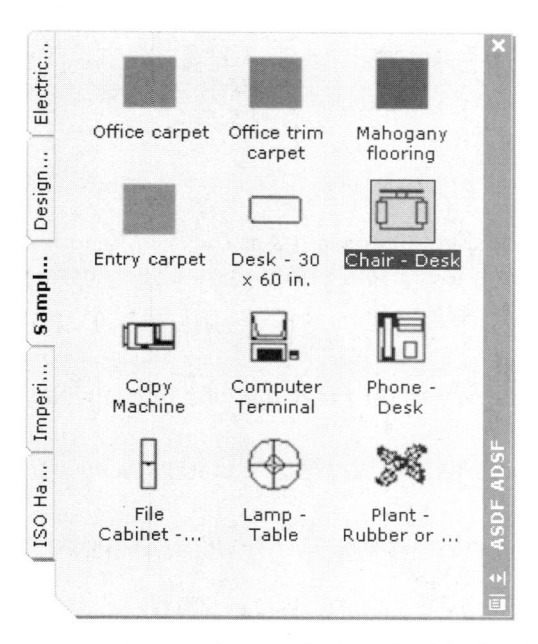

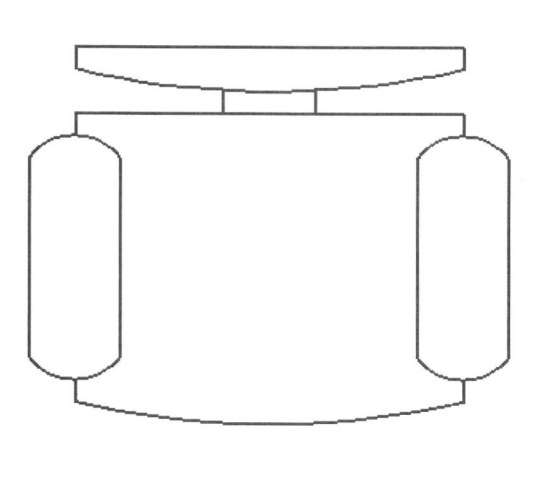

When you want control over placing the block, right-click the icon, and then select **Properties** from the shortcut menu; the Properties window is displayed, where you can make changes to the scale factor, rotation angle, and so on. The changes made are semi-permanent: they affect future insertions of the block until you again change the properties.

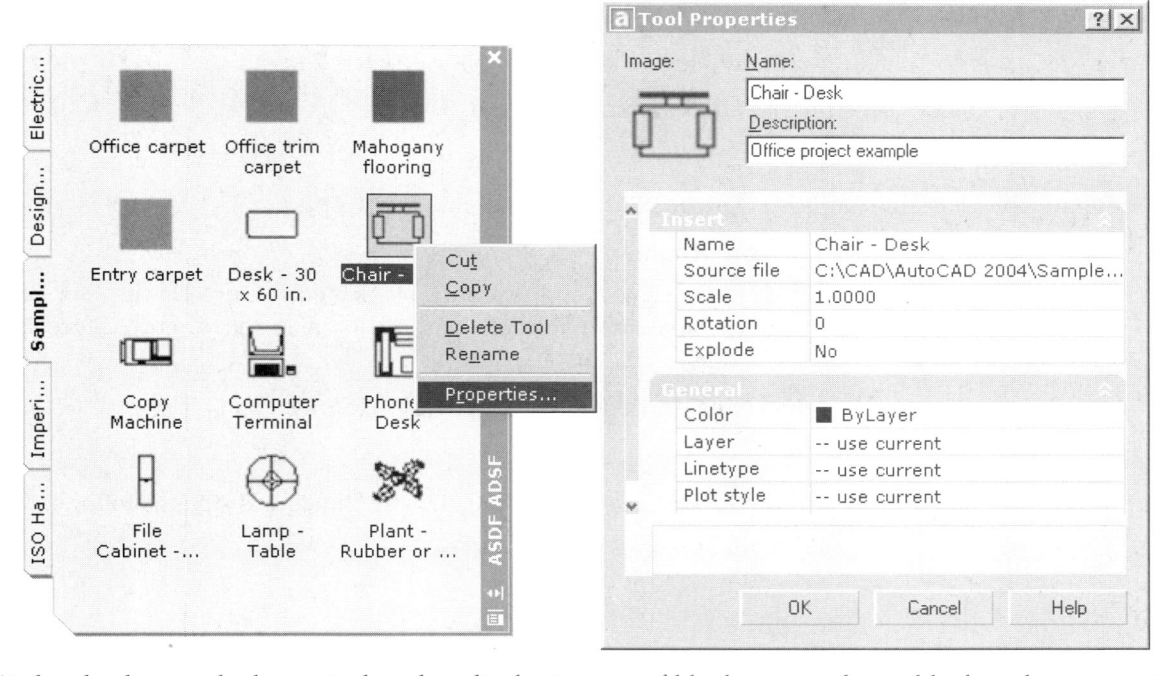

(A drawback to toolpalettes is that they display images of blocks, even when a block no longer exists. In that case, dragging the icon into the drawing results in nothing.)

CREATING AND INSERTING BLOCKS: ADDITIONAL METHODS

AutoCAD has several commands that are useful for working with blocks:

- **-INSERT** command places blocks using the command line.
- **BASE** command specifies the base point (origin) of the drawing.
- **WBLOCK** command exports blocks as *.dwg* files on disk.
- **MINSERT** command inserts blocks as rectangular arrays.
- **BLOCKICON** command creates icons of blocks.

Let's look at each command.

-Insert Command

AutoCAD also provides the **-INSERT** command, which displays its prompts at the command line. This version of the command is meant for scripts and for power users who prefer to use the keyboard (like myself). The command displays the following prompts:

> Command: **-insert**
>
> Enter block name or [?]: *(Enter a name, or type* **?** *to list the names of blocks already defined in the drawing.)*
>
> Specify insertion point or [Scale/X/Y/Z/Rotate/PScale/PX/PY/PZ/PRotate]: *(Pick a point, or enter an option.)*
>
> Enter X scale factor, specify opposite corner, or [Corner/XYZ] <1>: *(Press* ENTER, *or enter an option).*
>
> Enter Y scale factor <use X scale factor>: *(Press* ENTER, *or enter a scale factor).*
>
> Specify rotation angle <0>: *(Press* ENTER, *or enter an angle).*

Unlike the Insert dialog box, which displays all available options, the **-INSERT** command "hides" those same options. Here's how to access them:

Inserting Exploded Blocks

To insert an exploded block, prefix the block's name with an asterisk (*):

> Enter block name or [?]: ***blockname**

Redefining Blocks

To redefine a block definition, suffix the block's name with an equals sign (=):

> Enter block name or [?]: **blockname=**

Careful, because this forces AutoCAD to replace the existing block definition with the new one. The new definition will be applied to all block insertions of the same name in this drawing. AutoCAD then asks:

> Block "blockname" already exists. Redefine it? [Yes/No] <No>: **y**

Replacing Blocks with Drawings

To replace a block definition with an external .dwg drawing file, use the equals sign, as follows:

> Enter block name or [?]: **blockname=filename.dwg**

Base

As noted earlier, the Insert dialog box can insert another drawing into the current drawing as a block. The inserted drawing is normally inserted using coordinates 0,0 as its base point. If you wish to change the base point, use the **BASE** command on the drawing *before* inserting it.

> Command: **base**
>
> Specify base point <default>: *(Pick the point for the new base point.)*

Save the drawing to save the new base point, and then insert it in the other drawing.

WBlock

The **WBLOCK** command has the opposite of the **INSERT** command's ability to insert any *.dwg* drawing file as a block: **WBLOCK** exports all or part of the current drawing to be saved on disk as a *.dwg* file. This allows you to use blocks in other drawings, and is an alternative to the DesignCenter.

- From the **File** menu, choose **Export**. In the **Export Data** dialog box, choose "Block (*.dwg)" as from the **File of type** list box.
- At the 'Command:' prompt, enter the **wblock** command.
- Alternatively, enter the **w** alias at the keyboard.

> Command: **wblock** *(Press* ENTER.*)*

AutoCAD displays the Write Block dialog box, which looks very similar to the **BLOCK** command's dialog box. The primary difference is that you have a choice of what to write to disk in the Source area:

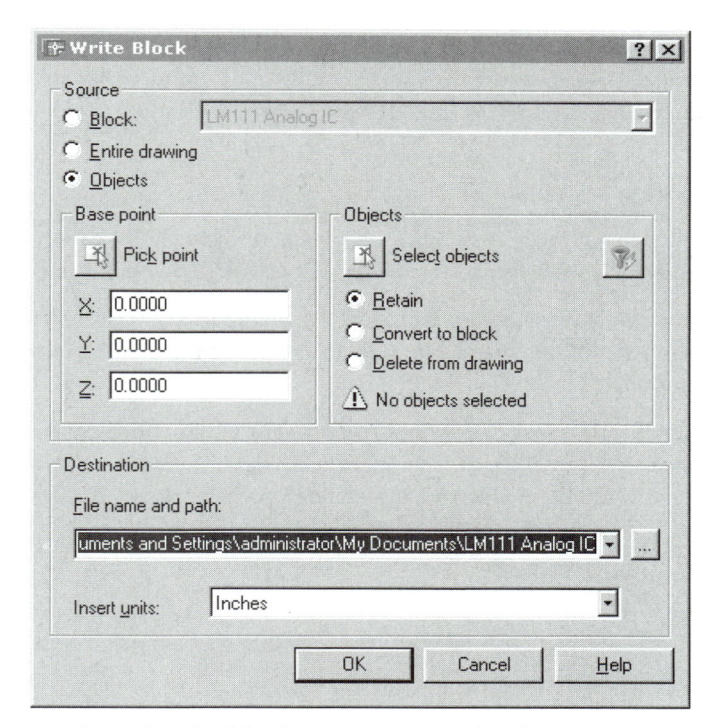

Block saves a block as a *.dwg* file; the block must exist in the drawing.

Entire Drawing saves the entire drawing with one exception: unused layers and blocks are not saved.

Objects saves the objects you select. You must specify the base point and select the objects.

In the Destination area, specify the file name and location (drive and folder.) When you change the file name extension from *.dwg* to *.dxf*, AutoCAD saves the drawing (or block) in DXF format. (DXF is short for "drawing interchange format.") With AutoCAD 2004, Autodesk removed **WBLOCK**'s ability to save the drawing as an *.xml* file (extended markup language).

MInsert

The **MINSERT** (short for "multiple insert") command combines the **INSERT** command and the rectangular options of the **ARRAY** command. AutoCAD doesn't use a dialog box, but displays all prompts at the command line. The sequence starts by issuing prompts in the same manner as the **-INSERT** command, followed by the prompts for constructing a rectangular array.

- At the 'Command:' prompt, enter the **minsert** command.

    Command: **minsert** *(Press ENTER.)*

AutoCAD displays prompts on the command lines:

> Enter block name or [?]: *(Enter a name, or type ? to list the names of blocks in the drawing.)*
>
> Specify insertion point or [Scale/X/Y/Z/Rotate/PScale/PX/PY/PZ/PRotate]: *(Pick a point.)*
>
> Enter X scale factor, specify opposite corner, or [Corner/XYZ] <1>: *(Press ENTER.)*
>
> Enter Y scale factor <use X scale factor>: *(Press ENTER.)*
>
> Specify rotation angle <0>: *(Press ENTER.)*
>
> Enter number of rows (---) <1>: *(Enter a number for the rows).*
>
> Enter number of columns (| | |) <1>: *(Enter a number for the columns).*
>
> Enter distance between rows or specify unit cell (---): *(Enter a distance between rows, or pick two points that show the row and column distance.)*
>
> Specify distance between columns (| | |): *(Enter a distance between columns.)*

The **MINSERT** command produces arrays that have many of the same properties as blocks, with some exceptions. Create arrays using **MINSERT** when you need to edit the array as a whole. For example, a seating arrangement consisting of several rows of chairs may need to be moved around for design purposes. Creating the arrangement with the **MINSERT** command allows you to move and rotate the seating as a whole.

The following qualities apply to minserts:

- The array reacts to editing commands as if it were one block. You cannot edit individual blocks making up the array. When you select one object to move or copy, the entire array is affected.

- You cannot "minsert" the block as exploded. If you need to do this, use the **INSERT** command followed by the **ARRAY** command. Nor can you explode the block into individual objects following insertion.

- When the initial block is inserted with a rotation angle, the entire array is rotated around the insertion point of the initial block (see figure below). This creates an array in which the original object appears to have been inserted at a standard zero angle, with the entire array rotated.

The illustration shows a block inserted at a 30-degree angle through the **MINSERT** command, with three rows and four columns.

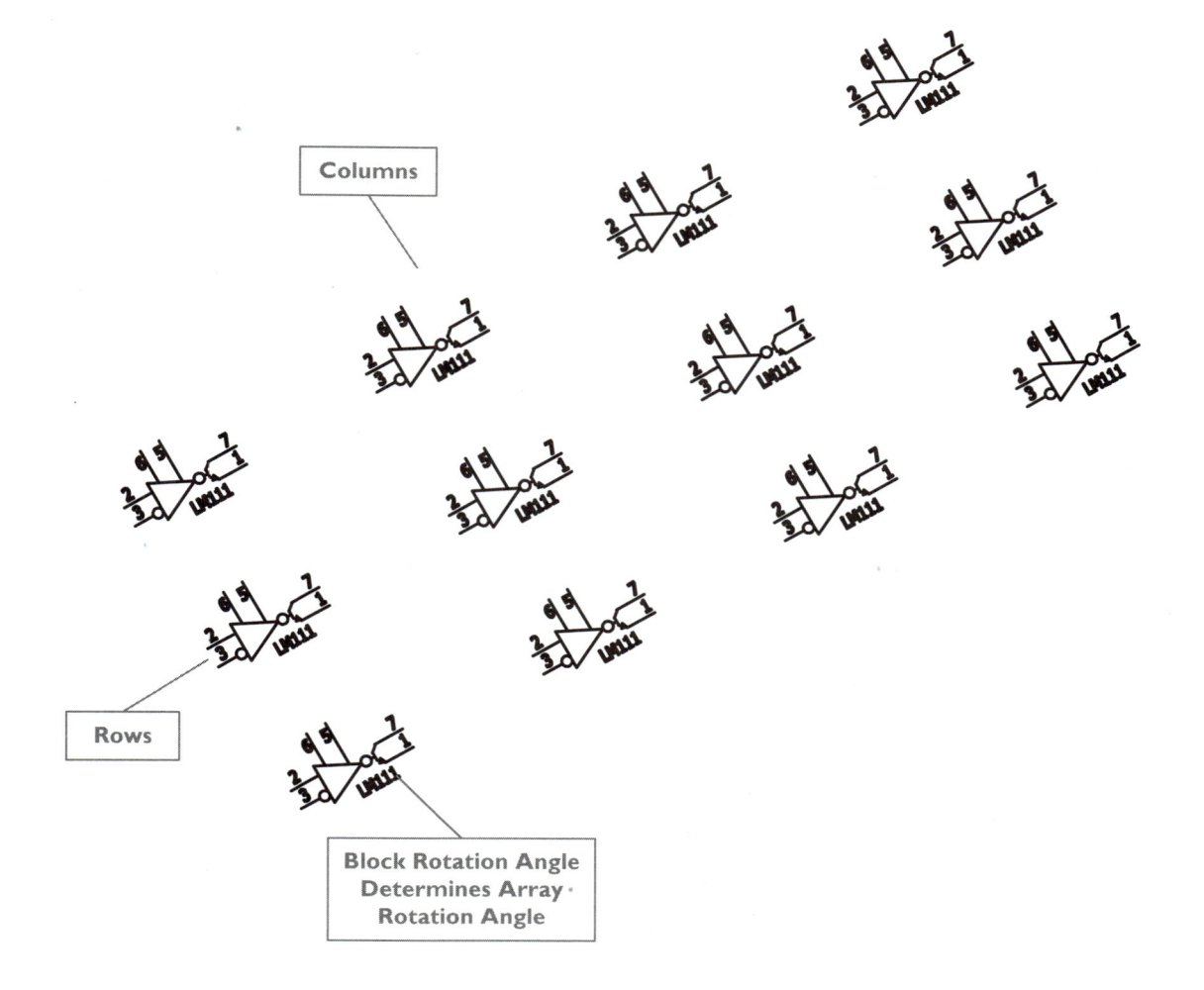

Columns

Rows

Block Rotation Angle Determines Array Rotation Angle

Note: To make a block immune to the **EXPLODE** command, insert it with the **MINSERT** command, and set the number of rows and columns to 1.

BlockIcon

In the Block Definition dialog box, you had the option to create a preview image of the block. This "icon" is used by the DesignCenter and Toolpalettes to display an image of the block.

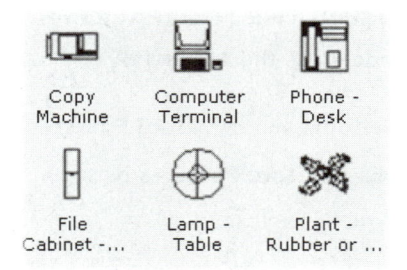

If you forget to ask for the icon, or have access to old blocks created without icons, you can use the **BLOCKICON** command to add icons to all blocks in the drawing.

Command: **blockicon**

Enter block names <*>: *(Press* ENTER *to iconize all blocks.)*

nnn blocks updated.

After the **BLOCKICON** command finishes, the drawing looks no different. In DesignCenter, however, blocks now display the preview image.

REDEFINING BLOCKS

If you have used many inserts of the same block in a drawing, redefining just one block changes all the blocks. This is an especially powerful feature: imagine being able to change 100 identical drawing parts in a single operation! Here's how:

1. Explode one of the block insertions. This action breaks the link to the block definition, and reduces the block to its constituent parts.
2. Edit the exploded block.
3. Use the **BLOCK** command to convert the edited parts back into a block. Give this "new" block the same name. You may select the name from the **Name** drop list.
4. Click **OK**. AutoCAD displays the following dialog box:

AutoCAD

? Chair - Desk is already defined.
Do you want to re-define it?

[Yes]   [No]

5. Click **Yes**. Notice that all other blocks of the same name change, including blocks inserted with MINSERT.

If the block is another drawing that was inserted whole, edit the original drawing, and then reinsert it as described above.

NEW IN 2004 — TOOLPALETTES

The **TOOLPALETTES** command displays a window that provides access to frequently-used content.

The palette is a centralized collection of blocks, and hatch and fill patterns that you use often. It replaces the need to hunt through DesignCenter, the Boundary Hatch dialog box, and the Insert block dialog box for specific blocks and hatch patterns.

In addition, tool palettes can be shared with other AutoCAD users, as well as the DesignCenter.

TUTORIAL: DISPLAYING AND USING THE TOOLPALETTE

1. To display the ToolPalette window, use the **ADCENTER** command with one of these methods:
   - From the menu bar, choose **Tools**, and then **Tool Palettes Window**.
   - On the standard toolbar, pick the **Tool Palettes** button.
   - On the keyboard, press **CTRL+3**.
   - At the 'Command:' prompt, enter the **toolpalettes** command.
   - Alternatively, enter the **tp** alias at the keyboard.

   Command: **toolpalettes** (Press ENTER.)

2. Notice the Tool Palettes window.

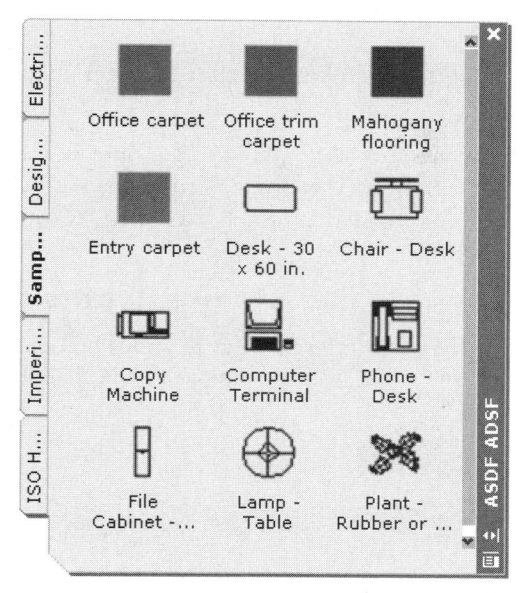

The *tabs* along the side access the palettes. Click a tab, such as **Imperial**, to view the associated palette.

Each palette contains one or more *tools*. A tool is a hatch pattern or a block; it cannot be any other AutoCAD object.

3. To place a block in the drawing, drag it from the palette into the drawing. As in dragging blocks from DesignCenter, AutoCAD places the block at the point where you let go of the mouse button. There are no prompts to answer; the block is scaled automatically to a preset value.
4. To place a hatch pattern or fill color, drag it from the palette into an object in the drawing. (AutoCAD ignores it if you try to place the hatch or fill in an empty part of the drawing.)
5. To turn off the Tool Palettes window, press **CTRL+3** or enter the **TOOLPALETTESCLOSE** command. As an alternative, click the **x** in the window's upper corner.

NAVIGATING THE TOOLPALETTES: ADDITIONAL METHODS

AutoCAD "hides" many of the Tool Palettes' options in shortcut menus, which differ depending on which part of the palette you right-click.

Right-click Commands: Tabs

Right-click the tab of the current palette (the one "on top") to see the following commands:

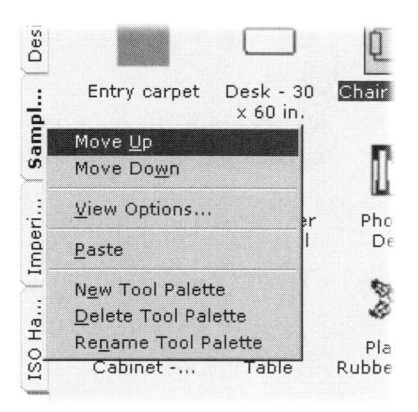

Move Up moves up the palette one position.

Move Down moves down the palette by one position.

View Options displays the View Options dialog box that lets you change the size and style of icons and text.

Paste pastes data from the Clipboard into the palette. This option is available only if the Clipboard contains appropriate data, such as an AutoCAD block. I can copy a block in DesignCenter for pasting in the tool palette, but not from AutoCAD itself, because in AutoCAD you never see the block definition; you only see insertions of it.

New Tool Palette creates a new blank palette, and prompts you to name the tab.

Delete Tool Palette warns you against deleting the palette, and then deletes it when you answer OK.

Rename Tool Palette allows you to rename a tab.

View Options Dialog Box

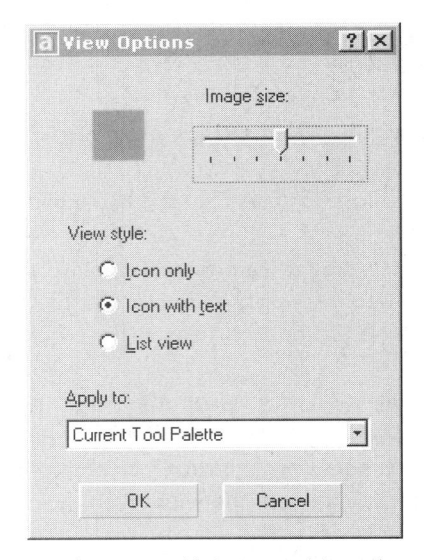

Image Size changes the size of icons from small (14 pixels) to large (54 pixels).

View Style changes the display of icons and text:

- **Icon only** displays icons only.
- **Icon with text** displays icons and text.
- **List view** displays small icons with text.

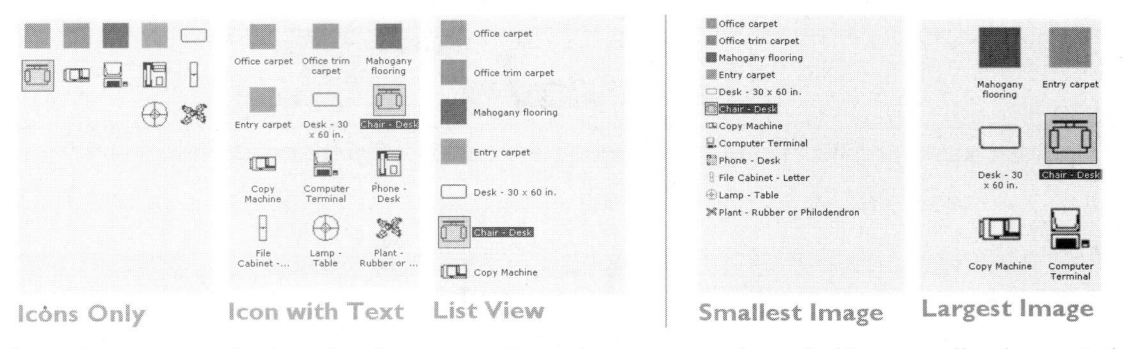

Apply to determines whether the changes apply to the current palette (tab), or to all palettes (tabs).

Right-click Commands: Icons

Right-click an icon (called "tools" by Autodesk) to get the following shortcut menu:

Cut removes the *tool* (a.k.a. icon — the hatch pattern or block) and places it in the Clipboard.

Copy copies the tool to the Clipboard.

Delete Tool removes the tool.

Rename renames the tool.

Properties displays the Properties dialog box for changing the properties of the hatch and block tools. This window is similar to the one displayed by the **PROPERTIES** command.

Tool Properties Dialog Box

The Tool Properties dialog box displays everything AutoCAD knows about the object. Shown below is the list of properties for the selected block. For example, you can change the name of the block, its description, scale, and whether it is inserted exploded.

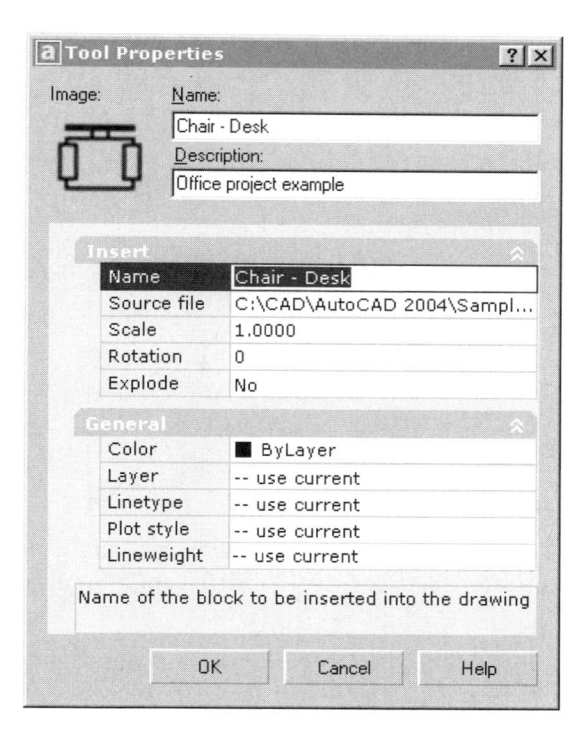

Properties shown in white can be changed; those in gray cannot.

Right-click Commands: Palettes

Right-click anywhere on an unused area of the palette — not on an icon or a tab — to get the following shortcut menu:

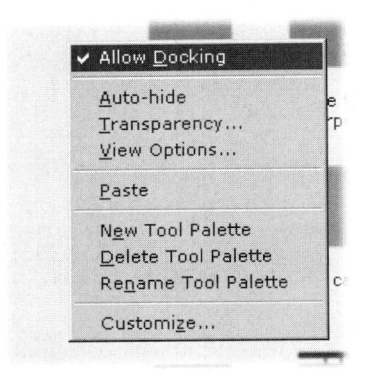

Allow Docking determines whether the tool palette docks against the side of the AutoCAD window when it get close enough.

Auto-Hide toggles whether the tool palette reduces itself to just the title bar when the cursor leaves the palette.

Transparency displays the Transparency dialog box for making the tool palette "see thru."

View Options displays the same dialog box as on the tab's shortcut menu.

Paste pastes the tool; command is available only when a tool has been copied or cut to the Clipboard.

New Tool Palette adds a new blank palette.

Delete Tool Palette removes the current palette.

Rename Tool Palette changes the name of the current palette.

Customize displays the Customize dialog box with the Tool Palettes tab.

Transparency Dialog Box

You can change the transparency of the tool palette:

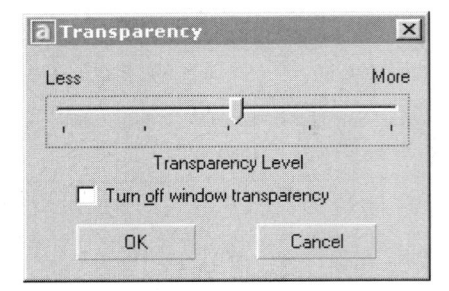

Moving the pointer towards **Less** makes the tool palettes more opaque, while moving it toward **More** make them more transparent. You can turn off transparency by clicking the **Turn off window transparency** check box.

Transparency lets you see both the Tool Palettes window and the drawing. Whether or not you use transparency is a matter of personal preference; I find transparency makes it harder to see both the window and the drawing, so I keep it turned off.

Here is what it looks like when the tool palettes are set to the maximum transparency:

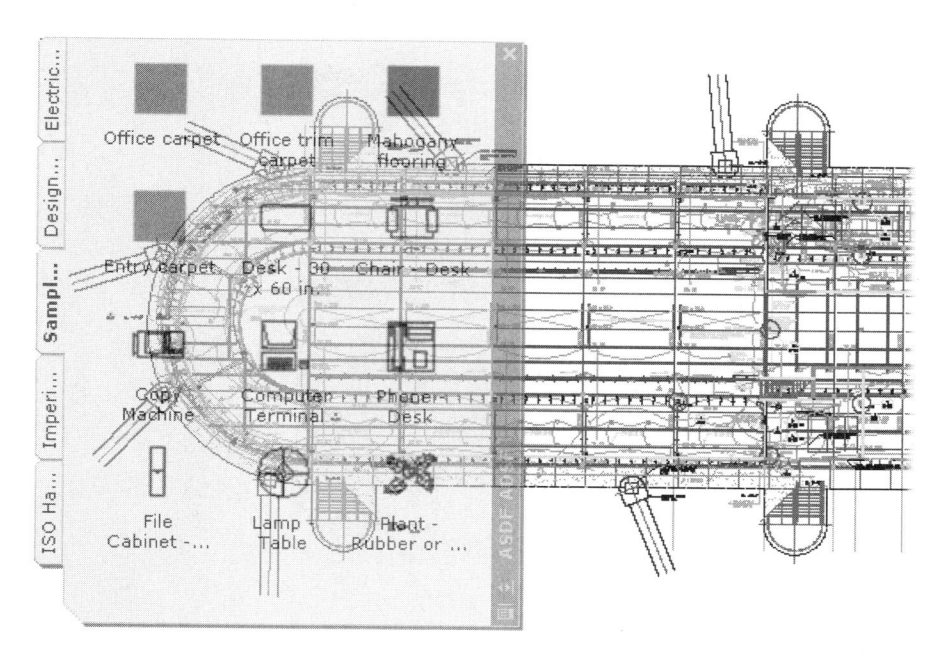

Customize Dialog Box

The Tool Palettes tab of the Customize dialog box lets you move the tabs up and down (through the **Up** and **Down** buttons). More importantly, it allows you to export the selected palette, so that it can be shared with other AutoCAD users.

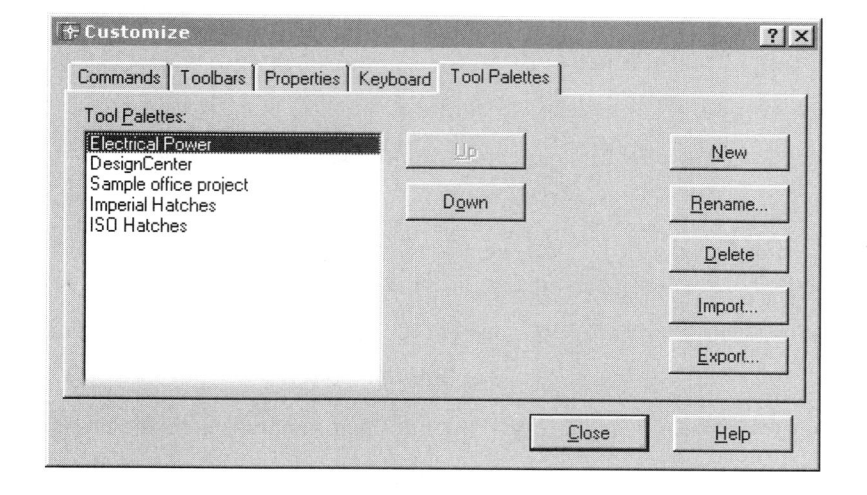

Right-click Commands: Titlebar

There is one more shortcut menu that comes up when you right-click the title bar — you actually customize the title bar:

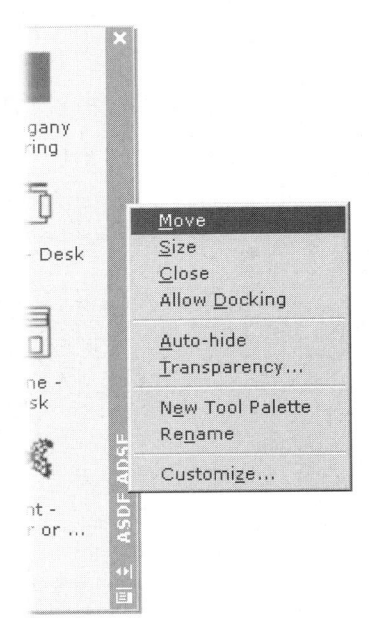

Many of the options are repeats of other shortcut menus.

Move displays the four-headed arrow cursor, but does nothing else. Because the window can be dragged by its title bar, the inclusion of this command seems redundant to me.

Size should resize the window, but it also does not work. As an alternative, drag the edges of the window to make it larger and smaller.

Close closes the window; alternatively, click the **x** on the title bar or press **CTRL+3**.

Rename changes the name on the title bar from the default of "Tool Palette."

Sharing Tool Palettes

In addition to dragging objects from DesignCenter into drawings, you can also drag them into the Tool Palettes. Here is a summary of the dragging operations you can perform:

- From DesignCenter into the Tool Palettes (but not from the Tool Palettes into DesignCenter).
- From the Tool Palettes into the drawing (but not from the drawing to the Tool Palettes).
- From DesignCenter into the drawing (but not from the drawing to the DesignCenter).

So that you can share tool palettes with other users, AutoCAD exports the content of the Tool Palette in a variant of XML with an extension of *.xtp*, which could be an acronym of either "eXport Tool Palette" or "Xml-format Tool Palette."

To export the Tool Palette, from the menu bar, select **Tools | Customize | Tool Palettes**, and then click the **Export** button. Notice that AutoCAD prompts you to save the palette as *.xtp* files. Use the **Import** button to read tool palettes created by others. (It's keen fun to share and trade palettes with your friends!)

⬛ BHATCH AND ✏️ HATCHEDIT

The **BHATCH** command places hatch patterns in closed areas, while the **HATCHEDIT** command edits the patterns.

Many times, parts and assemblies cannot be fully described by orthographic projection. It can be helpful to view the object as if it were cut apart, a view called a "section." Mechanical designers use sections to show interior details. Civil engineers detail profiles by showing sections along roadways and railways. Architects use sections through entire structures show how buildings are designed.

Crosshatching is used to show solid parts of the section. The spacing of the crosshatching should be relative to the scale of the section. The angle of the crosshatching should be oriented 45 degrees from the main lines of the cut area whenever possible. AutoCAD crosshatches sectional views easily. After drawing the section, apply the hatch pattern(s) to the cut areas with the **BHATCH** command.

Full Sections. A full section cuts across the entire object, and are usually cut through the larger axis.

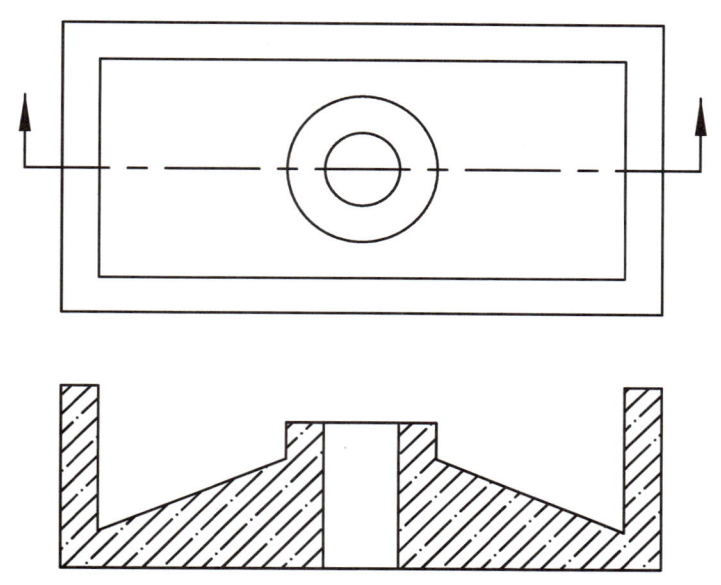

A section cut along the longer axis is cut along the longitudinal axis. If the section is cut along the minor axis, it is referred to as the latitudinal axis. Parts of the object that are "cut" are shown with crosshatching.

Revolved Sections. It is often helpful to view a section of a part of an object that is transposed on top of the point where the section was cut. Such a section is referred to as a revolved section.

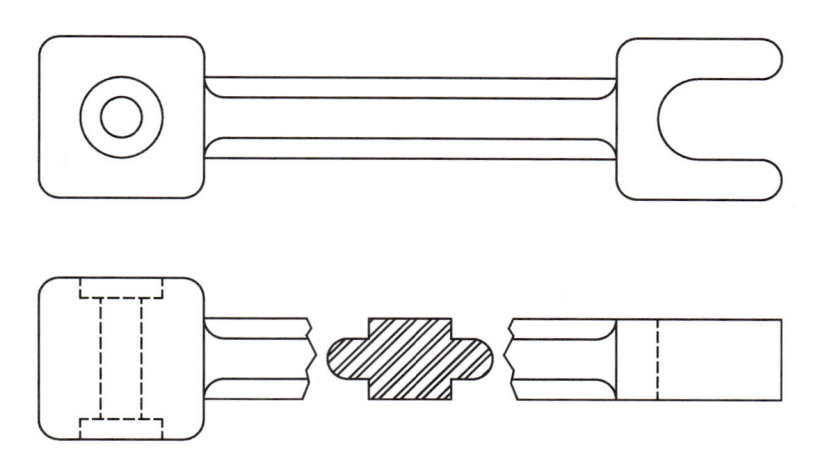

Removed Sections. A removed section is similar to a revolved section, except the section is not placed at the point where the section was cut.

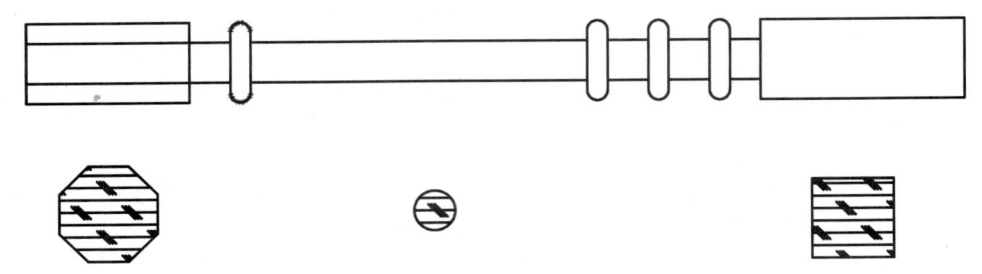

Offset Sections. Offset sections are cut along an uneven line. Offset sections should be used carefully; change the cutting plane only to show essential elements.

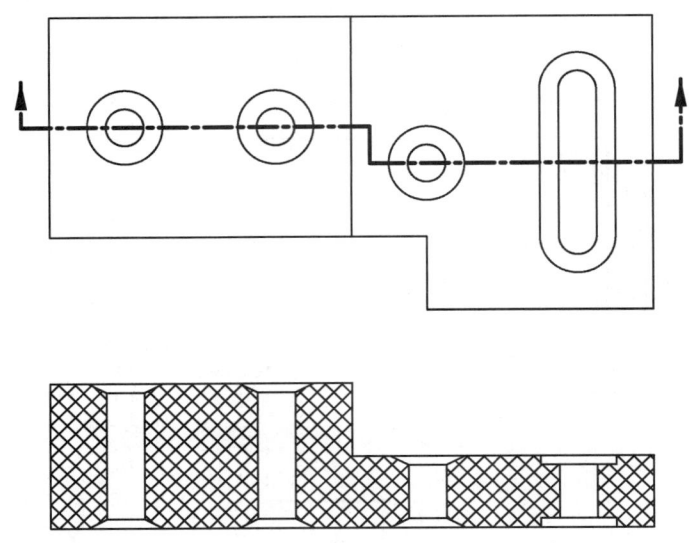

The **BHATCH** (short for "boundary hatch") command generates a boundary around an area, and then places an *associative* hatch pattern within that area. Associative hatch patterns automatically update themselves when you change the hatch boundary. If you prefer that the pattern not update itself, then you can place a nonassociative hatch.

Hatches differ from other objects in that the entire hatch is treated as one object. When you select one line of a hatch, the entire hatch is selected.

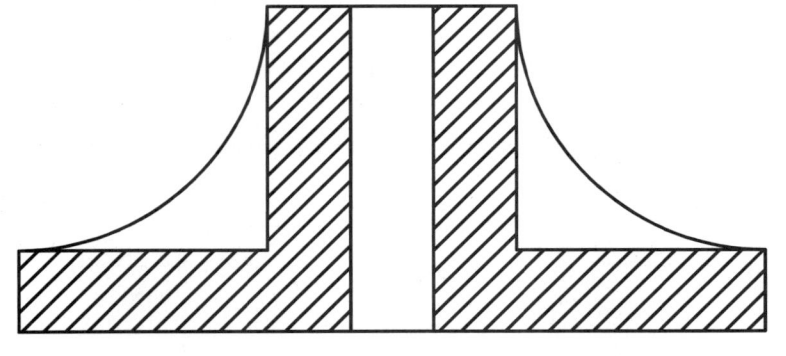

To place a hatch pattern in the drawing, AutoCAD needs to know at least two parameters: (1) the name of the hatch pattern; and (2) the area to be hatched. Optionally, you can also specify: (3) the scale; (4) rotation angle; and (5) style of area detection. You can also determine whether the boundary should be retained, and create simple custom patterns.

BASIC TUTORIAL: PLACING HATCH PATTERNS

1. To place hatch patterns in the drawing, use the **BHATCH** command with one of these methods:

   - From the menu bar, choose **Draw**, and then **Hatch**.
   - On the **Draw** toolbar, pick the **Hatch** button.
   - At the 'Command:' prompt, enter the **bhatch** command.
   - Alternatively, enter the aliases **h** or **bh** at the keyboard.

   Command: **bhatch** *(Press* ENTER.*)*

2. Notice the Boundary Hatch and Fill dialog box.

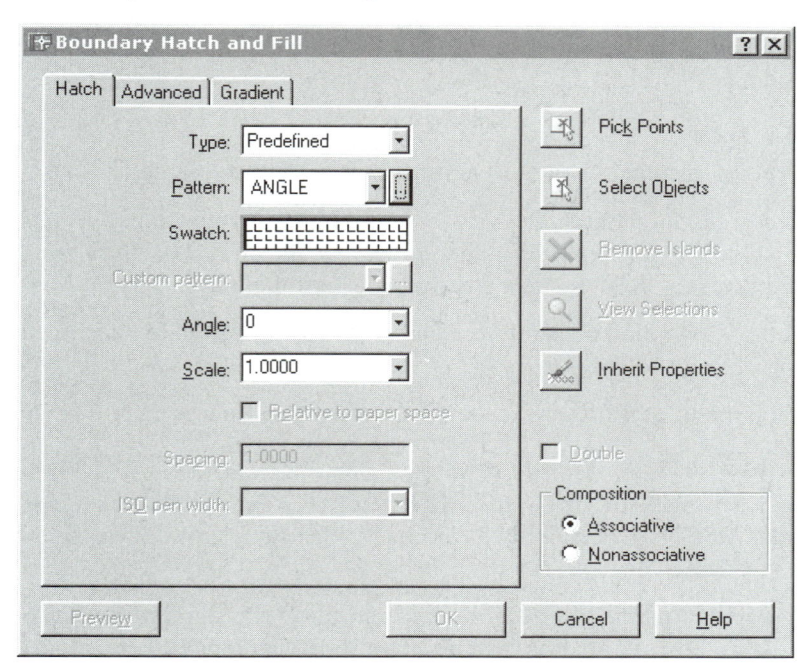

3. From the **Pattern** drop list, select a hatch pattern name.

   If you prefer to select the pattern visually, click the **...** button to display the Hatch Pattern Palette dialog box.

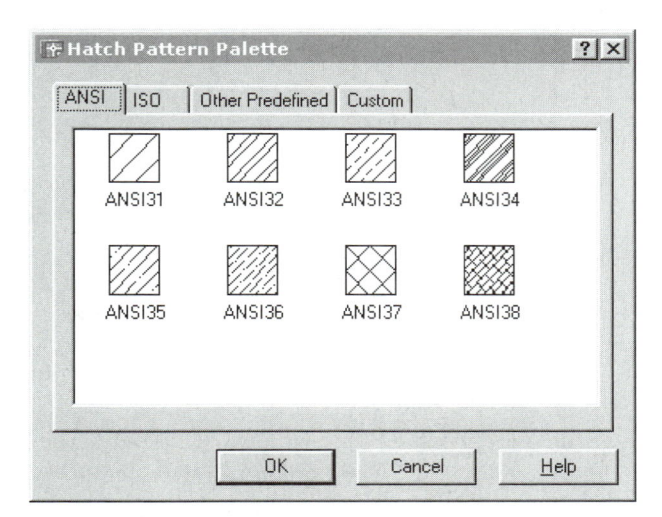

   Its tabs segregate patterns provided with AutoCAD into four groups: ANSI, ISO, Other Predefined, and Custom. Select a pattern, and then click **OK**. (For a solid color fill, select the Solid pattern found in the Other Predefined tab.)

4. Select the area to be hatched by clicking the **Pick Points** button. Notice that the dialog box disappears, and that AutoCAD prompts on the command line:

   Select internal point: *(Pick a point.)*

   Selecting everything... Selecting everything visible... Analyzing the selected data...

   Analyzing internal islands...

   Select internal point: *(Press* ENTER *to return to the dialog box.)*

5. Click **OK**. AutoCAD immediately hatches the area.

PLACING HATCH PATTERNS: ADDITIONAL METHODS

The Boundary Hatch and Fill dialog box has three tabs: **Quick**, **Advanced**, and **Gradient** (new to AutoCAD 2004), plus an area common to all tabs. This many options can be confusing, so let me step you through them.

Select a Type of Hatch

First, select the **Type** of pattern. AutoCAD supports three classes of patterns.

Predefined selects a predefined hatch pattern listed by the **Pattern** drop list. Hatch patterns provided with AutoCAD are stored in the *acad.pat* and *acadiso.pat* files. (ISO patterns permit a pen width to be assigned to the pattern; "ISO" is short for the International Organization of Standards.)

User Defined specifies a simple pattern of lines based on the angle and spacing. To create a unique pattern, apply linetypes to the hatching.

Custom selects patterns in *.pat* files typically provided by third party or in-house developers.

Select a Pattern Name

Next, select the pattern you want to use from the **Pattern** drop list. The default is made of diagonal lines, and is called "ANSI31." To select a different pattern, click the down arrow, and select from the available hatch patterns.

If you don't recognize the name, the alternative is to click the pattern next to **Swatch**. AutoCAD displays a tabbed dialog box of patterns grouped together. Select a pattern swatch, and choose **OK**.

Specify the Angle and Scale

Hatch patterns can be drawn at varying angles and scales. The hatch pattern samples shown in the dialog box have an angle of 0 degrees, even if the lines of the hatch are drawn at a different angle. To change the angle in the **Angle** text box, select one of the preset angles, or else use the keyboard to enter a different angle.

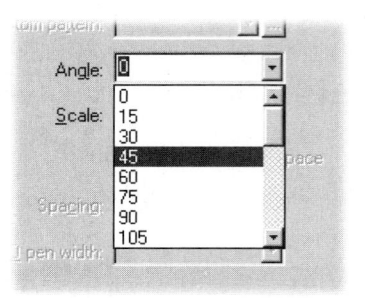

The scale factor multiplies the hatch pattern by a factor. Changing the value in **Scale** to 2, for example, doubles the spacing between lines, while changing the scale to 0.5 reduces the spacing to half-size. To change the scale in the **Scale** box, select one of the preset scales, or use the keyboard to enter another scale factor.

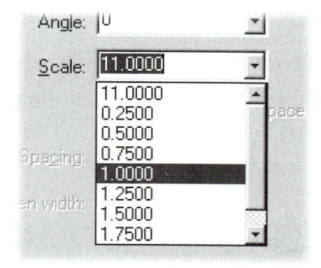

Scaling hatch patterns is the same as scaling text and linetypes: enter the inverse scale factor in the **Scale** box. For example, if the drawing is to be plotted at a scale of 1:100, then the hatch pattern must be applied at a scale of 100.

> **Note:** AutoCAD normally draws the hatch pattern using 0,0 as the starting point, even if the pattern does not physically pass through the origin (0,0). If it is critical that the hatch pattern start at a different location, such as a brick pattern on a wall, use the **Snap** command's **Rotate** option to change the base point.

Specifying User Defined Patterns

When you select **User Defined** as the hatch type, then you work with a single pattern consisting of simple, parallel lines. You specify three parameters only: the angle, the spacing between lines, and double hatching. (The angle option is the same as described earlier.)

The **Spacing** option lets you specify how far apart the parallel lines are drawn. It is similar to the scale option.

Turning on the **Double** check box causes AutoCAD to draw the pattern twice, the second time at 90 degrees to the first pattern, creating a crosshatch. The illustration below shows examples of using linetypes together with user-defined hatch patterns.

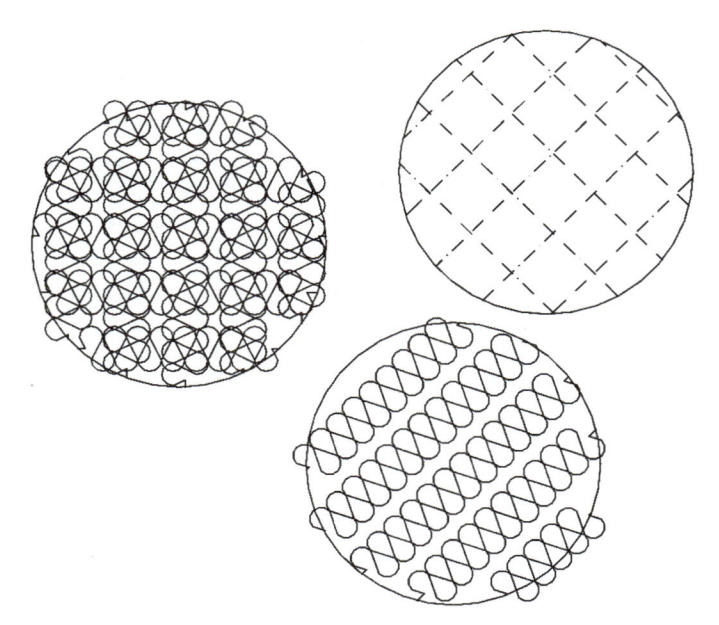

In the left circle, the Batting hatch pattern was used with double hatching; below it, with single hatching.

Selecting Gradient Fills
(New in 2004)

AutoCAD 2004 adds the ability to place *gradient* fills. These are colors that change in intensity (more white or more black), or change from one color to another. To place a gradient fill, select the **Gradient** tab of the Boundary Hatch and Fills dialog box:

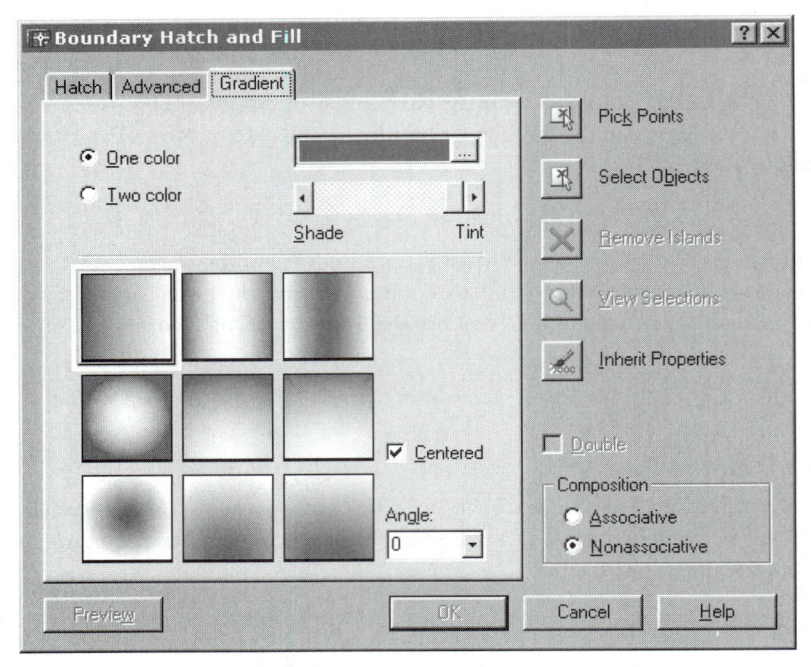

One Color means just one color is used with white or black to provide the second "color." To select the color, click the **...** button, which displays the Select Color dialog box. Pick a color, and then click **OK**.

Move the Shade-Tint slider to make the "second" color more black (Shade) or more white (Tint). Moving the slider to the center makes the gradient subtle.

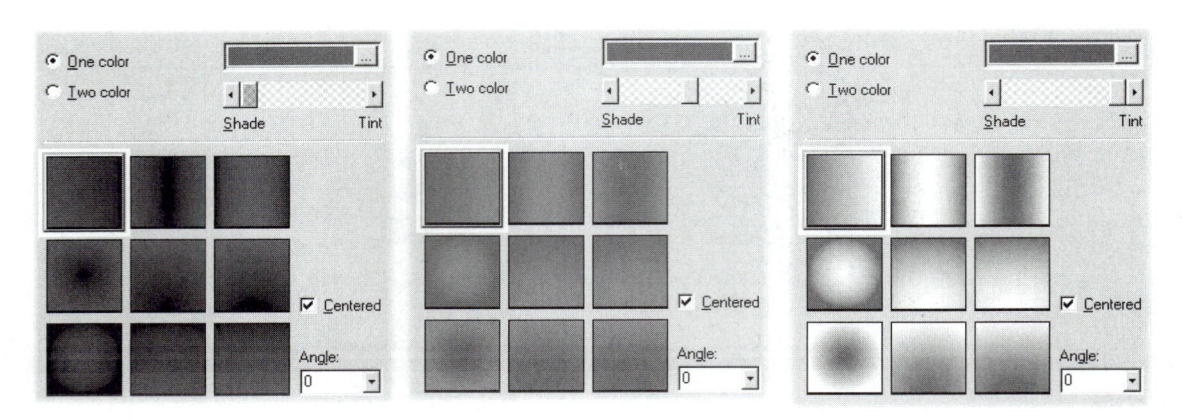

Color gradients to black.   Color with little gradient.   Color gradients to white.

Two Color creates a gradient from two colors. Pick the **...** button next to each color sample to select the colors. You don't have the slider to help you create subtle gradients; instead, select two similar colors.

Centered toggles the gradient between being centered in the fill area, or starting at the left edge.

Angle rotates the gradient inside the fill area.

Using gradients allows you to create 3D-like effects in 2D drawings, as the example below illustrates.

Selecting Other Hatch Parameters

There are a several other options in the **Quick** tab of the Boundary Hatch and Fill dialog box that are used less often. The **Relative to paper space** option is available only when you start the BHATCH command while in layout mode (paper space). It matches the hatch scale to the paper space scale.

The **ISO pen width** option is available only when you select an ISO hatch pattern. It allows you to specify the width of the lines making up the ISO pattern.

The **Composition** area lets you decide whether you want associative or non-associative hatch patterns. In almost all cases, you would keep the default of **Associative**. Associative patterns stretch when their boundaries are stretched. In addition, associative hatch patterns store data about themselves. This provides two advantages: you can copy hatch parameters through the **Inherit Properties** button, and change the parameters of an existing hatch.

Nonassociative hatch patterns do not update themselves, nor do they know their own parameters. The pattern is applied as a block. The advantage is that you can explode the block (reduce the pattern to its composite lines and points), and then edit them. You can use the EXPLODE command to turn an associative hatch into a non-associative one; once exploded, a non-associative hatch pattern cannot be converted to an associative; you would need to erase it, and then reapply the hatch.

Selecting the Hatch Boundary

A hatch must be contained within a boundary. The boundary is made up of one or more objects that form a closed polygon. Objects such as lines, polylines, circles, and arcs can form the boundary.

Before you can place a hatch, you must identify the objects that form the boundary. If the boundary is not closed, the hatch pattern leaks out. There are several ways to identify the boundary:

- Selecting the objects that make up the boundary.
- Selecting points within the hatch areas, and letting AutoCAD find the boundaries.
- Removing islands inside boundaries.

Defining the Boundary by Selecting Objects

You can specify the boundary for the hatch pattern by selecting objects that surround the area to be hatched. The illustration shows a shape constructed with four lines.

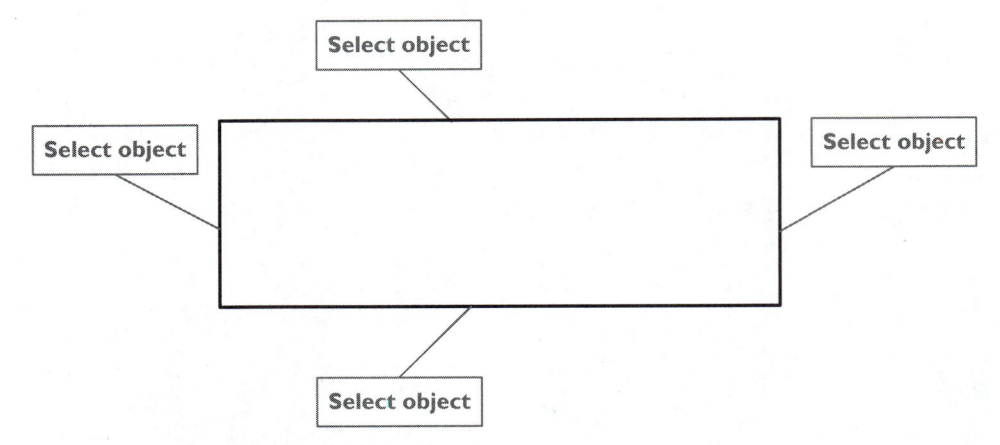

To place a hatch in the rectangle, choose the **Select Objects** button. The dialog box disappears from the screen; notice the command prompt area:

> Select objects: *(Pick one or more objects.)*
>
> Select objects: *(Press ENTER.)*

AutoCAD is prompting you to select the objects that will form the boundary. Place the pickbox over each object and select.

Note: Use the window or crossing option to select all the sides of the boundary in one operation.

When you are finished selecting objects, press ENTER. You have selected the objects that bound the area to be hatched.

To test the boundary, choose the **Preview** button. If the boundary is inadequate to contain the hatch pattern, AutoCAD reports on the command line:

> Unable to hatch the boundary.

Press ENTER or right-click to return to the Boundary Hatch and Fill dialog box. To review the objects selected for the boundary, choose **View Selections**. The dialog box again disappears, and AutoCAD highlights the boundary objects. If you need to, make changes to the boundary objects by choosing the **Select Objects** button to select and deselect objects.

Defining the Boundary by Selecting an Area

The simplest way to define the hatch area is to show AutoCAD the area by selecting a point within the area to be hatched.

Let's look at an example. The illustration shows an object with different areas that could be hatched.

[Pick Point]   [Resulting Hatch]

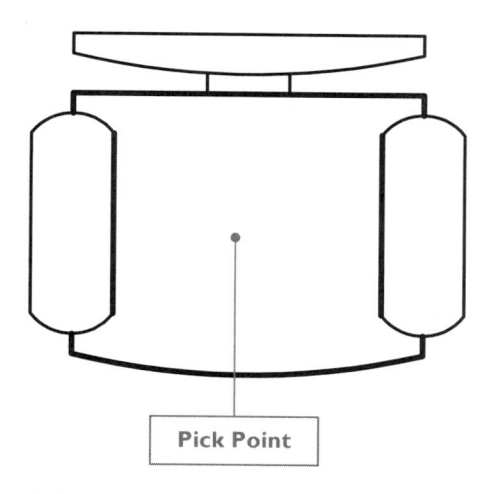

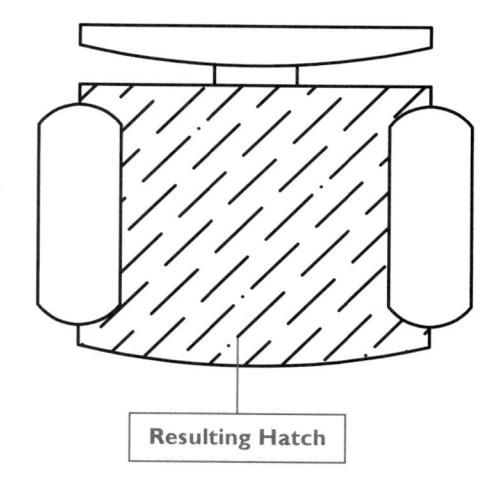

In the dialog box, choose the **Pick Points** button. When the dialog box disappears from the screen, place the crosshairs in the area.

> Select internal point: *(Pick point.)*

On the command line, AutoCAD reports:

> Selecting everything... Selecting everything visible... Analyzing internal islands...

You can think of those statements as AutoCAD mumbling to itself as it searches for the boundary. AutoCAD analyzes the area, looking for leakage. Unknown to you, AutoCAD then places a polyline boundary around the inside of the area. (You can place a boundary in any area with the **BOUNDARY** command, discussed later in this chapter.)

If you select a point in an area that is not contained, AutoCAD warns:

> Valid hatch boundary not found.

You must reconsider the area you want hatched; perhaps you need to draw another line to close off the area.

You can select more than one area. When you finish picking areas, press **ENTER** to return to the dialog box. Choose **Preview** to see how the hatch will turn out. When satisfied, choose **OK** and the hatch is completed.

Removing Islands

Sometimes the area to be hatched is not as straightforward as the rectangle hatched above. It is common for the hatch area to contain *islands*. An island is another closed object within the hatch boundary. In the illustration below, the two circles would be considered islands if the area around them were to be hatched.

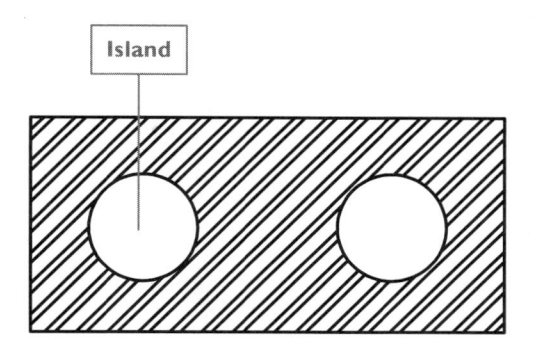

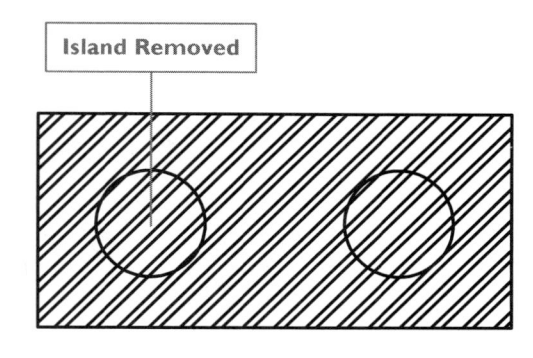

[Island]   [Island Removed]

AutoCAD assumes you do *not* want the islands hatched. If, however, you *do* want AutoCAD to draw the hatch pattern right through the islands, choose the **Remove Island** button. The dialog box disappears, and AutoCAD prompts you to pick the islands to be removed:

> Select island to remove: *(Pick object.)*

After you select an object, AutoCAD prompts you to pick another:

> <Select island to remove>/Undo:

Or, you can type **U** to deselect the object. When done removing island, press **ENTER**. Back in the dialog box, choose the **Preview** button to see the effect of having removed the islands.

Selecting Advanced Hatch Options

As you have read, the **Pick Points** button effectively automates the process of finding the hatch area. In some cases, the hatch area might be complicated. The **Advanced** tab of the Boundary Hatch dialog box provides additional parameters for determining the hatch area.

Let's look at each part of the dialog box, starting with island detection.

Island Detection Style

When the area to be hatched contain other objects, you can choose which get hatched. Island detection controls the hatching of complex areas. Let's look at how boundary styles work. The figure below illustrates a part with text.

Normal is the default style, which hatches inward from the outermost boundary, skips the next boundary, and hatches the next. Notice how the text is hatched around in the illustration. With this style hatching, an invisible window protects text from being obscured by the hatch pattern.

Outer style hatches only the outermost enclosed boundary. The hatch continues only until it reaches the first inner boundary, and continues no further. Text is not hatched.

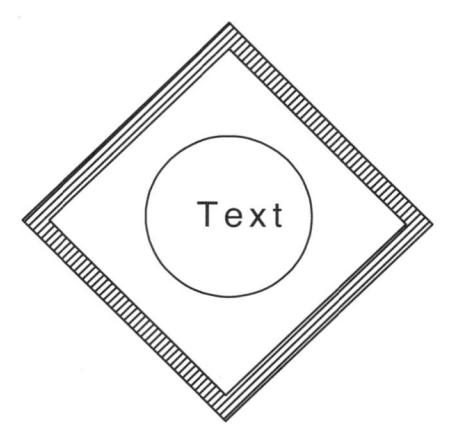

Ignore style hatches all areas defined by the outer boundary — with no exceptions. This style also hatches through text.

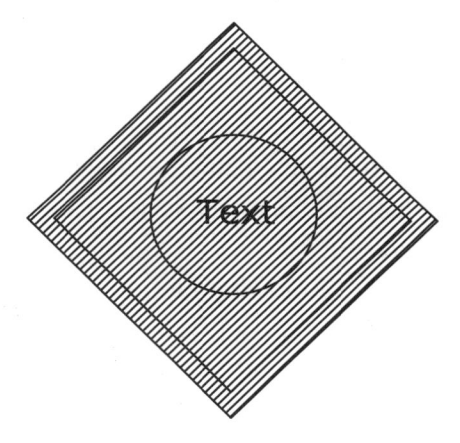

Object Type

The boundary is drawn as a polyline or a region object. When the **Retain Boundary** option is on, you can request that AutoCAD use a polyline for compatibility with older versions of AutoCAD; otherwise, use the **Region** option.

AutoCAD normally erases the boundary after the BHATCH command is finished. To retain the polyline boundary in the drawing, check the **Retain Boundaries** box.

Boundary Set

When you use the **Pick Points** option to define a boundary, AutoCAD analyzes all objects visible in the current viewport. You can, however, change the set of objects AutoCAD examines. In large drawings, reducing the set lets AutoCAD operate faster.

The **Current Viewport** option is the default, and examines all objects visible in the current viewport to create the boundary.

The **New** button prompts you to select objects from which to create the boundary set. (AutoCAD includes only objects that can be hatched.)

The **Existing Set** option creates the boundary from the objects selected with the **New** option. (You must use the **New** option *before* the **Existing Set** option.)

Island Detection Method

The **Flood** option includes islands as boundary objects.

The **Ray Casting** option runs an imaginary line from the point you pick to the nearest object. It then traces the boundary in the counter-clockwise direction; it excludes islands as boundary objects.

Pattern Alignment

There may be times when you want adjacent hatches to line up. AutoCAD guarantees alignment by using the 0,0 point as the origin for all hatches. This means that the hatches align properly. You can change the origin point by using the **SNAPBASE** system variable to change the base point.

> **Note**: Hatches can be handled more easily if they are put on their own layer. They can also be turned off and frozen to speed redraw time. Be sure that the layer linetype is continuous. Although the hatch pattern may contain dashed lines and dots, the linetype should be continuous to ensure a proper hatch.

CREATING HATCHES: COMMAND LINE

The **HATCH** command creates hatches after you enter options at the command line.

Command: **hatch**

Enter a pattern name or [?/Solid/User defined] <ANSI31>: *(Enter a pattern name or an option, or just plain press* ENTER.*)*

Specify a scale for the pattern <1.0000>: *(Enter a scale factor, or press* ENTER.*)*

Specify an angle for the pattern <0>: *(Enter an angle, or press* ENTER.*)*

Select objects to define hatch boundary or <direct hatch>,

Select objects: *(Select an object.)*

1 found Select objects: *(Press* ENTER *to exit the command.)*

The **?** option lists the name of hatch patterns loaded into AutoCAD. If you know letter(s) that the hatch pattern begins with, you can enter those few letters to narrow down the search, such as:

Enter a pattern name or [?/Solid/User defined] <ANSI31>: **?**

Enter pattern(s) to list <*>: **b***

This lists all hatch patterns that begin with 'b'. Pressing **ENTER** lists all hatch pattern names, like this:

```
AutoCAD Text Window - Drawing1.dwg
Edit
Enter pattern(s) to list <*>:
SOLID           -  Solid fill
ANGLE           -  Angle steel
ANSI31          -  ANSI Iron, Brick, Stone masonry
ANSI32          -  ANSI Steel
ANSI33          -  ANSI Bronze, Brass, Copper
ANSI34          -  ANSI Plastic, Rubber
ANSI35          -  ANSI Fire brick, Refractory material
ANSI36          -  ANSI Marble, Slate, Glass
ANSI37          -  ANSI Lead, Zinc, Magnesium, Sound/Heat/Elec Insulation
ANSI38          -  ANSI Aluminum
AR-B816         -  8x16 Block elevation stretcher bond
AR-B816C        -  8x16 Block elevation stretcher bond with mortar joints
AR-B88          -  8x8 Block elevation stretcher bond
AR-BRELM        -  Standard brick elevation english bond with mortar joints
AR-BRSTD        -  Standard brick elevation stretcher bond
AR-CONC         -  Random dot and stone pattern
AR-HBONE        -  Standard brick herringbone pattern @ 45 degrees
AR-PARQ1        -  2x12 Parquet flooring: pattern of 12x12
AR-RROOF        -  Roof shingle texture
AR-RSHKE        -  Roof wood shake texture
AR-SAND         -  Random dot pattern
Press ENTER to continue:
```

Press ENTER to continue the listing (which can go on for a long time), or else press ESC to return to the 'Command:' prompt.

Specifying the hatch style is "hidden" in this command: append the pattern name with a comma and a code for the style. For example:

> Enter a pattern name or [?/Solid/User defined] <ANSI31>: **ansi32,o**

The style codes are:

Style Code	Meaning
,i	Ignore
,n	Normal
,o	Outer
,?	List styles

The **Solid** option is a shortcut to the solid fill.

The User defined option take syou through the steps of defining the distance between lines and their angle:

> Specify angle for crosshatch lines <0>: *(Enter an angle, or press* ENTER.*)*
> Specify spacing between the lines <1.0000>: *(Enter a distance, or press* ENTER.*)*
> Double hatch area? [Yes/No] <N>: *(Type* **y** *or* **n**.*)*

The **direct hatch** option allows you to hatch areas not enclosed by a boundary. To access direct hatch, you have to press ENTER at this prompt:

> Select objects to define hatch boundary or <direct hatch>,
> Select objects: *(Press* ENTER.*)*

AutoCAD displays a new set of prompts, which guide you through creating the boundary for the hatch:

> Retain polyline boundary? [Yes/No] <N>: *(Type* **y** *or* **n**.*)*
> Specify start point: *(Pick a point.)*
> Specify next point or [Arc/Length/Undo]: *(Pick a point.)*
> Specify next point or [Arc/Close/Length/Undo]: *(Pick a point.)*
> Specify next point or [Arc/Close/Length/Undo]: *(Press* ENTER.*)*
> Specify start point for new boundary or <apply hatch>: *(Press* ENTER.*)*

AutoCAD fills the area with the hatch pattern.

TUTORIAL: CHANGING HATCH PATTERNS

1. To change hatch patterns in the drawing, use the **HATCHEDIT** command with one of these methods:
    - From the **Modify** menu, choose **Objects**, and then **Hatch**.
    - On the **Modify II** toolbar, pick the **Edit Hatch** button.
    - Double-click the associative hatch object.
    - At the 'Command:' prompt, enter the **hatchedit** command.
    - Alternatively, enter the **he** alias at the keyboard.

    > Command: **hatchedit** *(Press* ENTER.*)*

2. AutoCAD prompts you to select the hatch pattern to modify:

    > Select associative hatch object: *(Pick a hatch pattern.)*

Notice the Hatch Edit dialog box.

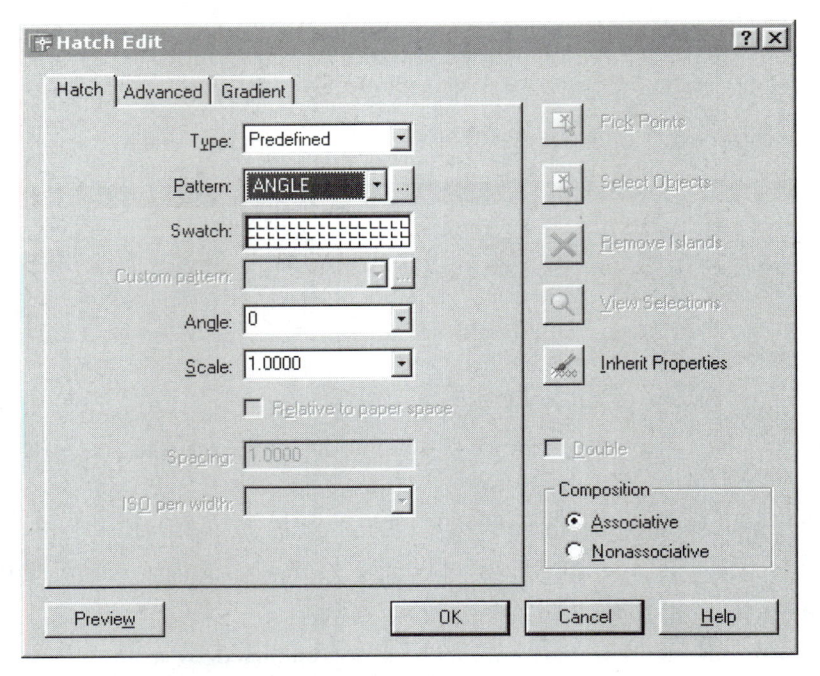

3. Make changes, such as the pattern name or scale.
4. Click **Preview** to see the effect of the changes. Notice that the dialog box disappears, and that AutoCAD prompts you:

   Pick or press Esc to return to dialog or <Right-click to accept hatch>:

5. To accept the changes, press **ENTER** or right-click the mouse.
   To return to the dialog box, press **ESC**, or left-click the mouse.

As an alternative, click the **Inherit Properties** button. This allows you to match the pattern to another in the drawing. AutoCAD temporarily dismisses the dialog box, and then prompts you:

   Select associative hatch object: *(Pick a hatch pattern.)*

AutoCAD copies the properties of the other hatch pattern, such as its name, rotation angle, and scale factor. When the Hatch Edit dialog box returns, click **OK**. The pattern changes to match.

BOUNDARY

The **BOUNDARY** command creates single regions from disparate areas.

Earlier, you read how AutoCAD creates a temporary boundary out of a polyline or region to hold the hatching. To work with boundaries independent of hatch patterns, use the **BOUNDARY** command. The boundary is selected in the same manner, and is constructed as a polyline or region. The difference is that no hatch is placed within the boundary.

> **Note:** The **BOUNDARY** command has an option that converts the outline of an object to a polyline; this allows you to select it more easily for editing.

TUTORIAL: CREATING BOUNDARIES

1. To place boundaries in the drawing, use the **BOUNDARY** command with one of these methods:
   - From the menu bar, choose **Draw**, and then choose **Boundary**.
   - At the 'Command:' prompt, enter the **boundary** command.
   - Alternatively, enter the aliases **bo** or **bpoly** (the command's original name, short for "boundary polyline") at the keyboard.

   Command: **boundary** *(Press ENTER.)*

2. Notice the Boundary Creation dialog box looks identical to the Advanced tab of the Hatch Pattern and Fill dialog box — except that numerous options are grayed out (are unavailable).

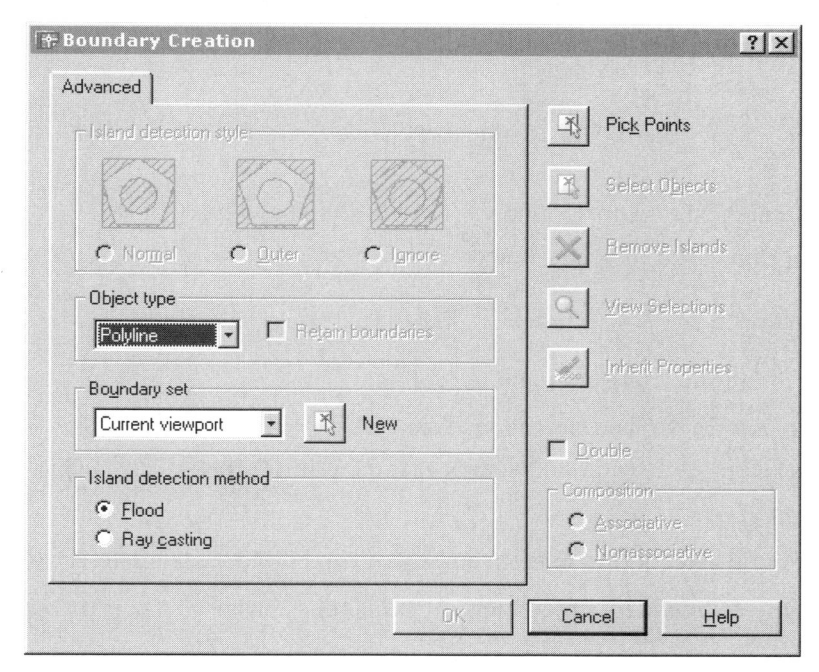

3. Select the object you want used for the boundary: a polyline or a region.
   Polyline boundaries are more easily edited, while region boundaries can be analyzed for properties, such as the centroid (geometrically-weighted center), using the **MASSPROP** command.

4. Click **Pick Points** to pick a point within the area to be bounded. If the area is enclosed, AutoCAD creates the boundary, makes the following report, and exits the command:

   BOUNDARY created 1 polyline

   Command:

   If AutoCAD finds the area is not closed, a dialog box complains in somewhat misleading terms, "Valid hatch boundary not found."

You probably won't notice the new boundary in the drawing, because AutoCAD traces over the objects defining the boundary. To edit it, such as to move or color it, use the "last" object selection, which selects the last-drawn object in the drawing.

CREATING BOUNDARIES: COMMAND LINE

The **-BOUNDARY** command creates boundaries after you specify them at the command line. If you don't need to change options, the command-line is faster than the dialog box. The command runs like this:

> Command: **-boundary**
> Specify internal point or [Advanced options]: *(Pick a point.)*
> Selecting everything... Selecting everything visible... Analyzing the selected data...
> Analyzing internal islands...
> Specify internal point or [Advanced options]: *(Press* ENTER.*)*
> BOUNDARY created 1 polyline

The **Advanced options** include **Boundary set**, **Island detection**, and **Object type**:

> Enter an option [Boundary set/Island detection/Object type]: *(Enter an option.)*

Boundary Set

The **Boundary set** option allows you to select which objects are considered for determining the boundary:

> Specify candidate set for boundary [New/Everything] <Everything>

The **New** option prompts you to select objects, while the **Everything** option includes all objects in the drawing.

Island Detection

The **Island detection** option toggles island detection:

> Do you want island detection? [Yes/No] <Y>: *(Enter an option.)*

When off, islands are ignored; when on, islands are taken into account.

Object Type

The **Object type** option selects the object from which the boundary is made, a region or a polyline:

> Enter type of boundary object [Region/Polyline] <Polyline>: *(Enter an option.)*

SELECT

The **SELECT** command allows you to select objects by their location. For example, you can select all objects that are inside a selection window, or all objects in the entire drawing. AutoCAD identifies selected objects by displaying them with *highlighting*, as if they were drawn with a dashed line. In the illustration below, the chair on the right has been selected.

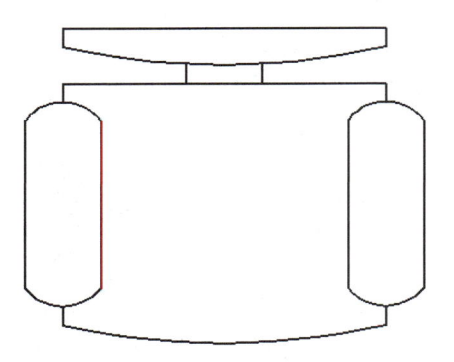

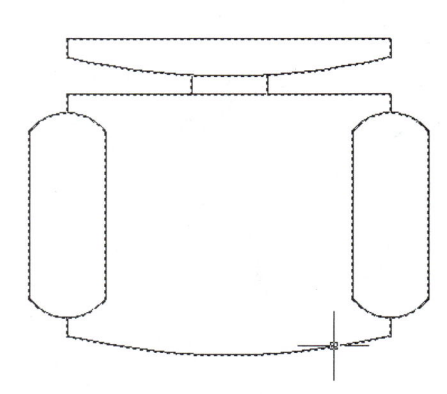

More commonly, objects are not selected with SELECT, but during other commands that present the "Select objects:" prompt. For example, you start the MOVE command, and then select the objects you wish to move.

In addition, AutoCAD allows you to select objects without the SELECT command or its options. With no command active, you pick one or more objects, and then AutoCAD highlights them, as well as displays *grips* (called "handles" in other software applications). You then edit the objects by manipulating their grips, until you press ESC to exit this direct editing mode.

TUTORIAL: SELECTING OBJECTS BY LOCATION

1. To select objects in the drawing, use the SELECT command with one of these methods:
   - At the 'Command:' prompt, enter the **select** command.
   - Alternatively, enter a selection option at any "Select objects" prompt.

   Command: **select** *(Press ENTER.)*

2. AutoCAD prompts you to select objects:

   Select objects: *(Select an object.)*

   1 found Select objects:

   Notice that AutoCAD highlights the objects you select, as well as keeps a running tally of the number of objects selected, as in "1 found."

3. You can keep selecting objects until you press ENTER to exit the command:

   Select objects: *(Select more objects.)*

   3 found Select objects: *(Press ENTER.)*

4. After pressing ENTER to exit the SELECT command, the highlighting disappears. The objects you selected are added to AutoCAD's *selection set*.

So, you may wonder, what good is the SELECT command? Not a lot, which is why most CAD operators never use it. You can, however, access the selection set with the **Previous** option of the SELECT command, or at any "Select objects:" prompt. In this way, the SELECT command is good for creating a selection set that might be used later by more than one command.

Note: The SELECT command fails to list its options. To force it to display options, enter **?**, as follows:

Command: **select**

Select objects: **?**

Invalid selection

Expects a point or

Window/Last/Crossing/BOX/ALL/Fence/WPolygon/CPolygon/Group/CLass/Add/Remove/Multiple/Previous/Undo/AUto/SIngle

Select objects: *(Enter an option.)*

Select Options

The select command's options have the following meaning:

Select Option	Meaning
Expects a point	Selects one object under the cursor.
Window	Selects all objects fully within a rectangle defined by two points.
Last	Selects the most recently created visible object.
Crossing	Selects all objects within and crossing a rectangle defined by two points. Crossing mode is indicated by a dashed rectangle.
BOX	Selects all objects within and/or crossing a rectangle specified by two points. If the selection rectangle is picked from right to left, performs Crossing selection; when selection rectangle is picked left to right, Window..
All	Selects all objects in the drawing, except those residing on frozen and locked layers.
Fence	Selects all objects crossing a selection line. The fence line can cross itself.
WPolygon	Selects all objects completely within a selection polygon, which can be any shape but cannot cross itself; AutoCAD closes polygon.
CPolygon	Selects all objects within and crossing a selection polygon, which can be any shape but cannot cross itself; AutoCAD closes polygon.
Group	Selects all objects comprising a named group.
CLass	Selects object classes defined by add-on software, such as Autodesk Map. *Classification* assigns properties, such as "road" and "river," to objects to make it easier to analyze drawings.
Add	Switches selection mode to Add, after being in Remove mode. Objects selected by any means listed above are added to the selection set.
Remove	Switches selection mode to Remove. Objects selected by any means listed above are removed from the selection set. As an alternative, you can hold down the SHIFT key to remove objects from the selection set.
Multiple	Selects objects without highlighting them. Also selects two intersecting objects when the intersecting point is selected twice.
Previous	Adds objects that were previously selected to the selection set. The Previous selection set is ignored when you switch between model and paper space.
Undo	Removes the object most recently added to the selection set.
Auto	Selects objects by three methods. Picking a point in a blank area of the drawing starts Window or Crossing mode, depending on how you to the cursor: • Right to left performs Crossing selection. • Left to right performs Window selection. • Picking an object selects it.
Single	Selects the first object(s) picked, and then does not repeat the "Select objects:" prompt.

That probably seems like too many options for you! It is useful to know about all of them, but in practice you probably use just a half dozen — All, Previous, Last, Window, Crossing, and point are the ones I use most frequently.

Note: A common problem is how to select overlapping objects. You may find that you repeatedly (and unsuccessfully) try to select one of two overlapping objects, and it's the "wrong" one that AutoCAD picks. The solution is to hold down the CTRL key, and then pick again. AutoCAD displays the prompt:

Select objects: <Cycle on> *(Pick again.)*

and then selects the other object the next time you pick.

Pick (Select Single Object)

To select a single object, click on it with the cursor. There is no need to enter any options at the "Select objects:" prompt:

Select objects: *(Pick an object.)*

In this mode, AutoCAD selects a single object. To select another object, pick again; to "unselect" the object, hold down the **SHIFT** key. To select from two or more overlapping objects, hold down the **CTRL** key. To select more than one object at a time, use one of the modes described next.

ALL (Select All Objects)

The **All** option selects all the objects in the drawing, except those on frozen and locked layers — most of the time. For some commands, such as **COPYCLIP**, the **All** option selects only those objects visible in the viewport.

> **Notes**: To select most objects, first select all objects with the **All** option, and then use the **Remove** (explanation follows) option to "deselect" the exceptions.
>
> As a shortcut, you can press the **CTRL+A** keys, which selects all objects in the drawing (subject to the restrictions noted above).

W (Select Within a Window)

One of the most common forms of object selection is the **W** option (short for "window"). It places a rectangle around the objects to be selected. You define the rectangle by picking its opposite corners, as follows:

    Select objects: w
    Specify first corner: (Pick point 1.)
    Specify opposite corner: (Pick point 2.)

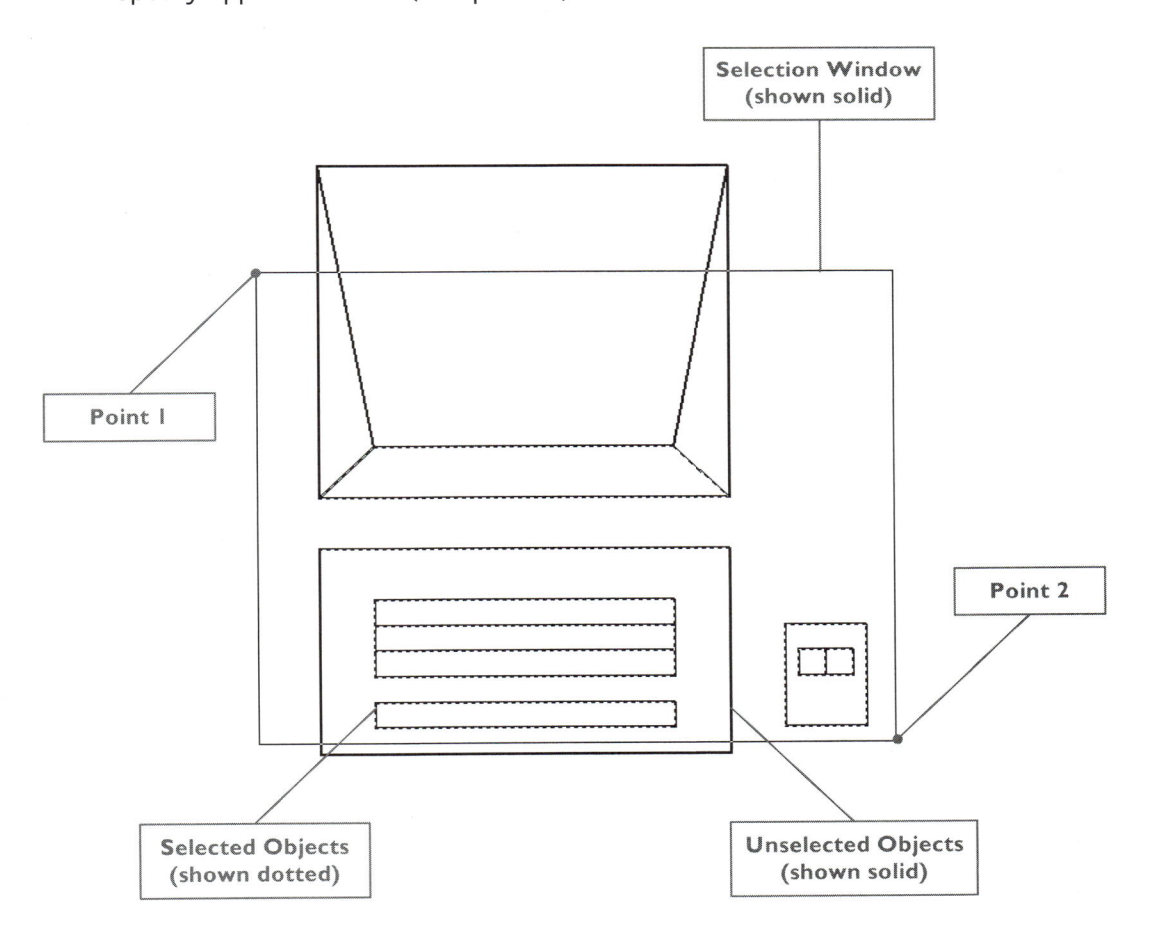

Objects *entirely* within the window rectangle are selected. If an object crosses the rectangle, it is not selected (see Crossing mode). Only objects currently visible on the screen are selected.

You can place the selection rectangle so that all parts of those objects you want to choose are contained in the windows, while those that you don't want are not entirely contained. With this method, you select objects in an area of your drawing where objects overlap.

C (Select with a Crossing Window)

The C option (short for "crossing") is similar to the Window option, with an important difference: objects crossing the rectangle are included in the selection set — as well as those objects entirely within the rectangle. Define the rectangle by picking its opposite corners.

    Select objects: **c**

    Specify first corner: (Pick point 1.)

    Specify opposite corner: (Pick point 2.)

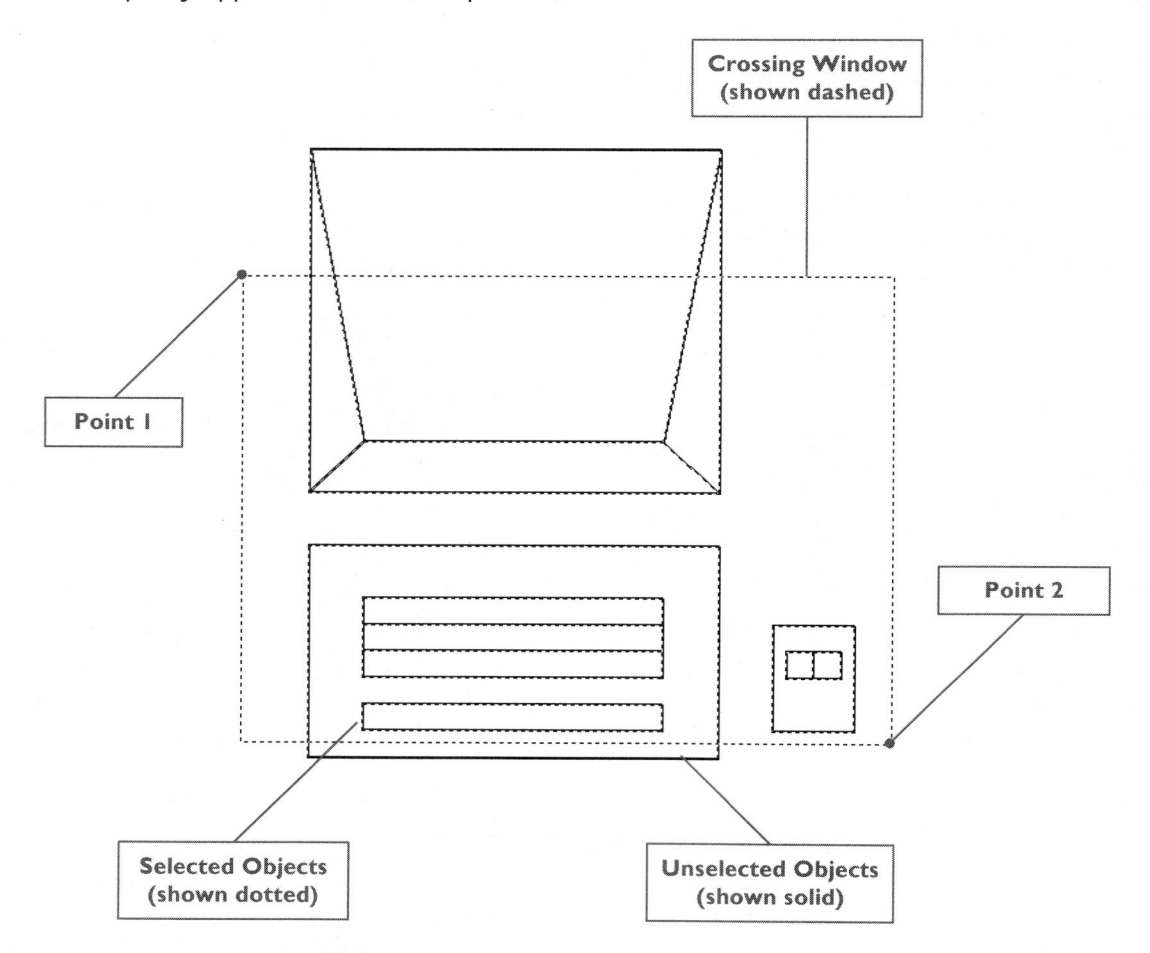

Notice that the crossing window is dashed. This distinguishes it from the Window rectangle, which is made of solid lines.

BOX (Select with a Box)

The **BOX** option allows you to use either the crossing or window rectangle to select objects. Define the rectangle by picking its opposite corners.

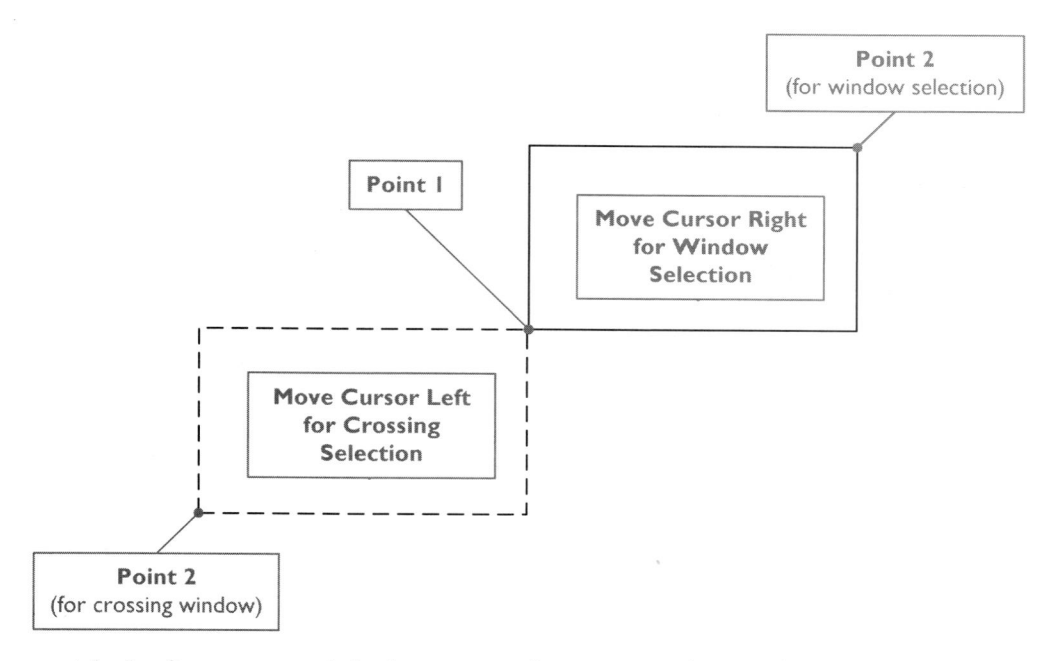

After you pick the first corner of the box, move the cursor either to the right or to the left. If you move to the right, the result is the Window selection: AutoCAD selects objects completely within the selection rectangle.

When you move to the left, the result is the crossing selection: AutoCAD selects objects within the rectangle, as well as those crossing it. In addition, the rectangle is drawn with dashed lines.

F (Select with a Fence)

The **F** option (short for "fence") uses a polyline to select objects. The fence is displayed as dashed lines; all objects crossing the fence are selected. You can construct as many fence segments as you wish, and can use the **Undo** option to undo a fence line segment.

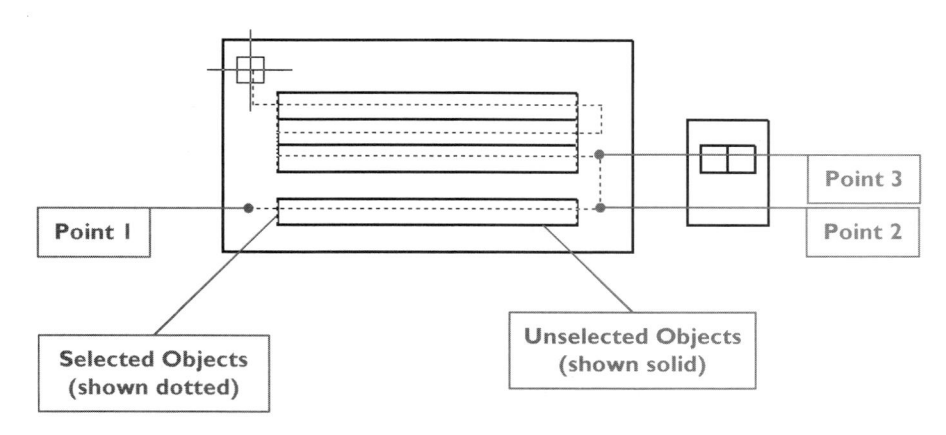

Select objects: f

First fence point: *(Pick point 1.)*

Specify endpoint of line or [Undo]: *(Pick point 2.)*

Specify endpoint of line or [Undo]: *(Pick point 3.)*

et cetera

AU (Select with the Automatic Option)

The **AU** option (short for "automatic") combines the pick and **Box** options. It is the default for most selection operations. After you enter **AU** in response to the "Select objects" prompt, you select a point with the pickbox. If an object is found, the selection is made.

If an object is not found, the selection point becomes the first corner of the Box option. Move the box to the right for Window, or to the left for Crossing. The **AU** option is excellent for advanced users who wish to reduce the number of modifier selections.

WP (Select with a Windowed Polygon)

The **WP** option (short for "windowed polygon") selects objects by placing a polygon window around them. The polygon window selects in the same manner as the Window option: all objects completely within the polygon are selected. Objects crossing, or outside the polygon, are not selected. The difference is that the **Window** option creates a selection rectangle, while the WP option creates a multisided window. Let's look at a sample command sequence:

> Select objects: **wp**
>
> First polygon point: *(Pick point 1.)*
>
> Specify endpoint of line or [Undo]: *(Pick point 2.)*
>
> Specify endpoint of line or [Undo]: *(Pick point 3.)*

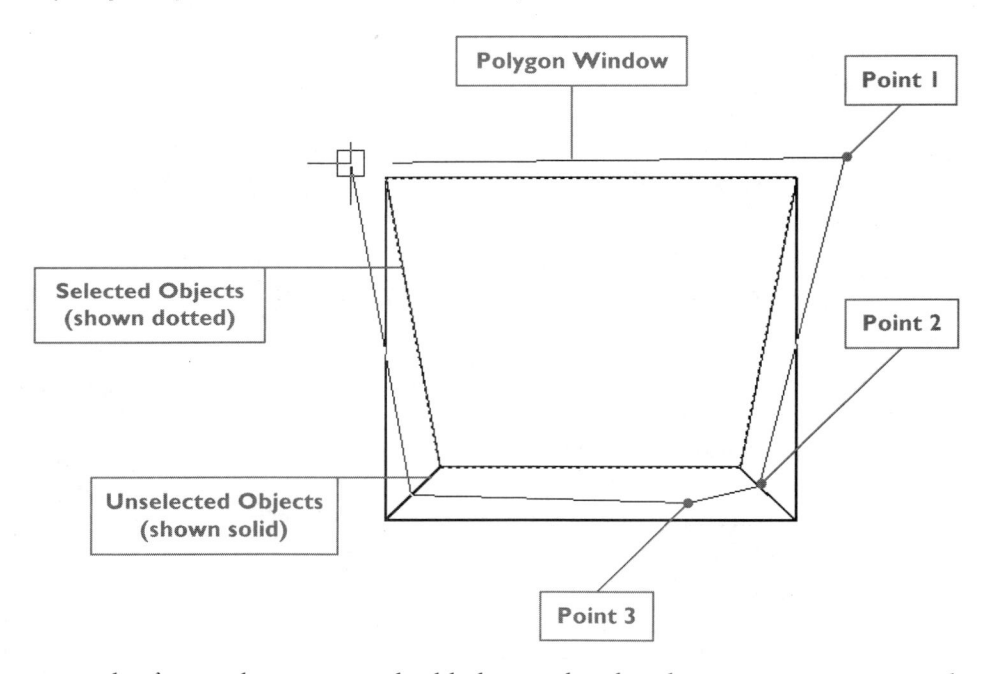

After you enter the first polygon point, build the window by placing one or more endpoints. The polygon rubber bands to the cursor intersection, always creating a closed polygon window.

Undo the last point entered by typing U in response to the prompt. Pressing ENTER closes the polygon window and completes the process.

Note that the polygon window must not cross itself or rest directly on a polygon object. If it does, AutoCAD warns "Invalid point, polygon segments cannot intersect," and refuses to place the vertex.

CP (Select with a Crossing Polygon)

The **CP** option (short for "crossing polygon") works in the same manner as the **WP** option, except that the polygon functions in the same manner as a crossing window. All objects within or crossing the polygon are selected.

The crossing polygon is displayed with dashed lines, similar to a crossing window.

SI (Select a Single Object)

The **SI** option (short for "single") forces AutoCAD to issue a single "Select objects:" prompt. (All other selection modes repeat the "Select objects:" prompt until you press **ENTER**.) You can use other selection options in conjunction with **SI** mode, such as the Crossing selection shown in this example:

> Select objects: **si**
> Select objects: **c**
> Specify first corner: *(Pick point 1.)*
> Specify opposite corner: *(Pick point 2.)*
> *nnn* found

The single option is useful for efficient single object selection in macros, because it does not require an **ENTER** to end the object selection process.

M (Select Through the Multiple Option)

Each time you select an object, AutoCAD scans the entire drawing to find the object. If you are selecting an object in a drawing that contains a large number of objects, there can be a noticeable delay.

The **M** option (short for "multiple") forces AutoCAD to scan the drawing just once. This results in shorter selection times. Press **ENTER** to finish the object selection and begin the scan. Note that AutoCAD does not highlight the objects selected until you press **ENTER** .

> Select objects: **m**
> Select objects: *(Pick three times.)*
> *nnn* selected, *nnn* found

Special Selection Modes

P (Select the Previous Selection Set)

The **P** option (short for "previous") uses the previous selection set. This very useful option allows you to perform several editing commands on the same set, without re-selecting them.

L (Select the Last Object)

The **L** option (short for "last") selects the last object drawn still visible on the screen. When the command is repeated, and **Last** is used a second time, AutoCAD chooses the same last object and reports "1 found (1 duplicate), 1 total." This option is useful for immediately editing an object just drawn.

G (Select the Group)

The **G** option (short for "group") adds the members of a named group to the selection set. (Groups are covered later in this chapter.) The option prompts:

> Enter group name: *(Enter a group name.)*

Changing Selected Items

You can add and remove objects from the group of selected objects by using modifiers. Modifiers must be entered after you select at least one object, and before you press **ENTER** to end object selection.

U (Undo the Selected Option)

The **U** option (short for "undo") removes the most recent addition to the set of selections. If the undo is repeated, you will step back through the selection set. This shortcut replaces the two-step process of using the Remove option, and then remembering the objects to pick.

R (Remove Objects from the Selection Set)

The **R** option (short for "remove") allows you to remove objects from the selection set. You can remove objects from the selection set by any object selection method.

>Select objects: **r**
>Remove objects: *(Pick an object.)*
>1 found, 1 removed, 2 total

Note: The **Remove** option is useful when you need to select a large number of objects with the exception of one or two objects located in the area. Select all the objects, then use the **R** option to remove the excess objects.

A (Add Objects to the Selection Set)

The **A** option (short for "add") adds objects to the set. The **Add** option is usually used after the **Remove** option. **Add** changes the prompt back to "Select objects:" so you may add objects to the selection set.

Cancelling the Selection Process.

Pressing ESC at any time cancels the selection process and removes the selected objects from the selection set. The prompt line returns to the 'Command:' prompt.

Pressing ENTER ends the selection process, and continues on with the editing command's other options.

Selecting Objects During Commands

To use the selection options at any "Select objects:" prompt, enter one of the abbreviations listed in the table above (abbreviations shown in uppercase letters). For example, to move all the objects within the selection rectangle:

>Command: **move**
>Select objects: **w**
>Specify first corner: *(Pick point 1.)*
>Specify opposite corner: *(Pick point 2.)*
>5 found Select objects: *(Press* ENTER *to end object selection.)*
>Specify base point or displacement: *(Pick a point.)*
>Specify second point of displacement or <use first point as displacement>: *(Pick another point.)*

Notice that the "Select objects:" prompt repeats until you press ENTER. This means you can keep selecting (and deselecting) objects until you are satisfied with the selection set.

DIRECT SELECTION AND GRIPS EDITING

AutoCAD provides a shortcut to selecting and editing objects. You can select objects without using commands. At the 'Command:' prompt, select an object. AutoCAD displays *grips* on the object.

Grips are small, colored squares that indicate where the object can be edited. For example, lines have grips at both ends and at the midpoint; circles have grips at the centerpoint and at the four quadrant points. The illustration below shows some objects and their grip locations.

Blocks are a special case. Normally, AutoCAD displays a single grip at the block's insertion point. If you wish to display grips on the objects within the block, select the **Enable grips within blocks** check box in the Options dialog box's Selection tab.

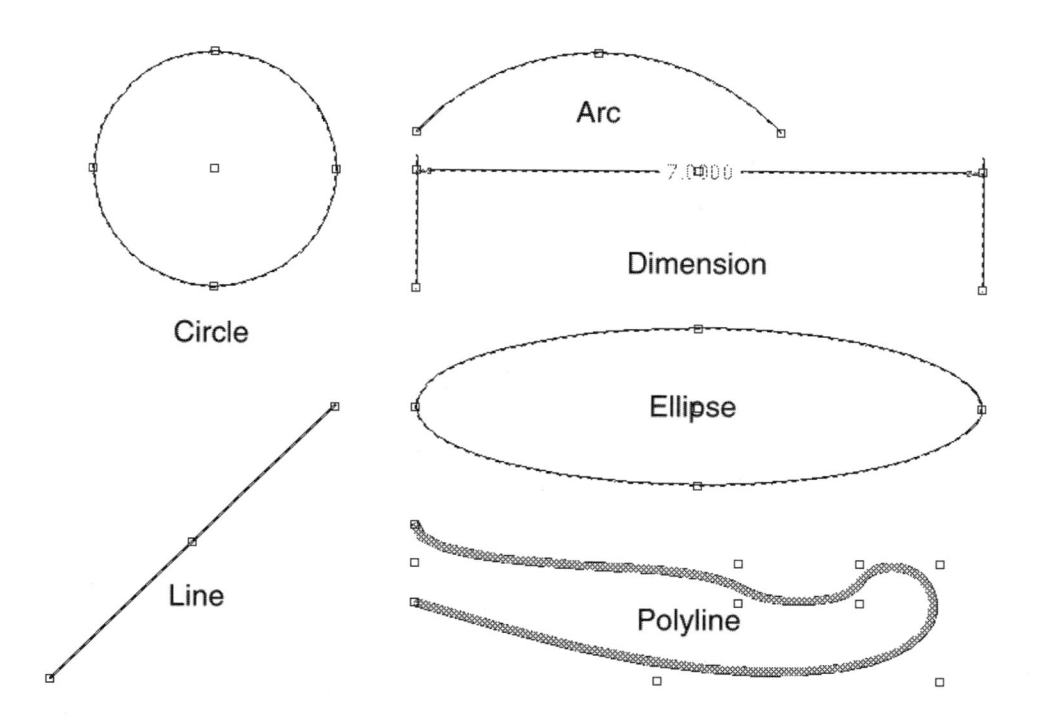

[Figures labeled: Circle, Arc, Dimension (7.0000), Ellipse, Line, Polyline]

Grip Color and Size

As a visual aid, grips are assigned colors and names:

Grip Color	Name	Meaning
Blue	Cold	Grip is not selected.
Green	Hover	Cursor is positioned over grip (new to AutoCAD 2004).
Red	Hot	Grip is selected, and object can be edited.

You can change the color and size of grips in the Selection tab of the Options dialog box. You can chose any color, but it makes sense to keep the ones AutoCAD assigned. The grip box size can range from 1 to 255 pixels.

Tutorial: Using Grips for Editing

The purposes of grips is to provide modeless editing free of commands, much like other drawing programs. Rather than entering commands and selecting objects, you select the objects first, and then change them.

1. Select any object for editing. The object is highlighted.
   Notice that blue (cold) grips are placed on the object. You can to select as many objects as you wish, by selecting more than once.
   If you find you cannot select more than one object, hold down the **SHIFT** key on the keyboard when selecting the grips.
2. After selecting the object to be edited, move the cursor over one of the blue grips as a base point for editing. As you get close to the it, notice that the cursor jumps to the grip, and the grip changes color to green (hover).
3. Click on the green grip. The grip box changes color to red (hot) to denote selection.
4. Edit the grip as described in Chapter 9; editing options include **Stetch**, **Mirror**, **Move**, **Copy**, **Rotate**, **Scale**, and **Mirror**. Right-click a red grip to see the editing options.
5. Press **ESC** once to clear the grips.

QSELECT

The **QSELECT** command allows you to select objects based on their properties, instead of on their location. For example, use the command to select all objects with hidden linetypes on a specific layer. Or, all circles with a radius larger than one inch.

TUTORIAL: SELECTING OBJECTS BY PROPERTIES

1. To select objects by their common properties, start the **QSELECT** command with one of these methods:
   - From the menu bar, choose **Tools**, and then **Quick Select**.
   - Right-click, and from the shortcut menu choose **Quick Select**.
   - At the 'Command:' prompt, enter the **qselect** command.

   Command: **qselect** *(Press ENTER.)*

2. Notice the **Quick Select** dialog box.

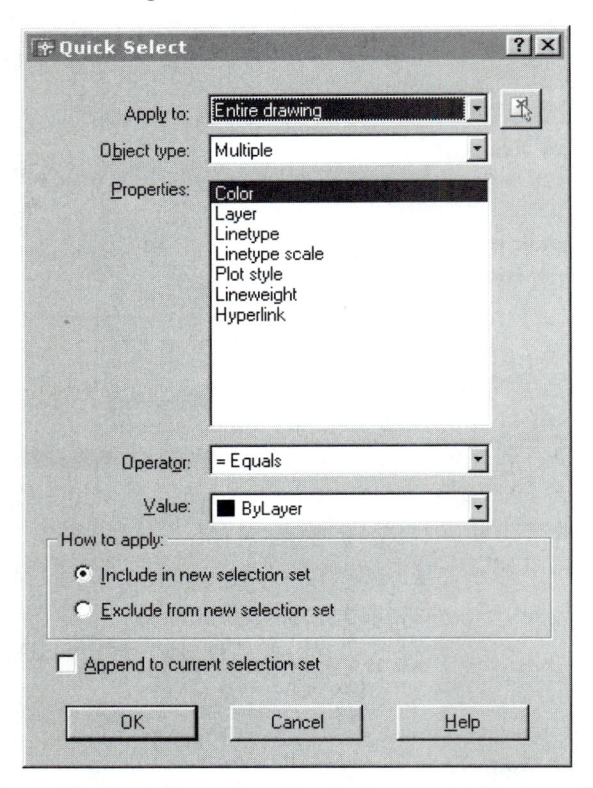

3. The selection can be made from the entire drawing or a subset of the drawing, called the "current selection."

   To select a subset, click the **Select Objects** button next to the **Apply to** drop list. AutoCAD prompts you:

   Select objects: *(Select one or more objects.)*

   Select objects: *(Press ENTER to return to the dialog box.)*

   Notice that the **Apply to** drop list now has two options: "Entire Drawing" and "Current Selection." Select one.

4. The **Object Type** drop list allows you to narrow down the selection to specific objects. Only the objects found in the drawing or the current selection are listed here. To include all objects, select "Multiple."

5. Narrow down your selection by picking one of the item in the **Properties** list. The type of object you select affects the list shown in the list, which contains all properties that AutoCAD can search for. You can select only one property from this list.
6. The **Operator** list lets you narrow the range AutoCAD searches for. Depending on the property you select, the choice of operator includes:

Operator	Meaning
=	Equals
<>	Not Equal To
>	Greater Than
<	Less Than
*	Wildcard Match (selects all).

   Use the ***Wildcard Match** operator with text fields that can be edited, such as the names of blocks.
7. The **Value** field goes with the Operator list: if AutoCAD can determine values, it lists them here; if not, you type the value.
8. Under **How to Apply**, you can have AutoCAD include or exclude from an existing selection set.
9. The **Append to Current Selection Set** check box determines whether these are added to (on) or replace (off) the current selection set.
10. Choose **OK** to exit the dialog box. AutoCAD highlights the objects selected, and reports the number selected:

    534 item(s) selected.

CONTROLLING THE SELECTION: ADDITIONAL METHODS

In addition to selection modes and direct selection, AutoCAD has these commands:

- **PICKBOX** changes the size of the square pick cursor.
- **DDSELECT** changes selection options; displays the Selection tab of the Options dialog box.
- **FILTER** selects objects based on their properties.
- **SELECTURL** highlights all objects that contain a hyperlink. See *Using AutoCAD: Advanced* for details on using hyperlinks in drawings.

Let's look at how these commands work.

Pickbox

When an editing command displays the "Select objects:" prompt, the cursor is supplemented by a small square called the "pickbox." When you place the pickbox over the object, and then click (press the left mouse button), AutoCAD scans the drawing, and selects the object under by the pickbox.

The size of the pickbox can be made larger and smaller. The Selection tab of the Options dialog box allows you to change the pickbox size; or, you just might prefer to use the **PICKBOX** system variable:

   Command: **pickbox**

   Enter new value for PICKBOX <3>: *(Enter a value between 0 and 50.)*

The size of the pickbox is measured in pixels. A large pickbox forces AutoCAD to scan through more objects, which can take longer amounts of time in complex drawings on slow computers.

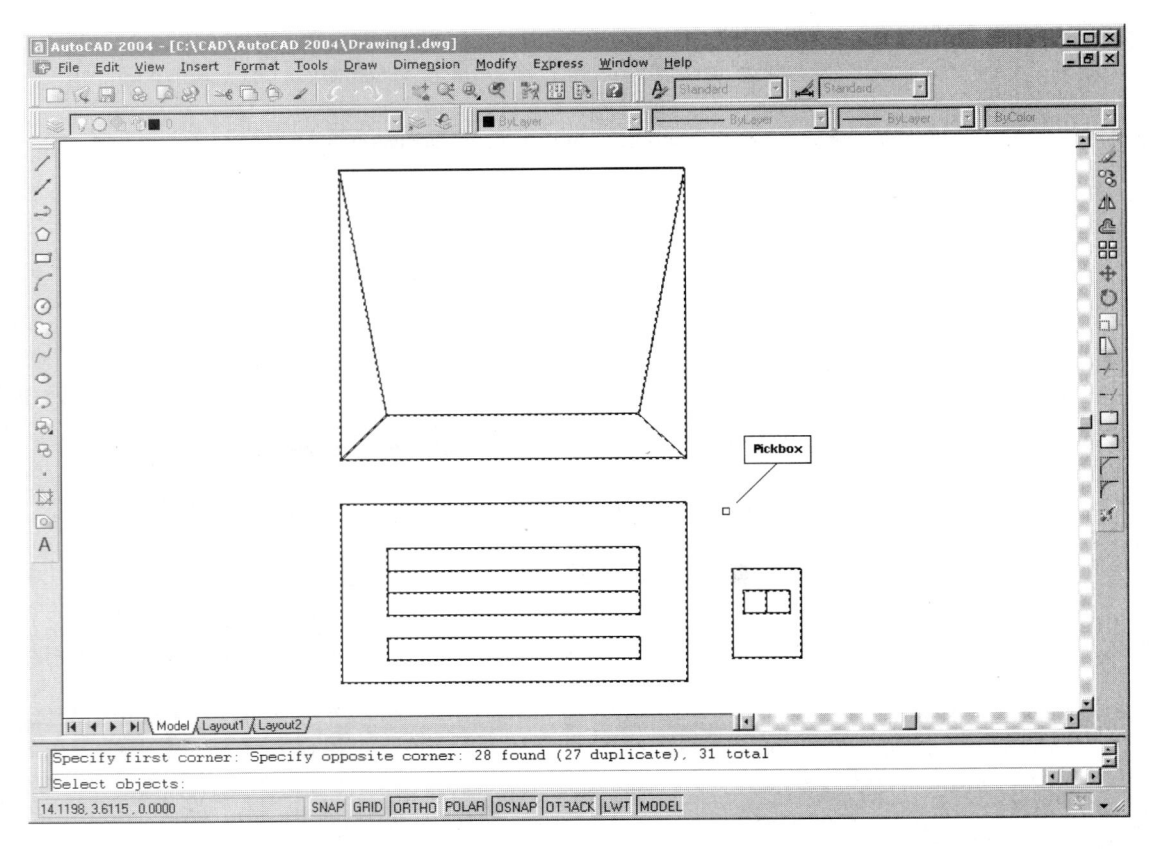

DdSelect

The DDSELECT command allows you to change selection modes. (The command is actually an alias for displaying the Selection tab of the Options dialog box.)

- From the **Tools** menu, choose **Options**, and then choose the **Selection** tab.
- Right-click, and from the shortcut menu choose **Options**.
- At the 'Command:' prompt, enter the **ddselect** command.

    Command: **ddselect** *(Press ENTER.)*

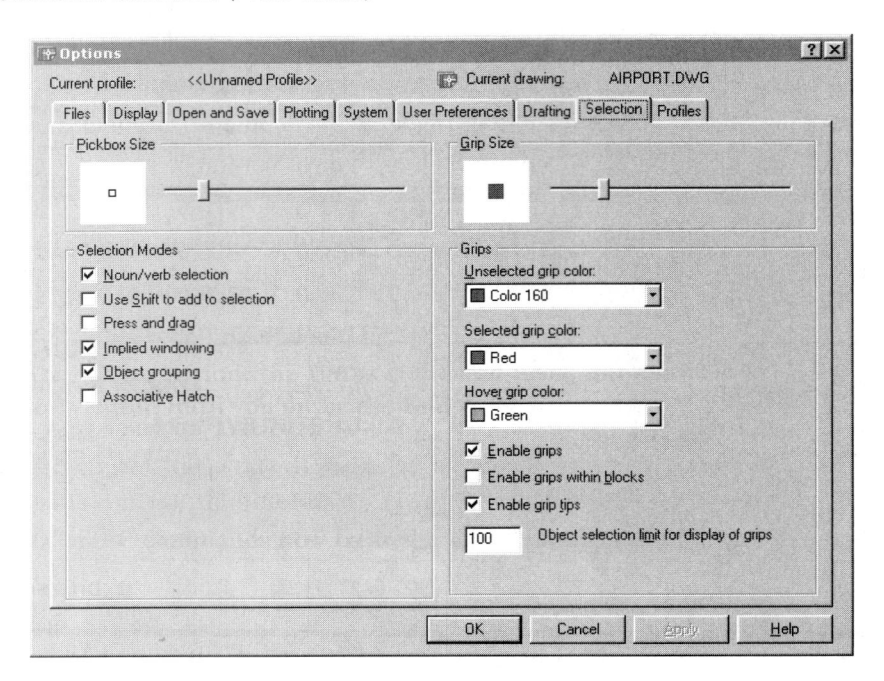

Let's look at each option.

Pickbox Size

The pickbox appears at the "Select objects:" prompt, and you can change its size.

Noun/Verb Selection

The "normal" way to edit drawings is to: (1) first select the editing command; and (2) then select the objects to modify. Technically, this is choosing the *verb* (the action represented by the edit command), and then the *noun* (the object of the action represented by the selection set).

As an alternative, AutoCAD allows you to reverse the procedure: (1) first select the object(s) to edit; and (2) then choose the editing command. This is called "noun/verb selection."

When the **Noun/Verb Selection** option is turned on (as it is by default), AutoCAD places the pickbox at the intersection of the crosshairs. The presence of the pickbox is your indication that you can select objects before one of these editing commands:

> ALIGN LIST ARRAY MIRROR BLOCK MOVE CHANGE PROPERTIES CHPROP
>
> ROTATE COPY SCALE DVIEW STRETCH ERASE WBLOCK EXPLODE

Here is the command sequence to erase a single object, for example:

> Command: *(Select the object.)*
> Command: **erase**
> 1 found

Use Shift to Add to Selection

When selecting, you normally choose the object(s), and then press ENTER when you are finished. As you select each object, it is automatically added to the selection set. To remove objects from the selection set, hold down the SHIFT key while picking them.

As an alternative, turn on the **Use Shift to Add to Selection** option. When this option is turned on, you must hold down the SHIFT key to add objects to the selection set; this is similar to the method used by many other Windows programs. When you make more than one selection without holding the SHIFT key, the previous selections are removed from the selection set.

Press and Drag

The traditional AutoCAD method for selecting objects through windowing is to: (1) click one corner; (2) move the cursor to the other corner; and (3) click the other corner.

As an alternative, turn on the **Press and Drag** option. You build a window by: (1) clicking one corner; and then (2) dragging (by holding down the mouse button) the cursor to the other corner. Like the Shift to Add option, this is the method used by most other Windows programs.

Implied Windowing

You previously learned how to use a selection window or a crossing window to select objects. All you have to do is enter either a **W** or a **C** to invoke the window mode.

The **Implied Windowing** option does this automatically (the option is on by default). You "imply" the window by clicking in an empty area of the drawing. When AutoCAD does not find an object within the area of the pickbox, it assumes that you want to use windowing. When you move the cursor to the right, Window selection mode is entered; move to the left, Crossing selection mode is entered.

Because the first point entered describes the first corner of the window, be sure to select a desirable position.

Object Grouping

When the **Object Grouping** option is turned on, AutoCAD selects the entire group when you pick one object in the group. When off, just the picked object is selected. (More on groups later in this chapter.)

As a shortcut, you can toggle object grouping mode with CTRL+SHIFT+A. Each time you press the key combination, AutoCAD comments:

Command: *(Press CTRL+SHIFT+A.)* <Group off>

"Group off" means objects are selected, while "Group on" means the entire group is selected:

Command: *(Press CTRL+SHIFT+A.)* <Group on>

Curiously, CTRL+H does the same thing. Technically, it toggles the PICKSTYLE system variable between 0 (individual objects selected from the group) and 1, the entire group is selected. PICKSTYLE can also have the values of 2 and 3 for toggling the selection of associative hatch patterns.

Associative Hatch

When the **Associative Hatch** option is turned on, AutoCAD selects the boundary object(s) when you pick an associative hatch pattern. When off, only the hatch is selected.

Filter

The FILTER command displays a dialog box that lets you select objects based on their properties. You can save the resulting selection set by name, and then access it at the "Select objects:" prompt.

- At the 'Command:' prompt, enter the **filter** command.
- Alternatively, enter the **f** alias at the keyboard.

Command: **filter** *(Press ENTER.)*

Because FILTER is a *transparent* command, it can be used during another command to filter the selection set:

Command: **erase**

Select objects: **'filter**

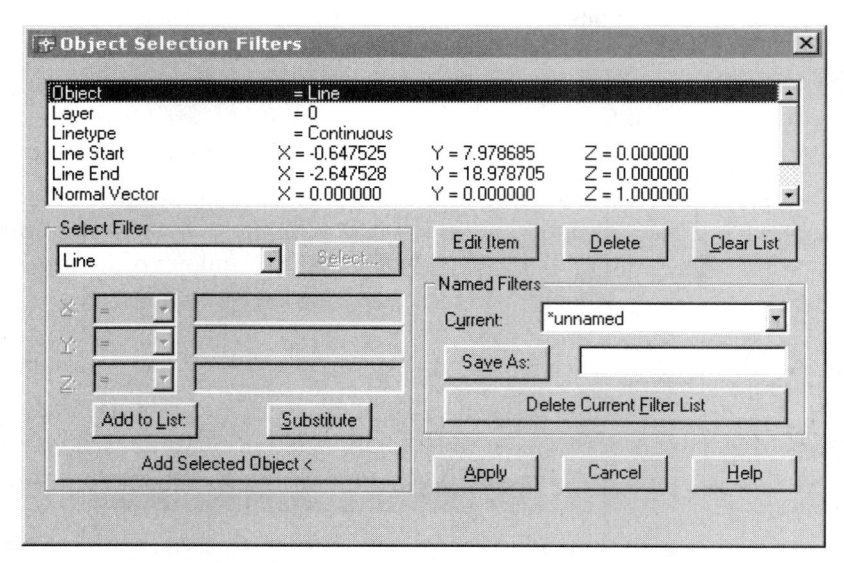

The selection set created by FILTER can be accessed via the **P** (previous) selection option. In the following tutorial, you erase all construction lines from a drawing:

1. Start the **ERASE** command:
   Command: **erase**
2. Then invoke the **FILTER** command transparently (prefix the command with the quotation mark):
   Select objects: **'filter**
   Notice that AutoCAD displays the Object Selection Filters dialog box.
3. In the **Select Filter** drop list, select "Xline."

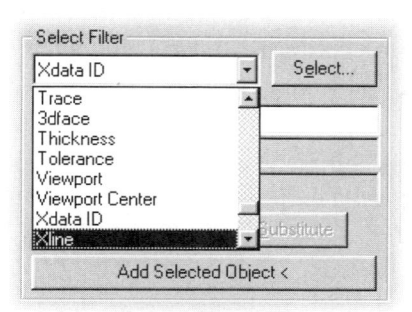

   Xline is near the end of the list. Here's a quick way to get to any item in an alphabetical list: after clicking on the drop list, press **x** on the keyboard. You are taken to the first word starting with "x".
4. Click the **Add to List** button. Notice that AutoCAD adds this text to the "list" (the large white area at the top of the dialog box):
   Object   =Xline

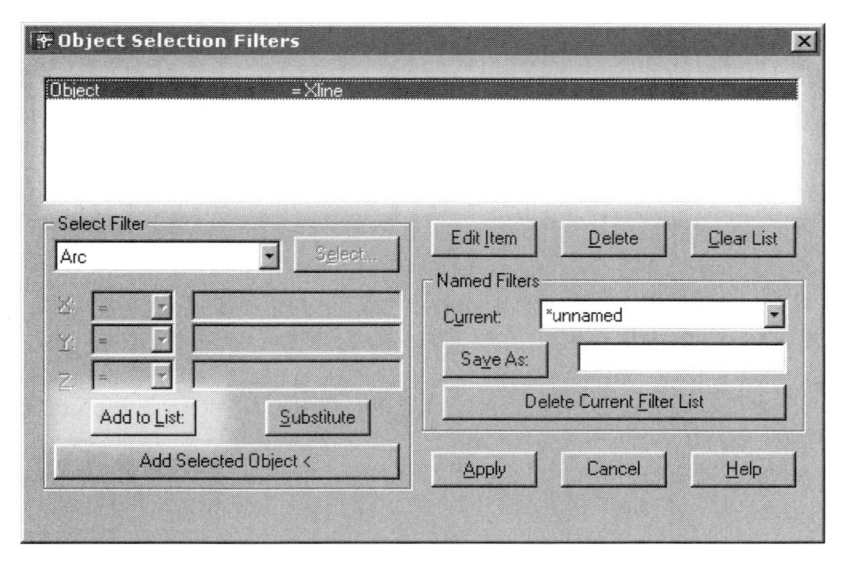

4. Click **Apply**. Notice that the dialog box disappears, and that the **FILTER** command displays prompts on the command line:
   Applying filter to selection.

   This means that AutoCAD will apply the filter (search for all xlines) to the objects you now select.
5. Specify that the filter should apply to the entire drawing with the All option:
   >>Select objects: **all**
   9405 found 9401 were filtered out.

   The double angle bracket (>>) is indicates AutoCAD is currently in a transparent command (the **ERASE** command continues on later).

6. Return to the **ERASE** command by exiting the **FILTER** command. Press **ENTER**:

    >>Select objects: *(Press* **ENTER** *to exit the* **FILTER** *command.)*

    Exiting filtered selection.

    Resuming ERASE command.

    AutoCAD returns to the **ERASE** command, and picks up the four objects selected by the **FILTER** command:

    Select objects: 4 found

7. Press **ENTER** to erase the four objects and exit the **ERASE** command:

    Select objects: *(Press* **ENTER** *to exit the* **ERASE** *command.)*

    AutoCAD uses the selection set created by the **FILTER** command to erase the xlines.

In addition to filtering objects, this command also filters x,y,z coordinates, such as the center point of circles. Other options become available in the dialog box as you select them. For example, when you select Elevation, the **X text** entry box allows you to specify the elevation using the following operators:

Operator	Meaning
<	Less than.
<=	Less than or equal to.
=	Equal to.
!=	Not equal to.
>	Greater than.
>=	Greater than or equal to.
*	All values.

Specifying elevation > 100, for example, means that all objects with an elevation greater than 100 will be added to the filtered selection set.

In addition, you can group filter sequences together using these operators:

Operator	Meaning
**Begin OR with **End OR	Include *any* of these items.
**Begin AND with **End AND	Include *all* of these items.
**Begin NOT with **End NOT	Include *none* of these items.
**Begin XOR with **End XOR	Include none of these items if one item is found.

You can save selection sets by name to an *.nfl* (short for "named filter") file on disk for use in other drawings or editing sessions.

GROUP

The **GROUP** command creates selectable groups of objects.

A selection set lasts only until a new selection set is created. When you select a circle, for example, and then later a line, AutoCAD "forgets" about the circle you selected first. (There is a work-around, but it involves writing code with the AutoLISP programming language.)

To overcome the limitation, AutoCAD lets you create "groups." Each group has a name and consists of any selection set of objects. Members of one group can be members of other groups. Unlike blocks, however, groups cannot be shared with other drawings.

TUTORIAL: CREATING GROUPS

1. To create groups, start the **GROUP** command with one of these methods:
   - At the 'Command:' prompt, enter the **group** command.
   - Alternatively, enter the **g** alias at the keyboard.

   Command: **group** *(Press ENTER.)*

2. Notice the **Object Grouping** dialog box.

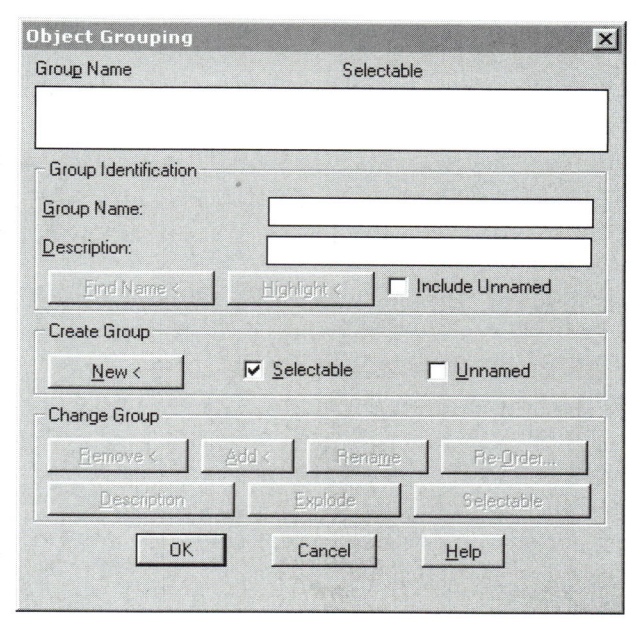

3. In creating a group, you first give it a descriptive name, such as "Linkage." The name can consist of up to 255 letters and numbers.

   Group Name: **Linkage**

4. Optionally, you can describe the group with a label of up to 448 characters:

   Description: **The left end of the linkage**

5. The next step is to select the objects that become part of the group. Choose the **New** button. AutoCAD dismisses the dialog box temporarily so that you can select objects in the drawing. (Use any object selection mode.)

   Select objects for grouping:

   Select objects: *(Pick one or more objects.)*

   5 found Select objects: *(Press ENTER to end object selection, and return to the dialog box.)*

6. Decide whether the group should be *selectable*:

   When selectable, selecting one member of the group selects the entire group.

   When not selectable, selecting a member of the group selects the member only.

   The **Selectable** setting is not crucial, because you can toggle selectable mode on and off at any time by pressing CTRL+SHIFT+A.

7. As you create groups, AutoCAD adds their name to the list at the top of the dialog box. When done creating groups, click **OK**.

> **Note:** The **Unnamed** check box determines whether the group is named. When selected (turned on), AutoCAD automatically assigns a name to the group. The first assigned name is ***A0**; the next is ***A1**, and so on. When **Unnamed** is not selected (turned off), you must give the group its name.

WORKING WITH GROUPS: ADDITIONAL METHODS

Once a group is created, AutoCAD treats all members of the group as a single object, much like a block. Select one member of the group, and AutoCAD selects the entire group; drag one member, and the entire group moves.

Unlike blocks, however, you can quickly work with the individual members. Press CTRL+SHIFT+A, and AutoCAD turns off group mode:

>Command: *(Press* CTRL+SHIFT+A.*)*

>\<Group off\>

Now you can select a single member of the group — not the entire group. To return to group mode, press CTRL+SHIFT+A again:

>Command: *(Press* CTRL+SHIFT+A.*)*

>\<Group on\>

Highlighting Groups

AutoCAD unfortunately has no easy way of identifying groups in drawings. (Curiously enough, AutoCAD LT has a different — and, in my opinion, better — user interface for handling groups.) The workaround is to return to the Object Grouping dialog box:

1. In the list under **Group Name**, select the name of a group. Notice that the **Highlight** button becomes available.
2. Choose the **Highlight** button. The dialog box disappears, and AutoCAD highlights the objects in the group.
3. Click the **Continue** button to return to the dialog box.

The method works in reverse: select an object to find out the name of its group(s).

4. Choose the **Find Name** button.
5. The dialog box disappears, and AutoCAD prompts you to pick an object:
   Pick a member of a group. *(Select one object).*
6. Select the object you are curious about, and AutoCAD displays a dialog box listing the name(s) of the group(s) to which the object belongs.

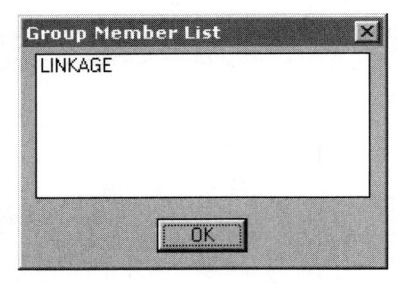

If the object is not a member of a group, AutoCAD complains:
   Not a group member.

Changing Groups

After the groups are created, you can change their descriptions, as well as the selection of objects in the group. You can make these changes by choosing the appropriate button in the Object Grouping dialog box:

- **Remove** removes objects from the group. AutoCAD temporarily dismisses the dialog box, and then highlights members of the group so that you can select the ones to remove.

Select objects to remove from group...

Remove objects: *(Select one or more objects.)*

Remove objects: *(Press* ENTER *to return to dialog box.)*

- **Add** adds objects to the group. AutoCAD again dismisses the dialog box, highlighting members of the group so that you don't accidently select those already members.

    Select objects to add to group...

    Select objects: *(Select one or more objects.)*

    Select objects: *(Press* ENTER *to return to dialog box.)*

- **Rename** changes the name of the group. Enter a new name in **Group Name** text box, and then choose the **Rename** button.

- **Re-order** changes the "order" of the objects in which they are handled for selection, tool paths, and so on. AutoCAD numbers the objects in the group as you add them, the first object being numbered 0. Other than reversing the order, the renumbering process is, unfortunately, painful to execute.

- **Description** changes the description of the group. Enter a new description in the **Description** text box, and then choose the **Description** button.

- **Explode** deletes the selected group from the drawing.

DRAWORDER

The **DRAWORDER** command determines the order in which overlapping objects are displayed. In the figure below, the text "Using AutoCAD 2004" is obscured by the rectangle. By changing the display order — the order in which AutoCAD redraws objects — you can make the text visible on top of the inconsiderately placed rectangle.

TUTORIAL: SPECIFYING THE DRAWORDER

1. To specify the display order of overlapping objects, start the **DRAWORDER** command with one of these methods:

    - From the menu bar, choose **Tools**, and then **Display Order**.

    - On the **Modify II** toolbar, pick the **Draworder** button.

    - At the 'Command:' prompt, enter the **draworder** command.

    - Alternatively, enter the **dr** alias at the keyboard.

        Command: **draworder** *(Press* ENTER.*)*

2. Select one or more objects:
    Select objects: *(Select one or more objects.)*
    1 found Select objects: *(Press* ENTER *to end object selection.)*
3. Specify the display order for the select objects(s)
    Enter object ordering option [Above object/Under object/Front/Back] <Back>: *(Press* ENTER.*)*
    Regenerating model.

CHANGING DRAWORDER: ADDITIONAL METHODS

The draworder command has four options for controlling the visual overlapping of objects. The first two listed below are useful when three or more objects overlap.

- **Above object** places the object visually on top of other selected objects.
- **Under object** places the object under other selected objects.
- **Front** places the object above all other objects.
- **Back** places the object below all other objects.

EXERCISES

1. In this exercise, you use some of the options of the **SELECT** command to selectively erase portions of a valve housing drawing.

    From the CD-ROM, open the *edit1.dwg* drawing.

    Use the **ERASE** command to delete some of the objects. From the **Modify** menu, select **Erase**.

    You should now see a pickbox on the screen. Place the pickbox over the bottom line of the part, as shown in the illustration, and click. The line should be highlighted.

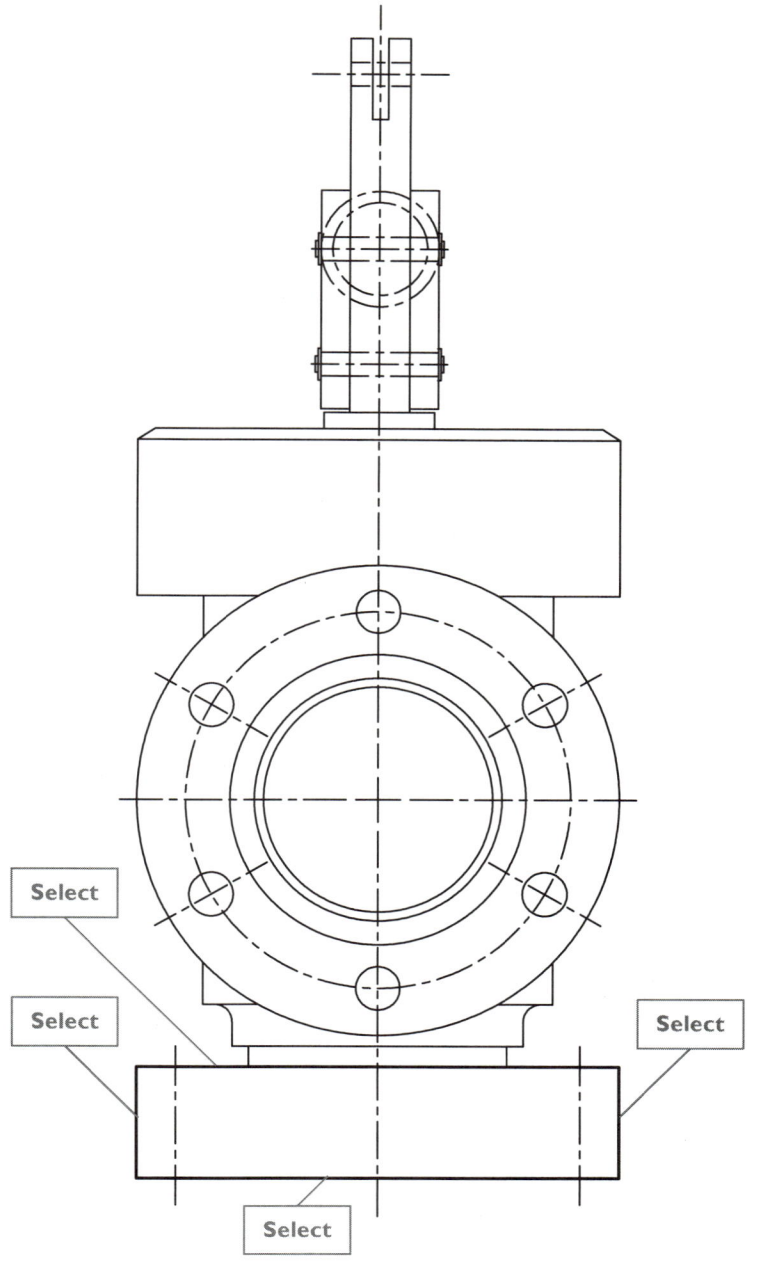

    One by one, select the remaining lines, as shown in the figure.

    Finally, press **ENTER**. The four lines are removed from the drawing.

2. Repeat the above exercise on the same drawing with the Window object selection mode. But first, use the **u** command to undo the erasure.

   Select **ERASE** again, and this time enter **w** in response to the "Select objects:" prompt. Refer to the figure for the points referenced in the following command sequence.

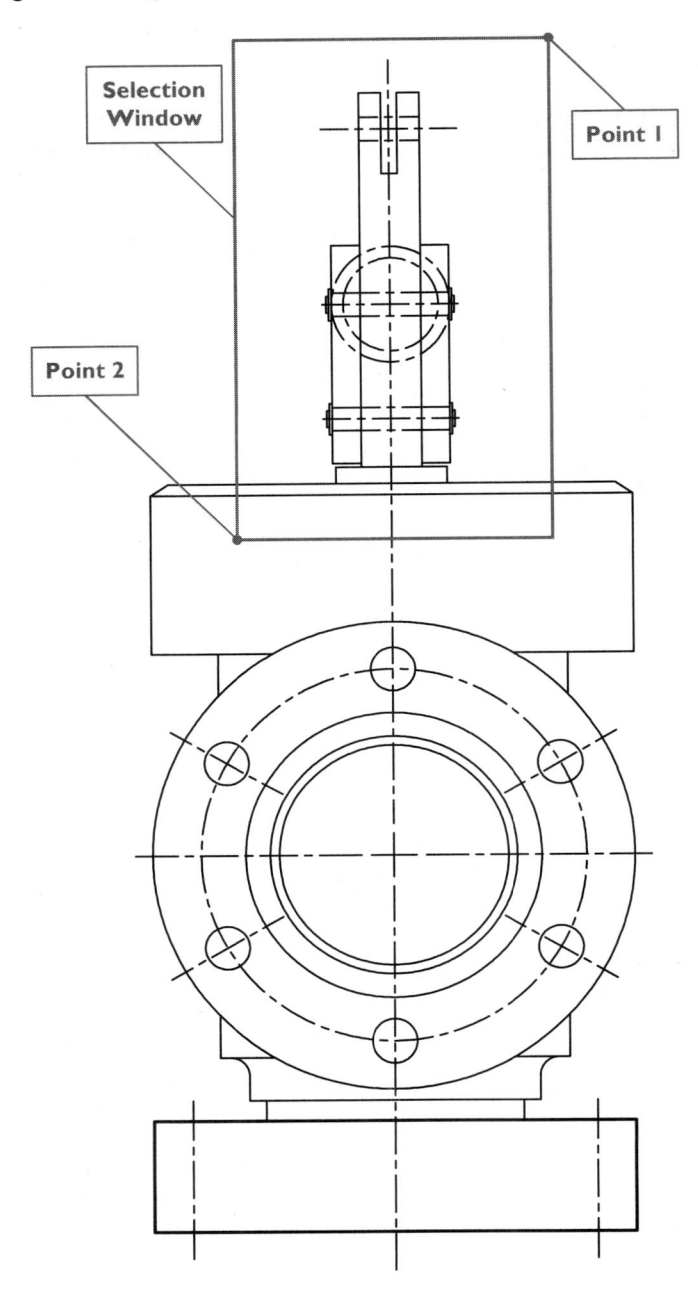

   Command: **erase**

   Select objects: **w**

   Specify first corner: *(Select point 1.)*

   Specify other corner: *(Select point 2.)*

   All items within the window are selected and highlighted. Notice that objects that extend outside the window area (but not wholly contained therein) are not selected.

   Select objects: *(Press* **ENTER**.*)*

   Press **ENTER** and the selected objects are deleted.

3. Repeat the above exercise on the same drawing with the Crossing window object selection mode. Again, undo the erasure with the **u** command.

   Select **ERASE** again, entering **C** (for "crossing") as the option. Refer to the following command sequence and the figure.

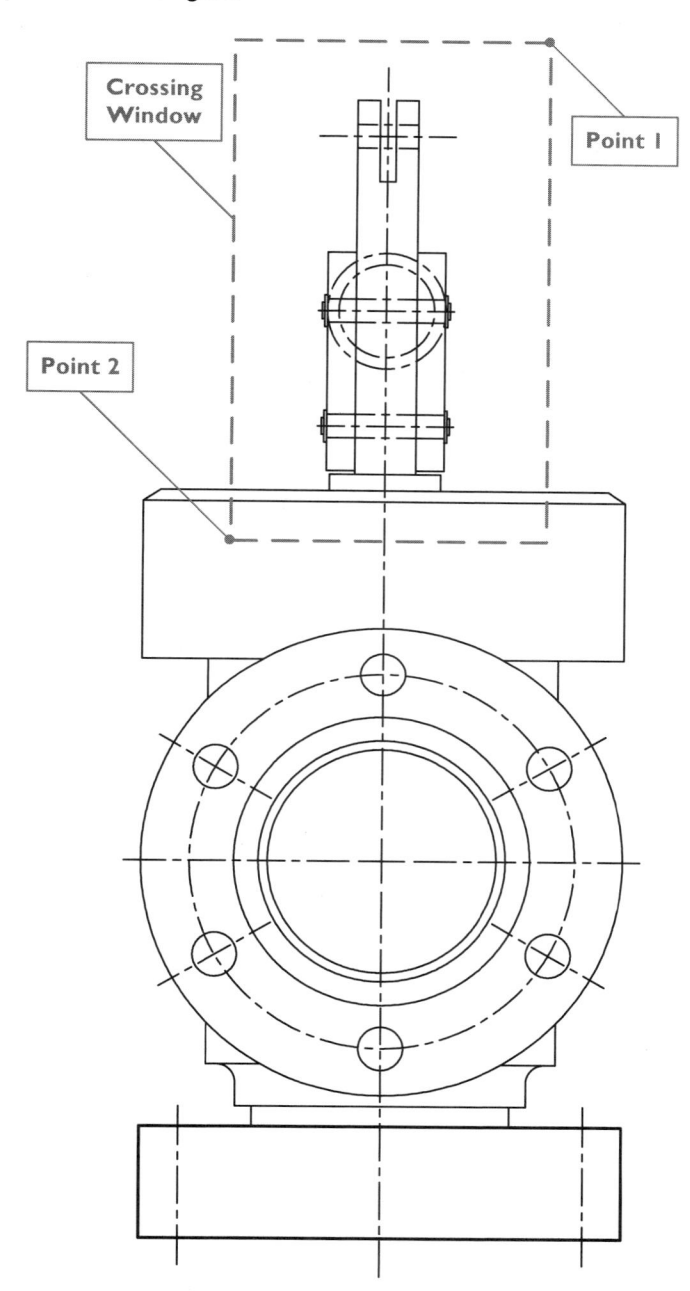

   Command: **erase**

   Select objects: **c**

   Select first corner: *(Select point 1.)*

   Select other corner: *(Select point 2.)*

   Notice that all the objects within and crossing the window are selected (as noted by the highlighting).

4. Continue from the above exercise, remove objects from the selection set.

   After you placed the crossing window in the previous exercise, AutoCAD asked you select more objects. This time, enter **R** for remove. The command sequence continues:

   Select objects: **r**

   Remove objects: *(Select one of the horizontal lines.)*

   1 found, 1 removed Remove objects: *(Select the other horizontal line.)*

   1 found, 1 removed Remove objects: *(Press ENTER.)*

   The objects you removed from the object selection set are not erased.

   Exit the drawing, discarding the changes you made so that the edits are not recorded.

5. In this exercise, you create a block from a drawing, and then insert the block in the drawing several times.

   From the CD-ROM, open the drawing named *office.dwg*.

   Start the **BLOCK** command, and then window the entire desk.

   Name the block "SDESK."

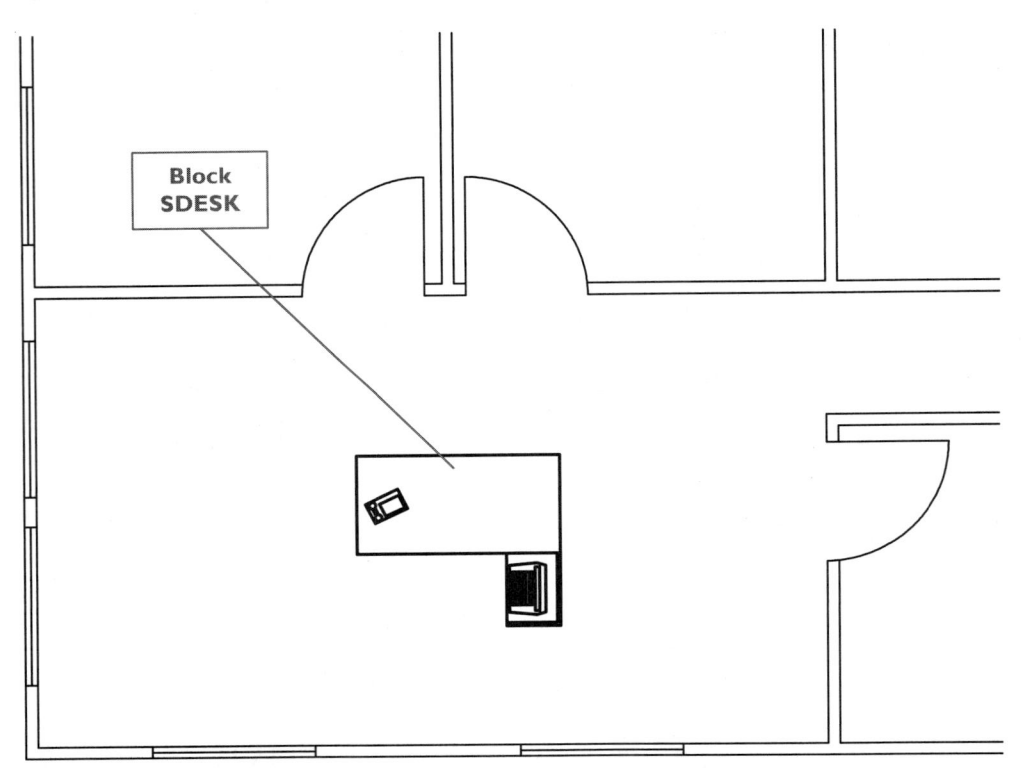

   The block is now stored with the drawing, and can be used as many times as you require. With the **INSERT** command, place the SDESK block in the drawing three times, using these scales and rotation angles:

   Scale = 1.0   Angle = 0 degrees
   Scale = 1.5   Angle = 90 degrees
   Scale = 0.75   Angle = 45 degrees

6. In this exercise, you use DesignCenter to place blocks in a drawing.
   From the CD-ROM, open the site plan drawing named *insert.dwg*. The landscape items are drawn for you and stored as blocks in the drawing.

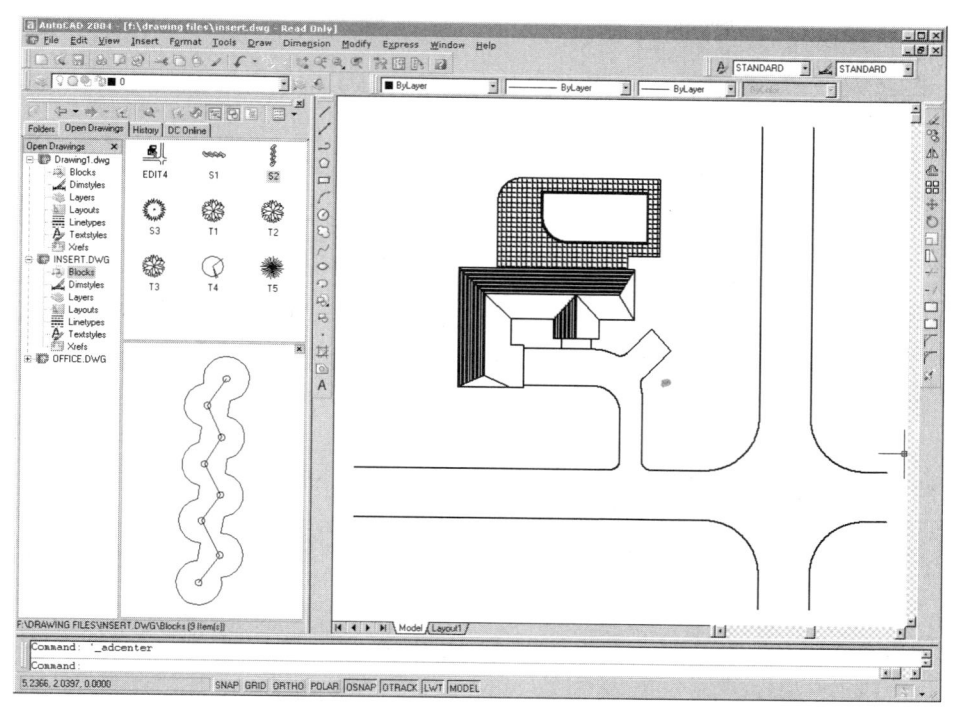

   Use **ADCENTER** to view the blocks, and then insert them in the drawing to create a landscaping design of your own.

7. In this exercise, you draw an electronic part called a "diode," and then insert it as an array of blocks.
   Using drawing commands, construct a diode as shown in the figure below.

   With the **BLOCK** command, select the diode, and name it DIODE; use the left end of the diode as the insertion point.
   You have now created the block DIODE in the drawing.
   Use the **MINSERT** command to place an array of diodes, as follows:

   Command: **minsert**

   Enter block name or ?: **diode**

   Specify insertion point or [Scale/X/Y/Z/Rotate/ PScale/PX/PY/PZ/ PRotate]: *(Pick a point.)*

   Enter X scale factor, specify opposite corner, or [Corner/XYZ] <1>: *(Press ENTER.)*

   Enter Y scale factor <use X scale factor>: *(Press ENTER.)*

   Specify rotation angle <0>: *(Press ENTER.)*

   Enter number of rows (—-) <1>: **4**

   Enter number of columns (| | |) <1>: **6**

   Enter distance between rows or specify unit cell (—-): *(Pick point 1.)*

   Specify opposite corner: *(Pick point 2.)*

You may need to use the **ZOOM All** command to see the completed array.

Notice the (- - -) and (||||) notations in the rows and columns prompts that make it easier to remember which way rows and columns operate.

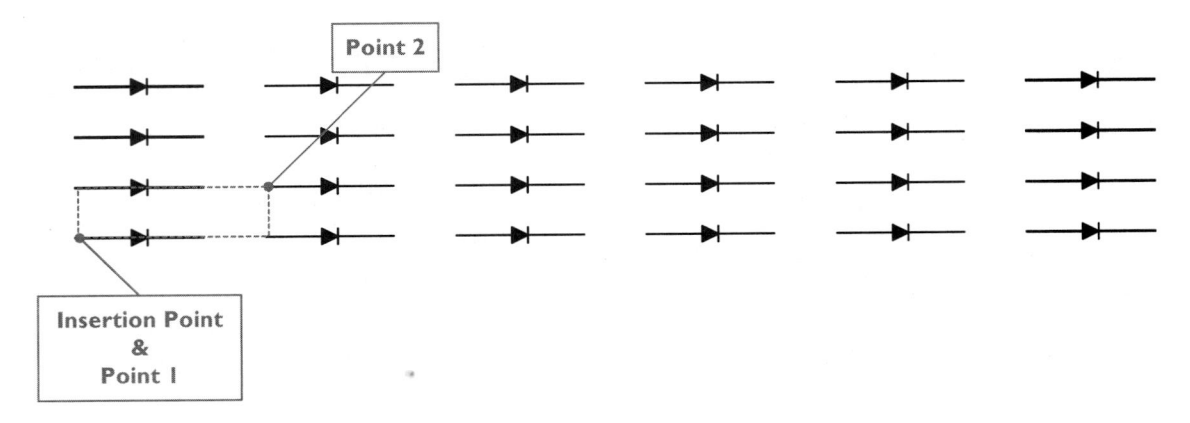

7. In this exercise, you hatch portions of a drawing of several hot air balloons..

   From the CD-ROM, open the drawing named *solids.dwg*.

   Use the **BHATCH** command to place the solid hatch pattern in a variety of colors in some of the balloon areas.

Now try placing a variety of hatch patterns in the same drawing.

8. In this exercise, you specify the area to be hatched with a pick point. The figure shows an object containing three areas that could be hatched.

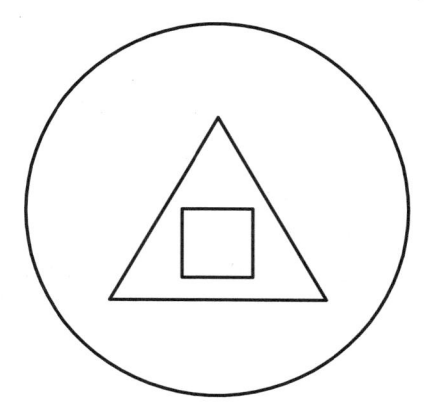

You need to hatch the square and circle, but not the triangle.

Draw the circle, triangle, and square using the **CIRCLE** and **POLYGON** commands.

From the **Draw** menu, select **Hatch**.

In the **Pattern** list box, select **ANSI31**.

Select the **Advanced** tab, and then ensure that **Normal** appears in the **Style** list box and that the **Island Detection** method has **Flood** selected.

Select the **Pick Points<** button, and then click inside the circle, but outside the triangle

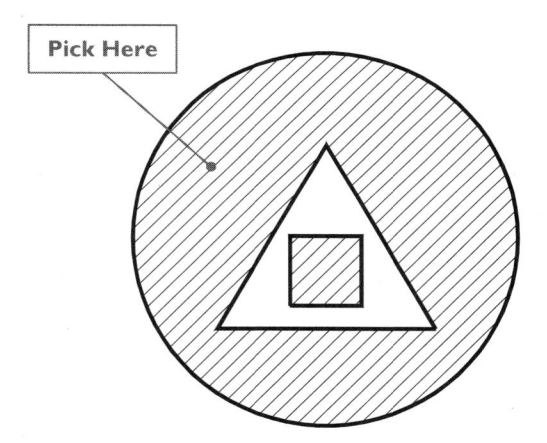

Right-click to return to the dialog box.

Choose the **Preview** button. Your hatch should look similar to the figure above. If not, make changes to the hatch options and/or your pick points.

When the preview hatch looks right, choose **OK** to apply the hatch pattern.

9. In this exercise, you use hatch styles to control the behavior of hatch patterns. From the CD-ROM, open the *hatch.dwg* drawing file.

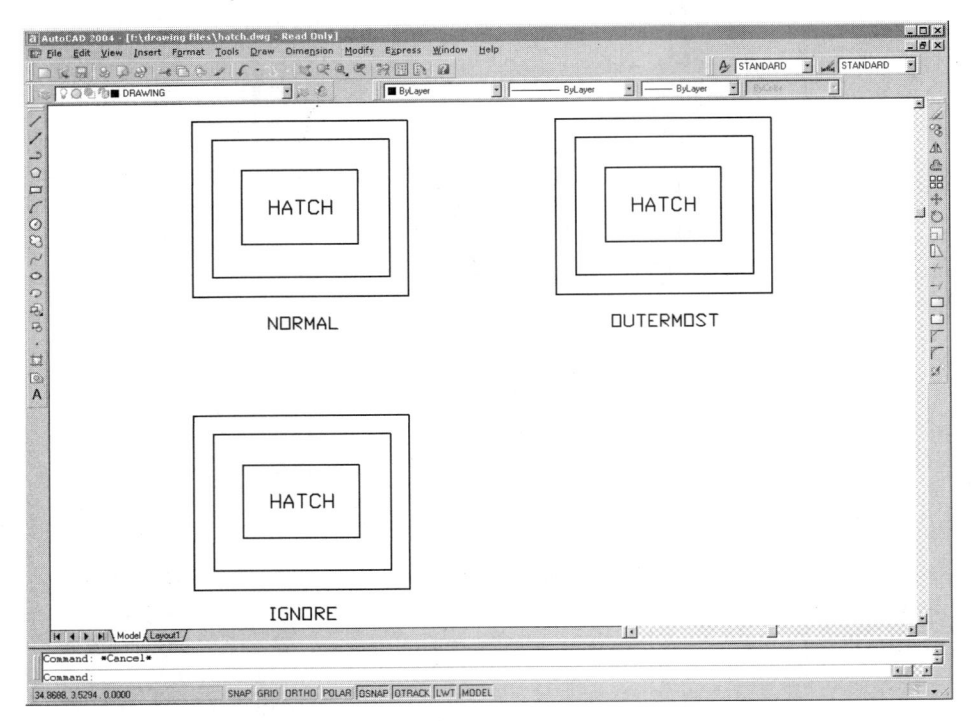

Start the **BHATCH** command, and set the following options:

Pattern	**ANSI31**
Scale	**1**
Angle	**0**
Island Detection Style	**Normal**
Select objects	**W**

Place the window around the object, and then press **ENTER**. Your drawing should look similar to the following illustration:

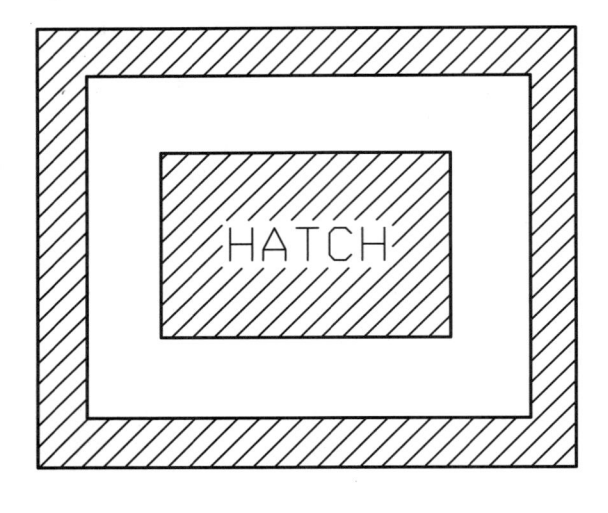

NORMAL

10. Repeat the above exercise on the same drawing, but use the **Outer** style.
    The drawing should look similar to the following illustration:

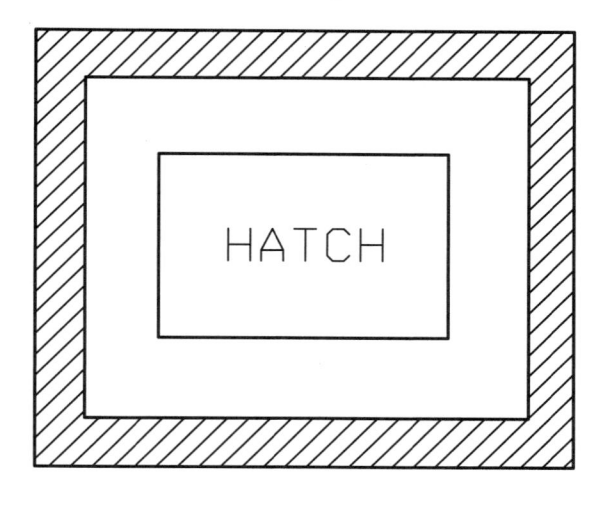

    OUTERMOST

11. Repeat the above exercise on the same drawing, but use the Ignore style of island detection.
    Notice that the hatch has ignored the boundaries and the text. Your drawing should look similar to the following illustration:

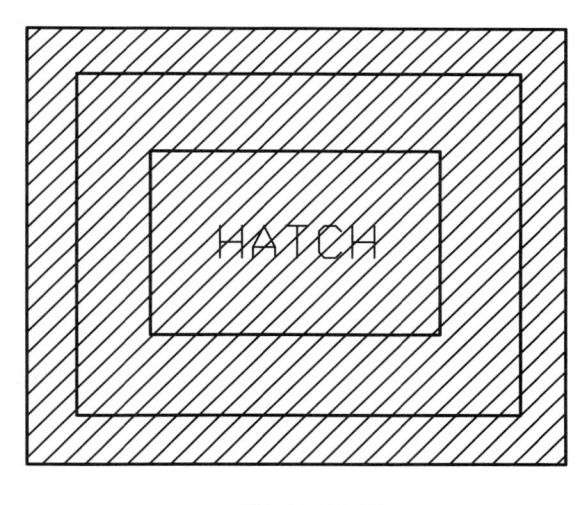

    IGNORE

CHAPTER REVIEW

1. Describe the purpose of the DesignCenter.
2. What are two ways to insert a block from DesignCenter into a drawing?
   a.
   b.
3. What is the purpose of the History button?
4. Which key do you hold down to select more than one item?
5. List four types of *.dwg* content displayed by DesignCenter:
   a.
   b.
   c.
   d.
6. Which shortcut keystroke toggles the display of the DesignCenter?
7. What is the purpose of the **ADCNAVIGATE** command?
8. What is the meaning of the message "Duplicate names ignored"?
9. Explain what the **BHATCH** command does.
10. If you did not want the solid filled areas in your drawing to plot, what could you do?
11. What is a hatch boundary?
12. What is a block?
13. What is the insertion base point of a block?
14. When placed in a drawing, how does a block handle its layer definitions?
15. What is a nested block?
16. How would you create a separate drawing file from an existing block?
17. How do you place a block in a drawing?
18. How can you place a block in another drawing?
19. When you place a block in the drawing, what is the insertion point?
20. Can you place one AutoCAD drawing in another AutoCAD drawing?
21. Name two advantages of using blocks.
22. What commands are combined to create the **MINSERT** command?
23. How would you change the origin point of a hatch?
24. When does a hatch pattern obscure text?
25. When an edit command is invoked, must a selection option, such as **SIngle** or **Window**, be entered before an object is selected?
26. When choosing a set of objects to edit, other than by using the **U** command, how can you alter an incorrect selection without starting over?
27. If you wish to edit an item that was not drawn last, but was the last item selected, would the **Last** selection option allow you to select the item?
28. How do you increase and decrease the size of the pickbox?
29. When entering **BOX** in response to the "Select objects:" prompt, how are you then allowed to choose objects?
30. What makes it evident that an item has been selected?
31. When a group of items is selected with the Window option, does an item become a part of the selection set as long as it is partially inside the window?
32. In the object selection process, how is the **BOX** option different from the **AUtomatic** option?
33. What is the purpose of the tool palette?
34. How do you add a block from a tool palette to the drawing?
35. Describe the purpose of the **BASE** command.

36. In what situation do you use the **BLOCKICON** command?
37. What kind of hatches can the **HATCHEDIT** command not change?
38. Can the **HATCHEDIT** command copy the properties of one hatch pattern and apply them to another?
39. What is the difference between the **SELECT** and **QSELECT** commands?
40. Can you change the size of the pick cursor?
41. What are the two keyboard shortcuts for toggling group mode?
42. What is the purpose of the **DRAWORDER** command?
43. Name two differences between blocks and groups:
    a.
    b.

CHAPTER 7

Changing Object Properties

As you create a drawing, you sometimes need to change the properties of objects, such as their color, weight, and linetype.

In some earlier chapters, you learned how to modify some objects, such as hatch patterns with the **HATCHEDIT** command and point styles with the **DDPDMODE** command. These commands are specific to certain objects. In this chapter, you learn how to change the properties of *all* objects, using these commands:

COLOR and **-COLOR** change the colors of objects.

LWEIGHT and **-LWEIGHT** change the display width of lines.

LINETYPE and **-LINETYPE** change the patterns of lines.

Properties toolbar interactively changes color, lineweight, and linetype.

LTSCALE and **CELTSCALE** change the scale factor of linetypes.

LAYER and **-LAYER** apply properties to all objects on layers.

AI_MOLC and **CLAYER** set the current layer.

LAYERP and **LAYERPMODE** set the layers to the previous state.

PROPERTIES and **CHPROP** change nearly all properties of objects.

MATCHPROP matches the properties of one object to other objects.

FINDING THE COMMANDS

On the **PROPERTIES** and **LAYERS** toolbars:

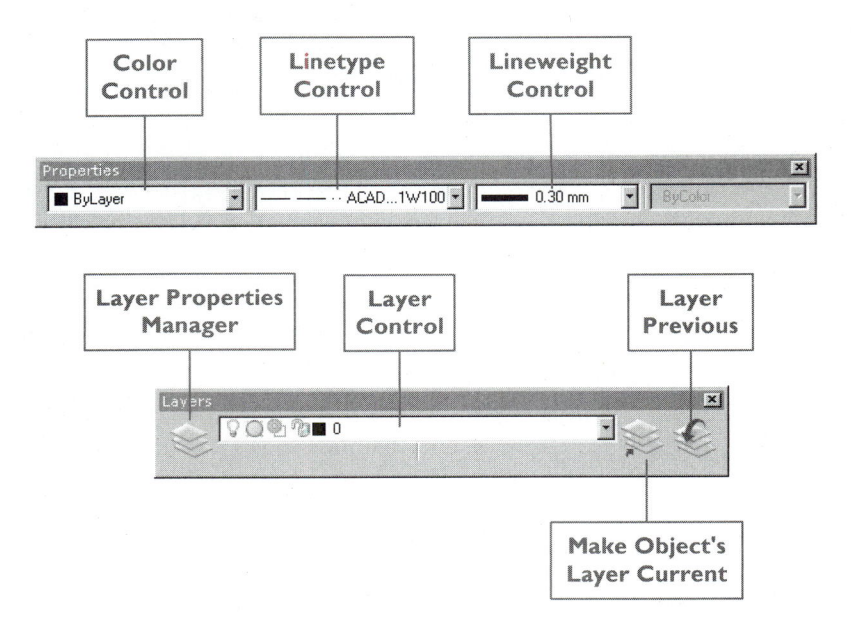

- Color Control
- Linetype Control
- Lineweight Control
- Layer Properties Manager
- Layer Control
- Layer Previous
- Make Object's Layer Current

On the **FORMAT** and **MODIFY** menus:

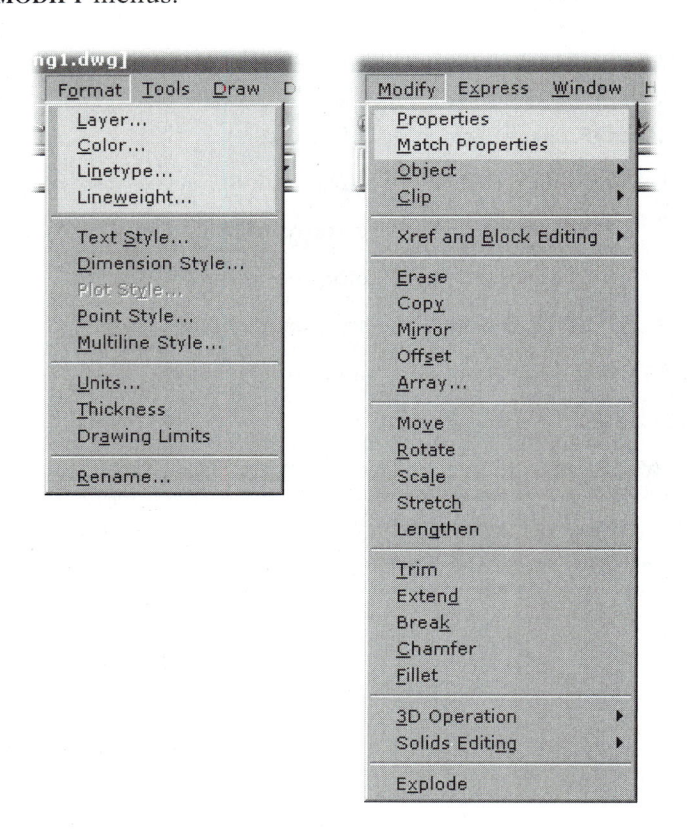

Format
- Layer...
- Color...
- Linetype...
- Lineweight...
- Text Style...
- Dimension Style...
- Plot Style...
- Point Style...
- Multiline Style...
- Units...
- Thickness
- Drawing Limits
- Rename...

Modify
- Properties
- Match Properties
- Object
- Clip
- Xref and Block Editing
- Erase
- Copy
- Mirror
- Offset
- Array...
- Move
- Rotate
- Scale
- Stretch
- Lengthen
- Trim
- Extend
- Break
- Chamfer
- Fillet
- 3D Operation
- Solids Editing
- Explode

COLOR

AutoCAD has a number of methods of setting and changing the color of objects:

- To *preset* the color for objects you plan to draw, the best method is to set the **Color** option of the **LAYER** command; another method is to use the **COLOR** command.

- To *change* the color of objects already in the drawing, the best method is to use the **Color Control** drop list on the **Properties** toolbar; another method is to use the **PROPERTIES** command.

The **COLOR** command sets the color for all *subsequently* drawn objects; the **LAYER** command sets the color for objects drawn on that layer. It is important to understand that the **COLOR** command overrides the colors set by the **LAYER** command. Thus, it is possible for layers to contain objects of different color, regardless of the color set for that layer.

Some drafters are upset that Autodesk introduced the **COLOR** command with AutoCAD v2.5, because they feel that layer settings should determine the colors of objects. Like this book's technical editor, drafters feel it is wrong to override the layer's color setting with the **COLOR** command; it reflects a preferred method of working with CAD, which you may come across it in your future place of work.

TUTORIAL: CHANGING COLORS

1. To change the color setting, start the **COLOR** command with one of these methods:
   - From the menu bar, choose **Format**, and then **Color**.
   - From the **Properties** toolbar, choose the **Color Control** drop list, and then **Select Color**.
   - At the 'Command:' prompt, enter the **color** command.
   - Alternatively, enter the aliases **col**, **colour** (the Britsh spelling), or **ddcolor** (the old name) at the 'Command:' prompt.

   Command: **color** *(Press* ENTER.*)*

2. In all cases, AutoCAD displays the Select Color dialog box.

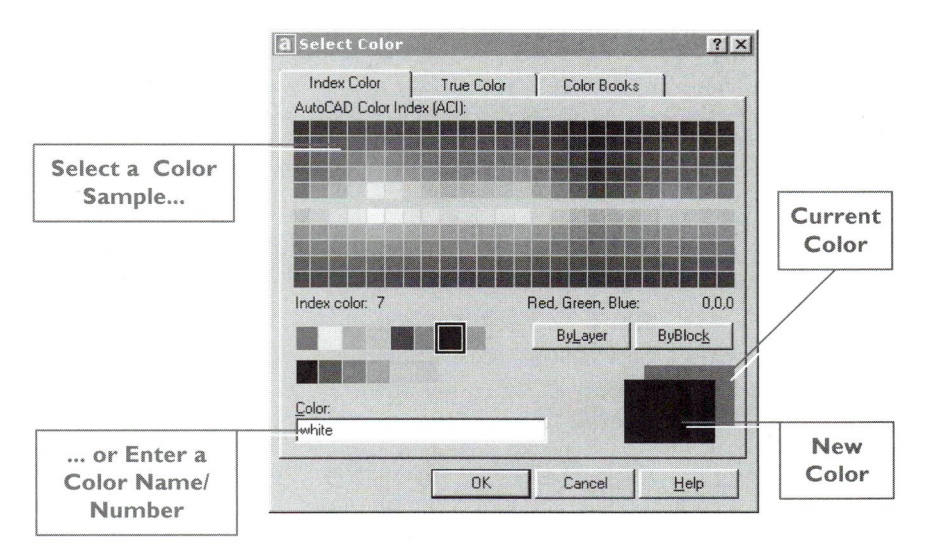

3. Select a color sample (one of the colored squares) or enter a color name/number.
4. Click **OK**. All the objects you now draw take on the new color — until you change color again.

Note: The Select Color dialog box is used with more than just the **COLOR** command. The dialog box is also used by many other commands for selecting colors, such as the Options dialog box.

WORKING WITH AUTOCAD COLORS

For many drafters, color is extremely important, because color is more than the hues and shades of objects. In older versions of AutoCAD, color was the only way to control pens in plotters. The color red, for example, was assigned to pen #1, which in turn could be the color red, or it could be a black pen with a width of 0.1". In AutoCAD 2004, you have the choice of using colors to control the plotter (the old method) or using styles (the new method). More details following in Chapter 17, "Plotting Drawings."

You specify colors by one of several methods: select a color sample; specify a number from 1 to 255; or enter a color name, such as "BLUE".

Color Numbers and Names

Of the 16.7 million colors available, AutoCAD identifies the first seven colors by name, and the first 255 colors by number. The color numbers are called the Autodesk Color Index, or ACI for short.

The color names and their associated numbers are:

Number	Name	Abbreviation
1	Red	R
2	Yellow	Y
3	Green	G
4	Cyan	C
5	Blue	B
6	Magenta	M
7	White/Black	W

(The abbreviations can be used with the **-COLOR** command at the command line.)

You may be wondering about *first* 255 colors? Before AutoCAD 2004, AutoCAD worked with just 255 colors, called the ACI (short for "AutoCAD Color Index"). These colors were selected by Autodesk, the colors you see in the Index Color tab of the Select Color dialog box. Colors 250 through 255 are shades of gray.

Background Color 0

Color 0 is reserved as the *background* color, the color displayed in the background of the drawing area. To change the background color, use the **Tools | Options | Display | Colors** command.

Typically, the background color is white or black (but can be any color). Because white lines would be invisible against a white background. AutoCAD automatically changes color 7 as necessary. When the background color is white (or another light color), AutoCAD changes color 7 to black; when the background color is black (or another dark color), AutoCAD changes color 7 to white.

So, this is one case in life when white is black, and black is white. Officially, however, color 7 is white, because AutoCAD's original background color was black. For this reason, you may hear and read of color 7 being referred to as "white" even though it appears black on your screen.

Colors ByLayer and ByBlock

In addition to "special" colors 0 and 7, AutoCAD has two more special "colors" called ByLayer and ByBlock. These handle the problem of what to call colors when they are controlled by layers and blocks.

ByLayer means that the color of the object is determined by the layer's color setting. Choosing ByLayer causes all subsequently drawn objects to inherit the layer's color; you return control of objects' colors to the layer's color setting.

ByBlock means that the objects in a block take on the color of the block. Choosing ByBlock causes all subsequent objects to be drawn in white until they are blocked. When the block is inserted, the objects inherit the color of the block insertion.

NEW IN 2004 True Color

AutoCAD 2004 increases the number of colors from 255 to millions. (Technically, it works with 24-bit color, which translates into 16.7 million hues.) Because it can be confusing to select one color out of millions, AutoCAD provides several methods to make your selection. The methods are called RGB, HSL, and color books.

By the way, *hue* refers to differences in color, while *shade* refers to differences in gray.

RGB

RGB is short for "red, green, blue." It is a system of selecting a specific color through numbers from 0 to 255. The numbers represent the amount of color, ranging from black (0) to full color (255).

As you may recall from elementary school physics, all colors of light can be represented by varying amounts of red, green, and blue. Yellow is made from mixing red and green, orange is made from full (255) red and half (127) green, and no (0) blue. The hue of blue used in this book is represented by R=38, G=133, and B=187.

To specify a color through RGB:

1. In the Select Color dialog box, select the True Color tab.
2. Under Color Model, select **RGB**.

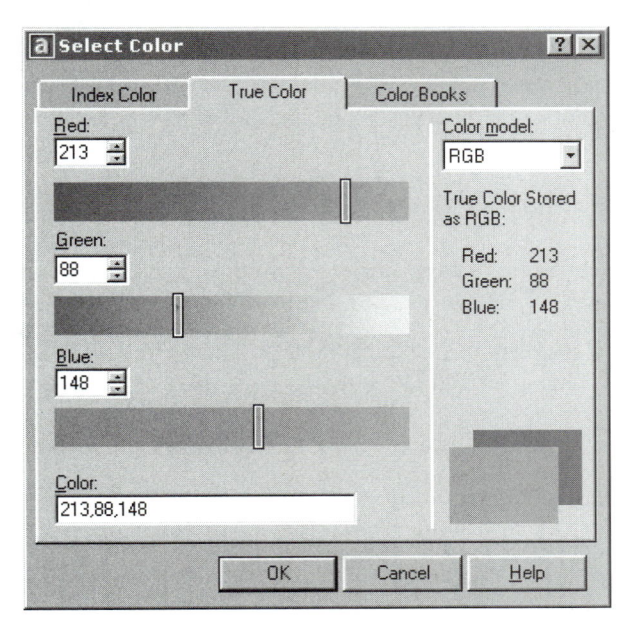

3. In the **Color** text box, enter a color triplet, or drag the three sliders to select a color to your liking.
4. Click **OK**.

HSL

HSL is short for "hue, saturation, luminance." It is another system for specifying colors in a range, but this time by varying the hue (color), saturation (amount of color, ranging from gray to full color), and luminance (brightness of color, ranging from darkened to bright).

Hue ranges from 0 to 360, while saturation and luminance range from 0 to 100 (per cent). The hue of 0 (and 360) is red, yellow is 60, green is 120, cyan is 180, blue is 240, and violet is 300. The HSL setting for the blue used in this book is H=202, S=66, and L=44.

To specify a color through HSL:

1. In the Select Color dialog box, select the True Color tab.
2. Under Color Model, select **HSL**.

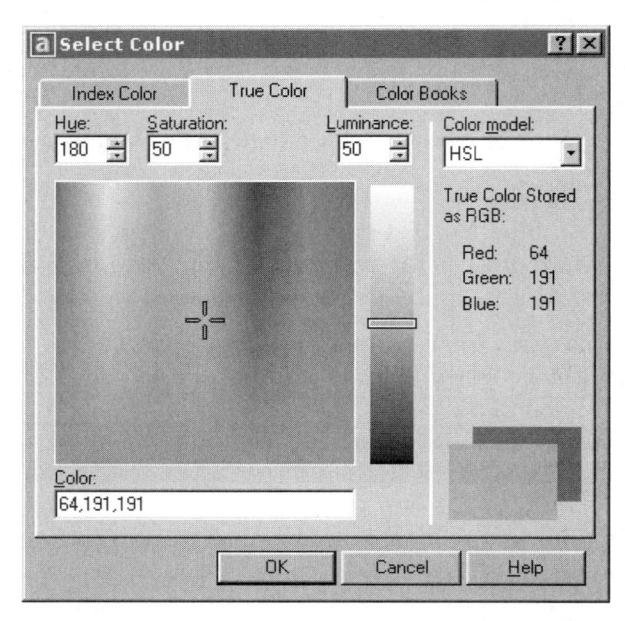

3. In the **Color** text box, enter a color triplet.
   As an alternative, move the cross slider left and right to select the hue; move it up and down to select saturation. Move the slider up and down to choose luminance.
4. Click **OK**.

Color Books

Using RGB or HSL, you can set a precise color in your drawings, such as the color of a client's logo. The publishing and design industries, however, tend to use a different system called Pantone.

The Pantone Color System (www.pantone.com) was designed in 1963 to specify color for graphic arts, textiles, and plastics, based on the assumption that colors are seen differently by individuals. Designers typically work with Pantone's fan-format book of standardized colors. This system is used primarily in North America.

The RAL color system (www.ral.de) was designed in 1927 to standardize colors by limiting the number of color gradations, first just 30 and now over 1,600. RAL (Reichs Ausschuß für Lieferbedingungen – German for "Imperial Committee for Supply Conditions") is administered by the German Institute for Quality Assurance and Labeling. This system is used primarily in Europe.

To specify a color book:

1. In the Select Color dialog box, select the Color Books tab.
2. Under **Color Book**, select a Pantone or RAL book name.

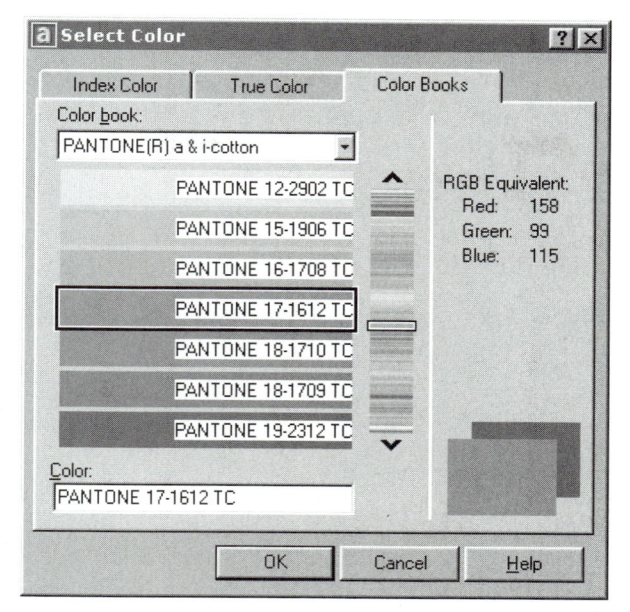

3. In the **Color** text box, enter a color book specification.

    As an alternative, move the slider up and down to choose a color group, and then pick a specific color name.

4. Click **OK**.

> **Note:** AutoCAD includes a drawing that shows all of its colors in the form of *color wheels*. In the \autocad 2004\samples folder, open *colorwh.dwg*. On the left, looking like a smoothly shaded donut, is the true color wheel. On the right is the original 256-color ACI color wheel. It is flanked by the named colors (to the left) and shades of gray (to the right).

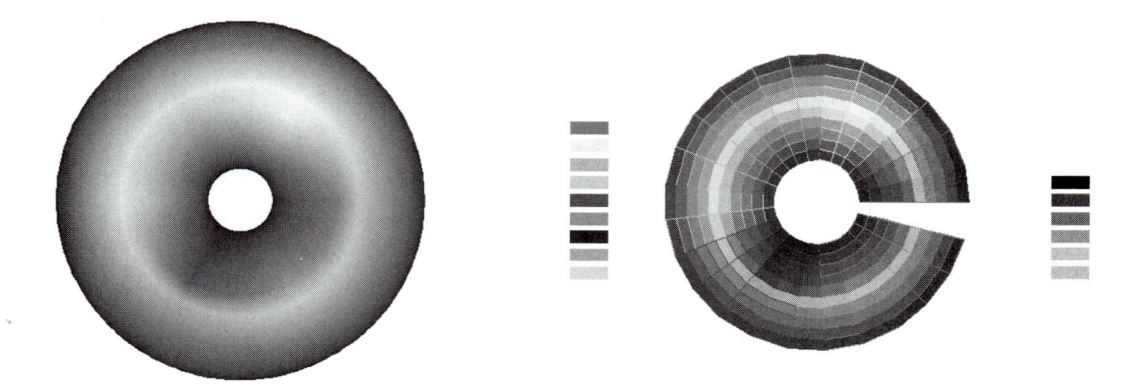

CHANGING COLORS: ADDITIONAL METHODS

AutoCAD provides several other ways of changing the color of objects:

- **-COLOR** command changes the working color at the command line.
- **Color Control** drop list on the Properties toolbar changes the color of selected objects.

Let's look at each option. (Later in this chapter, you also learn about changing color with the **PROPERTIES** and **CHPROP** commands.)

-Color

The **-COLOR** command changes the working color at the command line. Power users prefer this over the dialog box because it can be faster for changing colors. *Real* power users, however, always set all colors to ByLayer.

Command: **-color**

Enter default object color [Truecolor/COlorbook] <1 (red)>: *(Enter a color number, name, or abbreviation, or else enter an option.)*

The color names, numbers, and abbreviations were discussed earlier in this section.

The **Truecolor** option prompts you for a color specified by red, green, and blue. (The HSL method is not available).

Red,Green,Blue: **38,133,187**

The **COlorbook** option prompts you to enter a colorbook name:

Enter Color Book name: *(Enter **pantone** or **ral**.)*

Enter color name: *(Enter a color name, such as **11-0604 TC**.)*

Color Control

As an alternative to the color commands, you can use the **Color Control** in the **Properties** toolbar. The control affects color by two different methods, depending on the order in which you use it. To set the *working color* (subsequent objects take on the color), simply select a color from the drop list. To *change the color* of objects(s), first select the objects, and then a color from the drop list.

In the following tutorial, we look at both methods.

1. Start AutoCAD with a new drawing.
   Draw a few lines.
   Notice that they are colored white (or black, depending on the background color of your screen).
2. From the Properties toolbar, click on the **Color Control** drop list.
   It can be hard sometimes to figure out the location of this drop list, because it is not labeled — and because there are three drop lists that display "ByLayer." I look for the one with the color square.
   Another method for finding the correct drop list is to pause the cursor over one. After a second or two, AutoCAD displays a tooltip naming the drop list.

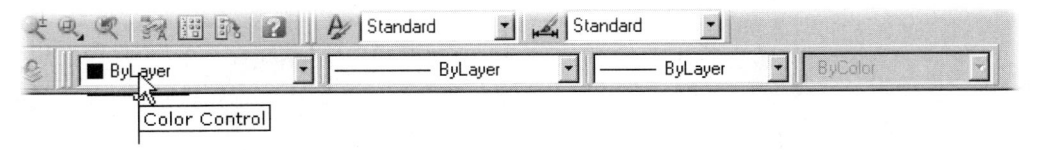

After clicking on the Color Control list box, it drops down to list ByBlock, ByLayer, and the first seven colors, red through white. (Notice that white has a black/white icon, indicating the ambiguity of the color name.)

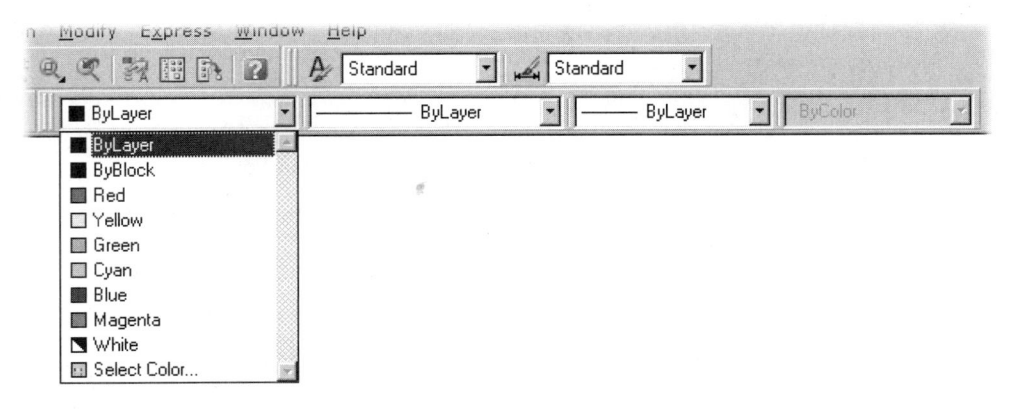

3. Select **Red**.
4. Draw some more lines. Notice that they are colored red.
5. Now, change the color of the red lines to blue: select the red lines by picking them. They become highlighted and show grips (squares).

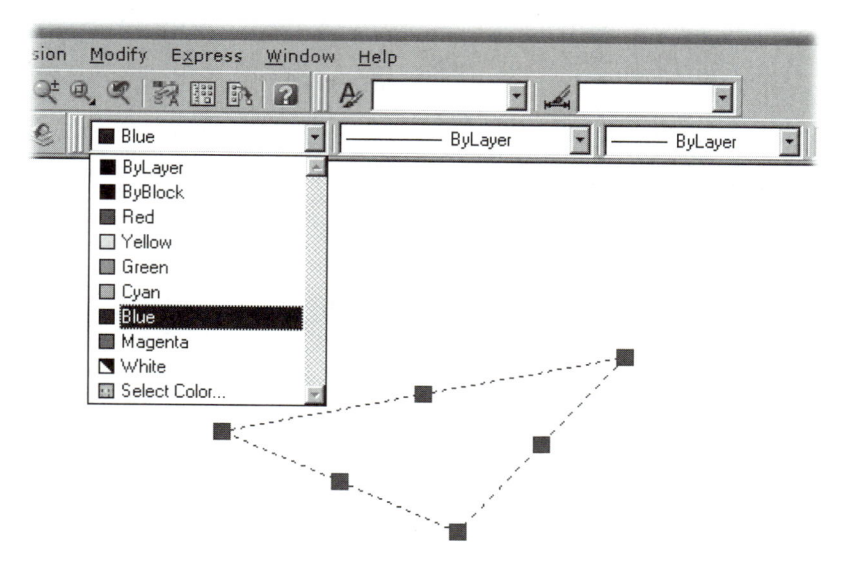

6. From the Color Control, select **Blue**. Notice that the lines turn blue.

   Press ESC to "unselect" the lines (remove the highlighting and grips).

LWEIGHT

The **LWEIGHT** command changes the apparent width of lines and all other objects.

In addition to filling areas with hatch patterns and colored fills, you can give objects a width (also called "weight"). Using a variety of weights (heavy and thin lines) helps make drawings clearer. Weights can be assigned to layers or to individual objects. In the figure below, the upper half of the drawing has lineweights turned off; in the lower half, lineweights are turned on.

In earlier versions of AutoCAD, only polylines and traces could have a width; all other objects were drawn one pixel wide, which means they are drawn as thin as the display or plotter allows. Widths were assigned at plot time based on the object's color. As of AutoCAD 2000, almost any object can have a width.

Lineweights range from 0.05 mm (0.002") to 2.11 mm (0.083"). AutoCAD displays the values in millimeters or inches. The lineweight of 0 is compatible with earlier versions of AutoCAD and displays as one pixel in model space. It is plotted at the thinnest width.

When on, the weight appears when both displayed and when plotted (with one exception: an object with lineweight 0.025 mm or less is displayed one pixel wide in model space). At plot time, objects with weight are plotted at the exact same widths. Objects copied to the Clipboard (through **CTRL+C**) retain their lineweight data when pasted (**CTRL+V**) back into AutoCAD, but retain their weight when pasted into many other Windows applications.

As with colors, some drafters prefer to define lineweight through layers, and do not approve of overriding the layer setting.

AutoCAD includes three named lineweights:

> **ByLayer** means that objects take on the lineweight defined by the layer.
>
> **ByBlock** means that objects take on their block's lineweight.
>
> **Default** displays the default value of 0.25 mm, or any other default value.

Note: Lineweights are meant as a visual aid. Do not use lineweights to represent the actual width of objects. If a printed circuit board trace is 0.05 inches wide, use a polyline with width of 0.05 inches. Do not draw a single line with a weight of 0.05 inches.

TUTORIAL: CHANGING LINEWEIGHT

1. In a new drawing, lineweights are not displayed. To display them, choose the **LWT** button on the status bar. Lineweights are on when the button looks depressed.
2. To change the lineweight, start the **LWEIGHT** command with one of these methods:
   - From the menu bar, choose **Format**, and then **Lineweight**.
   - On the status line, right-click **LWT**, and then choose **Settings** from the shortcut menu.
   - At the 'Command:' prompt, enter the **lweight** command.
   - Alternatively, enter the aliases **lw** or **lineweight** at the 'Command:' prompt.

   Command: **lweight** (Press ENTER.)

2. In all cases, AutoCAD displays the Lineweight Settings dialog box.

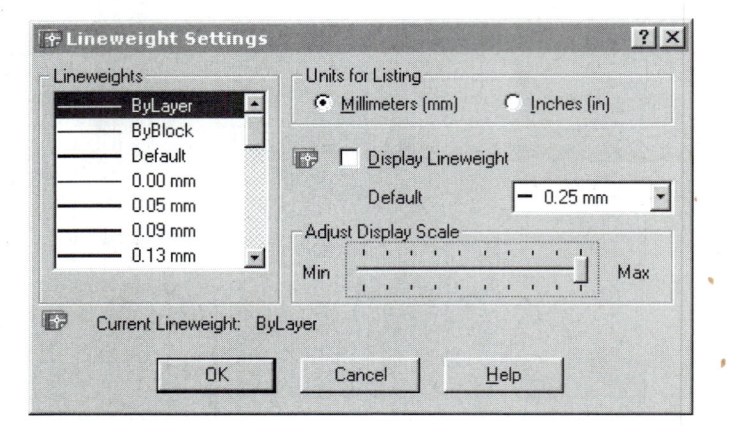

AUTOCAD'S DEFAULT LINEWEIGHTS

The table below shows the default lineweight values used by AutoCAD and their equivalent values for industry standards. There are 25.4 mm per inch; 72.72 points per inch.

Millimeters	Inches	Points	Pen size	ISO	DIN	JIS	ANSI
0.05	0.002						
0.09	0.003	1/4 pt.					
0.13	0.005				*		
0.15	0.006						
0.18	0.007	1/2 pt.	0000	*	*	*	
0.20	0.008						
0.25	0.010	3/4 pt.	000	*	*	*	
0.30	0.012		00				2H or H
0.35	0.014	1 pt.	0	*	*	*	
0.40	0.016						
0.50	0.020		1	*	*	*	
0.53	0.021	1-1/2 pt.					
0.60	0.024		2				H, F, or B
0.70	0.028	2-1/4 pt.	2-1/2	*	*	*	
0.80	0.031		3				
0.90	0.035						
1.00	0.039		3-1/2	*	*	*	
1.06	0.042	3 pt.					
1.20	0.047		4				
1.40	0.056			*	*	*	
1.58	0.062	4-1/4 pt.					
2.0	0.078			*	*		
2.11	0.083	6 pt.					

4. Under Lineweights is a list of weights available in AutoCAD. This is the same list displayed by the Lineweights Control list box on the Properties toolbar. You cannot specify custom lineweights.

   Select a lineweight; this becomes the default until you change the lineweight again.

5. Other items in the dialog box:

   Units for Listing selects between inches and millimeters (the default).

   Display Lineweight check box functions identically to the LWT button on the status bar.

   Default list box specifies the lineweight with which all objects are drawn, unless overridden by layer.

   Adjust Display Scale slider controls the display scale of lineweights in model space. This slider is beneficial if your computer displays AutoCAD on a high-resolution monitor, which tends to make lines look thinner. Experiment with adjusting the lineweight scale to see if you get a better display of different lineweights. As you move the slider, notice that the widths of lines in the Lineweights list change.

6. Choose **OK** to dismiss the dialog box.

CHANGING LINEWEIGHTS: ADDITIONAL METHODS

AutoCAD provides several other ways to change the lineweight of objects:

- **-LWEIGHT** command changes the working lineweight at the command line.
- **Lineweight Control** drop list on the Properties toolbar changes the lineweight of selected objects.

Let's look at each option. (Later in this chapter, you also learn about changing lineweight with the PROPERTIES and CHPROP commands.)

-LWeight

The -LWEIGHT command changes the lineweights and related options at the command line:

   Command: **-lweight**

   Current lineweight: 0.024"

   Enter default lineweight for new objects or [?]: *(Enter a value, such as **.042**, or type **?**.)*

The ? option lists the available lineweights:

   Enter default lineweight for new objects or [?]: **?**

   ByLayer ByBlock Default
   0.000" 0.002" 0.004" 0.005" 0.006" 0.007"
   0.008" 0.010" 0.012" 0.014" 0.016" 0.020"
   0.021" 0.024" 0.028" 0.031" 0.035" 0.039"
   0.042" 0.047" 0.055" 0.062" 0.079" 0.083"

To change the display between imperial and metric units, use the **LWUNITS** system variable, as follows:

   Command: **lwunits**

   Enter new value for LWUNITS <0>: **1**

A value of 0 (zero) causes AutoCAD to display lineweights in imperial units (inches), while a value of 1 displays units in metric (millimeters).

Lineweight Control

As an alternative to using commands and dialog boxes, you can change lineweights directly through the **Lineweight Control** drop list in the Properties toolbar. As with color, the control affects lineweights by two methods, depending on the order in which you use it. To set the working lineweight (subsequent objects take on the selected lineweight), simply select a lineweight from the drop list. To change the lineweight, first select the objects, and then a lineweight from the drop list.

In the following tutorial, we look at both methods.

1. Start AutoCAD with a new drawing, and then draw a few lines. Notice that they are drawn with thin looking lines.
2. From the Properties toolbar, click on the **Lineweight Control** drop list.
   Because there are three drop lists that display "ByLayer" it can be hard sometimes to figure out the location of this particular drop list. It is not labeled, so I look for the one with the slightly thicker line. Or, pause the cursor over a drop list, and wait for AutoCAD to displays a tooltip naming the drop list.

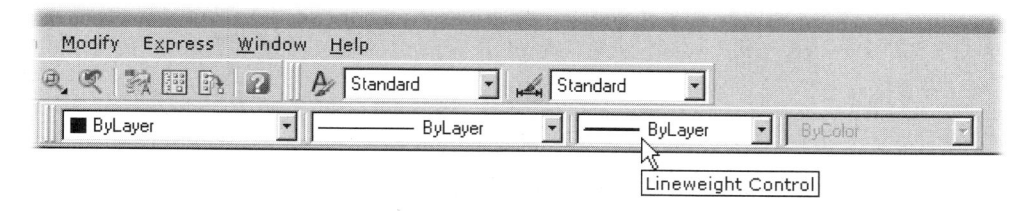

3. Select **0.30mm** or **0.012"**.
4. Draw some more lines. Notice that they are drawn thicker.
5. Now, change the weight of the lines, as follows: select the lines (they become highlighted and show grips).

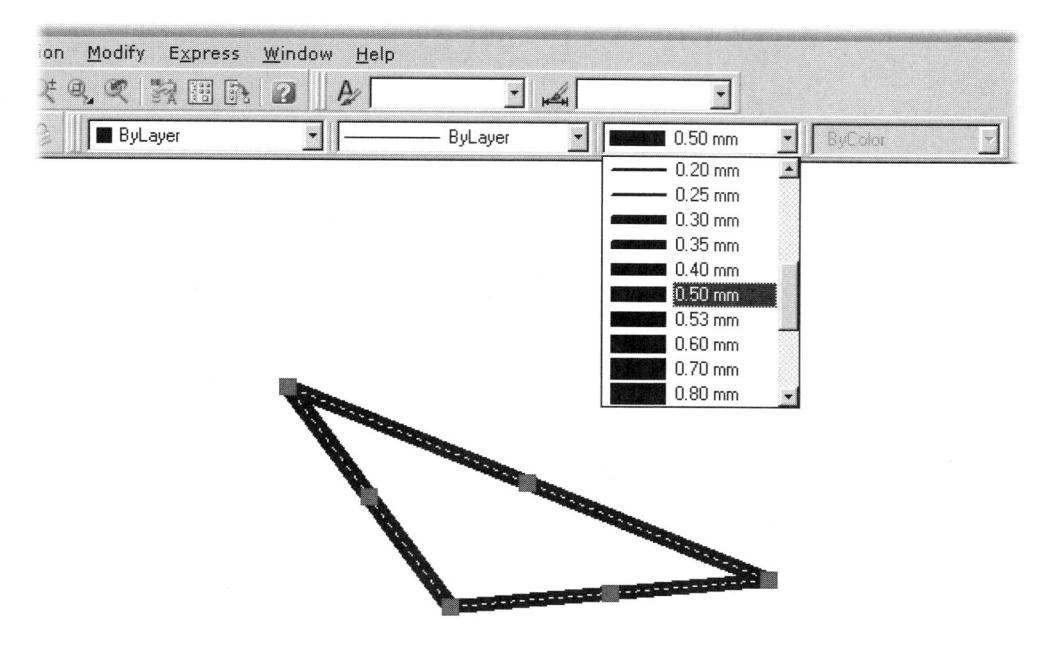

6. From the Lineweight Control drop list, select **0.050mm** or **0.020"**. Notice that the lines become thicker.
7. Press **ESC** to "unselect" the lines.

LINETYPE

The **LINETYPE** command changes the patterns of lines.

AutoCAD provides many linetypes; the figure on the next page illustrates the linetypes provided with AutoCAD. In addition, you can add custom linetypes.

AutoCAD uses these linetypes:

>**Simple** linetypes consist of line segments, dots, and gaps; these are the most commonly-used linetypes.

>**ISO** linetypes are similar to simple linetypes, except that they conform to ISO standards and can have a pen width assigned to them.

>**Complex** linetypes are like simple linetypes, but include characters and shapes, such as HW (to show a hot water line) and squiggles to show insulation.

To use a linetype in drawings takes two steps: (1) load the linetype into the drawing; and (2) select the linetype for use. Technically, AutoCAD stores linetype definitions in *acad.lin* and *acadiso.lin*, and the shapes for complex linetypes in *ltypeshp.shp* — all found in the *\support* folder.

Note: In drawings, solid lines represent the edges of the object. In AutoCAD, solid lines are referred to as "continuous lines."

Edges are hidden from view are shown in a linetype referred to as "hidden," a line constructed from a series of short line segments. The figure shows an object containing edges that are hidden in some views. These edges are defined with the hidden linetype.

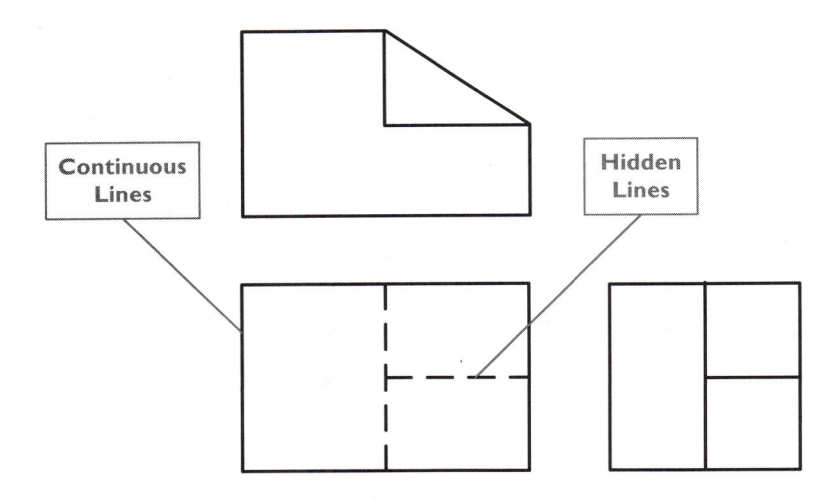

Linetype	Description
ACAD_ISO02W100	ISO dash __ __ __ __ __ __ __ __ __
ACAD_ISO03W100	ISO dash space __ __ __ __ __ __
ACAD_ISO04W100	ISO long-dash dot ____ . ____ . ____ . __
ACAD_ISO05W100	ISO long-dash double-dot ____ .. ____ .. ____
ACAD_ISO06W100	ISO long-dash triple-dot ____ ... ____ ... ____
ACAD_ISO07W100	ISO dot . . . .
ACAD_ISO08W100	ISO long-dash short-dash ____ __ ____ __
ACAD_ISO09W100	ISO long-dash double-short-dash ____ __ __
ACAD_ISO10W100	ISO dash dot __ . __ . __ . __ . __
ACAD_ISO11W100	ISO double-dash dot __ __ . __ __ . __ __
ACAD_ISO12W100	ISO dash double-dot __ . . __ . . __ . .
ACAD_ISO13W100	ISO double-dash double-dot __ __ . . __ __ . .
ACAD_ISO14W100	ISO dash triple-dot __ . . . __ . . . __ . . .
ACAD_ISO15W100	ISO double-dash triple-dot __ __ . . . __ __
BATTING	Batting SSSSSSSSSSSSSSSSSSSSSSSSSSSSSSSS
BORDER	Border __ __ . __ __ . __ __ . __ __ .
BORDER2	Border (.5x) __ __ . __ __ . __ __ .
BORDERX2	Border (2x) ____ ____ . ____ ____ .
CENTER	Center ____ __ ____ __ ____ __ ____
CENTER2	Center (.5x) ____ __ ____ __ ____ __
CENTERX2	Center (2x) ____ __ ____ __ ____
DASHDOT	Dash dot __ . __ . __ . __ . __ . __ .
DASHDOT2	Dash dot (.5x) _._._._._._._._._.
DASHDOTX2	Dash dot (2x) ____ . ____ . ____ . ____
DASHED	Dashed __ __ __ __ __ __ __ __ __
DASHED2	Dashed (.5x) _ _ _ _ _ _ _ _ _ _ _ _
DASHEDX2	Dashed (2x) ____ ____ ____ ____ ____
DIVIDE	Divide ____ . . ____ . . ____ . . ____
DIVIDE2	Divide (.5x) __ . . __ . . __ . . __ . .
DIVIDEX2	Divide (2x) ____ . . ____ . . ____ _
DOT	Dot . . . .
DOT2	Dot (.5x) ....
DOTX2	Dot (2x) . . . .
FENCELINE1	Fenceline circle ----0-----0----0----0---
FENCELINE2	Fenceline square ----[]-----[]----[]----[]----
GAS_LINE	Gas line ----GAS----GAS----GAS----GAS----GAS---
HIDDEN	Hidden __ __ __ __ __ __ __ __ __ __ __
HIDDEN2	Hidden (.5x) _ _ _ _ _ _ _ _ _ _ _ _
HIDDENX2	Hidden (2x) ____ ____ ____ ____ ____
HOT_WATER_SUPPLY	Hot water supply ---- HW ---- HW ---- HW ----
PHANTOM	Phantom _____ __ __ _____ __ __ _____
PHANTOM2	Phantom (.5x) ____ __ __ ____ __ __ ____
PHANTOMX2	Phantom (2x) _____ ____ ____ _
TRACKS	Tracks -|-|-|-|-|-|-|-|-|-|-|-|-|-|-|-|-|-|-|
ZIGZAG	Zig zag /\/\/\/\/\/\/\/\/\/\/\/\/\/

TUTORIAL: CHANGING LINETYPES

1. To change the linetype, start the **LINETYPE** command with one of these methods:
   - From the menu bar, choose **Format**, and then **Linetype**.
   - From the **Properties** toolbar, choose the **Linetype Control** drop list, and then **Other**.
   - At the 'Command:' prompt, enter the **linetype** command.
   - Alternatively, enter the aliases **lt**, **ltype**, or **ddltype** (the old name) at the 'Command:' prompt.

   Command: **linetype** *(Press ENTER.)*

2. In all cases, AutoCAD displays the Linetype Manager dialog box.

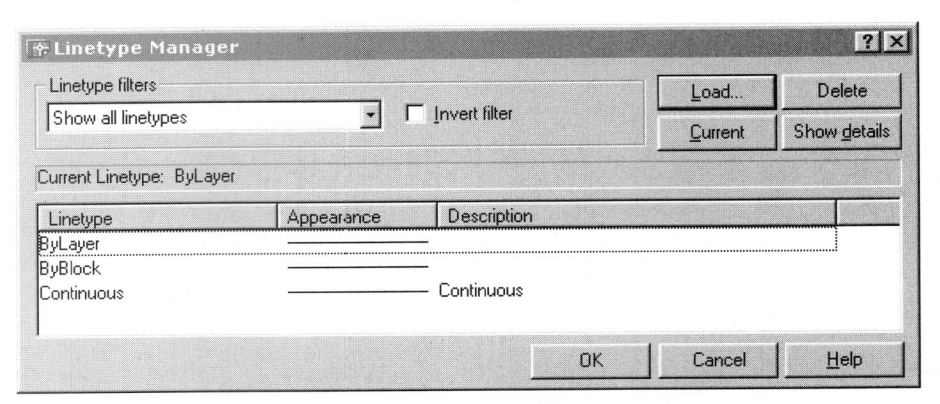

AutoCAD displays the Linetype Manager dialog box, which lists the linetypes currently loaded into the current drawing. As with colors and lineweights, there are two special linetypes:

ByLayer displays the linetype assigned by the layer.

ByBlock displays the linetype assigned by the block.

3. To load a linetype, choose the **Load** button. AutoCAD displays the Load or Reload Linetypes dialog box.

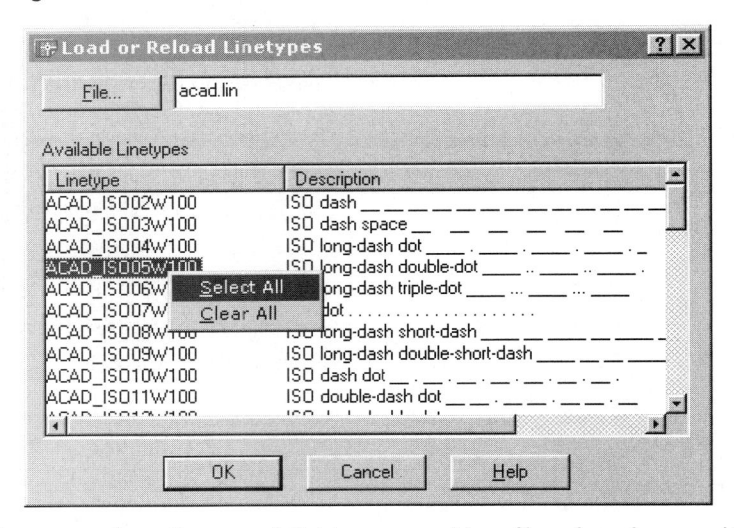

Unless you know you have linetype definitions stored in a file other than *acad.lin*, you can ignore the **File** button.

4. I find it easier to load all linetypes at once, and then later remove any that are unused.
   To do so, right-click on any linetype. Notice that AutoCAD displays a small pop-up menu with two selections.
   Choose **Select All,** and AutoCAD highlights all linetypes.
5. Choose **OK**, and AutoCAD loads all linetypes.
6. Back in the Linetype Manager dialog box, select a linetype, and then click **Current**.
7. Choose **OK**. From now on, all objects are drawn in this linetype — until you change it again.

CHANGING LINETYPES: ADDITIONAL METHODS

AutoCAD provides several other ways to handle linetypes:

- **-LINETYPE** command changes the working linetype at the command line.
- **LTSCALE** command changes the scale of linetypes.
- **Linetype Control** drop list on the Properties toolbar changes lineweights of selected objects.
- **DesignCenter** window allows you to share linetypes between drawings.

Let's look at each option. (Later you also learn about changing linetypes with the **PROPERTIES** and **CHPROP** commands.)

-Linetype

The **-LINETYPE** command loads and sets linetypes at the command line. Use the **Set** option to set the working linetype:

Command: **-linetype**

Enter an option [?/Create/Load/Set]: **s**

Specify linetype name or [?] <ByLayer>: *(Enter a linetype name, such as* **hidden**.*)*

Enter an option [?/Create/Load/Set]: *(Press* **ENTER** *to exit the command.)*

The **?** option prompts you to select a *.lin* file, and then lists the available linetypes. The display starts off like this:

Linetypes defined in file C:\Documents and Settings\username\Application
Data\Autodesk\AutoCAD 2004\R16.0\enu\Support\acad.lin:

Name	Description
"BORDER"	Border __ __ . __ __ . __ __ . __ __ . __ __ .
"BORDER2"	Border (.5x) _.__.__.__.__.__.__.__.__.
"BORDERX2"	Border (2x) ____ ____ . ____ ____ . ____
"CENTER"	Center ____ _ ____ _ ____ _ ____ _ ____ _ ____
"CENTER2"	Center (.5x) __ _ __ _ __ _ __ _ __ _ __

Press ENTER to continue: *(Press* **ENTER** *to see more, or else* **ESC** *to exit the listing.)*

The **Load** option is like the **Load** button in the linetypes dialog box: it selects a *.lin* file to load into the drawing. The **Create** option allows you to create custom linetypes by defining the pattern of segments, dots, and gaps, and their lengths.

LtScale

Because linetypes are constructed of line segments and dashes, the pattern can look too small or too large for your drawing. How do you tell when the scale is wrong? You don't see the pattern — lines look continuous. Like hatch patterns, linetypes need to be set to an appropriate scale; fortunately, the same scale factor applies to both. When you figure out the correct factor for hatch patterns, you can use the same one for linetypes (and text and dimensions, as it turns out).

Linetype scale = 1.0   *Linetype scale = 0.5*   *Linetype scale = 2.0*

The scale of linetypes is adjusted by the **LTSCALE** (short for "linetype scale") command:

> Command: **ltscale**
>
> Enter new linetype scale factor <1.0000>: *(Enter a value.)*

Entering a number larger than 1.0 results in longer line segments, while numbers smaller than 1.0 create shorter line segments.

> Regenerating model.

To display the new linetype scale correctly, AutoCAD automatically performs a regeneration. If a regeneration does not occur after you change the linetype scale, force a regeneration with the **REGEN** command.

If you change the scale, but linetypes still appear continuous, change **LTSCALE** to a different value. It is possible for the linetype scale to be too large or small to display. If lines take a longer time to draw, it is likely that the scale is too small.

Global and Object Linetype Scale

AutoCAD lets you specify the linetype scale in two ways: as a *global* factor for all objects in the drawing, and as a *local* factor for individual objects. While it is poor drafting practice to use multiple linetype scales in a drawing, the object scaling is particularly useful for complex linetypes, such as Batting and Hot Water Supply.

The global linetype scale is set with the **LTSCALE** command, as described above. To change the linetype scale of individual objects, use either the **PROPERTIES** command (as described later in this chapter) or the **CELTSCALE** system variable (short for "current entity linetype scale"). There is a difference between the two: The **PROPERTIES** command changes the linetype scale of objects that already exist in the drawing, while the **CELTSCALE** system variable changes the scale of objects you are about to draw.

CELTSCALE is a bit tricky to use, because it sets the linetype scale *relative* to the value set by the **LTSCALE** command setting. Here's an example: when **LTSCALE** is set to 3.0 but **CELTSCALE** is set to 0.75, objects will be drawn at a linetype scale of 2.25 (3.0 x 0.75 = 2.25).

Both global and object linetype scaling are also available as a "hidden" feature in the Linetype Manager dialog box. Click the **Details** button to reveal it:

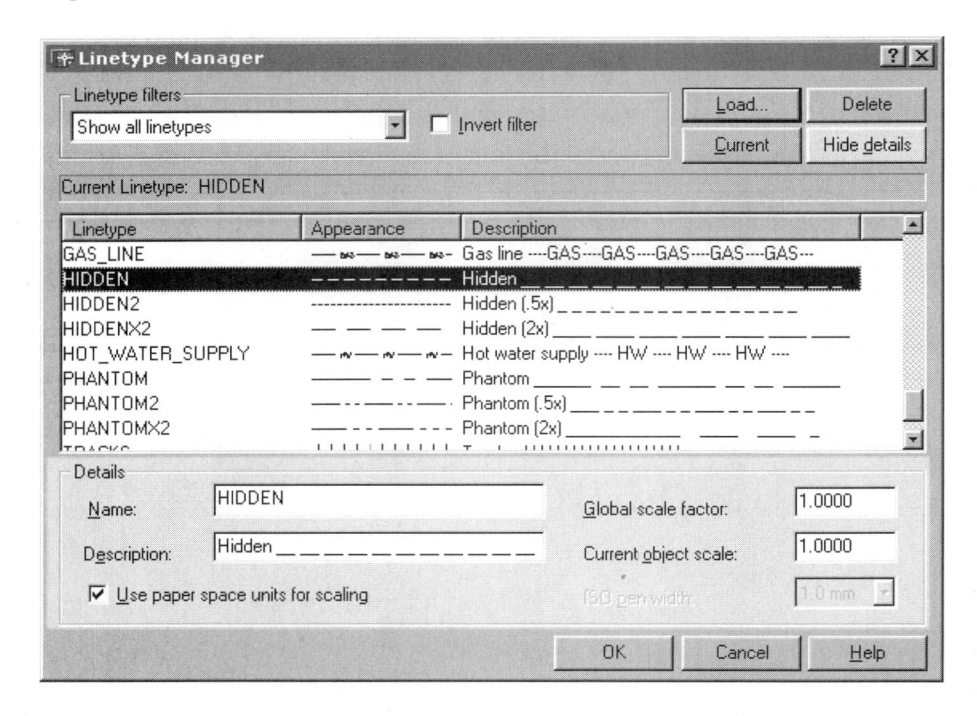

Global scale factor sets the linetype scale for all objects in the drawing. This scale factor takes effect with the next drawing regeneration.

Current object scale sets the linetype scale for subsequently-created objects.

ISO pen width selects from one of the ISO's pen widths, such as 1.00 mm.

Use paper space units for scaling scales linetypes the same in paper space and model space. More details may be found in *Using AutoCAD 2004: Advanced.*

Linetype Control

As with color and lineweights, you can change linetypes directly through the **Linetype Control** drop list on the Properties toolbar. The control affects linetypes by two methods, depending on the order in which you use it. To set the working lineweight (objects subsequently taking on the selected linetype), simply select a linetype from the drop list. To change the linetype of objects(s), first select the objects, and then select a linetype from the drop list.

In the following tutorial, we look at both methods.

1. Start AutoCAD with a new drawing, and then draw a few lines. Notice that they are drawn with continuous lines.
2. From the Properties toolbar, click on the **Linetype Control** drop list.
   You can find it by pausing the cursor over a drop list, and waiting for AutoCAD to display a tooltip naming the drop list.

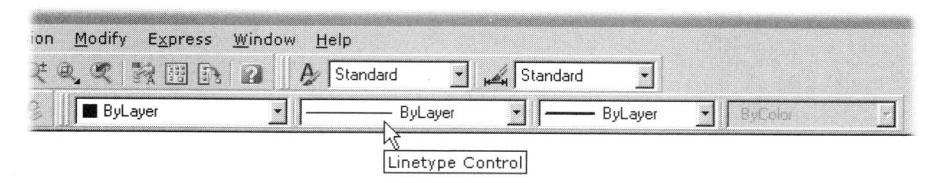

3. Select **Hidden**. (If the linetype name does not appear in the list, return to the earlier tutorial to learn how to load linetypes into drawings.)
4. Draw some more lines. Notice that they are drawn with dashed lines.
5. Now, change pattern of the lines: select the lines. They become highlighted and show grips.

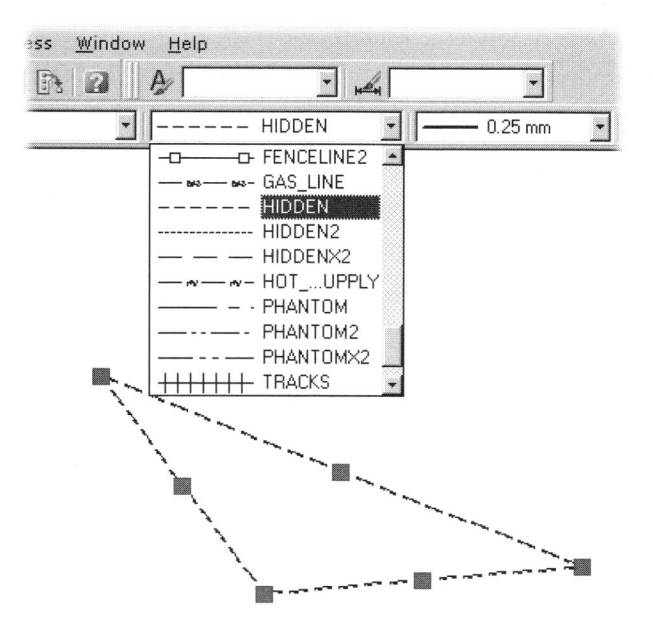

6. From the Linetype Control drop list, select **phantom**. Notice that the lines change, showing dashes and dots.
7. Press ESC to "unselect" the lines.

DesignCenter

If you (or someone else) create custom linetypes, AutoCAD has two ways to share them. One is to give each other a copy of the *.lin* file that stores the custom definitions. The other is to use DesignCenter to drag a copy of the linetype from one drawing to another.

LAYER

The **LAYER** command applies a common set of properties to all objects on layers.

Nearly all CAD drawings are constructed using *layers* that can be turned on and off, and changed. Traditional drafting techniques often include a method of drawing called "overlay drafting." This consists of sheets of transparent drafting media overlaid, so the drawings below show through. Both sheets are blueprinted together, resulting in a print that shows the design on both sheets.

The bottom sheet is typically referred to as the "base drawing." Each additional sheet places different items. For instance, when you prepare a set of floor plans, you typically draft separate drawings for the to-be-removed plan, the plumbing plan, the electrical plan, and so on. The floor plan can be thought of as the base drawing, with each discipline, such as electrical and plumbing, placed on overlay sheets.

Think of AutoCAD's layers as transparent sheets of glass stacked on top of each other. You draw different aspects of the drawing on each layer, yet are able to see through all the layers. The work appears as though it were one drawing.

AutoCAD goes one step further. You can turn on or off each layer, so that it is either visible or invisible, as well as change its properties. The properties of the layer affect all objects assigned to that layer, but can be overridden with the other commands in this chapter, such as **COLOR**, **LINETYPE**, and **PROPERTIES**.

Before you can add objects to a layer, you need to switch to the layer. The working layer is called the "current layer." AutoCAD can have only one layer current at a time. The current layer cannot be *frozen*. (Frozen layers are invisible and cannot be edited.)

TUTORIAL: CREATING LAYERS

1. To create layers, start the **LAYER** command with one of these methods:
   - From the menu bar, choose **Format**, and then **Layer**.
   - From the **Layers** toolbar, choose the **Layer Properties Manager** button.
   - At the 'Command:' prompt, enter the **layer** command.
   - Alternatively, enter the aliases **la** or **ddlmodes** (the old name for this command) at the 'Command:' prompt.

   Command: **layer** *(Press* ENTER.*)*

2. In all cases, AutoCAD displays the Layer Properties Manager dialog box.

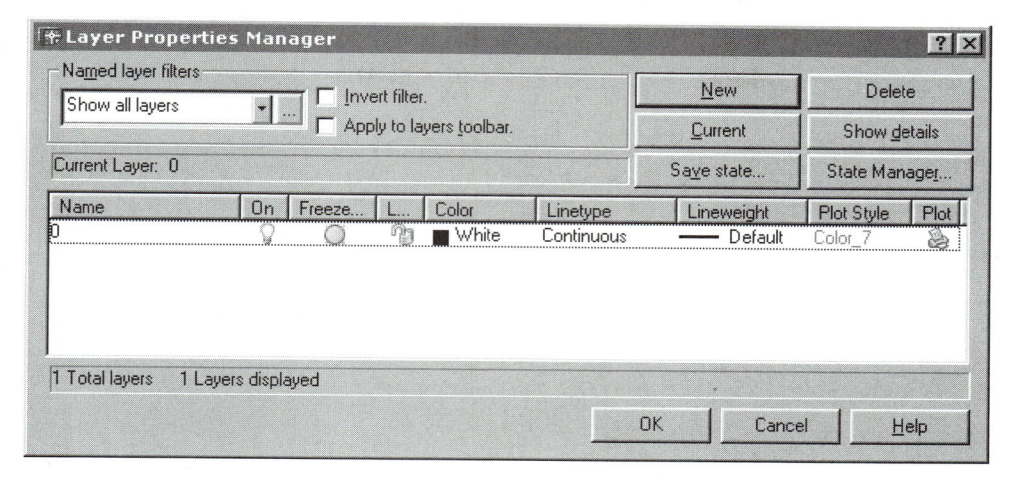

3. Click New.

   Notice that AutoCAD creates a new layer named "Layer1." The icons next to the name indicate that the layer is turned on, thawed, unlocked, colored white/black, has the continuous linetype, default lineweight, and will be plotted/printed. (More about these settings later.)

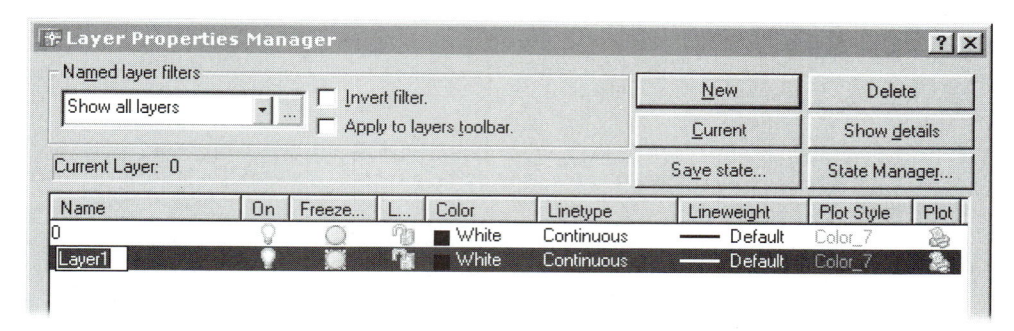

4. To change the layer name, edit the Layer1 text.
5. Click **Current** to make the layer current.
6. Click **OK** to dismiss the dialog box. Notice that the name also appears on the Layer toolbar.

> **Notes:** AutoCAD uses some layer names for its own purposes. Two layers you must become acquainted with are 0 and Layer Defpoints. Every new drawing contains one layer, called 0 (zero). This layer cannot be removed or renamed. Layer 0 has a special property for creating blocks, which you learn about elsewhere in this book.
>
> The first time you draw a dimension in a drawing, AutoCAD creates layer called "Defpoints" (short for "definition points"). This layer contains data that AutoCAD needs to keep its dimensions associative. The layer cannot be removed; it can be renamed, but then AutoCAD creates a new Defpoints layer with the next dimension. Anything you draw on this layer, accidentally or otherwise, AutoCAD does not plot. For this reason, some students become frustrated to find that part of their drawing won't plot, because they accidentally drew on layer Defpoints.
>
> Use the No Print toggle to make *any* layer non-printing.

Working with Layers

The Layer Properties Manager dialog box provides you with a fair degree of control over layers. Here's what you can do with them.

Turning Layers On or Off

When layers are *on*, their objects are seen and can be edited; the status is indicated by a yellow lightbulb icon in the **On** column. When *off*, the objects on that layer are not seen; the lightbulb icon turns blue-gray. To turn one or more layers on or off, first highlight the layer(s) by selecting their name(s), and then choosing the lightbulb icon. The "on" column immediately reflects the change.

> **Notes:** To select *all* layers in the layer dialog box, right-click on a layer name. AutoCAD displays a small pop-up menu. Choose **Select All** to highlight all layer names. Choose **Clear All** to deselect all layers.

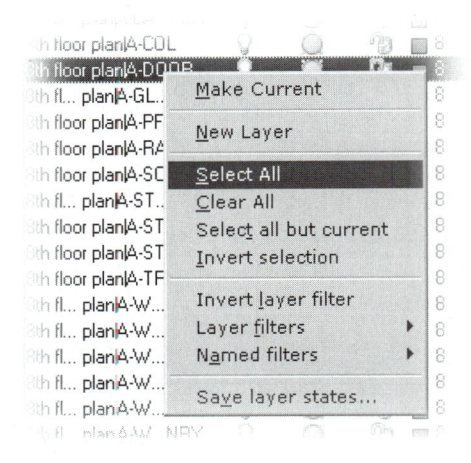

To change the width of the columns in the layer dialog box, grab the black bar separating column tiles and drag left or right.

Freezing and Thawing Layers

Frozen layers cannot be seen or edited; their status is shown by the snowflake icon in the **Freeze in All Viewpoints** column. *Thawed* layers can be seen and edited, just like *on* layers; the symbol for thawed layers is the sun icon. To freeze or thaw one or more layers, highlight the target layer(s), and then choose the sun or snowflake icon.

It is more efficient to freeze a layer than to turn it off. Objects on frozen layers are ignored by AutoCAD during regenerations and other compute-intensive operations.

Locking and Unlocking Layers

Objects on *locked* layers can be seen, but not edited; locked layers show a padlock icon in the Locked column. *Unlocked* layers are like *on* layer: their objects are seen and can be edited; an unlocked layer shows an open padlock. To lock or unlock layers, select the target layer(s) and then choose the padlock icons in the **Lock** column.

Setting the Layer Color

To set the color for objects on a layer, first select the layer names(s), and then click on the color square. AutoCAD displays the Select Color dialog box. Choose a color, and then click OK. From now on, all objects on the layer are displayed in that color, unless overridden by the **COLOR** command or by a block's properties.

Setting the Layer Linetype

When you select a linetype for a layer, it affects all the objects residing on that layer. First select the layer names(s) for which you wish to set a linetype, then choose the linetype. AutoCAD displays the Select Linetype dialog box. If necessary, load linetypes. Select a linetype and choose **OK**.

Setting the Layer Lineweight
You can specify the lineweight for all the objects residing on a layer. First, select the layer name(s) for which you wish to set a lineweight, and then choose the lineweight, such as Default. The Lineweight dialog box is displayed; select a lineweight, and then choose **OK**.

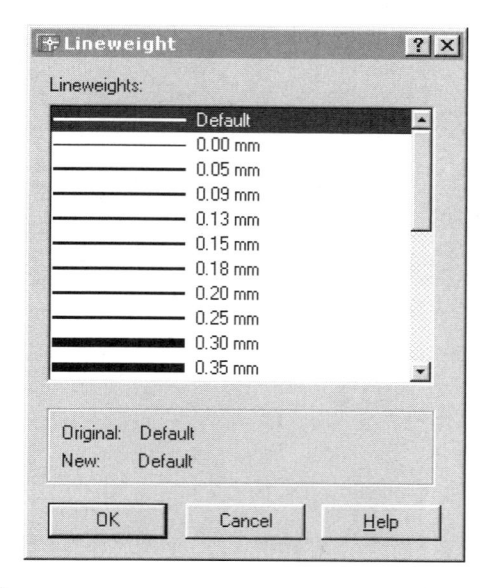

Setting the Layer Plot Style
Plot styles are available only when the feature is turned on in AutoCAD; otherwise, the plot style names are shown in gray and cannot be changed. Plot styles determine how all objects residing on a layer are plotted.

To change the plot style, select the layer names(s) for which you wish to set a plot style. Select a plot style from those available in the Select Plot Style dialog box, and then choose **OK**.

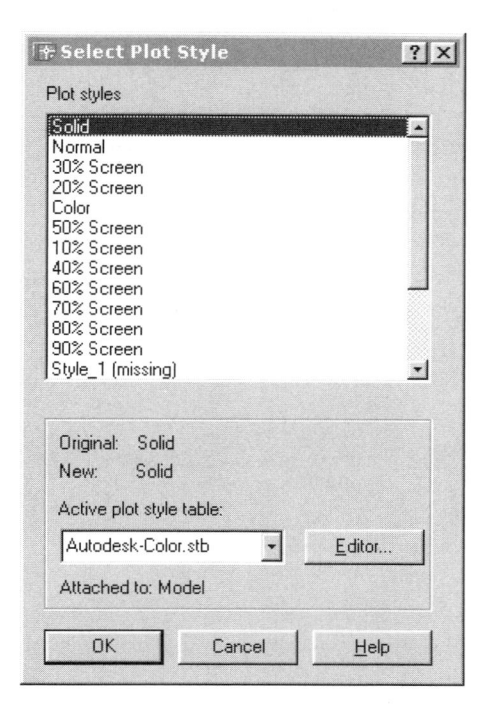

Setting the Layer Print Toggle
You can specify that some layers print, while others do not. Under the **Plot** column, choose the printer icon for the layer(s) you don't want to print or plot. To allow the layer to print, simply choose the icon a second time.

Show Details
The **Show Details** button expands the Layer Properties Control dialog box, showing details of each layer — in detail greater than just icons.

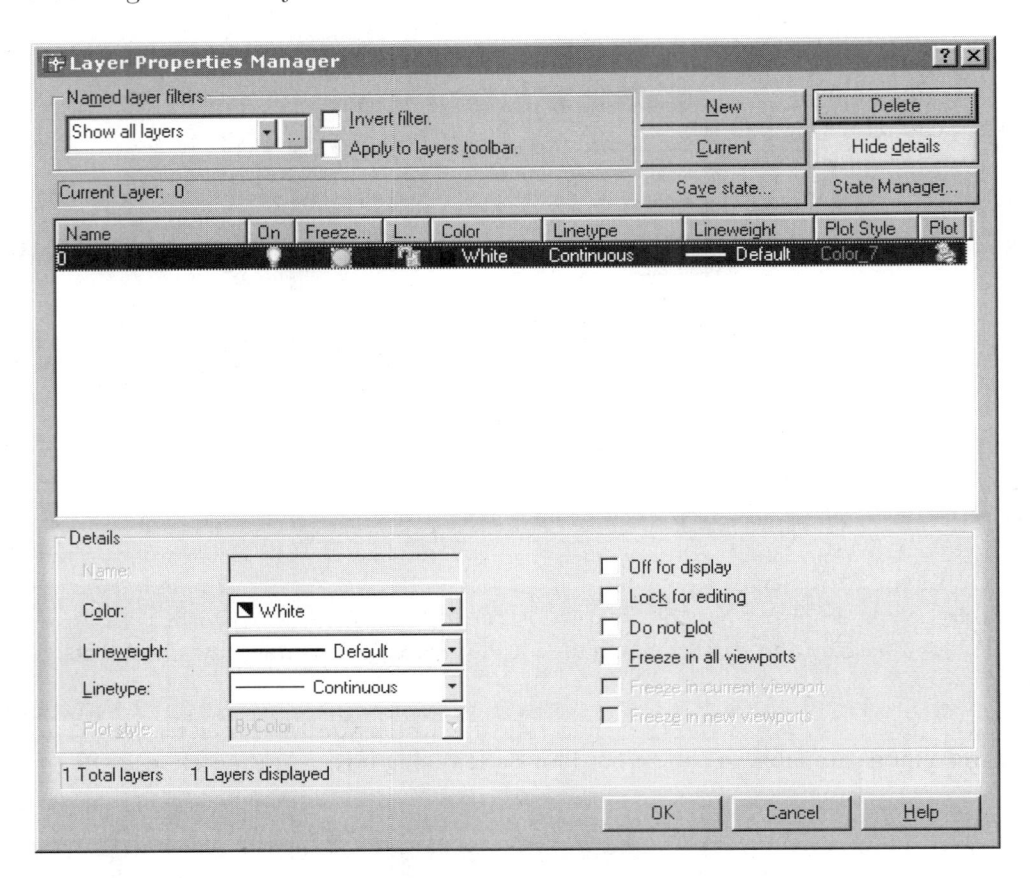

Layout Modes
In layout mode, two more columns appear in the layer dialog box: **Current VP Freeze** and **New VP Freeze**. To see these columns, switch to layout mode by selecting any tab except Model. Open the Layer Properties Manager dialog box, and scroll the layer listing all the way to the right.

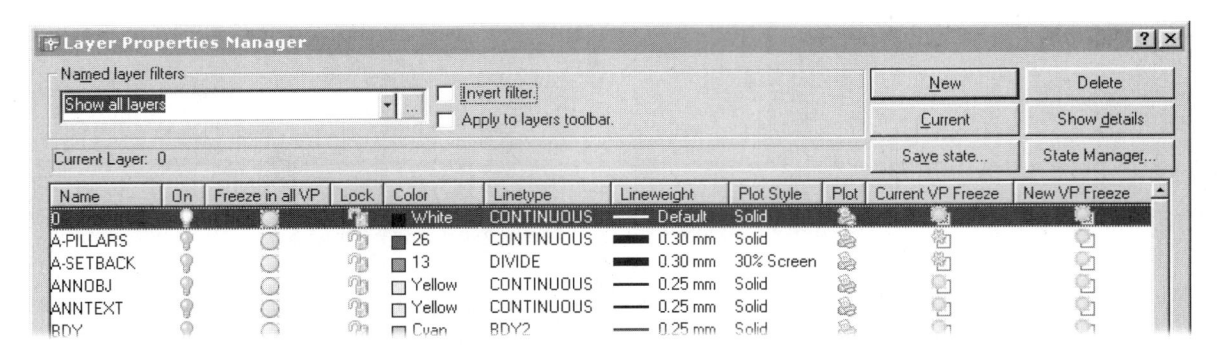

As described in *Using AutoCAD 2004: Advanced*, AutoCAD can work with tiled or floating viewports. You can freeze layers in floating viewports — something you cannot do in model view — to display different sets of layers in different viewports. AutoCAD has two controls over layer visibility in floating viewports: (1) freeze active viewport and (2) freeze new viewports.

The first option automatically freezes the specified layers in the current viewport. Select the layer name(s), and then choose the icon in the **Current VP Freeze** column. When the icon is a shining sun, the layer is thawed; when the icon is a snowflake, the layer is frozen. When the icon is gray, AutoCAD cannot display floating viewports, because the drawing is not in layout mode.

The second option freezes specified layers when a new viewport is created. Select the layer name(s), and then choose the icon in the **New VP Freeze** column. As before, when the icon is a shining sun, the layer is thawed; when the icon is a snowflake, the layer is frozen.

Selecting Layers

Some types of drawings may have many, many layer names. In theory, an AutoCAD drawing can have hundreds of thousands of layers; in practice, some drawings contain hundreds of layers. Working your way through that many layers becomes tedious. To help out, AutoCAD allows you to sort and to shorten the list.

You sort and shorten layers:

* By columns
* By the Show list
* By filters

Sort by Columns

The names of the columns — **Name**, **On**, **Freeze**, **Lock**, and so on — are actually buttons. Click once on the **Name** column button, and the column sorts in alphabetical (0 - 9 followed by A - Z) order; click a second time and the column sorts in reverse order (Z - A followed by 9 - 0). The figure below illustrates layer names sorted in reverse order and starting with Z:

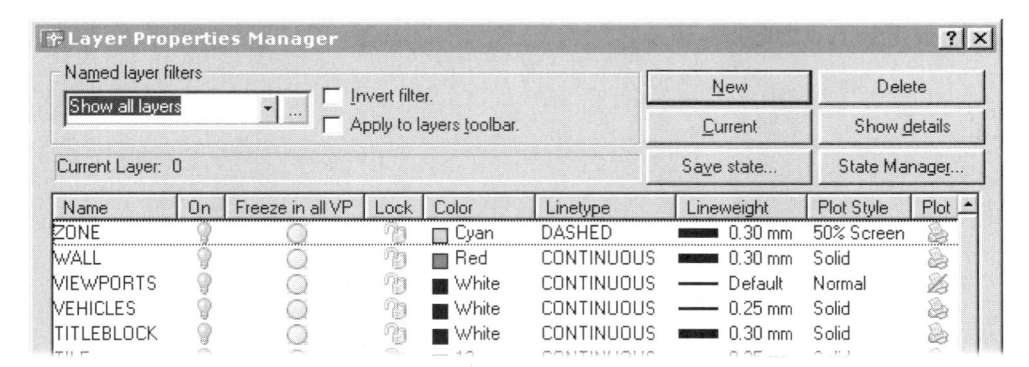

Choosing the **Color** column button sorts the colors by color number (red is color #1). Choosing a second time reverses the color numbers, starting with the highest color number used by drawing.

Choosing the **On**, **Freeze**, and other toggle column titles, sorts the layers by icons. Click once, and the column sorts by light bulb, sun, or open padlock. Click a second time, and the column reverses the sortation by dim bulb, snowflake, or closed padlock.

Show List

The drop list under **Named Layer Filters** lets you see logical groups of layers.

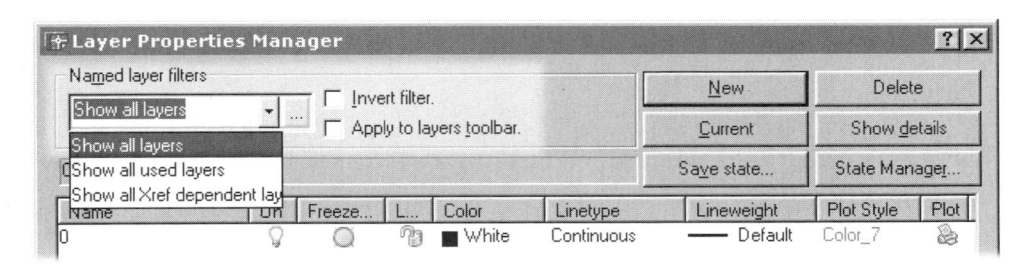

Your choices include:

Show all layers displays *all* layer names in the drawing; this is the default setting.

Show all used layers displays all layers in use, which means any layer containing at least one object.

Show all Xref dependent layers displays all layers found in externally-referenced drawings.

Invert filter displays the inverse of the above three options. For example, when **Invert Filter** is on and you select **All in Use**, the layer dialog box displays all layers *not* in use.

Apply to layers toolbar means that the names listed by the **Layers** toolbar will also be filtered.

... displays the Named Layer Filter dialog box, which allows even finer selection of layer names. The filter capabilities are used to display or suppress layers that are on or off, frozen or thawed, locked or unlocked, and frozen in the current viewport. You can also filter by name, color, and linetype. With these filters, you can use the wild-card characters of question mark (?) and asterisk (*) to designate multiple layers. For instance, you can filter the list so that only thawed layers are displayed.

Some of the filters shown above use drop lists to list either property, or both. For example, when you select the **Freeze/Thaw** drop list, the selections are Both, Frozen, and Thawed.

Other selections require you to enter the filter type in the text box. The default listing is an asterisk (*), which means that all layer names will be listed. Let's assume that you want to list all layers starting with the letters LEVEL and ending with any two characters. The filter would read:

    LEVEL??

This would list layers named LEVEL21, LEVEL_A, and LEVEL2C. It would not list 1LEVEL, LEV21, or 3RD_LEVEL, because the names don't match LEVEL.

You can reset the values so that all layer names are listed by choosing the **Reset** button.

Deleteing Layers

AutoCAD allows you to delete *empty* layers only; an empty layer is one that has no objects on it. As well, you cannot delete layer 0, DefPoints, and externally-referenced layers — even when empty.

Select one or more layer names, and choose **Delete**. If the layer cannot be erased, AutoCAD displays the following dialog box:

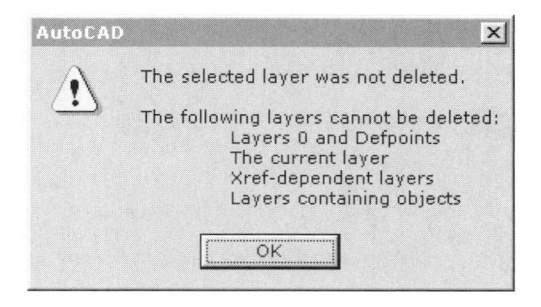

Save and Restore States

The **Save State** and **Restore State** buttons allow you save the current state of layers, and then restore the state at a later time. The *state* of layers includes names and properties, such as whether thawed or frozen, as well as colors and linetypes.

Once you save a layer state by name, you can edit the state, and even export states to a *.lay* file for sharing with others.

The **Save State** button displays the **Save Layer States** dialog box. Here you provide a name for the layer state, as well as determine which properties will be saved.

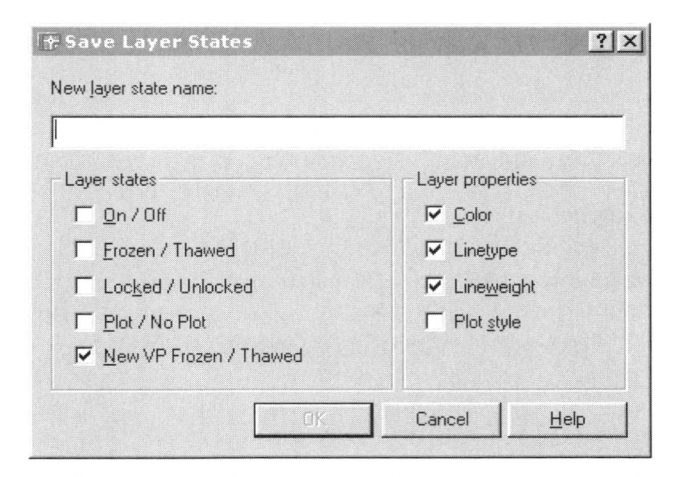

The **Restore State** button displays the Layer States Manager dialog box. Here you edit the states with a dialog box similar to that of the Save Layer States, as well as delete, rename, export, and import states.

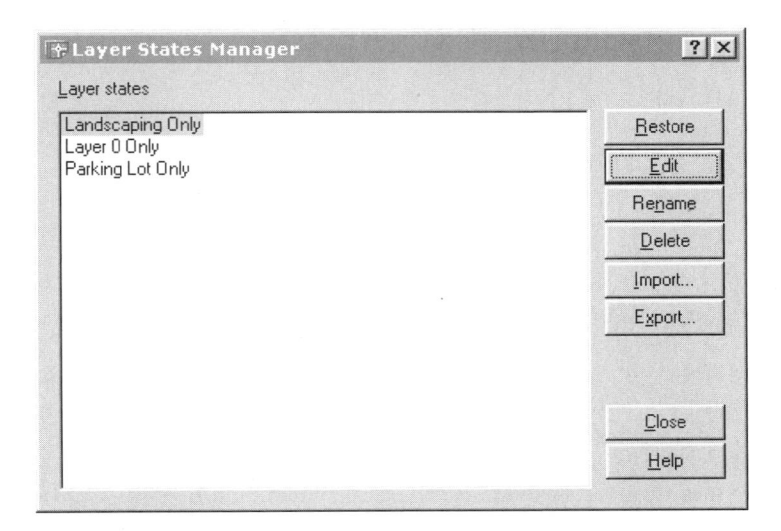

CONTROLLING LAYERS: ADDITIONAL METHODS

AutoCAD provides several other methods to control layers:

- **-LAYER** command controls layers at the command line.
- **Layers** toolbar controls layers from the toolbar.
- **CLAYER** system variable is a fast way to make a layer current.
- **AI_MOLC** command makes the selected object's layer current.
- **LAYERP** command restores the previous layer state.

Let's look at each of these methods.

-Layer

The **-LAYER** command creates and changes layers at the command line:

Command: -layer

Current layer: "0"

Enter an option

[?/Make/Set/New/ON/OFF/Color/Ltype/LWeight/Plot/PStyle/Freeze/Thaw/LOck/Unlock/stAte]: *(Enter an option.)*

Layers Toolbar

When the Layers toolbar is available (as it usually is), it lets you select the current layer without needing the **LAYER** command. In addition, it controls four layer properties.

Changing the Current Layer

To change the current layer, click the down arrow at the right end of the layer name on the toolbar. It drops down a list of layer names, along with icons signifying their status.

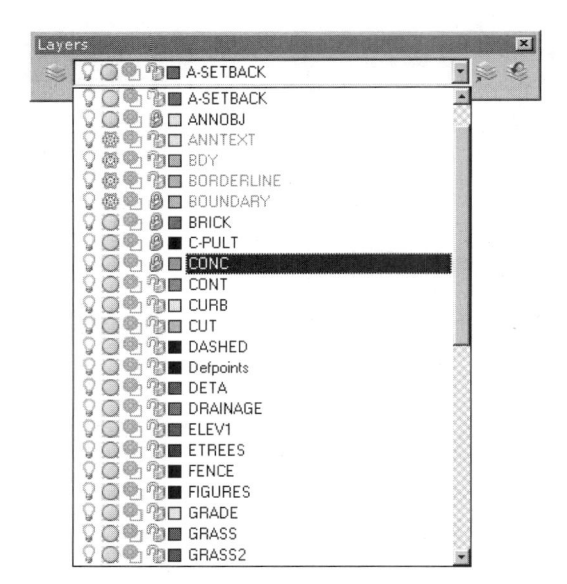

Select the name of a layer to make it current. It's that simple.

There is, unfortunately, one catch. You cannot make current any layer whose name is shown in gray, because it is frozen (and you can't see or work with frozen layers.)

Changing the Layers' Status

The drop-down box lists a quintet of icons beside each layer name. The colored square icon is the color assigned this layer. The other four icons each have two states:

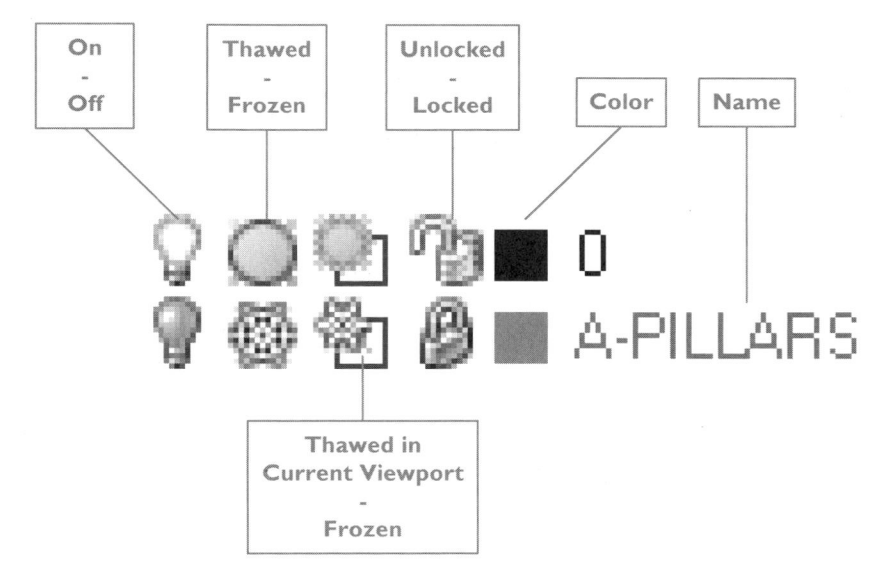

Lightbulb on or off: layer is on (default) or off.

Sun or **snowflake**: layer is thawed (default) or frozen; frozen layers also show their names in gray.

Sun or **snowflake on square**: layer is thawed (default) or frozen in the current viewport; this icon changes only when the drawing is in paper space (layout mode).

Padlock open or closed: layer is unlocked (default) or locked.

You change the status of each of the four icons simply by clicking them. Note that the thawed/frozen in current viewport icon changes only when the drawing is in layout mode (paper space).

Ai_Molc

An easy way to switch to another layer is with the Make Object's Layer Current button. Suppose you are interested in switching to the layer holding a green dotted line whose name you're not sure about. The AI_MOLC command switches you to the layer when you select the green dotted line. You use the feature as follows:

1. Choose the Make Object's Layer Current button. AutoCAD prompts you:
    Select object whose layer will become current:
2. Select the green dotted line. AutoCAD reports:
    layername is now the current layer.

Check the toolbar, and you will see that layer "layername" is now current.

CLayer

If you prefer the keyboard to dialog boxes, then CLAYER system variable (short for "current layer") is the fastest way to switch between layers, as follows:

Command: **clayer**

New value for CLAYER <default>: *(Enter the name of the layer, and then press* ENTER.*)*

AutoCAD immediately makes that layer current.

LayerP

The **LAYERP** command (short for "layer previous") undoes changes to layer settings, much like the **ZOOM Previous** command restores the previous view.

Command: **layerp**

Restored previous layer status.

The command cannot, however, undo the effect of renamed, deleted, purged, and newly created layers. When you rename a layer and change its properties, only the properties are changed back, not the name.

A related command is **LAYERPMODE** (short for "layer previous mode"). It toggles whether AutoCAD keeps track of changes to layers.

Command: **layerpmode**

Enter LAYERP mode <ON>: *(Enter* **ON** *or* **OFF**.)

When turned off, the **LAYERP** command does not work, and so AutoCAD reminds you:

Layer-Previous is disabled. Use LAYERPMODE to turn it on.

PROPERTIES

The **PROPERTIES** command displays a window that allows you to change (almost) *all* properties of objects. It's called a *window*, because you can continue working in AutoCAD without needing to close it, unlike dialog boxes.

The **Properties** window displays different sets of properties, depending on the object(s) you select: text, mtext, 3D face, 3D solid, multiline, arc, point, attribute definition, polyline, block insertion, ray, body, region, circle, shape, dimension, 2D solid, ellipse, spline, external reference, hatch, tolerance, image, trace, leader, viewport, line, and xline.

Each object has a somewhat different set of modifiable properties. For example, you can modify a line's endpoint X, Y, Z coordinates; you cannot, however, modify the coordinates of an arc's endpoints; instead, you alter its start and end angle.

TUTORIAL: CHANGING ALL PROPERTIES

1. To change all the properties of objects, start the **PROPERTIES** command with one of these methods:
   - From the **Tools** or **Modify** menu bars, choose **Properties**.
   - From the **Standard** toolbar, choose **Properties**.
   - At the keyboard, press **CTRL+1**.
   - At the 'Command:' prompt, enter the **properties** command.
   - Alternatively, enter the aliases **pr**, **props**, **mo**, (short for "modify"), **ch**, (short for "change"), **ddchprop**, or **ddmodify** (the old names in AutoCAD) at the 'Command:' prompt.

   Command: **properties** *(Press* ENTER.*)*

Changing Object Properties **341**

2. In all cases, AutoCAD displays this window:

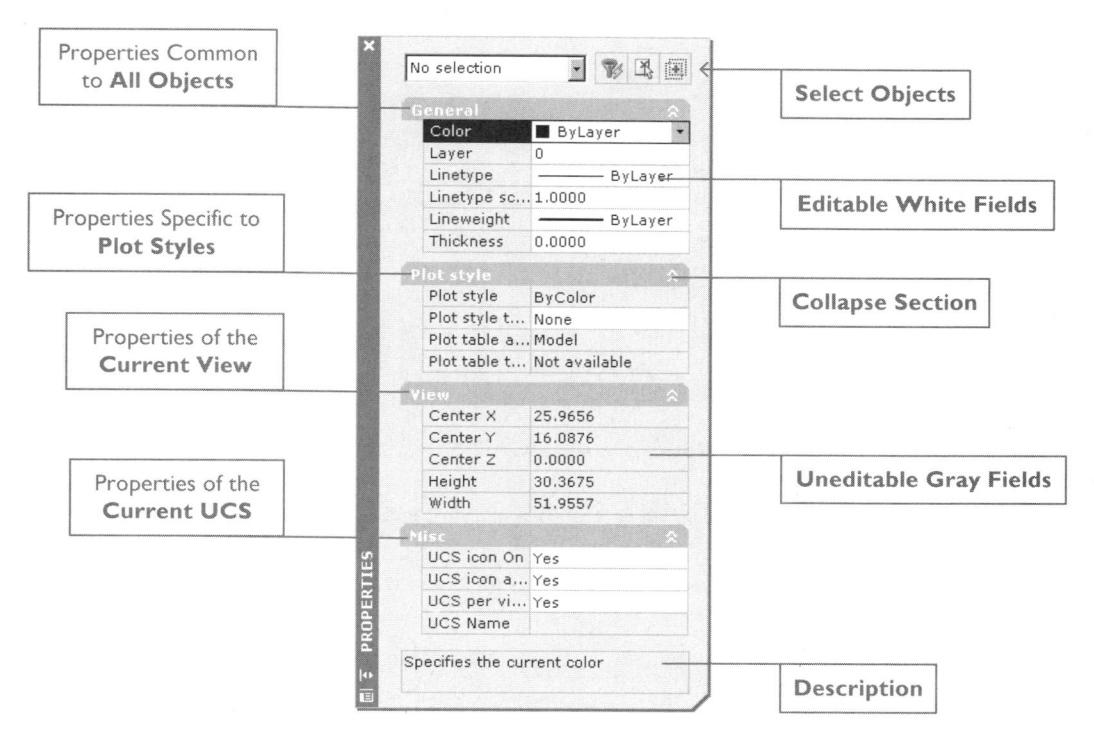

- Properties Common to **All Objects** → (General section)
- Properties Specific to **Plot Styles**
- Properties of the **Current View**
- Properties of the **Current UCS**
- Select Objects
- Editable White Fields
- Collapse Section
- Uneditable Gray Fields
- Description

3. Notice that there are two colors of field: white and gray. White means you can modify the property, while gray means you cannot.

   The effect of changes you make in this window depends on whether or not objects are selected — much like changing settings in the **Properties** toolbar. When no objects are selected (as shown above), making changes to properties affects objects drawn from now on; when objects are selected, changing the properties affects the selected objects.

   You can "collapse" sections by clicking the double-arrow button, making the window smaller.

4. Notice the section called "General" located in the upper part of the window. The General section includes these modifiable properties:

   Color. To change the color, select the color square. AutoCAD displays the standard color names.

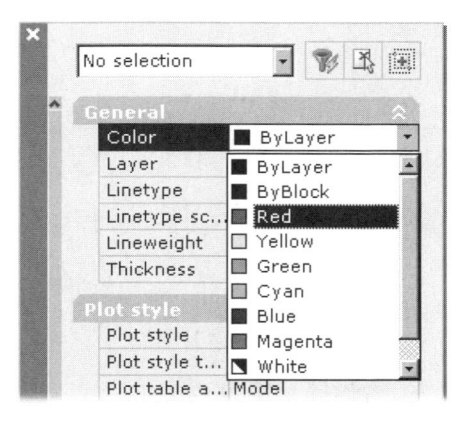

   To select other colors, select **Other**, and AutoCAD displays the Select Color dialog box. Select a color, choose **OK**, and return to the Properties window.

   Layer. To change the layer, select the **Layer** field. AutoCAD displays a list of the layer names defined in the drawing. Select a layer name.

Linetype. To change the linetype, select the **Linetype** field. AutoCAD displays a list of linetypes loaded into the drawing.

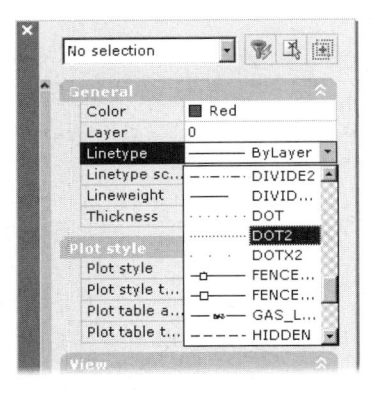

Select another linetype name. Notice that AutoCAD displays a sample. You cannot load additional linetypes from here.

Linetype Scale. To change the scale factor (if available), delete the linetype scale factor, and type a new number.

Lineweight. To change the lineweight, select the field, and then one of the preset lineweights.

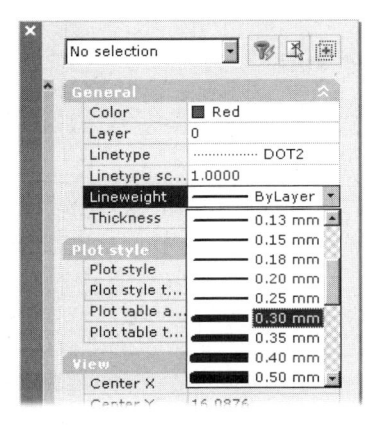

5. To select one or more objects, pick an object in the drawing, such as a line. Note that you do not dismiss the Properties window: simply move the cursor, and pick the object.
   (To select all objects in the drawing, press **CTRL+A**.)
   As an alternative, choose one of the three buttons at the top of the window:

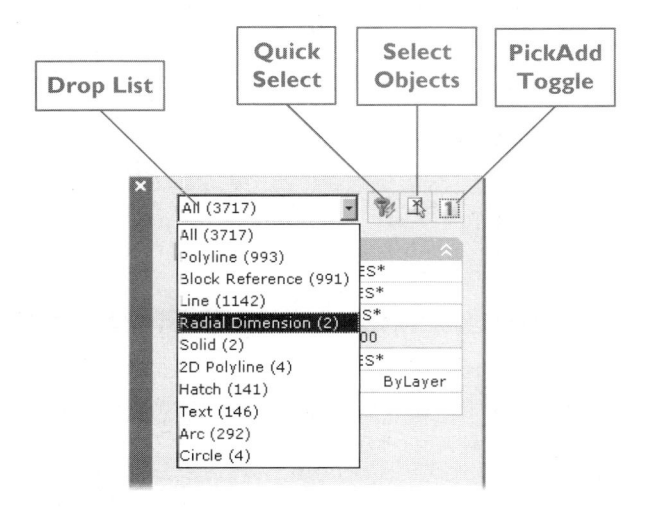

The drop list shows the variety (and, in parentheses, the number) of objects.

Quick Select displays the Quick Select dialog box, described in Chapter 6, "Drawing with Efficiency."

Select Objects causes AutoCAD to prompt you (unnecessarily):

Select objects: *(Select one or more objects.)*

Select objects: *(Press ENTER to end object selection.)*

Toggle value of PICKADD sysvar changes how you select objects. When the plus (**+**) sign appears on the button, you select additional objects by picking them; when the number 1 appears on the button, select additional objects by holding down the SHIFT key.

6. The Properties window changes to reflect the properties of the selected object. In the case of lines, you see its geometry, such as the x and y coordinates of the segment's start and end points.

    (When you pick two objects whose properties differ, AutoCAD displays *VARIES* as the property value.)

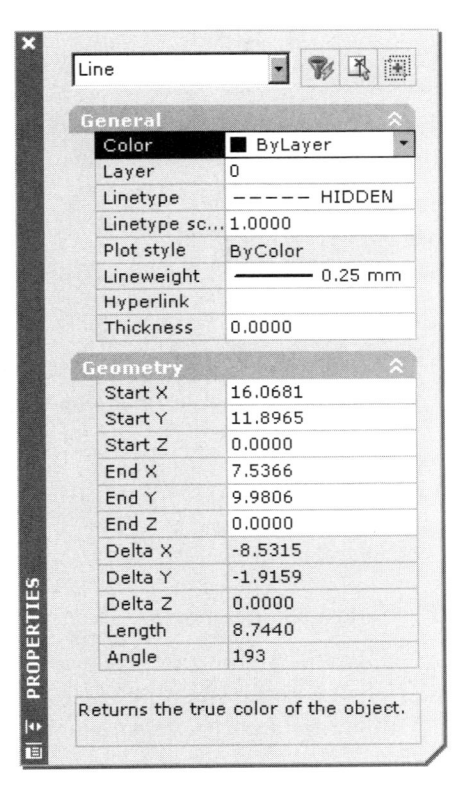

6. Click in the **Start X** field, and change the value to a different number. Notice how the line's endpoint moves.

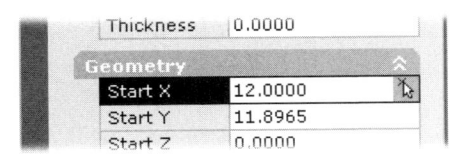

    As an alternative, click the arrow button. AutoCAD prompts you:

    Pick a point in the drawing: *(Move the cursor, and then pick a point.)*

    You may use object snaps to assist you in making an accurate pick.

7. When done, press **CTRL+1**, click the **x**, or enter the **PROPERTIESCLOSE** command.

> **Note**: You can easily display the Properties window by double-click (almost) any object in the drawing. Exceptions include text, which display the text editing window.

CHANGING PROPERTIES: ADDITIONAL METHODS

AutoCAD provides another way to change properties:

- **CHPROP** command changes properties at the command line.

ChProp

The **CHPROP** command prompts you to change the properties of selected objects at the command line:

Command: **chprop**

Select objects: *(Select one or more objects.)*

Select objects: *(Press ENTER to end object selection.)*

Enter property to change

[Color/LAyer/LType/ltScale/LWeight/Thickness/PLotstyle]: *(Enter an option.)*

Depending on the option you select, AutoCAD presents a different set of prompts that match the related command. For example, to change the color, AutoCAD prompts you in a manner similar to the -COLOR command:

Enter property to change

[Color/LAyer/LType/ltScale/LWeight/Thickness/PLotstyle]: **c**

New color [Truecolor/COlorbook] <BYLAYER>: *(Enter a color number, such as **2**.)*

Enter property to change

[Color/LAyer/LType/ltScale/LWeight/Thickness/PLotstyle]: *(Press ENTER to exit the command.)*

MATCHPROP

The **MATCHPROP** command "reads" the properties of one object and applies them to other objects.

At times, you may want a group of objects to match the properties of another object. For example, you may accidentally place several doors on the Landscape layer, instead of the Door layer. Or, you may realize that some lines drawn with hidden linetype should be in another linetype — whose name you don't recall.

You could use the Properties toolbar to change layers, colors, linetype, and so on, but it doesn't work well if you are not sure of the exact layer and linetype name. Complex drawings have hundreds of layer names, many of which look similar. One sample drawing provided with AutoCAD has the following layer names: ARCC, ARCCLR, ARCDIMR, ARCDSHR, ARCG, ARCM, ARCR, ARCRMNG, ARCRMNR, and ARCTXTG. And that's just the first ten! Similarly, linetypes can look confusingly similar.

The **MATCHPROP** command solves those problems. It lets you copy the properties from one object to a selection set of objects. You must, however, be sure to select objects in the correct order: (1) select the single object whose properties to copy; and then (2) select the object(s) to take on those properties.

TUTORIAL: MATCHING PROPERTIES

1. To match the properties of one object to other objects, start the **MATCHPROP** command with one of these methods:
   - From the menu bar, choose **Modify**, and then **Match Properties**.
   - From the **Standard**, toolbar choose **Match Properties**.
   - At the 'Command:' prompt, enter the **matchprop** command.
   - Alternatively, enter the aliases **ma** or **painter** (the old name in AutoCAD LT) at the 'Command:' prompt.

   Command: **matchprop** *(Press ENTER.)*

2. In all cases, AutoCAD prompts you at the command line:

   Select source object: *(Pick a single object.)*

   Notice that the cursor turns into a paintbrush.

3. AutoCAD lists the *active settings*, properties to be "picked up" by the paintbrush.
   Select the objects to whom the properties should be applied.

   Current active settings: Color Layer Ltype Ltscale Lineweight Thickness

   PlotStyle Text Dim Hatch Polyline Viewport

   Select destination object(s) or [Settings]: *(Pick one or more objects, or enter S.)*

4. You can continue selecting objects using any form of selection, such as windows, fences, and single picks.
   Press **ENTER** to exit the command.

   Select destination object(s) or [Settings]: *(Press ENTER.)*

MATCHING PROPERTIES: ADDITIONAL OPTION

The **MATCHPROP** command includes a single option that lets you decides which properties should be picked up and copied:

- **Settings** determines which properties are painted.

Settings

At the "Select destination object(s) or [Settings]:" prompt, enter **S** to display the Property Settings dialog box.

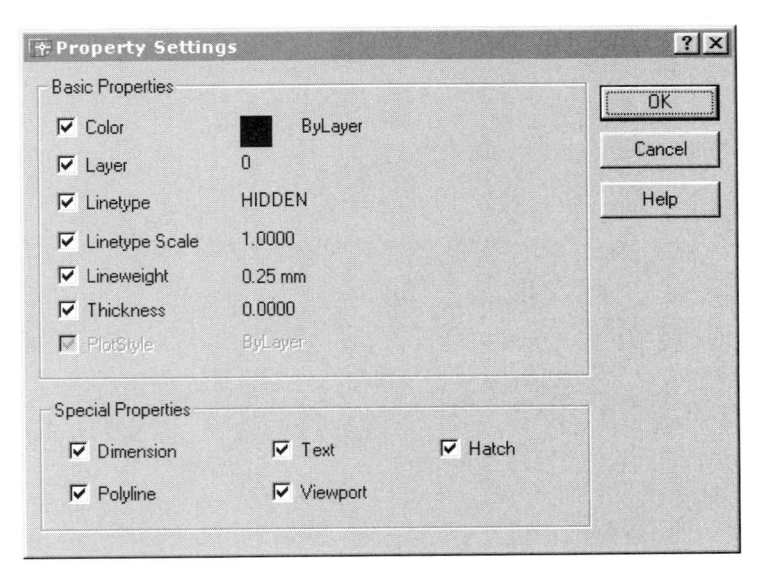

The properties listed by this dialog box are more extensive than those listed by the Properties toolbar.

Not all properties, however, work with all objects. For example, it makes no sense to match the hatch properties of text objects, since text cannot be hatched. (Thickness is the extrusion height of objects in the Z direction.)

Property	Objects Affected
Color	All except OLE objects.
Layer	All except OLE objects.
Linetype	All except attributes, hatches, multiline text, OLE objects, points, and viewports.
Linetype Scale	All objects except attributes, hatches, multiline text, OLE objects, points, and viewports.
Lineweight	All.
Thickness	Arcs, attributes, circles, lines, points, 2D polylines, regions, text, and traces.
Plot Style	All OLE objects; unavailable when PSTYLEPOLICY is 1.
Dimension	Dimension, leader, and tolerance objects; also paints dimension styles.
Polyline	Polylines; also paints width and linetype generation, but not fit/smooth, variable width, nor elevation.
Text	Text and multiline text only; also paints the text style.
Viewport	Viewports; also paints on/off, display locking, standard or custom scale, shade plot, snap, grid, as well as UCS icon visibility and location settings, but not clipping, UCS-per-viewport, or layer freeze/thaw state.
Hatch	Hatches; also paints the hatch pattern.

By default, all properties are on, meaning they will all be copied. When you turn off some properties, AutoCAD remembers this the next time you use MATCHPROP, which is why it has that list of "Current active settings." For example, if you turn off text, dimension, and hatch properties, next time you see a shorter list of active settings:

Command: **matchprop**
Select source object: (Pick.) Current active settings: Color Layer Ltype Text Dim Hatch
Select destination object(s) or [Settings]: *(Pick.)*

EXERCISES

1. In this exercise, you change the colors of objects.
   Open AutoCAD to start a new drawing.
   Use the **LAYER** command to set the color of layer 0 to cyan (light blue).
   Click **OK** to exit the dialog box.
   Draw some lines. Do the circles and boxes change color?

   Use the **COLOR** command to set the current color to yellow.
   Draw three circles. Did the **COLOR** command override the layer's color setting?
   Next, set the current color to red.
   Draw three boxes. Do they appear in the correct color?

   Start the **PROPERTIES** command, and then select all the objects on the screen.
   Select blue as the color. Did all the objects change to blue?

2. In this exercise, you create a layer, and then freeze and thaw it.
   Create a layer named "Mylayer."
   Set Mylayer as the current layer. Do you see the layer name in the toolbar at the top of the screen?
   Set the layer color to green, and then draw some objects on the layer.
   Using the Layer toolbar, freeze layer Mylayer, and then set the current layer to 0. Do the objects you drew disappear?
   Use the **-LAYER** command to thaw Mylayer. Do the objects reappear?

3. In this exercise, you create several layers, and then change their properties.
   With the **LAYER** command, create seven layers using these names:

       Landscape
       Roadways
       Hydro
       ToBeRemoved
       StormSewer
       CableTV
       Building

   Using the layer dialog box:
   Change the color of layer Landscape from white to green.
   Change the lineweight of layer Roadways from Default to 0.083" (2.11mm).
   Change the linetype of layer Hydro from Continuous to Gas_line.
   Change the ToBeRemoved layer from white to red, and from Continuous to Dashed linetype.
   Change the linetype of the CableTV layer from Continuous to Hidden, and change the lineweight to 0.020" (0.51mm).
   Make the CableTV layer current by choosing the **Current** button.
   Click **OK** to exit the layer dialog box.

   Draw some lines. Do they appear in hidden linetype? If the lines do not look thick, ensure the **LWT** button is depressed on the status line.
   Use the Layer toolbar to change to the ToBeRemoved layer, and then draw some lines. Do they appear red and dashed?

Again, use the Layer toolbar to change to the Building layer, and then draw some lines. Do they appear black and continuous?

4. In this exercise, you copy the properties from one object to all others in the drawing. Continue with the drawing you created in the previous exercise.
   Start the **MATCHPROP** command.

   Select source object: *(Select one of the red, dashed lines drawn on the CableTV layer.)*

   Select destination object(s) or [Settings]: **all**

   Select destination object(s) or [Settings]: *(Press ENTER.)*

   Do all the objects become red and dashed?

CHAPTER REVIEW

1. What do hidden lines show in a drawing?
2. Name two ways a linetype can be applied to objects in AutoCAD:
   a.
   b.
3. How do you control the length of individual segments in a dashed line?
4. How do you load all linetypes into a drawing?
5. What is the difference between *global* and *object* linetype scaling?
6. Which command allows you to change the names of layers to match those of another drawing?
7. Is it possible to save the current state of layers?
   If so, how?
8. Describe five properties that the **PROPERTIES** command changes.
9. What is meant when you change an object's color to "BYLAYER"?
10. Why would you use the **PROPERTIES** command instead of the Properties toolbar?
11. Can the Properties toolbar change objects? If so, how?
12. What is the **MATCHPROP** command used for?
13. Why would a CAD drafter use layers?
14. What layer option would you use if you wanted to create a new layer?
15. How do you turn on a frozen layer?
16. What is the difference between locking a layer and freezing a layer?
17. How do you obtain a listing of all the layers in your drawing?
18. Can you have objects of more than one color on the same layer? Explain.
19. What do the following layer symbols mean?
    Snowflake:
    Open lock:
    Lightbulb glowing:
    Printer:
    Colored square:
20. Name a benefit to using lineweights.
21. Is it possible to add custom linetypes to your drawing?
    Custom lineweights?
22. What is color is designated by R?
    What is color 7?
    What is a *color book*?
23. Describe what is meant by RGB in terms of colors.
24. Should you use lineweights to represent objects with width?
25. What is the purpose of the LWT button on the status bar?
26. Describe the steps to change the lineweight of a circle using the Properties toolbar:
    a.
    b.
27. Explain how the three classes of linetype differ:
    Simple:
    ISO:
    Complex:
28. Before using a linetype in a new drawing, what must you do first?
29. What is the name of the linetype used to show hidden edges?

30. **LTSCALE** is set to 0.5 and **CELTSCALE** is set to 5.0. At what scale is the next linetype drawn?
31. Under what condition can a layer be deleted?

    Can layer 0 be deleted?
32. Describe two ways to manage long lists of layer names:

    a.

    b.
33. When the drawing is in model space, why do the **Current VP Freeze** and **New VP Freeze** columns not appear in the layer dialog box?
34. What is the purpose of the **Show all used** layers option in the layers dialog box?
35. Can you assign a lineweight to layers?

    Assign a linetype?

    Assign a hatch pattern?
36. When a linetype is assigned to a layer, are all objects on that layer displayed with that linetype?

    If not, why not?
37. What do the following wildcard characters mean?

    *

    ?
38. What is the purpose of the **CLAYER** system variable?
39. Which keystroke shortcut displays the Properties window?
40. Describe how the **PICKADD** system variable affects object selection.
41. Can the Properties window be used to change the endpoints of line segments?

CHAPTER 8

Correcting Mistakes

As you create a drawing, you sometimes make mistakes. One way to correct them is through editing, which is the subject of the chapters following. Another method is to says "Oops!," and reverse the mistake by undoing it.

In all chapters until now, you have been learning how to create things. In this chapter, you learn how to un-create them — how to revert, reverse, and repair — using the following commands:

U and **UNDO** reverse the changes of most commands.

REDO and **MREDO** (new to AutoCAD 2004) reverse the effect of the undoing.

RENAME changes the names of linetypes, layers, and so on.

PURGE removes unused layers, blocks, and so on.

RECOVER and **AUDIT** attempt to fix corrupted drawings.

File Manager allows you to retrieve drawing backups.

QUIT and **CLOSE** discard editing changes made to drawings.

FINDING THE COMMANDS

On the **STANDARD** toolbar:

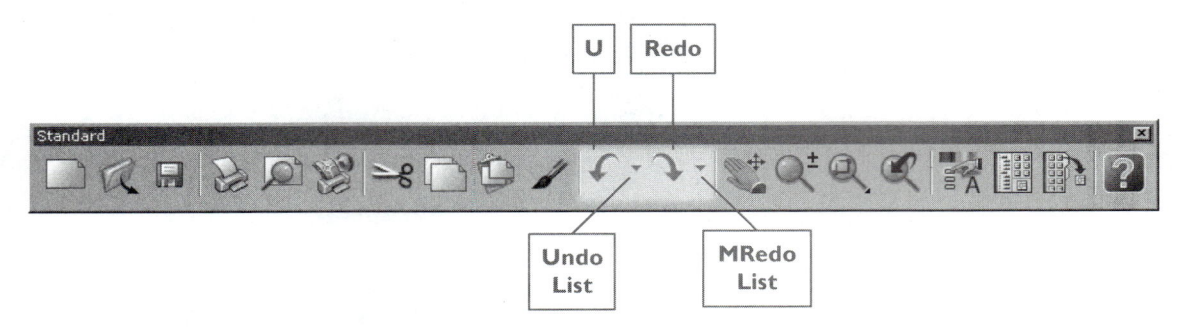

On the **FILE**, **EDIT**, and **FORMAT** menus:

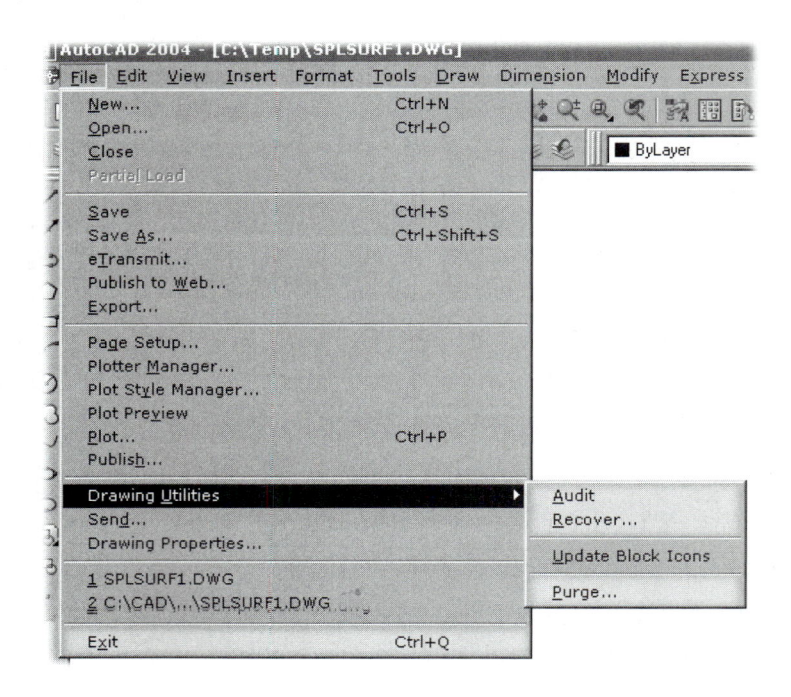

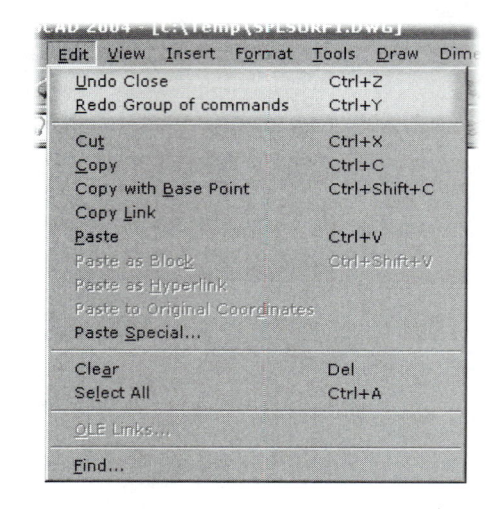

U AND UNDO

The **U** and **UNDO** commands reverse the effect of many (not all) commands.

- The **U** command undoes the last action; this is the "quick" command to use when just one or two changes need to be undone.
- The **UNDO** command provides many options for undoing the effects of commands; this is the "advanced" command when you want utter control.

AutoCAD allows you to undo your work, then to redo it. This is useful if you have just performed an operation that you wish to reverse — either a mistake, or a what-if scenario. After you undo an operation, you can use the **REDO** command once to reverse the undo.

The **U** command reverses the effect of the most recent command. You can execute a series of "undoes" to back up through a string of changes. (The U command is *not* an alias for the more complex **UNDO** command, although it functions identically to **UNDO 1**.)

Undoing a command restores the drawing to its state before the command was executed. For example, erase an object, and then execute the U command: the object is restored. When you scale an object, and then undo it, the object is scaled back to its original size. Undoing a just-completed **BLOCK** command restores the block, and deletes the block definition that was created, leaving the drawing exactly as it was before the block was inserted.

At the command prompt, the U command lists the command that is undone to alert you to the type of command that was affected.

TUTORIAL: UNDOING COMMANDS

1. To undo the effect of a command, start the **u** command with one of these methods:
   - From the menu bar, choose **Edit**, and then **Undo**.
   - At the keyboard, press **CTRL+Z**.
   - At the 'Command:' prompt, enter the **u** command.

   Command: **u** *(Press ENTER.)*

2. AutoCAD undoes the previous command, and then displays the name of the undone command:

   ERASE

   When you undo all the way back to the first command, AutoCAD reports:

   Everything has been undone

> **Note**: Several commands *cannot be undone*. **SAVE**, **PLOT**, and **WBLOCK**, for example, are unaffected, because AutoCAD cannot "unsave" or "unplot." If you attempt to use the U or UNDO command after these commands, the name of the command is displayed, but the command's action is not undone.
>
> **UNDO** and **U** have no effect on commands and system variables that open, save, export, and close files, change the arrangement of windows (as opposed to viewports), and redraw or regenerate (such as **REGEN**) the drawing.
>
> The two undo commands also have no effect on commands that took place before the drawing was opened. For example, you work on a drawing, save it, and close the drawing. When you open the drawing again, AutoCAD cannot undo the commands of the previous editing session.

UNDOING COMMANDS: ADDITIONAL METHODS

AutoCAD has several additional methods to undo the effect of commands:

- **UNDO** command provides control over the undo process at the command line.
- **Standard** toolbar provides a list of undo-able actions.
- **Undo** and **Previous** options in commands, and **Cancel** buttons in dialog boxes.

Let's look at each method.

Undo

The UNDO command provides fine control over the undo process.

Command: **undo**

Enter the number of operations to undo or [Auto/Control/BEgin/End/Mark/Back] <1>: *(Enter a number or an option.)*

As an alternative, you can click the **Undo** button on the Standard toolbar.

By default, the UNDO command operates like the U command: press **ENTER** at the "Enter the number of operations to undo" prompt, and AutoCAD undoes the last command (if possible). Thus, **UNDO 1 = U**.

Entering a number, such as **4**, undoes the last four commands. This is the same as pressing U four times.

Auto

The Auto option groups all the actions of a single command into a single undo. The **Auto** option is not available when the **Control** option is turned off.

Enter UNDO Auto mode [ON/OFF] <On>: *(Enter ON or OFF.)*

Control

The **Control** option limits the effect of the undo command:

Enter an UNDO control option [All/None/One] <All>: *(Enter an option.)*

The **All** option turns on the UNDO and U commands, and allows them to undo all the way to the beginning of the drawing session.

The **None** option turns off the UNDO and U commands, grays out the **Undo** button on the Standard toolbar, and discards the undo history. Use this option only if your computer is low on disk space.

The **One** option restricts the UNDO command to a single undo, and the **Auto**, **Begin**, and **Mark** options are unavailable.

BEgin/End

The **BEgin** option groups several commands into a *set*.

Command: **undo**

Enter the number of operations to undo or [Auto/Control/BEgin/End/Mark/Back] <1>: **be**

After entering the **BEgin** option, the UNDO command ends, but AutoCAD starts recording all the commands you now enter, until you enter the **End** option:

Command: **undo**

Enter the number of operations to undo or [Auto/Control/BEgin/End/Mark/Back] <1>: **e**

The UNDO, REDO, and U commands now treat the set of commands as a single undo:

Command: **u**

GROUP

Mark/Back

The **Mark** option places a marker in the undo collection:

>Command: **undo**
>
>Enter the number of operations to undo or [Auto/Control/BEgin/End/Mark/Back] <1>: **m**

The **Back** option undoes all commands back to the marker:

>Enter the number of operations to undo or [Auto/Control/BEgin/End/Mark/Back] <1>: **b**
>
>Mark encountered

You may place as many marks as you require; the **Back** option moves back through the undo collection one mark at a time, removing each mark.

> **Note:** This is a dangerous action: to undo everything in the drawing, use the **UNDO** command's **Back** option without marks:
>
> Command: **undo**
>
> Enter the number of operations to undo or [Auto/Control/BEgin/End/Mark/Back] <1>: **b**
>
> This will undo everything. OK? <Y> *(Enter* **Y**.*)*
>
> Everything has been undone

Standard Toolbar

As an alternative to the undo commands, you can use the **Undo** drop list on the Standard toolbar. Click the arrow next to **Undo**, and you see a list of commands.

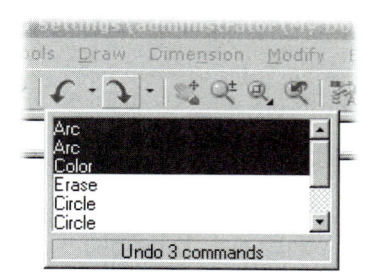

Move the cursor to the command back to which you want to undo. You cannot, however, selectively pick and choose noncontiguous commands.

Undo, Previous Options and Cancel Buttons

Several commands include methods of undoing operations within them.

Undo Option

Commands that execute more than one action often include an undo option. This allows you to reverse the effect of the last action without leaving the command. For example, the **LINE** command's **Undo** option erases the last-drawn segment. Other commands that have an undo option include **PLINE**, **PEDIT**, and **3DPOLY**.

Previous Option

Instead of an undo option, other commands have an equivalent option called "Previous." The **ZOOM** command's **Previous** option, for example, displays the previous view, whether created by the **ZOOM**, **PAN**, or **SHADEMODE** command. This option remembers up to ten previous views.

Other commands with a previous option include **SELECT** and **UCS**. One command is itself a "previous" command: **LAYERP** restores a previous layer state.

Cancel Buttons

Dialog boxes have a button labelled **Cancel**. You click this button after making changes to the dialog box, and then deciding you don't want the changes to take place.

REDO AND MREDO

The **REDO** and **MREDO** (short for "multiple redo") commands are the antidotes to the U and UNDO commands: they reverse the undoing.

- The **REDO** command redoes the last undo; this is the "quick" command to use when just one or two changes need to be reversed.
- The **MREDO** command provides a couple of options for redoing the effects of undo; this "advanced" command is also available as a drop list on the toolbar.

The **REDO** command must be used immediately after one of the undo commands.

TUTORIAL: REDOING UNDOES

1. To undo the effect of an undo command, start the **REDO** command with one of these methods:
   - From the menu bar, choose **Edit**, and then **Redo**.
   - At the keyboard, press **CTRL+Y**.
   - At the 'Command:' prompt, enter the **redo** command.

   Command: **redo** *(Press ENTER.)*

2. AutoCAD redoes the previous undo, and then displays the name of the redone command:
   ARC
   When you redo back to the first undo, AutoCAD reports:
   Everything has been redone

REDOING UNDOES: ADDITIONAL METHODS

AutoCAD has several additional methods to reverse the effect of undo commands:

- **MREDO** command provides control over the undo process at the command line.
- **Standard** toolbar provides a list of redo-able actions.

Let's look at each method.

MRedo
(NEW IN 2004)

The **MREDO** command provides addition control over the redo process.

Command: **mredo**
Enter number of actions or [All/Last]: *(Enter a number or an option.)*

Enter the number of undone commands you want reversed.

All

The **All** option reverses all previous undo actions.

Last

The **Last** option reverses the last undo action only; this is like the **REDO** command.

Standard Toolbar

As an alternative to the redo commands, you can use the **Redo** drop list on the Standard toolbar. Click the arrow next to **Redo**, and you see a list of commands.

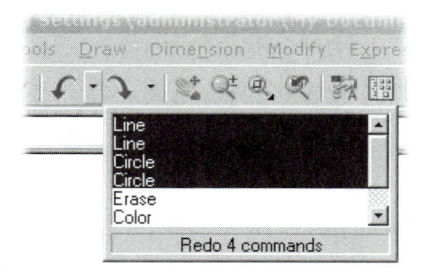

Move the cursor to the command to which you want to redo. You cannot pick and choose non-contiguous commands. When the **MRedo** button is gray, there are no actions to redo.

RENAME

The **RENAME** command changes the names of linetypes, layers, and so on. Very often, the related commands, such as **LINETYPE** and **LAYER**, include options for renaming them. The Rename dialog box, however, is a handy way for changing the names of many items at once.

If you are new to AutoCAD, you might not be familiar with the concept of *named objects*. Named objects are anything in the drawing that you name. These include:

| Blocks | Text styles | Dimension styles | Named user coordinate systems |
| Layers | Named views | Linetypes | Named viewports | Plot styles |

(The technical definition of a named object is any object that appears in the "tables" section of the drawing file.) Certain names cannot be changed, such as layer 0; these do not appear in the Rename dialog box. The Plot Styles option does not appear when plot styles are not enabled for the drawing.

TUTORIAL: RENAMING TABLES

1. To rename a table object, start the **RENAME** command with one of these methods:
   - From the menu bar, choose **Format**, and then **Rename**.
   - At the 'Command:' prompt, enter the **rename** command.
   - Alternatively, enter the **ren** alias.

   Command: **redo** *(Press* ENTER.*)*

2. In all cases, AutoCAD displays the Rename dialog box:

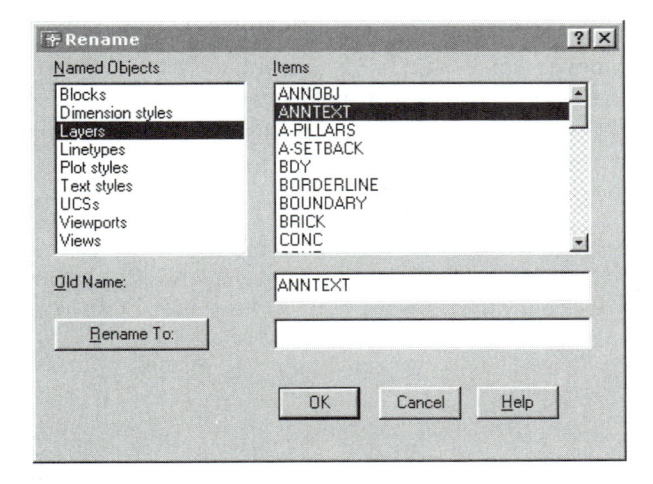

3. Under the **Named Objects** list, select a table name, such as Layers.

   Notice that AutoCAD displays a list of items for that name, such as all layer names.

   If there are no layers in the drawing that can be changed, the list remains blank.

4. Under **Items**, select a layer name.

   Notice it appears in the **Old Name** text box.

5. In the **Rename To** text box, enter a new name, and then click **Rename To**.

   Notice that the name changes in the list under Items.

6. Click **OK** to accept the changes.

RENAMING TABLES: ADDITIONAL METHODS

AutoCAD has one other method for renaming objects:

- **-RENAME** changes the names of tables at the command line.

Let's look at it.

-Rename

The -RENAME command provides addition control over the redo process.

   Command: **-rename**

   Enter object type to rename

   [Block/Dimstyle/LAyer/LType/Plotstyle/textStyle/Ucs/VIew/VPort]: *(Enter an option, such as* **la**.*)*

   Enter old layer name: *(Enter a valid layer name.)*

   Enter new layer name: *(Enter a new name for the layer.)*

You have to know the names of the table items to change ahead of time; when you enter the incorrect "old" name, AutoCAD complains:

   Cannot find layer "name".

When you enter names that cannot be changed, such as layer 0, AutoCAD complains:

   Cannot rename layer "0".

   Invalid

As with the dialog box, the **Plotstyle** option does not appear when plot styles have not been enabled for the drawing.

PURGE

The **PURGE** command removes unused layers, blocks, and so on from drawings.

Why would you want to remove named objects? The primary reason is that they clutter up drawings. In the course of creating a drawing, you sometimes create objects you never use, such as text styles, blocks, or dimension styles. For example, I find it better to load all linetypes, and then later purge the unused ones — this is a lot faster than loading linetype definitions one at a time as I need them.

In the old days of computing, when disk space was very limited, drafters purged drawings to reduce their file size. Today, with gigabytes of disk space being cheap like borsch, we no longer use that excuse. Instead, the clutter occurs in other ways, as in long lists of layer names.

You needn't worry that the **PURGE** command might erase items of importance. It's built-in safety guard prevents it from touching any named object in use. "In use" means, for example, layers with objects on them, or text styles used by text in the drawing.

In addition, **PURGE** does not touch named objects set as "current" and those found in a new, empty drawings:

Table Object	Names Untouched
Dimension Styles	Standard
Layers	0 and Defpoints
Linetypes	Continuous, Bylayer, and Byblock
Multilines	Standard
Blocks	Nested blocks and xrefs
Plot Styles	Normal
Text Styles	Standard

As an alternative to the **PURGE** command, you can remove unused named objects with the **Delete** button in certain dialog boxes. These include:

> **Layers** through the Layer Properties Manager dialog box (**LAYER** command).
>
> **Linetypes** through the Linetype Manger dialog box (**LINETYPE** command).
>
> **Styles** through the Text Style dialog box (**STYLE** command).

The Purge dialog box handles all these and, as an added bonus, deletes all unused named objects at once.

One last concept to discuss is the *nested purge*. This action is required for nested objects, such as nested blocks. That's where one block contains other blocks. The Purge dialog box removes nested items automatically, when that option is turned on.

> **Note:** You can use the **U** command to reverse the effect of the **PURGE** command, and return purged objects to the drawing.

TUTORIAL: PURGING UNUSED OBJECTS

1. To purge a drawing of unused objects, start the **PURGE** command with one of these methods:
   - From the **File** menu, choose **Drawing Utilities**, and then **Purge**.
   - At the 'Command:' prompt, enter the **purge** command.
   - Alternatively, enter the **pu** alias.

   Command: **u** *(Press* ENTER.*)*

2. In all cases, AutoCAD displays the Purge dialog box:

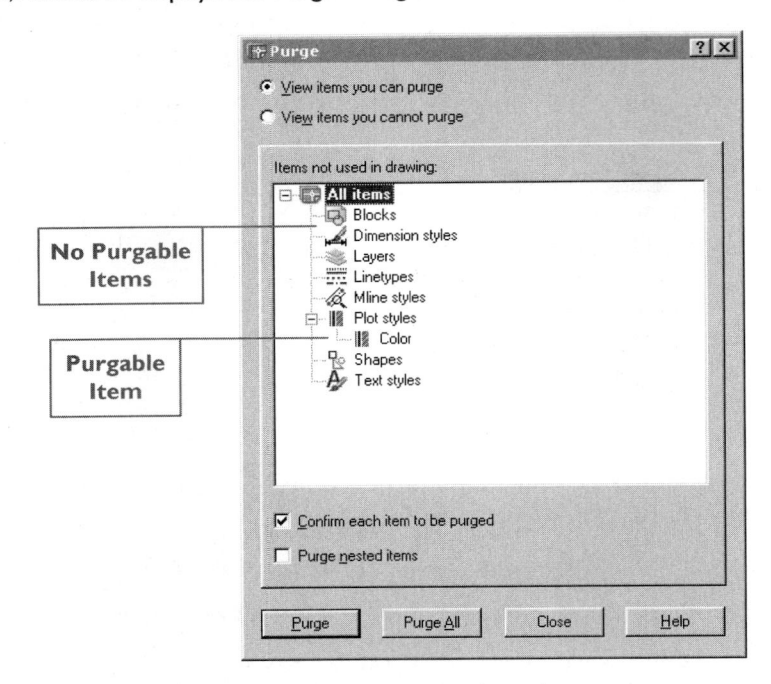

No Purgable Items

Purgable Item

3. Ensure the **View items you can purge** option is selected.
4. Look at the list under **Items not used in drawing**.
   Notice that some items have a plus sign next to them. This indicates there are purgable objects for that item.
   If there is no plus sign next to an item, there is nothing to purge.
5. Click the + sign.
   Notice that the list expands to show the purgable items.
6. Click **Purge All**.
   When the **Confirm each item to be purged** option is turned on, AutoCAD displays a dialog box asking your permission to purge:

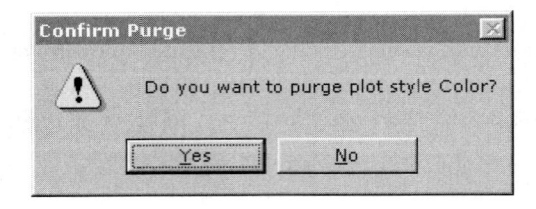

   Click **Yes**.
7. When done, click **Close**.

The dialog box's other radio button (the round buttons near the top of the dialog box) is labelled, "View items you cannot purge." It displays the inverse of the purgable list: named objects that cannot be purged, along with an explanation at the bottom of the dialog box. (You can right-click the explanation area, and copy the text to the Clipboard.)

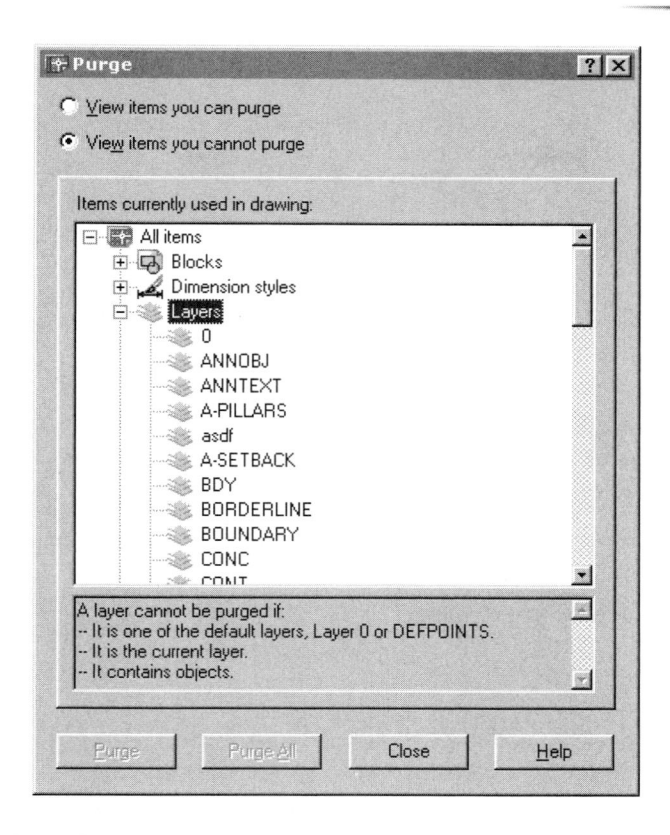

PURGING UNUSED OBJECTS: ADDITIONAL METHODS

AutoCAD has a couple of other methods for purging unused objects:

- **-PURGE** command executes the purge at the command line.
- **WBLOCK** command purges drawings when saved to disk.

-Purge

The **-PURGE** command provides AutoCAD's purge services at the command line. The fastest way to complete the purge process is to answer the prompts in this manner:

> Command: **-purge**
>
> Enter type of unused objects to purge
>
> [Blocks/Dimstyles/LAyers/LTypes/Plotstyles/SHapes/textSTyles/Mlinestyles/All]: **a**
>
> Enter name(s) to purge <*>: *(Press* ENTER.*)*
>
> Verify each name to be purged? [Yes/No] <Y>: **n**
>
> Deleting plotstyle "Color".
>
> 1 plotstyle deleted.

As an alternative, you can purge each object type and item individually. But if that's the case, it's better to use the dialog box.

WBlock

With the version of the PURGE command found in AutoCAD 2000 and earlier, you often needed to repeat the command, because AutoCAD would purge only one level of nesting at a time. Indeed, this was such a nuisance that power users employed a hidden feature of the WBLOCK command to cleanse the entire drawing. (With the Purge dialog box, the **Purge nested items** option cleans with nested objects automatically.)

To make WBLOCK purge drawings, you must use the **Entire drawing** option, and then save it to disk.

RECOVER AND AUDIT

The **RECOVER** command attempts to fix corrupted drawings, while **AUDIT** checks drawings for errors.

In a perfect world, everything works as planned. In our real world, however, we sometimes encounter difficulties. One day you will see a message from AutoCAD that reads:

> INTERNAL ERROR
>
> *(followed by a host of numbers)*

or

> FATAL ERROR.

This means that AutoCAD has encountered a problem severe enough that it cannot continue. You are usually given a choice of whether or not you wish to save the changes you have made since the last time you saved your work. The following message is displayed:

> AutoCAD cannot continue, but any changes to your drawing made up to the start of the last command can be saved.
>
> Do you want to save your changes? <Y>: **y**

If you enter "Y," AutoCAD attempts to write the changes to disk. If it is successful, AutoCAD displays the following message:

> DRAWING FILE SUCCESSFULLY SAVED

If the save is unsuccessful, the "INTERNAL ERROR" or "FATAL ERROR" messages are displayed again. When you see these ominous messages, you can wave good-bye to unsaved changes. In some severe cases, AutoCAD cannot recover the drawing, and reports:

> Unable to recover this drawing.

(Of course, as a good CAD operator, you save your work regularly, right? Recall the **SAVETIME** system variable discussed earlier in this book.) In AutoCAD 2004, Autodesk changed the drawing file format to do a better job at preventing corruption, and reduced the time between automatic saves from a dismal two hours down to a more reasonable ten minutes.

You can also attempt to recover damaged drawing files with the **RECOVER** command.

TUTORIAL: RECOVERING CORRUPT DRAWINGS

1. To attempt to recover a damaged drawing, start the **RECOVER** command with one of these methods:
   - From the **File** menu, choose **Drawing Utilities**, and then **Recover**.
   - At the 'Command:' prompt, enter the **recover** command.

   Command: **recover** *(Press* ENTER.*)*

2. AutoCAD displays the Select File dialog box.
   Select the drawing, and then click **Open**.

3. AutoCAD chatters as it works its way through the drawing, looking for errors:

   Drawing recovery.

   Drawing recovery log.

   Scanning completed.

   Validating objects in the handle table.

   Valid objects 519   Invalid objects 1

   Validating objects completed.

   Creating new ACAD_LAYOUT dictionary

```
Creating new ACAD_PLOTSTYLENAME dictionary
Creating new ACAD_PLOTSETTINGS dictionary
Creating new ACAD_MATERIAL dictionary
Creating new ACAD_COLOR dictionary16 error opening *Model_Space's layout.
Setting layout id to null.
16 error opening *Paper_Space's layout.
Setting layout id to null.
Used contingency data.
   Salvaged database from drawing.
Removed 2 unread objects from entity lists.
Opening a Release 13 format file.
 0     Blocks audited
Pass 1  491    objects audited
Pass 2  491    objects audited
Pass 3  500    objects audited
Total errors found 0 fixed 0
Regenerating model.
```

When AutoCAD detects a damaged drawing file when opened with the **OPEN** command, it performs the audit automatically. If the recovery is successful, the drawing is loaded; if not, the drawing is typically unrecoverable.

If the recovery is successful, and you save the drawing, you can load it normally the next time. When you exit the drawing without saving, the "repair" performed by AutoCAD is discarded.

RECOVERING CORRUPT DRAWINGS: ADDITIONAL METHODS

AutoCAD has one other methods for checking drawing integrity:

- **AUDIT** command executes the purge at the command line.

Audit

The **AUDIT** command looks for damage in drawings already loaded into AutoCAD. **RECOVER** checks drawings as they open; **AUDIT** checks them after they are open. This diagnostic tool also corrects damage to a file.

```
Command: audit
Fix any errors detected? [Yes/No] <N>: (Enter Y or N.)
```

If you answer "N" AutoCAD displays a report, but does not fix the errors.

When you answer "Y," AutoCAD displays the report and fixes errors.

In addition to providing the screen report, the **AUDIT** command can save the its report to a text file. This is controlled by the **AUDITCTL** system variable, where 0 means no file is produced (the default), and 1 means AutoCAD writes the *.adt* file. The file has the same name as the drawing, and is located in the same folder. You can read and print the audit file with any word processor.

FILE MANAGER

The file manager (or Explorer) included with Windows allows you to retrieve drawing backups. Whenever you ask AutoCAD to save the drawing, AutoCAD rename the previous *.dwg* file as a *.bak* file; when AutoCAD performs one of its automatic backups, a copy of the drawing is saved as a *.sv$* file. (This works, however, only if you asked AutoCAD to save backup files in the **Options** dialog box.)

Each time you use the **QSAVE** command, AutoCAD renames the current *.dwg* file with the *.bak* extension, and then saves the current state of the drawing as a new *.dwg* file.

When AutoCAD crashes, it tries to rename the current backup file so that it doesn't replace the previous backup file. AutoCAD renames it using the file extension of *.bk1*. If a file with such an extension already exists, it uses *.bk2* on through *.bkz*.

Should you need access to the previous version of the drawing, you can use Explorer to rename files with the *.bak* extension to the *.dwg* extension.

QUIT AND CLOSE

The **QUIT** command discards editing changes made to drawings, and exits AutoCAD. The **CLOSE** command does the same, but without exiting AutoCAD. If all else fails, quit the drawing, and AutoCAD discards all changes you've made.

TUTORIAL: DISCARDING DRAWINGS

1. To discard changes you've made to a drawing, start the **QUIT** command with one of these methods:
   - From the **File** menu, choose **Exit**.
   - At the keyboard, enter the **CTRL+Q** shortcut.
   - At the 'Command:' prompt, enter the **quit** command.
   - Alternatively, enter the **exit** alias.

   Command: **quit** *(Press* ENTER.*)*

2. In all cases, AutoCAD displays this dialog box.

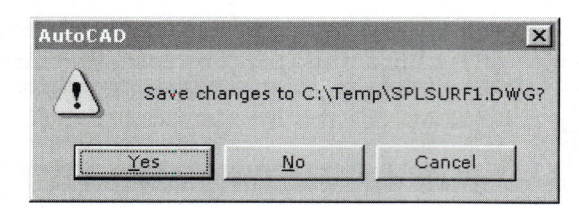

3. Click **No** to discard the changes, and exit AutoCAD.
   Click **Yes** if you've had a change of heart, and want to save the changes after all, and then exit AutoCAD.
   Click **Cancel** to *not* save changes *and* remain in AutoCAD.

To use the **CLOSE** command instead of **QUIT**, from the **Window** menu, select **Close**.

> **Notes:** To change a drawing, yet preserve it in its original form, use the **SAVEAS** command to save the changed drawing by another name, and then use **CLOSE** or **QUIT** to exit the drawing without saving changes.

EXERCISES

1. In this exercise, you work with the undo and redo commands.
   Start AutoCAD with a new drawing.
   Choose the **LINE** command, and then draw a line segment (press **ESC** to end the command).
   Enter **U** at the keyboard, and then press **ENTER**. Did the line segment disappear?

   Use the **LINE** command to draw several line segments.
   While the **LINE** command is still active, use its **Undo** option to remove one segment.
   End the **LINE** command, and then restart it.
   Draw more line segments.
   Use the **U** command to undo the lines drawn with the last **LINE** command. Which segments were undone?
   Press spacebar to repeat the **U** command. What happened?

   Use the **LINE** command to draw two line segments.
   Before entering the last point, press function key **F8** to turn on ortho mode. Notice the ORTHO button on the status line.
   Now use the U command to undo the sequence. Is the ortho mode on now?
   Press **F8** to turn on ortho mode.
   Now use the **LINE** command to draw a line segment.
   Next, use the **U** command to undo the sequence. Is ortho mode still turned on? What is the difference between this and the last sequence you performed?

   Use the **LINE** command to draw several line segments.
   Use the **U** command to undo the lines.
   Now use the **REDO** command. Did the lines reappear?
   Enter the **REDO** command again.
   What does the prompt line say? Why?

2. Use the **RECOVER** command to open the *splsurf1.dwg* drawing file (found on the CD-ROM). Read the report.
   How many invalid objects did AutoCAD find?

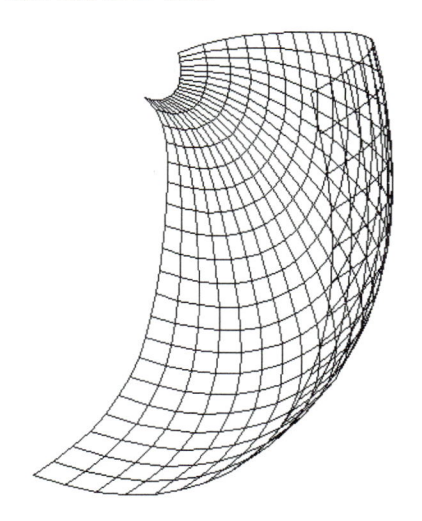

3. From the CD-ROM, open the *airport.dwg* file.
   Use the **PURGE** command to remove unused objects.
   Which items were removed?

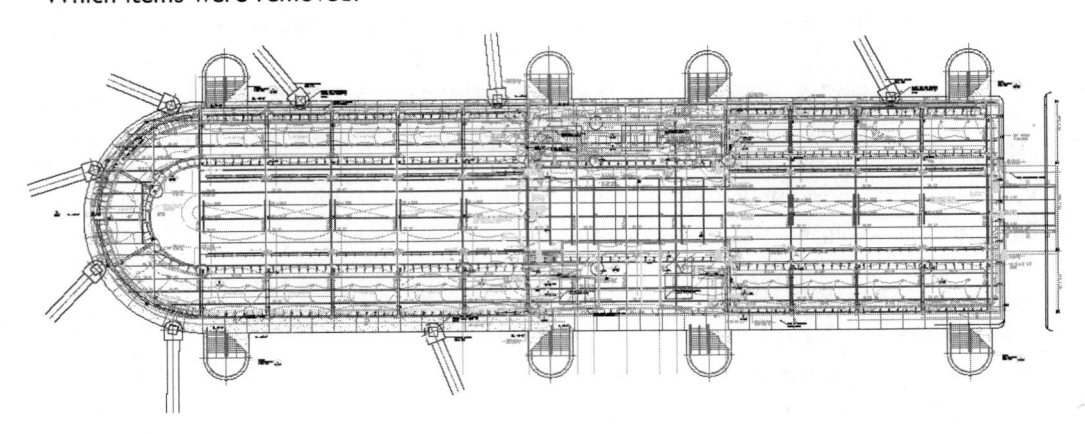

4. From the CD-ROM, copy the *forkift.bak* file to your computer's hard drive.
   Use Windows Explorer to rename the file to *forklift.dwg*.
   Open the drawing in AutoCAD.
   Did the drawing open normally?

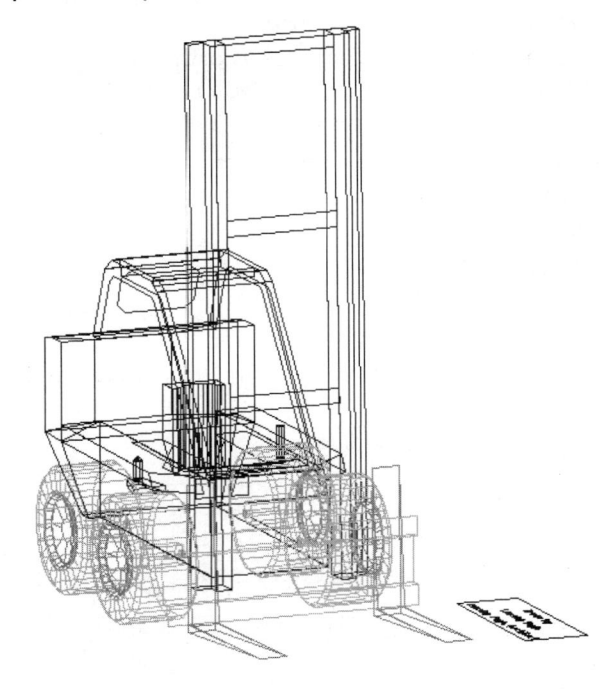

CHAPTER REVIEW

1. Can the **U** command be used to undo a sequence of commands?
2. Can the **U** command be entered from the menu bar as well as from the keyboard?
3. If you use the **Undo** option while in the LINE command, do all the segments disappear or is each segment stepped through backwards?
4. What is the AUDIT command used for?
   How does it differ from the RECOVER command?
5. What happens to the drawing if you use the UNDO command's **Back** option without setting any marks?
6. What is the keyboard shortcut for the following commands:

   UNDO

   REDO

   QUIT
7. What effect does the UNDO command have on the drawing after you use the SAVE command?
8. What happens to a block if you use the **U** command immediately after creating the block?
9. How far back can you use the undo commands?
10. Name a limitation to using the REDO command:
11. How does the MREDO command differ from the REDO command?
12. Describe the purpose of the RENAME command.
13. Can you rename layer 0?
14. Under what conditions are you unable to rename plot styles?
15. List two ways in which the PURGE command is useful:
    a.
    b.
16. When would a *nested purge* be required?
17. Name two alternatives to the PURGE command:
    a.
    b.
18. What does the message "Internal error" indicate?
19. Which command closes individual drawings?

CHAPTER 9

Direct Editing of Objects

AutoCAD traditionally used the command line for entering the names of commands and their options. That was the way computer software operated twenty years ago when AutoCAD was first developed. In the following two decades, programmers worked to make the user interface interactive: the first innovation was the menu bar, followed by the icon-bearing toolbar; more recently, programmers concentrated on getting software to allow direct manipulation of objects on the screen.

In Chapter 7 "Changing Object Properties" you gained experience in changing the properties of objects directly through the Properties toolbar. In this chapter, you learn to use AutoCAD's direct editing and drawing methods. They are:

Shortcut Menus provide direct access to appropriate commands.

Grips allow direct editing of objects.

Stretch, Move, Rotate, Scale, Copy, and **Mirror** comprise the grips editing options.

Direct Distance Entry allows "pen down" movement during commands.

TRACKING allows "pen up" movement during drawing and editing commands.

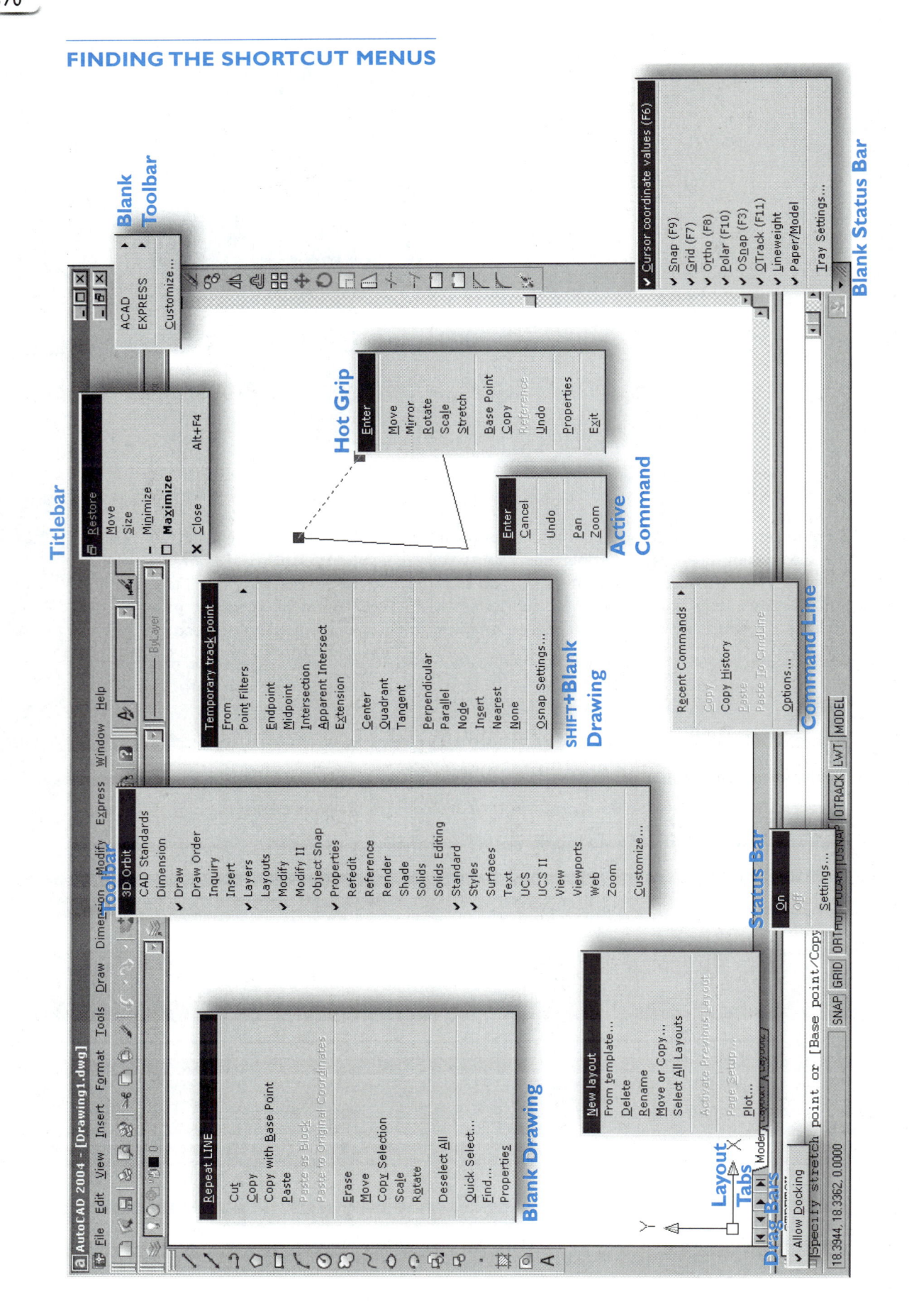

FINDING THE SHORTCUT MENUS

SHORTCUT MENUS

Shortcut menus provide direct access to commands appropriate to the context.

AutoCAD displays a shortcut menu almost every time you press the mouse's right button; these menus are also sometimes called "right-click menus" and "context menus." *Context* means that AutoCAD displays a different menu depending on where you press the right mouse button. The figure opposite illustrates the shortcut menus displayed in different areas of AutoCAD. On the top part of the window, there are shortcut menus for:

- Title bar
- Toolbars
- Blank toolbar area

In the drawing area, there are shortcut menus for:

- Blank drawing area
- **SHIFT**+blank drawing area
- Active commands
- Hot grips

And in the bottom part of AutoCAD, there are shortcut menus for:

- Layout tabs
- Command line
- Drag bar
- Status bar
- Blank status bar area

In addition, AutoCAD's text and other windows, such as Properties and DesignCenter have context-sensitive shortcut menus.

Commands available in shortcut menus are shown by black text. Unavailable commands are shown by gray text. Some commands have shortcut keystrokes associated with them; these are shown to the right of the command, such as Alt+F4.

A check mark next to a command means the option is turned on. Some shortcut menus have *submenus*, with further options; the presence of submenus is indicated by an arrowhead. Some commands display dialog boxes; this is indicated by the ellipsis (three dots ...).

TITLE BAR

Right-clicking the title bar displays the shortcut menu with commands for changing the AutoCAD window (the same menu is found with all other Windows software).

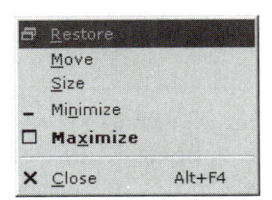

Restore windowizes AutoCAD, from either the maximized or minimized state.

Move moves the AutoCAD window within the computer monitor's screen.

Size changes the size of the AutoCAD window.

Minimize drops the AutoCAD window down to the Windows taskbar.

Maximize maximizes the AutoCAD window so that it fills the entire screen.

Close exits AutoCAD, after asking if you want to save changes for drawings.

TOOLBARS

Right-clicking the toolbar displays the shortcut menu with the names of toolbars available in AutoCAD. (Technically, the toolbar names are associated with a specific partial menu.) The check marks indicate the toolbars that are currently displayed. Select a toolbar name to toggle its display, on or off.

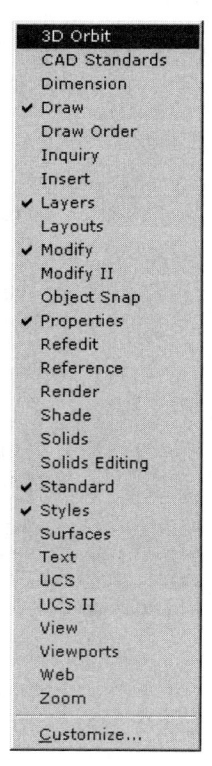

```
3D Orbit
CAD Standards
Dimension
✔ Draw
Draw Order
Inquiry
Insert
✔ Layers
Layouts
✔ Modify
Modify II
Object Snap
✔ Properties
Refedit
Reference
Render
Shade
Solids
Solids Editing
✔ Standard
✔ Styles
Surfaces
Text
UCS
UCS II
View
Viewports
Web
Zoom
Customize...
```

Customize displays the Customize dialog box.

BLANK TOOLBAR AREA

Right-clicking a blank part of the toolbar area displays the shortcut menu with the names of partial menus currently loaded. Each *partial menu* holds one or more toolbars, as well as items for the menu bar.

```
ACAD      ▶
EXPRESS   ▶
Customize...
```

The menu above shows two partial menus: Acad and Express. Acad is the standard menu provided with AutoCAD. Express is specific to the optional Express Tools package, which must be installed following the installation of AutoCAD 2004. AutoCAD allows you to write custom menus, which you can also load into AutoCAD. See *Using AutoCAD 2004: Advanced*.

Selecting the name of a partial menu displays a submenu of toolbar names.

DRAWING AREA

Right-clicking anywhere in the drawing area (while no command is active) displays the shortcut menu with the names of commands related to the Clipboard and viewing.

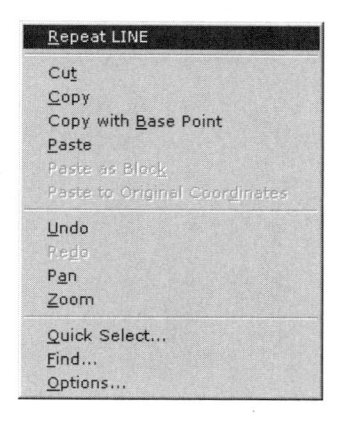

```
Repeat LINE
Cut
Copy
Copy with Base Point
Paste
Paste as Block
Paste to Original Coordinates

Undo
Redo
Pan
Zoom

Quick Select...
Find...
Options...
```

Repeat repeats the last command.

Cut cuts objects from the drawing, and sends them to the Clipboard.

Copy copies objects from the drawing, and sends them to the Clipboard.

Copy with Base Point copies objects to the Clipboard, after prompting for a base point.

Paste pastes objects from the Clipboard into the drawing.

Paste as Block pastes objects from the Clipboard in the drawing as a block; available only after using the Copy with Base Point command.

Paste to Original Coordinates pastes objects from the Clipboard into another drawing.

Undo reverses the effect of the last command.

Redo reverses the most-recent undo; available only when the previous command was U.

Pan enters real-time pan mode.

Zoom enters real-time zoom mode.

Quick Select displays the Quick Select dialog box for selecting objects based on their properties.

Find displays the Find and Replace dialog box for searching for text.

Options displays the Drafting tab of the Options dialog box.

SHIFT+BLANK DRAWING AREA

Holding down the **SHIFT** key (or the **CTRL** — it matters not) and then right-clicking anywhere in the drawing area displays the shortcut menu with the names of object snap modes and other drawing aids.

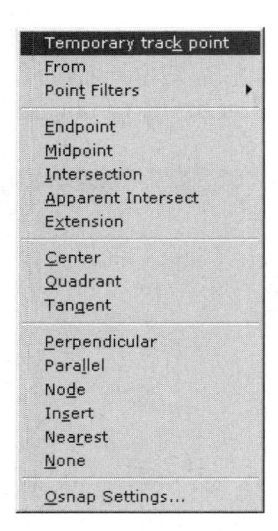

```
Temporary track point
From
Point Filters         ▶

Endpoint
Midpoint
Intersection
Apparent Intersect
Extension

Center
Quadrant
Tangent

Perpendicular
Parallel
Node
Insert
Nearest
None

Osnap Settings...
```

The object snaps listed in this shortcut menu are described fully in Chapter 5 "Drawing with Precision."

Point Filters displays a submenu of point filters, which are described in *Using AutoCAD 2004: Advanced*.

Osnap Settings displays the Object Snap tab of the Drafting Settings dialog box.

ACTIVE COMMANDS

Right-clicking during an active command displays a shortcut menu specific to the command, with options. The shortcut menu is different for every command, because each command has a different set of options.

In addition, the shortcut menu can vary during a command sequence, because different options are available at different times during the command. As an example, the figures below illustrate the two shortcut menus that appear during the LINE command:

Command: **line**

Specify first point: *(No shortcut menu available)*

Specify next point or [Undo]: *(Shortcut menu displays* **Undo** *option.)*

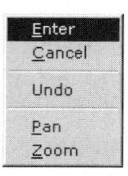

```
Enter
Cancel

Undo

Pan
Zoom
```

Specify next point or [Close/Undo]: *(Shortcut menu adds the* **Close** *option.)*

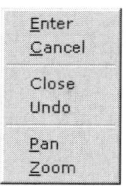

```
Enter
Cancel

Close
Undo

Pan
Zoom
```

Common to all active-command shortcut menus are these commands:

Enter is equivalent to pressing the ENTER key: it accepts default values, or ends the option or command, depending on the context.

Cancel is equivalent to pressing the ESC key: it cancels the command. In some cases, the command is halted; in other cases, the command ends repeating itself.

Pan and **Zoom** enter real-time pan and zoom modes.

HOT GRIPS

Right-clicking a hot grip displays the shortcut menu with the names of commands available during grip editing.

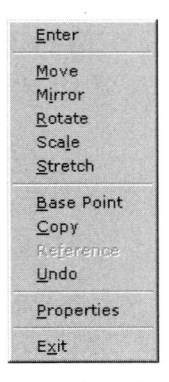

The commands are explained more fully later in this chapter.

LAYOUT TABS

Right-clicking any of the layout tabs displays the shortcut menu with the commands related to layouts and their tabs.

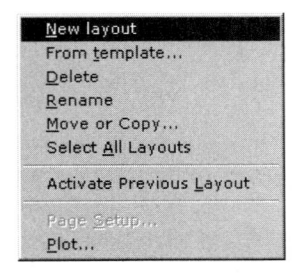

The commands are explained more fully in *Using AutoCAD 2004: Advanced*.

COMMAND LINE

Right-clicking the command line displays the shortcut menu with actions that affect the command line, as well as a submenu listing recently-used commands.

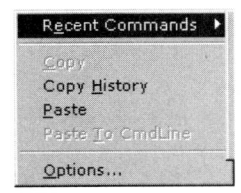

Recent Commands displays a submenu listing recently use commands. Select a command to re-execute it.

Copy copies selected text from the command line; this command is available only when text has been selected with the cursor.

Copy History copies all the text from the command line to the Clipboard.

Paste pastes text from the Clipboard into the command line. This command is available only when the Clipboard contains text. *Warning*: AutoCAD may react unintentionally to text pasted from another source.

Pasteto CmdLine pastes text to the command line, provided the Clipboard contains appropriate data.

Options displays the Display tab of the Options dialog box.

DRAG BAR

Right-clicking the drag bar at the end of the command-line window displays the shortest shortcut menu. The check mark indicates that the window will dock at the edge of AutoCAD, when dragged there.

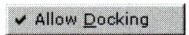

STATUS BAR

Right-clicking any button on the status bar displays shortcut menus specific to the button. Shown below is a typical menu:

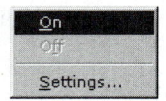

On turns on the button.

Off turns off the button.

Settings displays the appropriate dialog box.

BLANK STATUS BAR

Right-clicking a blank part of the status bar displays the shortcut menu with the names of the buttons available for the status bar. The check marks indicate the buttons that are currently displayed.

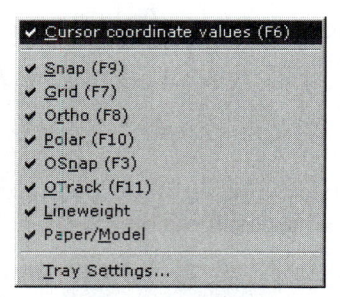

Tray Settings displays the Tray Settings dialog box.

TUTORIAL: CONTROLLING RIGHT-CLICK BEHAVIOR

1. To control the behavior of the right-click, display the Right Click Customization dialog box:
   - From the **Tools** menu, choose **Options**.
   - At the 'Command:' prompt, enter the **option** command.

- Alternatively, enter the **op** alias at the 'Command:' prompt.

    Command: **options (Press ENTER.)**

2. In all cases, AutoCAD displays the Options dialog box.
   Choose **User Preferences** tab, and then choose the **Right-click Customization** button. Notice the Right-click Customization dialog box.

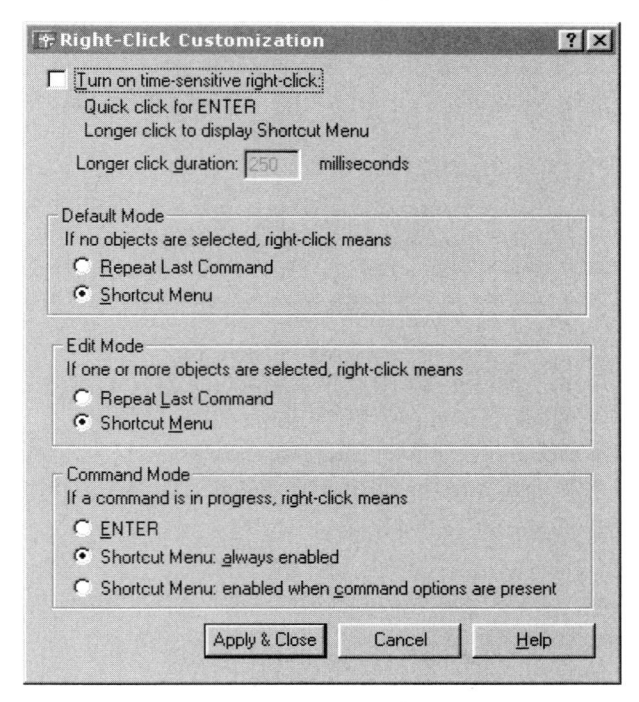

3. Make changes to the settings, and then click **Apply & Close**.
   Click **OK** to exit the Options dialog box.

The Right-Click Customization dialog box has the following options:

Turn on Time-Sensitive Right-Click

This option determines how AutoCAD reacts to right-click behavior; this option is normally turned off. When on, a short click acts like pressing the ENTER key. A long click displays the appropriate shortcut menu.

The difference between the short and long click is specified in milliseconds. The default value of 250 milliseconds is a quarter-second.

Default Mode

These two options determine the action when you right-click in the drawing area when no objects are selected and no commands are in progress.

Repeat Last Command repeats the last command.

Shortcut Menu displays the default shortcut menu (default).

Edit Mode

These two options determine the action when you right-click selected objects in grips mode (no command is active):

Repeat Last Command repeats the last command.

Shortcut Menu displays the edit (grips) shortcut menu (default).

Command Mode

These three options determine the action when you right-click in the drawing area while a command is active:

ENTER is the same as pressing **ENTER**; shortcut menus are disabled.

Shortcut Menu: Always Enabled displays shortcut menus applicable to the command in progress.

Shortcut Menu: Enabled When Command Options Are Present displays shortcut menus only when options are shown (enclosed in square brackets) on the command line. When the command displays no options as available, a right-click acts like pressing **ENTER** (default).

SHORTCUT MENUS: ADDITIONAL METHOD

The **SHORTCUTMENU** system variable determines whether the default, edit, and command-related shortcut menus are displayed in the drawing area.

ShortcutMenu	Meaning
0	Disables the default, edit, and command shortcut menus (this makes AutoCAD 2004 act like AutoCAD Release 14).
1	Displays shortcut menus when no command is active.
2	Displays shortcut menu for grips editing.
4	Displays shortcut menus during commands.
8	Displays shortcut menus during commands, but only when options are available at the command line.

To turn on more than one option, enter their sum. For example, the default is 11 (1 + 2 + 8):

    Command: **shortcutmenu**

    Enter new value for SHORTCUTMENU>: *(Enter a value, such as* **11**.*)*

GRIPS

Grips allow direct editing of objects.

In traditional AutoCAD usage, drafters enter an editing command, and then select the objects to edit (verb-noun). More recently, Autodesk added the reverse procedure: first select the objects, and then edit them (noun-verb). With no command active, you pick one or more objects. As feedback, AutoCAD highlights the selected objects by showing them with dashed lines, and displays *grips* (called "handles" in other software applications). You then edit the objects by manipulating their grips — until you press **ESC** to exit direct editing mode.

Grips are small, colored squares that indicate where the object can be edited. For example, lines have grips at both ends and at the midpoint; circles have grips at the centerpoint and at the four quadrant points.

A special case is blocks. Normally, AutoCAD displays a single grip at the block's insertion point. If you wish to display grips on the objects within the block, select the **Enable grips within blocks** check box in the Options dialog box's Selection tab.

Direct Editing of Objects 379

Circle

Arc

Dimension 7.0000

Ellipse

Line

Polyline

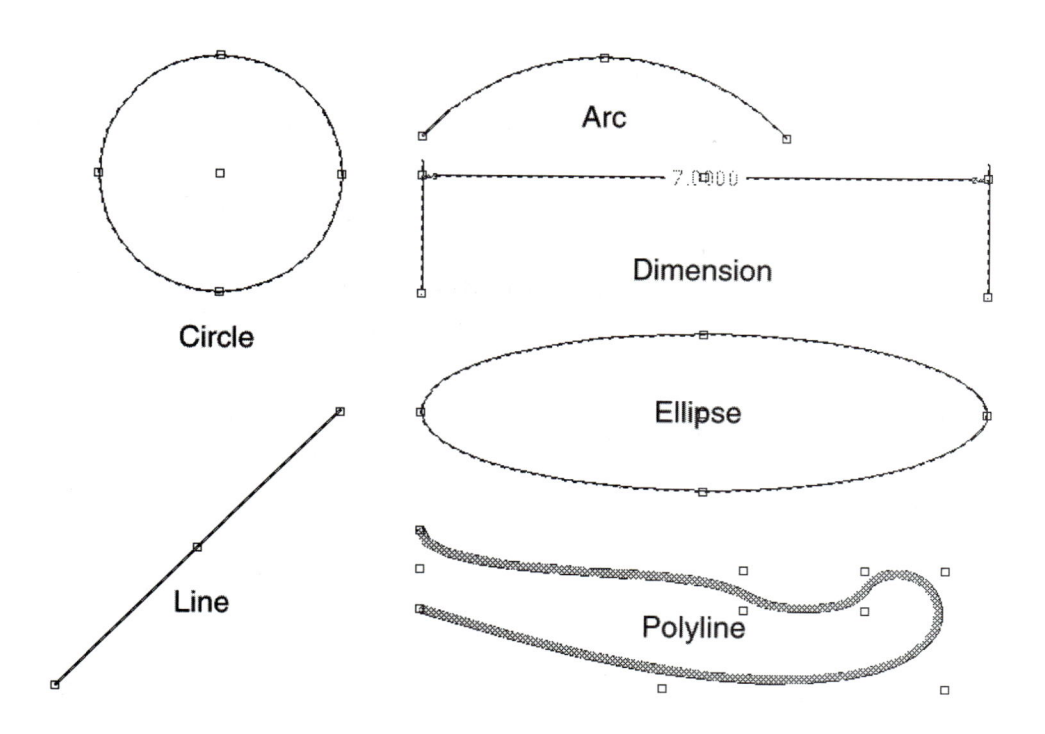

Note: For grip editing to work, the following criteria must be met:

Grip editing must be enabled. You can tell by the small selection box (called the *pickbox*) displayed at the intersection of the crosshairs when no command is active.

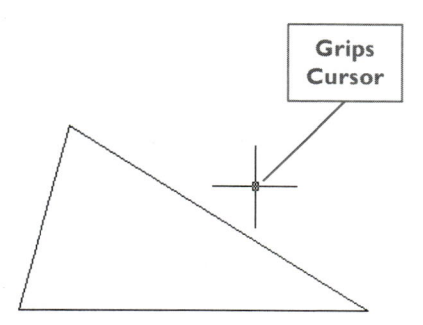

Grips Cursor

If not, go to the Selection tab of the Options dialog box, and then turn on these two options: **Enable Grips** and **Noun/Verb Selection**.

No commands can be active. You can tell by the command area being blank after the 'Command:' prompt.

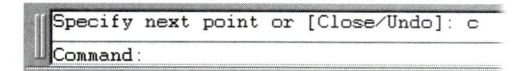

```
Specify next point or [Close/Undo]: c
Command:
```

If necessary, press **ESC** until the command area is blank.

GRIP COLOR AND SIZE

As a visual aid, grips are assigned colors and names:

Grip Color	Name	Meaning
Blue	Cold	Grip is not selected.
Green	Hover	Cursor is positioned over grip (new to AutoCAD 2004).
Red	Hot	Grip is selected, and object can be edited.

You can change the color and size of grips in the Selection tab of the Options dialog box. You can chose any color, but it makes sense to keep the ones AutoCAD assigned. The grip box size can range from 1 to 255 pixels.

TUTORIAL: DIRECT OBJECT SELECTION

1. Open a drawing in AutoCAD. Or, in a new drawing, place some objects.
2. To select objects in the drawing, pick them using one of the direct selection modes:
   - Selecting one object at a time:

     Command: *(Pick an object.)*

   Notice that AutoCAD highlights the object you selected, and displays one or more blue grips.

   To select additional objects, continue picking:

     Command: *(Pick another object.)*

   To "unselect" objects, hold down the **SHIFT** key while picking them. (If **PICKADD** mode is turned on, you need to hold down **SHIFT** to *add* objects.)

   - Selecting all objects in the drawing:

     Command: *(Press* CTRL+A.*)*

   AutoCAD highlights all objects in the drawing, with the exception of those on layers that are frozen or locked.

   - Selecting more than one object at a time:

     Command: *(Pick in a blank area of the drawing.)*

     Specify opposite corner: *(Move the cursor to form a selection window.)*

   Moving the cursor to the left forms a Crossing selection window (all objects within and crossing the window are selected), as illustrated below.

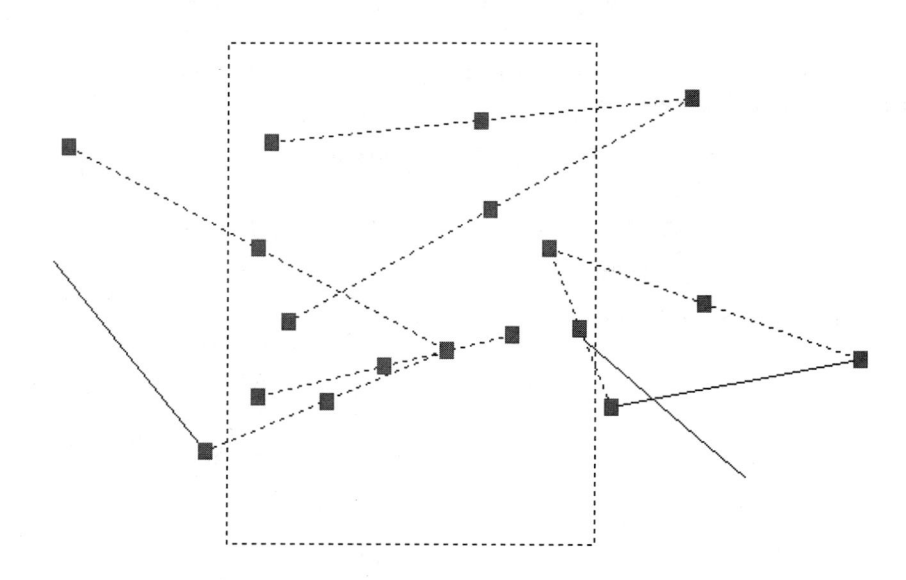

Moving to the right forms a Window selection window (only objects fully within the window are selected).

2. You can now edit the selected objects, as described in the sections following.
3. To unselect the objects, press **ESC** one or twice (until the highlighting and grips disappear).

GRIP EDITING OPTIONS

Stretch, **Move**, **Rotate**, **Scale**, **Copy**, and **Mirror** are the primary grip editing options.

After objects are selected, these options appear on the command line and the shortcut menu. In addition to these six, you can use other AutoCAD commands with grips: **ERASE**, **COPYCLIP**, **CUTCLIP**, and **PROPERTIES**. , as well **MOVE**, **SCALE** and **STRETCH** as commands. Other editing commands, such as **TRIM** and **OFFSET**, cannot be used with grips.

TUTORIAL: EDITING WITH GRIPS

1. Select any object in the drawing. The object is highlighted.

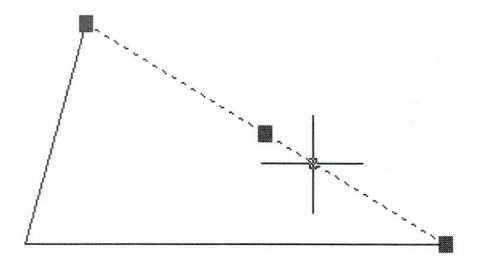

    Notice that one or more blue grips are placed on the object. (Blue grips are sometimes called "cold").

    You can to select as many objects as you wish by selecting more than once.

    If you find you cannot select more than one object, hold down the **SHIFT** key on the keyboard when selecting the grips.

2. Move the cursor over a blue grip.

    As the cursor gets close, notice that it jumps to the grip, and that the grip changes color to green (sometimes called "hover").

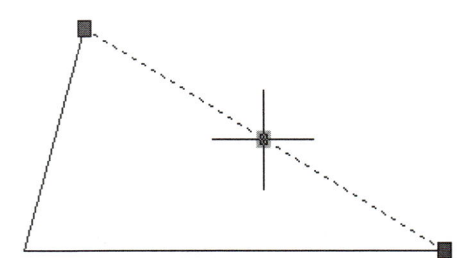

3. Click the green grip.

    Notice that the grip box changes color to red to denote selection (sometimes called "hot"). The hot grip is the base point from which some editing action, such as stretch and rotate, take place.

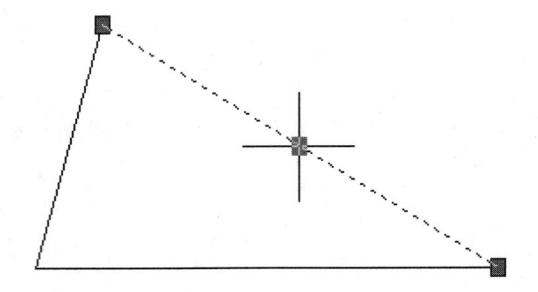

4. At the command line, AutoCAD prompts:

   ** STRETCH **

   Specify stretch point or [Base point/Copy/Undo/eXit]:

   Edit the grip as described below; right-click a red grip to see the editing options.
   Or, press the spacebar to see the other editing options: move, rotate, scale, and mirror.

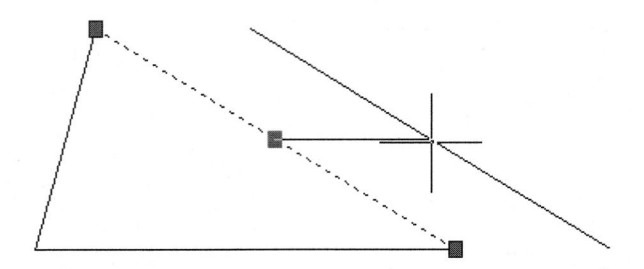

5. Press ESC once to clear the hot grip and selected objects.

GRIP EDITING MODES

When you select a single grip (so that it turns red), AutoCAD displays editing options on the command line. The selections cycle as you press **SPACEBAR** or **ENTER**. (Right-click to see a cursor menu listing the same options.)

   ** STRETCH **

   Specify stretch point or [Base point/Copy/Undo/ eXit]: *(Press spacebar.)*

   ** MOVE **

   Specify move point or [Base point/Copy/Undo/ eXit]: *(Press spacebar.)*

   ** ROTATE **

   Specify rotation angle or [Base point/Copy/Undo/Reference/eXit]: *(Press spacebar.)*

   ** SCALE **

   Specify scale factor or [Base point/Copy/Undo/Reference/eXit]: *(Press spacebar.)*

   ** MIRROR **

   Specify second point or [Base point/Copy/Undo/ eXit]: (Press spacebar.)

Press the spacebar until the desired edit option is listed, and then proceed with the command. For example, to resize the object, press the spacebar until **SCALE** appears, and then enter a scale factor at the "Specify scale factor:" prompt.

The selected grip point becomes the base point for editing. Alternatively, select one of the options listed on the command line. The commands and their options are covered later in this section.

GRIP EDITING OPTIONS

Each grip editing option has a set of suboptions in common. These are:

Base point

Enter **B** at the command line to specify base point other than the hot grip. AutoCAD prompts you:

> Specify base point: *(Pick a point.)*

The point you pick becomes the new base point.

Copy

Enter **C** to copy the selected object, leaving the original intact. The prompt changes to:

> ** STRETCH (multiple) **
>
> Specify stretch point or [Base point/Copy/Undo/eXit]: *(Pick a point.)*

The selected object is copied to the point you pick. The prompt repeats itself so that you can make multiple copies, as indicated by the word "(multiple)." Press **ENTER** or **ESC** to exit the command.

Undo

Enter **U** to undo the last operation.

Reference

Enter **R** to provide a reference for the **Scale** and **Rotate** modes. For Scale mode, AutoCAD prompts:

> Specify reference length <1.0000>: *(Enter a length, or pick two points.)*

Sometimes the reference option can be difficult to understand. Here is an example. Suppose you have an object that is 6 units in length, but you wish to resize it to 24 units in length. Select the **Reference** option, and then enter original length, as follows:

> Specify scale factor or [Base point/Copy/Undo/Reference/eXit]: **r**
>
> Reference length <1.0000>: **6**

AutoCAD then prompts for the new length.

> <New length>/Base point/Copy/Undo/Reference/eXit: **24**

AutoCAD calculates the scale factor, and then resizes the object.

For **Rotate** mode, AutoCAD prompts:

> Specify reference angle <0>: *(Enter an angle, or pick two points.)*

The reference angle is the angle at which the object is currently rotated. AutoCAD next prompts for the "New angle." The new angle is the angle you want to rotate the object.

eXit

Enter **X** or press **ESC** to exit grip editing.

GRIP EDITING OPTIONS

Let's look at each grip editing option.

Stretch

Stretch functions like the **STRETCH** command, except that AutoCAD uses the hot grip to determine the stretch results. Because only one object can have a hot grip, only one object at a time can be stretched or moved — unless two lines share a hot grip, as illustrated on the next page.

> ** STRETCH **
>
> Specify stretch point or [Base point/Copy/Undo/eXit]: *(Pick a point, or enter an option.)*

When you select the end grip of a line or arc, or quad grip of a circle, the object is *stretched*. Lines and arcs stretch longer or shorter, while circles change their diameter. The figure illustrates stretching these objects. (For clarity, the hot grips are shown as black squares, and the cold grips as white.) The triangle is being stretched smaller, while the circle is being stretch larger. The triangle is made of lines, so AutoCAD is stretching two lines at one, because of their common endpoints. The circle is not being stretched into an ellipse, as you might think from the word "stretch." Instead, its diameter is being stretched.

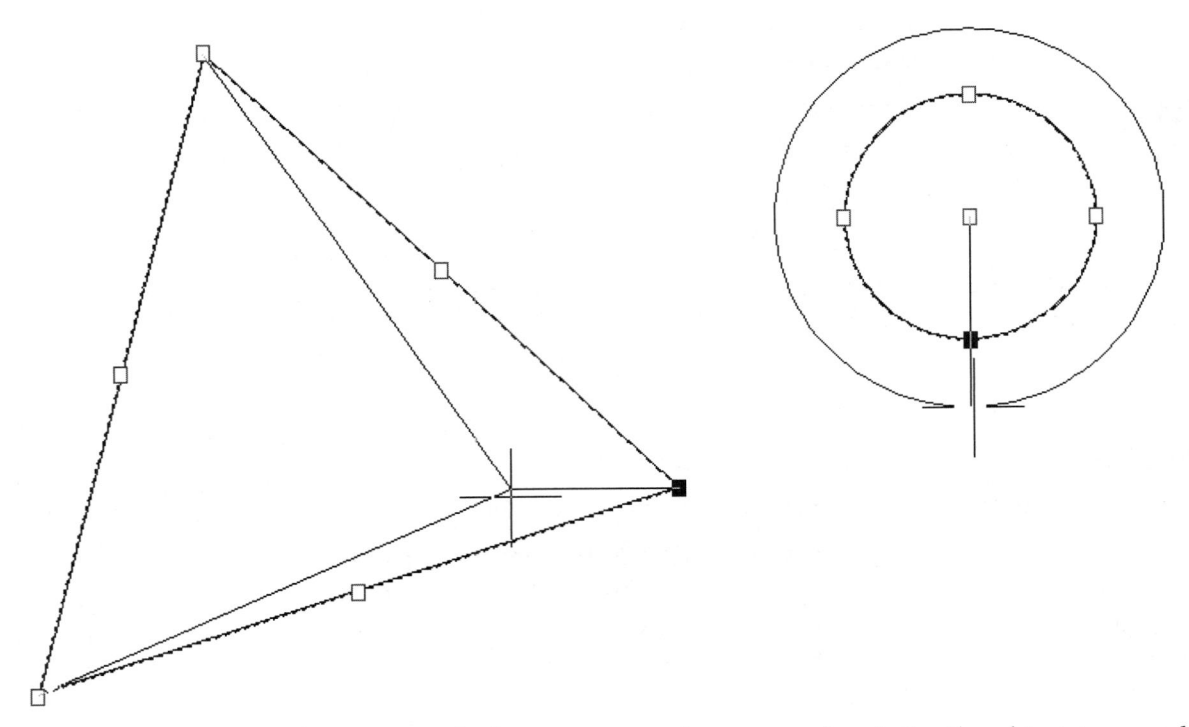

When you select the midpoint grip of a line, or arc or the center of a circle, the object is *moved*, but not stretched. The figure illustrates of the **Stretch** mode moving the selected objects. (Again, for clarity, the hot grips are shown as black squares, and the cold grips as white.) One line of the triangle is being moved, because the grip is at the center of line. The entire circle is being moved.

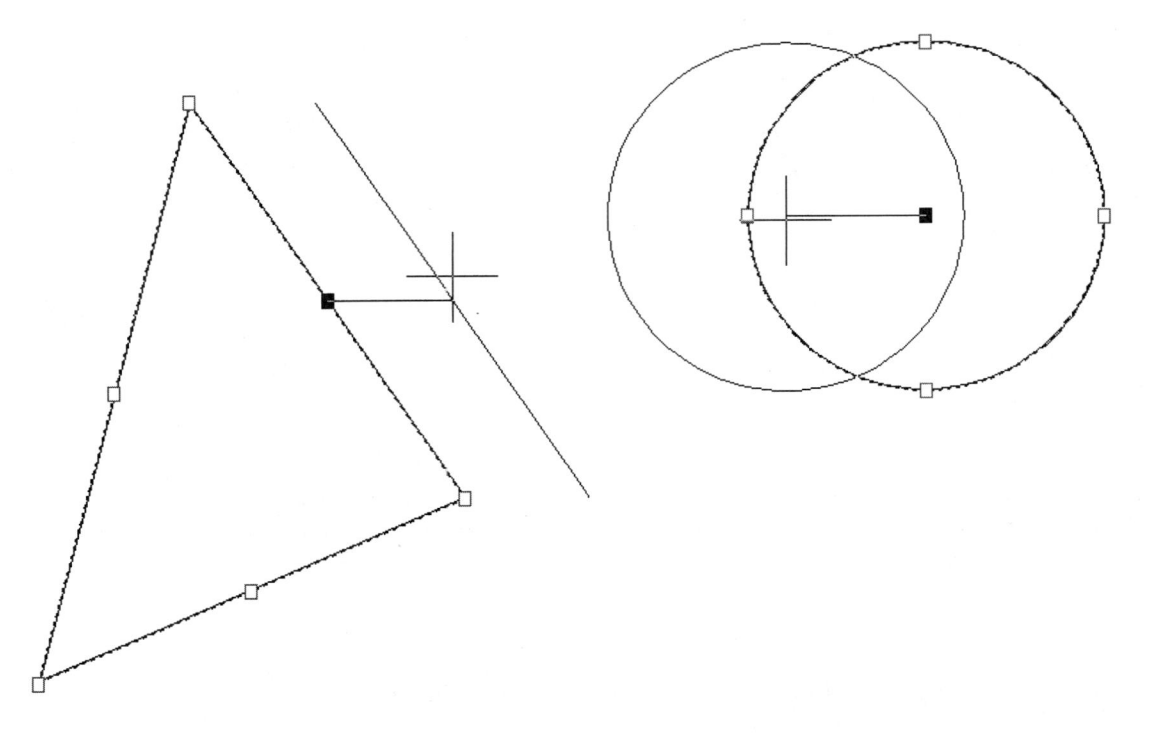

Move

Move moves the selected objects. All selected objects are moved, relative to the hot grip.

> ** MOVE **
>
> Specify move point or [Base point/Copy/Undo/eXit]: *(Pick a point, or enter an option.)*

Objects move by the distance from the base point to the point you pick. Recall that the base point is either the hot grip, or another point specified by the **Base point** option. You can either pick a point, or enter absolute or relative coordinates to specify the distance and direction of the move. For example, enter relative coordinates that specify a distance of 50 units at an angle of 45 degrees from the x-axis:

> Specify move point or [Base point/Copy/Undo/eXit]: **@50<45**

Rotate Mode

The **Rotate** option rotates the selected objects around the base point. All selected objects are rotated, relative to the hot grip.

> ** ROTATE **
>
> Specify rotation angle or [Base point/Copy/Undo/ Reference/eXit]: *(Pick a point, or enter an option.)*

Unless you use the **Base point** option to position the base point at a location other than a grip, the rotation occurs around the hot grip.

When rotating objects, you can dynamically set the rotation angle by moving the crosshairs, or by specifying a rotation in degrees.

Scale Mode

If you wish to resize the selected objects, cycle through the mode list until **Scale** mode is listed on the command line. All selected objects are resized, relative to the hot grip.

> .** SCALE **
>
> Specify scale factor or [Base point/Copy/Undo/ Reference/eXit]: *(Pick a point, or enter an option.)*

Scale mode resizes the selected object(s). The hot grip serves as the base point for resizing. Dynamically resize the objects by moving the crosshairs away from the base grip.

You can also resize the selected objects by entering a scale factor. A scale factor greater than one enlarges the object by that multiple. For example, a scale factor of 2.0 makes an object twice the size. Entering a decimal scale factor results in an object that is smaller than the original. For example, entering a scale factor of 0.5 shrinks the object to one-half the original size.

Mirror Mode

Mirror mode mirrors the selected object(s). All selected objects are mirrored relative to the hot grip. Objects are moved (not copied) about a *mirror* line created by the cursor. The first point of the mirror line is established by the hot grip (unless you change it with the **Base point** option.) AutoCAD prompts you for the second point of the mirror line:

> ** MIRROR **
>
> Specify second point or [Base point/Copy/Undo/ eXit]: *(Pick a point, or enter an option.)*

The figure illustrates the mirror line stretching between the hot grip and the crosshair cursor.

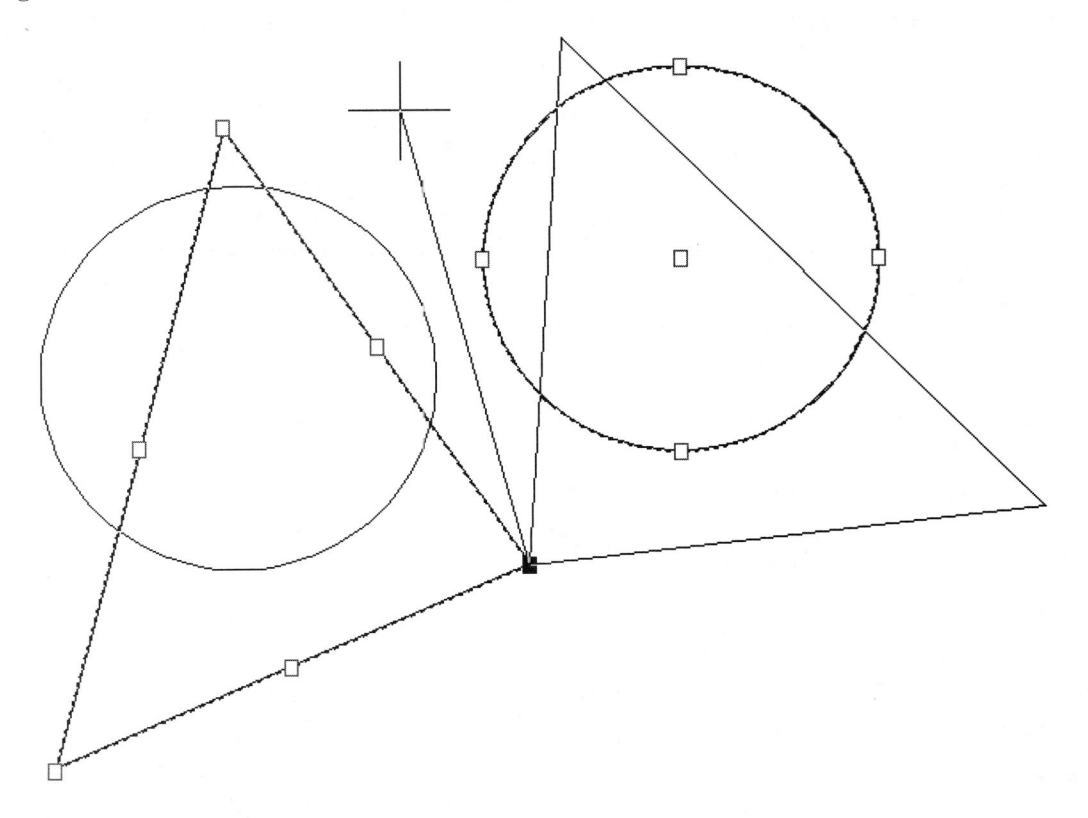

OTHER GRIPS COMMANDS

In addition to the commands and options listed at the command prompt, you can also use these commands when objects are highlighted with grips.

ERASE deletes the gripped objects from the drawing. As an alternative, you can press the **DEL** key. *Caution*: selecting all objects and then pressing **DEL** erases the entire drawing! Use **U** to recover.

COPYCLIP copies the gripped objects to the Clipboard; **CUTCLIP** cuts the objects. As alternatives, you can press **CTLR+C** and **CTRL+X**, respectively.

PROPERTIES displays the Properties window for the gripped objects.

MATCHPROP copies the properties of one gripped object; if more than one object is gripped, AutoCAD complains, "Only one entity can be selected as source object."

ARRAY creates rectangular and polar arrays of the gripped objects.

BLOCK creates a block of gripped objects.

EXPLODE explodes gripped objects; if they cannot be exploded, AutoCAD complains, "1 was not able to be exploded."

DIRECT DISTANCE ENTRY

Direct distance entry allows "pen down" movement during commands.

Direct distance entry is an alternative to entering polar or relative coordinates that rely on a distance and an angle. To show the angle, you move the mouse; then you type the distance. With ortho or polar modes turned on, direct distance entry is an efficient way to draw lines.

Direct distance entry can be used any time a command prompts you to specify a point, such as "Specify next point:".

TUTORIAL: DIRECT DISTANCE ENTRY

1. To draw a line 10 units long using direct distance entry, enter a drawing or editing command:

    Command: **line**

2. Before you can use direct distance entry, you must pick an initial point:

    Specify first point: *(Pick a point.)*

3. At the next prompt, move the mouse to indicate the angle, and then enter the distance at the keyboard:

    Specify next point: *(Move the mouse, and you see a rubber band line. At the keyboard, type* **10** *and press* ENTER.*)*

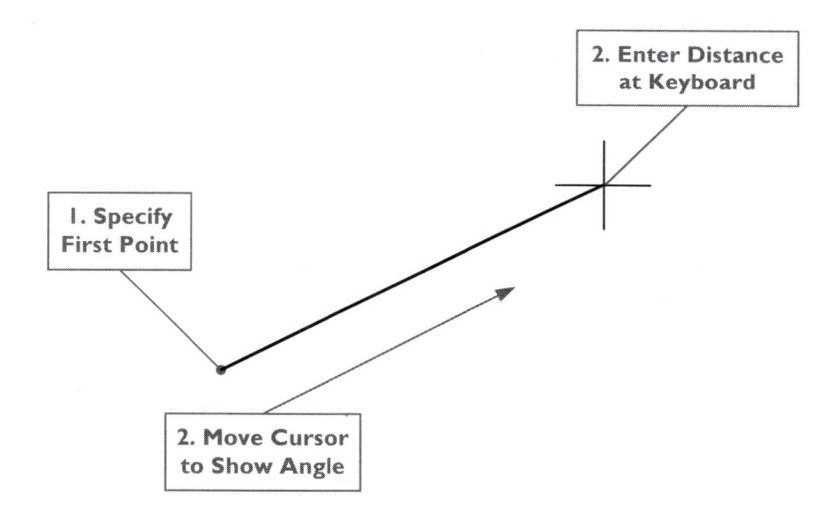

4. End the command:

    Specify next point: *(Press* ENTER*)*

AutoCAD draws a line segment ten units long in the direction you move the mouse. You can use direct distance entry for drawing polylines, arcs, multilines, and most other objects. Direct distance entry does not make sense with some drawing commands, like **CIRCLE** and **DONUT**.

Additionally, you can use direct distance entry with editing commands, such as **MOVE**, **STRETCH**, and **COPY**.

TRACKING

The Tracking modifier allows "pen up" movement during drawing and editing commands.

Whereas direct distance entry lets you draw relative distances, tracking lets you move relative distances within a command. Tracking is not a command, but a command option; you can use tracking only within another command.

Upon entering tracking mode, AutoCAD automatically switches to ortho mode, and then prompts for the "First tracking point:". (If polar mode is on, AutoCAD switches it off and turns on ortho mode.) Once you exit tracking mode, AutoCAD changes ortho and polar back to their original states.

Some commands keep prompting you for additional points, such as the **LINE** and **PLINE** commands. With these, you can go in and out of tracking mode as often as you like.

TUTORIAL: TRACKING

1. To employ tracking, enter a drawing or editing command:
   Command: **line**

2. At a "Specify first point:" or "Specify next point:" prompt, enter **tracking**, **track**, or **tk**.
   Specify first point: **tk**

3. AutoCAD enters tracking mode. Notice that ortho mode is turned on. (Look at the ORTHO button on the status bar.)
   Move the mouse a distance, and then click.
   First tracking point: *(Move the mouse and click.)*

4. This time, move the mouse, and then enter a distance.
   Next point (Press Enter to end tracking): *(Move the mouse in another direction, and then enter a distance, such as **2**.)*

5. Press **ENTER** to exit tracking mode, and resume drawing with the **LINE** command.
   If ortho mode was off when you started the line command, AutoCAD switches it off automatically upon exiting tracking mode.
   Next point (Press Enter to end tracking): *(Press* **ENTER**.*)*
   To point: **2,3**
   To point: **10,5**

6. Switch back to tracking mode, and enter distances as absolute x,y coordinates:
   To point: **tk**
   First tracking point: **5,10**

7. Exit tracking mode, draw another line, and exit the **LINE** command.
   Next point (Press Enter to end tracking): *(Press* **ENTER**.*)*
   To point: **3,2**
   To point: *(Press* **ENTER**.*)*

If you want tracking to move in angles other than 90 degrees, use the **Rotate** option of the **SNAP** command to change the angle. Since the **SNAP** command is transparent, you can change the tracking angle in the middle of tracking, as follows:

   Command: **line**
   Specify first point: *(Pick.)*
   Specify next point: **tk**

First tracking point: *(Move cursor and pick a point.)*
Next point (Press Enter to end tracking): **'snap**
>>Specify snap spacing or [ON/OFF/Aspect/Rotate/Style/Type] <0.5000>: **r**
>>Specify base point <0.0000,0.0000>: *(Press ENTER to accept default.)*
>>Specify rotation angle <0>: **45**
Resuming LINE command.
Next point (Press Enter to end tracking): *(Move cursor and pick a point.)*
Next point (Press Enter to end tracking): *(Press ENTER to end tracking.)*
Specify next point: *(Pick a point.)*
Specify next point: *(Press ENTER to exit command)*

Note: Tracking normally assumes you want to switch direction each time you use it. For example, if you first move north (or south), AutoCAD assumes you next want to move east (or west).

It is not easy, however, to back up or track forward in the same direction. For example, if you track north and then want to track north some more, you find the cursor wanting to move east or west, but not north or south.

To make the tracking cursor continue in the same direction, you move it back to its most recent starting point. Then you can move in the direction you want.

EXERCISES

1. In the following exercise, you practice using shortcut menus to control AutoCAD and execute commands.

   Right-click a blank part of the status bar, and then turn off the **PAPER/MODEL** button. Does the button disappear? Repeat to return the button.

   Right-click the **ORTHO** button, and then select **Settings**. Which dialog box appears? Click **OK** to dismiss the dialog box.

   Right-click the **LWT** button, and then select **Settings**. Does a different dialog box appear? Click **OK** to dismiss the dialog box.

   Draw a circle. When you are done, right-click the drawing area. Do you see **Repeat CIRCLE** at the top of the shortcut menu?

   Right-click again, and select **Undo**. Does the circle disappear?

2. In this exercise, you use grips to edit objects in the drawing.
   From the CD-ROM, open the *17_35.dwg* file, a drawing of an angle.

   Select a circle, and then click on the center grip.
   Drag the grip to another place in the drawing. Does the circle move?

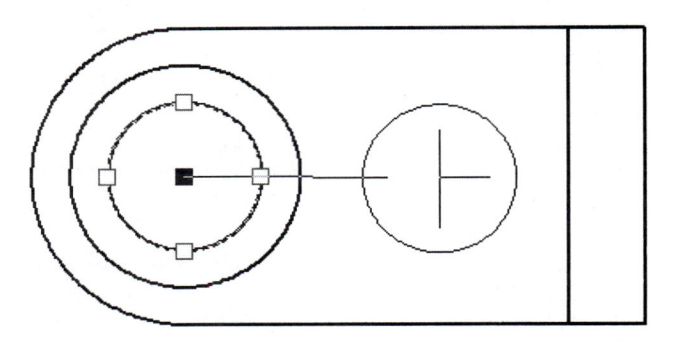

   Now click one of the four quadrant grips.
   Drag the grip. Does the circle become larger?

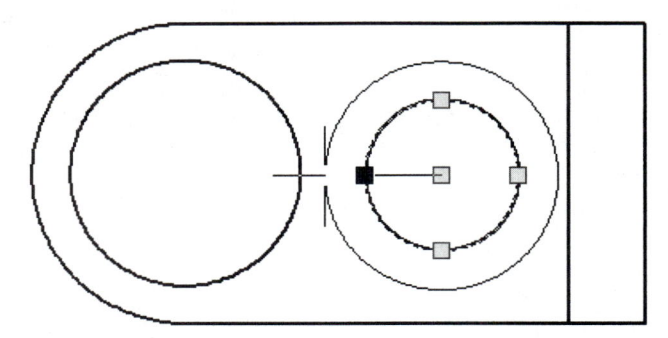

   Press **ESC** to remove the grips.

Select the entire drawing, and then click on any grip.
Press the spacebar to see ** MOVE ** on the command line.
Drag the grip. Do all parts of the drawing move?

Press the spacebar again to see ** ROTATE ** on the command line.
Drag the grip. Does the drawing rotate around the grip?

Enter **b** to access the base point option:
   Specify base point: *(Pick another point.)*
Move the cursor. Does the drawing now rotate about the new point?

Press the spacebar again to see ** SCALE ** on the command line.
Drag the grip. Does the drawing change its size?

Press the spacebar again to see ** MIRROR ** on the command line.
Drag the grip. Do you see a mirrored copy?

3. Use the **LINE** command with direct distance entry to draw the object defined by the following relative coordinates:
   - Point 1: 0,0
   - Point 2: @3,0
   - Point 3: @0,1
   - Point 4: @–2,0
   - Point 5: @0,2
   - Point 6: @–1,0
   - Point 7: 0,0

4. Use the **PLINE** command with direct distance entry to draw the object defined by the following relative coordinates:
   - Point 1: 0,0
   - Point 2: @4<0
   - Point 3: @4<90
   - Point 4: @4<180
   - Point 4: @4<270

5. Use direct distance entry and tracking to draw the following fuse link. Each square represents one unit.

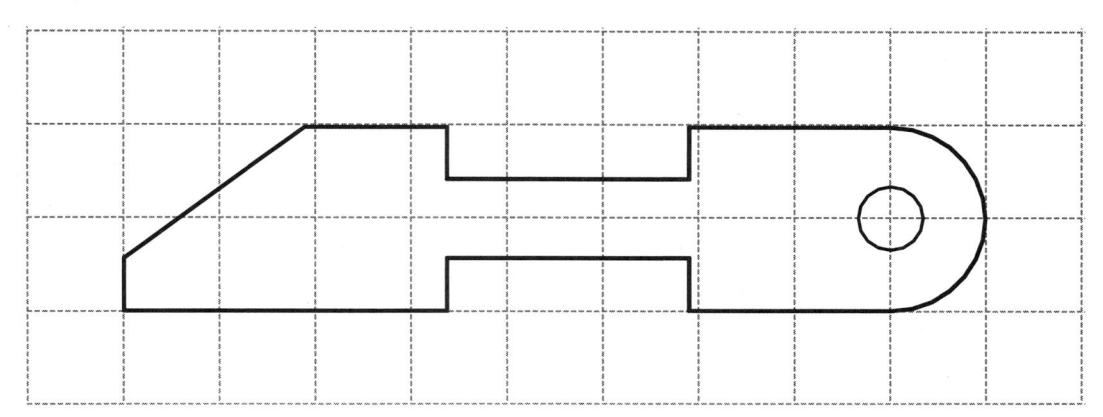

6. Use direct distance entry and tracking to draw the following cylinder. Each square represents 2 units.

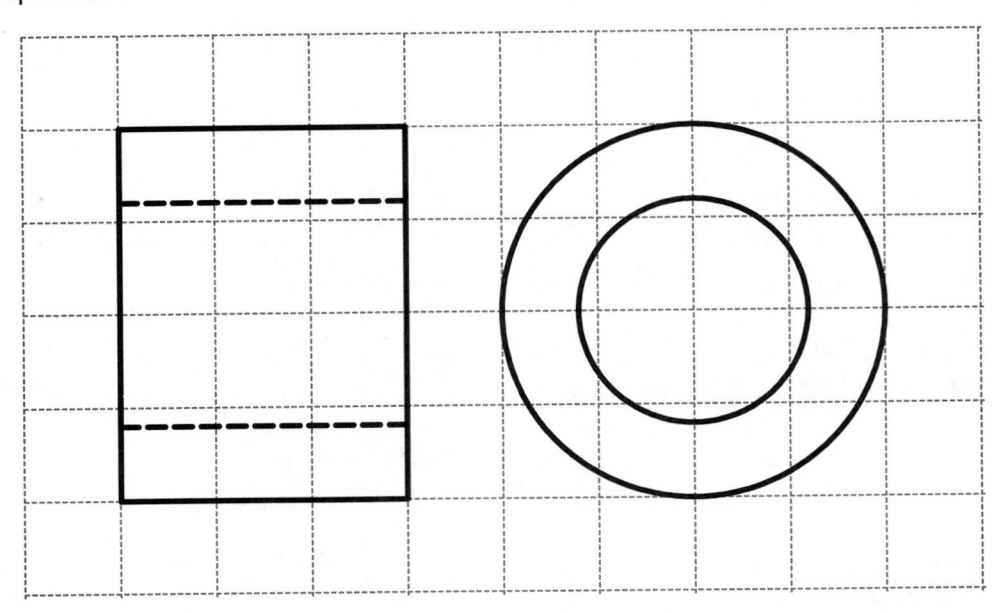

7. Use direct distance entry and tracking to draw the following baseplate. Each square represents a half-unit.

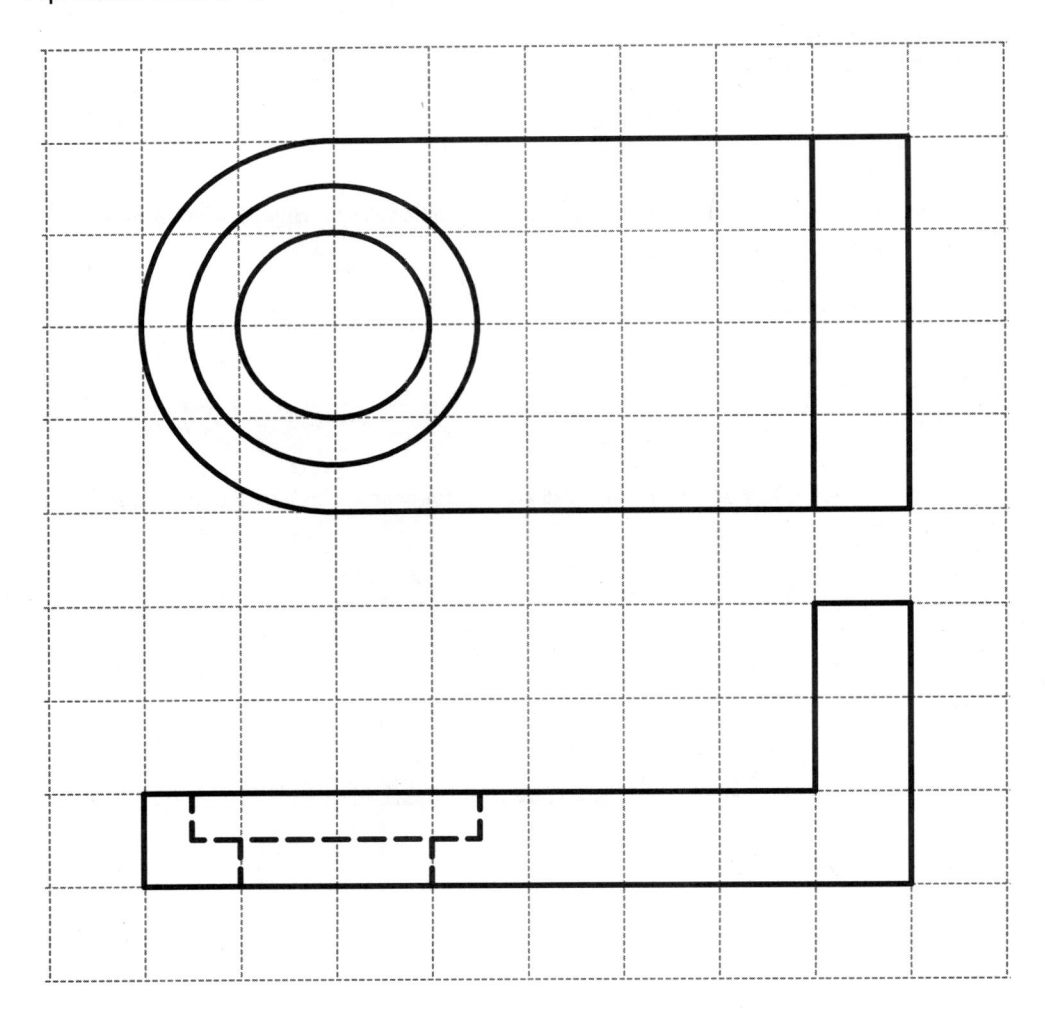

8. Use direct distance entry and tracking to draw the following wrench. Each square represents one unit.

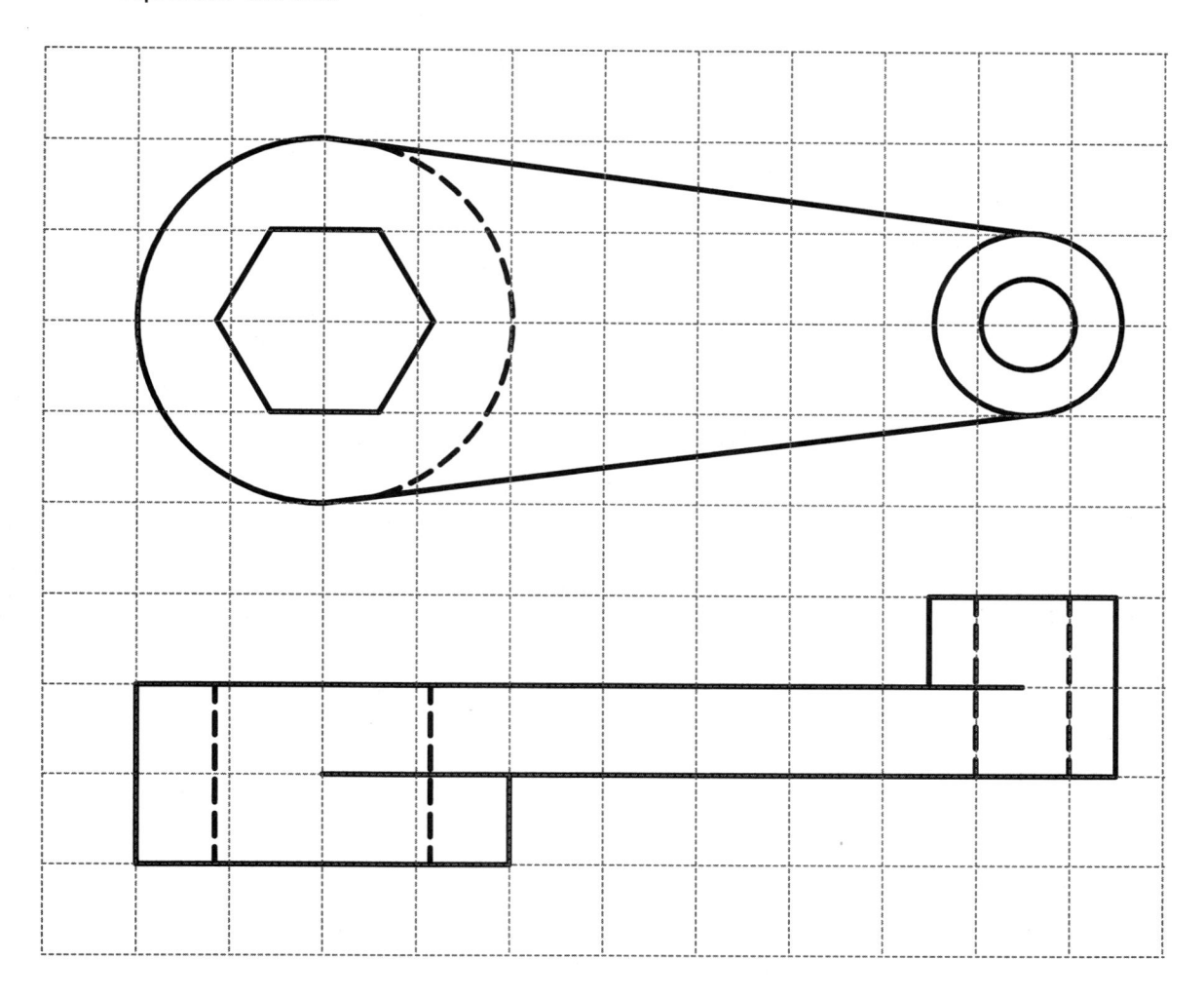

CHAPTER REVIEW

1. What is the difference between a hot and a cold grip?
2. What is the default colors for the following grips:
   Hot:
   Cold:
   Hover:
3. Name three editing operations you can perform with grips.
4. How do you deselect gripped objects?
5. What is the primary difference between tracking and direct distance entry?
6. How do you access shortcut menus?
7. Which shortcut menu do you see when holding down the **SHIFT** key in a blank part of the drawing?
8. Is the same shortcut menu shown during all commands?
9. What is the primary purpose of grips editing?
10. Name two conditions under which grips editing can take place:
    a.
    b.
11. Where are grips located on a line?
    On a circle?
    On a block?
12. Can more than one object be selected for simultaneous grips editing?
13. Using grips, can more than one object be stretched at a time?
14. Describe the purpose of the **Base point** option in grips editing.
15. What happens when you drag a circle by its centerpoint grip?
    What happens when you drag a circle by its quad grip?
16. Can direct distance entry be used with editing commands?
17. What is the alias for tracking?
18. Name the mode that AutoCAD automatically turns on when you enter tracking mode?
19. Other than the standard six grips editing commands, name three other editing commands that work with gripped objects:
    a.
    b.
    c.
20. Can you use the **BLOCK** command with gripped objects?

CHAPTER 10

Constructing Objects

Drawing the same objects over and over was a tedious part of hand drafting. In Chapter 7 "Drawing with Efficiency," you learned to insert blocks as one method for quickly placing parts in a drawing. In some circumstances, blocks are not the best method. This chapter introduces other means of making copies — mirrored copies, parallel offsets, and arrays of copies. After completing this chapter, you will be able to use AutoCAD's commands for constructing objects from existing objects. They are:

COPY makes one or more identical copies of objects.

MIRROR make mirrored copies.

MIRRORTXT determines whether text is mirrored

OFFSET makes parallel copies.

OFFSETDIST and **OFFSETGAPTYPE** preset parameters for offsetting objects.

MEASURE and **DIVIDE** place copies of points and blocks along objects.

ARRAY, **-ARRAY** and **3DARRAY** construct linear, rectangular, and polar copies.

FILLET and **CHAMFER** create rounded and angled corners.

REVCLOUD creates revision clouds (new to AutoCAD 2004).

RMLIN and **-RMLIN** insert redline markup .rml files from Volo View.

FINDING THE COMMANDS

On the MODIFY and DRAW toolbars:

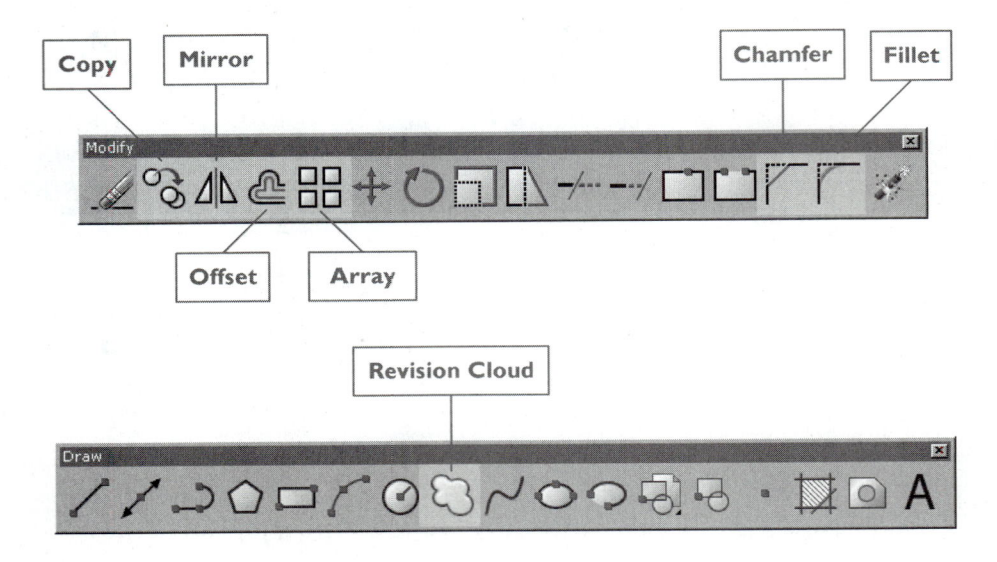

- Copy
- Mirror
- Chamfer
- Fillet
- Offset
- Array
- Revision Cloud

On the MODIFY and DRAW menus:

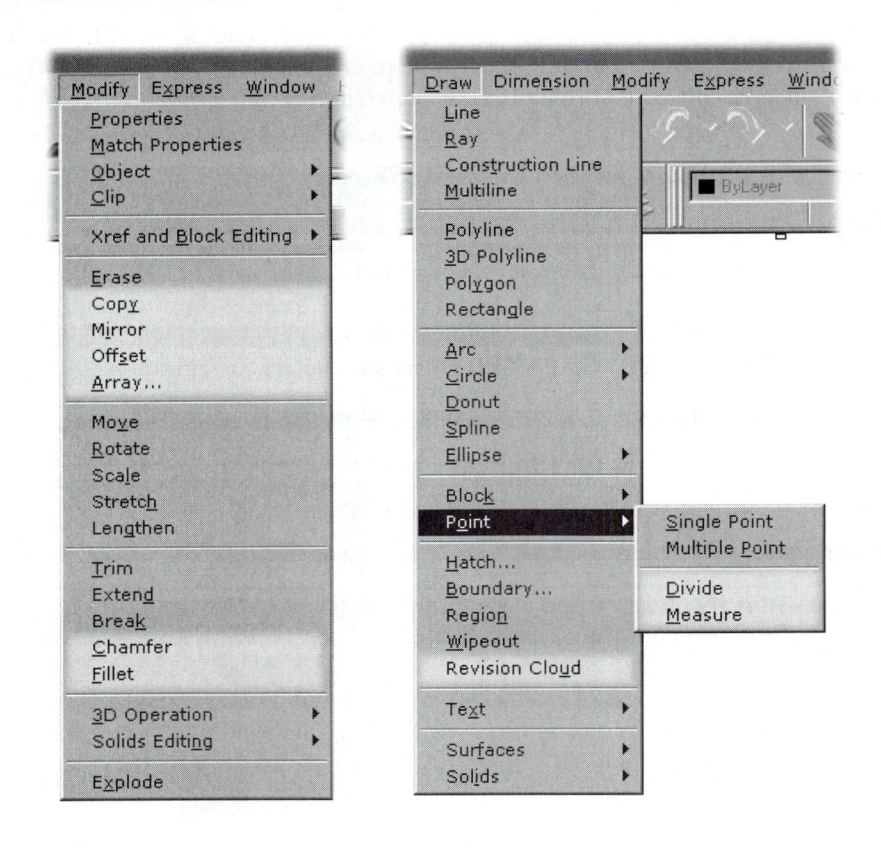

COPY

The COPY command makes one or more copies of objects.

The copies are identical to the original; all that changes is their location in the drawing. This command works within drawings; to copy objects to other drawings, use the COPYCLIP command.

TUTORIAL: MAKING A COPY

1. To make a copy of one or more objects, start the **COPY** command:
   - From the menu bar, choose **Modify**, and then **Copy**.
   - From the **Modify** toolbar, choose the **Copy** button.
   - At the 'Command:' prompt, enter the **copy** command.
   - Alternatively, enter the aliases **co** or **cp** at the 'Command:' prompt.

   Command: **copy** *(Press ENTER.)*

2. In all cases, AutoCAD prompts you to select the objects you want copied:

   Select objects: *(Pick one or more objects.)*

   Select objects: *(Press ENTER to end object selection.)*

3. Identify the point from which the displacement is measured:

   Specify base point or displacement, or [Multiple]: *(Pick point 1.)*

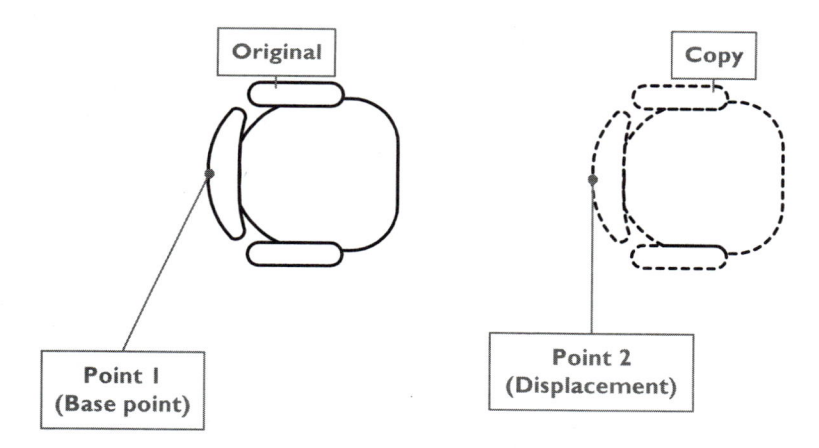

4. And identify the location for the copied object:

   Specify second point of displacement or <use first point as displacement>: *(Pick point 2.)*

After you use any of AutoCAD's many object selection modes to choose the object(s), the prompt asks for the *displacement*: this is the distance from the original object (known as the *base point*) to the location of the new object.

You can choose any point you like for the base point, but some points make more sense than others. Sometimes, it can use useful to use an object snap, such as INTersection or INSertion point, to pick the base point precisely. To copy the objects vertically or horizontally, turn on ortho mode.

AutoCAD has several ways to determine the displacement, which you learn about in the next tutorial.

MAKING COPIES: ADDITIONAL METHODS

In addition to making one copy at a time, the COPY command's **Multiple** option allows you repeatedly to make copies. This option is an alternative to the ARRAY command (discussed later in this chapter) for when you want an array of copies.

- **Multiple** makes multiple copies.
- **Displacement** specifies the distance between original and copies.

Let's look at each option.

Multiple

The **Multiple** option allows you to place more than one copy during the command. The "Second point of displacement" prompt repeats until you cancel the command. The copy originates from the originally selected object, using the base point you first selected.

1. Start the **COPY** command, and then select objects to copy:

    Command: **copy**

    Select objects: *(Select one or more objects.)*

    Select objects: *(Press ENTER.)*

2. Specify the **Multiple** option, and then pick a base point:

    Specify base point or displacement, or [Multiple]: **m**

    Specify base point: *(Pick a point.)*

3. Repeatedly pick points for the copies, and then press **ENTER** to end the command:

    Specify second point of displacement or <use first point as displacement>: *(Pick a point.)*

    Specify second point of displacement or <use first point as displacement>: *(Pick another point.)*

    Specify second point of displacement or <use first point as displacement>: *(Press ENTER.)*

Displacement

The displacement options provide you with several methods to specify the distance between the original and the copies. When AutoCAD prompts, "Specify base point or displacement," this is a hint of what is to come. *Specify base point* means you enter x,y coordinates (such as 4,5) *or* pick a point in the drawing. Both methods can be used as the base point, but the x,y coordinates could be interpreted differently by AutoCAD; *displacement* is the hint that the next prompt interprets these, depending on your actions.

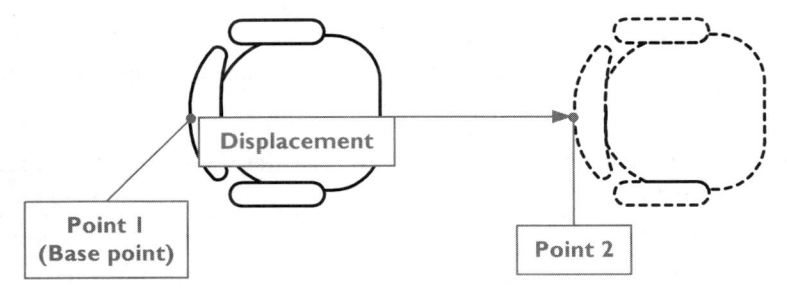

The second prompt, "Specify second point of displacement or <use first point as displacement>" requires that you perform one of three actions. *Specify second point of displacement* means that you should enter another x,y coordinate or pick another point. Both actions place the copy at a distance that AutoCAD calculates from the two sets of x,y coordinates or pick points. You can, of course, mix and match coordinate entry and pick points.

It's not clear from the prompt, but AutoCAD wants you to press **ENTER** at *<use first point as displacement>*. It then interprets the coordinates you entered at the earlier prompt (such as 4,5) as relative distances. In this case, there is no need to use the @ prefix. AutoCAD places the copy 4 units right and 5 units up from the original.

This option can, unfortunately, have an unexpected result, because, as this book's technical editor Bill Fane once said, pressing ENTER sometimes makes the copied objects end up near Hawaii, not a problem, he notes, if you live in Maui. This problem occurs when you use the mouse to pick the first point, and then just press ENTER for the second point; AutoCAD interprets the second point as 0,0. Use the ZOOM Extents command to find the "missing" copy.

1. Start the COPY command, and then select objects to copy:

    Command: **copy**

    Select objects: *(Select one or more objects.)*

    Select objects: *(Press ENTER.)*

2. Enter coordinates for the base point:

    Specify base point or displacement, or [Multiple]: **4,5**

3. Press ENTER to place the copy by a relative distance and end the command:

    Specify second point of displacement or <use first point as displacement>: *(Press ENTER.)*

> **Notes:** You may enter x,y coordinates for 2D displacement, or x,y,z coordinates for 3D displacement. In addition, you can use direct distance entry to specify the displacement.
>
> Being familiar with AutoCAD's many object selection modes is crucial for working efficiently with the commands in this (and the next) chapter. Review the SELECT command in Chapter 6 "Drawing with Efficiency."

MIRROR

The **MIRROR** command makes mirrored copies.

The command saves time when you are drawing symmetrical objects. Draw a half, or one-fourth, of an object, and then construct the other parts by mirroring.

You have the option to retain or delete the original objects, as well as to decide whether text should be mirrored or not. The *mirror line* is the line about which the objects are mirrored.

Other drawing programs use the phrases "flip horizontal" and "flip vertical" in place of mirror.

TUTORIAL: MAKING MIRRORED COPIES

1. To make mirrored copies of one or more objects, start the MIRROR command:
    - From the menu bar, choose **Modify**, and then **Mirror**.
    - From the **Modify** toolbar, choose the **Mirror** button.
    - At the 'Command:' prompt, enter the **mirror** command.
    - Alternatively, enter the **mi** alias at the 'Command:' prompt.

    Command: **mirror** *(Press ENTER.)*

2. In all cases, AutoCAD prompts you to select the objects you want mirrored:

    Select objects: *(Pick one or more objects.)*

    Select objects: *(Press ENTER to end object selection.)*

3. Identify the points that define the mirror line:
     Specify first point of mirror line: *(Pick point 1.)*
     Specify second point of mirror line: *(Pick point 2.)*

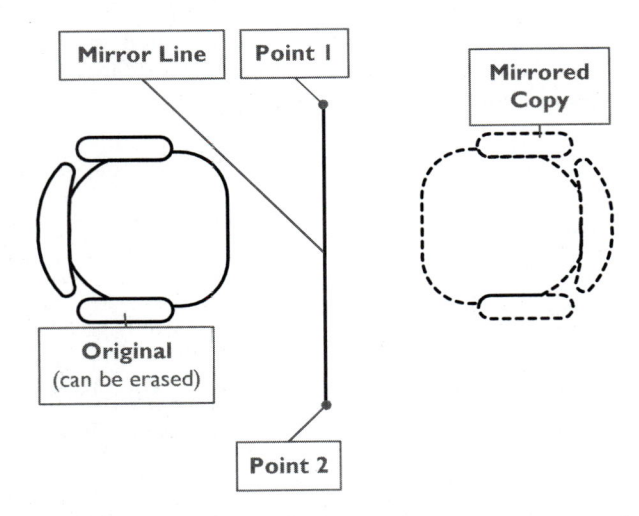

4. Decide whether you want the original objects(s) erased:
     Delete source objects? [Yes/No] <N>: *(Enter Y or N.)*

Note: To make the mirror line absolutely horizontal or vertical, turn on ortho more. Polar mode is handy for making mirrored copies at specific angles.

MAKING MIRRORED COPIES: ADDITIONAL METHODS

Whether text should also be mirrored has to be decided by you with the help of a system variable.

- **MIRRTEXT** system variable decides whether mirrored text is mirrored.

MirrText

Sometimes objects contain text. The dilemma is whether to mirror the text, which makes it read backwards — or not mirror, which makes it read normally. The **MIRRTEXT** system variable determines this, as follows:

   Command: **mirrtext**
   New value for MIRRTEXT <1>: *(Enter 1 or 0.)*

Entering **0** produces non-mirrored text, which is the default setting. Enter **1** if you prefer text to be mirrored.

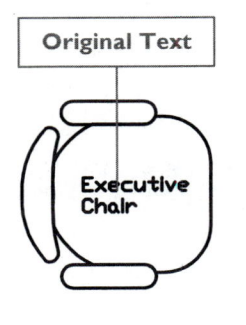

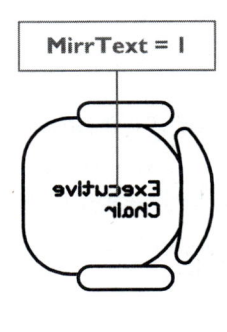

OFFSET

The **OFFSET** command makes parallel copies.

This command constructs copies parallel to objects; AutoCAD limits you to making offset copies of one object at a time. When the objects have curves or are closed, the copies become larger or smaller, depending on whether they are on the inside or outside radius.

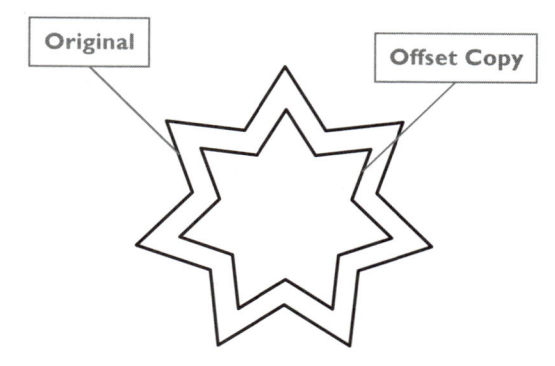

AutoCAD offsets lines, arcs, circles, ellipses, elliptical arcs, polylines (2D only), splines, rays, and xlines. Sometimes, unexpected results happen, as illustrated below. The dashed lines are copies offset from the original polyline (shown as the heavy line).

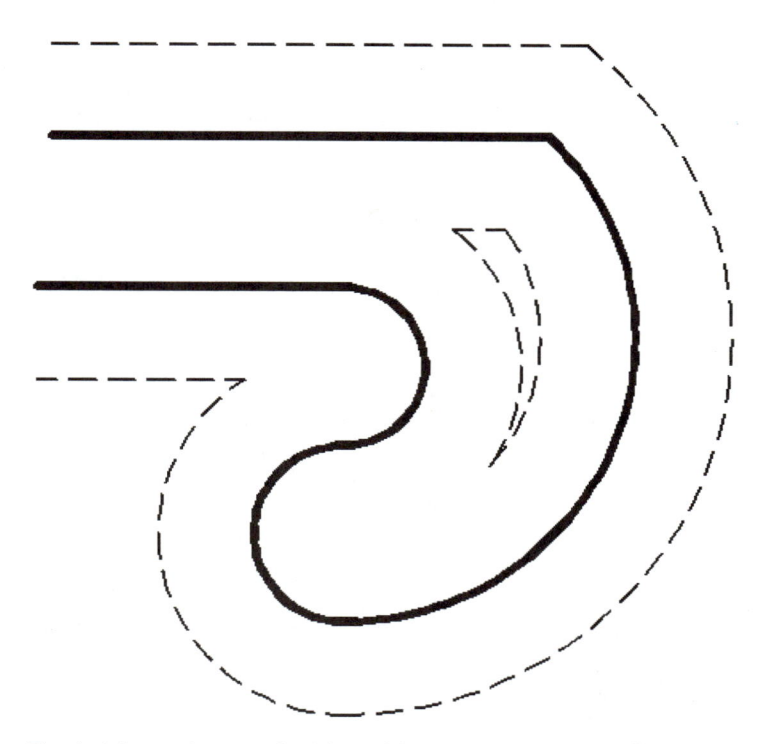

The thick line is the original polyline, while the dashed lines are offset copies.

To create an offset copy, AutoCAD needs to know three pieces of information, in this order: (1) the offset distance; (2) the object to offset; and (3) the side on which to place the offset copy. It may seem counter-intuitive *first* to specify the distance and *then* select the object, but "that's the way the Mercedes bends," as the driver said after his automobile accident.

TUTORIAL: MAKING OFFSET COPIES

1. To make offset copies of an object, start the **OFFSET** command:
   - From the menu bar, choose **Modify**, and then **Offset**.
   - From the **Modify** toolbar, choose the **Offset** button.
   - At the 'Command:' prompt, enter the **offset** command.
   - Alternatively, enter the **o** alias at the 'Command:' prompt.

   Command: **offset** *(Press ENTER.)*

2. In all cases, AutoCAD prompts you to specify the offset distance:
   Specify offset distance or [Through] <Through>: *(Enter a distance.)*

3. Select the object to offset:
   Select object to offset or <exit>: *(Select one object.)*

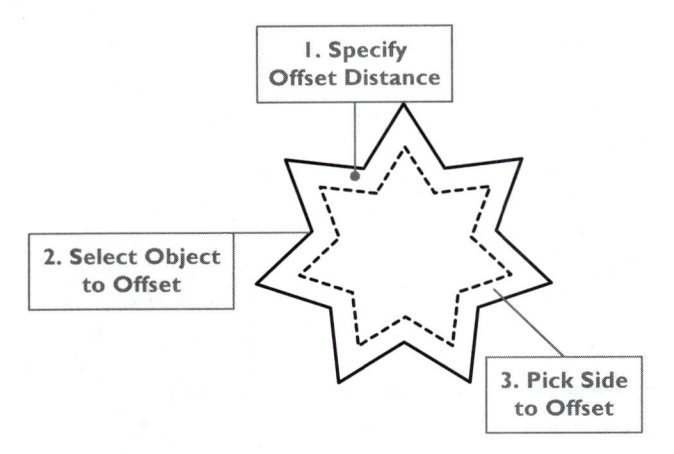

   1. Specify Offset Distance
   2. Select Object to Offset
   3. Pick Side to Offset

4. Pick the side on which the offset should be placed:
   Specify point on side to offset: *(Pick point 3.)*

5. Press enter to exit the command:
   Select object to offset or <exit>: *(Press ENTER.)*

The "Select object to offset:" prompt repeats to allow you to offset as often as you wish, but just one object at a time. Press **ENTER** to terminate the command.

For the offset distance, you can pick two point on the screen, or enter a number representing the distance. If you get the message, "That object is not parallel with the UCS," this means that the direction of the object's Z axis was not parallel to the current user coordinate system.

MAKING OFFSET COPIES: ADDITIONAL METHODS

The **OFFSET** command's **Through** option lets you specify a point through which the copy is offset. In addition, two system variables let you preset parameters.

- **Through** option combines the distance and side options.
- **OFFSETDIST** system variable presets the offset distance.
- **OFFSETGAPTYPE** system variable determines how polyline gaps are handled.

Let's look at each.

Through

The **Through** option constructs the offset copy "through" a point. In effect, it combines the "Specify offset distance" and "Side to offset" options.

1. Draw a circle.
2. Start the **OFFSET** command, and then enter the **Through** option.

    Command: **offset**

    Specify offset distance or [Through]: **t**

3. Select the circle, and then pick the point through which the copy should be offset:

    Select object to offset or <exit>: *(Pick the circle.)*

    Specify through point: *(Pick point 1.)*

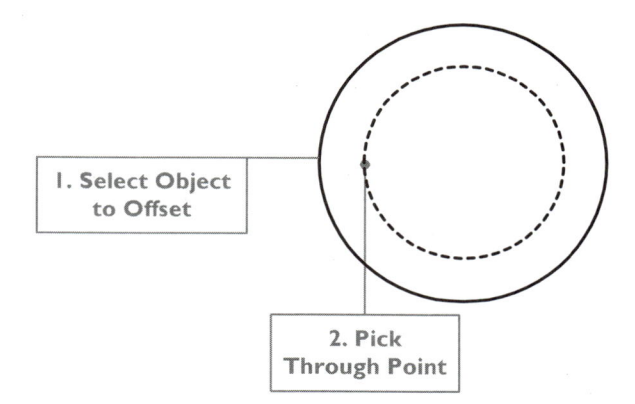

[Figure: Circle with dashed offset copy inside. Labels: "1. Select Object to Offset" and "2. Pick Through Point"]

4. Continue making offset copies, or press **ENTER** to end the command.

    Select object to offset or <exit>: *(Press ENTER.)*

OffsetDist

With the **OFFSETDIST** system variable, you preset the offset distance. As well, you can use it to preselect the **Through** option. Then, during the **OFFSET** command, you need only press **ENTER** at the "Specify offset distance" prompt.

Command: **offsetdist**

Enter new value for OFFSETDIST <1.00>: *(Enter a distance, positive or negative, such as* **2.54** *or* **-1.***)*

Enter a distance, and that becomes the default offset distance next time you use the **OFFSET** command. The prompt looks like this (the change is emphasized in blue):

Specify offset distance or [Through] <**2.5400**>: *(Press ENTER.)*

If, on the other hand, you enter a negative number, such as **-1**, then the **OFFSET** command sets the **Through** option as the default:

Specify offset distance or [Through] <**Through**>: *(Press ENTER.)*

OffsetGapType

Offsetting polylines can be tricky. For this reason, AutoCAD includes the **OFFSETGAPTYPE** system variable to help you decide how potential gaps between polyline segments should be handled. The choices are extending lines, creating fillets (arcs), or creating bevels (chamfers):

Command: **offsetgaptype**

Enter new value for OFFSETGAPTYPE <0>: *(Enter a number between 0 and 2.)*

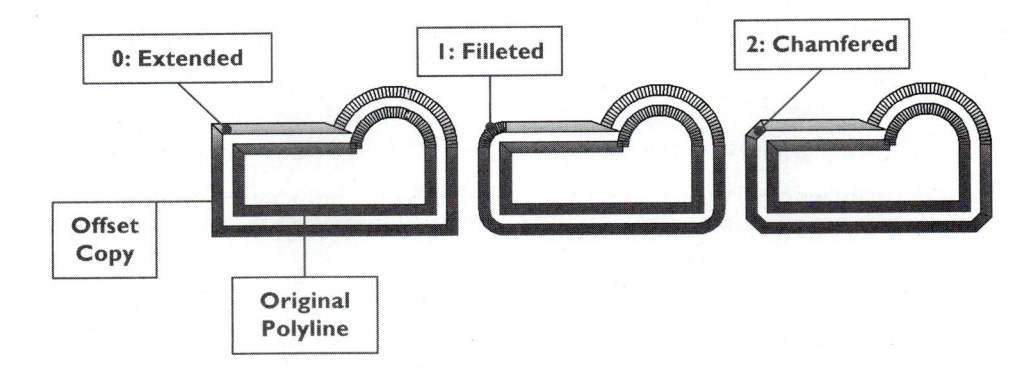

OffsetGapType	Comment
0	Gaps filled with extended line segments (default).
1	Gaps filled with filleted segments, creating arcs.
2	Gaps filled with chamfered line segments, creating beveled edges.

DIVIDE AND MEASURE

The **DIVIDE** and **MEASURE** commands place copies of points and blocks along objects.

DIVIDE divides an object into an equal number of parts, while **MEASURE** works with a specific distance. So, *number* of segments versus *length* of segment. Both commands place either blocks or points along the object.

The **MEASURE** command has nothing to do with measuring distances or lengths; for those, use the **DIST** and **LIST** commands.

Like the **OFFSET** command, dividing and measuring works with just one object at a time. Use the cursor to select a single object, because you cannot use Window, Crossing, or Last selection modes. In addition, AutoCAD is limited to working with lines, arcs, circles, splines, and polylines; picking a different object results in the complaint:

Cannot divide that object. * Invalid*

AutoCAD does not place a point or block at the start or end of open objects.

> **Note:** After using the **DIVIDE** and **MEASURE** commands to place points on an object, you can snap to the points with the **NODe** object snap.

TUTORIAL: DIVIDING OBJECTS

1. To divide an object into equal parts, start the **DIVIDE** command:
   - From the **Draw** menu, choose **Point**, and then **Divide**.
   - At the 'Command:' prompt, enter the **divide** command.
   - Alternatively, enter the **div** alias at the 'Command:' prompt.

   Command: **divide** *(Press ENTER.)*

2. In all cases, AutoCAD prompts you to select the single object to divide:
   Select object to divide: *(Select one object.)*
3. Specify the number of division:
   Enter the number of segments or [Block]: *(Enter a number between 2 and 32767.)*

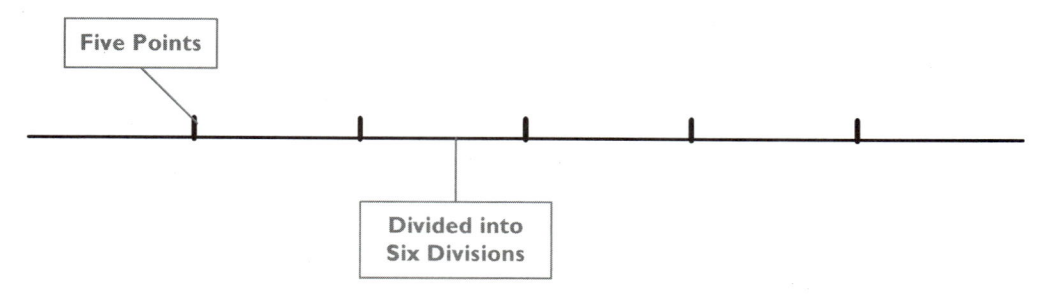

[Figure: Line divided into six divisions with five points marked]

The **DIVIDE** command results in evenly spaced points, and with one fewer points than you would expect: divide a line by *six*, and AutoCAD places *five* points, creating *six* divisions. (Count the number of points and divisions on the line illustrated above.) The points and blocks that are placed are independent of the line, and you can move and erase them at will.

The **MEASURE** command operates similarly to the **DIVIDE** command, the difference being that you specify the length of segment by which to space the points.

TUTORIAL: "MEASURING" OBJECTS

1. To place points along an object at specific distances, start the **MEASURE** command:
   - From the **Draw** menu, choose **Point**, and then **Measure**.
   - At the 'Command:' prompt, enter the **measure** command.
   - Alternatively, enter the **me** alias at the 'Command:' prompt.

   Command: **measure** *(Press ENTER.)*

2. In all cases, AutoCAD prompts you to select the single object to divide:
   Select object to measure: *(Select one object.)*
3. Specify the number of division:
   Specify the length of segment or [Block]: *(Enter a number between 2 and 32767.)*

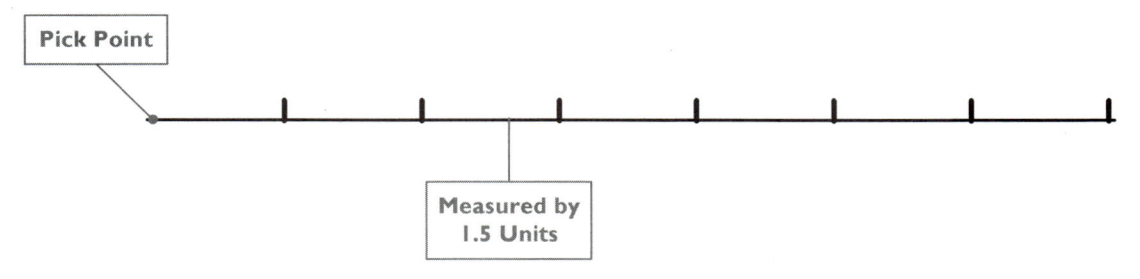

[Figure: Line measured by 1.5 units from pick point]

This may sound arcane, but the **MEASURE** command needs a *starting point*, which depends on the object. As the figure above illustrates, measurement does not result in the even distribution of points as does division.

For open objects — lines, arcs, splines and open polylines — the starting point is the endpoint closest to your pick point; for closed polylines, it's the point where you began drawing the polyline. For circles, it is at the current snap angle, which is usually 0 (at the circle's 3 o'clock point); the measurement is made in the counterclockwise direction.

DIVIDING AND MEASURING: ADDITIONAL METHODS

The commands' **Block** option lets you specify a block to place along the object, in place of a point. And, since points tend to be invisible, the **PDMODE** and **PDSIZE** system variables (accessed by **DDPTYPE**) are useful.

- **Block** option places blocks along the object.
- **DDPTYPE** command changes the look of the points.

Let's look at each.

Block

The **Block** option places blocks (symbols) along the object. The block must already exist in the drawing. The option operates identically for both commands:

1. Start either command. Enter the **b** option, and then the name of a block:

    Enter the number of segments or [Block]: **b**

    Enter name of block to insert: *(Type name.)*

2. Decide whether you want the block aligned with the object (Y) or at its own orientation (N):

    Align block with object? [Yes/No] <Y>: *(Enter **Y** or **N**.)*

3. Continue on with the command.

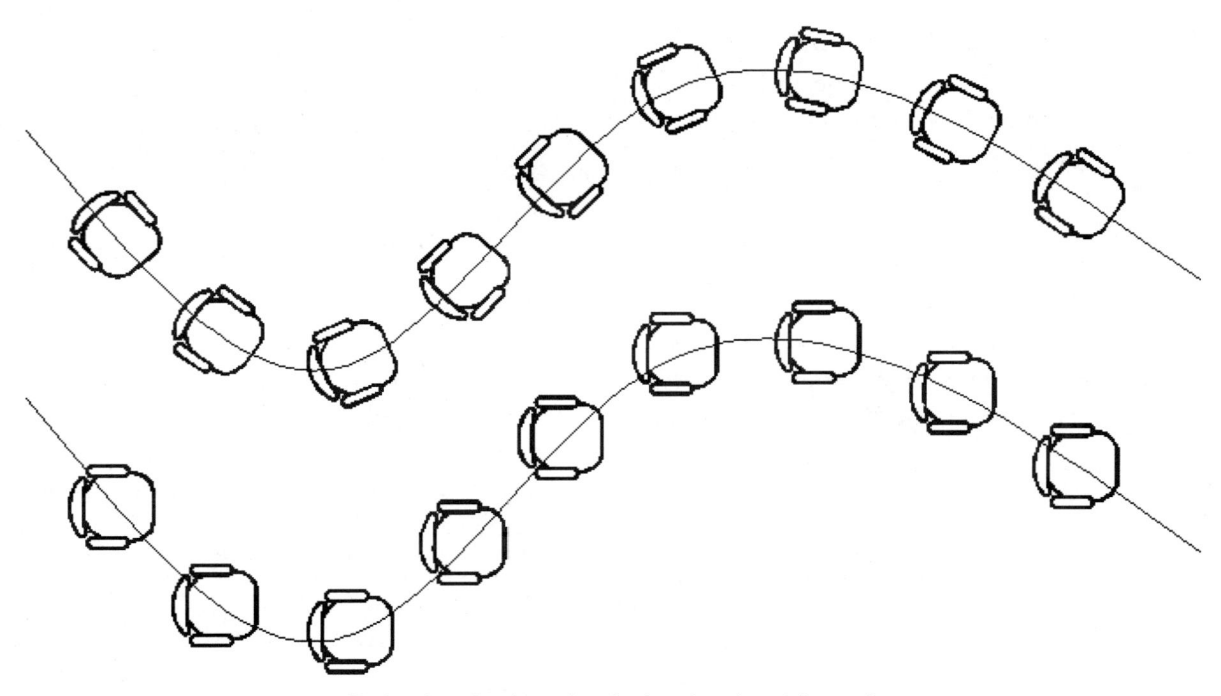

Blocks aligned with a spline (top) and unaligned (bottom).

"Yes" means the inserted block turns with the divided object, such as arc, circle, or spline. "No" means the block is always oriented in the same direction. If you are unsure of the block's name or even of its existence in the drawing, use the DesignCenter to help you.

DdPType

Recall from Chapter 4 "Drawing with Basic Objects" that the **DDPTYPE** command changes the look and size of points. Points are normally invisible (for all intents and purposes), so it may be useful to change their size. The figure below illustrates the before-and-after difference.

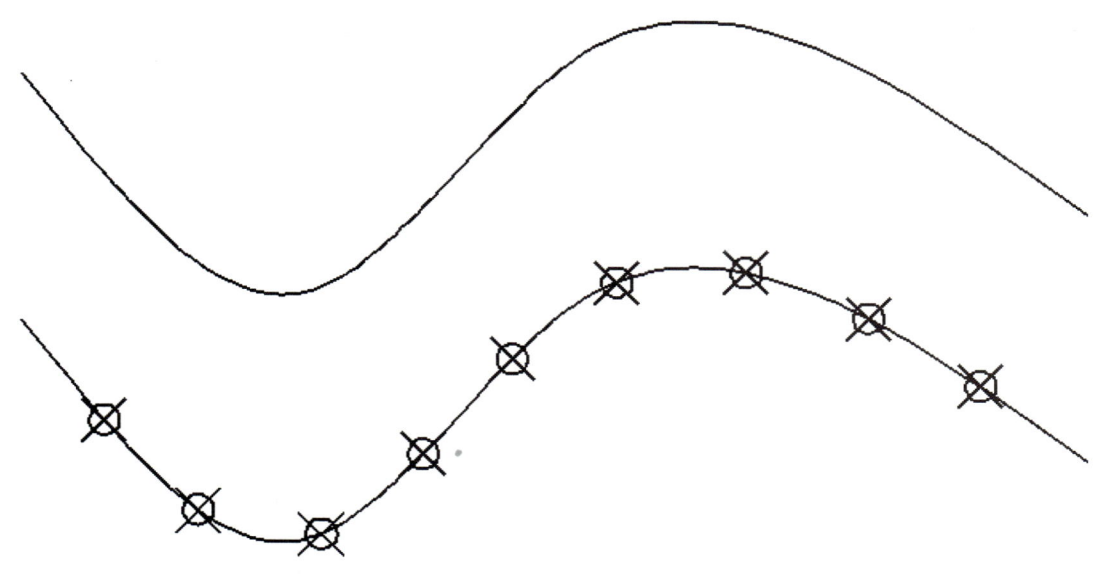

Spline divided by points of normal size (top), and enlarged (bottom).

ARRAY

The **ARRAY** command makes evenly-spaced copies in linear, rectangular, and round patterns.

There are times when you want to place multiple copies of an object in a pattern. Consider the rows of seats in a movie theater, or the columns of parking spaces at a shopping center. If you were using traditional drafting techniques, you would draw each one separately.

Earlier in the chapter, you saw how the **COPY** command (with its **Multiple** option) and the **DIVIDE** and **MEASURE** commands can place many copies of objects in the drawing. The **ARRAY** command, however, proves superior for making copies in precise rows, columns, matrices, circles, and semicircles. After you array an object, each copy can be edited separately — unlike the similar **MINSERT** command. For placing many copies in random places, **COPY Multiple** is better.

The largest number of rows and columns you can enter is 32,676; the smallest number is 1. Because a 32676 x 32676-array creates one billion elements, AutoCAD limits the total number to 100,000 elements — otherwise your computer system overloads. Where does the seemingly arbitrary value of 32,767 come from? I'll give you hints: it's an integer, and it's 2 to the power 15.

You can change the upper limit to another value between 100 and 10,000,000 with the **MAXARRAY** system registry variable. At the command prompt, enter "MaxArray" exactly as shown:

    Command: **(setenv "MaxArray" "1000")**
    "1000"

To create an array, AutoCAD needs to know: (1) the type of array, rectangular or polar; (2) the object(s) you plan to array, which must already exist in the drawing; and (3) the parameters of the array.

> **Note**: Sometimes it's difficult to distinguish between *rows* and *columns*. Rows go side to side, while columns go up and down. Still puzzled? Look at the preview window, or click the **Preview** button for a sneak peak.

TUTORIAL: LINEAR AND RECTANGULAR ARRAYS

A linear array copies objects in the horizontal direction (in a row or the x direction) or the vertical direction (in a column or the y direction). A rectangular array copies the objects in a rectangular pattern made up of rows and columns.

1. To make an array of copies, start the **ARRAY** command:
   - From the menu bar, choose **Modify**, and then **Array**.
   - From the **Modify** toolbar, choose **Array**.
   - At the 'Command:' prompt, enter the **array** command.
   - Alternatively, enter the **ar** alias at the 'Command:' prompt.

   Command: **array** *(Press ENTER.)*

2. In all cases, AutoCAD displays the Array dialog box:

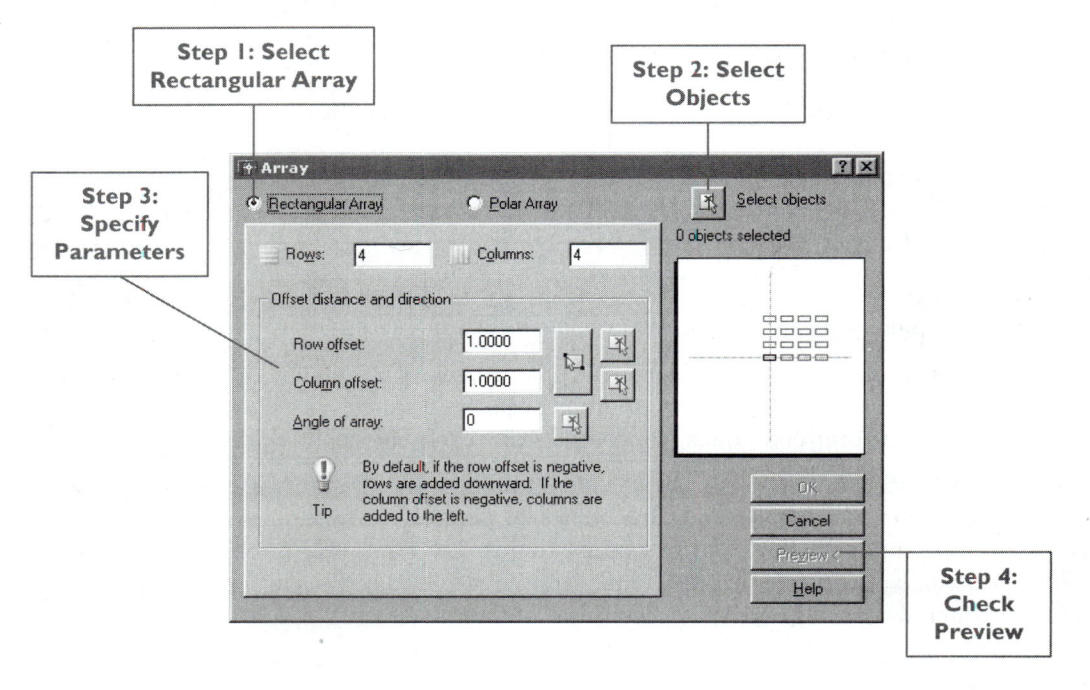

   Step 1: Select Rectangular Array
   Step 2: Select Objects
   Step 3: Specify Parameters
   Step 4: Check Preview

   Notice the dialog box has two radio buttons (the round buttons at the top of the dialog box) that determine the type of array:
   - **Rectangular** displays the options for creating linear and rectangular arrays.
   - **Polar** displays options for creating polar (circular and semicircular) arrays.

3. Choose **Select Objects** to select the object(s) to array. The dialog box disappears, and this prompt appears:

   Select objects: *(Select one or more objects.)*
   Select objects: *(Press ENTER to return to the dialog box.)*

   Note: You can avoid this step by selecting the objects before starting the **ARRAY** command.

4. In the **Rows** and **Columns** text boxes, specify the number of copies to made in rows and columns. When you enter a 1 for both, AutoCAD complains,

   Only one element; nothing to do.

   because a 1x1 array consists of the original object only.

5. The **Row Offset** and **Column Offset** options measure the distance between the elements of the array. You can enter a specific distance, or click the adjacent buttons to select the distances in the drawing:

   The **Pick Both Offsets** button clears the dialog box, and AutoCAD prompts you at the command line:

   > Specify unit cell: *(Pick a point.)*
   >
   > Other corner: *(Pick another point.)*

   Pick two points, creating a rectangle that specifies the row and column distance between elements in the array. After you pick the second point, the dialog box returns.

   The **Pick Row Offset** and **Pick Column Offset** buttons clear the dialog box, and AutoCAD prompts you at the command line:

   > Specify the distance between rows (or columns): *(Specify a distance.)*

   Enter a distance, or pick two points that specify the distance between row (or column) elements in the array. After picking the second point, the dialog box returns.

   Note: AutoCAD normally creates rows to the right, and columns upwards. To have AutoCAD draw the array elements in the other direction, enter negative values for the row and column offsets.
   If you use the mouse to pick two points, and the second point is below or to the left of the first, then AutoCAD automatically sets negative values.

6. **Angle of array** tilts the rectangular array at an angle; note that the object itself does not tilt, but the elements of the array are staggered by the angle. Entering an angle of 180 degrees draws the array downward. (To select the angle in the drawing, click the **Pick Angle of Array** button.)

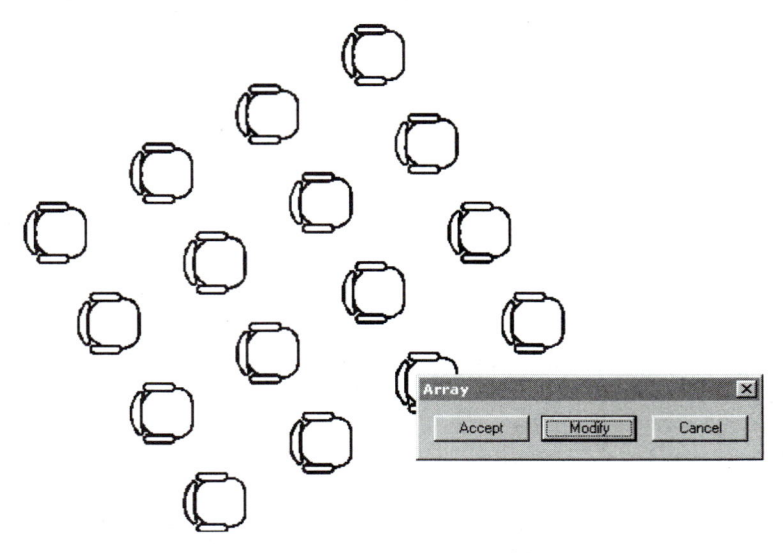

   A 4x4 rectangular array at an angle of 30 degrees.

7. Click **Preview** to see what the array will look like. Notice the dialog box with its three buttons:

   Accept keeps the array, and exits the **ARRAY** command.

   Modify returns to the Array dialog box.

   Cancel doesn't keep the array (leaves things the way they were), and exits the Array command.

8. Don't like the look of the preview? Click **Modify**, change parameters, and then click **Preview** until you're satisfied.

   Finally, click **Accept**.

TUTORIAL: POLAR AND SEMICIRCULAR ARRAYS

Polar arrays arrange objects in circular patterns. To construct a polar array, you must define the angle between the items (from center to center, not actually between) and either the number of items or degrees to fill. You have the option of rotating (or not rotating) each object as it is arrayed.

1. Start the **ARRAY** command.
2. Click the **Polar Array** radio button. Now let's take a look at the options for creating a round array.

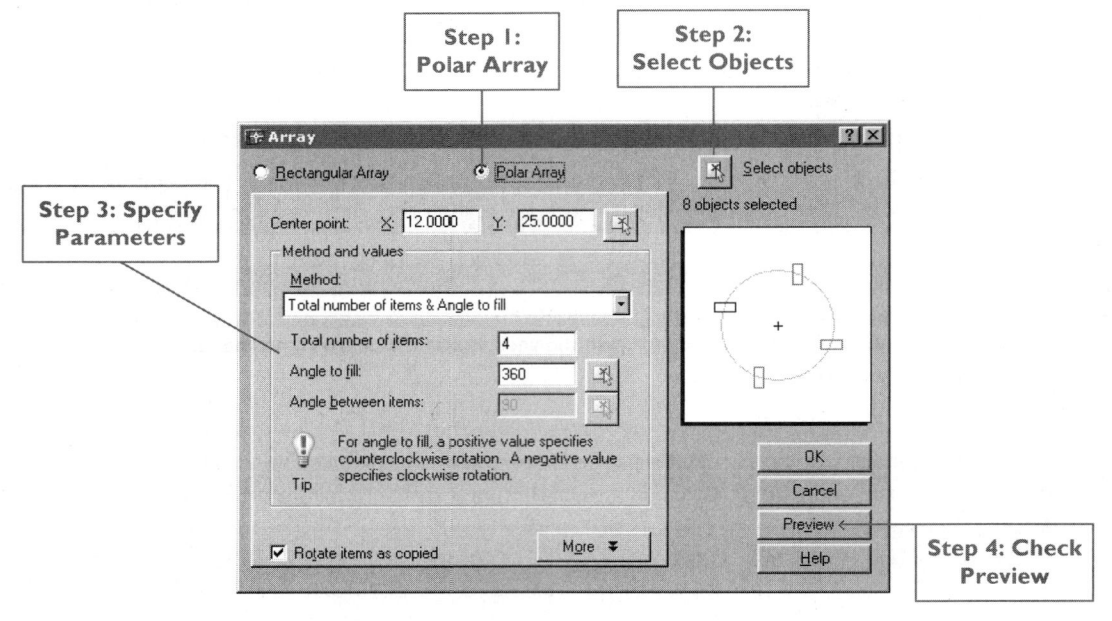

3. Select the **Center point** of the polar array; think of it as being the same as the center of a circle. AutoCAD automatically picks a point for itself (using a method I have never figured out; the technical editor agrees: "It's totally random!"), but you can change it. Enter new x,y-values or click the **Pick Center Point** button.
4. To create a polar array, you must provide data for any two of the following options. The Method drop list lets you pick the pair of options from the three possible choices.
   - **Total Number of Items** specifies the number of elements in the array. AutoCAD draws them to fit the polar route.
   - **Angle to Fill** changes the angle. By default, AutoCAD creates a 360-degree polar array, but you can create an arc array, instead. For example, specifying 270 degrees gives you three-quarters of a polar array.
   - **Angle Between Items** specifies the angle between the elements of the array.

5. Decide whether you want AutoCAD to **Rotate Items as Copied**. When turned on (the check mark appears), AutoCAD makes sure the items "face the center."
6. Click the **More** button to see more options for polar arrays:

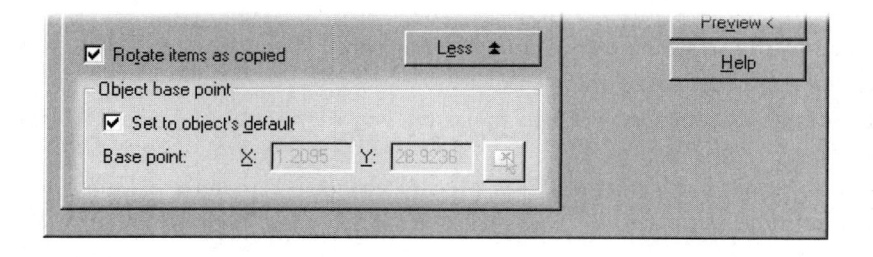

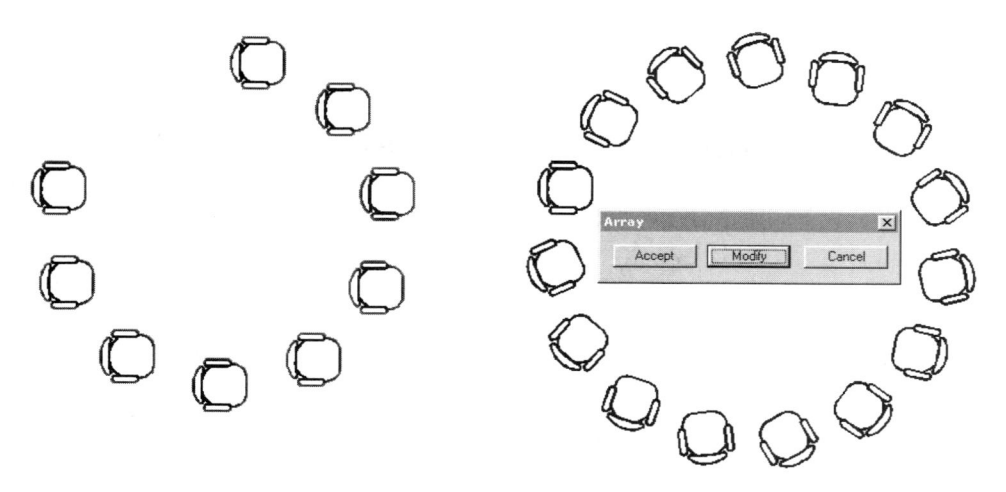

At left: 270-degree polar array with un-rotated objects.
At right: 360-degree polar array with rotated objects.

The **Object base point** options are tricky to grok, but essentially you get to pick where AutoCAD measures the distance from the object to the center of the polar circle. AutoCAD uses the following default values, which change depending on the type of object being arrayed:

Object	Default Base Point
Lines, polylines, donuts, 3D polylines, rays, splines	Starting point.
Arcs, circles, ellipses	Center point.
Polygons, rectangles	First corner drawn.
Xlines	Midpoint.
Blocks, mtext, text	Insertion point.
Regions	Grip point.

7. Click **Select objects** to pick one or more objects to array.
   Click **Preview** to see what the array will look like.

CONSTRUCTING ARRAYS: ADDITIONAL METHODS

The -ARRAY command creates arrays at the command line, while **3DARRAY** creates them in 3D space.

- **-ARRAY** command creates arrays at the command line.
- **3DARRAY** command creates three-dimensional arrays.

Let's look at each.

-Array

The -ARRAY command produces arrays after you enter options at the command line. To produce a rectangular array, use the **R** option, as follows:

Command: **-array**
Select objects: *(Select one or more objects.)*
Select objects: *(Press* ENTER *to end object selection.)*
Enter the type of array [Rectangular/Polar] <P>: **r**
Enter the number of rows (---) <1>: *(Enter the number of rows, such as* **4.**)

Enter the number of columns (| | |) <1> *(Enter the number of columns, such as **3**.)*

Enter the distance between rows or specify unit cell (---): *(Enter the distance between rows, such as **2.5**.)*

Specify the distance between columns (| | |): *(Enter the distance between columns, such as **3.3**.)*

To rotate the rectangular array, change the angle of the SNAPANG system variable before starting the -ARRAY command:

Command: **snapang**

Enter new value for SNAPANG <0>: *(Enter a new angle, such as **30**.)*

To produce a polar array, use the **P** option, as follows:

Command: **-array**

Select objects: *(Select one or more objects.)*

Select objects: *(Press ENTER to end object selection.)*

Enter the type of array [Rectangular/Polar] <P>: **p**

Specify center point of array or [Base]: *(Pick a point.)*

Enter the number of items in the array: *(Enter a number, such as **16**.)*

Specify the angle to fill (+=ccw, -=cw) <360>: *(Press ENTER, or enter an angle.)*

Rotate arrayed objects? [Yes/No] <Y>: *(Enter **Y** or **N**.)*

The options are similar to those available in the dialog box. Entering a negative angle draws the polar array clockwise; do not enter an angle of 0 degrees.

3DArray

The **3DARRAY** command creates the array in 3D space.

The **Rectangular** option arrays the objects in the x, y, and z directions — called rows, columns, and levels. The prompts are similar to the -ARRAY command, but with two added for the z direction (illustrated by ...):

Enter the number of levels (...) <1>: *(Enter a number, such as **3**.)*

Specify the distance between levels (...): *(Enter a distance, such as **1.28**.)*

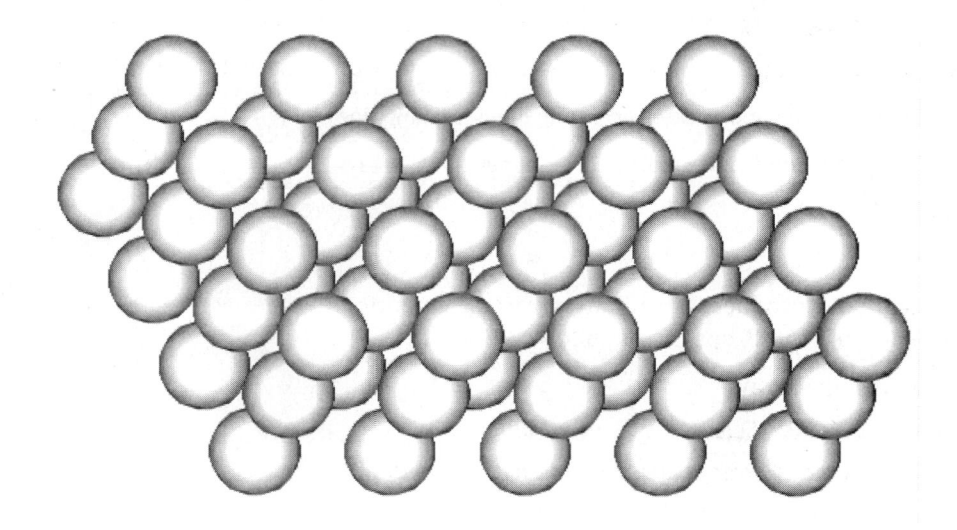

Looking very molecular, this 3D rectangular array was created with spheres, and then rendered.

The **Polar** option isn't three-dimensional. Instead, it creates a circular array at a specified angle in 3D space. The prompts are identical to that of the -ARRAY command, until the end. The *axis of rotation* allows you to tilt the polar array in 3D space:

> Specify center point of array: *(This pick point also specifies the first point for the axis.)*
> Specify second point on axis of rotation: *(Pick another point.)*

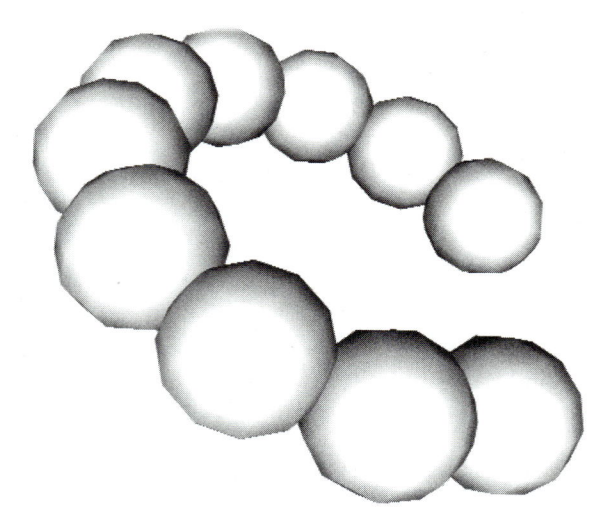

Looking like a girl's bracelet, this 3D polar array was created with spheres, and then rendered in perspective mode.

FILLET AND CHAMFER

The **FILLET** and **CHAMFER** commands create rounded and angled corners, respectively.

The **FILLET** command connects two lines or polylines with a perfect intersection, or with an arc of specified radius. Fillets can also connect two circles, two arcs, a line and a circle, a line and an arc, or a circle and an arc.

The two objects need not need touch to be filleted — this includes parallel lines. This allows you to intersect two non-touching lines; in the case of parallel lines, a 180-degree arc is drawn between their ends.

TUTORIAL: FILLETING OBJECTS

1. To fillet a pair of objects, start the **FILLET** command:
   - From the menu bar, choose **Modify**, and then **Fillet**.
   - From the **Modify** toolbar, choose the **Fillet** button.
   - At the 'Command:' prompt, enter the **fillet** command.
   - Alternatively, enter the **f** alias at the 'Command:' prompt.

   Command: **fillet** *(Press* ENTER.*)*

2. In all cases, AutoCAD first displays the current fillet settings, and then asks you to select objects:

    Current settings: Mode = TRIM, Radius = 0.0000
    Select first object or [Polyline/Radius/Trim/mUltiple]: *(Pick object 1.)*
    Select second object: *(Pick object 2.)*

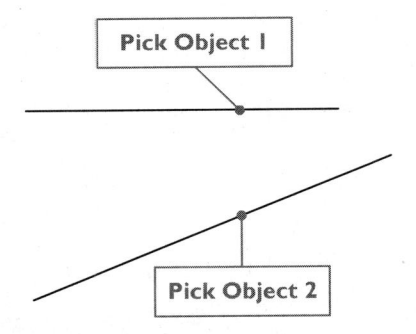

3. And the command is done! Because the radius was set to 0 (the fresh-off-the-distribution-CD default value), AutoCAD created a clean intersection, without the arc you might have been expecting.

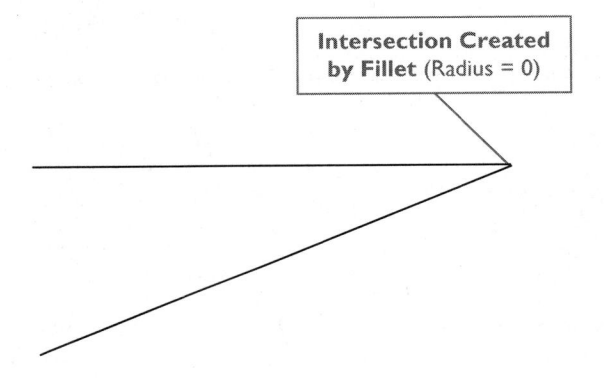

DIFFERENT FILLET RESULTS FOR DIFFERENT OBJECTS

Depending on the objects involved, the fillet can differ.

Lines

Two lines are trimmed back (or extended, as necessary), so that an arc fits between them. A zero-radius fillet connects two lines with a perfect intersection.

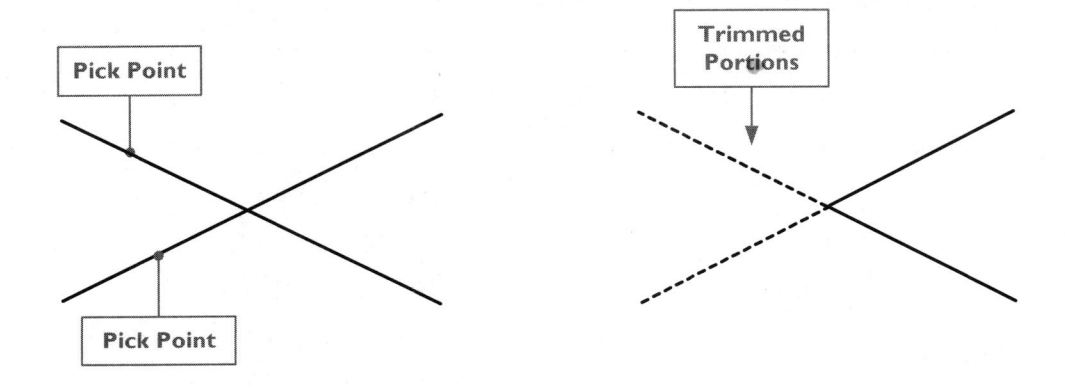

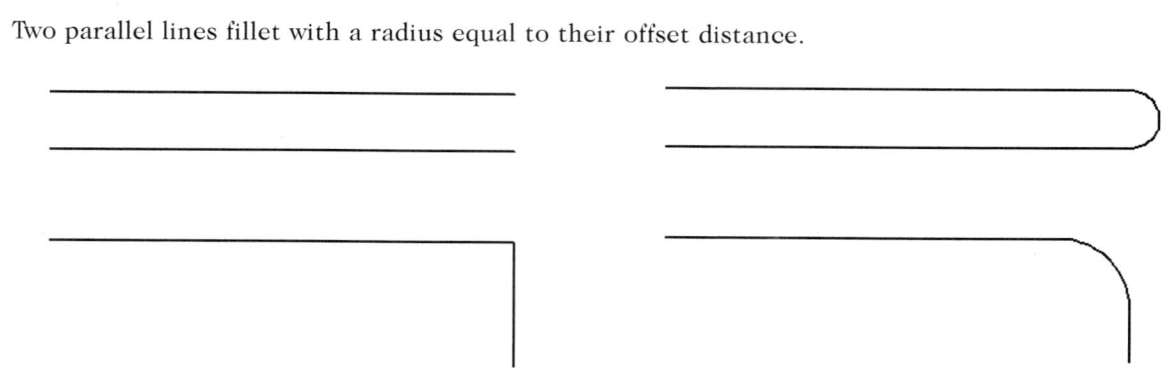

Two parallel lines fillet with a radius equal to their offset distance.

Two pairs of lines before and after being filleted with radius = 1.

Polylines

You can fillet an entire polyline in one operation when you select the **P** option (short for "polyline"). The fillet radius is placed at all vertices of the polyline. If arcs exist at any intersections, they are changed to the new fillet radius. Note that the fillet is applied to one continuous polyline.

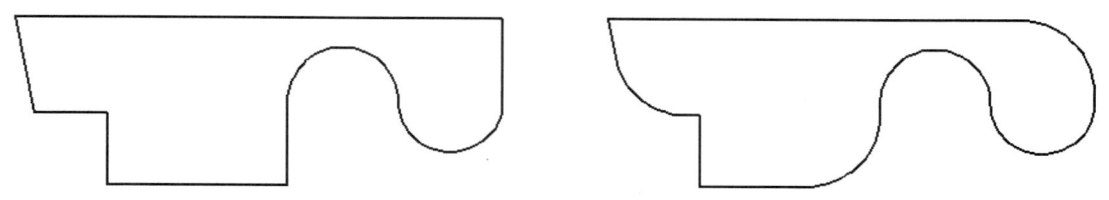

*A polyline filleted with the **Polyline** option; notice that two vertices were too short to be filleted with radius = 1.*

If the fillet radius is too large for a line or polyline segment, AutoCAD does not apply the fillet:

    Select 2D polyline: *(Select a single polyline.)*

    4 lines were filleted

    3 were too short

When a line and a polyline are filleted, all three objects (the line, the fillet arc, and the polyline) are converted to a single polyline.

Arcs and Circles

Lines, arcs, and circles can also be filleted. When you fillet such objects, however, there are often several possible fillet combinations possible. You specify the type of fillet by the points picked when you select the objects. AutoCAD attempts to fillet the endpoint closest to the selection point.

The figure illustrates several combinations between a line and arc. Observe the placement of the points used to pick the objects in the middle row, and the resulting fillet (shown in the bottom row).

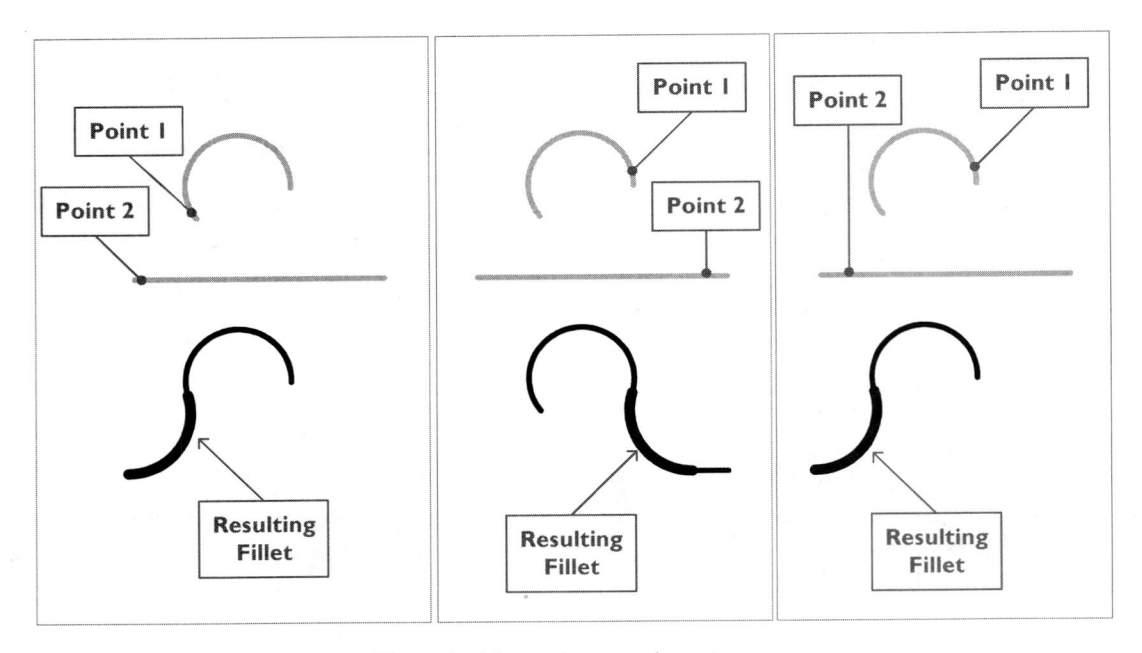

*The original line and arc are shown in gray;
the resulting fillet is shown by the heavy arcs.*

As with lines and arcs, the result of filleting two circles depends on the location of the two points you use to select the circles. The figure illustrates three possible combinations, each using different selection points.

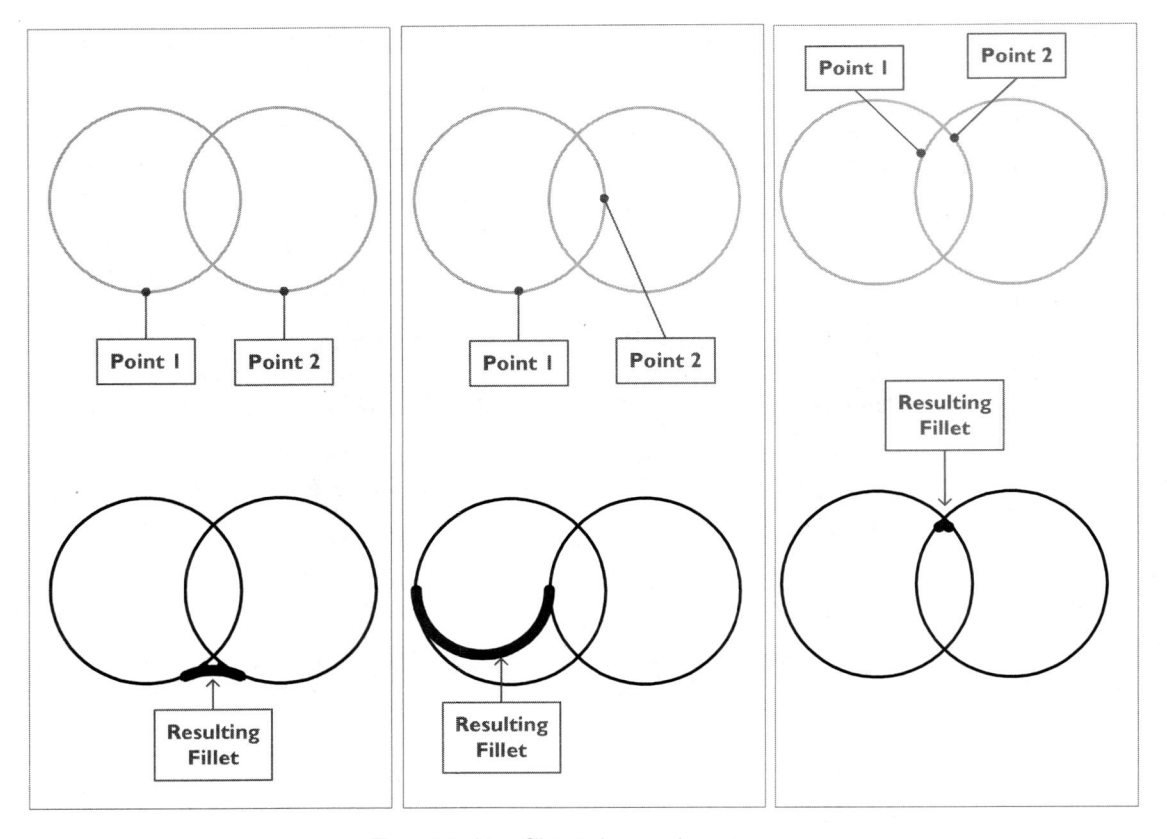

*The original pre-fillet circles are shown in gray;
the resulting fillet is shown by the heavy arcs.*

Notes: When you select two objects for filleting, and get an undesirable result, use the **U** command to undo the fillet. Try to respecify the points closer to the endpoints you want to fillet.

If you have several filleted corners to draw, construct your intersections at right angles, and then fillet each later. This allows you to continue the **LINE** command without interruptions, and requires fewer commands.

Changing an arc radius by fillet is cleaner and easier than erasing the old arc and cutting in a new one. Let AutoCAD do the work for you!

You "clean up" line intersections by setting the fillet radius to zero and filleting the intersections.

CONSTRUCTING FILLETS: ADDITIONAL METHODS

The command has several option for special cases:

- **Polyline** option treats polylines differently.
- **Radius** option changes the radius of the three-dimensional arrays.
- **Trim** option determines what happens to the leftover bits.
- **mUltiple** option continues the command to fillet additional objects (new to AutoCAD 2004).

Let's look at each.

Polyline

The **Polyline** option fillets all vertices of a single polyline. When you fillet a polyline without using this option, AutoCAD expects that you want to place a fillet between two adjacent segments. At the "Select first object" prompt, select a single polyline.

Select 2D polyline: *(Select a single polyline.)*

Radius

The **Radius** option determines the radius of the fillet arc. When set to 0, the command ensures that the two lines match precisely; an arc is not created.

Specify fillet radius <1.0000>: *(Enter a radius.)*

If the fillet radius is too large for a line, AutoCAD complains:

Radius is too large
Invalid

In that case, use a radius smaller than the shortest line.

Trim

The **Trim** option determines what happens to the trimmed bits. When on (the default), the command trims away the selected edges, up to the fillet arc's endpoint. When off, no trim occurs

Enter Trim mode option [Trim/No trim] <Trim>: *(Enter T or N.)*

mUltiple

The **mUltiple** option repeats the command until you press **ESC**:

Select first object or [Polyline/Radius/Trim/mUltiple]: *(Press ESC to exit command.)*

CHAMFER POINTS

The **CHAMFER** command trims segments from the ends of two lines or polylines, and then draws a straight line or polyline segment between them. The distance to be trimmed from each segment can be different or the same. The two objects do not have to intersect, but they must be capable of intersecting if there were extended. Unlike fillets, parallel lines cannot be chamfered.

Chamfering only works with line segments, such as lines, 2D polylines, and traces. It does not work with arc segments, such as arcs, circles, and ellipses.

TUTORIAL: CHAMFERING OBJECTS

1. To fillet a pair of objects, start the **CHAMFER** command:
   - From the menu bar, choose **Modify**, and then **Chamfer**.
   - From the **Modify** toolbar, choose the **Chamfer** button.
   - At the 'Command:' prompt, enter the **chamfer** command.
   - Alternatively, enter the **cha** alias at the 'Command:' prompt.

   Command: **chamfer** *(Press ENTER.)*

2. In all cases, AutoCAD first displays the current chamfer settings:
   (TRIM mode) Current chamfer Dist1 = 0.0000, Dist2 = 0.0000

3. Before selecting objects, change the chamfer distance from the current setting of 0:
   Select first line or [Polyline/Distance/Angle/Trim/Method/mUltiple]: **d**
   Specify first chamfer distance <0.0000>: *(Enter a value, such as .5.)*

4. Chamfers work with two distances, which can be different:
   Specify second chamfer distance <0.5000>: *(Enter another value, such as .25.)*

5. Now pick the two lines to chamfer:
   Select first line or [Polyline/Distance/Angle/Trim/Method/mUltiple]: *(Pick line 1.)*
   Select second line: *(Pick line 2.)*

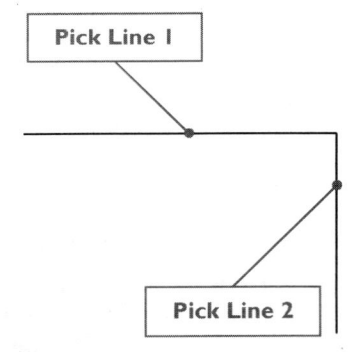

6. AutoCAD creates a chamfer at the intersection.

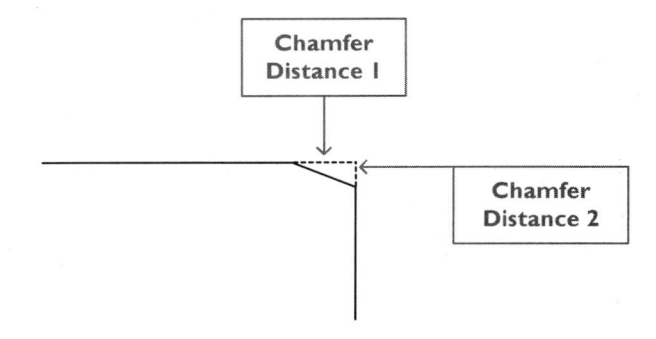

You specify the amount to be trimmed by entering either a numerical value, or by showing AutoCAD the distance using two points on the screen.

CONSTRUCTING CHAMFERS: ADDITIONAL METHODS

The command has several other options for special cases:

- **Polyline** option chamfers a single polyline.
- **Angle** option specifies the chamfer angle.
- **Trim** option determines whether end pieces are saved.
- **Method** option specifies the chamfer method.
- **mUltiple** option continues the command to fillet additional objects (new to AutoCAD 2004).

Let's look at each.

Polyline

Like filleting, chamfering a polyline is different from chamfering a pair of lines. If you had used the **PLINE** command's **Close** option to finish the polyline, AutoCAD chamfers *all* corners of the polyline; if not, the final vertex is not chamfered.

Select 2D polyline: *(Pick a polyline.)*

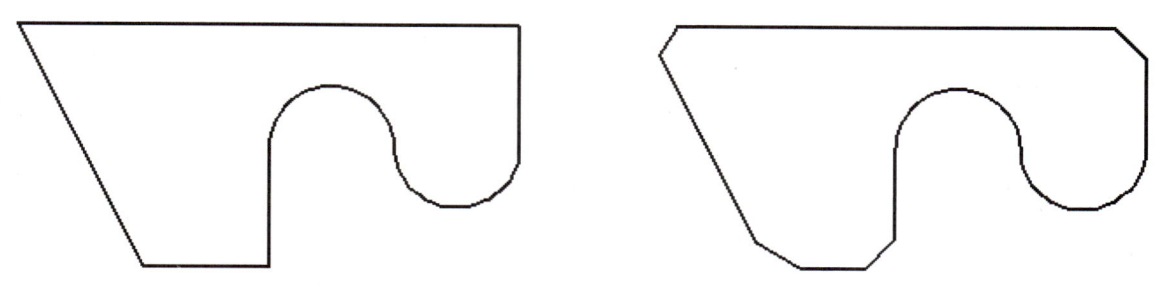

*A polyline chamfered with the **Polyline** option; notice that arc vertices are not chamfered.*

If the polyline contains arcs, they are not chamfered. If some parts of the polyline do not chamfer, it could be that the segments are too short or are parallel to each other. In that case, AutoCAD warns: "*n* were too short."

Angle

As an alternative to specifying a chamfer by two distances, AutoCAD lets you specify a distance and an angle.

Specify chamfer length on the first line <1.0000>: *(Enter the distance.)*

Specify chamfer angle from the first line <0>: *(Enter the angle.)*

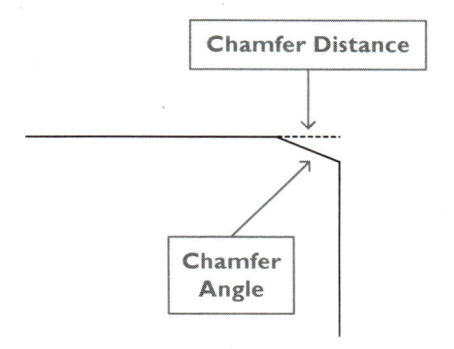

Trim

Normally, the CHAMFER command erases the line segments not needed after the chamfer. The **Trim** option, however, determines whether you keep the excess lines.

Enter Trim mode option [Trim/No trim] <Trim>: (Enter T or N.)

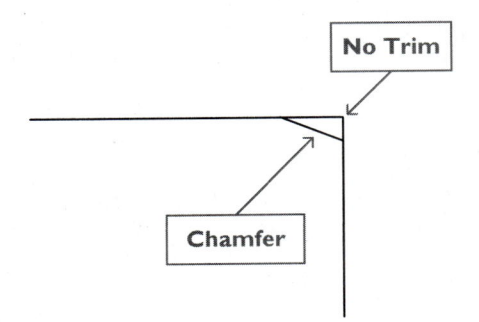

Method

The default method specifies two distances. If you prefer the distance-angle method, the **Method** option lets you change the default:

Enter trim method [Distance/Angle] <Angle>: (Enter D or A.)

mUltiple
(NEW IN 2004)

The **mUltiple** option repeats the command until you press ESC:

Select first object or [Polyline/Radius/Trim/mUltiple]: (Press ESC to exit command.)

REVCLOUD
(NEW IN 2004)

The REVCLOUD command creates revision clouds.

Revision clouds are often used to highlight areas in drawings that require attention, such as a revision or a potential error. Revision clouds are sometimes called "markups" or "redlines," because they were often drawn with red pencils to stand out in drawings.

AutoCAD's REVCLOUD command creates revision clouds, or converts other objects into revision clouds — specifically circles, ellipses, closed polylines, and closed splines. Because you cannot invoke transparent zooms and pans (other than through the mouse wheel) during the command, ensure you can see the entire area before starting.

TUTORIAL: REDLINING OBJECTS

1. To markup drawings, start the **REVCLOUD** command:
   - From the menu bar, choose **Draw**, then choose **Revision Cloud**.
   - From the **Draw** toolbar choose the **Revcloud** button.
   - At the 'Command:' prompt, enter the **revcloud** command.

   Command: **revcloud** *(Press ENTER.)*

2. AutoCAD displays the current settings, and then prompts you to start drawing the revision cloud:

   Minimum arc length: 0.5000   Maximum arc length: 0.5000

   Specify start point or [Arc length/Object] <Object>: *(Pick a point.)*

3. Move your cursor. Notice that AutoCAD automatically creates the cloud pattern. As an alternative, you can define the size of arcs by picking points.

   Guide crosshairs along cloud path...

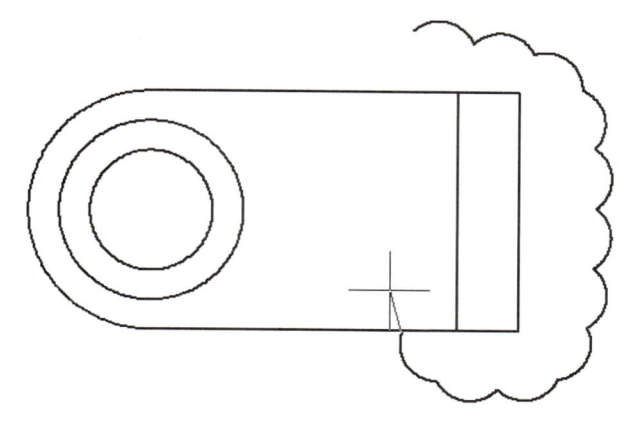

4. When the cursor is close to the start point, AutoCAD automatically closes the cloud: Revision cloud finished.

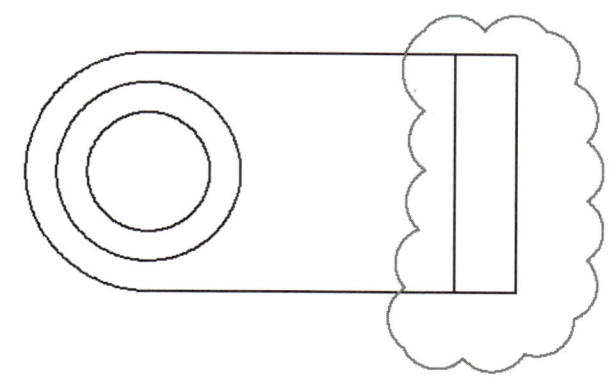

Note: Before starting **REVCLOUD**, switch to a layer set to red. That makes it easier to turn on and off the display of the revision cloud, and makes it traditional red, to boot.

After placing the revision cloud, you can edit it like other objects in the drawing — through coping, grips editing, erasing, and so on.

REDLINING OBJECTS: ADDITIONAL METHODS

To help you in redlining, AutoCAD has these options and commands:

- **Arc length** option defines the size of the arcs making up the clouds.
- **Object** option converts existing objects into revision clouds.
- **RMLIN** command imports *.rml* redline markup files from Volo View.

Let's look at each.

Arc length

The **Arc length** option determines the size of the arcs making up the clouds, which are scale-dependent, like linetypes and hatch patterns. **REVCLOUD** saves the arc length as a factor of the **DIMSCALE** system variable, so that new clouds, drawn after the dimension scale changes, still look the right size when the scale factor changes.

By giving the minimum and maximum lengths different values, AutoCAD draws clouds with random-size arcs. AutoCAD limits the maximum arc length to three times the minimum length.

Specify minimum length of arc <0.5000>: *(Enter a value.)*

Specify maximum length of arc <0.5000>: *(Enter a value.)*

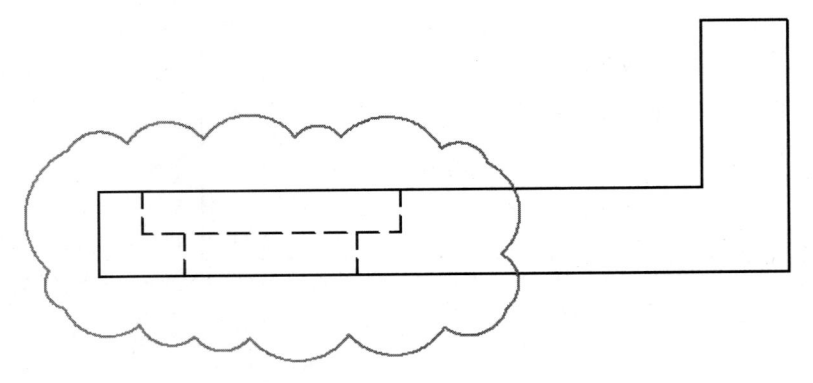

This revision cloud has arcs of varying sizes.

Object

The **Object** option lets you convert existing objects into revision clouds.

Select object: *(Select one object.)*

Reverse direction [Yes/No] <No>: *(Enter* **Y** *or* **N**.*)*

Revision cloud finished.

Top: The original polyline object.
Middle: Object converted to revision cloud.
Bottom: Revision cloud with reversed arcs.

You can convert closed polylines, circles, ellipses, and closed splines to revision clouds. After applying this option, what happens to the original object? That depends on the setting of the **DELOBJ** system variable. If set to 1 (the default), the original is erased; when set to 0, the original object stays in the drawing.

Command: **delobj**

Enter new value for DELOBJ <1>: *(Enter **1** or **0**.)*

RMLin

The **RMLIN** and **-RMLIN** commands insert *.rml* redline markup files created by Volo View.

You use Autodesk's Volo View software to view and print AutoCAD *.dwg* drawings and *.dxf* interchange files. In addition, the software allows for "redline markups" — comments and diagrams, such as rectangles, callouts, and revision clouds. Volo View provides a lower-cost method for more people to access AutoCAD drawings.

Comments made on drawings in Volo View are saved in *.rml* files. When imported into AutoCAD with the **RMLIN** command, the markups are placed on the red, locked layer called _MARKUP_, which AutoCAD creates automatically. Additional *.rml* imports are merged onto the same layer. (Technically, *.rml* is short for "redline markup language," because it is based on XML, the extended markup language.)

TUTORIAL: INSERTING REDLINES FROM VOLO VIEW

1. To insert an *.rml* file from Volo View, start the **RMLIN** command:
   - From the menu bar, choose **Insert**, and then **Markup**.
   - At the 'Command:' prompt, enter the **rmlin** command.

   Command: **rmlin** *(Press ENTER.)*

2. In either case, AutoCAD displays the Insert Markup dialog box.
   Select an *.rml* file, and then click **Open**.

3. AutoCAD displays the Map Markup dialog box, whose purpose is to map layouts stored in the *.rml* file with layouts in the current drawing.

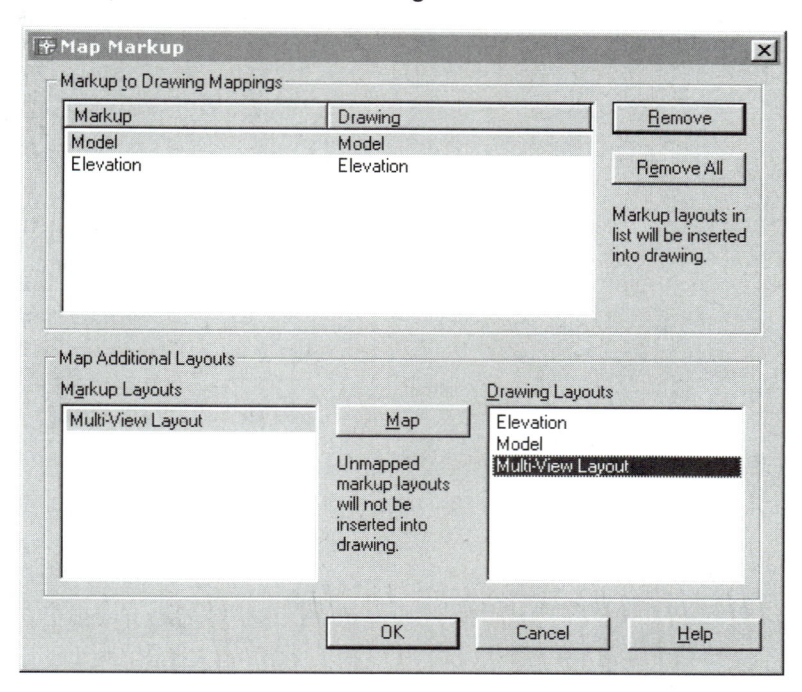

The **Markup to Drawing Mappings** section lists how markup layouts are mapped to drawing layouts. The **Remove** button removes the selected mapping (not the layout).

4. To change the layout mapping, click **Remove All**.

   Select a layout name in Markup Layouts, and then a matching layout under Drawing Layouts. Click **Map**, and repeat for all other layouts.

> **Note:** This dialog box lists the layout name of Default when there are no layouts in the *.rml* file.

5. Click **OK**, and AutoCAD adds the redlines to the drawing.

The **-RMLIN** command displays its prompts at the command line:

   Command: **-rmlin**
   Enter name of RML file: *(Enter a .rml file name, or type ~ for dialog box.)*
   Enter an option [Single/Multiple/sAme] <Single>: *(Enter an option.)*
   Enter a markup layout name or [?] <Model>: *(Enter the name of a layout, or type ? for a list of layouts.)*
   Enter a drawing layout name <Model>: *(Enter layout name to match.)*

The **Single** option displays prompts to map one markup layout to one drawing layout. The **Multiple** option displays prompts to map more than one markup layout to drawing layouts. The **Same** option automatically maps markup layouts to drawing layouts of the same name.

EXERCISES

1. From the CD-ROM, open the *edit3.dwg* drawing file, an architectural elevation.

   Use the **COPY** command to copy the windows from the left side to the right side. Then, copy all the windows on the lower level (including those you just copied) to the upper level. When copying, use the object selection options you think will work best.

2. From the CD-ROM, open the *edit4.dwg* drawing file, a landscaping plan.

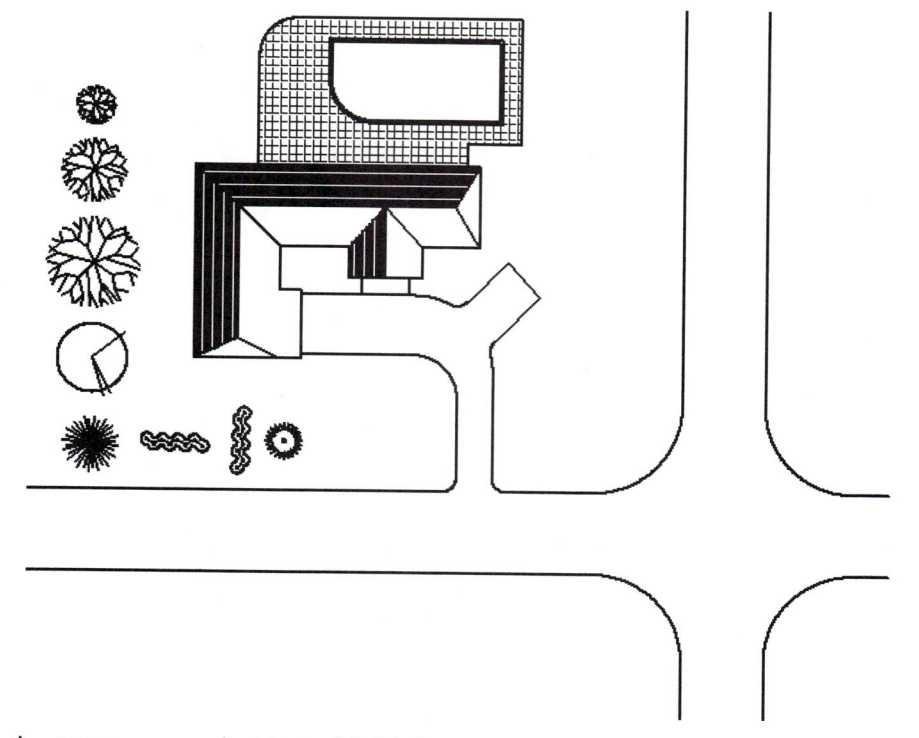

   Use the **COPY** command with the **Multiple** option to copy the landscaping blocks (trees, shrubbery, and so on) and to create a landscape scheme.

3. Connect two lines with fillets of varying radius.
   First, draw lines similar to those in the illustration.

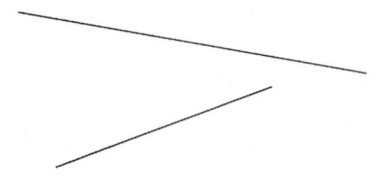

   a. Set the fillet radius to zero, and then apply the fillet to the two lines.
      Do the two lines now connect in a perfect intersection?
   b. Continue by connecting the same two lines with a fillet of radius of 0.15.

4. From the CD-ROM, open the *edit6.dwg* drawing file, a practice drawing.
   Fillet the objects to achieve the results shown on right side of the figure.

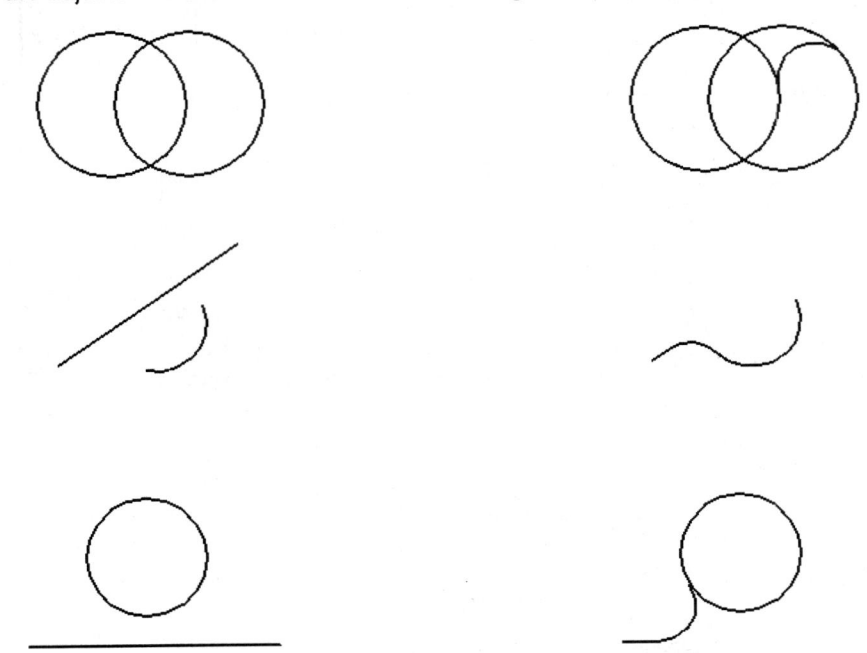

5. Suppose that you have been directed to add another circle to your drawing, which is identical to an existing circle.
   Draw a circle, and then use the **COPY** command to place a second circle.
   Is the second circle an exact copy of the first?

6. In this exercise, you create a rectangular array.
   Start a new drawing.
   As a visual aid, turn on the grid with a value of one.
   Draw a circle with the center point located at 1,1 and a radius of one.
   Select the circle, and then start the **ARRAY** command.
   In the dialog box, enter the following options:

Type of array	**Rectangular**
Rows	**3**
Columns	**5**
Row offset	**2**
Column offset	**2**

Your array should look like the following illustration:

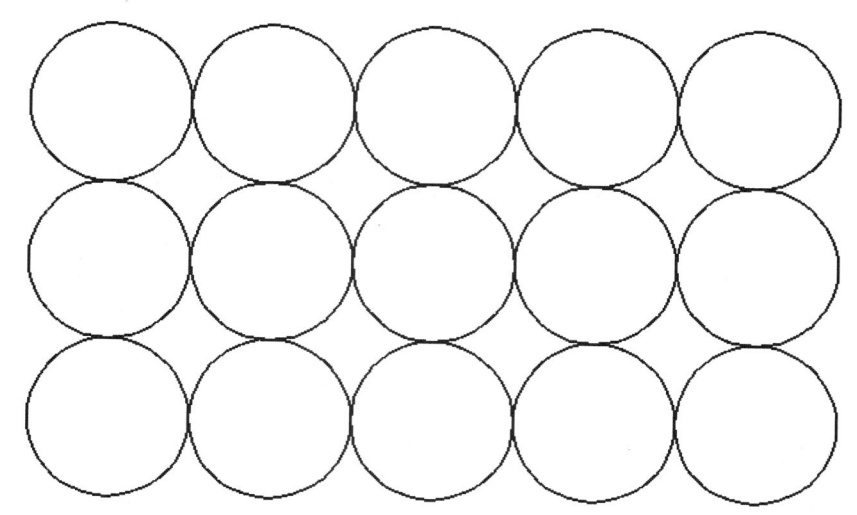

Save the drawing with the name *array.dwg*.

7. Start another drawing to create a polar array.
   Draw a square with sides of one unit each.
   Start the **ARRAY** command, and enter these options:

Select objects	*Select the square*
Type of array	**Polar**
Center point	**2,2**
Total number of items	**4**
Angle to fill	**270**
Rotate items as copied	**off**

   The polar array should look like the one in the figure.
   Save the drawing with the name *parray.dwg*.

8. From the CD-ROM, open the *mirror1.dwg* drawing file, a floor plan.
   Suppose that you are designing a house, and you want to reverse the layout of the bathroom.

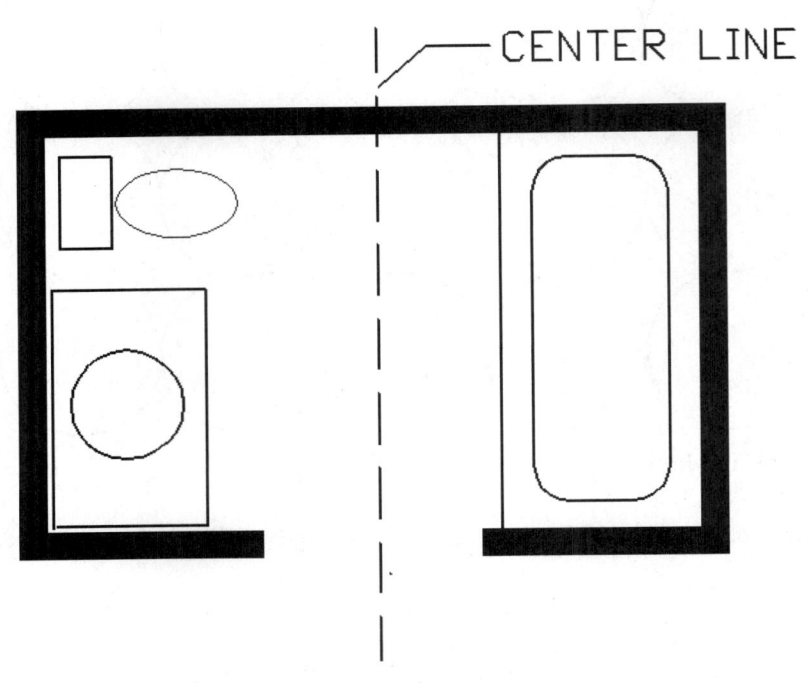

   Reverse the room using the **MIRROR** command.
   Hint: turn on ortho mode (**F8**), and delete the source objects.

9. In this exercise, the **PDMODE** system variable allows you to see points placed by the **DIVIDE** command.
   Start a new drawing.
   Draw a circle: center the circle in the viewport, and use a radius of 3.
   Start the **DIVIDE** command, and use 8 for the number of segments.
   Do you see any difference to the circle?
   Set the **PDMODE** system variable to 34.
   If necessary, use the **REGEN** command to make the new point style visible.

10. Draw a symbol similar to the one in the figure.
    Which two commands make it easier to draw the symbol?

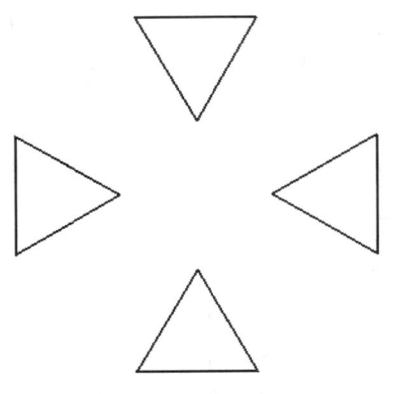

    Turn the symbol into a block, and name it "Symbol."
    a. Draw a circle on the screen, and then start the **DIVIDE** command.
       Use 8 segments, along with the **Block** option, with blocks aligned with the object.

b. Repeat the exercise, but this time don't rotate the block.

11. Let's "measure" an object.
    Draw a horizontal line 6 units in length.
    Keep the point mode setting of 34 to make the points visible.
    With the **MEASURE** command, specify a segment length of 1.0.

12. In this exercise, you use the **ELLIPSE** command to draw a can.
    Draw one ellipse with the **ELLIPSE** command. Don't worry about the size.
    Next, use the **COPY** command to copy the ellipse to create the top of the can.
    Hint: Use ortho mode to align the copy perfectly with the original ellipse.
    Now use the **LINE** command to draw lines between the two ellipses, as shown in the figure.
    Hint: Use QUADrant object snap to capture the outer quadrants of the ellipses.

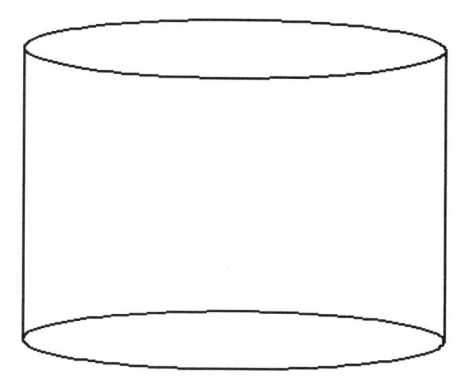

13. In this exercise, you use the **OFFSET** command to help draw a cityscape plan.
    From the CD-ROM, open the *offset.dwg* drawing file.

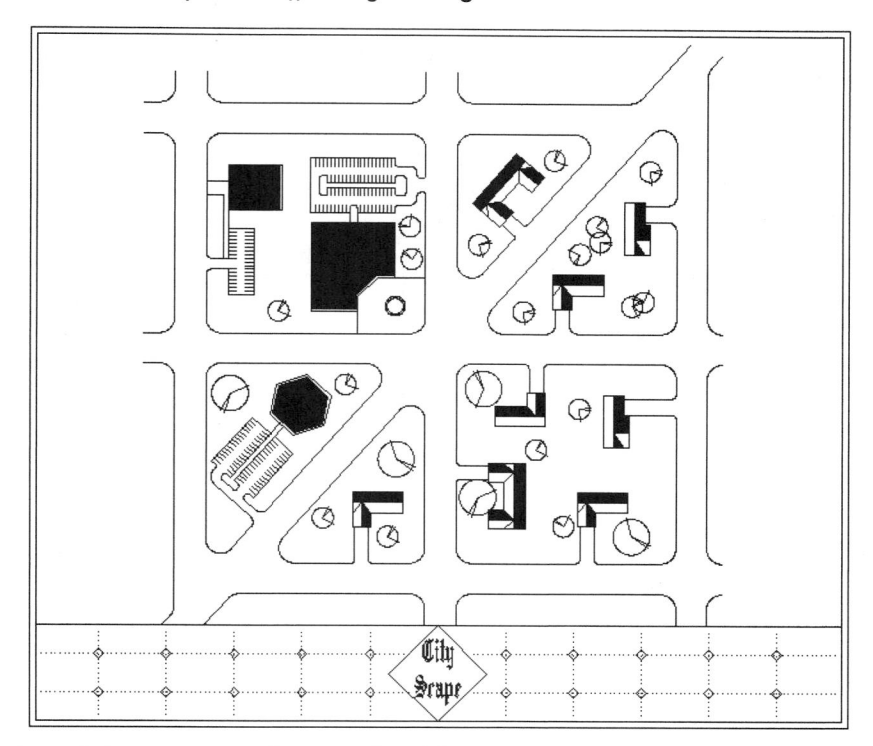

    The edges of the streets are drawn with polylines.
    Use the **OFFSET** command's **Through** option to offset the curbs by 6".

14. Draw the object shown in the figure below.
    Use the **CHAMFER** command to bevel all its corners.

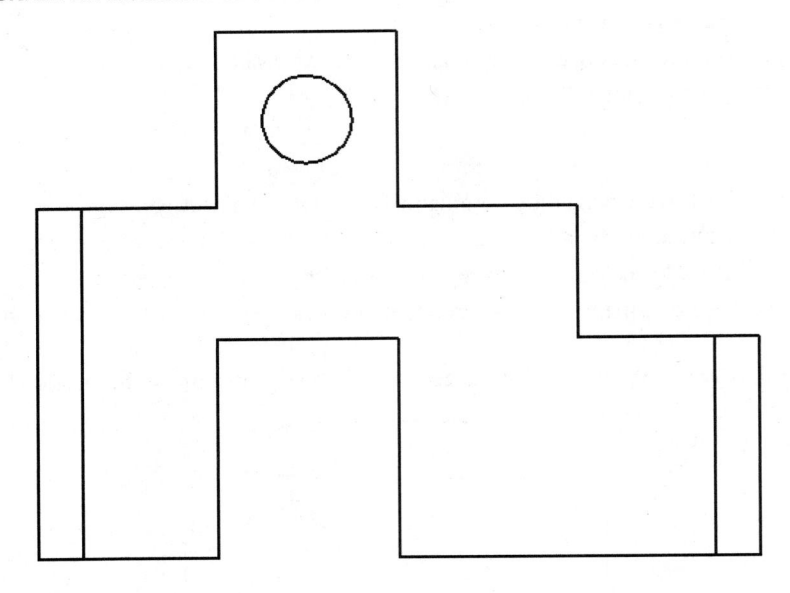

15. From the CD-ROM, open the *edit5.dwg* drawing, a piping diagram.
    Use the **REVCLOUD** command to place revision clouds around the two unfinished areas, as illustrated by the figure.

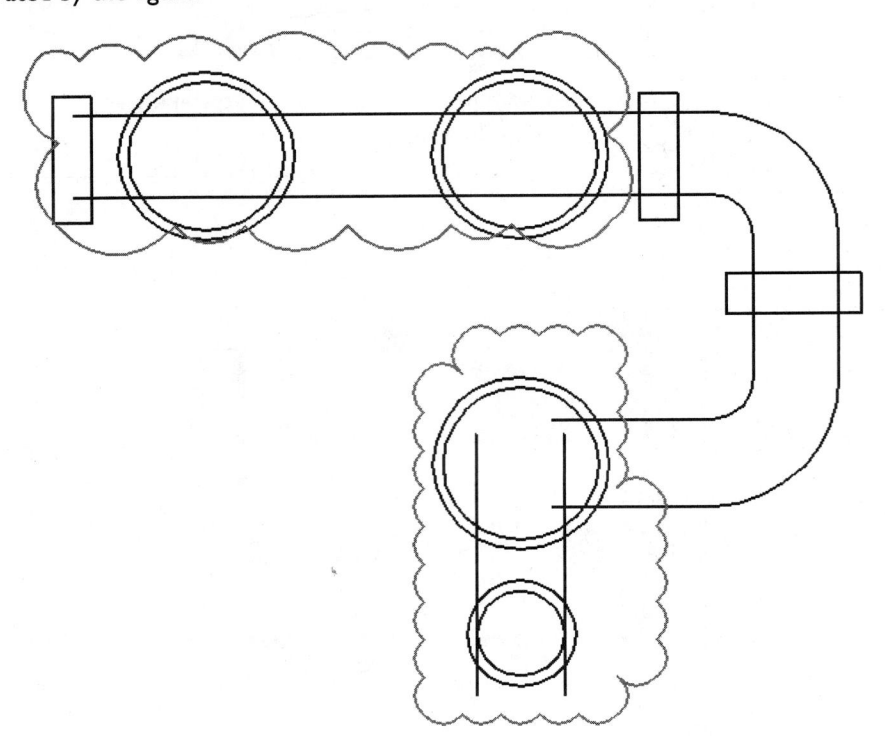

16. Draw the profiles of the aluminum extrusions diagrammed and dimensioned below. All dimension in inches.

   a. Angle.

A	B	C	D
4.000	1.000	0.155	0.125

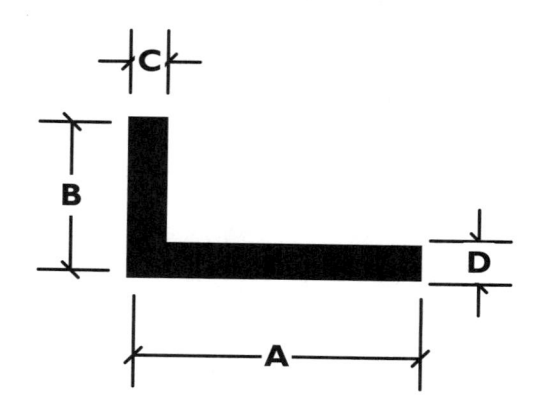

   b. Square tube.

A	T
4.000	0.145

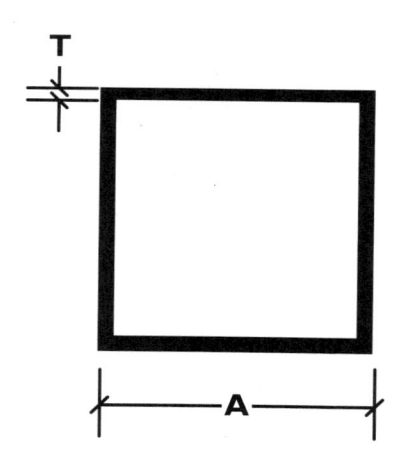

   c. Rectangular tube with rounded corners.

A	B	T	R1	R2
2.250	1.750	0.125	0.125	0.125

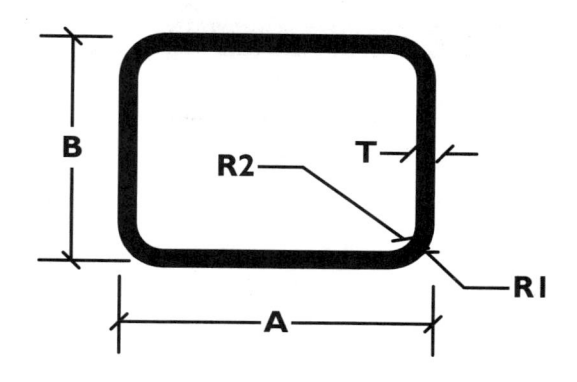

d. Rectangular bar with rounded corners.

A	B	R
3.250	0.375	0.030

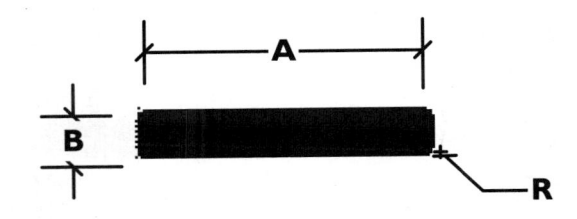

e. Tee.

A	B	T1	T2	R
2.000	2.000	0.125	0.125	0.015

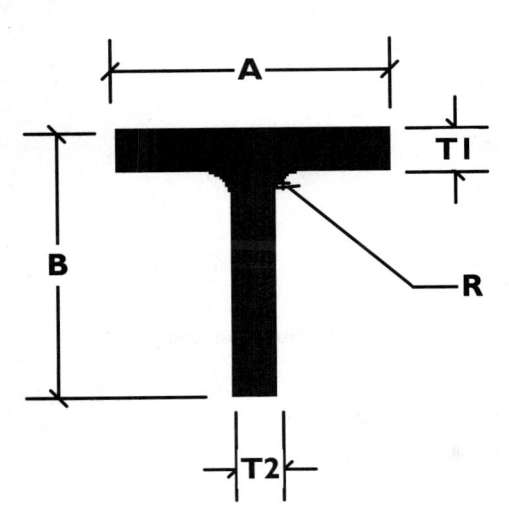

f. Channel with rounded corners.

A	B	C	D	R
3.000	1.500	0.375	0.375	0.375

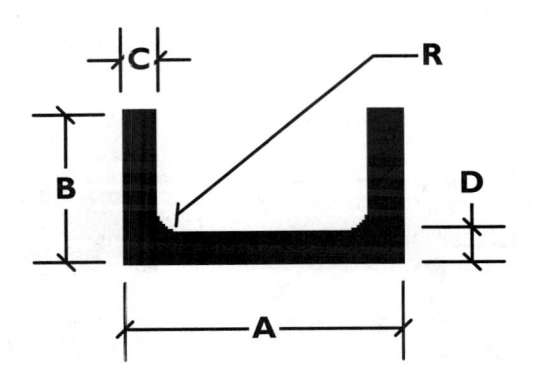

17. Draw the outline and foldlines of the diecut carboard packaging that holds a tube of toothpaste. Cutlines are shown in black; fold lines in white. All dimensions in mm. Print the drawing, and the cut and fold along the lines to recreate the carboard box.

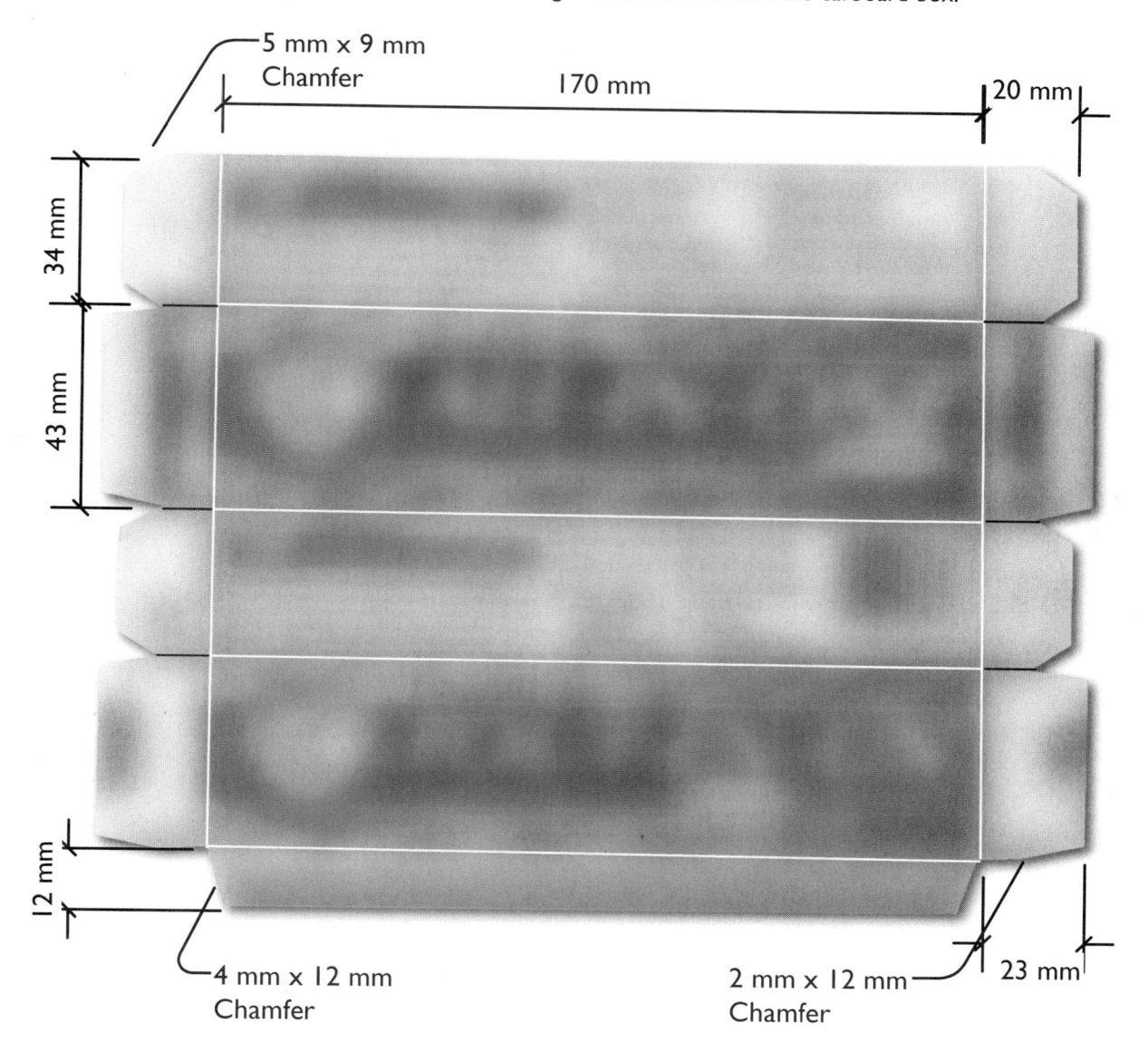

5 mm x 9 mm Chamfer

170 mm

20 mm

34 mm

43 mm

12 mm

4 mm x 12 mm Chamfer

2 mm x 12 mm Chamfer

23 mm

CHAPTER REVIEW

1. What two changes can a fillet make to an intersection?
2. When using the **Multiple** option of the **COPY** command, is each copy relative to the point of the last copy made, or the first base point entered?
3. Can objects of different types be filleted (such as a line and an arc), or must they be alike?
4. What are the three types of arrays?
5. Describe the function of the **MIRROR** command.
6. Which system variable controls whether text is mirrored?
7. What is the mirror line?
8. What objects can you use with the **MEASURE** command?
6. How many objects can be offset at one time?
7. Describe the purpose of the **Through** option of the **OFFSET** command?
8. Draw an example of using the **OFFSET** option of **Through**.
9. What is the purpose of the **CHAMFER** command?
10. What is the procedure for setting **CHAMFER** distances?
11. Describe the purpose of the **ARRAY** command.
12. Explain how the **-ARRAY** command differs from the **ARRAY** command.
13. Can the **ARRAY** command can draw arrays at an angle?
14. Under what condition does the **ARRAY** dialog box's **OK** button stay grayed (and unavailable)?
15. To construct an array in the -x direction, you enter a negative value in which option?
16. What could happen to your computer if you construct too large an array?
17. Define *displacement*, as used by the **COPY** command.
18. Which mode helps you copy objects precisely horizontally and vertically?
19. What are the aliases for the following commands:
    COPY
    MIRROR
    OFFSET
    FILLET
20. What problem can occur when the **OFFSET** command is applied to curved objects?
21. How does the **OFFSETGAPTYPE** system variable affect polylines?
22. Does the **MEASURE** command measure the lengths of objects?
23. What two kinds of objects does the **DIVIDE** command place along objects?
24. When you specify 5 segments for the **DIVIDE** command, how many objects does it place?
25. Can parallel lines be filleted?
26. Describe the purpose of the **FILLET** and **CHAMFER** commands' **Trim** option.
27. What is the difference between a fillet and a chamfer?
28. Can all vertices of a polyline be chamfered at once? If so, how?
29. Describe the purpose of the **REVCLOUD** command.
30. What is another name for "redlines"?
31. How does the **DELOBJ** system variable affect the **Object** option of the **REVCLOUD** command?
32. Which software creates *.rml* files?
33. Onto which layer does AutoCAD place imported *.rml* files?
34. What is the purpose of the **RMLIN** command's mapping function?

CHAPTER 11

Additional Editing Options

In previous chapters, you learned numerous methods of creating and changing objects. In Chapter 9, "Direct Editing of Objects," for instance, you changed objects directly through grips editing. In this chapter, you learn AutoCAD's other editing commands, including some that may already be familiar to you from grips editing. This chapter summarizes the remaining editing commands used on basic objects in two-dimensional drafting. They are:

- **ERASE** deletes objects from drawings.
- **OOPS** brings them back.
- **BREAK** removes portions of objects.
- **TRIM, EXTEND,** and **LENGTHEN** change the length of open objects.
- **STRETCH** makes portions of objects larger and smaller.
- **MOVE** moves objects in the drawing.
- **ROTATE** rotates objects.
- **SCALE** changes the size of objects.
- **CHANGE** changes the size, properties, and other characteristics of objects.
- **EXPLODE** and **XPLODE** reduce complex objects into their simplest form.
- **PEDIT** edits polylines.
- **SPLFRAME, SPLINETYPE,** and **SPLINESEGS** control the look of polyline splines.

FINDING THE COMMANDS

On the MODIFY and MODIFY II toolbars:

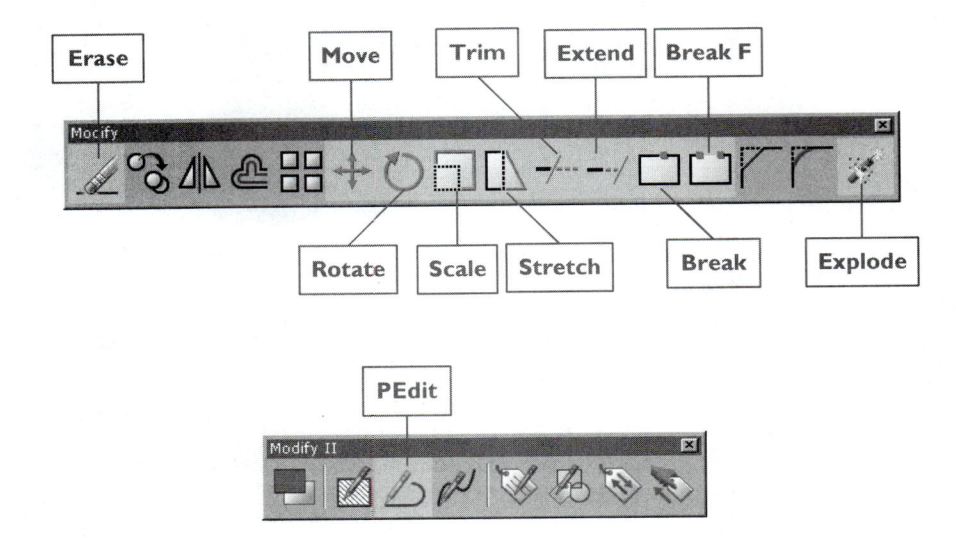

On the MODIFY menu:

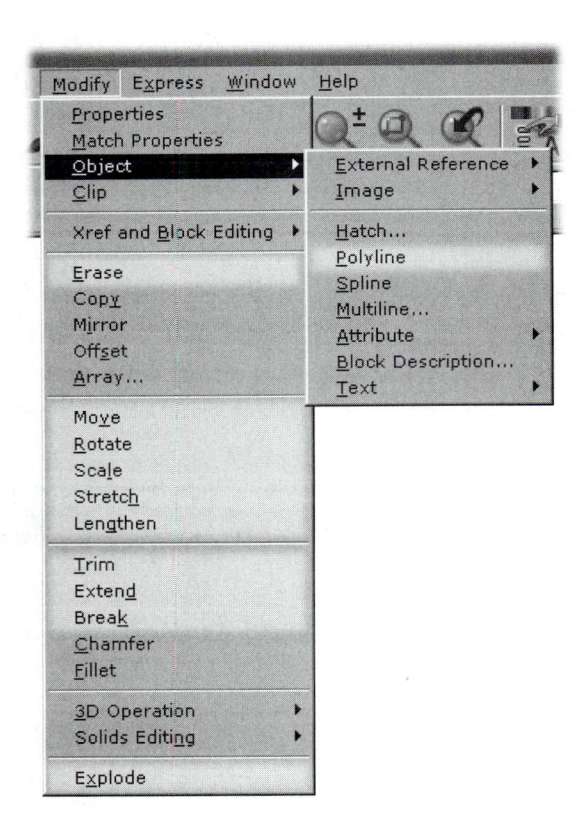

ERASE

The **ERASE** command deletes objects from drawings.

TUTORIAL: ERASING OBJECTS

1. To erase one or more objects, start the **ERASE** command:
   - From the menu bar, choose **Modify**, and then **Erase**.
   - From the **Modify** toolbar, choose the **Erase** button.
   - At the 'Command:' prompt, enter the **erase** command.
   - Alternatively, enter the **e** alias at the 'Command:' prompt.

   Command: **erase** *(Press ENTER.)*

2. In all cases, AutoCAD prompts you to select the objects you want erased:

   Select objects: *(Pick one or more objects.)*

   Select objects: *(Press ENTER to end object selection.)*

   Notice that the selected objects disappear from the drawing.

At the "Select objects:" prompt, the crosshair cursor changes to the pickbox for selecting one object at a time (the point method). If you wish to use one of the other object selection options (such as Window), enter the option at the "Select objects:" prompt.

It may be necessary to use the **REDRAW** command to refresh the screen following erasure.

RECOVERING ERASED OBJECTS

AutoCAD has two commands for restoring erased objects:

- **OOPS** returns the last erased objects.
- **U** undoes the last operation, including erasure.

Let's look at each.

Oops

The **OOPS** command restores the objects that were last erased from the drawing.

   Command: **oops**

This command also works in conjunction with the **BLOCK** and **-BLOCK** commands, when the original objects are erased in making the block.

U

If you find yourself unable to oops back erased objects, remember that you can undo the erase with the **U** command.

> **Notes:** To erase the object you just drew, enter **ERASE** with the **Last** option. A peculiarity is that AutoCAD only erases the last-drawn object if it is visible on the screen.
>
> Similarly, to erase the previous selection set, use **ERASE Previous**. You can erase the entire drawing with the **ERASE All** command. To do the same using keyboard shortcuts, press **CTRL+A** and then **DEL**.

BREAK

The **BREAK** command removes portions of objects.

To "break" an object, select two points on it. The portion of the object between the two points is erased. Unlike some other editing commands, **BREAK** operates on only one object at a time.

The command removes a portion of almost any object, and shorten open objects — depending on the place you pick. Objects that can be broken are: arcs, circles, ellipses, elliptical arcs, lines, polylines, rays, splines, and xlines.

BREAK can remove an object entirely: snap to one end of a line for the first point, and then snap to the other endpoint for the second point. The entire line disappears from the drawing. Use **U** to recover, if necessary.

BREAK is also useful for "cracking" objects: click the same point twice, and **BREAK** creates two segments that touch.

TUTORIAL: BREAKING OBJECTS

1. To remove a portion of an object, start the **BREAK** command:
   - From the menu bar, choose **Modify**, and then **Break**.
   - From the **Modify** toolbar, choose the **Break** button.
   - At the 'Command:' prompt, enter the **break** command.
   - Alternatively, enter the **b** alias at the 'Command:' prompt.

   Command: **break** *(Press* ENTER.*)*

2. In all cases, AutoCAD prompts you to select the single object you want broken:

   Select object: *(Pick one object at point 1.)*

   The point you pick is crucial, because it becomes the first point of the break.

3. Pick a second point, the other end of the break:

   Specify second break point or [First point]: *(Pick point 2.)*

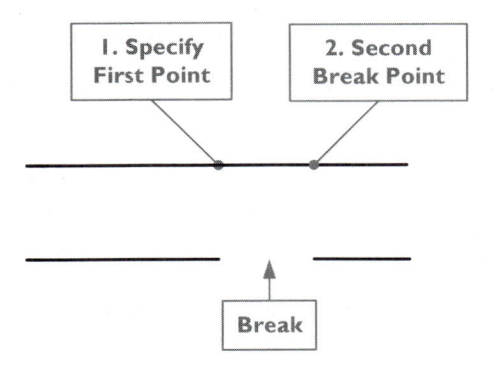

BREAKING OBJECTS: ADDITIONAL METHODS

By default, the first point you select on the object becomes the *first break point*. To redefine the first break point, enter **F** in response to the prompt. You then select another point to be the first break point. Redefining the first point is useful when the drawing is crowded, or when the break occurs at an intersection where pointing to the object at the first break point might result in the wrong object being selected.

- **First** defines the first pick point.

Let's look at this option.

First

The **First** option allows you to redefine the first pick point, providing greater control over the segment being broken out.

> Command: **break**
> Select object: *(Pick object.)*
> Specify second break point or [First point]: **f**
> Specify first break point: *(Pick point 1.)*
> Specify second break point: *(Pick point 2.)*

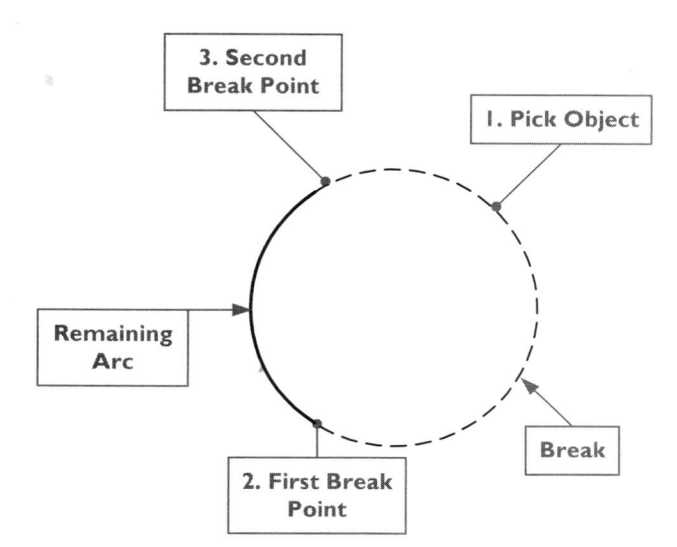

Note: The first tip I ever learned from technical editor Bill Fane was that sometimes it is easier to create an arc by drawing a circle, and then using the **BREAK** command to remove the unwanted portion.

The **BREAK** command affects objects in different ways:

Object	Break Action
Arcs, Lines	Removes the portion of the line or arc between the pick points.
	When one point is on the line, but the other point is off the end of the line, the line is "trimmed back" to the first break point.
Circles	Breaks into an arc.
	Unwanted piece is determined by going counterclockwise from the first point to the second point.
Ellipses	Same manner as circles, except that its results in an elliptical arc.
Splines	Same manner as lines and arcs.
Traces	Same manner as lines, except that the new endpoints are trimmed square.
Polylines	Cuts square, like traces.
	Breaking a closed polyline creates an open polyline.
Viewports	Cannot be broken.
Xlines	Become rays.

TRIM, EXTEND, AND LENGTHEN

The **TRIM**, **EXTEND**, and **LENGTHEN** commands change the length of open objects.

The **TRIM** command shortens objects by defining other objects as *cutting edges*; any portion of the object beyond the cutting line is "cut off."

AutoCAD needs to know two pieces of information for trimming: (1) the object(s) to be used as the cutting edges; and (2) the portion of the object to be removed.

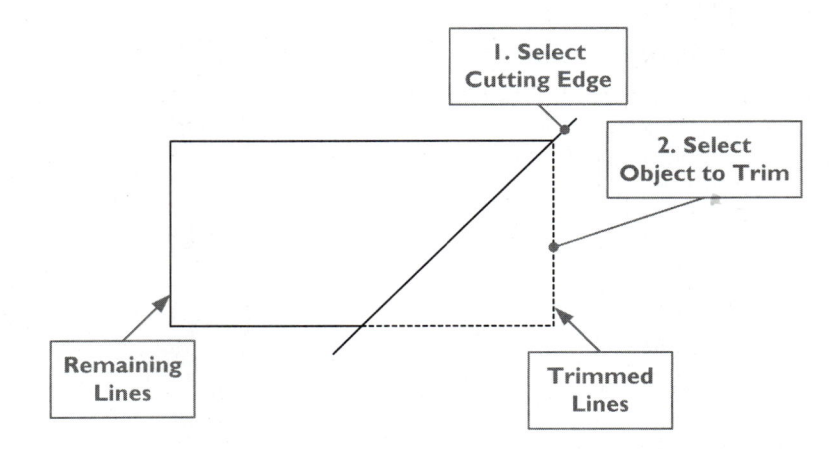

Cutting edges are any combination of arcs, circles, ellipses, lines, viewports, rays, regions, splines, text, xlines, 2D and 3D polylines (polylines with non-zero width trim to the centerline of the polyline). A quick method is to select *all* objects as cutting edges: at the "Select objects" prompt, press **ENTER**, and AutoCAD selects all objects in the drawing, including those not in the current view.

The part that is picked is the part that is trimmed. An undocumented method of quickly selecting many objects to trim is the **Fence** selection mode.

TUTORIAL: SHORTENING OBJECTS

1. To shorten one or more objects, start the **TRIM** command:
   * From the menu bar, choose **Modify**, and then **Trim**.
   * From the **Modify** toolbar, choose the **Trim** button.
   * At the 'Command:' prompt, enter the **trim** command.
   * Alternatively, enter the **tr** alias at the 'Command:' prompt.

   Command: **trim** *(Press ENTER.)*

2. In all cases, AutoCAD reports the current trim settings, and then prompts you to select cutting edges:

   Current settings: Projection=UCS Edge=None

   Select cutting edges ...

   Select objects: *(Pick one or more objects as cutting edges.)*

   Select objects: *(Press ENTER to end object selection.)*

3. Select the object to be trimmed. *Warning!* AutoCAD trims the portion you select.

   Select object to trim or shift-select to extend or [Project/Edge/Undo]: *(Pick object to be trimmed.)*

4. AutoCAD repeatedly prompts you to select additional objects to trim. When done trimming, press **ENTER**.

Select object to trim or shift-select to extend or [Project/Edge/Undo]: *(Press **ENTER** to exit the command.)*

When you select objects that cannot serve as cutting edges, such as blocks and dimensions, AutoCAD displays the message:

No edges selected.

If an object cannot be trimmed, AutoCAD displays:

Cannot TRIM this object

If the object to be trimmed does not intersect a cutting edge, AutoCAD complains:

Object does not intersect an edge.

Trimming Closed Objects

To trim closed objects, such as circles and polygons, they must be intersected with two cutting edges, such as with a line drawn through two points on a circle's circumference. If only one cutting edge intersects the closed object, AutoCAD complains:

Circle must intersect twice.

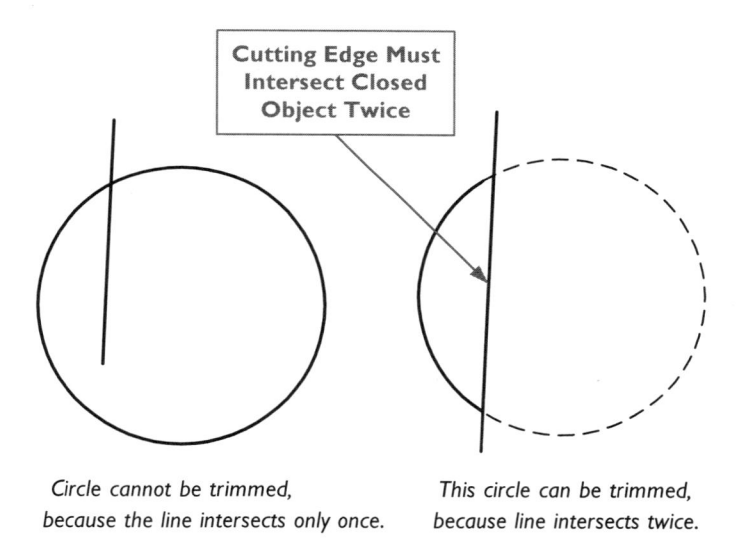

Circle cannot be trimmed, because the line intersects only once.

This circle can be trimmed, because line intersects twice.

Trimming Polylines

Polylines are trimmed at the intersection of the centerline of the polyline and the cutting edge. The trim is a square edge. Therefore, if the cutting edge intersects a polyline of non-zero width at an angle, the square-edged end may protrude beyond the cutting edge.

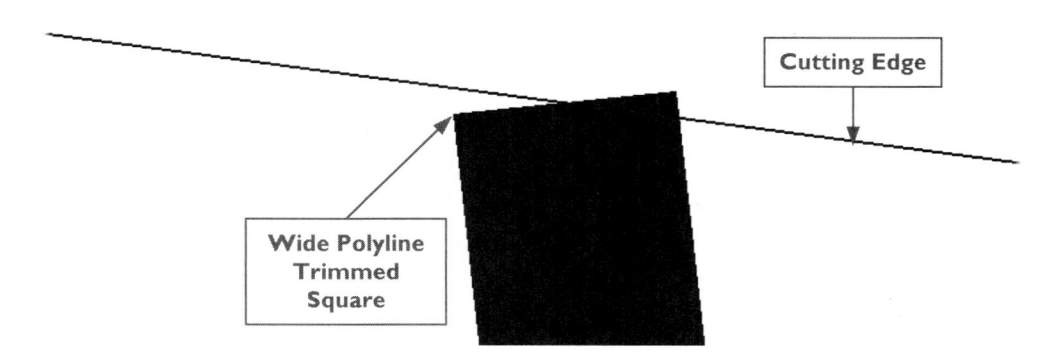

Notes: The **TRIM** and **EXTEND** commands can switch roles:
To extend an object during the **TRIM** command, hold down the **SHIFT** key while selecting the object. Similarly, to trim an object during the **EXTEND** command, hold down the **SHIFT** key while selecting the object.

TRIMMING OBJECTS: ADDITIONAL METHODS

The TRIM command has several options:

- **Project** determines trimming in 3D space.
- **Edge** toggles actual and implied cutting edges in 3D space.
- **Undo** undoes the last trim.
- **SHIFT-select** extends the object to the cutting line, instead of trimming.

Let's look at each option.

Project

The **Project** option is used in 3D drafting to specify the projection AutoCAD uses when it trims objects.

Select object to trim or [Project/Edge/Undo]: **p**
Enter a projection option [None/Ucs/View]: *(Enter an option.)*

The sub-options are:

None specifies no projection; AutoCAD trims only objects that actually intersect with cutting edges.

UCS projects the objects onto the x,y-plane of the current UCS (user-defined coordinate system). AutoCAD trims objects, even if they do not intersect with the cutting edge in 3D space—in effect, flattening 3D objects into the 2D plane.

View specifies a projection along the current view direction. AutoCAD trims objects that look as if they should be trimmed from your viewpoint, even if they don't physically intersect.

Edge

The **Edge** option is also used in 3D drafting, and determines whether objects are trimmed at an actual cutting edge, or at an implied cutting edge:

Select object to trim or [Project/Edge/Undo]: **e**
Enter an implied edge extension mode [Extend/No extend]: *(Enter an option.)*

The options are:

Extend causes AutoCAD to project the cutting edge so that it intersects the object(s).

No extend trims objects at cutting edges that physically intersect; cutting edges are not extended.

Undo

The **Undo** option undoes the last trim.

Extend (SHIFT-select)

When you hold down the **SHIFT** key at the "Select object to trim or shift-select to extend" prompt, AutoCAD extends the object to the cutting line, instead of trimming — in effect, reversing the **TRIM** command. The complimentary option is available in the **EXTEND** command.

EXTEND

The **EXTEND** command extends objects in a drawing to meet a boundary object. It functions very much like the **TRIM** command. Instead of a cutting edge, this command uses *boundary* objects — the point to which selected objects are extended.

AutoCAD needs to know two things: (1) the object(s) to be used as boundary objects; and (2) the object to extend.

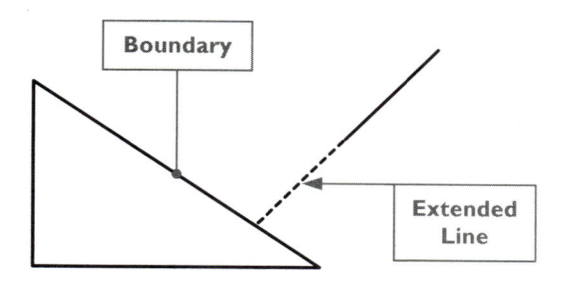

Boundary objects can be lines, arcs, circles, polylines, ellipses, splines, rays, xlines, regions, blocks, and viewports (in paper space). If several boundary edges are selected, the objects are extended to the first boundary encountered. If none can be met, AutoCAD complains:

> No edges selected.

Generally, closed objects cannot be extended. Specifically: text, splines, 3D solids, xlines. mutlilines, ellipses, donuts, hatches, revclouds, regions, 3D surfaces, and points (dimensions can be extended). If object cannot be extended, AutoCAD displays:

> Cannot EXTEND this object.

Take care! To tell AutoCAD which end of the object to extend, you must point to the object close to the end to be extended.

TUTORIAL: EXTENDING OBJECTS

1. To extend one or more objects, start the **EXTEN** command:
   * From the menu bar, choose **Modify**, and then **Extend**.
   * From the **Modify** toolbar, choose the **Extend** button.
   * At the 'Command:' prompt, enter the **extend** command.
   * Alternatively, enter the **ex** alias at the 'Command:' prompt.

   Command: **extend** *(Press ENTER.)*

2. In all cases, AutoCAD reports the current extend settings, and then prompts you to select boundary edges:

   Current settings: Projection=UCS Edge=None

   Select boundary edges ...

   Select objects: *(Pick one or more objects as boundaries.)*

   Select objects: *(Press ENTER to end object selection.)*

3. Select the object to be extended. AutoCAD extends the end that you select.

   Select object to extend or shift-select to trim or [Project/Edge/Undo]: *(Pick object to be extended.)*

4. AutoCAD repeatedly prompts you to select additional objects to extend. When done trimming, press **ENTER**:

   Select object to trim or shift-select to extend or [Project/Edge/Undo]: *(Press ENTER to exit command.)*

Extending Polylines

Polylines of non-zero width are extended until the centerline meets the boundary object; similarly, objects extend to the centerline of the polylines. Only open polylines can be extended. If you attempt to extend a closed polyline, AutoCAD complains:

> Cannot extend a closed polyline.

If a wide polyline and the boundary intersect at an angle, a portion of the square end of the polyline may protrude over the boundary. Extending a tapered polyline adjusts the length of the segment that tapers — the taper extends over the longer length.

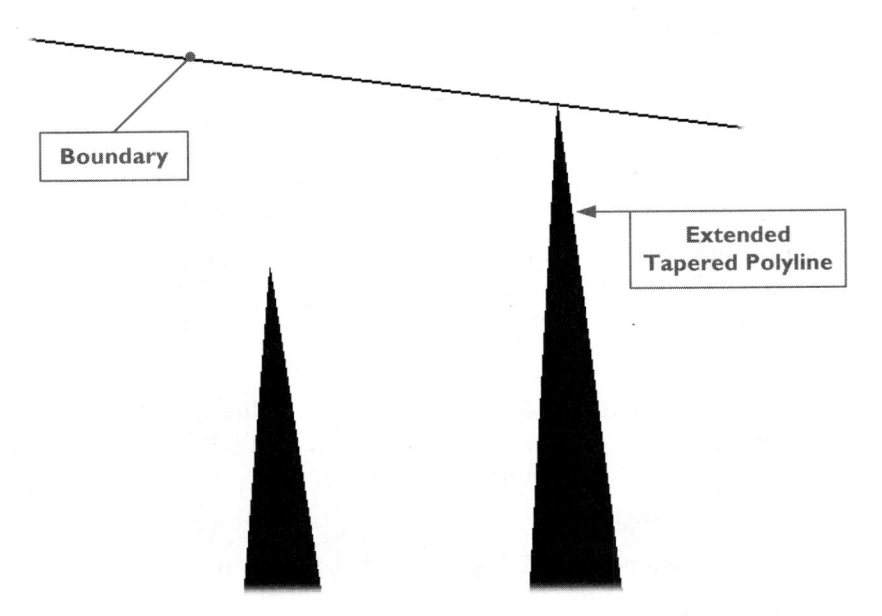

When a tapered polyline is extended to a boundary, its taper is extended as well.

EXTEND OPTIONS

The **EXTEND** command's options are identical to those of **TRIM**'s.

LENGTHEN

The **LENGTHEN** command is a faster version of **TRIM** and **EXTEND**: it changes the length of open objects, making them longer or shorter; it does not work with closed objects. **LENGTHEN** is faster because you do not need cutting edges or boundaries: you just point in the drawing for the change to occur, or specify a percentage change numerically.

This command works with lines, arcs, open polylines, elliptical arcs, and open splines.

TUTORIAL: LENGTHENING OBJECTS

1. To change the length of an open object, start the **LENGTHEN** command:
   * From the menu bar, choose **Modify**, and then **Lengthen**.
   * At the 'Command:' prompt, enter the **lengthen** command.
   * Alternatively, enter the **len** alias at the 'Command:' prompt.

   Command: **lengthen** *(Press ENTER.)*

2. In all cases, AutoCAD prompts you to select an object. AutoCAD reports the length of the object:

   Select an object or [DElta/Percent/Total/DYnamic]: *(Select one open object.)*
   Current length: 24.6278

3. AutoCAD repeats the prompt. It wants you to select an option:

    Select an object or [DElta/Percent/Total/DYnamic]: *(Enter an option, such as **dy**.)*

4. Curiously, AutoCAD has forgotten which object you selected; you need to select it a second time — or, you can select a different object:

    Select an object to change or [Undo]: *(Select one open object.)*

5. As you move the cursor, notice that AutoCAD ghosts in the lengthened line. When satisfied with the new length, click.

    Specify new end point: *(Move the cursor, and then click.)*

6. AutoCAD repeats the prompt. When done lengthening, press **ENTER**:

    Select an object to change or [Undo]: *(Press **ENTER** to exit the command.)*

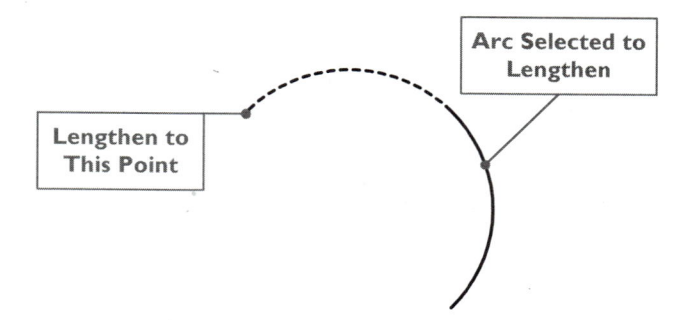

Note: When you select an object at the initial "Select an object:" prompt, AutoCAD reports on its length, as follows: *Current length: 10.2580.* For an arc, AutoCAD also reports its angle:

Current length: 8.4353, included angle: 192.

When you select a closed object, such as a circle or closed polyline, AutoCAD complains, "This object has no length definition." You cannot lengthen dimensions, hatch patterns, and splines.

LENGTHENING OBJECTS: ADDITIONAL METHODS

The **LENGTHEN** command changes the length of open objects by four methods:

- **Delta** by an incremental amount.
- **Percent** by a percentage.
- **Total** to the total amount.
- **DYnamic** by cursor movement.

Let's look at how each option.

Delta

The **DElta** option changes the length by adding the indicated amount to the object. Enter a negative number to shorten the object. As an alternative to typing a value, you can indicate the amount by picking two points anywhere in the drawing.

    Command: **lengthen**
    Select an object or [DElta/Percent/Total/DYnamic]: **de**
    Enter delta length or [Angle] <0.0000>: *(Enter a number, or pick two points on screen.)*
    Select an object to change or [Undo]: *(Select object.)*
    Select an object to change or [Undo]: *(Press **ESC** to exit the command.)*

Note that this command lengthens the open object at the end you select. To lengthen both ends of an object, select it twice: once at one end, and again at the other.

The **Angle** option changes the angle of a selected arc or polyarc. The command adds length to the end of the arc you select:

> Enter delta angle <0>: *(Type an angle, or pick two points.)*
>
> Select an object to change or [Undo]: *(Select the arc.)*
>
> Select an object to change or [Undo]: *(Press* ESC *to exit the command.)*

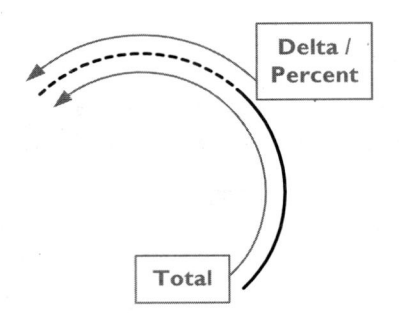

The arc is lengthened by the angle you specify. When you select two points, AutoCAD uses the angle of the rubber-band line, not the length of the line.

The angle you specify must be small enough for the arc not to total 360 degrees. If the angle is too large, AutoCAD curtly informs you: "Invalid angle."

Percent

The **Percent** option changes the length by a percentage of the object. For example, entering **25** shortens a line to 25 percent of its original length, while entering **200** doubles its length.

> Select an object or [DElta/Percent/Total/DYnamic]: **p**
>
> Enter percentage length <100.0000>: *(Enter a percentage, such as* **25**.*)*
>
> Select an object to change or [Undo]: *(Select the object.)*
>
> Select an object to change or [Undo]: *(Press* ESC *to exit the command.)*

Total

The **Total** option changes the length of a line to an absolute length. For example, a value of **5** changes the line to a length of 5.0 units, no matter its existing length.

> Select an object or [DElta/Percent/Total/DYnamic]: **t**
>
> Specify total length or [Angle] <1.0000>: *(Enter a value, such as* **5**.*)*
>
> Select an object to change or [Undo]: *(Select the object.)*
>
> Select an object to change or [Undo]: *(Press* ESC *to exit the command.)*

For arcs, this option changes the length to an absolute circumference. The **Angle** option changes the length of an arc to the included specified angle.

Dynamic

The **DYnamic** option visually changes the length of open objects. Notice that this option reverses the order: (1) first you select the object to lengthen, and (2) then you specify the length.

> Select an object or [DElta/Percent/Total/DYnamic]: **dy**
>
> Select an object to change or [Undo]: *(Select the object.)*
>
> Specify new end point: *(Pick a point.)*
>
> Select an object to change or [Undo]: *(Press* ESC *to exit the command.)*

After you select the object, the length of the object changes as you move the cursor. You can use object snap modes to make the dynamic lengthening more accurate.

STRETCH

The **STRETCH** command makes portions of objects larger and smaller, *and* moves selected objects, while retaining their connections to other objects. As you will see, it is one of the most useful editing commands in AutoCAD — second only to the **UNDO** command. The command works with lines, arcs, elliptical arcs, solids, traces, rays, splines, and polylines.

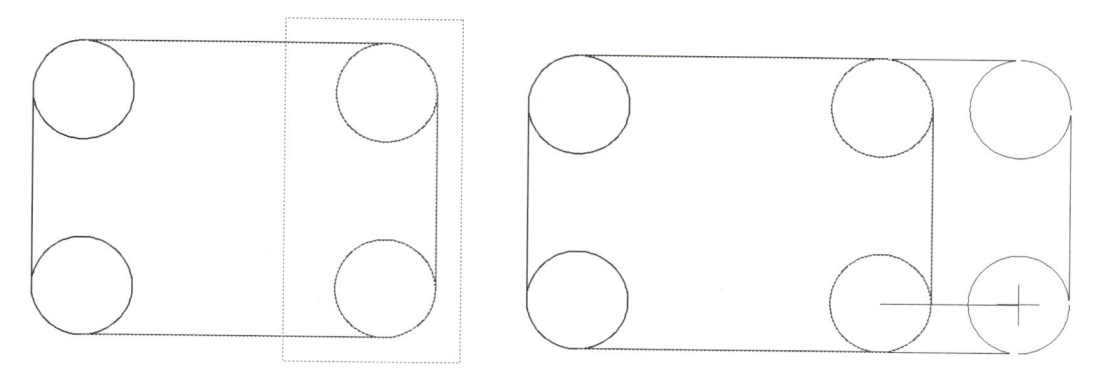

Objects within the selection window are moved; those crossing the window are stretched.

STRETCHING — THE RULES

You must understand the rules associated with the **STRETCH** command to execute it properly:

Rule 1: Crossing Selection. You can use any object selection mode, but at least one must be **Crossing** or **CPolygon**. If you do not use a *windowed* selection, AutoCAD complains, "You must select a crossing or polygon window to stretch." When more than one windowed selection mode is used to select objects, the **STRETCH** command uses the last window. The window must include at least one vertex or endpoint of the object.

Rule 2: Move vs. Stretch. If the object is completely inside the selection, it will be moved rather than stretched. Objects — specifically arcs, elliptical arcs, lines, polyline segments, solids, rays, traces, and splines — entirely within the selection window are moved just like the **MOVE** command. The endpoints within the selection window are moved, while those outside remain fixed.

Arcs are stretched like lines, except that the arc's center, start, and endpoints are adjusted so the distance from the midpoint of the chord to the arc is constant. Polylines are handled by their individual segments; the polyline's width, tangent, and curve fitting are not affected.

Rule 3: Definition Points. Some objects, such as circles and blocks, cannot be stretched; they are either moved or left alone by the **STRETCH** command, depending on their *definition point*. If this point lies inside the selection window, the object is moved; if outside, it is not affected.

Object	Definition Point
Point	Center of the point.
Circle	Center point of the circle.
Block	Insertion point.
Text	Leftmost point of the text line.

TUTORIAL: STRETCHING OBJECTS

1. To stretch one or more objects, start the **STRETCH** command:
   - From the menu bar, choose **Modify**, and then **Stretch**.
   - From the **Modify** toolbar, choose the **Stretch** button.
   - At the 'Command:' prompt, enter the **stretch** command.

- Alternatively, enter the **s** alias at the 'Command:' prompt.

    Command: **stretch** *(Press* ENTER.*)*

2. In all cases, AutoCAD prompts you to select the objects using a crossing selection mode. If you do not enter a selection option, such as Crossing, AutoCAD defaults to AUtomatic.

    Select objects to stretch by crossing-window or crossing-polygon...

    Select objects: **c**

    Specify first corner: *(Pick a point.)*

    Specify opposite corner: *(Pick another point.)*

    Select objects: *(Press* ENTER *to end object selection.)*

3. Identify the points that measure the displacement:

    Specify base point or displacement: *(Pick a point, using an object snap mode if necessary.)*

    Specify second point of displacement: *(Pick the destination point.)*

For the displacement, you can enter Cartesian, polar, cylindrical, or spherical coordinates — without the @ prefix, because AutoCAD assumes relative coordinates.

An alternative to the **STRETCH** command is grips editing, as described in Chapter 9.

MOVE

The **MOVE** command moves objects in the drawing.

It is similar to the **COPY** command in that AutoCAD needs to know two things: (1) the objects to move; and (2) the displacement — the distance to move (see Chapter 10, "Constructing Objects").

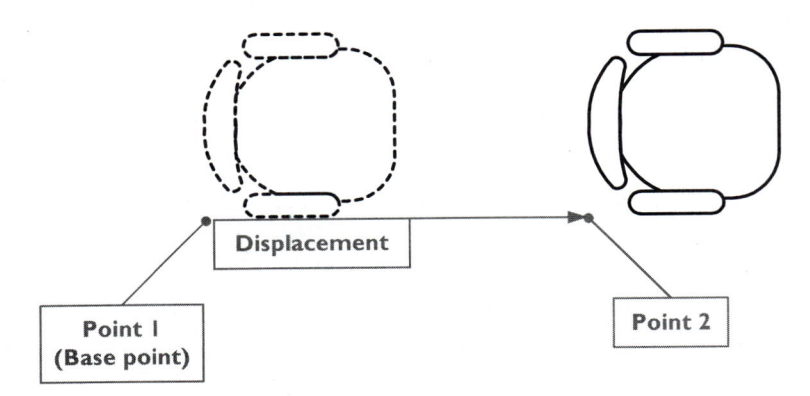

Displacement

Point 1 (Base point)

Point 2

TUTORIAL: MOVING OBJECTS

1. To move one or more objects, start the **MOVE** command:
    - From the menu bar, choose **Modify**, and then **Move**.
    - From the **Modify** toolbar, choose the **Move** button.
    - At the 'Command:' prompt, enter the **move** command.
    - Alternatively, enter the **m** alias at the 'Command:' prompt.

    Command: **move** *(Press* ENTER.*)*

2. In all cases, AutoCAD prompts you to select the objects to move:

    Select objects: *(Select one or more objects.)*

Select objects: *(Press ENTER to end object selection.)*
3. Identify the point from which the displacement is measured:
Specify base point or displacement: *(Pick point 1.)*
4. And pick the second point:
Specify second point of displacement or <use first point as displacement>: *(Pick point 1.)*

The first point need not be on the object to be moved; using a corner point or another convenient point of reference on the object makes the displacement easier to visualize. I find that object snap and ortho mode help make the move more precise.

An alternative to the **MOVE** command is grips editing, as described in Chapter 9.

ROTATE

The **ROTATE** command turns the orientation of objects about a base point. The *base point* is the point about which the selected objects are rotated.

You can specify the rotation angle three ways: by entering an angle; by dragging the angle; or by choosing a reference angle. (The similar sounding **REVOLVE** command is meant for creating 3D solid models.)

TUTORIAL: ROTATING OBJECTS

1. To turn one or more objects, start the **ROTATE** command:
   - From the menu bar, choose **Modify**, and then **Rotate** .
   - From the **Modify** toolbar, choose the **Rotate** button.
   - At the 'Command:' prompt, enter the **rotate** command.
   - Alternatively, enter the **ro** alias at the 'Command:' prompt.

   Command: **rotate** *(Press ENTER.)*

2. In all cases, AutoCAD indicates the current rotation settings, and then prompts you to select the objects you want rotated:
   Current positive angle in UCS: ANGDIR=counterclockwise ANGBASE=0
   Select objects: *(Pick one or more objects.)*
   Select objects: *(Press ENTER to end object selection.)*

1. Select Objects

2. Pick Basepoint

3. Specify Rotation Angle

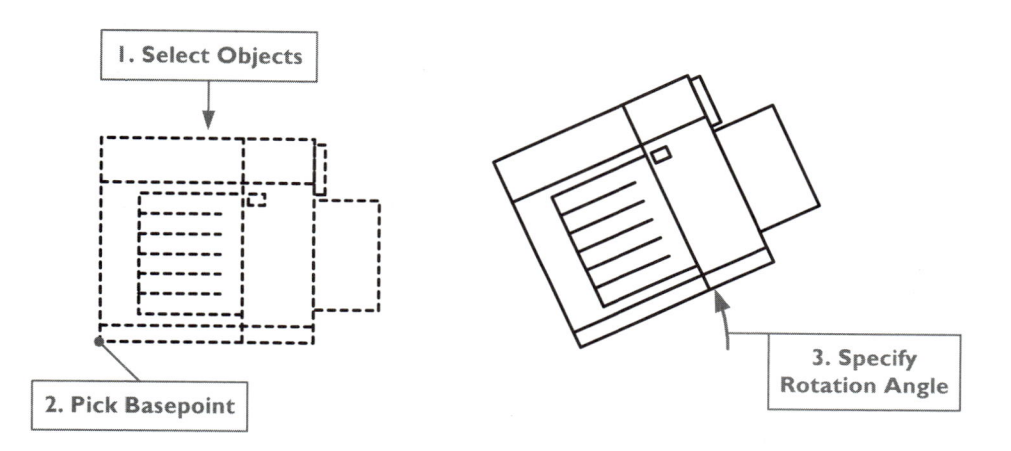

3. Identify the point about which the objects are rotated:
   Specify base point: *(Pick a point, such as the lower left corner.)*
4. Specify the angle of rotation:
   Specify rotation angle or [Reference]: *(Enter an angle, such as **45**.)*

You can specify a simple angle, or change one angle to another. For example, if an object is currently oriented at 58 degrees, and you wish to rotate the object to 26 degrees, rotate the object the difference of –32 degrees.

Positive angles rotate counterclockwise; negative angles, clockwise.

ROTATING OBJECTS: ADDITIONAL METHODS

The **ROTATE** command has two alternative methods for specifying the angle. In addition, rotation is affected by two system variables.

- **Reference** rotates relative to another angle.
- **Dragging** rotates the object in real-time.
- **ANGDIR** system variable specifies the direction of positive angles.
- **ANGBASE** system variable specifies the direction of 0 degrees.

Let's look at each.

Reference

The **Reference** option is used when you need to align the object to another. You don't need to know the angle of the other object, but object snaps are most useful. In this example, we want Desk B to be at the same slant as Desk A.

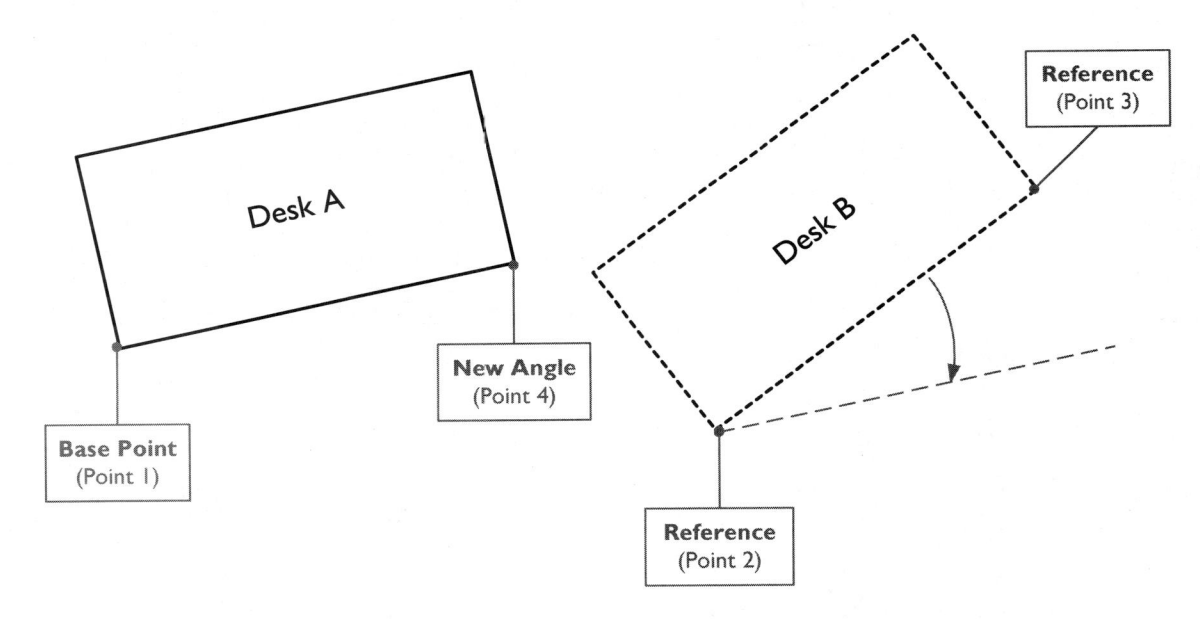

Specify base point: **end**
of *(Pick point 1.)*
Specify rotation angle or [Reference]: **r**
Specify the reference angle <0>: **end**
of *(Pick point 2.)*
Second point: : **end**

of *(Pick point 3.)*

Specify the new angle: **end**

of *(Pick point 4.)*

Dragging

To rotate the objects in real-time, move the cursor in response to the "Specify rotation angle or [Reference]:" prompt. Keep an eye on the status line as it reports the x,y,z coordinates and angle. Click to fix the object at the location shown on the screen.

AngBase

The **ANGBASE** system variable sets the direction of 0 degrees. In AutoCAD, 0 degrees can point in any direction; the default is at 3 o'clock or east, in the direction of the positive x-axis. To change the position of zero degrees, enter a new angle, as follows:

Command: **angbase**

Enter new value for ANGBASE <0>: *(Enter an angle, such as* **90**.*)*

Entering 90, for example, rotates 0 degrees by 90 degrees, so that it now points to 12 o'clock, or north.

AngDir

The **ANGDIR** system variable specifies the direction of positive angles. It takes on two values: 0 means positive angles are measured counterclockwise, while 1 means they are measured clockwise.

Command: **angdir**

Enter new value for ANGDIR <0>: *(Enter* **0** *or* **1**.*)*

Notes: Both **ANGBASE** and **ANGDIR** are usually not changed, but is it is helpful to know about them in case your drafting discipline requires that angles be represented differently, or AutoCAD isn't responding to angle input as you expect.

An alternative to the **ROTATE** command is grips editing, as described in Chapter 9.

SCALE

The **SCALE** command changes the size of objects. Contrary to what the name's suggests, this command does not change the *scale* of objects in the drawing — but their *size*. The x, y, and z directions of the objects are changed equally.

AutoCAD needs to know three things: (1) the objects to be resized; (2) the base point from which the objects are resized; and (3) the amount to resize.

TUTORIAL: CHANGING SIZE

1. To resize one or more objects, start the **SCALE** command:
   - From the menu bar, choose **Modify**, and then **Scale**.
   - From the **Modify** toolbar, choose the **Scale** button.
   - At the 'Command:' prompt, enter the **scale** command.
   - Alternatively, enter the **sc** alias at the 'Command:' prompt.

   Command: **scale** *(Press* ENTER.*)*

2. In all cases, AutoCAD prompts you to select the objects you want resized:

    Select objects: *(Pick one or more objects.)*

    Select objects: *(Press ENTER to end object selection.)*

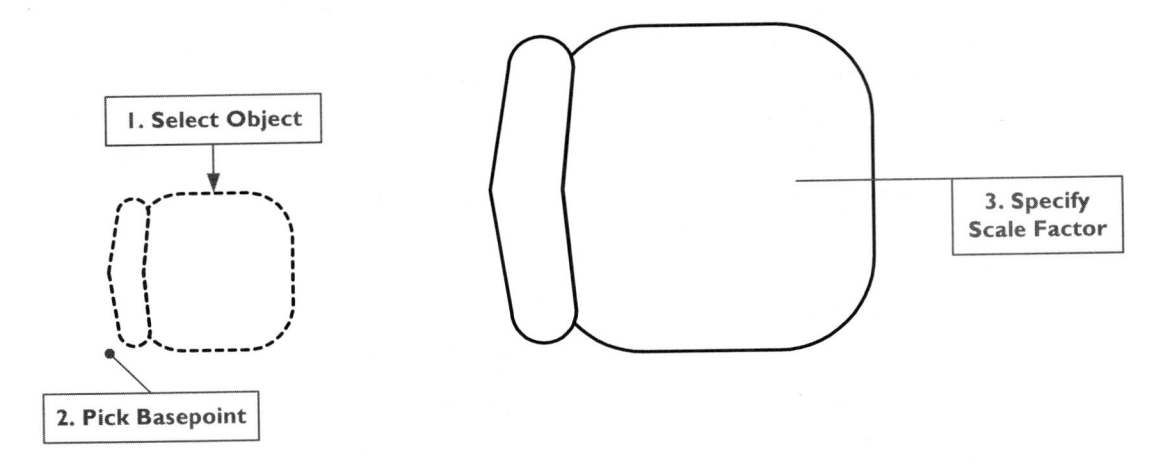

3. Identify the point about which the objects are rotated:

    Specify base point: *(Pick a point, such as the lower left corner.)*

4. Specify the resize factor:

    Specify scale factor or [Reference]: *(Enter a factor, such as* **2** *or* **0.5**.*)*

The base point can be specified anywhere in the drawing. This point remains stationary, and the object is resized from that point.

You can specify the size by a numerical factor, by dragging, or by referencing a known length, and then entering a new length. Entering a decimal factor makes objects smaller: a factor of 0.25 makes an object 25 percent of the original size. Factors larger than 1.0 increase the size of the objects. For example, entering 2.0 makes the object twice the original size in the direction of all three axes.

CHANGING SIZE: ADDITIONAL METHODS

The **SCALE** command has one option:

- **Reference** changes the size relative to other distances.

Let's look at it.

Reference

The **Reference** option adjusts objects to a "correct" size. This is particularly useful when you scan a drawing, and then bring the raster image into AutoCAD with the **IMAGE** command. To trace over the scanned image, you first have to change the image to the right size: you achieve this with the **Reference** option. All you need is the true length of just one line in the scanned image.

1. At the "Select objects:" prompt, select the raster image:

    Command: **scale**

    Select objects: *(Pick the edge of the raster image.)*

    Select objects: *(Press ENTER.)*

2. For the base point, select the lower left corner of the raster image.

    Specify base point: *(Pick.)*

3. Specify the **Reference** option:

    Specify scale factor or [Reference]: **r**

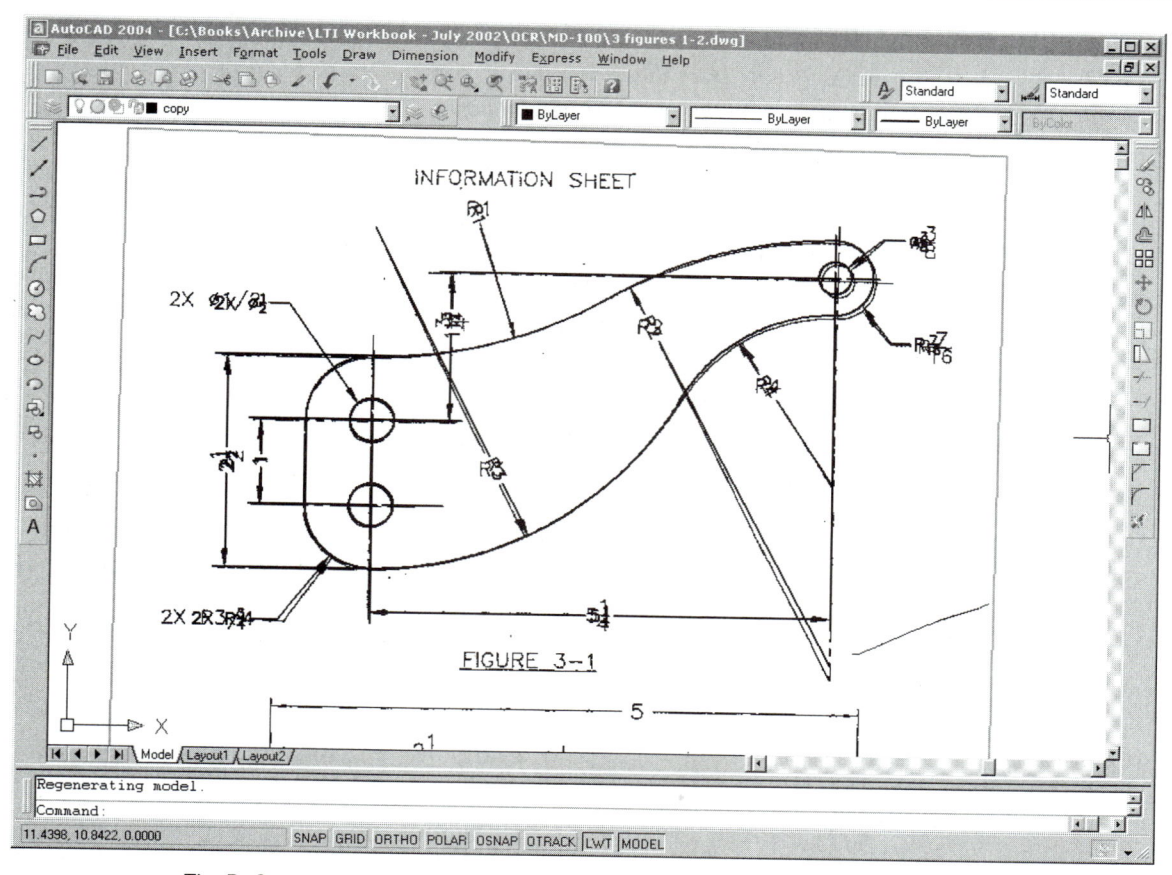

*The **Reference** option of the SCALE and ROTATE commands correct the size and skew
of scanned images before copying the old drawing in AutoCAD.*

4. Select the end of the known line in the raster image.

   Specify reference length <1>: *(Pick one end of the raster line.)*

5. Select the other end of the raster line; this allows AutoCAD to measure the distance of the line:

   Specify second point: *(Pick other end of raster line.)*

6. For the new length, enter the value of the known dimension.

   Specify new length: *(Enter value.)*

AutoCAD resizes the scanned image, making it the correct size. Use the **ROTATE** command to correct a skewed image. (If the scanned image is warped, you need to correct it with "rubber sheeting," a feature not available in AutoCAD.)

The **SCALE** command is also useful for converting an Imperial drawing to metric, and vice versa. An alternative to the **SCALE** command is grips editing, as described in Chapter 9.

CHANGE

The **CHANGE** command changes the size, properties, and other characteristic of objects. It is similar to, but more advanced than, **CHPROP**; but **PROPERTIES** is the easiest and most versatile of all (See Chapter 7, "Changing Object Properties.")

TUTORIAL: CHANGING OBJECTS

1. To change one or more objects, start the **CHANGE** command:
   - At the 'Command:' prompt, enter the **change** command.
   - Alternatively, enter the **-ch** alias at the 'Command:' prompt.

   Command: **change** *(Press ENTER.)*

2. In all cases, AutoCAD prompts you to select the complex objects you want to change:

   Select objects: *(Pick one or more objects.)*

   Select objects: *(Press ENTER to end object selection.)*

3. The **Properties** options are somewhat different from the **CHPROP** command:

   Specify change point or [Properties]: *(Pick a point.)*

The **CHANGE** command's **Elev** option changes the elevation of objects, an option lacking in **CHPROP**.

Although the Properties window is handier than this command-line command, **CHANGE** has some unique options, because the **Change Point** option reacts differently, depending on the object selected.

CHANGING OBJECTS: ADDITIONAL METHODS

Here's how the **CHANGE** command's Change Point option handles different objects:

- **Lines** change their endpoints.
- **Circles** change their radius.
- **Blocks** change their insertion point and rotation angle.
- **Text** changes their properties.
- **Attribute Text** changes their properties.

Let's look at each.

Lines

The **Change Point** option moves the endpoints of lines closest to the pick point.

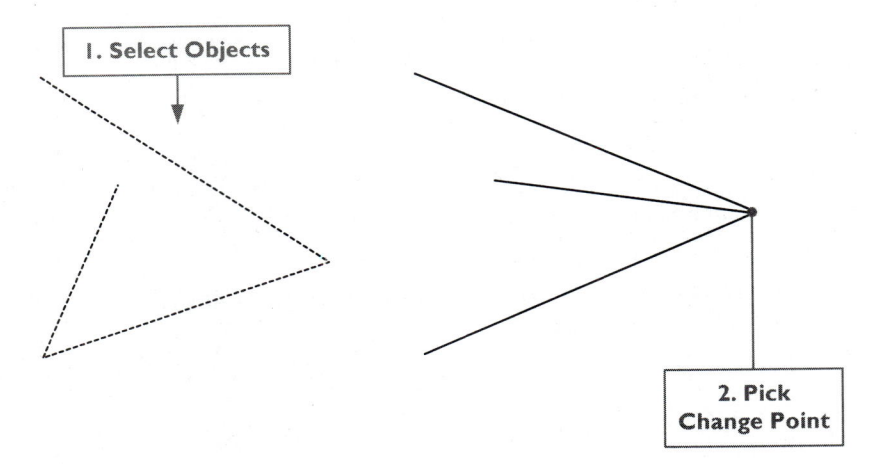

When ortho mode is on, AutoCAD makes the lines parallel to either the x or y axis, depending on which is closest. This is a quick way to straighten out a bunch of crooked lines, and is not available in any other AutoCAD command.

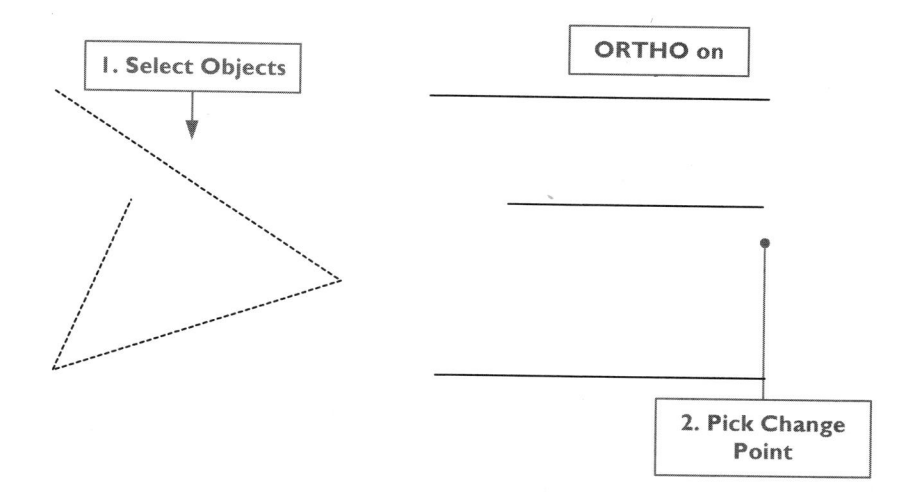

Circles

The **Change Point** option changes the circle's radius. When more than one circle is selected, AutoCAD repeats the prompt for each circle:

    Specify new circle radius <no change>: *(Enter a new radius, or press* ENTER.*)*

Press ENTER to keep the circle's radius.

Blocks

The **Change Point** option changes the location and rotation of blocks.

    Specify new block insertion point: *(Pick a new insertion point, or press* ENTER.*)*
    Specify new block rotation angle <current>: *(Enter a new angle, or press* ENTER.*)*

Press ENTER to keep each option in place.

Text

The **Change Point** option changes the position of the text, as well as the text's other properties. (Until the **PROPERTIES** command's predecessor, **DDMODIFY**, came along, this was the only way to change text properties in a drawing.)

    Specify new text insertion point <no change>: *(Pick a new insertion point, or press* ENTER.*)*
    Enter new text style <current>: *(Enter a different text style, or press* ENTER.*)*
    Specify new height <current>: *(Enter a new height, or press* ENTER.*)*
    Specify new rotation angle <current>: *(Enter a new angle, or press* ENTER.*)*
    Enter new text <current>: *(Enter a new line of text, or press* ENTER.*)*

Press ENTER to keep each option as is. If you select more than one line of text, AutoCAD repeats the prompts for the next one.

Attribute Definitions

The **Change Point** option changes the text and properties of attributes, unless not part of a block.

    Specify new text insertion point: *(Pick a new insertion point, or press* ENTER.*)*
    Enter new text style <current>: *(Enter a different text style, or press* ENTER.*)*
    Specify new height <current>: *(Enter a new height, or press* ENTER.*)*
    Specify new rotation angle <current>: *(Enter a new angle, or press* ENTER.*)*
    Enter new text <current>: *(Enter new text, or press* ENTER.*)*
    Enter new tag <current>: *(Enter a new tag, or press* ENTER.*)*
    Enter new prompt <current>: *(Enter a new prompt, or press* ENTER.*)*

Enter new default value <current>: *(Enter a new default value, or press* **ENTER**.*)*

Press **ENTER** to keep each option in place.

EXPLODE AND XPLODE

The **EXPLODE** and **XPLODE** commands reduce compound objects to their simplest form.

The commands "break down" blocks, polylines, and other compound objects into basic lines, arcs, and text. Some objects must be exploded several times before they are finally reduced to vector primitives.

TUTORIAL: EXPLODING COMPOUND OBJECTS

1. To explode one or more compound objects, start the **EXPLODE** command:
   - From the menu bar, choose **Modify**, and then **Explode**.
   - From the **Modify** toolbar, choose the **Copy** button.
   - At the 'Command:' prompt, enter the **explode** command.
   - Alternatively, enter the **x** alias at the 'Command:' prompt.

   Command: **explode** *(Press* **ENTER**.*)*

2. In all cases, AutoCAD prompts you to select the compound objects you want explode:

   Select objects: *(Pick one or more objects.)*

   Select objects: *(Press* **ENTER** *to end object selection.)*

The objects are exploded, but you might not see any difference. Different compound objects react differently to being exploded.

Arcs

Arcs do not explode, unless they are part of a nonuniformly scaled block. When the block is exploded, the arc becomes an elliptical arc.

Note: When a block is inserted with different scale factors in the x, y, and/or z directions, it is called a "nonuniformly scaled block." When exploded, the results may be different from what you expect. Arcs, for example, are converted to elliptical arcs to maintain their distortion.

Sometimes nonuniformly scaled blocks contain objects that cannot be exploded, such as bodies, 3D solids, and regions. In that case, AutoCAD places them in an *anonymous* block with the ***E** prefix in the block's name.

Circle

Circles do not explode, unless they are part of a nonuniformly scaled block. When the block is exploded, the circle becomes an ellipse.

Blocks

Blocks explode into the lines and other objects originally used to define them. Nested blocks must be exploded a second time. Attribute text is deleted; attribute definitions are redisplayed.

Xrefs and blocks inserted with the **MINSERT** command cannot be exploded.

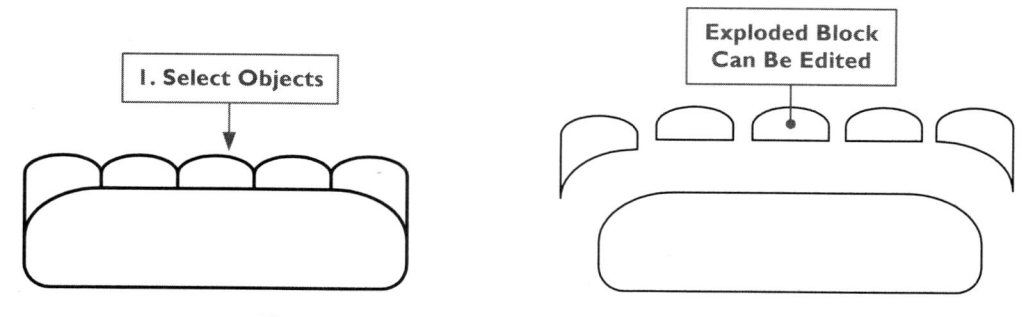

*Block before being exploded (at left), and after (at right).
Parts have been pulled apart for clarity.*

Polylines

Polylines explode into lines and arcs. Width and tangent information is discarded; the resulting lines and arcs follow the centerline of the old polyline. If the exploded polyline has segments of width, AutoCAD displays the message:

> Exploding this polyline has lost width information.
> The UNDO command will restore it.

The new lines and arcs are placed on the same layer as the polyline, and inherit the same color.

*At left: a polyline with variable width before being exploded.
At right: the polyline after being exploded; lines and arcs have been pulled apart for clarity.*

3D Polylines

Three-dimensional polylines explode into line segments. Linetypes are retained.

Dimensions

Associative dimensions explode into basic objects (lines, polyline arrowheads, and text), which are placed on the same layer as they original dimensions, and inherit that layer's properties.

Leaders

Leaders explode into lines and multiline text; the arrowheads explode into 2D solids. Depending on how the leader was constructed, resulting objects could also include splines, block inserts of arrowheads and annotation blocks, and tolerance objects. More than one explode may be required.

Multiline Text

Multiline text explodes into single-line text.

Multilines

Multilines explode into lines and arcs.

Polyface Meshes

Single-vertex polyface meshes explode into point objects. Two-vertex meshes are exploded into lines. Three-vertex meshes explode into 3D faces.

3D Solids

3D solids exploded into bodies, which can be further exploded into regions. Regions, in turn, explode into lines, arcs, and circles.

XPLODE

The XPLODE command controls what happens when compound objects are exploded.

TUTORIAL: CONTROLLED EXPLOSIONS

1. To explode compound objects with control, start the **XPLODE** command:
   - At the 'Command:' prompt, enter the **xplode** command.

   Command: **xplode** *(Press ENTER.)*

2. In all cases, AutoCAD prompts you to select the compound objects you want exploded:
   Select objects to XPlode.
   Select objects: *(Pick one or more objects.)*
   Select objects: *(Press ENTER to end object selection.)*

3. Enter an option:
   Enter an option
   [All/Color/LAyer/LType/LWeight/Inherit from parent block/Explode] <Explode>: *(Enter an option.)*

4. If you had selected more than one object, an additional prompt appears:
   Enter an option [Individually/Globally] <Globally>: *(Enter I or G.)*

The **Global** option applies the previous options (color, layer, and so on) to all selected objects. The **Individually** option applies the previous options to each object, one at a time.

All

The **All** option prompts you to specify the color, linetype, lineweight, and layer on which the exploded objects should land. The prompts are the same as those for the four following options.

Color

The **Color** option sets the color of exploded objects:

[Red/Yellow/Green/Cyan/Blue/Magenta/White/BYLayer/BYBlock/Truecolor/Colorbook] <BYLAYER>: *(Enter a color name or number.)*

Layer

The **Layer** option sets the layer of the exploded objects:

Enter new layer name for exploded objects <current>: *(Enter the name of an existing layer.)*

LType

The **LType** option sets the linetype of the exploded objects:

Enter new linetype name for exploded objects <BYLAYER>: *(Enter the name of a linetype.)*

Inherit from Parent Block

The **Inherit from parent block** option sets the color, linetype, lineweight, and layer of the exploded objects to that of originating block.

Explode

The **Explode** option reduces compound objects in the same manner as the EXPLODE command.

PEDIT

The **PEDIT** command (short for "polyline edit") edits polylines.

This command also edits 3D polylines drawn with the **3DPOLY** command, and 3D polyfaces drawn with AutoCAD's 3D surfacing commands.

TUTORIAL: EDITING POLYLINES

1. To edit one or more polylines, start the **PEDIT** command:
   - From the **Modify** menu bar, choose **Objects**, and then **Polyline**.
   - From the **Modify II** toolbar, choose the **Edit Polyline** button.
   - At the 'Command:' prompt, enter the **pedit** command.
   - Alternatively, enter the **pe** alias at the 'Command:' prompt.

   Command: **pedit** *(Press ENTER.)*

2. In all cases, AutoCAD prompts you to select the polyline you want edited:

   Select polyline or [Multiple]: *(Pick a polyline.)*

3. Many options are listed:

   Enter an option [Close/Join/Width/Edit vertex/Fit/Spline/Decurve/Ltype gen/Undo]: *(Enter an option, or press ESC to exit the command.)*

As a reminder, here are the parts of a polyline:

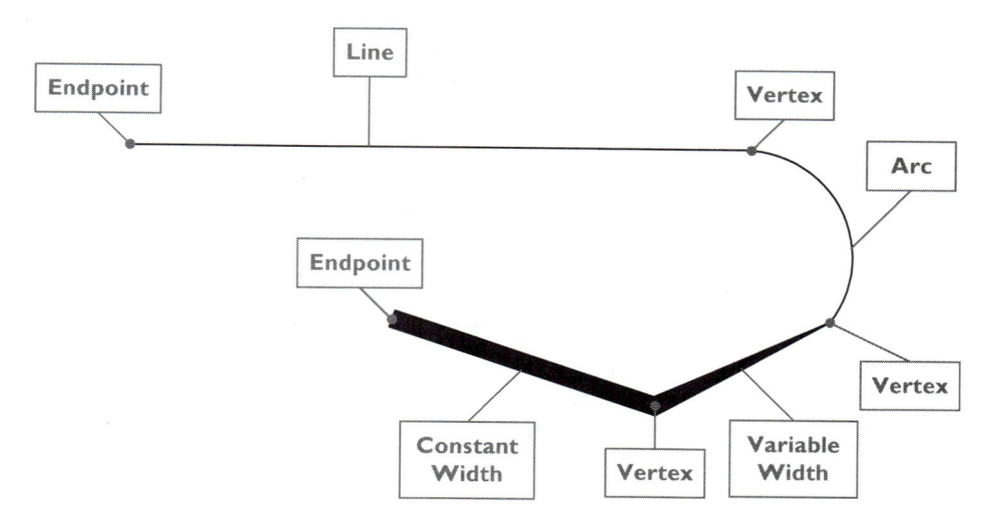

Close/Open

The **Close** option closes open polylines. It connects the last point to the first point of the polyline. This is like the **PLINE** command's **Close** option, except it allows you to close the polyline after exiting **PLINE**.

The **Open** option replaces the **Close** option when the polyline is closed. It opens closed polylines by removing the last segment created by the **Close** option. If the polyline looks closed, but was not closed with **Close**, the **Open** option has no effect.

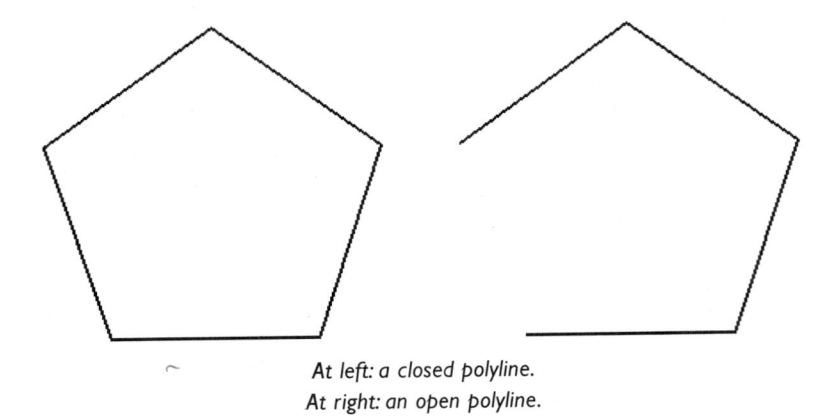

At left: a closed polyline.
At right: an open polyline.

Join

The **Join** option converts and connects polyline and non-polyline objects (lines and arcs only) to the original polyline. AutoCAD prompts:

> Select objects: *(Select one or more lines, arcs, and polylines.)*

The objects you select become part of the original polyline; if the polyline was curve-fitted, it is first decurved.

AutoCAD determines which arcs and lines share common endpoints with the original polyline, and then merges them into that polyline. To join successfully, the objects must touch, or be within a *fuzz factor* distance, as described later in this chapter. If endpoints are too far away to join, use the **FILLET** or **CHANGE** command to extend them to a perfect match. For alternatives, see the **Multiple** and **Jointype** options later in this chapter.

Width

The **Width** option changes the width for the entire polyline. Unlike the **PLINE** command's **Width** option, you can't have tapers or different widths for each segment; the width is applied uniformly to all segments of the polyline. AutoCAD prompts:

> Specify new width for all segments: *(Enter a value.)*

You can type the width at the keyboard, or else pick two points in the drawing. The figure illustrates the effect of this option:

At left: a polyline with varying widths.
*At right: the polyline after applying the **Width** option.*

Edit vertex

The **Edit vertex** options edits the vertices of a polyline. (Recall that the vertex is the connection between polyline segments, as well as the two endpoints of open polylines.) The name, "edit vertex," is somewhat misleading, because it also edits the segments between pairs of vertices.

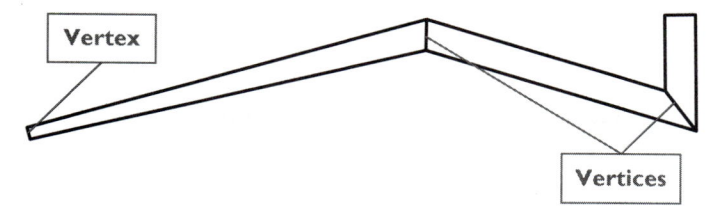

To help keep track of which vertex is being edited, AutoCAD displays a marker in the shape of an X. Press **N** and **P** to move the marker to the next and previous vertex.

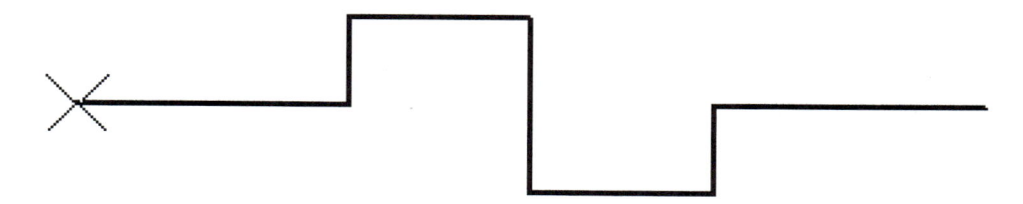

*The **X** marks the vertex.*
*Pressing **N** and **P** moves the marker from vertex to vertex.*

A second marker, an arrow, is displayed when you work with the **Tangent** option.

Command: **pedit**

Select polyline or [Multiple]: *(Pick a polyline.)*

Enter an option [Close/Join/Width/Edit vertex/Fit/Spline/Decurve/Ltype gen/Undo]: **e**

[Next/Previous/Break/Insert/Move/Regen/Straighten/Tangent/Width/eXit]: *(Enter an option.)*

Note: Quite frankly, the **Edit vertex** options are a pain to use. In many cases, you may find it much easier to edit the polyline using grips.

The many sub-options are:

Next/Previous

The **Next** option moves the x-marker to the next vertex, while the **Previous** option moves it to the previous vertex.

Break

The **Break** option removes one or more segments between two vertices. The first vertex (on one side of the break) is at the x-marker when you enter this option. The other vertex, on the other side of the break, you mark using the following options:

Enter an option [Next/Previous/Go/eXit]: *(Enter an option.)*

Press **N** and **P** to move the x-marker to another vertex, and then press **G** to effect the break. Press **X** if you decide to exit this mode without breaking the polyline.

The entire segment is removed; to remove a portion of a segment, use the **BREAK** command. This option does not work under two conditions: the same vertex is marked twice, so that no segment is selected for removal; or the polyline's two endpoints are marked, in effect removing the entire polyline.

Insert

The **Insert** option adds a vertex to the polyline at your pick point. AutoCAD prompts:

> Specify location for new vertex: *(Pick a point anywhere in the drawing.)*

AutoCAD redraws the polyline so that it includes the new vertex.

Move

The **Move** option moves the marked vertex. AutoCAD prompts:

> Specify new location for marked vertex: *(Pick a point anywhere in the drawing.)*

Regen

The **Regen** option regenerates the polyline, without needing to exit the command. It is required after applying the **Width** option.

Straighten

The **Straighten** option removes vertices, line segments, and arcs between two marked vertices, and replaces them with a single segment. It works like the **Break** option: AutoCAD remembers the x-marked vertex before you enter this option, and then you move the x-marker to another vertex.

> Enter an option [Next/Previous/Go/eXit]: *(Enter an option.)*

Press **N** and **P** to move the x-marker to another vertex, and the press **G** to straighten. Press **X** if you decide to exit this mode without changing the polyline.

Tangent

The **Tangent** option attaches a tangent direction to the marked vertex for use later with the **Fit** and **Spline** options. AutoCAD prompts:

> Specify direction of vertex tangent: *(Pick a point, or enter an angle.)*

AutoCAD displays an arrow pointing in the direction of tangency.

Width

The **Width** option changes the starting and ending widths for the segment immediately following the marked vertex. AutoCAD prompts:

> Specify starting width for next segment: *(Enter a width, or pick two points.)*
>
> Specify ending width for next segment: *(Enter a width, or pick two points.)*

Use the **Regen** option to see the new width.

Exit

The **Exit** option exits **Edit vertex** mode, and returns to **PEDIT**'s original prompt line.

Fit

The **Fit** option constructs smooth curves from the vertices in the polyline. The curve consists of arcs joined at vertices, as illustrated by the figure. Notice how each segment has at least one arc associated with it. AutoCAD inserts extra vertices and arcs, where necessary; in the figure, two segments have two arcs each.

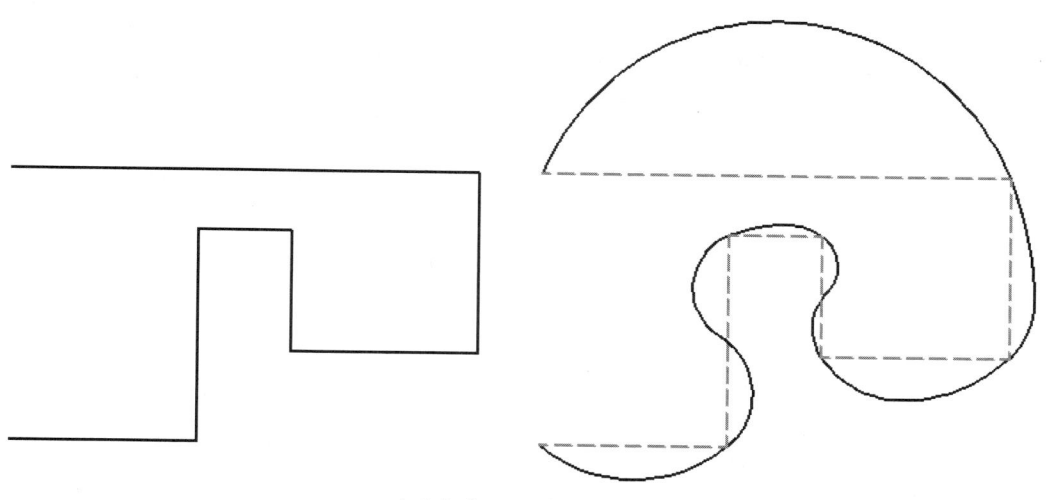

At left: the original polyline.
*At right: the polyline after applying the **Fit** curve option;*
the original polyline is shown overlaid by gray dashed lines.

Spline

The **Spline** option creates a spline curve from the polyline, using the vertices as the control points to approximate a B-spline curve. This is not a true B-spline, as constructed by the **SPLINE** command, but only approximates it.

When the original polyline also contains arcs, they are first converted to straight segments. When the original polyline also has widths, they are tapered from the first vertex to the last.

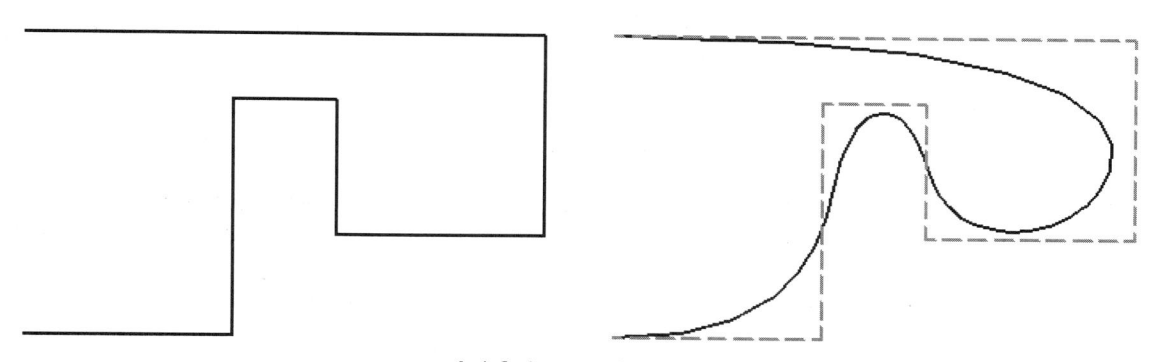

At left: the original polyline.
*At right: the polyline after applying the **Spline** option; the original polyline is shown overlaid by gray dashed lines.*

Decurve

The **Decurve** option negates the effect of the two curve options, **Fit curve** and **Spline**, by straightening all segments of the polyline, as well as removing extra vertices that may have been added.

Ltype gen

The **Ltype gen** option (short for "Linetype generation") controls the generation of linetypes through the vertices of the polyline. Normally, AutoCAD generates the dashes, dots, and gaps for each segment of the polyline; the pattern starts and stop at each vertex.

Enter polyline linetype generation option [ON/OFF] <Off>: *(Enter **ON** or **OFF**.)*

When the option is turned on, the polyline is treated as a single segment, resulting in a uniform linetype pattern along the entire length of the polyline. In most cases, you probably want this option turned on.

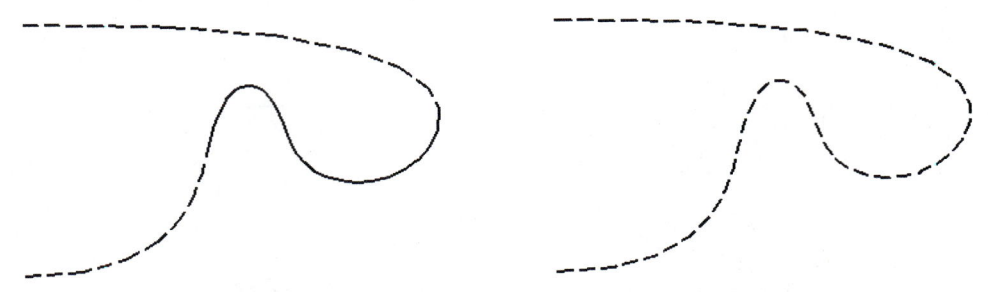

*At left: a polyline with **Ltype gen** turned off; part of the polyline lacks the line pattern.*
At right: the polyline after turning on the option; all of the polyline is patterned..

Note: Linetype generation does not work on polylines with tapered segments.

Undo

The **Undo** option reverses the last **PEDIT** operation without having to exit the command to access the U command.

EDITING POLYLINES: ADDITIONAL METHODS

Of the **PEDIT** command's many options, some are not obvious, and some are accessed through system variables:

- **Select** option selects a line or arc.
- **Multiple** option selects more than one polyline.
- **Jointype** option determines how multiple objects are joined into a single polyline.
- **SPLFRAME** system variable toggles the spline frame.
- **SPLINETYPE** and **SPLINESEGS** system variables determine the spline's type and quality.

Let's look at each.

Select

The "Select polyline" prompt is misleading, because you can also select a non-polyline object — specifically a line or an arc. In that case, AutoCAD notices:

Command: **pedit**

Select polyline or [Multiple]: *(Pick a line or arc.)*

Object selected is not a polyline.

Do you want to turn it into one? <Y>: *(Enter **Y** or **N**.)*

When you respond with "Y", the object is converted to a polyline. If you had selected an object other than a line or arc, the conversion fails, and AutoCAD complains:

Object selected is not a polyline.

Mutiple

You can select more than one polyline to edit, but only by entering the **Multiple** option, as follows:

> Command: **pedit**
>
> Select polyline or [Multiple]: **m**
>
> Select objects: *(Pick one or more polylines.)*
>
> Select objects: *(Press ENTER to end polyline selection.)*

Jointype

The **Join** option merges lines, arcs, and polylines into a single polyline, but they all need to be touching. When you use the **Multiple** option at the start of the PEDIT command, however, then the objects *need not touch!* AutoCAD instead displays the following prompt:

> Join Type = Extend
>
> Enter fuzz distance or [Jointype]<0.0000>: *(Enter a distance, or type J.)*

The *fuzz distance* determines how far apart the objects can be, and still be joined to the original polyline. The **Jointype** option specifies how distant objects should join:

> Enter a vertex editing option
>
> Enter join type [Extend/Add/Both] <Extend>: *(Enter an option.)*

The **Extend** option joins the selected objects by extending and trimming them to fit.

The **Add** option joins the selected objects by bridging them with a straight segment.

The **Both** option joins the selected objects by extending or trimming; if that is not possible, AutoCAD adds the straight segment.

When you attempt to join two non-touching polylines, AutoCAD complains:

> 0 segments added to polyline

No matter which option you pick, fit-curved and splined polylines lose their curvature, and are converted back to the original frame.

SplFrame

The **SPLFRAME** system variable toggles the visibility of the *spline frame*. The original polyline from which the spline curve was created with the PEDIT command's **Spline** option.

The frame is normally turned off, because there isn't much need to see it; it just clutters the drawing. To see the frame, turn on the system variable, and follow that with a drawing regeneration:

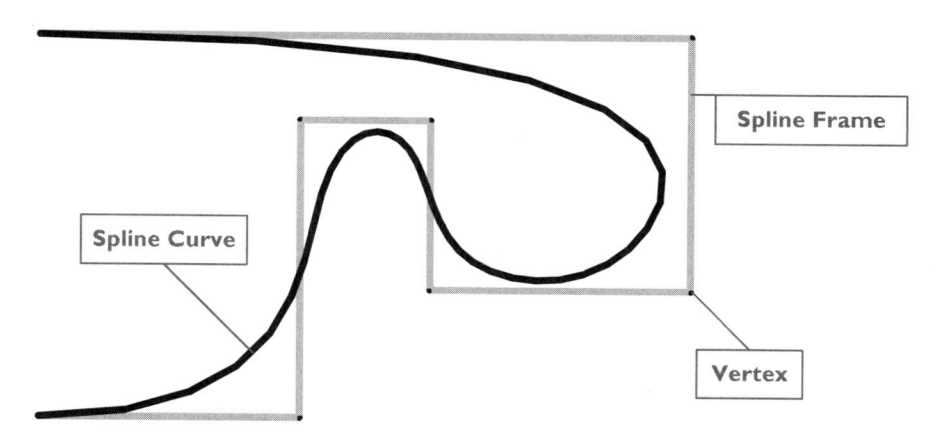

In black: a splined polyline.
In gray: the original polyline (its frame).

Command: **splframe**

Enter new value for SPLFRAME <0>: **1**

Command: **regen**

Editing commands act differently with the spline and its frame:

Editing Command	Comment
COPY, ERASE, MIRROR, MOVE, ROTATE, SCALE	Affects both spline curve and frame, whether frame is visible or not.
EXTEND	Adds a vertex to the frame where it intersects the boundary.
BREAK, EXPLODE, TRIM	Deletes the frame, and generates a new spline.
OFFSET	Copies the spline.
STRETCH	After spline and frame are stretched, refits the spline.
DIVIDE, MEASURE, HATCH, FILLET, CHAMFER, and AREA's **Object** option	Affects the spline curve only, not the frame.

SplineType

The **SPLINETYPE** system variable gives you the choice of quadratic and cubic spline-fit polylines: a value of 5 approximates quadratic B-splines, while a value of 6 approximates cubic B-splines (values of 1 through 4 have no meaning). Technically, a *B-spline* is a generalization of the Bézier curve; the quadratic B-spline produces a tighter curve than does the cubic B-spline.

Command: **splinetype**

Enter new value for SPLINETYPE <6>: *(Enter 5 or 6.)*

The setting is not retroactive; it comes into effect the next time you convert a polyline into a spline.

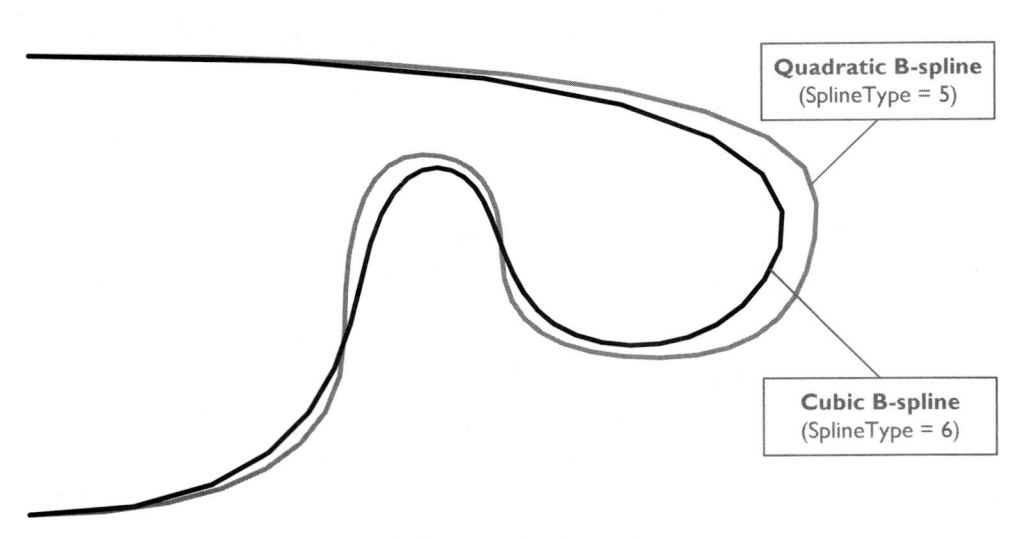

In black: a cubic B-spline polyline.
In gray: a quadratic B-spline.

SplineSegs

The **SPLINESEGS** system variable (short for "spline segments") controls the *apparent* smoothness of spline and fit-curves, apparent because AutoCAD approximates the spline curve with straight line segments. The default value is 8 segments between vertices; higher values construct more segments.

Command: **splinesegs**

Enter new value for SPLINESEGS <8>: *(Enter a value.)*

Settings **SPLINESEGS** to a negative value forces AutoCAD to change the spline to a fit-curve — the same as using the **PEDIT** command's **Fit** option.

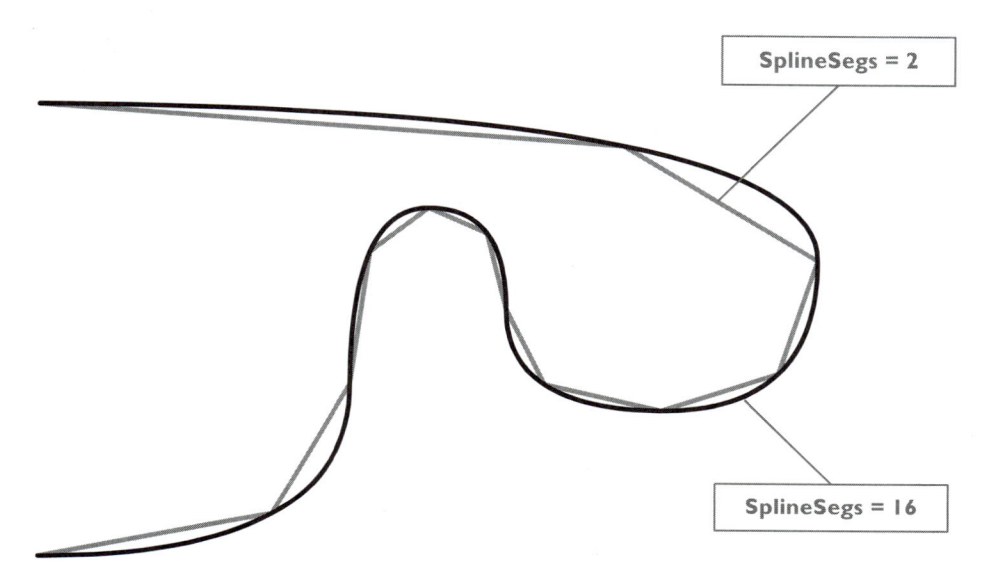

In black: 16 segments per vertex.
In gray: 2 segments per vertex.

Note: The **SPLFRAME** system variable applies to all polylines in the drawing; AutoCAD cannot display the frames of selected polylines.

In contrast, the **SPLINETYPE** and **SPLINESEGS** system variables apply to polylines on individually.

EXERCISES

1. In this exercise, you erase objects from a drawing, and then bring them back.
   From the CD-ROM, open the *edit1.dwg* drawing file, a valve housing.
   Use the **ERASE** command to delete some of the objects.
   Next, issue the **OOPS** command. Did the objects return?

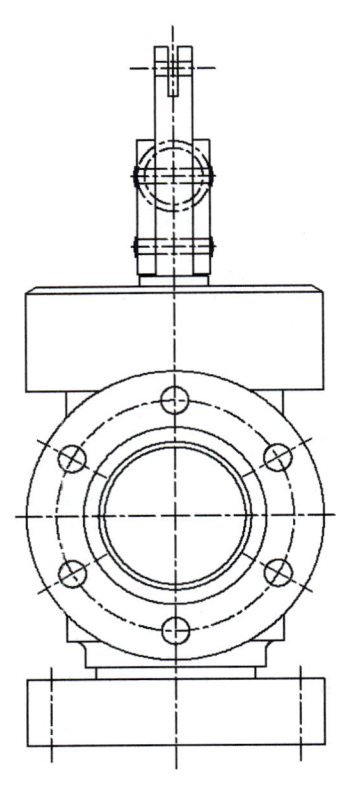

   Repeat the **ERASE** command with the **All** option. Did everything disappear?
   This time, use the **U** command to bring back the drawing.

2. In this exercise, you assemble a jigsaw puzzle.
   From the CD-ROM, open the *edit2.dwg* file, a drawing of puzzle pieces.
   Use the **MOVE** command to move the pieces into position, leaving a small space between them. The most effective method is to move the pieces roughly into position, then zoom in, and finely position the pieces using object snap.
   You may also want to try using grips editing to move pieces into place.

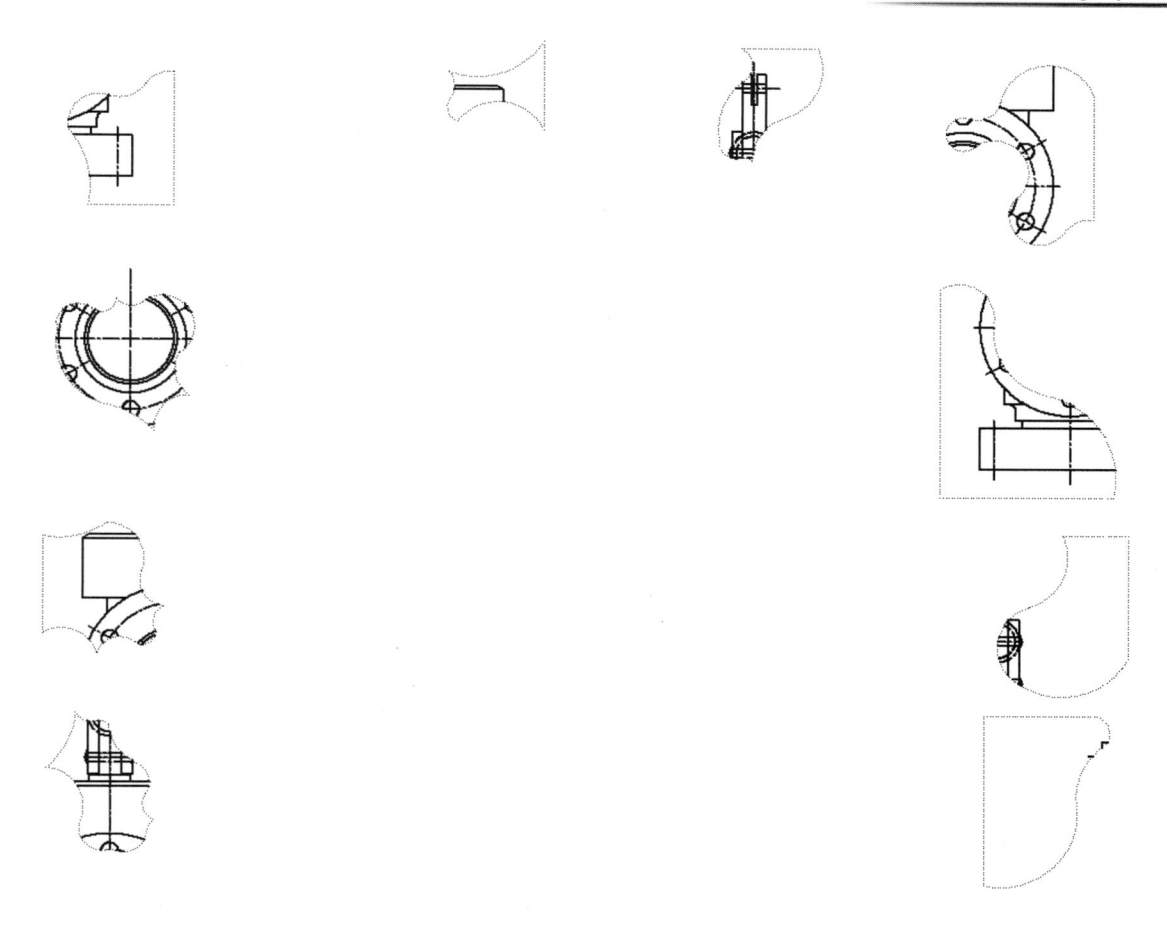

3. In this exercise, you practice breaking and trimming lines.
   From the CD-ROM, open the *edit5.dwg* drawing file, a piping diagram.

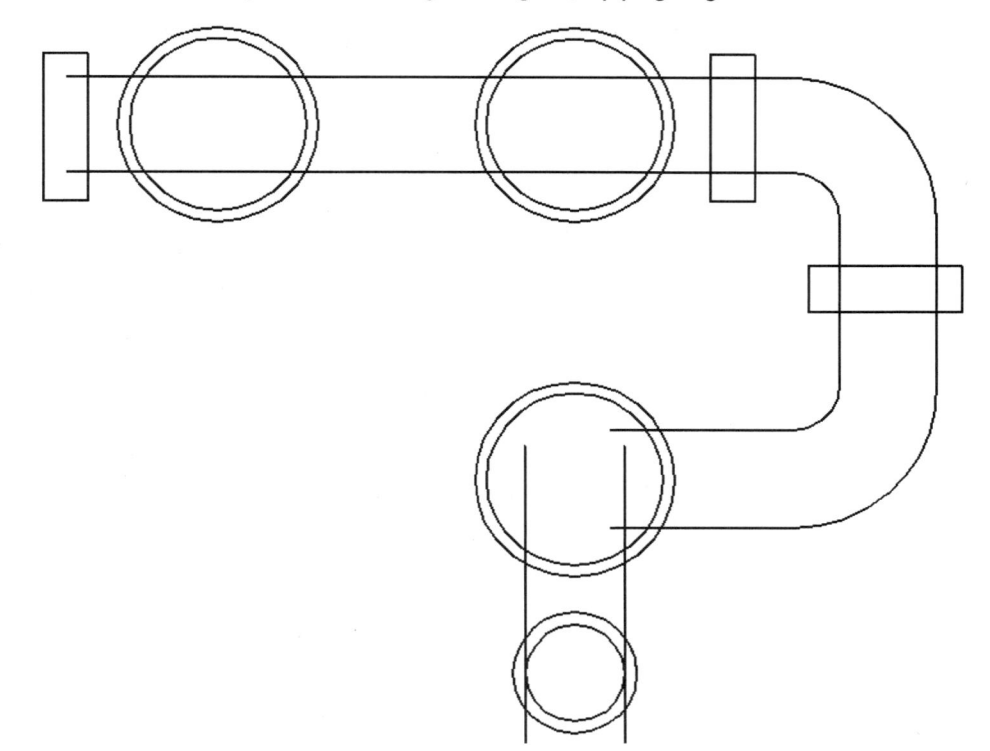

Use the **BREAK** command to break each of the objects in the drawing, achieving the result shown.

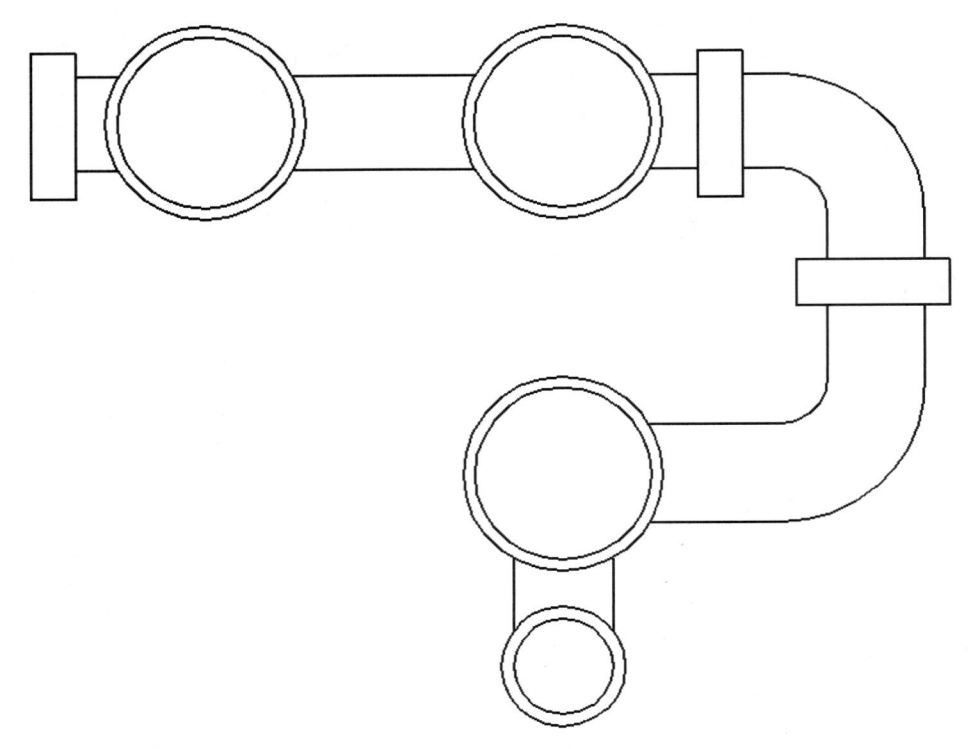

Quit the drawing — don't save your work!

Open *edit5.dwg* again; this time use the **TRIM** command to produce the same result as shown above.

Here's a timesaving tip: at the "Select cutting edges" prompt, just press **ENTER**; this is a "hidden" option that selects all objects in the drawing.

You may need to use the **ERASE** command to clean up. Which command did you find easier — **BREAK** or **TRIM**?

4. In this exercise, you work with several editing commands.

   From the CD-ROM, open the *edit7.dwg* drawing file, a piece of sheet metal.

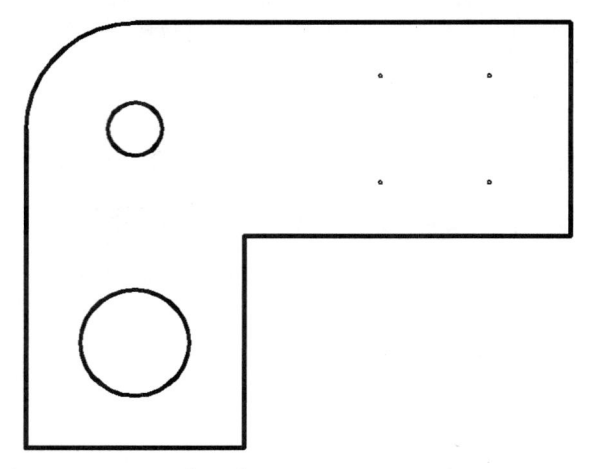

   Suppose that a design change is initiated, and you are instructed to remove the lower circle from the drawing.

   From the **Modify** menu, use the **Erase** command.

Now, erase the four points on the object by using a window selection.

Your boss changes his mind.

Put back the four points you just erased.

After erasing the lower circle, your boss decides the remaining circle should be moved down by 2.0 units.

Use CENter object snap, ortho mode, and direct distance entry to move it accurately.

After reviewing the drawing, your boss's boss feels the sheet metal needs two holes.

Add another circle, identical to the remaining circle. Use grips editing to copy the circle.

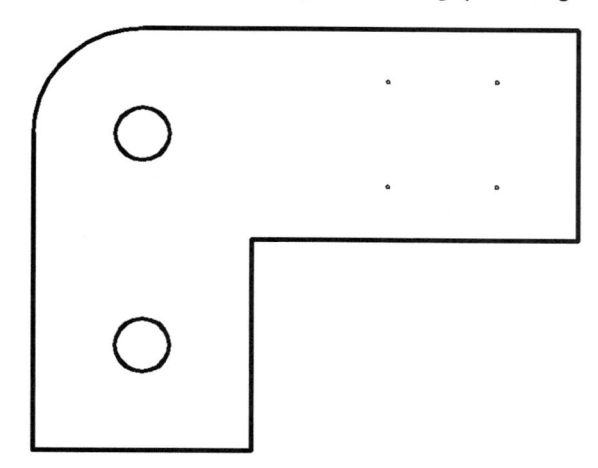

You learn from the product design department that the sheet metal is too large. A notch needs to be taken out.

Turn on snap mode, and set the snap spacing to 0.25. Use the **BREAK** and **LINE** commands to draw the notch.

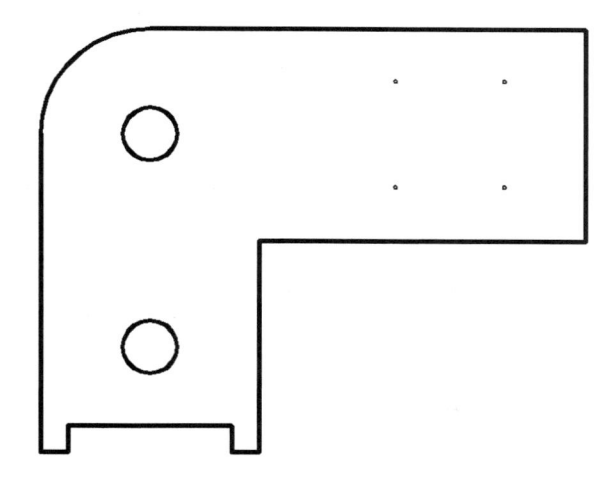

The product safety department reviews the drawing, and feels the sheet metal may be hazardous to children under three years of age. You are asked to round off two corners. Add a radius of 0.5 to each of the two right corners.

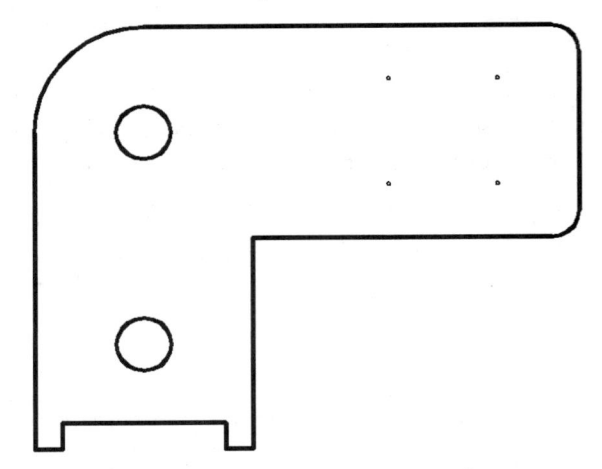

At last, all departments are satisfied. While the scenario in this exercise may seem silly, you will probably experience such changes to your drawings in the workplace.

Save the drawing.

5. In this exercise, you trim objects.
   Draw four intersecting lines about 6 units long and 2 units apart, as illustrated by the figure.
   Which command quickly makes the second, parallel line?

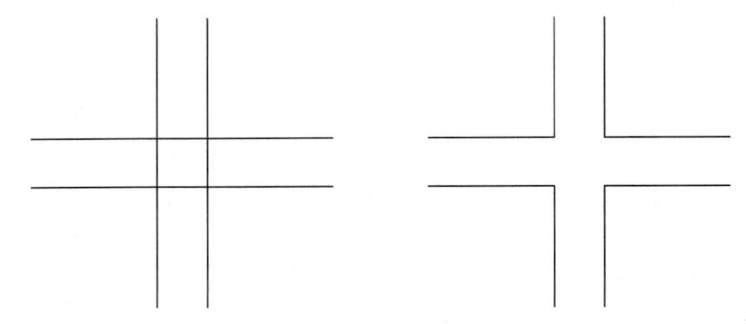

   At left: four lines before trimming.
   At right: after trimming.

   Use the **TRIM** command to trim the intersections. Remember to use the trick of responding **all** to the "Select cutting edges" prompt.

6. In this exercise, you extend objects.
   Draw two vertical parallel lines 6 units long and 6 units apart.
   Between them, draw one horizontal line 4 units long, as illustrated by the figure.

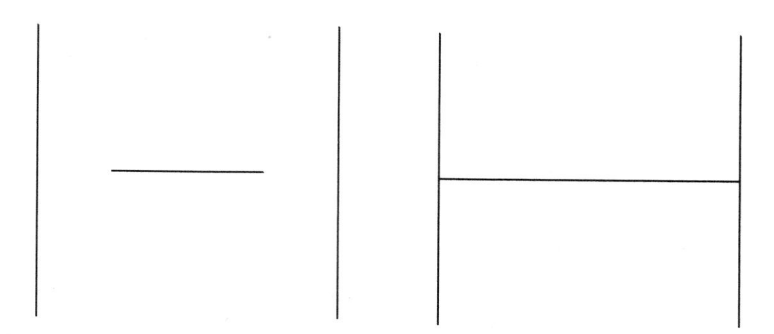

At left: one line before extending.
At right: after extending both ends.

Extend the horizontal line to the vertical lines with the **EXTEND** command.

To extend the other end (remember, chose both vertical lines as boundaries), respond to the repeating "Select object to extend:" prompt by selecting a point at the other end of the horizontal line.

7. In this exercise, you rotate objects.

   Draw the arrow symbol illustrated by the figure.

   Using the **ROTATE** command, turn the arrow by 45 degrees about the center of the arrow's base. Which object snap mode helps you find the midpoint of a line?

   Using grips editing, turn the arrow a further 90 degrees.

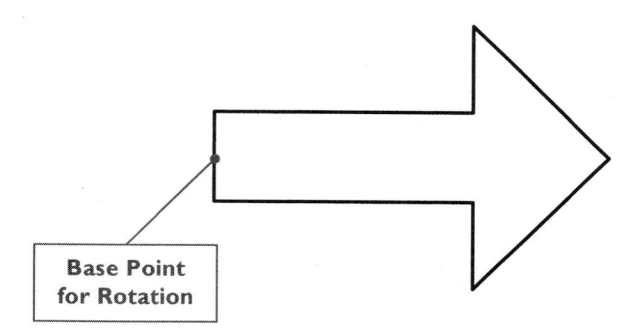

   Draw a cam similar to the one illustrated by the figure.

   Use Window object selection to rotate the entire cam by -45 degrees.

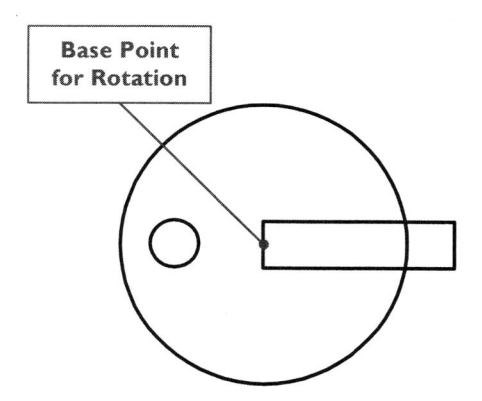

8. In this exercise, you resize objects.

   With the **POLYGON** command, draw the triangle illustrated by the figure. The edge is 2 units long.

   Use the **SCALE** command to double the size of the triangle.

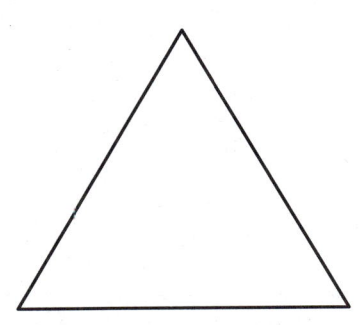

   Repeat the command, but this time use the **Reference** option to further double the size of the triangle.

   Specify scale factor or [Reference]: **r**

   Specify reference length <1>: **2**

   Specify second point: **4**

   Did AutoCAD resize all the lines making up the triangle?

9. In this exercise, you "stretch" a window to move it within the wall — without editing the wall.

   From the CD-ROM, open the *stretch1.dwg* file, a drawing of a wall cross section.

   Use the **STRETCH** command to move the window along the wall.
   Which selection mode must you use?

   Did the wall stretch?

10. From the CD-ROM, open the *stretch2.dwg* file, a drawing of a pencil.

    Use the **STRETCH** command to make the pencil shorter.

11. In this exercise, you lengthen and shorten objects with a different command.

    Draw an arc, using these parameters:

Start point	**2,5**
Second point	**0,3**
Endpoint	**2,1.5**

    Start the **LENGTHEN** command, and select the arc.
    What is its length and included angle?

Use the **DElta** option to add 2 units to the length.

Use the **Total** option to change the arc to 5 inches.
Did it become longer or shorter?

Use the **Percent** option to change the arc to 100%.
Did the arc change?

12. In this exercise, you change objects.
    Ensure ortho mode is turned off.
    Draw four lines, randomly, as illustrated by the figure.

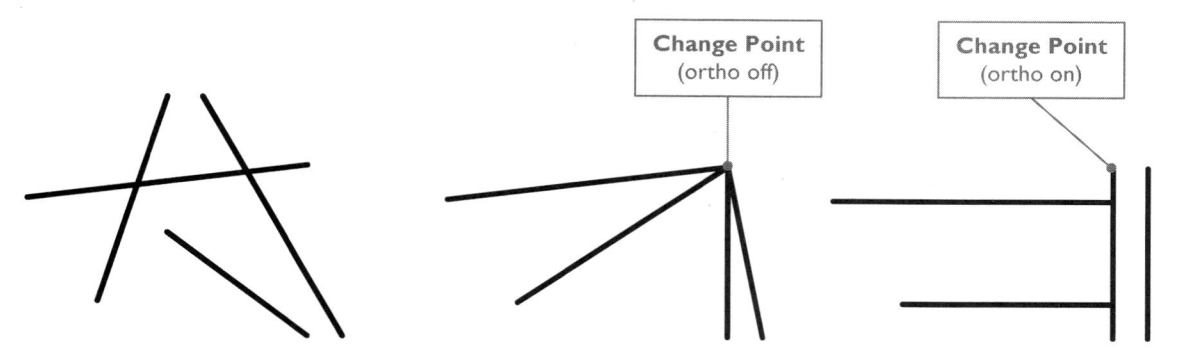

    Change Point (ortho off)

    Change Point (ortho on)

    Use the **CHANGE** command to select the lines, and then use the **Change Point** option to give the lines a common endpoint.
    Now use the **CHANGE** command's **Properties** option to change the lines to red.

13. In this exercise, you explode objects.
    From the CD-ROM, open the *edit2.dwg* drawing file, the drawing of the puzzle.
    Select one of the puzzle pieces. What does the single grip tell you?
    With the piece still selected, use the **EXPLODE** command. Does the piece look different?
    Select the piece again. Has the number of grips changed? What does this tell you?
    Use the **EXPLODE** command a second time. What message does AutoCAD give you?

    Select a different, unexploded piece.
    Start the **XPLODE** command, and specify a color of red. What happens to the puzzle piece?

14. In this exercise, you edit polylines.
    From the CD-ROM, open the *offset.dwg* drawing file, the drawing of the small town.
    Select any of the polylines that define the edges of streets.
    Use the **PEDIT** command's **Width** option to change the width to 0.1 units.
    Exit the command, and start the **PROPERTIES** command.
    Select *all* the polylines in the drawing by clicking the **Quick Select** button (funnel with the lightening strike) and entering these parameters:

Apply to	**Entire drawing**
Object type	**Polyline**
Properties	**Layer**
Operator	**=**
Value	**0**

    After you click **OK**, AutoCAD highlights all polylines.

In the **Geometry** section of the **Properties** window, change *VARIES* to 0.1 units.

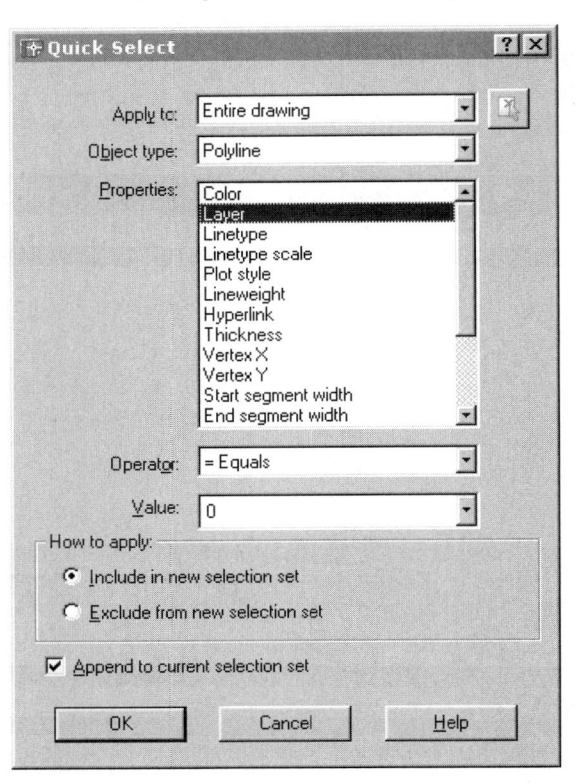

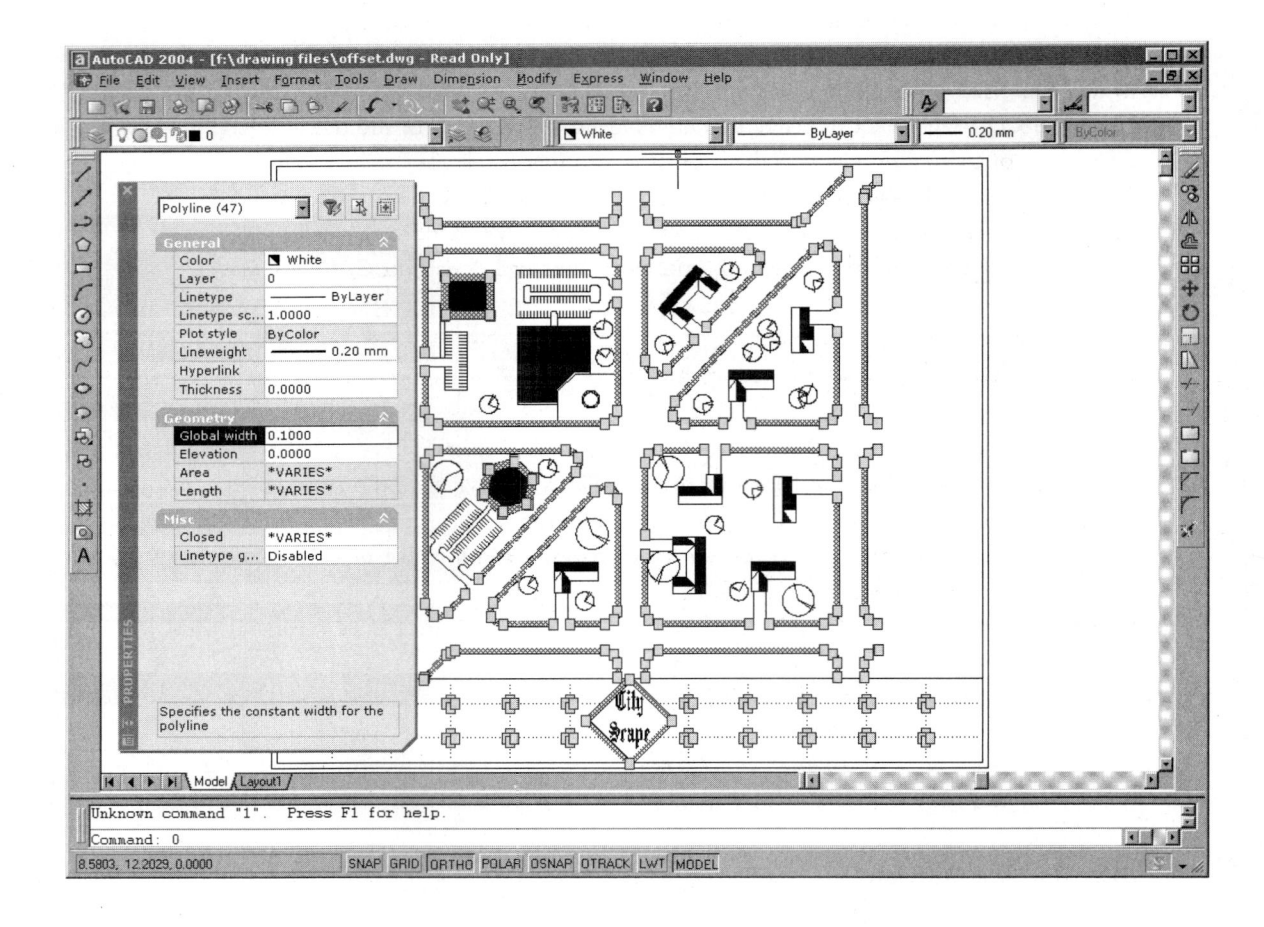

Press **ESC** to remove the grips and highlighting. The drawing should look like the figure below.

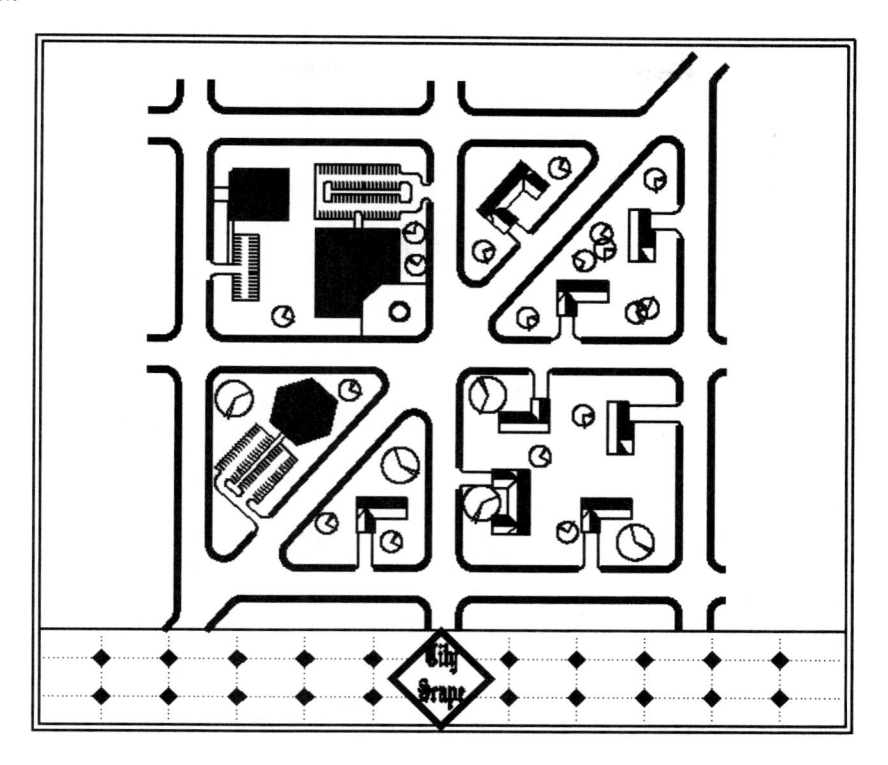

CHAPTER REVIEW

1. What is the difference between a hot and a cold grip?
2. When moving an object, must the base point be on the selected object(s)?
3. What objects are affected by the **BREAK** command?
4. In breaking an object, what happens if you do not enter **F** for selection of the first point?
5. Which commands restore objects just erased?
6. What happens when you explode a block?
7. Describe what happens when you explode a polyline.
8. What is a *cutting edge*?
9. At what point is a polyline trimmed?
10. What is the special requirement before a circle can be trimmed?
11. Name the borderline used by the **EXTEND** command.
12. Describe three ways to rotate an object.
    a.
    b.
    c.
13. Can an object be resized differently in the x and y axes with the **SCALE** command?
14. What value changes objects to half their original size? <u>0.5</u>
15. When you select objects with three window operations, which operation determines the stretch?
16. Name the command that explicitly edits polylines?
17. Which parts of the polyline are the *vertices*?
18. How do you edit polyline vertices?
19. What types of vertex editing can you perform?
20. List the aliases for the following commands:
    TRIM
    BREAK
    EXTEND
    PEDIT
21. When can the **OOPS** command be used?
22. Is it possible to erase all objects in the drawing?
23. Can the **BREAK** command shorten objects?
24. What does a circle become after applying the **BREAK** command?
    An ellipse?
    Xlines?
25. Describe how the **LENGTHEN** command differs from the **TRIM** and **EXTEND** commands.
26. Can the **TRIM** command extend open objects?
    If so, how?
27. Can the **EXTEND** command extend close objects?
    If so, how?
28. Do blocks and dimensions work as cutting edges?
29. In what kind of drafting are the **TRIM** command's **Project** and **Edges** options used?
30. List the two things AutoCAD needs to know before extending objects:
    a.
    b.
31. Do blocks work as boundary edges?
32. Can circles be extended?

33. Is it acceptable to select all objects as cutting edges?
34. Can arcs be lengthened with the **LENGTHEN** command?
    Shortened?
35. Describe how the **LENGTHEN** command's **Total** and **Delta** options differ.
36. When does the **STRETCH** command only move objects?
37. Can objects be moved using grips editing?
38. What does the **ROTATE** command rotate objects about?
39. Where does 0 degrees point, by default?
    Can the direction be changed?
    If so, how?
40. In which direction does AutoCAD measure negative angles, by default?
41. What three pieces of information does AutoCAD need to know before resizing objects?
    a.
    b.
    c.
42. Describe how the **CHANGE** command's **Change Point** option changes lines:
    Ortho mode turned off:
    Ortho mode turned on:
43. What is the difference between the **EXPLODE** and **XPLODE** commands?
44. How are arcs in non-uniformly scaled blocks exploded?
45. Under what condition must blocks be exploded more than once?
46. What are polylines exploded into?
47. Can you specify the color of exploded objects?
    If so, how?
48. Describe how AutoCAD reacts when you select lines and arcs with the **PEDIT** command?
49. Can lines and arcs that don't touch be turned into a single polyline?
    If so, how?
50. During polyline vertex editing, which keystrokes move the x marker from vertex to vertex?
51. How does the **Straighten** option affect a polyline?
52. What is the difference between the **Fit** and **Spline** options of the **PEDIT** command?
53. Are the splines created by **PEDIT** true splines?
54. Explain why the **Ltype gen** option is important?
55. What is the purpose of the **SPLFRAME** system variable?
56. Can the type of polyline spline be changed?
    If so, how?

CHAPTER 12

Placing and Editing Text

Text is used in many parts of drawings: for identifying drawings through title blocks; labeling drawings through callouts; lists of parts; and including paragraphs of text that warn and explain. Traditionally, placing text by hand in a paper drawing was laborious. In contrast, placing text in an AutoCAD drawing is easy. This chapter shows you how to place text in drawings, and then how to change the text and its properties. The commands are:

TEXT and **-TEXT** place text in drawings, one line at a time.

MTEXT and **-MTEXT** place paragraphs of formatted text (updated in AutoCAD 2004).

QLEADER and **LEADER** place callouts in drawings.

DDEDIT edits text.

STYLE and **-STYLE** define named text styles based on fonts.

SPELL checks the drawing for unfamiliar words.

FIND searches and optionally replaces text.

WIPEOUT places blank areas under text for legibility (new in AutoCAD 2004).

QTEXT displays lines of text as rectangles.

SCALETEXT and **JUSTIFYTEXT** change the size and justification of text.

COMPILE converts PostScript font files for use in drawings.

FINDING THE COMMANDS

On the TEXT, STYLES, and DIMENSION toolbars:

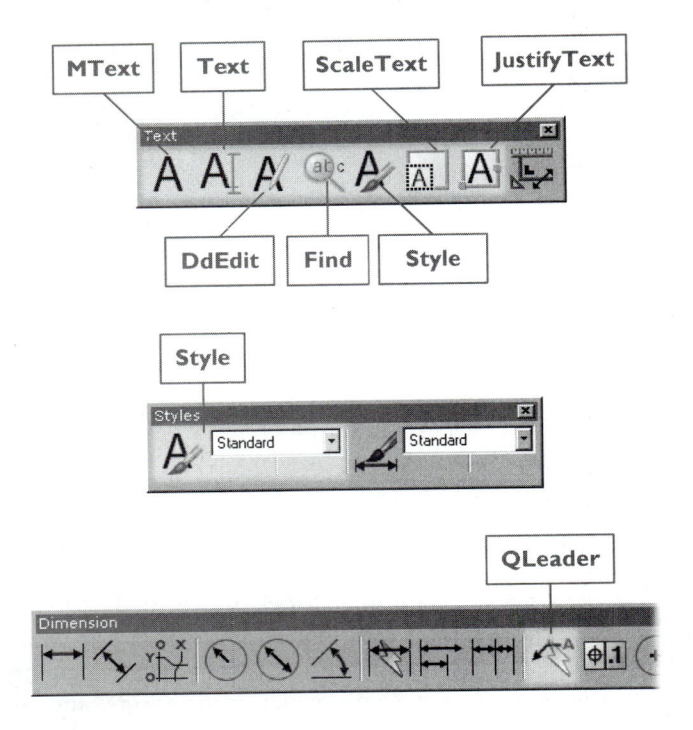

- MText
- Text
- ScaleText
- JustifyText
- DdEdit
- Find
- Style

Style

QLeader

On the EDIT and FORMAT menus:

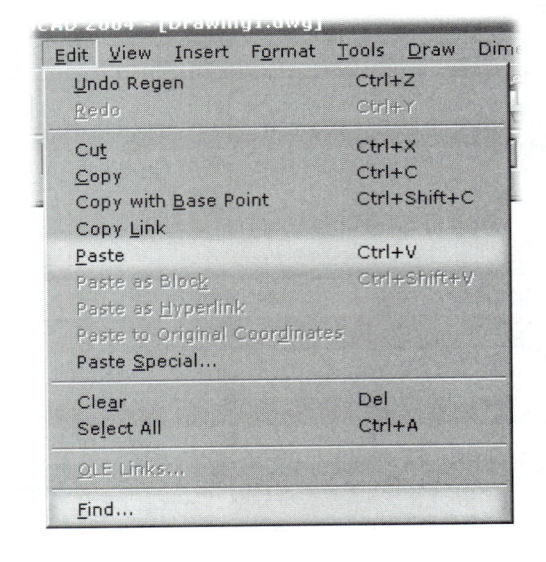

On the DRAW and DIMENSION menus:

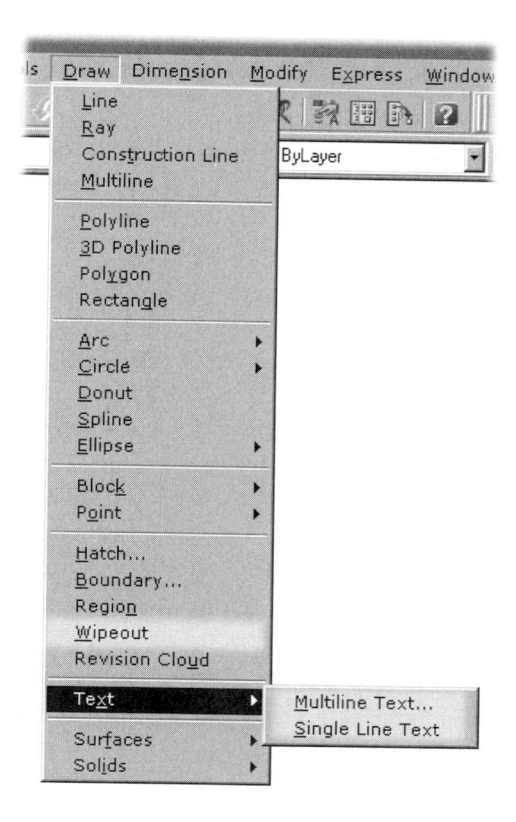

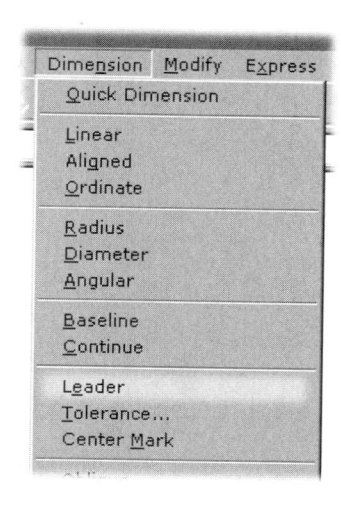

On the MODIFY menu:

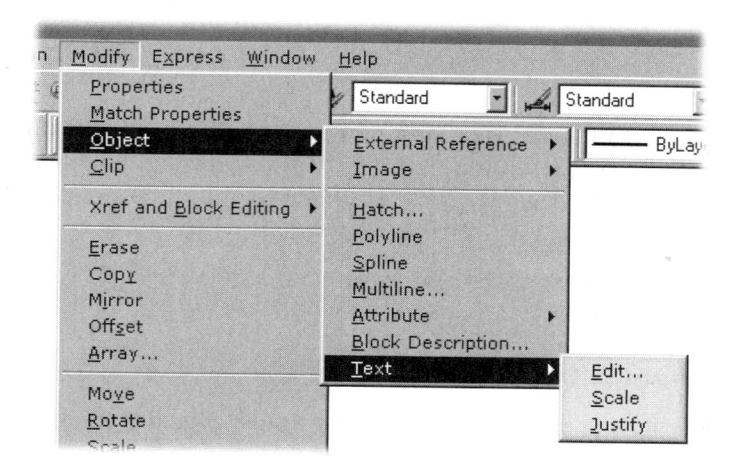

A| TEXT

The **TEXT** command places text, one line at a time, while the related **MTEXT** command places paragraphs of text, as described later in this chapter.

The figure illustrates the many uses of text in an architectural drawing. Can you spot the notes, general notes, leaders, dimensions, and title block (shown enlarged on the next page)?

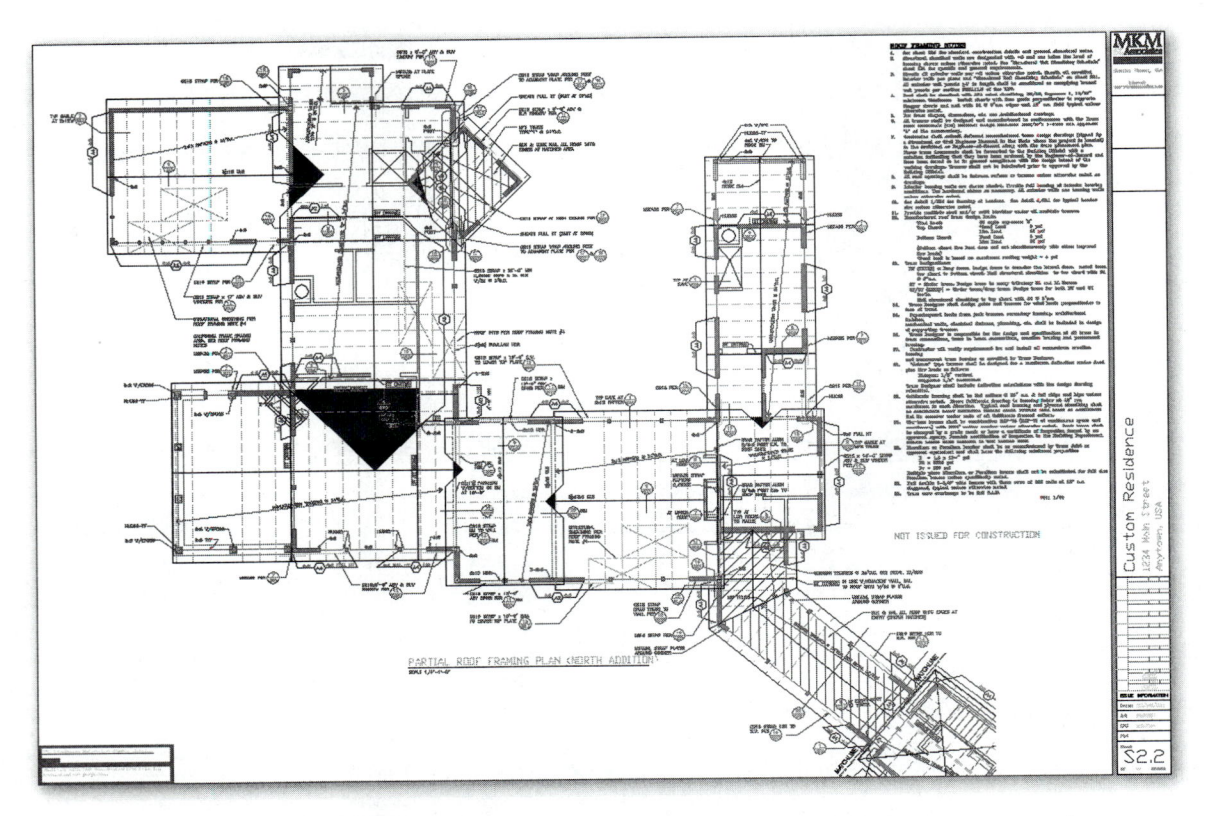

The use of text in drafting is usually governed by standards. Many North American industries use the American National Standards Institute (ANSI) style of letters and numbers. European companies use ISO lettering (International Organization for Standardization). Other companies and countries set standards of their own.

The most important aspect of lettering is that it be clear and concise. In general:

- **Headings** should be 3/16-inch high.
- **Note text** should be 1/8-inch high.
- **Text** typically should be left-justified (each line of text aligned at its left edge).

The default font in every new drawing is the rather ugly **Txt** font. The clear font often used by drafters is called **Simplex** or **RomanS** by AutoCAD.

At left: AutoCAD's default Txt font.
At right: AutoCAD's Simplex font preferred by many drafters.

Fonts are the "design" of text letters. Caslon, for example, is the name of the font used for this paragraph, while GillSans is used for headings and tutorial text in this book. Arial, TimesRoman, `Courier`, and *Script* are examples of other fonts.

AutoCAD can use any TrueType font (*.ttf* files) found on your computer, its own font format called "SHX" (*.shx* files), as well as PostScript fonts (*.pfb* files) after conversion.

Many (but not all) fonts can be stretched, compressed, obliqued, mirrored, or drawn in a vertical stack. Text is affected by color and lineweight, but not linetype.

TUTORIAL: PLACING TEXT

1. To place single lines of text, start the **TEXT** command:
   - From the **Draw** menu, choose **Text**, and then **Single Line Text**.
   - From the **Text** toolbar, choose the **Single Line Text** button.
   - At the 'Command:' prompt, enter the **text** command.
   - Alternatively, enter the aliases **dt** or **dtext** at the 'Command:' prompt.

   Command: **text** *(Press* ENTER.*)*

2. In all cases, AutoCAD reports the current settings, and then prompts you for the start point:

   Current text style: "Standard" Text height: 0.2000

   Specify start point of text or [Justify/Style]: *(Pick a point, or specify x,y coordinates.)*

3. Specify the height and rotation angle:

   Specify height <0.2000>: *(Enter a height, or press* ENTER.*)*

   Specify rotation angle of text <0>: *(Enter an angle, or press* ENTER.*)*

4. Enter a line of text, and then press **ENTER**.

   Enter text: *(Enter text, and then press* ENTER.*)*

5. Enter additional lines of text, which AutoCAD places under the previous line.

   If you wish to place text elsewhere in the drawing, move the cursor and click; AutoCAD places the next line of text at that point.

   Enter text: *(Move cursor, and click. Enter text, and then press* ENTER.*)*

5. Press **ENTER** a second time to exit the command:

   Enter text: *(Press* ENTER *to exit the command.)*

Menu selections, tablet picks, and command options are "locked out" during the **TEXT** command's "Enter text:" prompt. Only keyboard entry is permitted while placing text.

Press **BACKSPACE** to erase one character at a time; this works for text placed on previous lines; it does not work, however, for text placed during the previous **TEXT** command.

When you use a text alignment other than starting point, the text does not appear to be properly aligned until you finish the command by pressing **ENTER** twice.

As you type text, AutoCAD displays the *I-beam cursor*, called that because it looks like the letter I. (Earlier releases of AutoCAD used a rectangle cursor.) The cursor has three horizontal lines that represent the top of uppercase letters, the text baseline, and the bottom of descenders. (A *descender* is the part of lowercase letters that hangs below the baseline — g, j, p, q, and y.) Thus, the total height of the I-beam represents the height of uppercase characters in the current text style + the depth of lowercase descenders.

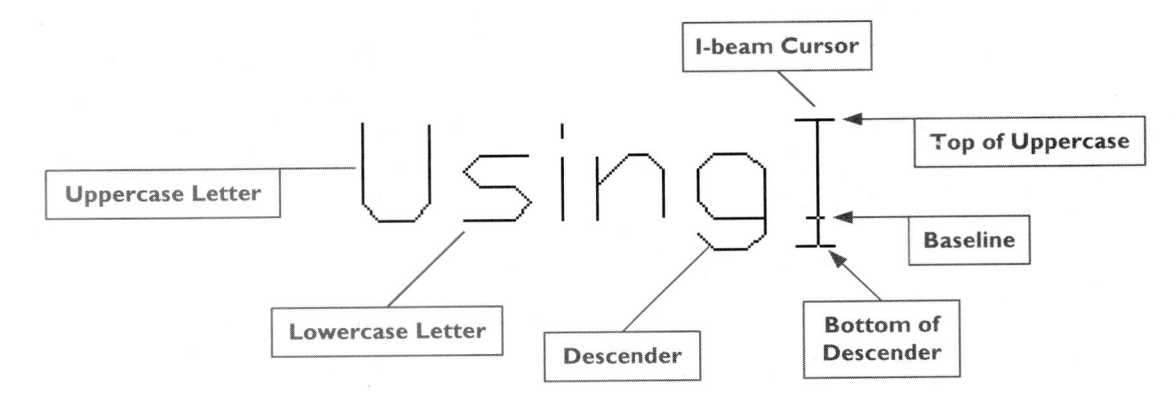

If you cancel the **TEXT** command at any time during the operation, all text placed during that operation is erased, not just the current line.

The next time you use the **TEXT** command, the last text string is highlighted (shown in dashed lines).

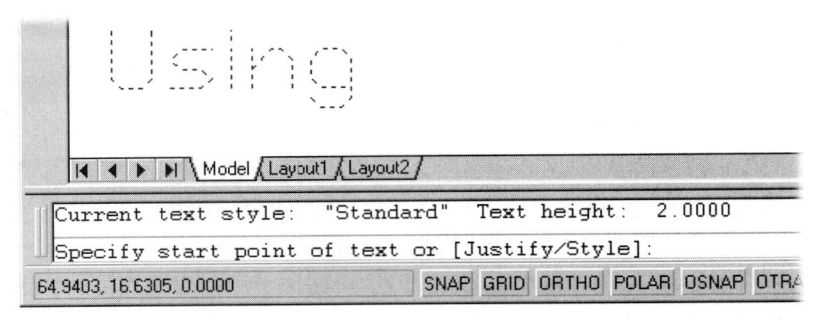

AutoCAD does this for a reason: when you press **ENTER** at the "Specify start point of text:" prompt, the I-beam cursor is placed on the next line after that last text, as though you had not exited **TEXT**.

Command: *(Press ENTER to repeat the TEXT command.)*

TEXT Current text style: "Standard" Text height: 2.0000

Specify start point of text or [Justify/Style]: *(Press ENTER to continue text below the last line.)*

This is useful for placing additional lines of text. Notice that AutoCAD skips the height and rotation prompts, because it assumes you want them to be the same as before.

PLACING TEXT: ADDITIONAL METHODS

The **TEXT** command allows you to control text justification and style, as listed by the options below. In addition, the **-TEXT** command handles non-text characters in a unique manner.

- **Justify** specifies a text justification mode.
- **Style** selects a predefined text style.
- **Height** specifies the height of the text.
- **Rotation angle** specifies the angle at which the line of text is rotated.
- **Control Codes** add formatting and special characters.
- **-TEXT** evaluates AutoLISP expressions during text placement.

Let's look at each.

Justify

The **Justify** option specifies the text alignment. The default is *left-justified*, where lines of text start at a common left edge. When you respond to the prompt "Specify start point of text" by picking a point, or typing x,y coordinates, this becomes the starting point for the text string.

To see a list of all text justification options, respond with a **J** (short for "justify") at the "Specify start point of text or [Justify/Style]:" prompt. AutoCAD prompts for the alignment options.

Specify start point of text or [Justify/Style]: **j**
Enter an option [Align/Fit/Center/Middle/Right/ TL/TC/TR/ML/MC/MR/BL/BC/BR]:

Text can be justified (or aligned) both vertically and horizontally. AutoCAD uses one- and two-letter abbreviations to designate each alignment option. (Historically, the one-letter options were present with the original AutoCAD from 20 years ago. The two-letter options were added more recently.) One letter describes vertical alignments, and the other horizontal alignments. For example, TL is the top-left alignment. The figures below illustrate the alignment modes.

Justify	Meaning
Start point	Equivalent to BL.
Align	Fits text between two points, and adjusts height appropriately.
Fit	Fits text between two point at a specific height.
Center	Equivalent to BC.
Middle	Equivalent to MC.
Right	Equivalent to BR.

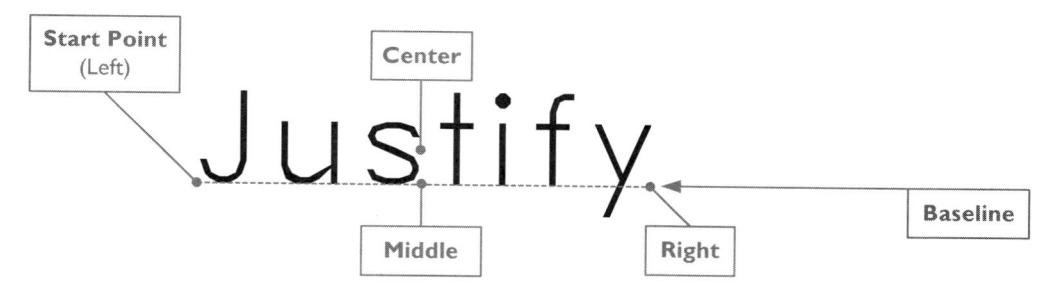

TL	Top left.
TC	Top center.
TR	Top right.
ML	Middle left.
MC	Middle center.
MR	Middle right.
BL	Bottom left.
BC	Bottom center.
BR	Bottom right.

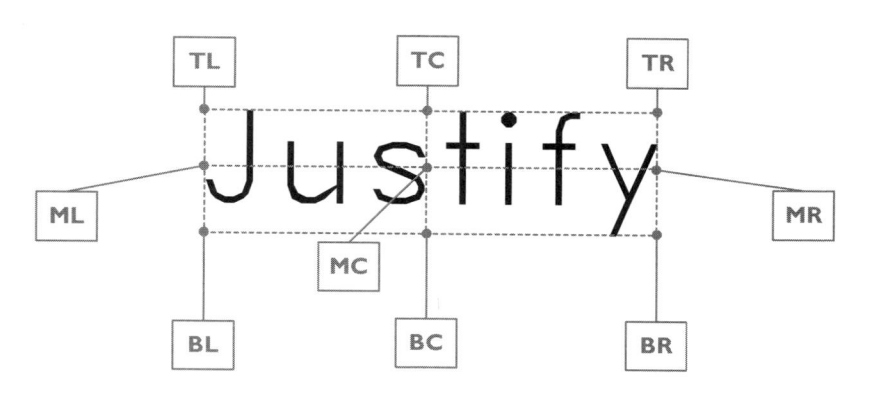

Middle-aligned text is often used in title blocks. Right-justified text is often used to the left of a leader line.

If you know the alignment you want, it is not necessary to enter the **J** option; simply enter the one- or two-letters abbreviation for the alignment. For example, to right-justify text:

>Specify start point of text or [Justify/Style]: **r**

Two justification modes operate differently from the others, because they require two pick points: **Align** and **Fit**.

Align

The **Align** option requires you to select two points that the text will fit between. AutoCAD adjusts the text height so that the baseline of the text fits perfectly between the two points. Note that the two points can be placed at any angle in relation to each other.

>Command: **text**
>Current text style: "Standard" Text height: 0.2000
>Specify start point of text or [Justify/Style]: **a**
>Specify first endpoint of text baseline: *(Pick point 1.)*
>Specify second endpoint of text baseline: *(Pick points.)*
>Enter text: *(Type text.)*
>Enter text: *(Press* ENTER *to exit command.)*

There are no "height" and "rotation angle" prompts, because the two pick points determine the height of text and the angle of the text line. When you pick the two points in reverse order, AutoCAD draws the text upside down.

Fit

The **Fit** option is similar to the **Aligned** option: it prompts you for two points between which to place the text. The option also prompts for a text height. AutoCAD fits the text between the two points, but draws the text at the height you specify.

>Specify start point of text or [Justify/Style]: **f**
>Specify first endpoint of text baseline: *(Pick point 1.)*
>Specify second endpoint of text baseline: *(Pick point 2.)*
>Specify height <0.2>: *(Enter a height.)*
>Enter text: *(Type text.)*
>Enter text: *(Press* ENTER *to exit command.)*

Fit Justification

Examples of text fitted between two points, but assigned different heights.

When you pick the second point to the left of the first, AutoCAD draws the text upside down. Fitted text is often used in constrained areas, for example when labeling a small closet. Text is squished or stretched horizontally to fit between the two points.

Style

The **Style** option selects a predefined text style. *You must first create the text style before you can use it.* The exception is the **Standard** style, which is included in each new drawing.

Styles (covered in detail later in this chapter) include parameters such as font name, boldface, height, and oblique angle.

> **Note:** The **Style** option of the TEXT command and the STYLE command are not the same. The **Style** *option* selects a style, while the STYLE *command* creates the style.

Height

Text can be drawn at any height — even as tall as the orbit of Pluto. (The technical editor notes that text can be as tall as 9.49×10^{94}, which is considerably taller than the known universe.) The default, however, is a miserable 0.2 units.

Specify height <0.2000>: *(Enter a height such as **7.5**, or press ENTER.)*

Once the text height is set, it remains the default until changed. There is one condition where the "Specify height" prompt does not appear: when the height is predefined in the style.

Like hatch patterns and linetypes, text must be scaled appropriately for the drawing. When you sketch a picture of your house on a piece of paper, for instance, you draw the house small enough to fit the paper (there might be a sheet of paper big enough to draw the house full size, but it would be hard to find!). Because *scaling* is so important for text, a review may be in order.

In the sketch of my 50-foot house (at left), I drew it 4 1/4" long. I had to draw the house 140 times smaller than its actual size: 4.25" = 50' (recall that the units must be the same, so 50' x 12 = 600"):

$$\frac{4.25"}{600"}$$

equals 1:140 (roughly). The 1:140 is called the "scale factor." The text I wrote on the paper, however, is *full size*, which happens to measure 1/8" tall in the sketch.

In AutoCAD, however, scaling is done the other way around. Instead of drawing the house smaller to fit the paper, I draw it full size: the 50-foot house is drawn 50 feet long; the scale is 1:1.

Text, however, can't be drawn 1:1 — it has to be 140 times larger. The standard in drafting is to draw text 1/8" tall for "normal-sized" text. When I place text in my house drawing, I specify a height of 17.5" tall (=1/8" x 140). This may seem to you much too tall (nearly one-and-a half feet tall!) but trust me: when plotted, it looks exactly right.

The figure below illustrates text placed at several heights — some too small, some just right, and some too large.

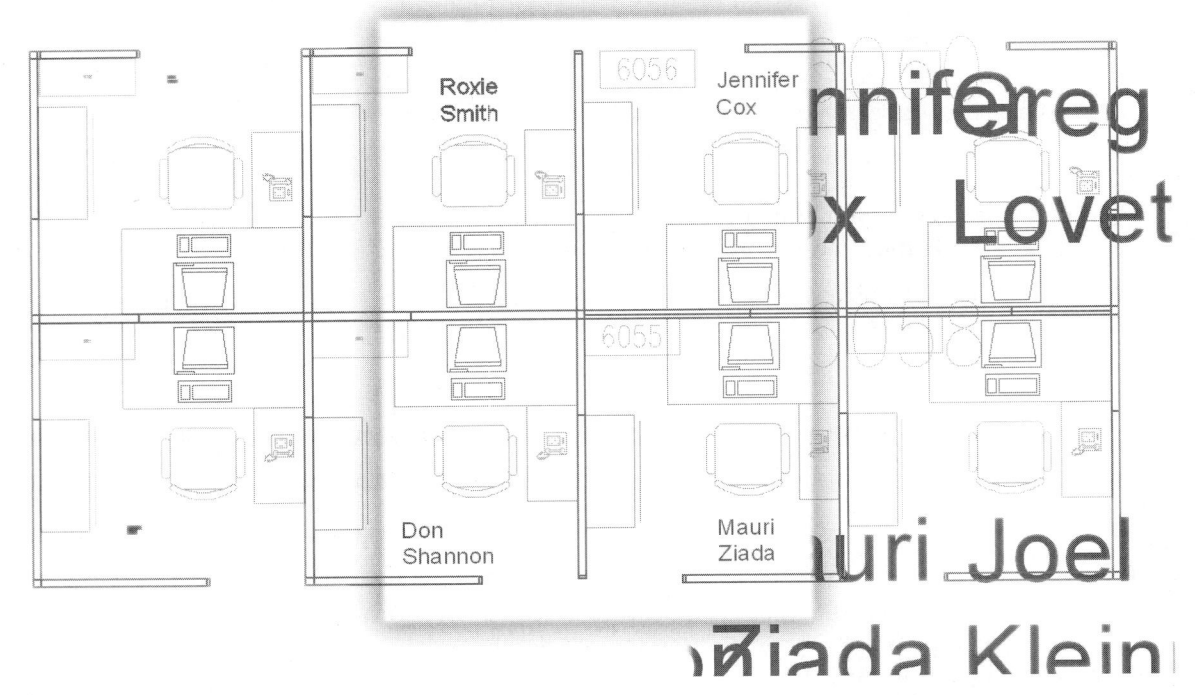

At left: text height of 1" is too small.
Center: text height of 6" is just right.
At right: text height of 2' is too tall.

Rotation Angle

Lines of text can be placed at any angle in drawings. You can specify the angle at which the text is drawn by designating the angle.

Specify rotation angle of text <0>: *(Enter an angle, or pick two points.)*

Press **ENTER** to accept the default angle, 0 degrees in this case. The text is rotated about the *insertion point*, which varies depending on the justification mode.

Once the text angle is set, the angle remains the default angle until it is changed. The figure illustrates text placed at several angles.

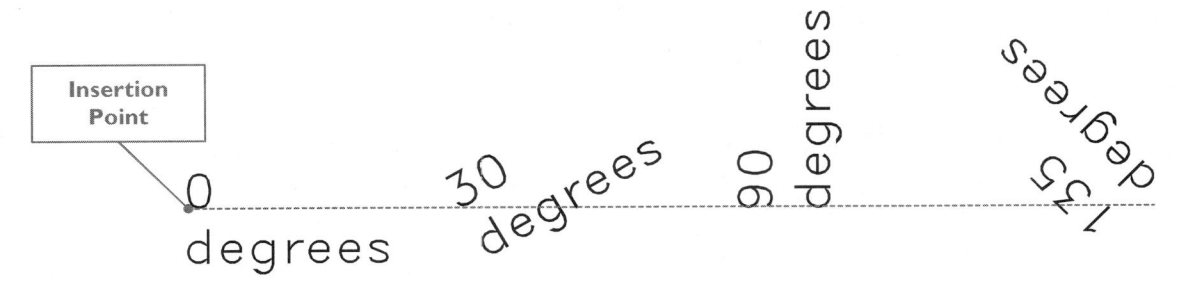

The angle is affected by the **ANGBASE** and **ANGDIR** system variables, as are other angles in AutoCAD.

Control Codes

AutoCAD allows you to add *metacharacters* to the text, called "control codes." A metacharacter is text that means something other than itself. (The copy editor notes that *all* text means something other than itself. We'll ignore that philosophical argument for now.) Use control codes to specify underlined text, and to include symbols, such as the degree and plus/minus symbols, and other text notations.

With the TEXT command, this is accomplished with a code that start with two percent characters: %%. The following table lists the control codes and their function:

Control Code	Meaning
%%c	Diameter symbol Ø
%%d	Degree symbol °
%%o	Overline
%%%	Percent symbol %
%%p	Plus-minus symbol ±
%%u	Underline
%%nnn	ASCII character *nnn*

Let's look at an example. The following text string:

Enter text: **If the piece is fired at 400%%dF for %%utwenty%%u hours,**

Enter text: **it will achieve %%p95%%% strength.**

is placed by AutoCAD as:

> If the piece is fired at 400°F for twenty hours, it will achieve ±95% strength.

Notice that you turn on and off the underscore (a.k.a. underlining) by typing %%u twice — once to turn on, and once to turn off; if you don't turn off underscoring, AutoCAD automatically turns it off at the end of the line.

You can "overlap" symbols, using, for instance, the degree symbol between the underscore symbols.

When you use text control codes, the codes are initially displayed instead of the effect. For example, when you enter:

Enter text: **This is %%uunderlined%%u.**

the text codes %%u are shown in the drawing as you enter the text:

> This is %%uuderlined%%u.

After you exit the TEXT command, AutoCAD redraws the text, showing the effect:

> This is underlined.

ASCII Characters

ASCII characters refer to the ASCII (short for "American Standard Code for Information Interchange") character set. This character set assigns number codes to symbols. To use one a symbol in text, enter two percent signs (%%) followed by the ASCII character code. For example, to place a tilde (~) in your text, enter %%126.

-Text

The **-TEXT** command is exactly like **TEXT**, except that it obeys the setting of the **TEXTEVAL** system variable.

TextEval	Meaning
0	Text strings are read literally: text is text.
1	Text prefixed with (or ! is evaluated as an AutoLISP expression.

Change **TEXTEVAL** to 1...

> Command: **texteval**
>
> Enter new value for TEXTEVAL <0>: *(Enter* **1**.*)*

... and then enter an AutoLISP expression:

> Command: **-text**
>
> Current text style: "STANDARD" Text height: 0.2"
>
> Specify start point of text or [Justify/Style]: *(Pick a point.)*
>
> Specify height <0'-6">: *(Press* ENTER.*)*
>
> Specify rotation angle of text <0.00>: *(Press* ENTER.*)*
>
> Enter text: **(+ 2 3)**
>
> 5

AutoCAD evaluates the AutoLISP expression, (+ 2 3), and places the result, 5, as text in the drawing.

> **Note:** The **TEXT** command is efficient for quickly placing strings of text at different positions in the drawing. (*String* is another way of saying "a line of text.") **MTEXT** is best for paragraphs of text, and text that contains formatting, such as **boldface** and color.

MTEXT

The **MTEXT** command places paragraphs of text, with flexible formatting.

While **TEXT** places text in the drawing one line at a time, the **MTEXT** command (short for "multiline text") creates paragraphs of text. It fits the text into an invisible boundary that you define (the boundary is not printed or plotted). It permits more text enhancements than **TEXT**, such as varying heights and colors, and stacked fractions.

The figure below illustrates an example of paragraph text.

```
ROOF FRAMING NOTES
 1.   See sheet SD1 for standard construction details and general structural notes.
 2.   Structural sheathed walls are designated with ~G and are below the level of
      framing shown unless otherwise noted. See "Structural Wall Sheathing Schedule"
      sheet SD1 for specific and general requirements.
 3.   Sheath all exterior walls per ~R unless otherwise noted. Sheath all specified
      interior walls per plans and "Structural Wall Sheathing Schedule" on sheet SD1.
      All exterior wall panels ≥4' in length shall be considered as complying braced
      wall panels per section 2320.11.3 of the UBC.
 4.   Roof shall be sheathed with APA rated sheathing, 32/16, Exposure 1, 15/32"
      minimum thickness.  Install sheets with face grain perpendicular to supports.
      Stagger sheets and nail with 8d @ 6"o.c. edges and 12" o.c. field typical unless
      otherwise noted.
 5.   For truss shapes, dimensions, etc. see Architectural drawings.
 6.   All trusses shall be designed and manufactured in conformance with the Truss
```

TUTORIAL: PLACING PARAGRAPH TEXT

1. To place paragraphs of text in drawings, start the **MTEXT** command:
   - From the **Draw** menu, choose **Text**, and then **Muliline Text**.
   - From the **Text** toolbar, choose the **Multiline Text** button.
   - At the 'Command:' prompt, enter the **mtext** command.
   - Alternatively, enter the aliases **t** or **mt** at the 'Command:' prompt.

   Command: **mtext** *(Press ENTER.)*

2. In all cases, AutoCAD reports the current settings, and then prompts you to pick two corners of the bounding box :

   Current text style: "Standard"   Text height: 0.2000

   Specify first corner: *(Pick point 1.)*

   Specify opposite corner or [Height/Justify/Line spacing/Rotation/Style/Width]: *(Pick point 2.)*

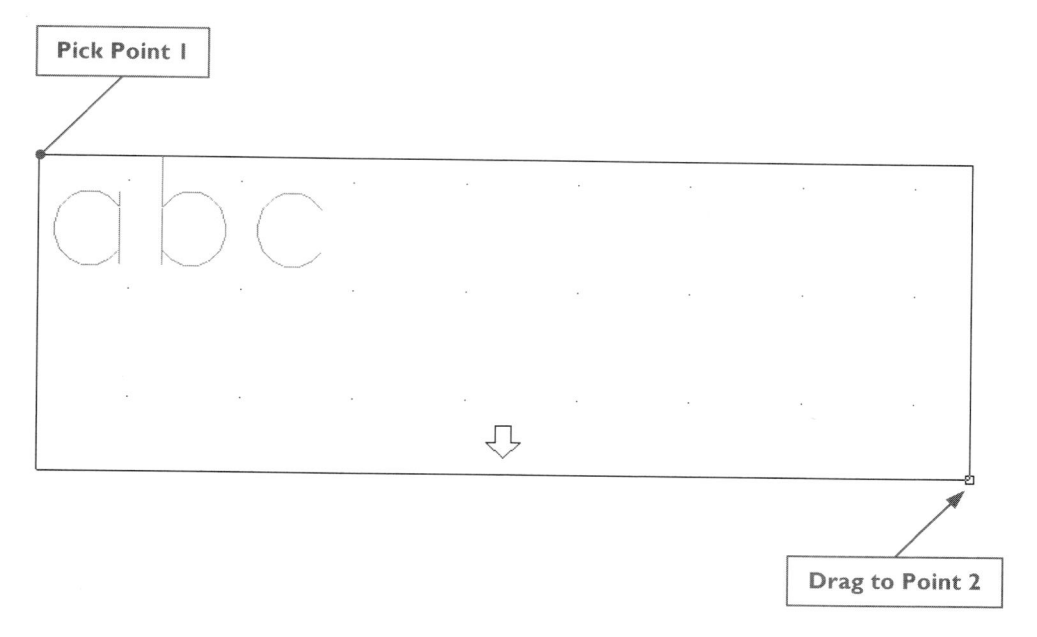

3. The text editor appears. It consists of two parts: the Text Formatting toolbar, and the text entry area topped by the tab bar.

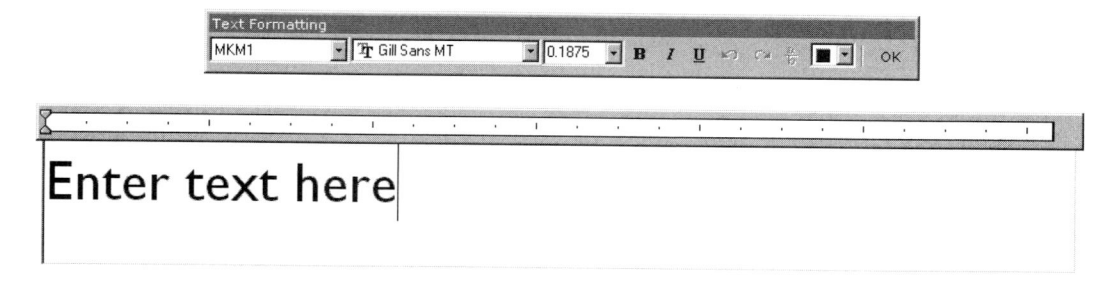

4. Type text into the text entry area.

   To format the text, select it by highlighting (drag the cursor over text you want formatted). From the Text Fomatting toolbar, select the format options.

   Additional options are hidden in a shortcut menu that you access by right-clicking the text entry area.

5. When done, click **OK**. Notice that AutoCAD places the text in the drawing.

To exit the mtext editor without saving changes, press ESC.

Note: The mtext editor displays diameter symbols as **%%c** and nonbreaking spaces as hollow rectangles. To insert the Euro symbol, press **CTRL+ALT+E**; if that doesn't work, turn on NumLock, hold down the **ALT** key, and then type **0128** on your keyboard's numeric keypad.

The Ä (Euro currency) symbol is included with all AutoCAD fonts shipping since Release 2000.

Text Formatting Toolbar

The Text Formatting toolbar allows you to apply a style to the text, as well as override the style with other properties, such as font and color.

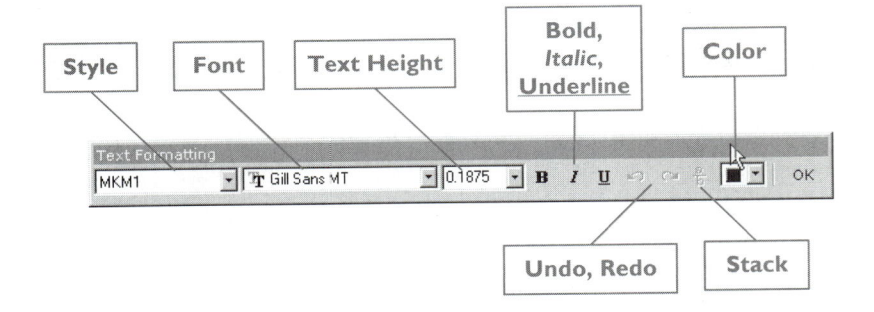

Style

Select a style name from the drop list. Only those text styles defined in the drawing are listed. The style applies to all text in the boundary box; you cannot selectively apply styles. You can, however, override the styles with the Font, Text Height, and other properties listed on the Text Formatting toolbar.

Font

Select a font name from the drop list. The fonts listed are those installed on your computer. The list probably includes all the TrueType fonts included with Windows, as well as fonts installed by AutoCAD. TrueType fonts have a small TT logo in front of them; AutoCAD fonts are prefixed by the Autodesk logo. (Printer fonts, PostScript fonts, and other fonts are not listed because AutoCAD does not support them directly.)

You can use several fonts with mtext, so you must highlight the text before selecting a font. (To highlight text, click and drag.)

To select all text, right-click the text and choose **Select All** from the pop-up menu. You can then change to all UPPERCASE or all lowercase.

Text Height

If you want some (or all) text at a different height, highlight the text, and then type a text height in the list box.

Bold

To **boldface** text, highlight the text, and then choose the **B** button. Choose the button a second time to "unboldface" the text. When the **B** on the button is gray, in the current font cannot be boldfaced — usually a problem only with certain AutoCAD fonts. Pressing **CTRL+B** performs the same function.

Italic

To *italicize* text, highlight then text, and then choose the **I** button. When the **I** button is gray, the text cannot be italicized. As an alternative, you can press **CTRL+I**.

Underline

To underline text, highlight the text, and then choose the **U** button. All fonts can be underlined; CTRL+U is the keystroke shortcut.

Undo

Choose the **Undo** button to reverse the last operation. Or press CTRL+Z.

Redo

Choose the **Redo** button to "undo" the undo. Or press CTRL+Y.

Stack

The **a/b** button stacks fractions. Instead of using a side-by-side format, like 11/32, this button places the 11 over the 32. You use it like this:

1. Type the fraction, such as 11/32.
2. Highlight the entire fraction.
3. Choose the **a/b** button.

AutoCAD replaces the numbers with a stacked fraction. Repeat steps 2 and 3 to "unstack" stacked fractions.

This button is not limited to numbers: AutoCAD stacks any combination of text, numbers, and symbols between spaces — whatever is in front of the slash goes on top; whatever is behind the slash goes underneath.

Also, this button is not limited to slash symbols. The carat (^) and the pound or hash mark (#) symbols perform different kinds of stacking:

Slash (/) stacks with a horizontal bar.   15/16 $\frac{15}{16}$

Carat (^) stacks without a bar (tolerance style).   12^12 $\frac{12}{12}$

Pound (#) stacks with a diagonal bar.   12#12 $^{12}\!/_{12}$

When the cursor comes next to stacked text, AutoCAD displays the following dialog box for controlling the "autostack" feature:

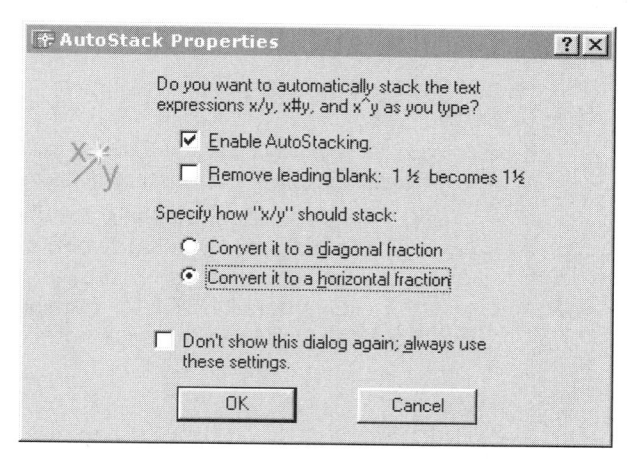

Enable AutoStacking automatically stacks numbers on either side of the /, #, and ^ characters.

Remove Leading Blank removes blanks between whole numbers and stacked fractions. This option is available only when AutoStacking is turned on.

Convert It to a Diagonal Fraction creates stacked numbers with the diagonal slash, regardless of the stack character (/, #, and ^).

Convert It to a Horizontal Fraction creates stacked numbers with the diagonal slash, regardless of the stack character.

> **Note:** When drawings with diagonal stacked fractions are opened in AutoCAD Release 14 or earlier, they are converted to horizontal fractions, but are restored when opened in AutoCAD 2000 or later.

Color

Highlight a portion of the text, and select a color from the **Color** list box. You can choose Bylayer, Byblock, the seven basic AutoCAD colors, and AutoCAD's other 16.7 million colors and color books.

Text Entry Area

The text entry area consists of a rectangle that defines the width of the mtext boundary box. It is topped by a tab bar that sets tabs, indents, and margins. You can apply different indent and tab settings to every paragraph.

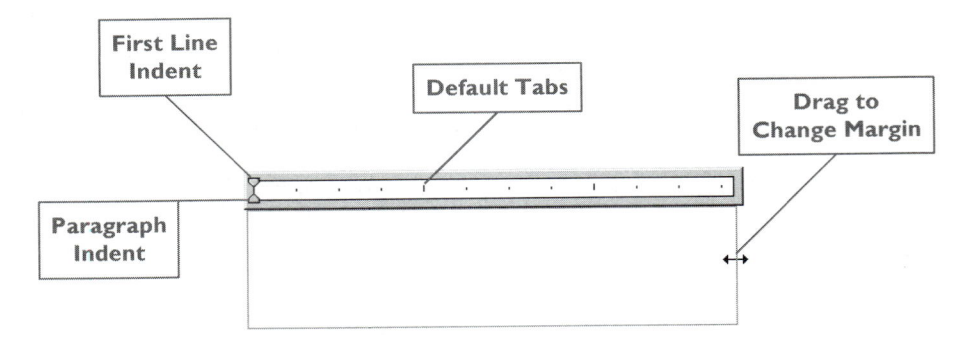

First Line Indent

You can move the *first line indent* arrow along the tab bar: by dragging it with the cursor. This shows how far in the first line of each paragraph should be indented. You can also create *hanging indents* when this indent is further left than the paragraph indent.

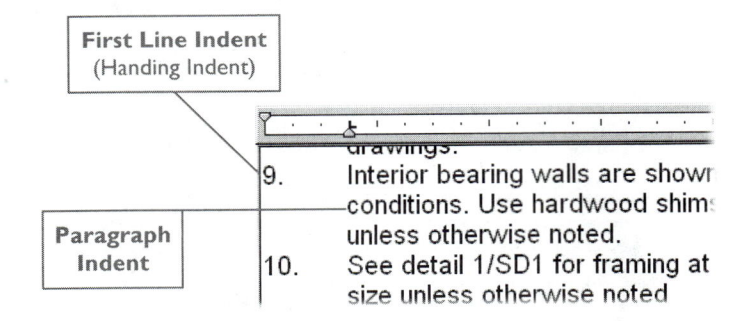

Paragraph Indent

Similarly, the paragraph indent shows how far the entire paragraph should be indented. Only the left indent can be modified; to change the right indent, drag the boundary box margin.

Tabs

Click the tab bar to set tabs; existing tabs can be dragged back and forth along the tab bar.

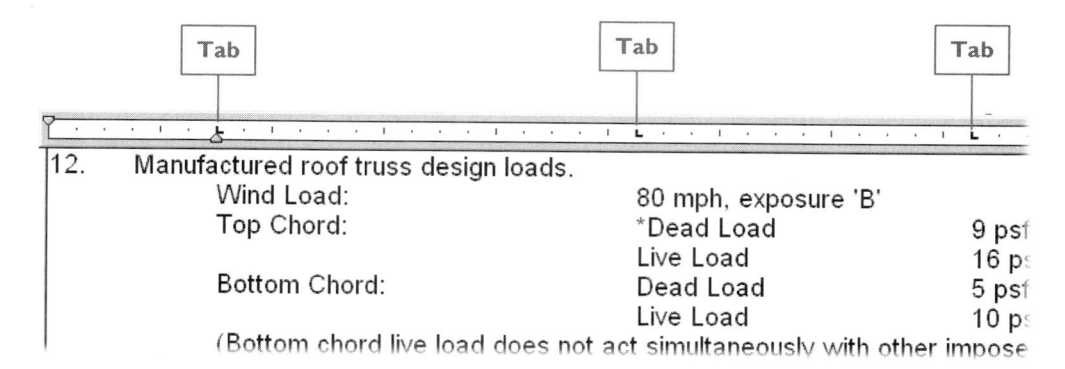

Right-clicking the tab bar reveals "hidden" options.

Indents and Tabs

The **Indents and Tabs** option displays a dialog box:

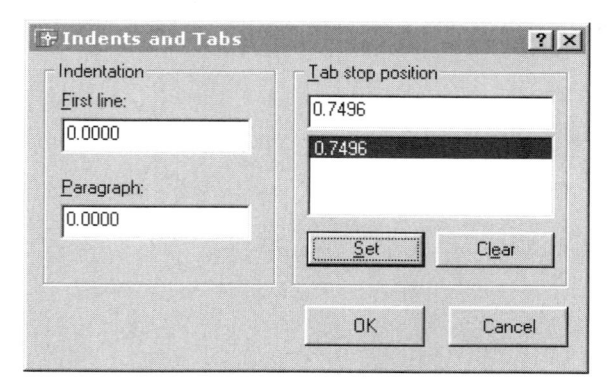

The **Indentation** section allows you to enter precise distances to indent the first line and the paragraph (i.e. every line). The indentation is available only for the left edge of the paragraph. As an alternative, you can drag the two arrows on the tab bar: the top arrow handles first line indentation, the bottom arrow the paragraph indentation.

The **Tab Stop Position** section lets you enter exact tab stops. *Tab stops* line up columns of data. As an alternative, you can click on tab bar to set tabs interactively.

Set Mtext Width

The **Set Mtext Width** option displays a dialog box that allows you to enter a precise width for the boundary box. As an alternative, you can drag the right edge of the boundary box, making it wider and narrower.

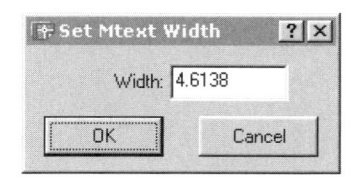

MText Shortcut Menu

Right-clicking the text entry area displays a shortcut menu that reveals many "hidden" options.

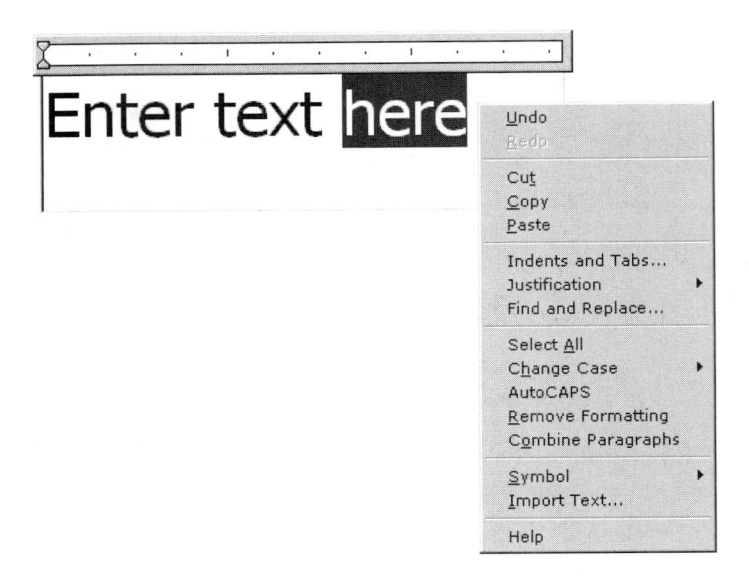

The first five options on the shortcut menu are standard: **Undo** (CTRL+Z), **Redo** (CTRL+Y), **Cut** (CTRL+X), **Copy** (CTRL+C), and **Paste** (CTRL+V).

Indents and Tabs

The **Indents and Tabs** option displays the dialog box described earlier.

Justification

The **Justification** option displays a submenu listing the justification modes described earlier.

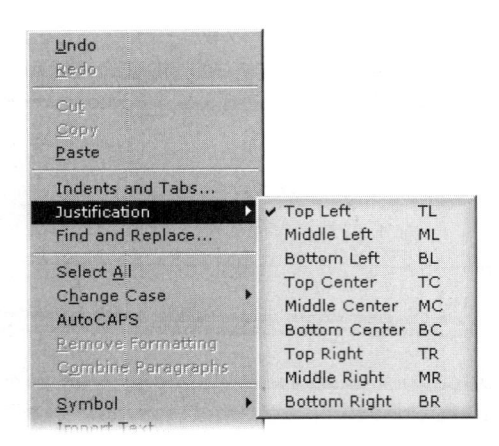

Find and Replace

The **Find and Replace** option displays the dialog box for searching for and optionally replacing text.

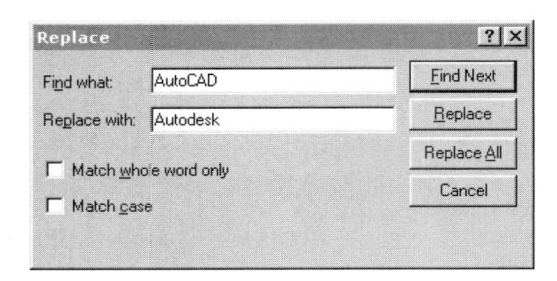

Select All

The **Select All** option selects all the text in the editor. Pressing CTRL+A does the same thing, and doesn't require opening this menu.

Change Case

The **Change Case** option changes selected text to all UPPERCASE or all lowercase. There are a number of keyboard shortcuts:

> **CTRL+SHIFT+A** toggles selected text between all uppercase and all lowercase.
>
> **CTRL+SHIFT+U** converts selected text to all uppercase; cannot be converted back to lowercase.
>
> **CTRL+SHIFT+L** converts selected text to all lowercase; cannot be converted back to uppercase.

AutoCAPS

The **AutoCAPS** toggle forces the computer's CapsLock mode, so that all typed text and imported text is displayed in uppercase characters. If the CapsLock light won't go off on your keyboard, it's because this option is still turned on.

Remove Formatting

The **Remove Formatting** option removes bold, italic, and underline from the selected text; does not affect other formatting, such as font and alignment changes.

Combine Paragraphs

The **Combine Paragraphs** option groups selected paragraphs into a single paragraph.

Stack

The **Stack** option operates identically to the icon on the Text Formatting toolbar: it toggles stacking of selected text on either side of a stack character. When stacked text is selected, this option changes to **Unstack**.

Properties

The **Properties** option is available only when stacked text is selected.

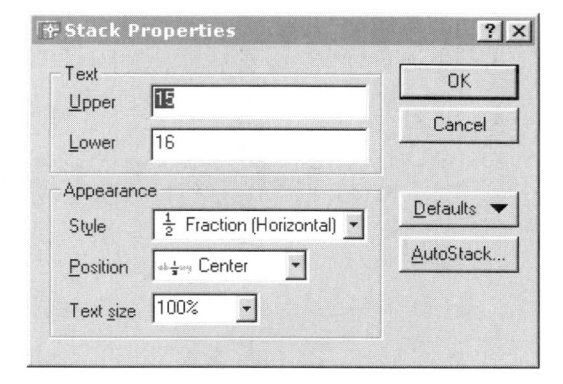

Upper changes the text in the upper half of stacked fractions; text need not be a number.

Lower changes the text in the lower part.

Style selects a format for stacked text: **Tolerance** (no dividing line), **Fraction** (horizontal line), or **Fraction** (diagonal line).

Position aligns the factions: **Top** (top of fraction aligns with top of line), **Center** (center of fraction aligns with center of the text line), or **Bottom** (bottom of fraction aligns with text baseline).

Text Size specifies the size of the stacked text as a percentage, from 25 to 125 percent.

Defaults saves the settings as the new default, or restores the previous default.

AutoStack displays the AutoStack Properties dialog box described earlier.

Symbol

The **Symbol** option displays a submenu listing common drafting symbols.

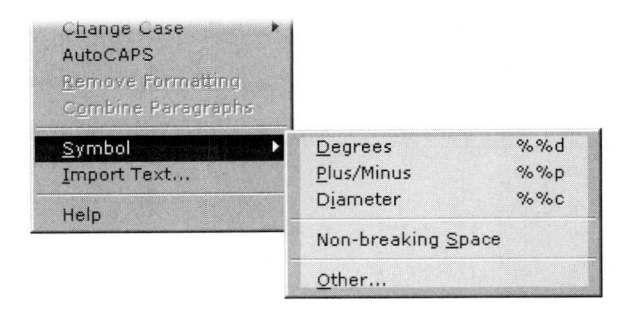

The three basic symbols are the most commonly used in drafting: degree, plus–minus, and diameter.

A fourth symbol is invisible, and is called the "non-breaking space." When you place it between two words, it prevents AutoCAD from using that space for word wrapping. For example, if your text includes the phrase "six inches," you might want a non-breaking space between the "six" and "inches" to keep the two words together.

Other opens the Windows Character Map dialog box, which contains all the characters available for the font. Inserting a symbol requires the rather awkward process of: (1) selecting the symbol; (2) clicking **Select**; (3) clicking **Copy**; (4) clicking x to close the dialog box; (5) right-clicking in the mtext editor, and then selecting **Paste**.

To insert the Euro symbol, press CTRL+ALT+E.

Import Text

The **Import Text** option displays a dialog box for selecting documents saved in plain text (a.k.a. ASCII format) or RTF (rich text format). You cannot import text saved in a word processing formats, such as Write (WRI), Word (DOC), or WordPerfect (WP) files. Attempting to import these formats results in this warning:

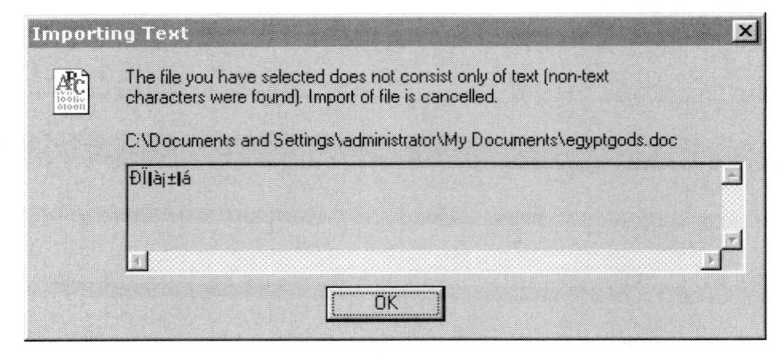

AutoCAD can import files up to 32KB in size, although a warning dialog box incorrectly states the limit is 16KB. Excel spreadsheets imported through RTF format are truncated at 72 rows, unless created in Office 2002 with service pack 2 installed. Text color is set to the current color; some (but not all) other formatting is preserved, such as font names and sizes, but not justification or columns.

Note: As ab alternative to selecting commands from the toolbar and shortcut menus, the following keystroke shortcuts also work in the mtext editor:

Shortcut	Meaning
TAB	Moves cursor to the next tab stop.
CTRL+A	Selects all text.
CTRL+B	**Boldface** toggle.
CTRL+I	*Italicize* toggle.
CTRL+U	Underline toggle.
Case Conversion:	
CTRL+SHIFT+A	Toggles between all UPPERCASE and all lowercase.
CTRL+SHIFT+U	Converts selected text to all UPPERCASE.
CTRL+SHIFT+L	Converts selected text to all lowercase.
Justification:	
CTRL+L	Left-justifies selected text.
CTRL+E	Centers selected text.
CTRL+R	Right-justifies selected text.
Line Spacing:	
CTRL+1	Single-spaces lines.
CTRL+5	1.5-spaces lines.
CTRL+2	Double-spaces lines.
Font Sizing:	
CTRL+SHIFT+,	(*comma*) Reduces font size.
CTRL+SHIFT+.	(*period*) Increases font size.
CTRL+=	Superscript toggle.
CTRL+SHIFT+=	Subscript toggle.
Copy and Paste:	
CTRL+C	Copies selected text to the Clipboard.
CTRL+X	Cuts text from the editor and send it to the Clipboard.
CTRL+V	Pastes text from the Clipboard.
Undo and Redo:	
CTRL+Y	Redoes last undo.
CTRL+Z	Undoes last action.
Symbols:	
CTRL+ALT+E	Inserts Euro symbol.
ALT+SHIFT+X	Converts selected text to ASCII number (A becomes 41, and so on).

MText Control Codes

In addition to the %% control codes discussed earlier for the TEXT command, MTEXT has a host of its own control codes. These codes are used by AutoCAD to format mtext; in addition, you can use them to format text with a text editor or word processor external to AutoCAD. (You may be able to write a set of macros that converts the word processor's codes to those recognized by AutoCAD.)

Mtext codes consist of a backslash symbol followed by a character. Codes are *case-sensitive*, which means you must use an uppercase **P** and not a lowercase **p**, for example. The table shows the codes and their meanings.

Control Code	Meaning
\~	Nonbreaking space.
\\	Backslash.
\{	Opening brace; encloses multiple control codes.
\}	Closing brace.
\A*n*;	Sets vertical alignment of text in boundary box:
	0 = bottom alignment.
	1 = center alignment.
	2 = top alignment.
\C*n*;	Sets color of text:
	1 = red.
	2 = yellow.
	3 = green.
	4 = cyan.
	5 = blue.
	6 = magenta.
	7 = white.
\F*x*;	Changes to font file name *x*.
\H*n*;	Changes text height to *n* units.
\L	Underline.
\l	Turns off underline.
\M+*nnn*;	Multibyte shape number *nnn*.
\O	Turns on overline.
\o	Turns off overline.
\P	End of paragraph.
\Q*n*;	Changes obliquing angle to *n*.
\S*n*^*m*;	Stacks character *n* over *m*.
\T*n*;	Changes tracking to *n*; for example, \T3; is wide spacing between characters.
\U+*nnn*;	Places Unicode character *nnn*.
\W*n*;	Changes width factor (width of characters) to *n*.

Let's look at some examples. Normally, MTEXT automatically wraps text to fit the boundary box. To force a line break in the text, use the \P code (short for "paragraph"). AutoCAD reads the following line:

> This is the first line of text\P and the second line.

and places it in the drawing as:

> This is the first line of text
>
> and the second line.

To force text to remain together (hard spaces), use the \~ code:

> This\~is\~text\~that\~must\~remain\~together.

To underline text, use the \L and \l codes:

> This is \Lunderlined\l text.

Use the curly brackets, { and }, to specify a change in text for a specific part. For example, to write the following warning in white (color #7) and red (color #1) text:

> {\C7;Do not} {\C1;change} {\C7;this drawing.}

Notice how the semicolon is required when the mtext code consists of more than a single character:

> \P does not require the semicolon.
>
> \C7; requires the semicolon.

PLACING PARAGRAPH TEXT: ADDITIONAL METHODS

The **MTEXT** command has many options. As well, there is the command-line version of the command, and a system variable that selects the text editor. Each time you respond to an option (except the **Width** option), AutoCAD redisplays the MTEXT prompt until you pick the opposite corner:

> Specify opposite corner or [Height/Justify/Line spacing/Rotation/Style/Width]:

- **Height** specifies the height of text.
- **Justify** specifies the justification of text in the boundary box.
- **Line spacing** specifies the spacing between lines of text.
- **Rotation** rotates the boundary box.
- **Style** selects the text style.
- **Width** specifies the width of the boundary box.
- **-MTEXT** command accepts text entry at the command-line.
- **MTEXTED** system variable selects the mtext editor.

Let's look at each.

Height

The **Height** option changes the height of the text used by the mtext editor. It prompts you:

> Specify height <0.2000>: *(Type a number or indicate a height.)*

Justify

The **Justify** option selects the justification and positioning of text within the boundary box. AutoCAD prompts:

> Enter justification [TL/TC/TR/ML/MC/MR/BL/BC/BR] <TL>: *(Enter an option, or press* **ENTER***.)*

The justification options are the same as for the **MTEXT** command, except that in this case, there are two areas where justification applies: (1) text; and (2) flow.

Text justification is left, center, or right, relative to the left and right boundaries of the rectangle. *Flow justification* positions the block of text top, middle, and bottom relative to the top and bottom boundaries of the rectangle. The abbreviations have the following meaning:

Linespacing

The **Linespacing** option changes the spacing between lines of text, sometimes called the "interline spacing" or "leading." The option has two methods for specifying the spacing: **At least** and **Exactly**. The **At least** option specifies the minimum distance between lines, while **Exactly** specifies the precise distance between lines of text.

> Enter line spacing type [At least/Exactly] <At least>: **a**
>
> Enter line spacing factor or distance <1x>: *(Enter a value, including the* **x** *suffix.)*

The value you enter is a multiplier of the standard line spacing distance, which is defined in the font.

Rotation

The **Rotation** option specifies the rotation angle of the boundary box.

> Specify rotation angle <default>: *(Enter an angle, or pick two points.)*

This option rotates the entire block of text. For example, specify 90 degrees to place text sideways at the edge of a drawing.

Note: When you use the mouse to show the rotation angle, AutoCAD calculates the angle as follows:

- The *start* of the angle is the X-axis.
- The *end* of the angle is the line anchored by the "Specify first corner:" pick; the other corner is now in line with the indicated angle.

Style

The **Style** option selects a text style to use for the multiline text.

> Enter style name (or ?) <Standard>: *(Enter a name, or press ? for a list of style names.)*

When you enter ? in response to this prompt, AutoCAD displays the names of text styles defined in the drawing. You create new text styles with the STYLE command (described later in this chapter) or with a style borrowed from another drawing, using the **DesignCenter**.

Width

The **Width** option specifies the width of the boundary box.

> Specify width: *(Enter a value, or pick a point to show the width.)*

A width of 0 (zero) has special meaning. AutoCAD draws the multiline text as one long line (no word wrap), as if there were no bounding box at all.

-MText

The -MTEXT command operates identically to the MTEXT command, except that it prompts you to enter text at the command line.

> Command: **-mtext**
>
> Current text style: "ROMAND" Text height: 3/16"
>
> Specify first corner: *(Pick a point.)*
>
> Specify opposite corner or [Height/Justify/Line spacing/Rotation/Style/Width]: *(Pick another point, or enter an option.)*
>
> MText: *(Enter text.)*
>
> MText: *(Press ENTER to exit the command.)*

At the "MText:" prompt, you can type lines of text; as an alternative, you can right-click and select **Paste**: text in the Clipboard is pasted to the command line and placed in the drawing. The drawback to this technique is that the text cannot contain blank lines, which AutoCAD misinterprets as its cue to exit the -MTEXT command.

MTextEd

The MTEXTED system variable defines the text editor that should be used to edit mtext.

> Command: **mtexted**
>
> Enter new value for MTEXTED, or . for none <"Internal">: *(Enter a name.)*

MTextEd	Meaning
.	Uses the default mtext editor (does not disable mtext editor.)
Internal	Uses the default mtext editor.
:lisped	Uses an AutoLISP-based editor.
notepad.exe	Uses Notepad as the editor.

As an alternative, you can specify any word processor, but you must include the full path and executable name. For example, to use the Atlantis word processor:

Command: **mtexted**

Enter new value for MTEXTED, or . for none <"Internal">: **"c:\program files\atlantis\atlantis.exe"**

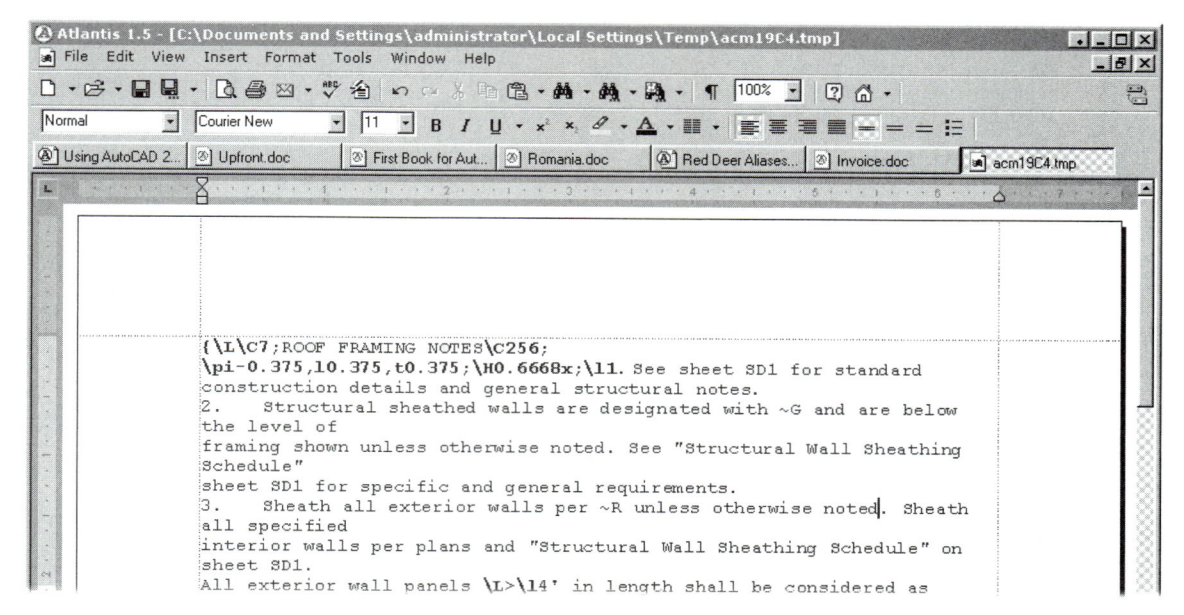

The drawback to external text editors and word processors is that they do not understand mtext format codes, and so display them literally, which can be an advantage if you want to examine the code.

The **:lisped** option works only with lines of text with fewer than 80 characters.

QLEADER AND LEADER

The **QLEADER** and **LEADER** commands place *callouts* (a.k.a leaders) in drawings.

The difference between the two commands is that **QLEADER** (short for "quick leader") includes a Settings dialog box for fashioning the leader's properties, whereas the **LEADER** command's options are entered at the command-line. The figure illustrates examples of leaders.

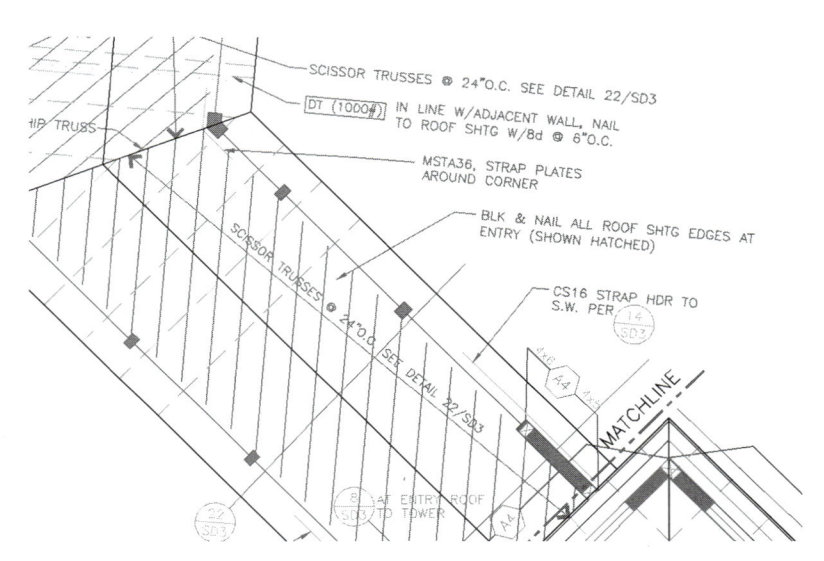

TUTORIAL: PLACING LEADERS

1. To place leaders in drawings, start the **QLEADER** command:
   - From the menu bar, choose **Dimension**, and then **Leader**.
   - From the **Dimension** toolbar, choose the **Quick Leader** button.
   - At the 'Command:' prompt, enter the **qleader** command.
   - Alternatively, enter the **le** alias at the 'Command:' prompt.

   Command: **qleader** *(Press ENTER.)*

2. In all cases, AutoCAD prompts you for the first point, which is where the tip of the arrowhead is placed:

   Specify first leader point, or [Settings] <Settings>: *(Pick point 1.)*

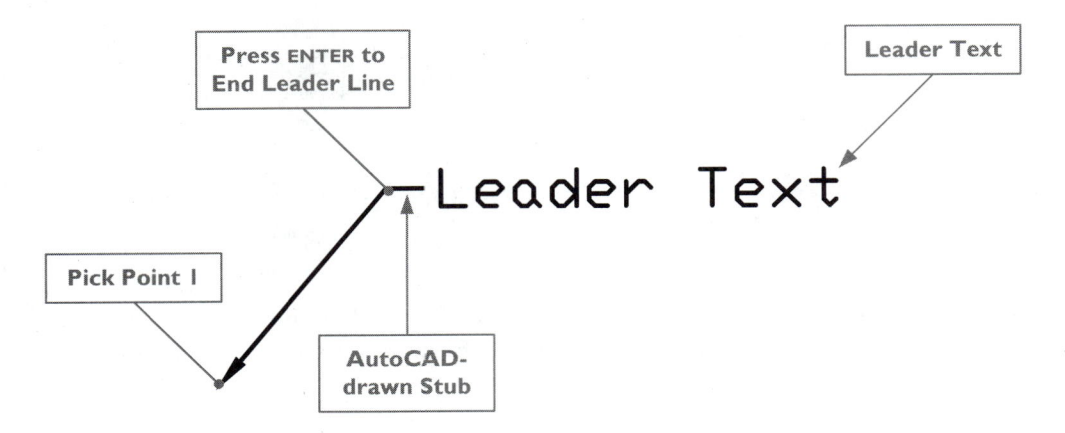

3. Continue picking points. Each pick indicates a vertex in the leader line.
   Press **ENTER** to end the leader line. AutoCAD later draws a short horizontal stub automatically.

   Specify next point: *((Press ENTER to end the leader line.)*

4. If the text needs to be constrained in width, enter the width here; otherwise, enter 0 for unconstrained width:

   Specify text width <0.0000>: *(Press ENTER.)*

5. Enter the text, and then press **ENTER**.
   Press **ENTER** twice in a row to exit the command.

   Enter first line of annotation text <Mtext>: *(Enter text.)*

   Enter next line of annotation text: *(Press ENTER to exit command.)*

If the leader's start point touches an object, it is *associated* with the object. This means that when you move the object, the arrowhead moves with the object, although the leader text stays in place. (Associativity is available only when associative dimensioning is turned on through the **DIMASSOC** system variable — which is the case in most drawings.)

Many of the leader's properties, such as line color and arrowhead scale, are determined by system variables related to dimensioning, all of which start with **DIM**. To modify these properties, use the **DIMSTYLE** command, as described in the next chapter.

You can use grips editing to modify the leader's position. The figures illustrates some possibilities.

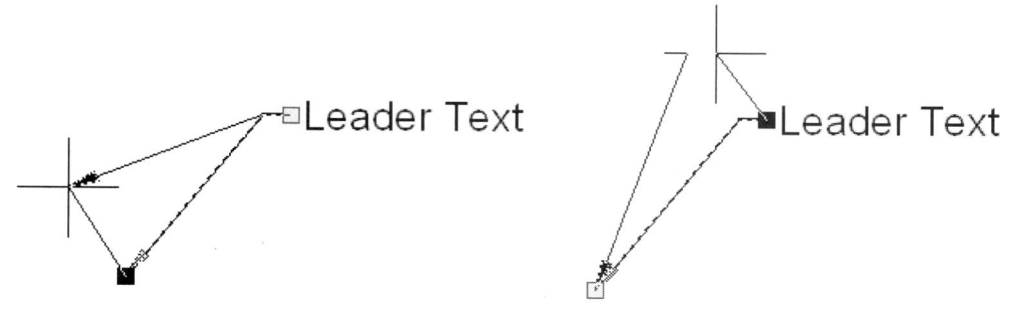

To edit the leader text, double-click the text (not the leader line), and AutoCAD displays the familiar mtext editor.

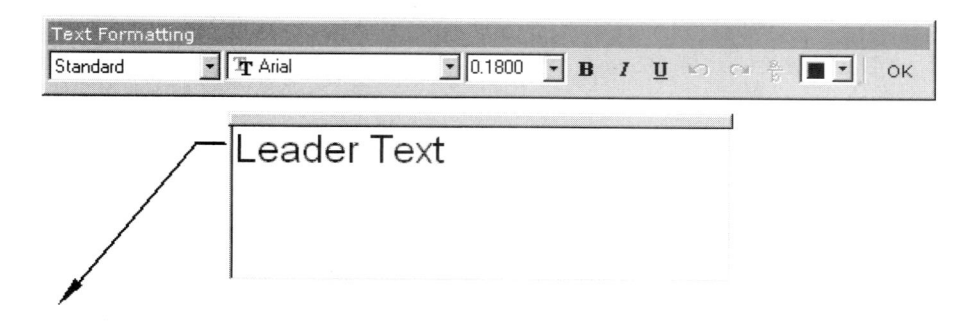

PLACING LEADERS: ADDITIONAL METHODS

The **QLEADER** command has these options:

- **Settings** specifies the characteristics of leaders.
- **LEADER** operates at the command line.

Let's look at each.

Settings

The **Settings** option displays a dialog box for specifying the look and operation of leaders.

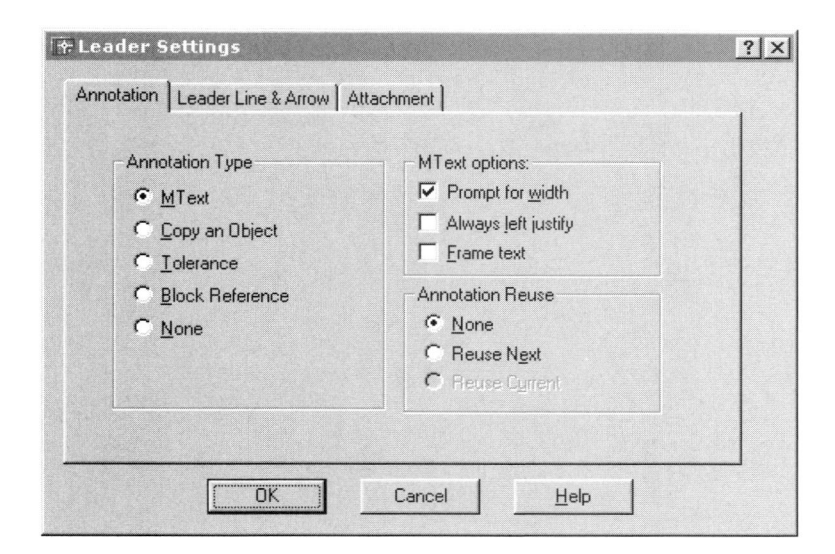

Annotation Type

The options you select under **Annotation Type** affect the prompt displayed by the **QLEADER** command.

MText (default) changes the prompts to create leader text from mtext.

> Specify text width <0.0000>: *(Enter a width, or press* ENTER.*)*
>
> Enter first line of annotation text <Mtext>: *(Enter text.)*
>
> Enter next line of annotation text: *(Press* ENTER *to end the command.)*

Copy an Object changes the prompt to select multiline text (including other leader text), single-line text, a tolerance object, or block to be copied as the leader annotation:

> Select an object to copy: *(Select one object.)*

Tolerance displays the Tolerance dialog box; see Chapter 15, "Geometric Dimensioning and Tolerancing."

Block Reference changes the prompt for inserting a block:

> Enter block name or [?]: *(Enter the name of a block)*
>
> Specify insertion point or [Scale/X/Y/Z/Rotate/PScale/PX/PY/PZ/PRotate]: *(Pick a point.)*
>
> Enter X scale factor, specify opposite corner, or [Corner/XYZ] <1>: *(Enter a scale factor.)*
>
> Enter Y scale factor <use X scale factor>: *(Press* ENTER.*)*
>
> Specify rotation angle <0>: *(Enter an angle.)*

None removes the prompt for leader text and other annotation types; it draws just the arrowhead and leader line.

MText Options

The **MText Options** are available only when the **MText** option was selected (see above).

Prompt for Width (default = on) toggles the display of the prompt for the width of the mtext leader text:

> Specify text width <0.0000>: *(Enter a width, or press* ENTER.*)*

Always Left Justify (default = off) toggles left-justification of mtext, regardless of leader location. AutoCAD normally adjusts the text justification based on the leader line's orientation.

When this option is turned on, multiple lines of text are left-justified when the leader angles left, as illustrated by the figure below. When this option is turned off, multiple lines of text are right-justified.

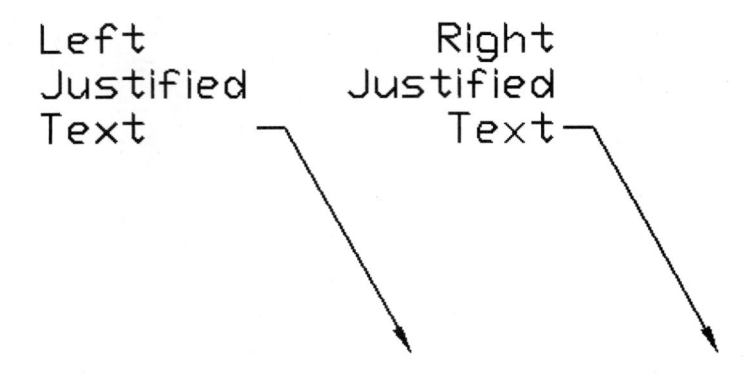

Frame Text (default = off) toggles the addition of a rectangular frame around the mtext annotation.

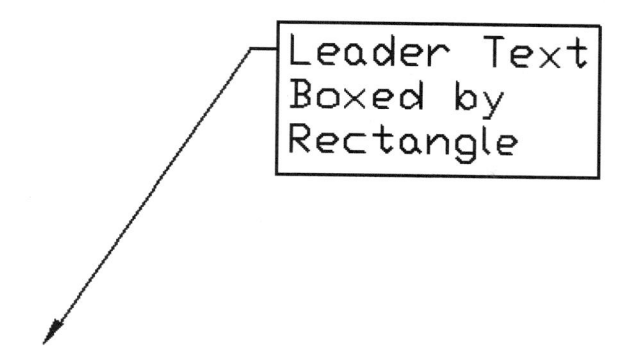

```
Leader Text
Boxed by
Rectangle
```

Annotation Reuse

The options under **Annotation Reuse** determine whether the leader text or other annotation is reused by subsequent **QLEADER** commands.

None (default) causes AutoCAD to prompt you for an annotation, as described above.

Reuse Next uses the next annotation you create for subsequent leaders. There is no prompt for mtext, blocks, tolerances, or copying of objects.

Reuse Current reuses the current annotation for future leaders.

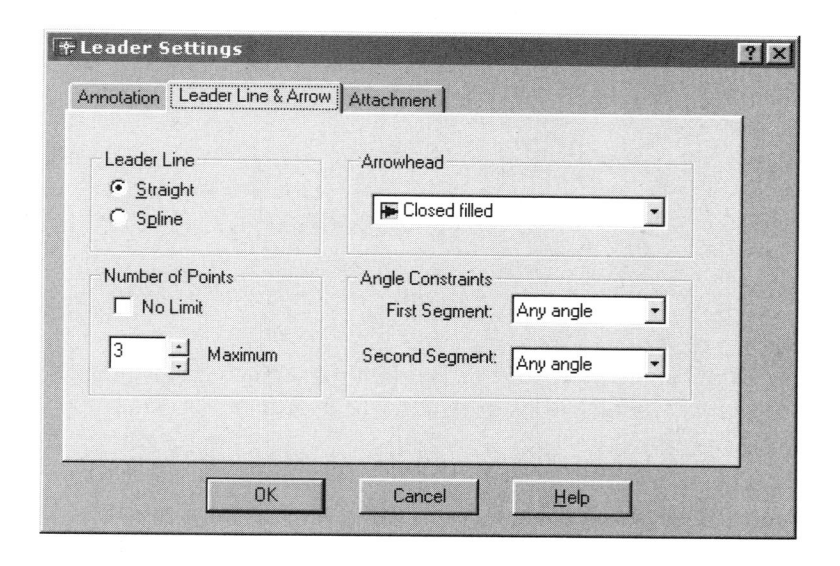

Leader Line

Straight draws the leader line out of straight line segments.

Spline draws the leader from a spline object, resulting in a curved leader.

```
Splined
Leader
```

Number of Points

No Limit means that the QLEADER command prompts you to "Specify next point:" until you get tired and press ENTER.

Maximum means that the command stops prompting you after a set number, such as 3. You can always stop earlier by pressing ENTER. You must set the number to one more than the number of leader segments you want to create: 3 means that you are prompted for two segments; recall that AutoCAD draws the third segment automatically. The minimum maximum is 2; maximum is 999!

Arrowhead

Select one of the arrowheads shown by the drop list:

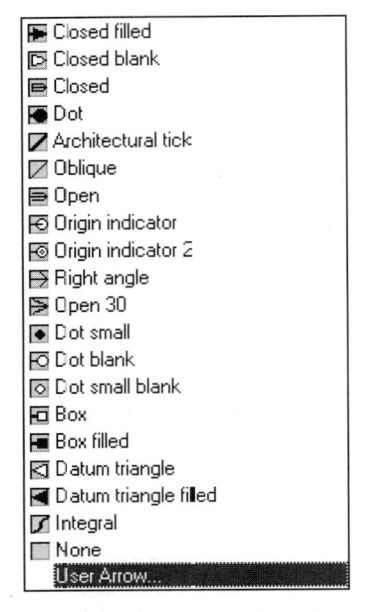

User Arrow lists blocks in the drawing, of which one can be selected for use as the leader's arrowhead.

Angle Constraints

First Segment specifies the angle of constraint for the first segment. "Any angle" means the segment is not constrained; otherwise, select from Horizontal (0 degrees), 90, 45, 30, or 15 degrees.

Second Segment specifies the angle-of-constraint for the second leader segment.

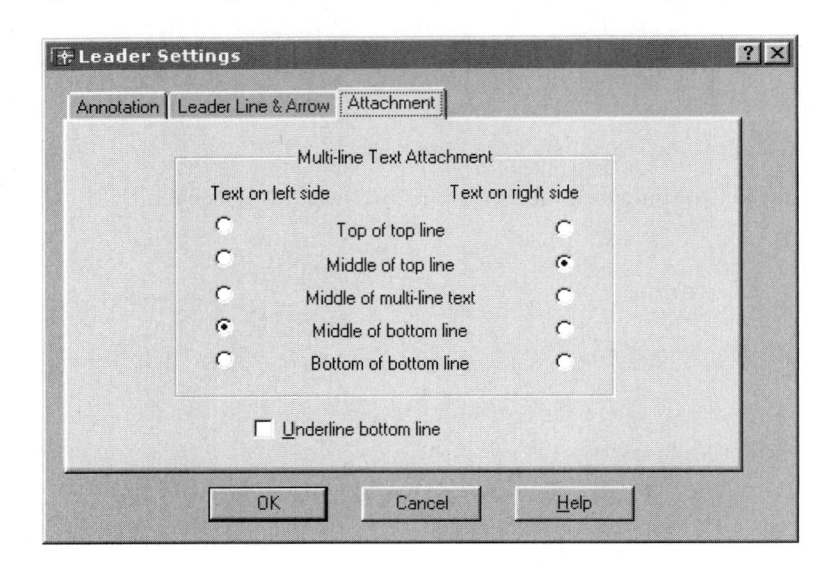

Multiline Text Attachment

The **Multiline Text Attachment** options are available only when the **Mtext** option is selected on the Annotation tab. These options align the text with the leader line, and can be set differently for left- and right-justified text.

Top of Top Line aligns the leader line with the top of the top mtext line.

Middle of Top Line aligns the leader line with the middle of the top mtext line.

Middle of Multiline Text aligns the leader line with the middle of the mtext.

Middle of Bottom Line aligns the leader line with the middle of the bottom mtext line.

Bottom of Bottom Line aligns the leader line with the bottom of the bottom mtext line.

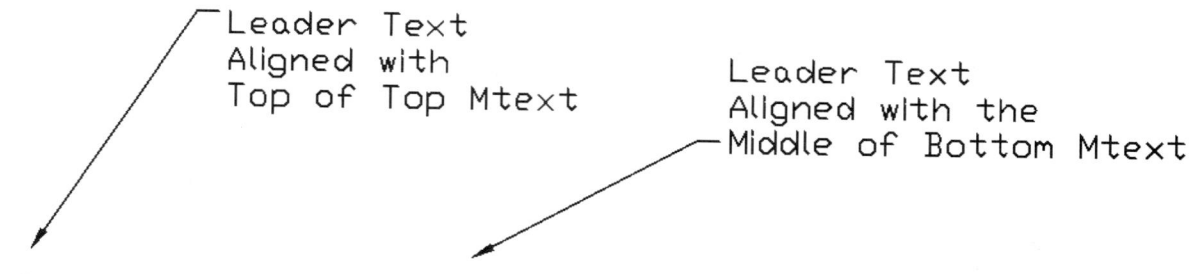

If none of the above options works for you, you may have selected the wrong side: select an option for the correct side.

Underline Bottom Line attaches the leader line to the bottom of the mtext, and underlines the last mtext line.

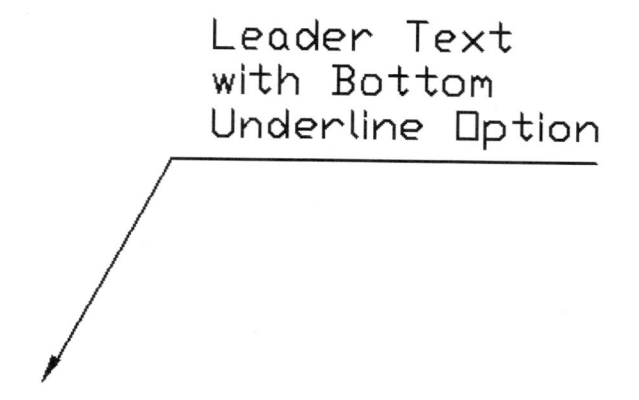

Leader

The **LEADER** command displays prompts at the command line.

    Command: **leader**

    Specify leader start point: *(Pick a point.)*

    Specify next point: *(Pick another point.)*

    Specify next point or [Annotation/Format/Undo] <Annotation>: *(Enter an option.)*

The **Annotation** and **Format** options are similar to those of **QLEADER**; the **Undo** option undoes the last action.

DDEDIT

The **DDEDIT** command edits text.

This one command handles all forms of AutoCAD text: single-line text, mtext, dimensions, and attributes. It displays a different dialog box for each type of text. (**DDEDIT** is short for "dynamic dialog editor," an old reference to dialog boxes.)

TUTORIAL: EDITING TEXT

1. To edit text in drawings, start the **DDEDIT** command:
   - From the **Modify** menu, choose **Object**, choose **Text**, and then **Edit**.
   - From the **Edit** toolbar, choose the **Edit Text** button.
   - At the 'Command:' prompt, enter the **ddedit** command.
   - Alternatively, enter the **ed** alias at the 'Command:' prompt.

   Command: **ddedit** *(Press ENTER.)*

2. In all cases, AutoCAD prompts you to select the text:

   Select an annotation object or [Undo]: *(Select a text object.)*

   AutoCAD displays different dialog boxes, depending on the text you selected:
   - Single-line text placed with the **TEXT** command: the **Edit Text** dialog box.
   - Multi-line text placed with the **MTEXT** command: the **Text Formatting** dialog box.
   - Leaders and dimension text placed with dimensioning commands: the **Text Formatting** dialog box.
   - Attribute definitions placed with the **ATTDEF** command: the **Edit Attribute Definition** dialog box.

> **Notes:** As an alternative to entering the **DDEDIT** command, you can double-click single-line and multiline text, and attribute definitions. AutoCAD automatically displays the appropriate editor. **MTEDIT** is an alias for displaying the mtext editor. Double-clicking leaders and dimension text displays the Properties window, which can also be used to edit text.
>
> Attributes placed with the **INSERT** command are edited with a separate command, **ATTEDIT**.

Single-line Text Editor

When you select single-line text placed by **TEXT**, AutoCAD displays the Edit Text dialog box:

Make your changes, such as deleting and adding text, and then choose the **OK** button. AutoCAD returns to the prompt:

Select an annotation object or [Undo]: *(Select another line of text, or press ESC to exit the command.)*

Multiline and Leader Text Editor

NEW IN 2004

When you select leaders, dimensions, or text created by the **MTEXT** command, AutoCAD displays the Text Formatting bar — identical to that displayed by **MTEXT**. (This was called the **Multiline Text Editor** dialog box in earlier releases, and has been greatly changed in AutoCAD 2004.)

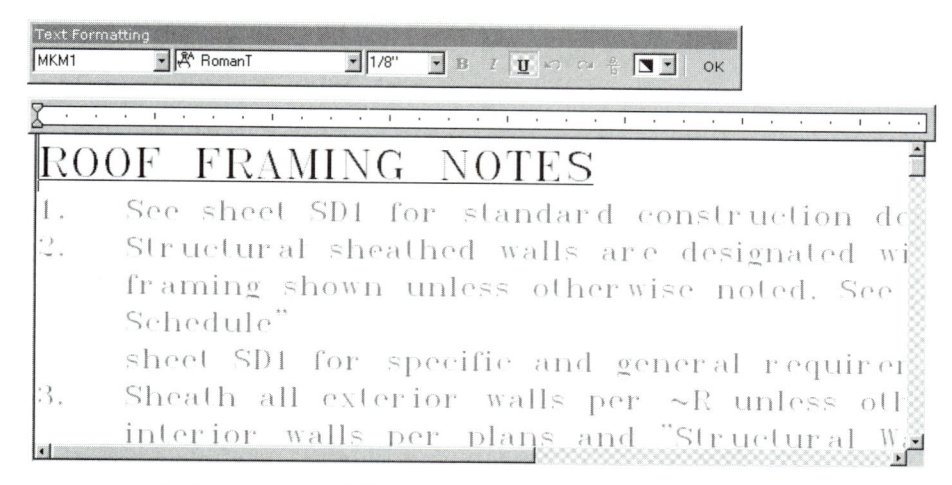

Make your changes, and choose the **OK** button. AutoCAD returns to the prompt:

> Select an annotation object or [Undo]: *(Select another paragraph of text, or press* **ESC** *to exit the command.)*

Attribute Text Editor

When you select attribute definitions created by the **ATTDEF** command, AutoCAD displays the Edit Attribute Definition dialog box.

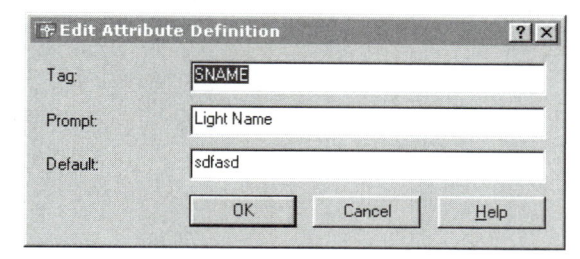

Make your changes to the tag, prompt, and default. Choose the **OK** button. Attributes are discussed in *Using AutoCAD 2004: Advanced*.

EDITING TEXT: ADDITIONAL METHODS

The **DDEDIT** command has but a single option — undo; an alternative editor is the **PROPERTIES** command.

- **Undo** option undoes changes made to the text.
- **PROPERTIES** command changes all properties of text.

Let's look at each.

Undo

The **Undo** option undoes changes made to text while still in the **DDEDIT** command:

> Select an annotation object or [Undo]: **u**

Properties

The **Properties** window is able to edit text, as well as many other aspects of the text — color, layer, insertion point, and so on. The only property that cannot be changed is the font; that must be changed with the **STYLE** command. Select the text, and then right-click: from the shortcut menu, select **Properties**.

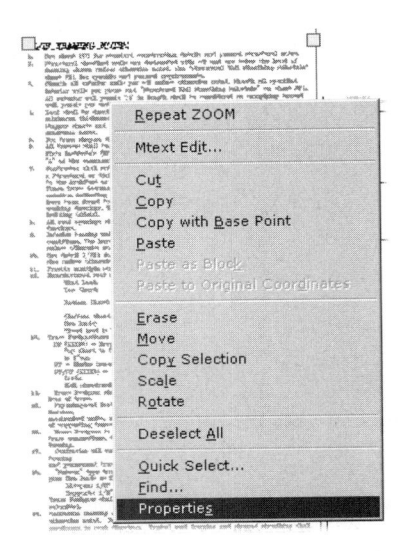

AutoCAD displays the Properties window. The Contents area displays the text of the text object. The strange characters — such as {\L\C7; — are mtext formatting codes. If the text is paragraph text, choose the [...] button, which causes AutoCAD to load the familiar mtext editor.

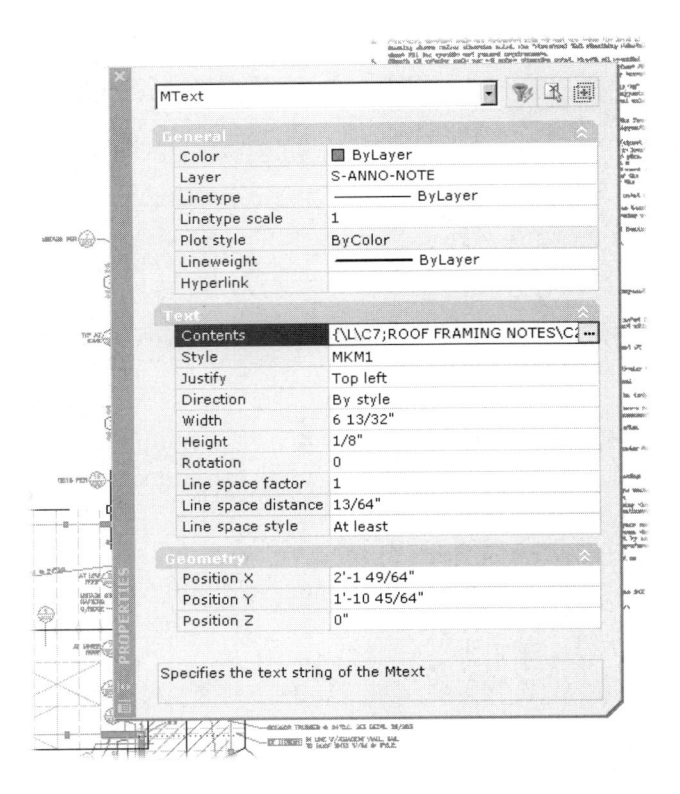

STYLE

The **STYLE** command defines named text styles based on fonts.

A *style* is a collection of properties applied to the font of your choice, such as slanted text, or perhaps condensed text. Styles are often used in word processing and desktop publishing. The idea is to preset most text properties, so that you don't need to do it each time with each line of text.

Some properties cannot, however, be set by styles, such as color and layer.

Styles are used by the **TEXT** and **MTEXT** commands, as well as by attributes and dimensions. Styles are stored by a name of your choice. Text styles are composed of the following information, although not all fonts support all properties:

- **Style name** describes the style — can be up to 255 characters in length.
- **Font file** is associated with the style. The font can be either *.ttf* (TrueType) or *.shx* (AutoCAD) format. PostScript *.pfb* font files can be used after conversion to *.shx* with the **COMPILE** command.
- **Font style** selects regular, boldface, or italicized text.
- **Height** specifies the height of the text. Normally, this is set to 0, which means that the **TEXT** command prompts you for the height. Entering a height here means the command doesn't bug you later.
- **Width factor** specifies a multiplier making the text wider or narrower. A width of 1 is standard; a width factor of 0.5 produces text at one half the width.
- **Obliquing angle** that determines the slant of the text. A positive angle is a forward slant; a negative angle a backward slant.
- **Backwards** mirrors the text.
- **Upside down** places the text upside down.
- **Vertical** places the text vertically.

When you change the style, all text assigned that style also changes.

TUTORIAL: DEFINING TEXT STYLES

1. To define one or more text styles, start the **STYLE** command:
   - From the menu bar, choose **Format**, and then **Text Style**.
   - From the **Text** toolbar, choose the **Text Style** button.
   - From the **Styles** toolbar choose the **Text Style Manager** button.
   - At the 'Command:' prompt, enter the **style** command.
   - Alternatively, enter the aliases **st** or **ddstyle** (the old name) at the 'Command:' prompt.

   Command: **style** *(Press* ENTER.*)*

2. In all cases, AutoCAD displays the dialog box:

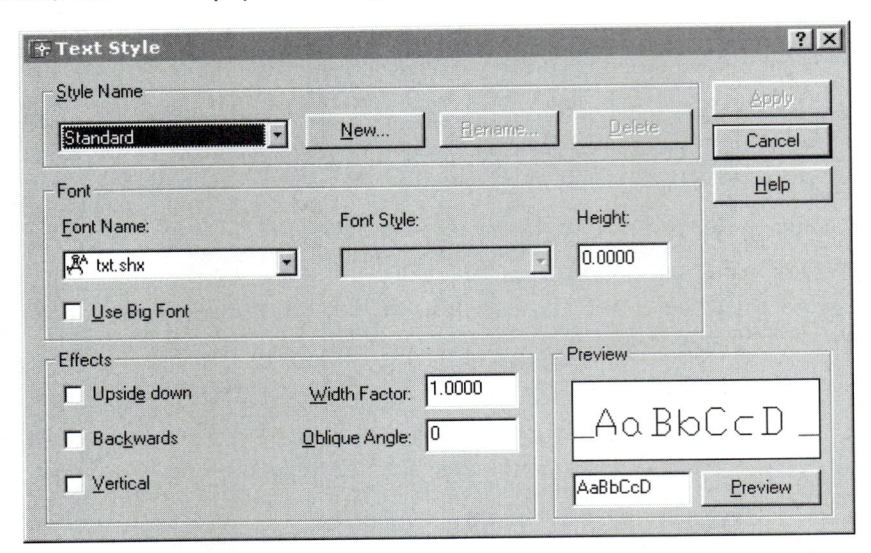

The **Style Name** list box provides the names of styles already in the drawing. Use this list to change the characteristics of an existing style.

Options that do no apply to a font or style are grayed out.

3. Click **New**.

AutoCAD displays the New Text Style dialog box.

Change the default name "style1" to anything else — up to 255 characters — and then choose **OK**. Notice that AutoCAD adds the name to the **Style Name** list.

The **Rename** button lets you rename the selected style; you cannot rename the Standard text style.

The **Delete** button erase the selected style from the drawing — but only if it is unused. If used by text, AutoCAD complains, "Style is in use, can't be deleted."

4. From the **Font Name** drop list, select a font. There are many font files available for AutoCAD, which comes with an large selection: specialty fonts for map making (symbols instead of letters), cursive writing, and for other applications.

Note: The font file should be appropriate for the application. Most mechanical applications use the ANSI type lettering, often called "Leroy." The Simplex font (also known as "RomanS" — roman, single stroke) closely approximates this style. Architectural drawings typically use "hand-lettered" fonts, such as City Blueprint. Engineering applications often use the Simplex font, since it closely resembles a traditional font constructed with a lettering template.

Multistroke fonts use a code, with "S" denoting single stroke, "D" duplex or double stroke, "C" complex (also double stroke), and "T" triple stroke.

5. Select any of the following options, if available (not grayed out):

From the **Font Style** drop list, select Regular, **Boldface**, *Italic*, or ***Bold Italic***. Not all font styles are available for all fonts.

In the **Height** text box, enter a height, which is measured from the baseline to the top of uppercase letters. You may first need to determine the scale of the drawing before deciding the text height. In fact, you may find it preferable to leave the height at the default of 0. This gives you the flexibility to change the height as the text is being placed. This applies particularly if you are not sure of the scale until the end of the drawing process.

The **Use Big Font** option refers to a class of AutoCAD fonts that handle more than 256 characters, primarily for Chinese and other languages with thousands of characters.

The **Width Factor** determines the width of characters. A width factor of 1.0 is "standard." A decimal value, such as 0.85, draws text narrower by 15%, which is useful for condensed text. Values greater than 1 create expanded text.

0.5 Width Factor

2.0 Width Factor

Oblique Angle is the slant applied to characters. A zero obliquing angle draws text that is "straight up." Positive angles, such as 15 degrees, draw text that slants forward, while negative angles draw backward slants. Valid values are between –85 and 85 degrees.

15° Oblique

-15° Oblique

Backwards means the text is drawn in reverse. Backwards text is useful for drawings plotted on the back side of clear media, such as the casting and injection-molding industries.

txeT esreveR

Upside Down draws text upside down.

nwoD edispU txeT

Vertical draws text vertically, which is *not* the same as rotating text by 90 degrees. Text is drawn so that each letter is vertical and the text string itself is vertical. TrueType and some AutoCAD fonts cannot be drawn vertically.

V T
e e
r x
t t
i
c
a
l

Note: Obliquing, underscoring, or overscoring should not be used with vertical text, since the result will look incorrect.

6. After applying effects, such as a width factor, choose the **Preview** button to see the result.
7. Click **OK**.

USING TEXT STYLES

The difference between fonts and styles is sometimes confusing. You cannot use fonts directly in AutoCAD; you can only use fonts through styles (the exception is in the mtext editor). Styles modify the properties of a font (such as its height, width, and slant). One font can be used by many styles; each style can only specify one font.

You must create a style to use a font. Then, in the **TEXT** and **MTEXT** commands, you specify the style to use for the text you place. Each line or paragraph of text is assigned a style; you may change the style with the **PROPERTIES** command. When the style changes, the look of the text changes. To change text globally, change the style with the **STYLE** command.

Often, drafting offices create a standard selection of styles, each of which is used for a particular type of text in drawing. For example, one style is used for notes, another for titles, and others for the different parts of the title block. Using styles ensures consistency across drawings, no matter who drafter.

Styles Toolbar *(NEW IN 2004)*

An alternative method to applying styles is with the Styles toolbar.

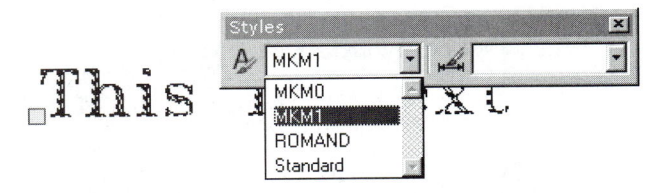

Select one or more lines of text created with the **TEXT** command, and then select a style name from the style droplist. AutoCAD changes the look of the text to match the style.

The **Text Style Manager** button displays the Text Style dialog box.

DEFINING STYLES: ADDITIONAL METHODS

The **-STYLE** command defines styles at the command prompt. In addition, several system variables affect the look and quality of TrueType fonts.

- **-STYLE** command defines styles at the command prompt.
- **TEXTFILL** system variable toggles fill of TrueType fonts for plots.
- **TEXTQLTY** system variable adjusts quality of TrueType fonts for plots.
- **FONTALT** system variable specifies font substitution.

Let's look at each.

-Style

The **-STYLE** command prompts you to create text styles at the command line:

> Command: **-style**
>
> Enter name of text style or [?]: *(Enter a name, type **?**, or press* ENTER.*)*

Press ENTER to redefine an existing font. AutoCAD prompts:

> Existing style. Full font name = current:
>
> Specify full font name or font file name <TTF or SHX>: *(Enter the name of a .ttf or .shx font file.)*

To specify the name of a Big Font, enter a *.shx* file name, a comma, and the Big Font file name.

Enter ~ to display the Select Font File dialog box.

Enter ? to list the names of text styles in the drawing.

The remaining prompts are similar to the options in the Style dialog box:

> Specify height of text: *(Enter a height, or enter **0**.)*
> Specify width factor: *(Enter a factor.)*
> Specify obliquing angle: *(Enter an angle between -85 and 85 degrees.)*
> Display text backwards? [Yes/No] <N>: *(Enter **Y** or **N**.)*
> Display text upside-down? [Yes/No] <N>: *(Enter **Y** or **N**.)*
> Vertical? <N>: *(Enter **Y** or **N**.)*

TextFill

The **TEXTFILL** system variable toggles the fill of TrueType fonts for plots. When on (the default), fonts are filled; when off, only the outline of the font is plotted. This system variable affects only the plotting of fonts; it does not affect the display of fonts.

> Command: **textfill**
>
> Enter new value for TEXTFILL <1>: **0**

Outline Font

TextQlty

The **TEXTQLTY** system variable specifies the resolution of TrueType font outlines (technically, the tessellation fineness) when plotted: 0 means the text is not smoothed; 100 means maximum smoothness. The difference is seen only when drawings are plotted; this system variable does not affect the display.

> Command: **textqlty**
>
> Enter new value for TEXTQLTYL <50>: **100**

FontAlt

The **FONTALT** system variable specifies the name of a font to use when other fonts cannot be found.

Often, different computers have different collections of installed fonts. When you receive a drawing from another AutoCAD drafter, the drawing may use fonts not found on your computer. So that text doesn't go missing, AutoCAD substitutes the *simplex.shx* font for missing fonts. (If the Simplex font cannot be found, AutoCAD displays the Alternate Font dialog box, so that you can select another font file.)

Command: **fontalt**

Enter new value for FONTALT, or . for none <"simplex">: *(Enter the name of a TrueType or AutoCAD font.)*

For missing TrueType fonts in Release 14 drawings, AutoCAD substitutes another font without warning you, which can be annoying. On my computer, AutoCAD substitutes the inappropriate Absalom font, which makes drawing texts look like this:

New paragraph with unstacked 11/32 and stacked fractions 13/64.

SPELL

The **SPELL** command checks the drawing for words unfamiliar to AutoCAD.

TUTORIAL: CHECKING SPELLING

1. To check the spelling of text, start the **SPELL** command:
   - From the menu bar, choose **Tools**, and then **Spell**.
   - At the 'Command:' prompt, enter the **spell** command.
   - Alternatively, enter the **sp** alias at the 'Command:' prompt.

   Command: **spell** *(Press ENTER.)*

2. In all cases, AutoCAD prompts you to select objects.
   I find the easiest method is to specify "all", because then I don't have to select lines of text individually, and AutoCAD filters out non-text objects automatically.

   Select objects: **all**

   2030 found 1 was an external reference.

   Select objects: *(Press ENTER to end object selection.)*

3. AutoCAD displays the Check Spelling dialog box:

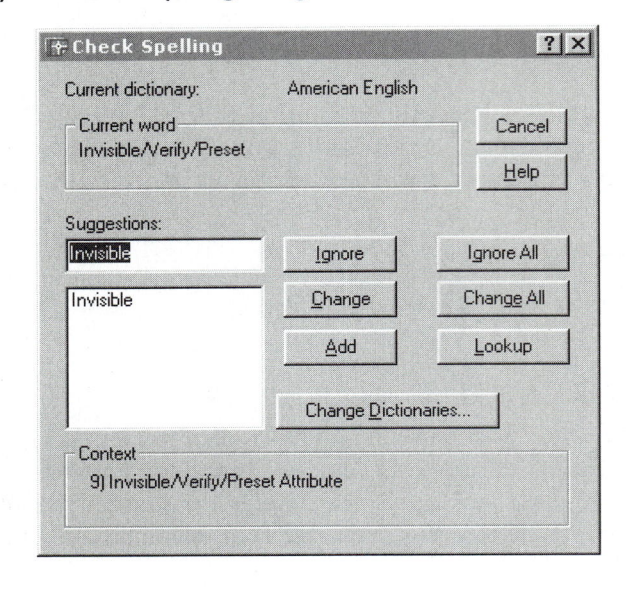

4. Confirm the spelling of words AutoCAD does not recognize.
5. When done, AutoCAD reports: "Spelling check complete."

This command works with words placed or imported by the **TEXT**, **-TEXT**, **MTEXT**, **-MTEXT**, **LEADER**, **QLEADER**, and **ATTDEF** commands. When it comes to attributes, only values are spell checked, not tags. Because '**SPELL** can be run as a transparent command, you can use it while other commands are active.

FIND

The **FIND** command searches for, and optionally, replaces text.

TUTORIAL: FINDING TEXT

1. To find and replace text, start the **FIND** command:
   - From the menu bar, choose **Edit**, and then **Find**.
   - From the **Text** toolbar, choose the **Find and Replace** button.
   - At the 'Command:' prompt, enter the **find** command.

   Command: **find** *(Press* ENTER.*)*

2. In all cases, AutoCAD displays the Find and Replace dialog box.

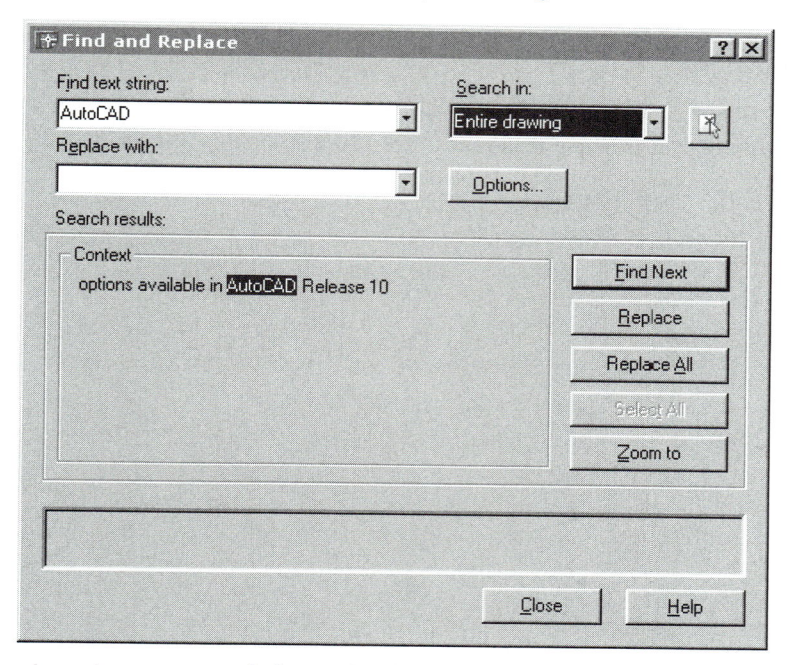

3. Enter a word or phrase to search for in the drawing.
4. Optionally, enter a replacement word or string.
5. Click **Find Next** to find the first instance of the phrase.
   If necessary, click **Replace** or **Replace All**.
6. When done, click **Close**.

The **U** command reverses changes made by this command.

NEW IN 2004: WIPEOUT

The **WIPEOUT** command places blank areas in drawings to make text and other objects more legible in cluttered areas.

The **BHATCH** and **HATCH** commands normally avoid the problem by hatching around text. If you place text after the hatch, however, you can either use **HATCHEDIT** to "clean up" around the text, or else use the **WIPEOUT** command.

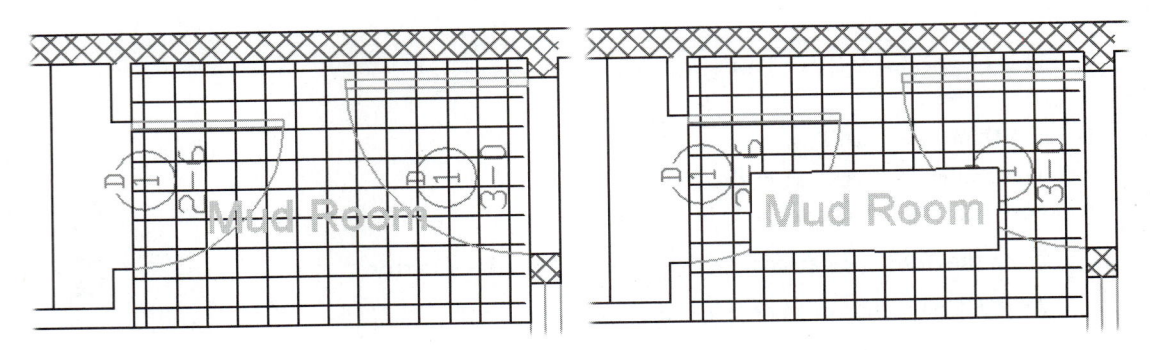

At left: text is obscured by the hatch pattern.
At right: text is legible over the wipeout polygon.

TUTORIAL: WIPING OUT OBJECTS

1. To wipe out objects, start the **WIPEOUT** command:
   - From the menu bar, choose **Draw**, and then **Wipeout**.
   - At the 'Command:' prompt, enter the **wipeout** command.

   Command: **wipeout** (Press ENTER.)

2. In all cases, AutoCAD prompts you to pick points to define the wipeout area:

   Specify first point or [Frames/Polyline] <Polyline>: *(Pick a point.)*

   Specify next point: *(Pick a point.)*

   Specify next point or [Undo]: *(Pick another point.)*

   Specify next point or [Close/Undo]: **c**

The **WIPEOUT** command responds to snap mode and to object snap modes, but not to ortho mode. The workaround is to draw a closed polyline with ortho mode, and then convert the polyline to a wipeout object. Wipeouts are raster images drawn in the background color, usually white or black. To plot drawings with whiteouts, your computer must be connected to a raster-capable plotter using the ADI v4.3 driver or the Windows printer driver.

WIPING OUT OBJECTS: ADDITIONAL METHODS

The **WIPEOUT** command has these options:

- **Frames** toggles the display of the wipeout's frame.
- **Polyline** converts a closed polyline into a frame.
- **Undo** undoes the last segment.
- **Close** closes the polygon.

Let's look at how each option works.

Frames

The **Frames** option toggles the display of the wipeout's border. When the frame is off, wipeouts cannot be selected.

>Enter mode [ON/OFF] <ON>: **off**

Polyline

The **Polyline** option converts a closed polyline into a frame. The polyline must contain line segments only (no arcs), and have zero width.

>Select a closed polyline: *(Select a single polyline.)*
>
>Erase polyline? [Yes/No] <No>: *(Enter* **Y** *or* **N**.*)*

Not erasing leaves the polyline in place, providing the wipeout with a border; erasing the polyline leaves the wipeout area without a border.

Undo

The **Undo** option undoes the last segment.

Close

The **Close** option closes the polygon.

QTEXT

The **QTEXT** command displays lines of text as rectangles.

QTEXT uses rectangular boxes to represent the height and length of the text. The change to text boxes does not take place until the next regeneration; therefore you should follow with the **REGEN** command. The figure illustrates before and after applying **QTEXT/REGEN** to a drawing:

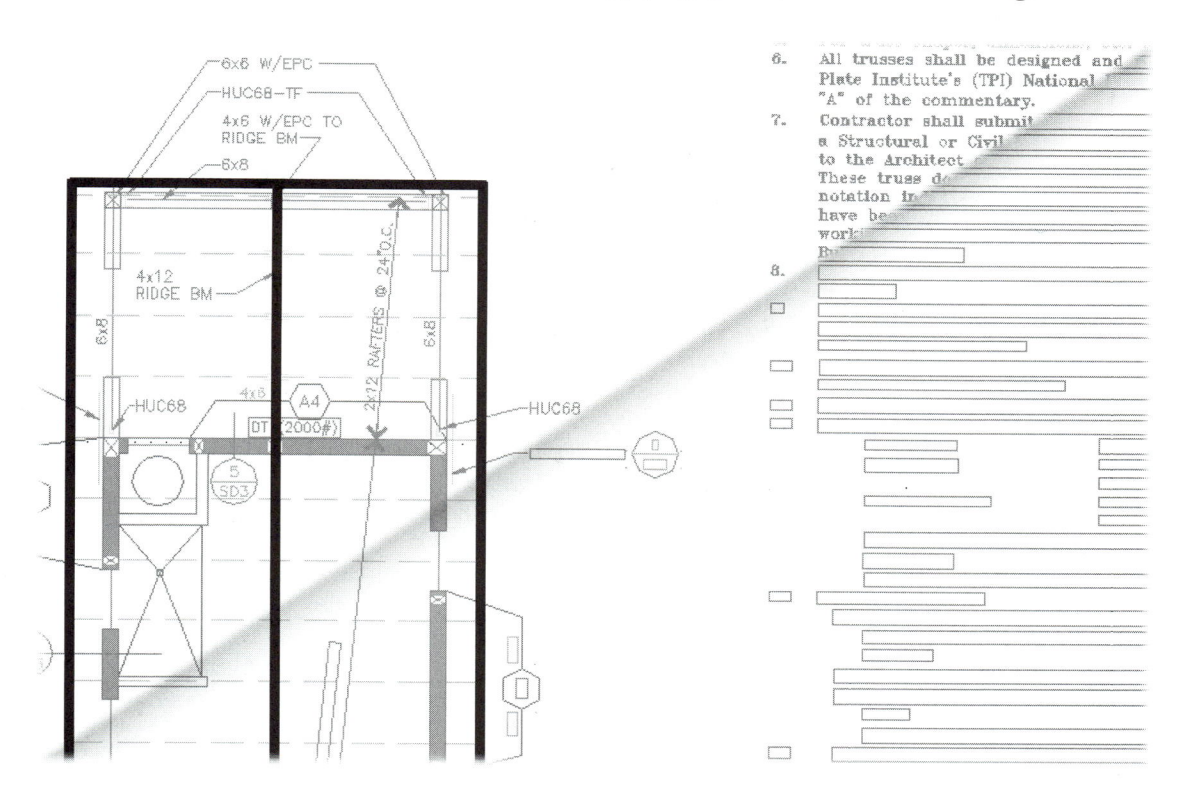

This command, which is short for "quick text," replaces text with rectangles of the same size. Being a transparent command, 'QTEXT can be used during other commands.

Historically, this command was useful in the early days of AutoCAD, when computers were very slow. Back then, turning off text helped speed up the redraw and regeneration times. QTEXT is still helpful today for speedier plots of drawings.

TUTORIAL: QUICKENING TEXT

1. To display text as rectangles, start the **QTEXT** command:
   - At the 'Command:' prompt, enter the **qtext** command.

   Command: **qtext** *(Press ENTER.)*

2. AutoCAD displays the tersest prompt of all commands :
   ON/OFF <OFF>: **on**

3. To see the rectangles, follow with the **REGEN** command:
   Command: **regen**

Even after a regeneration, some text may still appear normal. That's because it lies in another space, either model or paper space. Switch to model space or to a layout to force the regeneration. The figure below illustrates the problem:

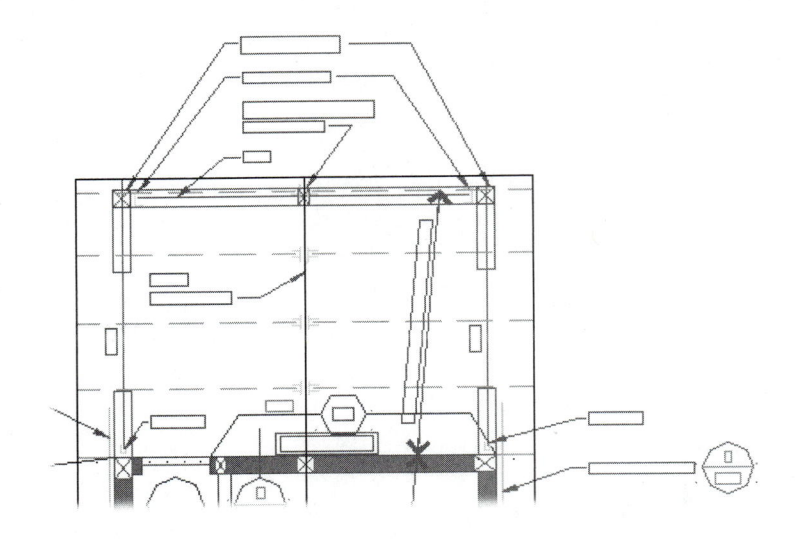

Notes: When a drawing contains a large amount of text, several techniques can be used to minimize the time required to display and plot the text:.

Place the text in the drawing last. This eliminates having to regenerate text while you are constructing the drawing elements.

Create a special layer for text. Freeze the layer when you are not using the text.

Use the **QTEXT** command if you want the text placements shown, but still need fast regenerations.

Plot check drawings with the **QTEXT** command turned on. The text plots as rectangles, greatly reducing plot time.

Whenever possible, avoid multistroke "fancy" fonts or TrueType fonts. These fonts regenerate and plot more slowly than single stroke fonts. If you wish to use fancy fonts, the quick text option reduces the regeneration time.

Use a simple font while working, and then use the style command to change to a fancier font for plotting. The drawback, the technical editor notes, is that the character spacing and other parameters may change.

SCALETEXT AND JUSTIFYTEXT

The **SCALETEXT** and **JUSTIFYTEXT** command change the size and justification of text.

After placing text in the drawing, you may find it necessary to change the height of the text, or its justification. The **PROPERTIES** command allows you to change these, but may not do it accurately.

AutoCAD provides the **SCALETEXT** command to change the height (scale) of text. It makes selected text larger or smaller. The advantage of this command over **SCALE** is that it resizes multiple text objects without changing their location; the **SCALE** command scales everything, including location, relative to the base point.

And the **JUSTIFYTEXT** command changes the justification of text. It can change, for example, left-justified text to right-justified. The size of justification of mtext can be changed with the mtext editor through the **MTEDIT** command.

TUTORIAL: RESIZING TEXT

1. To resize text, start the **SCALETEXT** command:
   - From the **Modify** menu, choose **Object**, choose **Text**, and then **Scale**.
   - From the **Text** toolbar, choose the **Scale Text** button.
   - At the 'Command:' prompt, enter the **scaletext** command.

   Command: **scaletext** *(Press ENTER.)*

2. In all cases, AutoCAD prompts you to select text.
   Select objects: *(Pick one or more lines of text.)*
   Select objects *(Press ENTER to end object selection.)*

3. Pick a base point, or select a justification point:
   Enter a base point option for scaling [Existing/Align/Fit/Center/Middle/Right/TL/TC/TR/ML/MC/MR/BL/BC/BR] <Existing>: *(Enter an option, or press ENTER.)*

4. Specify the new height:
   Specify new height or [Match object/Scale factor] <3/16">: *(Enter a height, or an option.)*

When AutoCAD prompts you to select objects, you can select a mix of text and non-text objects; AutoCAD filters out the non-text objects automatically.

The **SCALETEXT** command changes the height of selected text relative to a *base point*. The base point can be the existing insertion point or one of the many text justification points. By default, the base point is the existing insertion point.

The **Match Object** option matches the height to that of another text object:

   Select a text object with the desired height: *(Pick another text object.)*
   Height=3/16"

The **Scale Factor** option scales the text by a factor. A factor larger than 1 enlarges the text, while a value under 1 reduces it.

   Specify scale factor or [Reference]: *(Enter a factor, such as 2.)*

The **Reference** option is identical to that of the **SCALE** command: the text is scaled relative to another size.

TUTORIAL: REJUSTIFYING TEXT

1. To change the justification of text, start the **JUSTIFYTEXT** command:
   - From the **Modify** menu, choose **Object**, choose **Text**, and then **Justify**.
   - From the **Text** toolbar, choose the **Justify Text** button.
   - At the 'Command:' prompt, enter the **justifytext** command.

   Command: **justifytext** *(Press ENTER.)*

2. In all cases, AutoCAD prompts you to select text.

   Select objects: *(Pick one or more lines of text.)*

   Select objects *(Press ENTER to end object selection.)*

3. AutoCAD prompts you to select a justification option; the option you select will override the justification of the existing text.

   Enter a justification option [Existing/Align/Fit/Center/Middle/Right/ TL/TC/TR/ML/MC/MR/ BL/BC/BR] <Existing>: *(Press ENTER.)*

The same rules apply for **JUSTIFYTEXT** and **SCALETEXT** regarding object selection. Following both commands, the U command will undo any damage.

COMPILE

The **COMPILE** command converts *.shp* and *.pfb* files into *.shx* files.

Many kinds of PostScript fonts are defined by *.pfb* files. Some older versions of AutoCAD directly supported PostScript fonts, which are the *de facto* standard in desktop publishing and the popular Acrobat Reader software — not surprisingly, since all three were defined by Adobe. When Microsoft licensed the competitive TrueType font technology from Apple Computer, it quickly became popular with Windows users. Autodesk replaced support for PostScript with TrueType in more recent releases of AutoCAD; PostScript files can still be used in drawings, but only after conversion with the **COMPILE** command.

AutoCAD uses *.shx* files to define its own font format, as well as shapes, an older but more efficient version of blocks. (Technically, *.shp* files are the source code for compiled *.shx* files.)

TrueType fonts cannot be converted, nor is there is need to.

TUTORIAL: COMPILING POSTSCRIPT FONTS

1. To use PostScript fonts in drawings, start the **COMPILE** command:
   - At the 'Command:' prompt, enter the **compile** command.

   Command: **compile** *(Press ENTER.)*

2. AutoCAD displays the Select Shape or Font File dialog box.
   In **Files of Type,** select "PostScript Fonts (*.pfb)".
   Select a *.pfb* file, and then click **Open**.

Placing and Editing Text **527**

![Select Shape or Font File dialog box showing Fonts folder with cob.pfb, cobo.pfb, com.pfb, coo.pfb files]

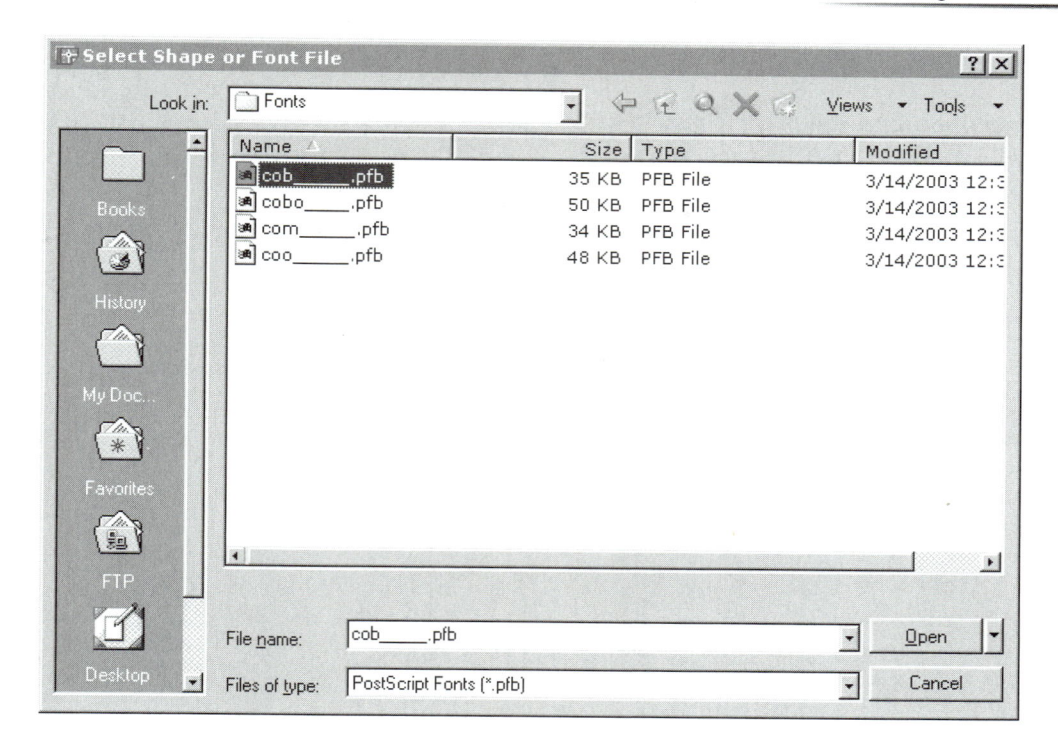

3. AutoCAD converts the font file:

    Compiling shape/font description file

    Compilation successful. Output file

    C:\Adobe\Acrobat\Distillr\Data\Fonts\coo_____.shx contains 40891 bytes.

4. Use Widows Explorer (or File Manager) to copy the *.shx* files from the folder in which they were converted, to the *\autocad 2004\fonts* folder — otherwise AutoCAD cannot find them.

5. Exit AutoCAD, and then restart the software. This forces AutoCAD to update its list of fonts.

You can now use the converted PostScript fonts in the drawing. As illustrated by the figure, the fonts appear unfilled; they are outlined only. This can be an advantage or a disadvantage, depending on your needs. A definite advantage, however, is that converted PostScript fonts display and plot faster than equivalent TrueType fonts.

PostScript Font

Note: Most fonts are *proportionally spaced*, which means the letter **i** takes up less width than the letter **w**. This makes it easier to read the text.

AutoCAD includes a font meant especially for columns of text, where it is important that the text line up vertically. The *monotxt.shx* font is designed so that every letter takes up the same width. A sample is shown below:

```
       braced wall panels per section 2320.11.3.
    3. Install sheets with face grain perpendicular to suports.
       Stagger sheets and nail with 8d @ 6'o.c. edges and 12' o.c. field typicl
       unless otherwise noted.
    4. For truss shapes, dimensions, and so on, see the Architectural drawings.
    5. All trusses shall be designed and manufactured in confimance with the
```

EXERCISES

1. In this exercise, you practice placing text.
   Start AutoCAD with a new drawing.
   With the **TEXT** command, place following lines of text:
   THIS IS THE FIRST LINE OF TEXT
   THIS IS THE SECOND & CENTERED LINE OF TEXT

2. In this tutorial, you place text centered on a point.
   The text "Part A" is to be centered both vertically and horizontally on a selected point.
   Use the **TEXT** command to place the text.
   Which justification mode do you use?

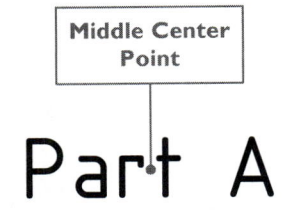

   Middle Center Point

   Part A

3. In this exercise, you practice placing text at a variety of alignments.
   The figure illustrates several text strings. The placement point is marked with a solid dot.
   Use the **TEXT** command to place the text as shown.

   Kitchen   Radius
             AutoCAD
   Drill-thru   .Title
   Capacitor   Isometric View
           Section A-A

4. In this exercise, you practice placing text at a variety of angles.
   Using the **TEXT** command, place the following line of text at angles of:
   THIS IS ANGLED TEXT
   a. 0 degrees.
   b. 45 degrees.
   c. 90 degrees.
   d. 135 degrees.
   e. 180 degrees.
   Use the same base point for each line.

5. In this exercise, use control codes to construct the following text string in the drawing.
   The story entitled MY LIFE is ±50% true.

6. In this exercise, you practice using mtext.
   Start a new drawing, and use the **MEXT** command to place several lines of text in the drawing, such as your name and address.

   Your First and Last Name

   1234 First Avenue

   Anytown, BC

   V8C 1T2 Canada

   Click the **OK** button to exit the mtext editor.

7. Continuing from the previous exercise, double-click the mtext.
   Does the mtext editor reappear?
   Make the following changes to parts of the text:

   Font: **Times New Roman**
   Color: **Red**

   Click the **OK** button to exit the mtext editor.
   Did the changes come into effect?

8. Continuing from the previous exercise, turn on the **QTEXT** command.
   Did the text change its look?
   If no, which command did you forget to use?

9. In this exercise, you import text from another file.
   If necessary, create a brief text file in Notepad, the text editor included with every version of Windows.
   Save the file in *.txt* format.
   Import the text file into your drawing with the **mtext** command.

10. Create five different styles that represent very different text appearances, such as wide text, slanted, and so on.
    Which command creates styles?
    Place examples of each style in the drawing.

11. Using only the **MTEXT** command, design your own business card with at least two fonts, three font variations, and two colors.
    Do not design a logo — just position the text. Include at least your name, address, and Internet information.
    Standard business cards are rectangles that measure 3.5" wide by 2" tall.
    Which command is useful for drawing rectangles?
    How many business cards can you fit onto a vertical A-size sheet?

12. Use a combination of text and drawing commands to design a logo for an engineering office. The logos often include variations on the owner's names.

13. In this and the next exercise, you practice drawing leaders.
    Draw a rectangle.
    Attach a leader to each of the four corners.
    Use the following text for each leader:
    > North East Corner
    > North West Corner
    > South East Corner
    > South West Corner

14. Continuing from the previous exercise, place a splined leader pointing to the center of the rectangle.
    Change the arrowhead to a dot.
    Use the following annotation:
    > Parcel of Land.

15. Continuing from the previous exercise, double-click the annotation of the splined leader, and change the text to:
    > Disputed Parcel of Land.
    > Not to be Subdivided.

16. In this and the following exercise, you correct and change text in drawings.
    From the CD-ROM, open the *spelling.dwg* drawing file, a drawing that contains text with spelling errors.

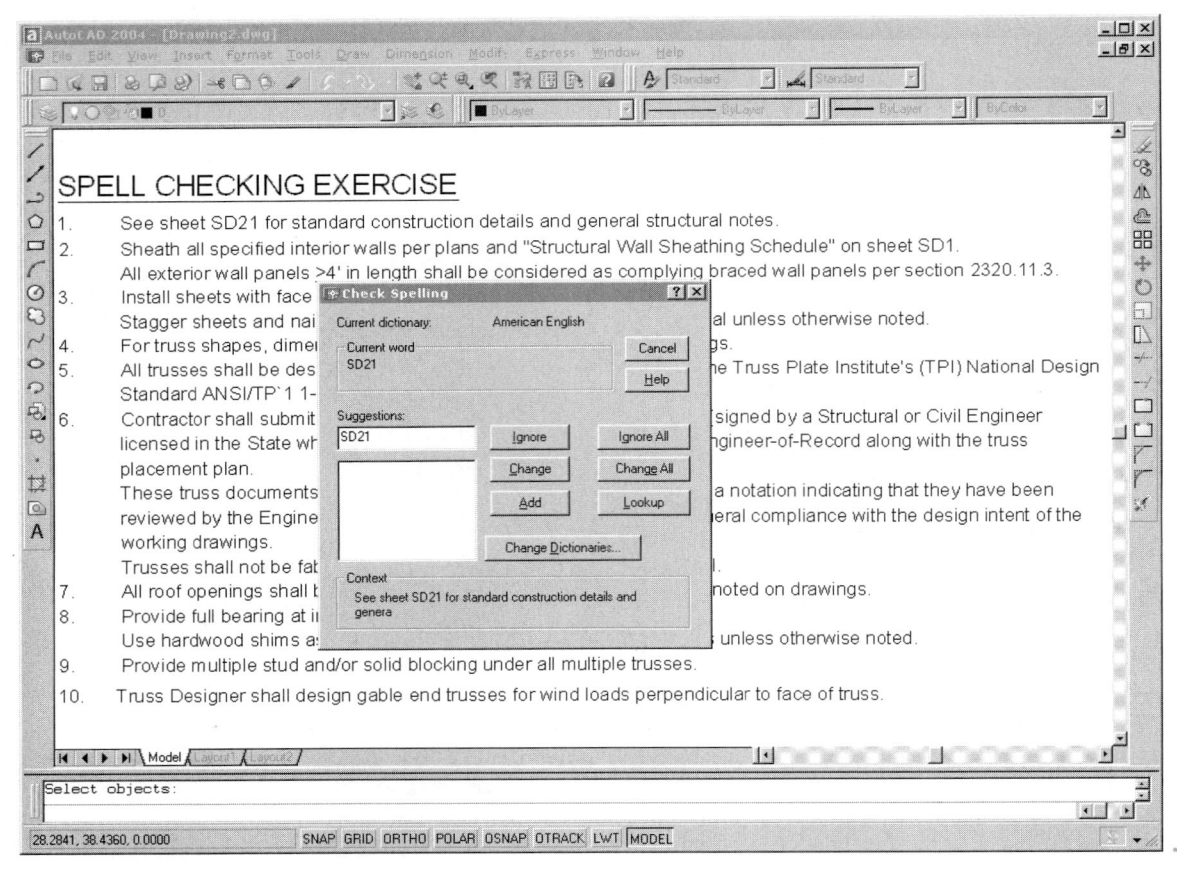

    Correct the spelling.
    How many mistakes does AutoCAD's **SPELL** command find?
    How many *real* spelling mistakes are there?

17. Continuing with the same drawing, find and replace the following phrases:

Find	Replace With
sheet SD1	sheet AA-01
Building Official	local Building Official

    Which command do you use to find and replace words in drawings?

18. From the CD-ROM, open the *insert.dwg* drawing file, a house plan.

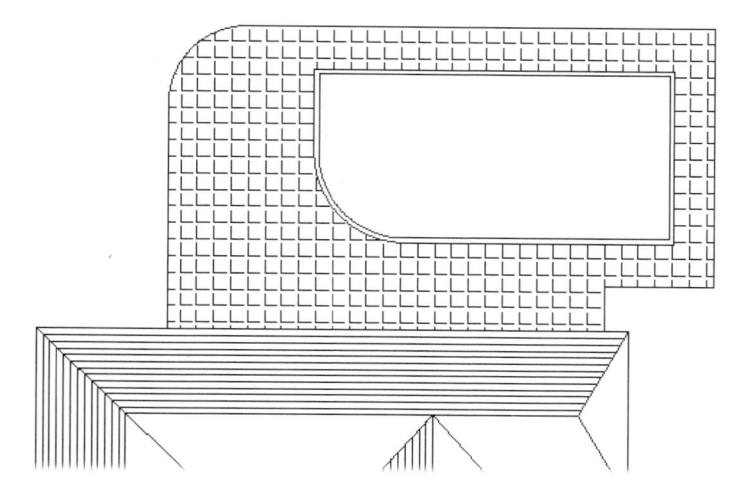

    Use the **WIPEOUT** command to place a blank rectangle on the patio area at the rear of the house.

    Over the wipeout, place the following note:

    Interlocking Belgium Bricks

    Use **Fit** justification to make the text fit within the wipeout rectangle.

19. Continuing with the same drawing, use the **SCALETEXT** command to reduce the size of the text you added in the previous exercise.

20. Continuing with the previous exercise, use the **JUSTIFYTEXT** command to change the justification to **Center**. How does the text change?

CHAPTER REVIEW

1. Name three ways you can see a list of the text styles stored in the drawing.
   a.
   b.
   c.
2. What are fonts?
   How do styles differ from fonts?
3. What four text properties are altered by the **STYLE** command?
   a.
   b.
   c.
   d.
4. There are several methods of placing text. List three of them, with a brief explanation of each:
   a.
   b.
   c.
5. When an underscore or overscore is added to text, is the entire string altered, or can words be treated separately?
   Explain.
6. When AutoCAD asks for a text height, is a numerical entry required?
   Explain.
7. List five alignment modes available through the **TEXT** command.
   a.
   b.
   c.
   d.
   e.
8. Can text be rotated at any angle?
   How does the oblique angle differ from the text angle? rs
9. How do the **TEXT** and **-TEXT** commands differ?
10. How can the text height be altered each time you enter text without redefining the style?
11. How can you compress and expand the text width?
12. Which formats of text can AutoCAD import into drawings?
13. What height should note text be in drawings?
14. Name four areas where text is used in drawings:
    a.
    b.
    c.
    d.
15. Can AutoCAD use *any* TrueType font found on your computer?
16. What happens when you open a drawing that contains a font not found on your computer?
17. Which command is better for placing lines of text at many locations over the drawing: **TEXT** or **MTEXT**?

18. List the command associated with each alias:
    dt
    t
    ed
    st
19. Can you access the menu while entering text? *no*
20. What do the three horizontal lines of the I-beam cursor represent?
    a.
    b.
    c.
21. Which one of the following letters has a descender? a, B, d, j, L.
22. Explain the meaning of the following justification codes:
    TL
    MC
    BR
    TR
23. Is BR justification the same as the Right justification?
    Explain.
24. Describe the one difference between Align and Fit justification modes.
25. What height should text be for a drawing scaled at:
    1:100
    1" = 50'
    1:1
    1" = 6"
    Show your work for each calculation.
26. Why is it wrong to use 1/8"-high text in an A-size drawing of a house?
27. When might you place rotated text in drawings?
28. Explain the meaning of the following control codes:
    %%u
    %%%
    %%d
    %%c
29. If you fail to turn off underlining, what happens on the next line of text?
30. What does *string* mean?
31. If the Euro symbol is not on your keyboard, how would you enter it?
32. Can you use more than one font with the **TEXT** command?
    With the **MTEXT** command?
33. Explain the meaning of the following mtext editor keyboard shortcuts:
    CTRL+B
    CTRL+I
    CTRL+A
    CTRL+SHIFT+A
34. How does AutoCAD stack text on either side of these characters?
    /
    ^
    #

35. Can different colors be applied to text placed with the **TEXT** command?
    With the **MTEXT** command?
36. Explain the meaning of the following zeros:
    Text Height = 0
    Mtext Width = 0
37. Can only numbers be used for stacked text?
38. Is it possible to use a text editor or word processor other than AutoCAD's built-in mtext editor?
    If so, how?
39. List the three primary parts of a leader:
    a.
    b.
    c.
40. What is the quickest way to edit the text of a leader?
    To edit the position of leader line?
41. Can leaders use only one kind of arrowhead?
42. Name three kinds of leader annotation:
    a.
    b.
    c.
43. What does a splined leader line look like?
44. Can leaders be attached to objects?
45. List three ways to edit text:
    a.
    b.
    c.
46. Is a different command needed to edit text created by the **TEXT** and **MTEXT** commands?
47. Does a different editor display for text created by the **TEXT** and **MTEXT** commands?
48. Can you have more than one text style in drawings?
49. What happens when you apply a different style to a line of text?
50. What happens to text when you make changes to a style?
51. Which font most closely approximates the Leroy lettering guides? s
52. Can TrueType fonts have the vertical style?
53. Name one advantage to using styles in drawings.
54. What is the purpose of the **TEXTFILL** and **TEXTQLTY** system variables?
    When do they come into effect?
55. Can the **SPELL** command be used for text placed by commands other than **TEXT** and **MTEXT**?
56. Explain a benefit of the **WIPEOUT** command.
57. Which command turns lines of text into rectangles?
58. Describe the purpose of the **SCALETEXT** command.
    And the **JUSTIFYTEXT** command.
59. How would you change the size of text in a drawing being converted from imperial to metric units?
60. Can PostScript fonts be used in AutoCAD 2004 drawings?

CHAPTER 13

Placing Dimensions

Dimension play an important role in defining the size, location, angle, and other attributes of objects in your drawings. Dimensions show the length and width of objects, the diameter of holes, and the angles of sloped parts. This chapter describes how to place dimensions in the drawing using these commands:

- **DIMHORIZONTAL, DIMVERTICAL,** and **DIMROTATED** place horizontal, vertical, and rotated dimensions.
- **DIMLINEAR** places horizontal, vertical, and rotated dimensions.
- **DIMALIGNED** places dimensions aligned with objects.
- **DIMBASELINE** and **DIMCONTINUE** add baseline and continuous dimensions.
- **QDIM** generates a variety of continuous dimensions.
- **DIMRADIUS** and **DIMDIAMETER** place radial and diameter dimensions.
- **DIMCENTER** places center marks on arcs and circles.
- **DIMANGULAR** places angular dimensions.

FINDING THE COMMANDS

On the **DIMENSION** toolbar:

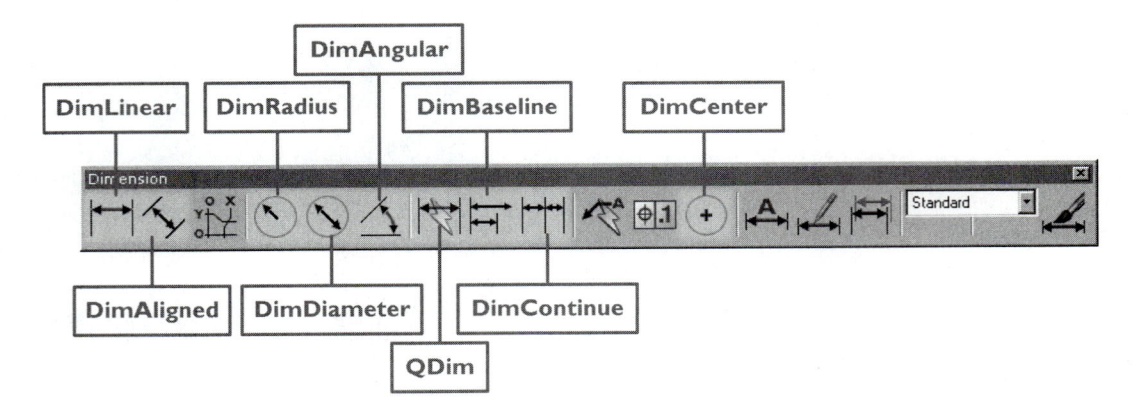

[Dimension toolbar with labeled buttons: DimLinear, DimRadius, DimAngular, DimBaseline, DimCenter, DimAligned, DimDiameter, QDim, DimContinue]

On the **DIMENSION** menu:

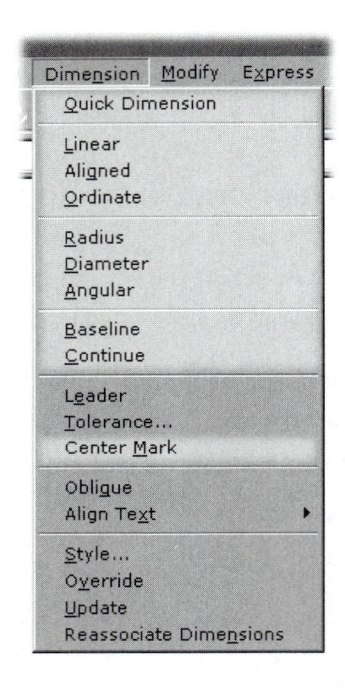

[Dimension menu showing: Quick Dimension; Linear, Aligned, Ordinate; Radius, Diameter, Angular; Baseline, Continue; Leader, Tolerance..., Center Mark; Oblique, Align Text; Style..., Override, Update, Reassociate Dimensions]

Placing Dimensions 537

INTRODUCTION TO DIMENSIONS

Dimensions in drawings eliminate the need to use scale rulers and protractors to work out measurements and angles.

Typically, the most important parts of the drawing are dimensioned, not every detail. In particular, overall sizes and items subject to interpretation should be dimensioned. In addition, dimensions should not interfere with the drawing.

The figure below illustrates metric dimensions in a sample drawing provided with AutoCAD 2004.

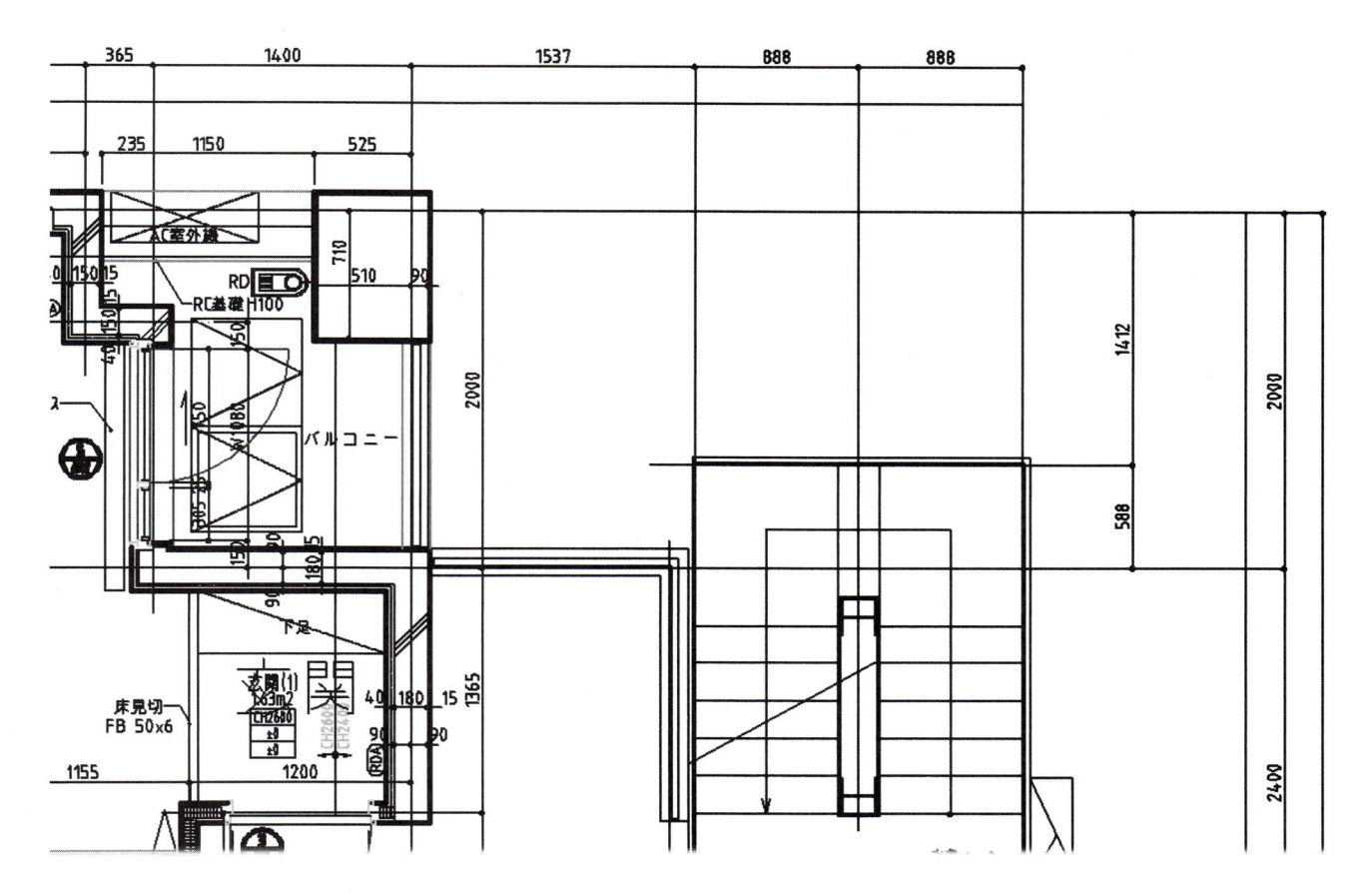

THE PARTS OF DIMENSIONS

Dimensions consist typically of these parts: a dimension line with an arrowhead at each end, a pair of extension lines, and text.

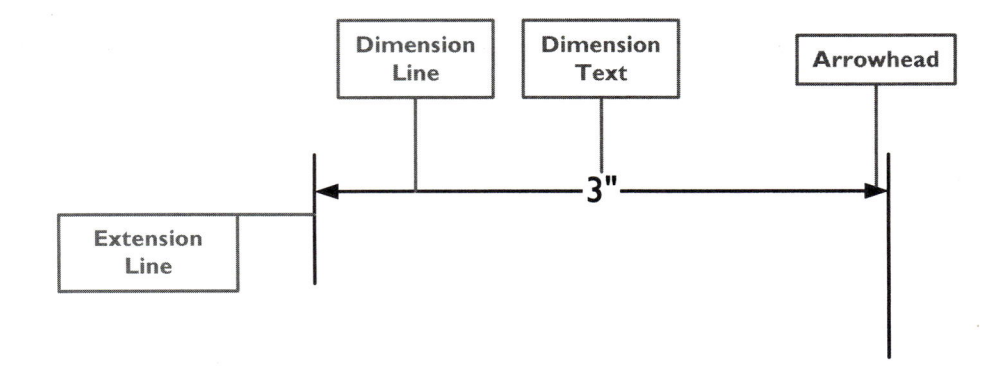

Dimension Line

The *dimension line* is the line with the arrows or "ticks" at each end. It indicates distances or angles being measured. To report the distance or angle, text is place in or over the dimension line. If there is too little room between the extension lines, the dimension line is not drawn.

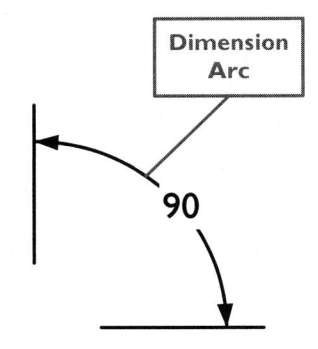

When you measure angles, the dimension line is an arc instead of a straight line.

Extension Lines

The *extension lines* (sometimes called "witness lines") indicate the points being dimensioned. There is usually an extension line at either end of the dimension line. They are usually perpendicular to the dimension line, but need not be. Sometimes, the first or second extension line is not drawn, and in rare cases, neither is drawn.

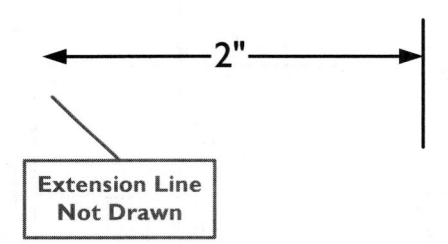

Almost all dimension commands start by prompting you to pick two points: these become the endpoints for extension lines.

AutoCAD offsets the extension lines from the pick points by a distance of 0.0625 units to mimic standard drafting practice. (The offset distance is called "offset from origin" and can be changed in the Dimension Style Manager dialog box.)

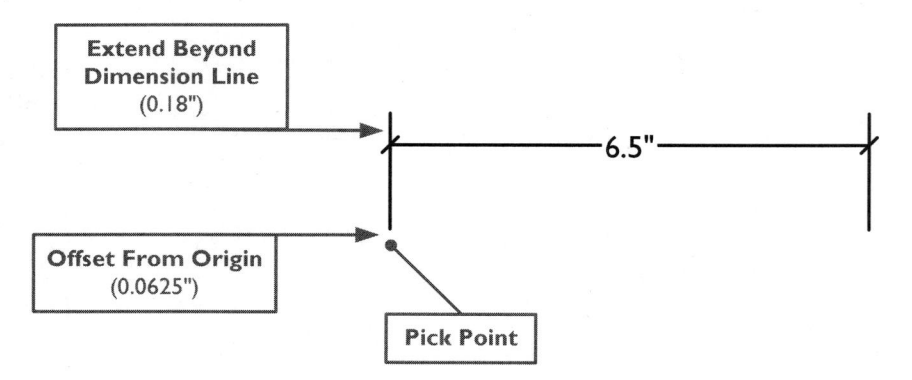

In addition, AutoCAD draws the extension lines a short distance beyond the dimension line. Again, this is standard drafting practice. The default distance is 0.18 units, but can be changed in the Dimension Styles Manager dialog box.

Arrowheads

The *arrowheads* are placed at either end of the dimension line, and point to the extension lines. If there is too little room between the extension lines, the arrowheads are placed outside. The Dimension Style Manager dialog box lets you specify whether arrowheads or text should be moved outside first.

The arrowheads can be replaced with tick marks and other symbols. AutoCAD includes many different arrowheads, and you can define your own.

The figure at right illustrates the arrowhead styles included with AutoCAD; to create your own arrowhead, you first define it as a block.

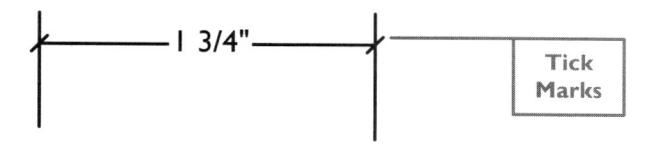

- Closed filled
- Closed blank
- Closed
- Dot
- Architectural tick
- Oblique
- Open
- Origin indicator
- Origin indicator 2
- Right angle
- Open 30
- Dot small
- Dot blank
- Dot small blank
- Box
- Box filled
- Datum triangle
- Datum triangle filled
- Integral
- None

The length of the arrowhead varies, depending on the scale of the drawing. Generally, arrowheads are 1/8 inch in length when used in small drawings, and 3/16 inch in larger drawings. Arrowheads that are too large or small prove either distracting or difficult to read. Arrowheads are usually drawn with a 1:3 aspect ratio: the arrow is three times longer than it is wide. Other arrowheads, such as tick marks and circles, are 1:1.

Dimension Text

The dimension *text* appears in or over the dimension line, depending on the standard. If there is too little room between the extension lines, the text is placed outside, or even some distance away, and then referenced with a leader line.

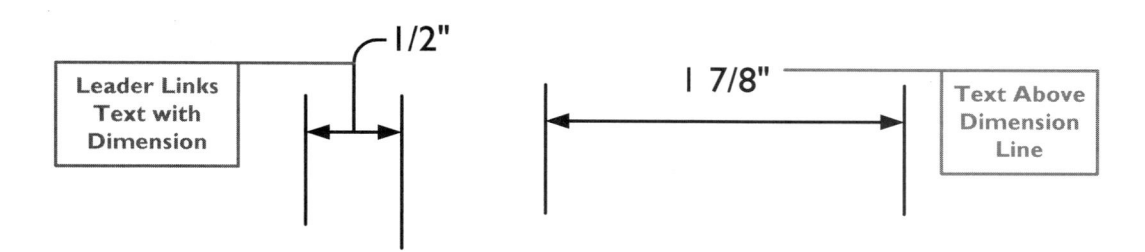

Text can be above, below, or in the dimension lines. It can be centered, left justified, or right justified on the dimension line. It can be horizontal, vertical, or rotated. Horizontal, centered text is said to be in the "home" position. The location of dimension text depends on the discipline. In architectural and structural drawings, the dimension is placed on top of the dimension line. In mechanical drawings, it is usually placed within the dimension line.

When distances are stipulated, the text designating feet and inches is separated by a dash, as in 5'-4". If there are no inches, a zero is used: 6'-0". If dimensions are stipulated strictly in inches, such as 72", use the inch mark to avoid confusion; in mechanical design, inch marks are not used.

Fractions are given either as common fractions, such as 1/2, 3/4, and so on, or as decimal fractions, such as 0.50 and 0.75. Normally, inches and common fractions are stipulated without a dash between them. In CAD, however, many text fonts do not have "stacked" fractions. When fractions must be constructed from standard numerical text, use a dash to avoid confusion, such as 3-1/2".

Typically you let AutoCAD measure the distance and determine the text; you can, however, override it and enter your own text. Dimension text does not follow the text style set by the **STYLE** command, nor the format set by the **UNITS** command. Instead, the style, units, and precision are determined by the dimension style.

Tolerances

Dimension *tolerances* are plus and minus amounts appended to the dimension text. AutoCAD can add the tolerances automatically, if you request. Typically, you specify the plus and minus amounts, which can be equal or unequal.

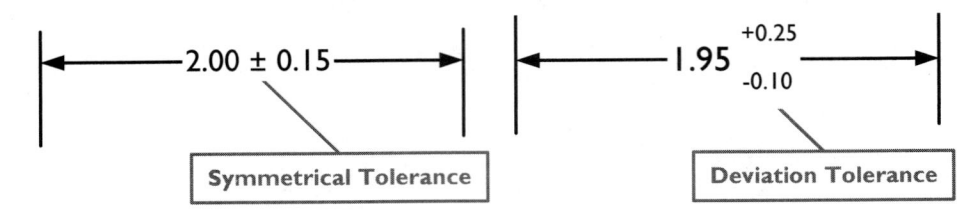

|←———2.00 ± 0.15———→| |←———1.95 +0.25 / -0.10———→|

Symmetrical Tolerance   **Deviation Tolerance**

If the tolerances are equal ("symmetrical"), they are drawn with a plus/minus symbol. If unequal ("deviation"), tolerances are drawn one above the other: above for the plus amount, below for minus.

Limits

Instead of showing dimension tolerances, the tolerance can be applied to the text — added to and subtracted from the dimension text. These are called *limits*. The example below is a measurement of 4.0000 unit, with a tolerance of ±0.15.

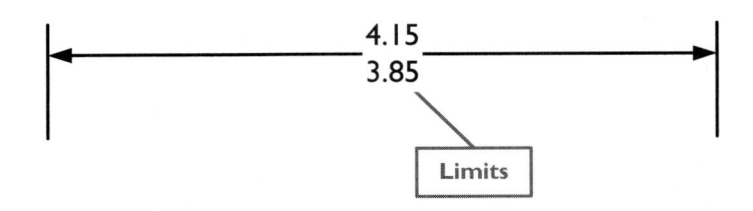

|←———— 4.15 / 3.85 ————→|

Limits

Alternate Units

Alternate units show two forms of measurement, such as English and metric, on the same dimension line. The second set is usually shown in square brackets.

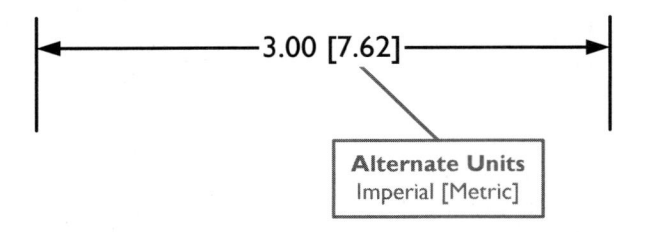

|←————3.00 [7.62]————→|

Alternate Units
Imperial [Metric]

DIMENSIONING OBJECTS

AutoCAD can place dimensions between any two points in the drawing. In addition, AutoCAD has a *direct dimensioning* mode, where it automatically dimensions specific objects: lines and polyline segments, arcs and polyarcs, circles, vertices, and single points.

Lines and Polyline Segments

AutoCAD dimensions lines and polyline segments with a single pick. The dimension is complete, consisting of dimension and extension lines, arrowheads, and text. This is accomplished typically with the **DIMLINEAR** and **DIMALIGNED** commands.

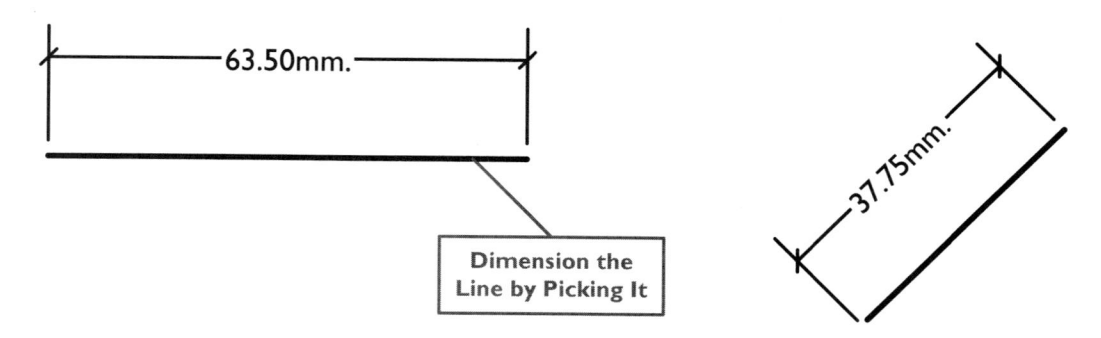

Arcs, Circles, and Polyarcs

An arc is dimensioned as a radius. The leader line should be placed at an angle, avoiding horizontal or vertical placement. The letter "R," designating a radius dimension, should prefix the dimension text, such as R2.125.

If a circle is dimensioned by its diameter, the dimension text is prefixed by the diameter symbol, such as Ø4.25.

An object constructed of several arcs is dimensioned in two stages: (1) locating the center of the arcs with horizontal or vertical dimensions; and (2) showing their radii with radius dimensioning.

AutoCAD dimensions arcs and polyline arcs with a single pick. The resulting dimension depends on the command you use. The **DIMLINEAR** and **DIMALIGNED** commands place dimension and extension lines, arrowheads, and text.

The **DIMRADIUS** and **DIMDIAMETER** commands place dimension lines or leaders, arrowheads, and text.

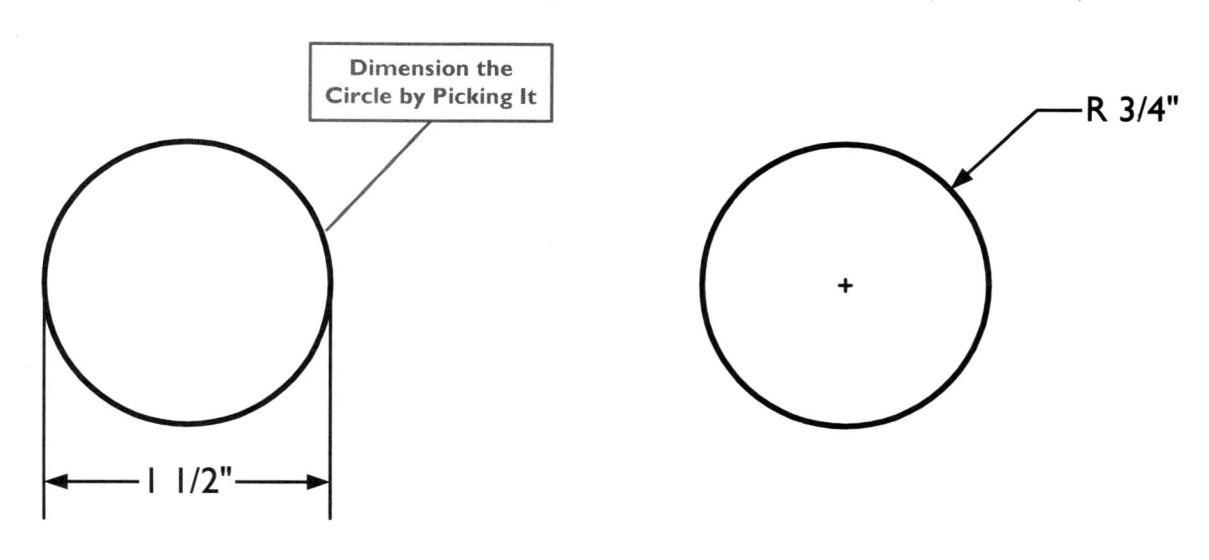

Wedges and Cylinders

Wedges are dimensioned in two views. The three distances dimensioned are the length, width, and height. Cylinders are dimensioned for diameter and height. The diameter is typically dimensioned in the non-circular view. If a drill-through is dimensioned, it is described by a diameter leader.

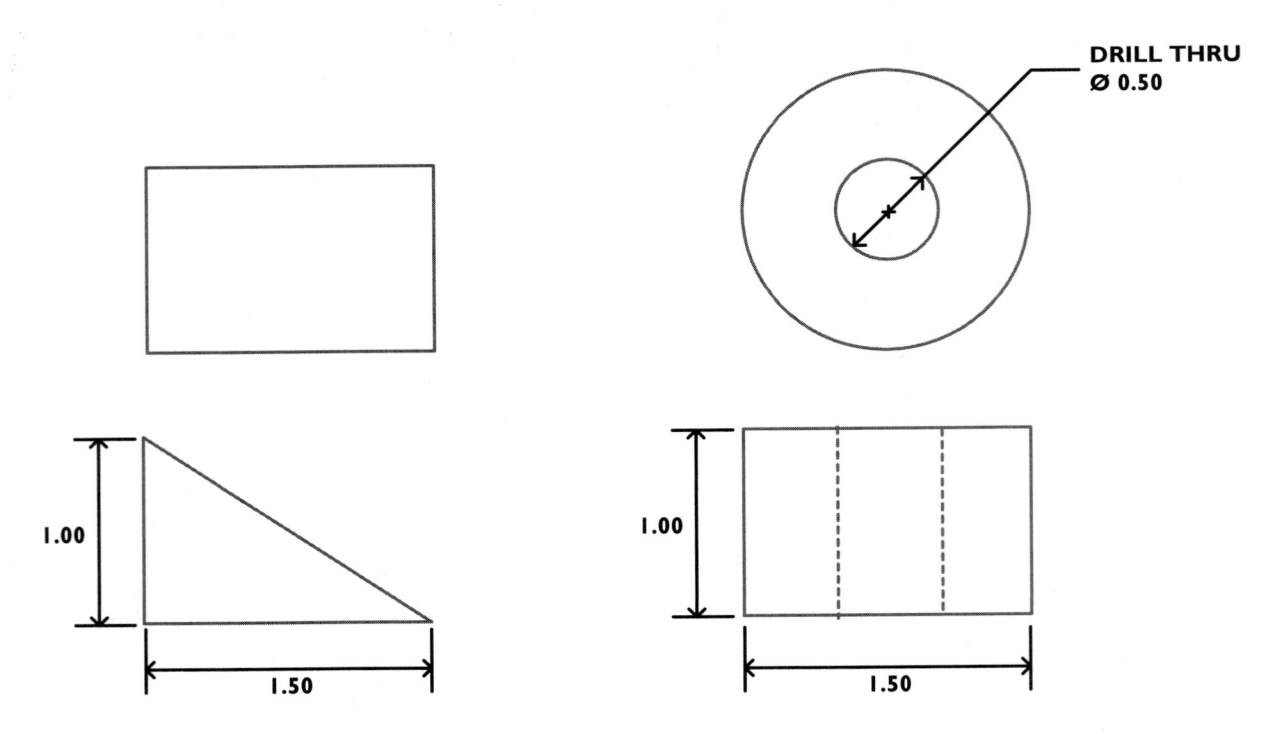

DRILL THRU
Ø 0.50

1.00

1.50

1.00

1.50

Cones and Pyramids

Cones are dimensioned at the diameter and the height. Some conical shapes, such as the truncated cone, require two diameter dimensions. Pyramids are dimensioned like cones.

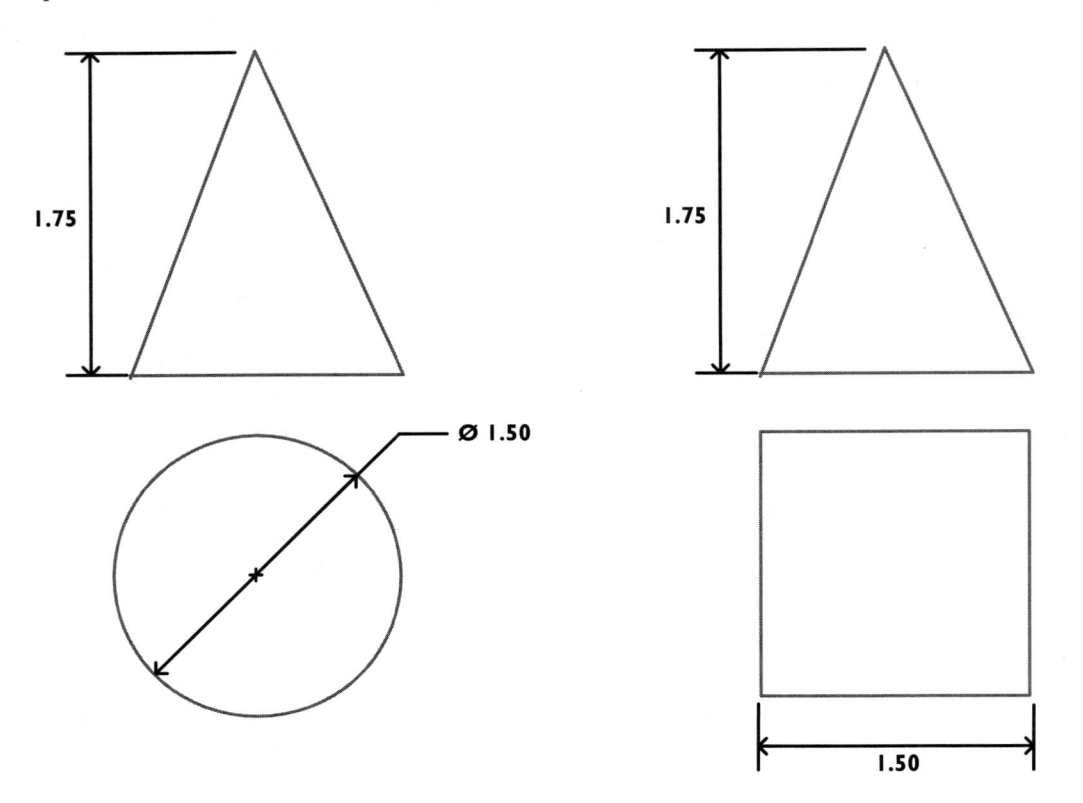

1.75

1.75

Ø 1.50

1.50

Holes

Holes can be drilled, reamed, bored, punched, or cored. It is preferable to dimension the hole by notes, giving the diameter, operation, and (if there is more than one hole) the number. The operation describes such techniques as counter-bored, reamed, and countersunk.

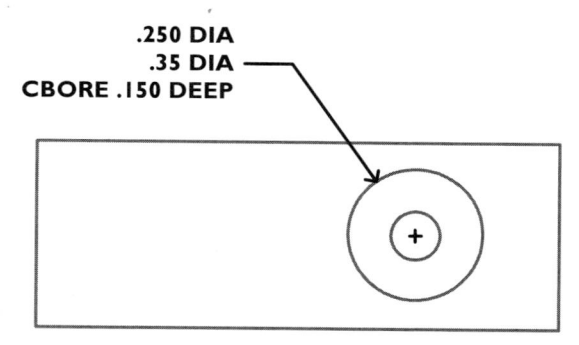

Standards dictate that drill sizes be designated as decimal fractions.

Whenever possible, point the dimension leader to the hole in the circular view. Holes made up of several diameters can be dimensioned in their section.

Vertices

AutoCAD measures the angle of lines, arcs, circles, and polyline arcs with a single pick using the **DIMANGULAR** command.

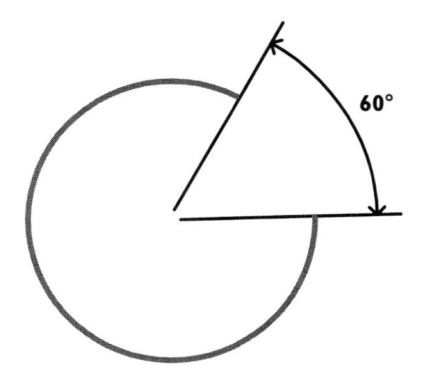

Single Points

To "dimension" a single point, you typically use the **LEADER**, **QLEADER**, or **DIMORDINATE** commands. Leaders are typically pieces of text that point at a spot in the drawing through lines. Ordinate dimensions measure the x and y coordinates of objects relative to a base point.

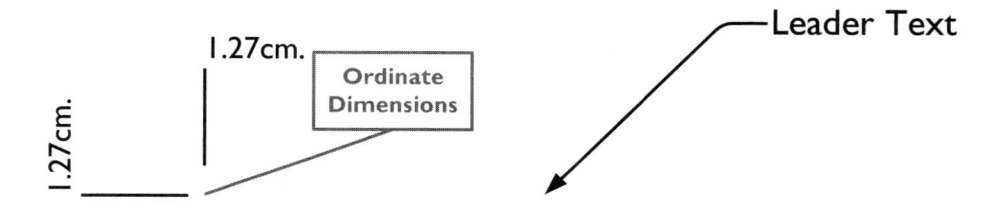

For drawing leaders, see Chapter 12 "Placing and Editing Text"; for ordinates, see Chapter 15, "Geometric Dimensioning and Tolerancing."

Chamfers and Tapers

A chamfer is an angled surface applied to an edge. Use a leader to dimension chamfers of 45 degrees, with the leader text designating the angle and one (or two) linear distances.

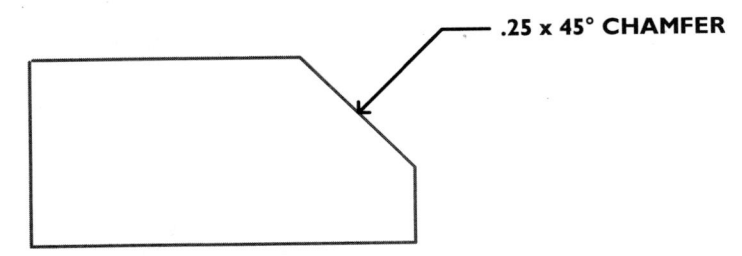

If the chamfer is not 45 degrees, dimensions showing the angle and the linear distances describe the part.

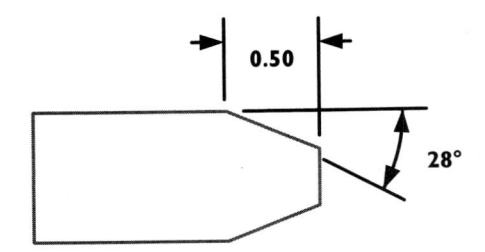

A taper can be described as the surface of a *cone frustum* (cone with the top sliced off). Dimension tapers by giving the diameters of both ends and the rate of taper, given as the distance of taper per foot.

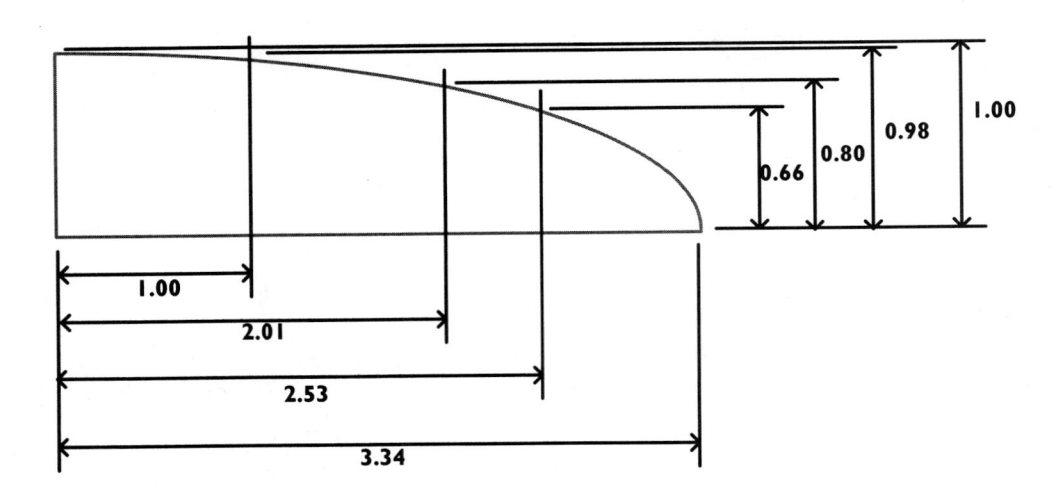

DIMENSION MODES

AutoCAD constructs dimensioning *semi-automatically*. All it needs from you is some basic information, such as what to dimension, or where to start and end the dimension. Then it constructs the dimension for you: it measures the distance or angle, and then draws in the dimension.

AutoCAD's dimensions are *associative*. That means they automatically update themselves when the associated objects are moved and stretched. This is a very powerful feature and a great time-saver.

Dimensions, however, do not need to be associative; AutoCAD works with three types of associativity:

Fully associative is the "normal" mode, where AutoCAD attaches the ends of the extension lines to geometry of the object. Move the object, and the dimension moves with it; stretch the object, and the dimension text updates automatically.

Partly associative is an early attempt at associativity. The extension lines are attached to *dimension points* placed on layer Defpoints (short for "definition points"). When stretching or moving objects, you have to include the defpoint; otherwise, the dimension does not update correctly.

Non-associative means the dimension is not attached to objects in any way.

Dimensions are placed as if they were blocks: the lines, arrows, and text act as a single object. You can use the **EXPLODE** command to break apart dimensions.

> **Note:** When AutoCAD reports "Non-associative dimension created," this means that AutoCAD was not able to attach the dimension to an object.

Dimension Variables and Styles

Dimension variables determine how the dimensions are drawn; they are system variables specific to dimensions, sometimes called "dimvars" for short. Some variables store values, such as the color of the dimension line and the look of the arrowhead; others are toggles that turn values on and off, such as whether the first or second extension lines should be displayed.

Just as styles determine the look of text, *dimension styles* determine the look of dimensions. Dimension styles ("dimstyles" for short) are created and modified with the **STYLE** command, which collects together all the settings of dimension variables. Dimvars and dimstyles are discussed in Chapter 14, "Editing Dimensions."

Dimension Standards

Because there are so many ways of drawing dimensions, some industries and countries have defined standards. AutoCAD includes some of these standards with its *.dwt* template drawings, including ANSI (American), ISO (international), DIN (German), JIS (Japanese), and Gb (Chinese).

The standards organizations make the dimension standards available to members at a cost <www.nationalcadstandard.org>, or sometimes make them freely available by posting standards at Web sites <www.cadinfo.net/editorial/archdim1.htm>.

Object Snaps

Use object snap modes for accurate dimensioning. For example, the ENDpoint and INTersection osnaps are useful for capturing the ends and intersections of lines, while the CENter and QUAdrant osnaps help with dimensions of arcs and circles.

DIMHORIZONTAL

The **DIMHORIZONTAL** command places horizontal dimensions.

This command restricts the dimension line to the horizontal. To locate the extension lines, you select an object (limited to line, arc, circle, and polyline) or pick two points. Another pick point is required to locate the dimension line.

While the **DIMLINEAR** command is more flexible (it combines horizontal, vertical, and rotated dimensioning), the **DIMHORIZONTAL** command may be easier to understand for users new to AutoCAD.

TUTORIAL: PLACING DIMENSIONS HORIZONTALLY

1. To place a horizontal dimension, start the **DIMHORIZONTAL** command:
    - At the 'Command:' prompt, enter the **dimhorizontal** command.

    Command: **dimhorizontal** *(Press ENTER.)*

2. AutoCAD prompts you to select the object you want dimensioned:
    Specify first extension line origin or <select object>: *(Press ENTER.)*
    Select object to dimension: *(Pick one object.)*

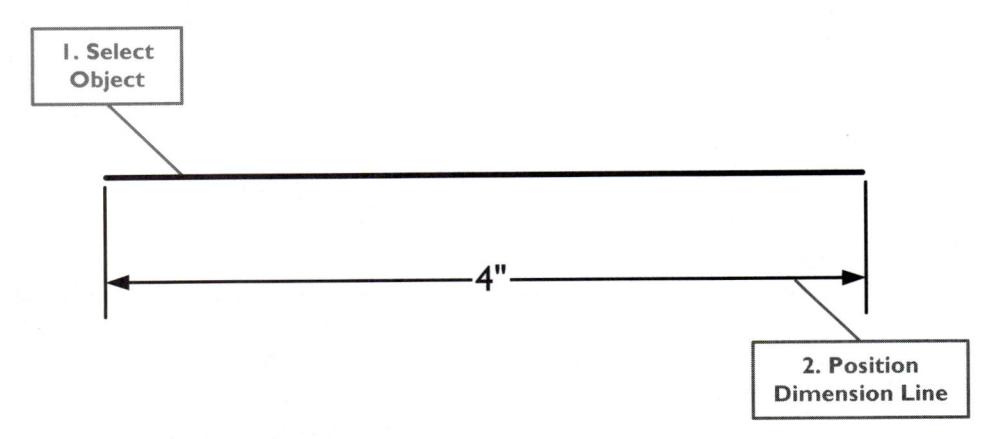

3. Pick a point to place the dimension line:
    Specify dimension line location or [Mtext/Text/Angle]: *(Pick a point to place the dimension line.)*

4. Notice that AutoCAD reports the length measurement at the command prompt, and draws the horizontal dimension.
    Dimension text = 4"

HORIZONTAL DIMENSIONS: ADDITIONAL METHODS

In addition to dimensioning objects, the **DIMHORIZONTAL** command has options to place horizontal dimensions between any points, as well as to change the dimension text.

- **Extension** places horizontal dimensions between two points.
- **MText** displays the mtext editor.
- **Text** prompts you to edit the dimension text.
- **Angle** rotates the dimension text.

Let's look at each option.

Extension

The **Extension** options prompts for two points, which are the endpoints of the two extension lines. The horizontal dimension is placed between them.

> Command: **dimhorizontal**
> Specify first extension line origin or <select object>: *(Pick point 1.)*
> Specify second extension line origin: *(Pick point 2.)*
> Specify dimension line location or [Mtext/Text/Angle]: *(Pick point 3.)*
> Dimension text = 4"

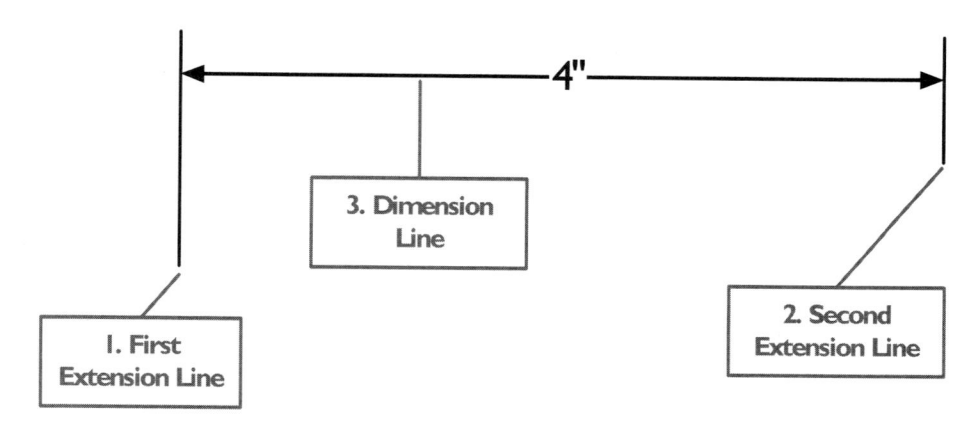

MText

The **MText** option displays the mtext editor, allowing you to edit the dimension text. The editor should be familiar to you from Chapter 12, "Placing and Editing Text."

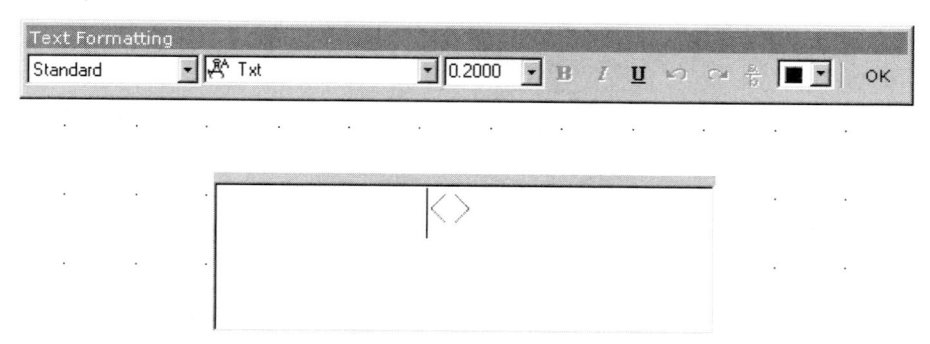

The unfamiliar symbol is <>, the double angle bracket. It may seem odd to you, but this is an important shorthand notation. AutoCAD uses it to indicate the *default* dimension text. Here are some examples of how you can work with the double angle brackets:

- Replace <> with other text, and AutoCAD shows the new text on the dimension line.

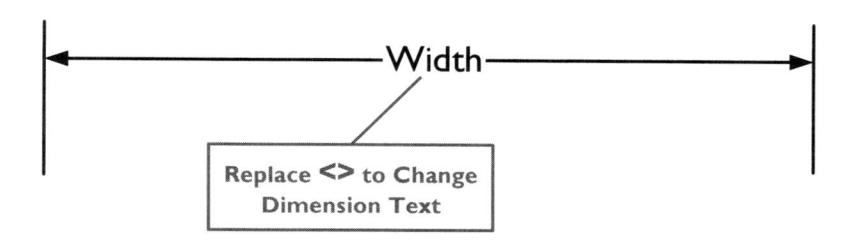

- Add text on either side of <>, and AutoCAD shows the added text on either side of the dimension text.

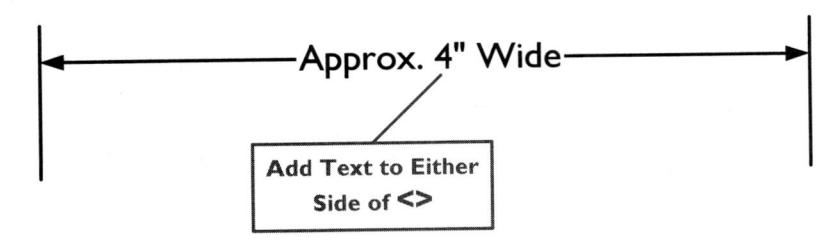

Approx. 4" Wide
(Add Text to Either Side of <>)

- Insert text between the brackets, and AutoCAD displays the brackets and the text.

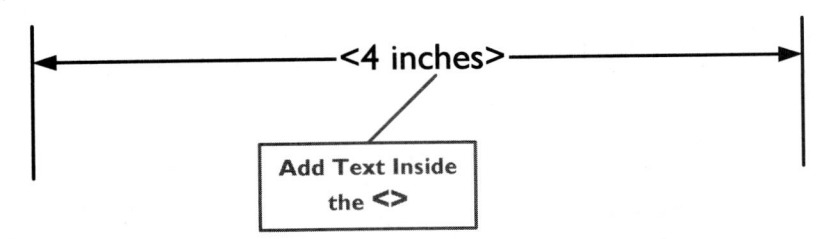

<4 inches>
(Add Text Inside the <>)

- Erase <> and AutoCAD erases the dimension text.

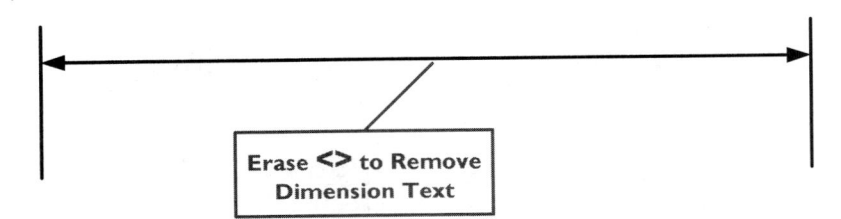

(Erase <> to Remove Dimension Text)

You can insert the %%d (°), %%p (±), %%c (Ø) metacharacters to specify special characters.

Text

The **Text** option prompts you to edit the dimension text at the command prompt.

> Enter dimension text <3.5000>: **The distance is <> inches**

Although the < > is not shown explicitly, you can type the angle brackets with the text you enter. The result is the same as with the **MText** option: replace, add, insert, and erase the < > marker.

Angle

The **Angle** option rotates the dimension text about its center point.

> Specify angle of dimension text: *(Enter an angle, such as **45**.)*

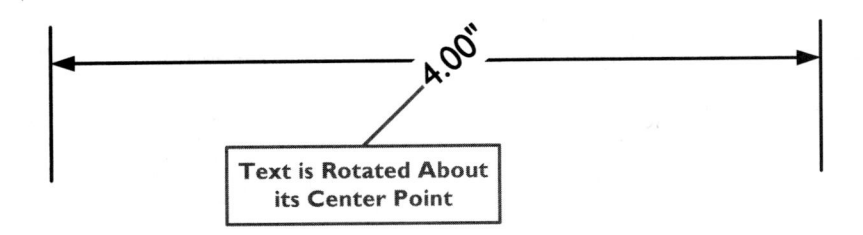

4.00"
(Text is Rotated About its Center Point)

DIMVERTICAL AND DIMROTATED

The **DIMVERTICAL** command places vertical dimensions, while the **DIMROTATED** command places dimensions at an angle.

These two commands operate identically to DIMHORIZONTAL, except that **DIMVERTICAL** forces the dimension line to be vertical, while **DIMROTATED** asks you for a rotation angle, and then draws the dimension line at that angle.

TUTORIAL: PLACING DIMENSIONS VERTICALLY

1. To place a vertical dimension, start the **DIMVERTICAL** command:
   - At the 'Command:' prompt, enter the **dimvertical** command.

   Command: **dimvertical** *(Press ENTER.)*

2. AutoCAD prompts you to select the object you want dimensioned:
   Specify first extension line origin or <select object>: *(Press ENTER.)*
   Select object to dimension: *(Pick one object.)*

3. Pick a point to place the dimension line:
   Specify dimension line location or [Mtext/Text/Angle]: *(Pick a point to place the dimension line.)*

   Dimension text = 10

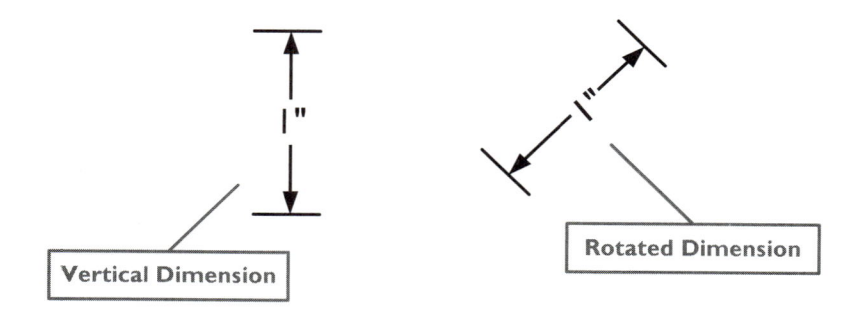

TUTORIAL: PLACING DIMENSIONS AT AN ANGLE

1. To place a rotated dimension, start the **DIMROTATED** command:
   - At the 'Command:' prompt, enter the **dimrotated** command.

   Command: **dimrotated** *(Press ENTER.)*

2. AutoCAD asks for the dimension's angle:
   Specify angle of dimension line <0>: *(Enter an angle, such as **45**, or show the angle by picking two points in the drawing.)*

3. AutoCAD prompts you to pick the endpoints of the extension lines:
   Specify first extension line origin or <select object>: *(Pick a point.)*
   Specify second extension line origin: *(Pick another point.)*

4. Pick a point to place the dimension line:
   Specify dimension line location or [Mtext/Text/Angle]: *(Pick a point to place the dimension line.)*

   Dimension text = 1"

DIMLINEAR

The **DIMLINEAR** command places horizontal, vertical, and rotated dimensions.

After specifying the location of the extension lines, AutoCAD determines whether to draw a horizontal or vertical dimension depending on where you place the dimension line:

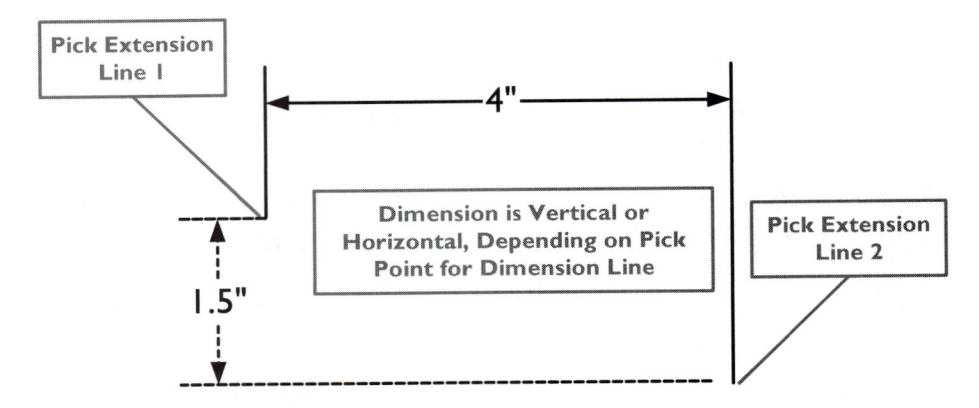

This command also applies dimensioning directly onto objects, and more has options than the dimension commands described earlier.

TUTORIAL: PLACING LINEAR DIMENSIONS

1. To place a linear dimension, start the **DIMLINEAR** command:
   - From the menu bar, choose **Dimension**, and then **Linear**
   - From the **Dimension** toolbar, choose the **Linear Dimension** button.
   - At the 'Command:' prompt, enter the **dimlinear** command.
   - Alternatively, enter the aliases **dli** or **dimlin** at the 'Command:' prompt.

   Command: **dimlinear** (Press ENTER.)

2. In all cases, AutoCAD prompts you to select the object you want dimensioned:
   Specify first extension line origin or <select object>: (Press ENTER.)
   Select object to dimension: (Pick one object.)

3. Pick a point to place the dimension line:
   Specify dimension line location or
   [Mtext/Text/Angle/Horizontal/Vertical/Rotated]: (Pick a point to place the dimension line horizontally or vertically.)
   Dimension text = 4"

LINEAR DIMENSIONS: ADDITIONAL METHODS

The **DIMLINEAR** command has the same options as **DIMHORIZONTAL**:

- **Extension** option places horizontal dimensions between two points.
- **MText** option displays the mtext editor.
- **Text** option prompts you to edit the dimension text.
- **Angle** option rotates the dimension text.

See the **DIMHORIZONTAL** command for more information on the four options listed above. In addition, there are these options:

- **Horizontal** option forces the dimension line horizontally.
- **Vertical** option forces the dimension line vertically.
- **Rotated** option rotates the dimension line.

Let's look at the three new options.

Horizontal and Vertical

The **Horizontal** option forces the dimension line horizontally, no matter where you locate the dimension line. This makes it operate like **DIMHORIZONTAL**. Similarly, the **Vertical** option forces the dimension line vertically, just like the **DIMRVERTICAL** command.

Rotated

The **Rotated** option rotates the dimension line, just like the **DIMROTATED** command. Enter **r** at the prompt, and AutoCAD asks for the angle:

>Specify dimension line location or
>[Mtext/Text/Angle/Horizontal/Vertical/Rotated]: **r**
>Specify angle of dimension line <0>: **34**

You can enter different forms of angle measurement by including their units designation. For example, to enter the angle in grads, include the **g** suffix:

>Specify angle of dimension line <0>: **38g**

Or to enter the angle in radians, include the **r** suffix:

>Specify angle of dimension line <0>: **0.59r**

Or pick two points to show the angle:

>Specify angle of dimension text: *(Pick point 1.)*
>Specify second point: *(Pick point 1.)*

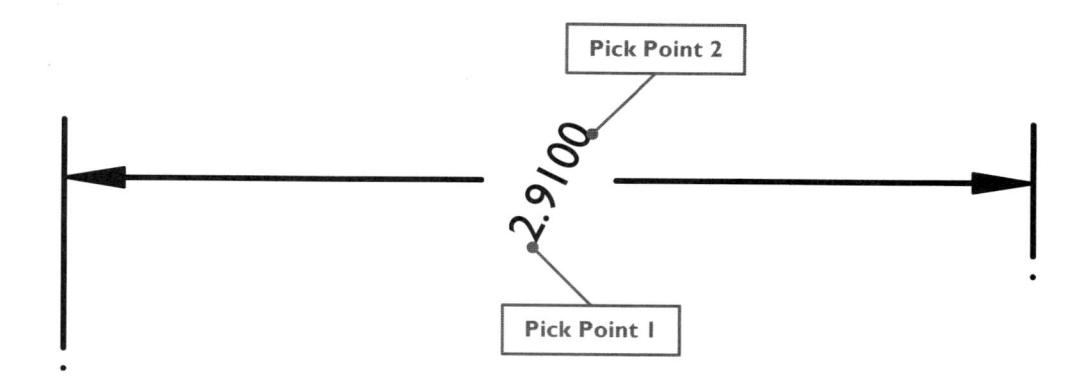

DIMALIGNED

The **DIMALIGNED** command places dimensions aligned to two points.

This command is like the **DIMLINEAR** command, except that the pick points for the extension lines determine the angle of the dimension line.

When you employ direct dimensioning, **DIMALIGNED** reacts differently, depending on the object you select:

Lines, Arcs, and Polylines — endpoints of lines, arcs, and polylines determine the angle of the dimension line; measures the length of line and polyline segments, and the chord length of arcs.

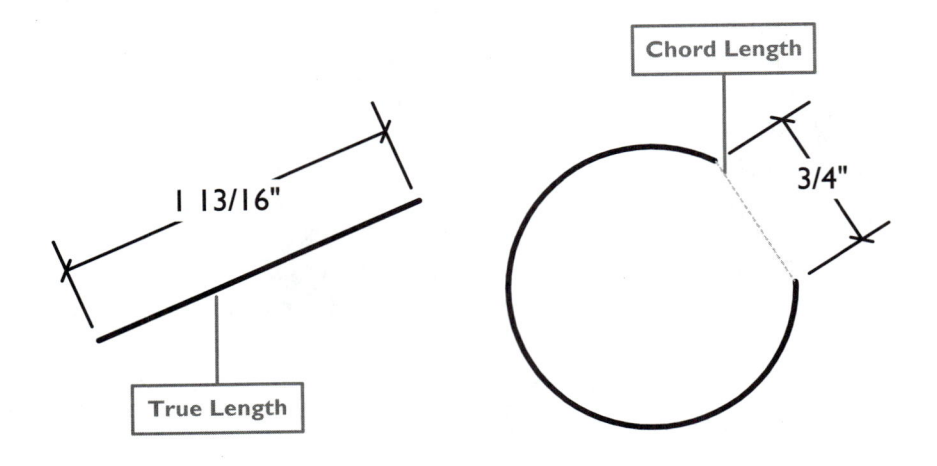

Circles — pick points on circles' determine angle of dimension line; measures the circles' diameter.

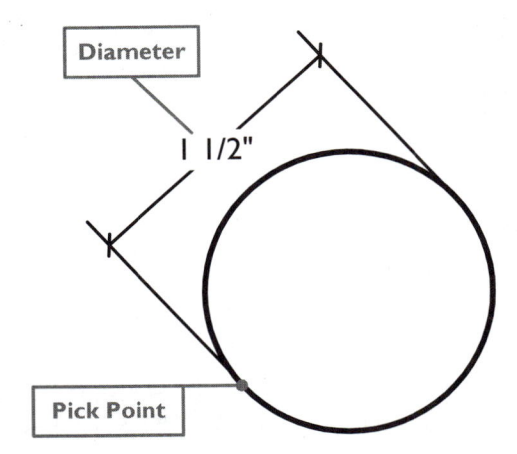

TUTORIAL: PLACING ALIGNED DIMENSIONS

1. To place a dimension aligned to two points, start the **DIMALIGNED** command:
   - From the menu bar, choose **Dimension**, and then **Aligned**
   - From the **Dimension** toolbar, choose the **Aligned Dimension** button.
   - At the 'Command:' prompt, enter the **dimaligned** command.
   - Alternatively, enter the aliases **dal** or **dimali** at the 'Command:' prompt.

   Command: **dimaligned** (Press ENTER.)

2. In all cases, AutoCAD prompts you to select the object you want dimensioned:
   Specify first extension line origin or <select object>: *(Press ENTER.)*
   Select object to dimension: *(Pick one object.)*
3. Pick a point to place the dimension line:
   Specify dimension line location or
   [Mtext/Text/Angle]: *(Pick a point to place the dimension line.)*
   Dimension text = 10.0

This command's **Extension line**, **Mtext**, **Text**, and **Angle** options operate identically to that of the **DIMHORIZONTAL** command.

DIMBASELINE AND DIMCONTINUE

The **DIMBASELINE** command (also called "parallel dimensioning") continues dimensions from baselines, while the **DIMCONTINUE** command (also called "chain dimensioning") continues dimensions from the previous dimension.

Baseline dimensions tend to stack over one another, while continuous dimension tend to be located next to each other. The stack distance is 0.38, but can be changed. Continuous dimensions share common extension lines at one end.

Both commands continue from the last placed dimension, which can be linear, angular, or ordinate. If no dimension was created during the current drafting session, AutoCAD prompts you to select a dimension as the base. You cannot use either of these commands until at least one other dimension has been placed in the drawing. These commands do not work with leaders, radial, or diameter dimensions.

TUTORIAL: PLACING ADDITIONAL DIMENSIONS FROM A BASELINE

1. To place baseline dimensions, start the **DIMBASELINE** command:
   - From the menu bar, choose **Dimension**, and then **Baseline**
   - From the **Dimension** toolbar, choose the **Baseline Dimension** button.
   - At the 'Command:' prompt, enter the **dimbaseline** command.
   - Alternatively, enter the aliases **dba** or **dimbase** at the 'Command:' prompt.

   Command: **dimbaseline** *(Press ENTER.)*
2. If necessary, AutoCAD prompts you to select a dimension to use as the base:
   Select base dimension: *(Pick a linear, angular, or ordinate dimension.)*
3. AutoCAD prompts you to pick the endpoint for the next dimension:
   Specify a second extension line origin or [Undo/Select] <Select>: *(Pick a point.)*
   Dimension text = 10.0

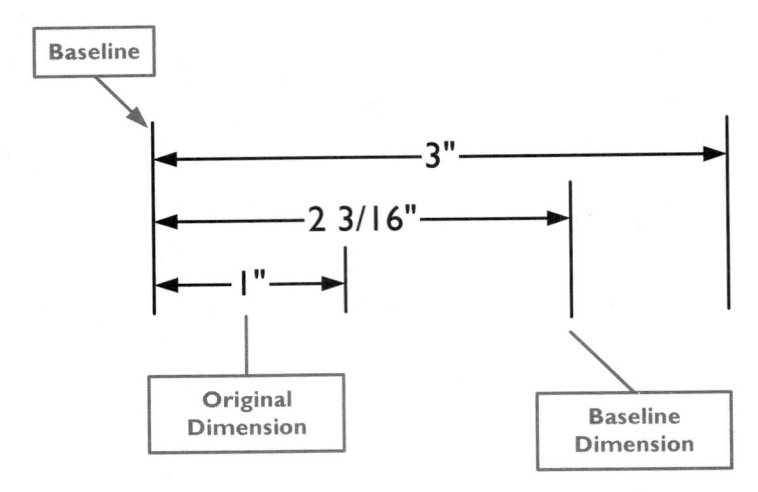

Notice that the baseline dimension continues from the first extension line of the previous dimension.

4. AutoCAD repeats the prompt for the next dimension:

    Specify a second extension line origin or [Undo/Select] <Select>: *(Pick another point.)*

5. Press **ESC** to exit the command:

    Specify a second extension line origin or [Undo/Select] <Select>: *(Press* **ESC**.*)*

TUTORIAL: CONTINUING DIMENSIONS

The **DIMCONTINUE** command is identical to **DIMBASELINE**, except that continuous dimensions are placed next to each other.

1. To continue dimensioning, start the **DIMCONTINUE** command:
   - From the menu bar, choose **Dimension**, then choose **Continue**
   - From the **Dimension** toolbar, choose the **Continue Dimension** button.
   - At the 'Command:' prompt, enter the **dimcontinue** command.
   - Alternatively, enter the aliases **dco** or **dimcont** at the 'Command:' prompt.

    Command: **dimcontinue** *(Press* ENTER.*)*

2. If necessary, AutoCAD prompts you to select a dimension to use as the base:

    Select base dimension: *(Pick a linear, angular, or ordinate dimension.)*

3. AutoCAD prompts you to pick the endpoint for the next dimension:

    Specify a second extension line origin or [Undo/Select] <Select>: *(Pick a point.)*

    Dimension text = 10.0

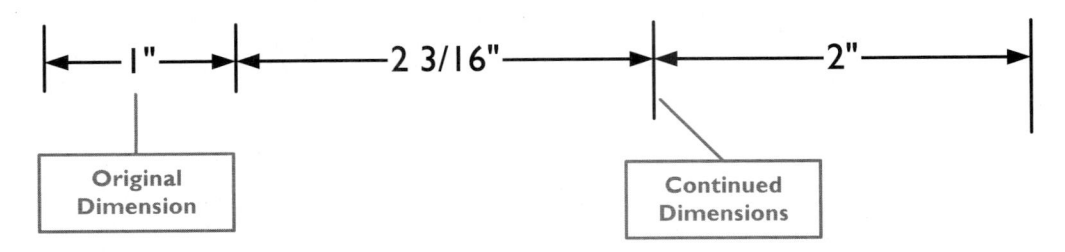

Notice that the dimension continues from the second extension line of the previous dimension.

4. AutoCAD repeats the prompt for the next dimension:

   Specify a second extension line origin or [Undo/Select] <Select>: *(Pick another point.)*

5. Press **ESC** to exit the command:

   Specify a second extension line origin or [Undo/Select] <Select>: *(Press ESC.)*

CONTINUED DIMENSIONS: ADDITIONAL METHODS

The **DIMBASELINE** and **DIMCONTINUE** commands have the following options:

- **Undo** option undoes the last dimension placed.
- **Select** option selects another dimension to continue from.

Let's look at both.

Undo

The **Undo** option undoes the last placed dimension. Because these two commands repeat, this option is handy for undoing a dimension you did not mean to place.

Select

The **Select** option selects another dimension to continue from. AutoCAD prompts:

   Select continued dimension: *(Pick an extension line.)*

By picking an extension line, you determine the direction in which the dimension continues — to the left or to the right — assuming the dimension is horizontal.

QDIM

The **QDIM** command places baseline and continuous style dimensions in one fell swoop.

It is like the **DIMCONTINUE** and **DIMBASELINE** commands, but more powerful: it places continued radial dimensions, applies the continued dimensions all at once, and has a greater number of options. On the negative side, it is more complex to use than the two previous commands.

The figure below illustrates radius (at left) and continuous dimensions generated by this command.

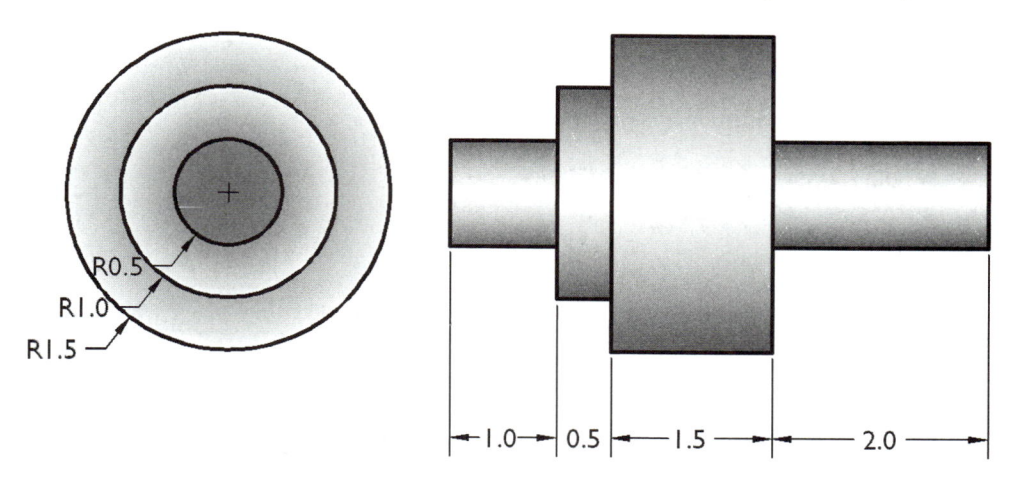

CHANGING DIMENSION TYPE

Perhaps the most powerful aspect of **QDIM** is its ability to change the dimension type. For instance, you might dimension an object in staggered mode, but some time later change your preference to ordinate mode. Restart the command, select the same object, and then specify ordinate mode. Rather than adding ordinate dimensions, AutoCAD changes the staggered dimensions to the new style.

TUTORIAL: QUICK CONTINUOUS DIMENSIONS

1. To place continuous-style dimension quickly, start the **QDIM** command:
   - From the menu bar, choose **Dimension**, and then **Quick Dimension.**
   - From the **Dimension** toolbar, choose the **Quick Dimension** button.
   - At the 'Command:' prompt, enter the **qdim** command.

   Command: **qdim** *(Press ENTER.)*

2. In all cases, AutoCAD reports the dimensioning priority, and then prompts you to select the geometry to dimension:

   Associative dimension priority = Endpoint
   Select geometry to dimension: *(Select one or more objects.)*
   Select geometry to dimension: *(Press ENTER to end object selection.)*

   Unlike other dimensioning commands, you may select one or more objects.

3. Pick a point to define the position of the dimension line:

   Specify dimension line position, or
   [Continuous/Staggered/Baseline/Ordinate/Radius/Diameter/datumPoint/Edit/seTtings]
      <Radius>: *(Pick a point.)*

   Depending on the objects you select, AutoCAD attempts to make a best guess at how to dimension them.

QUICK DIMENSIONS: ADDITIONAL METHODS

The **QDIM** command boasts a host of options to force the type of continuous dimension, as well as specify settings:

- **Continuous** forces continued dimensions.
- **Staggered** places staggered dimensions.
- **Baseline** places baseline dimensions.
- **Ordinate** places ordinate dimensions.
- **Radius** places radial dimensions.
- **Diameter** places diameter dimensions.
- **datumPoint** specifies the start point for baseline and ordinate dimensions.
- **Edit** edits continuous dimensions.
- **seTtings** selects the object snap priority.

Let's look at all the options.

Continuous

The **Continuous** option forces continued dimensions, like those created by **DIMCONTINUE**.

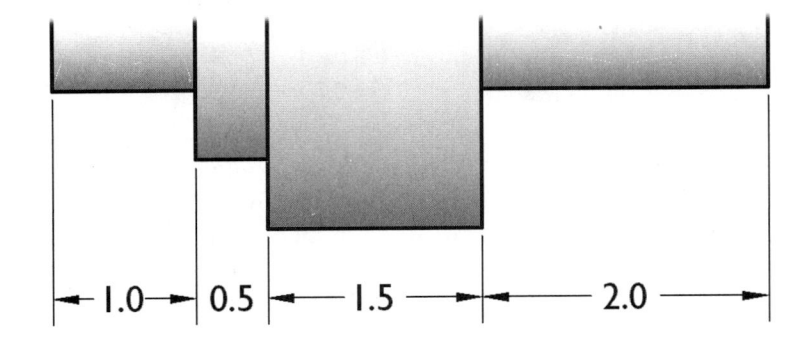

Staggered

The **Staggered** option forces staggered dimensions, which are like stacked continuous dimensions. This option, however, requires an even number of dimensionable objects. As illustrated in the figure below, segment can be left out, unfortunately, and are not dimensioned.

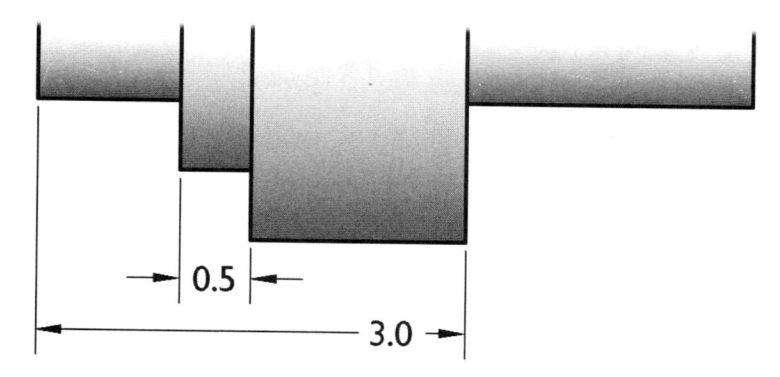

Baseline

The **Baseline** option forces baseline dimensions, like those created by the **DIMBASELINE** command. The figure below shows the dimensions starting at the left end of the flywheel; you can change the start point with the **Datum point** option, as described later.

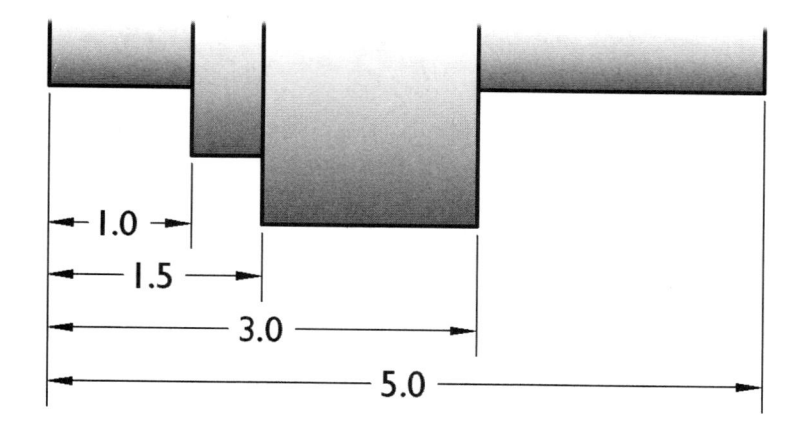

Ordinate

The **Ordinate** option forces ordinate dimensions, like those created by the **DIMORDINATE** command, with two differences. **DIMORDINATE** prevents ordinate dimensions from ending on top of each other, while **QDIM** does not. **DIMORDINATE** measures relative to the UCS; this command measures relative to a datum point.

The figure illustrates x-ordinate dimensions starting with a value of 5.0, which is the distance from the origin (0,0). You can change the origin point with the **Datum point** option.

To create y-ordinate dimensions, move the cursor to the left or right side of the object during the "Specify dimension line position" prompt.

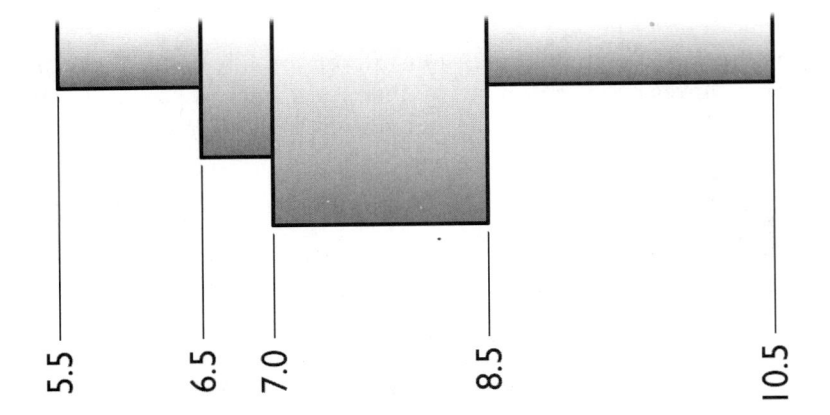

Radius

The **Radius** option forces radius dimensions, and works only with circle and arcs. Selecting other objects causes AutoCAD to complain, "No arcs or circles are currently selected." The radial dimensions are placed at the point you pick at the "Specify dimension line position" prompt.

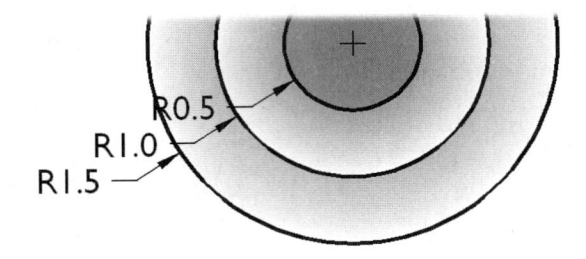

Diameter

The **Diameter** option forces diameter dimensions. It operates identically to the **Radius** option. In both cases, it's a good idea to avoid placing dimensions horizontally, because they tend to bunch up.

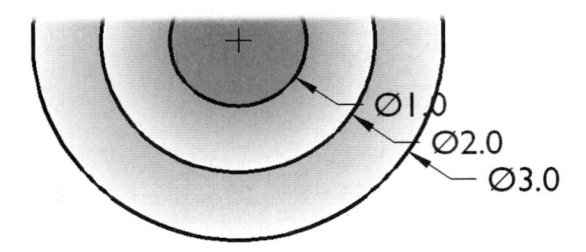

datumPoint

The **datumPoint** option prompts you to select a new base point from which to measure baseline and ordinate dimensions.

> Select new datum point: *(Pick a point.)*

Edit

The **Edit** option enters editing mode, and prompts you to add and remove dimension markers. These markers show up as an **x** at the endpoints of lines, and the center points of circles and arcs.

> Indicate dimension point to remove, or [Add/eXit] <eXit>: *(Enter an option.)*

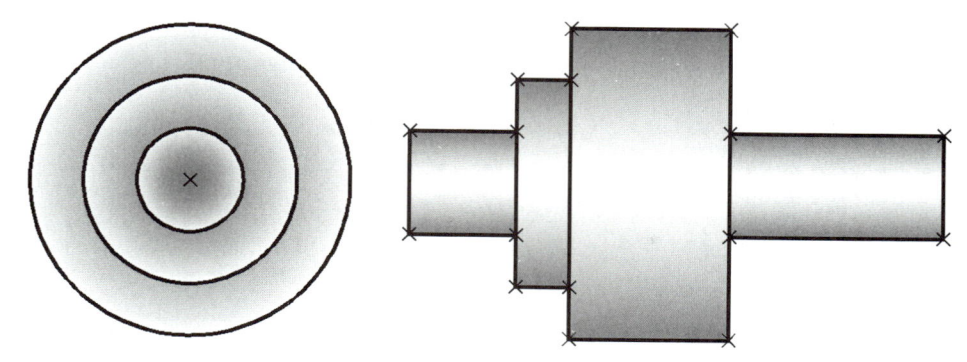

Remove removes markers from objects. To remove, pick a marker with the cursor. AutoCAD reports, "One dimension point removed."

Add adds markers to the drawing. To add, pick a point anywhere in the drawing; it need not be on an object. AutoCAD reports, "One dimension point added."

eXit exits **Edit** mode, and returns to the "Specify dimension line position" prompt.

seTtings

The **seTtings** options doesn't set much, except whether endpoints or intersections have priority, and even then I am not convinced either setting makes any difference. AutoCAD displays the following prompt:

Associative dimension priority [Endpoint/Intersection]: *(Enter E or I.)*

DIMRADIUS AND DIMDIAMETER

The **DIMRADIUS** command places radius dimensions, while the **DIMDIAMETER** command places diameter dimensions.

These commands dimension circles, arcs, and *polyarcs* (arcs that are parts of polylines). They do not dimension other circular objects, including ellipses, splines, and regions.

The dimensions created by these two commands are more like leaders than the two-extension-line dimensions you have seen so far in this chapter. They don't prompt for a point, but immediately ask you to select a circle or an arc.

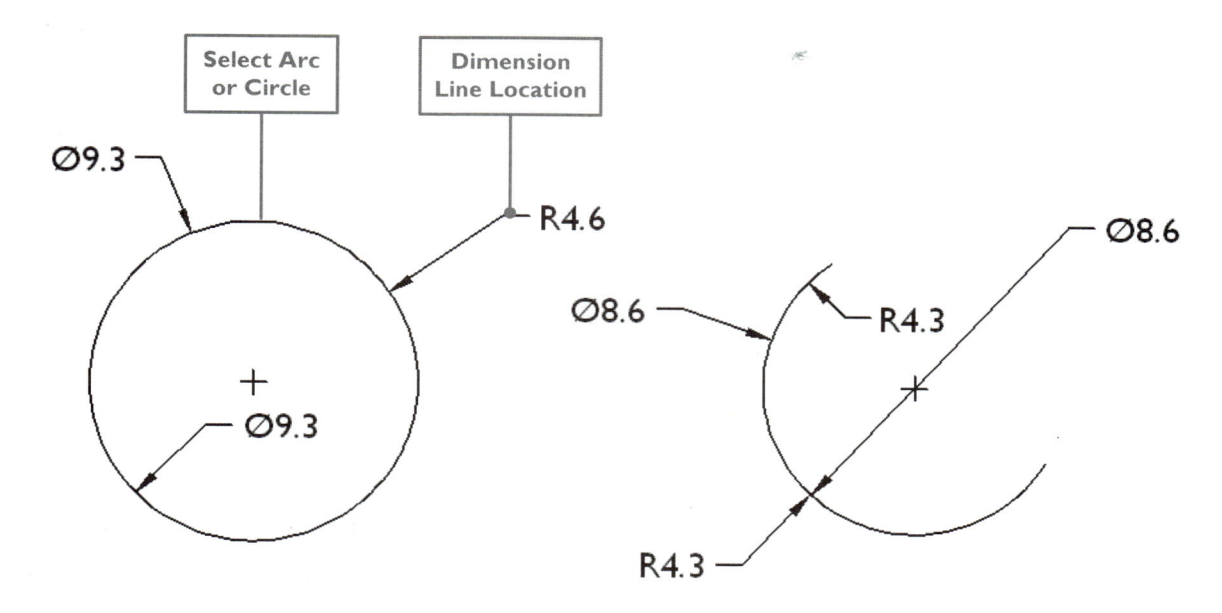

The second prompt asks you to "Specify dimension line location." Although this sounds like a single item (the dimension line), the point you pick determines several things at once, which can be difficult for new users:

- Angle of the leader line.
- Length of the leader line.
- Location of the dimension text.
- Placement of the center mark.

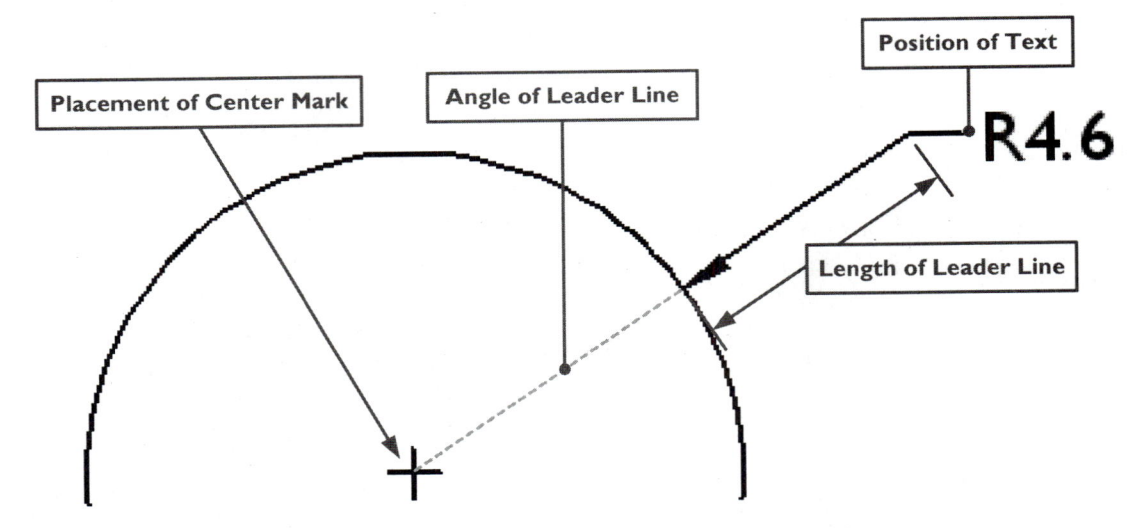

TUTORIAL: PLACING RADIAL DIMENSIONS

1. To dimension a radius, start the **DIMRADIUS** command:
   - From the menu bar, choose **Dimension**, and then **Radius**
   - From the **Dimension** toolbar, choose the **Radius Dimension** button.
   - At the 'Command:' prompt, enter the **dimradius** command.
   - Alternatively, enter the aliases **dra** or **dimrad** at the 'Command:' prompt.

   Command: **dimradius** *(Press ENTER.)*

2. In all cases, AutoCAD prompts you to select the arc or circle you want dimensioned:
   Select arc or circle: *(Pick one arc, circle, or polyarc.)*
   Dimension text = 10.0

3. Pick a point to place the dimension leader:
   Specify dimension line location or [Mtext/Text/Angle]: *(Pick a point to position the dimension leader.)*

The **DIMDIAMETER** command operates identically:

Command: **dimdiameter**

Select arc or circle: *(Pick one arc, circle, or polyarc.)*

Dimension text = 10.0

Specify dimension line location or [Mtext/Text/Angle]: *(Pick a point to position the dimension.)*

The **Mtext**, **Text**, and **Angle** options are the same as described earlier in this chapter under the **DIMHORIZONTAL** command.

DIMCENTER

The **DIMCENTER** command places center marks and lines on arcs and circles.

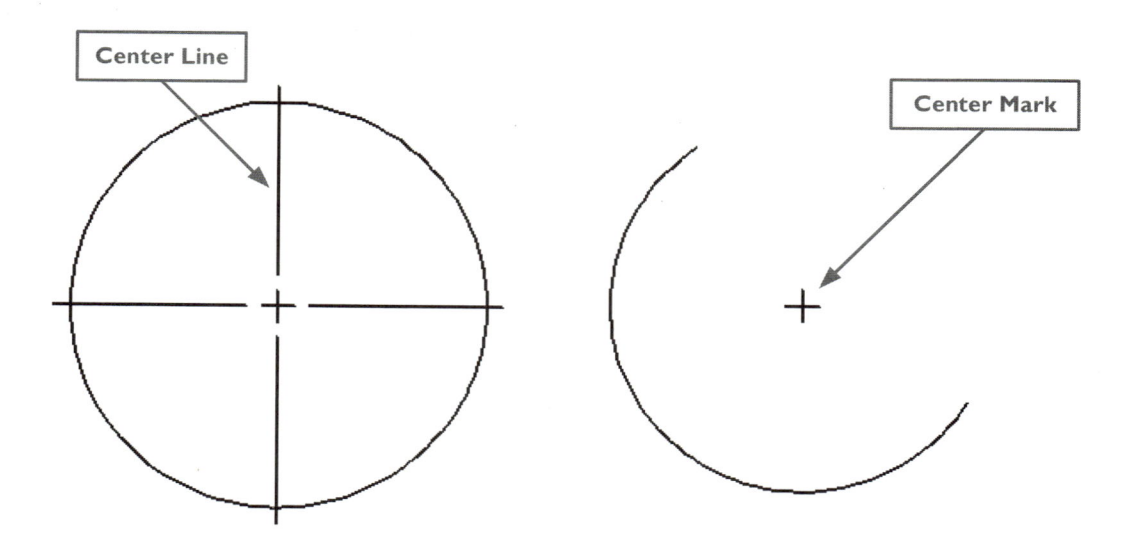

TUTORIAL: PLACING CENTER MARKS

1. To mark the center of arcs and circles, start the **DIMCENTER** command:
   - From the menu bar, choose **Dimension**, and then **Center Mark**.
   - From the **Dimension** toolbar, choose the **Center Mark** button.
   - At the 'Command:' prompt, enter the **dimcenter** command.
   - Alternatively, enter the **dce** alias at the 'Command:' prompt.

   Command: **dimcenter** *(Press ENTER.)*

2. In all cases, AutoCAD prompts you to select the arc or circle you want dimensioned:

   Select arc or circle: *(Pick one arc, circle, or polyarc.)*

   AutoCAD places the center mark.

Through the **DIMCEN** system variable, you change the look of the center mark. This variable also affects the center marks created by the **DIMRADIUS** and **DIMDIAMETER** commands, although in their cases, center marks are drawn only when the dimension is placed outside the circle or arc.

DimCen	Meaning
0	Draws neither center lines nor center marks.
<0	Draws center lines and center marks.
>0	Draws center marks only.

The default value is 0.09, which means DIMCENTER draws each half of the center mark 0.09 units long. When the value is -0.09, AutoCAD leaves a gap of 0.09 between the mark and the line, and extends the line 0.09 beyond the circumference of the circle or arc.

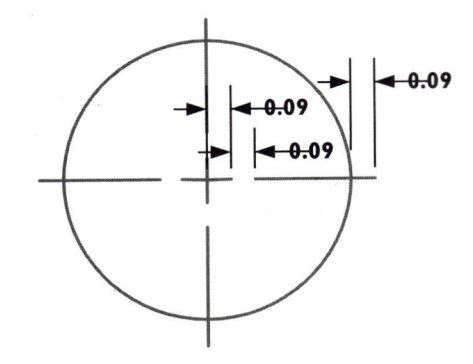

To draw center lines as well, change the value of DIMCEN to -0.09 (the negative value), as follows:

Command: **dimcen**

Enter new value for DIMCEN <0.0900>: **-0.09**

DIMANGULAR

The **DIMANGULAR** command dimensions angles, arcs, circles, and lines.

Unlike the other dimensioning commands in this chapter, this command displays angles only — not lengths or diameters. The dimension requires a *vertex* and some means of determining the angle. The result depends on the object selected:

Two Lines — vertex is the real or apparent intersection of the two lines; angle is measured between the two lines.

Arc — vertex is the arc's center point; angle is measured between two endpoints.

Circle — vertex is the circle's center point; angle is measured between the "Select circle" pick and the next pick point.

Three Points — vertex is the first point picked; the angle is measured between the next two pick points.

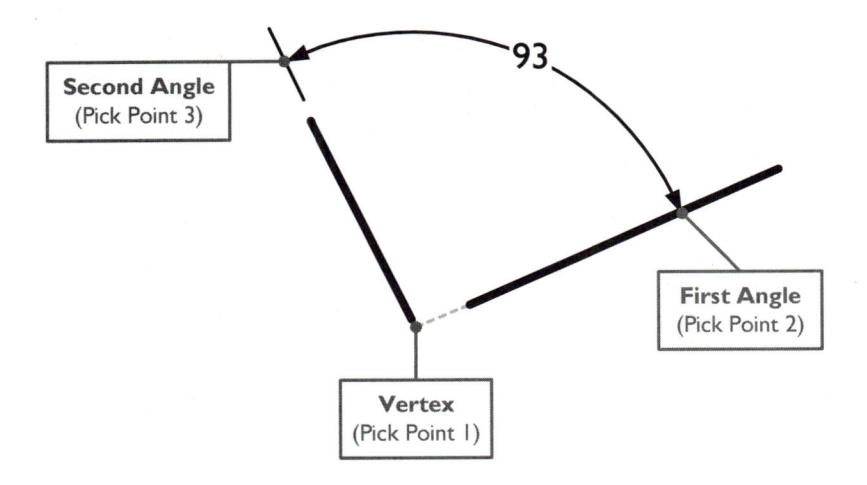

For intersecting lines, **DIMANGULAR** draws four possible angles, each supplementary pair adding up to 180 degrees. In the figure below, the supplementary pairs are shown in the same color.

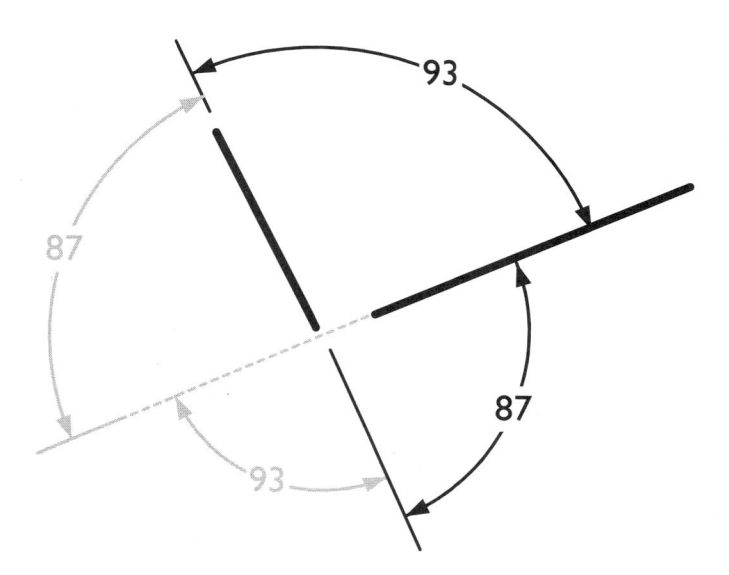

In the case of arcs and circles, two angles are possible: the *minor arc* (shown in black, below) and the *major arc* (shown in gray). Together they add up to 360 degrees.

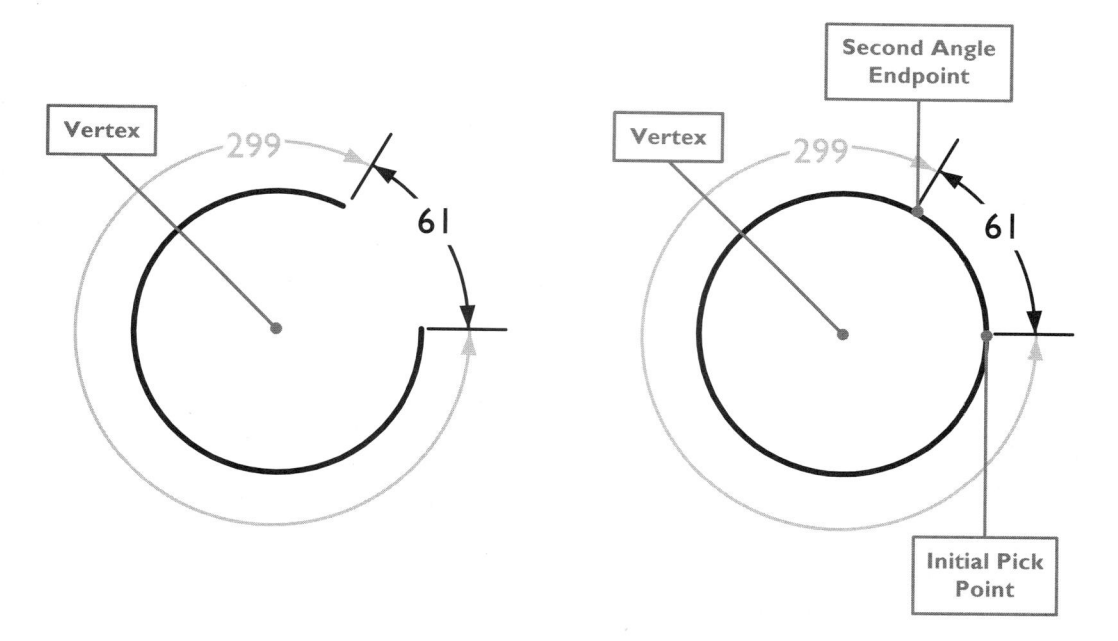

TUTORIAL: PLACING ANGULAR DIMENSIONS

1. To dimension an angle, start the **DIMANGULAR** command:
   * From the menu bar, choose **Dimension**, and then **Angular**.
   * From the **Dimension** toolbar, choose the **Angular Dimension** button.
   * At the 'Command:' prompt, enter the **dimangular** command.
   * Alternatively, enter the aliases **dan** or **dimang** at the 'Command:' prompt.

   Command: **dimangular** *(Press ENTER.)*

2. In all cases, AutoCAD prompts you to select the object you want dimensioned:
   Select arc, circle, line, or <specify vertex>: *(Press* ENTER. *for the vertex option)*
3. Pick three points to specify the angle:
   Specify angle vertex: *(Pick a vertex at point 1.)*
   Specify first angle endpoint: *(Pick point 2.)*
   Specify second angle endpoint: *(Pick point 3.)*
   Non-associative dimension created.
4. Pick a point to locate the dimension:
   Specify dimension arc line location or [Mtext/Text/Angle]: *(Pick a point to position the dimension.)*
   Dimension text = 90

The **Mtext**, **Text**, and **Angle** options are the same as described earlier in this chapter under the DIMHORIZONTAL command.

TUTORIA: DIMENSIONING

In the following exercise, you dimension the fuse link below with some of the command learned in this chapter. As you work through the tutorial, refer to the numbered points in the figure below.

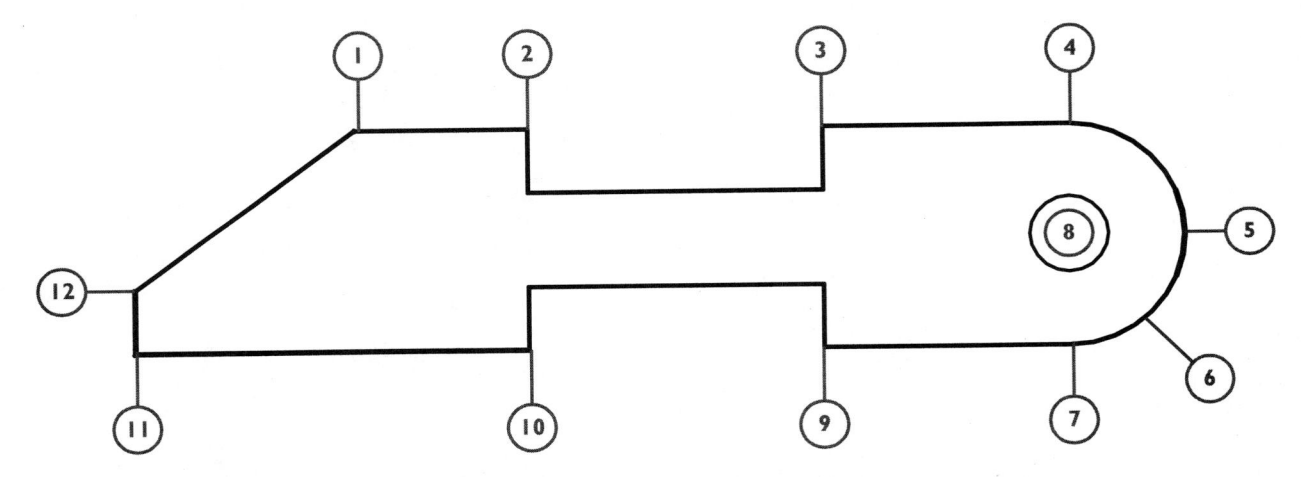

1. From the companion CD-ROM, open the *dimen.dwg* file, a drawing of a fuse link.
   This drawing has layers, dimension styles, and object snaps preset for you:
   Check the Layer Properties toolbar to ensure layer "Dimensions" is current.
   Check on the status bar that OSNAP is turned on, and check the object snap settings.

2. Start by placing a linear dimension along the top of the fuse link.
   From the **Dimension** menu, select **Linear**.
   Use the **Select Object** option to select the line between points 1 and 2.
   Locate the dimension line roughly 0.5 units above.

3. Continue dimensioning with the **DIMCONTINUE** command.
   Use INTersection object snap at points 3 and 4.

Complete the dimension at point 5 with the QUAdrant object snap.
Remember to press ESC to exit the command.
Do the new dimension lines continue at the same height as the first dimensions?

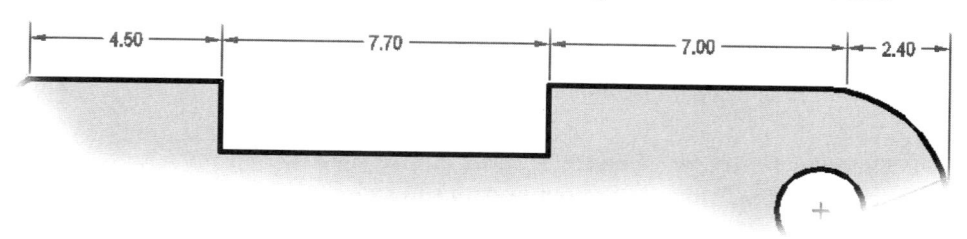

4. Now start a linear dimension along the bottom of the fuse link with the **DIMLINEAR** command.
   Place extension lines at points 11 and 10.
   Does INTersection object snap help you?

5. Switch to baseline dimensioning. From the **Dimension** menu, choose **Baseline**.
   Pick points 9 and 7 as the second extension line origins.
   Complete the command by picking point 5 with QUAdrant object snap.

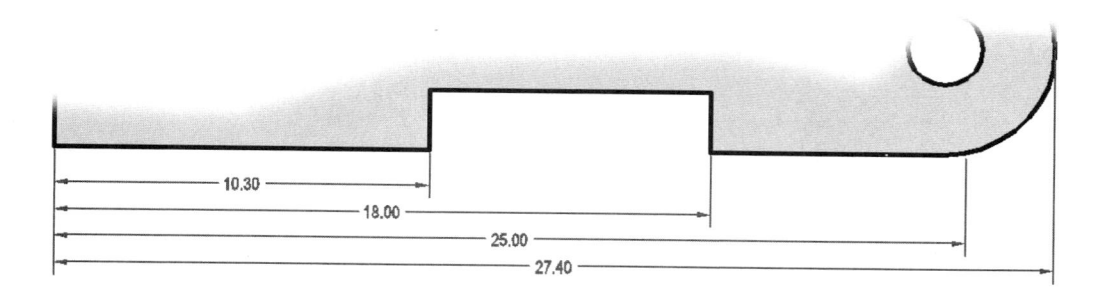

6. Place a vertical dimension by entering the **DIMLIN** alias.
   Choose point 11 as the first extension line origin, and point 12 as the second.
   If you make a mistake, use the **Undo** command to eliminate the last dimension.

7. Dimension the angle at the left side with the **DIMALIGNED** command.
   Use the **Select Object** option to select the angled line between points 12 and 1.

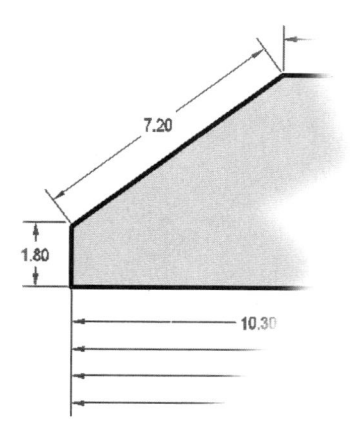

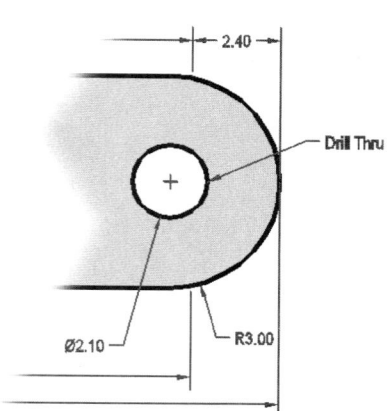

8. Select **Diameter** from the **Dimension** menu.
   Dimension the circle at point 8.
   Make sure the leader line extends outside the fuse link.

9. Dimension the arc with **DIMRADIUS**.
   Pick point 6, and then place dimension leader outside the fuse link.

10. Construct a leader by selecting **Leader** from the **Dimension** menu. (See Chapter 12, "Placing and Editing Text," for more about leaders.)
    Start the leader at the circle (point 8), and end it outside the object.
    For the leader text, enter "Drill Thru."
    Remember to press **ENTER** to exit the **QLEADER** command.

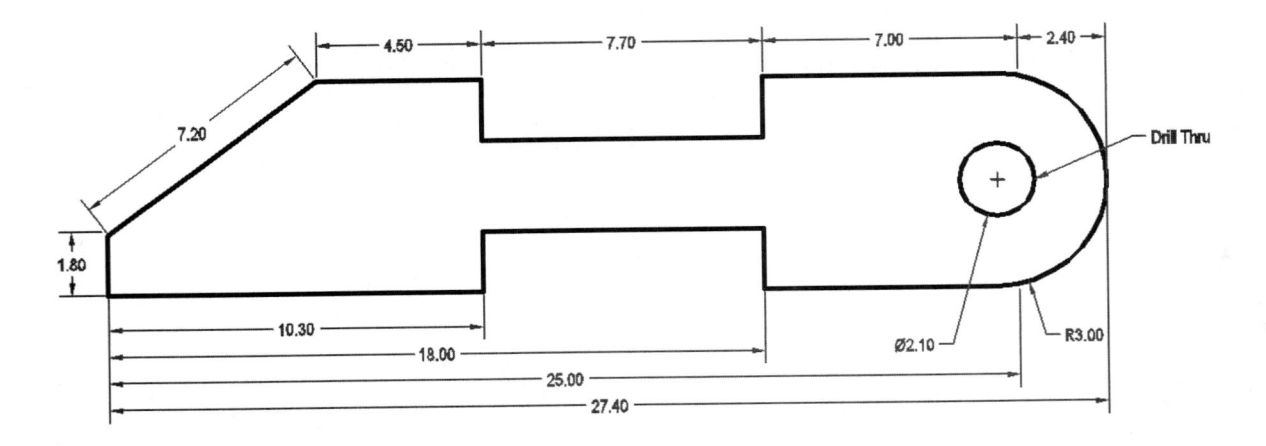

11. Save the completed drawing as *dimen.dwg*.

EXERCISES

1. Start a new drawing.
   With the **POLYGON** command, draw a square approximately one-third the size of the screen.
   From the **Dimension** menu, select **Linear**. Use object snap INTersection to capture the lower left and lower right corners of the square.
   Place the dimension line below the bottom line of the square.

2. Continuing from the previous exercise, repeat the command for the upper line of the square, with a change. Instead of picking the corners, dimension the line by selecting it.
   Is the dimension text the same for the lower and upper lines?

3. Continuing, select **Linear** from the **Dimension** menu, and then dimension one of the vertical sides of the box.
   Is the vertical dimension the same as the horizontal dimension?

4. Continuing, dimension the remaining vertical line.
   At the "Specify dimension line location or [Mtext/Text/Angle/Horizontal/Vertical/Rotated]" prompt, enter **m**.
   Does the mtext editor dialog box appear?
   Type your name.
   Do you recall the purpose of the < > characters?
   Choose **OK** when finished.
   What appears along the dimension line?
   Save the drawing as *square1-4.dwg*.

5. Start a new drawing.
   Draw a scalene triangle.
   Use the **DIMALIGNED** command to dimension the three angled sides.
   Does AutoCAD draw the dimension lines parallel to the sides?

6. Draw two lines that are not parallel, similar to those shown below.
   Place an angular dimension on the two lines.

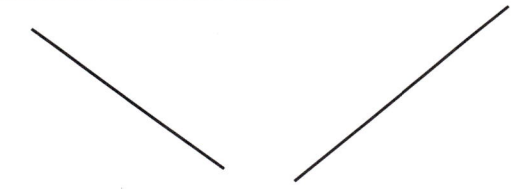

   Repeat the command to draw the supplemental angle.

7. Draw an arc and a circle.
   Dimension the minor and major arcs with the **DIMANGULAR** command.
   Save the drawing as *dimen5-7.dwg*.

8. Start a new drawing.
   Draw a diagonal line, similar to the ones in exercise #6.
   Use the following commands to dimension the line three times: horizontally, vertically, and rotated.
   a. **DIMHORIZONTAL**
   b. **DIMVERTICAL**
   c. **DIMROTATED**

9. Undo the three dimensions you drew in the previous exercise.
   Use the **DIMLINEAR** command to draw the same three orientations of dimension.
   Which commands did you find easier to use?
   Save the drawing as *linear8-9.dwg*.

10. Start a new drawing.
    Draw an arc and a circle on the screen, and then dimension each with the **DIMDIAMETER** command.
    Practice placing diameter dimensions inside and outside the circle, and at different locations around the arc.

11. Repeat exercise #9 with the **DIMRAD** command.
    Save the drawing as *diarad10-11.dwg*.

12. From the companion CD-ROM, open the *qdim.dwg* drawing of a flywheel.
    Dimension the two views with the **QDIM** command.
    When complete, save the drawing as *flywheel12.dwg*.

13. Draw the following base plates, and then place the dimensions at the locations shown.
    a. Save the drawing as *baseplate13-a.dwg*.

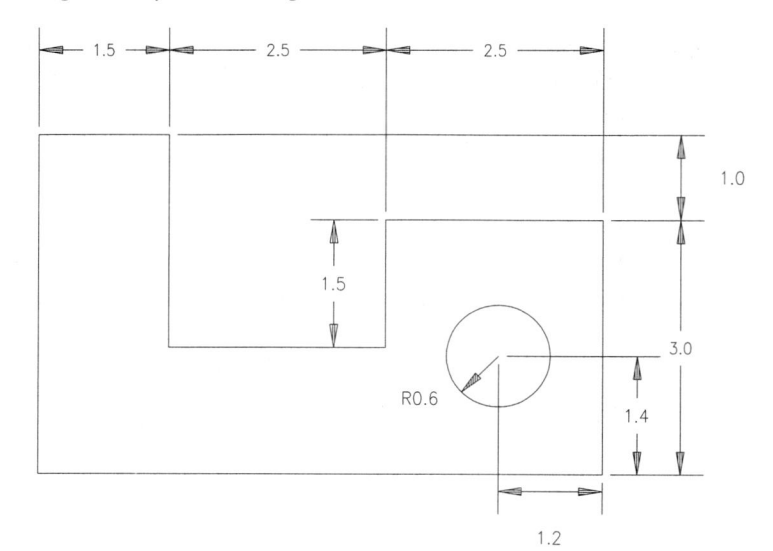

b. Save the drawing as *baseplate13-b.dwg*.

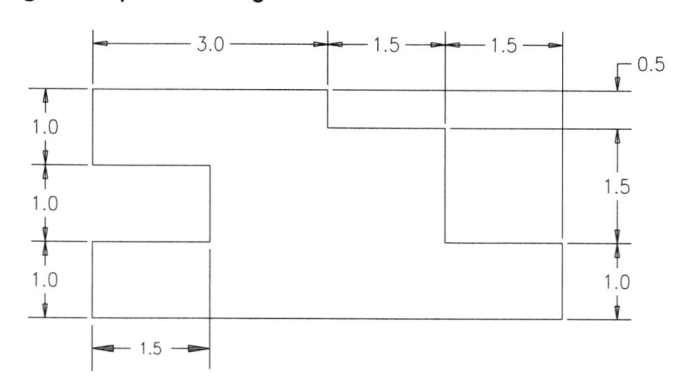

c. Save the drawing as *baseplate13-c.dwg*.

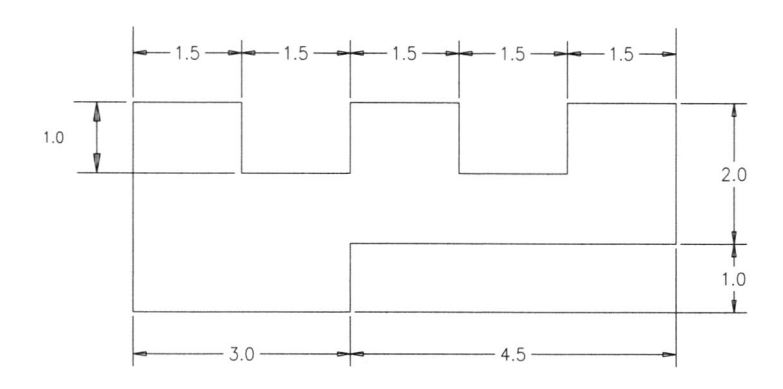

14. Draw the two views of the clamp, and then use the dimension commands to place the vertical dimensions at the locations shown. Save the drawing as *clamp14.dwg*.

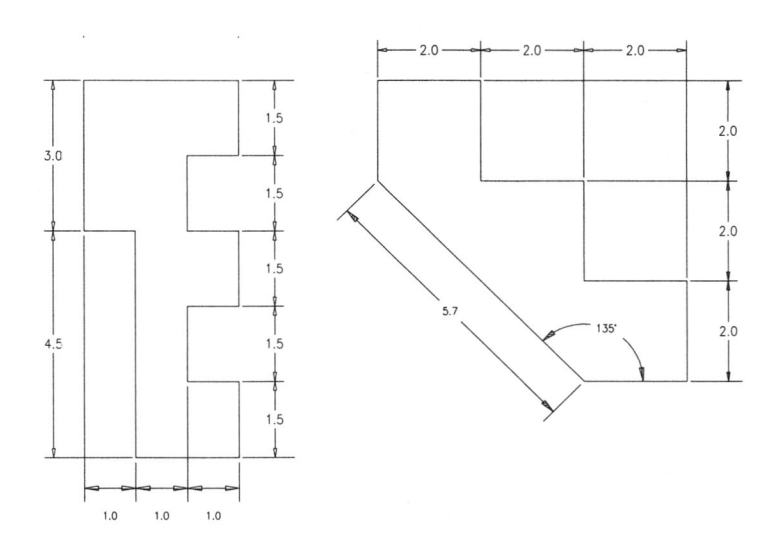

15. Construct the drawing of the corner shelf, and then place all the dimensions as shown. Save the drawing as *shelf15.dwg*.

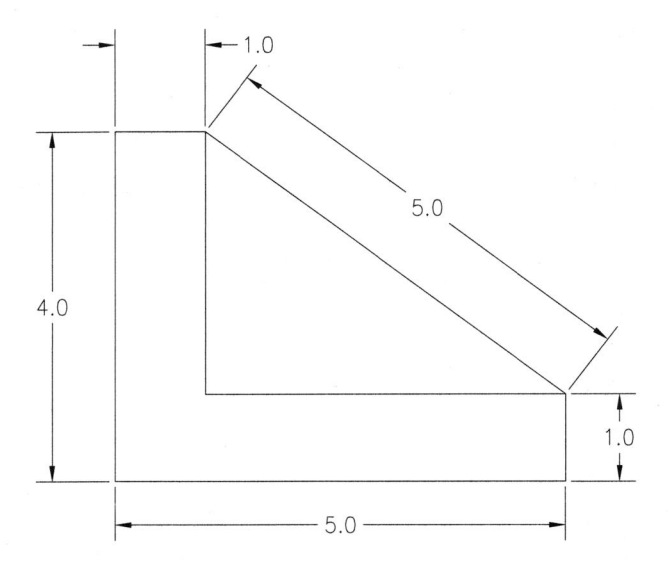

16. Draw the following 5.5" x 3.5" object, and then use radius dimensioning to specify the two 1.5" fillets. Save the drawing as *radial16.dwg*.

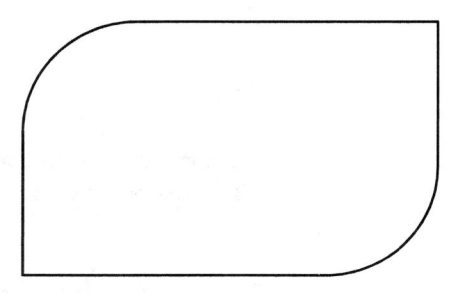

17. Construct the drawing of the corner shelf, and then place all the dimensions as shown. Save the drawing as *shelf17.dwg*.

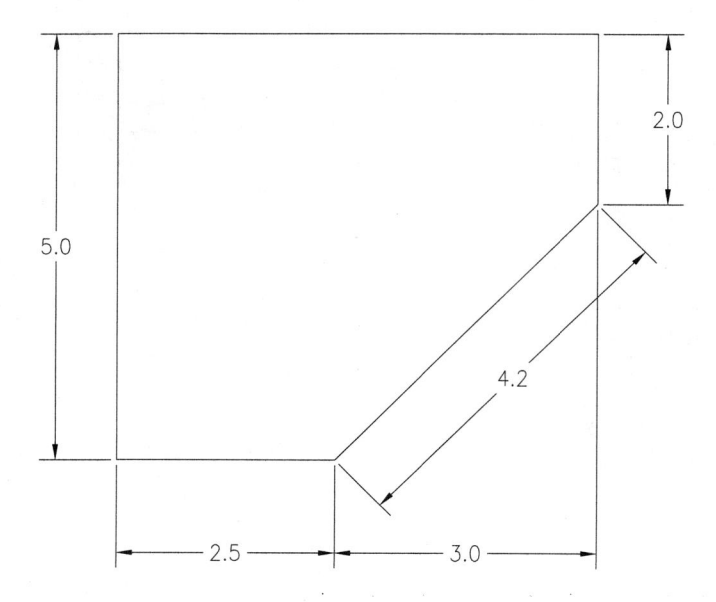

18. Construct the drawing of the jig, and then place the baseline dimensions. Save the drawing as *jig18.dwg*.

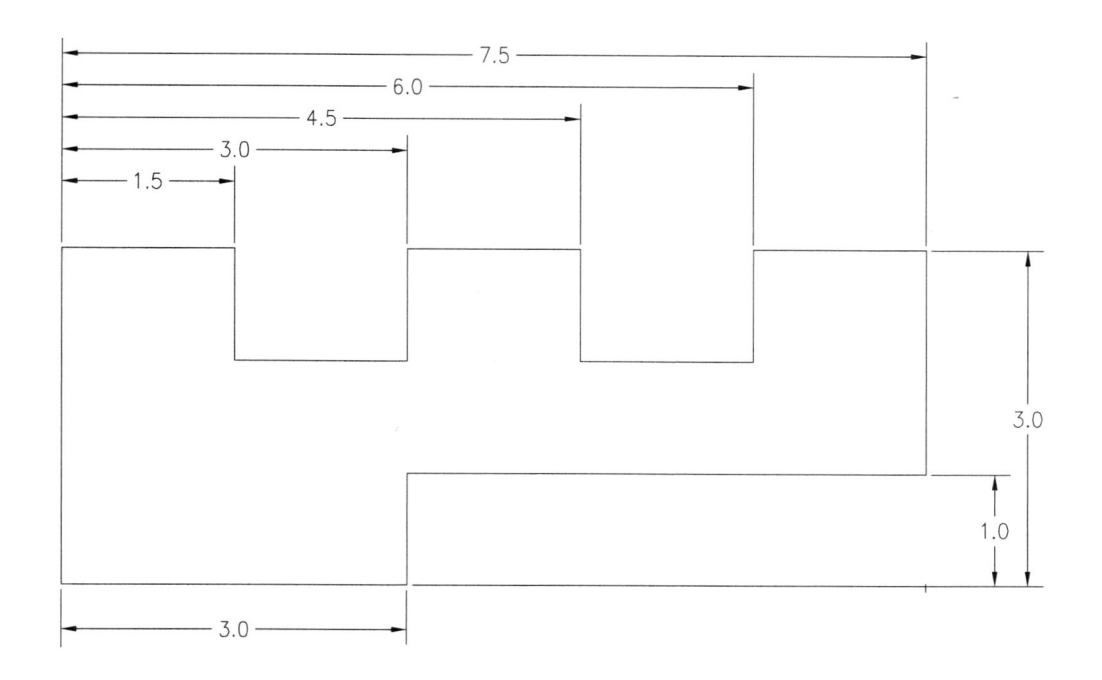

19. From the companion CD-ROM, open the *15-47.dwg* of a gasket. Dimension the drawing with the **DIMRAD** and **DIMDIA** commands. When complete, save the drawing.

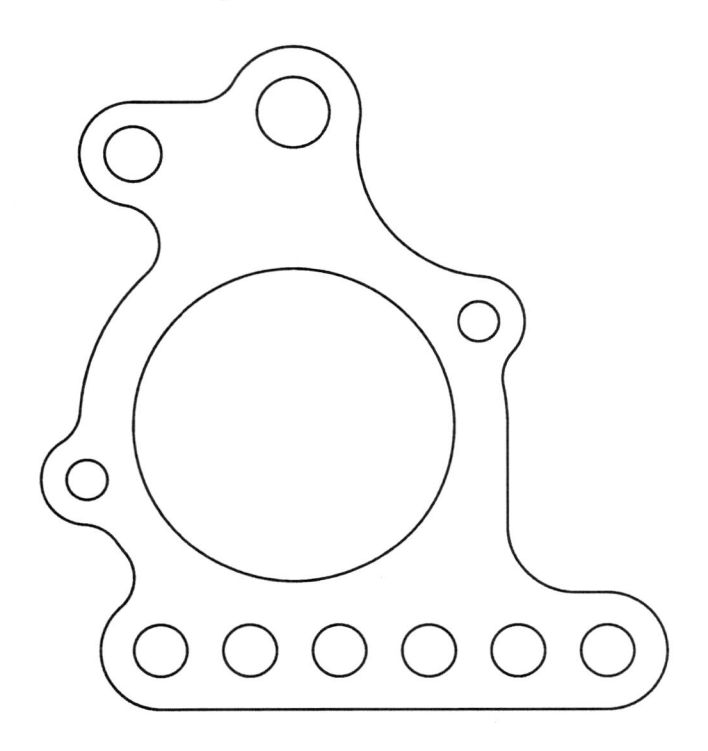

20. From the companion CD-ROM, open the following drawings, and then dimension the gaskets.

   a. Save as *5-43.dwg*

   b. Save as *15-44.dwg*

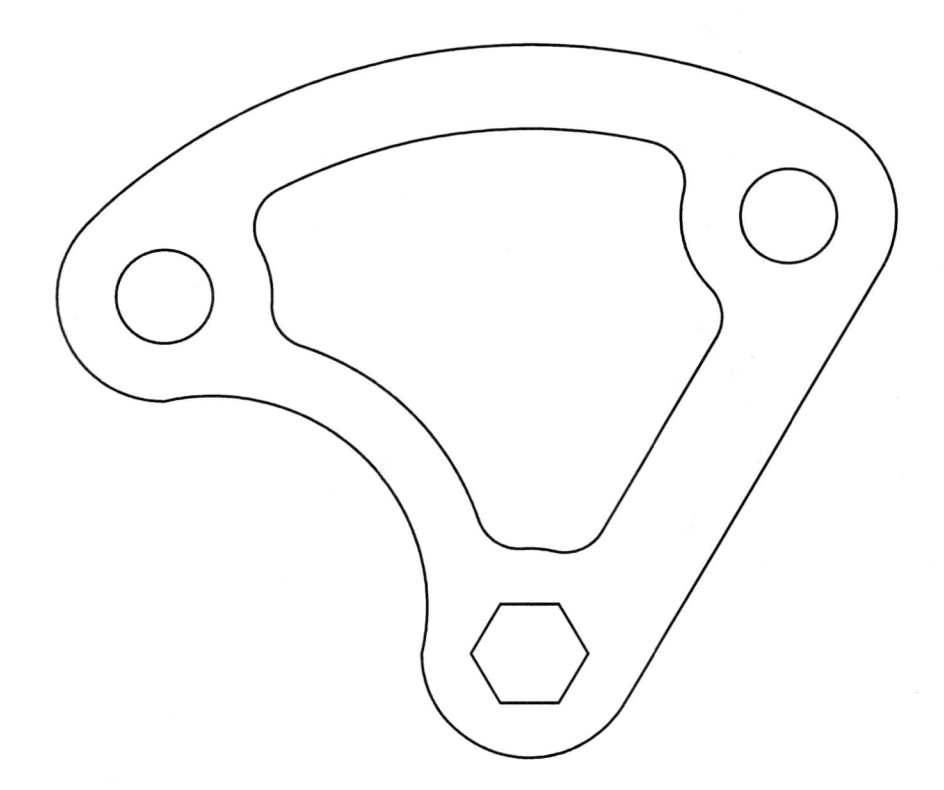

Placing Dimensions **573**

c. Save as *15-45.dwg*

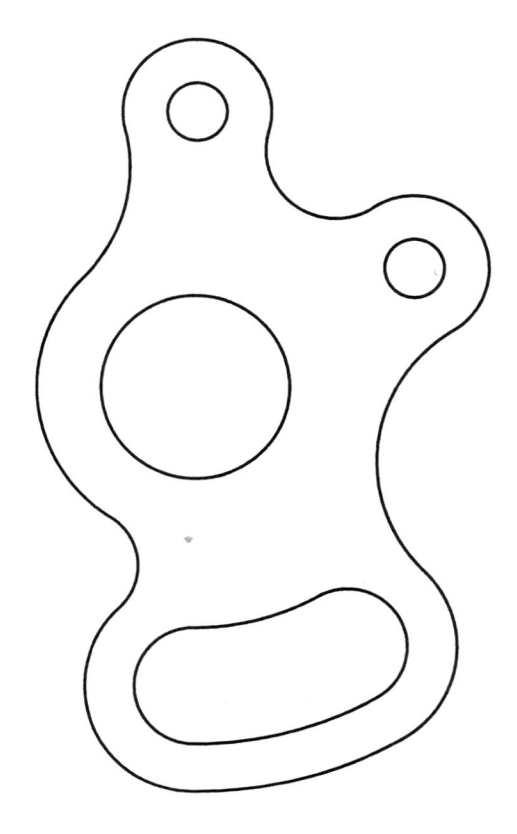

d. Save as *15-46.dwg*

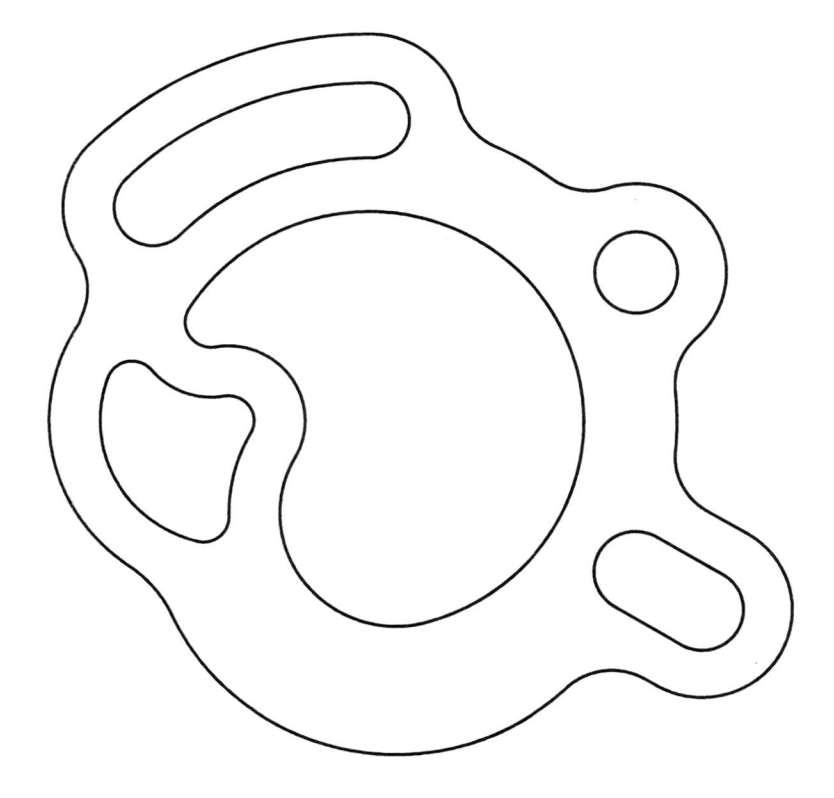

21. From the CD-ROM, open the following drawings. Dimension each drawing, and then save your work.

   a. *17_19.dwg:* Attachment for a three-point hitch.

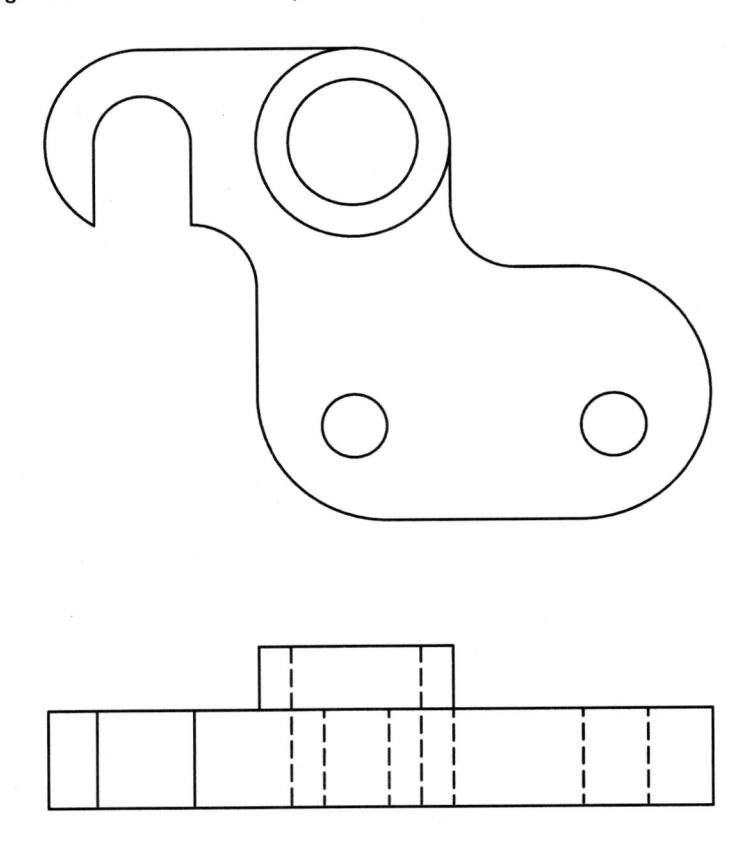

   b. *17_20.dwg:* Flag pole stabilizer.

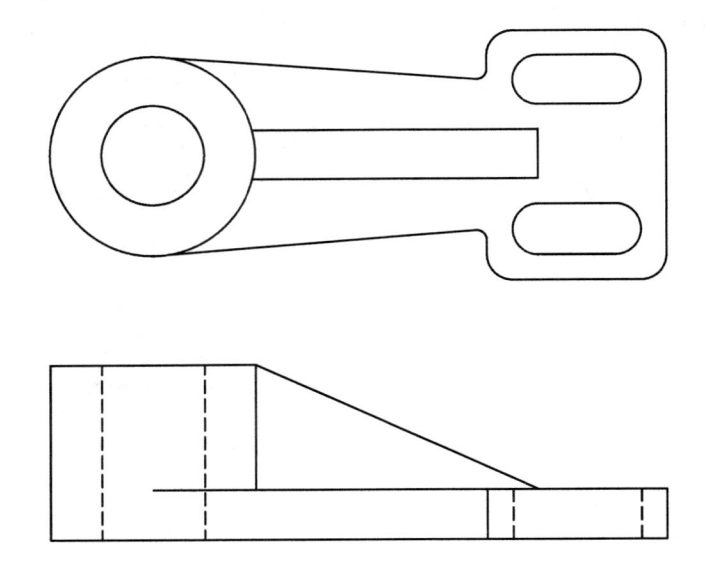

c. *17_21.dwg:* Reciprocating linkage.

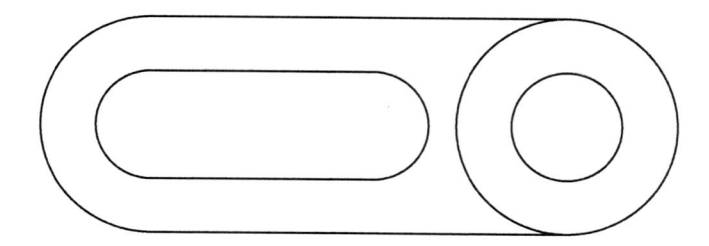

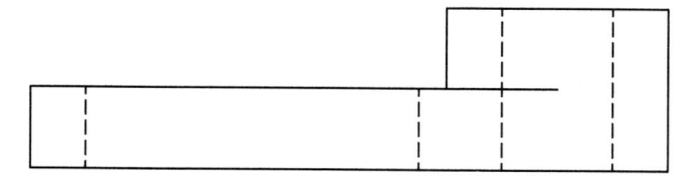

d. *17_22.dwg:* Socket linkage.

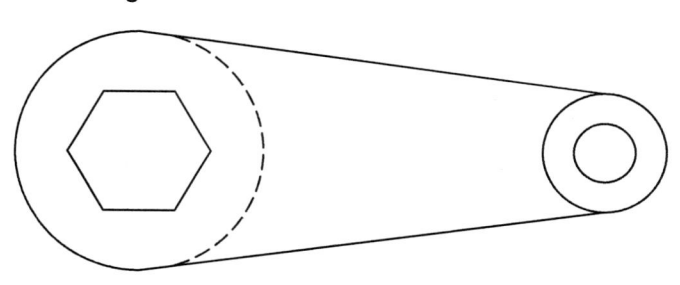

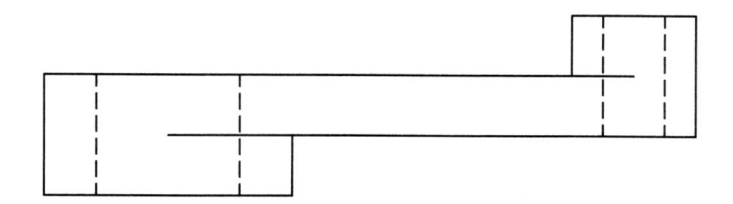

e. *17_23.dwg:* Attachment plate.

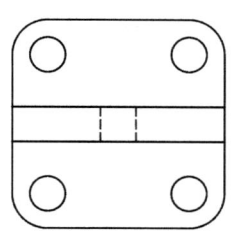

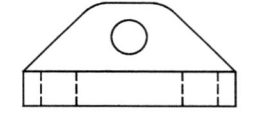

f. *17_24.dwg:* **Hitch yoke.**

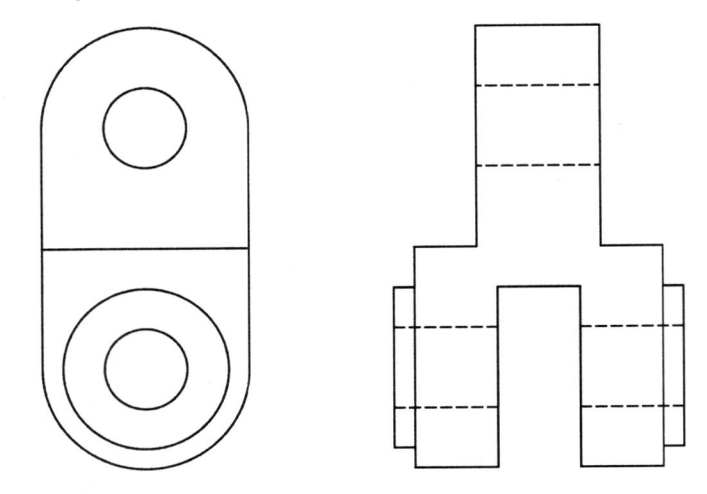

g. *17_25.dwg:* **Hitch linkage.**

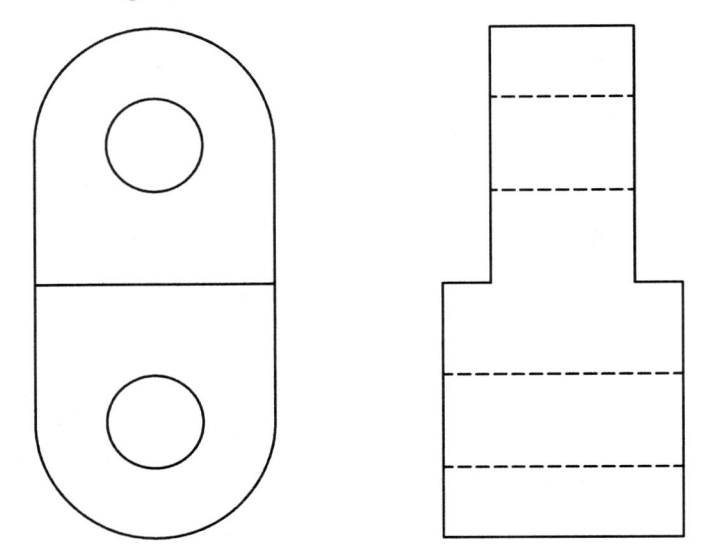

h. *17_26.dwg:* **Axle suppport.**

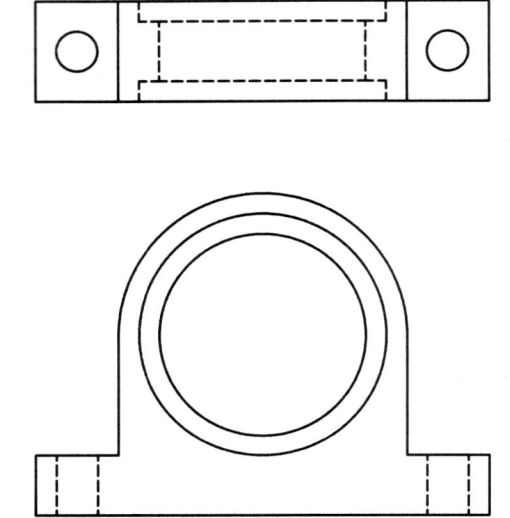

i. *17_27.dwg:* Cone of silence. (Technical editor comments: "Yes! Maxwell Smart lives!" Copy editor comments: "Sorry about that, Chief.")

j. *17_28.dwg:* Bushing.

k. *17_29.dwg:* Wedge.

l. *17_30.dwg:* Concrete footing.

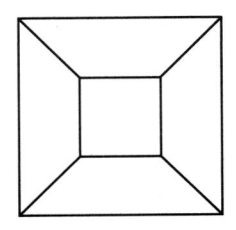

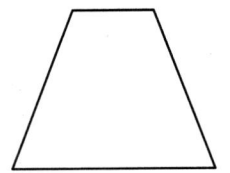

m. *17_31.dwg:* Heavy-duty axle support.

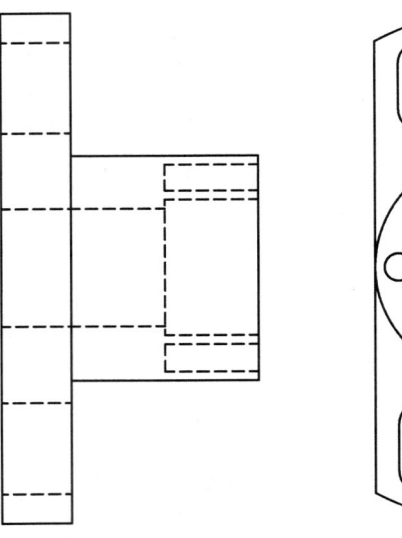

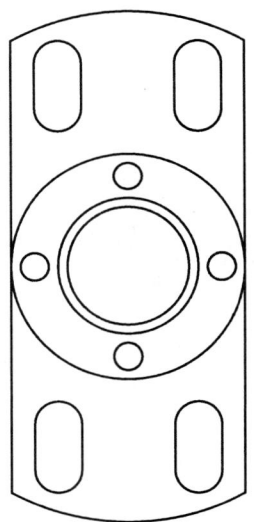

n. *14_32.dwg:* Paintbrush handle.

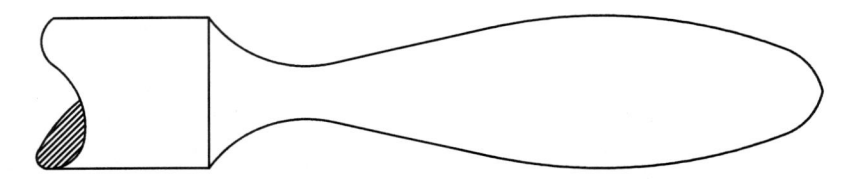

o. *17_33.dwg:* Spacer.

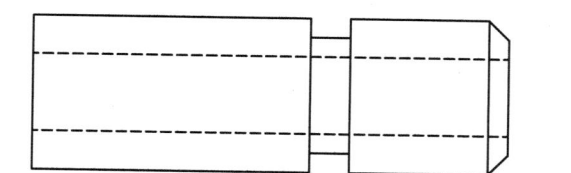

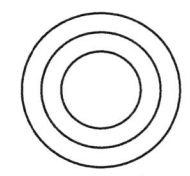

p. *17_34.dwg:* Base plate.

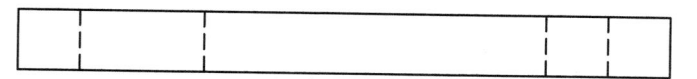

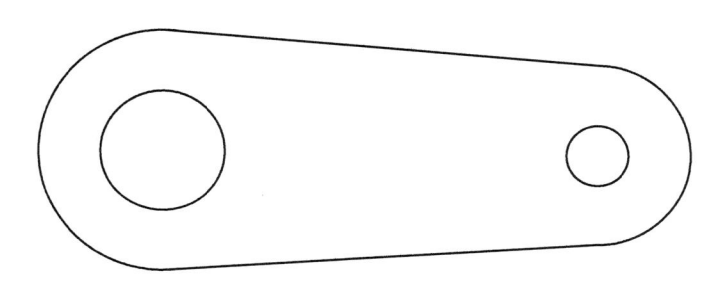

q. *17_35.dwg:* Angled baseplate.

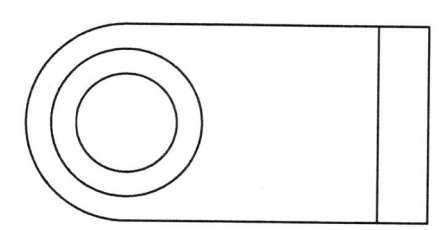

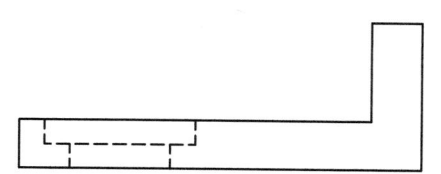

CHAPTER REVIEW

1. A dimension measures _____.
2. Should extension lines touch the object they reference?
3. Where should dimensions be placed, whenever possible?
4. Leaders indicate _____.
5. On which menu do you find the dimensioning commands?
6. Should you attempt to dimension every detail?
7. Under what conditions are dimension lines not drawn?
8. What is another name for "witness lines"?
9. How are the endpoints of extension lines determined?
10. Must a dimension always have two extension lines?
11. Sketch and label the parts of measurement dimensions. Include the following parts:
    Dimension line
    Arrowheads
    Extension lines
    Text
12. Sketch and label the parts of a radial dimension. Include the following parts:
    Leader line
    Arrowhead
    Text
    Radius Symbol
13. Sketch and label the parts of an extension line:
    Extension line
    Offset
    Extension beyond dimension line
14. List the types of linear dimensions that the **DIMLINEAR** command draws.
    a.
    b.
    c.
15. Explain the function of the **DIMDIAMETER** command?
16. What is the purpose of the **QDIM** command?
17. Draw an example of a baseline dimension:
18. Must arrowheads always be arrows?
19. What is the aspect ratio of arrowhead?
20. List three places where dimension text can be placed.

21. Must dimension text always be horizontal?
22. Write out an example of each of the following terms:
    Deviation Tolerance
    Symmetrical Tolerance
    Limits
    Alternate Units
23. Explain why *direct dimensioning* is beneficial.
    Can any object be dimensioned directly?
24. Which dimensioning commands do you use most often with arcs and circles?
25. What is the purpose of the **DIMANGULAR** command?
26. Describe the meaning of "fully associative dimensioning."
27. Are all dimensions associative?
    Give an example.
28. How is the look of dimensions controlled?
29. Where in AutoCAD might you find dimension standards?
30. Name three object snaps useful for dimensioning:
    a.
    b.
    c.
31. Write out the names of commands that can draw dimensions horizontally.
32. Explain the meaning of the following options available in many dimensioning commands:
    Angle
    Mtext
    Text
33. What is the <> symbol meant for?
34. A dimension reads 2.5. Write out the text string that changes the dimension to read:
    a. "Adjust to 2.5m"
    b. "Turn by 2.5° increments."
    c. "Temperature range is 2.5°±0.5°"
35. Which part of an arc does the **DIMALIGNED** command measure?
    Of a circle?
    Of a line?
36. How do these commands differ?
    DIMBASELINE
    DIMCONTINUE
37. Can you apply the **DIMCONTINUE** command to a diameter dimension?
38. Describe the steps to create y-ordinate dimensions with **QDIM**.
39. Can you dimension ellipses and splines with the **DIMRADIUS** command?
40. Write out the command for each alias:
    dra
    dco
    dli
    dimali
41. Explain the purpose of the center mark.
    Describe the difference between a center mark and a center line?
    How do you instruct AutoCAD to draw one or the other?

42. What angle does the **DIMANGULAR** command measure for the following objects:
    Circle
    Two Lines
    Arc
43. How many possible angles can **DIMANGULAR** draw on:
    Circles?
    Pair of lines?

CHAPTER 14

Editing Dimensions

You often need to change existing dimensions. This chapter describes how to edit the position and properties of dimensions. After completing this chapter, you will have an understanding of these commands:

- **Grips** edits dimensions and leaders directly.
- **DIMEDIT** slants extension lines.
- **DIMTEDIT** relocates dimension text.
- **DDEDIT** changes dimension text with the mtext editor.
- **AIDIM** and **AI_DIM** prefix a group of "hidden" text editing commands.
- **Dimension Variables** specify the look and position of dimension elements.
- **DIMSTYLE** and **-DIMSTYLE** create and modify dimension styles.
- **Styles** toolbar updates selected dimensions with new dimstyles.
- **DIMOVERRIDE** applies changed dimvars to selected dimensions.

FINDING THE COMMANDS

On the **DIMENSION** toolbar:

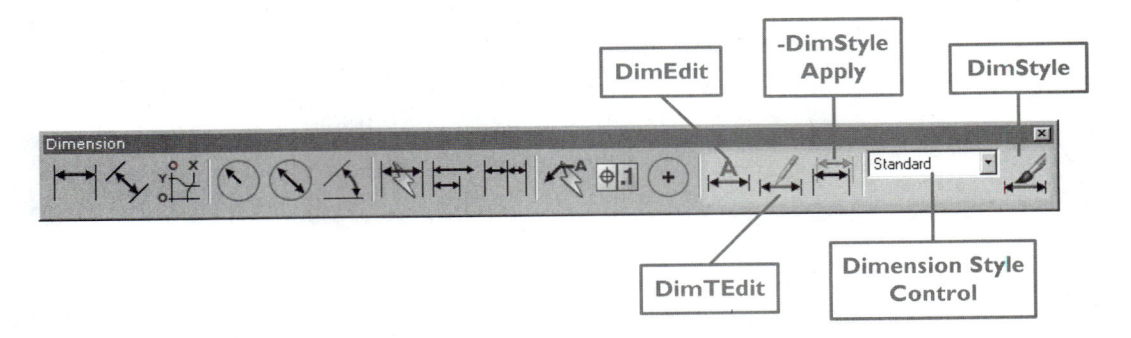

- DimEdit
- -DimStyle Apply
- DimStyle
- DimTEdit
- Dimension Style Control

On the **STYLES** and **TEXT** toolbars:

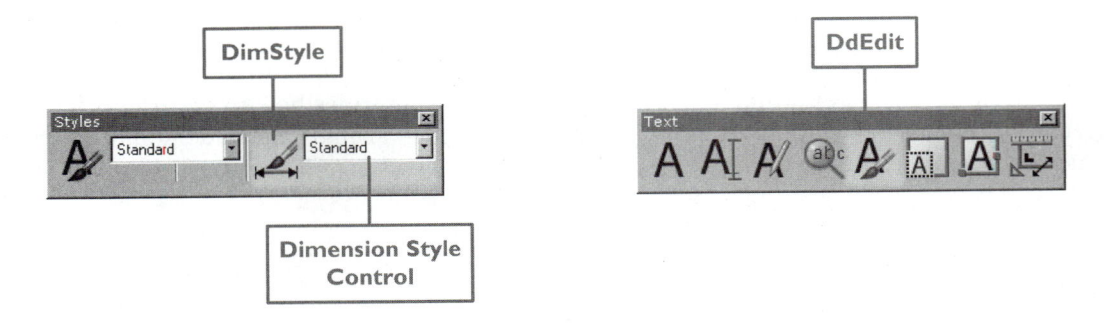

- DimStyle
- Dimension Style Control
- DdEdit

On the **FORMAT** and **DIMENSION** menus:

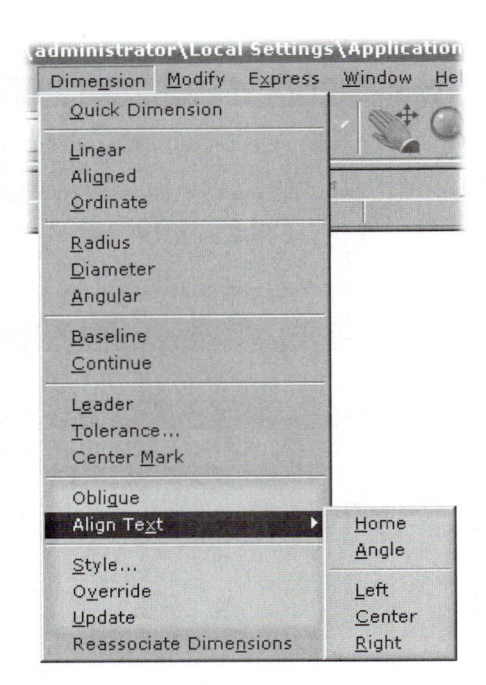

GRIPS EDITING

Grips allow you to edit dimensions and leaders directly — without needing to learn editing commands.

Every associative dimension has three or more grips. The grips relocate the dimension line, text, extension lines, and so on. The tricky part, however, is in understanding which grip performs what action. This section describes the action of each grip. To use grips to move and rotate, refer to Chapter 9, "Direct Editing of Objects."

Associative dimensions "stick" to the objects they dimension. When you edit the object, the dimension changes as well. Move the object, and the dimension moves along; stretch the object, and the dimension updates.

TUTORIAL: EDITING DIMENSIONED OBJECTS

1. Draw, and then dimension an object using the **Select Object** option.
   Command: **dimlinear**
   Specify first extension line origin or <select object>: *(Press ENTER.)*
   Select object to dimension: *(Pick an object, such as the line.)*
   Specify dimension line location or
   [Mtext/Text/Angle/Horizontal/Vertical/Rotated]: *(Pick a point.)*
   Dimension text = 3.00
2. Select the object (not the dimension!). Notice the grips.
   Select one of the grips, and then stretch the object.
   Notice that the associated dimensions update themselves, as illustrated by the figure below.

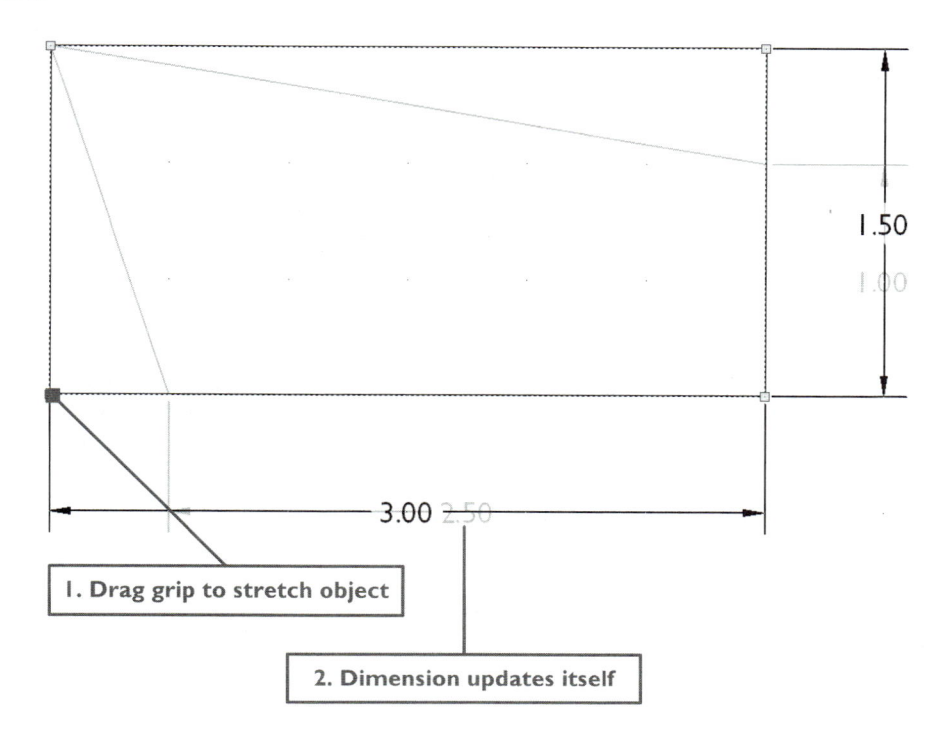

1. Drag grip to stretch object

2. Dimension updates itself

3. Select the **Move** option, and then move the object.
   Notice that the dimensions move along.
4. Press **ESC** to end grips editing.

GRIPS EDITING OPTIONS

Grips perform six editing operations:

Stretch changes the position of dimension parts, such as the extension line or text, while keeping the rest of the dimension in place. The dimension remains partially associated with its object. The dimension text is updated, if the stretching changes the length of the dimension line. Each part of the dimension reacts differently to stretching, as detailed later.

Copy copies the dimension, subject to the peculiarities of the selected grip.

Move moves the entire dimension, but disassociates it from its object.

Rotate rotates the entire dimension about the selected grip, but disassociates it from its object.

Scale resizes the dimension, using the selected grip as the base point; the dimension remains associated with its object. The dimension text is updated to reflect the new size. As with copying, scaling is subject to the peculiarities of the selected grip.

Mirror does not work with dimensions; dimensions are rotated, not mirrored, about the selected grip. The "mirrored" dimension is disassociated from its object.

EDITING LEADERS WITH GRIPS

Identical leaders are drawn with the **LEADER** and **QLEADER** commands; the only difference is the location of the grip on the text, which is immaterial for grips editing.

Leaders have three grips for changing the position of the arrowhead end of the leader, the vertices, and the leader endpoint. The leader text is placed independently, and thus is edited separately from the leader line.

Stretch is detailed below.

Copy makes copies of the leader line, independent of the text.

Move moves the leader line.

Rotate rotates the leader line about the selected grip.

Scale resizes the leader along the axis of the line.

Mirror mirrors the leader line, and erases the original.

Arrowhead

Relocate the arrowhead end of the leader line by dragging its grip:

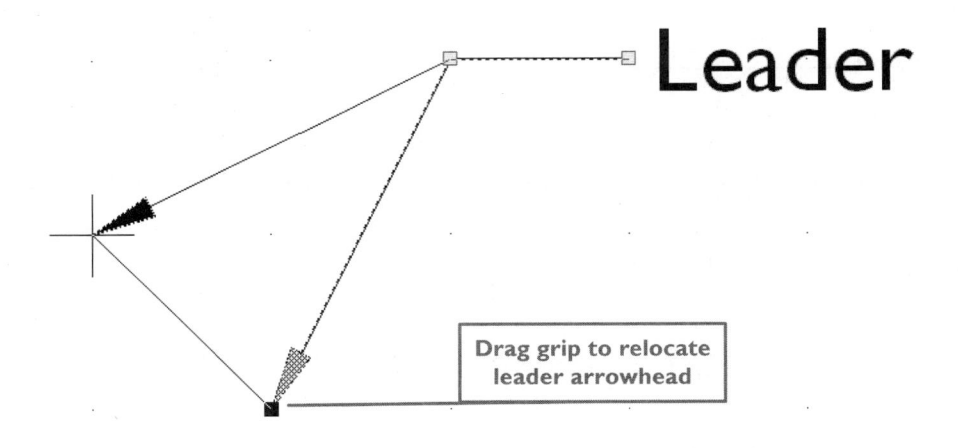

Vertex

There are grips located at each vertex along the leader line. The vertex is relocated by dragging its grip:

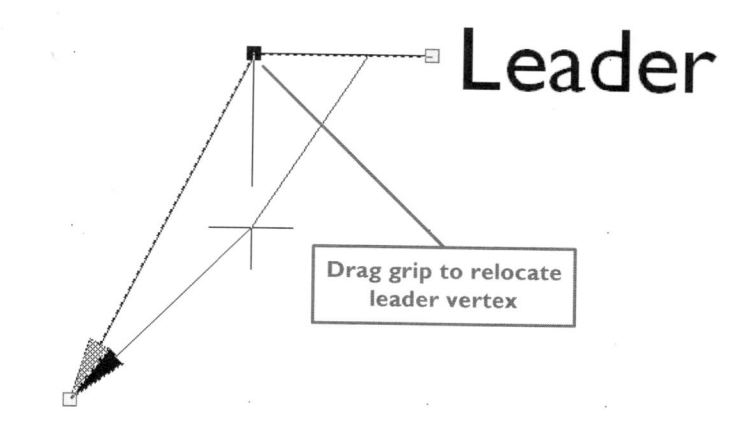

Drag grip to relocate leader vertex

Endpoint

The endpoint of the leader line is the last point you pick; it is relocated by moving its grip. (Moving the leader line does not move the text.)

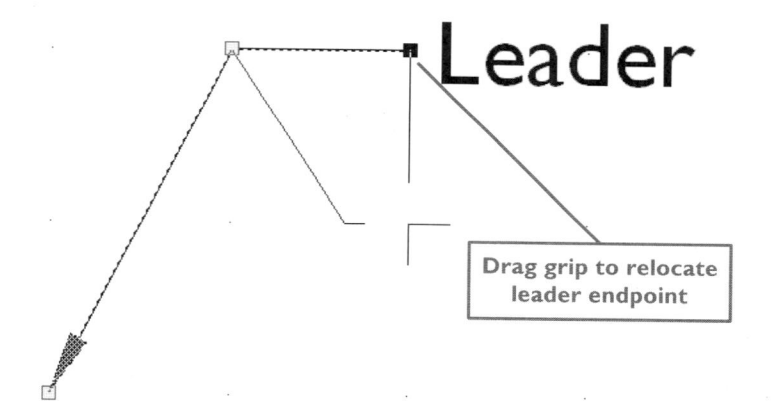

Drag grip to relocate leader endpoint

Text

The text is independent of the leader line; you must select it separately from the leader line. The text is relocated by dragging its grip. (Moving the text brings the leader line along with it.)

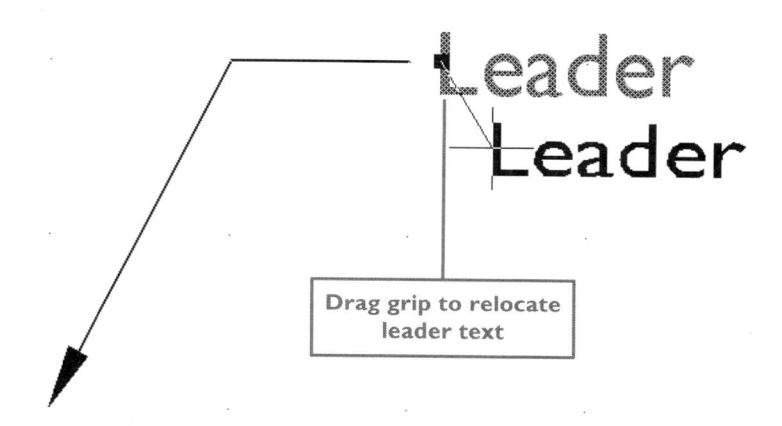

Drag grip to relocate leader text

GRIPS EDITING LINEAR AND ALIGNED DIMENSIONS

Linear dimensions are drawn by many dimensioning commands, including **DIMLINEAR**, **DIMALIGNED**, and **DIMROTATED**. Grips editing is identical for dimensions created by all these commands, whether horizontal, vertical, or rotated.

Linear dimensions have five grips: one at either end of the dimension line, one at the end each extension line, and one on the text.

Stretch is detailed below.

Copy copies the dimension, subject to the grip selected:

- Extension line grips constrain copying in the direction of the dimension line.
- Dimension line grips constrain copying in the direction of the extension lines.
- Text grip constrains copying in the direction of the extension lines, but also moves the location of the text.

Move moves the entire dimension freely, but disassociates it from its object.

Rotate rotates the dimension about the selected grip.

Scale resizes the dimension, subject to the grip selected:

- Extension line grips resize the dimension relative to the selected grip.
- Dimension line grips constrain resizing in the direction of the dimension line, relative to the selected grip.
- Text grip constrains resizing in the direction of the dimension line, but resizes symmetrically. (Both extension lines move in mirrored fashion.)

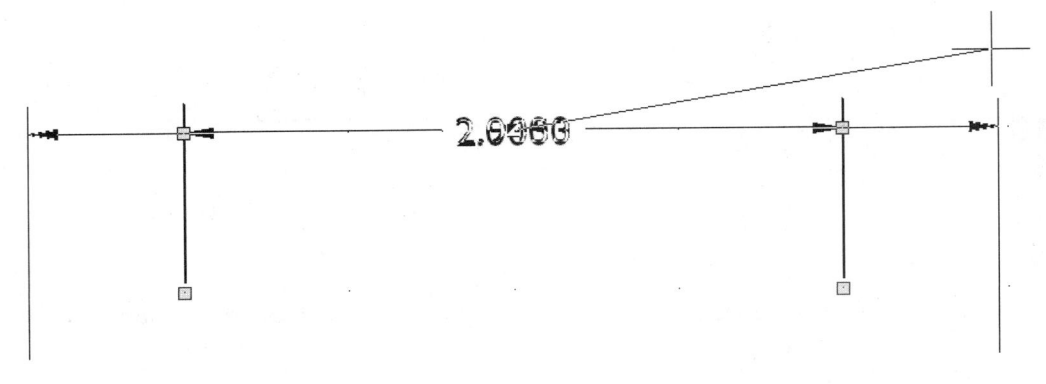

Mirror mirrors the dimension about the selected grip; the original is erased. Text is not mirrored, ignoring the setting of the **MIRRTEXT** system variable.

Dimension Line

There is a grip located at each end of the dimension line. The two grips perform the identical task: moving the dimension line closer and text to and further away from the object.

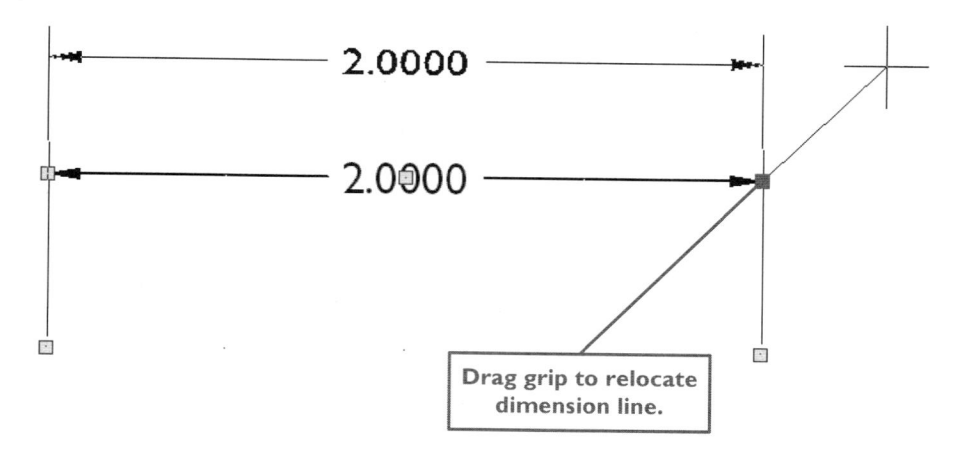

Drag grip to relocate dimension line.

Extension Lines

Each extension line has a grip at its endpoint performing two functions: changing the length of the extension line, and changing the width of the dimension line, thus updating the dimension text. Turning on ortho mode constrains movement of either the extension or the dimension line.

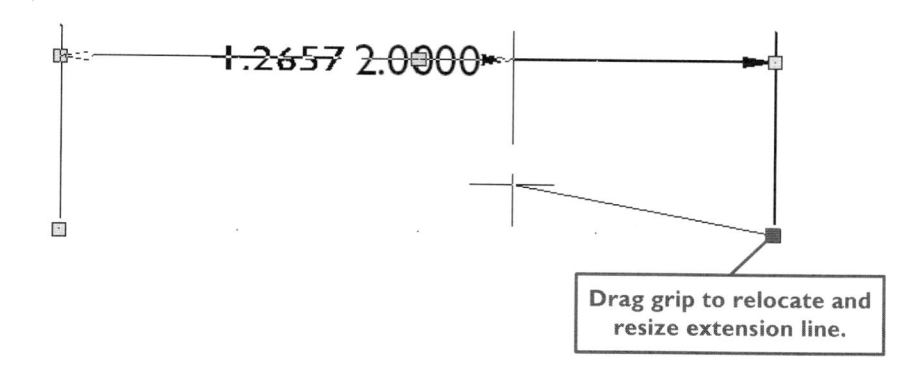

Drag grip to relocate and resize extension line.

Text

The text has a single grip at its center performing two functions similar to the extension line grips: changing the position of the text line along the dimension line, and changing the location of the dimension line. Turning on ortho mode constrains movement of the text and the dimension line.

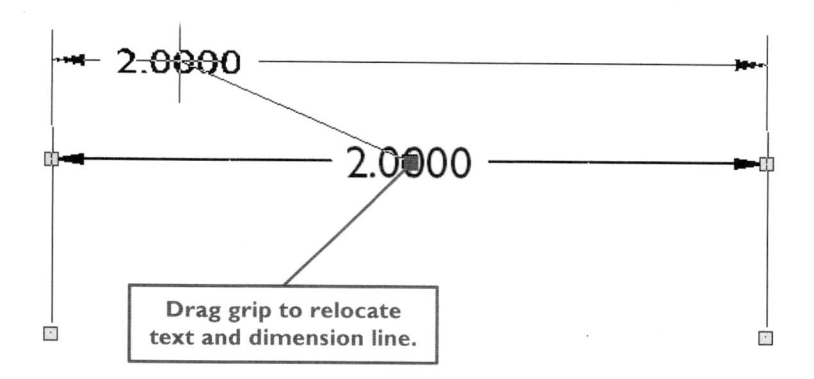

Drag grip to relocate text and dimension line.

GRIPS EDITING ANGULAR DIMENSIONS

Angular dimensions are drawn by the **DIMANGULAR** command, and have five grips: one at the vertex, one on the dimension arc, one at the end of each extension line, and one on the text.

Stretch is detailed below.

Copy copies the dimension, subject to the grip selected:

- Extension line grips constrain copying in the direction of the dimension line.
- Dimension line grips constrain copying in the direction of the extension lines.
- Text grip constrains copying in the direction of the extension lines, but also moves the location of the text.

Move moves the entire dimension freely, no matter which grip is selected. The dimension is disassociated from its object.

Rotate rotates the dimension about the selected grip.

Scale resizes the dimension, using the selected grip as the base point.

Mirror mirrors the dimension about the selected grip; the original is erased. Text is not mirrored, ignoring the **MIRRTEXT** system variable.

Dimension Arc

Stretching the grip attached to the dimension arc moves it and the text in and out from the vertex.

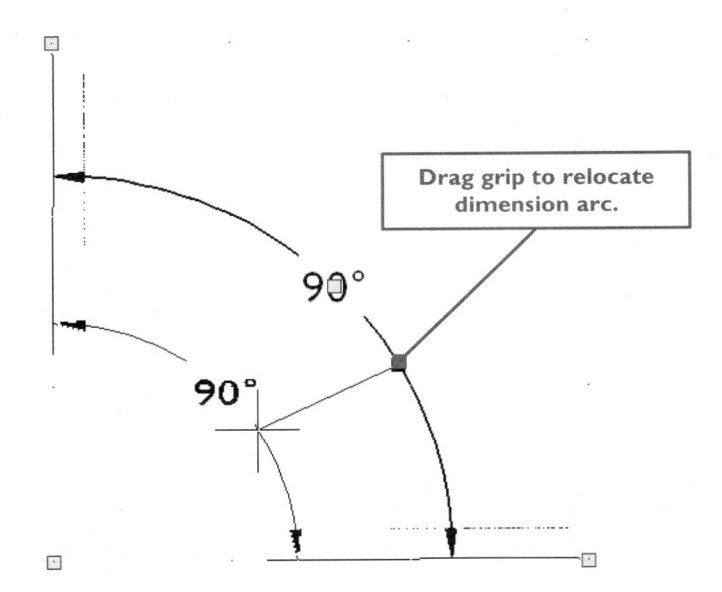

Text

Stretching the text grip performs two functions simultaneously: moves the text along the dimension arc, and moves the dimension arc (as described above).

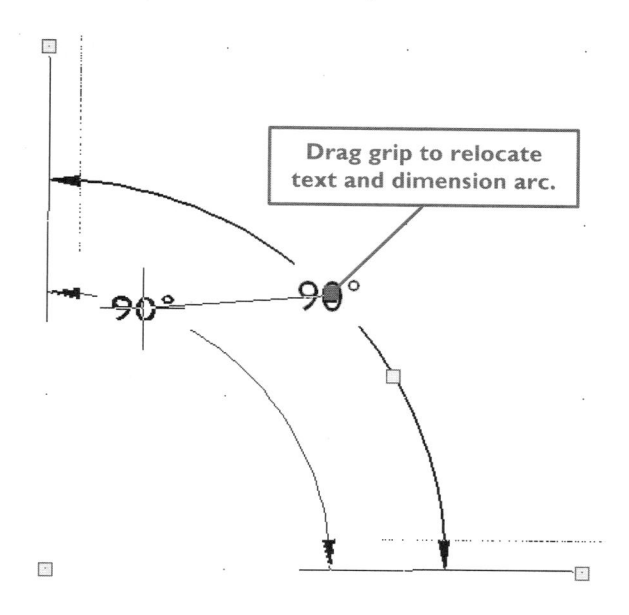

Drag grip to relocate text and dimension arc.

Extension Lines

Stretching the grips at the ends of the extension lines also performs two functions simultaneously: changes the length of the extension line, and moves the extension line. Moving the extension line changes the angle, and so the text is also updated.

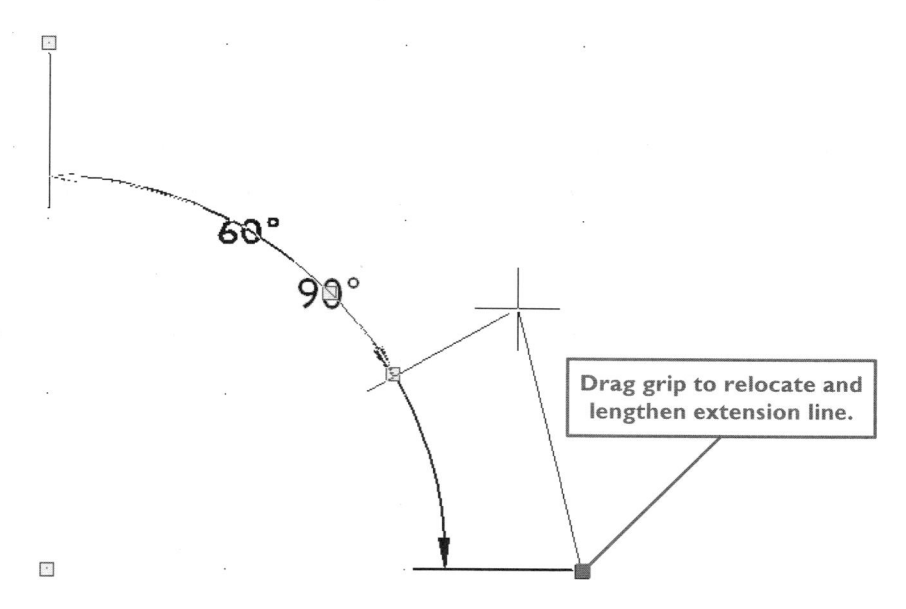

Drag grip to relocate and lengthen extension line.

Vertex

Stretching the grip at vertex is an exercise in mind bending. The text grip and all other non-grip parts of the angular dimension change — or rather, bend: as you move the vertex grip, the extension lines and dimension arc change their lengths and angles, but are constrained by the extension line and dimension arc grips remaining fixed. As the angle changes, the text is also updated.

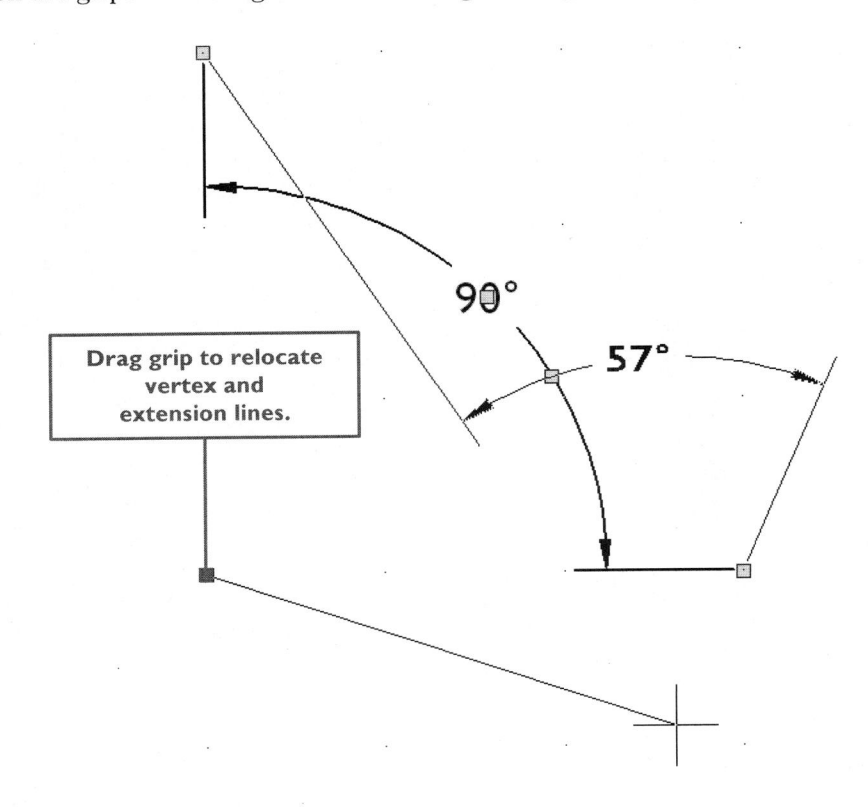

EDITING DIAMETER AND RADIUS DIMENSIONS

The radial dimensions are drawn by the **DIMDIAMETER** and **DIMRADIUS** commands. These commands operate on circles, arcs, and polyarcs only.

Each dimension has three grips: one where the dimension attaches to the circle or arc, one at the diameter or radius point, and one in the center of the text.

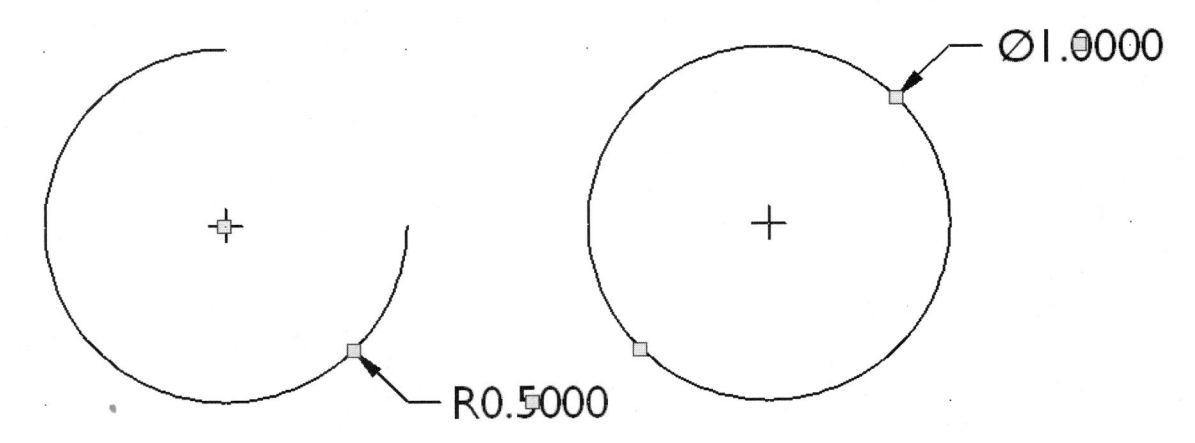

At left: The grips found on radius dimensions.
At right: The grips found on diameter dimensions.

Stretch is detailed below.

Copy copies the dimension, subject to the grip selected:

- Extension line grips constrain copying in the direction of the dimension line.
- Dimension line grips constrain copying in the direction of the extension lines.
- Text grip constrains copying in the direction of the extension lines, but also moves the location of the text.

Move moves the entire dimension freely, no matter which grip is selected. The dimension is disassociated from its object.

Rotate rotates the dimension about the leader and text grips. When the center point grip is selected, the radius dimension rotates about the circle. The text and leader stubs maintain their horizontal orientation.

Scale resizes the dimensions, using the selected grip as the base point.

Mirror mirrors the dimension about the selected grip; the original is erased. Mirroring with the radius dimension's center point grip rotates the dimension about the circle. Text maintains its horizontal orientation and is not mirrored, ignoring the MIRRTEXT system variable.

On Circumference

The radius dimension has one grip located on the circumference of circles and arcs, while diameter dimensions have two. Drag the grip to relocate the dimension around the circle or arc; the text maintains its orientation, and the leader, its length.

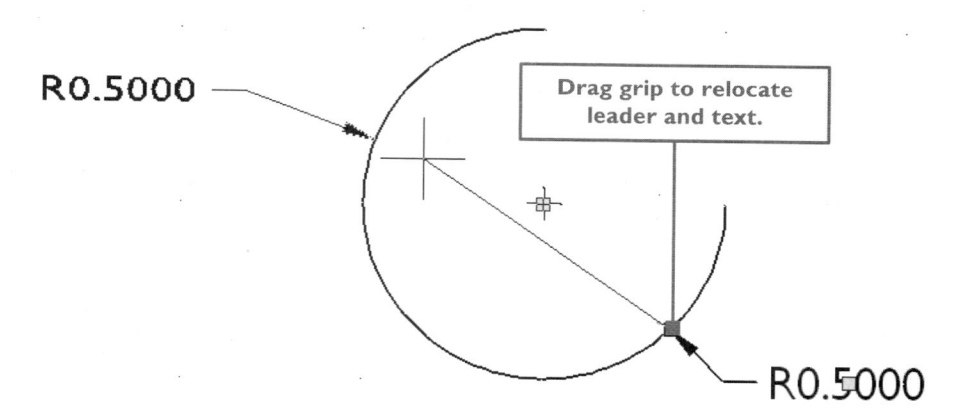

Center Mark

Only the radius dimension has a grip at the center mark. Drag the grip to relocate the leader away from the arc or circle; the dimension text stays in place, roughly.

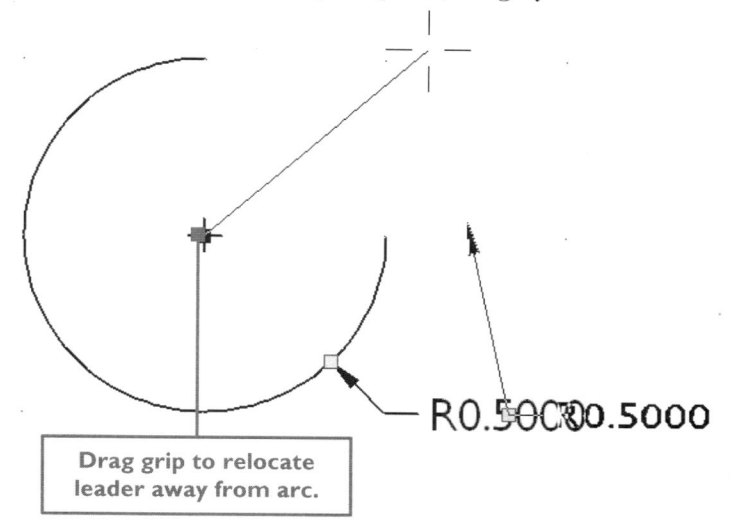

Text

Both the radius and diameter dimensions have a grip located in the center of the dimension text. Drag the grip to relocate the dimension around the circle or arc; the text maintains its orientation, but the leader changes its length.

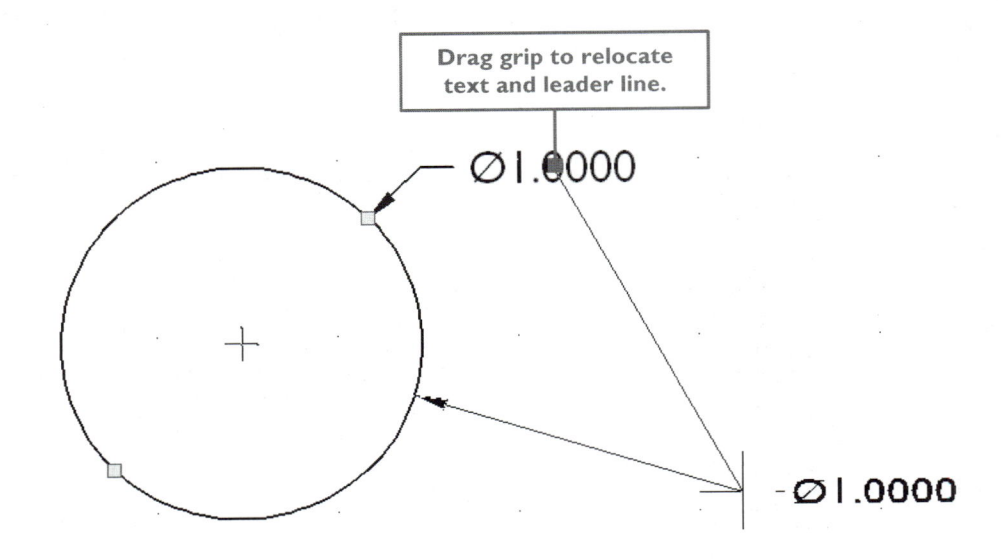

Drag grip to relocate text and leader line.

DIMEDIT AND DIMTEDIT

The **DIMEDIT** and **DIMTEDIT** commands edit extension lines and dimension text.

Extension lines can be *obliqued* with the **DIMEDIT** command, so that they are not longer at right angles to the dimension line. Obliqued (slanted) extension lines are required by isometric drawings, and are useful in difficult dimensioning situations, such dimensioning along shallow curves. Obliquing cannot be applied to leader dimensions, including diameter, radius, and ordinates.

Dimension text can be edited with the mtext editor, as well as rotated, relocated, and justified along the dimension line — left, center, and right. After all the changes, both commands can be used to "home" the text by sending it back to its default location: centered on the dimension line.

One command name is short for "dimension edit," while the other is short for "dimension text edit," yet both edit dimension text. The toolbar icons are unhelpful, **DIMTEDIT** showing a pencil, implying the dimension itself can be edited. The two commands also overlap each other's text editing options; the difference can be confusing to understand. The table below compares and contrasts the abilities of each command; identical operations are shown in **boldface**:

Editing Operation	*DimEdit*	*DimTEdit*
Extension Lines:		
Apply obliquing angle	☑	...
Dimension Text:		
Edit with mtext editor	☑	...
Rotate	☑	☑
Move to default location (home)	☑	☑
Drag text to new location	...	☑
Center text on dimension line	...	☑
Right- and left-justify text on dimline	...	☑

It would make more sense if **DIMTEDIT** handled *all* editing of dimension text, while **DIMEDIT** handled extension lines, as well as other functions ignored by AutoCAD, such as relocating and removing arrowheads. But they don't, and so we carry on.

Here's how to keep them straight:

> **DimEdit** obliques extension lines (ignore its other capabilities).
>
> **DimTEdit** changes the position of dimension text.
>
> **DdEdit** edits dimension text with the mtext editor.

TUTORIAL: OBLIQUING EXTENSION LINES

1. To change the angle of extension lines, start the **DIMEDIT** command:
   - From the menu bar, choose **Dimension**, and then **Oblique**.
   - From the **Dimension** toolbar, choose the **Dimension Edit** button.
   - At the 'Command:' prompt, enter the **dimedit** command.
   - Alternatively, enter the aliases **ded** or **dimed** at the 'Command:' prompt.

   Command: **dimedit** *(Press* ENTER.*)*

2. In all cases, AutoCAD lists the options available. Enter **o** for the oblique option:

   Enter type of dimension editing [Home/New/Rotate/Oblique] <Home>: **o**

3. Select the dimensions whose extensions lines to slant.
   You can enter **all**, and AutoCAD filters out non-dimension objects.
   Select objects: *(Pick one or more dimensions.)*
   Select objects: *(Press* ENTER *to end object selection.)*

4. Specify the angle of slant between 0 and 360 degrees:
   - To slant extension lines for isometric drawings, enter **30**.
     Enter obliquing angle (press ENTER for none): **30**
   - To slant in the other direction, enter negative angles, such as **-30**.
   - To keep the extension lines at their current angle (no change), press **ENTER**.
   - To return obliqued extension lines to perpendicular, enter **0** or **90**.
   - Or, you can pick two points to show AutoCAD the angle.

   AutoCAD obliques the extension lines of selected dimensions.

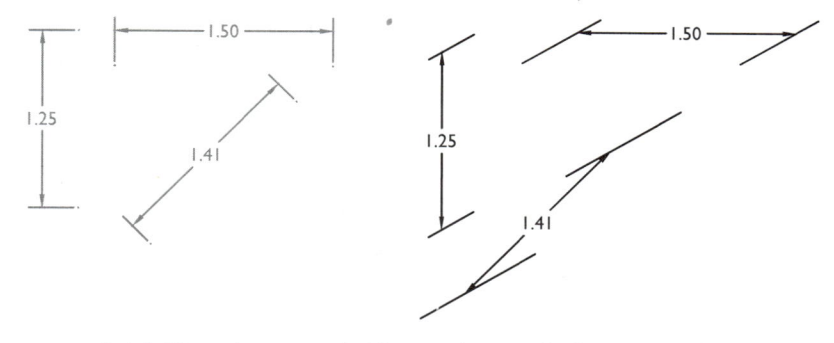

   At left: Dimensions created with normal, perpendicular extension lines.
   At right: Dimensions with extension lines obliqued by 30 degrees.

> **Note:** Be careful with obliquing angles that are very close to the angle of the dimension (1.0 degrees for horizontal dimensions, 89.0 degrees for vertical dimensions, and so on). They stretch the dimension into an unrecognizable jagged line. You can always use the **U** command to reverse undesirable obliquing.

TUTORIAL: REPOSITIONING DIMENSION TEXT

1. To change the position of dimension text, start the **DIMTEDIT** command:
   - From the menu bar, choose **Dimension**, and then **Align Text**.
   - From the **Dimension** toolbar, choose the **Dimension Text Edit** button.
   - At the 'Command:' prompt, enter the **dimtedit** command.
   - Alternatively, enter the **dimted** alias at the 'Command:' prompt.

   Command: **dimtedit** *(Press ENTER.)*

2. AutoCAD prompts you to select one dimension:

   Select dimension: *(Select one dimension.)*

3. Enter an option to relocate the text:

   Specify new location for dimension text or [Left/Right/Center/Home/Angle]: *(Move cursor, or enter an option.)*

   - **Specify new location** dynamically relocates the text along the dimension line, and at the same time moves the dimension line perpendicular to the extension lines.

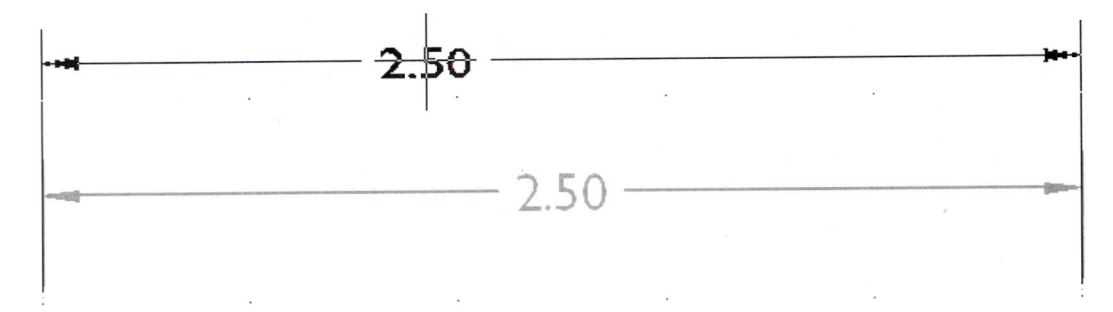

   - **Left** moves the text to the left end of the dimension line.
   - **Right** moves the text to the right end of the dimension line.
   - **Center** centers the text on the dimension lines.
   - **Home** returns the text to its original position.
   - **Angle** rotates text about its center point. AutoCAD prompts you:

   Specify angle for dimension text: *(Enter an angle, such as **30**.)*

   Angled text is used with isometric dimensions.

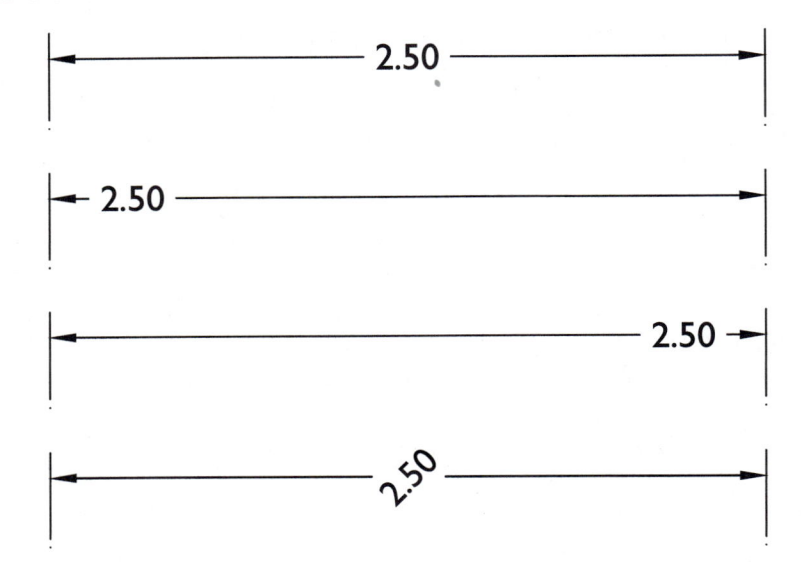

   Top: Dimension text in home and center positions.
   Below: Text in left, right, and rotated positions.

TUTORIAL: EDITING DIMENSION TEXT

1. To edit dimension text, start the **DDEDIT** command:
   - From the **Modify** menu, choose **Object**, then **Text**, and then **Edit**.
   - From the **Text** toolbar, choose the **Edit Text** button.
   - At the 'Command:' prompt, enter the **ddedit** command.
   - Alternatively, enter the **ed** alias at the 'Command:' prompt.

   Command: **ddedit** *(Press ENTER.)*

2. In all cases, AutoCAD prompts you to select the text to edit.
   Select a single dimension:

   Select an annotation object or [Undo]: *(Select a dimension or leader text.)*

3. AutoCAD opens the mtext editor with the <> symbol, indicating the dimension text.

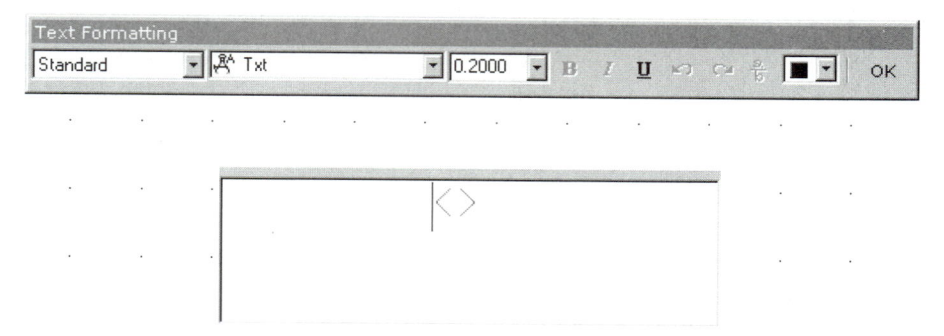

   Recall from the previous chapter that you can edit the <> symbol in several ways:
   - **Erase** to remove the text from the dimension.
   - **Replace** to change the dimension text.
   - **Add** text in front of and behind to create prefixes and suffixes.
   - **Enter** text between the angle brackets to show the brackets with text.

4. When done editing the text, click **OK**.
5. AutoCAD repeats the prompt.
   Select another dimension, or any other text object.
   When done editing, press **ENTER**.

   Select an annotation object or [Undo]: *(Press ENTER to exit command.)*

AIDIM AND AI_DIM

AutoCAD has a group of "hidden" commands for editing dimension text. These commands are meant for use in menu and toolbar macros, but are also handy for making quick fixes to mucked-up dimensions. The commands are shortcuts to options within other commands and dimension variables:

Command	Shortcut For	Comment
AiDimPrec	DIMDEC	Selectively changes precision of fractional and decimal text.
AiDimStyle	DIMSTYLE	Saves and applies up to six dimension styles.
AiDimTextMove	DIMTMOVE	Relocates text with optional leader.
Ai_Dim_TextAbove	DIMTAD	Moves text above dimension line for JIS compliance.
Ai_Dim_TextCenter	DIMTEDIT C	Centers dimension text.
Ai_Dim_TextHome	DIMTEDIT H	Returns text to home position.

The AI- prefix indicates custom commands written by Autodesk Incorporated. Why do some commands start with AI and some with AI_? My guess is that they were written by two different programmers. There is a difference between the two: those that start with AI have multiple options, while those with AI_ perform a single operation.

AIDIMPREC

The **AIDIMPREC** command retroactively and selectively changes the precision of fractional and decimal dimension text and angles of all dimensions, except leaders.

This command is more convenient than its official alternative, changing the **DIMDEC** system variable, followed by the **-DIMSTYLE** command's **Apply** option. Decimal text and angles are rounded to the nearest decimal place, while fractional text is rounded to the nearest fraction.

AiDimPrec	Decimal Units	Fractional Units
0	0	1"
1	0.0	1/2"
2	0.00	1/4"
3	0.000	1/8"
4	0.0000	1/16"
5	0.00000	1/32"
6	0.000000	1/64"

TUTORIAL: CHANGING DISPLAY PRECISION

1. To change the precision of dimension text, start the **AIDIMPREC** command:
    Command: **aidimprec** *(Press ENTER.)*
2. AutoCAD prompts you to specify a precision:
    Enter option [0/1/2/3/4/5/6] <4>: *(Enter the number of decimal places.)*
3. Select one or more dimension objects.
    Enter all to select everything in the drawing; AutoCAD filters out non-dimension objects.
    This command does not change the precision of leader objects.
    Select objects: *(Pick one or more dimensions.)*
    Select objects: *(Press ENTER to end object selection.)*
    AutoCAD changes the display precision of the selected dimensions.

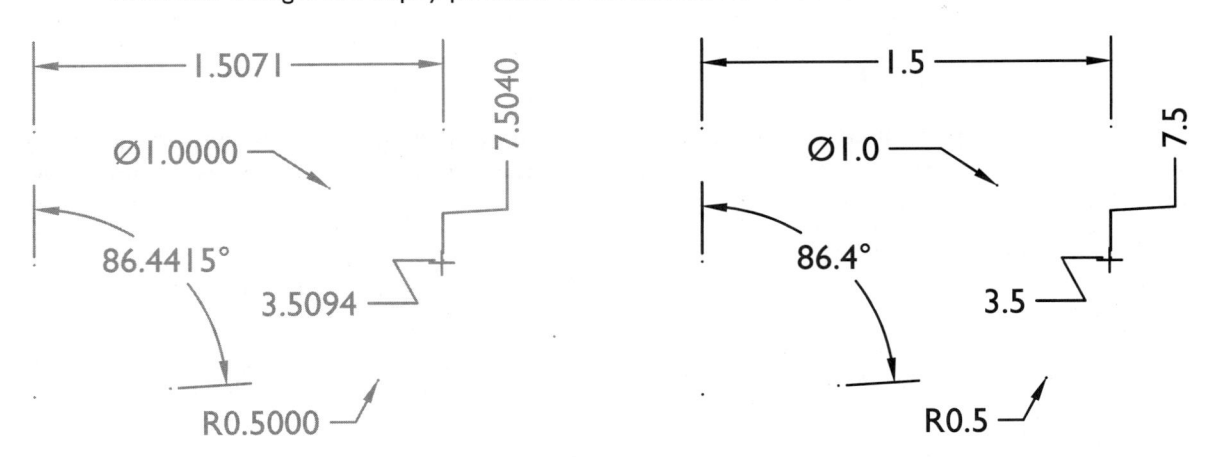

At left: Dimension text displayed to four decimal places.
At right: Precision reduced to one decimal place.

The change is not permanent; AutoCAD remembers the actual measurements and angles — which is why reference is made to "display precision." You can reapply the command to increase and decrease the precision displayed by the dimensions.

Note: Because the **AIDIMPREC** command rounds off dimension text, it can create false measurement. For instance, if a dimension measures 3.4375", setting **AIDIMPREC** to 0 rounds down to 3". Similarly, a measurement of 2.5" rounds up to 3".

AIDIMTEXTMOVE

The AIDIMTEXTMOVE command relocates text, either with or without moving the dimension line, and optionally adds a leader.

This command is more convenient than its official alternative, changing the DIMTMOVE system variable, followed by the DIMOVERRIDE command. AIDIMTEXTMOVE has these options:

AiDimTextMove	DimTMove	Comment
0	0	Moves text within dimension line.
1	1	Adds a leader to the moved text.
2	2	Moves text anywhere without leader line; also moves dimension line (default).

Top: Original dimension with text centered in dimension line.
Below: Text position modified by AIDIMTEXTMOVE = 0, 1, and 2, respectively.

TUTORIAL: MOVING DIMENSION TEXT

1. To move the dimension text, start the **AIDIMTEXTMOVE** command:
   Command: **aidimtextmove** *(Press ENTER.)*
2. AutoCAD prompts you to specify a precision:
   Enter option [0/1/2] <2>: *(Enter an option.)*

3. Select one dimension.

   Although the command allows you to select more than one dimension, it operates on the first-selected dimension only.

   Select objects: *(Pick one dimension.)*

   Select objects: *(Press ENTER to end object selection.)*

4. Move the cursor to relocate the dimension text.

The following trio of commands quickly relocate dimension text to above, centered, and in the dimension line. These commands operate on just one dimension at a time, unfortunately.

This command is more convenient than changing the **DIMTAD** system variable, followed by the **-DIMSTYLE** command's **Apply** option.

Command	DimTAD	Comment
Ai_Dim_TextAbove	3	Moves text above dimension line.
Ai_Dim_TextCenter	0	Centers text vertically.
Ai_Dim_TextHome	2	Centers text horizontally and vertically.

Ai_Dim_TextAbove

The **AI_DIM_TEXTABOVE** command moves text above dimension line, which is helpful for making drawings JIS compliant.

   Command: **ai_dim_textabove**

   Select objects: *(Pick one dimension.)*

   Select objects: *(Press ENTER to end object selection.)*

Ai_Dim_TextCenter

The **AI_DIM_CENTER** command centers text along dimension line. If the text is above or below the dimension line, it keeps that vertical position, but is centered horizontally.

   Command: **ai_dim_textcenter**

   Select objects: *(Pick one dimension.)*

   Select objects: *(Press ENTER to end object selection.)*

Ai_Dim_TextHome

The **AI_DIM_HOME** command centers the dimension text on the side of the dimension line farthest from the defining points.

   Command: **ai_dim_texthome**

   Select objects: *(Pick one dimension.)*

   Select objects: *(Press ENTER to end object selection.)*

DIMENSION VARIABLES AND STYLES

How dimensions look and act depend on the settings of *dimension variables* (called "dimvars" for short). The variables control properties, such as whether dimension text is placed within or above the dimension line, the use of arrowheads or tick marks, the color of the extension lines, and the overall scale. (Technically, dimvars are no different from system variables.)

Dimvar names all start with "dim" to make it easy to identify them. The remainder of the names are very strange, making them hard to decode. Who could guess that **DIMSOXD** means "suppress the drawing of dimension lines outside the extension lines." The "soxd" is shorthand for "suppress outside extension dimension (line)." Another puzzler is **DIMTOFL**, which means "dimension line is drawn between the extension lines even when the text is placed outside." I suppose "tofl" could be short for "text outside forced line."

Other dimvars have names that are easier to decode, such as **DIMSCALE** for dimension scale and **DIMTXT** for the text style.

Like system variables, dimvars can be toggle (turn values on and off), and hold numeric values, or other data. The advantage to dimvars is that they control almost every aspect of a dimension; the problem with dimvars is that there so many of them! In total, there are 70 dimvars, two-thirds of which are dedicated to formatting the dimension text.

After setting the values of dimvars, you can store them in a *style*. Creating dimension styles (or "dimstyle" for short) is similar to creating text styles. In both cases, you specify options, and then save them by name. New AutoCAD drawings based on the *acad.dwt* template drawing have a text style named "Standard," and a dimension style called "Standard." (The similarly of the names is coincidence; they are two different things.)

CONTROLLING DIMVARS

AutoCAD's dimension variables are set through two methods: using the Dimension Style Manager dialog box, or by entering dimvar names at the command prompt. For example, to set the scale for dimensions, it can be faster to enter the **DIMVAR** at the command prompt than to hunt it down in the multi-tabbed dialog box:

Command: **dimscale**

New value for DIMSCALE <1.0>: *(Enter a scale factor, such as **2.5**.)*

The scale is not retroactive, however, but applies to dimensions drawn from now on. To change the scale of existing dimensions in the drawing, use the **DIMSCALE** dialog box's **Override** option.

Two dimvars are not set by the dialog box: **DIMSHO** (toggles whether dimensions are updated while dragged) and **DIMASO** (determines whether dimensions are drawn as associative or non-associative). Both have a value of 1, which means they are turned on. You can turn them off, but there isn't any good reason to do so. Historically, both dimvars exist only for compatibility with old versions of AutoCAD. When computers were slow, generating highlighted images of dragged dimensions slowed the computer even more; this is no longer an issue.

SOURCES OF DIMSTYLES

When you create a new drawing based on template drawings, it contains a specific dimstyle. AutoCAD includes a number of *.dwt* template drawings that meet standards from Germany, Japan, China, and United States, as well as the international metric standard. For example, the *ISO A0 title block.dwt* template contains a dimstyle named "ISO" that conforms to the dimensioning standards of the International Organization for Standardization.

Template drawings contain the dimension styles, shown in the following table:

Template File	Dimstyle Name	Comment
acad.dwt	Standard	Default setting for dimension variables.
ansi*.dwt	Standard	American National Standards Institute.
din*.dwt	DIN	German industrial standard.
gb_*.dwt	GB-5	Chinese standard.
iso*.dwt	ISO-25	International Organization for Standardization.
jis*.dwt	JIS	Japanese Industrial Standard.

Differences Between International Standards

The table below lists the differences between the Standard and international dimstyles. Differences are highlighted in **boldface**.

Variable	Description	Standard	DIN	GB-5	JIS	ISO-25
DimAltD	Alternate precision	2	2	**4**	2	**3**
DimAltF	Alt scale factor	25.4	**0.04**	**0.04**	**0.04**	**0.04**
DimAltTD	Alt tolerance precision	2	2	**4**	2	**3**
DimAltU	Alternate units	2	**8**	**8**	2	2
DimASZ	Arrow size	0.18	**2.5**	**5**	**2.5**	**2.5**
DimCen	Center mark size	0.09	**2.5**	**5**	**0**	**2.5**
DimDec	Precision	4	4	**2**	4	**2**
DimDLI	Dim line spacing	0.38	**3.75**	**7.5**	**7**	**3.75**
DimDSsp	Decimal separator	.	.	.	.	**,**
DimExE	Ext line extension	0.18	**1.25**	**2.5**	**1**	**1.25**
DimExO	Extension line offset	0.06	**0.63**	**1**	**1**	**0.63**
DimGap	Text offset	0.09	**0.63**	**1.25**	**0**	**0.63**
DimLUnit	Length units	2	**6**	2	2	2
DimSOXD	No dim lines outside	Off	Off	Off	**On**	Off
DimTAD	Text position vertical	0	**1**	**1**	**1**	**1**
DimTDec	Tolerance precision	4	4	**2**	4	**2**
DimTFac	Tol text scale factor	1	1	**0.7**	1	1
DimTIH	Text inside align	On	On	**Off**	**Off**	**Off**
DimTIX	Text inside	Off	Off	**On**	**On**	Off
DimTOFL	Dim line forced	Off	**On**	**On**	**On**	**On**
DimTOH	Text outside align	On	**Off**	**Off**	**Off**	**Off**
DimTOLJ	Tolerance pos vert	1	1	**0**	1	**0**
DimTVP	Text vert position	0	**1**	0	0	0
DimTxt	Text height	0.18	**2.5**	**5**	**2.5**	**2.5**
DimTZIN	Tol zero suppression	0	0	0	0	**8**
DimZIN	Zero suppression	0	**8**	**8**	**8**	**8**

Comparing Dimvars

The **DIMSTYLE** command's **Compare** button opens a dialog box that lets you compare dimension variables of two dimension styles. The **Copy** button copies the data to the Clipboard in tab-delimited ASCII format. You can then paste the data into a spreadsheet or other document (which is how the table above was constructed).

Dimstyles from Other Sources

You can use the DesignCenter to copy dimension styles from other drawings. In the tree view of DesignCenter, open the **Dimstyles** item of the drawing from which you wish to copy the dimstyles. Right-click the dimstyle name, and then select Add Dimstyle(s) from the cursor menu.

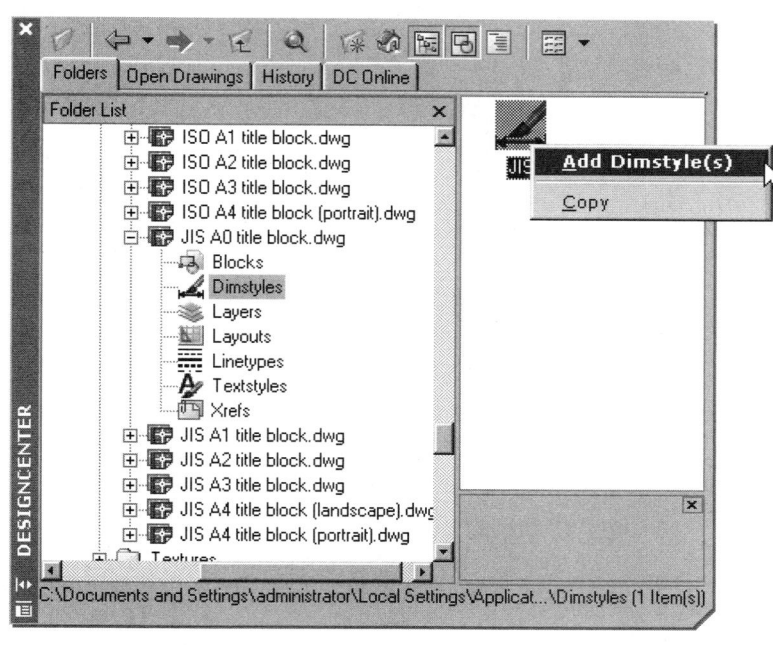

To check that the dimstyle is added to the drawing, click the Dimstyle Control list box on the Styles toolbar.

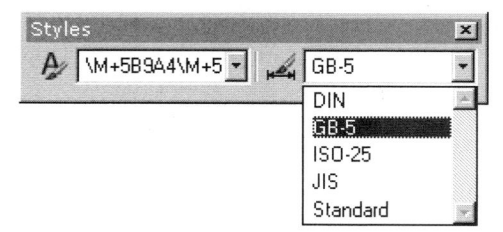

AutoCAD 2004 has several additional methods of sharing dimension standards between drawings. These include:

Express Tools includes the **DIMEX** and **DIMIM** commands for exporting and importing files that contain dimension styles. From the **Express** menu, select **Dimension**, and then **DimStyle Export** or **DimStyle Import**. AutoCAD displays the dialog boxes shown below.

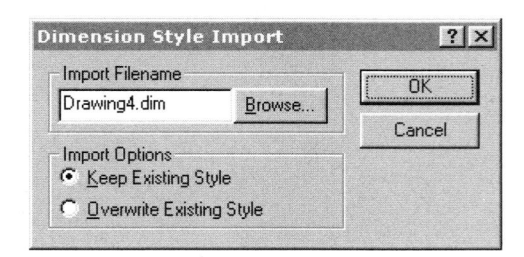

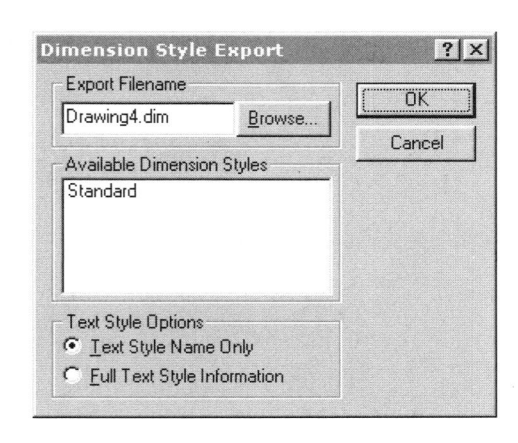

When working with externally-referenced drawings, you can use the **XBIND** command to *bind* (add in) dimension styles found in attached drawings.

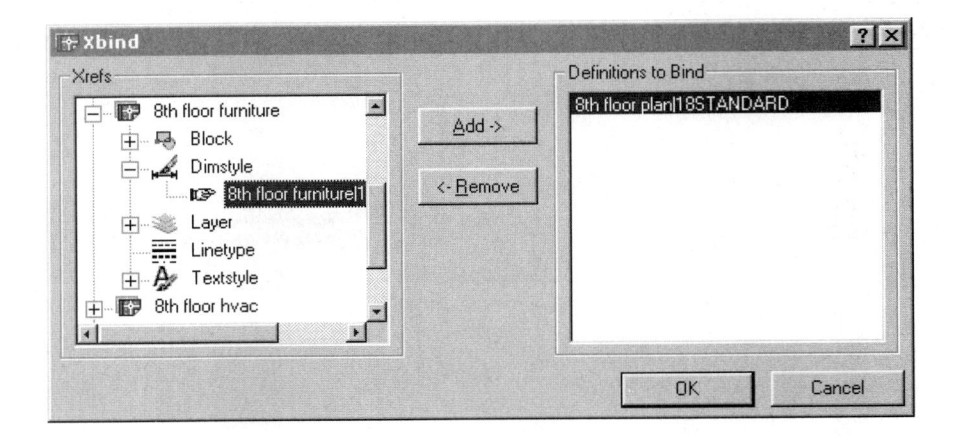

SUMMARY OF DIMENSION VARIABLES

General

DimAso toggles dimensions between associative and non-associative.

DimAssoc determines how dimensions are created: exploded, attached to defpoints, or attached to objects.

DimScale specifies the overall scale factor for dimensions.

DimSho updates dimensions while dragging.

DimStyle specifies the current dimension style as set by the **DIMSTYLE** command (note that the system variable and the command share the same name).

Dimension Line

Extension and Offset

DimDLE specifies how far the dimension line extends past the extension lines.

DimDLI specifies the offset for baseline dimensions.

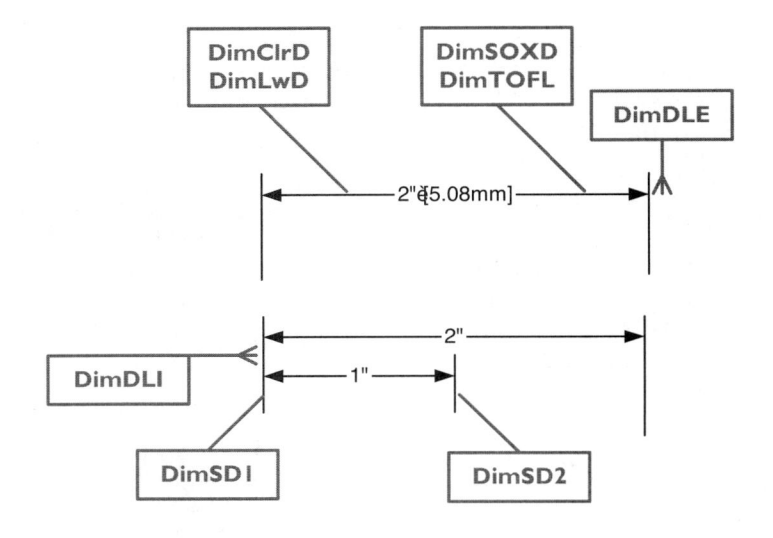

Suppression and Position

DimSD1 suppresses the first dimension line.

DimSD2 suppresses the second dimension line.

DimSOXD suppresses dimension lines outside extension lines.

DimTOFL draws dimension lines inside extension lines.

Color and Lineweight

DimClrD specifies the color of dimension lines.

DimLwD specifies the lineweight for dimension lines.

Extension Lines

Color and Lineweight

DimClrE specifies the color of extension lines.

DimLwE specifies the lineweight of extension lines.

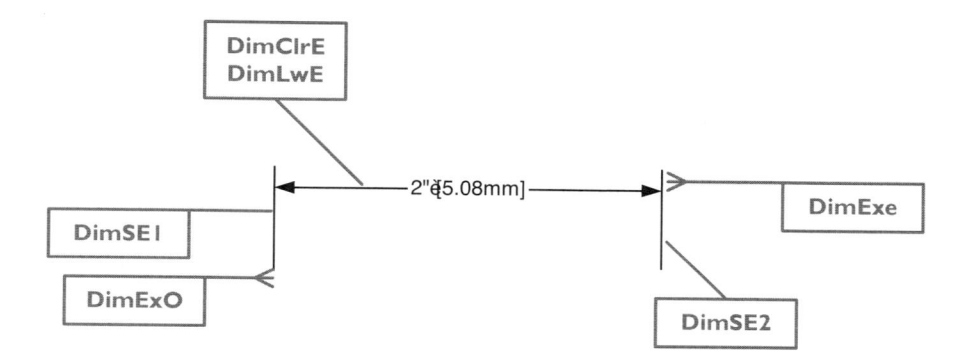

Extension and Offset

DimExe extends the extension lines above the dimension line.

DimExO offsets the extension line from the origin.

Suppression and Position

DimSE1 suppresses the first extension line.

DimSE2 suppresses second extension line.

Arrowheads

Size and Fit

DimASz determines the length of arrowhead blocks.

DimTSz specifies the size of tick strokes.

DimAtFit determines how text and arrowheads are fitted between extension lines.

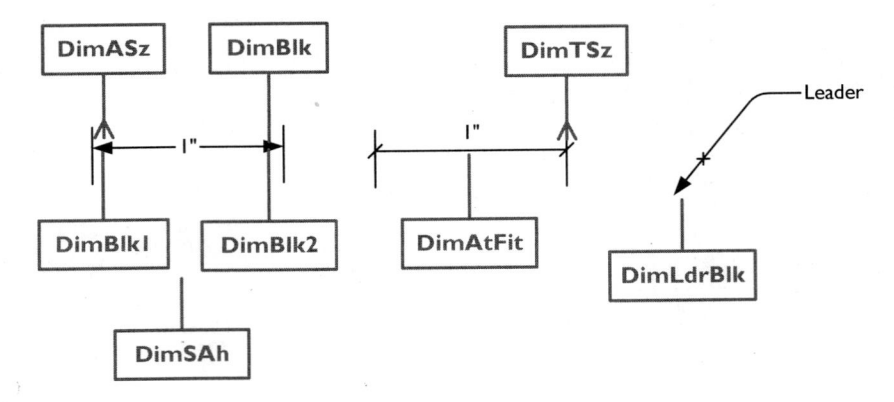

Names of Blocks

DimBlk names the arrowhead block.

DimBlk1 names the first arrowhead block.

DimBlk2 names the second arrowhead block.

DimLdrBlk names the leader arrowhead.

DimSAh determines whether separate arrowhead blocks are used.

Center Marks

DimCen specifies the mark size and line.

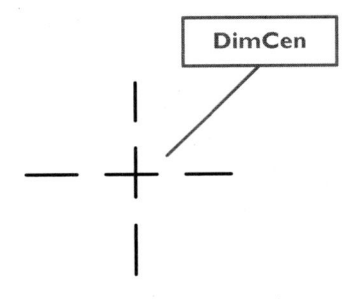

Text

Color and Format

DimClrT specifies the color of the text.

DimGap determines the gap between the dimension line and text.

DimTxSty specifies the text style.

DimTxt stores the text height.

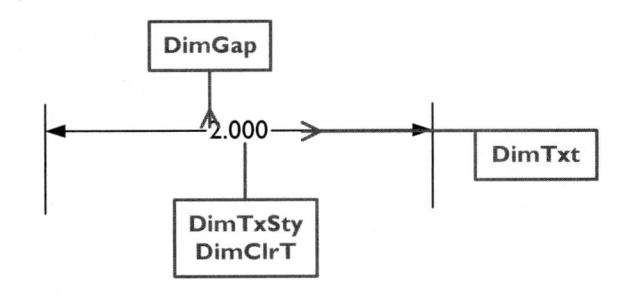

Units, Scale, and Precision

DimLUnit specifies the format of linear units.

DimPost specifies prefixes and suffixes of dimension text.

DimLFac is the linear unit scale factor.

DimDSep specifies the decimal separator.

DimRnd rounds distances.

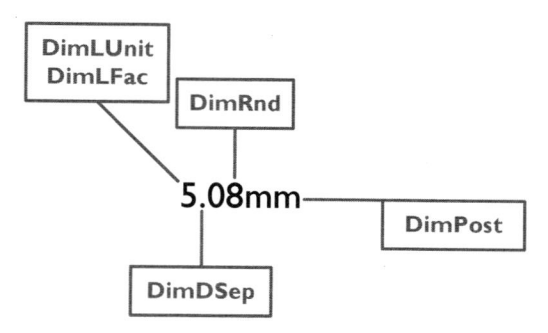

DimFrac determines how fractions are stacked.

DimZIN suppresses zeroes in feet-inches units.

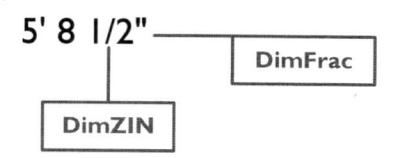

DimADec specifies teh precision of angular dimensions.

DimAUnit specifies the format for angular dimensions.

DimAZin controls how zeros are suppressed in angular dimensions.

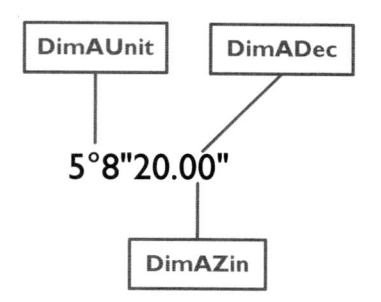

Justification

DimJust justifies horizontal text.

DimTAD vertically positions dimension text.

DimTVP specifies the vertical text position.

DimTIH determines whether text inside extensions is horizontal.

DimTIX forces text between extension lines.

DimTMove determines how dimension text is relocated.

DimTOH determines whether text outside extension lines is horizontal.

DimUPT toggles whether user positions dimension line and/or text.

Alternate Text

DimAlt toggles alternate units.

DimAltU determines the format of alternate units.

DimAltD specifies decimal places of alternate units.

DimAltF specifies the scale factor of alternate units.

DimAltRnd rounds off alternate units.

DimAltZ suppresses zeros in alternate units.

DimAPost determines the prefixes and suffixes for alternate text.

Limits

DimLim toggles dimension limits.

Tolerance Text

DimTol toggles whether tolerances are drawn.

DimTZin suppresses zeros in tolerances.

DimTFac scales the tolerance text height.

DimTolJ justifies tolerance text vertically.

DimTP specifies the plus tolerance value.

DimTM specifies the minus tolerance value.

Primary Tolerance

DimDec specifies the decimal places for the primary tolerance.

DimTDec specifies the decimal places for primary tolerance units.

Alternate Tolerance

DimAltTD specifies decimal places for tolerance alternate units.

DimAltTZ suppresses zeros in alternate tolerance units.

ALPHABETICAL LISTING OF DIMSTYLES

The table below summarizes AutoCAD's 70 dimension variables, their default value in the *acad.dwt* template, and their optional values.

DimVar	Default	Settings and Options
A		
DimADec	0	Angular dimension precision:
		-1 Use DimDec setting (default).
		0 Zero decimal places (minimum).
		8 Eight decimal places (maximum).
DimAlt	Off	Alternate units:
		On Enabled.
		Off Disabled.
DimAltD	2	Alternate unit decimal places.
DimAltF	25.4000	Alternate unit scale factor.
DimAltRnd	0.0000	Rounding factor of alternate units.
DimAltTD	2	Tolerance alternate unit decimal places.
DimAltTZ	0	Alternate tolerance units zeros:
		0 Zeros not suppressed.
		1 All zeros suppressed.
		2 Include 0 feet, but suppress 0 inches.
		3 Includes 0 inches, but suppress 0 feet.
		4 Suppresses leading zeros.
		8 Suppresses trailing zeros.
DimAltU	2	Alternate units:
		1 Scientific.
		2 Decimal.
		3 Engineering.
		4 Architectural; stacked.
		5 Fractional; stacked.
		6 Architectural.
		7 Fractional.
		8 Windows desktop units setting.
DimAltZ	0	Zero suppression for alternate units:
		0 Suppress 0 ft and 0 in.
		1 Include 0 ft and 0 in.
		2 Include 0 ft; suppress 0 in.
		3 Suppress 0 ft; include 0 in.
		4 Suppress leading 0 in decimal dims.
		8 Suppress trailing 0 in decimal dims.
		12 Suppress leading and trailing zeroes.
DimAPost	" "	Prefix and suffix for alternate text.
DimAso	On	Toggle associative dimensions:
		On Dimensions are associative.
		Off Dimensions are not associative.
DimAssoc	2	Controls creation of dimensions:
		0 Dimension elements are exploded.
		1 Single dimension object, attached to defpoints.
		2 Single dimension object, attached to geometric objects.
DimASz	0.1800	Arrowhead length.

DimVar	Default	Settings and Options

DimAtFit	3	When insufficient space between extension lines, dimension text and arrows are fitted:
		  0 Text and arrows outside extension lines.
		  1 Arrows first outside, then text.
		  2 Text first outside, then arrows.
		  3 Either text or arrows, whichever fits better.

DimAUnit	0	Angular dimension format:
		  0 Decimal degrees.
		  1 Degrees.Minutes.Seconds.
		  2 Grad.
		  3 Radian.
		  4 Surveyor units.

DimAZin	0	Suppress zeros in angular dimensions:
		  0 Display all leading and trailing zeros.
		  1 Suppress 0 in front of decimal.
		  2 Suppress trailing zeros behind decimal.
		  3 Suppress zeros in front and behind the decimal.

B

DimVar	Default	Settings and Options

DimBlk	""	Arrowhead block name:
		  Architectural tick: "Archtick"
		  Box filled: "Boxfilled"
		  Box: "Boxblank"
		  Closed blank: "Closedblank"
		  Closed filled: "" (default)
		  Closed: "Closed"
		  Datum triangle filled: "Datumfilled"
		  Datum triangle: "Datumblank"
		  Dot blanked: "Dotblank"
		  Dot small: "Dotsmall"
		  Dot: "Dot"
		  Integral: "Integral"
		  None: "None"
		  Oblique: "Oblique"
		  Open 30: "Open30"
		  Open: "Open"
		  Origin indication: "Origin"
		  Right-angle: "Open90"

DimBlk1	""	Name of first arrowhead's block; uses same list of names as under DimBlk. . Return to default arrowhead.
DimBlk2	""	Name of second arrowhead's block; uses same list of names as under DimBlk. . Return to default arrowhead.

C

DimVar	Default	Settings and Options

DimCen	0.0900	Center mark size:
		  -n Draws center lines.
		  0 No center mark or lines drawn.
		  +n Draws center marks of length n.

DimVar	Default	Settings and Options
DimClrD	0	Dimension line color:
		0 BYBLOCK (default).
		1 Red.
		...
		255 Dark gray.
		256 BYLAYER.
DimClrE	0	Extension line and leader color.
DimClrT	0	Dimension text color.
D		
DimDec	4	Primary tolerance decimal places.
DimDLE	0.0000	Dimension line extension.
DimDLI	0.3800	Dimension line continuation increment.
DimDSep	"."	Decimal separator (must be a single character).
E		
DimExe	0.1800	Extension above dimension line.
DimExO	0.0625	Extension line origin offset.
F		
DimFrac	0	Fraction format when DimLUnit is set to 4 or 5:
		0 Horizontal.
		1 Diagonal.
		2 Not stacked.
G		
DimGap	0.0900	Gap from dimension line to text.
J		
DimJust	0	Horizontal text positioning:
		0 Center justify.
		1 Next to first extension line.
		2 Next to second extension line.
		3 Above first extension line.
		4 Above second extension line.
L		
DimLdrBlk	" "	Block name for leader arrowhead; uses same name as DimBlock.
		. Return to default.
DimLFac	1.0000	Linear unit scale factor.
DimLim	Off	Generate dimension limits.
DimLUnit	2	Dimension units (except angular); replaces DimUnit:
		1 Scientific.
		2 Decimal.
		3 Engineering.
		4 Architectural.
		5 Fractional.
		6 Windows desktop.
DimLwD	-2	Dimension line lineweight; valid values are BYLAYER, BYBLOCK, or an integer multiple of 0.01 mm.

DimVar	Default	Settings and Options
DimLwE	-2	Extension lineweight; valid values are BYLAYER, BYBLOCK, or an integer multiple of 0.01 mm.

P

DimPost	""	Default prefix or suffix for dimension text (maximum 13 characters): " " No suffix. <>mm Millimeter suffix. <>Å Angstrom suffix.

R

DimRnd	0.0000	Rounding value for dimension distances.

S

DimSAh	Off	Separate arrowhead blocks: Off Use arrowhead defined by DimBlk. On Use arrowheads defined by DimBlk1 and DimBlk2.
DimScale	1.0000	Overall scale factor for dimensions: 0 Value is computed from the scale between current model space viewport and paperspace. >0 Scales text and arrowheads.
DimSD1	Off	Suppress first dimension line: On First dimension line is suppressed. Off Not suppressed.
DimSD2	Off	Suppress second dimension line: On Second dimension line is suppressed. Off Not suppressed.
DimSE1	Off	Suppress the first extension line: On First extension line is suppressed. Off Not suppressed.
DimSE2	Off	Suppress the second extension line: On Second extension line is suppressed. Off Not suppressed.
DimSho	On	Update dimensions while dragging: On Dimensions are updated during drag. Off Dimensions are updated after drag.
DimSOXD	Off	Suppress dimension lines outside extension lines: On Dimension lines not drawn outside extension lines. Off Are drawn outside extension lines.
DimStyle	"STANDARD"	Name of the current dimension style.

T

DimTAD	0	Vertical position of dimension text: 0 Centered between extension lines. 1 Above dimension line, except when dimension line not horizontal and DimTIH = 1. 2 On side of dimension line farthest from the defining points. 3 Conforms to JIS.
DimTDec	4	Primary tolerance decimal places.
DimTFac	1.0000	Tolerance text height scaling factor.

DimVar	Default	Settings and Options

DimTIH	On	Text inside extensions is horizontal:
		    Off Text aligned with dimension line.
		    On Text is horizontal.

DimTIX	Off	Place text inside extensions:
		    Off Pace text inside extension lines, if room.
		    On Force text between the extension lines.

DimTM	0.0000	Minus tolerance.

DimTMove	0	Determines how dimension text is moved:
		    0 Dimension line moves with text.
		    1 Adds a leader when text is moved.
		    2 Text moves anywhere; no leader.

DimTOFL	Off	Force line inside extension lines:
		    Off Dimension lines not drawn when arrowheads are outside.
		    On Dimension lines drawn, even when arrowheads are outside.

DimTOH	On	Text outside extension lines:
		    Off Text aligned with dimension line.
		    On Text is horizontal.

DimTol	Off	Generate dimension tolerances:
		    Off Tolerances not drawn.
		    On Tolerances are drawn.

DimTolJ	1	Tolerance vertical justification:
		    0 Bottom.
		    1 Middle.
		    2 Top.

DimTP	0.0000	Plus tolerance.

DimTSz	0.0000	Size of oblique tick strokes:
		    0 Arrowheads.
		    >0 Oblique strokes.

DimTVP	0.0000	Text vertical position when DimTAD=0:
		    1 Turns DimTAD on.
		    >-0.7 or <0.7 Dimension line is split for text.

DimTxSty	"STANDARD"	Dimension text style.
DimTxt	0.1800	Text height.

DimTZin	0	Tolerance zero suppression:
		    0 Suppress 0 ft and 0 in.
		    1 Include 0 ft and 0 in.
		    2 Include 0 ft; suppress 0 in.
		    3 Suppress 0 ft; include 0 in.
		    4 Suppress leading 0 in decimal dim.
		    8 Suppress trailing 0 in decimal dim.
		    12 Suppress leading and trailing zeroes.

U

DimVar	Default	Settings and Options

DimUPT	Off	User-positioned text:
		    Off Cursor positions dimension line.
		    On Cursor also positions text.

DimVar	Default	Settings and Options
Z		
DimZIN	0	Suppression of 0 in feet-inches units:
		0 Suppress 0 ft and 0 in.
		1 Include 0 ft and 0 in.
		2 Include 0 ft; suppress 0 in.
		3 Suppress 0 ft; include 0 in.
		4 Suppress leading 0 in decimal dim.
		8 Suppress trailing 0 in decimal dim.
		12 Suppress leading and trailing zeroes.

Two dimvars are considered obsolete by Autodesk, but remain in AutoCAD for compatibility reasons:

Obsolete DimVar	Default	Comments
DimFit	3	Use DimATfit and DimTMove instead.
DimUnit	2	Replaced by DimLUnit and DimFrac.

DIMSTYLE

The **DIMSTYLE** command displays a dialog box for creating and modifying named dimension styles.

The Dimension Style Manager consists of an initial dialog box, which leads to several others. They control almost all dimension variables, which in turn affect the look of dimensions in your drawings. You can create many dimstyles, each with its own name. This means you can create many customized sets of dimensions — one for every drawing, if need be.

The dialog box is called a "manager," because it lets you manage dimension styles in the drawing. You can create new styles, override and modify existing styles, compare the differences between two styles, and delete styles.

A preview window provides a graphical overview of the current dimstyle.

Every new drawing contains at least one dimension style, named "STANDARD" in Imperial drawings, "ISO-25" in metric drawings, and so on. For this reason, you typically see a single dimstyle name the first time you use the **DIMSTYLE** command.

There are several ways to obtain additional dimension styles. You can get dimstyles by creating your own, opening a template drawing, loading them with the **DIMIM** command, using the **XBIND** command on an externally-referenced drawing, and copying dimstyles through the DesignCenter. You cannot, however, set an xref'ed dimstyle as current.

TUTORIAL: WORKING WITH DIMENSION STYLES

1. To create and change dimension styles, start the **DIMSTYLE** command:
   - From the menu bar, choose **Format**, and then **Dimension Style**.
   - From the **Dimension** or **Styles** toolbars, choose the **Dimension Style** button.
   - At the 'Command:' prompt, enter the **dimstyle** command.
   - Alternatively, enter the aliases **d**, **dst**, **dimsty**, or **ddim** (the command's old name) at the 'Command:' prompt.

   Command: **dimstyle** (Press ENTER.)

In all cases, AutoCAD displays the Dimension Style Manager dialog box.

Setting the Current Dimstyle

To set a dimension style as active, select a dimstyle name from the list under **Styles**, and then choose the **Set Current** button. (The current dimension style name is stored in dimvar **DIMSTYLE**.)

New to AutoCAD 2004 is the Styles toolbar, which provides drop lists of dimension and text styles in the drawing. To make a dimstyle current, select it from the drop list.

Renaming and Deleting Dimstyles

It is not immediately apparent how to rename or delete a dimension style in the Dimension Style Manager. Select a dimension style name in the **Styles** list. Right-click to display the cursor menu, and then choose **Rename** to rename the dimstyle, or choose **Delete** to erase the dimstyle from the drawing. (Styles in use cannot be erased.) The "Standard" dimstyle can be renamed, but not deleted. Xref'ed dimstyles cannot be renamed or deleted.

As alternatives, use the **RENAME** command to rename dimstyles, and the **PURGE** command to remove unused ones.

CREATING NEW DIMSTYLES

The **New** button leads to the Create New Dimension Style dialog box. Creating a new style is as easy as typing a name, and then choosing **OK**. There are, however, some nuances of which you should be aware.

The standard method is to copy an existing style, make changes, and then save the result. For this reason, AutoCAD displays "Copy of Standard" (or the current style name) in the **New Style Name** text field.

The **Start With** text field lets you select which dimstyle to copy — provided the drawing contains two or more dimstyles. This allows you to start with an existing dimstyle (such as the DIM, JIS, Gb, and ISO styles stored in the template drawings) and modify it according to your needs.

The **Use for** drop-down list lets you apply changes to all or a limited group of dimensions:

- All dimensions
- Linear dimensions
- Angular dimensions
- Ordinate dimensions
- Radius dimensions
- Diameter dimensions
- Leaders and tolerance

TUTORIAL: CREATING NEW DIMENSION STYLES

1. To create dimension styles, start the **DIMSTYLE** command:
2. In the Dimension Style Manager dialog box, choose the **New** button.

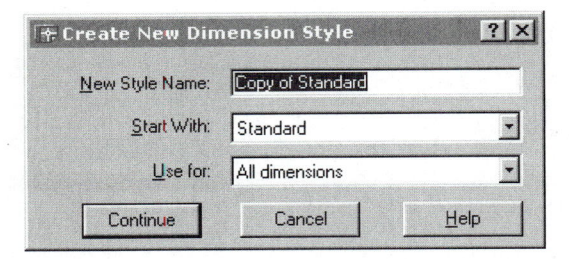

3. From the **Start With** drop list, select an existing dimension style. Most likely, your choice is "Standard," since most drawings contain that one dimension style.
4. Decide whether the changes apply to all dimensions, or to a subgroup. If a subgroup, select it from the **Use for** drop list.
5. In **New Style Name** field, type a name for the style. The name can be any descriptive name up to 255 characters long.
6. Choose **Continue**. Notice that the new style name appears in the Styles list.
7. Now choose the **Modify** button to make the changes to the dimension style.

MODIFYING THE DIMSTYLE

The **Modify** button leads to the Modify Dimension Style dialog box, which contains several tab views listing most of the options that affect dimensions. Here you modify the values to create new dimstyles. The options are discussed in detail later in this chapter.

(The **Override** button leads to the Override Current Dimension Style dialog box, which looks identical to the Modify dialog box, but is meant to *override* the settings of a current dimstyle. The difference between modifying and overriding is discussed in greater detail later in this chapter.)

Xref'ed dimstyles cannot be modified or overridden.

The **Modify** button leads to a group of tabbed dialog boxes that change the settings of dimensions. Each tab contains a preview window that gives you a rough idea of the effect of your changes on the look of the dimensions. Here is an overview of the tabs:

- **Lines and Arrows** specifies properties for dimension lines, extension lines, arrowheads, and center marks.
- **Text** specifies format, placement, and alignment of dimension text.
- **Fit** specifies placement of dimension text, arrowheads, leader lines, and dimension line.
- **Primary Units** specifies format of primary dimension units, as well as sets the prefix and suffix of dimension text.
- **Alternate Units** specifies format of units, angles, dimension, and scale of alternate measurement units.
- **Tolerances** specifies format of dimension text tolerances. (This tab is *not* for formatting Tolerance dimensions.)

LINES AND ARROWS

The **Lines and Arrows** tab specifies the properties for dimension lines, extension lines, arrowheads, and center marks.

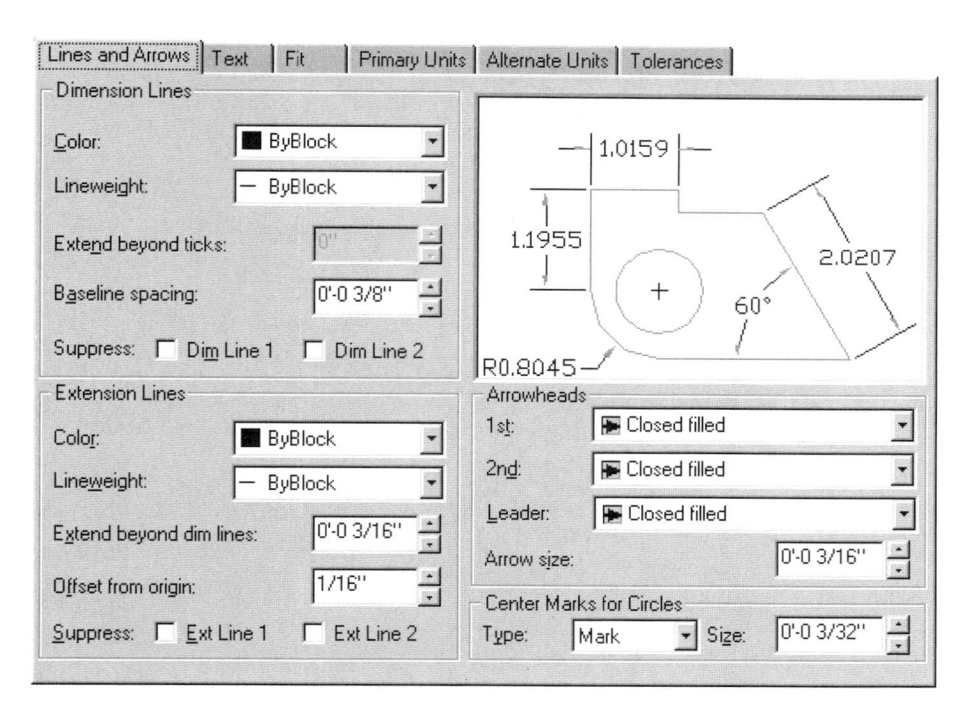

Dimension Lines

The **Color** drop list change the color of dimension lines. The default color is Byblock, because dimensions are considered blocks. "ByBlock" means that dimension lines take on the color of the entire dimension, which is Bylayer (meaning the dimension takes on the color defined by its layer). Although you can choose another color, it is best to place dimensions on layers that specify the color of the dimension. The color is stored in the dimension variable named **DIMCLRD** (short for "DIMension CoLoR Dimension").

To see the effect of changing a variable in the Preview box, select the color blue. Notice that the dimension line color changes from black to blue.

To change the lineweight of the dimension line, choose the drop-down list next to **Lineweight** and select a predefined width. As with colors, it is better to specify the lineweight via the layer, rather than to change the lineweight here. The dimension line lineweight is stored in dimension variable **DIMLWD**.

Extend beyond ticks sets the distance that extension lines extend beyond the dimension line (stored in **DIMDLE**). This option is available only when tick marks are used in place of arrowheads.

Baseline spacing determines the distance between dimension lines when they are automatically stacked by the **DIMBASELINE** and **QDIM** commands. The distance is stored in dimvar **DIMDLI**.

AutoCAD can suppress either or both dimension lines. To do so, select the **Suppress** check boxes to suppress **Dim Line 1** (stored in dimvar **DIMSD1**) or **Dim Line 2** (dimvar **DIMSD2**).

Extension Lines

Color and **Lineweight**: As with dimension lines, you can specify the color (stored in dimvar **DIMCLRE**) and lineweight (**DIMLWE**) of extension lines.

Extend beyond dim lines is the distance the extension line protrudes beyond the dimension line (stored in **DIMEXE**).

Offset from origin sets the distance the extension line begins away from the pick point or the object being dimensioned (**DIMEXO**).

Suppress: You can suppress the first (**DIMSE1**) or second (**DIMSE1**) or both extension lines. (Historically, this was an issue for pen plotters. When multiple extension lines were drawn on top of each other with felt pens, the paper eventually wore through.)

Arrowheads

1st, **2nd**, and **Leader**: Choose from 20 predefined arrowhead styles (stored in **DIMBLK**), or select an arrowhead that you created previously with the **BLOCK** command. In addition to the standard arrowhead, AutoCAD includes arrowheads such as the tick, the dot, open arrowhead, open dot, right-angle arrowhead, or no head at all; see figure at right.

You can specify a different arrowhead at each end of the dimension line (**DIMBLK1** and **DIMBLK2**) and for the end of a leader dimension (**DIMLDRBLK**). When you use a different arrowhead at each end, dimvar **DIMSAH** is set to 1. The User Arrow (customized arrowhead) is defined by the creation of an object scaled to unit size, and then saved as a named block. The name is stored in **DIMBLK**.

Arrow size is the distance from the left to right end; for custom arrowheads, AutoCAD scales the unit block to size (0.18 units, by default). The value is stored in **DIMASZ**.

- Closed filled
- Closed blank
- Closed
- Dot
- Architectural tick
- Oblique
- Open
- Origin indicator
- Origin indicator 2
- Right angle
- Open 30
- Dot small
- Dot blank
- Dot small blank
- Box
- Box filled
- Datum triangle
- Datum triangle filled
- Integral
- None

TUTORIAL: CREATING CUSTOM ARROWHEADS

1. Start AutoCAD with a new drawing.
2. Set the grid to 1.0, or use xlines to demarcate a one square-inch area, because the arrowhead must fit within the one square inch.

3. Draw the arrowhead with the tip pointing West.
   The arrowhead attaches to the dimension line on the East (see figure below).

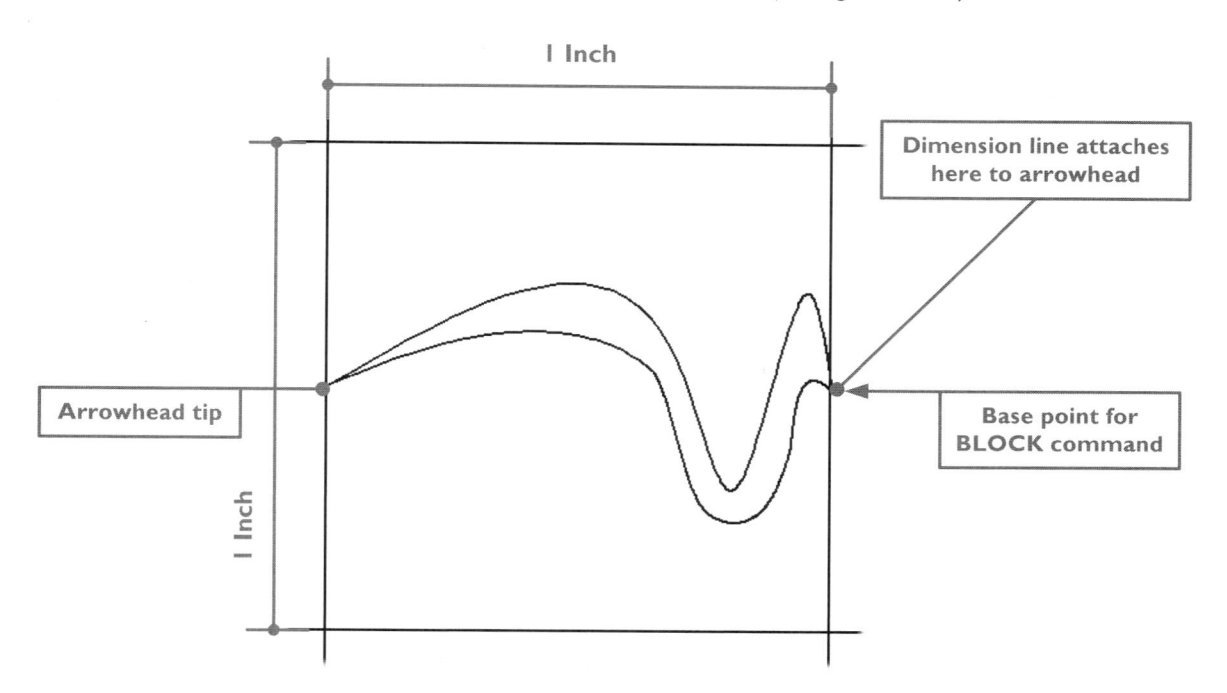

Custom arrowhead designed with the SPLINE and OFFSET commands.

4. Use the **BLOCK** command to turn the arrowhead into a block.
   Give the block a meaningful name, such as "spiral."
   Important! For the **Base point**, pick the East end of the arrowhead, where it attaches to the dimension line, as illustrated above.
5. Start the **DIMSTYLE** command, and then click **Modify**.
   Select the **Lines and Arrows** tab.
   Under Arrowheads, in the **1st** drop list, select **User Arrow**. 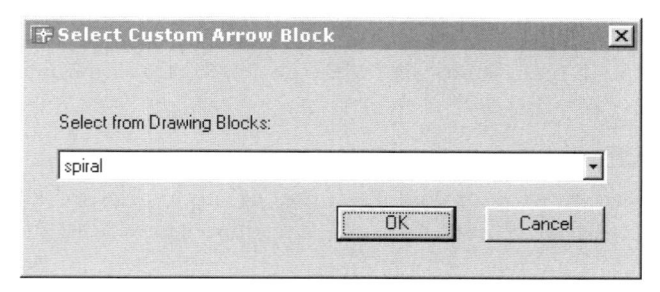
6. The Select Custom Arrow Block dialog box lists all the blocks defined in the drawing.
   Select the name of the block you created earlier, and then click **OK**.

Notice that AutoCAD automatically fills in the **2nd** drop list with the same name, and that the preview image shows the custom arrowheads.

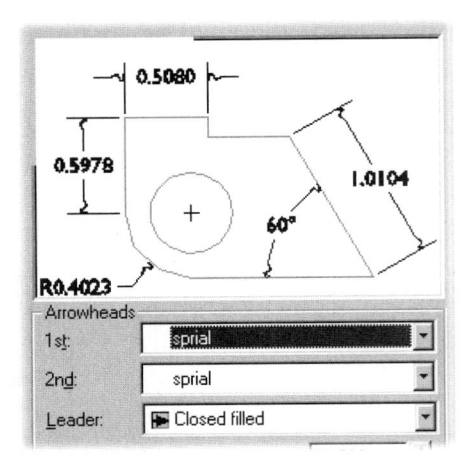

7. Click **OK** to exit the Lines and Arrow tab.
   Choose the **Set Current** button, and then click **Close** to close the dialog box.
8. Try drawing dimensions with your custom arrowhead.

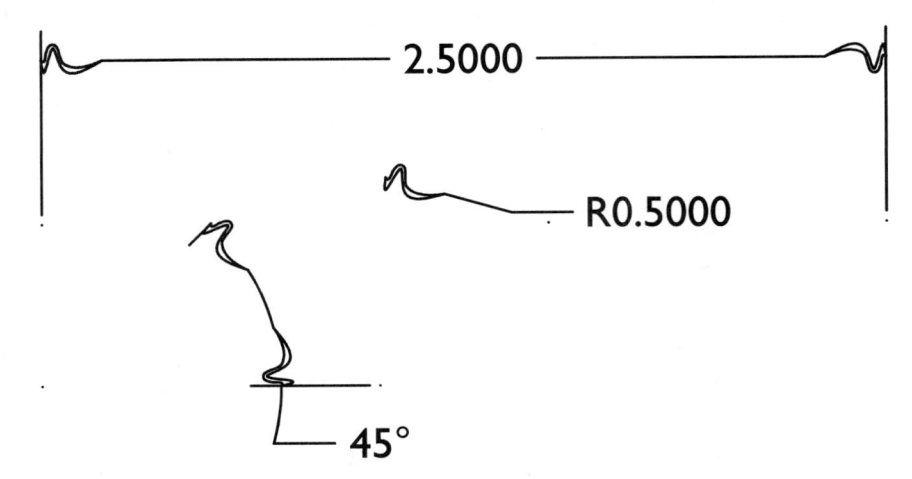

If the arrowhead is too small, increase its size with the DIMSTYLE command's **Arrow size** option.

Center Marks for Circles

AutoCAD lets you mark the centers of circles and arcs with the DIMCENTER command, which places center marks, center marks with extending lines, or no marks at all. The DIMRADIUS and DIMDIAMETER commands also place center marks. The value is stored in DIMCEN.

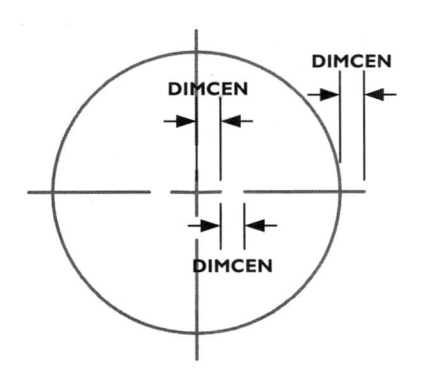

From the **Type** drop list, select **None**, **Mark**, or **Line**.

Size specifies the size of the mark.

TEXT

The **Text** tab specifies the format, placement, and alignment of dimension text.

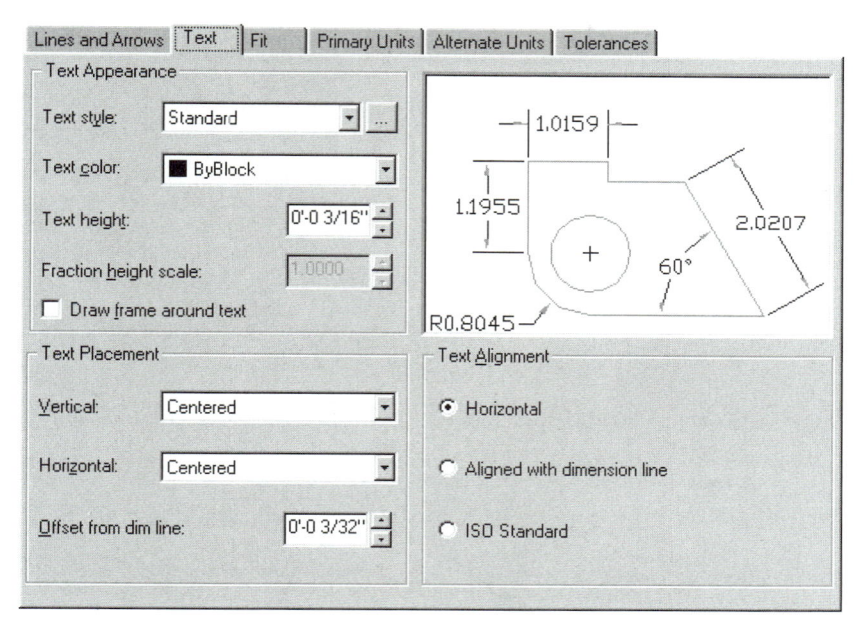

Text Appearance

Text Style selects a style previously defined by the STYLE command, or imported from other drawings. The default is Standard. Choose the ... button to display the **Text Style** dialog box, which lets you create or modify a text style. The text style name is stored in dimvar DIMTXSTY.

Text Color determines the color of the dimension text (DIMCLRT).

Text Height operates like the height in text styles: 0 means use the same height as defined in the text style. A height other than 0 overrides the text style height (DIMTXT).

Fraction Height Scale determines the size of fractional dimension text, as a fraction of regular text (DIMTFAC).

Draw frame around text check box draws a box around the dimension text (DIMGAP < 0). The distance is set further down the dialog box with the **Offset from dim line** option.

Text Placement

Vertical drop list box specifies the vertical placement of dimension text (DIMTAD):

- **Centered** centers dimension text between extension lines and on the dimension line (default).
- **Above** places text above the dimension line.
- **Outside** places text on the side farthest away from the extension line pick points.
- **JIS** places text according to the Japanese Industrial Standards for dimensions, above the dimension line.

Horizontal drop list specifies the horizontal placement of dimension text (DIMJUST).

- **Centered** centers text between the extension lines and on the dimension line (the default).
- **At Ext Line 1** left-justifies text against first extension line.

- **At Ext Line 2** right-justifies text against the second extension line.
- **Over Ext Line 1** positions text vertically over the first extension line.
- **Over Ext Line 2** positions text vertically over the second extension line.

Offset from dim line specifies the gap between dimension the line and text (**DIMGAP**).

Text Alignment

Text Alignment determines the orientation of dimension text when inside and outside the extension lines (**DIMTIH** and **DIMTOH**).

- **Horizontal** forces text to be horizontal, whether or not it fits inside the extension lines.
- **Aligned with dimension line** forces text to align with the dimension line.
- **ISO Standard**:
  - When text fits inside extension lines, text aligns with the dimension line.
  - When text is outside extension lines, text is drawn horizontally.

FIT

The **Fit** tab determines the placement of dimension text, arrowheads, leader lines, and dimension lines.

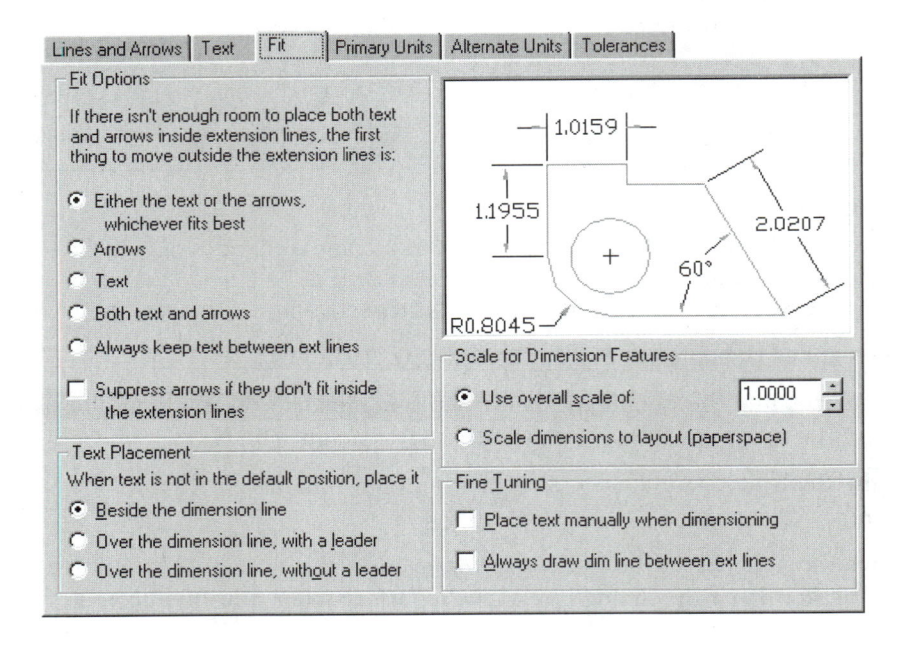

Fit Options

The **Fit Options** section controls where text and arrowheads are placed when the distance between extension lines is too narrow (**DIMATFIT**).

- **Either the text or the arrows, whichever fits best** places the elements where there is room.
- **Arrows** fits only the arrowheads between extension lines, when space is available.
- **Text** fits the dimension text between extension lines, while arrowheads are placed outside — when space is lacking.
- **Both text and arrows** forces both text and arrows between extension lines.
- **Always keep text between ext lines** always keeps text between the extension lines (**DIMTIX**).
- **Suppress arrows if they don't fit inside the extension lines** draws no arrowheads in that case (**DIMSOXD**).

Text Placement

When AutoCAD cannot place the dimension text normally, you have these options (**DIMTMOVE**):

- **Beside the dimension line** places text beside the dimension line.
- **Over the dimension line, with a leader** draws a leader line between the dimension line and the text when there isn't enough room for the text.
- **Over the dimension line, without a leader** draws no leader.

Scale for Dimension Features

Use overall scale of is an important area of the dialog box, because it controls the scale of text and arrowheads when dimensions are placed in the drawing (**DMSCALE**). Entering a value of 2, for example, doubles the size of arrowheads and text; the length of the dimension and extension lines are unaffected. To change the width of dimension and extension lines, use the **Lineweight** option in the Lines and Arrows tab.

Scale dimensions to layout (paperspace) scales dimensions to a factor based on the scale between model and paper space (**DMSCALE = 0**).

Fine Tuning

Place text manually when dimensioning determines whether dimension commands prompt you for the position of the dimension text, allowing you to change the text position, if necessary (**DIMUPT**).

Always draw dim line between ext lines forces AutoCAD always to draw the dimension lines inside the extension lines (**DIMTOFL**). The arrowheads and text are placed outside, if there isn't enough room.

PRIMARY UNITS

The **Primary Units** tab sets the format of primary dimension units, as well as set the prefix and suffix of dimension text.

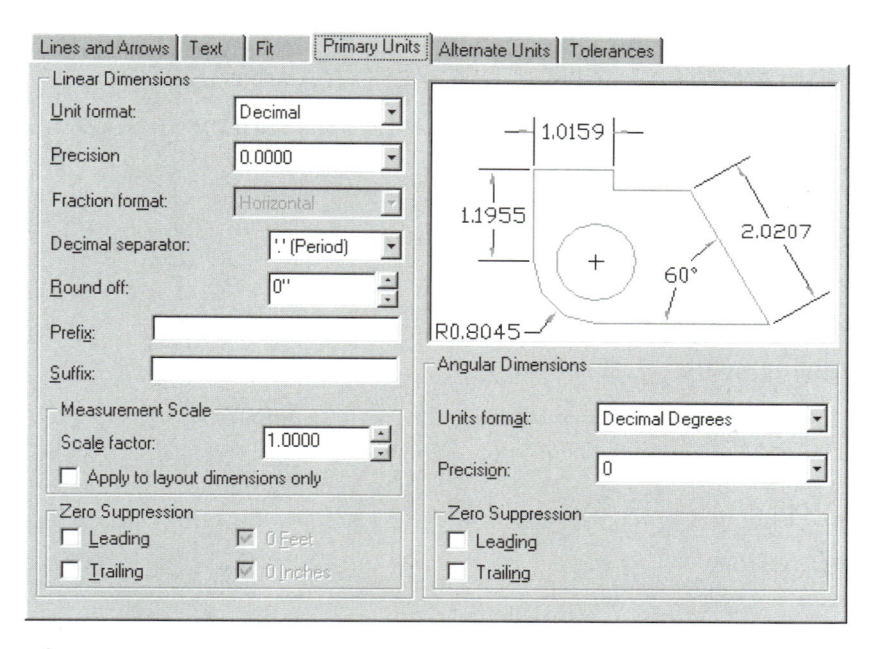

Linear Dimensions

In this section you specify the dimensioning units, and whether the dimension text has a prefix or suffix. This section has features similar to those of the **UNITS** command. Whereas the **UNITS** controls the units used by AutoCAD for everything in the drawing, the Primary Units tab overrides them. The values you select here apply only to the dimension text.

Units format specifies format: decimal (default), architectural, engineering, fractional, and scientific units (**DIMLUIT**).

Precision specifies the number of decimal places to be displayed. The actual measured value does not change.

Fraction format stacks fractions horizontally, diagonally, or not at all (**DIMFRAC**). This option is available only when you select a unit with fractions, such as architectural.

Decimal separator specifies the decimal separator, other than period (.) in decimal units. This setting is typically used in countries outside North America, some of which use the comma (,) to separate units from fractions (**DIMDSEP**). This option is available only when you select a unit with decimals, such as decimal.

Round off specifies how decimals are rounded, such as to the nearest 0.5; note that this setting is different from Precision, which *truncates* the display of decimal places (**DIMRND**). The actual measured value does not change.

Prefix and **Suffix** provide room for any alphanumeric values, which are then added in front of and behind the dimension text. For example, to prefix every dimension with the word Verify, enter that in the Prefix box. To suffix every dimension with "(TYPICAL)," enter that in the Suffix box (**DIMPOST**).

Measurement Scale

Scale Factor sets the scale factor that is multiplied to linear dimensions (**DIMLFAC**). By entering 25.4, for example, imperial dimensions are converted to millimeters (metric units).

Apply to layout dimensions only applies the linear scale factor to dimensions created in layouts (paper space). When turned on, the scale factor is stored as a negative value in **DIMLFAC**.

Zero Suppression

Leading and **Trailing** specifies whether leading and trailing zeros and the zero feet or inches are suppressed (**DIMZIN**).

Angular Dimensions

Units format specifies the format of angular dimensions (**DIMAUNIT**).

Precision specifies the number of decimal places or fractions (**DIMADEC**). Select from:

    Decimal degrees (*DdD.dddd*, the default).

    Degrees/ Minutes/Seconds (*DD.MMSSdd*).

    Grads (*DDg*).

    Radians (*DDr*).

    Surveyor units (*NDDE*).

Zero Suppression suppresses leading and trailing zeros (**DIMAZIN**).

ALTERNATE UNITS

The **Alternate Units** tab selects the format of units, angles, dimensions, and scale of alternate measurement units. The options are the same as for the Primary Units tab.

AutoCAD allows you to place dimensions with double units: a primary unit plus a second or *alternate* units. The alternate units appear in square brackets. This is particularly useful for drawings that must show imperial and metric units.

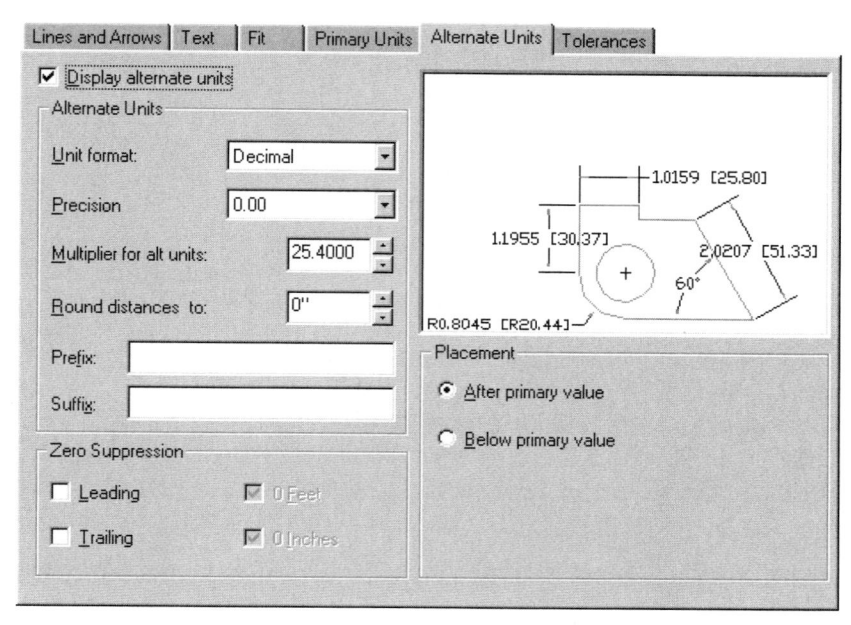

Display alternate units check box turns on alternate units (**DIMALT**); when off, all options are grayed out, meaning they are unavailable.

This tab is similar to the Primary Units tab, with the exception of the Placement option. Options selected in this tab are stored in the following dimvars:

DIMALTD specifies alternate unit decimal places.

DIMALTTD specifies alternate tolerance unit decimal places.

DIMALTRND specifies the rounding of alternate units.

DIMALTTZ specifies the alternate-tolerance-units zero suppression.

DIMALTU specifies the alternate units format, except for angular dimensions.

DIMALTZ specifies the alternate unit zero suppression.

Placement

The placement options determine where the alternate units are placed: **After** or **Below** the primary value (**DIMAPOST**).

TOLERANCES

The **Tolerances** tab controls the format of dimension text tolerances (and is not for formatting dimensions created with the **TOLERANCE** command).

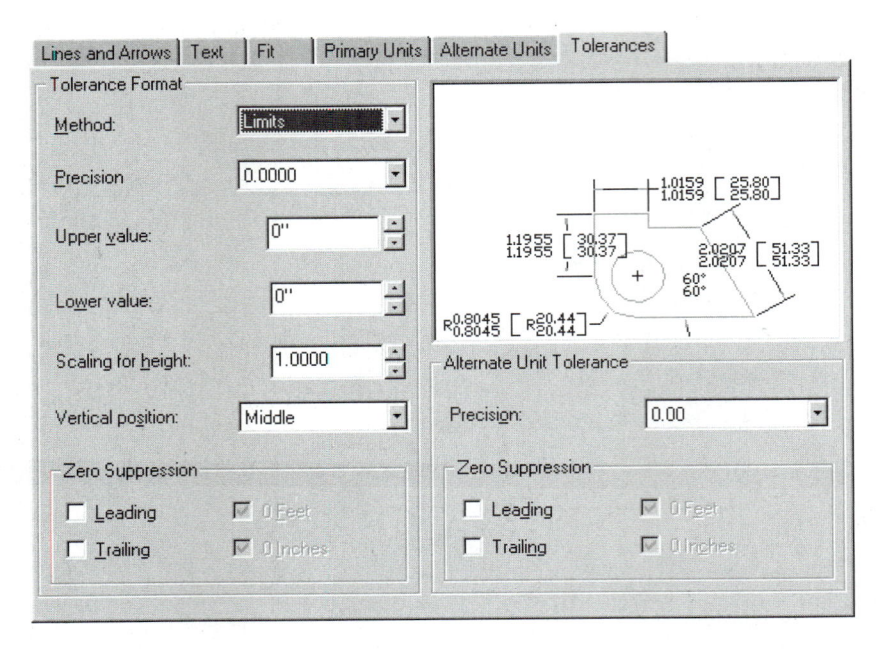

Tolerance Format

The **Method** drop list offers five styles of tolerance text (controlled by DIMTOL and DIMLIM).

- **None** places no tolerance text (default); most options are grayed out in this tab.
- **Symmetrical** adds a single plus/minus tolerance.
- **Deviation** adds a plus and a minus tolerance.
- **Limits** draws two dimensions.
- **Basic** draws a box around the dimension text (distance between text and box is stored in **DIMGAP**).

Dimvar **DIMTOL** appends the tolerances to the dimension text. **DIMLIM** displays dimension text as limits; turning on **DIMLIM** turns off **DIMTOL**.

Precision determines the number of decimal places.

Upper Value and **Lower Value** specify the upper or plus value (DIMTP), and the lower or minus tolerance value (DIMTM).

Scaling for height value determines the size of the tolerance text relative to the main dimension text (DIMTFAC). A value of 1 indicates each character of tolerance (or limits) text is the same size as the primary dimension text. A scale of 0.5 reduces the text by half.

Vertical position determines the relative placement of tolerance text (DIMTOLJ), whether aligned to the top, middle, or bottom of the primary dimension text.

Zero Suppression options are identical to those found in the Primary Units tab.

Alternate Unit Tolerance specifies the number of decimal places for the tolerance values in the alternate units of a dimension (DIMALTTD).

MODIFYING AND OVERRIDING DIMSTYLES

You create a dimension style for your drawings, and then use it to draw dimensions. Sometimes, however, you might realize that you need to change the style. Perhaps the client wants dimension lines thicker than the extension lines, or a particular extension line needs to be removed, or perhaps some dimensions need to be drawn with smaller text.

Some of these changes are *global*, which means they apply to every dimension in the drawing, the thicker dimension lines, for instance. Other changes are *local*, which means they apply to selected dimensions, the extension line to be removed, for example.

Some changes are *retroactive*, which means they apply to all dimensions already drawn. Other changes are *temporary*, which means they apply to dimensions not yet drawn.

Here is how AutoCAD handles these four cases:

To apply *global changes retroactively*, modify the dimension style with the **DIMSTYLE** command's **Modify** button. It modifies the dimension style permanently (well, at least until you use modify it again). Make a change in a dimension style, and AutoCAD retroactively changes all dimensions in the drawing.

To apply *global changes temporarily*, modify the dimension style with the **DIMSTYLE** command's **Override** button, which creates a "sub-style." The next dimension you draw reflects the change in dimension sytle.

To apply a *different dimension style* to selected dimensions, use the -**DIMSTYLE** command's **Apply** option. This command prompts you to select dimensions, and then applies a dimstyle to them — a local modifications.

To *change the dimension variable* setting of selected dimensions, use the **DIMOVERIDE** command. This command asks you for the name of a dimension variable, the new value, and then prompts you to select the dimensions to which the local override is applied.

> **Note**: The difference between *modifying* and *overriding* dimension style:
> **Modify** — modifies retroactively and globally updates all existing dimensions created using that style.
> **Override** — changes apply only to new dimensions created after the override is applied. It applies overrides to the current dimension style only.

In the following tutorial, you learn how to change the style of all dimensions in the drawing at once.

TUTORIAL: GLOBAL RETROACTIVE DIMSTYLE CHANGES

1. Start a new drawing, and draw some dimensions.
2. To change the style of all dimensions in the drawing, start the **DIMSTYLE** command:
3. In the dialog box, click **Modify**.

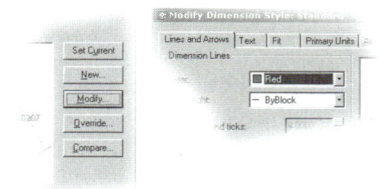

4. In the Lines and Arrows tab, change the dimension line **Color** to red.
   Notice that the preview shows the dimension lines changed to red.
5. Click **OK**, and then click **Close**.
   Notice that all dimensions in the drawing now have red dimension lines.

In the next tutorial, you learn how to override the dimension style temporarily, so that new dimensions take on the changed style; existing dimensions remain unaffected.

TUTORIAL: GLOBAL TEMPORARY DIMSTYLE CHANGES

1. Continue with the drawing from the previous tutorial.
2. Start the **DIMSTYLE** command, and then choose the **Override** button.
3. In the Lines and Arrows tab, change the dimension line color to blue.

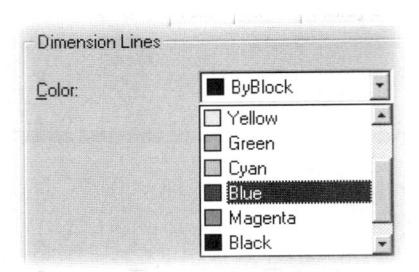

4. Click **OK**.
   Notice that the Standard dimstyle has a sub-style called "<style override>." It is highlighted, meaning that it is the current style. (After exiting this dialog box, any dimension you draw from now on has blue dimension lines.)
5. In the **Description** area, notice the change to the blue dimension lines: "Standard+Dim line color = 5 (blue)".

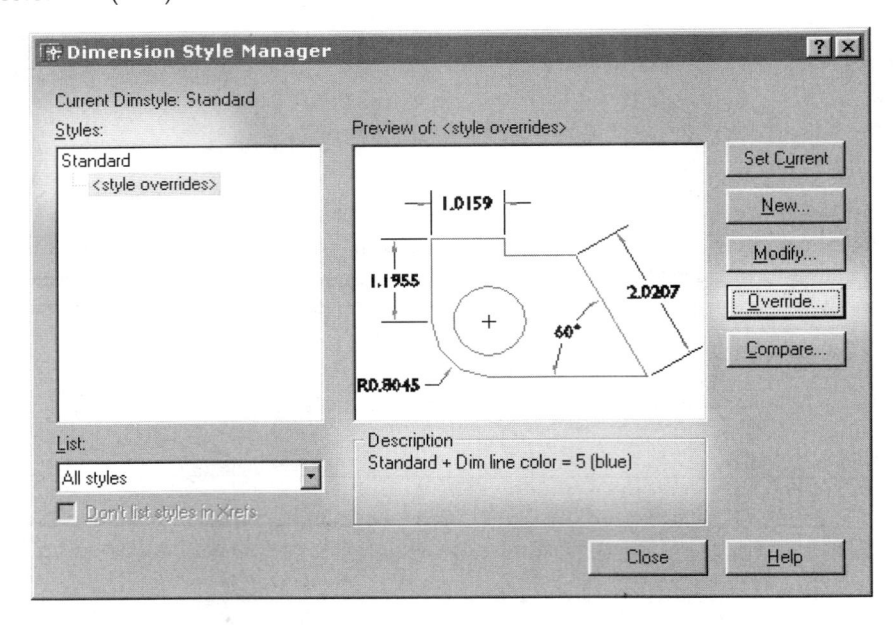

6. Click **Close**.
   Notice that none of the existing dimensions in the drawing change.
7. Draw a linear dimension with the **DIMLINEAR** command.
   Notice that its dimension line is blue.

Overrides are not, unfortunately, listed by the droplist of the **Styles** toolbar.

> **Note:** You can turn overrides into permanent dimension styles. To turn the override into a *new* dimstyle, right-click "<style override>" (an AutoCAD-generated name), and then select **Rename** from the shortcut menu. Change the name to something meaningful, like "Blue Dimlines."
>
> To make the override the current dimstyle, right-click "<style override>", and then select **Save to current style** from the shortcut menu. Notice that the override disappears, and that the Standard style now shows blue dimension lines. Click **Close**, and all dimensions change to blue dimlines. (This is equivalent to having used the **Modify** button in the first place.)

The dimension style dialog box is lacking, however. It cannot apply changed dimension styles to individual dimensions. That responsibility lies with the **Style** toolbar, the DIMOVERIDE commands, and the Properties window.

STYLE TOOLBAR AND DIMOVERRIDE

The **Style** toolbar applies a dimension style to selected dimensions, while the DIMOVERRIDE command applies changed dimension variables to selected dimensions.

TUTORIAL: LOCAL DIMSTYLE CHANGES

For this tutorial, you need a drawing with at least two dimension styles. If necessary, create a second style, as described earlier.

1. To apply a dimstyle to just a few dimensions, select them in the drawing.
   Notice that they are highlighted.
2. From the Styles toolbar, click the awkwardly-named **Dim Style Control** list box.
   Notice the list of dimension style names.

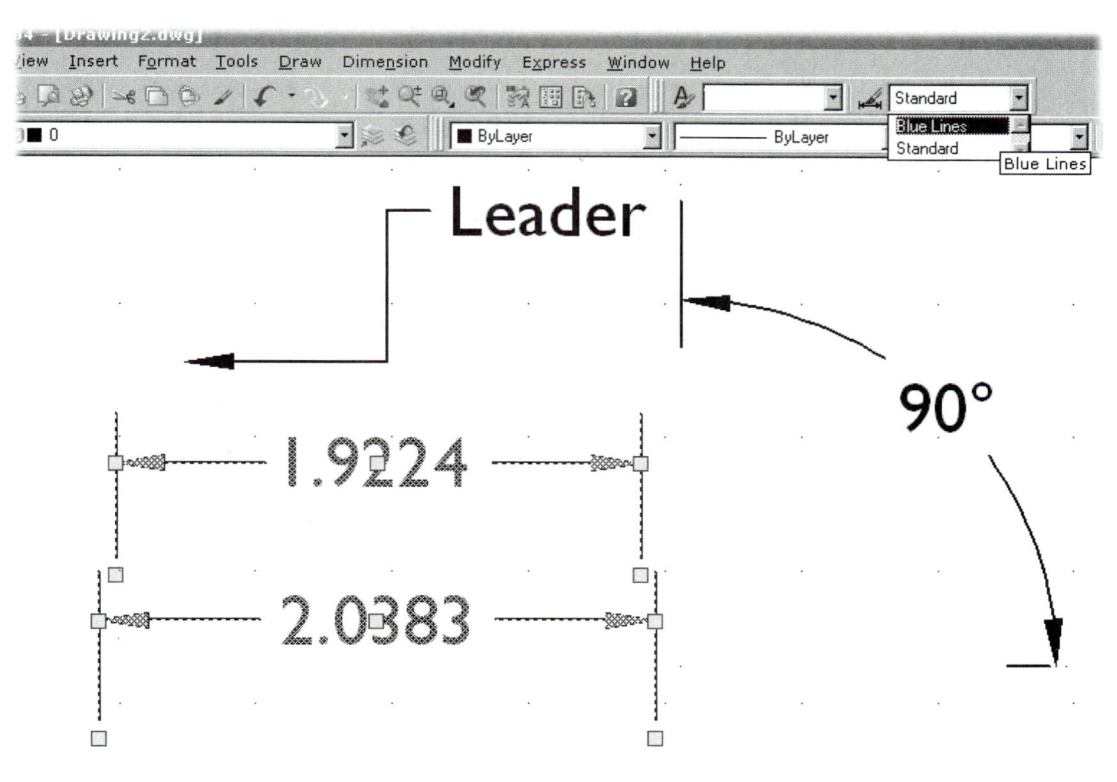

3. Select a different style name.
   Notice that the selected dimensions change their style.

4. Press ESC to remove the highlighting from dimensions.

In the next tutorial, you change the color of one dimension's extension line to yellow.

TUTORIAL: LOCAL DIMVAR OVERRIDES

For this tutorial, you need a drawing with at least two dimensions.
1. To change a dimvar for selected dimensions, start the DIMOVERRIDE command.
   - From the menu bar, choose **Dimension**, and then **Override**.
   - At the 'Command:' prompt, enter the **dimoverride** command.
   - Alternatively, enter the aliases **dov** or **dimover** at the 'Command:' prompt.

   Command: **dimoverride** (Press ENTER.)

2. Enter the name of the dimvar, DIMCLRE, which sets the color of extension lines:
   Enter dimension variable name to override or [Clear overrides]: (Enter the name of a dimension variable, such as **dimclre**.)

3. Enter the new value for the dimvar, yellow:
   Enter new value for dimension variable <BYBLOCK>: **yellow**

4. The prompt repeats so that you can override other dimvars.
   Enter dimension variable name to override: (Press ENTER to continue.)

5. Select the dimensions to change:
   Select objects: (Select a dimension.)
   Select objects: (Press ENTER to end object selection.)

   Notice that the extension lines change to yellow.

Note: You can override dimvars *during* dimensioning commands. In the following example, the dimension line color is changed to green during the DIMALIGNED command:

Command: **dimali**
Specify first extension line origin or <select object>: **dimclrd**
Enter new value for dimension variable <byblock>: **green**
Specify first extension line origin <select object>: (Continue with DIMALIGNED command.)

Dimension lines drawn from now on are green, until you override again.

Clearing Local Overrides

If you change your mind, you can clear the local override you applied in the tutorial above. Start the DIMOVERRIDE command, and enter the **Clear override** option:

Command: **dimoverride**
Enter dimension variable name to override or [Clear overrides]: **c**
Select objects: (Select the dimensions to return to normal.)
Select objects: (Press ENTER to end object selection.)

The dimension returns to its previous look, because the override is cancelled.

EXERCISES

1. In the following exercises, you modify the Standard dimension style.
   Start AutoCAD with a new drawing.
   Draw one each of the following dimensions — linear, aligned, diameter, and leader — similar to the figure below.
   Which commands did you need to draw the dimensions?

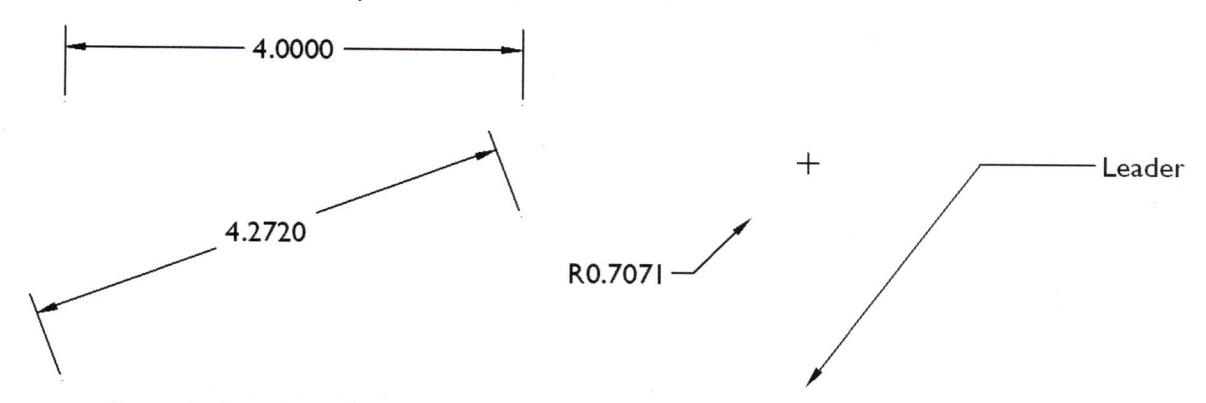

   ```
   |←———— 4.0000 ————→|
   ```

   4.2720

   R0.7071

   Leader

   Open the Dimension Style Manager dialog box.
   Which dimstyle is listed there?
   Modify the dimstyle: change the color of extension lines to red.
   When you exit the dialog box, what happens?

   Return to the Dimension Style Manager dialog box.
   Override the dimstyle: change the color of text to blue.
   When you exit the dialog box, what happens?
   Draw another linear dimension. Is it different from the others?

   Return to the Dimension Style Manager dialog box.
   Convert the override into a dimstyle called "Blue Text."
   When you exit the dialog box, what happens?

   Use the **Styles** toolbar to apply the Blue Text dimstyle to the aligned dimension. Does it change?
   Repeat, applying the dimstyle to the leader. Does it change?

2. Continuing with the drawing from above, create a new dimension style.
   Create the new dimstyle with the following settings:

   Extension line:
       Color   **8**
   Text:
       Font   **Times New Roman**
       Color   **155**
       Vertical Placement   **JIS**
   Arrowheads:
       Both   **Architectural Tick**

   Save style by the name of "JayGray."
   Apply the style to all dimensions in the drawing.
   Save the drawing as *jaygray2.dwg*.

3. Continuing with the drawing from the previous exercise, use grips editing on the different grips of each dimension.
   Do not save the drawing.

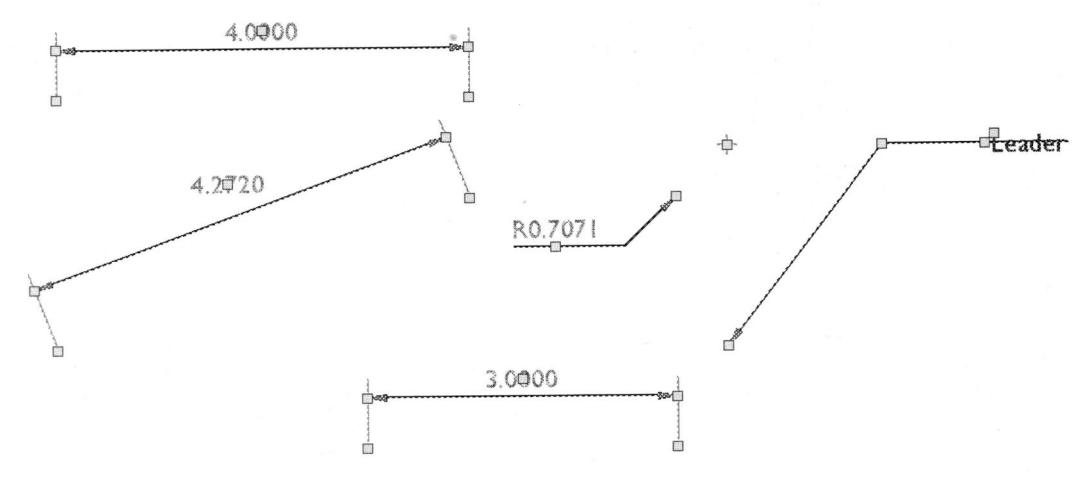

5. Start a new drawing.
   Open the DesignCenter, and then import dimension styles from the following drawings:
       JIS A1 title block.dwg
       DIN A2 title block.dwg
       ISO A3 title block.dwg

6. From AutoCAD 2004's \sample folder, open the *Taisei Detail Plan.dwg* file, a drawing of an office.
   Select a dimension, and edit its text

CHAPTER REVIEW

1. Which command is used to create, edit, and delete dimension styles?
2. What is the difference between modifying and overriding a dimension?
3. Can the color and lineweight of dimension lines be set independently of extension lines?
4. Can the extension lines be suppressed independently of each other?
5. Is it better practice to use dimension styles or layers to set the color and lineweight of dimension lines?
6. Can you define your own arrowhead for use by dimensions?
7. When the distance between extension lines is too tight, where does AutoCAD place the dimension text?
8. Describe some situations where you might need to edit dimensions.
9. What are alternative units?
10. When you move an object, do associative dimensions move also?
11. Describe how the following grips editing commands affect dimensions:
    Stretch
    Copy
    Move
    Rotate
    Scale
    Mirror
12. What happens to the text when a leader is grips edited?
13. Can the extension lines be grips-edited independently of the dimension line?
14. Explain the meaning of the following abbreviations:
    dimvar
    dimstyle
    dimline
15. What is the "T" in **DIMTEDIT** short for?
16. Describe the function of the **DIMEDIT** command's **Oblique** option. When is it useful?
17. How do you edit the wording of the dimension text?
    The properties of dimension text?
    The position of dimension text?
18. Briefly explain the difference among these three similar-sounding dimension-editing commands:
    DDEDIT
    DIMEDIT
    DDIMTEDIT
19. If you make a mistake editing a dimension, which command changes back the dimension?
20. Can dimensions be scaled independently of the drawing?
21. What is the danger in rounding dimension text?
22. When AutoCAD rounds off dimension values, is the change permanent?
23. Round the following numbers to the nearest unit:
    2.5
    1.01
    4.923
    10.00
    6.489

24. Where does the JIS dimension standard expect text to be placed?
25. Describe the purpose of *dimension variables*.
26. Does every drawing have a dimension style?
27. How do you change the scale of dimensions relative to the rest of the drawing?
28. List two sources of dimension styles:

    a.

    b.
29. If you saw the following dimension styles in a drawing, what would be their source?

    DIN

    ISO-25

    JIS

    GB-5
30. Briefly describe the purpose of the following dimvars. (For help, look up the tables in this chapter.)

    DIMCLRE

    DIMSE1

    DIMJUST

    DIMCLRT

    DIMAFIT
31. Find the name of the dimvar that does the following task. (For help, look up the tables in this chapter.)

    Size of the center mark

    Enable alternate units

    Length of the arrowhead

    Linear dimension units

    Control suppression of zeros
32. Why does the Standard dimension style set **DIMALTF** (alternate scale factor) to 25.4?

    And why does the ISO-25 dimstyle set the same dimvar to 0.04?
33. The DIN dimstyle sets **DIMASZ** to 2.5, but the GB-5 dimstyle sets it to 5.0. What difference does that make in the drawing?
34. What is the command that corresponds to the alias?

    ddim

    dimted

    ded

    dov

    d
35. Can you rename dimension styles?

    Purge dimension styles?
36. Can a drawing have more than one dimension style?

    Can a dimension have two different arrowheads?
37. Can you create your own arrowheads for dimensions?
38. Under what condition do the arrowheads and text appear outside the extension lines?
39. What is the purpose of *zero suppression*?

    Why might you want to use zero suppression?
40. When are alternative units used?

    How are alternative units shown?

41. Briefly describe the sequence to change the style of all dimensions in the drawing retroactively.
42. Briefly describe the sequence to change the style of dimensions selectively.
43. Can dimvars be changed during a dimensioning command?

CHAPTER 15

Geometric Dimensions and Tolerances

Ordinate dimensions and GD&T (geometric dimensioning and tolerancing) are typically used for machined parts — metal parts drilled with holes, slots, and contours — often in conjunction with CNC programming (computer numerical control of machinery). Tolerancing informs the machine operator about the finish of the part. In this chapter, you learn these commands:

- **DIMORDINATE** and **QDIM** draw ordinate dimensions.
- **DIMCONTINUE** draws additional ordinates
- **UCS** relocates the datum, and changes the leader angle.
- **TOLERANCE** and **QLEADER** construct GD&T symbols.

FINDING THE COMMANDS

On the **DIMENSION** toolbar:

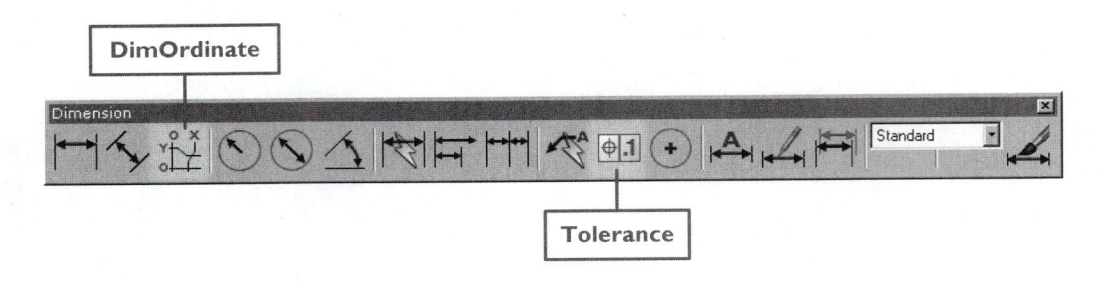

On the **DIMENSION** menu:

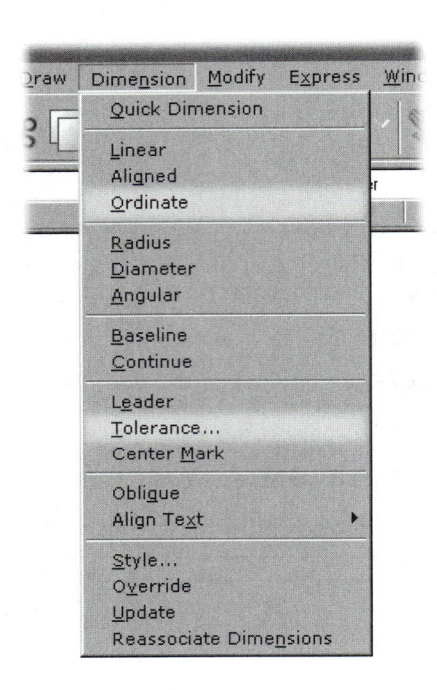

DIMORDINATE

The **DIMORDINATE** command places ordinate dimensions relative to a base point.

Ordinate dimensions (also called "datum dimensions") display X and Y coordinates at the end of a leader. They measure the perpendicular distance from the origin (also called the "datum") to features in drawings, commonly corners and centers of holes. When indicating hole centers, the Center linetype is applied to the leader line.

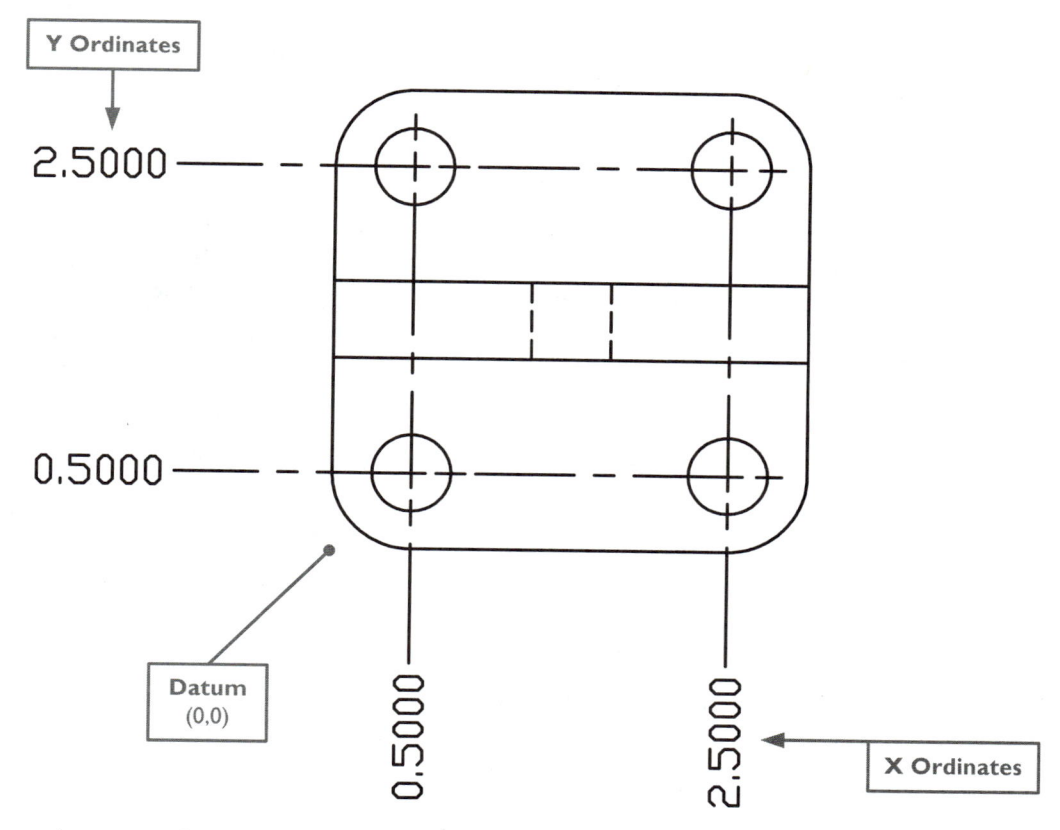

Ordinate dimension have just one number: the x distance or the y distance from the datum:

- X distance measured along the x axis, and specified by the x ordinate.
- Y distance measured along the y axis, and specified by the y ordinate.

AutoCAD's **DIMORDINATE** command has two methods of placing x and y ordinate dimensions: specifying the **Xdatum** or **Ydatum** options, or determine the correct ordinate automatically from the cursor position. AutoCAD measures the distance from the cursor location to the origin, and ghosts whichever ordinate has the farther distance.

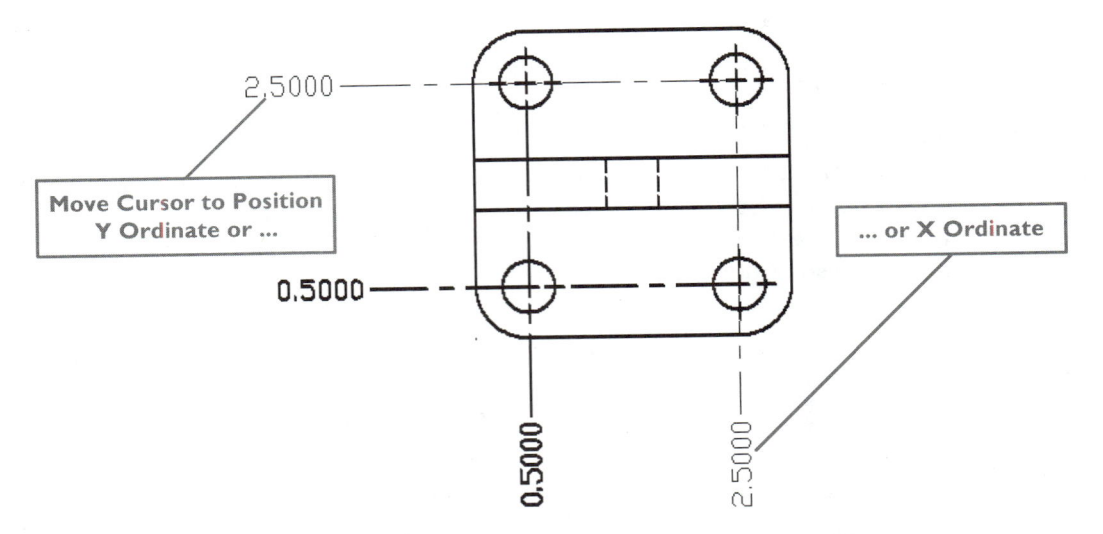

[Figure: Drawing of a hinge-like part with ordinate dimensions. Callouts: "Move Cursor to Position Y Ordinate or ..." pointing to 2.5000 label; "... or X Ordinate" pointing to vertical ordinate dimensions. Labels shown: 2.5000, 0.5000 (horizontal leaders on left); 0.5000, 2.5000 (vertical leaders on bottom).]

Ordinate dimensions ignore text orientation defined in the dimension style. The ordinate text aligns vertically or horizontally with the leader, unless you override it with the **Angle** option. In addition, you can override the calculated ordinate with your own text, through the **Text** and **MText** options.

Once an ordinate dimension is placed in the drawing, the DIMCONTINUE and DIMBASELINE commands quickly places additional dimensions. The two command operate identically: DIMBASELINE does not stack ordinates, as it does for linear and angular dimensions.

> **Note:** You will find it useful to place ordinate dimensions accurately with object snaps, such as CENter to find the centers of holes. Turning on ortho mode helps keep the leaders straight.

TUTORIAL: PLACING ORDINATE DIMENSIONS

1. To place x and y ordinate dimensions, start the **DIMORDINATE** command:
   - From the menu bar, choose **Dimension**, and then **Ordinate**.
   - From the **Dimension** toolbar, choose the **Dimension Ordinate** button.
   - At the 'Command:' prompt, enter the **dimordinate** command.
   - Alternatively, enter the aliases **dor** or **dimord** at the 'Command:' prompt.

   Command: **dimordinate** (Press ENTER.)

2. In all cases, AutoCAD prompts you to select the feature you want dimensioned:
   Specify feature location: (Pick a point in the drawing.)

3. Move the cursor to located the ordinate text:
   Specify leader endpoint or [Xdatum/Ydatum/Mtext/Text/Angle]: (Move the cursor, and then pick a point.)

   AutoCAD reports the distance it calculated from the datum (origin).
   Dimension text = 2.5000

When you pick an object at the "Specify feature location" prompt, AutoCAD associates the ordinate dimension with the coordinate — not the object. Move the object, and the ordinate dimension doesn't change (unless you use grips editing); move the dimension, and the datum distance updates automatically.

The dimensions are calculated relative to the origin of the current UCS, as described later in this section — and not relative to the drawing's origin (0,0). If it seems that ordinates are drawn relative to the drawing's origin, it's because it happens to coincide with the base point for the UCS.

The value of the coordinates is always positive, even when you draw ordinates "below" the x axis of the UCS.

PLACING ORDINATES: ADDITIONAL METHODS

The **DIMORDINATE** command has the following options:

- **Xdatum** specifies the x-ordinate.
- **Ydatum** specifies the x-ordinate.
- **Mtext** opens the mtext editor.
- **Text** prompts for text at the command line.
- **Angle** specifies the angle of the text.

Let's look at each option.

Xdatum

The **Xdatum** option forces the ordinate in the x-direction. AutoCAD repeats the prompt:

Specify leader endpoint or [Xdatum/Ydatum/Mtext/Text/Angle]: **x**

Specify leader endpoint or [Xdatum/Ydatum/Mtext/Text/Angle]: *(Move the cursor, and then pick a point.)*

Ydatum

The **Ydatum** option forces the ordinate in the y-direction.

Mtext

The **Mtext** option displays the mtext editor. The **<>** symbol indicates the default ordinate text. You can use the Text Formatting toolbar to apply styles, fonts, and formatting. In the text edit window, you can add prefix and suffix text to the ordinate.

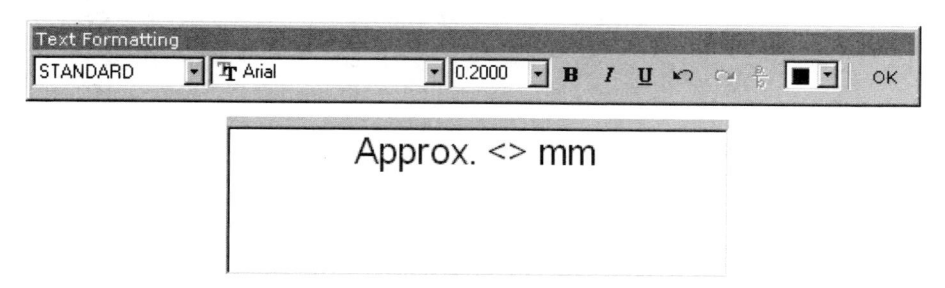

Click **OK** to exit the mtext editor. AutoCAD repeats the prompt:

Specify leader endpoint or [Xdatum/Ydatum/Mtext/Text/Angle]: *(Move the cursor, and then pick a point.)*

Text

The **Text** option prompts you to change the ordinate text:

Enter dimension text <0.7500>: *(Type text, such as **Approx. <> mm**, and then press* ENTER.*)*

Use the double angle brackets (**<>**) as a placeholder for AutoCAD's calculated ordinate distance. After pressing ENTER, AutoCAD repeats the prompt:

Specify leader endpoint or [Xdatum/Ydatum/Mtext/Text/Angle]: *(Move the cursor, and then pick a point.)*

Angle

The **Angle** option rotates the ordinate text:

> Specify angle of dimension text: *(Enter an angle, such as **45**.)*

After you press ENTER, AutoCAD repeats the prompt:

> Specify leader endpoint or [Xdatum/Ydatum/Mtext/Text/Angle]: *(Move the cursor, and then pick a point.)*

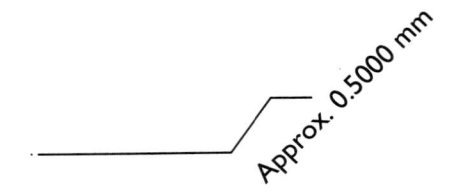

*An ordinate dimension modified by the **Text** and **Angle** options.*

DIMCONTINUE

The **DIMCONTINUE** command draws additional ordinate dimensions of the same type — helpfully avoiding the pair of prompts displayed by **DIMORDINATE**. For example, if you placed an x ordinate dimension, **DIMCONTINUE** places additional x ordinates.

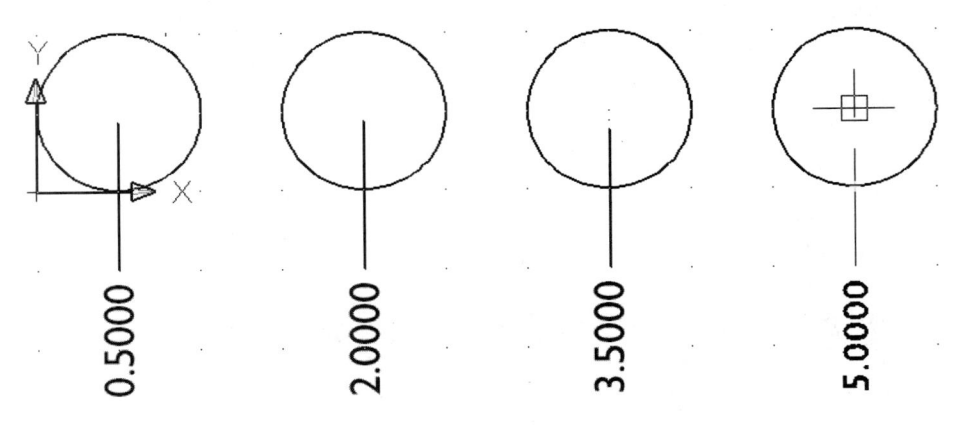

Command: **dimcontinue**
Specify feature location or [Undo/Select] <Select>: *(Pick feature.)*
Dimension text = 2.0000
Specify feature location or [Undo/Select] <Select>: *(Pick feature.)*
Dimension text = 3.5000
Specify feature location or [Undo/Select] <Select>: *(Pick feature.)*
Dimension text = 5.0000
Specify feature location or [Undo/Select] <Select>: *(Press ENTER to exit command.)*

UCS ORIGIN AND 3POINT

The **UCS** command defines the base point from which ordinate dimensions are measured; the **Origin** option moves the base point. This is a shortcut not documented by Autodesk. You typically move the UCS origin to the lower-left corner of the part, a procedure called "datum shift." (Recall that the **BASE** command defines the origin of the drawing, while the **SNAPBASE** system variable controls the origin of snap, grid markings, and hatch patterns.)

To see the effect of the origin on a drawing containing ordinate dimensions, use the **ZOOM Extents** command. Instead of the objects filling the screen, as you would expect, the objects may appear in the upper right corner of the drawing. This happens because AutoCAD takes into account the origin of the ordinate dimensions when calculating the extents of the drawing.

In addition, AutoCAD uses the axes of the UCS to draw the leaders orthogonally. Rotate the axes with the **3point** option, and the ordinate dimensions are drawn at an angle.

TUTORIAL: CHANGING THE ORDINATE BASE POINT

Ensure the UCS icon is turned on for this tutorial.

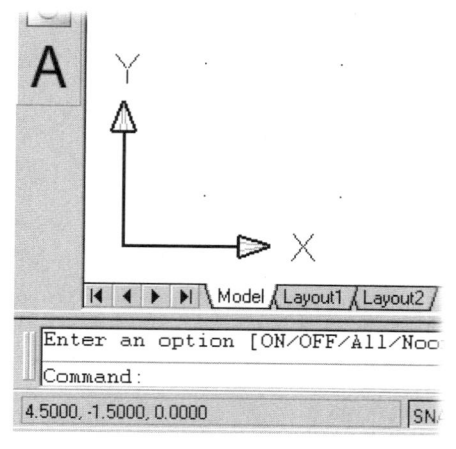

If you do not see the icon, turn it on, as follows:
Command: **uscicon**

Enter an option [ON/OFF/All/Noorigin/ORigin/Properties] <OFF>: **on**

Repeat the command to force the icon to appear at the origin:
Command: *(Press spacebar.)*

Enter an option [ON/OFF/All/Noorigin/ORigin/Properties] <OFF>: **or**

1. To see the effect of the UCS origin command, first draw an x-ordinate at 2,2, as follows:
   Command: **dimordinate**
   Specify feature location: **2,2**
   Specify leader endpoint or [Xdatum/Ydatum/Mtext/Text/Angle]: **x**
   Specify leader endpoint or [Xdatum/Ydatum/Mtext/Text/Angle]: *(Pick a point.)*
   Dimension text = 2.0000

2. Start the **ucs** command, and then specify the **Origin** option:
   Command: **ucs**
   Enter an option [New/Move/orthoGraphic/Prev/Restore/Save/Del/Apply/?/World] <World>: **o**

3. Move the UCS origin to 1,1:
   Specify new origin point <0,0,0>: **1,1**

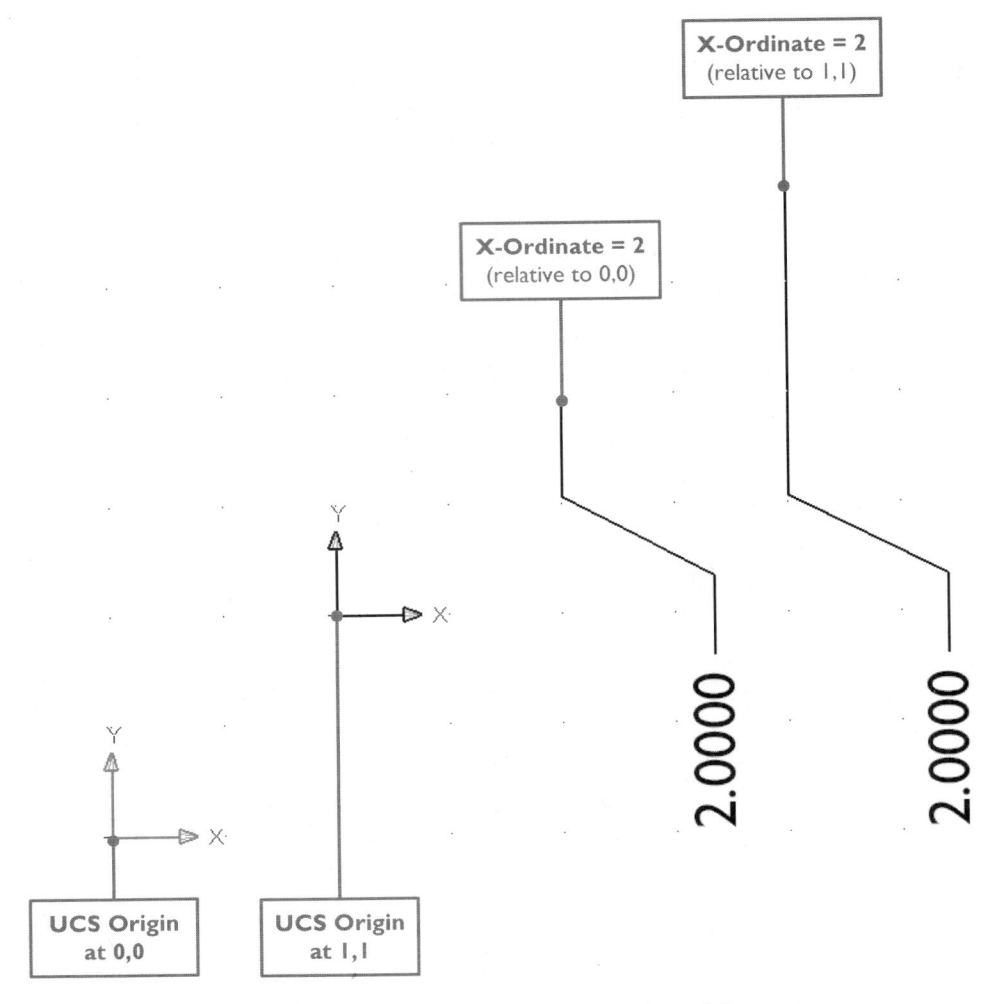

X-Ordinate = 2
(relative to 1,1)

X-Ordinate = 2
(relative to 0,0)

2.0000

2.0000

UCS Origin at 0,0

UCS Origin at 1,1

4. Draw another x-ordinate at 2,2 and notice that it appears in a different location.
   Command: **dimord**
   Specify feature location: **2,2**
   Specify leader endpoint or [Xdatum/Ydatum/Mtext/Text/Angle]: **x**
   Specify leader endpoint or [Xdatum/Ydatum/Mtext/Text/Angle]: *(Pick a point.)*
   Dimension text = 2.0000

TUTORIAL: ROTATING THE ORDINATE

1. Ensure the UCS icon is turned on for this tutorial.
2. Start the **ucs** command, and then specify the **3point** option:
   Command: **ucs**
   Enter an option [New/Move/orthoGraphic/Prev/Restore/Save/Del/Apply/?/World] <World>: **3point**
3. You can keep the same UCS origin (datum), or specify a new datum:
   Specify new origin point <0,0,0>: *(Press ENTER to keep datum.)*

Geometric Dimensions and Tolerances 645

4. Pick a point to align the x axis at an angle. You can pick a point, or enter coordinates.
   To rotate 45 degrees, for example, enter polar coordinates: d<a.
      Specify point on positive portion of X-axis <1.0,0.0,0.0>: *(Pick a point, or enter polar coordinates, such as* **1.0<45**.*)*
5. Press **ENTER** to keep the y axis perpendicular to the x axis:
      Specify point on positive-Y portion of the UCS XY plane <-0.7,0.7,0.0>: *(Press* **ENTER**.*)*

Notice that several things change in the drawing. The UCS icon rotates, but so do the crosshair cursor, grid, snap, and the angle for ortho mode (if you have them turned on). This is because the UCS rotates the entire coordinate system.

Coordinates may appear in an unexpected location, as you find out with the final step in this tutorial. You learn more about UCSs in Volume 2 of this book, *Using AutoCAD 2004: Advanced*, because they are primarily used for three-dimensional design.

6. Draw another x-ordinate at 2,2 and notice that it appears in a different location.
      Command: **dimord**
      Specify feature location: **2,2**
      Specify leader endpoint or [Xdatum/Ydatum/Mtext/Text/Angle]: **x**
      Specify leader endpoint or [Xdatum/Ydatum/Mtext/Text/Angle]: *(Pick a point.)*
      Dimension text = 2.0000

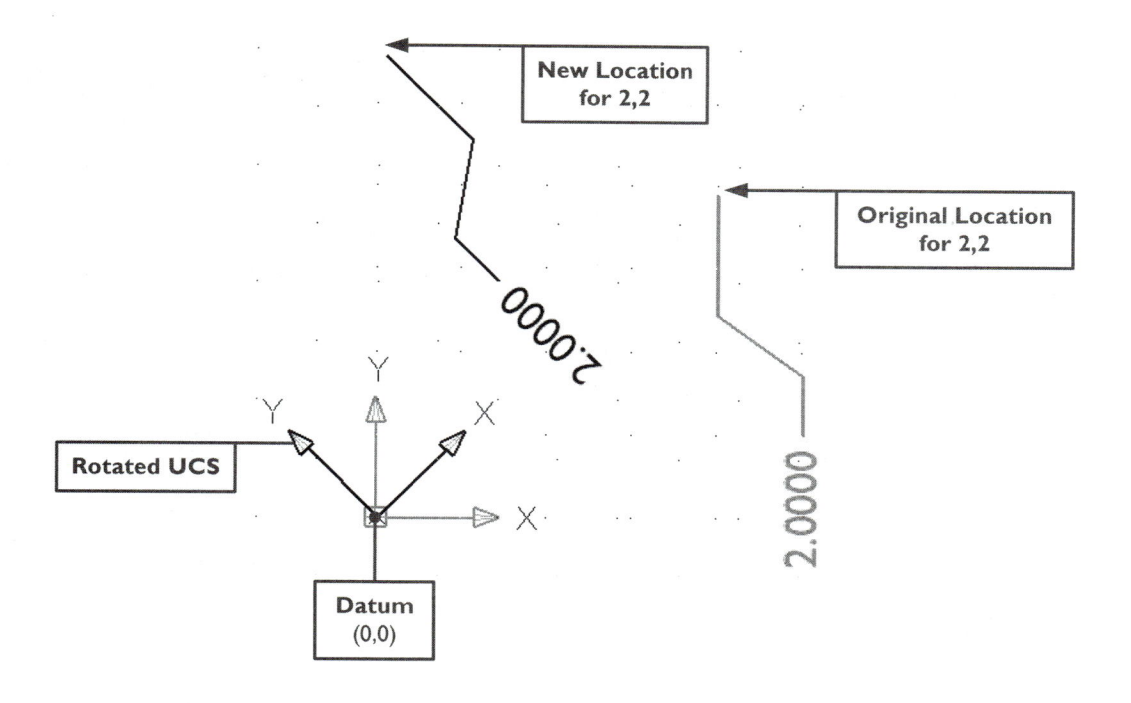

Note: To return the UCS icon back to "normal" — located at the drawing origin and unrotated — use the **UCS World** command:

   Command: **ucs**
   Enter an option [New/ ... /?/World] <World>: **w**

QDIM

The **QDIM** command places ordinate dimensions.

You may prefer using it, because it combines the **DIMORDINATE**, **DIMCONTINUE**, and **UCS Origin** commands.

TUTORIAL: ORDINATE DIMENSIONS WITH QDIM

1. Start the **QDIM** command:
   Command: **qdim**
2. Select the objects to dimensions:
   Associative dimension priority = Endpoint
   Select geometry to dimension: *(Select one or more objects.)*
   Select geometry to dimension: *(Press ENTER to end object selection.)*
3. Position the datum at the lower-left corner of the objects:
   Specify dimension line position, or
   [Continuous/Staggered/Baseline/Ordinate/Radius/Diameter/datumPoint/Edit/seTtings
   ] <Continuous>: **p**
   Select new datum point: *(Pick a point.)*

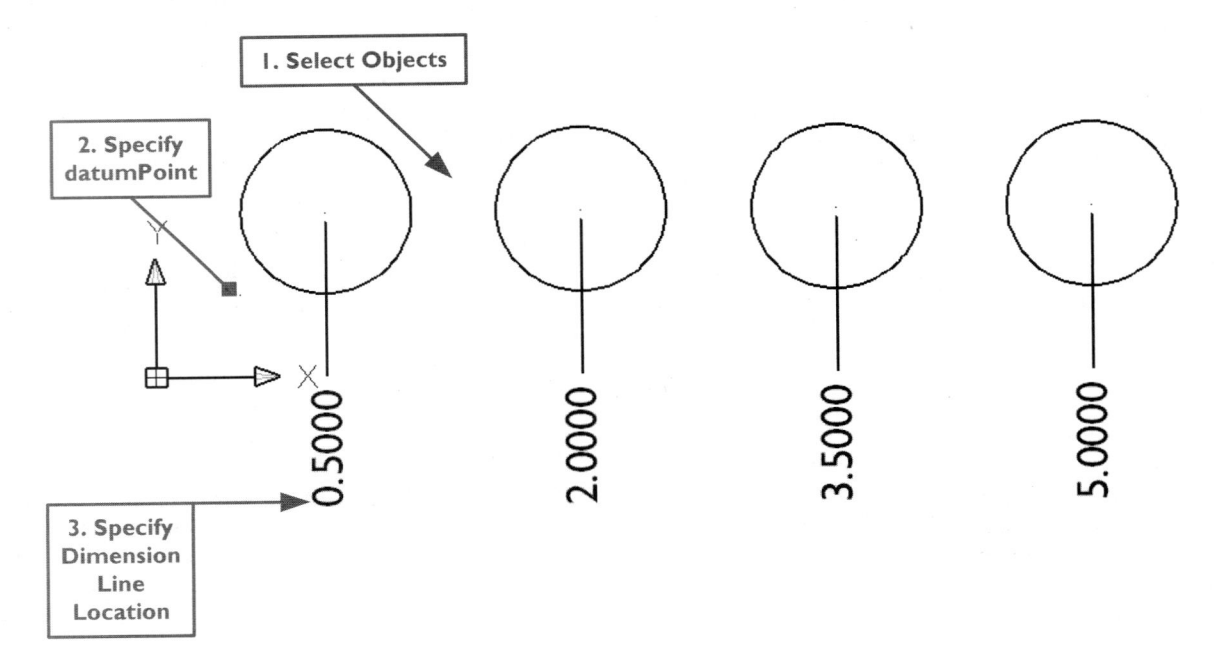

4. Specify ordinate dimensioning:
   Specify dimension line position, or
   [Continuous/Staggered/Baseline/Ordinate/Radius/Diameter/datumPoint/Edit/seTtings] <Continuous>: **o**
5. Specify the dimension line location:
   Specify dimension line position, or
   [Continuous/Staggered/Baseline/Ordinate/Radius/Diameter/datumPoint/Edit/seTtings] <Continuous>: *(Pick a point.)*

EDITING ORDINATE DIMENSIONS

Once ordinate dimensions are in place, you can edit them with grips and other commands that affect dimensions — with limitations.

Grips Editing

When you select ordinate dimensions, they display four grips: one at either end of the leader line, one on the text, and one at the datum (UCS origin).

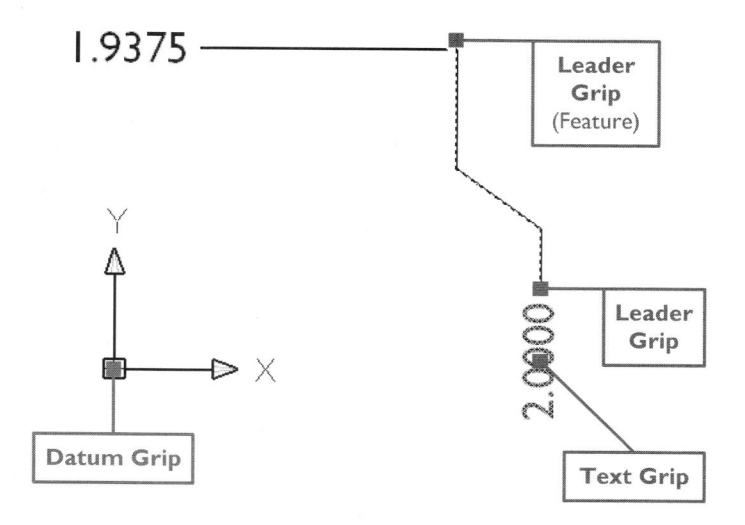

Leader Line

The leader line has four grips, one at each end — one at the datum, one on the text, and one at each end of the leader line. When you move the feature-end grip, the other end and the text remain in place; the text updates automatically.

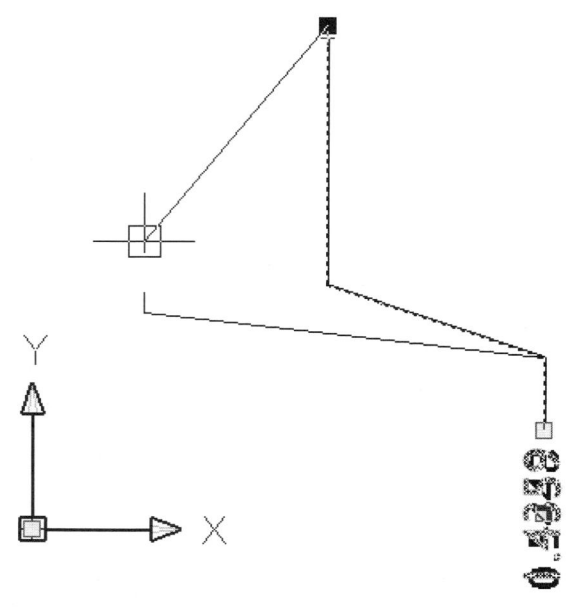

When you move the text-end grip, the text move with it; the other end remains in place. Unlike regular leaders, ordinate text is attached to its leader!

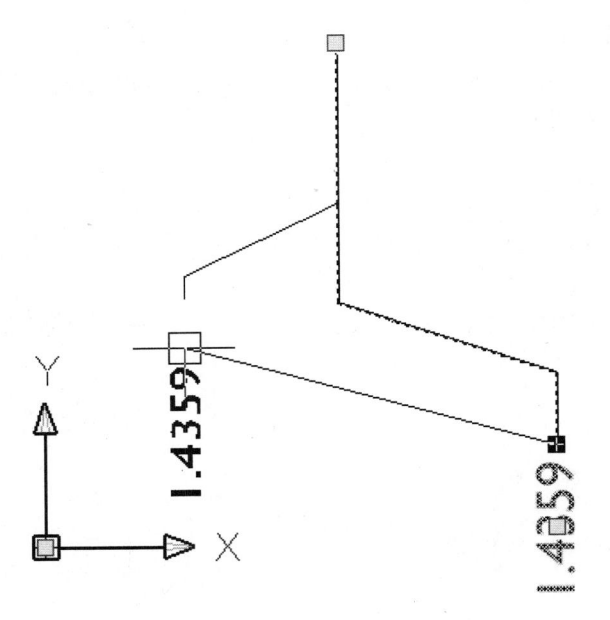

Text
The text grip is located in the center of the ordinate text. It moves the text and leader in exactly the same manner as the leader grip.

Datum
The datum grip is initially located at the UCS origin. When you move it, the datum changes for the selected dimension, but the UCS icon remains in place. The text updates its coordinates, and the dimension is no longer associative.

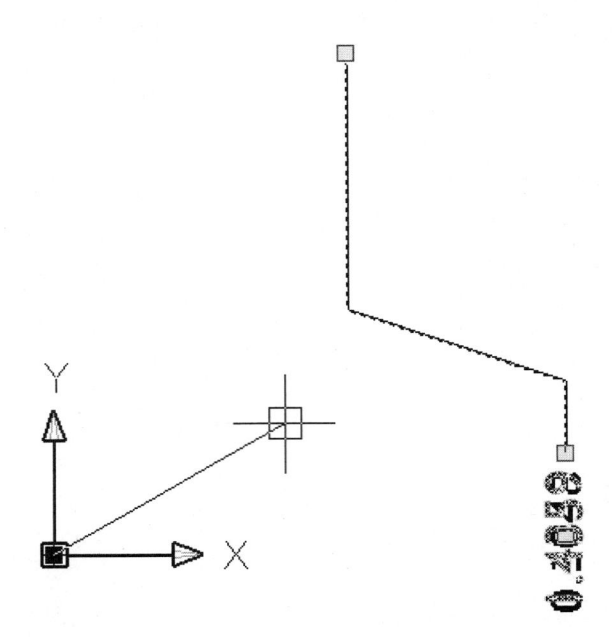

Editing Commands

The **DIMEDIT** command rotates and homes the ordinate dimension text, but does not oblique the leader lines.

The **DIMTEDIT** command relocates the ordinate text, as well as rotates and homes it. The Left, Right, and Center options have no effect on the text.

To edit the properties, double click the ordinate dimension. AutoCAD displays the Properties window (shown at right).

TOLERANCE

The **TOLERANCE** command displays a dialog box for selecting tolerance symbols.

Use the **TOLERANCE** command to create leaderless frames, and the **QLEADER** command to attach the frames to leader. Geometric tolerances are not associated with geometric objects. After placing the tolerance frame, you can edit the symbols by double-clicking: AutoCAD displays the Geometric Tolerance dialog box. The frame can be edited by most editing commands, as well as with grips.

GEOMETRIC DIMENSIONING AND TOLERANCES (GD&T)

Drafters use geometric tolerance symbols to show machinists the acceptable deviations of *form*, *profile*, *orientation*, *location*, and *runout* of features. Symbols are used because they reduce the need for notes describing complex geometry requirements. The symbols provided by AutoCAD are based on the ASME Y14.5M – 1994 standard.

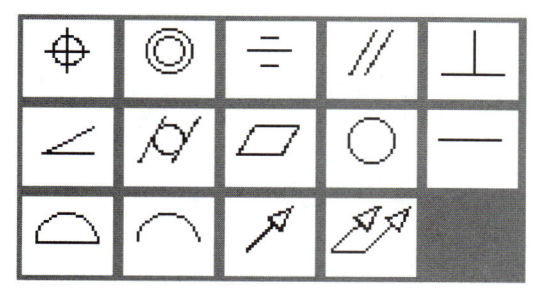

Geometric characteristic symbols

At left: Tolerance diameter symbol.
At right: Projected tolerance zone symbol.

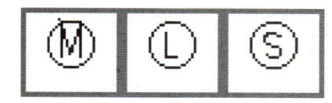

Material condition symbols

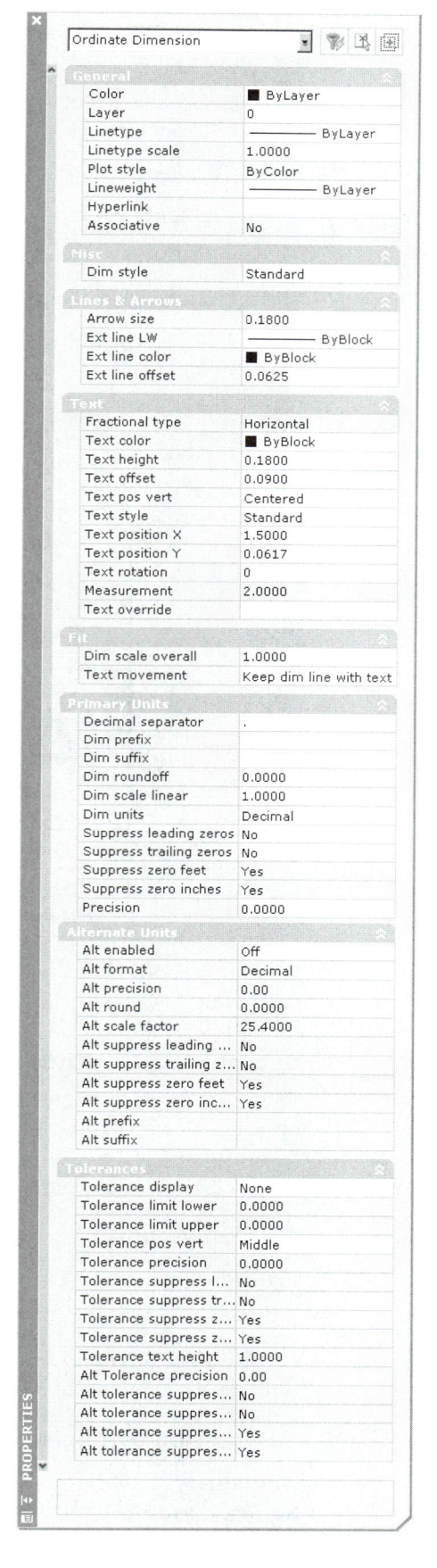

Tolerance symbols are defined in *gdt.shx*. Some GD&T symbols are missing from AutoCAD, such as the round datum target.

The symbols are placed in a *feature control frame*. The frame makes it easier to read the symbols, because it separates them into categories:

1. Geometric characteristic.
2a. Tolerance zone.
2b. Tolerance zone modifiers (MMC, LMC, or RFS).
3. Datum reference, and datum reference modifiers (optional).

The parts of a typical frame are shown below:

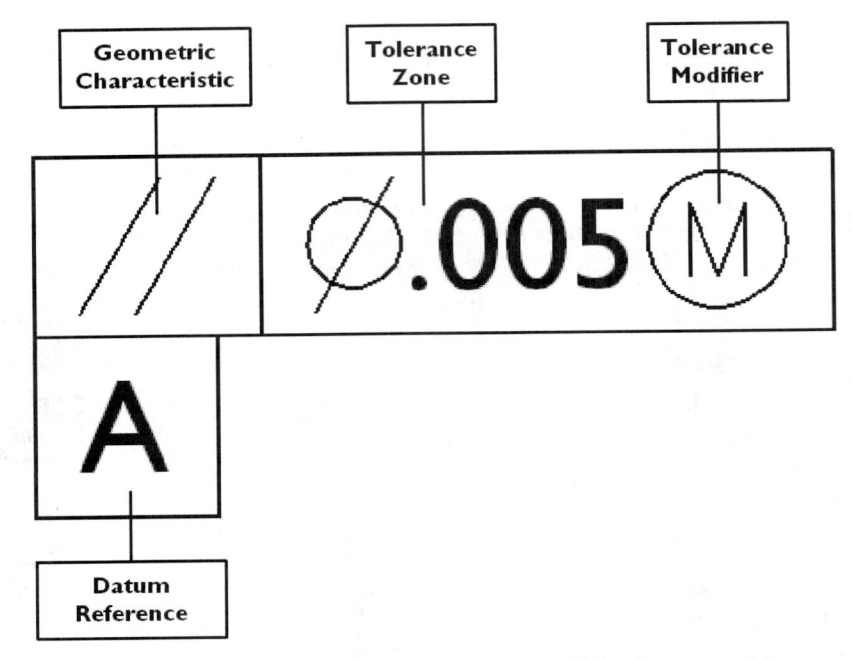

The frame shows that the part A needs to be machined to parallel tolerance of diameter 0.005 inches, maximum.

GEOMETRIC CHARACTERISTIC SYMBOLS

The geometric characteristic identifies the ideal geometry, such as a perpendicular surface or a cylinder. As with ordinate dimensions, *datums* in tolerances are the origin from which the geometry is established. The *datum target* is a line or area on parts that establishes datums. See www.engineersedge.com/gdt.htm for additional GD&T definitions.

Profile Symbols

Profile of a Line
Entire length of a feature must lie between two parallel zone lines.

Profile of a Surface
Entire surface must lie between envelope surfaces separated by the tolerance zone.

Orientation Symbols

Angularity
Surfaces, axes, and center planes must lie between two parallel plates sloped at a specified angle.

Perpendicularity
Surfaces, axes, median planes, and lines are exactly 90 degrees to the datum plane or axis.

Parallelism
All points on the surface or axis are equidistant from the reference datum.

Location Symbols

True Position
Specifies zone of tolerance for center, axis, and center planes.

Symmetry
Features are symmetric about the center plane of the datum.

Concentricity
Specifies cylindrical tolerance zone.

Runout Symbols

Circular Runout
Circular elements must be within the runout tolerance (full 360-degree rotation about the datum axis).

Total Runout
Surface elements across the entire surface must be within runout tolerance.

Form Symbols

Straightness
Surface or axis is a straight line, within the tolerance.

Flatness
Entire 3D surface must be flat, within the tolerance.

Circularity
All points on a surface of revolution, such as cylinders, spheres, and cones, are equidistant from the axis of the center, within the tolerance.

Cylindricity
All points on the surface of revolution, such as cylinders, are the same distance from a common axis, within the tolerance.

ADDITIONAL SYMBOLS

Tolerance Symbols
The *tolerance* is the difference between the minium and maximum limits.

No Symbol
No tolerance symbol means the tolerance is linear.

Tolerance Diameter
This symbol means the tolerance refers to a diameter measurement.

Projected Tolerance Zone
Specifies the perpendicularity and mating clearance for holes in which pins, studs, and screws will be inserted.

Material Characteristics Symbols

The *material characteristics* indicates the maximum, minimum, or required adherence to the tolerance.

Maximum Material Condition (MMC)

Feature contains the maximum material within the stated limits. Examples include the maximum shaft diameter and the minimum hole diameter.

Least Material Condition (LMC)

Feature contains the least amount of material within the stated limits; reverse of MMC. Examples include minimum shaft diameter and maximum hole diameter.

Regardless of Feature Size (RFS)

Geometric tolerance or datum reference applies at any increment of size within its tolerance.

SYMBOL USAGE

The following table shows appropriate uses for the symbols

Geometric Characteristic	Surface	Size	MMC	LMC	P
Form:					
Straightness	☑	☑	☑	☑	
Flatness	☑				
Circularity	☑				
Cylindricity	☑				
Orientation:					
Perpendicularity	☑	☑	☑	☑	☑
Angularity	☑	☑	☑	☑	☑
Parallelism	☑	☑	☑	☑	☑
Location:					
Positional Tolerance		☑	☑	☑	☑
Concentricity		☑			
Symmetry		☑			
Runout:					
Circular Runout	☑	☑			
Total Runout	☑	☑			
Profile:					
Profile of a Line	☑		☑	☑	
Profile of a Surface	☑		☑	☑	

TUTORIAL: PLACING TOLERANCE SYMBOLS

1. To make place a tolerance frame, start the **TOLERANCE** command:
   - From the menu bar, choose **Dimension**, and then **Tolerance**.
   - From the **Dimension** toolbar, choose the **Tolerance** button.
   - At the 'Command:' prompt, enter the **tolerance** command.
   - Alternatively, enter the **tol** alias at the 'Command:' prompt.

   Command: **tolerance** *(Press* ENTER.*)*

   In all cases, AutoCAD displays the Geometric Tolerance dialog box:

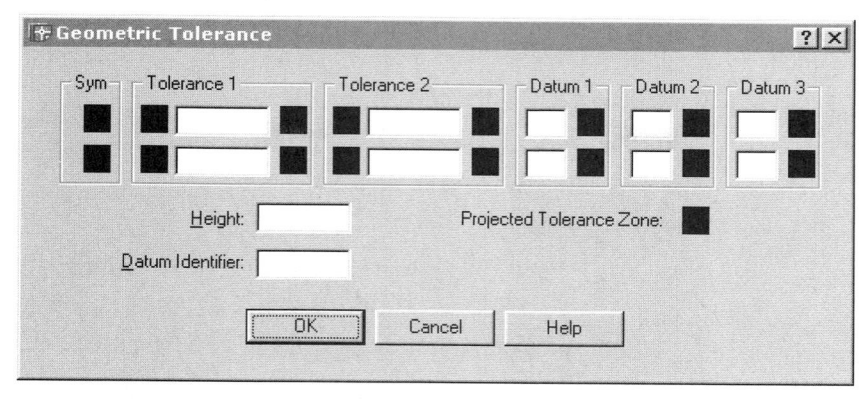

2. Under **Sym**, click the black square. AutoCAD displays the Symbol dialog box.

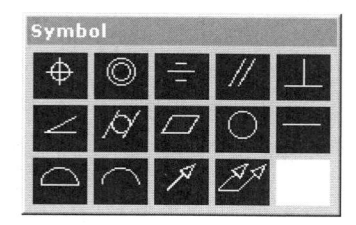

3. Select a Geometric Characteristic symbol. To select none, click the white square, or press **ESC**. Notice that the dialog box disappears, and AutoCAD fills in the symbol.
4. If you want the diameter tolerance symbol, click the first black square under **Tolerance 1**. To remove the symbol, click the square a second time.
5. Enter the tolerance value in the white text entry rectangle.
6. To add a material condition symbol, click the second black square under **Tolerance 1**. AutoCAD displays the Material Condition dialog box.

7. Select a material condition symbol.
    To select none, click the white square, or press **ESC**.
    Notice that the dialog box disappears, and AutoCAD fills in the symbol.

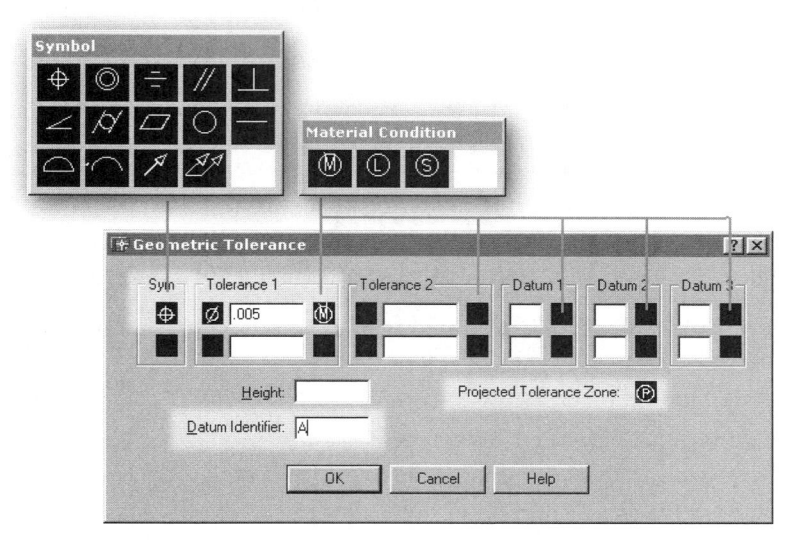

8. If necessary, fill in values and symbols for **Tolerance 2**, **Datum 1**, **2**, and **3**., as well as a height and a datum identifier.

9. Click **Projected Tolerance Zone** to fill in the **P** symbol.
10. When done, click **OK**.

    AutoCAD prompts:

    Select tolerance location: *(Pick a point.)*

    Pick a point in the drawing to place the tolerance frame.
    Using object snap modes may be helpful.

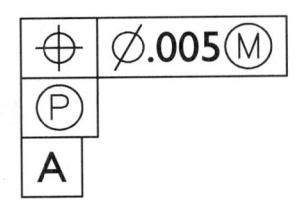

QLEADER

The **QLEADER** command includes an option to place tolerance marks.

TUTORIAL: PLACING TOLERANCES WITH QLEADER

1. To place tolerance frames at the end of leaders, start the **QLEADER** command, and then select the Settings option:

    Command: **qleader**

    Specify first leader point, or [Settings] <Settings>: **s**

2. In the Settings dialog box, select **Tolerance** in the Annotation tab.

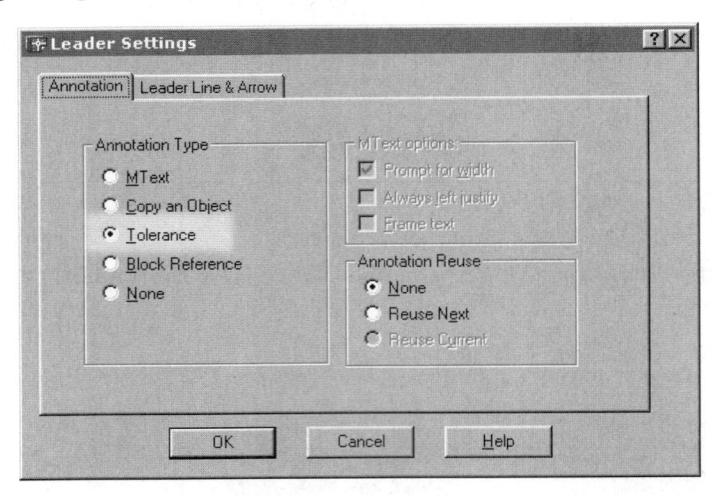

3. Make any other changes you wish in the dialog box, and then click **OK**.
4. AutoCAD prompts you to pick points for the leader's endpoint and vertices:

    Specify first leader point, or [Settings] <Settings>: *(Pick a point.)*

    Specify next point: *(Pick another point.)*

5. Press **ENTER** to end the leader.

    Specify next point: *(Press ENTER to end leader creation.)*

    AutoCAD displays the Geometric Tolerance dialog box.

6. Enter the tolerance symbols and values, and then click **OK**.

    AutoCAD places the leadered frame in the drawing.

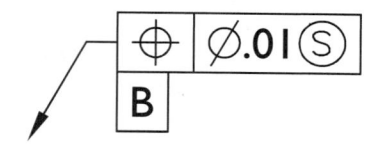

The LEADER command is less convenient, because you have to specify tolerances each time you use it:

> Command: **leader**
>
> Specify leader start point: *(Pick a point.)*
>
> Specify next point: *(Pick another point.)*
>
> Specify next point or [Annotation/Format/Undo] <Annotation>: *(Press* ENTER *to end leader line.)*
>
> Enter first line of annotation text or <options>: *(Press* ENTER *for options.)*
>
> Enter an annotation option [Tolerance/Copy/Block/None/Mtext] <Mtext>: **t**

EXERCISES

1. In the following exercise, you prepare a drawing for ordinate dimensioning.
   From the CD-ROM, open the *positioningplate.dwg* file, a drawing of a positioning plate.

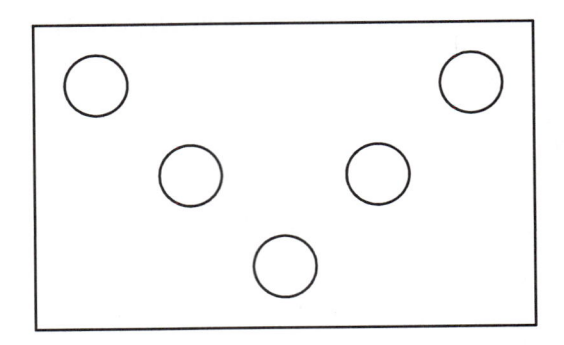

   Create a new layer with the following properties:
Name	**Ordinate**
Color	**Blue**
Linetype	**Center**

   Make the layer current.
   Change the value of **DIMCEN** to -0.09, and then use the **DIMCENTER** command to place center marks at the center of each circle.
   Relocate the UCS origin to the lower left corner of the plate.
   Dimension each hole with x and y ordinate dimensions. There should be eight dimensions in all.
   Save the drawing as *ordinate1.dwg*.

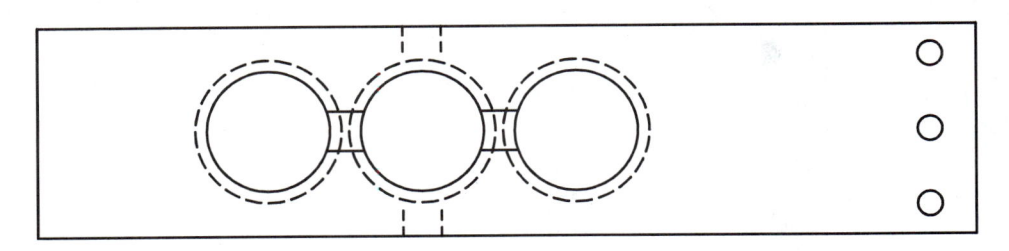

2. From the CD-ROM, open the *bar.dwg* file, a drawing of a mounting bar provided as part of a sample drawing included with AutoCAD.
   Dimension each hole with x and y ordinate dimensions.
   How many dimensions did you place?
   Save the drawing.

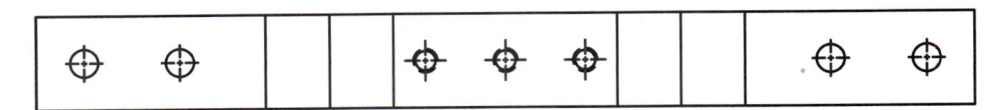

3. From the CD-ROM, open the *brace.dwg* file, a drawing of a locating brace.
   Dimension each hole with x and y ordinate dimensions.
   Save the drawing.

4. From the CD-ROM, open the *mount.dwg* file, a drawing of a motor mounting plate. Using the **QDIM** command, dimension the holes with x and y ordinate dimensions. Save the drawing.

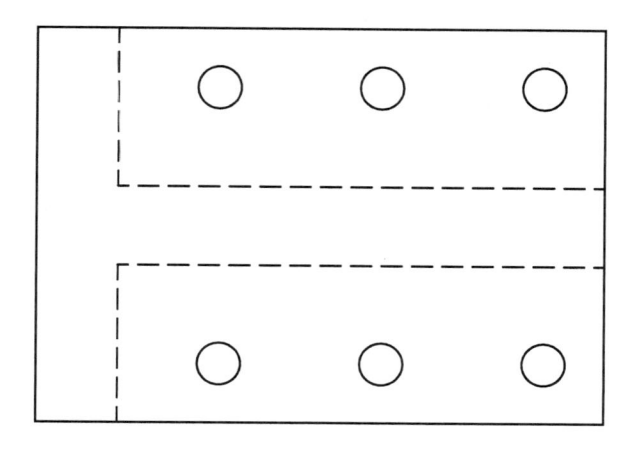

5. In the following exercises, you create tolerance frames.
   Open a new drawing.
   Use the **TOLERANCE** command to recreate the following tolerance frames:
   a.

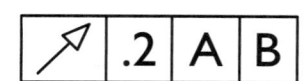

   | ⌖ | Ø1 Ⓜ | A |

   b.

   | ↗ | .2 | A | B |

   c.

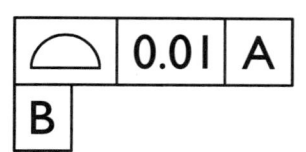

   | ⌒ | 0.01 | A |
   | B |

   d.

   | ⌖ | Ø0.5 Ⓜ | A Ⓜ | B |

6. Use the **QLEADER** command to recreate the following tolerance frames:
   a.

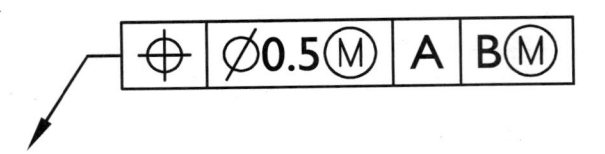

   | ⌖ | ⌀0.5Ⓜ | A | BⓂ |

   b.

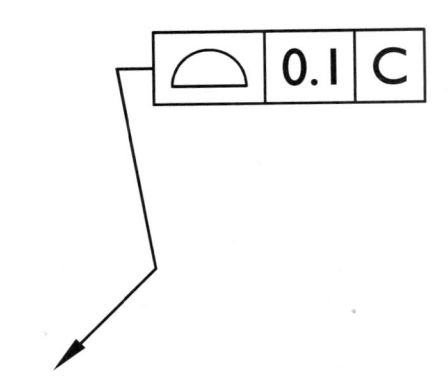

   | ⌒ | 0.1 | C |

   c.

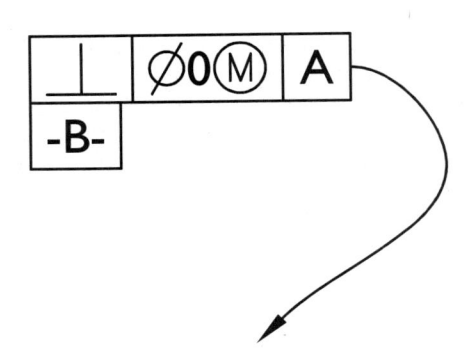

   | ⊥ | ⌀0Ⓜ | A |
   | -B- |

CHAPTER REVIEW

1. On which menu do you find the tolerance command?
2. Explain the purpose of the dimordinate command.
3. What is another name for ordinate dimensions?
   For the origin?
4. What do ordinate dimensions measure?
5. List two methods for moving the datum:
   a.
   b.
6. Describe how the dimcontinue command is useful for placing ordinate dimensions.
7. How many ordinate dimensions are needed for the holes in the following drawing:
   In the x direction?
   In the y direction?
   Sketch the locations of the x and y ordinate leader lines.

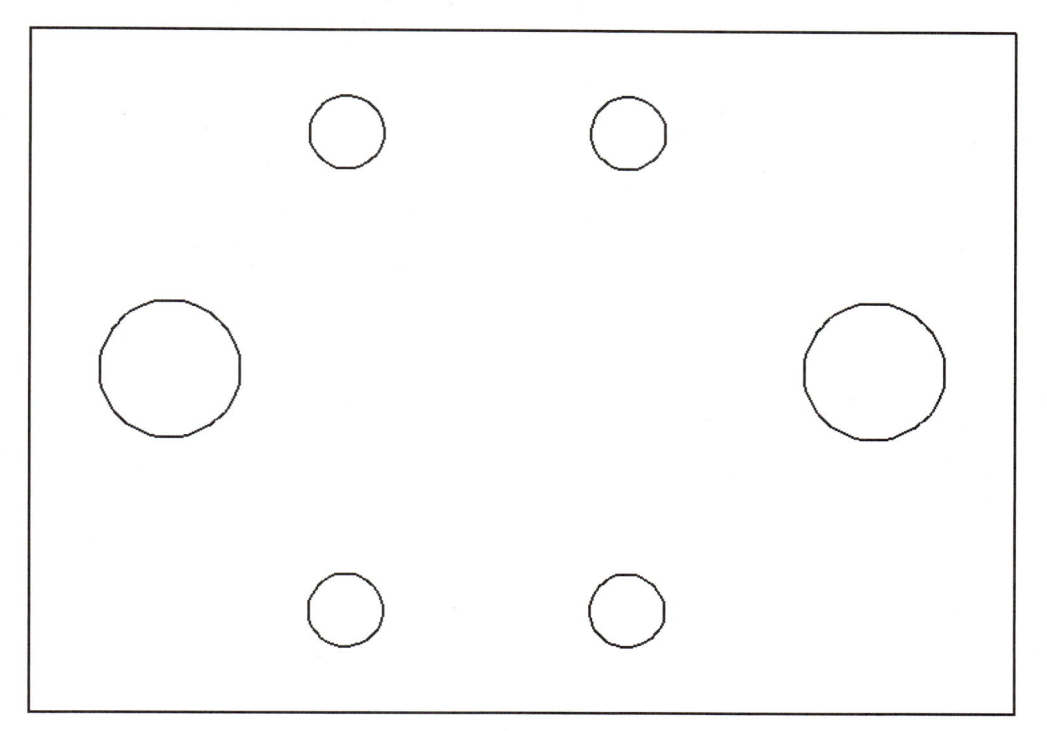

8. Why do you need to move the datum?
9. Two of the following commands relocate the datum for ordinate dimensions.
   Which two are correct?
   snapbase
   base
   qdim datumPoint
   ucs Origin
10. Can the leaders of ordinate dimensions be drawn at an angle?
    If so, how?

11. Can the text of ordinate dimensions be drawn at an angle?
    If so, how?
12. Name the command (and option) that return the datum to "normal."
13. Why is the qdim command a good alternative to the dimordinate command?
14. What is GD&T short for?
15. Label the parts of the tolerance frame:

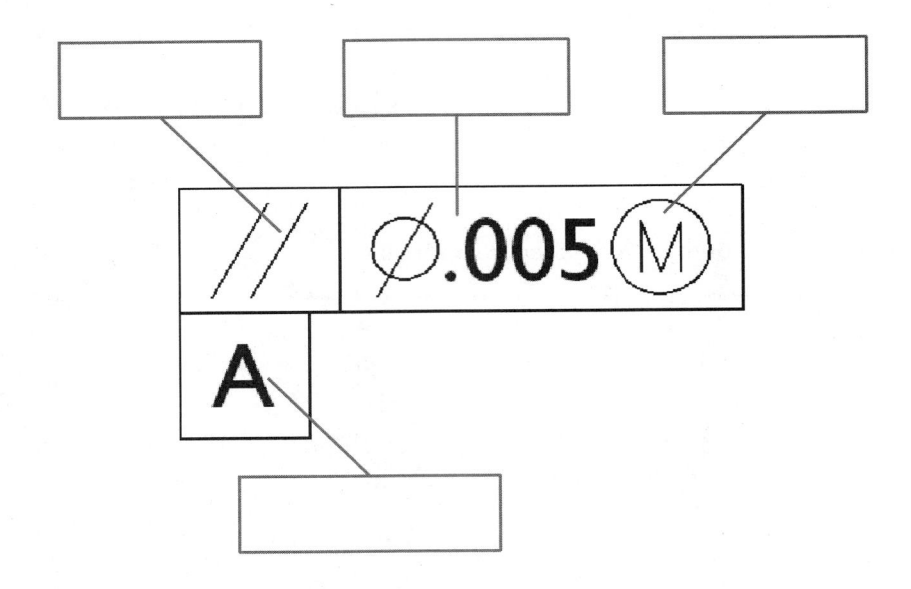

16. Decode the following abbreviations:
    RFS
    LMC
    MMC
    P

CHAPTER 16

Working with Layouts

Up until now, you have been drawing and editing drawings in "model" space. (Model space is where the model is drawn.) Preparing a drawing for plotting means working with *layouts*. This is where you set up drawings for plotting/printing. In this chapter, you find the following commands:

- **TILEMODE** switches the drawing between model and layout tabs.
- **ZOOM Xp** scales models relative to layouts.
- **LAYER Current VP Freeze** freezes layers in selected viewports.
- **LAYOUTWIZARD** steps through the layout creation process.
- **LAYOUT** creates and modifies layouts.
- **VIEWPORTS** creates rectangular viewports.
- **-VPORTS** and **VPCLIP** create and change polygonal viewports.
- **PSLTSCALE** scales model-space linetypes relative to layouts.
- **SPACETRANS** scales model-space distances relative to layouts.

FINDING THE COMMANDS

On the **LAYOUTS** and **VIEWPORTS** toolbars:

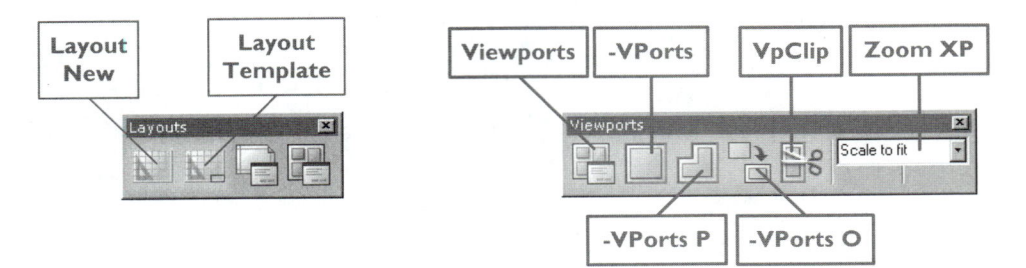

On the **VIEW** and **INSERT** menus:

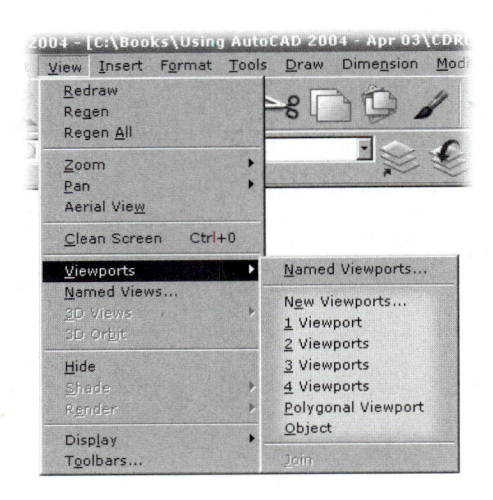

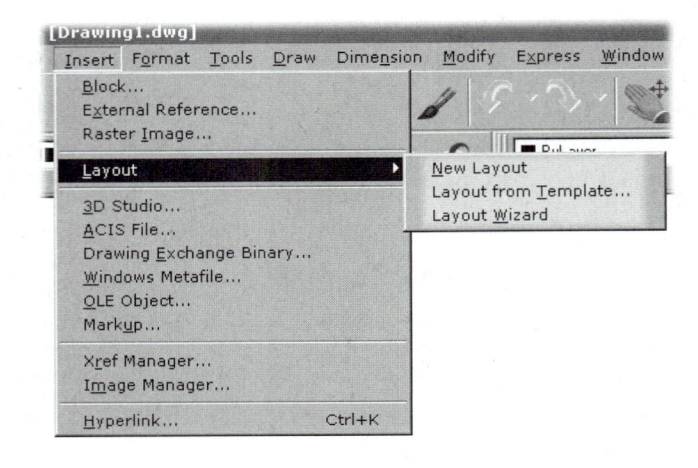

On the layout tabs:

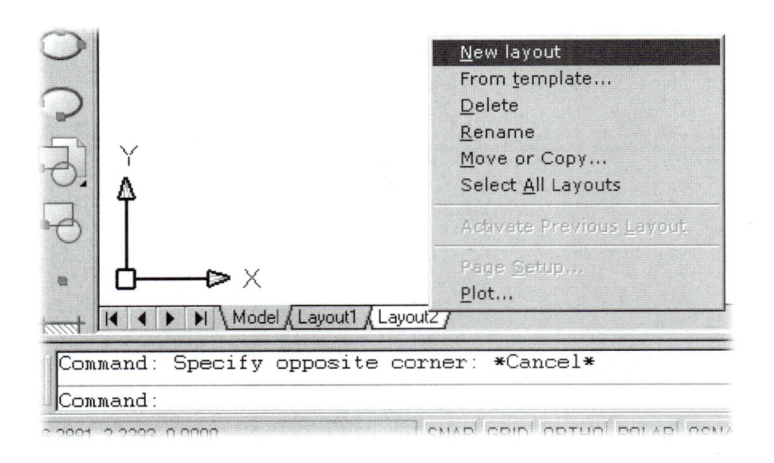

TILEMODE

The **TILEMODE** system variable switches AutoCAD between model space and layouts.

Historically, **TILEMODE** was the original method of switching between model and *paper space*, as layouts were originally called. With the introduction of layout tabs in AutoCAD 2000, the need for entering **TILEMODE** as a "command" disappeared. (I include the system variable here to point out that when clicking on the model or layout tabs, AutoCAD executes **TILEMODE**.)

ABOUT LAYOUTS

AutoCAD provides two drafting environments: Model space and layouts. Model space is where you draw the model full scale, 1:1. Recall that you had to calculate the scale factor for text, linetypes, and hatch patterns so that they would not appear too small when printed.

Layouts solve the scaling problem. Here is where you arrange the model as if it were on a sheet of paper. Now the paper is full size (1:1), so you don't need to figure out the scale factor for text, linetypes, and hatch patterns — they are all drawn full size.

Instead, the model is "scaled" to fit the paper. You don't have to do the scaling; instead, you create *viewports* in the layout. These act like windows into model space; you simply use the **ZOOM** command to scale the model to a specific scale factor.

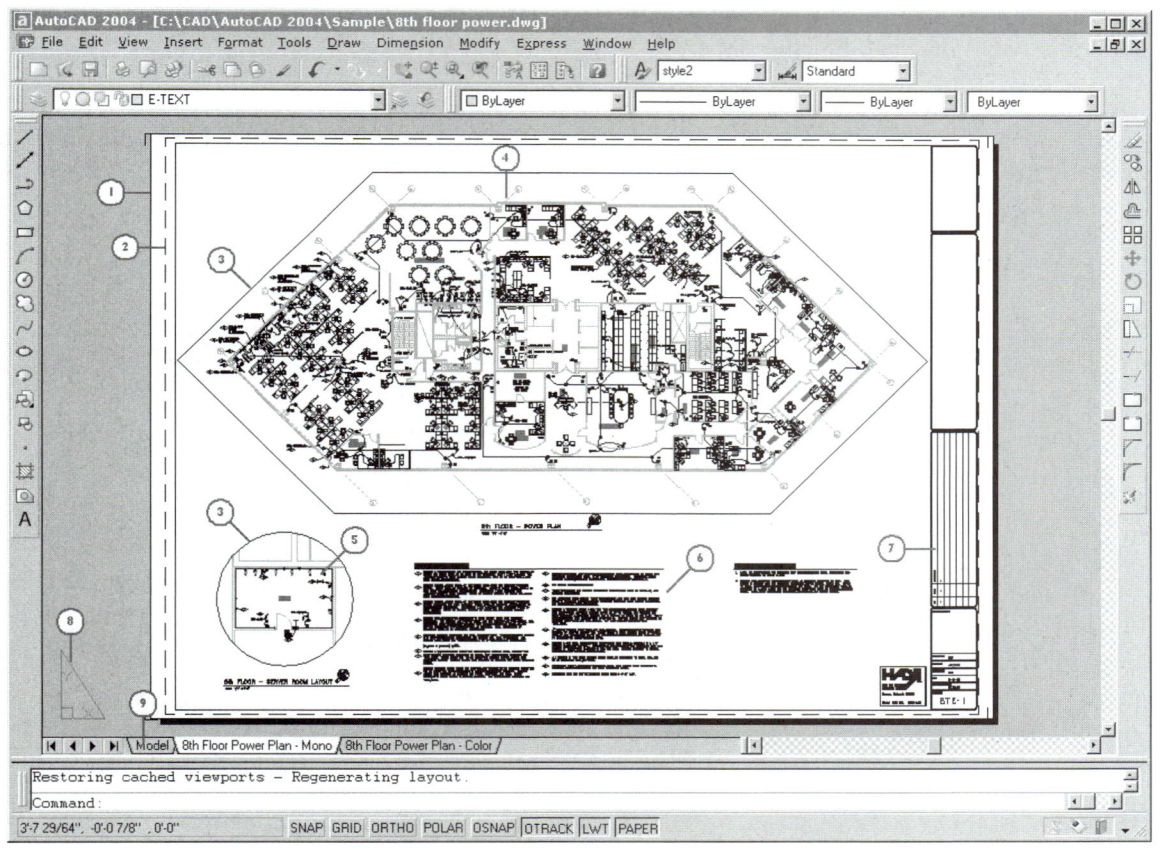

1: Edge of paper.   2: Printer margin.
3: Viewports.   4: Full drawing of model.   5: Detail view of model.
6: Text placed in layout.   7: Title block placed in layout.
8: Paper space UCS icon.   9: Model and layout tabs.

A single drawing has one model space, and one or more layouts. New drawings start with layouts given the generic names Layout1 and Layout2. Each layout can represent a drawing sheet. The layouts show the overall plan, details, and reference drawings. Each layout can be on a different paper size, and be assigned a different printer. You can think of layouts as interactive plot previews.

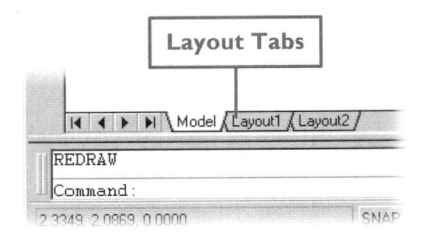

Each layout has at least one viewport, which shows some or all of the model. Some layouts have two or more viewports, each showing details. (AutoCAD can display the contents of up to 64 viewports.) Each viewport can show the model at a different scale factor, and in 3D drawings, show different views, such as the top, side, and front. (Layouts themselves are strictly 2D.)

Each layout typically has a drawing border and title block. (The viewports fit inside the drawing border.) You can add dimensions and text. A command is available that automatically translates linetype scales between the model scale factor and the layout.

You can create new layouts, import layouts from other drawings, rename and delete layouts, and change the order in which layouts appear. Similarly, you can create new viewports, resize and delete viewports, toggle layer visibility, and assign plot styles to viewports. While layouts are always rectangular, viewports can be any shape.

Exploring Layouts

To experience layouts, open the *8th floor.dwg* file from AutoCAD's *sample* folder, a floor plan drawing provided with AutoCAD. If necessary, click the Model tab. You see the entire model, but without any notes, drawing borders, and title blocks.

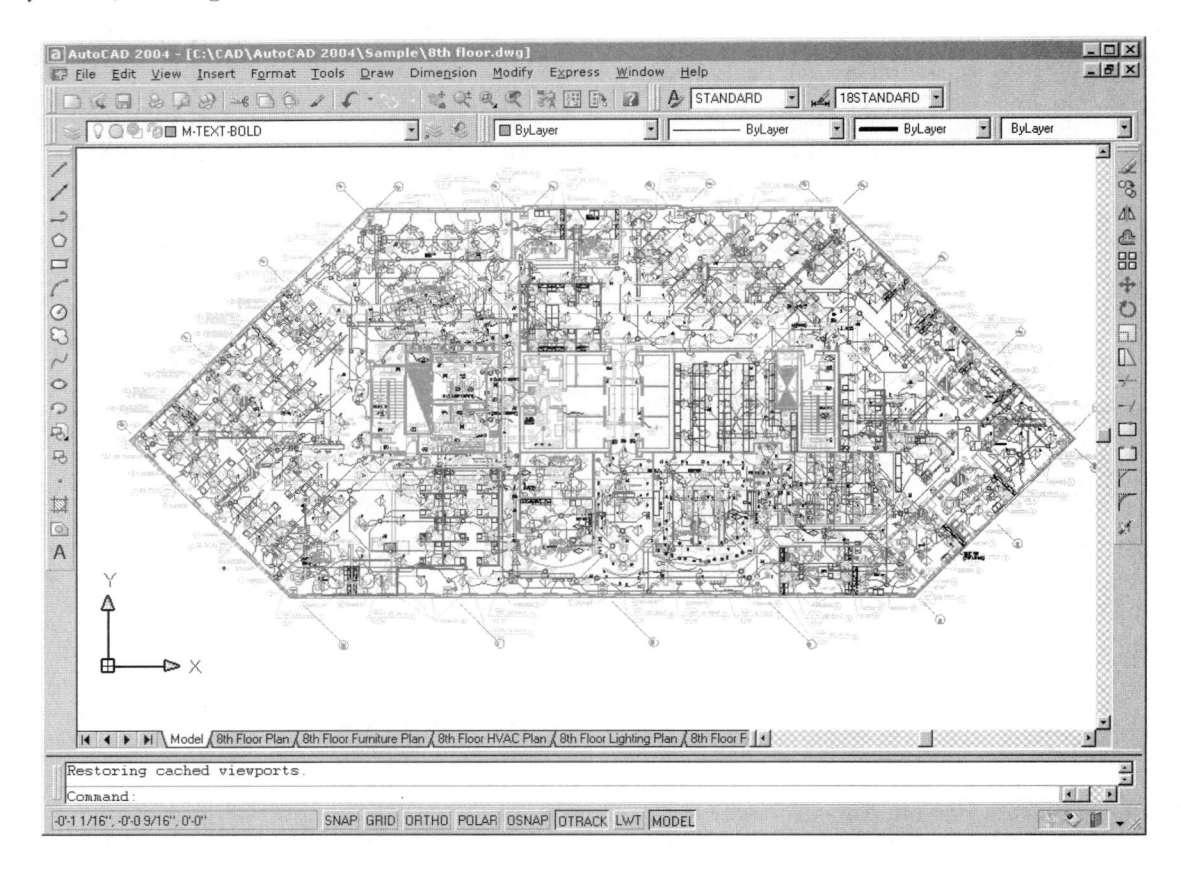

Click the **8th Floor Plan** tab. (This is equivalent to setting the TILEMODE system variable to 0. Technically, "tile mode" refers to the tiled viewports found in the Model tab.) The view changes dramatically. You see just the floor plan of the model, which is surrounded by the title block and drawing border. This illustrates how layers can be frozen to show selective details in layouts.

The drawing is on a white rectangle, which represents the paper. In this case, the selected paper size is Architectural E1, 30" x 42". The dashed line represent the printer margin, the unprintable area at the edges of the paper.

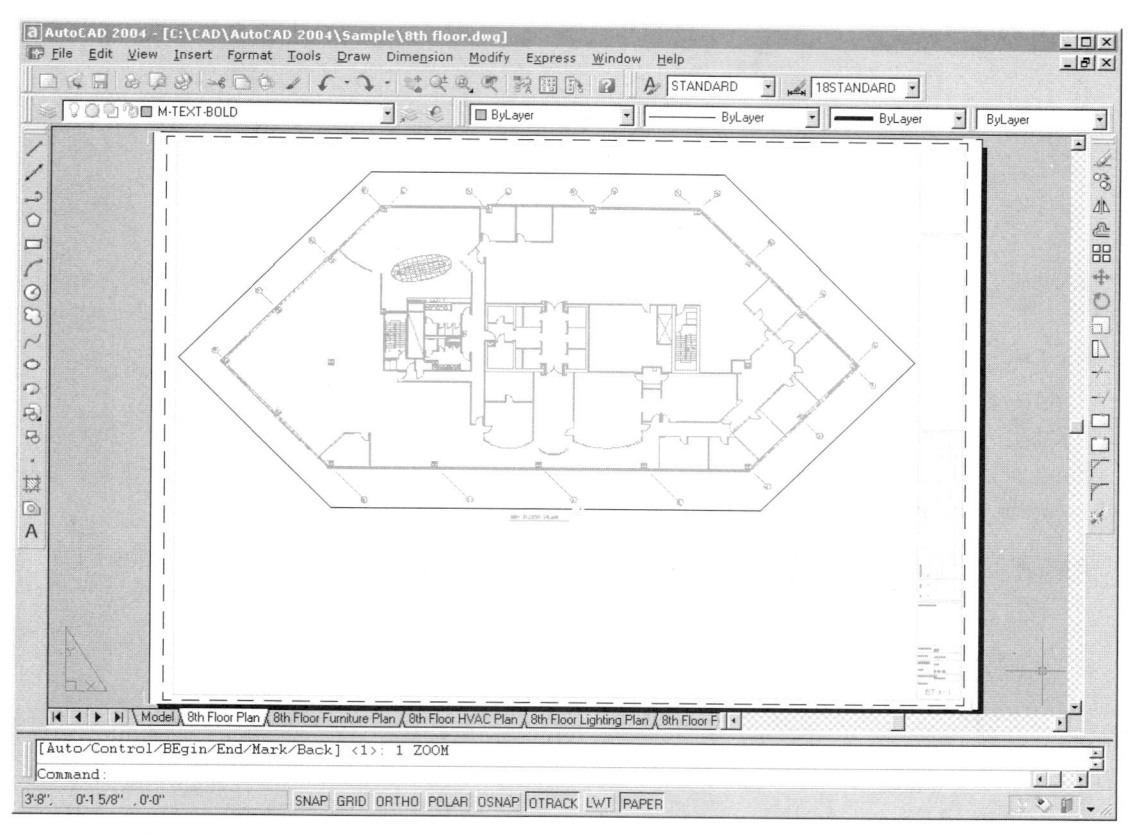

Click the **8th Furniture Plan** tab. The view changes again. In this layout, layers have been turned on to show the furniture — desks, chairs, and so on.

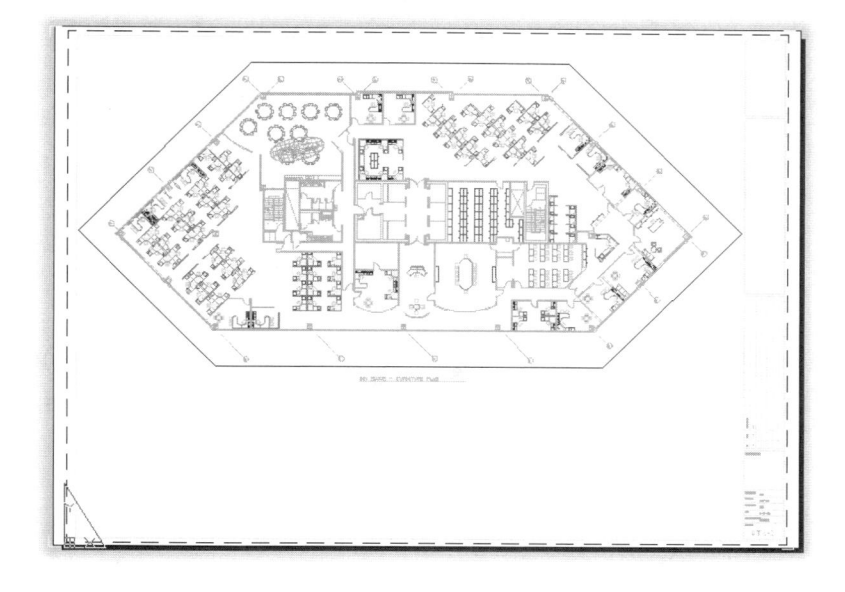

Click on the other layout tabs, working your way through the HVAC plan (heating, ventilation, air conditioning), lighting plan, power plan, and plumbing plan. As an alternative to clicking the tabs with the cursor, you can also press **CTRL+PGUP** and **CTRL+PGDN** to switch between layouts.

When drawings have many layouts, an additional set of controls becomes useful. The "VCR" buttons take you to the next and previous layouts, as well as to the first and last layouts.

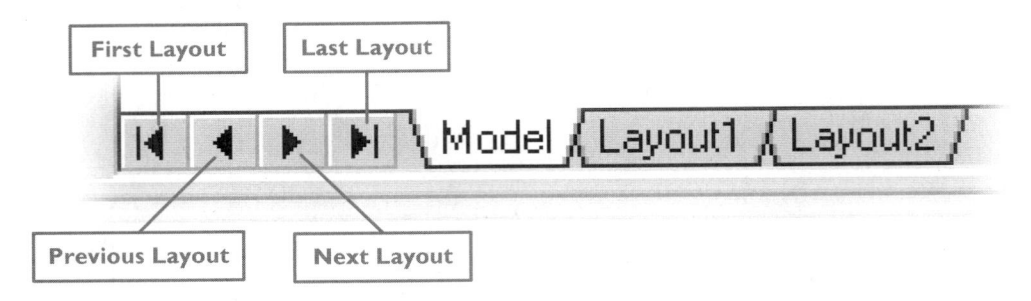

TUTORIAL: CREATING LAYOUTS

It can be disconcerting to create a layout for the first time. This tutorial takes you through the necessary steps.

1. From the CD-ROM, open the *layout.dwg* file, a streetscape drawing.
2. Click the **Layout1** tab.
   Notice that AutoCAD displays the Page Setup dialog box.
3. In the **Layout Name** text box, enter a name for the layout, such as "Cityscape."

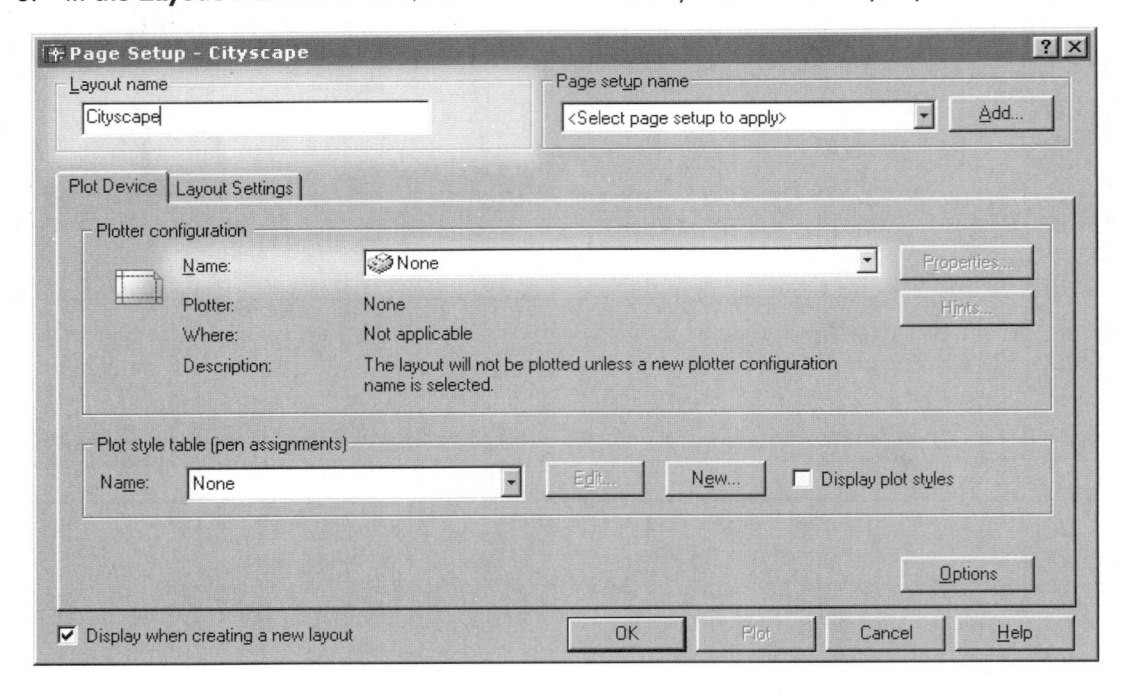

4. If you know the name of the printer that will be plotting this drawing, select it now. (You can always change the printer later.) In the Plot Device tab, select a printer from the **Name** drop list.
5. If you know the size of paper the drawing will be plotted on, select it now:
   Click the **Layout Settings** tab.
   From the **Paper Size** drop list, select the paper size. If you are not sure, select a size that matches 8.5 x 11.0 inches, such as Letter or ANSI A — a size that almost every printer supports.

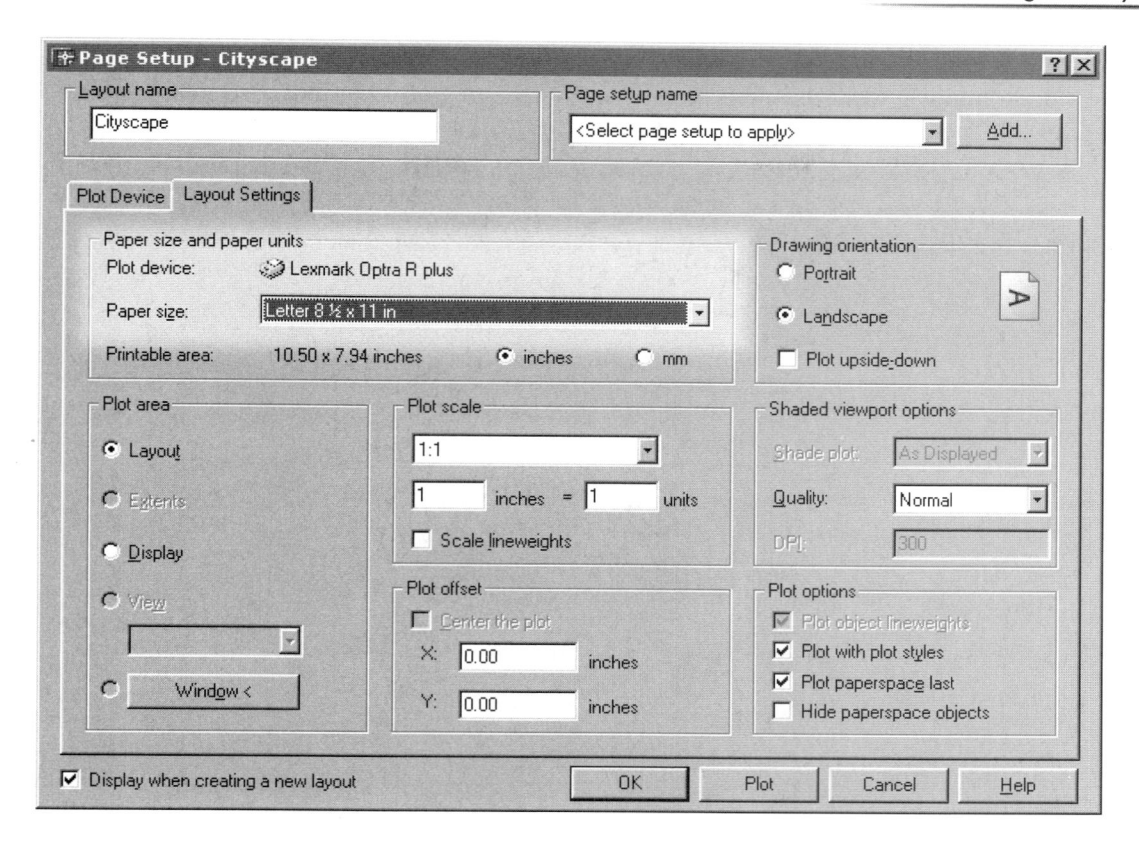

6. Click **OK**.

   Notice that the cityscape drawing appears on a "sheet of paper" surrounded by two rectangles. The inner rectangle is the viewport, the dashed rectangle is the margin. Notice, too, that the layout tab has changed its name from "Layout1" to "Cityscape."

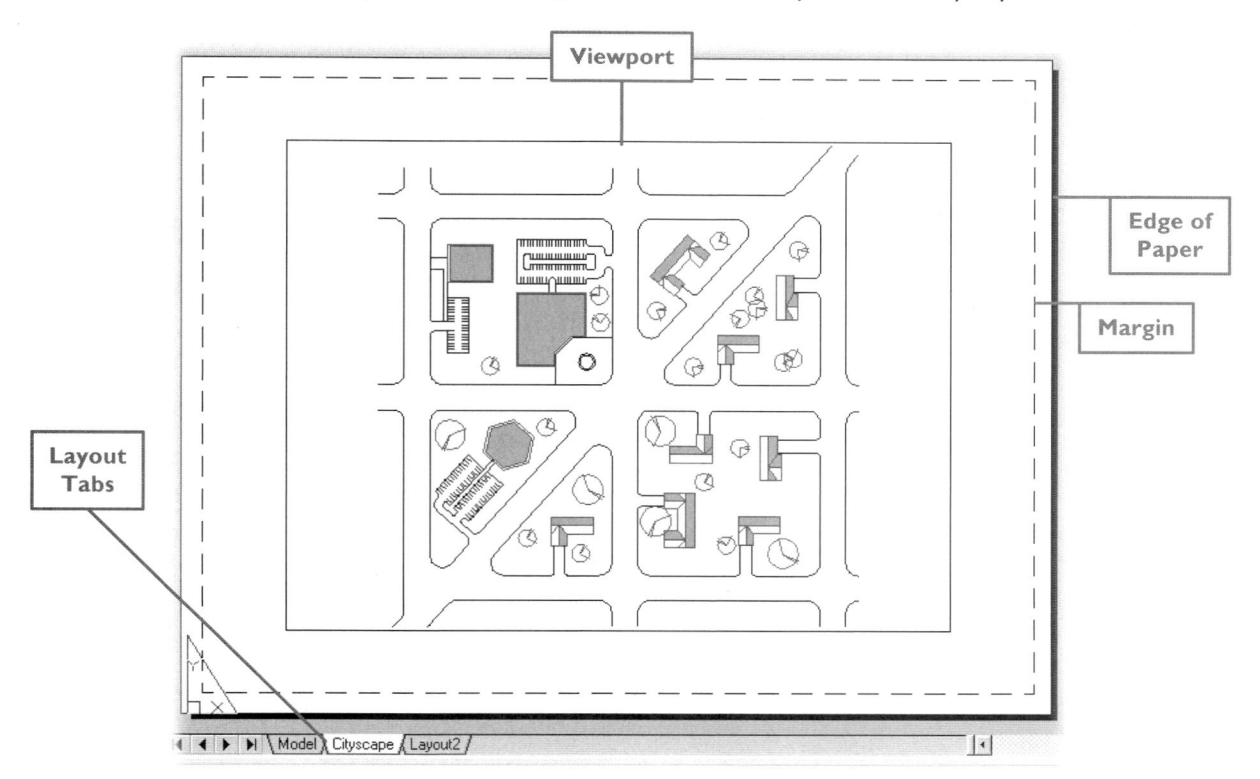

TUTORIAL: WORKING IN PAPER MODE

In paper mode, you can work only with the paper, not the model. "Working with the paper" includes drawing on the page, manipulating the viewports, and plotting.

1. Try selecting the model. Notice that you can't, because the layout is in "paper" mode. (On the status bar, the indicator reads PAPER.)
2. Click the viewport border. Notice that you can select it, unlike the model.

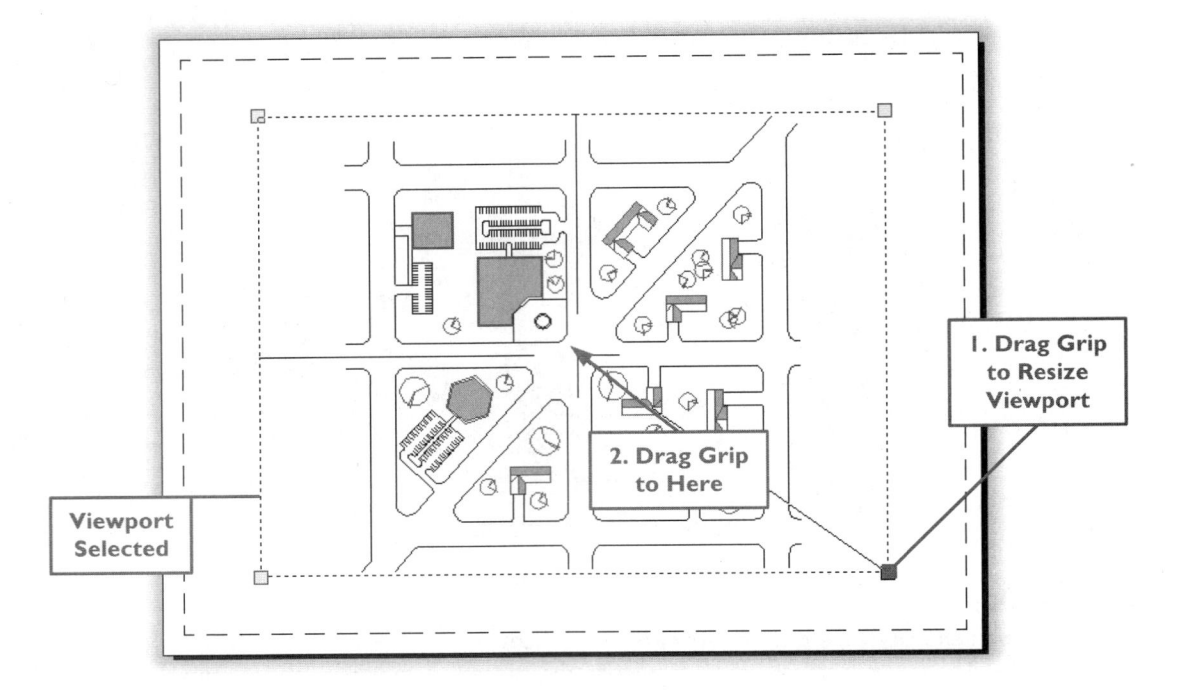

3. Grab one of the grips, and drag the viewport smaller. Notice what happens to the model: it has become "cropped" — part of the model has been cut off visually. This shows how you can selectively show details of the model in layouts. The layout need not show the entire model.

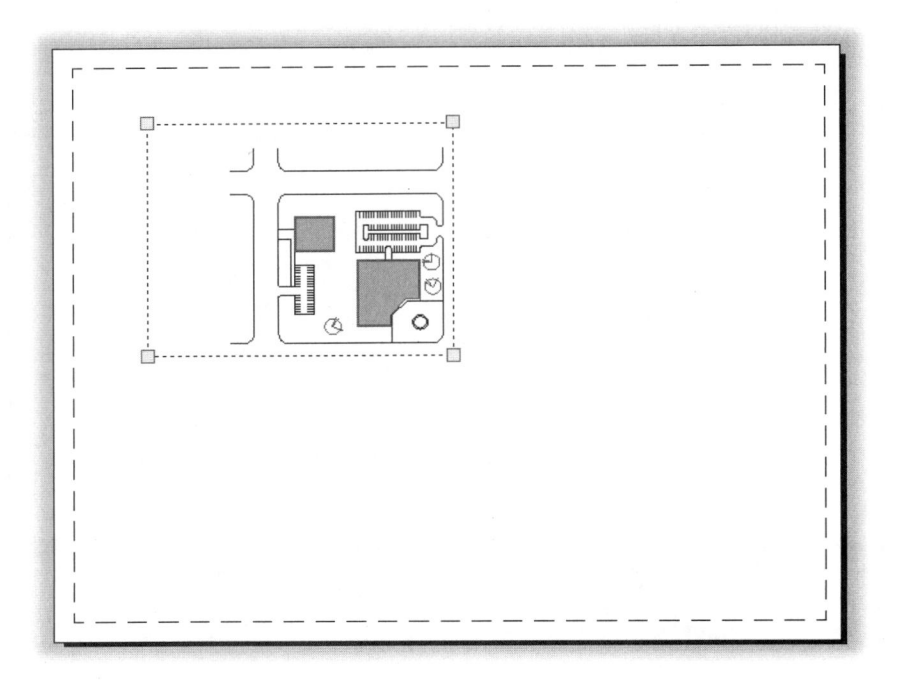

4. To move the viewport, grab one of its edges, and drag to its new position. Notice that the cropped model view moves with the viewport.

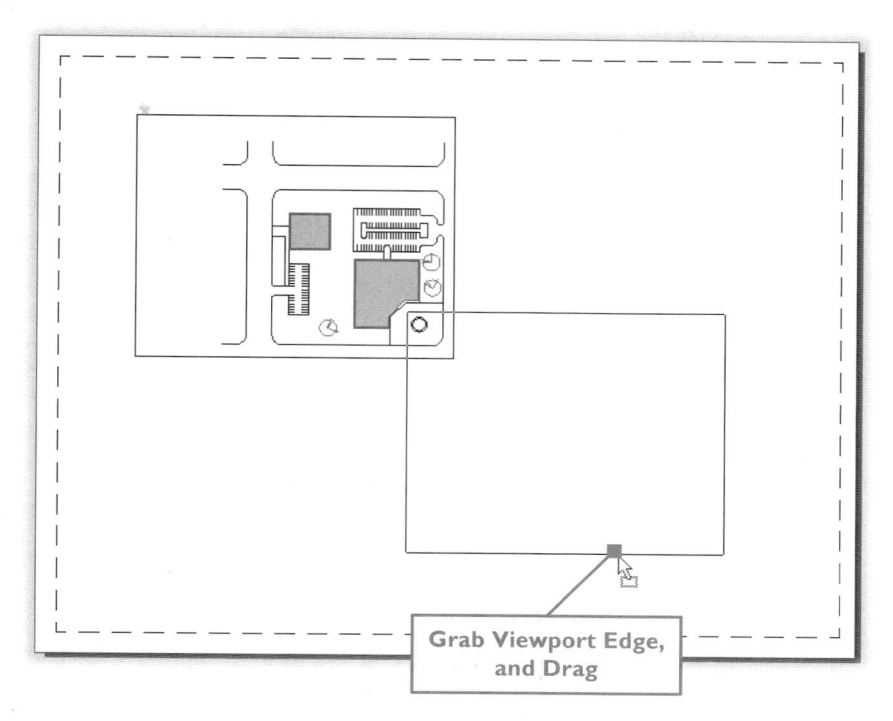

Grab Viewport Edge, and Drag

5. You can apply all of the grips editing commands on viewports. You can stretch, move, copy, scale (resize), rotate, and mirror viewports. Use the **COPY** command, for example, to make copies of viewports. Notice that the copy of the viewport contains exactly the same model view as the original. As the figure below illustrates, viewports can overlap.

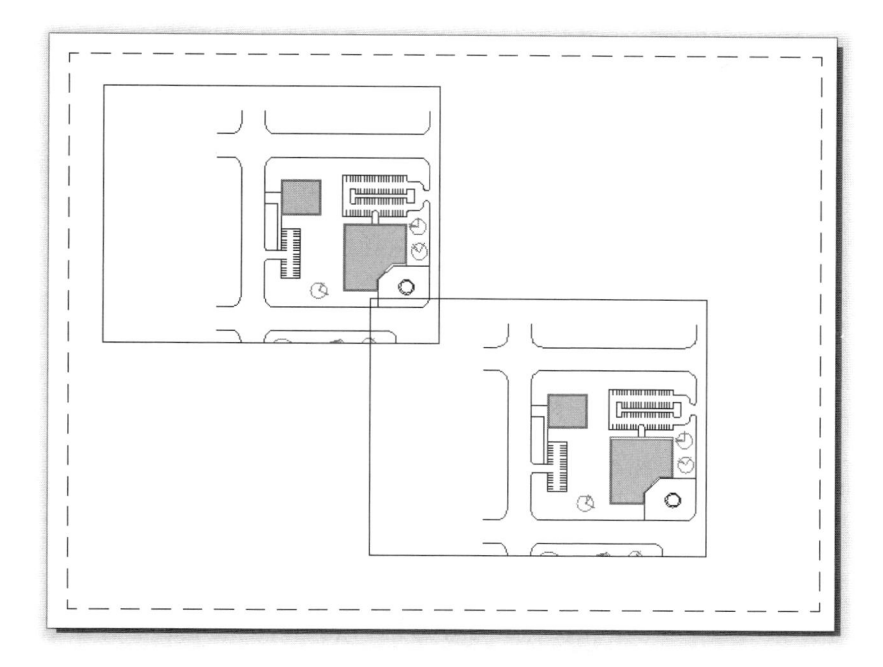

In addition, you can use the **ERASE** command to delete viewports. You can change the color of viewport borders, but you cannot apply linetypes or lineweight.

You cannot apply other editing commands to viewports. Commands like **OFFSET**, **TRIM**, **EXPLODE**, and **FILLET** don't work.

TUTORIAL: HIDING THE VIEWPORT BORDER

You can place viewports on layers, and then freeze the layer. The viewport border is no longer visible, but the contents of the viewport are. Use this trick to hide the viewport border:

1. With the **LAYER** command, create a new layer called "VPorts."
2. With the **PROPERTIES** command, select the viewport borders, and change the layer to VPorts.

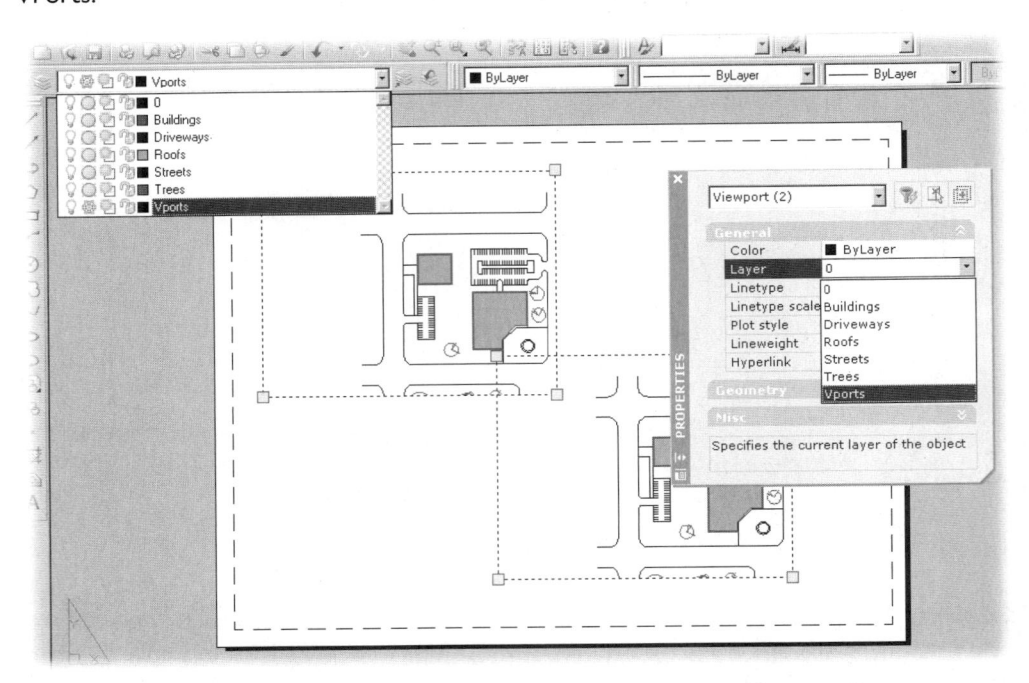

3. From the **Layer** toolbar, click the **Layer Control** drop list, and freeze layer VPorts. Notice that the viewport border disappears, but that the model view remains.

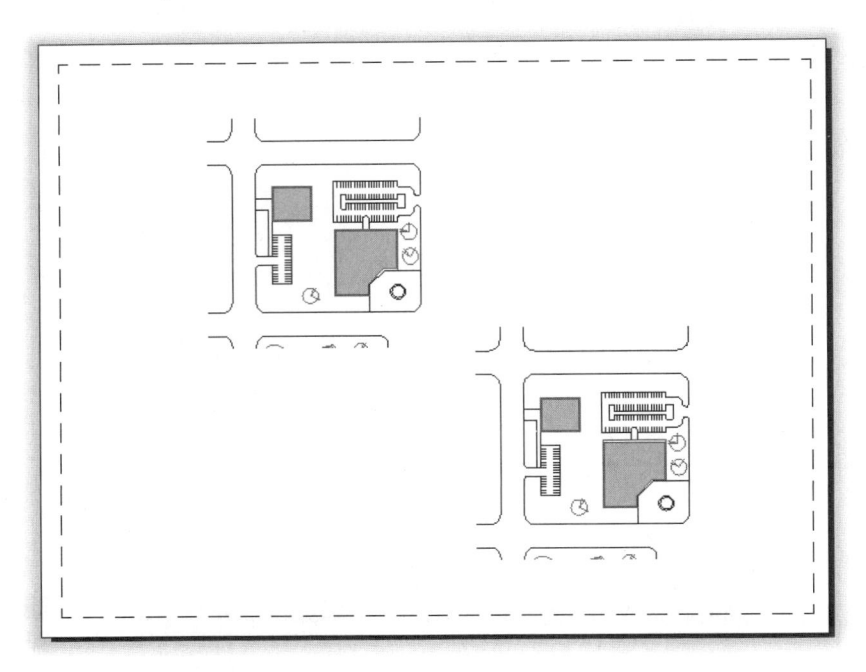

4. Thaw layer VPorts to bring back the viewport borders.

TUTORIAL: SWITCHING FROM PAPER TO MODEL MODE

Until now, you have been manipulating the viewports. Let's begin working with the model inside the viewports.

1. On the status bar, click PAPER.

   The button changes to MODEL. This means that you are now working with the model instead of the paper.

2. Notice that one of the viewports has a heavier border. This is the *active* viewport, the one in which you can manipulate the model.

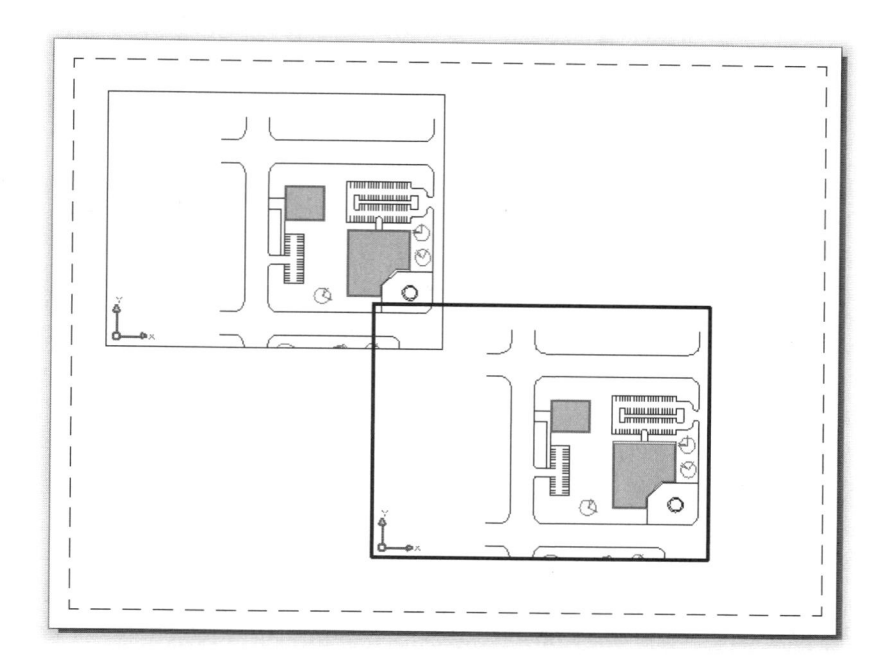

   To make another viewport active, simply click on it. Alternatively, you can press **CTRL+R** to cycle through the viewports.

3. Select an object in the model. Notice that the grips and highlight show in both viewports. This proves that you are seeing two images of the same model.

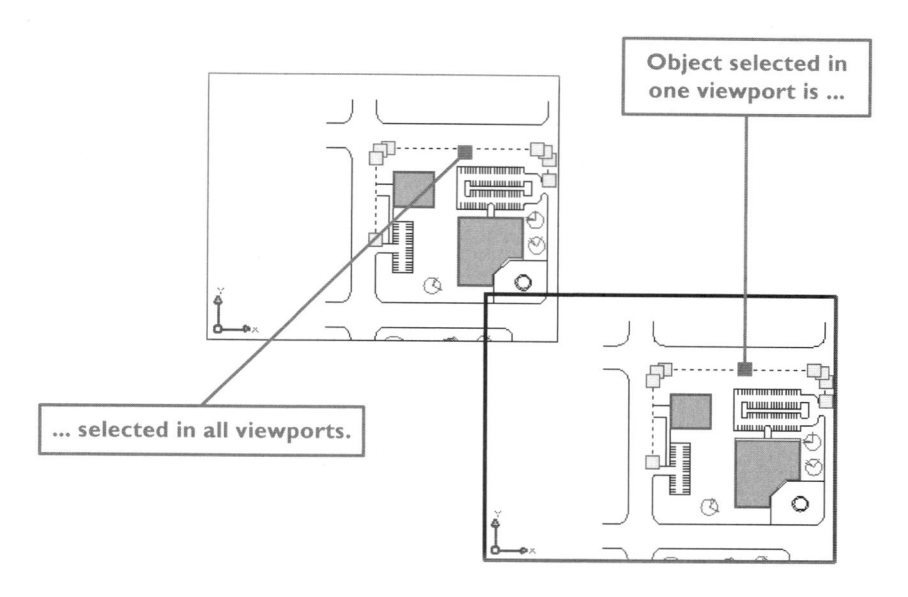

TUTORIAL: SCALING MODELS IN VIEWPORTS

While objects are not independent in viewports, their visibility is. AutoCAD's two primary tools are **ZOOM** and **LAYER**. In model mode, AutoCAD commands apply only to the active viewport; the other viewports are ignored.

1. The **ZOOM** command changes the size of the model in the viewport.
   Enter the **ZOOM** command, and then use the **Extents** option. You should see the entire model in the active viewport.

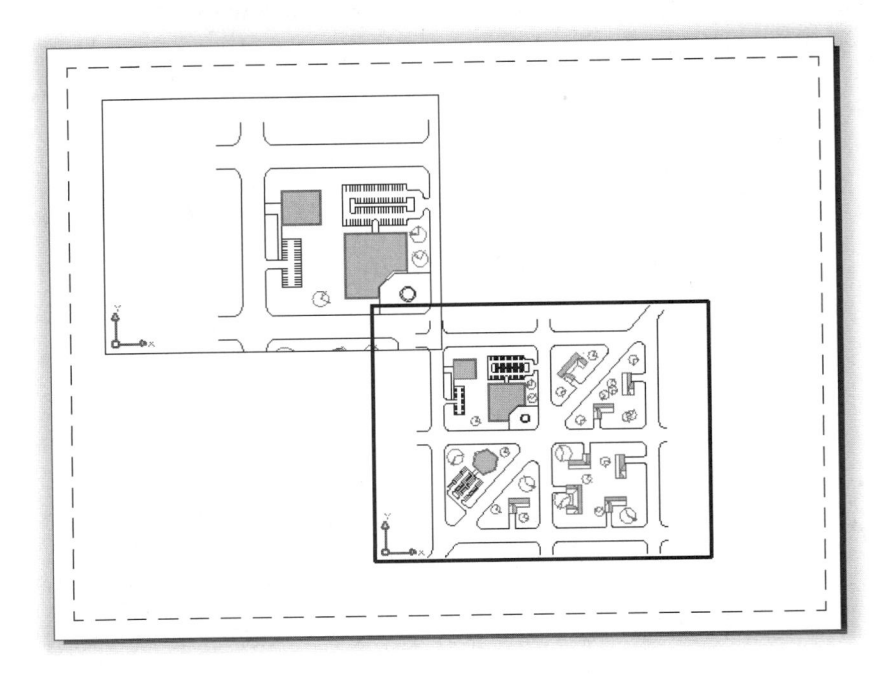

2. The **ZOOM** command's **XP** option is meant for use only in layout tabs. This option scales the model view relative to the layout (a.k.a. paper space). This is important: *You use ZOOM XP to scale models in layout viewports.*

   City plans are scaled typically at large scale factors, such as 1:1000, 1:2000, 1:5000, and so on. This model is approximately 470 feet across, while the viewport is about 4.7 inches across. Doing the math shows that the nearest standard scale factor is 1:2000:

   $$(470' \times 12 \text{ inches/foot}) \div 4.7" = 1200$$

   The scale factor of 1:2000 is the zoom factor used with the **XP** suffix:

   Command: **zoom**

   Specify corner of window, enter a scale factor (nX or nXP), or

   [All/Center/Dynamic/Extents/Previous/Scale/Window] <real time>: **1/2000xp**

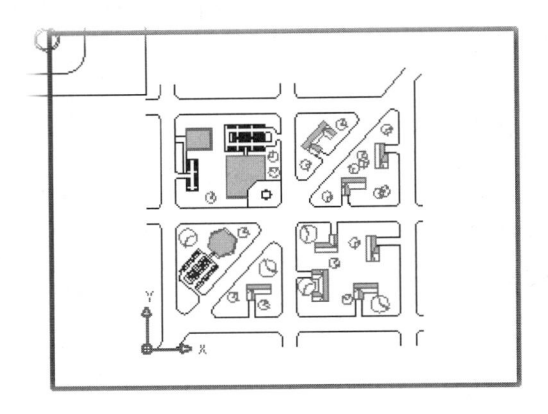

The layout now has two views of the model: a large one and a small one. This illustrates that viewports can have independent views of the model.

3. As an alternative to the **ZOOM XP**, command, you can use the **PROPERTIES** command to set the model scale.

   Switch back to PAPER mode.

   Double-click the viewport border.

   In the **Misc** section of the Properties window, click **Custom Scale**.

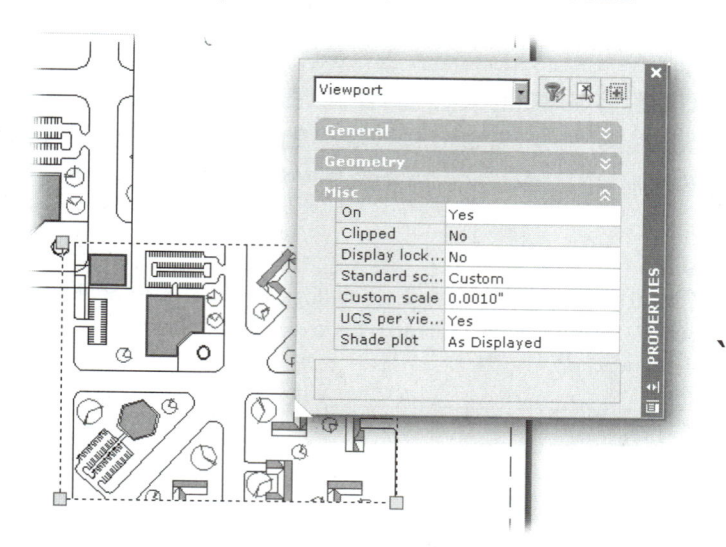

Enter a new scale factor, such as **0.0001**, and then press **ENTER**.

Notice that the size of the model changes in the viewport.

Note: You can use the **PAN** command to position the model inside the viewport.

TUTORIAL: SELECTIVELY DISPLAY DETAILS IN VIEWPORTS

The **LAYER** command allows you selectively to freeze layers in viewports. The **Current VP Freeze** option allows one viewport to show some details, and while other viewports show others.

1. In MODEL mode, select a viewport.
2. With the **LAYER** command, open the Layer Properties Manager dialog box.

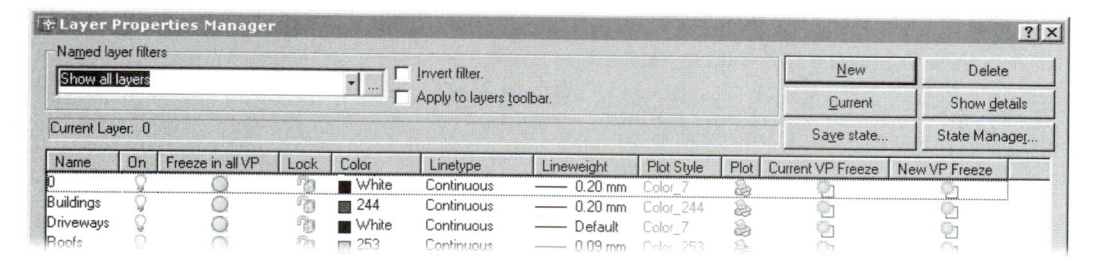

At the far right end are two columns you have not yet used:

- **Current VP Freeze** freezes the selected layers in the current viewport (VP).

- **New VP Freeze** freezes the selected layers when a new viewport is created.

These two options have no effect in model tab, or in other layouts and viewports. The **Freeze in all VP** option, with which you are already familiar, freezes the layers in *all* viewports, whether in model or layout tabs, in paper mode or in model mode.

3. Hold down the **CTRL** key, and then select the Buildings, Roofs, and Trees layers.
4. Let go of the **CTRL** key, and then click a highlighted icon in the Current VP Freeze column. Notice that the suns turn to snowflakes.

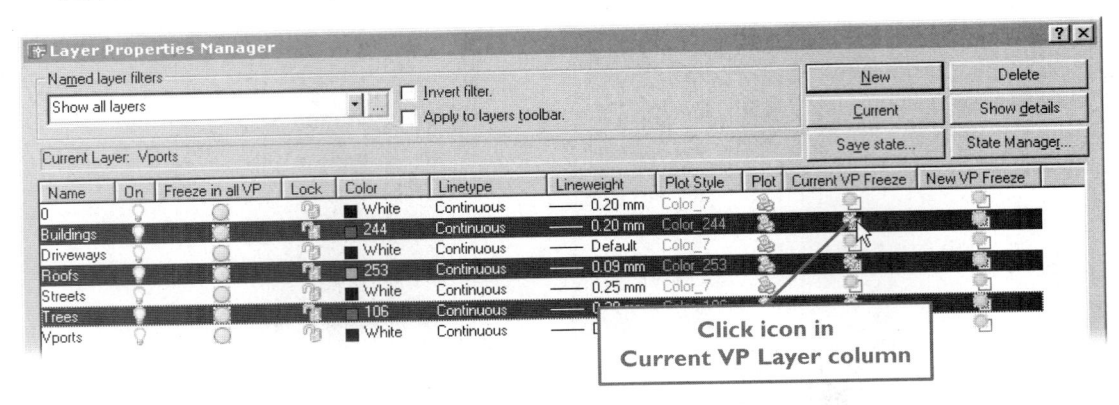

5. Choose the **OK** button, and notice that the buildings, roofs, and trees disappear from the viewport. The other viewport is unaffected. This shows that viewports can freeze layers independently of each other.

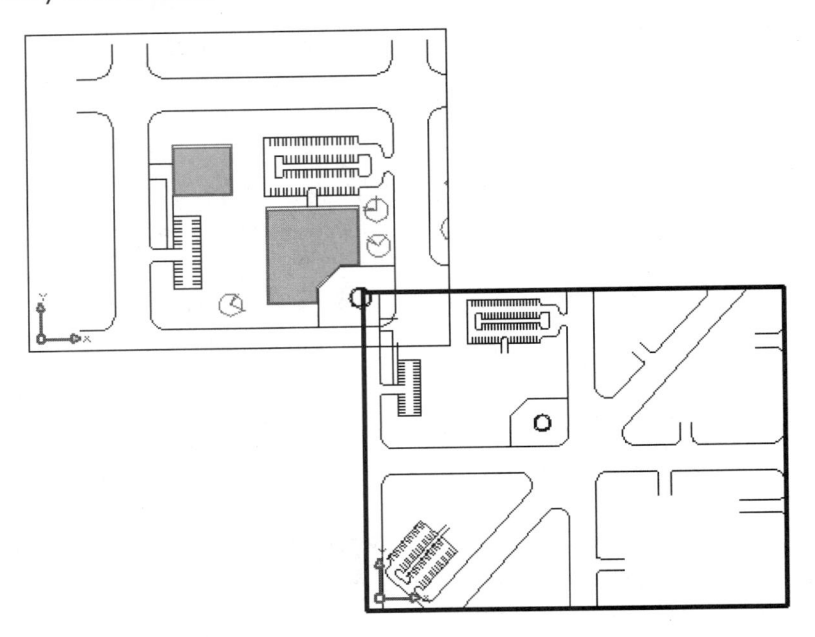

TUTORIAL: INSERTING TITLE BLOCKS

Layout mode is where you place scale-dependent objects, such as drawing notes and title blocks. In this tutorial, you add a title block and border to the drawing.

1. Ensure that Cityscape layout is in **PAPER** mode.
2. Create a new layer called "Titleblock," and then make it current.
3. Open DesignCenter (press **CTRL+2**).
4. In the Folder List, go to the folder holding the template drawings.
   In AutoCAD 2004, Autodesk relocated the *.dwt* template drawings from the handy *template* folder (now empty) to the more inconvenient (from the use standpoint) folder:

   c:\documents and settings*username*\local settings\autodesk\autocad\r16.0\enu\template

   Replace *username* with the name by which you log in to Windows.

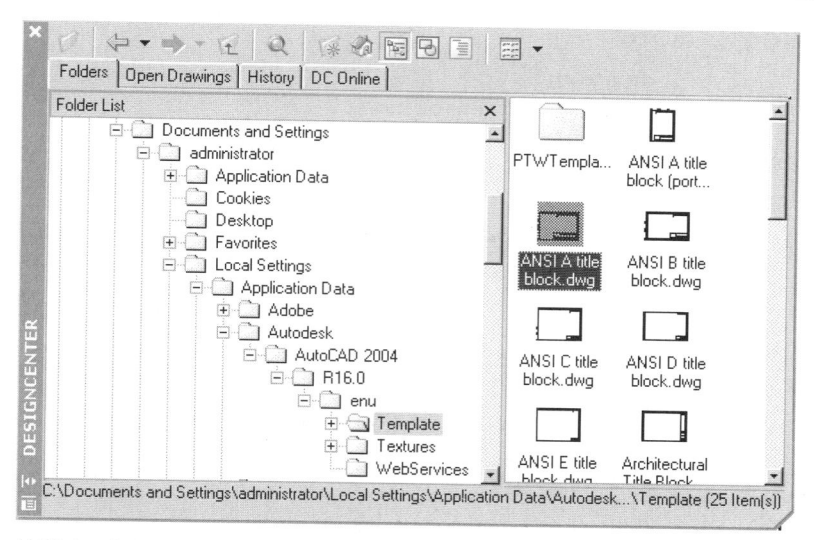

5. Drag the *ANSI A title block.dwg* file from DesignCenter into the drawing.
6. When AutoCAD prompts you for the insertion point, pick the lower-left corner of the margin.

    Command: _-INSERT Enter block name or [?] <A$C27FF03C9>: "C:\CAD\AutoCAD

    2004\UserDataCache\Template\ANSI A title block.dwg"

    Specify insertion point or [Scale/X/Y/Z/Rotate/PScale/PX/PY/PZ/PRotate]: *(Pick the lower left corner of the margin.)*

7. Press **ENTER** at the remaining prompts.

    Enter X scale factor, specify opposite corner, or [Corner/XYZ] <1>: *(Press* **ENTER** *to keep scale factor at 1.0.)*

    Enter Y scale factor <use X scale factor>: *(Press* **ENTER** *to keep scale factor at 1.0.)*

    Specify rotation angle <0>: *(Press* **ENTER** *to keep angle at 1.0.)*

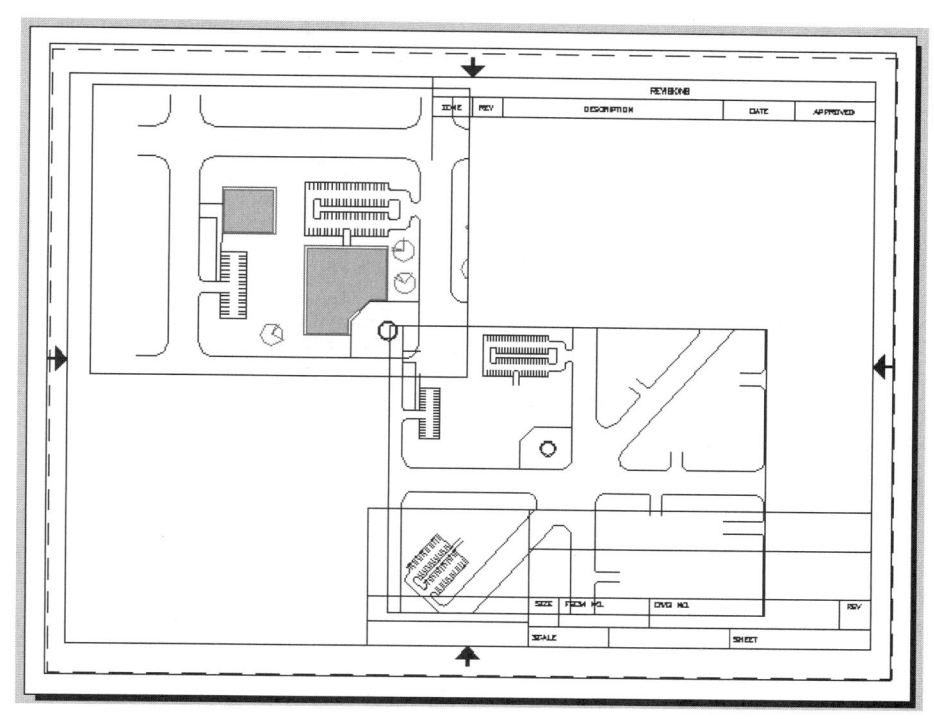

8. You may need to move the viewports so that they do not interfere with the title block or border.

TUTORIAL: USING TILEMODE

At the start of this chapter, I mentioned the TILEMODE system variable, which has been replaced by the layout tabs. For completeness, here is how to use it:

Command: **tilemode**
Enter new value for TILEMODE <0>: **1**

A value of **1** returns to the model tab, while a value of **0** returns to the last active layout tab.

LAYOUTWIZARD

The LAYOUTWIZARD command takes you through the steps for creating new layouts.

TUTORIAL: MANAGING LAYOUTS

1. To manage viewports, start the **LAYOUT** command:
   - From the **Insert** menu, choose **Layouts**, and then **Layout Wizard**.
   - In the **Layouts** toolbar, choose **New Layout**.
   - At the 'Command:' prompt, enter the **layoutwizard** command.

   Command: **layoutwizard** (Press ENTER.)

2. In all cases, AutoCAD displays the first dialog box:

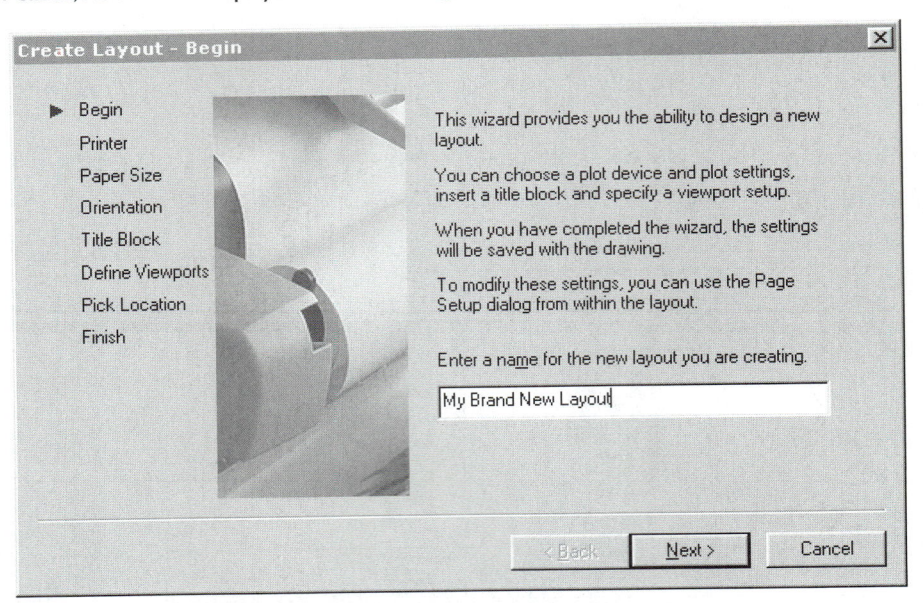

Enter a name for the layout, which will appear on the tab.
Click **Next**.

3. Select a printer from the list of Windows system printers and AutoCAD HDI printer drivers. (You can select a different printer later, if need be.)
   Click **Next**.

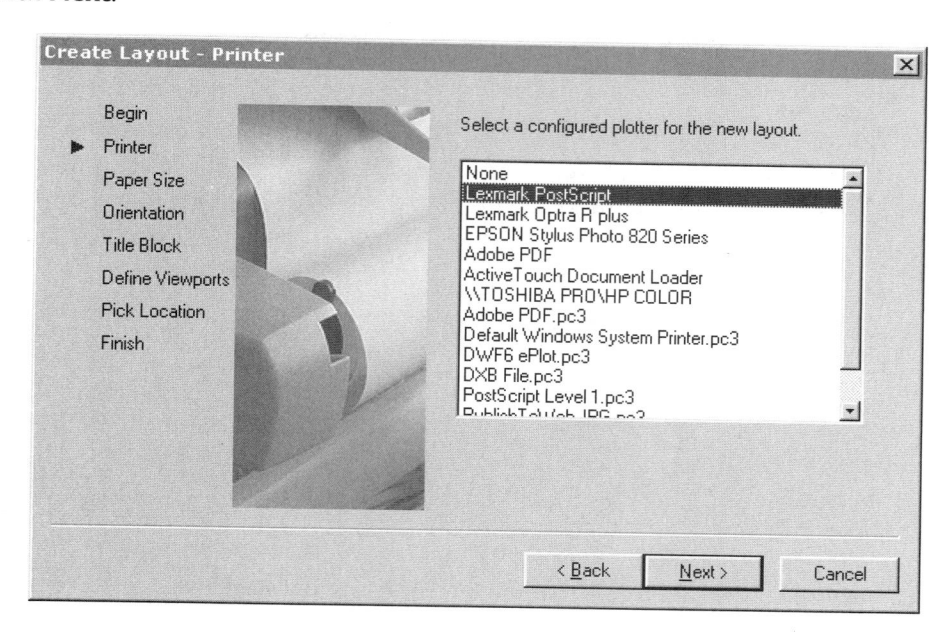

4. Select the drawing units — inches or mm.

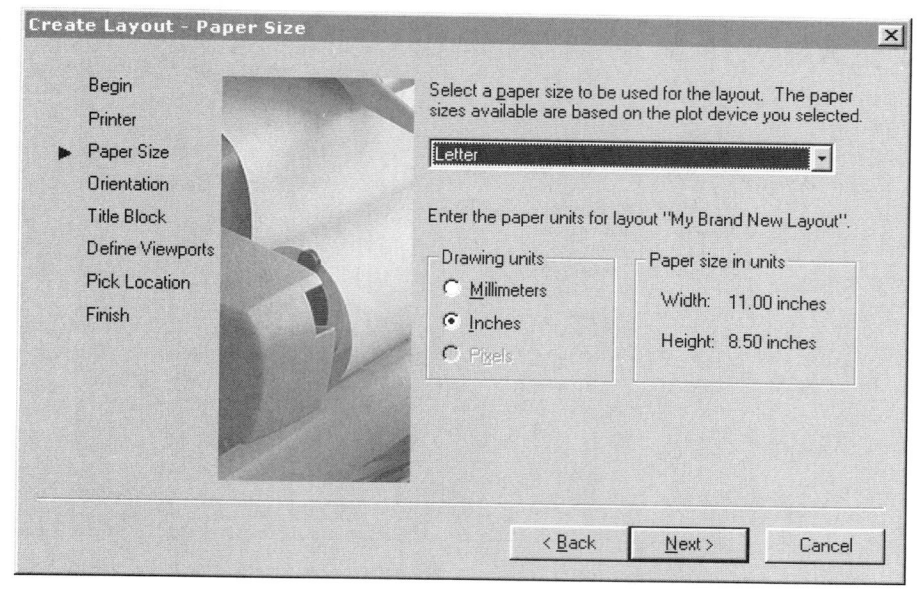

5. Select a paper size from the drop list. Only those sizes supported by the printer are listed.
   Click **Next**.

6. Select an orientation for the paper — landscape or portrait.

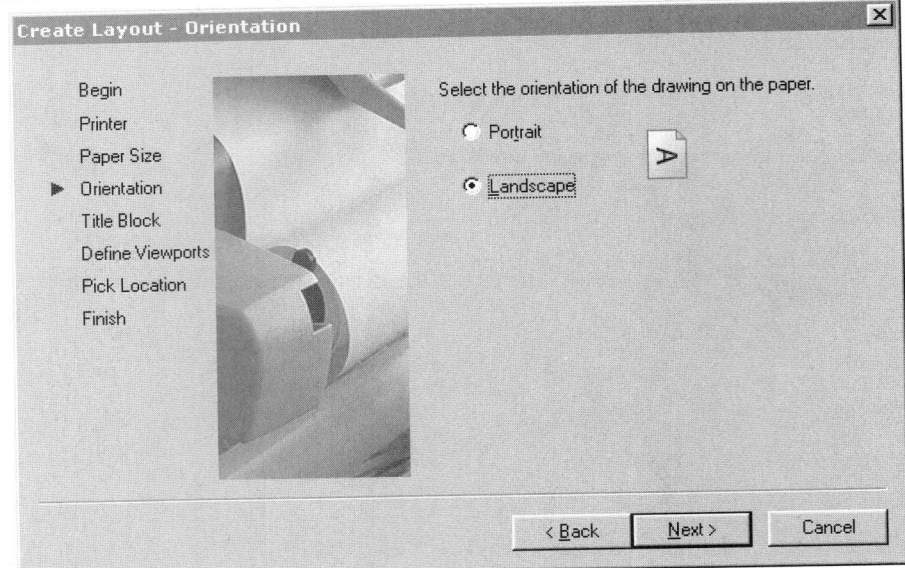

7. Select a drawing border/title block from the list. For A-size drawings, make sure the orientation of the border matches the orientation of the page you selected in the previous setup.

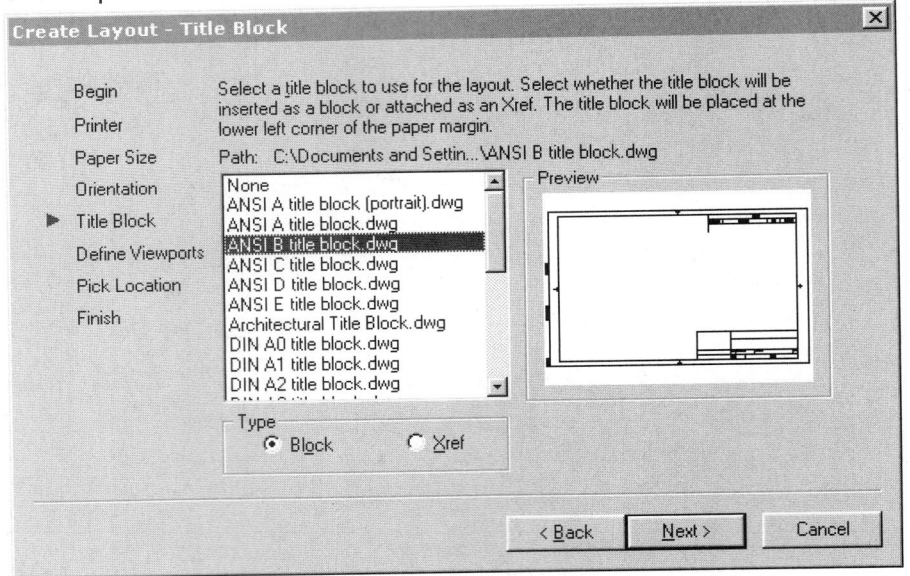

8. Decide whether you want the drawing border to be inserted as a block, or attached as an externally-referenced drawing (xref). If you are not sure, select **Block**.

Type	Pros	Cons
Block	Border and title block are part of the drawing, making it complete and easier to transport, such as by email.	Drawing size is larger. Blocks cannot be as readily updated as xrefs.
Xref	Drawing size is smaller. All xref'ed title blocks can be easily changed.	When sending the drawing to another office or client, you must ensure the xref is packaged with the drawing file.

9. Select the number and style of viewports. If you are not sure, select **Single**.

Viewport Setup	Comments
None	No viewport is created.
Single	One viewport is created.
Std. 3D Engineering Views	Four viewports are created, and the viewpoint adjusted to show the standard engineering views of 3D drawings: front, side, top, and isometric.
Array	*Rows* x *columns* array of viewports is created.

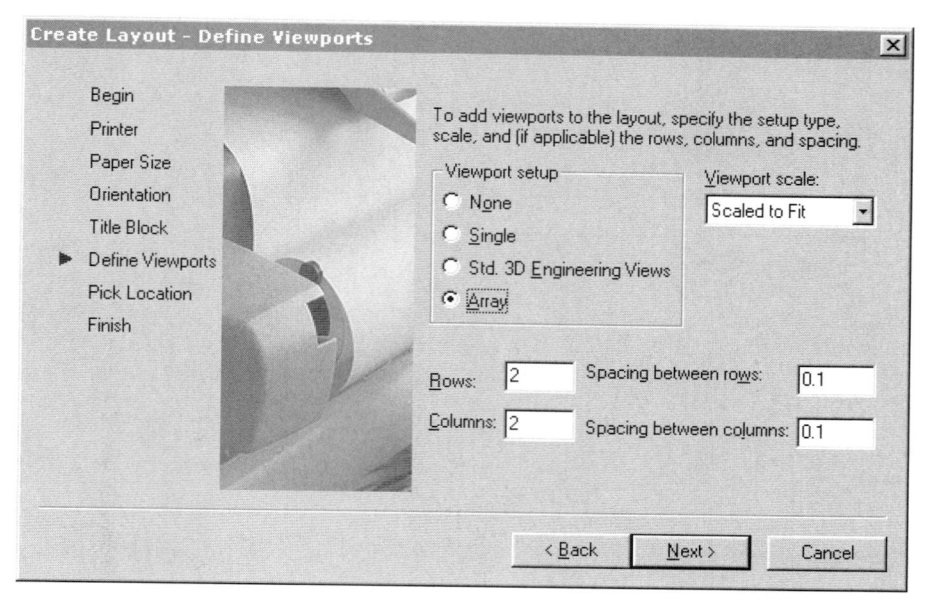

10. Select the viewport scale, which ranges from 100:1 to 1:100, and from 1/128"=1' to 1'=1'. If you are not sure of the scale, select **Scaled to Fit**. (You can change it later by double-clicking the viewport border, and changing the value of **Custom Scale**.)

11. Click **Select Location** to position the viewport(s).

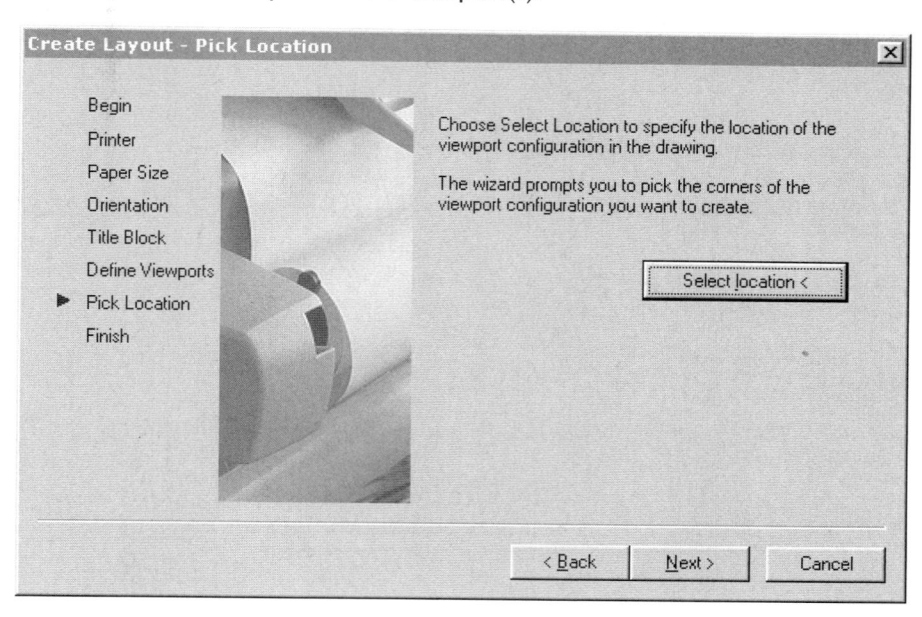

12. AutoCAD prompts you at the command line:

    Specify first corner: *(Pick a point.)*

    Specify opposite corner: *(Pick another point.)*

    Pick two points to form a rectangle. AutoCAD fits the viewport(s) to the rectangle.
13. Click **Finish**.

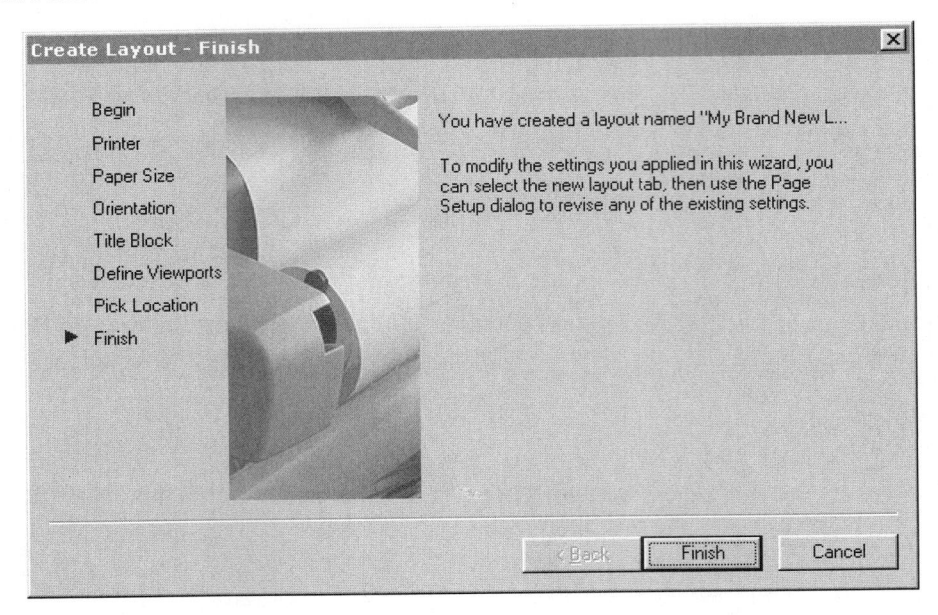

AutoCAD creates the viewport, which allows the model to show through into the layout.

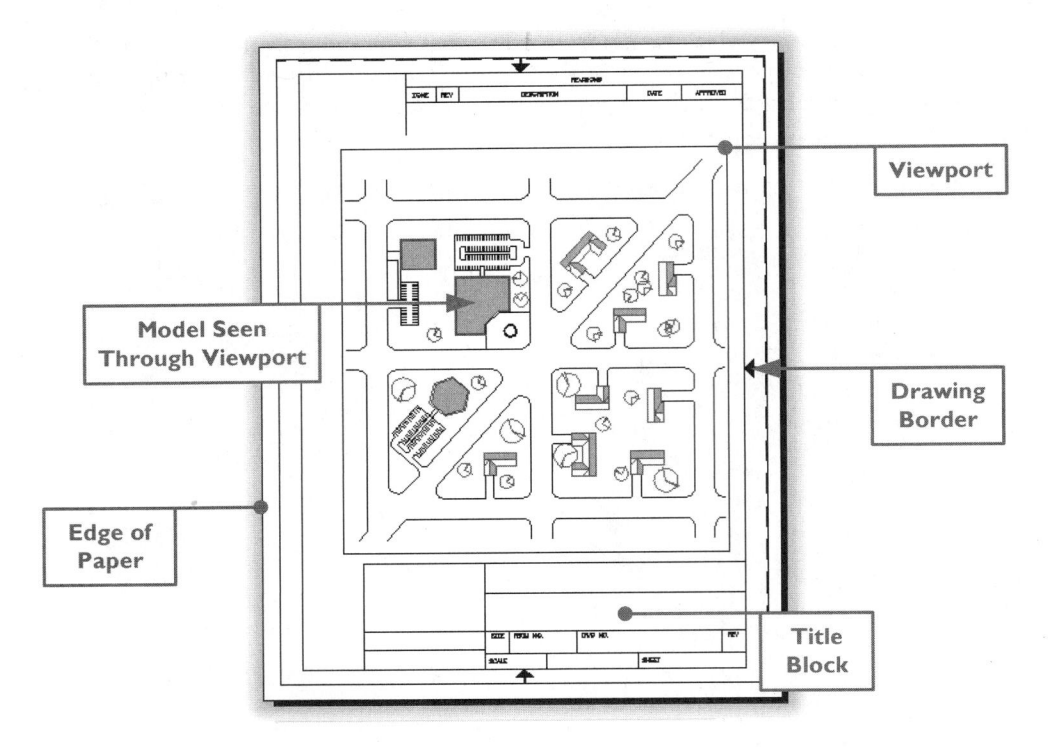

LAYOUT

The **LAYOUT** command manages layouts.

TUTORIAL: MANAGING LAYOUTS

1. To manage viewports, start the **LAYOUT** command:
   - From the **Insert** menu, choose **Layouts**.
   - In the **Layouts** toolbar, choose **New Layout**.
   - Right-click any layout tab:

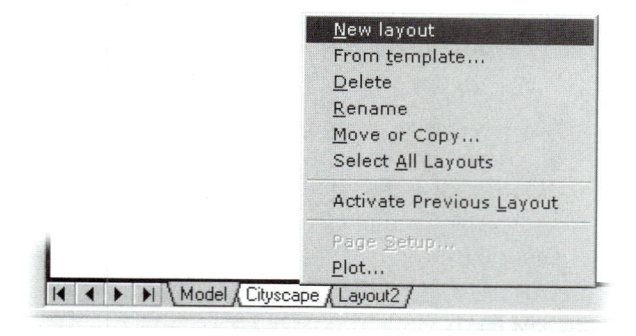

   - At the 'Command:' prompt, enter the **layout** command.
   - As an alternative, enter the **lo** alias.

   Command: **layout** *(Press ENTER.)*

2. In all cases, AutoCAD displays the prompts at the command line:
   Enter layout option [Copy/Delete/New/Template/Rename/SAveas/Set/?] <set>:

3. Enter an option, as described below:

Layout	Comment
Copy	Copies the selected layout.
	Enter name of layout to copy:
Delete	Removes the selected layout. (The Model tab cannot be deleted).
	Enter name of layout to delete:
New	Creates new layouts. Names can be up to 255 characters long, but only a maximum of 31 are displayed on the tab, fewer if there is less room.
	Enter name of new layout:
Template	Creates new layouts based on *.dwt*, *.dwg*, and *.dxf* files. Displays the Insert Layout(s) dialog box, and then inserts all objects and layouts into the drawing.

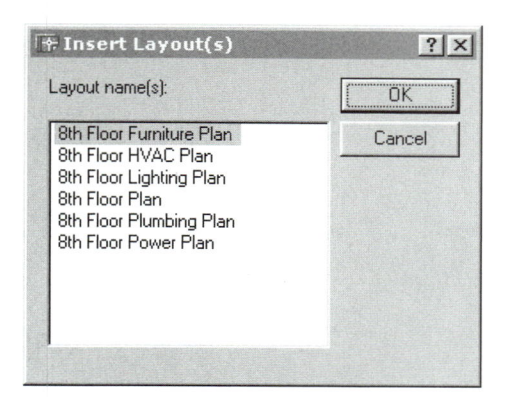

Layout (cont'd)	*Comment*
Rename	Renames layouts.
	Enter name of layout to rename:
	Enter new layout name:
Saveas	Saves layouts as *.dwt* template files.
	Enter layout to save to template:
Set	Makes the selected layout \current.
	Enter layout to make current:
?	List the names of layouts.

VIEWPORTS

The **VIEWPORTS** command creates and merges rectangular viewports.

When you switch to a layout for the first time, AutoCAD creates a single viewport for you, which you can copy, move, and resize. To create a new, rectangular viewport, use the **VIEWPORTS** dialog box. To create non-rectangular viewports, or convert objects into viewports, use **-VPORTS** at the command line. Both commands work in model space and in layouts, but non-rectangular viewports are limited to layouts. The **VIEWPORTS** command needs to know two pieces of information: (1) the number of viewports; and (2) the location of the viewports.

TUTORIAL: CREATING RECTANGULAR VIEWPORTS

1. To create one or more rectangular viewports, start the **VIEWPORTS** command:
   - From the **View** menu, choose **Viewports**, and then **New Viewport**.
   - From the **Viewports** toolbar, choose the **Display Viewports Dialog** button.
   - At the 'Command:' prompt, enter the **viewports** command.
   - Alternatively, enter the **vports** alias at the 'Command:' prompt.

   Command: **viewports** (Press ENTER.)

   In all cases, AutoCAD displays the Viewports dialog box.

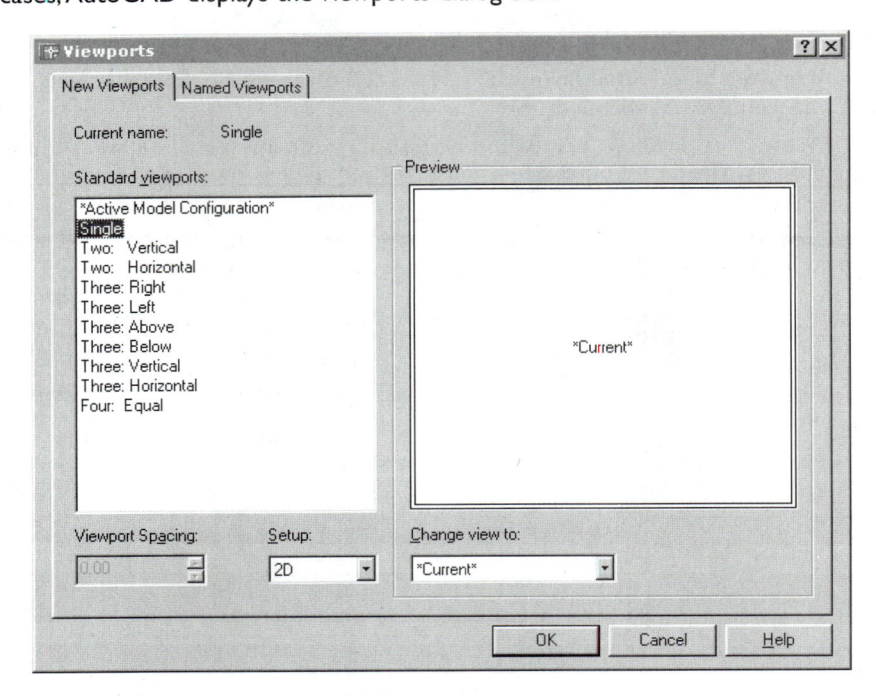

2. Select a style of viewport:

Standard Viewport	Style	Standard Viewport	Style
Single		Three Above	
Two Vertical		Three Below	
Two Horizontal		Three Vertical	
Three Right		Three Horizontal	
Three Left		Four Equal	

3. Click **OK**. AutoCAD prompts you for the location of the viewports:

    Specify first corner or [Fit] <Fit>: *(Press ENTER to fit the viewports to the display area.)*

    Press **ENTER**, and AutoCAD fits the viewports to the display area. Or, you can pick two points (forming a rectangle), and AutoCAD fits the viewports.

> **Note:** Viewports created in layouts are called "floating," because you can move them about. Each viewport is independent of the others. Polygonal (non-rectangular) viewports and viewports converted from objects can only be created in layouts.
>
> Viewports created in model space are called "tiled" viewports, because they stick together as tightly as tiles. They cannot be moved or copied. You can merge and erase them only with the **-VPORTS** command.

-VPORTS AND VPCLIP

The **-VPORTS** command creates non-rectangular viewports, while the **VPCLIP** command converts rectangular viewports into clipped viewports.

You are not stuck with just rectangular viewports. AutoCAD can create viewports that are polygonal (non rectangular) and circular. These are also known as "clipped viewports," because they are meant to hide a portion of the model.

-VPORTS converts these objects into viewports: closed polylines, circles, ellipses, splines, and regions. The polyline can be made of line and arc segments, and may intersect itself.

TUTORIAL: CREATING POLYGONAL VIEWPORTS

1. To create polygonal viewports, ensure AutoCAD is displaying a layout; this action cannot be carried out in model space.
2. Start the **-VPORTS** command:
   - From the **View** menu, choose **Viewports**, and then **Polygonal Viewport**.
   - At the 'Command:' prompt, enter the **-vports** command.

   Command: **-vports** *(Press ENTER.)*

3. In all cases, AutoCAD displays the prompts at the command line. Enter **P** to draw the outline of a polygonal viewport:

   Specify corner of viewport or
   [ON/OFF/Fit/Shadeplot/Lock/Object/Polygonal/Restore/2/3/4] <Fit>: **p**

4. Pick a point to start.

   Specify start point: *(Pick a point.)*

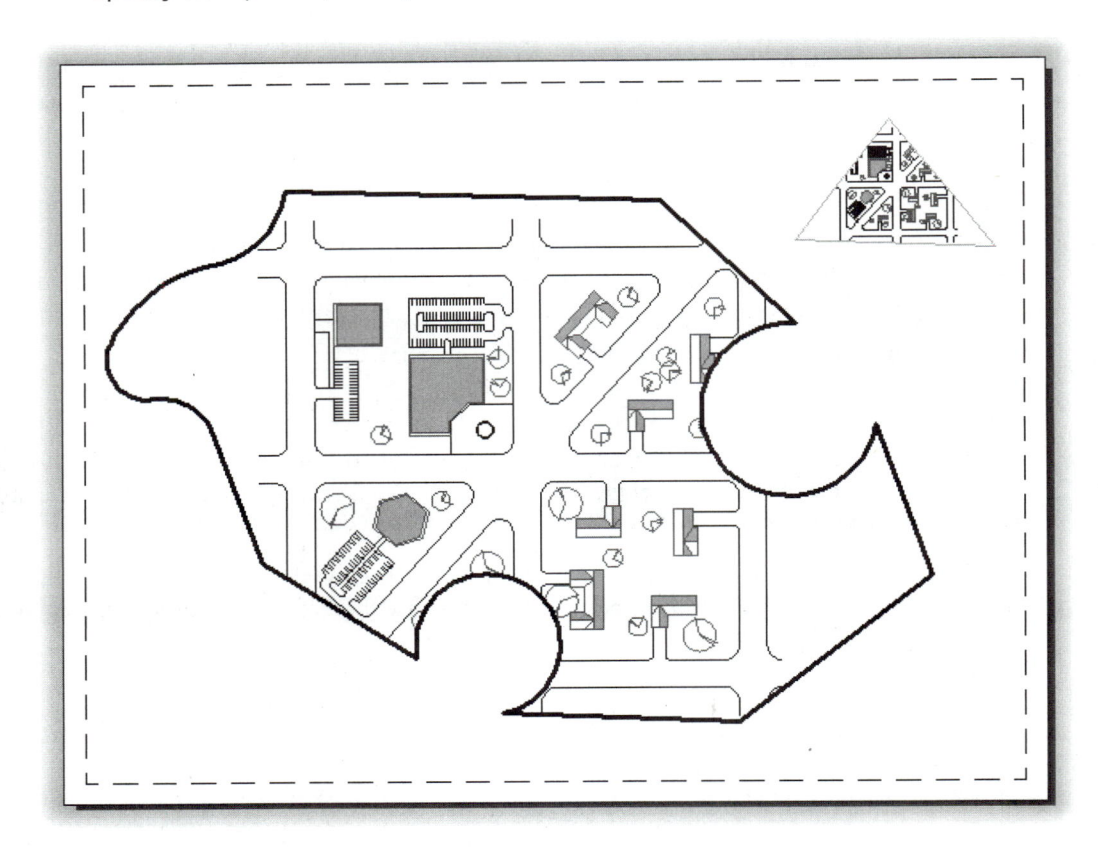

5. The prompts that follow are exactly like those of the **PLINE** command. You can draw line segments and arcs. You must pick a minimum of three points, and then close the polyline.

The figure opposite illustrates polygonal viewports. At left is one made with line and arc segments; in the upper right, one made with the minimum of three line segments.

TUTORIAL: CREATING VIEWPORTS FROM OBJECTS

1. To convert objects into viewports, ensure AutoCAD is displaying a layout; this action cannot be carried out in model space.
2. Draw a polyline, circle, ellipse, spline, or region. If drawing a polyline or spline, use the **Close** option to ensure the object is closed. The figure below illustrates a polyline drawn with the **PLINE** command, and then smoothed with the **PEDIT** command.

3. Start the **-VPORTS** command:
   - From the **View** menu, choose **Viewports**, and then **Objects**.
   - From the **Viewports** toolbar, choose the **Convert Object to Viewport** button.
   - At the 'Command:' prompt, enter the **-vports** command.

   Command: **-vports** *(Press ENTER.)*

4. Enter **O** to convert a closed object into a viewport:
   Specify corner of viewport or
   [ON/OFF/Fit/Shadeplot/Lock/Object/Polygonal/Restore/2/3/4] <Fit>: **p**
5. Pick the object to convert.
   Select object to clip viewport: *(Select a closed object.)*
   AutoCAD converts the object into a viewport, and the model shows through.

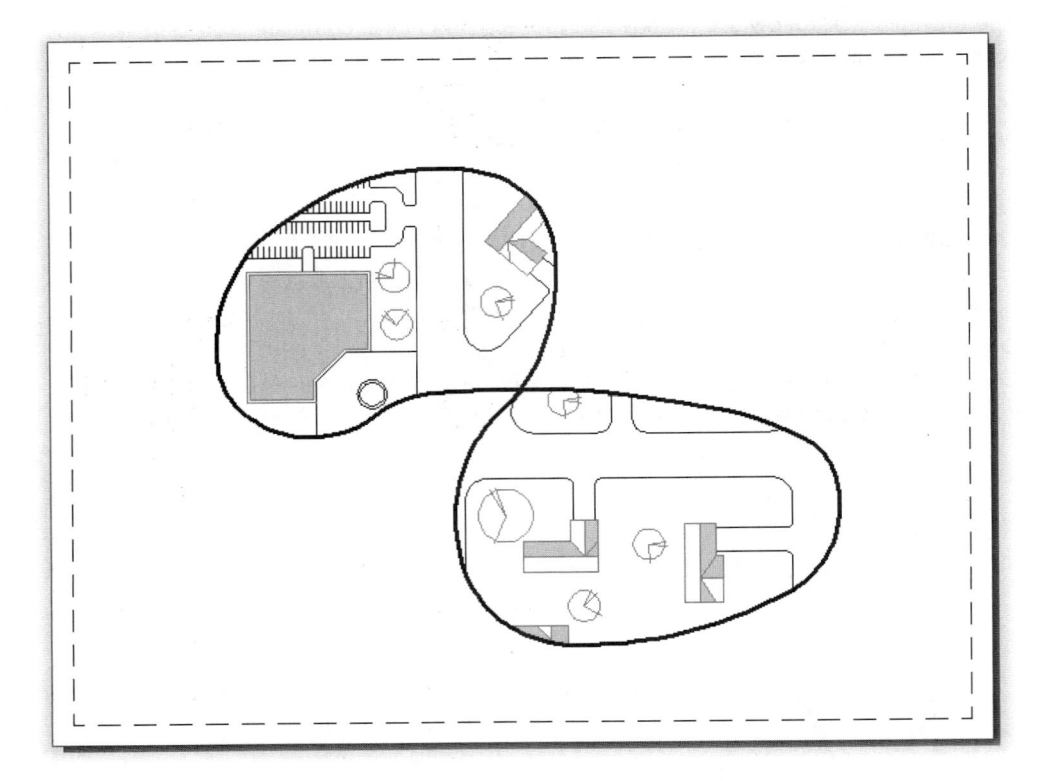

Once the viewport is in place, you can use grips editing to change the border, as illustrated by the figure below.

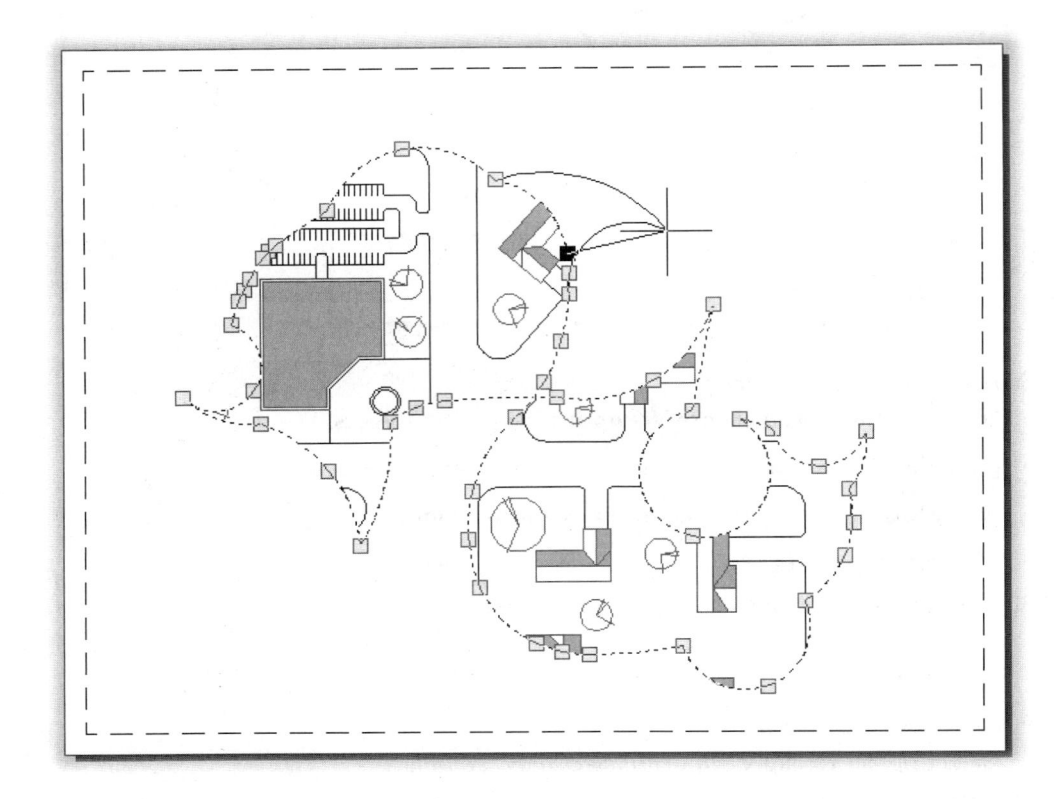

The **VPCLIP** command operates in a manner similar to **-VPORTS**: it clips rectangular viewports with polyline or objects. In addition, it clips 'clipped' viewports. Unlike **-VPORTS**, it can convert the clipped viewport back into a rectangular one.

Command: **vpclip** *(Press ENTER.)*

Select viewport to clip: *(Pick a viewport.)*

Select clipping object or [Polygonal] <Polygonal>:

If you select a viewport previously clipped by this command, the **Delete** option also shows. The three options are:

VpClip Option	Comment
Clipping object	Selects a closed object to be converted into a viewport boundary.
Polygonal	Picks points to designate line and arc segments for the viewport boundary.
Delete	Deletes the clipped viewport, and restores the rectangular viewport.

PSLTSCALE

The **PSLTSCALE** system variable matches linetype scaling to viewport scaling, so that linetypes drawn in the model appear the correct size (short for "paper space line type scale").

Command: **psltscale**

Enter new value for PSLTSCALE <1>: **0**

When set to **1** (the default), the viewport scale factor (set by **ZOOM Xp** or the viewport properties) controls linetype scaling. The lengths of dashes and gaps are based on paper space units — for objects drawn in model space and on layout. The advantage is that viewports can have different zoom levels (scale factors), yet the linetypes look the same.

When set to **0**, the linetype scale in model space is independent of layout scale factors.

Note: After changing the value of **PSLTSCALE**, you need to use the **REGENALL** command to update linetype scaling in all viewports.

TUTORIAL: MAKING LINETYPE SCALES UNIFORM

1. From the CD-ROM, open the *psltscale.dwg* file, a drawing of objects with several linetypes. In this drawing, **PSLTSCALE** is turned off (set to **0**). Notice that the linetypes in the two viewports have different scales, because the viewports are of different scales.

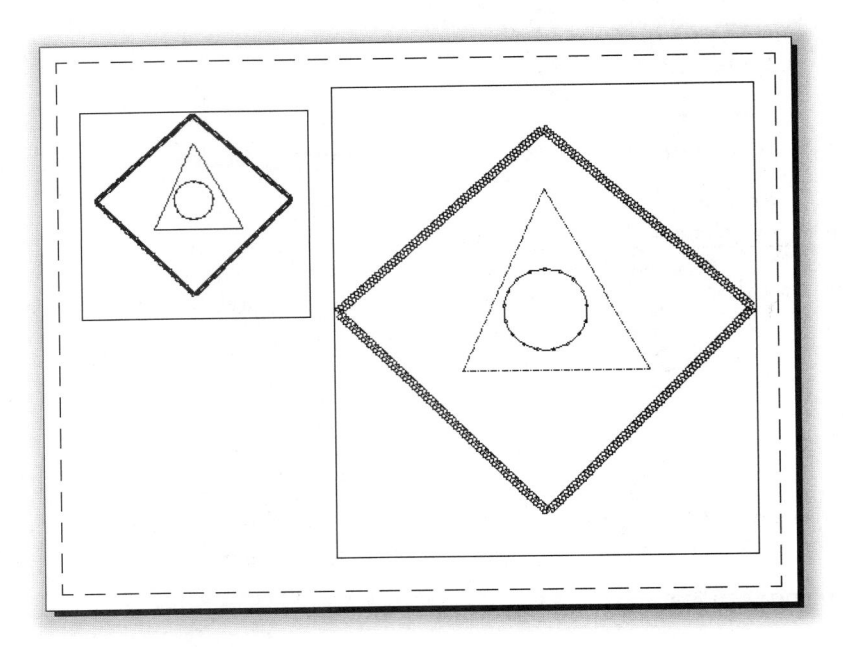

2. To make linetype scales the same in all layout viewports, start the **PSLTSCALE** system variable:

    Command: **psltscale** *(Press ENTER.)*

3. AutoCAD prompts you to change the value:

    Enter new value for PSLTSCALE <0>: **1**

4. The change to **PSLTSCALE** has no effect until you regenerate all the viewports:

    Command: **regenall** *(Press ENTER.)*

    Notice that the linetypes have the same scale factor in both viewports.

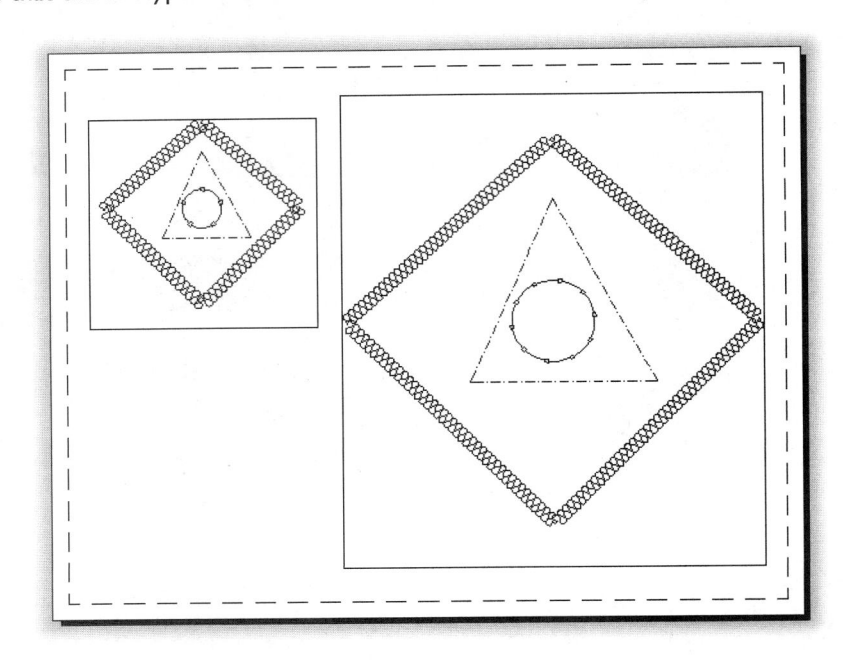

5. Change the linetype scale:

   Command: **ltscale**

   Enter new linetype scale factor <0.2500>: **.75**

   Regenerating layout.

   Regenerating model.

The linetype scale changes in both viewport, after AutoCAD automatically regenerates them.

'SPACETRANS

The **SPACETRANS** command converts distances between model space units and paper space units (short for "space translation").

This command automatically determines the height of text being placed model mode of a layout, but should be scaled appropriately for paper mode. The command is meant to be used transparently, during other commands, when AutoCAD asks for a height, length, or distance. In addition, the command may only be used in a layout — either in PAPER or MODEL mode; using it in model space results in the complaint:

   ** Command not allowed in Model Tab **

To specify the height of text that matches a given height in a layout, ensure AutoCAD is displaying the layout, and then enter the following:

   Command: **text**

   Specify start point of text or [Justify/Style]: *(Pick a point.)*

   Specify height <0.2>: **'spacetrans**

You can enter a fraction, and AutoCAD calculates it. For example, 1/8" (0.125) is a typical height for text in drawings.

   >>Specify paper space distance <1.000>: **1/8**

Press **ENTER**, and AutoCAD resumes the **TEXT** command, displaying the calculated height (0.125 x viewport scale factor):

   Resuming TEXT command.

   Specify height <0.2000>: 0.075783521943615

Continue with the command's other prompts:

   Specify rotation angle of text <0>: *(Press ENTER.)*

   Enter text: **SpaceTrans to the Rescue!** *(Press ENTER.)*

   Enter text: *(Press ENTER.)*

The text is placed in the layout at a size of 0.075... — instead of 0.125. When the viewport scale changes, the text also changes size, depending on the mode in which it was placed:

   When the text is placed in PAPER mode, it:

   - Is not visible in model space.
   - Does not change size when viewport scale changes.
   - Issues the following prompts (note the word "model"):
     Select a viewport: *(Select a viewport border.)*

     Specify model space distance <1.0>: *(Enter a value.)*

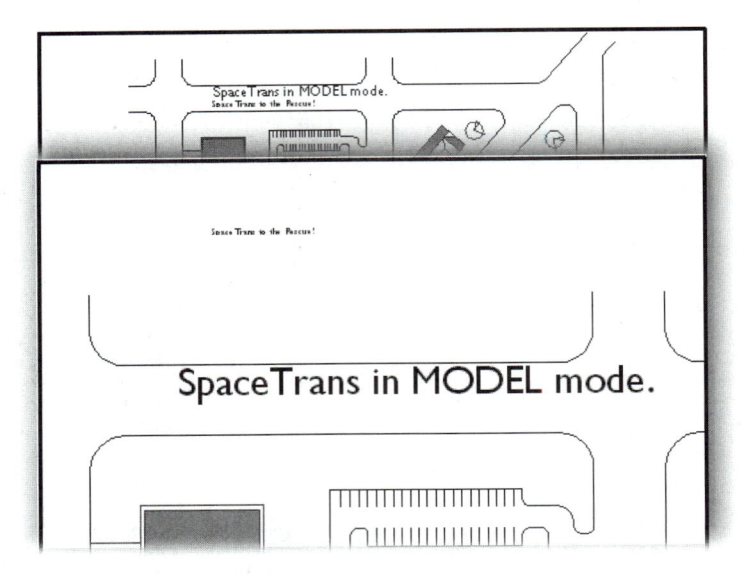

When the text is placed in MODEL mode, it:

- Is visible in model space.
- Changes size when viewport scale changes.
- Issues the following single prompt (note the word "paper"):
  Specify paper space distance <1.000>: *(Enter a distance.)*

Notes: Use **SPACETRANS** to create notes in model space that appear at the correct height in the layout, and are plotted at the correct size.

Use it together with the **SCALETEXT** command to change existing text. Start the **SCALETEXT** command, and then enter **'SPACETRANS** to specify the new height. AutoCAD scales the text correctly.

SPACETRANS cannot be used with the **MTEXT** command. Instead use it as a calculator at the 'Command:' prompt to find the scaled text height, and then enter the height manually in the mtext editor:

Command: **spacetrans**
Specify model space distance <0.1250>: **1.0**
0.663080708963838

EXERCISES

1. From the CD-ROM, open the *grader.dwg* file, the 2D drawing of a grader.
   Click the **Layout1** tab, and create two viewports on an A0- or E-size sheet of paper:
   One viewport shows the entire grader, without the drawing border.
   Second viewport shows a detail of the air intake.
   The result should look similar to the figure below. (All figure in these exercises are courtesy of Autodesk, Inc.)
   Save the drawing as *layout1.dwg*.

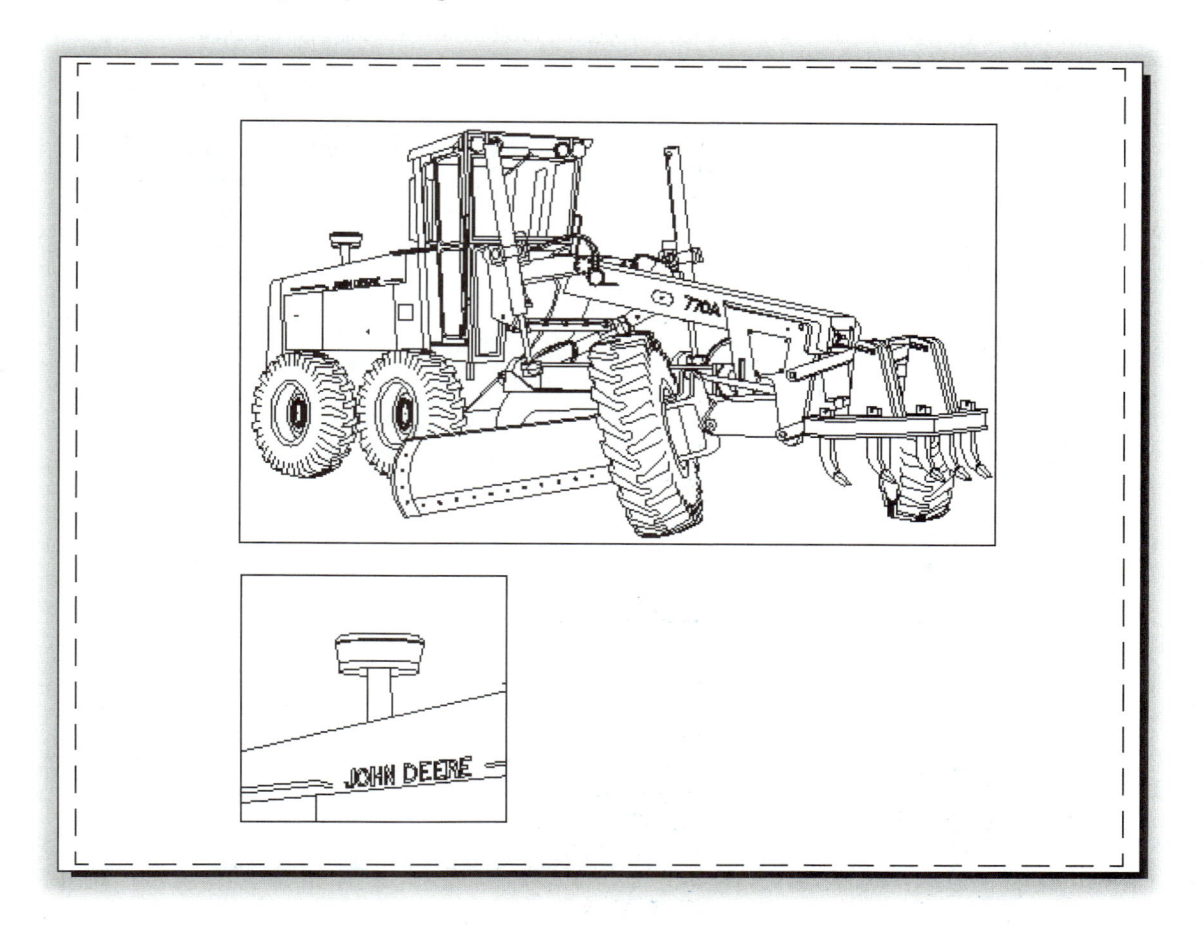

2. Start a new drawing, and then use the Layout Wizard to create a viewport with the following options:

Name	**Practice Layout**
Plotter	**DWF 6**
Paper Size	**ANSI B**
Drawing Units	**Inches**
Orientation	**Landscape**
Title Block	**ANSI B**
Viewport Setup	**Single**
Viewport Scale	**Scaled to Fit**

   Save the result as *layout2.dwg*.

3. From the CD-ROM, open the *langer.dwg* file, a drawing detailing sump pumps.
   Create four layout tabs of the following names, which match the names of the details.
   Pump Connection Detail
   Branch Pipe Support Detail
   Duplex Ejector Pumps
   Air Handler Unit Hanging Detail

   In each layout, place the namesake detail.
   What is the viewport scale factor?
   Save the drawing as *layout3.dwg*. The result should look similar to the figure below.

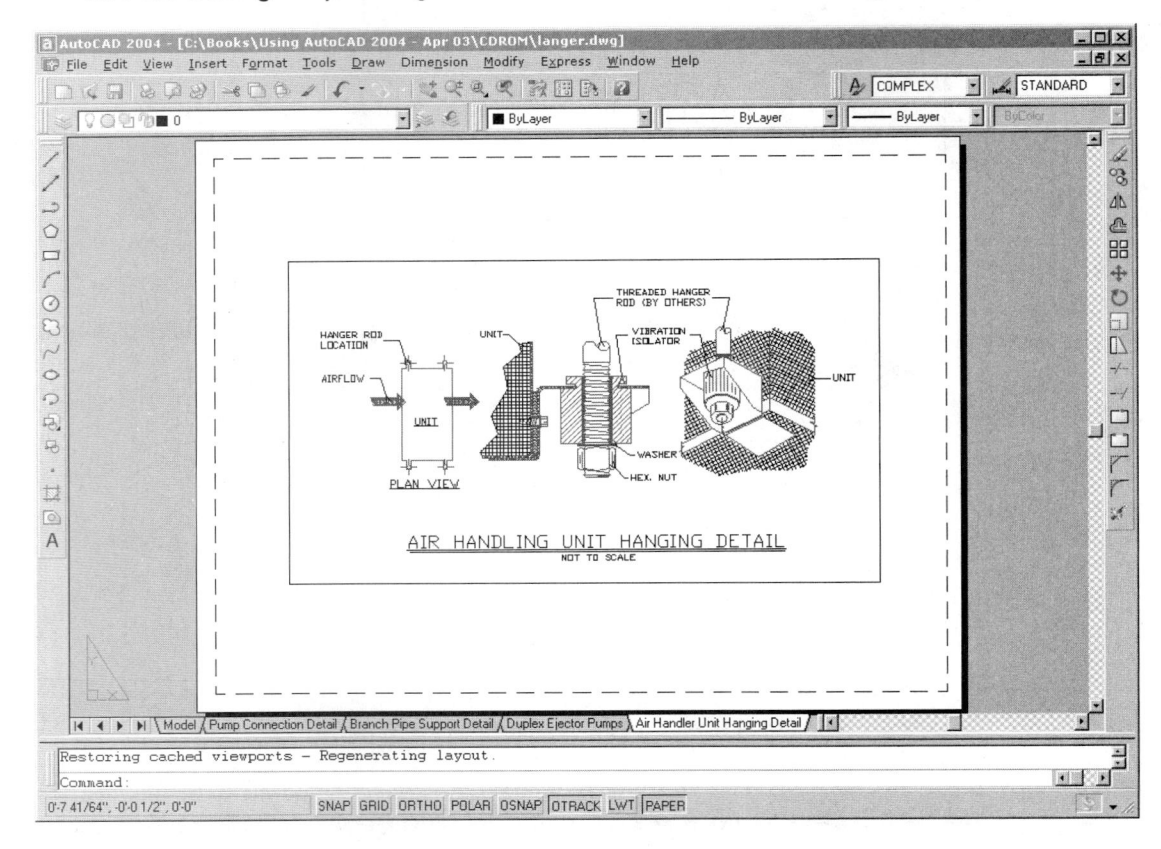

4. Use the **VIEWPORTS** command to create viewports with the following arrangements:
   a.

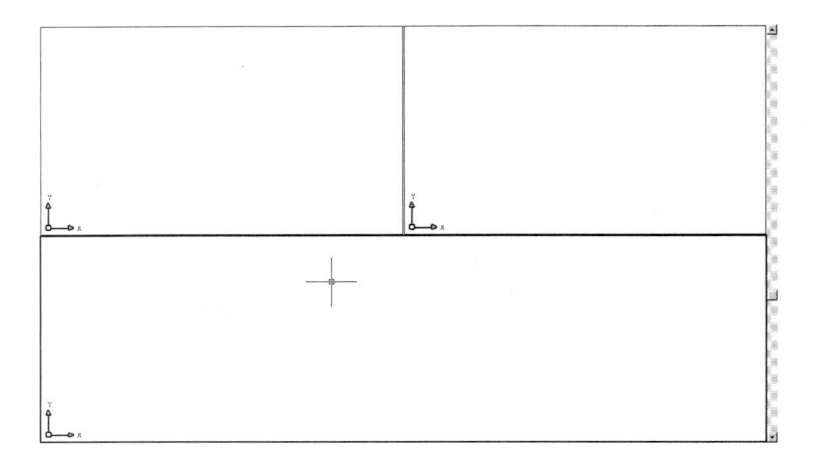

b.

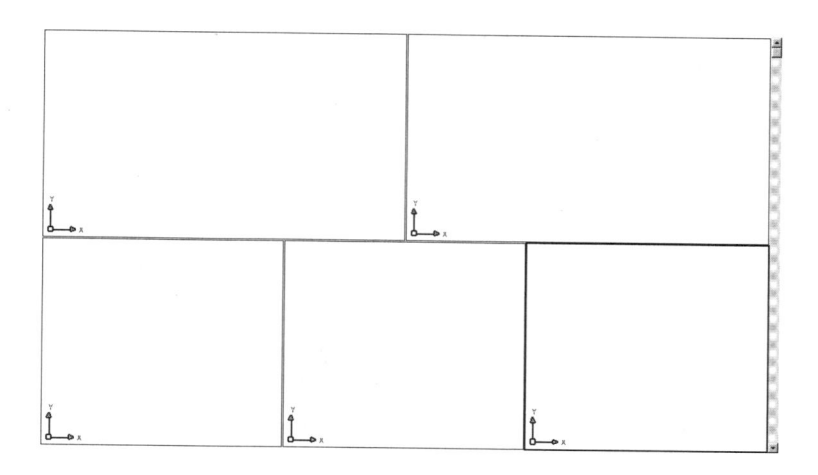

c.

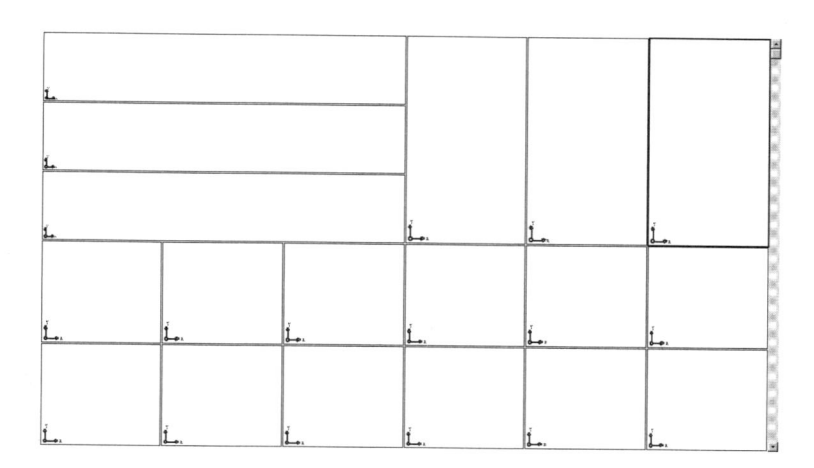

5. From the CD-ROM, open the *wright.dwg* file, a floor plan and elevation of a house designed by architect Frank Lloyd Wright.

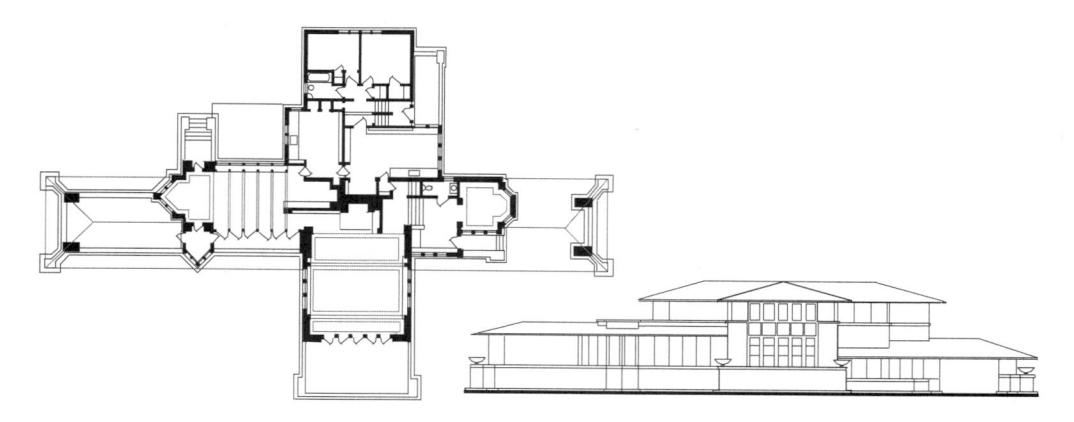

6. Create a pair of clipped viewports that show the two views independently, as shown below.
   Save the drawing as *layouts5.dwg*.

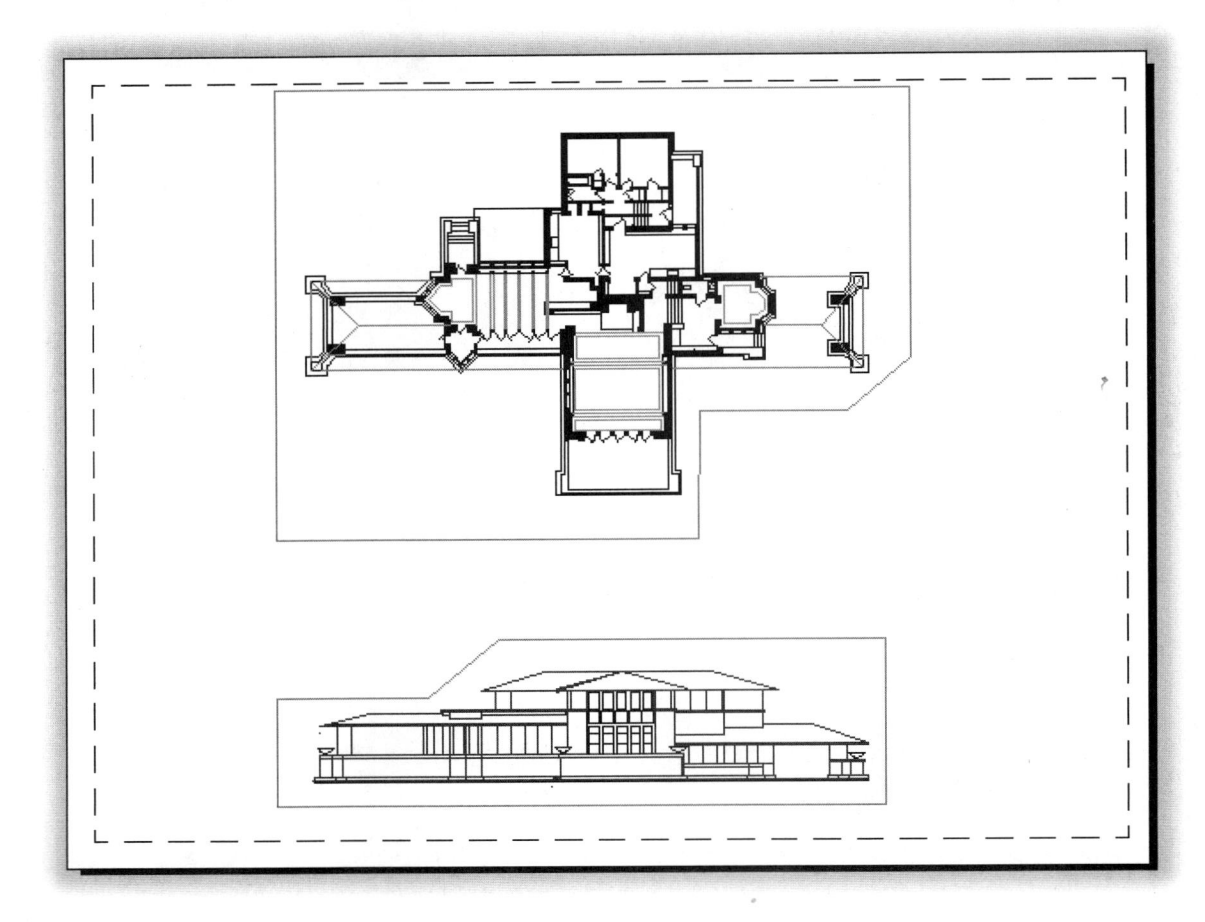

CHAPTER REVIEW

1. Explain the purpose of layouts.
2. Describe two ways to switch between model and layout mode:
   a.
   b.
3. In which tab are the following elements drawn full size?
   The model
   Title block
4. What does the layout represent?
5. Can a drawing have more than one layout?
6. How do you see the model in layouts?
7. Can viewports be modified?
   Can viewports be copied?
8. Name the parts of the layout:

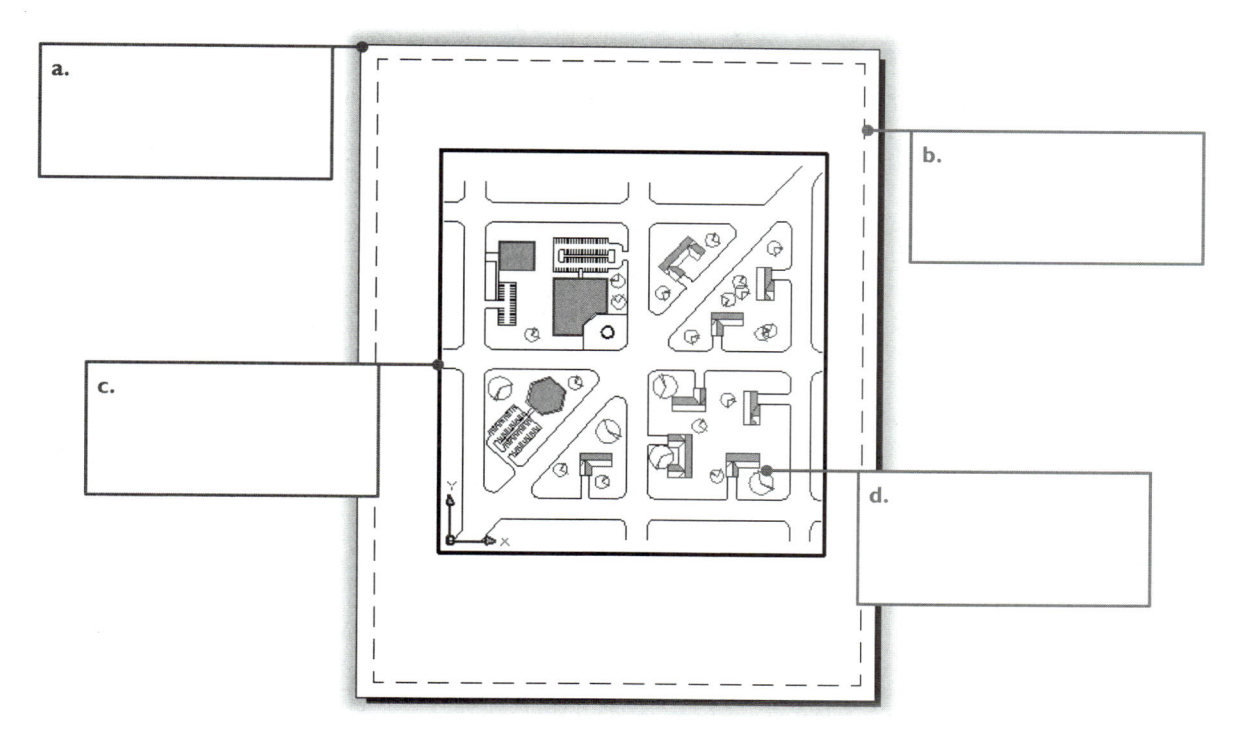

   a.
   b.
   c.
   d.
9. A layout has two viewports, of which one has a heavier border.
   What is the significance of the heavy border?
   What mode is AutoCAD in?
10. Explain the function of the following keystrokes:
    CTRL+R
    CTRL+PGDN

11. What kind of drafting can you do when the layout is in:
    PAPER
    MODEL
12. Describe what happens when you resize a viewport in a layout?
    When you copy a viewport.
    When you select an object in one viewport.
13. Describe how to hide the viewport border.
14. What are the differences between the two viewports shown below?
    Explain a possible reason for the differences.

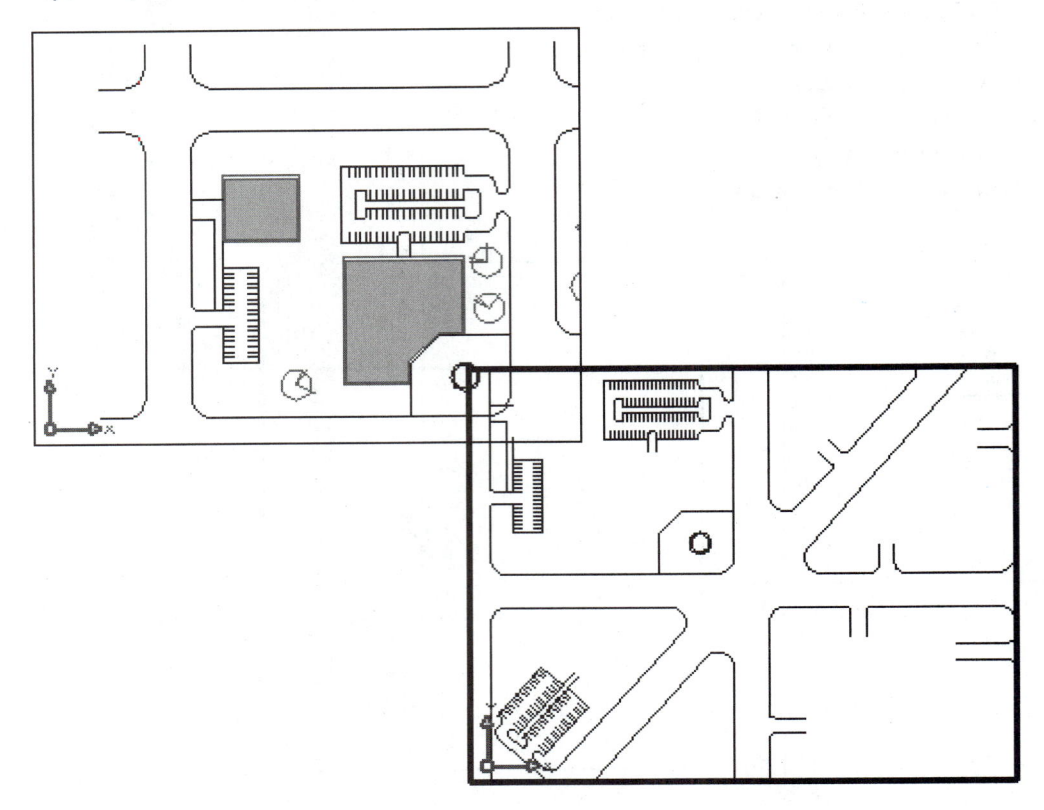

15. What are two ways to scale models in layouts?
    a.
    b.
16. What happens when you double-click a viewport?
17. Calculate the scale factor for the following viewport:
    Drawing of automobile = 12', bumper to bumper length
    Width of viewport = 10"
18. How would you move the model in the viewport?
19. Explain the difference between the **LAYER** command's **Current VP Freeze** and **New VP Freeze** options.
20. What does the snowflake mean in the Layer Properties Manager dialog box?
21. In which mode would you insert a title block?
    PAPER mode of layout tabs.
    MODEL mode of layout tabs.
    Model tab.

22. What advantage does the **VIEWPORTS** command have over the **-VPORTS** command?
    -VPORTS command over **VIEWPORTS** ?
23. Must viewports be rectangular?
    If no, in what way?
    If yes, why?
24. Describe the difference between:
    Tiled viewports
    Floating viewports
25. Briefly explain the function of the **PSLTSCALE** command.
26. When does **PSLTSCALE** take effect?
27. When is the **SPACETRANS** command commonly used?
28. What scale factor does **SPACETRANS** calculate?

CHAPTER 17

Plotting Drawings

The end product of most CAD drafting is often plotted on paper. In this chapter, you learn to plot drawings with these commands:

- **PAGESETUP** prepares drawings for plotting.
- **PREVIEW** previews drawings before plotting.
- **PLOT** plots drawings.
- **PLOTSTAMP** stamps plots with information about drawings and plotting.
- **PUBLISH** creates and plots drawing sets (new to AutoCAD 2004).
- **PLOTTERMANAGER** creates and edits plotter configurations.
- **STYLESMANAGER** creates and edits plot style tables.
- **CONVERTPSTYLES** and **CONVERTCTB** convert color-based and named plot style tables.
- **BATCHPLT** plots drawings unattended.

FINDING THE COMMANDS

On the **STANDARD** toolbar:

On the **FILE** menu:

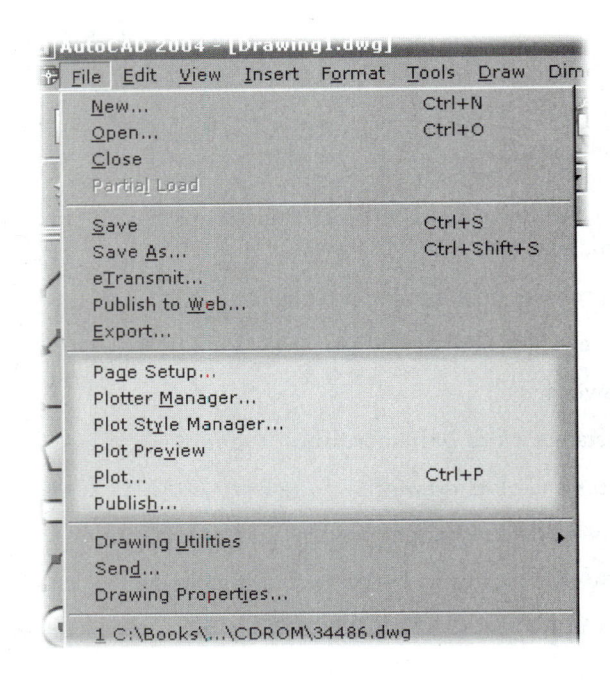

PAGESETUP

The **PAGESETUP** command prepares drawings for plotting.

Before AutoCAD can plot a drawing, you must assign a plotter to the drawing file. You can do this as part of the initial set up for the drawing, saving the plotter assignment to your template drawings. Otherwise, you must select a plotter the first time you plot the drawing. With **PAGESETUP** or **PLOT**, you see the dialog box shown below.

TUTORIAL: PREPARING FOR PLOTTING

1. To prepare drawings for plotting, start the **PAGESETUP** command:
   - From the menu bar, choose **File**, and then **Page Setup**.
   - At the 'Command:' prompt, enter the **pagesetup** command.

   Command: **pagesetup** *(Press ENTER.)*

2. AutoCAD displays the Page Setup dialog box.

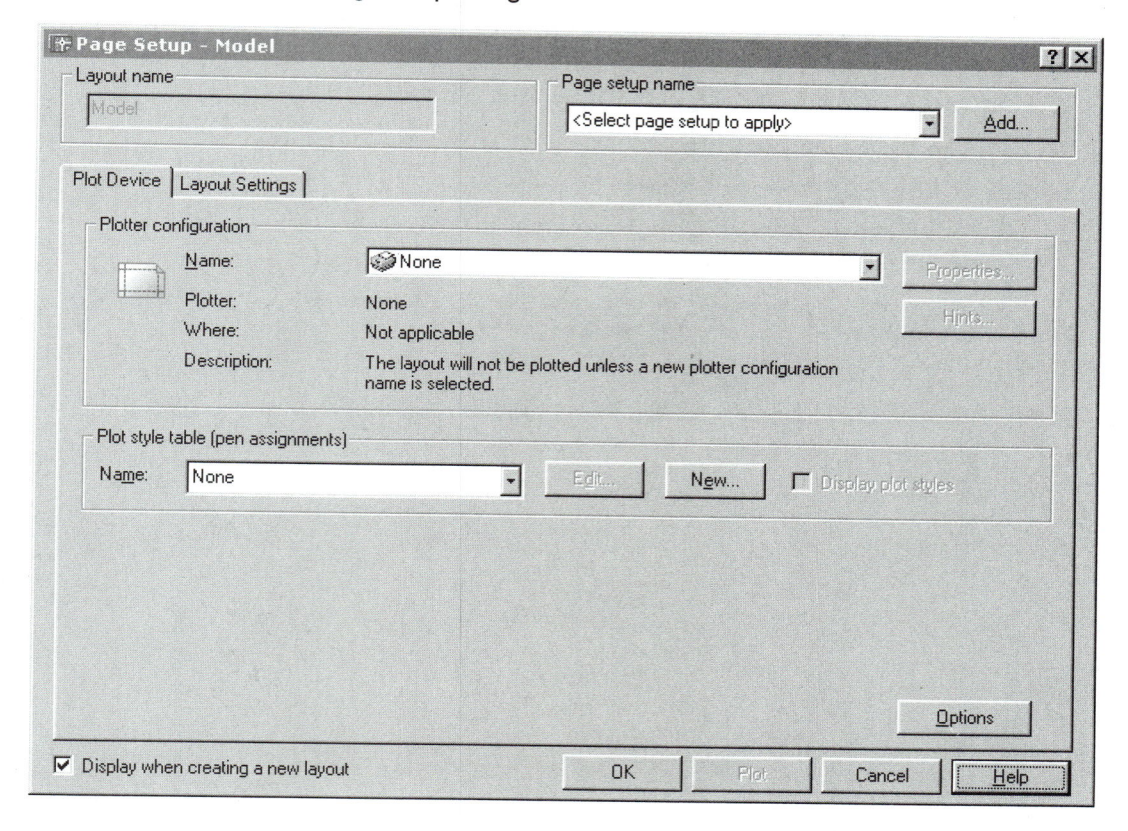

3. In the Plotter Configuration area, select a printer from the **Name** drop list.
4. Click **OK**.
   Your drawing is now ready to plot.

The **Properties** button is discussed with the **PLOTTERMANAGER** command. Pen style tables are discussed with the **STYLESMANAGER** command. The **Layout Settings** tab is discussed with the **PLOT** command.

The **Display When Creating a New Layout** option displays this dialog box each time a new layout tab is created.

PREVIEW

The **PREVIEW** command previews drawings before plotting them.

Previewing is important, because it ensures the drawing will be plotted as you expect. The preview shows whether all drawing elements will appear on the paper correctly. It also provides a visual check of plot settings, such as whether the drawing is centered correctly. This saves you time and money because previewing is faster than plotting and costs nothing.

The preview depends on the printer configuration defined by the **PAGESETUP** command. If you have not yet selected a printer for the drawing, you will be prompted to do so after entering the **PREVIEW** command.

TUTORIAL: PREVIEWING PLOTS

1. To preview drawings before plotting, start the **PREVIEW** command:
   - From the menu bar, choose **File**, and then **Preview**.
   - From the **Standard** toolbar, choose the **Preview** button.
   - At the 'Command:' prompt, enter the **preview** command.
   - Alternatively, enter the **pre** alias at the 'Command:' prompt.

   Command: **preview** (Press ENTER.)

2. In all cases, AutoCAD displays the preview window.
   (If a plotter has not yet been assigned to the drawing, then AutoCAD first displays the Page Setup dialog box.)

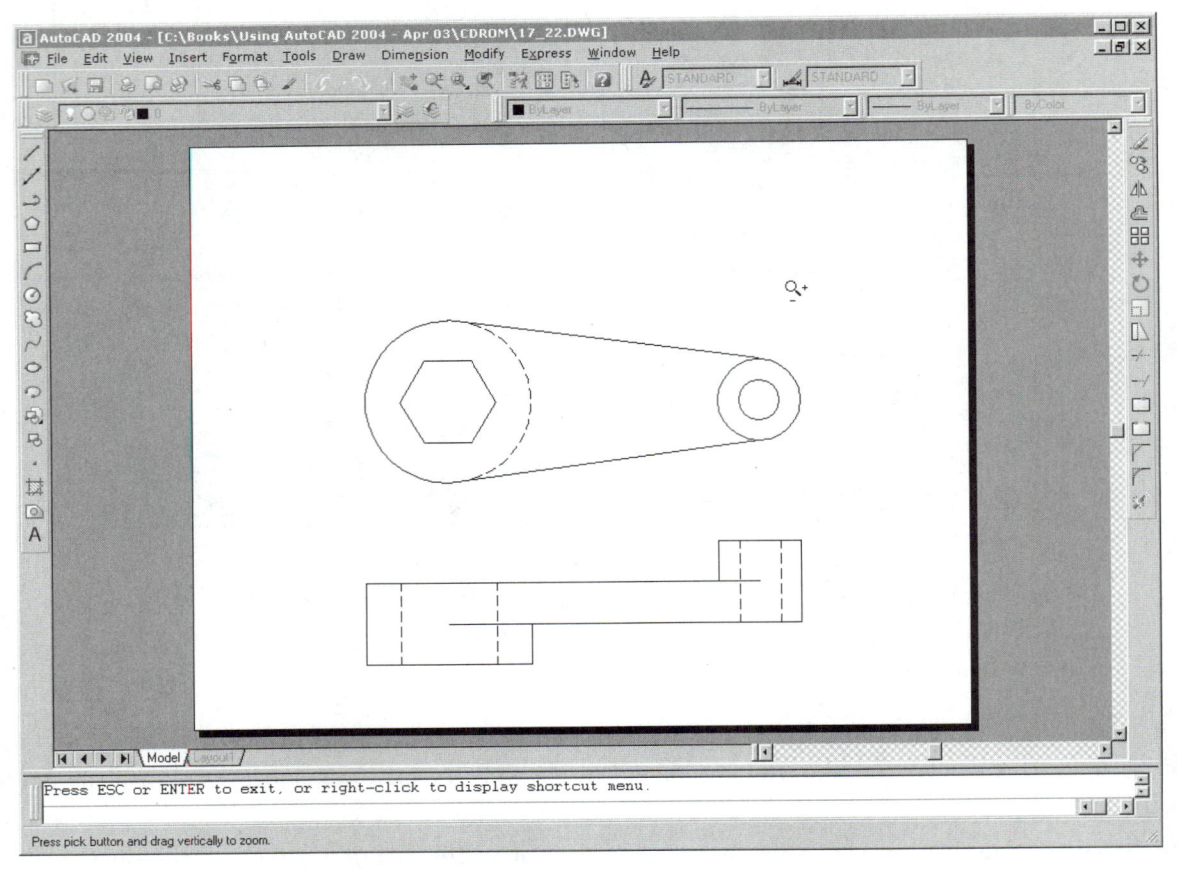

3. Press **ESC** to return to the drawing.

PREVIEWING THE PLOT: ADDITIONAL METHODS

The preview shows a white rectangle that represents the paper. The cursor looks like a magnifying glass. As you drag the cursor up and down, the preview zooms in and out. Right-click the "paper" to display a shortcut menu.

Option	Cursor	Meaning
Exit		Exits preview.
Plot		Plots the drawing.
Pan	🖐	Switches from zoom to pan mode.
Zoom	🔍+	Switches from pan to zoom mode.
Zoom Window	▸▫	Zooms into a windowed area.
Zoom Original		Returns to normal view.

When you select **Pan**, the cursor changes to a hand. Hold down the left mouse button and the view moves (pans) as you move around the mouse. The **Zoom Window** option operates identically to the regular ZOOM command's **Window** option. The **Zoom Original** option returns the original view, much like the ZOOM **Extents** command.

PLOT

The **PLOT** command displays the Plot dialog box, and then plots the drawing.

The dialog box consists of two tabs: Plot Device and Plot Settings. In the **Plot Device** tab, you select the plotter and the plot style table (more on plot styles later in this chapter). Think of this tab as "*where* to plot."

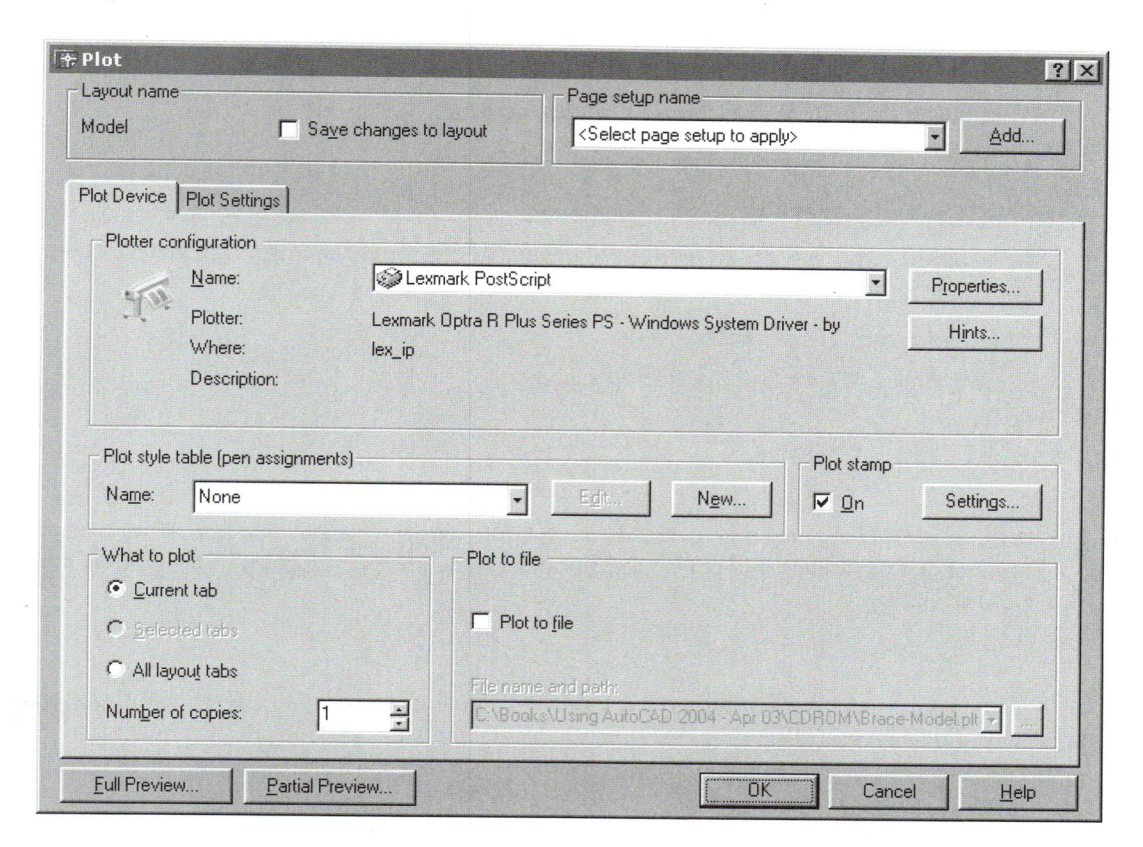

In the **Plot Settings** tab, you select the plot scale, media size, orientation, and other parameters. Think of this tab as "*what* to plot."

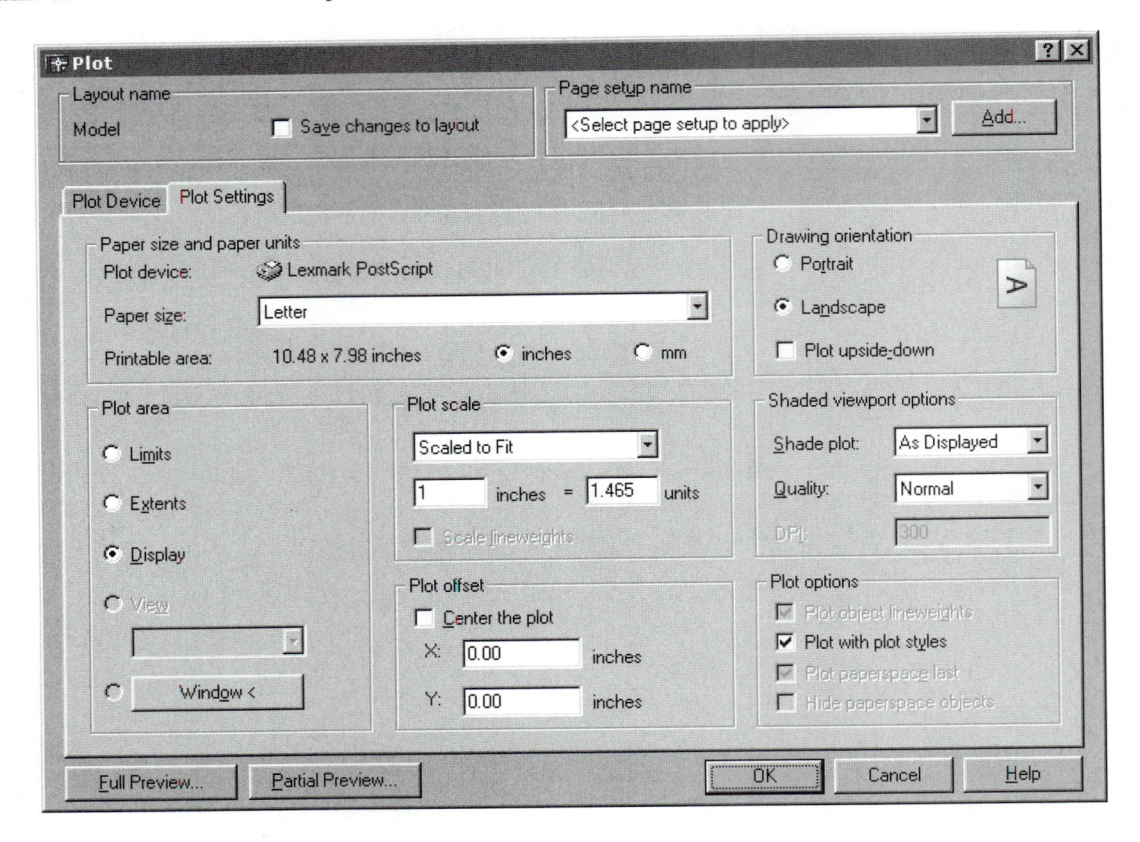

TUTORIAL: PLOTTING DRAWINGS

1. To plot drawings, start the **PLOT** command:
   - From the menu bar, choose **File**, and then **Plot**.
   - From the **Standard** toolbar, choose the **Plot** button.
   - At the 'Command:' prompt, enter the **plot** command.
   - On the keyboard, press the **CTRL+P** shortcut.
   - Alternatively, enter the aliases **print** or **dwfout** (an old command integrated into plotting) at the 'Command:' prompt.

   Command: **plot** *(Press* ENTER.*)*

   In all cases, AutoCAD displays the Plot dialog box.
2. Click **OK**.
   AutoCAD plots the drawing.

   Note: AutoCAD does not plot layers that are frozen or have the No Plot property. As well, AutoCAD does not plot the Defpoints layer. If part of your drawing does not plot, it could be because you drew on the Defpoints layer.

The Plot dialog box has many options, and it can be confusing to navigate. Here are the steps you need to take to plot successfully.

Step 1: Select a Plotter/Printer

AutoCAD stores the settings of many plotters and printers. For example, your computer may have two different printers available for plotting. The **Plotter Configuration** list box of the Plot Device tab contains the list of printers available to you — local and networked — as well as special drivers for plotting to file.

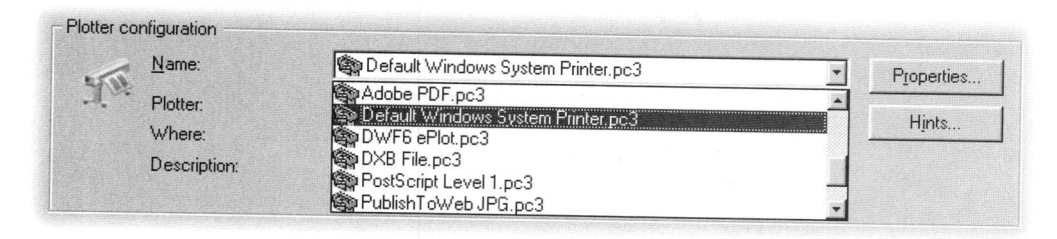

From the **Name** drop list, select a printer or plotter. If the printer or plotter you want is not listed here, read about the **PLOTTERMANAGER** command later in this chapter.

To change properties specific to the printer, such as its resolution or color management, choose the **Properties** button.

Step 2: Select a Plot Style

Plotters are capable of plotting with *pens*. Historically, "pen" derives from an earlier age when plotters plotted with actual pens — most commonly felt pens, but also ball point pens, technical ink pens, and even pencils. The term has carried over to today's non-pen plotters (laser and inkjet), and refers to varying widths, colors, and shades of gray.

Earlier releases of AutoCAD assigned the colors of objects to pens. For example, all objects colored red (either by layer or by entity) would be plotted by a specific pen; those colored blue should be plotted with another pen, and so on. The pens didn't need to contain red or blue ink; typically they contained black ink, and had tips of varying widths, such as 0.1" or 0.05".

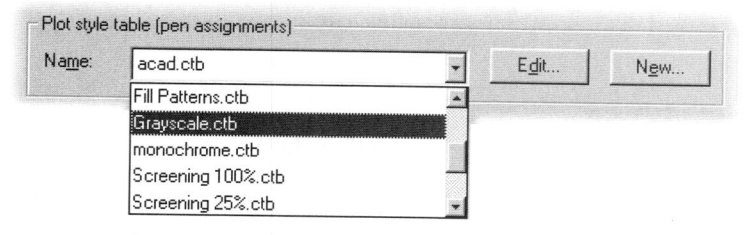

Select one of the preassigned plot styles from the list. *Plot styles* assign plotter-specific properties to layers and objects, such as widths, colors, line-end capping, and patterns. You learn more about plot styles with the **STYLESMANAGER** command later in this chapter.

Step 3: Select the Layout to Plot

You can specify one or more layouts to plot. The choices are contained in the **What to Plot** area of the dialog box, together with the number of copies.

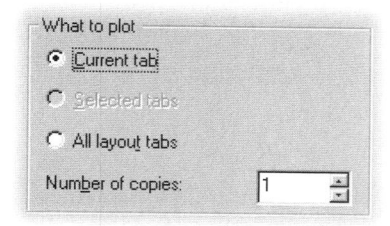

Step 4: Select the Media Size
Select the **Plot Settings** tab.

The **Paper Size and Paper Units** area specifies the size of paper (a.k.a. media). AutoCAD knows the media sizes that the selected printer can work with. In addition, AutoCAD reports the **Printable Area**, which is the size of the paper minus the *margins* (the unprintable edges).

Make your selection from the **Paper Size** list box.

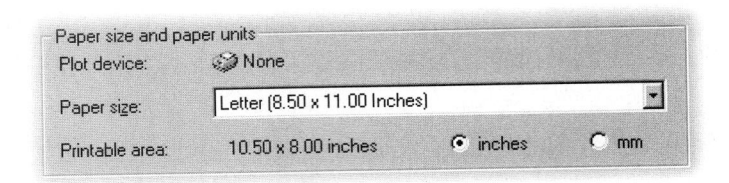

The **inches** and **mm** options toggle between Imperial and metric units; the options do not affect the selection of paper sizes.

Step 5: Select the Orientation
Larger CAD drawings are usually plotted in *landscape* mode, with the long edge of the page laid horizontally. To do this, choose the **Landscape** button in the **Drawing Orientation** area.

Smaller drawings are often printed in *portrait* mode, where the long edge is upright. Choose the **Portrait** button.

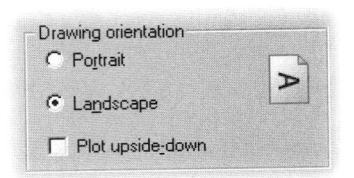

The **Plot upside-down** option is handy if you stuck the paper in the printer with the title block the wrong way around.

Step 6: Select the Plot Area
In most cases, you plot the entire drawing, also called the "extents" of the drawing. In other cases, you may want to plot a specific area of the drawing. AutoCAD offers the following options:

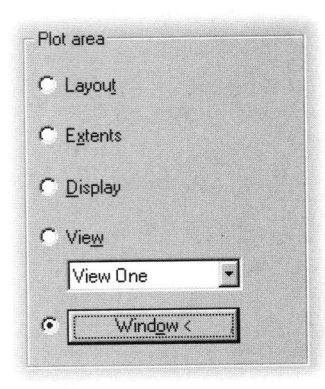

Plot Area	Comments
Limits	Plots the limits of a drawing as set by the LIMITS command.
Display	Plots the current view of the drawing — the "what you see is what you get" plot.
Extents	Plots the extents, a rectangle that encompasses every part of the drawing containing objects; same as performing a ZOOM **Extents**, and then plotting with the **Display** option.
View	Plots named views; the drawing must contain at least one named view (created with the VIEW command). Click **View**, and then select a view name.

| Window | Plots a rectangular area identified by two windowed picks. After clicking **Window**, you are prompted to pick the two corners of a rectangle: |

Specify first corner: *(Pick a point or type X,Y coordinates.)*

Specify opposite corner: *(Pick another point or type X,Y coordinates.)*

Step 7: Select Plot Scale

To plot the drawing at a specific scale, select one of the predefined scale factors provided in the **Plot Scale** list box. These range from 1:1 to 1:100 and 1/128"=1'.

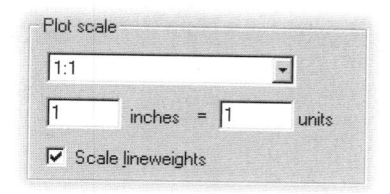

For draft plots, select **Scaled to Fit**. Together with a plot area of **Extents**, this ensures your entire drawing fits whatever size of media you select.

A third choice is **Custom**, where you specify the scale factor. Specify the number of *inches* (or millimeters) on the paper to match the number of drawing *units* to be plotted. For example, a scale of 1" = 8'-0" means that 1" of paper contains 8'-0" (or 96") of drawing. To set the custom scale correctly, enter:

- Inches   1"
- Units   9' 6"

(This is the same as a scale of 1/8" = 1'-0". The example assumes that the drawing units are set to architectural units.)

If you are working in metric, choose the **mm** radio button in the **Paper size and paper units** section. AutoCAD provides metric conversion: type the number of inches in the **inches** text box, and then choose the **mm** button.

Step 9: Preview the Plot

You should preview the plot to see the area of the paper on which the plot appears. AutoCAD's plot preview allows two types of preview: partial and full.

Notes: Use plot previews to discover plotting results before committing time, paper, and pens to a drawing that may be set up incorrectly.

If the **Full Preview** and **Partial Preview** buttons are grayed out, you have not yet selected a plotter for the layout.

Full Preview
The *full* preview shows the drawing as it would appears as a final plot on paper. See the PREVIEW command earlier in this chapter.

Partial Preview
The *partial* preview does not show the drawing, but rather the position of the plot on the paper. Historically, this option was more desirable than full preview, because slower computers took a long time to generate full previews.

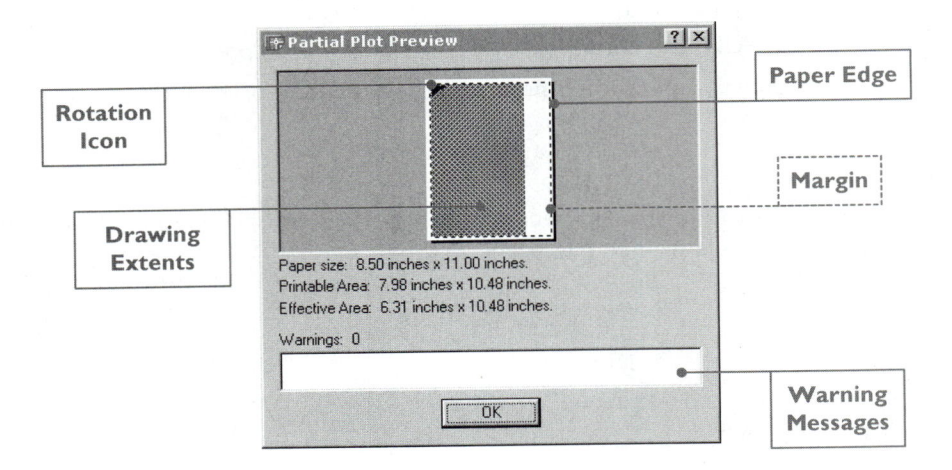

The paper is shown by the white rectangle.

The margin (printable area of the paper) is shown by the black dashed rectangle.

The area covered by the drawing is shown by the blue hatched rectangle. When the drawing area matches the margins, alternating red and blue dashed lines are shown.

The red triangle is the *rotation icon*, which represents the lower left corner of the drawing as positioned on the screen. When the drawing is rotated 90 degrees (such as for landscape mode shown above), the icon appears at the upper left.

When the plot origin is offset so that the drawing extends beyond the margins, AutoCAD displays a green line along the clipped edge, and warns:

    Effective area clipped to display image.

    Plotting area exceeds maximum.

Step 10: Save the Settings
If you plan to use the same plot parameters again, it makes sense to save them. In the **Page Setup Name** section (near the top of the Plot dialog box), choose the **Add** button.

AutoCAD displays the User Defined Page Setups dialog box.

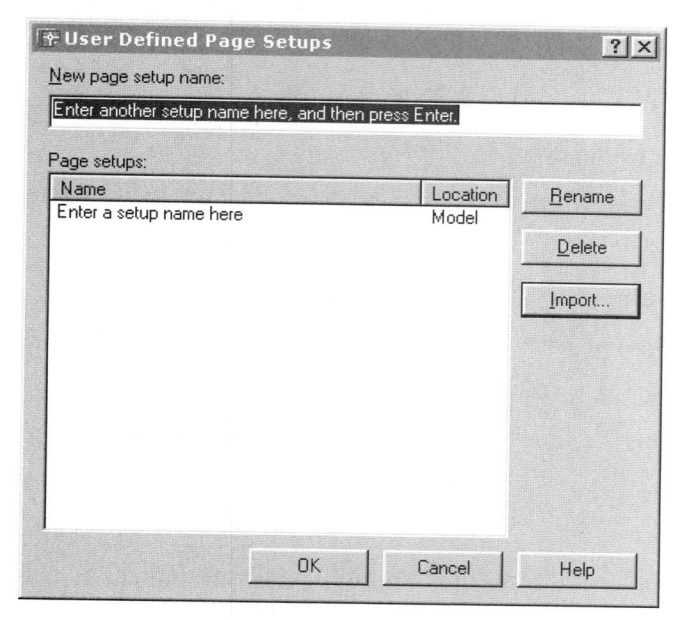

Enter a descriptive name, and choose **OK**. As an alternative, you can click **Import**, and then select a drawing file containing plot setups.

Back in the Plot dialog box, select a saved plot setup from the **Page setup name** list box.

Step 11: Plot the Drawing

When you have completed all these steps, make sure the plotter is ready, and choose the **OK** button at the bottom of the Plot dialog box. (Click **Cancel** to exit the dialog box, losing the changes you made.)

AutoCAD displays a dialog box indicating its progress plotting each layout.

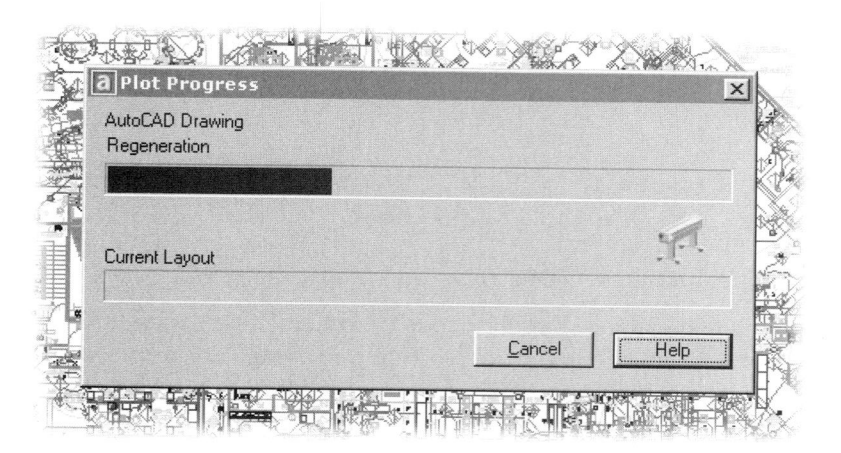

PLOTTING DRAWINGS: ADDITIONAL METHODS

The **PLOT** command provides fine control over how the plot is created, as these options indicate:

In the **Plot Device** tab:

- **Plot Stamp** stamps the plot with information about the plot.
- **Plot to File** saves drawings as plot files.
- **AutoSpool** generates plot files intercepted by other software programs.

In the **Plot Settings** tab:

- **Plot Offset** centers or offsets the plot on the page.
- **Shaded Viewport** removes hidden lines or creates rendered plots.
- **Plot Options** lists miscellaneous options.

Let's look at each option; the **Plot Stamp** option is discussed later in this chapter.

In the **Plot Device** tab:

Plot to File

The **Plot to File** option saves the drawing to a file. The *plot file* can be read by plotters, or imported into other software. For example, many brands of plotters and graphics programs can import files created by HPGL plotter drivers, usually meant for Hewlett-Packard brand plotters.

To create plot files:

1. Start the **PLOT** command.
2. In the Plot Device tab's **Plot to File** section, select the **Plot To File** check box.

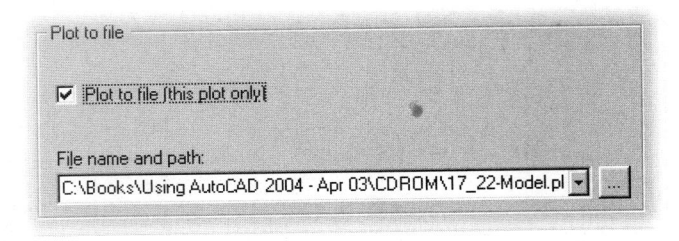

3. Enter a file name, or choose the **...** (browse) button to select a name from the file dialog box. (Choosing the **Browse the Web** button lets you save the file to a location on the Internet.) If you do not include an extension with the file name, AutoCAD appends *.plt* (short for "plot").
4. Click **OK**.

AutoSpool

The **AutoSpool** option lets AutoCAD *spool* the plot.

Spooling is a technique (short for "simultaneous peripheral operations online") that speeds up printing jobs. When printing a document or plotting a drawing, the print data is sent to a file on disk; shortly thereafter, the spooling software starts up automatically, and sends the print data from the file to the printer. It is faster to save a file to disk than to print, so the application finishes the print job faster. (Spooling is also known as *buffering*, and in Windows it is handled by the Print Manager.)

AutoCAD allows you to use independent software for plot spooling. Before using the software, however, you must set up AutoCAD:

1. Install and configure the spooler software according to the vendor's instructions.
2. AutoCAD needs to know the name of the spooler, as well as the folder in which to place the spool files.

   From the **Tools** menu, select **Options**, and then the **Files** tab:
   - Open **Print File, Spooler, and Prolog Section Names**, and in **Print Spool Executable**, specify the spooler program name.
   - Open **Printer Support File Path**, and in **Print Spool File Location**, specify the folder name.

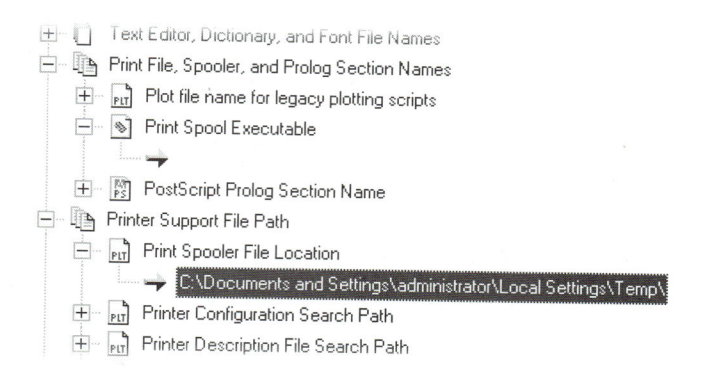

   Choose **OK**.
3. Set up the plotter by selecting **File | Plotter Manager**, and then working through the wizard.
4. Start the **PLOT** command, and then select the plotter from the **Plotter Configuration** drop list.

In the **Plot Settings** tab:

Plot Offset

The plot origin is at the lower left corner of the media for most plotters, just as it is for AutoCAD drawings. Some plotters have their origin at the center of the media; AutoCAD normally adjusts for that.

Sometimes you need to shift the plot on the media, for example to avoid a preprinted title block. Use the **Plot Offset** section of the Plot dialog box:

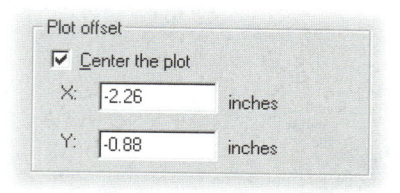

- Positive values shift the plot to the right and up.
- Negative values shift the plot to the left and down.

In most cases, however, you will probably choose the **Center the plot** option to have AutoCAD center the plot on the media.

> **Note**: Changing the plot's offset may result in a *clipped* plot, where part of the drawing is not plotted, because it extends beyond the edge of the media's margin.

Shaded Viewport
New in 2004

The **Shaded Viewport** option controls how model space is plotted. (This option is new to AutoCAD 2004.)

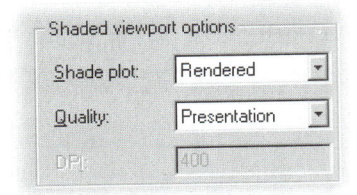

Shade Plot	Comment
As Displayed	Plots objects as displayed in model space. If shaded with the SHADE command or rendered with the RENDER command, then the drawing is plotted that way.
Wireframe	Plots objects in wireframe, regardless of the display mode in model space.
Hidden	Plots objects with hidden lines removed, regardless of display mode.
Rendered	Plots objects rendered, regardless of display mode.

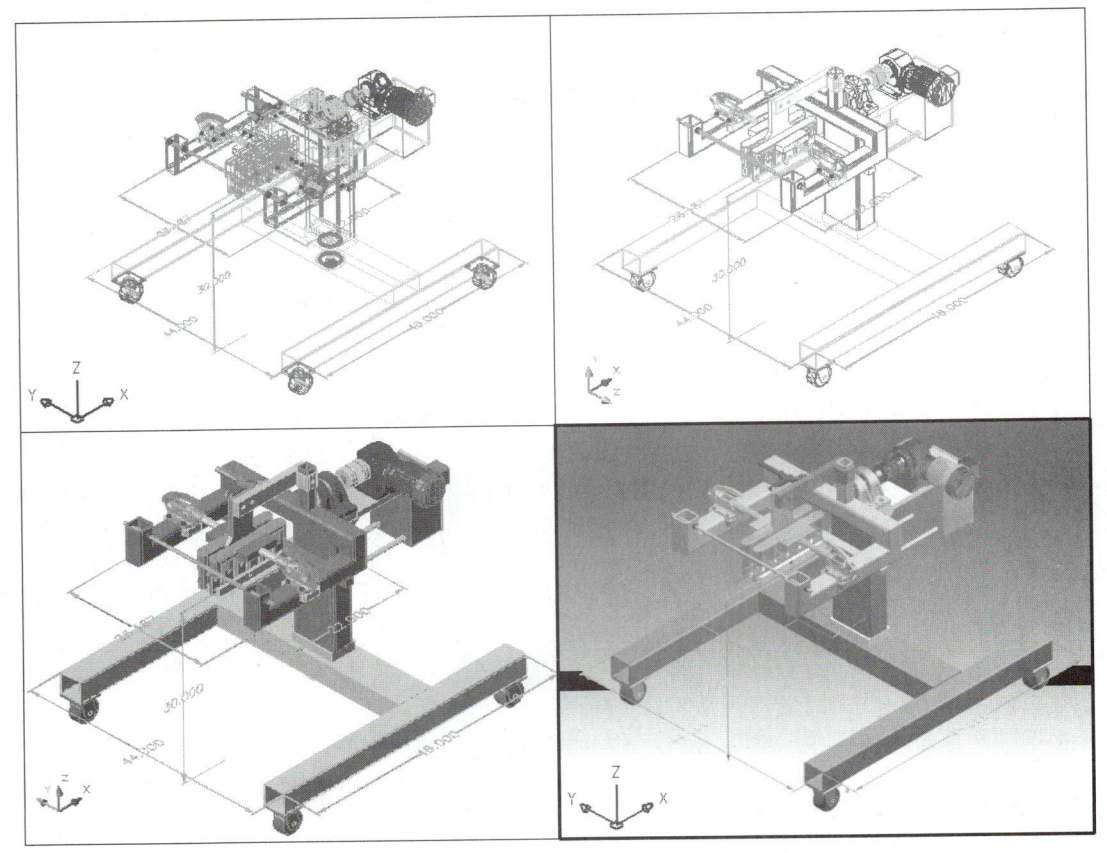

Top left: 3D drawing displayed as wireframe.
Bottom left: 3D drawing shaded.
Top right: 3D drawing with hidden lines removed.
Bottom right: 3D drawing fully rendered.

Notes: Two-dimensional drawings cannot be rendered, shaded, or have hidden lines removed. You can see the effect of shading and hidden-line removal on 3D drawings in the plot preview window. Note that removing hidden lines from complex drawings can take time and slow the plotting.

The **Quality** drop list specifies the resolution for shaded and rendered plots, because they are raster images. (Wireframe and hidden-line removed plots are vectors.)

Quality	Meaning
Draft	Plotted as wireframes.
Preview	Plotted at 150 dpi.
Normal	Plotted at 300 dpi.
Presentation	Plotted at 600 dpi.
Maximum	Plotted at output device's current resolution.
Custom	Plotted at the resolution specified in the DPI text box

The **Shaded Viewport** option applies to plots made from model space only; it is not available when plotting layouts. To plot a layout viewport with hidden lines removed or rendered

1. Exit the **PLOT** command.
2. In the layout, double-click the viewport border.
3. In the **Misc** section of the Properties window, select a plotting mode from the **Shade plot** drop list.

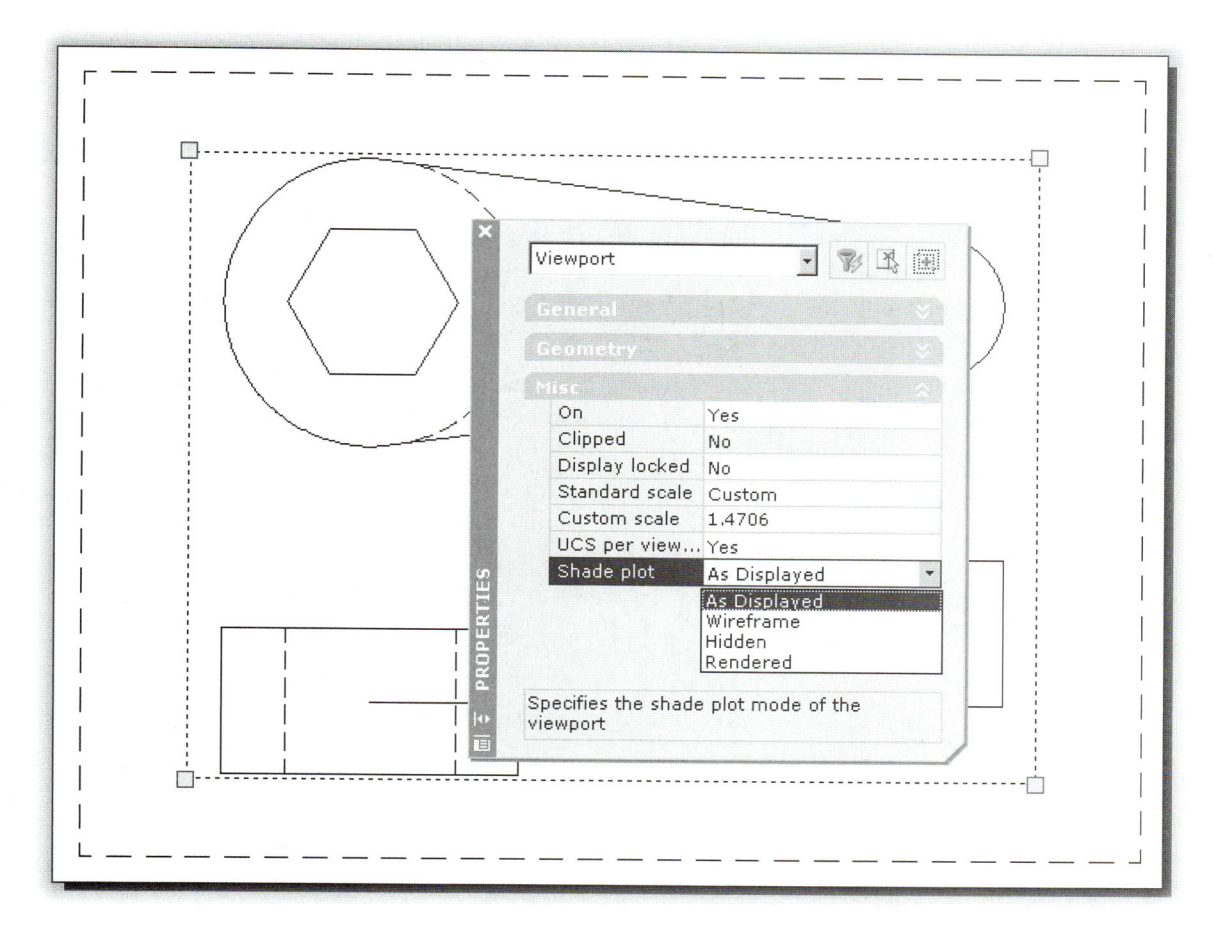

Plot Options
The **Plot Options** area lists miscellaneous options.

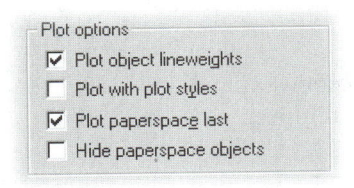

Plot Object Lineweights and **Plot with Plot Styles**
The **Plot with Lineweights** and **Plot with Plot Styles** options are toggles:

- **Plot with Lineweights** plots the drawing with lineweights, if lineweights are turned on, and if lineweights have been assigned to layers and objects.

- **Plot with Plot Styles** plots the drawing with plot styles, again only if defined and turned on.

Plot Paperspace Last
AutoCAD normally plots paper space objects before plotting model space objects. This option reverses the order. This option is not available when plotting from the model tab.

Hide Paperspace Objects
When turned on, this option determines if hidden-line removal applies to objects in paperspace. This option is not available when plotting from the model tab.

PLOTSTAMP

The **PLOTSTAMP** command stamps the plot with information about the drawing and plot.

When you plot sets of drawings, it sometimes becomes difficult to determine which set was plotted most recently — or even which drawing file produced the plot. To identify each plot, apply a *plot stamp*. The plot stamp is one or two lines of text that list the drawing name, date and time plotted, the plotter device, plot scale, name of the computer that generated the plot, and so on. In addition, you may specify two custom pieces of data.

The plot stamp is usually placed along the edge of the drawing, but you can change the location, as well as the font and size of text. The plotstamp data can be saved to a file, which is useful for billing clients, seeing who is hogging the plotter, and so on.

You access plot stamping with the **PLOTSTAMP** command, or from the **Plot Device** tab of the Plot dialog box.

TUTORIAL: STAMPING PLOTS

1. To apply a stamp to the plotted drawing, start the **PLOTSTAMP** command:
   - In the Plot Device tab of Plot dialog box, choose the **Plot Stamp** options.
   - At the 'Command:' prompt, enter the **plotstamp** command.
   - Alternatively, enter the aliases **ddplotstamp** (the command's old name) at the 'Command:' prompt.

   Command: **plotstamp** *(Press ENTER.)*

   In all cases, AutoCAD displays the Plot Stamp dialog box.

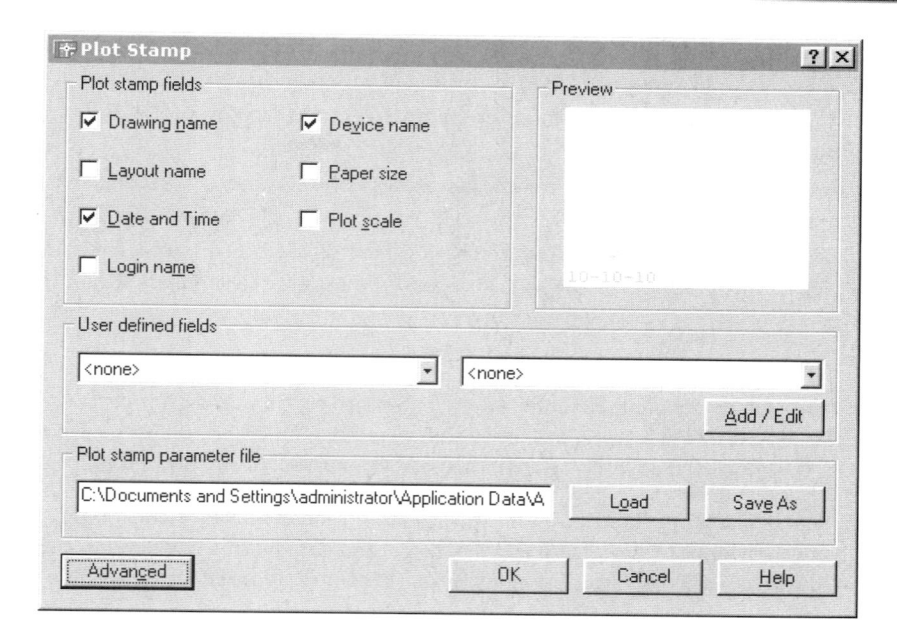

2. In the **Plot Stamp Fields** area, select the text you wish stamped on the plot:

Plot Stamp Fields	Example	Comments
Drawing Name	C:\17_22.DWG	Full path and file name of the drawing.
Layout Name	Model	Layout name; "Model" if plotted from model space.
Date and Time	6/26/2003 10:24:42 AM	Date and time of the plot; format is determined by the Regional Settings dialog box of the Control Panel.
Login Name	Administrator	Windows login name, as stored in the LOGINNAME sysvar.
Device Name	Lexmark Optra R+	Name of the plotting device.
Paper Size	Letter 8 ½ x 11 in	Size of the paper.
Plot Scale	1:0.8543125	Plot scale factor; an unusual scale factor, such as shown in the example, means the drawing was scaled to fit the margins.

3. To define your own fields, click the **Add/Edit** button. AutoCAD displays the User Defined Fields dialog box.

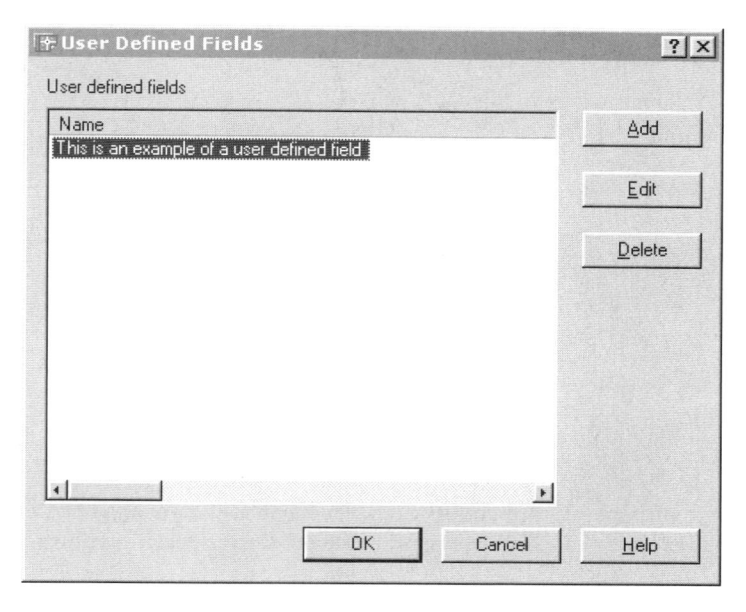

Click **Add**, and then enter text. You can add as many user-defined fields as you wish, but AutoCAD includes a maximum of two per plot stamp. When done, click **OK**.

To include a user-defined field with the plot stamp, select them from the drop lists under **User defined** fields.

4. Click **Save As** to save the settings to file. This allows you to reuse the settings with other drawings, or swap with friends.

   AutoCAD displays the Plotstamp Parameter File Name dialog box. Enter a file name, and then click **Save**. AutoCAD saves the data in a *.pss* (plot stamp parameter) file.

5. Click **OK**.

Notes: When the options of the Plot Stamp dialog box are grayed out, this means that the *inches.pss* or *mm.pss* files in AutoCAD's *\support* folder are set to read-only. To change the setting with Windows Explorer: (1) right-click the file; (2) select Properties; (3) uncheck Read-only; and (4) click **OK**.

The plot stamp data is not saved with the drawing, but is generated with each plot. The plot stamp's orientation is previewed by the Plot Stamp dialog box, and not with the **PREVIEW** command. Plot stamps are plotted with color 7 (on raster plotters) or pen 7 (on pen plotters).

STAMPING PLOTS: ADVANCED OPTIONS

The Advanced Options dialog box provides additional options for locating the stamp on the plot. In the Plot Stamp dialog box, click **Advanced**.

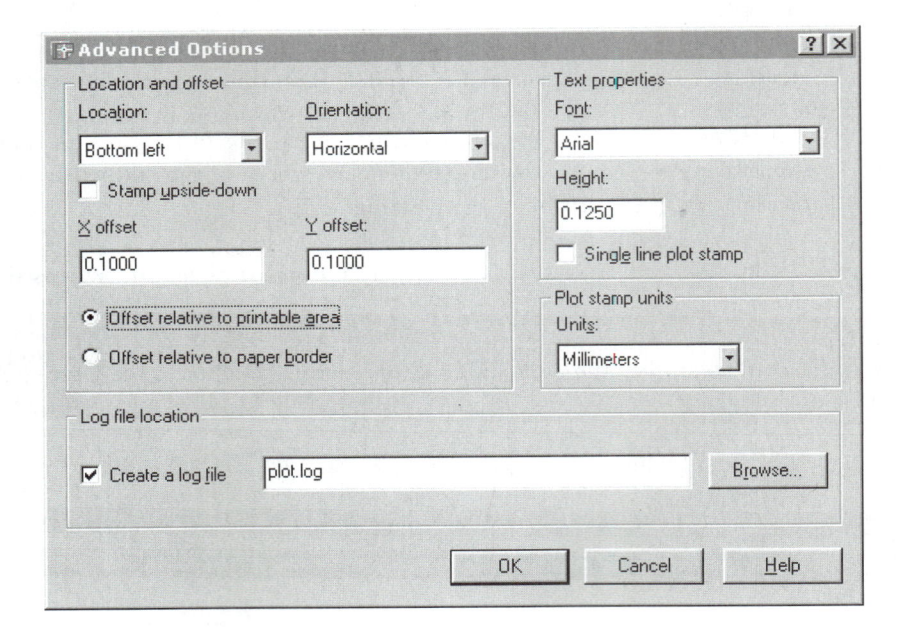

The Location and Offset area positions the stamp on the plot. The preview, unfortunately, is not located in this dialog box, so you need to go through a cycle: change settings, click **OK** to get out of this dialog box, check the preview, and return.

The **Stamp upside-down** option is useful, because then the stamp is confused for other text in the drawing. Also, some filing systems work better with upside-down plot stamps.

Note: Too large of an offset value positions the plot stamp text beyond the plotter's printable area, which may cause the text to be cut off. To prevent this, use the **Offset relative to printable area** option.

The **Text Properties** section selects the font and size of text. I recommend a narrow font, such as Arial Narrow or Future Condensed, to ensure all the plot stamp text fits the page. The default size, 0.2", may be too large; change it to 0.125" or smaller.

Plot stamp units defines the offset and height numbers used by this dialog box. Select from inches, millimeters, and pixels.

The **Log File Location** area lets you specify the name and folder for the plot logging file. The *.log* file is in ASCII format, and contains exactly the same information as stamped on the plot:

C:\17_22.DWG,Model,6/26/2003 12:34:48 PM,Administrator,Lexmark Optra R plus,
Letter 8 ½ x 11 in,1:0.854315,

Click **OK** to exit the dialog box.

PUBLISH

NEW IN 2004

The **PUBLISH** command creates drawing sets.

A *drawing set* consists of one or more drawings and layouts that are plotted, in order, at one time; alternatively, the drawing set can be a *.dwf* file or a *.plt* plot file, which is sent later to printers. Engineering and architectural offices often use drawing sets to combine all drawings belonging to a project.

This command allows you to select the drawing files and layouts, and then reorder, rename, or copy them. (This command does also plots drawings.) Lists of drawing sets can be saved as *.dsd* (drawing set description) files for later reuse.

TUTORIAL: PUBLISHING DRAWING SETS

1. To set create a drawing set, start the **PUBLISH** command:
   - From the menu bar, choose **File**, and then **Publish**.
   - From the **Standard** toolbar, choose the **Publish** button.
   - At the 'Command:' prompt, enter the **publish** command.

   Command: **publish** *(Press* ENTER.*)*

   In all cases, AutoCAD displays the Publish Drawing Sheets dialog box.

2. Initially, the dialog box lists the names of the current drawing(s) and layout(s) open in AutoCAD.

   Click **Add Sheets** to add more drawings. AutoCAD displays the Select Drawings dialog box.

   Select one or more drawings in the manner of the OPEN command, and then click **Select**.

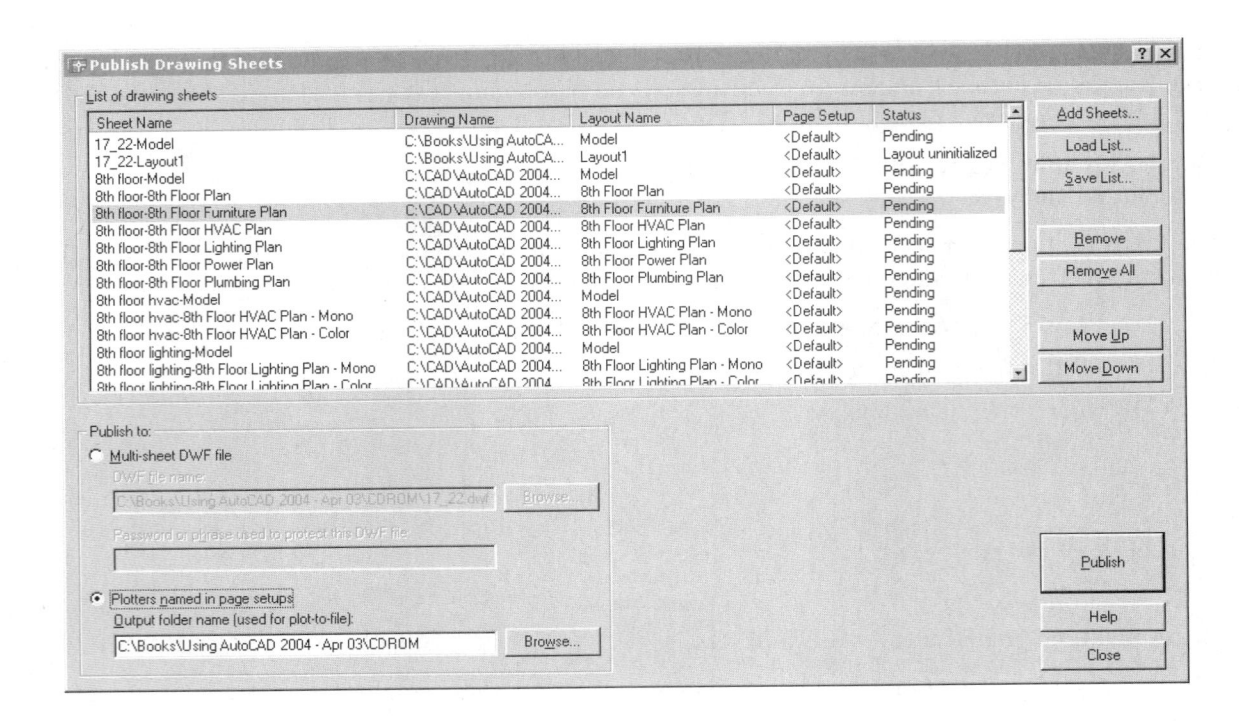

AutoCAD adds the drawings and their layouts to the list:
- **Model** indicates the model tab (model space view).
- *Layout names* indicates the layout tabs (paper space view).

If the layouts already exist in the list, AutoCAD prompts you to change their names:

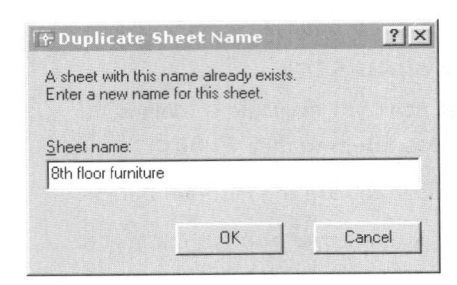

Change the name, and then click **OK**.

If you click **Cancel**, the layout is not added to the list.

3. To remove layouts from the list, select one or more, and then click **Remove**.
   - Hold down the SHIFT key to select a range of layout names.
   - Hold down the CTRL key to select nonconsecutive layout names.

   AutoCAD removes the selected layouts without asking if you're sure.

4. Drawing sets are usually plotted in a specific order. Use the **Move Up** and **Move Down** buttons to change the order of the layouts.

5. The drawing sets can be "published" to the printer, or saved as a *.dwf* file for transmittal by email. Make the selection in the Publish To section:
   - **Multi-sheet DWF file** saves all of the layouts in a single *.dwf* file. You have the option of protecting the file with a password to stymie unauthorized viewing.
   - **Plotters name in page setups** sends all of the layouts to plotter(s). As the option name indicates, the layouts might not end up being plotted by the printer you expect.

Prior to pressing **Publish**, ensure each layout is set up for the correct printer. This complexity is necessary, because drawing sets sometimes need to be plotted to a variety of printers.

6. Click **Save List** to save the list of layouts for later reuse.
7. Click **Publish**.

    AutoCAD loads each drawing, and then generates the plot or the *.dwg* file. A dialog box notes its progress.

    If a drawing or layout cannot be found, AutoCAD skips it and carries on publishing the next layout. If errors are found, AutoCAD generates a *.csv* (comma separated value) log file, which you can read with a spreadsheet program.

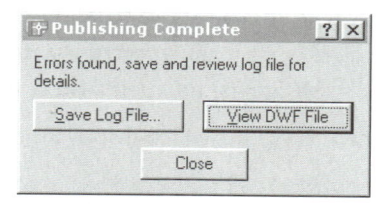

When done, AutoCAD asks if you want to view the *.dwf* file in the Express Viewer software provided with AutoCAD 2004.

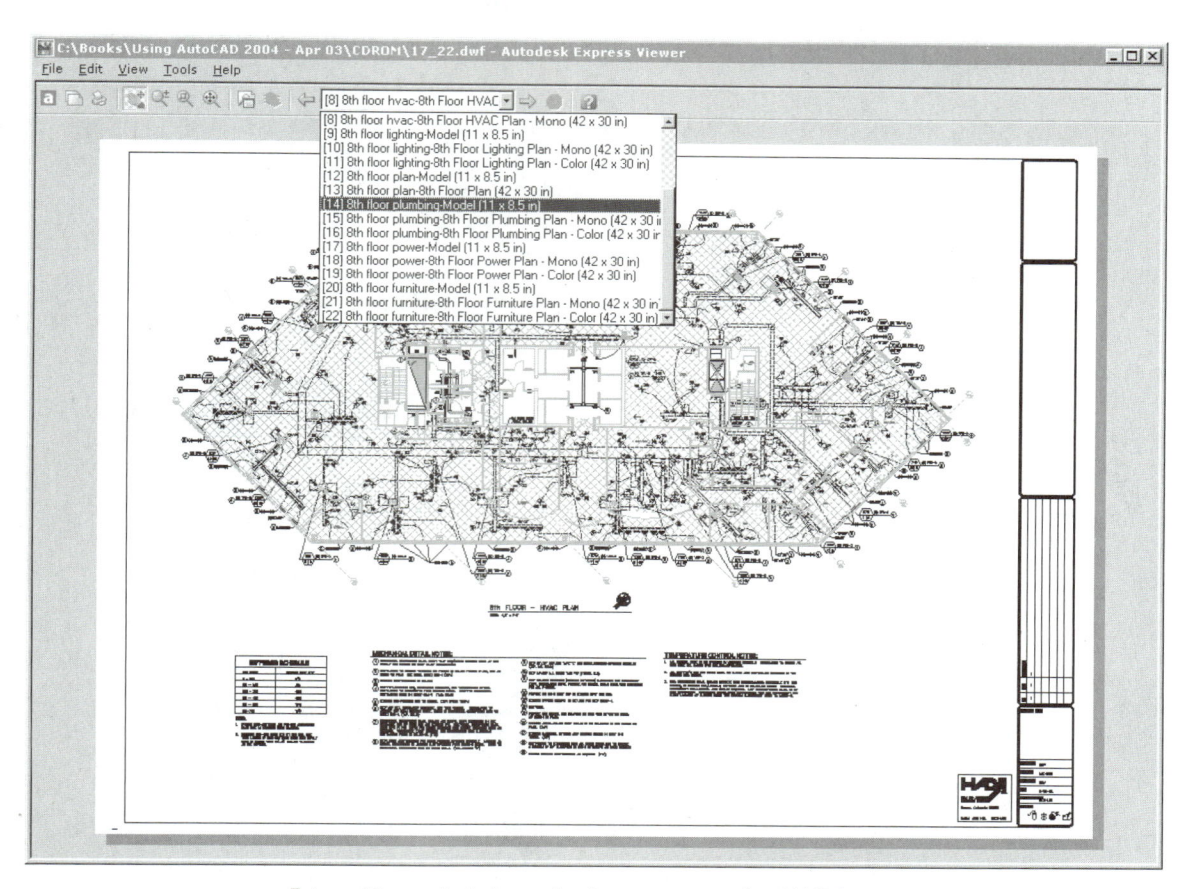

Express Viewer displaying a drawing set generated in DWF format.

PLOTTERMANAGER

The **PLOTTERMANAGER** command creates and edits plotter configurations.

AutoCAD plots to any printer connected to your computer. This includes system printers found in any office, and large-format plotters, often used by engineering and architectural offices for creating D- and E-size plots (roughly three to four feet across).

The problem is that the device drivers provided by Microsoft are not accurate or flexible enough for AutoCAD. For this reason, Autodesk includes its own set of improved drivers, known as HDI (short for "HEIDI Device Interface;" HEIDI is short for "HOOPS Extended Immediate-mode Drawing Interface;" HOOPS is short for "Hierarchical Object-Oriented Picture System" – whew!).

To plot drawings with HDI drivers, first run the Plotter Manager to create configurations that specify in great detail how plotters print drawings.

TUTORIAL: CREATING PLOTTER CONFIGURATIONS

1. To create configurations for plotters, start the **PLOTTERMANAGER** command:
   - From the **Tools** menu, choose **Wizards**, and then **Add A Plotter**.
2. AutoCAD starts the *addplwiz.exe* program, which displays the Introduction Page.

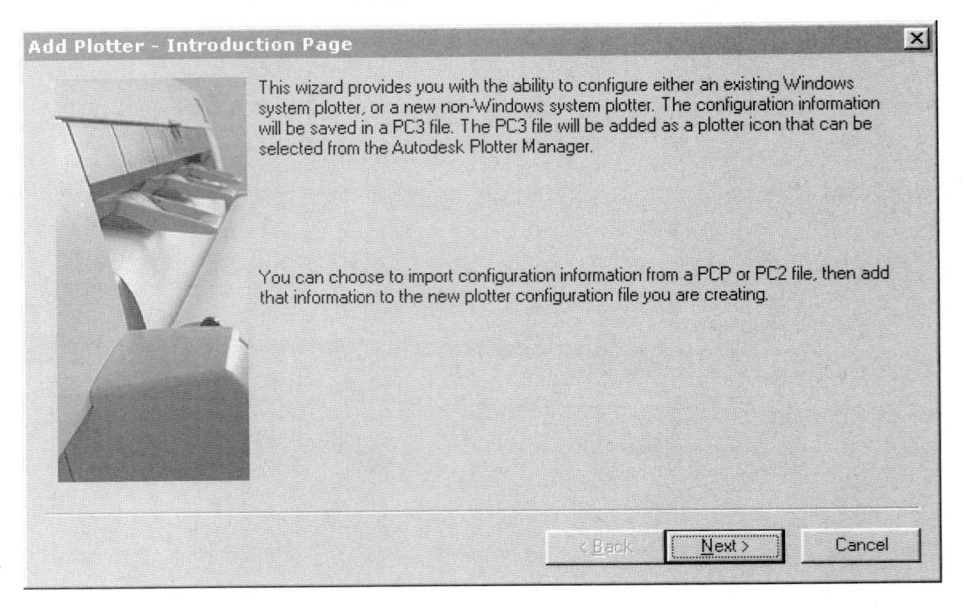

3. Choose **Next**.
4. In the Begin page, you set up AutoCAD with one of these styles of printer:

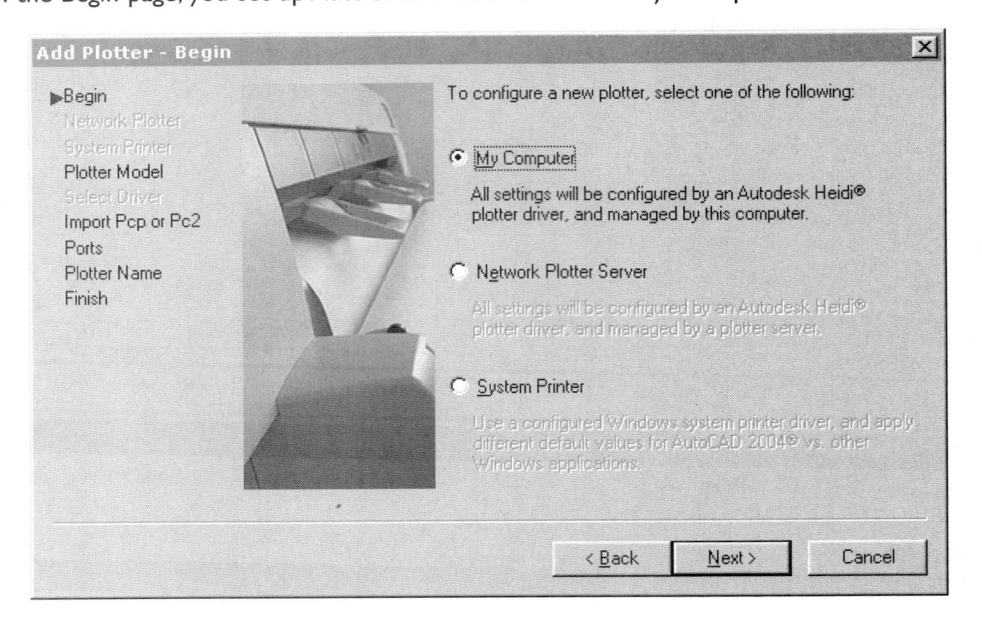

- **My Computer** are local printers and plotters controlled by AutoCAD's plotter drivers.
- **Network Plotter Server** are printers located on the network, also controlled by AutoCAD's plotter drivers.
- **System Printer** are local and network printers controlled by Windows printer drivers.

Unless you have a reason to do otherwise, choose **My Computer,** and then click **Next**.

5. AutoCAD displays a list of printer and plotter drivers provided by Autodesk.

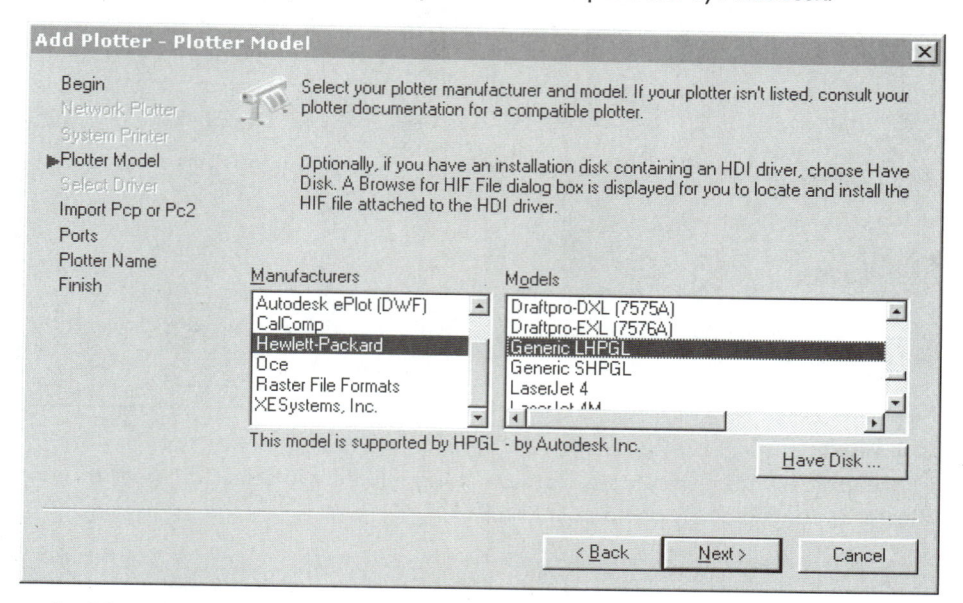

From the **Manufacturers** list, select the brand name of the plotter.

From the **Models** list, select the specific model number.

If your new plotter comes with AutoCAD-specific drivers on a diskette or CD-ROM, choose the **Have Disk** button.

(If you had selected Network Plotter Server, AutoCAD would have prompted you for the network location of the server. If System Printer, AutoCAD would have prompted you to select one.)

Choose **Next**.

Note: If you do not see the name of your plotter's manufacturer, check the documentation. Often it lists brand names and model numbers of compatible plotters. If you cannot find this information, try selecting Adobe for PostScript printers, and Hewlett-Packard for large-format inkjet plotters.

To plot the drawing in a raster format, select **Raster File Formats**, and then a specific format:

Format	Color Depths	File Extension
Independent JPEG Group JFIF	Gray, RGB	.jpg
Portable Network Graphics PNG	Bitonal, gray, indexed, RGB, RGBA	.png
TIFF Uncompressed	Bitonal, indexed, gray, RGB, RGBA	.tif
TIFF Compressed	Bitonal, indexed, gray, RGB, RGBA	.tif
CALS MIL-R-28002A Type 1	Bitonal	.cal
Dimensional CALS Type 1	Bitonal	.cal
MS-Windows BMP Uncompressed	Bitonal, gray, indexed, RGB	.bmp
TrueVision TGA 2.0 Uncompressed	Indexed, gray, RGB, RGBA	.tga
Z-Soft PC Paintbrush PCX	Indexed, RGB	.pcx

6. The Import PCP or PC2 page is important only if you created plotter configuration files with AutoCAD Release 13 (*.pcp*) and 14 (*.pc2*). Here, you can import those files for use with AutoCAD 2000 (as *.pc3* files).

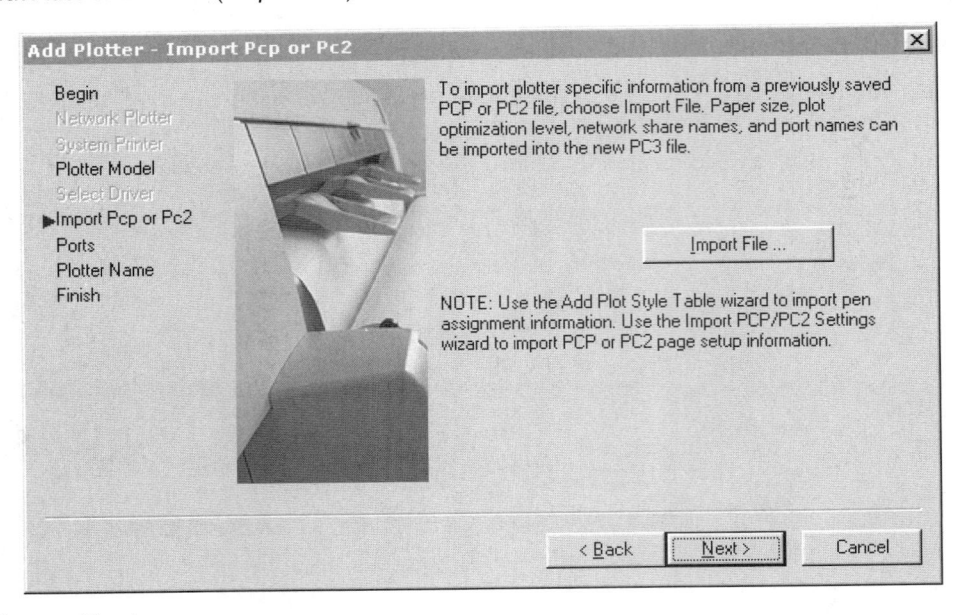

Choose **Next**.

7. The Ports page is most difficult step for some users. Here you select the port to which the plotter is connected — or no port at all.

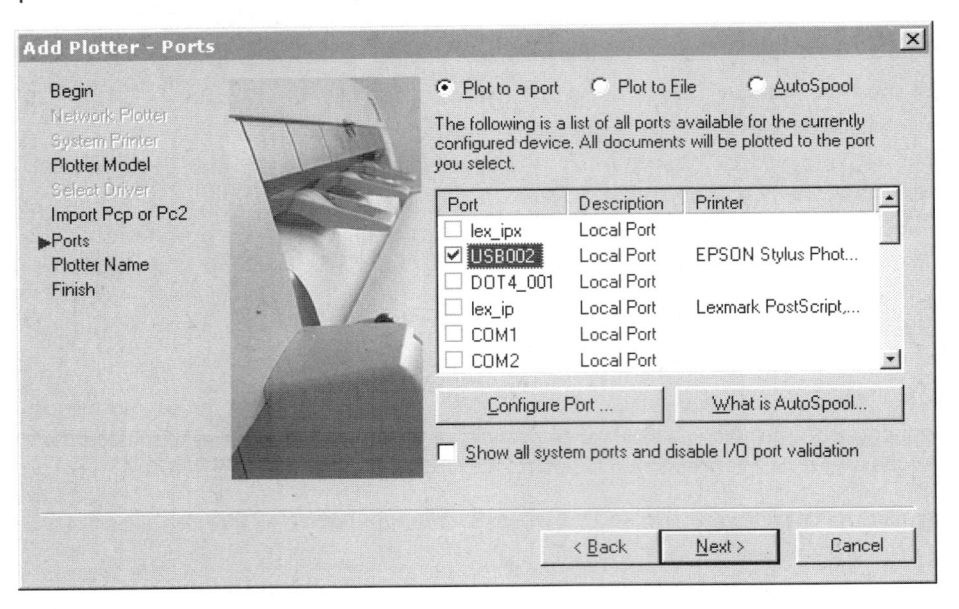

Your choices are:

- **Plot to a Port** — AutoCAD sends the drawing to the plotter through a local port or network port, such as parallel ports (designated LPT), serial ports (designated COM), or USB ports. Select this option for drawings plotted by printers and plotters.

- **Plot to File** — AutoCAD sends the drawing to a file on disk. Select this option to save drawing to disk as a file in both plotter and raster formats.

- **AutoSpool** — AutoCAD sends the drawing to a file in a specified folder (defined by AutoCAD's Options dialog box), where another program processes the file. Select this option only if you know what you are doing. (If you need to ask, then you don't know.)

8. Some ports have further options. If necessary, choose the **Configure Port** button:

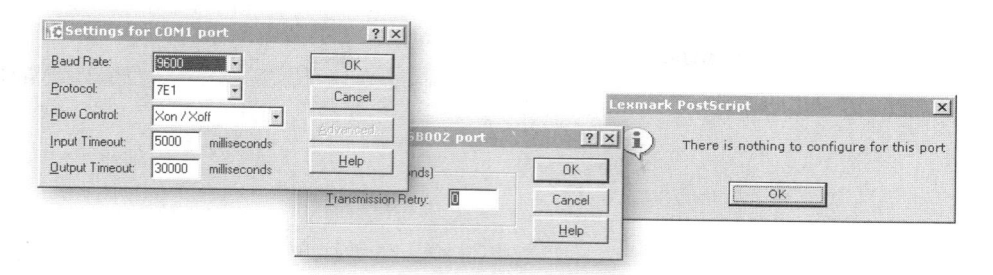

- **Serial** ports specify communications settings.
- **Parallel** and **USB** ports specify the transmission retry time.
- **Network** ports specify nothing.

Change settings, and then click **OK**.
Choose **Next**.

9. In the Plotter Name page, give the plotter configuration a name.

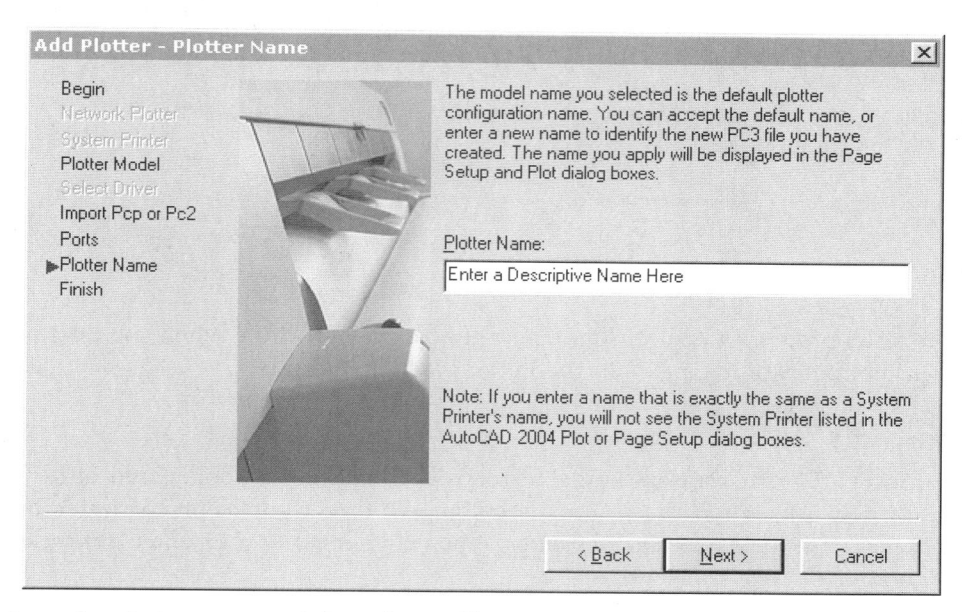

Enter a descriptive name, and then choose **Next**.

10. The Finish page has three buttons:
    - **Edit Plotter Configuration** — Displays AutoCAD's Plotter Configuration Editor dialog box, which allows you to specify options, such as media source, type of paper, type of graphics, and initialization strings.
    - **Calibrate Plotter** — calibrates the plotter. This allows you to confirm that a, say, ten-inch line is indeed plotted 10.0000000 inches long.
    - **Finish** — completes the plotter configuration.

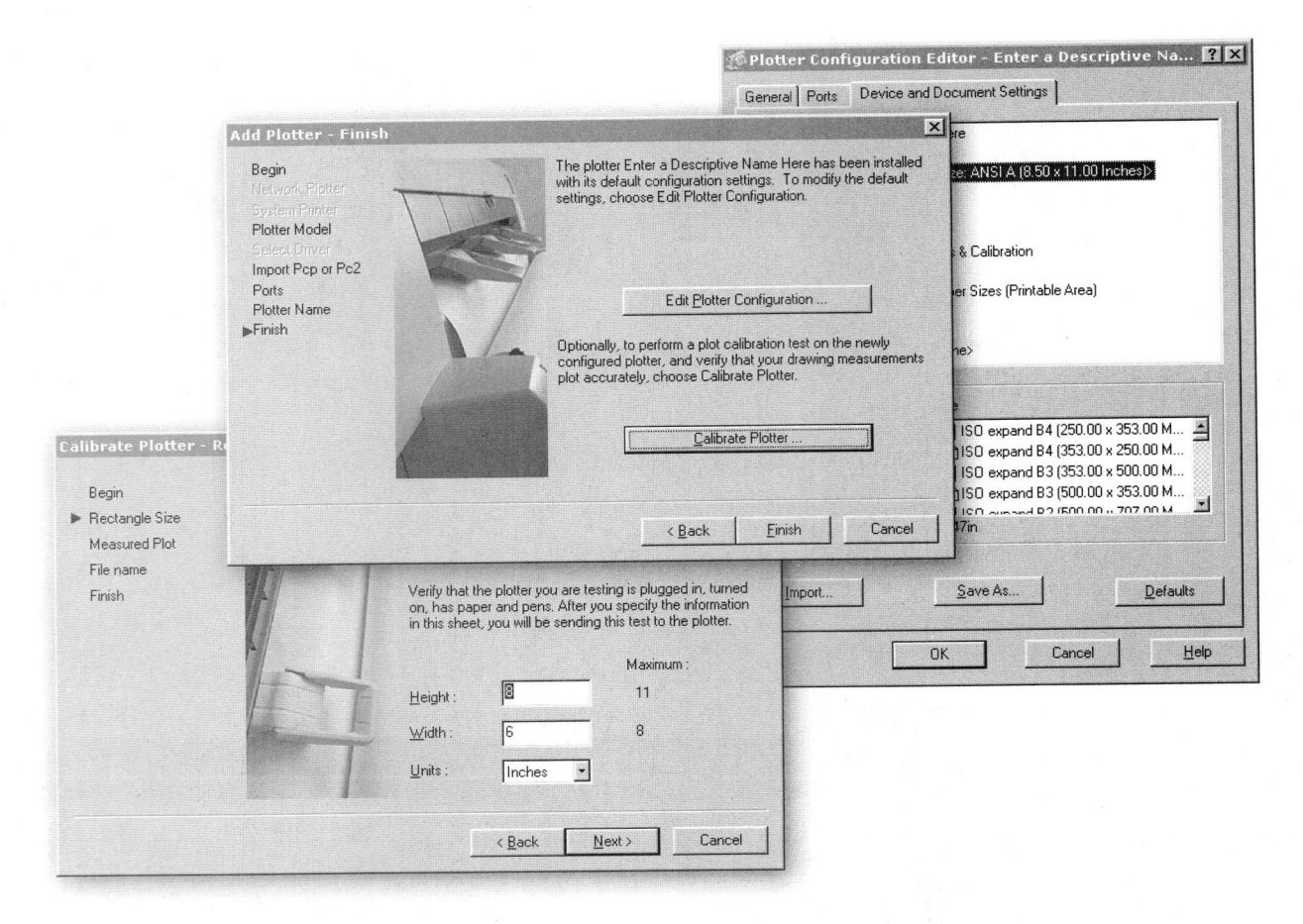

11. Choose **Finish**.

This completes the process of configuring a new plotter for AutoCAD. When you next use the **PLOT** command, this configuration appears in the list of available plotters.

ABOUT PRINTERS

You can use AutoCAD with printers available to your computer — connected either directly, or indirectly via a network. When the printer is connected directly to your computer by a parallel, serial, or USB port, it is known as a *local printer*. When connected to your office network, it is known as a *network printer*.

Local and Network Printer Connections

Printers are usually connected to computers through the USB (universal serial port) connections. Computer can have up to 128 USB ports, although two to eight is common. To support older printers, computers also have one or two parallel ports; serial ports are used only for the oldest printers and pen plotters, because they operate slowly and are difficult to configure.

Network printers are connected *directly* or *indirectly* to the network. If connected directly, the printer contains its own network card; if indirectly, the printer is connected to a computer (through parallel, serial, or USB ports), and the computer is connected to the network.

Computers connected to the network can print to any network printer — provided the computers have been given permission to access the network printer. This works in reverse, too: you can give other networked computers permission to use your computer's printer.

Differences Between Local and Network Printers

The primary difference between local and network printers is how Windows sees them. During printer setup, you need to tell Windows whether the printer is Local or Network printer, so that it knows whether to search for it on your computer's local ports, or along the network.

Another difference is that local printers are typically more available. A network printer might be inaccessible, because the network is down, or because too many other computers are sending it files.

The advantage to networking printers is that everyone in the office can share all printers. If the printer attached to your computer breaks down (through mechanical failure, lack of paper or ink, and so on), you can easily access another printer.

Nontraditional Printers

Windows also works with nontraditional printers, the most common today being Adobe Acrobat PDF (short for "portable document format"). If Acrobat or Distiller are installed on your computer, you can create PDF files of drawings by plotting with the Acrobat printer driver.

Less popular now is the fax. Most computers have fax capability included with the modems. Windows lets you fax from any software program, including AutoCAD. The process is as simple as selecting the fax as the printer. The drawback to faxing is that AutoCAD is typically limited to squeezing large drawings onto A- or A4-sizes of paper (roughly 8" x 11").

System Printers

AutoCAD 2004 works "out of the box" with any printer connected to your computer and your network. That's because AutoCAD checks for all *system printers* registered with Windows. (System printers refer to all local and network printers recognized by Windows.)

To see the list of system printers, choose the **Start** button on the task bar, and then select **Settings | Printers**. Windows opens the Printers window, which lists your computer's system printers.

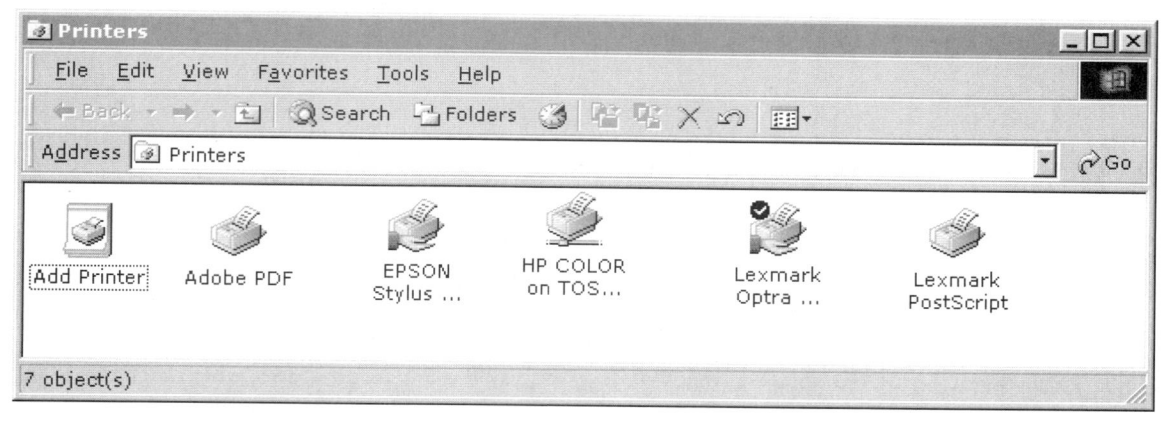

The figure illustrates the Printers window for the author's computer. From left to right, the icons indicate:

Add A Printer — Double-click this icon to add a new printer to your computer. You only need to do this when Windows does not automatically recognize a new printer or plotter added to your computer. (The automatic recognition is called "Plug and Play" — and nicknamed "plug and pray" because success is uncertain).

Adobe PDF — The unadorned printer icon indicates a local printer.

EPSON Stylus — The hand icon represents a local printer that others on the network can access. To allow others to access your computer's printer(s) over the network, right-click the printer icon, and then select **Sharing**. In the **Properties | Sharing** dialog box, select the **Shared As** radio button, and then choose **OK**.

HP COLOR on TOS — Notice the "wires" attached to the bottom of the icon. These indicate a network printer. This printer is attached to another computer on the network.

Lexmark Optra — The check mark on this icon indicates the default printer. AutoCAD and other Windows programs use this printer, unless you specify another. To select a different default printer, right-click another of the other printer icons, and then select **Default Printer** from the menu.

To change the properties of system printers, right-click the icon, and then select **Properties**. The content of the Properties dialog box varies, depending on the printer's capabilities. Commonly, though, you can set the default resolution, paper size and source, color management, and so on.

IMPORTING OLDER CONFIGURATION FILES

As Autodesk added more capabilities to plotting, it also provided a way to import older plotter configuration files into new releases of AutoCAD. The table below, taken from an AutoCAD dialog box, summarizes the paths that configuration files can take:

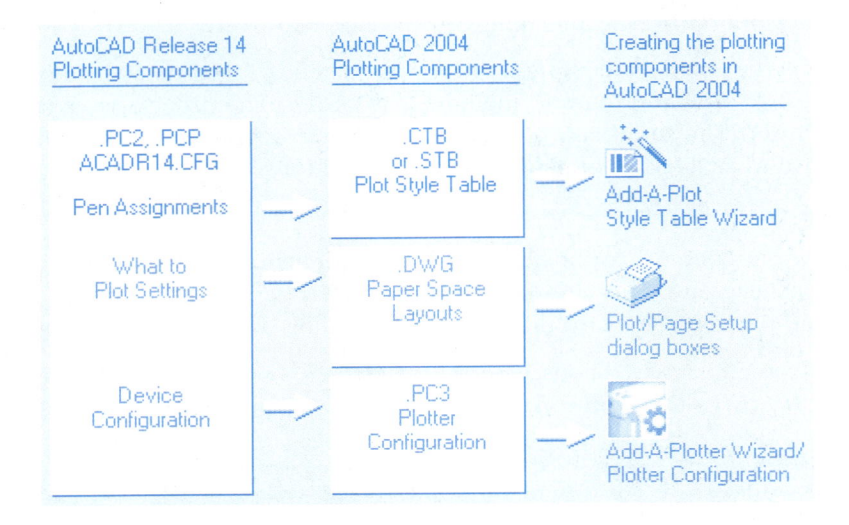

TUTORIAL: EDITING PLOTTER CONFIGURATIONS

1. To edit a plotter configuration, start the **PLOTTERMANAGER** command:
   - From the menu bar, choose **File**, and then **Plotter Manager**.
   - At the 'Command:' prompt, enter the **plottermanager** command.

     Command: **plottermanager** (Press ENTER.)

   In all cases, AutoCAD requests that Windows display the Plotters window, which lists all HDI plotter configurations as *.pc3* files.

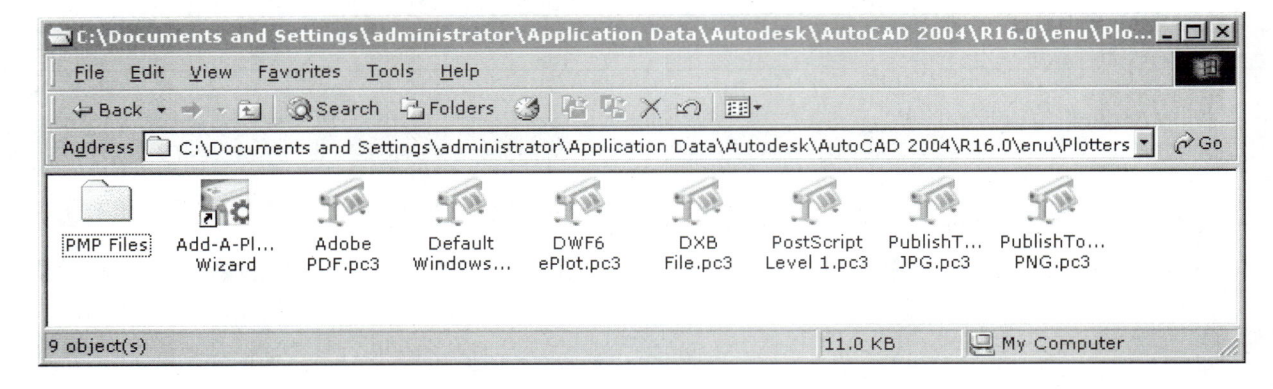

2. Double-click any .pc3 icon, except *Default Windows System Printer.pc3*, because it must be edited through Windows.
3. Windows displays the Plotter Configuration Editor dialog box.
   Click the **Device and Settings** tab.

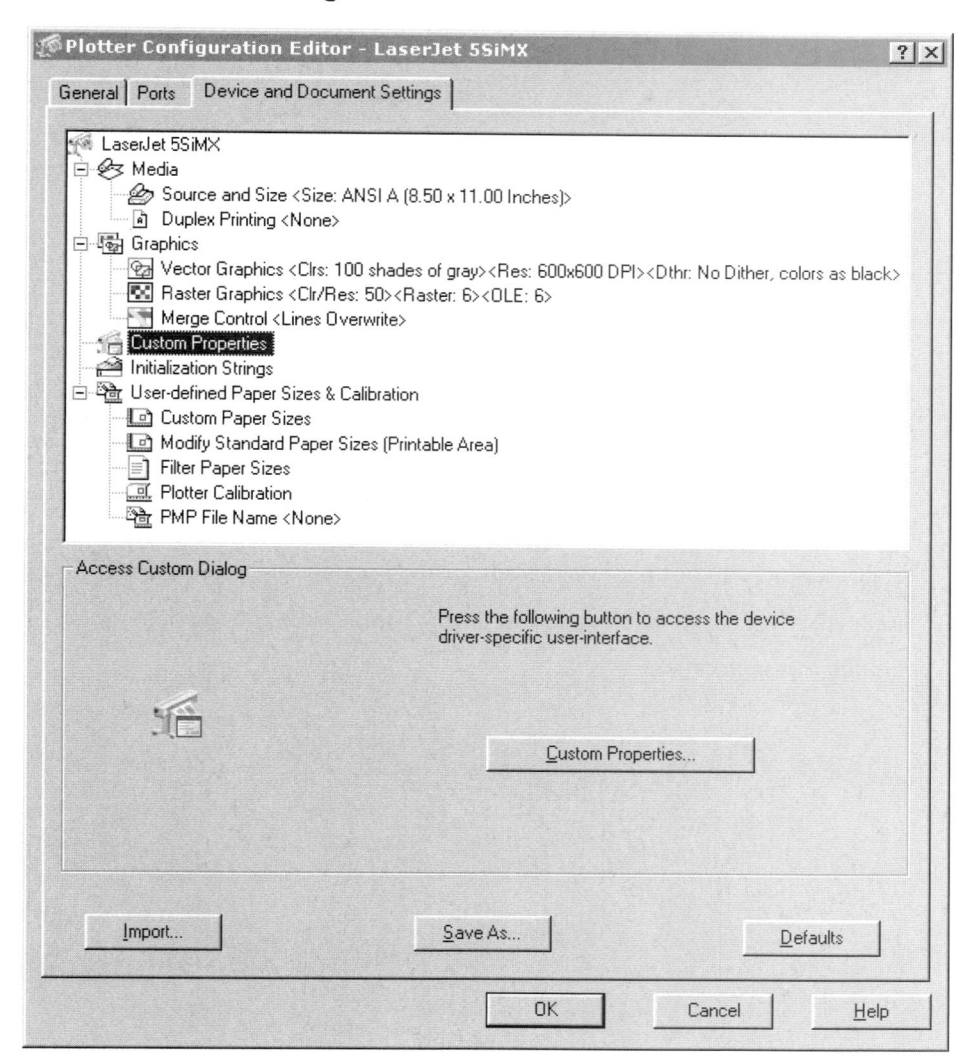

The available settings vary, depending on the capabilities of the plotter. (Shown below are settings for the HP LaserJet 5SiMX printer.)

Some settings are unavailable when the plotter does not support them. Other settings are handled through the **Custom Properties** button.

Selected options are noted in angle brackets, <like this>. When a change is made to a setting, AutoCAD and places a red check mark in front of the it

4. Changes the settings, and then press **OK**.

THE PLOTTER CONFIGURATION EDITOR OPTIONS

The Plotter Configuration Editor lists features and capabilities that AutoCAD takes advantage of. Not all settings listed below are available for all printers and plotters. Some features must be set through Windows: select the **Custom Properties** node, and then click the **Custom Properties** button.

Media

The **Media** option specifies the paper source and destination, as well as the size and type of paper.

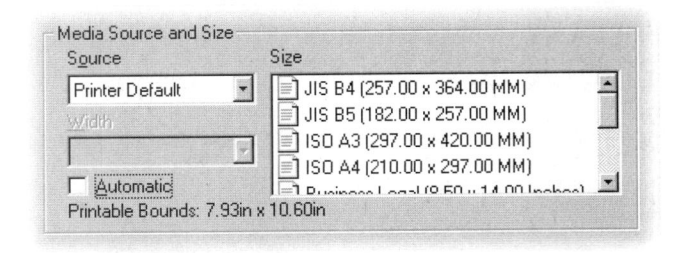

Source specifies the location of the paper. Examples include trays, sheet feed, and roll feed.

Width specifies the width of the paper; applies only to roll-fed sources.

Automatic means that the printer determines the paper source.

Size lists the standard and custom paper sizes the printer is capable of handling.

Printable Bounds lists the margin measurement for the selected paper size.

Media Type specifies the type of paper.

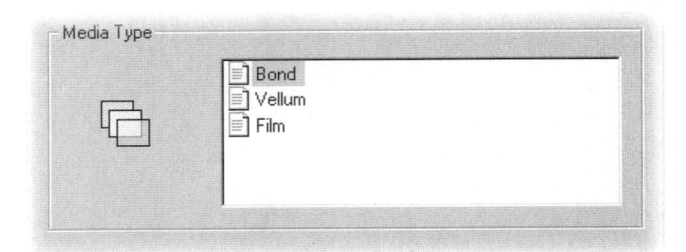

Duplex Printing

The **Duplex Printing** option determines whether the paper is printed on both sides.

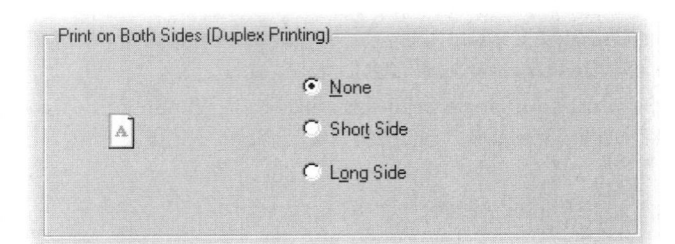

None turns off double-sided printing.

Short Side sets the binding margin on the short edge.

Long Side sets the binding margin on the long edge.

Media Destination
The **Media Destination** option handles paper activities after the plot, such as cutting, collating, and stapling.

Physical Pen Configuration
The **Physical Pen Configuration** option is for pen plotters only. It controls the actions of each pen.

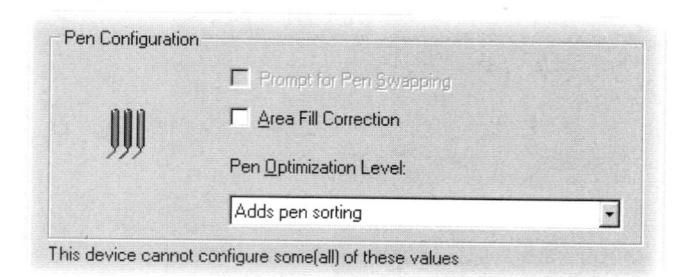

Prompt for Pen Swapping forces AutoCAD to pause the plot so that you can change pens.

Area Fill Correction forces AutoCAD to plot filled areas by half-a-pen width narrower, so as not to draw filled areas too wide.

Pen Optimization Level optimizes pen motion to reduce total plotting time. This setting is not effective with slow computers and slow plotters.

Physical Pen Characteristics allows you to specify the color, width, and manufacturer-recommended top speed of each pen (in inches or millimeters per second).

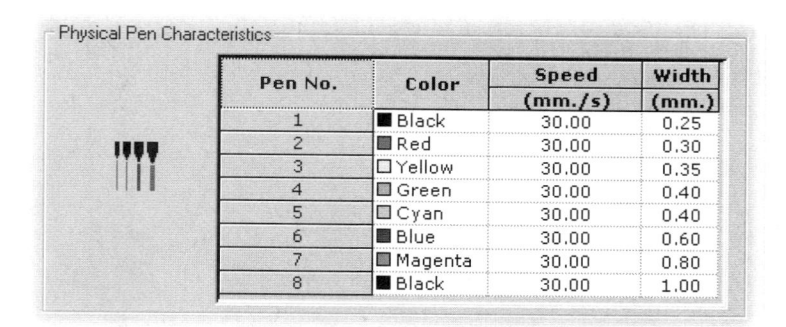

Pen No.	Color	Speed (mm./s)	Width (mm.)
1	Black	30.00	0.25
2	Red	30.00	0.30
3	Yellow	30.00	0.35
4	Green	30.00	0.40
5	Cyan	30.00	0.40
6	Blue	30.00	0.60
7	Magenta	30.00	0.80
8	Black	30.00	1.00

Graphics
The **Graphics** options specify options for vector and raster graphics and TrueType fonts. The settings here vary depending on the capabilities of each printer and plotter.

Installed Memory reports to AutoCAD the added memory (RAM or hard disk) residing in the printer.

Total Installed Memory reports the total amount of memory installed in the printer.

Vector Graphics specifies options for color depth (number of colors), resolution (dpi), and dithering (simulating a color by mixing two colors). Some printers trade off fewer colors in exchange for higher resolution.

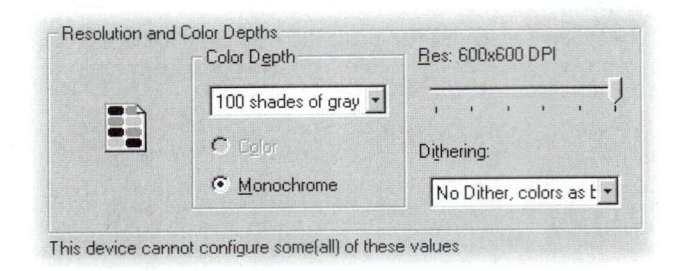

The **Raster Graphics** option trades off plotting speed for output quality (available only for plotters that don't use pens).

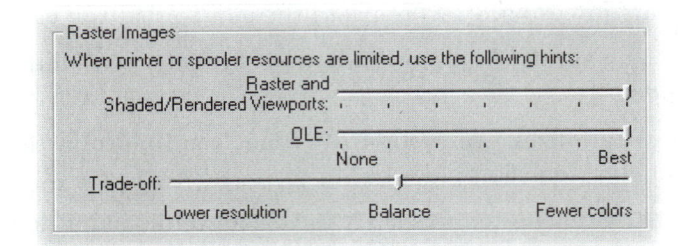

The **TrueType Text** option specifies whether to plot TrueType text as graphic images (slower, but guarantees the look of the font) or as vector outlines (faster, but may not print with correct font).

The **Merge Control** option specifies the look of crossing lines. This option is not valid when printing all colors as black, when using PostScript printers, and when your printer does not support merge control.

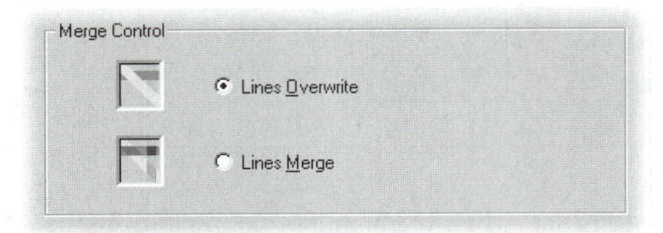

- **Lines Overwrite** — only topmost lines are visible at intersections.
- **Lines Merge** — colors of crossing lines are merged.

Custom Properties

The **Custom Properties** option displays the **Custom Properties** button. Click the button to view dialog boxes by Windows that control the printer.

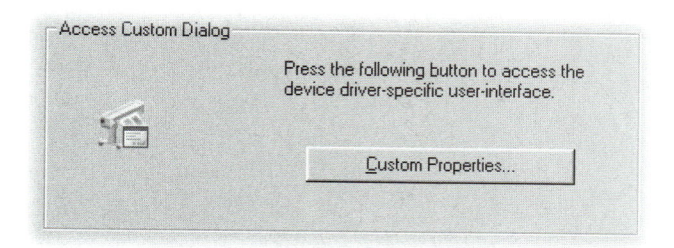

Initialization Strings

The **Initialization Strings** option sets control codes for non-system printers, including those for pre-initialization, post-initialization, and termination. These codes allow you to use plotters and printers not supported entirely by AutoCAD.

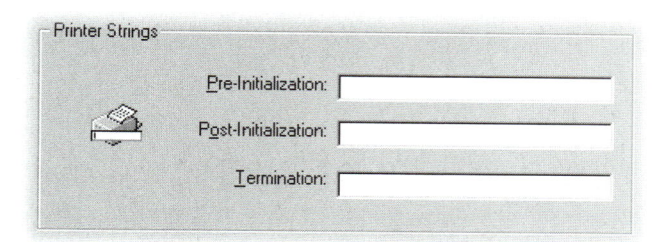

Note: Use the backslash (\ ) to emulate escape strings. For example, \27 is sent as the escape character, and \10 as the line-feed character.

User-Defined Paper Sizes and Calibration

The **User-Defined Paper Sizes and Calibration** option allows you to specify custom paper sizes (those not recognized by AutoCAD) and calibrates the printer. To help you define custom paper sizes, AutoCAD runs the Custom Paper Size wizard.

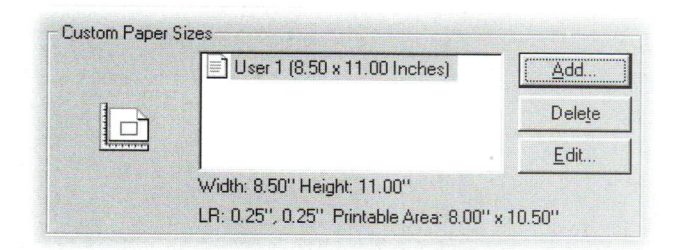

Note: Calibration ensures drawings are plotted accurately. Autodesk recommends, "If your plotter provides a calibration utility, it is recommended that you use it instead of the AutoCAD utility."

Following printer calibration, AutoCAD stores the calibration data in *.pmp* (plotter model parameter) files. The *.pmp* file must be attached to the printer's *.pc3* file, unless it was created during the Calibration stage of the Add-a-Plotter wizard.

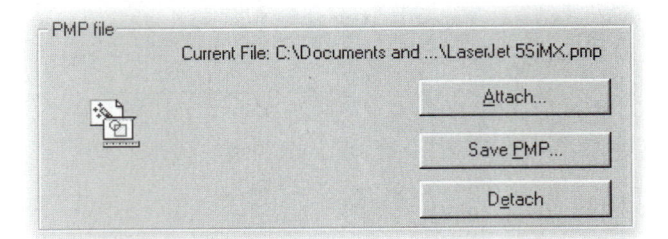

Save As

To save the settings, click the **Save As** button. You can provide the *.pc3* file to share with other AutoCAD users, who can read the file through the **Import** button.

STYLESMANAGER

The **STYLESMANAGER** command creates and edits plot style tables.

Plot style *tables* are collections of plot styles assigned to layouts and the model tab. *Plot styles* control the printing properties of objects. Depending on the capabilities of the printer, plot styles control color or grayscale, dithering and screening, pen number and virtual pens, linetype and lineweight, line end and join styles, and fill style.

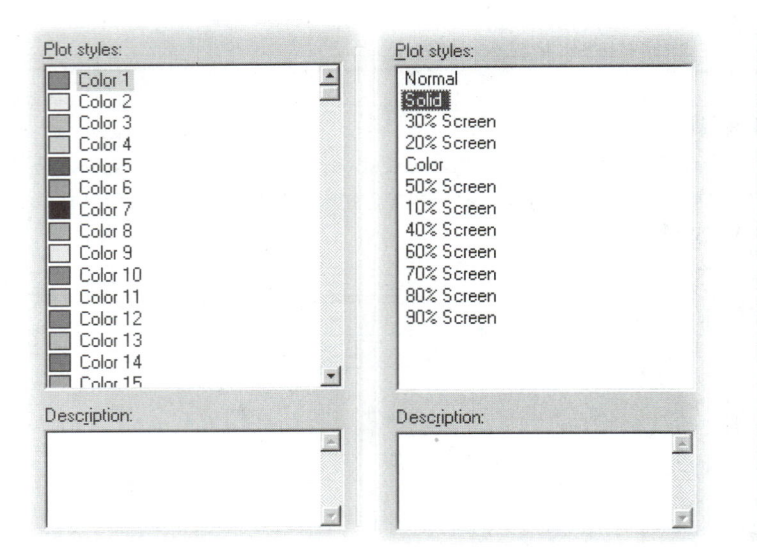

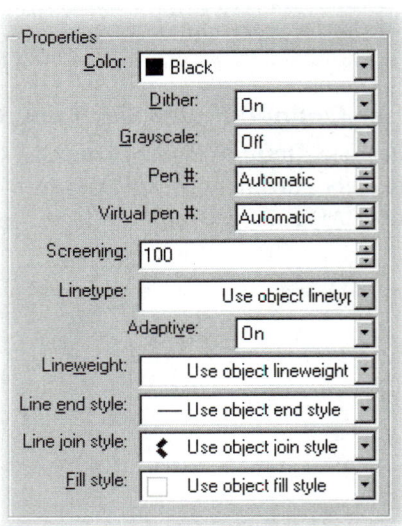

Left: Color-based plot style names are limited to the 256 ACI colors.
Center: Named plot styles are unlimited.
Right: The plot properties that can be set for each plot style.

AutoCAD works with two types of plot style table: color-dependent and named.

Color-dependent plot style tables (*.ctb* files) assign plot styles by the color of the layer or the objects, if ByLayer color has been overridden. (Prior to AutoCAD 2000, this was the only way to control printing. During the old **PLOT** command, the CAD operator assigned AutoCAD colors to pen numbers. For example, blue may have been assigned pen #3. During plotting, AutoCAD instructed the plotter to plot all blue-colored objects with whatever pen was in pen holder #3. Pens could be assigned linetypes, colors, and widths, depending on the capabilities of the plotter.)

Color-dependent plot style tables have exactly 256 plot styles, one for each ACI color (AutoCAD Index Color). AutoCAD cannot accurately plot true colors (16.7 million) and color books (Pantone and Ral) with these style tables; if you must, however, use the **Use Entity Color** option to plot with the ACI color nearest to the true color.

Named plot style tables (*.stb* files) contain named plot styles assigned to layers and objects. This style of plot style is much more flexible than color-dependent styles, because it is not limited to controlling objects by their color. Instead, each and every layer and object can have its own plot style. You can assign a different named STB to each layout in the drawing.

> **Note**: If you are not sure whether a drawing uses color-dependent or named plot styles, look at the **Plot Style Control** drop list on the Properties toolbar. If it is grayed out, color-dependent styles are in effect; if not, named plot styles are in effect.

Left: Grayed-out drop list indicates color-based plot styles are in effect.
Right: Named plot styles are in effect.

CONVERTCTB AND CONVERTPSTYLES

You can switch drawings between color-dependent and named plot style tables with the CONVERTPSTYLES command (short for "convert plot styles").

- **From color-dependent** to named — removes CTB color-dependent plot style tables from layouts, and replaces them with named ones. Color-dependent plot style tables should be first converted using these steps:

    Step 1: **CONVERTCTB** command converts color-dependent plot style tables to named tables. Select a *.ctb* file from the dialog box, and then specify the name for the *.stb* file. The 256 converted plot styles are given generic names: Style1, Style2, and so on.

    Step 2: **STYLESMANAGER** command renames the generic plot styles. Styles should be renamed *before* being attached to layouts.

    Step 3: **CONVERTPSTYLES** command switches the drawing to named plot styles. The command has no options, but displays a warning:

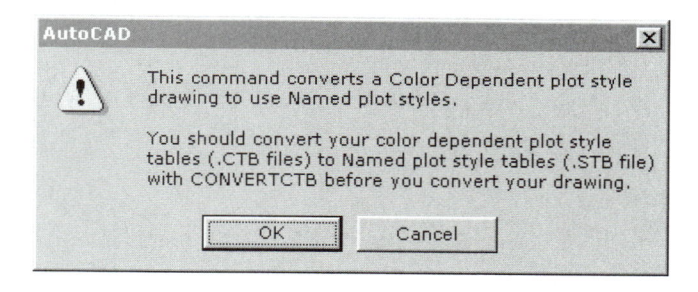

Click **OK**, and a moment later AutoCAD confirms the conversion:

Drawing converted from Named plot style mode to Color Dependent mode.

- **From named** to color-dependent — any STB plot style names assigned to layers and objects in the drawing are erased. Using the **CONVERTPSTYLES** command results in a message similar to that shown above, along with the confirmation message:

Drawing converted from Color Dependent plot style mode to Named mode.

Although it appears that the UNDO command reverses the conversion, named plot styles are labeled "missing" and are not returned.

Setting Default Plot Styles

In almost all cases, your drawings should use named plot styles, because they have more benefits than color-based ones.

Color-based Tables	Named Tables
256 plot styles.	Only as many plot styles as your drawings need. *Benefit*: fewer plot styles to deal with.
Generic style names, such as "Color1" and "Color2."	Styles can be given descriptive names, such as "50% Screening." *Benefit*: easier to understand style names.
Object colors controls plots.	Object color independent of plotting. *Benefit*: more versatile use of colors in drawings.
Single plot style for all layouts.	Every layout can have its own plot style. *Benefit*: drawing can use layouts for different plotters.

Despite the drawbacks to color-based plot styles, they remain the default setting for new drawings. To change the default plot style table for new drawings, use the **OPTIONS** command, and then select the **Plotting** tab.

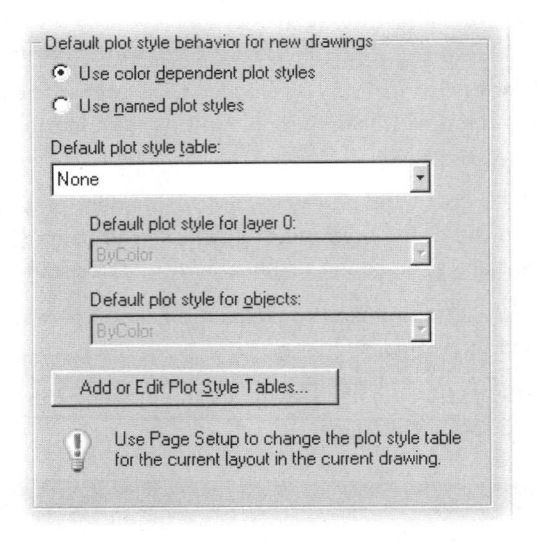

Select color or named plot style, and then a default table. Click **OK** to exit the dialog box, and then save the template file.

Assigning Plot Styles

Named plot styles can be assigned to objects, layers, and layouts. Like colors and linetypes, plot styles can be **ByLayer** and **ByBlock**. *ByLayer* means objects take on the plot style assigned to the layer; change the plot style, and all objects on the layer change. ByBLock means objects in a block take on the plot style assigned to the block. In addition, a plot style can be **Default**, which means objects take on the plot style assigned to the layout.

Objects

To assign a plot style to one or more objects, select them in the drawing, and then select the named plot style from the **Plot Style Control** drop list on the Properties toolbar. Click **Other** to see a list of all plot styles in the table.

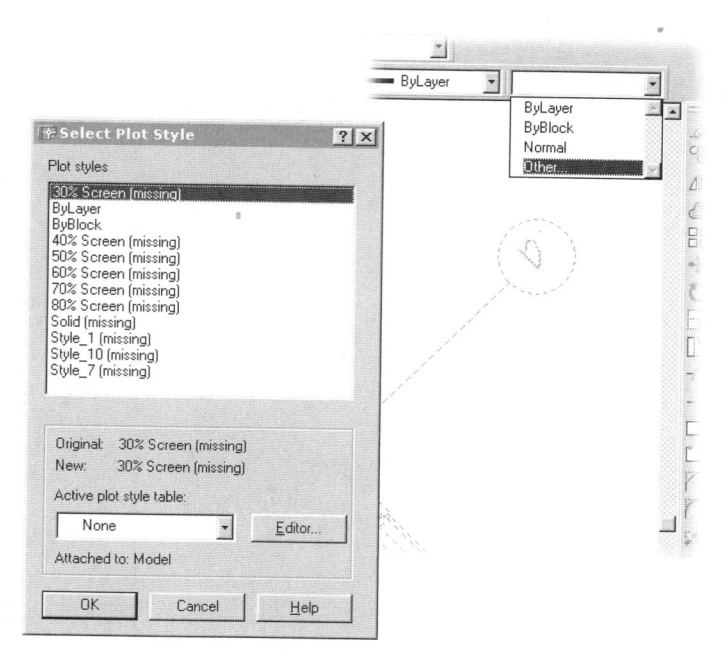

Alternatively, you can assign plot styles to objects with the Properties window.

Layers

To assign a plot style to layers, start the **LAYER** command.

In the dialog box, select the layer, and then select a named plot style from the **Plot Style** column. Repeat for other layers.

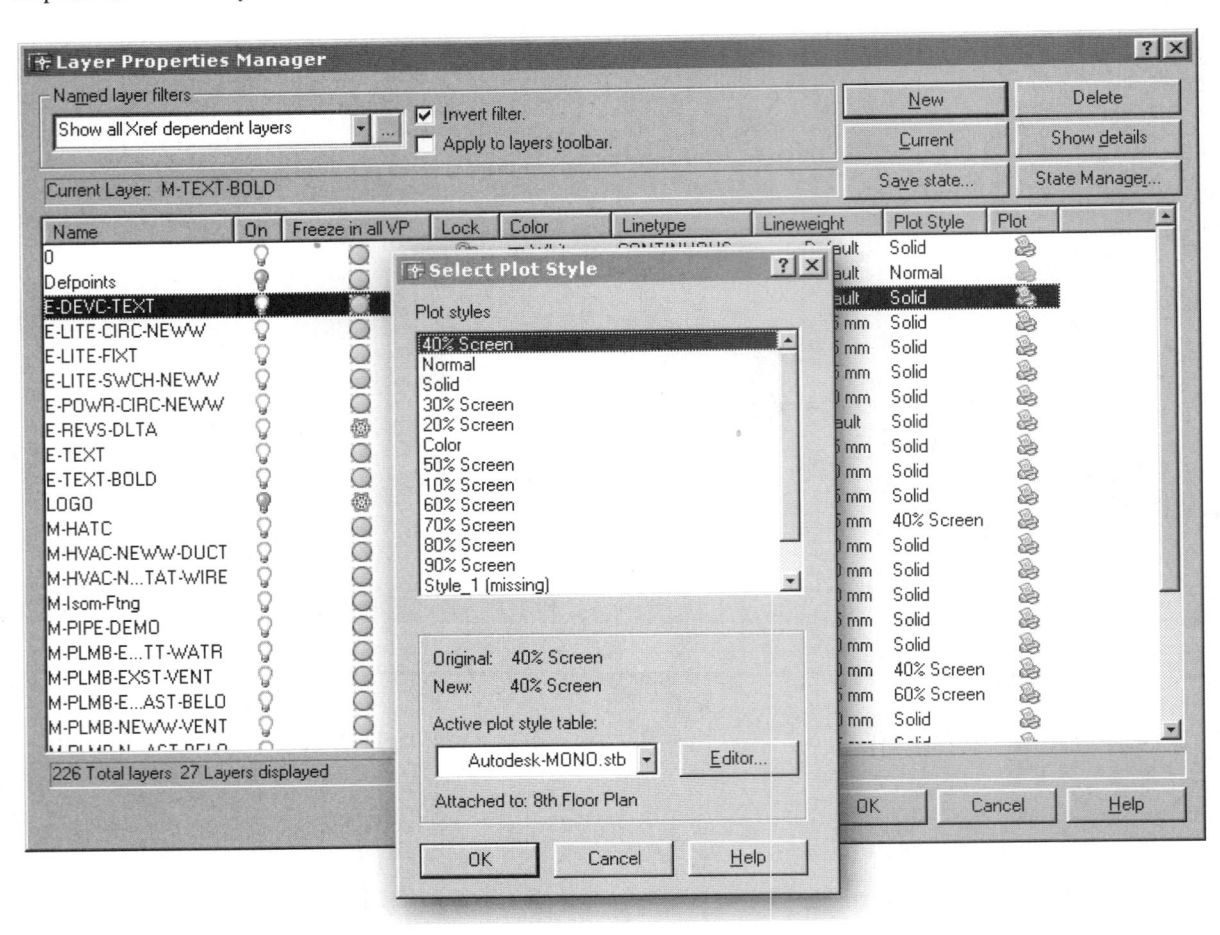

Click **OK** to exit the dialog box. All objects assigned to that layer now take on the same plot style.

Layouts

To assign a plot style to layouts, select the Model tab, and then start the **PAGESETUP** command. In the **Plot Device** tab, select a plot style table from the **Name** drop list.

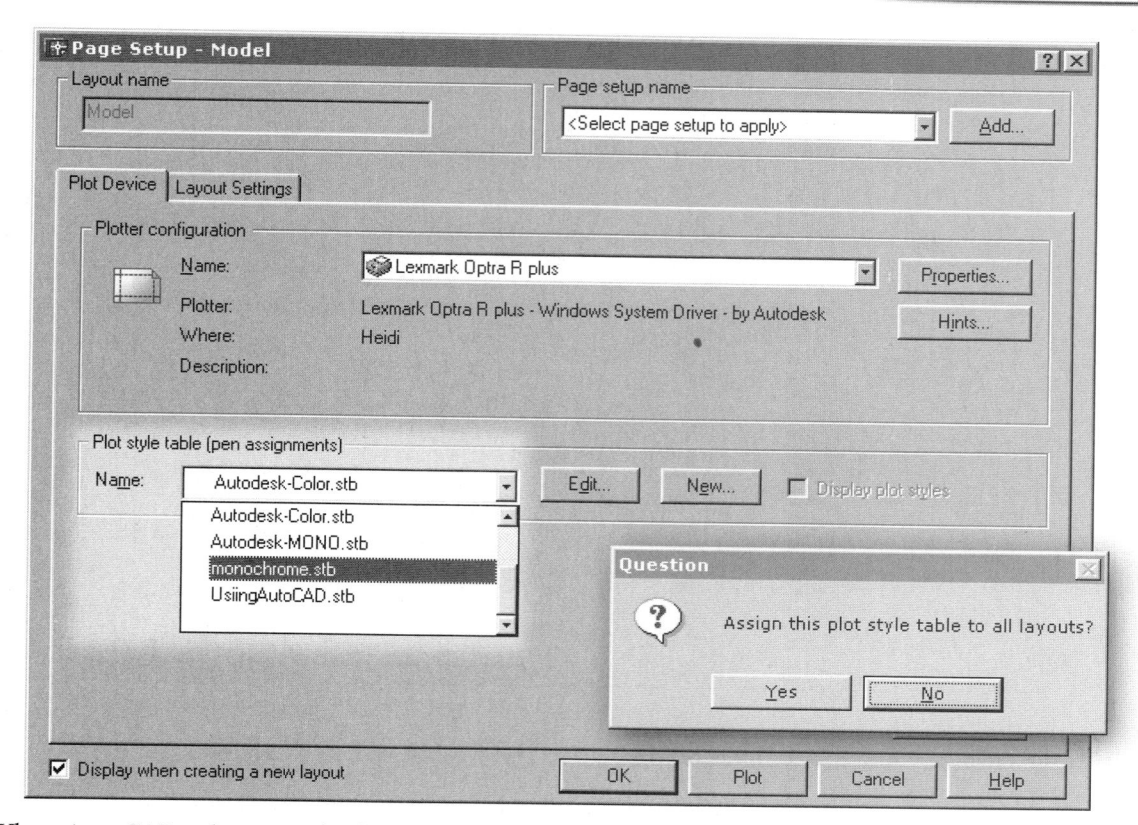

When AutoCAD asks to apply the table to all layouts, answer **No**. Click **OK** to exit the dialog box.

Select a layout tab. AutoCAD automatically displays the Page Setup dialog box again. Notice that the **Plot style table** is "None." Select a table from the Name drop list, and then turn on the **Display plot styles** option.

Click **OK** to exit the dialog box. Repeat for each layout tab.

TUTORIAL: CREATING PLOT STYLES

1. To create new plot styles, start the **stylesmanager** command:
   - From the menu bar, choose **File**, and then **Plot Style Manager**, and then select **Add-A-Plot Style Table Wizard**.
   - From **Tools** menu, choose **Wizard**, and then **Add Plot Style Table**.
   - At the 'Command:' prompt, enter the **stylesmanager** command, and then select **Add-A-Plot Style Table Wizard**.

   Command: **stylesmanager**(Press ENTER.)

   In all cases, AutoCAD runs the Add Plot Style Table "wizard," which leads you through the steps for creating a new plot style table.

2. Click **Next**.

3. In the Begin dialog box, AutoCAD asks for the kind of plot style table to create. Select **Start from Scratch**, and then click **Next**.

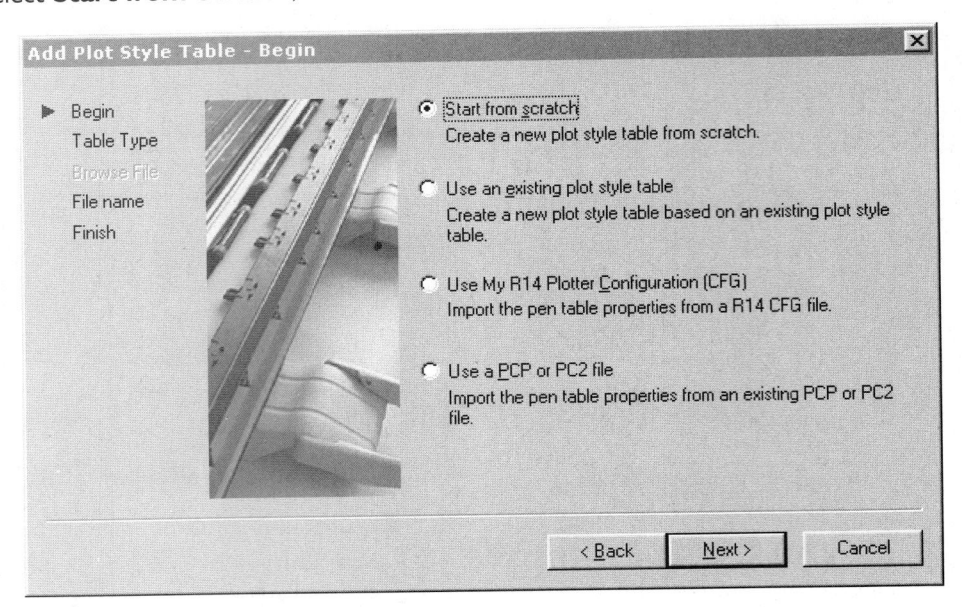

3. In the Table Type dialog box, select **Named Plot Style Table**, and then click **Next**.

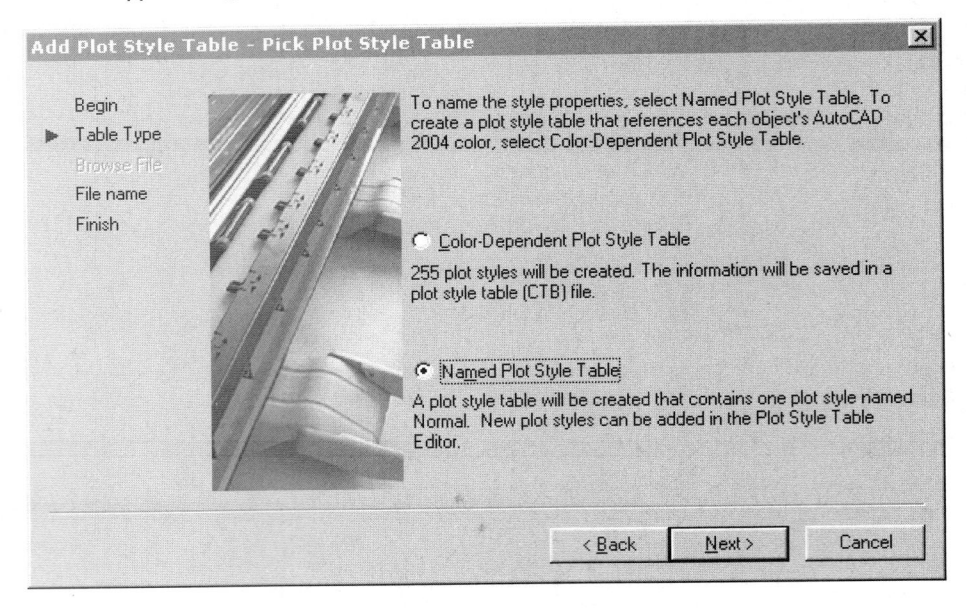

5. In the File Name dialog box, enter a descriptive name for the style table, and then click **Next**.

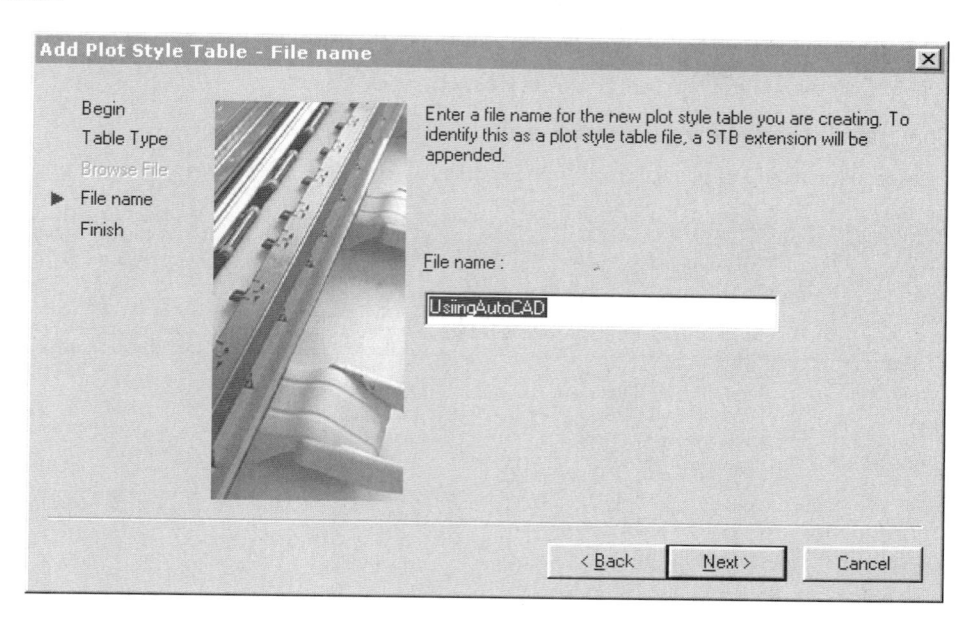

6. In the Finish dialog box, click **Finish**.

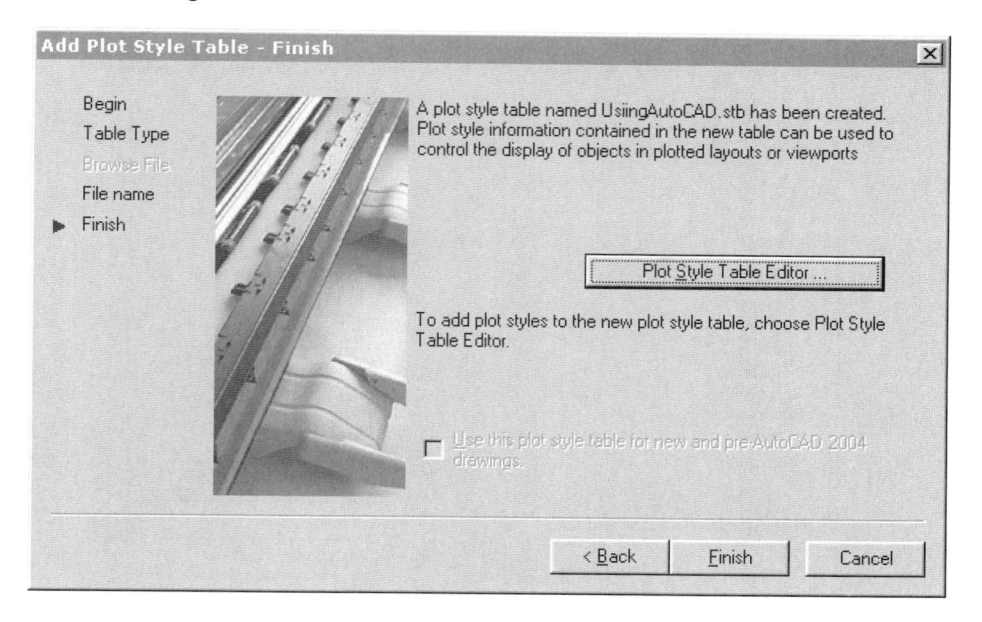

AutoCAD adds the *.stb* style table file to its collection stored in the *\plot styles* folder.

TUTORIAL: EDITING PLOT STYLES

1. To edit plot styles, start the **stylesmanager** command:
   - From the menu bar, choose **File**, and then **Plot Style Manager**.
   - At the 'Command:' prompt, enter the **stylesmanager** command.

   Command: **stylesmanager***(Press* ENTER.*)*

   In all cases, AutoCAD requests that Windows display the Plot Styles window.

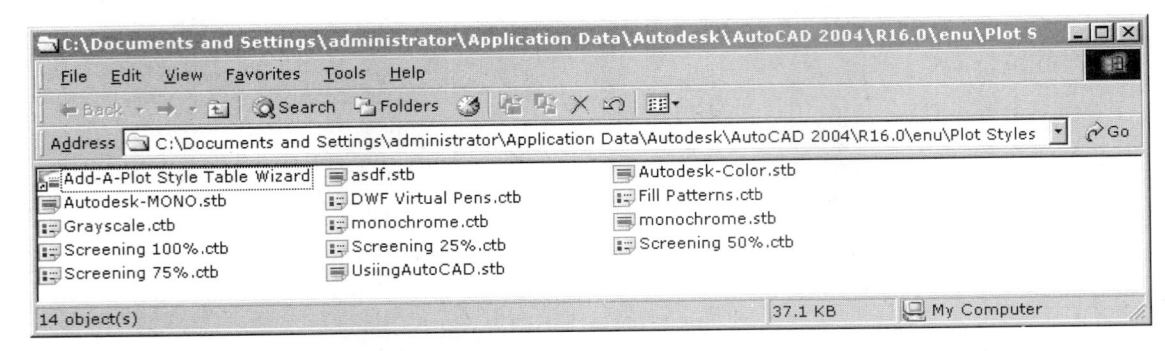

2. Double-click a plot style table file:
   - CTB: color-based plot style table files.
   - STB: named plot style table files.

   Notice that AutoCAD opens the Plot Style Table Editor.

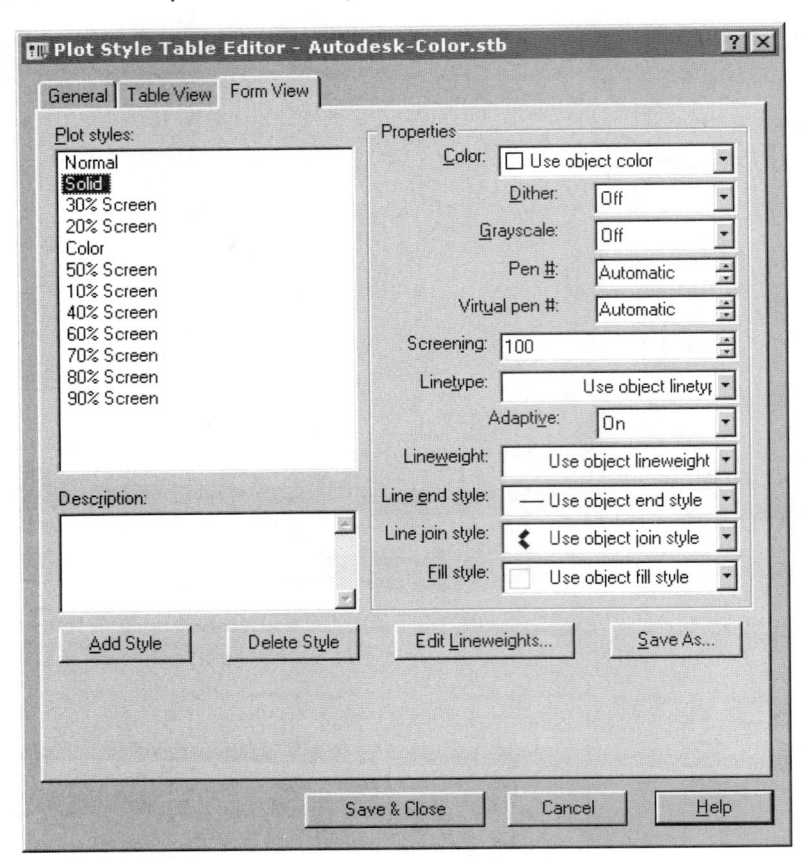

3. Click the **Form View** tab, which lists plot style names on the left, and the associated plot style properties on the right.

Plotting Drawings 741

4. Under **Plot styles**, select a plot style name — other than Normal. Notice the plot style properties at the right.

   Edit the properties, as required.

   Note: The first plot style in every named plot style table is called "NORMAL." It lists the object's default properties, so that you can see its properties with no plot style applied. The NORMAL style cannot be editor or removed.

5. To add a plot style, click **Add Style**.

   Give the style a name, and then edit its properties. (Color-based tables cannot have plot styles added to, or deleted from them.)

Plot Style Property	Comment
Color	Plotted color of objects; plot color overrides object color. *Default:* **Use Object Color**.
Dither	Toggles dithering to approximate colors with dot patterns. *Default:* **Off** for *.stb* files; **On** for *.ctb* files.
Grayscale	Converts object colors to grayscale during printing. *Default:* **Off**
Pen #	Specifies pen #; ranges from 0 to 32 (available for pen plotters only). *Default:* **Automatic** (pen #0).
Virtual Pen #	Simulates pen plotters for non-pen plotters; ranges from 0 to 255. *Default:* **Automatic** (pen #0).
Screening	Color intensity; ranges from 0 (white) to 100. Selecting a value other than 100 turns on Dithering. *Default:* **100**.
Linetype	Overrides object linetype with selected linetype. *Default:* **Use Object Linetype**.
Adaptive	Adjusts scale of the linetype to complete the pattern. Turn off if linetype scale is crucial. *Default:* **On**.
Lineweight	Overrides object lineweight with selected lineweight. *Default:* **Use Object Lineweight**.
Line End Style	Assigns a style to the end of object lines: Butt, Square, Round, and Diamond. *Default:* **Use Object End Style**.
Line Join Style	Assigns a style to the intersection of object lines: Miter, Bevel, Round, and Diamond. *Default:* **Use Object Join Style**.
Fill Style	Overrides object fill style with the following: Solid, Crosshatch, Diamonds, Horizontal Bars, Slant Left, Slant Right, Square Dots, Checkerboard, and Vertical Bar. *Default:* **Use Object Fill Style**.

6. To delete a style, select it, and then click **Delete Style**.
7. To save the plot style table, click **Save As**, and then give the table file a name.
8. When done, click **Save and Close**.

BATCHPLT

The **BATCHPLT** command plots drawings unattended.

Batch plotting is an alternative method for plotting drawings with AutoCAD. "Batching" means lining up more than one drawing for plotting. Many plotters are capable of producing more than one plot unattended, because they have a sheet feeder or continuous rolls of paper. (Historically, pen plotters had to be fed one piece of paper at a time, making batch plotting impossible.) Thus, if you have many drawings to plot, it can be more efficient to let the *batchplt.exe* program, AutoCAD, and your plotter do the work overnight, rather than tie up the plotter all day.

AutoCAD includes the BatchPlt utility (short for "batch plot"), which runs externally to AutoCAD; it is not a command. You use BatchPlt to put together a list of drawings; it then launches AutoCAD to go open and plot each drawing on the list. BatchPlt lets you specify a different plotter and setup for each drawing through *.pc3* (plotter control) files.

TUTORIAL: BATCH PLOTTING

To create a list of drawings for plotting, follow these steps:

1. To set up a batch of drawings to plot, start the **BATCHPLT** program:
   - In Windows Explorer, double-click **batchplt.exe** in the *autocad 2004* folder.
   - At the 'Command:' prompt, enter the **shell** command:

     Command: **shell** *(Press* ENTER.*)*

     OS Command: **batchplt** *(Press* ENTER.*)*

   Windows starts another session of AutoCAD, and then starts the AutoCAD Batch Plot Utility program.

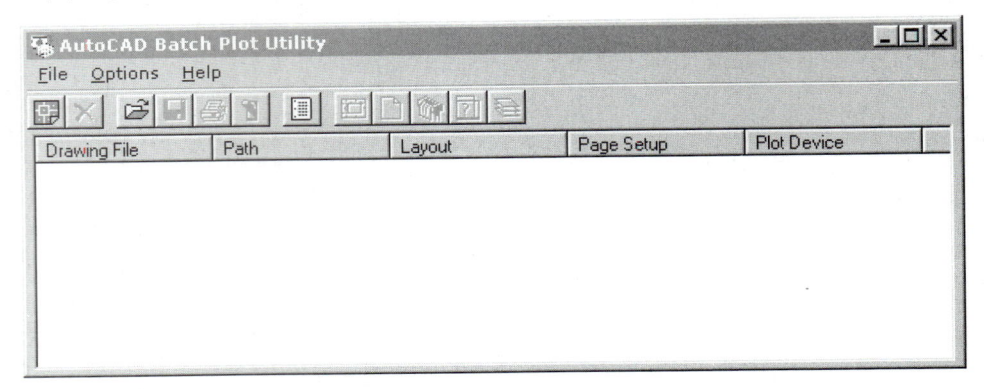

   Note: The AutoCAD opened with Batchplt must remain open; otherwise Batchplt crashes, because it relies on communicating with AutoCAD to access data in drawings.

2. From the **File** menu, select **Add Drawing**.
   Select the drawings to plot; you can select more than one file by holding down the CTRL key while selecting names.

3. To change the plotter configuration, right-click a drawing name. The cursor menu displays five options:

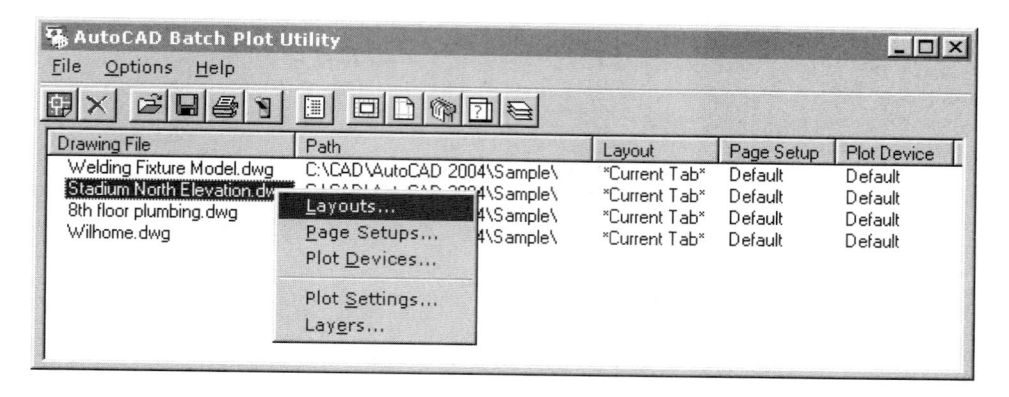

- **Layouts** selects predefined layouts from the Layouts dialog box.

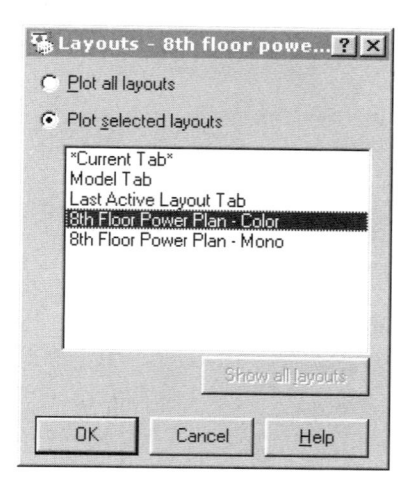

- **Page Setups** selects predefined page setups from the Page Setups dialog box.

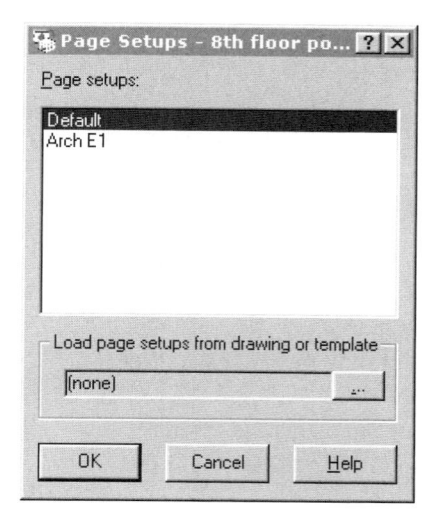

- **Plot Devices** selects plotters from the Plot Devices dialog box. (These are the same plotters as listed by the Plot dialog box inside AutoCAD.)

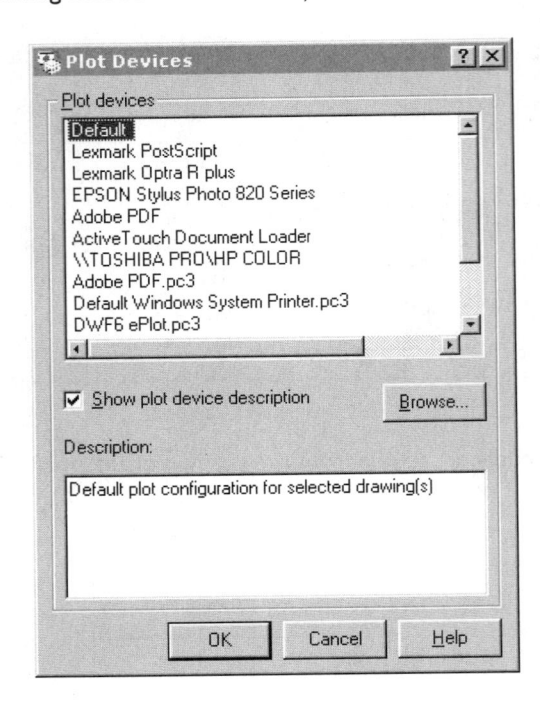

- **Plot Settings** selects plot settings from an abbreviated version of the Plot dialog box.

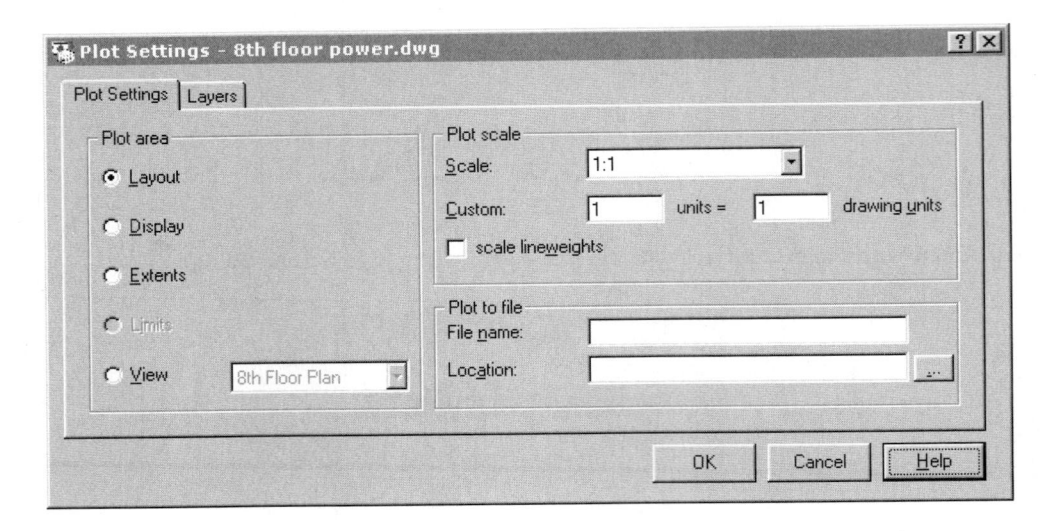

- **Layers** toggles the display of layers.

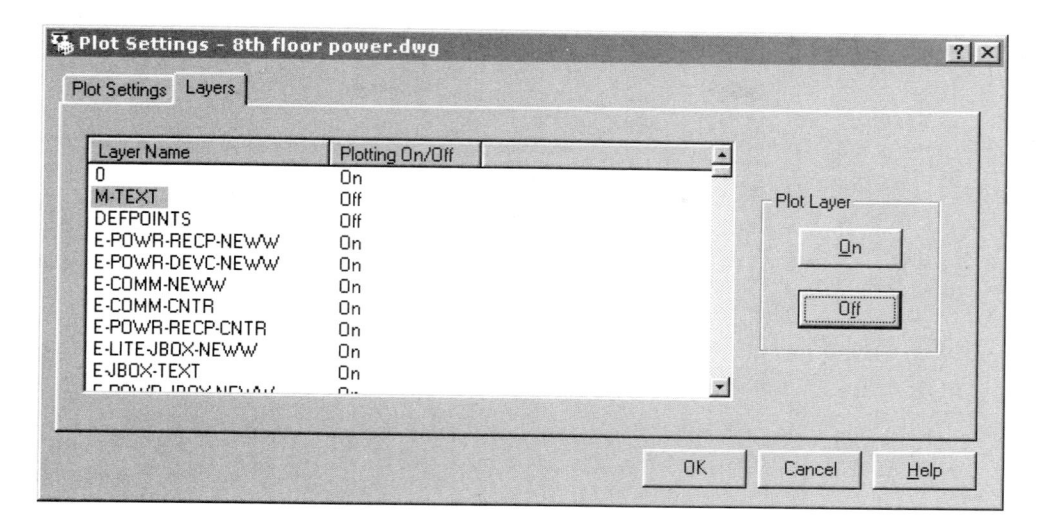

4. Check that the plotter is turned on, online, and filled with paper.
   If you wish, select **File | Plot Test**, which performs a "dry run" on all drawings. (This loads each drawing into AutoCAD, but does not plot it.)
5. Finally, select **File | Plot**.
   As each drawing is processed by AutoCAD and sent to the plotter, the Batch Plot Progress dialog box reports the progress of the batch plotting.

When you exit Batchplt, the associated session of AutoCAD also exits.

EXERCISES

1. From the CD-ROM, open the *edit4.dwg* file, a landscape drawing.

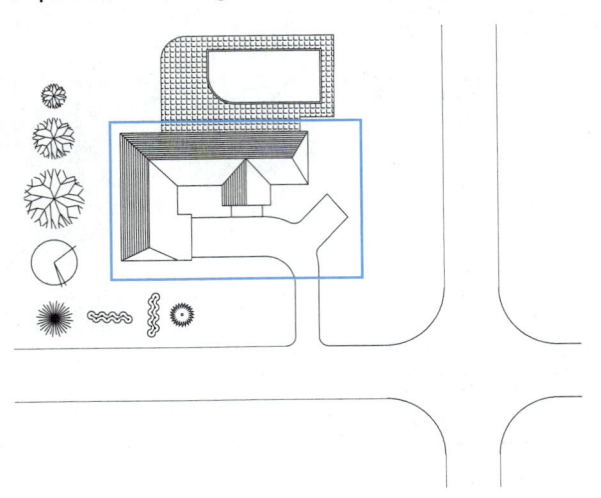

   Start the **PLOT** command, and assign a printer, if necessary.
   Plot the drawing at a scale factor of "Scaled to Fit."
   Does the drawing fit the page?

2. Use a ruler to measure the length of the house's longest side. Assume the length represents 50'.
   At what scale was the house plotted?

3. Plot the drawing again, rotating the plot through landscape or portrait orientation.

4. Using the same drawing, set the plot scale factor to 1:1.
   Preview the drawing.
   Is the drawing larger or smaller?

5. Plot the drawing again, but this time use the **Window** option, and window the area shown above by the blue rectangle.

6. Use the **PLOTSTAMP** command to attach a stamp to the plot.

7. Use the **PUBLISH** command to create a drawing set of the three drawings list below:
   15_43.dwg
   15_44.dwg
   15_45.dwg

8. Use the same three drawings to perform a batch plot.

9. Create a plotter configuration for the printer attached to your computer.
   Name the configuration "UsingAutoCAD."

10. Create a plotter configuration for creating compressed TIFF files with the **PLOT** command.
    Test the configuration by plotting a drawing, and then viewing the *.tif* file with AutoCAD's **REPLAY** command.

11. Calibrate the printer attached to your computer.
    How inaccurate was it before calibration?

12. Create a named plot style table with the following plot styles:
    50% screening.
    Checkerboard fill.
    Diamond join style.
    0.02 lineweight.
    Attach the plot style table to a drawing, and then plot it.

CHAPTER REVIEW

1. What is the purpose of plotting?
2. What part of the drawing is plotted when plotted to the extents of the drawing?
3. Describe how to plot only a portion of drawings?
4. How do you rotate the plot on the paper?
5. Why would you plot drawings to file?
6. Can you plot a drawing without first assigning a plotter or printer?
7. Explain the benefit to using the preview command.
8. What is the name of the Plot dialog box's tab that matches the following description:
   What to plot.
   Where to plot.
9. Does AutoCAD plot objects on frozen layers? <u>no</u>
10. What kinds of plotters and printers does AutoCAD work with?
11. How large is a drawing plotted when the scale factor is Scaled to Fit?
    1:1?
    1/128" = 1?
12. Name an advantage to using Full Preview.
    To using Partial Preview.
13. Explain how *spooling* helps the performance of the computer.
14. When might you want to offset the drawing on the paper?
    What is the drawback of offsetting?
15. Can AutoCAD create rendered plots?
    Can 2D drawings be rendered?
16. How do you instruct AutoCAD to plot with hidden lines removed in model space?
    In a layouts?
17. Describe how plot stamps are useful.
18. When a plot stamp indicates the scale factor is 1:0.97531864, what might this indicate?
19. What is a drawing set?
    When might you want to use it?
20. Must all layouts in a drawing set come from the same drawing?
21. Explain the purpose of the **PUBLISH** command.
    Of the **PLOTTERMANAGER** command.
22. Can AutoCAD plot to printers located on networks?
    Connected via USB ports?
23. Explain the meaning of the parts of the printer icon:

    a.
    b.
    c.

24. What is the purpose of the .*pc3* file?
    The .*stb* file?
    The .*ctb* file?
25. Can AutoCAD print on both side of the paper?

26. Explain the purpose of the **STYLESMANAGER** command.
27. Of the two lists of plot style names shown below, which are color-based names and which are named styles?

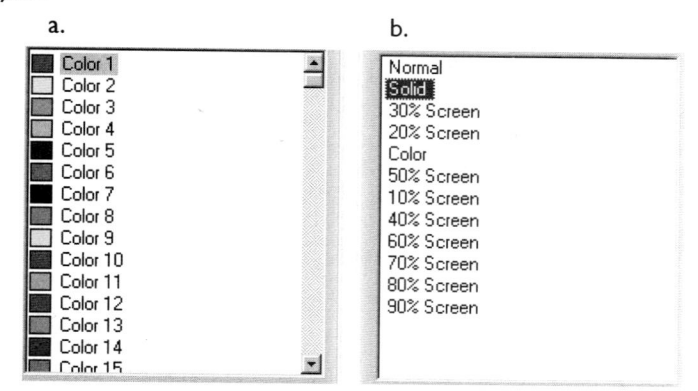

          a.   b.

    a.
    b.
28. Describe two advantages of named plot styles over color-based ones:
    a.
    b.
29. Can named plot styles be applied to layers?
Layouts?
Views?
30. Explain the purpose of the Batchplt program.

CHAPTER 18

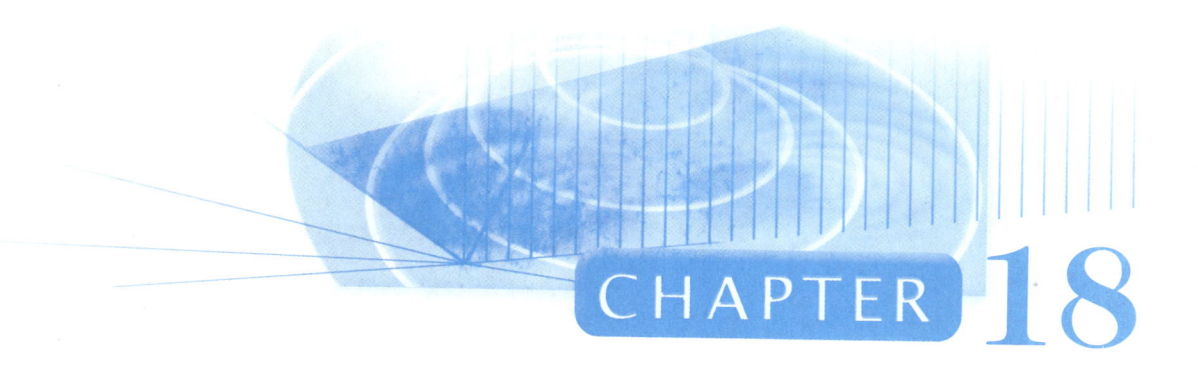

Reporting on Drawings

A benefit of drawing with CAD is that you can obtain information from the drawing. AutoCAD's inquiry commands provide the following kinds of information:

- **ID** reports the x,y,z coordinates of single points.
- **DIST** reports the distance and angle between two points.
- **AREA** reports the area and perimeter of areas.
- **MASSPROP** reports the area, perimeter, centroid, and so on of boundaries.
- **LIST**, **DBLIST**, and **PROPERTIES** report information about objects.
- **DWGPROPS** reports and stores information about the drawing.
- **TIME** reports on time spent editing the drawing.
- **STATUS** reports on the state of the drawing.
- **SETVAR** reports the settings of system variables.

FINDING THE COMMANDS

On the **INQUIRY** toolbar:

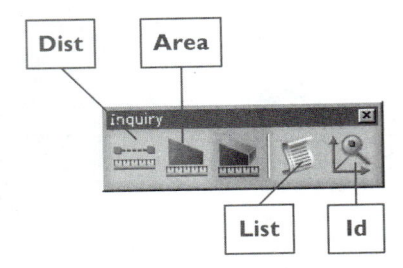

On the **TOOLS** menu:

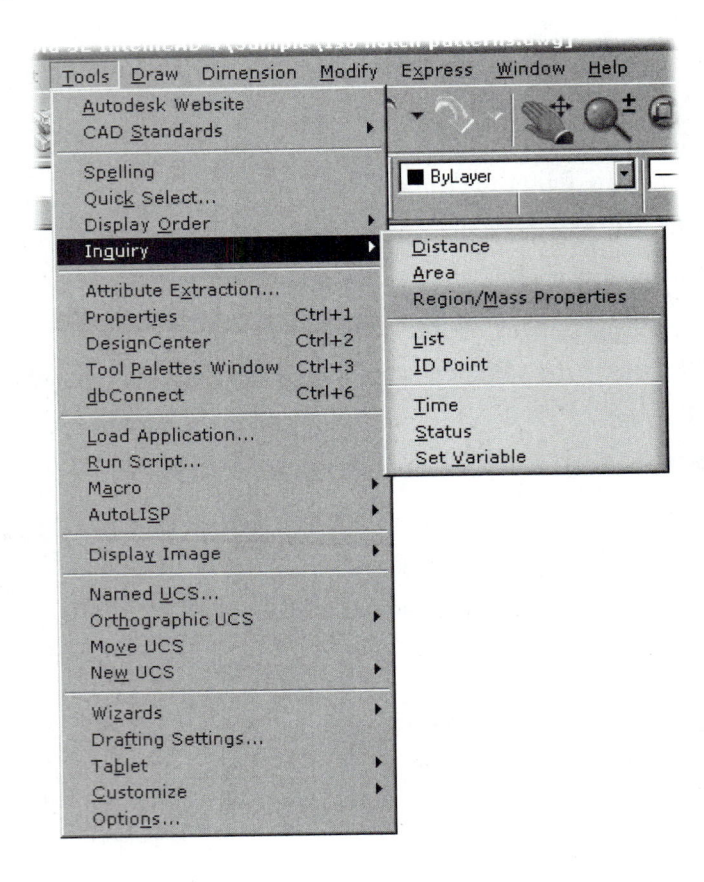

ID

The **ID** command reports the x,y,z coordinates of single points. (It is the shortest command name in AutoCAD; the longest is **HYPERLINKOPTIONS**.)

This command is transparent, so it can be used during other commands.

TUTORIAL: REPORTING POINT COORDINATES

1. To find the coordinates of a point, start the **ID** command:
   * From the **Tools** menu, choose **Inquiry**, and then **ID Point**.
   * From the **Inquiry** toolbar, choose the **Locate Point** button.
   * At the 'Command:' prompt, enter the **id** command.

   Command: **id** *(Press* ENTER.*)*

2. In all cases, AutoCAD prompts you to pick a point in the drawing.
   To capture the point of a geometric feature accurately, use an object snap mode, such as ENDpoint or CENter.

   Specify point: *(Pick a point.)*

   AutoCAD reports the x, y, z coordinates of the point:
   X = 13.3006   Y = 4.4596   Z = 0.0000

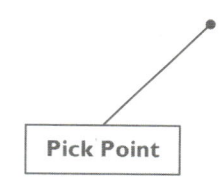

Pick Point

AutoCAD stores the coordinates in the **LASTPOINT** system variable, which you can access in the next command by entering @ at the prompt. Here is an example with the **LINE** command:

   Command: **line**

   Specify first point: **@**

   Specify next point or [Undo]: *(Pick a point.)*

And the **CIRCLE** command:

   Command: **circle**

   Specify center point for circle or [3P/2P/Ttr (tan tan radius)]: **@**

   Specify radius of circle or [Diameter]: *(Pick a point.)*

The coordinates are stored in **LASTPOINT** only until the next "Specify point" prompt. As soon as you pick another point, or enter coordinates at the keyboard, AutoCAD stores the new coordinates in the **LASTPOINT** system variable.

REPORTING POINT COORDINATES: ADDITIONAL METHODS

In addition reporting the 3D coordinates of points you pick, the **ID** command can show you coordinates through blip marks.

* **BLIPMODE** command turns on blips.
* **Marking points** shows the location of entered coordinates.

Let's look at how they work.

Blipmode

The **BLIPMODE** command toggles the display of blips in the drawing. *Blips* are small + markers that appear any time you pick a point in the drawing. AutoCAD normally keeps blip mode turned off.

Command: **blipmode**
Enter mode [ON/OFF] <OFF>: **on**

Now when you use the **ID** command (and almost any other command), each screen pick leaves behind the blip marker.

Command: **id**
Specify point: *(Pick a point.)*
X = 7.0   Y = 10.0   Z = 0.0

The size or shape of blip marks cannot be changed. Blip marks are like grid marks: they do not plot. Unlike grid marks, however, blips are not permanent. The next redraw or regeneration command erases them.

Command: **redraw**

Marking Points

The **ID** command also operates in reverse: specify a coordinate, and it draws a blip mark at that point:

Command: **id**
Specify point: **7.9,10.2**
X = 7.9   Y = 10.2   Z = 0.0

DIST

The **DIST** command reports the distance and angle between two points.

TUTORIAL: REPORTING DISTANCES AND ANGLES

1. To find the distance and angle between two points, start the **DIST** command:
   - From the **Tools** menu, choose **Inquiry**, and then **Distance**.
   - From the **Inquiry** toolbar choose the **Distance** button.
   - At the 'Command:' prompt, enter the **dist** command.
   - Alternatively, enter the **di** alias at the 'Command:' prompt.

   Command: **dist** *(Press ENTER.)*

2. In all cases, AutoCAD prompts you to pick two points in the drawing. To capture the points accurately, use an object snap mode.

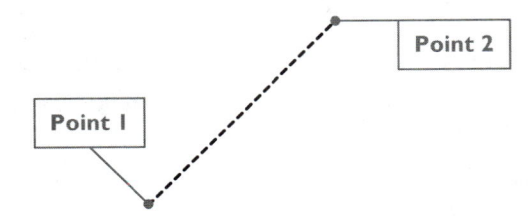

Specify first point: *(Pick point 1.)*
Specify second point: *(Pick point 2.)*

AutoCAD reports the distance and angles between the points in several formats:

Distance = 0.7071, Angle in XY Plane = 45, Angle from XY Plane = 0

Delta X = 0.5000, Delta Y = 0.5000, Delta Z = 0.0000

For 2D drafting, you need only the x- and y-related information, as illustrated by the figures:

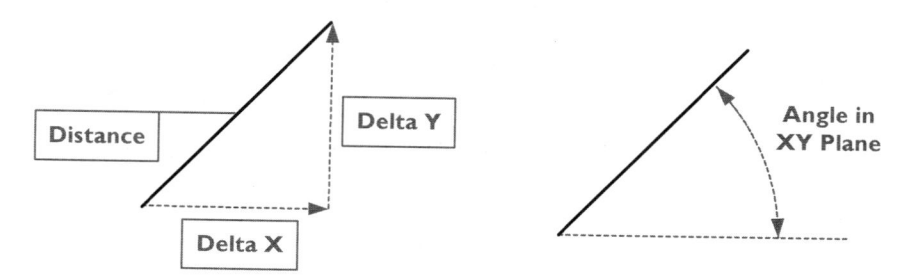

At left: the distance, delta X and delta Y reported by the DIST command.
At right: the angle in the XY plane.

AutoCAD stores the distance in the **DISTANCE** system variable. You can use the **DIST** command to show distances by specifying relative or polar coordinates, with blip mode turned on:

Command: **dist**

Specify first point: *(Pick first point.)*

Specify second point: **@10,0**

Use the PERpendicular object snap to find the shortest distance between two objects, such as the line and circle illustrated below:

Command: **dist**

Specify first point: **per**

to *(Pick a point on the line.)*

When there are more than two objects in the drawing, AutoCAD does not find the perpendicular on the line until you pick the next point. An ellipsis ... appears next to the perpendicular icon to indicate this deferment.

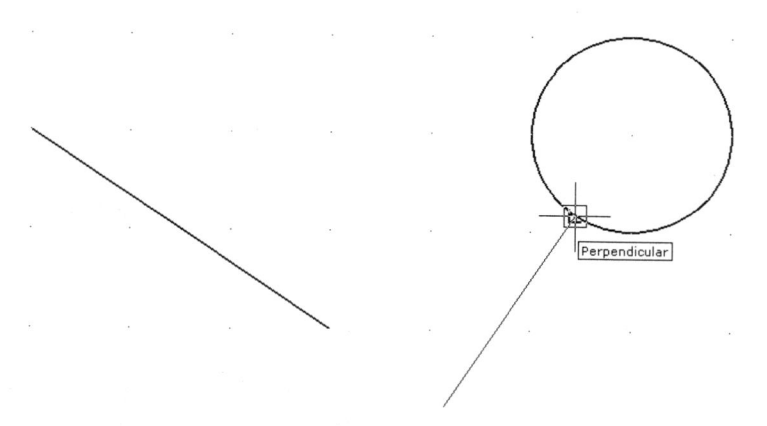

Specify second point: **per**

to *(Pick a point on the circle.)*

AutoCAD measures the shortest distance between the two objects; in the figure above, AutoCAD has "extended" the line segment to find the point of perpendicularity.

AREA

The AREA command reports the area and perimeter of areas.

This command measures the area and perimeter of closed objects, such as polygons and circles, as well as arbitrary areas of picked points. Areas can be added and subtracted to arrive at a total.

TUTORIAL: REPORTING AREAS AND PERIMETERS

1. To find the area and perimeters of objects, start the **AREA** command:
   - From the **Tools** menu, choose **Inquiry**, and then **Area**.
   - From the **Inquiry** toolbar, choose the **Area** button.
   - At the 'Command:' prompt, enter the **area** command.
   - Alternatively, enter the **aa** alias at the 'Command:' prompt.

   Command: **area** (Press ENTER.)

2. In all cases, AutoCAD prompts you to pick corner points. These are the vertices of an imaginary polygon.
   To capture the points accurately, use object snap modes.

   Specify first corner point or [Object/Add/Subtract]: *(Pick point 1.)*
   Specify next corner point or press ENTER for total: *(Pick point 2.)*
   ...
   Specify next corner point or press ENTER for total: *(Pick point 5.)*

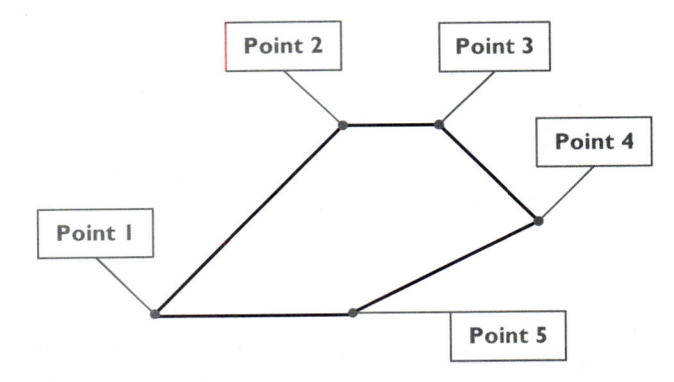

3. Press **ENTER** to end the measurement. (It is not necessary for the last pick to match the first one; AutoCAD automatically "closes" the polygon for you.)

   Specify next corner point or press ENTER for total: *(Press ENTER to end the command.)*

   AutoCAD reports the area and perimeter:
   Area = 2.0316, Perimeter = 5.9220

AutoCAD stores the area and perimeter measurements in the **AREA** and **PERIMETER** system variables. AREA is one of a few system variables with the same name as a related command. To access the value stored in the system variable, you must use the **SETVAR** command, because entering AREA at the 'Command:' prompt executes the command.

   Command: **setvar**
   Enter variable name or [?]: **area**
   AREA = 2.0316 (read only)

REPORTING AREA: ADDITIONAL METHODS

AutoCAD can add and subtract areas from the total, as well as find the area of circles, ellipses, splines, polylines, polygons, regions, and 3D solids.

- **Add** option adds another area to the total.
- **Subtract** option subtracts another area from the total.
- **Object** option finds the area of closed objects.

Let's look at each option.

Add

The **Add** option adds an area to the total. You can pick points or select objects to add.

> Command: **area**
> Specify first corner point or [Object/Add/Subtract]: **a**
> Specify next corner point or press ENTER for total (ADD mode): *(Pick a point.)*
> et cetera

The "(ADD mode)" message reminds you that AutoCAD is adding areas.

Subtract

The **Subtract** option removes an area from the total. You can pick points or select an object to remove. You can find negative areas by immediately going into **Subtract** mode.

> Command: **area**
> Specify first corner point or [Object/Add/Subtract]: **s**
> Specify next corner point or press ENTER for total (SUBTRACT mode): *(Pick a point.)*
> et cetera

The "(SUBTRACT mode)" message reminds you that AutoCAD is subtracting areas.

Object

The **Object** option finds the area and perimeter of selected objects: circles, ellipses, open and closed splines, open and closed polylines, polygons, regions, and 3D solids.

Objects	Measurement
Circles	Area and circumference.
Ellipses, polygons, regions	Area and perimeter.
Closed polylines and closed splines	Area and perimeter.
Open polylines and open splines	Area and length.
3D solids	Area of all faces.

An open polyline (or spline) has a shorter "perimeter" than a closed polyline of the same area.

Even though the "Select objects" prompt suggests you can select more than object at a time, you are limited to just one. To add the area of additional objects, use the **Add** option.

> Command: **area**
> Specify first corner point or [Object/Add/Subtract]: **o**
> Select objects: *(Select one object.)*
> Area = 0.9085, Perimeter = 3.5000

MASSPROP

The **MASSPROP** command reports the area, perimeter, centroid, and so on of boundaries.

The boundary must be made from a single region object. You can use the **SUBTRACT** command to remove *islands* from the region.

TUTORIAL: REPORTING INFORMATION ABOUT REGIONS

1. To find out information about regions, start the **MASSPROP** command:
   - From the **Tools** menu, choose **Inquiry**, and then **Region/Mass Properties**.
   - From the **Inquiry** toolbar, choose the **Mass Properties** button.
   - At the 'Command:' prompt, enter the **massprop** command.

   Command: **massprop** *(Press ENTER.)*

2. In all cases, AutoCAD prompts you to select objects. While you can select more than one object, it is better to pick just one, because of the amount of information generated.

   Select objects: *(Select region objects.)*

   Select objects: *(Press ENTER to end object selection.)*

   AutoCAD switches to the Text window, and displays the report:

   ```
   ---------------- REGIONS ----------------
   Area:                 1.7211
   Perimeter:            14.3193
   Bounding box:         X: 1.2616  --  3.7327
                         Y: 0.9440  --  3.0999
   Centroid:             X: 2.6010
                         Y: 2.2224
   Moments of inertia:   X: 9.0893
                         Y: 12.3840
   Product of inertia:   XY: 9.9040
   Radii of gyration:    X: 2.2981
                         Y: 2.6825
   Principal moments and X-Y directions about centroid:
                         I: 0.5772 along [0.9653 -0.2612]
                         J: 0.7529 along [0.2612 0.9653]
   ```

3. AutoCAD asks if you wish to save the data to a file.

   Write analysis to a file? [Yes/No] <N>: **y**

4. If you answer **Y**, AutoCAD displays the Create Mass and Area Properties File dialog box. Enter a file name, and then click **Save**. AutoCAD creates the *.mpr* mass properties report file.

5. Press **F2** to return to the drawing window.

LIST AND DBLIST

The **LIST** command reports information about objects, while **DBLIST** reports on every object in the drawing (short for "database listing").

TUTORIAL: REPORTING INFORMATION ABOUT OBJECTS

1. To find out information about objects, start the **LIST** command:
   - From the **Tools** menu, choose **Inquiry**, and then **List**.
   - From the **Inquiry** toolbar, choose the **List** button.
   - At the 'Command:' prompt, enter the **list** command.
   - Alternatively, enter the **ls** alias at the 'Command:' prompt.

   Command: **list** *(Press ENTER.)*

2. In all cases, AutoCAD prompts you to select objects. While you can select more than one object, it is better to pick just one, because of the amount of information generated.

   Select objects: *(Select one object.)*

   Select objects: *(Press ENTER to end object selection.)*

   AutoCAD reports on the object in the Text window.

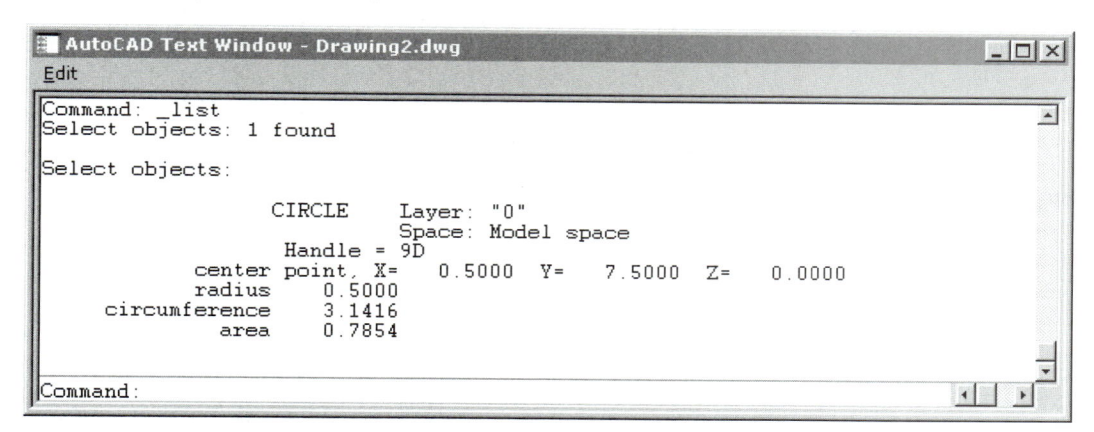

```
AutoCAD Text Window - Drawing2.dwg
Edit
Command: _list
Select objects: 1 found

Select objects:
                  CIRCLE    Layer: "0"
                            Space: Model space
                  Handle = 9D
        center point, X=   0.5000  Y=   7.5000  Z=   0.0000
              radius   0.5000
       circumference   3.1416
                area   0.7854

Command:
```

AutoCAD reports the following information:

Line	Information Reported
1	Type of object.
	Layer the object is on.
2	Space the object is in,: model or paper space (layouts).
3	Handle (an identification number assigned by AutoCAD.)
4	X, y, z coordinates of the object's location.
Other	Additional geometric information specific to the object.
	Area and perimeter or length.
	Color, linetype, and lineweight, if not set to BYLAYER.
	Thickness, if not zero.

The information displayed by the list is not reported uniformly for different objects. For example, the circles report their area before their location, while ellipses report area later. The geometric information also varies for each object, as illustrated by the table below. For complex objects, the listing can get very long, because AutoCAD reports on every segment.

Object	Information Reported
Point	Point.
Line	From point, To point, Length, Angle in XY Plane, Delta X, Delta Y, Delta Z.
Circle	Center point, Radius.
Ellipse	Center, Major Axis, Minor Axis, Radius Ratio.
Arc	Center point, Radius, Start angle, End angle, Length.
Polyline	Constant width (or Starting width and Ending width), Area, Perimeter.
Polyarc	Bulge, Center, Radius, Start angle, End angle.
Spline	Area, Circumference, Order, Properties, Parametric Range, Number of control points, Control Points, Number of fit points, User Data, Start Tangent, End Tangent.

You can select the text, and then use **CTRL+C** to copy it to the Clipboard for pasting into other documents. Press **F2** to return to the drawing window.

Note: The **LIST** command is more useful than **AREA** for individual objects, because it provides more information. For example, **AREA** reports the area and circumference of a circle, but **LIST** also reports its radius.

TUTORIAL: REPORTING INFORMATION ABOUT ALL OBJECTS

1. To find out information about all objects in the drawing, start the **DBLIST** command:
   Command: **dblist** *(Press* **ENTER**.*)*

2. AutoCAD switches to the Text window, and reports on every object in the drawing. Every screenfull, the report pauses.
   Press ENTER to continue: *(Press* **ENTER**.*)*

3. The report is very long, even for small drawings. To end it, press **ESC**:
   Press ENTER to continue: *(Press* **ESC**.*)*

Press **F2** to return to the drawing window.

PROPERTIES

The **PROPERTIES** command reports on objects in a format that easier to digest and more complete than **LIST** and **DBLIST**. Compare the data for the circle shown at right, with that listed on the opposite page.

When you select all objects in the drawing, the Property window reports the total number. The drawback is that the data listed by the Properties window cannot be copied to the Clipboard. For more information about this command, see Chapter 7, "Changing Object Properties."

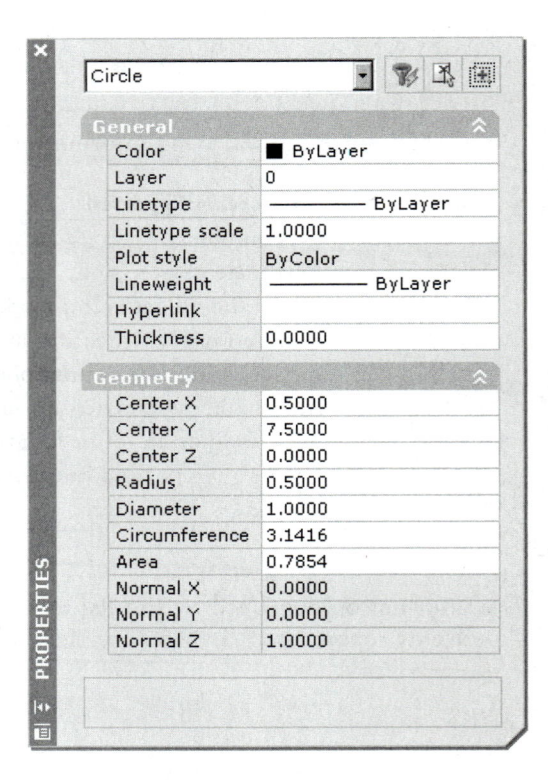

DWGPROPS

The **DWGPROPS** command reports and stores information about the drawing file.

This command displays a dialog box, into which you can add data, such as the name of the drafter, and general comments. This data can be searched with the Windows Explorer **FIND** command.

TUTORIAL: REPORTING INFORMATION ABOUT OBJECTS

1. To find out information about objects, start the **DWGPROPS** command:
    - From the **File** menu, choose **Drawing Properties**.
    - At the 'Command:' prompt, enter the **dwgprops** command.

    Command: **dwgprops** *(Press ENTER.)*

2. In all cases, AutoCAD displays the Properties dialog box.

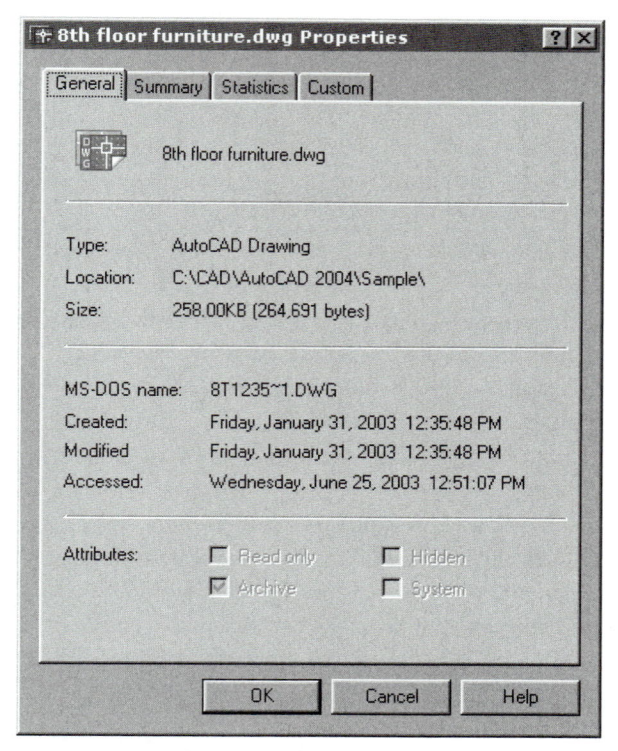

    The **General** tab shows information about the drawing file's name, location, size, and dates. The file attributes can only be changed by Explorer's **Property** option.
    If the drawing has not yet been saved, then most fields are blank.

3. Click the **Summary** tab.
    You can add data to the fields, such as your name, the subject of the drawing, and general comments.
    Title can be different from the drawing's file name: some use the project name, for example.
    Keywords are useful when drawings are searched by database programs.
    Hyperlink Base is the base URL address used by relative hyperlinks in the drawing.

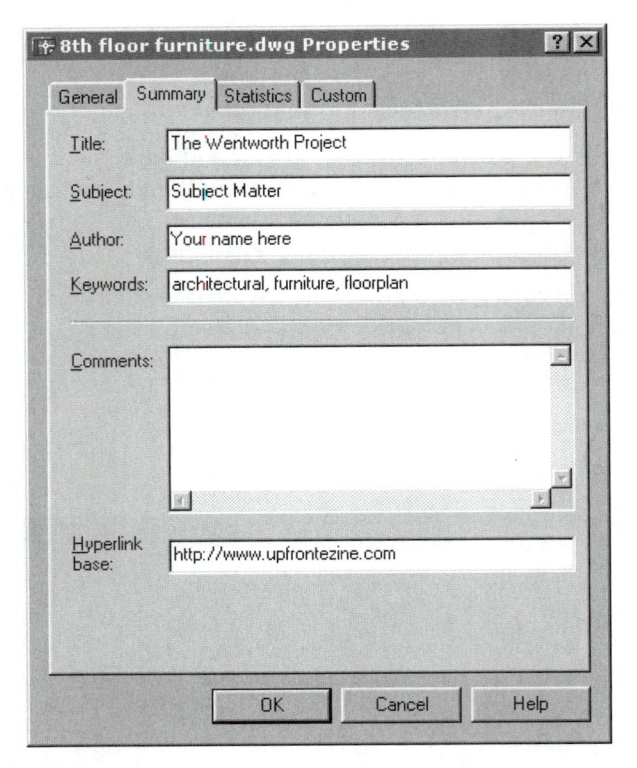

4. Click the **Statistic** tab.

    Created is the date and time the drawing was first created with the **NEW** command; the date and time are retrieved from the **TDCREATE** system variable.

    Last Modified is the date and time the drawing was last edited or had objects added, retrieved from the **TDUPDATE** system variable.

    Last Saved By is the user's login name, retrieved from **LOGINNAME** system variable, which in turn is obtained from Windows.

    Revision Number is the AutoCAD software revision number.

    Total Editing Time is the time the drawing has been open. The name is misleading, because it includes time when the drawing is displayed by AutoCAD without being edited. The value is retrieved from the **TDINDWG** system variable.

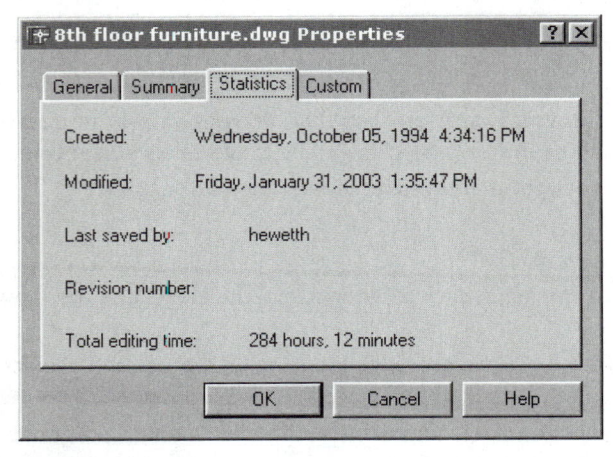

    Note. If the drawing was last saved by a non-Autodesk software product, an additional message appears in this tab:

    This file may have been last saved by a program other than Autodesk software.

    The value of 0 is stored from **DWGCHECK** system variable.

6. Click the **Custom** tab.

   You can enter text in up to ten fields, along with a value for each. The custom fields are searchable by the **Find** option of DesignCenter.

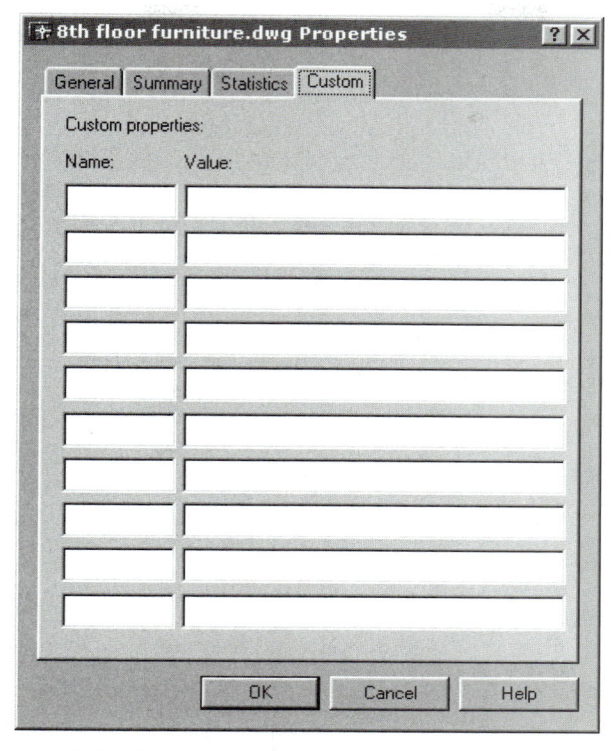

7. Click **OK** to close the dialog box.

'TIME

The **TIME** command reports on time spent editing the drawing, and provides timer functions.

This command can be used transparently, in the middle of other commands.

TUTORIAL: REPORTING ON TIME

1. To find out time information about the drawing, start the **TIME** command:
   * From the **Tools** menu, choose **Inquiry**, and then **Time**.
   * At the 'Command:' prompt, enter the **time** command.

   Command: **time** *(Press* ENTER.*)*

   AutoCAD reports time-related data:
   Command: **time**
   Current time: Wednesday, June 25, 2003 at 2:53:15:890 PM
   Times for this drawing:
     Created: Wednesday, June 25, 2003 at 1:04:52:812 AM
     Last updated: Wednesday, June 25, 2003 at 1:04:52:812 AM
     Total editing time: 0 days 13:48:23.203
     Elapsed timer (on):   0 days 13:48:23.047
     Next automatic save in: <no modifications yet>

2. Press **ENTER** to exit the command:

    Enter option [Display/ON/OFF/Reset]: *(Press* **ENTER**.*)*

The accuracy of the time depends on your computer's clock. The meaning of the data reported by this command is as follows:

Time	Meaning
Current Time	Current date and 24-hour time to the nearest millisecond.
Created	Date and time drawing was created.
Last Updated	Latest use of the SAVE and QSAVE commands.
Total Editing Time	Cumulative time drawing has open in AutoCAD, excluding plotting time, and sessions when drawing was exited without saving.
Elapsed Timer	Stopwatch-like timer.
Next Automatic Save In	Minutes remaining until the next automatic save.

REPORTING ON TIME: ADDITIONAL METHODS

The **TIME** command includes options to redisplay its timings, and act as a stopwatch.

- **Display** option redisplays the time report.
- **ON** option starts the timer.
- **OFF** option turns off the timer.
- **Reset** option resets the timer.

Let's look at each.

Display

The **Display** option redisplays the time report, with updated times.

ON, OFF, and Reset

The **ON** option starts the timer. The **OFF** option stops the time, and the **Reset** option sets the timer back to 0 days and 0 time (0 hours, 0 minutes, and 0 seconds).

STATUS

The **STATUS** command reports on the state of the drawing. The **DIM: STATUS** command reports the values of dimension variables.

    Command: **status**

AutoCAD switches to the Text screen, and reports the information shown below. You can copy the information to the Clipboard by selecting the text, and then pressing **CTRL+C**. To return to the drawing window, press **F2**.

Status	Meaning
Model Space limits; Paper Space limits	X,y coordinates stored in the LIMMIN and LIMMAX system variables. Off indicates limits checking is turned off (LIMCHECK).
Model Space uses; Paper Space uses	X,y coordinates of the lower-left and upper-right extents of objects. Over indicates drawing extents exceed the drawing limits.
Display shows; Insertion base is	X,y coordinates of the lower-left and upper-right corners of the current display. X,y,z coordinates stored in system variable INSBASE.

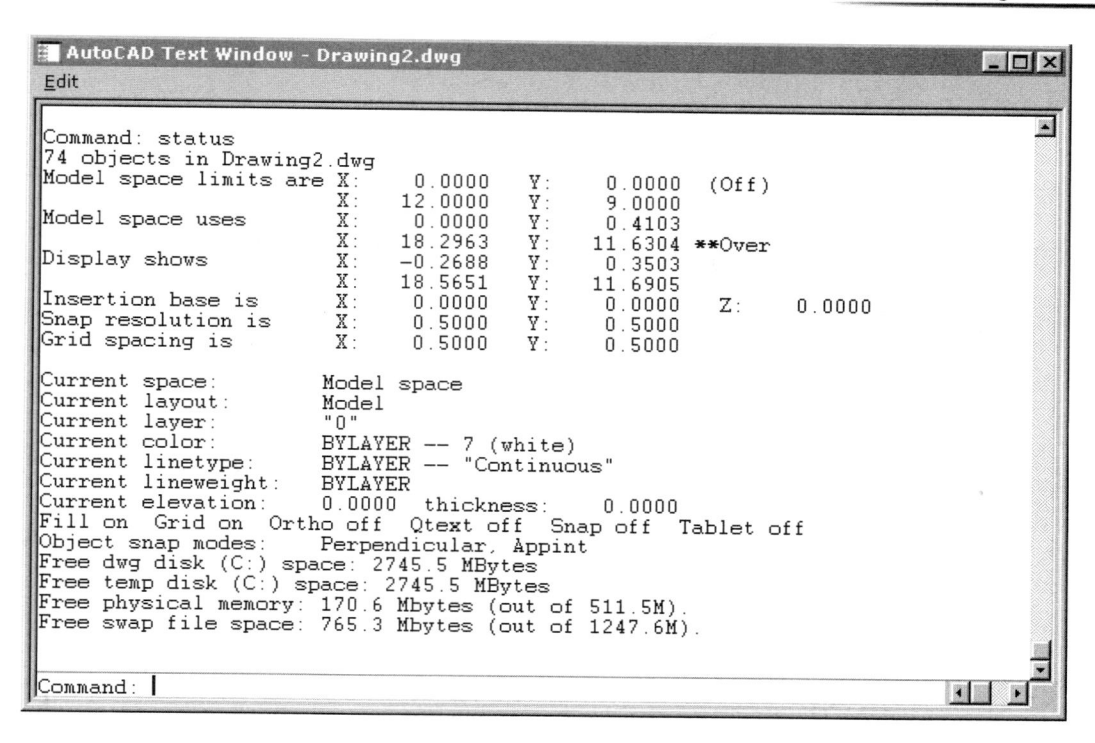

```
AutoCAD Text Window - Drawing2.dwg
Edit

Command: status
74 objects in Drawing2.dwg
Model space limits are X:     0.0000   Y:    0.0000   (Off)
                       X:    12.0000   Y:    9.0000
Model space uses       X:     0.0000   Y:    0.4103
                       X:    18.2963   Y:   11.6304  **Over
Display shows          X:    -0.2688   Y:    0.3503
                       X:    18.5651   Y:   11.6905
Insertion base is      X:     0.0000   Y:    0.0000   Z:    0.0000
Snap resolution is     X:     0.5000   Y:    0.5000
Grid spacing is        X:     0.5000   Y:    0.5000

Current space:          Model space
Current layout:         Model
Current layer:          "0"
Current color:          BYLAYER -- 7 (white)
Current linetype:       BYLAYER -- "Continuous"
Current lineweight:     BYLAYER
Current elevation:      0.0000   thickness:   0.0000
Fill on  Grid on  Ortho off  Qtext off  Snap off  Tablet off
Object snap modes:      Perpendicular, Appint
Free dwg disk (C:) space: 2745.5 MBytes
Free temp disk (C:) space: 2745.5 MBytes
Free physical memory: 170.6 Mbytes (out of 511.5M).
Free swap file space: 765.3 Mbytes (out of 1247.6M).

Command:
```

Status (cont'd)	*Meaning*
Snap resolution is; Grid spacing is	Snap and grid settings, as stored in the SNAPUNIT and GRIDUNIT system variables.
Current space	Whether model or paper space is current.
Current layout	Name of the current layout.
Current layer; Current color; Current linetype; Current lineweight; Current plot style; Current elevation; Thickness	Values for the layer name, color, linetype name, elevation, and thickness, as stored in system variables CLAYER, CECOLOR, CELTYPE, LWEIGHT, CPLOTSTYLE, ELEVATION, and THICKNESS.
Fill, Grid, Ortho; QTet, Snap, Tablet	Toggle settings (on or off) for the fill, grid, ortho, qtext, snap, and tablet modes, as stored in the system variables FILLMODE, GRIDMODE, ORTHOMODE, TEXTMODE, SNAPMODE, and TABMODE.
Object Snap modes	Current object snap modes, as stored in the system variable OSMODE.
Free dwg disk (C:) space	Hard disk space free on the specified drive.
Free temp disk (C:) space	Hard disk space free on the drive specified for AutoCAD's temporary files.
Free Physical Memory	Amount of free RAM memory, and total RAM memory.
Free Swap File Space	Amount of free swap file space, and total swap file size.

You can also use the **STATUS** command at the 'Dim:' prompt, as follows:

>Command: **dim**
>
>Dim: **status**

AutoCAD flips to the Text window, and displays the value of all dimension variables.

```
AutoCAD Text Window - Drawing2.dwg
Edit

Command: dim

Dim: status
DIMASO       Off                    Create dimension objects
DIMSTYLE     Standard               Current dimension style (read-only)
DIMADEC      0                      Angular decimal places
DIMALT       Off                    Alternate units selected
DIMALTD      2                      Alternate unit decimal places
DIMALTF      25.4000                Alternate unit scale factor
DIMALTRND    0.0000                 Alternate units rounding value
DIMALTTD     2                      Alternate tolerance decimal places
DIMALTTZ     0                      Alternate tolerance zero suppression
DIMALTU      2                      Alternate units
DIMALTZ      0                      Alternate unit zero suppression
DIMAPOST                            Prefix and suffix for alternate text
DIMASZ       0.1800                 Arrow size
DIMATFIT     3                      Arrow and text fit
DIMAUNIT     0                      Angular unit format
DIMAZIN      0                      Angular zero supression
DIMBLK       ClosedFilled           Arrow block name
DIMBLK1      ClosedFilled           First arrow block name
DIMBLK2      ClosedFilled           Second arrow block name
DIMCEN       0.0900                 Center mark size
DIMCLRD      BYBLOCK                Dimension line and leader color
DIMCLRE      BYBLOCK                Extension line color
Dim:
```

You can copy the information to the Clipboard by selecting the text, and then pressing **CTRL+C**. To return to the drawing window, press **F2**. Enter **EXIT** to return to the 'Command:' prompt:

>Dim: **exit**
>
>Command:

'SETVAR

The **SETVAR** command reports the settings of system variables (short for "set variables").

System variables store the state of the drawing, and control certain commands. You can change the value of most system variables, thus changing the state of the drawing (snap, grid, and so on) and the action of commands (mirrored text, current layer, and so on). Some system variables cannot be changed. They are called "read-only."

This command can be used transparently, during other commands.

You have used this command many times in this and previous chapters, without realizing it. Historically, system variables could only be changed with the **SETVAR** command. More recently, Autodesk made it possible to enter system variables at the 'Command:' prompt. Thus, the only time you need to use this command is when you need to examine a list of system variables, or when the name of the system variable is the same as the command.

All of AutoCAD's system variables are listed in Appendix C of this book.

TUTORIAL: LISTING SYSTEM VARIABLES

1. To list the values of system variables, start the **SETVAR** command:

- From the **Tools** menu, choose **Inquiry**, and then **Set Variable**.
- From the **Modify** toolbar, choose the **Copy** button.
- At the 'Command:' prompt, enter the **setvar** command.
- Alternatively, enter the **set** alias at the 'Command:' prompt.

   Command: **setvar** *(Press ENTER.)*

2. In all cases, AutoCAD prompts you to enter the name of a system variable. Enter the ? option:

   Enter variable name or [?] <AREA>: **?**

3. AutoCAD asks which variables you wish to see. Use wildcard characters:

   * — means all variables.
   a* — means all variables starting with the letter "a."
   grid* — means all variables starting with the word "grid."
   ? — means a single character.

   Enter variable(s) to list <*>: *(Press ENTER to list all variables.)*

   AutoCAD flips to the Text window, and displays the value of all system variables.

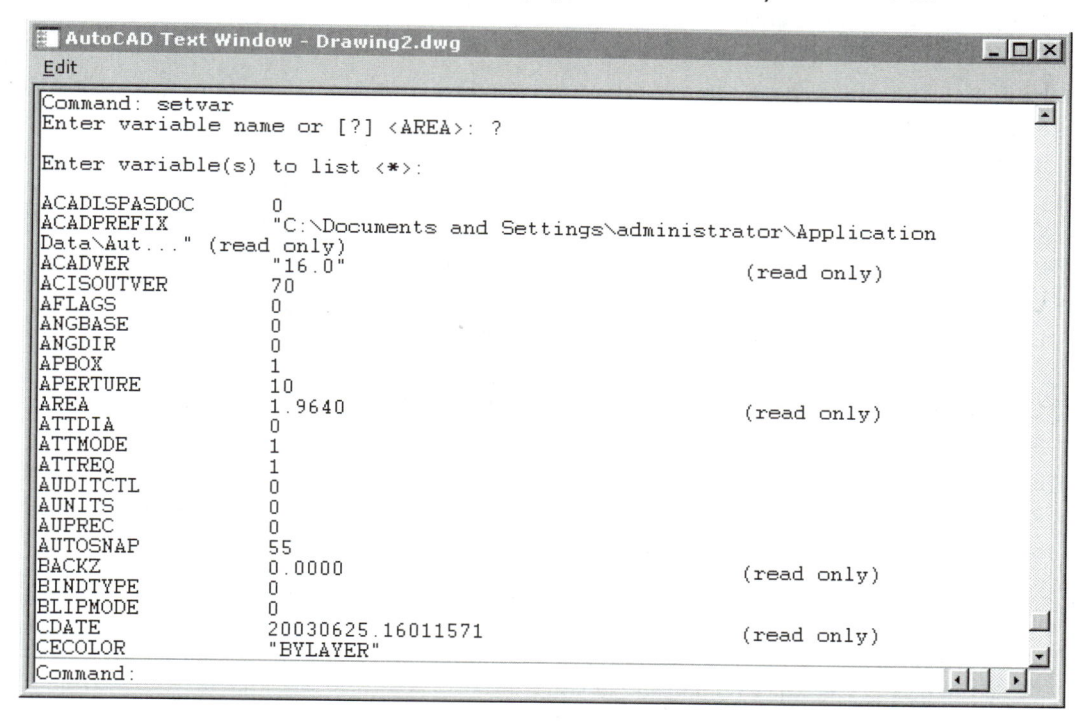

```
AutoCAD Text Window - Drawing2.dwg
Edit

Command: setvar
Enter variable name or [?] <AREA>: ?

Enter variable(s) to list <*>:

ACADLSPASDOC       0
ACADPREFIX         "C:\Documents and Settings\administrator\Application
Data\Aut..." (read only)
ACADVER            "16.0"                              (read only)
ACISOUTVER         70
AFLAGS             0
ANGBASE            0
ANGDIR             0
APBOX              1
APERTURE           10
AREA               1.9640                              (read only)
ATTDIA             0
ATTMODE            1
ATTREQ             1
AUDITCTL           0
AUNITS             0
AUPREC             0
AUTOSNAP           55
BACKZ              0.0000                              (read only)
BINDTYPE           0
BLIPMODE           0
CDATE              20030625.16011571                   (read only)
CECOLOR            "BYLAYER"
Command:
```

4. With over 400 system variables, the listing is very long. Press **ESC** to end it early. To return to the drawing window, press **F2**.

EXERCISES

1. Start AutoCAD with a new drawing. Use this drawing for all the exercises in this chapter.
   Draw a line with endpoints at 2,2 and 6,6.
   With the **ID** command and ENDpoint object snap, select the first endpoint of the line.
   What coordinates does the command return?
   Repeat for the other end. Are the coordinates what you expect?

2. Draw a line, circle, and arc on the screen.
   Select the line with the **LIST** command.
   What is the line's angle in the XY Plane?
   Select the circle with **LIST** command.
   What is the circle's radius?
   Select the arc with **LIST** command.
   What is the arc's starting angle?

3. Continue with the drawing from exercise #2.
   Select the line with the **PROPERTIES** command.
   Deselect the line, and then select the arc
   How does the data displayed by the Properties window change?.
   Select all three objects. Does the Properties window display a smaller or larger amount of information?

4. Continue with the same drawing, and enter the **DBLIST** command.
   What do you notice about the information listed by AutoCAD?
   Use the **COPY** command's **Multiple** option to copy the objects three times.
   Use the **DBLIST** command again, and notice the length of the listing.

5. Continuing with the same drawing, use the **DIST** command with ENDpoint object snap.
   Select two endpoints of one of the lines on the screen.
   Is the length the same as reported by the **LIST** command?
   Repeat the **DIST** command, but use PERpendicular object snap to find the distance between a line and a circle.

6. Draw two rectangles of different sizes with the **RECTANG** command.
   Use the **AREA** command to determine the area of one rectangle.
   Which option of the **AREA** command did you use?
   Repeat the same procedure on the second rectangle. Add the two areas together, and obtain the total area.
   Use the **AREA** command, but this time use the **Add** option to find the area of both rectangles.
   Does AutoCAD arrive at the same answer as you?

7. From the CD-ROM, open the *area.dwg* file, a drawing of geometric shapes.
   Find the area of each object, and then write down the figures.
   Using the **Add** option, find the total area of all objects.

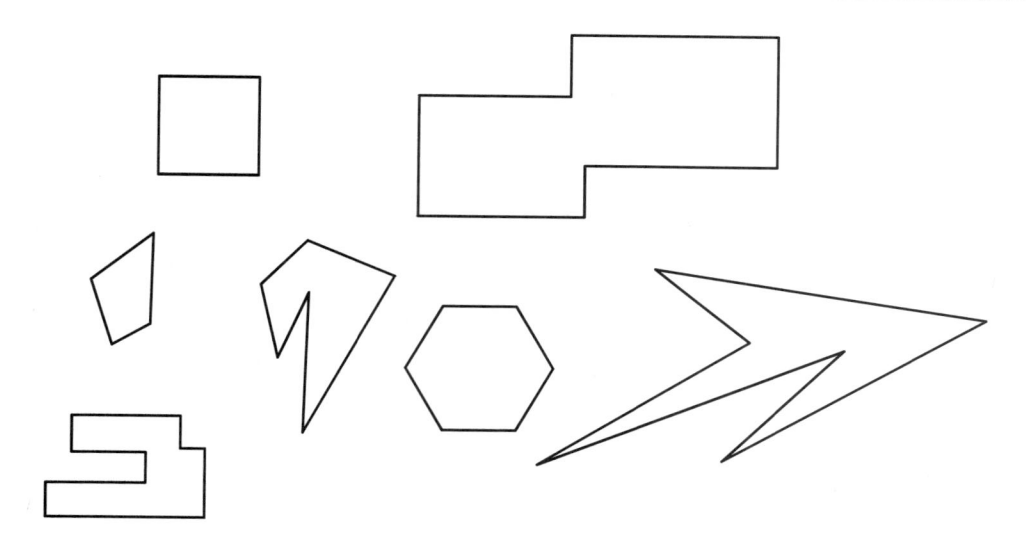

8. Use the **TIME** command to check the current time in the drawing.
   Is the time reported by AutoCAD the same as on your wristwatch?
   The same as shown on your computer?

9. Use the elapsed timer option of the **TIME** command to determine the time needed to draw compete the following exercise.
   Before starting, write down the time you estimate it will take you to complete exercise #10.

10. Draw the baseplate shown below.
    Use the **AREA** command to find the area.
    Hints: Use the **RECTANGLE** command with its **Fillet** option. Remember to use the **Subtract** option to remove the holes from the total.

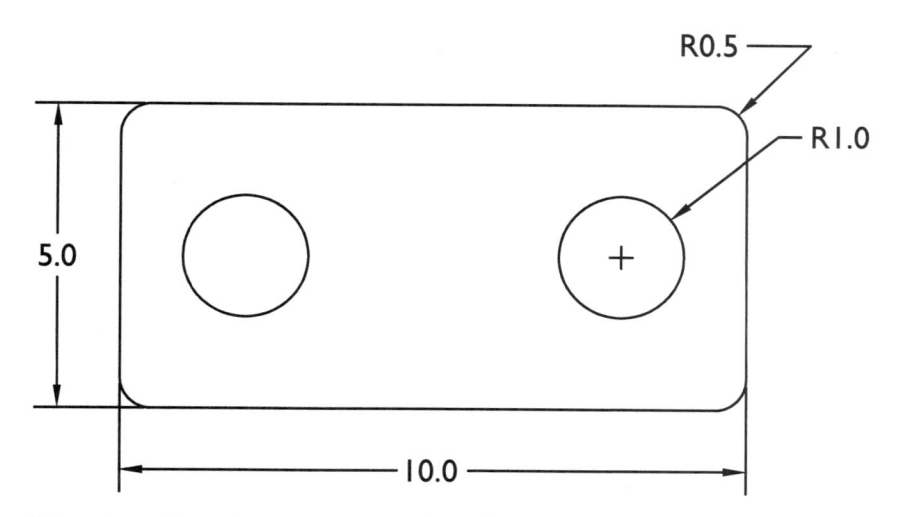

    Turn off the timer. How close was your estimate?

11. From the CD-ROM, open the *34486.dwg* file, a drawing of the author's 50' x 25' basement floor plan. Each unit in the drawing represents a foot.

What is the total square footage of the basement?
Create a new layer called "Areas."
Draw rectangles that match each room.
Find the area of each room:
  Storage Room
  Laundry Room
  Bathroom
  Mud Room
  Guest Room
  Bedroom
  Closet
  Storage
  Office
What is the area of the hallways, including stairs?

12. Start a new drawing with *acad.dwt*.
    Find the value of the following system variables:
    OFFSETGAPTYPE
    LASTPOINT
    AREA
    MAXSORT
    If you do not understand the meaning of a system variable, refer to Appendix C of this book.

13. From the CD-ROM, open the *15_44.dwg* file, a drawing of a gasket.
Ensure the DELOBJ system variable is set to 1.
Start the BOUNDARY command, and set object type to **Region**. Click **Pick Points**, and select a point inside the gasket. How many regions is the gasket?

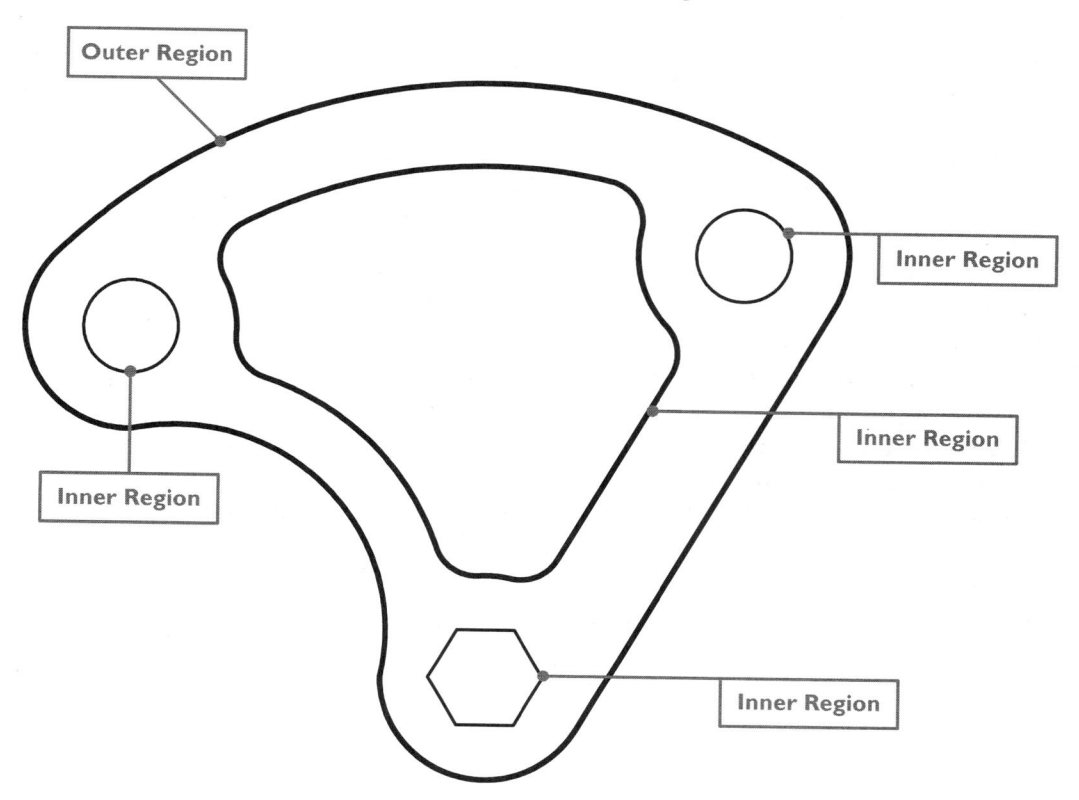

Outer Region

Inner Region

Inner Region

Inner Region

Inner Region

Start the SUBTRACT command, and then select the outermost region. When prompted "Select solids and regions to subtract," select the four inner regions (also called "islands").

Apply the MASSPROP command to the gasket, and then save the data to an .mpr file.
Write down the values from the report:
  Area: _____
  Perimeter: _____
  Bounding box: X: _____   Y: _____
  Centroid: X: _____   Y: _____
  Moments of inertia: X: _____   Y: _____
  Product of inertia: XY: _____
  Radii of gyration: X: _____   Y: _____
  Principal moments and X-Y directions about centroid: I: _____
                                                         J: _____

Use the POINT command to place a point at the centroid of the gasket.

13. From the CD-ROM, open the *15_43.dwg* file, and find the centroid of the gasket.

CHAPTER REVIEW

1. The **ID** command performs which two kinds of actions?
   a.
   b.
2. Where are the **ID** coordinates stored?
   How do you access them in other drawing and editing commands?
3. What happens when blipmode is turned on?
4. What is the difference among the **PROPERTIES**, **LIST**, and the **DBLIST** commands?
5. What are the six distances returned by the **DIST** command?
6. What option under **AREA** calculates the area of a circle?
7. After using the **AREA** and **LIST** commands on the circle, which command do you find more useful?
8. How do you determine the last time a drawing was updated?
9. Can the **AREA** command return the length of an open polyline?
10. Why would you want to subtract areas?
11. How do you stop the report generated by **DBLIST** ? press **ESC**
12. Describe a purpose for using the **DWGPROPS** dialog box.
13. List two reasons why the Total Editing Time might be inaccurate for drawings:
    a.
    b.
14. Can you change the value of all system variables?

APPENDIX A

Computers and Windows

The Windows operating system is the foundation on which all other programs run. Windows loads automatically when you start the computer. You might notice operating system messages and version numbers when your computer starts. AutoCAD 2004 operates only on Windows 2000 and XP.

HARDWARE OF AN AUTOCAD SYSTEM

Computers are the central part of CAD systems; *peripherals* connect to the computer, as discussed later in this chapter. The five primary categories of computer are:

Notebook computers are the smallest practical computer for running AutoCAD. They are about the size of a notebook (hence the name), which makes them easily transportable. Notebook computers are generally more expensive and somewhat less powerful than the fastest desktop computers; all of today's notebook computers, however, are powerful enough to run AutoCAD.

Desktop computers are the type you most commonly see on (or under) a desk. These versatile machines are sometimes referred to as "personal" computers because they are primarily designed for use by one person. Today's desktop computers are often connected to a network of many computers, which allows files and some programs to be used by more than one operator at a time.

Personal computers usually consist of a case that contains the central processing unit; memory; one or more disk drives; and additional adapters, such as for sound and video. The monitor, keyboard, mouse, and output devices (such as printers) are external to the case.

Workstations are larger, faster, and more expensive than desktop computers. They typically contain one or more faster processors, much more memory, and much larger storage space. This class of computer is meant for running larger and more sophisticated programs than personal computers typically handle. For example, the special effects of Hollywood movies are created on workstations; some specific brands of CAD software runs best on workstations.

Mainframe computers are the largest type of computer. They are capable of processing a large amount of data, and are used by governments and companies who handle large amounts of data, such as those involved in credit card and database transactions. These computers are not typically used for running CAD software any more.

Palm computers are the smallest of computers, fitting in the palm of your hand. Typically, you input data by writing with a stylus on the computer's screen; handwriting recognition software translates your writing into text and graphics.

Palm computers can be used to view and edit CAD drawings, although AutoCAD itself is not (yet) available. The software shown on the PalmOS computer at right is meant for use at the job site; the ZiPCAD data is imported into AutoCAD through *.dxf* files.

COMPONENTS OF COMPUTERS

Computers are made up of several essential parts. Let's look at some of them.

System Board

The system board (sometimes referred to as the "motherboard") is a fiberglass board that holds most of the computer's chips and expansion cards. The CPU (central processing unit), RAM memory chips, ROM (read-only memory), disc controllers, and other parts are mounted on this board.

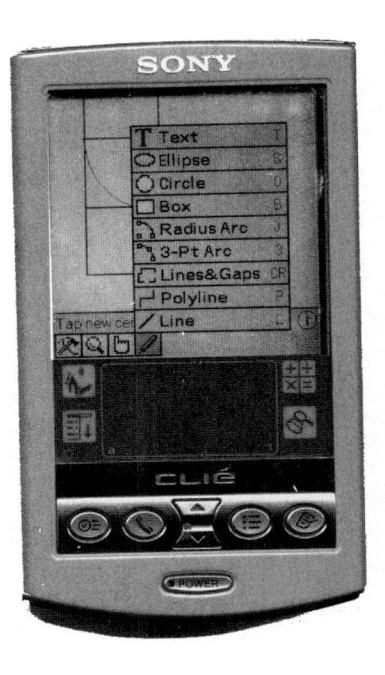

In addition, the board contains slots into which you mount expansion boards, such as video capture cards and drive adapters. There are four types of expansion slots:

- **ISA** (industry standard architecture) is the oldest type of expansion slot. Some computers may still have some of these slots for older adapter cards.

- **PCI** (peripheral connect interface) is the most common slot in today's computers found in both PC and Macintosh computers.

- **AGP** (advanced graphics port) is designed for high-speed data transfer between the CPU and graphics boards.

- **PC Card** or PCMCIA (personal computer memory card interface adapter) is designed for notebook computers. The adapter card looks like a thick credit card, and might contain a modem, network interface, or even a disk drive.

Central Processing Unit

The central processing unit (or CPU) is the "brain" of the computer. Here the software is processed. Then the CPU sends instructions to the graphics board, printer, plotter, and other peripherals. With few exceptions, all information passes through the central processing unit.

Physically, the central processor is a computer chip mounted on a large heat sink, and attached to the system board. Because of the heat generated by the CPU, it is common to have a heat sink (which looks like a series of black fins) and a small fan mounted on the chip. Today's high-speed CPUs are often packaged with cache memory, which helps speed up calculations.

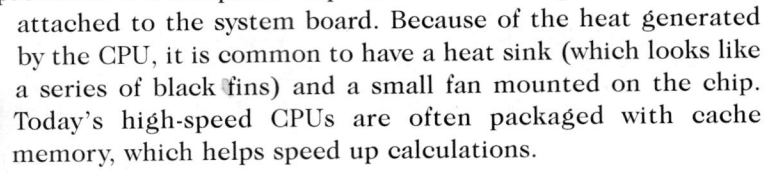

CPUs are also located in other parts of the computer system. Your computer's graphics board has its own dedicated CPU for processing graphical images. Your printer and plotter have CPUs for processing print data.

Intel, AMD, Motorola, and others manufacture CPUs. Today, the Pentium line of processors is most commonly used for AutoCAD.

Memory

Computer memory can be divided into two categories: ROM (read-only memory) and RAM (random access memory). ROM memory is contained in preprogrammed chips mounted on the motherboard. ROM stores basic sets of commands for the computer, such as the instructions for starting up the computer.

Random access memory (RAM) is what most people mean when they refer to "computer memory." It temporarily stores information in your computer. It is *temporary*, because all data in RAM is lost when the computer is turned off.

Software programs (such as AutoCAD) require a certa amount of RAM to run the program. This amount is given in kilobytes. A common number might be "256MB," meaning 256 *megabytes*. A megabyte is 1,024 *kilobytes*; a kilobyte is 1,024 *bytes*. You might see these numbers rounded, such as to 1,000 bytes in a kilobyte.

Disk Drives

Disk drives store large amounts of data, such as the Windows operating system, application programs, drawings, and other documents. A disk drive is either non-removable (*fixed*) or *removable*.

In almost all cases, fixed drives are hard drives. Today's hard drives hold 120 GB (gigabytes) or more of data. A *gigabyte* is one thousand megabytes.

Removable drives include diskettes, CD-ROMs, large capacity "diskettes," such as Iomega's Zip and Jaz drives, and even tape drives. The less-commonly-used 3-1/2" diskettes hold small amounts of data; their capacity is typically limited to 1.44 MB. Some computers support diskettes that hold 2.88 MB of data.

CD-ROM drives (short for "compact disc, read-only memory") are the most common method for distributing software today. Your copy of AutoCAD probably arrived on a CD-ROM. This disc can be read, but not written to. This makes the data secure from accidental erasure and attack from computer viruses.

CD-R discs can be written to (called "burning a CD") but not erased. These allow you to store up to 650 MB of data. CD-RW discs can be written to and erased.

While not as common because of their expense, DVD (short for "digital versatile disc") discs hold 4.7GB of data. DVD can only be read; DVD-R can be written to (recorded); and DVD-RW can be written to and erased.

Tape drives were the original backup medium, because they were relatively cheap. They suffer, however, from access slow speed. Today's tape drives can store 40GB or more data on tapes similar to those used by 8mm video cameras.

DISK CARE

All of your work is recorded to disk, so caring for the disk is very important. If a disk is damaged, you might lose your work! Frequent backing-up (copying files to a second disk, tape, or recordable CD-R) and proper handling of disks minimize the possibility of file loss.

Hard disks are installed inside the computer. This eliminates the danger of improper handling, but does not prevent damage. A hard drive can be damaged by shock. If you move the computer, be sure the power is off and move it gently. Today's hard drives have self-parking heads that, upon shutdown, move the head to a sector that does not have stored data. When the hard drive is on, do not move or tilt the computer. This is particularly important with notebook computers.

Floppy disks are also subject to damage from rough handling. The 3-1/2" diskettes and Zip disks are protected by a sliding door. The purpose of this door is defeated if you open it to look at and touch the magnetic material. Touching the disk surface leaves oil from the skin on the disk, possibly making it unreadable.

Dust and smoke too can leave particles on the disk surface that may prevent the drive from reading the disk properly. Heat and cold can cause the disk material to expand or contract, also causing problems. And magnets scramble the data on the disk's tracks. Spilling a liquid onto a disk leaves a residue; when the liquid is hot or cold, it can cause temperature damage, such as melting.

CD-ROMs and DVDs are relatively sturdy forms of data storage, but should also be handled with care. Keep the disk in its case, which protects the surface from which the laser reads data. Avoid touching the surface of the discs; handle them by their center and edges.

Write-Protecting Data

Normally, a disk can be both read by the computer and written to (new data can be saved on it). But you can write-protect the disk to preserve what is already stored on it, which allows the disk to be read by the computer, but not written to.

The 3-1/2" disk has a small slide switch in one corner of the disk. Sliding the switch alternates between write-protecting and write-enabling. CD-ROMs and DVDs cannot be written to, so the data is always safe.

Some CD-R drives can erase data from CD-R and DVD-R discs, while all DVD-RW and CD-RW discs have data erased and added. These discs have no write-protect switch, so it is possible to lose data.

PERIPHERAL HARDWARE

Peripheral hardware consists of add-on devices that perform specific functions. A properly-equipped CAD station consists of several peripheral devices.

PLOTTERS

Plotters produce "hard" copies of your drawings on paper. The two primary types of plotter today are inkjet plotters and laser printers. Some offices still have older pen plotters.

Inkjet Plotters

The most common plotter today is the inkjet plotter, which works by sending an electrical current to the print head. The electricity heats up ink in the printhead, causing it to squirt out the end of a nozzle and hit the paper.

In a typical print head, there are 48 nozzles lined up vertically. Large-format inkjet plotters have as many as 512 nozzles. For color plots, there are at least four heads: one each for cyan (light blue), magenta (pink), yellow, and black. By combining these four colors using a process called dithering, an inkjet plotter can produce color plots containing 256 or more colors. For higher quality output, some inkjet plotters have seven or more color cartridges.

Although a typical resolution is about 360 dots per inch (dpi), some inkjet plotters are capable of 2880 dpi. Inkjet plotters can use normal paper, but specially-coated paper results in brighter colors and a cleaner print. Desktop inkjet printers produce plots up to B-size (11" x 17"), while floor models products plots up to E-size (36" x 48").

Laser Printers

The laser printer creates a print using a process similar to photocopying. The image of the drawing is "painted" on a drum by a laser beam. This creates an electrostatic charge on the drum. The drum turns, picking up black toner particles, which stick to the charged portions of the drum. The toner particles are transferred to the paper, which passes over a hot fuser, which melts the toner onto the paper.

The laser-copy process produces sharp, clear prints. Although some laser printers have excellent graphics capabilities, they are usually restricted to a maximum of B-size. Resolution is typically 600 dpi, but can be 1200 dpi or higher.

Pen Plotters

Pen plotters are the oldest way of producing plots from CAD drawings. The pen plotter uses technical ink pens to draw on vellum, Mylar, or other suitable media. Pen plotters often use other types of pens, such as ballpoint, pencil, and markers.

The pen plotter is controlled by signals sent by the computer. Precise stepper motors control both the side-to-side pen movement and back-and-forth paper movement. A solenoid raises and lowers the pen. The accuracy (resolution) of the plotter is determined by the increment by which the motors move.

MONITORS

Monitors are used to view the work in progress. A monitor is sometimes referred to as a "display device" or "CRT" (cathode ray tube). As prices come down, flat-panel monitors are becoming more popular. These use the same LCD (liquid crystal display) technology found in notebook computers. CAD operators prefer monitors that measure 19 inches (diagonally) or larger, because they can see more of the drawing at a time.

The quality of the image on the monitor is determined by its resolution, which depends on the number of dots (pixels) displayed on the screen. These pixels make up the image. A typical 1024 x 768-resolution display contains 1,024 pixels horizontally and 768 pixels vertically. Many professional CAD users prefer higher resolution displays, such as 1280 x 1024 or higher.

The device driver, the operating system, and the graphics board contained in the computer must be capable of displaying the higher resolution. And the monitor must be matched to the graphics board.

INPUT DEVICES

Just as a word processing program requires a keyboard to input the individual letters, numbers, and symbols, a CAD program requires an input device to create and manipulate drawing elements. Although many programs allow input from the keyboard arrow keys, an input device speeds up the drawing process. The most common input devices for CAD drawing are the mouse, digitizer, and scanner.

Mouse

The mouse is an input device used for pointing and positioning the cursor. The name comes from its appearance. A mouse can be used for command input and screen interaction; it cannot digitize drawings. There are primarily two types: mechanical and optical.

A *mechanical* mouse has a ball under the housing. The rolling ball transmits its relative movement to the computer through a wire. An *optical* mouse uses optical technology to sense movement over almost any surface — except glass table tops.

Today's mouse devices come with one button (Macintosh), two buttons (Microsoft), three or four buttons (Logitech). Several buttons are better, because you can use them for additional functions; as well, AutoCAD supports functions on up to 15 buttons. For example, you can make the middle button of a three-button mouse act like a double-click, reducing the number of times you click the button. Some mouse devices come with a wheel, which allows you quickly to zoom and pan with AutoCAD.

Digitizers

The digitizer is an electronic input device that transmits the absolute X and Y location of a cursor resting on a sensitized pad. The digitizer can be used in two ways: as a pointing device to move a point around the screen, or as a tracing device to copy drawings into the computer in scale and proper proportion. Points are located on the pad by means of a stylus (similar in appearance to a pencil), or a puck (alternately referred to as a cursor). Digitizers use a fine grid of wires sandwiched between glass layers. The cursor is then moved across the pad, and the relative location is read and transmitted to the computer.

In "tablet" mode, the digitizing pad is calibrated to the actual absolute coordinates of the drawing. When used as a pointing device, the tablet is not calibrated.

Digitizers come in several sizes. You can use small pads to digitize drawings larger than the pad surface by moving the drawing on the pad and recalibrating. This can become annoying, however, when you frequently work with large-scale drawings. In that case, the expense of a large-format digitizer (36" x 24") may be justified.

Scanner

The scanner is another type of input device. It is used to "read" paper drawings and maps into the computer. The scanner works by shining a bright light at the paper. A head made of many CCDs (charge couple devices, the same technology used in digital cameras) measures the amount of reflected light. A bright area is the paper, while a dark area is part of a line.

The scanner sends data to the computer, where it can be displayed as a raster image by AutoCAD or converted to a vector file by specialized software. Scanners are available in a variety of sizes, ranging from very small units that read business cards to E-size scanners that read large engineering drawings.

DRIVES AND FILES

Windows names the computer's disk drives with a letter and colon, such as A: and C:. In all Windows computers, drives A: and B: are disk drives for removable media, usually 3-1/2" floppy disks. Drives C: and above are usually for hard drives, network drives, and additional removable drives. For example, E: is often, but not always, the CD-ROM or DVD removable drive.

Network drives are disk drives of other computers that you can access over a network. Your computer must be connected to the other computers through a network cable or a wireless network connection. By being networked, you and your co-workers can share drawings and other files. The networked version of AutoCAD allows a single copy to be used by two or more operators, depending on the terms of the network license.

Removable drives are any disk drives in which the medium can be removed, including floppy diskettes, CD-ROM discs, removable hard drives, Zip disks, and tape cartridges. When reading or writing using removable disks, you must be careful that the correct disk is inserted.

The position of the drives in the computer can differ according to the computer cabinet design. The figure below illustrates typical locations.

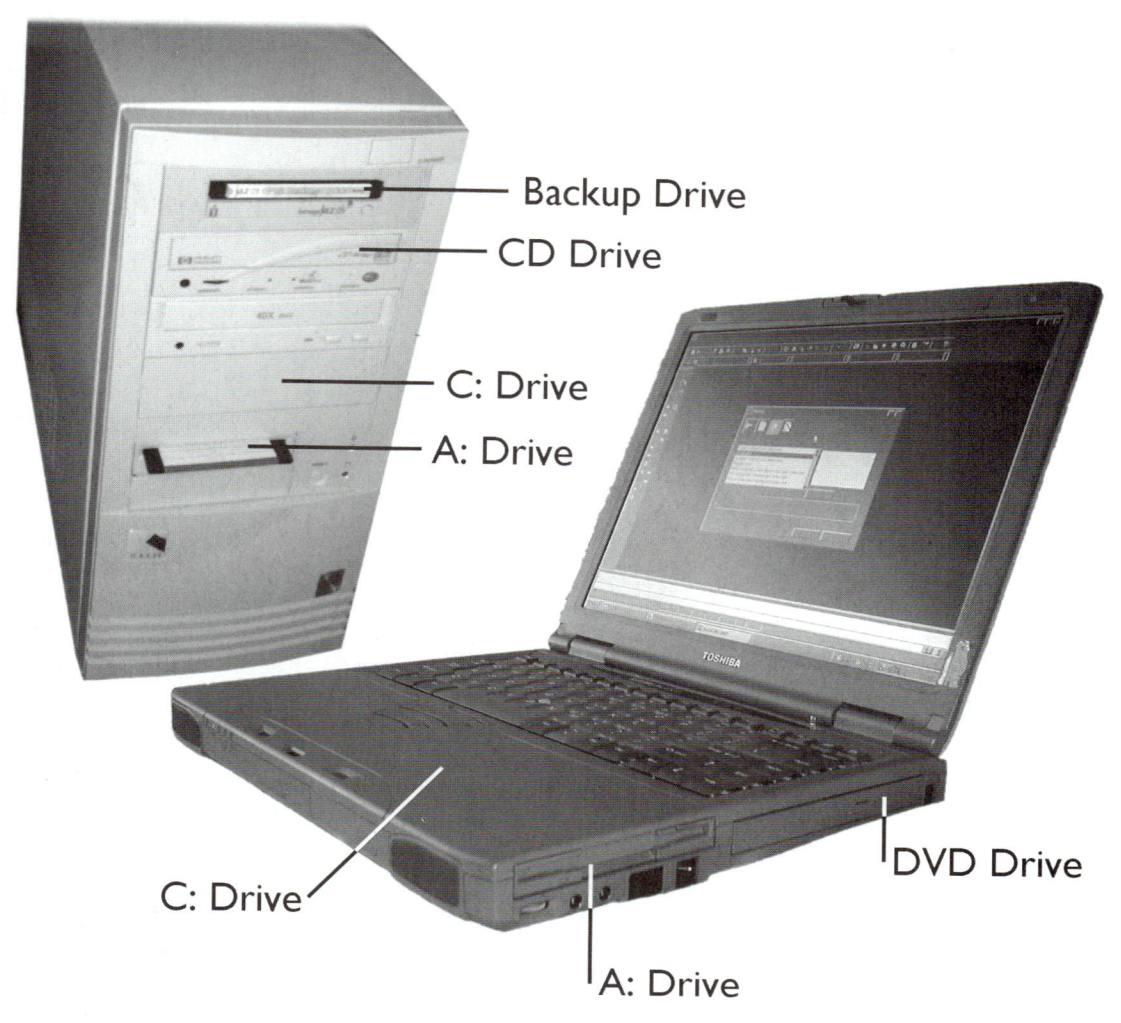

The location of disk drives in typical desktop and notebook computers.

FILE NAMES AND EXTENSIONS

Files can have names up to 255 letters and characters. A dot (.) followed by a three-letter code is usually added to the end of file names. This is called the *file extension*, and denotes the type of file. For example, an AutoCAD drawing file has a *.dwg* extension. Correct notation includes the file name, followed by a period, then the file extension. For example, an AutoCAD drawing named *widget* is stored as *widget.dwg*. Many programs add their own file extension code, while some require you to specify the extension.

(Display of file extensions are normally turned off by Windows. Since it is useful to know a file's extension, you can turn on. In Windows Explorer, select **View | Folder Options**. In the dialog box, click the **View** tab, then uncheck the "Hide extensions for known file types" option.)

Older versions of operating systems, such as Windows 3.1 and DOS, limit file names to eight characters. When a file name of nine or more characters is copied to a computer running an older OS, the file name is truncated (chopped off) to six characters and ~1 is added. For example, *houseplan.dwg* becomes *housep~1.dwg*.

Wild-Card Characters

In file dialog boxes, some Windows functions, and certain AutoCAD commands, you can use *wildcard* characters to specify files. The two most common wildcard characters are the question mark (?) and the asterisk (*).

The question mark fills in for any single character. By using one or more question marks, you represent one or more characters. When you enter

car??

you would refer to names, such as:

car   cars   carts   card2

The asterisk represents all characters. For example, when you enter:

*.dwg

you refer to *all* files with the *.dwg* file extension, such as:

floorplan.dwg   my drawing 1.dwg   8th floor.dwg

Alternately, when you type

floorplan.*

you refer to all files named "Floorplan" regardless of their file extension (or even to those with no extension):

floorplan.dwg   floorplan.bak   floorplan.dwf   floorplan.

DISPLAYING FILES

There are several ways to display and manipulate files in Windows 2000 and XP.

Windows Explorer

Windows Explorer is the "official" method to view files in Windows. Explorer, not to be confused with Internet Explorer, lets you view files in all folders on all drives, including on computers networked to yours.

File Manager

The File Manager is the old method of viewing files from Windows v3.x and NT v3.x, with a drawback and an advantage over Explorer. The drawback is that File Manager is limited to displaying eight-character file names; file names longer than eight characters are truncated to six characters and given a ~1 suffix. The advantage is that File Manager can sort files by file extension, something Explorer cannot do; the equivalent in Explorer is to sort files by Type, which is not as useful for the power user, because it is more useful to know file extensions than Type descriptions.

File-Related Dialog Boxes

Almost all of AutoCAD's file-related dialog boxes — such as **Open**, **Save As** and **Import** — let you view and manipulate files. This is a handy shortcut that obviates the need to switch to Explorer.

DOS Session

The oldest method is to start a DOS session, and then type in commands — such as DIR, TREE, COPY, MOVE, and DELETE — together with parameters, such as /s, *.*, and a: . This is good for power users, because Windows Explorer cannot perform some file-related tasks, such as renaming all files in a folder, or copying a file to another name in the same folder.

WINDOWS EXPLORER

Windows Explorer (not to be confused with Internet Explorer) lets you view and manipulate files on your computer's drives and all networked drives that you have permission to access. You can change the way Explorer looks, but the most useful configuration has two panes, as illustrated by the figure below.

Viewing Files

In Windows Explorer, the left *pane* displays the *tree* of drives (including networked drives) and folders (the new name for subdirectories). This is called the "Folder view." To see more of the tree, use the scroll bar. The + (plus) sign beside a folder means it has subfolders. Click + to open the folder; the + changes to a – (minus) sign. Click – to close the folder.

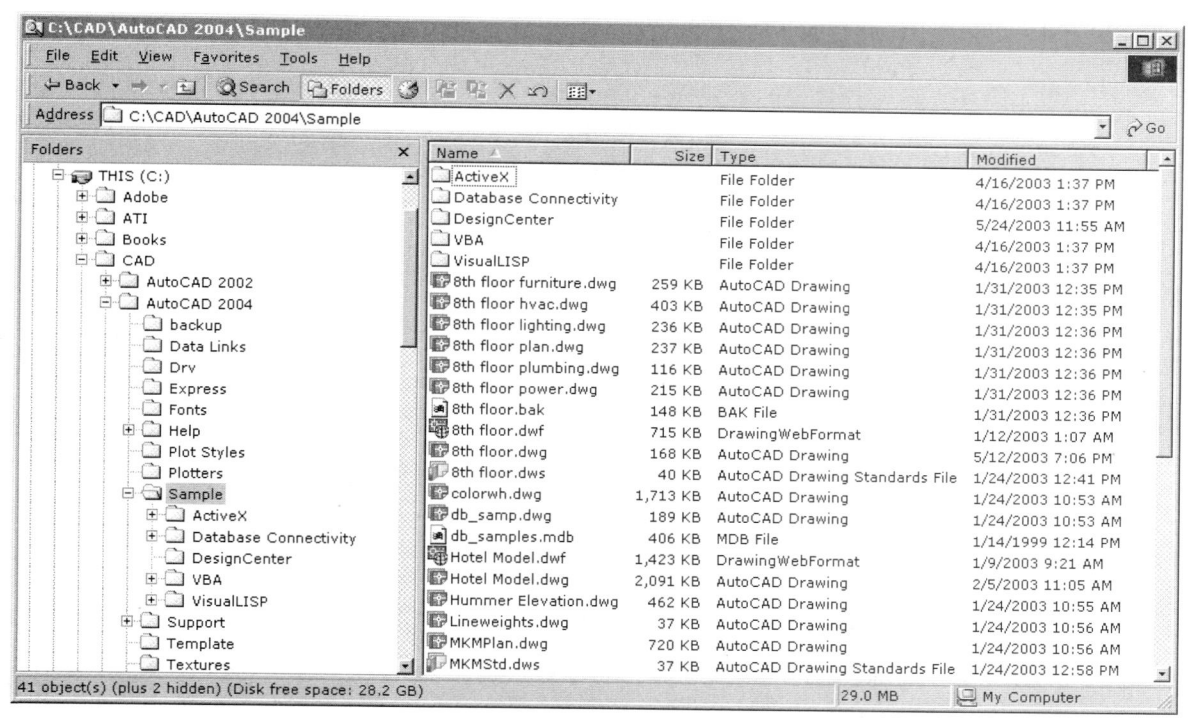

Windows Explorer can take on different configurations.
This configuration shows two panes and file details.

In the right pane, Windows Explorer displays the files and subfolders contained in the folder currently selected in the left window pane. This is called the Contents view. You change the sorting order of the **Name**, **Size**, **Type**, and **Modified** by clicking on the sort bar. (To see the sort bar, set the **View** to display file **Details**.) Click a second time and Explorer displays the list in reverse order, which is useful for displaying files sorted alphabetically or by size, either largest first or smallest first.

Windows Explorer also lends itself to displaying files and folders in different views, each giving you more detail. The View item on the menu bar lets you switch between large icons, small icons, list, detail, and thumbnail (new to AutoCAD 2004), as illustrated by the figure below.

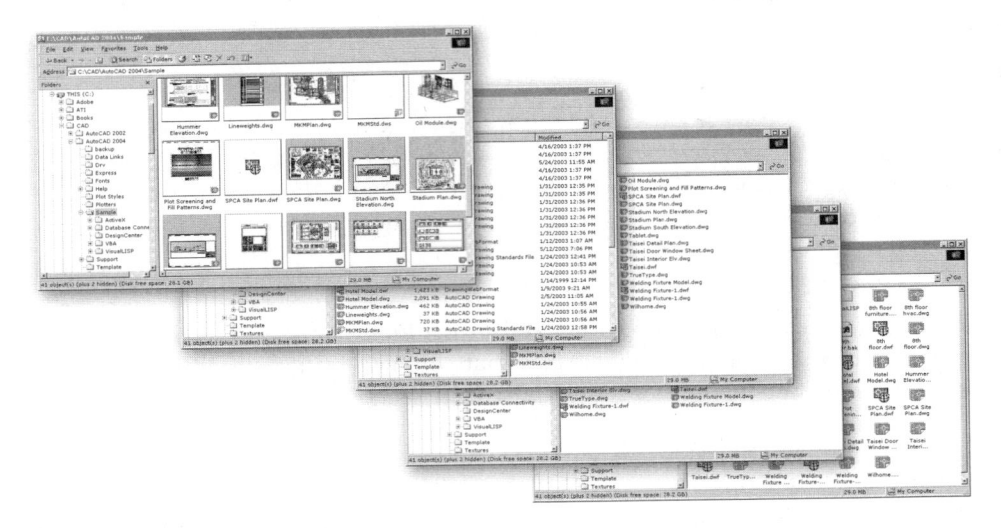

Manipulating Files

When you right-click on any file or folder, you can change its name and perform other tasks. (Alternatively, use the **File** and **Edit** items on the menu bar.) When you right-click, Windows Explorer displays a shortcut menu with the following options, which differ depending on whether you right-click a folder or file:

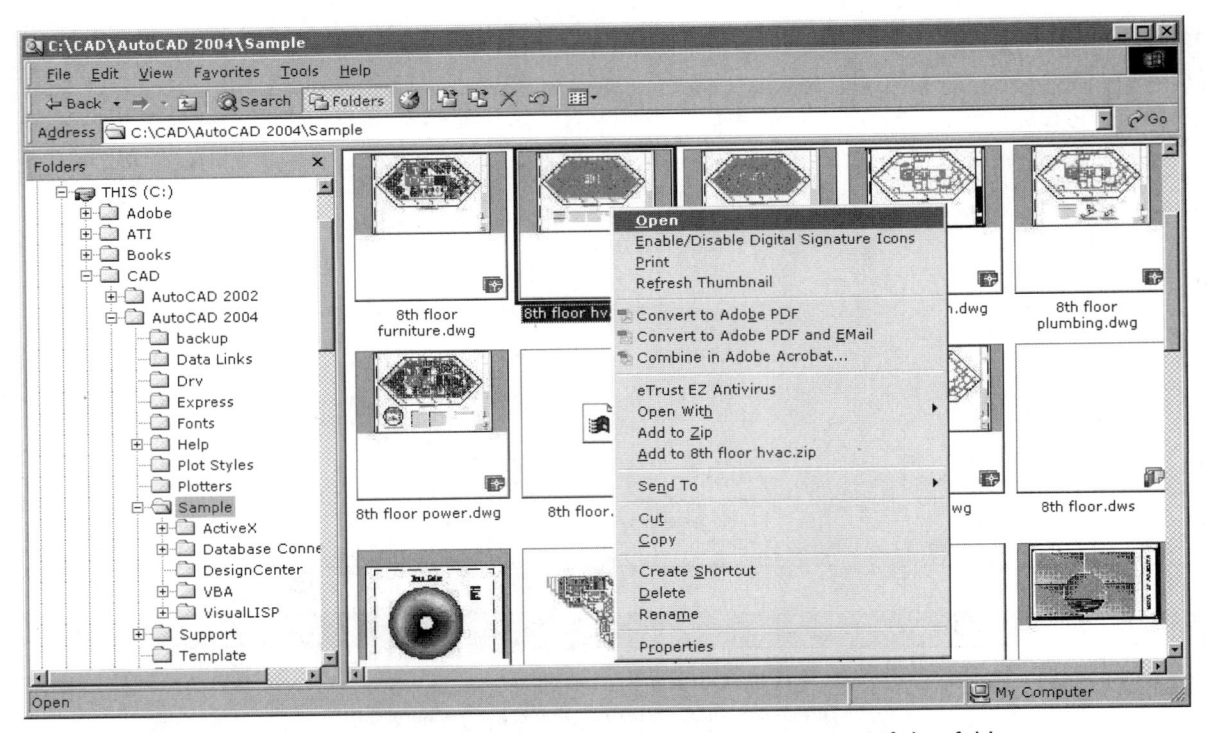

Open (folders) brings up another window displaying the contents of that folder.

Open (files) opens the file with the associated program. For example, a *.lsp* (AutoLISP source code) file is opened by Notepad. An *.exe* file, such as *acad.exe*, starts to run.

Explore (folders only) displays the contents of the folder in the pane at right.

Find (folders only) displays the Find: All Files dialog box, which lets you find files anywhere on the computer and (if connected) the network.

Sharing (folders only) allows you to give others on a network access to drives and folders. You can set three types of permission: (1) full read-write; (2) read only; and (3) access with password.

Send to sends the folder or file to one of several destinations, such as to the floppy drive or the Briefcase folder. Although the printer is listed, you cannot send folders to the printer; when you send a file to the printer, Windows Explorer first opens the associated application.

Create Shortcut places *shortcuts* as icons on the Windows desktop (shortcuts are small files that point to the original file's location). Shortcuts are excellent for accessing files without having to wander through the folders to locate them each time.

Delete erases folders and files. Windows first asks, "Are you sure?" Deleted files and folders are stored in the Recycle Bin for a limited time, from which they can be recovered, if on the hard drive. On floppy diskettes and Zip disks, the files are gone for good.

Rename changes the names of folders and files. As an alternative, click twice (slowly, do not double-click) on the name and Explorer lets you change the name directly.

Properties sets the attributes for the folder or file, including archive, read-only, hidden, and system.

Copying And Moving Files

Copying or moving a file using Windows Explorer takes three steps:

1. In the Folders pane (on the left side), make sure you can see the name of the folder you will be copying the file to. If you cannot see it, scroll the list to see more folders or open the appropriate closed folder by clicking the + sign.
2. In the Contents pane (on the right side), drag the file to the folder. (Drag means to click on the file and move the cursor without letting go of the mouse button until you reach its destination.)
3. To copy the file, hold down the **CTRL** key while dragging the file. As a reminder that you are copying the file, Windows Explorer displays a small + sign near the dragged icon.

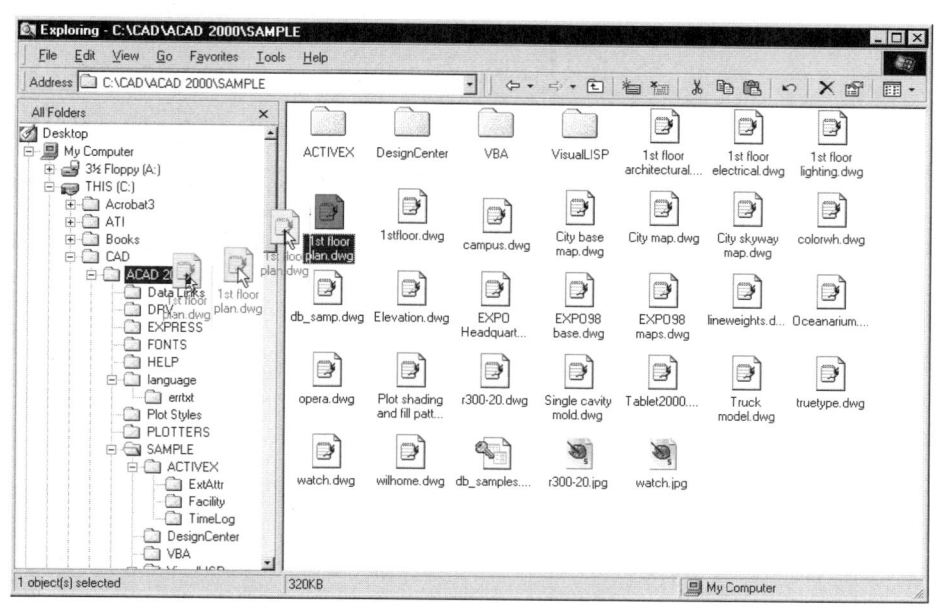

If you drag the file on top of an application, the application will attempt to open the file. For example, dragging a *.dwg* file onto the *acad.exe* icon causes AutoCAD to launch with that drawing.

AUTOCAD FILE DIALOG BOXES

Windows Explorer is the official way to view and manipulate files in Windows. You can, however, perform some of these functions from within AutoCAD's file-related dialog boxes, which makes sense, because when saving and opening files you probably want to work with them. Those dialog boxes are displayed when you use the Open, SaveAs, Export, and other file-related commands.

AutoCAD's File dialog box provides the following functions:

Switch Views. Choose the **Views** item to see the files listed in different views, such as a compact listing or with details (size, type, and modified date-time). Click the column headers to sort in alphabetical order; click a second time to sort in reverse order.

Select a Different Folder. Select a different folder (subdirectory) by several methods: (1) the **Look in** list box lets you move directly to another folder; (2) single-click the **Up One Level** icon to move up the folder tree one folder at a time; double-click the folder to move down into it; or (3) click the **Back** button to return to the previous folder.

Create a New Folder. Click the **Create New Folder** icon to create a new folder in the current folder. As an alternative, press **ALT+5**.

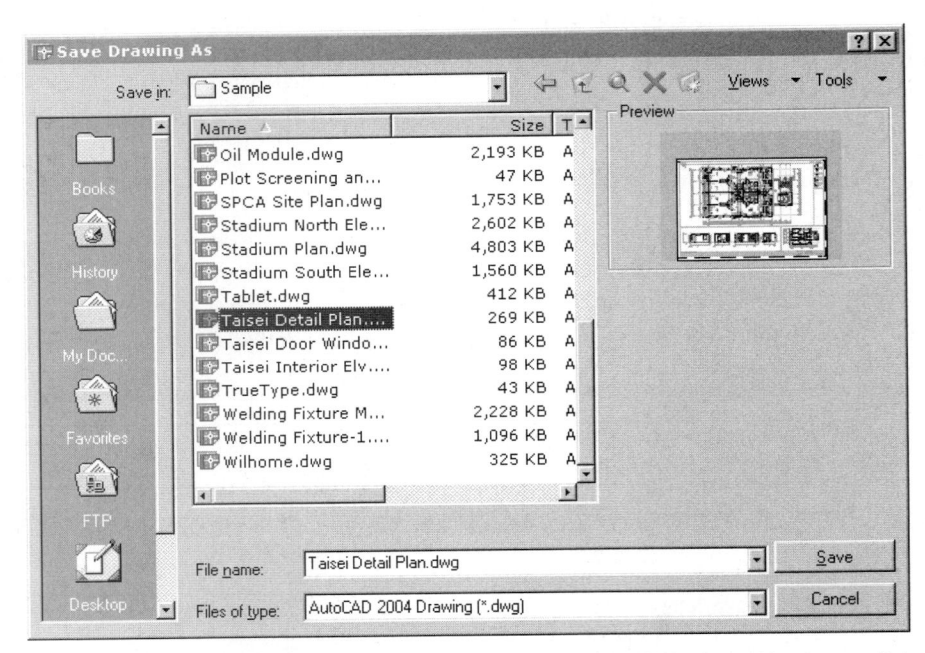

Delete a File. Select the file name and press the **DEL** key. Windows displays a dialog box that asks, "Are you sure you want to send 'file name' to the Recycle Bin?" When you respond **Yes**, the file is not erased but moved to the Recycle Bin folder. If you delete a file by accident, you can go to the Recycle Bin and retrieve the file — if on the hard drive. Floppy diskettes and Zip disks erase the files permanently.

Rename a File. Click twice (slowly) on the file name, and then type a new name.

Move or Copy a File. Drag the file to another folder. Hold down the **CTRL** key to copy the file; hold down **SHIFT** to move a file.

Context-Sensitive Menu. Right-click a file or folder to bring up the same context-sensitive menu as seen in Windows Explorer.

Open or save the File. Double-click a file name to load it without choosing **OK**.

Places List. Located on the left hand side, the Places list provides convenient access to frequently used folders. To add a folder, drag it from the file list into the Places list.

AUTOCAD FILE EXTENSIONS

AutoCAD uses many different types of files to support your drafting work. The contents of a file are often described by the file name extension. For example, linetypes are stored in files that end with .lin, while drawings are stored in files with an extension of *.dwg*. Many of the file extensions used by AutoCAD are listed here.

Drawing Files

Extension	Description
.$ac	Temporary files created by AutoCAD
.bak	Backup drawing files
.dwf	Drawing Web format files
.dwg	AutoCAD drawing files
.dws	CAD standards files
.dwt	Drawing template files
.dxb	AutoCAD binary drawing interchange files
.dxf	AutoCAD drawing interchange files

AutoCAD Program Files

Extension	Description
.arx	ObjectARx (AutoCAD Runtime eXtension) program files
.dll	Dynamic link libraries
.exe	Executable files, such as AutoCAD itself
.lsp	AutoLISP program files

Support Files

Extension	Description
.cfg	Configuration files
.chm	HTML format help files
.cus	Custom dictionary files
.dct	Dictionary files
.err	Error log files
.fmp	Font mapping files
.hlp	Windows-format help files
.lin	Linetype definition files
.lli	Landscape libraries
.log	Log file created by the Logfileon commands
.mli	Rendering material librarys
.mln	Multiline library files
.mnc	Compile menu files
.mnd	Uncompiled menu file containing macross
.mnl	AutoLISP routines used by AutoCAD menus
.mns	AutoCAD-generated menu source files
.mnu	Menu source files
.msg	Message files
.pat	Hatch pattern definition files
.pgp	Program parameters files for external commands and command aliases
.scr	Script files
.shp	Shape and font definition files
.shx	Compiled shape and font files

Plotting Support Files

Extension	Description
.ctb	Color-table based plot parmaeter files
.pcp	Plot configuration parameters files for AutoCAD Release 14
.pc2	Plot configuration parameters files for AutoCAD 2000
.pc3	Plot configuration parameters files for 2000i, 2002, and 2004
.plt	Plot files
.stb	Style-table based plot parameter files

Import-Export Files

Extension	Description
.3ds	3D Studio files (import and export)
.bmp	Windows raster files (device-independent bitmap)
.cdf	Comma delimited files (created by AttExt)
.dxx	DXF files (created by AttExt)
.pcx	Raster format files
.png	Portable Network Graphics raster files
.sat	ACIS solid object files (Save As Text)
.slb	Slide library files
.sld	Slide files
.stl	Solid object stereo-lithography files (solids modeling)
.tif	Raster format files (Tagged image file format)
.tga	Raster format files (Targa)
.txt	Space delimited files (created by AttExt)
.wmf	Windows metafile files
.xls	Excel spreadsheet files (created by EAttExt)
.xml	DesignXML format files

LISP and ObjectARX Programming Files

Extension	Description
.cpp	ObjectARX source code files
.dcl	Dialog control language descriptions of dialog boxes
.dvb	Visual Basic for Applications program files
.def	ObjectARX definition files
.fas	AutoLISP fast load programs
.frm	VBA form definitions
.h	Definitions for ADS and ObjectARX functions
.lib	ObjectARX function libraries
.mak	ObjectARX make files
.rx	Lists of ObjectARX applications that load automatically
.tlb	ActiveX Automation type libraries
.unt	Unit definition files
.vlx	Compiled Visual LISP files.

Miscellaneous Files

Extension	Description
.css	Cascading style sheet files
.html	HTML files
.htt	HTML template files
.ini	Initialization files
.js	JavaScript files
.map	Maps displayed by Light command
.xls	XML style sheet
.xml	Extended markup language files
.xmx	External messages files

REMOVING TEMPORARY FILES

While it is working, AutoCAD creates "temporary" files. When you exit AutoCAD, these temporary files are erased. If, however, your computer crashes, these files might be left on the hard disk. You can erase temporary files, but only when AutoCAD is not running. Never erase files with these extensions while AutoCAD is still running:

.$a   .ac$ .dwk   .dwl   .ef$ .sv$ .swr

These files can be erased to free up disk space when AutoCAD is not running.

APPENDIX B

AutoCAD Commands, Aliases, and Keyboard Shortcuts

AUTOCAD 2004 COMMANDS

The following commands are documented by Autodesk as available in AutoCAD 2004. Commands prefixed with ' (apostrophe) are *transparent* — they can be executed during another command. Commands prefixed with - (hyphen) display their prompts at the command line only, and not through a dialog box. Shortcut keystokes are indicated as **CTRL+F4**.

The icon indicates the command is new in AutoCAD 2004.

A complete reference guide to all of AutoCAD 2004's commands is found in *The Illustrated AutoCAD 2004 Quick Reference* by author Ralph Grabowski, available through Autodesk Press.

Command	Description
A	
'About	Displays an AutoCAD information dialog box that includes version and serial numbers.
AcisIn	Imports ASCII-format ACIS files into the drawing, and then creates 3D solids, 2D regions, or body objects.
AcisOut	Exports AutoCAD 3D solids, 2D regions, or bodies as ASCII-format ACIS files (file extension *.sat*).
AdcClose	CTRL+2: Closes the DesignCenter window.
AdCenter	CTRL+2: Opens DesignCenter; manages AutoCAD content.
AdcNavigate	Directs DesignCenter to the file name, folder, or network path you specify.
Align	Uses three pairs of 3D points to move and rotate (align) 3D objects.
AmeConvert	Converts drawings made with AME v2.0 and v2.1 into ACIS solid models.
'Aperture	Adjusts the size of the target box used with object snap.
AppLoad	Displays a dialog box that lets you list AutoLISP, Visual Basic, and ObjectARX program names for easy loading into AutoCAD.
Arc	Draws arcs by a variety of methods.
Area	Computes the area and perimeter of polygonal shapes.
Array	Makes multiple copies of objects.
Arx	Loads and unloads ObjectARX programs. Also displays the names of ObjectARX program command names.
Assist	Displays the real-time help window.

Command	Description
AttDef	Creates attribute definitions.
'AttDisp	Controls whether attributes are displayed.
AttEdit	Edits attributes.
AttExt	Extracts attribute data from drawings, and writes them to files for use with other programs.
AttReDef	Assigns existing attributes to new blocks, and new attributes to existing blocks.
AttSync	Synchronizes changed attributes with all blocks.
Audit	Diagnoses and corrects errors in drawing files.

B

Command	Description
Background	Sets up backgrounds for rendered scenes; can be solid colors, gradient shades, images in BMP, PCX, Targa, JPEG, or TIFF formats, or the current AutoCAD wireframe views.
'Base	Specifies the origin for inserting one drawing into another.
BAttMan	Edits all aspects of attributes in a block; short for Block Attribute Manager.
BHatch	Fills an automatically-defined boundary with hatch patterns, solid colors, and gradient fills; previews and adjusts patterns without starting over.
'Blipmode	Toggles display of marker blips.
Block	Creates symbols from groups of objects.
BlockIcon	Generates preview images for blocks created with AutoCAD Release 14 and earlier.
BmpOut	Exports selected objects from the current viewport to raster BMP files.
Boundary	Draws closed boundary polylines.
Box	Creates 3D solid boxes andr cubes (solid modeling command).
Break	Erases parts of objects, breaks objects in two.
Browser	Launches your computer's default Web browser with the URL you specify.

C

Command	Description
'Cal	Runs a geometry calculator that evaluates integer, real, and vector expressions.
Camera	Sets the camera and target locations.
Chamfer	Trims intersecting lines, connecting them a chamfer.
Change	Permits modification of an object's characteristics.
CheckStandards	Compares the settings of layers, linetypes, text styles, and dimension styles with those set in another drawing.
ChProp	Changes properties (linetype, color, and so on) of objects.
Circle	Draws circles by a variety of methods.
CleanScreenOn *(New in 2004)*	CTRL+0: Maximizes the drawing area by turning off toolbars, title bar, and window borders.
CleanScreenOff *(New in 2004)*	CTRL+0: Turns on toolbars, title bar, and window borders.
Close	CTRL+F4: Closes the current drawing.
CloseAll	Closes all open drawings; keeps AutoCAD open.
'Color	Sets new colors for subsequently-drawn objects.
Compile	Compiles shapes and .shp and .pfb font files.
Cone	Creates 3D cones (a solid modeling command).
Convert	Converts 2D polylines and associative hatches in pre-AutoCAD Release 14 drawings to the "lightweight" format to save on memory and disk space.
ConvertCTB	Converts drawings from plot styles to color-based tables.

Command	Description
ConvertPStyles	Converts drawings from color-based tables to plot styles.
Copy	Copies selected objects.
CopyBase	CTRL+SHIFT+C: Copies objects with a specified base point.
CopyClip	CTRL+C: Copies selected objects to the Clipboard in several formats.
CopyHist	Copies Text window text to the Clipboard.
CopyLink	Copies all objects in the current viewport to the Clipboard in several formats.
Customize	Customizes the toolbar and keyboard shortcuts.
CutClip	CTRL+X: Cuts selected objects from the drawing to the Clipboard in several formats.
Cylinder	Creates 3D cylinders (a solid modeling command).

D

Command	Description
DbcClose	Closes the dbConnect Manager.
DbConnect	CTRL+6: Connects objects in drawings with tables in external database files.
DblClkEdit	Toggles double-click editing.
DbList	Provides information about all objects in drawings.
DdEdit	Edits text, paragraph text, attribute text, and dimension text.
'DdPType	Specifies the style and size of points.
DdVPoint	Sets 3D viewpoints.
'Delay	Creates a delay between operations in a script file.
DetachURL	Removes hyperlinks from objects.

DIMENSIONS

Command	Description
Dim	Specifies semi-automatic dimensioning capabilities.
Dim1	Executes a single AutoCAD Release 12-style dimension command.
DimAligned	Draws linear dimensions aligned to objects.
DimAngular	Draws angular dimensions.
DimBaseline	Draws linear, angular, and ordinate dimensions that continue from baselines.
DimCenter	Draws center marks on circles and arcs.
DimContinue	Draws linear, angular, and ordinate dimensions that continue from the last dimension.
DimDiameter	Draws diameter dimensions on circles and arcs.
DimDisassociate	Removes associativity from dimensions.
DimEdit	Edits the text and extension lines of associative dimensions.
DimLinear	Draws linear dimensions.
DimOrdinate	Draws ordinate dimensions in the X and Y directions.
DimOverride	Overrides current dimension variables to change the look of selected dimensions.
DimRadius	Draws radial dimensions for circles and arcs.
DimReassociate	Associates dimensions with objects.
DimRegen	Updates associative dimensions.
DimStyle	Creates, names, modifies, and applies named dimension styles.
DimTEdit	Moves and rotates text in dimensions.
'Dist	Computes the distance between two points.
Divide	Divides objects into an equal number of parts, and places specified blocks or point objects at the division points.
Donut	Constructs solid filled circles and doughnuts.
'Dragmode	Toggles display of dragged objects.
DrawOrder	Changes the order in which objects are displayed: selected objects and images are placed above or below other objects.

Command	Description
'DSettings	Specifies drawing settings for snap, grid, polar, and object snap tracking.
'DsViewer	Opens the Aerial View window.
DView	Displays 3D views dynamically.
DwgProps	Sets and displays the properties of the current drawing.
DxbIn	Creates binary drawing interchange files.

E

Command	Description
EAttEdit	Enhanced attribute editor.
EAttExt	Enhanced attribute extraction.
Edge	Changes the visibility of 3D face edges.
EdgeSurf	Draws edge-defined surfaces.
'Elev	Sets current elevation and thickness.
Ellipse	Constructs ellipses and elliptical arcs.
Erase	Removes objects from drawings.
eTransmit	Packages the drawing and related files for transmission by email or courier.
Explode	Breaks down blocks into individual objects; breaks down polylines into lines and arcs.
Export	Exports drawings in several file formats.
Extend	Extends objects to meet boundary objects.
Extrude	Extrudes 2D closed objects into 3D solid objects (a solid modeling command).

F

Command	Description
'Fill	Toggles the display of solid fills.
Fillet	Connects two lines with an arc.
'Filter	Creates selection sets of objects based on their properties.
Find	Finds and replaces text.
Fog	Adds fog or depth effects to a rendering.

G

Command	Description
GoToURL	Links to hyperlinks
'GraphScr	F2: Switches to the drawing window from the Text window.
'Grid	F7 and CTRL+G: Displays grid of specified spacing.
Group	Creates named selection sets of objects (CTRL+SHIFT+A toggles group selection style).

H

Command	Description
Hatch	Performs hatching at the command prompt.
HatchEdit	Edits associative hatch patterns.
'Help	? and F1: Displays a list of AutoCAD commands with detailed information.
Hide	Removes hidden lines from the currently-displayed view.
HlSettings *(New in 2004)*	Adjusts the display settings for hidden-line removal.
HyperLink	CTRL+K: Attaches hyperlinks to objects, or modifies existing hyperlinks.
HyperLinkOptions	Controls the visibility of the hyperlink cursor and the display of hyperlink tooltips.

I

Command	Description
'Id	Describes the position of a point in x,y,z coordinates.
Image	Controls the insertion of raster images with an xref-like dialog box.
ImageAdjust	Controls the brightness, contrast, and fading of raster images.
ImageAttach	Attaches raster images to the current drawing.
ImageClip	Places rectangular or irregular clipping boundaries around images.
ImageFrame	Toggles the display of the image frames.

Appendix B: AutoCAD Commands, Aliases, and Keyboard Shortcuts **793**

Command	Description
ImageQuality	Controls the display quality of images.
Import	Imports a variety of file formats into drawings.
Insert	Inserts blocks and other drawings into the current drawing.
InsertObj	Inserts objects generated by another Windows application.
Interfere	Determines the interference of two or more 3D solids (a solid modeling command).
Intersect	Creates a 3D solid or 2D region from the intersection of two or more 3D solids or 2D regions (a solid modeling command).
'Isoplane	F5 and CTRL+E: Switches to the next isoplane.

J

Command	Description
JpegOut	Exports views as JPEG files. *(New in 2004)*
JustifyText	Changes the justification of text.

L

Command	Description
'Layer	Creates and changes layers; toggles the state of layers; assigns linetypes, lineweights, plot styles, colors, and other properties to layers.
LayerP	Displays the previous layer state.
LayerPMode	Toggles the availability of the LayerP command.
Layout	Creates a new layout and renames, copies, saves, or deletes existing layouts.
LayoutWizard	Designates page and plot settings for new layouts.
LayTrans	Translates layer names from one space to another.
Leader	Draws leader dimensions.
Lengthen	Lengthens or shortens open objects.
Light	Creates, names, places, and deletes "lights" used by the Render command.
'Limits	Sets drawing boundaries.
Line	Draws straight line segments.
'Linetype	Lists, creates, and modifies linetype definitions; loads them for use in drawings.
List	Displays database information for selected objects.
Load	Loads shape files into drawings.
LogFileOff	Closes the .log keyboard logging file.
LogFileOn	Writes the text of the 'Command:' prompt area to .log log file.
LsEdit	Edits landscape objects.
LsLib	Accesses libraries of landscape objects.
LsNew	Places landscape items in drawings.
'LtScale	Specifies the scale for all linetypes in drawings.
LWeight	Sets the current lineweight, lineweight display options, and lineweight units.

M

Command	Description
MassProp	Calculates and displays the mass properties of 3D solids and 2D regions (a solid modeling command).
'MatchProp	Copies properties from one object to other objects.
MatLib	Imports material-look definitions; used by the Render command.
Measure	Places points or blocks at specified distances along objects.
Menu	Loads menus of AutoCAD commands into the menu area.
MenuLoad	Loads partial menu files.
MenuUnLoad	Unloads partial menu files.
MInsert	Makes arrays of inserted blocks.
Mirror	Creates mirror images of objects.
Mirror3D	Creates mirror images of objects rotated about a plane.
MlEdit	Edit multilines.

Command	Description
MLine	Draws multiple parallel lines (up to 16).
MlStyle	Defines named mline styles, including color, linetype, and endcapping.
Model	Switches from layout tabs to Model tab.
Move	Moves objects from one location to another.
NEW IN 2004 MRedo	Redoes more than one undo operation.
MSlide	Creates *.sld* slide files of the current display.
MSpace	Switches to model space.
MText	Places formatted paragraph text inside a rectangular boundary.
Multiple	Repeats commands.
MView	Creates and manipulates viewports in paper space.
MvSetup	Sets up new drawings.

N

Command	Description
New	CTRL+N: Creates new drawings.

O

Command	Description
Offset	Constructs parallel copies of objects.
OleLinks	Controls objects linked to drawings.
OleScale	Displays the OLE Properties dialog box.
Oops	Restores objects accidentally erased by the previous command.
Open	CTRL+O: Opens existing drawings.
Options	Customizes AutoCAD's settings.
'Ortho	F8 and CTRL+L: Forces lines to be drawn orthogonally, or as set by the snap rotation angle.
'OSnap	F3 and CTRL+F: Locates geometric points of objects.

P

Command	Description
PageSetup	Specifies the layout page, plotting device, paper size, and settings for new layouts.
'Pan	Moves the view.
PartiaLoad	Loads additional geometry into partially-opened drawings.
-PartialOpen	Loads geometry from a selected view or layer into drawings.
PasteBlock	CTRL+SHIFT+V: Pastes copied block into drawings.
PasteClip	CTRL+V: Pastes objects from the Clipboard into the upper left corner of drawings.
PasteOrig	Pastes copied objects from the Clipboard into new drawings using the coordinates from the original.
PasteSpec	Provides control over the format of objects pasted from the Clipboard.
PcInWizard	Imports *.pcp* and *.pc2* configuration files of plot settings.
PEdit	Edits polylines and polyface objects.
PFace	Constructs polygon meshes defined by the location of vertices.
Plan	Returns to the plan view of the current UCS.
PLine	Creates connected lines, arcs, and splines of specified width.
Plot	CTRL+P: Plots drawing to printers and plotters.
PlotStamp	Add information about the drawing to the edge of the plot
PlotStyle	Sets plot styles for new objects, or assigns plot styles to selected objects.
PlotterManager	Launches the Add-a-Plotter wizard and the Plotter Configuration Editor.
NEW IN 2004 PngOut	Exports views as PNG files (portable network graphics format).
Point	Draws points.
Polygon	Draws regular polygons with a specified number of sides.

Command	Description
Preview	Provides a Windows-like plot preview.
Properties	CTRL+1: Controls properties of existing objects.
PropertiesClose	CTRL+1: Closes the Properties window.
PSetUpIn	Imports user-defined page setups into new drawing layouts.
PSpace	Switches to paper space (layout mode).
Publish *(NEW IN 2004)*	Plots one or more drawings as a drawing set, or exports them in DWF format.
PublishToWeb	Creates a Web page from one or more drawings in DWF, JPEG, or PNG formats.
Purge	Deletes unused blocks, layers, linetypes, and so on.

Q

Command	Description
QDim	Creates continuous dimensions quickly.
QLeader	Creates leaders and leader annotation quickly.
QNew *(NEW IN 2004)*	Starts new drawings based on template files.
QSave	CTRL+S: Saves drawings without requesting a file name.
QSelect	Creates selection sets based on filtering criteria.
QText	Redraws text as rectangles with the next regeneration.
Quit	ALT+F4 and CTRL+Q: Exits AutoCAD.

R

Command	Description
Ray	Draws semi-infinite construction lines.
Recover	Attempts recovery of corrupted or damaged files.
Rectang	Draws rectangles.
Redefine	Restores AutoCAD's definition of a command.
Redo	CTRL+Y: Restores the operations changed by the previous Undo command.
'Redraw	Cleans up the display of the current viewport.
'RedrawAll	Performs redraw in all viewports.
RefClose	Saves or discards changes made during in-place editing of xrefs and blocks.
RefEdit	Selects references for editing.
RefSet	Adds and removes objects from a working set during in-place editing of references.
Regen	Regenerates the drawing in the current viewport.
RegenAll	Regenerates all viewports.
'RegenAuto	Controls whether drawings are regenerated automatically.
Region	Creates 2D region objects from existing closed objects (a solid modeling command).
Reinit	Reinitializes the I/O ports, digitizer, display, plotter, and the *acad.pgp* file.
Rename	Changes the name of blocks, linetypes, layers, text styles, views, and so on.
Render	Creates renderings of 3D objects.
RendScr	Creates renderings on computers with a single, nonwindowing display configured for full-screen rendering.
Replay	Displays BMP, TGA, and TIFF raster files.
'Resume	Continues playing back a script file that had been interrupted by the ESC key.
RevCloud *(NEW IN 2004)*	Draws revision clouds.
Revolve	Creates 3D solids by revolving 2D closed objects around an axis (a solid modeling command).
RevSurf	Draws a revolved surface (a solid modeling command).
RMat	Defines, loads, creates, attaches, detaches, and modifies material-look definitions; used by the Render command.

Command	Description
RMLin	Imports *.rml* redline markup files created by Volo View.
Rotate	Rotates objects about specified center points.
Rotate3D	Rotates objects about a 3D axis.
RPref	Set preferences for renderings.
RScript	Restarts scripts.
RuleSurf	Draws ruled surfaces.

S

Command	Description
Save & SaveAs	CTRL+SHIFT+S: Saves the current drawing by a specified name.
SaveImg	Saves the current rendering in BMP, TGA, or TIFF formats.
Scale	Changes the size of objects equally in the x, y, z directions.
ScaleText	Resizes text.
Scene	Creates, modifies, and deletes named scenes; used by the Render command.
'Script	Runs script files in AutoCAD.
Section	Creates 2D regions from 3D solids by intersecting a plane through the solid (a solid modeling command).
SecurityOptions *(new in 2004)*	Sets up passwords and digital signatures for drawings.
Select	Preselects objects to be edited; CTRL+A selects all objects.
SetiDropHandler *(new in 2004)*	Specifies how i-drop objects are treated when dragged into drawings from Web sites.
SetUV	Controls how raster images are mapped onto objects.
'SetVar	Views and changes AutoCAD's system variables.
ShadeMode	Shades 3D objects in the current viewport.
Shape	Places shapes from shape files into drawings.
Shell	Runs other programs outside of AutoCAD.
ShowMat	Reports the material definition assigned to selected objects.
SigValidate *(new in 2004)*	Displays digital signature information in drawings.
Sketch	Allows freehand sketching.
Slice	Slices 3D solids with a planes (a solid modeling command).
'Snap	F9 and CTRL+B: Toggles snap mode on or off, changes the snap resolution, sets spacing for the X- and Y-axis, rotates the grid, and sets isometric mode.
SolDraw	Creates 2D profiles and sections of 3D solid models in viewports created with the SolView command (a solid modeling command).
Solid	Draws filled triangles and rectangles.
SolidEdit	Edits faces and edges of 3D solid objects (a solid modeling command).
SolProf	Creates profile images of 3D solid models (a solid modeling command).
SolView	Creates viewports in paper space of orthogonal multi- and sectional view drawings of 3D solid model (a solid modeling command).
SpaceTrans	Converts length values between model and paper space.
Spell	Checks the spelling of text in the drawing.
Sphere	Draws 3D spheres (a solid modeling command).
Spline	Draws NURBS (spline) curve (a solid modeling command).
SplinEdit	Edits splines.
Standards	Compares CAD standards between two drawings.
Stats	Lists information about the current state of rendering.
'Status	Displays information about the current drawing.
StlOut	Exports 3D solids to a *.stl* files, in ASCII or binary format, for use with stereolithography (a solid modeling command).

Command	Description
Stretch	Moves selected objects while keeping connections to other objects unchanged.
'Style	Creates and modifies text styles.
StylesManager	Displays the Plot Style Manager.
Subtract	Creates new 3D solids and 2D regions by subtracting one object from a second object (a solid modeling command).
SysWindows	CTRL+TAB: Controls the size and position of windows.

T

Command	Description
Tablet	F4: Aligns digitizers with existing drawing coordinates; operates only when a digitizer is connected.
TabSurf	Draws tabulated surfaces.
Text	Places text in the drawing.
'TextScr	F2: Displays the Text window.
TifOut *(new in 2004)*	Exports views as TIFF files (tagged image file format).
'Time	Keeps track of time.
Tolerance	Selects tolerance symbols.
-Toolbar	Controls the display of toolboxes.
ToolPalettes *(new in 2004)*	Opens the Tool Palettes window.
ToolPalettesClose *(new in 2004)*	Closes the Tool Palettes window.
Torus	Draws doughnut-shaped 3D solids (a solid modeling command).
Trace	Draws lines with width.
Transparency	Toggles the background of a bilevel image transparent or opaque.
TraySettings *(new in 2004)*	Specifies options for commands operating from the tray (right end of status bar).
'TreeStat	Displays information on the spatial index.
Trim	Trims objects by defining other objects as cutting edges.

U

Command	Description
U	CTRL+Z: Undoes the effect of commands.
UCS	Creates and manipulates user-defined coordinate systems.
UCSicon	Controls the display of the UCS icon.
UcsMan	Manages user-defined coordinate systems.
Undefine	Disables commands.
Undo	Undoes several commands in a single operation.
Union	Creates new 3D solids and 2D regions from two solids or regions (a solid modeling command).
'Units	Selects the display format and precision of units and angles.

V

Command	Description
VbaIDE	ALT+F8: Launches the Visual Basic Editor.
VbaLoad	Loads VBA projects into AutoCAD.
VbaMan	Loads, unloads, saves, creates, embeds, and extracts VBA projects.
VbaRun	ALT+F11: Runs VBA macros.
VbaStmt	Executes VBA statements at the command prompt.
VbaUnload	Unloads global VBA projects.
View	Saves the display as a view; displays named views.
ViewRes	Controls the fast zoom mode and resolution for circle and arc regenerations.

Command	Description
VLisp	Launches the Visual LISP interactive development environment.
VpClip	Clips viewport objects.
VpLayer	Controls the independent visibility of layers in viewports.
VPoint	Sets the viewpoint from which to view 3D drawings.
VPorts	CTRL+R: Sets the number and configuration of viewports.
VSlide	Displays *.sld* slide files created with MSlide.

W

Command	Description
WBlock	Writes objects to a drawing files.
Wedge	Creates 3D solid wedges (a solid modeling command).
WhoHas	Displays ownership information for opened drawing files.
WmfIn	Imports *.wmf* files into drawings as blocks.
WmfOpts	Controls how *.wmf* files are imported.
WmfOut	Exports drawings as *.wmf* files.

X

Command	Description
XAttach	Attaches externally-referenced drawing files to the drawing.
XBind	Binds externally referenced drawings; converts them to blocks.
XClip	Defines clipping boundaries; sets the front and back clipping planes.
XLine	Draws infinite construction lines.
XOpen *(New in 2004)*	Opens externally-referenced drawings in independent windows.
Xplode	Breaks compound objects into component objects, with user control.
Xref	Places externally-referenced drawings into drawings.

Z

Command	Description
'Zoom	Increases and decreases the viewing size of drawings.

3

Command	Description
3D	Draws 3D surface objects of polygon meshes (boxes, cones, dishes, domes, meshes, pyramids, spheres, tori, and wedges).
3dArray	Creates 3D arrays.
3dClip	Switches to interactive 3D view, and opens the Adjust Clipping Planes window.
3dConfig	Configures the 3D graphics system from the command line.
3dCOrbit	Switches to interactive 3D view, and set objects into continuous motion.
3dDistance	Switches to interactive 3D view, and makes objects appear closer or farther away.
3dFace	Creates 3D faces.
3dMesh	Draws 3D meshes.
3dOrbit	Controls the interactive 3D viewing.
3dOrbitCtr *(New in 2004)*	Centers the view.
3dPan	Invokes interactive 3D view to drag the view horizontally and vertically.
3dPoly	Draws 3D polylines.
3dsIn	Imports 3D Studio geometry and rendering data.
3dsOut	Exports AutoCAD geometry and rendering data in *.3ds* format.
3dSwivel	Switches to interactive 3D view, simulating the effect of turning the camera.
3dZoom	Switches to interactive 3D view to zoom in and out.

COMMAND ALIASES

Aliases are shortened versions of command names. They are defined in the *acad.ppg* file.

Command	Aliases
A	
AdCenter	Dc, Dcenter, Adc, Content
Align	Al
AppLoad	Ap
Arc	A
Area	Aa
Array	Ar
-Array	-Ar
AttDef	Att, Ddattdef
-AttDef	-Att
AttEdit	Ate, Ddatte
-AttEdit	-Ate, Atte
AttExt	Ddattext
B	
BHatch	H, Bh
Block	B, Bmake, Bmod, Acadblockdialog
-Block	-B
Boundary	Bo, Bpoly
-Boundary	-Bo
Break	Br
C	
Chamfer	Cha
Change	-Ch
CheckStandards	Chk
Circle	C
Color	Col, Colour, Ddcolor
Copy	Cp, Co
D	
Dbconnect	Ase, Aad, Dbc, Aex, Asq, Ali, Aro
Ddedit	Ed
Ddgrips	Gr
Ddvpoint	Vp
DIMENSIONS	
DimAligned	Dal, Dimali
DimAngular	Dan, Dimang
DimBaseline	Dba, Dimbase
DimCenter	Dce
DimContinue	Dco, Dimcont
DimDiameter	Ddi, Dimdia
DimDisassociate	Dda
DimEdit	Dimed, Ded

Command	Aliases
DimLinear	Dli, Dimlin
DimOrdinate	Dor, Dimord
DimOverride	Dov, Dimover
DimRadius	Dra, Dimrad
DimReassociate	Dre
DimStyle	D, Dimsty, Dst, Ddim
DimTEdit	Dimted
Dist	Di
Divide	Div
Donut	Doughnut
Draworder	Dr
DSettings	Se, Ds, Ddrmodes
DsViewer	Av
DView	Dv
E	
Ellipse	El
Erase	E
Explode	X
Export	Exp
Extend	Ex
Extrude	Ext
F	
Fillet	F
Filter	Fi
G	
Group	G
-Group	-G
H	
Hatch	-H
Hatchedit	He
Hide	Hi
I	
Image	Im
-Image	-Im
ImageAdjust	Iad
ImageAttach	Iat
ImageClip	Icl
Import	Imp
Insert	I, Inserturl, Ddinsert
-Insert	-I

Command	Aliases
Insertobj	Io
Interfere	Inf
Intersect	In

L

Command	Aliases
Layer	Ddlmodes, La
-Layer	-La
-Layout	Lo
Leader	Lead
Lengthen	Len
Line	L
Linetype	Ltype, Lt, Ddltype
-Linetype	-Ltype, -Lt
List	Ls, Li
LtScale	Lts
Lweight	Lineweight, Lw

M

Command	Aliases
MatchProp	Ma, Painter
Measure	Me
Mirror	Mi
MLine	Ml
Move	M
MSpace	Ms
MText	T, Mt
-MText	-T
MView	Mv

O

Command	Aliases
Offset	O
Open	Openurl
Options	Op, Preferences
OSnap	Os, Ddosnap
-OSnap	-Os

P

Command	Aliases
Pan	P
-Pan	-P
-PartialOpen	Partialopen
PasteSpec	Pa
PEdit	Pe
PLine	Pl
Plot	Print, Dwfout
PlotStamp	Ddplotstamp
Point	Po
Polygon	Pol
Preview	Pre
Properties	Props, Pr, Mo, Ch, Ddchprop, Ddmodify
PropertiesClose	Prclose
PSpace	Ps

Command	Aliases
PublishToWeb	Ptw
Purge	Pu
-Purge	-Pu

Q

Command	Aliases
QLeader	Le
Quit	Exit

R

Command	Aliases
Rectang	Rec, Rectangle
Redraw	R
RedrawAll	Ra
Regen	Re
RegenAll	Rea
Region	Reg
Rename	Ren
-Rename	-Ren
Render	Rr
Revolve	Rev
Rotate	Ro
RPref	Rpr

S

Command	Aliases
Save	Saveurl
Scale	Sc
Script	Scr
Section	Sec
SetVar	Set
ShadeMode	Sha, Shade
Slice	Sl
Snap	Sn
Solid	So
Spell	Sp
Spline	Spl
SplinEdit	Spe
Standards	Sta
Stretch	S
Style	St, Ddstyle
Subtract	Su

T

Command	Aliases
Tablet	Ta
Text	Dt, Dtext
Thickness	Th
Tilemode	Ti, Tm
Tolerance	Tol
Toolbar	To
ToolPalettes	Tp
Torus	Tor
Trim	Tr

Command	Aliases		Command	Aliases
U			**X**	
UcsMan	Uc, Dducs, Dducsp		XAttach	Xa
Union	Uni		XBind	Xb
Units	Un, Ddunits		-XBind	-Xb
-Units	-Un		XClip	Xc
V			XLine	Xl
View	V, Ddview		XRef	Xr
-View	-V		-XRef	-Xr
VPoint	-Vp		**Z**	
VPorts	Viewports		Zoom	Z
W			**3**	
Wblock	W, Acadwblockdialog		3dArray	3a
-Wblock	-W		3dFace	3f
Wedge	We		3dOrbit	3do, Orbit
			3dPoly	3p

KEYBOARD SHORTCUTS

CONTROL KEYS

Function keys can be customized with the **Tools | Customize | Keyboard** command.

Ctrl-key	Meaning
CTRL+0 *(New in 2004)*	Toggles clean screen.
CTRL+1	Displays the Properties window.
CTRL+2	Opens the AutoCAD DesignCenter window.
CTRL+6	Launches dbConnect.
CTRL+A	Selects all objects in the drawing.
CTRL+SHIFT+A	Toggles group selection mode.
CTRL+B	Turns snap mode on or off.
CTRL+C	Copies selected objects to the Clipboard.
CTRL+SH+C *(New in 2004)*	Copies selected objects with a base point to the Clipboard.
CTRL+D	Changes the coordinate display mode.
CTRL+E	Switches to the next isoplane.
CTRL+F	Toggles object snap on and off.
CTRL+G	Turns the grid on and off.
CTRL+H	Toggles pickstyle mode).
CTRL+K	Creates a hyperlink.
CTRL+L	Turns ortho mode on and off.
CTRL+N	Starts a new drawing.
CTRL+O	Opens a drawing.
CTRL+P	Prints the drawing.
CTRL+Q *(New in 2004)*	Quits AutoCAD.
CTRL+R	Switches to the next viewport.
CTRL+S	Saves the drawing.
CTRL+SH+S *(New in 2004)*	Displays the Save Drawing As dialog box.
CTRL+T	Toggles tablet mode.
CTRL+V	Pastes from the Clipboard into the drawing or to the command prompt area.
CTRL+SH+V *(New in 2004)*	Pastes with an insertion point.
CTRL+X	Cuts selected objects to the Clipboard.
CTRL+Y	Performs the REDO command.
CTRL+Z	Performs the U command.
CTRL+TAB	Switches to the next drawing.

COMMAND LINE KEYSTROKES

These keystrokes are used in the command-prompt area and the Text window.

Keystroke	Meaning
left arrow	Moves the cursor one character to the left.
right arrow	Moves the cursor to the right.
HOME	Moves the cursor to the beginning of the line of command text.
END	Moves the cursor to the end of the line.
DEL	Deletes the character to the right of the cursor.
BACKSPACE	Deletes the character to the left of the cursor.
INS	Switches between insert and typeover modes.
up arrow	Displays the previous line in the command history.
down arrow	Displays the next line in the command history.
PGUP	Displays the previous screen of command text.
PGDN	Displays the next screen of command text.
CTRL+V	Pastes text from the Clipboard into the command line.
ESC	Cancels the current command.

FUNCTION KEYS

Function keys can be customized with the **Tools | Customize | Keyboard** command.

Function Key	Meaning
F1	Calls up the help window.
F2	Toggles between the graphics and text windows.
F3	Toggles object snap on and off.
F4	Toggles tablet mode on and off; you must first calibrate the tablet before toggling tablet mode.
ALT+F4	Exits AutoCAD.
CTRL+F4	Closes the current drawing.
F5	Switches to the next isometric plane when in iso mode; the planes are displayed in order of left, top, right, and then repeated.
F6	Toggles the screen coordinate display on and off.
F7	Toggles grid display on and off.
F8	Toggles ortho mode on and off.
ALT+F8	Displays the Macros dialog box.
F9	Toggles snap mode on and off.
F10	Toggles polar tracking on and off.
F11	Toggles object snap tracking on and off.
ALT+F11	Starts Visual Basic for Applications editor.

ALTERNATE KEYS

The menu-related Alt keys can be customized in the *acad.mnu* file.

Alt-key	Meaning
ALT+TAB	Switches to the next application.
ALT+-	(dash) Accesses window control menu.
ALT+D	Accesses the Draw menu.
ALT+E	Accesses the Edit menu.
ALT+F	Accesses the File menu.
ALT+H	Accesses the Help menu.
ALT+I	Accesses the Insert menu.
ALT+M	Accesses the Modify menu.
ALT+N	Accesses the Dimension menu.
ALT+O	Accesses the Format menu.
ALT+T	Accesses the Tools menu.
ALT+V	Accesses the View menu.
ALT+W	Accesses the Window menu.
ALT+X	Accesses the Express menu.

MOUSE AND DIGITIZER BUTTONS

The mouse and digitizer buttons can be customized in the *acad.mnu* file.

Button #	Mouse Button	Meaning
...	Wheel	Zooms or pans.
1	Left	Selects objects.
3	Center	Displays object snap menu.
2	Right	Displays shortcut menus.
4		Cancels command.
5		Toggles snap mode.
6		Toggles orthographic mode.
7		Toggles grid display.
8		Toggles coordinate display.
9		Switches isometric plane.
10		Toggles tablet mode.
11		Not defined.
12		Not defined.
13		Not defined.
14		Not defined.
15		Not defined.
SHIFT+1	SHIFT+Left	Toggles cycle mode.
SHIFT+2	SHIFT+Left	Displays object snap shortcut menu.
CTRL+2	CTRL+Right	Displays object snap shortcut men

MTEXT EDITOR SHORTCUT KEYS

Shortcut	Meaning
TAB	Moves cursor to the next tab stop.
CTRL+A	Selects all text.
CTRL+B	**Boldface** toggle.
CTRL+I	*Italicize* toggle.
CTRL+U	Underline toggle.

Case Conversion:

CTRL+SHIFT+A	Toggles between all UPPERCASE and all lowercase.
CTRL+SHIFT+U	Converts selected text to all UPPERCASE.
CTRL+SHIFT+L	Converts selected text to all lowercase.

Justification:

CTRL+L	Left-justifies selected text.
CTRL+E	Centers selected text.
CTRL+R	Right-justifies selected text.

Line Spacing:

CTRL+1	Single-spaces lines.
CTRL+5	1.5-spaces lines.
CTRL+2	Double-spaces lines.

Font Sizing:

CTRL+SHIFT+,	(*comma*) Reduces font size.
CTRL+SHIFT+.	(*period*) Increases font size.
CTRL+=	Superscript toggle.
CTRL+SHIFT+=	Subscript toggle.

Copy and Paste:

CTRL+C	Copies selected text to the Clipboard.
CTRL+X	Cuts text from the editor and send it to the Clipboard.
CTRL+V	Pastes text from the Clipboard.

Undo and Redo:

CTRL+Y	Redoes last undo.
CTRL+Z	Undoes last action.

Symbols:

CTRL+ALT+E	Inserts Euro symbol.
ALT+SHIFT+X	Converts selected text to ASCII number (A becomes 41, and so on).

MTEXT CONTROL CODES

Control Code	Meaning
\~	Nonbreaking space.
\\	Backslash.
\{	Opening brac; encloses multiple controls codes.
\}	Closing brace.
\A$n$	Set vertical alignment of text in boundary box:
	0 = bottom alignment.
	1 = center alignment.
	2 = top alignment.
\C$n$	Sets color of text:
	1 = red.
	2 = yellow.
	3 = green.
	4 = cyan.
	5 = blue.
	6 = magenta.
	7 = white.
\F$x$;	Changes to font file name x.
\H$n$;	Changes text height to n units.
\L	Underline.
\l	Turns off underline.
\M+nnn	Multibyte shape number nnn.
\O	Turns on overline.
\o	Turns off overline.
\P	End of paragraph.
\Q$n$;	Changes obliquing angle to n.
\S$n$^m	Stacks character n over m.
\T$n$;	Changes tracking to n; for example, \T3; is wide spacing bewteen characters.
\U+nnn	Places Unicode character nnn.
\W$n$;	Changes width factor (width of characters) to n.

APPENDIX C

AutoCAD System Variables

AutoCAD stores information about its current state, the drawing and the operating system in over 400 *system variables*. Those variables help programmers — who often work with menu macros and AutoLISP — to determine the state of the AutoCAD system.

CONVENTIONS

The following pages list all documented system variables, plus several more not documented by Autodesk. The listing uses the following conventions:

Bold   System variable is documented in AutoCAD 2004.

Italicized   System variable is not listed by the **SETVAR** command or Autodesk documentation.

⌨   System variable must be accessed via the **SETVAR** command.

2004   System variable is new to AutoCAD 2004.

COLUMN HEADINGS

Default   Default value, as set in the *acad.dwg* prototype drawing.

R/O   Read-only; cannot be changed by the user or by a program.

Loc   Location where the value of the system variable is saved:

Location	Meaning
ACAD	Set by AutoCAD.
DWG	Saved in current drawing.
REG	Saved globally in Windows registry.
...	Not saved.

Variable	Default	R/O	Loc	Meaning
_PkSer	varies	R/O	ACAD	Software package serial number, such as "117-69999999".
_Server	0	R/O	REG	Network authorization code.
_VerNum	varies	R/O	REG	Internal program build number, such as "T.0.98".

A

Variable	Default	R/O	Loc	Meaning
AcadLspAsDoc	0	...	REG	*acad.lsp* is loaded into:
				0 Just the first drawing.
				1 Every drawing.
AcadPrefix	varies	R/O	...	Path spec'd by ACAD environment variable, such as *d:\acad 2004\support; d:\acad 2004\fonts*.
AcadVer	"16.0"	R/O	...	AutoCAD version number.
AcisOutVer	70	...	...	ACIS version number, such as 15, 16, 17, 18, 20, 21, 30, 40, or 70.
AdcState	0	R/O	...	Specifies if **DesignCenter** is active.
AFlags	0	...	...	Attribute display code:
				0 No mode specified.
				1 Invisible.
				2 Constant.
				4 Verify.
				8 Preset.
AngBase	0	...	DWG	Direction of zero degrees relative to UCS
AngDir	0	...	DWG	Rotation of angles:
				0 Clockwise
				1 Counterclockwise
ApBox	0	...	REG	AutoSnap aperture box cursor:
				0 Off.
				1 On.
Aperture	10	...	REG	Object snap aperture in pixels:
				1 Minimum size.
				50 Maximum size.
Area	0.0000	R/O	...	Area measured by the last **Area**, **List**, or **Dblist** commands.
AttDia	0	...	DWG	Attribute entry interface:
				0 Command-line prompts.
				1 Dialog box.
AttMode	1	...	DWG	Display of attributes:
				0 Off.
				1 Normal.
				2 On.
AttReq	1	...	REG	Attribute values during insertion are:
				0 Default values.
				1 Prompt for values.

Variable	Default	R/O	Loc	Meaning
AuditCtl	0	...	REG	Determines creation of *.adt* audit log file:
				0 File not created.
				1 *.adt* file created.
AUnits	0	...	DWG	Mode of angular units:
				0 Decimal degrees.
				1 Degrees-minutes-seconds.
				2 Grads.
				3 Radians.
				4 Surveyor's units.
AUPrec	0	...	DWG	Decimal places displayed by angles.
AutoSnap	63	...	REG	Controls AutoSnap display; sum of:
				0 Turns off all AutoSnap features.
				1 Turns on marker.
				2 Turns on SnapTip.
				4 Turns on magnetic cursor.
				8 Turns on polar tracking.
				16 Turns on object snap tracking.
				32 Turns on tooltips for polar tracking and object snap tracking.
AuxStat	0	...	DWG	*-32768* *Minimum value.*
				32767 *Maximum value.*
AxisUnit	*0.0000*	...	DWG	*Obsolete system variable.*
B				
BackZ	0.0000	R/O	DWG	Back clipping plane offset.
BindType	0	...	...	When binding xrefs or editings xref, converts xref names from:
				0 **xref|name** to **xref0name**.
				1 **xref|name** to **name**.
BlipMode	0	...	REG	Display of blip marks:
				0 Off.
				1 On.
C				
CDate	varies	R/O	...	Current date and time in the format YyyyMmDd.HhMmSsDd, such as 20010503.18082328
CeColor	"BYLAYER"	...	DWG	Current color.
CeLtScale	1.0000	...	DWG	Current linetype scaling factor.
CeLType	"BYLAYER"	...	DWG	Current linetype.
CeLWeight	-1	...	DWG	Current lineweight in millimeters; valid values are 0, 5, 9, 13, 15, 18, 20, 25, 30, 35, 40, 50, 53, 60, 70, 80, 90, 100, 106, 120, 140, 158, 200, and 211, plus the following:
				-1 BYLAYER.
				-2 BYBLOCK.
				-3 DEFAULT as defined by **LwDdefault**.

Variable	Default	R/O	Loc	Meaning
ChamferA	0.5000	...	DWG	First chamfer distance.
ChamferB	0.5000	...	DWG	Second chamfer distance.
ChamferC	1.0000	...	DWG	Chamfer length.
ChamferD	0	...	DWG	Chamfer angle.
ChamMode	0	...	...	Chamfer input mode:
				0 Chamfer by two lengths.
				1 Chamfer by length and angle.
CircleRad	0.0000	...	...	Most-recent circle radius.
CLayer	"0"	...	DWG	Current layer name.
CmdActive	1	R/O	...	Type of current command; sum of:
				1 Regular command active.
				2 Transparent command active.
				4 Script file active.
				8 Dialog box active.
				16 Dynamic data exchange active.
				32 AutoLISP active.
				64 ObjectARX command active.
CmdDia	1	...	REG	Formerly determined whether the **PLOT** command displayed at the command line prompt or via a dialog box; no longer has an effect as of AutoCAD 2000; replaced by **PLQUIET**.
CmdEcho	1	...	...	AutoLISP command display:
				0 No command echoing.
				1 Command echoing.
CmdNames	varies	R/O	...	Current command, such as "SETVAR".
CMLJust	0	...	DWG	Multiline justification mode:
				0 Top.
				1 Middle.
				2 Bottom.
CMLScale	1.0000	...	DWG	Scales width of multiline:
				-n Flips offsets of multiline.
				0 Collapses to single line.
				1 Default.
				n Scales by a factor of *n*.
CMLStyle	"STANDARD"	...	DWG	Current multiline style name.
Compass	0	...	...	Toggles display of the 3D compass:
				0 Off.
				1 On.
Coords	1	...	DWG	Coordinate display style:
				0 Updated by screen picks.
				1 Continuous display.
				2 Polar display upon request.

Variable	Default	R/O	Loc	Meaning
CPlotStyle	"ByColor"	...	DWG	Current plot style; values defined by AutoCAD are "ByLayer", "ByBlock", "Normal", and "User Defined".
CProfile	"<<Unnamed Profile>>"	R/O	REG	Current user profile.
CTab	"Model"	R/O	DWG	Current layout tab.
CursorSize	5	...	REG	Cursor size, in percent of viewport:
				1 Minimum size.
				100 Full viewport.
CVPort	2	...	DWG	Current viewport number:
				2 Minimum (default).

D

Variable	Default	R/O	Loc	Meaning
Date	*varies*	R/O	...	Current date in Julian format, such as 2448860.54043252
DbcState (NEW IN 2004)	0	R/O	DWG	Specifies if dbConnect Manager is active.
DBMod	4	R/O	...	Drawing modified; sum of:
				0 No modification since last save.
				1 Object database modified.
				2 Symbol table modified.
				4 Database variable modified.
				8 Window modified.
				16 View modified.
DctCust	"d:\acad 2004\support\sample.cus"	...	REG	Name of custom spelling dictionary.
DctMain	"enu"	...	REG	Code for spelling dictionary:
				ca Catalan.
				cs Czech.
				da Danish.
				de German; sharp 's'.
				ded German; double 's'.
				ena English; Australian.
				ens English; British 'ise'.
				enu English; American.
				enz English; British 'ize'.
				es Spanish; unaccented capitals.
				esa Spanish; accented capitals.
				fi Finish.
				fr French; unaccented capitals.
				fra French; accented capitals.
				it Italian.
				nl Dutch; primary.
				nls Dutch; secondary.
				no Norwegian; Bokmal.
				non Norwegian; Nynorsk.

Variable	Default	R/O	Loc	Meaning
				pt Portuguese; Iberian.
				ptb Portuguese; Brazilian.
				ru Russian; infrequent 'io'.
				rui Russian; frequent 'io'.
				sv Swedish.
DefLPlStyle	"ByColor"	R/O	REG	Default plot style for layer.
DefPlStyle	"ByColor"	R/O	REG	Default plot style for new objects.
DelObj	1	...	REG	Toggle source objects deletion:
				0 Objects deleted.
				1 Objects retained.
DemandLoad	3	...	REG	When drawing contains proxy objects:
				0 Demand loading turned off.
				1 Load application when drawing opened.
				2 Load application at first command.
				3 Load app when drawing opened or at first command.
DiaStat	1	R/O	...	User exited dialog box by clicking:
				0 **Cancel** button.
				1 **OK** button.

Dimension Variables

Variable	Default	R/O	Loc	Meaning
DimADec	0	...	DWG	Angular dimension precision:
				-1 Use **DimDec** setting (default).
				0 Zero decimal places (minimum).
				8 Eight decimal places (maximum).
DimAlt	Off	...	DWG	Alternate units:
				On Enabled.
				Off Disabled.
DimAltD	2	...	DWG	Alternate unit decimal places.
DimAltF	25.4000	...	DWG	Alternate unit scale factor.
DimAltRnd	0.0000	...	DWG	Rounding factor of alternate units.
DimAltTD	2	...	DWG	Tolerance alternate unit decimal places.
DimAltTZ	0	...	DWG	Alternate tolerance units zeros:
				0 Zeros not suppressed.
				1 All zeros suppressed.
				2 Includes 0 feet, but suppress 0 inches.
				3 Includes 0 inches, but suppress 0 feet.
				4 Suppresses leading zeros.
				8 Suppresses trailing zeros.

Variable	Default	R/O	Loc	Meaning
DimAltU	2	...	DWG	Alternate units:
				1 Scientific.
				2 Decimal.
				3 Engineering.
				4 Architectural; stacked.
				5 Fractional; stacked.
				6 Architectural.
				7 Fractional.
				8 Windows desktop units setting.
DimAltZ	0		DWG	Zero suppression for alternate units:
				0 Suppress 0 ft and 0 in.
				1 Includes 0 ft and 0 in.
				2 Includes 0 ft; suppress 0 in.
				3 Suppress 0 ft; include 0 in.
				4 Suppress leading 0 in dec dims.
				8 Suppress trailing 0 in dec dims.
				12 Suppress leading and trailing zeroes.
DimAPost	""	...	DWG	Prefix and suffix for alternate text.
DimAso	On	...	DWG	Toggle associative dimensions:
				On Dimensions are associative.
				Off Dimensions are not associative.
DimAssoc	2	...	DWG	Controls creation of dimensions:
				0 Dimension elements are exploded.
				1 Single dimension object, attached to defpoints.
				2 Single dimension object, attached to geometric objects.
DimASz	0.1800	...	DWG	Arrowhead length.
DimAtFit	3	...	DWG	When insufficient space between extension lines, dimension text and arrows are fitted:
				0 Text and arrows outside extension lines.
				1 Arrows first outside, then text.
				2 Text first outside, then arrows.
				3 Either text or arrows, whichever fits better.
DimAUnit	0	...	DWG	Angular dimension format:
				0 Decimal degrees.
				1 Degrees.Minutes.Seconds.
				2 Grad.
				3 Radian.
				4 Surveyor units.

Variable	Default	R/O	Loc	Meaning
DimAZin	0	...	DWG	Supress zeros in angular dimensions:

    0 Display all leading and trailing zeros.
    1 Suppress 0 in front of decimal.
    2 Suppress trailing zeros behind decimal.
    3 Suppress zeros in front and behind the decimal.

Variable	Default	R/O	Loc	Meaning
DimBlk	""	R/O	DWG	Arrowhead block name:

Architectural tick:	"_Archtick"
Box filled:	"_Boxfilled"
Box:	"_Boxblank"
Closed blank:	"_Closedblank"
Closed filled:	"" (default)
Closed:	"_Closed"
Datum triangle filled:	"_Datumfilled"
Datum triangle:	"_Datumblank"
Dot blanked:	"_Dotblank"
Dot small:	"_Dotsmall"
Dot:	"_Dot"
Integral:	"_Integral"
None:	"_None"
Oblique:	"_Oblique"
Open 30:	"_Open30"
Open:	"_Open"
Origin indication:	"_Origin"
Right-angle:	"_Open90"

- Closed filled
- Closed blank
- Closed
- Dot
- Architectural tick
- Oblique
- Open
- Origin indicator
- Origin indicator 2
- Right angle
- Open 30
- Dot small
- Dot blank
- Dot small blank
- Box
- Box filled
- Datum triangle
- Datum triangle filled
- Integral
- None

Variable	Default	R/O	Loc	Meaning
DimBlk1	""	R/O	DWG	Name of first arrowhead's block; uses same list of names as under **DIMBLK**.

    . Displays default arrowhead.

Variable	Default	R/O	Loc	Meaning
DimBlk2	""	R/O	DWG	Name of second arrowhead's block.
DimCen	0.0900	...	DWG	Center mark size:

    -n Draws center lines.
    0 No center mark or lines drawn.
    +n Draws center marks of length *n*.

Variable	Default	R/O	Loc	Meaning
DimClrD	0	...	DWG	Dimension line color:

    0 BYBLOCK (default)
    1 Red.
    ...
    255 Dark gray.
    256 BYLAYER.

Variable	Default	R/O	Loc	Meaning
DimClrE	0	...	DWG	Extension line and leader color.
DimClrT	0	...	DWG	Dimension text color.
DimDec	4	...	DWG	Primary tolerance decimal places.

Variable	Default	R/O	Loc	Meaning
DimDLE	0.0000	...	DWG	Dimension line extension.
DimDLI	0.3800	...	DWG	Dimension line continuation increment.
DimDSep	"."	...	DWG	Decimal separator (must be a single character).
DimExe	0.1800	...	DWG	Extension above dimension line.
DimExO	0.0625	...	DWG	Extension line origin offset.
~~DimFit~~	3	...	DWG	*Obsolete: Autodesk recommends use of **DimATfit** and **DimTMove** instead.*
DimFrac	0	...	DWG	Fraction format when **DimLUnit** is set to 4 or 5: **0** Horizontal. **1** Diagonal. **2** Not stacked.
DimGap	0.0900	...	DWG	Gap from dimension line to text.
DimJust	0	...	DWG	Horizontal text positioning: **0** Center justify. **1** Next to first extension line. **2** Next to second extension line. **3** Above first extension line. **4** Above second extension line.
DimLdrBlk	""	...	DWG	Block name for leader arrowhead; uses same name as DIMBLOCK. **.** Displays default arrowhead.
DimLFac	1.0000	...	DWG	Linear unit scale factor.
DimLim	Off	...	DWG	Generate dimension limits.
DimLUnit	2	...	DWG	Dimension units (except angular); replaces DIMUNIT: **1** Scientific. **2** Decimal. **3** Engineering. **4** Architectural. **5** Fractional. **6** Windows desktop.
DimLwD	-2	...	DWG	Dimension line lineweight; valid values are BYLAYER, BYBLOCK, or an integer multiple of 0.01mm.
DimLwE	-2	...	DWG	Extension lineweight; valid values are BYLAYER, BYBLOCK, or an integer multiple of 0.01mm.
DimPost	""	...	DWG	Default prefix or suffix for dimension text (maximum 13 characters): **""** No suffix. **<>mm** Millimeter suffix. **<>Å** Angstrom suffix.
DimRnd	0.0000	...	DWG	Rounding value for dimension distances.
DimSAh	Off	...	DWG	Separate arrowhead blocks: **Off** Use arrowhead defined by DIMBLOCK. **On** Use arrowheads defined by DIMBLOCK1 and DIMBLOCK2.

Variable	Default	R/O	Loc	Meaning
DimScale	1.0000	...	DWG	Overall scale factor for dimensions:
				0 Value is computed from the scale between current modelspace viewport and paperspace.
				>0 Scales text and arrowheads.
DimSD1	Off	...	DWG	Suppress first dimension line:
				On First dimension line is suppressed.
				Off Not suppressed.
DimSD2	Off	...	DWG	Suppress second dimension line:
				On Second dimension line is suppressed.
				Off Not suppressed.
DimSE1	Off	...	DWG	Suppress the first extension line:
				On First extension line is suppressed.
				Off Not suppressed.
DimSE2	Off	...	DWG	Suppress the second extension line:
				On Second extension line is suppressed.
				Off Not suppressed.
DimSho	On	...	DWG	Update dimensions while dragging:
				On Dimensions are updated during drag.
				Off Dimensions are updated after drag.
DimSOXD	Off	...	DWG	Suppress dimension lines outside extension lines:
				On Dimension lines not drawn outside extension lines.
				Off Are drawn outside extension lines.
DimStyle	"STANDARD"	R/O	DWG	Current dimension style.
DimTAD	0	...	DWG	Vertical position of dimension text:
				0 Centered between extension lines.
				1 Above dimension line, except when dimension line not horizontal and DIMTIH = 1.
				2 On side of dimension line farthest from the defining points.
				3 Conforms to JIS.
DimTDec	4	...	DWG	Primary tolerance decimal places.
DimTFac	1.0000	...	DWG	Tolerance text height scaling factor.
DimTIH	On	...	DWG	Text inside extensions is horizontal:
				Off Text aligned with dimension line.
				On Text is horizontal.
DimTIX	Off	...	DWG	Place text inside extensions:
				Off Text placed inside extension lines, if room.
				On Force text between the extension lines.
DimTM	0.0000	...	DWG	Minus tolerance.
DimTMove	0	...	DWG	Determines how dimension text is moved:
				0 Dimension line moves with text.
				1 Adds a leader when text is moved.
				2 Text moves anywhere; no leader.

Variable	Default	R/O	Loc	Meaning
DimTOFL	Off	...	DWG	Force line inside extension lines:
				Off Dimension lines not drawn when arrowheads are outside.
				On Dimension lines drawn, even when arrowheads are outside.
DimTOH	On	...	DWG	Text outside extension lines:
				Off Text aligned with dimension line.
				On Text is horizontal.
DimTol	Off	...	DWG	Generate dimension tolerances:
				Off Tolerances not drawn.
				On Tolerances are drawn.
DimTolJ	1	...	DWG	Tolerance vertical justification:
				0 Bottom.
				1 Middle.
				2 Top.
DimTP	0.0000	...	DWG	Plus tolerance.
DimTSz	0.0000	...	DWG	Size of oblique tick strokes:
				0 Arrowheads.
				>0 Oblique strokes.
DimTVP	0.0000	...	DWG	Text vertical position when **DIMTAD**=0:
				1 Turns on **DIMTAD**.
				>-0.7 or **<0.7** Dimension line is split for text.
DimTxSty	"STANDARD"	...	DWG	Dimension text style.
DimTxt	0.1800	...	DWG	Text height.
DimTZin	0	...	DWG	Tolerance zero suppression:
				0 Suppress 0 ft and 0 in.
				1 Include 0 ft and 0 in.
				2 Include 0 ft; suppress 0 in.
				3 Suppress 0 ft; include 0 in.
				4 Suppress leading 0 in decimal dim.
				8 Suppress trailing 0 in decimal dim.
				12 Suppress leading and trailing zeroes.
~~DimUnit~~	2	...	DWG	*Obsolete; replaced by* **DIMLUNIT** *and* **DIMFRAC**.
DimUPT	Off	...	DWG	User-positioned text:
				Off Cursor positions dimension line
				On Cursor also positions text
DimZIN	0	...	DWG	Suppression of 0 in feet-inches units:
				0 Suppress 0 ft and 0 in.
				1 Include 0 ft and 0 in.
				2 Include 0 ft; suppress 0 in.
				3 Suppress 0 ft; include 0 in.
				4 Suppress leading 0 in decimal dim.
				8 Suppress trailing 0 in decimal dim.
				12 Suppress leading and trailing zeroes.

Variable	Default	R/O	Loc	Meaning
DispSilh	0	...	DWG	Silhouette display of 3D solids:
				0 Off.
				1 On.
Distance	0.0000	R/O	...	Distance measured by last DIST command.
DonutId	0.5000	...	...	Inside radius of donut.
DonutOd	1.0000	...	...	Outside radius of donut.
DragMode	2	...	REG	Drag mode:
				0 No drag.
				1 On if requested.
				2 Automatic.
DragP1	10	...	REG	Regen drag display.
DragP2	25	...	REG	Fast drag display.
DwgCheck	0	...	REG	Toggles checking for problems in drawing and if it was edited by software other than AutoCAD:
				0 Supresses dialog box.
				1 Displays warning dialog box.
				2 Potential problems notified at the command line after opening the drawing.
				3 Potential problems and editing by non-AutoCAD notified at the command line after opening the drawing.
DwgCodePage	varies	R/O	DWG	Drawing code page, such as "ANSI_1252"
DwgName	varies	R/O	...	Current drawing filename, such as "*drawing1.dwg*".
DwgPrefix	varies	R/O	...	Drawing's drive and folder, such as "*d:\acad 2004*".
DwgTitled	0	R/O	...	Drawing has file name:
				0 "*drawing1.dwg*".
				1 User-assigned name.
E				
EdgeMode	0	...	REG	Toggle edge mode for TRIM and EXTEND commands:
				0 No extension.
				1 Extends cutting edge.
Elevation	0.0000	...	DWG	Current elevation, relative to current UCS.
EntExts	*1*	...	...	*Controls how drawing extents are calculated:*
				0 Extents calculated every time; slows down AutoCAD, but uses less memory.
				1 Extents of every object is cached as a two-byte value (default).
				2 Extents of every object is cached as a four-byte value (fastest, but uses more memory).
EntMods	*0*	R/O	...	*Increments by one each time an object is modified to indicate that an object has been modified since the drawing was opened; value ranges from 0 to 4.29497E9.*
ErrNo	0	...	...	Error number from AutoLISP, ADS, and ObjectArx applications.

Appendix C: AutoCAD System Variables **819**

Variable	Default	R/O	Loc	Meaning
Expert	0	...	...	Suppresses the displays of prompts:

    0 Normal prompts

    1 "About to regen, proceed?" and "Really want to turn the current layer off?"

    2 "Block already defined. Redefine it?" and "A drawing with this name already exists. Overwrite it?"

    3 LINETYPE command messages.

    4 UCS **Save** and VPORTS **Save**.

    5 DIMSTYLE **Save** and DIMOVERRIDE.

Variable	Default	R/O	Loc	Meaning
ExplMode	1	...	...	Toggle whether EXPLODE and XPLODE commands explode non-uniformly scaled blocks:

    0 Does not explode.

    1 Explodes.

Variable	Default	R/O	Loc	Meaning
ExtMax	-1.0E+20, -1.0E+20, -1.0E+20	R/O	DWG	Upper-right coordinate of drawing extents.
ExtMin	1.0E+20, 1.0E+20, 1.0E+20	R/O	DWG	Lower-left coordinate of drawing extents.
ExtNames	1	...	DWG	Format of named objects:

    0 Names are limited to 31 characters, and can include A - Z, 0 - 9, dollar ($), underscore (_), and hyphen (-).

    1 Names are limited to 255 characters, and can include A - Z, 0 - 9, spaces, and any characters not used by Windows or AutoCAD for special purposes.

F

Variable	Default	R/O	Loc	Meaning
FaceTRatio	0	...	...	Controls the aspect ratio of facets on rounded 3D bodies:

    0 Creates an *n* by 1 mesh.

    1 Creates an *n* by *m* mesh.

Variable	Default	R/O	Loc	Meaning
FaceTRres	0.0	...	DWG	Adjusts smoothness of shaded and hidden-line objects:

    0.01 Minimum value.

    10.0 Maximum value.

Variable	Default	R/O	Loc	Meaning
FileDia	1	...	REG	User interface:

    0 Command-line prompts.

    1 Dialog boxes, when available.

Variable	Default	R/O	Loc	Meaning
FilletRad	0.5000	...	DWG	Current fillet radius.
FillMode	1	...	DWG	Fill of solid objects:

    0 Off.

    1 On.

Variable	Default	R/O	Loc	Meaning
Flatland	*0*	R/O	...	*Obsolete system variable.*
FontAlt	"*simplex.shx*"	...	REG	Name for substituted font.
FontMap	"*acad.fmp*"	...	REG	Name of font mapping file.
Force_Paging	*0*	...	...	**0** *Minimum (default).*

    4.29497E9 *Maximum.*

Variable	Default	R/O	Loc	Meaning
FrontZ	0.0000	R/O	DWG	Front clipping plane offset.
FullOpen	1	R/O	...	Drawing is:
				0 Partially loaded.
				1 Fully open.

G

Variable	Default	R/O	Loc	Meaning
GfAng	0	...	...	Angle of gradient fill; 0 to 360 degrees.
GfClr1	"RGB 000,000,255"	...	...	First gradient color in RGB format.
GfClr2	"RGB 255,255,153"	...	...	Second gradient color in RGB format.
GfClrLum	1.0	...	...	Level of gradient in one-color gradients:
				0 Shade (mixed with black).
				1 Tint (mixed with white).
GfClrState	1	...	...	Specifies type of gradient fill:
				0 Two-color.
				1 One-color.
GfName	1	...	...	Specifies style of gradient fill:
				1 Linear.
				2 Cylindrical.
				3 Inverted cylindrical.
				4 Spherical.
				5 Inverted spherical.
				6 Hemispherical.
				7 Inverted hemispherical.
				8 Curved.
				9 Inverted curved.
GfAShift	0	...	...	Specifies the origin of the gradient fill:
				0 Centered.
				1 Shifted up and left.
GlobCheck	0	...	...	Reports statistics on dialog boxes:
				-1 Turn off local language.
				0 Turned off.
				1 Warns if larger than 640x400.
				2 Also reports size in pixels.
				3 Additional info.
GridMode	0	...	DWG	Display of grid:
				0 Off.
				1 On.
GridUnit	0.5000,0.5000	...	DWG	X,y-spacing of grid.
GripBlock	0	...	REG	Display of grips in blocks:
				0 At insertion point.
				1 At all objects within block.

Variable	Default	R/O	Loc	Meaning
GripColor	160	...	REG	Color of unselected grips:
				1 Minimum color number; red.
				160 Default color; blue.
				255 Maximum color number.
GripHot	1	...	REG	Color of selected grips:
				1 Default color, red.
				255 Maximum color number.
GripHover (NEW IN 2004)	3	...	REG	Grip fill color when cursor hovers.
GripObjLimit (NEW IN 2004)	100	...	REG	Grips not displayed when more than this number:
				1 Minimum
				32767 Maximum.
Grips	1	...	REG	Display of grips:
				0 Off.
				1 On.
GripSize	3	...	REG	Size of grip box, in pixels:
				1 Minimum size.
				255 Maximum size.
GripTips (NEW IN 2004)	1	...	REG	Determines if grip tips are displayed when the cursor hovers over custom objects:
				0 Off.
				1 On

H

Variable	Default	R/O	Loc	Meaning
HaloGap	0	...	DWG	Distance to shorten a haloed line; specified as the percentage of 1.0 units.
Handles	1	R/O	...	Obsolete system variable.
HidePrecision	0	...	...	Controls the precision of hide calculations:
				0 Single precision, less accurate, faster.
				1 Double precision, more accurate, but slower.
HideText	0	...	...	Determines whether text is hidden during the HIDE command:
				0 Text is not hidden nor hides other objects, unless text object has thickness.
				1 Text is hidden and hides other objects.
Highlight	1	...	...	Object selection highlighting:
				0 Disabled.
				1 Enabled.
HPAng	0	...	...	Current hatch pattern angle.
HPAssoc	0	...	REG	Determines if hatches are associative:
				0 Not associative.
				1 Associative.

Variable	Default	R/O	Loc	Meaning
HPBound	1	...	REG	Object created by **BHATCH** and **BOUNDARY** commands:
				0 Region.
				1 Polyline.
HPDouble	0	...	...	Double hatching:
				0 Disabled.
				1 Enabled.
HPName	"ANSI31"	...	...	Current hatch pattern name
				"" No default.
				. Set no default.
HPScale	1.0000	...	...	Current hatch scale factor; cannot be zero.
HPSpace	1.0000	...	...	Current spacing of user-defined hatching; cannot be zero.
HyperlinkBase	""	...	DWG	Path for relative hyperlinks.

I

Variable	Default	R/O	Loc	Meaning
ImageHlt	0	...	REG	When a raster image is selected:
				0 Image frame is highlighted.
				1 Entire image is highlighted.
IndexCtl	0	...	DWG	Creates layer and spatial indices:
				0 No indices created.
				1 Layer index created.
				2 Spatial index created.
				3 Both indices created.
InetLocation	"www.autodesk.com"	...	REG	Default browser URL.
InsBase	0.0000,0.0000,0.0000	...	DWG	Insertion base point relative to the current UCS for **INSERT**.
InsName	""	...	...	Current block name:
				. Set to no default.
InsUnits	1	...	...	Drawing units of block dragged into drawing from DesignCenter:
				0 Unitless.
				1 Inches.
				2 Feet.
				3 Miles.
				4 Millimeters.
				5 Centimeters.
				6 Meters.
				7 Kilometers.
				8 Microinches.
				9 Mils.
				10 Yards.
				11 Angstroms.
				12 Nanometers.
				13 Microns.

Variable	Default	R/O	Loc	Meaning
				14 Decimeters.
				15 Decameters.
				16 Hectometers.
				17 Gigameters.
				18 Astronomical Units.
				19 Light Years.
				20 Parsecs.
InsUnitsDefSource	1	...	REG	Source drawing units value; ranges from 0 to 20; see above.
InsUnitsDefTarget	1	...	REG	Target drawing units value; ranges from 0 to 20.
IntersectionDisplay *(New in 2004)*	0	...	DWG	Determines whether polylines are drawn by the HIDE command at intersections of 3D surfaces:
				0 Off.
				1 On.
ISaveBak	1	...	REG	Controls whether .*bak* file is created:
				0 No file created.
				1 .*bak* backup file created.
ISavePercent	50	...	REG	Percentage of waste in .*dwg* file before cleanup occurs:
				0 Every save is a full save.
IsoLines	4	...	DWG	Isolines on 3D solids:
				0 No isolines; minimum.
				16 Good-looking.
				2,047 Maximum.

L

Variable	Default	R/O	Loc	Meaning
LastAngle	0	R/O	...	Ending angle of last-drawn arc.
LastPoint	*varies*	...	...	Last-entered point, such as 15,9,56.
LastPrompt	""	R/O	...	Last string on the command line; includes user input.
LazyLoad	*0*	...	...	*Toggle: 0 or 1.*
LayoutRegenCtl	2	...	REG	Controls display list for layouts:
				0 Display list is regenerated with each tab change.
				1 Display list is saved for model tab and last layout tab.
				2 Display list is saved for all tabs.
LensLength	50.0000	R/O	DWG	Perspective view lens length, in mm.
LimCheck	0	...	DWG	Drawing limits checking:
				0 Disabled.
				1 Enabled.
LimMax	12.0000,9.0000	...	DWG	Upper right drawing limits.
LimMin	0.0000,0.0000	...	DWG	Lower left drawing limits.

Variable	Default	R/O	Loc	Meaning
LispInit	1	...	REG	AutoLISP functions and variables are:
				0 Preserved from drawing to drawing.
				1 Valid in current drawing only.
Locale	"enu"	R/O	...	ISO language code.
LocalRootPrefix (NEW IN 2004)	"\documents and Settings\administrator\local settings\appli..."	R/O	REG	Path to folder holding local customizable files.
LogFileMode	0	...	REG	Text window written to *.log* file:
				0 No.
				1 Yes.
LogFileName	"Drawing1_1_1_0000.log"	R/O	DWG	Filename and path for *.log* file.
LogFilePath	"d:\autocad 2004\"	...	REG	Path for the *.log* file.
LogInName	""	R/O	...	User's login name; max = 30 chars.
LTScale	1.0000	...	DWG	Current linetype scale factor; cannot be 0.
LUnits	2	...	DWG	Linear units mode:
				1 Scientific.
				2 Decimal.
				3 Engineering.
				4 Architectural.
				5 Fractional.
LUPrec	4	...	DWG	Decimal places of linear units.
LwDefault	25	...	REG	Default lineweight, in millimeters; must be one of the following values: 0, 5, 9, 13, 15, 18, 20, 25, 30, 35, 40, 50, 53, 60, 70, 80, 90, 100, 106, 120, 140, 158, 200, or 211.
LwDisplay	0	...	DWG	Toggles whether lineweight is displayed; setting is saved separately for Model space and each layout tab.
				0 Not displayed.
				1 Displayed.
LwUnits	1	...	REG	Determines units for lineweight:
				0 Inches.
				1 Millimeters.

M

Variable	Default	R/O	Loc	Meaning
MacroTrace	*0*	...	...	*Diesel debug mode:*
				0 Off.
				1 On.
MaxActVP	64	...	...	Maximum viewports to regenerate:
				2 Minimum.
				64 Maximum (increased from 48 in R14).
MaxObjMem	*0*	...	...	*Maximum number of objects in memory; object pager is turned off when value = 0, <0, or 2,147,483,647.*
MaxSort	1000	...	REG	Maximum names sorted alphabetically.

Variable	Default	R/O	Loc	Meaning
MButtonPan	1	...	REG	Determines behavior of wheelmouse:
				0 As defined by AutoCAD menu file.
				1 Pans when dragging with wheel.
MeasureInit	0	...	REG	Drawing units:
				0 English.
				1 Metric.
Measurement	0	...	DWG	Drawing units (overrides **MEASUREINIT**):
				0 English.
				1 Metric.
MenuCtl	1	...	REG	Submenu display:
				0 Only with menu picks.
				1 Also with keyboard entry.
MenuEcho	0	...		Menu and prompt echoing; sum of:
				0 Display all prompts.
				1 Suppress menu echoing.
				2 Suppress system prompts.
				4 Disable **^P** toggle.
				8 Display all input-output strings.
MenuName	"acad"	R/O	REG	Current menu filename.
MirrText	1	...	DWG	Text handling during **Mirror** command:
				0 Retain text orientation.
				1 Mirror text.
ModeMacro	""	...	...	Invoke Diesel programming language.
MTextEd	"Internal"	...	REG	Name of the mtext editor:
				. Use default editor.
				0 Cancel the editing operation.
				-1 Use the secondary editor.
				"blank" Mtext internal editor.
				"Internal" Mtext internal editor.
				"Notepad" Windows Notepad editor.
				":lisped" Built-in AutoLISP function.
				string Name of editor; must be less than 256 characters long and use this syntax:
				:AutoLISPtextEditorFunction#TextEditor
MTextFixed (NEW IN 2004)	0	...	REG	Specifies mtext editor appearance:
				0 Displays mtext editor and text at same size and position as object being edited.
				1 Displays mtext editor at the same location as last used; uses fixed height text.
MTJigStrings (NEW IN 2004)	"abc"	...	REG	Sample text displayed by mtext editor; maximum 10 letters; enter . for no text.

Variable	Default	R/O	Loc	Meaning

MyDocumentsPrefix *(New in 2004)*

"C:\Documents and Settings\administrator\My Documents"
    R/O   REG   Path to the *my documents* folder of the currently logged-in user.

N

NodeName   "AC$"   R/O   REG   Name of network node; range is one to three characters.

NoMutt   0   ...   ...   Suppresses the display of message (a.k.a. muttering) during scripts, LISP, macros:

    0 Display prompt, as normal.
    1 Suppress muttering.

O

ObscureColor   0   ...   DWG   Color of objects obscured by HIDE command:

    0 Invisible
    1 - 255 Color number.

ObscureLtype   0   ...   DWG   Linetype of objects obscured by HIDE command:

    0 Invisible.
    1 Solid.
    2 Dashed.
    3 Dotted.
    4 Short dash.
    5 Medium dash.
    6 Long dash.
    7 Double short dash.
    8 Double medium dash.
    9 Double long dash.
    10 Medium long dash.
    11 Sparse dot.

OffsetDist   1.0000   ...   ...   Current offset distance:

    <0 Offsets through a specified point.
    >0 Default offset distance.

OffsetGapType   0   ...   REG   Determines how to reconnect polyline when individual segments are offset:

    0 Extend segments to fill gap.
    1 Fill gap with fillet (arc segment).
    2 Fill gap with chamfer (line segment).

OleHide   0   ...   REG   Display and plotting of OLE objects:

    0 All OLE objects visible.
    1 Visible in paper space only.
    2 Visible in model space only.
    3 Not visible.

Variable	Default	R/O	Loc	Meaning
OleQuality	1	...	REG	Quality of display and plotting of embedded OLE objects:
				0 Line art quality.
				1 Text quality.
				2 Graphics quality.
				3 Photograph quality.
				4 High quality photograph.
OleStartup	0	...	DWG	Loads OLE source application to improves plot quality:
				0 Do not load OLE source application.
				1 Load OLE source app when plotting.
OrthoMode	0	...	DWG	Orthographic mode:
				0 Off.
				1 On.
OSMode	4133	...	REG	Current object snap mode; sum of:
				0 NONe.
				1 ENDpoint.
				2 MIDpoint.
				4 CENter.
				8 NODe.
				16 QUAdrant.
				32 INTersection.
				64 INSertion.
				128 PERpendicular.
				256 TANgent.
				512 NEArest.
				1024 QUIck.
				2048 APPint.
				4096 EXTension.
				8192 PARallel.
				16383 All modes on.
				16384 Object snap turned off via **OSNAP** on the status bar.
OSnapCoord	2	...	REG	Keyboard overrides object snap:
				0 Object snap override keyboard.
				1 Keyboard overrides object snap.
				2 Keyboard overrides object snap, except in script.

P

PaletteOpaque (New in 2004)

	0	...	REG	Determines if palettes can be made transparent:
				0 Turned off by user.
				1 Turned on by user.
				2 Unavailable, but turned on by user.
				3 Unavailable, and turned off by user.

Variable	Default	R/O	Loc	Meaning
PaperUpdate	0	...	REG	Determines how AutoCAD plots a layout with paper size different from plotter's default:
				0 Displays a warning dialog box.
				1 Changes paper size to that of the plotter configuration file.
PDMode	0	...	DWG	Point display mode; sum of:
				0 Dot.
				1 No display.
				2 +-symbol.
				3 x-symbol.
				4 Short line.
				32 Circle.
				64 Square.
PDSize	0.0000	...	DWG	Point display size, in pixels:
				>0 Absolute size.
				0 5% of drawing area height.
				<0 Percentage of viewport size.
PEdit Accept *(New in 2004)*	0	...	REG	Suppresses display of the PEDIT command's "Object selected is not a polyline. Do you want to turn it into one? <Y>" prompt.
PEllipse	0	...	DWG	Toggle ELLIPSE creation:
				0 True ellipse.
				1 Polyline arcs.
Perimeter	0.0000	R/O	...	Perimeter calculated by the last AREA, DBLIST, and LIST commands.
PFaceVMax	4	R/O	...	Maximum vertices per 3D face.
PHandle	0	...	ACAD	*Ranges from 0 to 4.29497E9.*
PickAdd	1	...	REG	Effect of SHIFT key on selection set:
				0 Adds to selection set.
				1 Removes from selection set.
PickAuto	1	...	REG	Selection set mode:
				0 Single pick mode.
				1 Automatic windowing and crossing.
PickBox	3	...	REG	Object selection pickbox size, in pixels:
				0 Minimum size.
				50 Maximum size.
PickDrag	0	...	REG	Selection window mode:
				0 Pick two corners.
				1 Pick a corner; drag to second corner.
PickFirst	1	...	REG	Command-selection mode:
				0 Enter command first.
				1 Select objects first.

Variable	Default	R/O	Loc	Meaning
PickStyle	1		REG	Include groups and associative hatches in selection:
				0 Neither included.
				1 Include groups.
				2 Include associative hatches.
				3 Include both.
Platform	"Microsoft Windows NT Version 5.0 (x86)"	R/O	...	AutoCAD platform (name of the operating system).
PLineGen	0	...	DWG	Polyline linetype generation:
				0 From vertex to vertex.
				1 From end to end.
PLineType	2	...	REG	Automatic conversion and creation of 2D polylines by **PLINE**:
				0 Not converted; old-format polylines created.
				1 Not converted; optimized polylines created.
				2 Polylines in older drawings are converted on open; **PLINE** creates optimized polylines with lwpolyline object.
PLineWid	0.0000	...	DWG	Current polyline width.
PlotRotMode	1	...	DWG	Orientation of plots:
				0 Lower left = 0,0.
				1 Lower left plotter area = lower left of media.
				2 X, y-origin offsets calculated relative to the rotated origin position.
PlQuiet	0	...	REG	Toggles display during batch plotting and scripts (replaces **CMDDIA**):
				0 Plot dialog boxes and nonfatal errors are displayed.
				1 Nonfatal errors are logged; plot dialog boxes are not displayed.
PolarAddAng	""	...	REG	Contains a list of up to 10 user-defined polar angles; each angle can be up to 25 characters long, each separated with a semicolon (;). For example: 0;15;22.5;45.
PolarAng	90	...	REG	Specifies the increment of polar angle; contrary to Autodesk documentation, you may specify any angle.
PolarDist	0.000	...	REG	The polar snap increment when **SNAPSTYL** is set to 1 (isometric).
PolarMode	0	...	REG	Settings for polar and object snap tracking; sum of:
				0 Measure polar angles based on current UCS (absolute), track orthogonally; don't use additional polar tracking angles; and acquire object tracking points automatically.
				1 Measure polar angles from selected objects (relative).
				2 Use polar tracking settings in object snap tracking.
				4 Use additional polar tracking angles (via **POLARANG**).
				8 Press **SHIFT** to acquire object snap tracking points.
PolySides	4	...	...	Current number of polygon sides:
				3 Minimum sides.
				1024 Maximum sides.
Popups	1	R/O	...	Display driver support of AUI:
				0 Not available.
				1 Available.

Variable	Default	R/O	Loc	Meaning
Product	"AutoCAD"	R/O	ACAD	Name of the software.
Program	"acad"	R/O	ACAD	Name of the software's executable file.
ProjectName	""	...	DWG	Project name of current drawing; searches for xref and image files.
ProjMode	1	...	REG	Projection mode for **TRIM** and **EXTEND** commands:
				0 No projection.
				1 Project to x,y-plane of current UCS.
				2 Project to view plane.
ProxyGraphics	1	...	DWG	Proxy image saved in the drawing:
				0 Not saved; displays bounding box.
				1 Image saved with drawing.
ProxyNotice	1	...	REG	Display warning message:
				0 No.
				1 Yes.
ProxyShow	1	...	REG	Display of proxy objects:
				0 Not displayed.
				1 All displayed.
				2 Bounding box displayed.
ProxyWebSearch	0	...	REG	Object enablers are checked:
				0 AutoCAD does not check for object enablers.
				1 AutoCAD checks for object enablers if an Internet connection is present.
PsLtScale	1	...	DWG	Paper space linetype scaling:
				0 Use model space scale factor.
				1 Use viewport scale factor.
PsProlog	""	...	REG	*PostScript prologue filename.*
PsQuality	75	...	REG	*Resolution of PostScript display, in pixels:*
				<0 *Display as outlines; no fill.*
				0 *Not displayed.*
				>0 *Display filled.*
PStyleMode	1	...	REG	Toggles the plot color matching mode of the drawing:
				0 Use named plot style tables.
				1 Use color-dependent plot style tables.
PStylePolicy	1	...	REG	Determines whether object color is associated with its plot style:
				0 Color and plot style not associated.
				1 Object's plot style is associated with its color.
PsVpScale	0	...	...	Sets the view scale factor (ratio of units in paper space to units in newly created model space viewports) for newly-created viewports:
				0 Scaled to fit.
PUcsBase	""	...	DWG	Name of UCS defining the origin and orientation of orthographic UCS settings in paper space only.

Variable	Default	R/O	Loc	Meaning

Q

QAFlags 0 ... ... Quality assurance flags:

    0 Turned off.

    1 The ^C metacharacters in a menu macro cancels grips, just as if user pressed ESC.

    2 Long text screen listings do not pause.

    4 Error and warning messages are displayed at the command line, instead of in dialog boxes.

    128 Screen picks are accepted via the AutoLISP (command) function.

QTextMode 0 ... DWG Quick text mode:

    0 Off.

    1 On.

R

RasterPreview 1 R/O REG Preview image:

    0 None saved.

    1 Saved in *.bmp* format.

RefEditName "" ... ... The reference filename when it is in reference-editing mode.

RegenMode 1 ... DWG Regeneration mode:

    0 Regen with each view change.

    1 Regen only when required.

Re-Init 0 ... ... Reinitialize I/O devices; sum of:

    1 Digitizer port.

    2 Plotter port.

    4 Digitizer.

    8 Plotter.

    16 Reload PGP file.

RememberFolders 1 ... REG Controls path search method:

    0 Path specified in desktop AutoCAD icon is default for file dialog boxes.

    1 Last path specified by each file dialog box is remembered.

ReportError (NEW IN 2004) 1 ... REG Determines if AutoCAD sends an error report to Autodesk:

    0 No error report created.

    1 Error report is generated and can be sent to Autodesk.

RoamableRootPrefix (NEW IN 2004) "d:\documents and settings\administrator\application aata\aut..." R/O REG Path to root folder where roamable customized files are located.

RTDisplay 1 ... REG Raster display during real-time zoom and pan:

    0 Display the entire raster image.

    1 Display raster outline only.

Variable	Default	R/O	Loc	Meaning
S				
SaveFile	"*varies*.sv$"	R/O	REG	Automatic save filename.
SaveFilePath	"temp\"	...	REG	Path for automatic save files.
SaveName	""	R/O	...	Drawing save-as filename.
SaveTime	10	...	REG	Automatic save interval, in minutes:
				0 Disable auto save.
ScreenBoxes	0	R/O	ACAD	Maximum number of screen menu items:
				0 Screen menu turned off.
ScreenMode	3	R/O	...	State of AutoCAD display screen; sum of:
				0 Text screen.
				1 Graphics screen.
				2 Dual-screen display.
ScreenSize	*varies*	R/O	...	Current viewport size, in pixels, such as 719.0000,381.0000.
SDI	0	...	REG	Toggles multiple-document interface (SDI is "single document interface"):
				0 Turns on MDI.
				1 Turns off MDI (only one drawing may be loaded in AutoCAD).
				2 MDI disabled for apps that cannot support MDI; read-only.
				3 MDI disabled for apps that cannot support MDI, even when SDI set to 1; R/O.
ShadEdge	3	...	DWG	**SHADE** style:
				0 Shade faces; 256-color shading.
				1 Shade faces; edges background color.
				2 Hidden-line removal.
				3 16-color shading.
ShadeDif	70	...	DWG	Percent of diffuse to ambient light:
				0 Minimum.
				100 Maximum.
ShortcutMenu	11	...	REG	Toggles availability of shortcut menus; sum of:
				0 Disables all default, edit, and command shortcut menus.
				1 Enables default shortcut menus.
				2 Enables edit shortcut menus.
				4 Enables command shortcut menus whenever command is active.
				8 Enables command shortcut menus only when command options are available at the command line.
ShpName	""	...	...	Current shape name:
				. Set to no default.
SigWarn (NEW IN 2004)	1	...	REG	Determines whether a warning is displayed when a file is opened with a digital signature.
SketchInc	0.1000	...	DWG	**SKETCH** command's recording increment.

Variable	Default	R/O	Loc	Meaning
SkPoly	0	...	DWG	Sketch line mode:
				0 Record as lines.
				1 Record as a polyline.
SnapAng	0	...	DWG	Current rotation angle for snap and grid.
SnapBase	0.0000,0.0000	...	DWG	Current origin for snap and grid.
SnapIsoPair	0	...	DWG	Current isometric drawing plane:
				0 Left isoplane.
				1 Top isoplane.
				2 Right isoplane.
SnapMode	0	...	DWG	Snap mode:
				0 Off.
				1 On.
SnapStyl	0	...	DWG	Snap style:
				0 Normal.
				1 Isometric.
SnapType	0	...	REG	Toggles between standard or polar snap for the current viewport:
				0 Standard snap.
				1 Polar snap.
SnapUnit	0.5000,0.5000	...	DWG	X,y-spacing for snap.
SolidCheck	1	...	...	Toggles solid validation:
				0 Off.
				1 On.
SortEnts	127	...	DWG	Object display sort order; sum of:
				0 Off.
				1 Object selection.
				2 Object snap.
				4 Off.
				8 Slide generation.
				16 Regeneration.
				32 Plot.
				64 Off
SplFrame	0	...	DWG	Polyline and mesh display:
				0 Polyline control frame not displayed; display polygon fit mesh; 3D faces invisible edges not displayed
				1 Polyline control frame displayed; display polygon defining mesh; 3D faces invisible edges displayed.
SplineSegs	8	...	DWG	Number of line segments that define a splined polyline.
SplineType	6	...	DWG	Spline curve type:
				5 Quadratic Bezier spline.
				6 Cubic Bezier spline.

Variable	Default	R/O	Loc	Meaning
StandardsViolation (NEW IN 2004)	2	...	REG	Displayed when CAD standards are violated:
				0 No alerts displayed.
				1 Alerts displayed when CAD standard violated.
				2 Displays icon on status bar when file is opened with CAD standards, and when non-standard objects are created.
Startup (NEW IN 2004)	0	...	reg	Determines which dialog box is displayed by the **NEW** and **QNEW** commands:
				0 Displays Select Template dialog box.
				1 Displays Startup and Create New Drawing dialog box.
SurfTab1	6	...	DWG	Density of surfaces and meshes:
				5 Minimum.
				32766 Maximum.
SurfTab2	6	...	DWG	Density of surfaces and meshes:
				2 Minimum.
				32766 Maximum.
SurfType	6	...	DWG	Pedit surface smoothing:
				5 Quadratic Bezier spline.
				6 Cubic Bezier spline.
				8 Bezier surface.
SurfU	6	...	DWG	Surface density in m-direction:
				2 Minimum.
				200 Maximum.
SurfV	6	...	DWG	Surface density in n-direction:
				2 Minimum.
				200 Maximum.
SysCodePage	"ANSI_1252"	R/O	...	System code page.

T

Variable	Default	R/O	Loc	Meaning
TabMode	0	...	...	Tablet mode:
				0 Off.
				1 On.
Target	0.0000,0.0000,0.0000	R/O	DWG	Target in current viewport.
TDCreate	varies	R/O	DWG	Time and date drawing created, such as 2448860.54014699.
TDInDwg	varies	R/O	DWG	Duration drawing loaded, such as 0.00040625.
TDuCreate	varies	R/O	DWG	The universal time and date the drawing was created, such as 2451318.67772165.
TDUpdate	varies	R/O	DWG	Time and date of last update, such as 2448860.54014699.
TDUsrTimer	varies	R/O	DWG	Time elapsed by user-timer, such as 0.00040694.
TDuUpdate	varies	R/O	DWG	The universal time and date of the last save, such as 2451318.67772165.
TempPrefix	"\temp"	R/O	...	Path for temporary files.

Variable	Default	R/O	Loc	Meaning
TextEval	0	...	...	Interpretation of text input:
				0 Literal text.
				1 Read (and ! as AutoLISP code.
TextFill	1	...	REG	Toggle fill of TrueType fonts during plotting:
				0 Outline text.
				1 Filled text.
TextQlty	50	...	DWG	Resolution of TrueType fonts during plotting:
				0 Minimum resolution.
				100 Maximum resolution.
TextSize	0.2000	...	DWG	Current height of text.
TextStyle	"Standard"	...	DWG	Current name of text style.
Thickness	0.0000	...	DWG	Current object thickness.
TileMode	1	...	DWG	View mode:
				0 Display layout tab.
				1 Display model tab.
ToolTips	1	...	REG	Display tooltips:
				0 Off.
				1 On.
TpState	0	R/O	...	Determines if Tool Palettes window is open:
				0 Closed.
				1 Open.
TraceWid	0.0500	...	DWG	Current width of traces.
TrackPath	0	...	REG	Determines display of polar and object snap tracking alignment paths:
				0 Displays object snap tracking path across the entire viewport.
				1 Displays object snap tracking path between the alignment point and "From point" to cursor location.
				2 Turns off polar tracking path.
				3 Turns off polar and object snap tracking paths.
TrayIcons	1	...	REG	Determines if the tray is displayed on the status bar.
TrayNotify	1	...	REG	Determines whether service notifications are displayed by the tray.
TrayTimeout	5	...	REG	Specifies the length of time (in seconds) that tray notificaitons are displayed:
				0 Minimum
				5 Default
				10 Maximum
TreeDepth	3020	...	DWG	Maximum branch depth in *xxyy* format:
				xx Model-space nodes.
				yy Paper-space nodes.
				>0 3D drawing.
				<0 2D drawing.

Variable	Default	R/O	Loc	Meaning
TreeMax	10000000	...	REG	Limits memory consumption during drawing regeneration.
TrimMode	1	...	REG	Trim toggle for CHAMFER and FILLET commands:
				0 Leave selected edges in place.
				1 Trim selected edges.
TSpaceFac	1.0000	...	...	Mtext line spacing distance; measured as a factor of text height; valid values range from 0.25 to 4.0.
TSpaceType	1	...	...	Type of mtext line spacing:
				1 At Least: adjust line spacing based on the height of the tallest character in a line of mtext.
				2 Exactly: use the specified line spacing; ignores character height.
TStackAlign	1	...	DWG	Vertical alignment of stacked text.
				0 Bottom aligned.
				1 Center aligned.
				2 Top aligned.
TStackSize	70	...	DWG	Sizes stacked text as a percentage of the selected text height:
				1 Minimum %.
				127 Maximum %.

U

Variable	Default	R/O	Loc	Meaning
UcsAxisAng	90	...	REG	Default angle for rotating the UCS around an axes (via the **UCS** command using the **X**, **Y**, or **Z** options; valid values are limited to: 5, 10, 15, 18, 22.5, 30, 45, 90, or 180.
UcsBase	""	...	DWG	Name of the UCS that defines the origin and orientation of orthographic UCS settings.
UcsFollow	0	...	DWG	New UCS views:
				0 No change.
				1 Automatic display of plan view.
UcsIcon	3	...	DWG	Display of UCS icon:
				0 Off.
				1 On.
				2 Display at UCS origin, if possible.
				3 On, and displayed at origin.
UcsName	""	R/O	DWG	Name of current UCS view:
				"" Current UCS is unnamed.
UcsOrg	0.0000,0.0000,0.0000	R/O	DWG	Origin of current UCS relative to WCS.
UcsOrtho	1	...	REG	Determines whether the related orthographic UCS setting is restored automatically:
				0 UCS setting remains unchanged when orthographic view is restored.
				1 Related orthographic UCS setting is restored automatically when an orthographic view is restored.
UcsView	1	...	REG	Determines whether the current UCS is saved with a named view:
				0 Not saved.
				1 Saved.

Appendix C: AutoCAD System Variables

Variable	Default	R/O	Loc	Meaning
UcsVp	1	...	DWG	Determines whether the UCS in active viewports remains fixed (locked) or changes (unlocked) to match the UCS of the current viewport: **0** Unlocked. **1** Locked.
UcsXDir	1.0000,0.0000,0.0000	R/O	DWG	X-direction of current UCS relative to WCS.
UcsYDir	0.0000,1.0000,0.0000	R/O	DWG	Y-direction of current UCS relative to WCS.
UndoCtl	5	R/O	...	State of undo; sum of: **0** Undo disabled. **1** Undo enabled. **2** Undo limited to one command. **4** Auto-group mode. **8** Group currently active.
UndoMarks	0	R/O	...	Current number of undo marks.
UnitMode	0	...	DWG	Units display: **0** As set by UNITS command. **1** As entered by user.
UserI1 thru **UserI5**	0	...	DWG	Five user-definable integer variables.
UserR1 thru **UserR5**	0.0000	...	DWG	Five user-definable real variables.
UserS1 thru **UserS5**	""	...	...	Five user-definable string variables.

V

Variable	Default	R/O	Loc	Meaning
ViewCtr	varies	R/O	DWG	X,y,z-coordinate of center of current view, such as 6.2433,4.5000,0.0000.
ViewDir	varies	R/O	DWG	Current view direction relative to UCS, such as 0.0000,0.0000,1.0000.
ViewMode	0	R/O	DWG	Current view mode; sum of: **0** Normal view. **1** Perspective mode on. **2** Front clipping on. **4** Back clipping on. **8** UCS-follow on. **16** Front clip not at eye.
ViewSize	9.0000	R/O	DWG	Height of current view.
ViewTwist	0	R/O	DWG	Twist angle of current view.
VisRetain	1	...	DWG	Determines xref drawing's layer settings — on-off, freeze-thaw, color, and linetype: **0** Xref layer settings in the current drawing takes precedence for xref-dependent layers. **1** Settings for xref-dependent layers take precedence over the xref layer definition in the current drawing.

Variable	Default	R/O	Loc	Meaning
VSMax	*varies*	R/O	DWG	Upper-right corner of virtual screen, such as 37.4600,27.0000,0.0000.
VSMin	*varies*	R/O	DWG	Lower-left corner of virtual screen, such as -24.9734,-18.0000,0.0000.

W

Variable	Default	R/O	Loc	Meaning
WhipArc	0	...	REG	Display of circlular objects:
				0 Display as connected vectors.
				1 Display as true circles and arcs.
WhipThread	3	...	REG	Controls multithreaded processing on two CPUs (if present) during drawing redraw and regeneration:
				0 Single-threaded calculations.
				1 Regenerations multi-threaded.
				2 Redraws multi-threaded.
				3 Regens and redraws multi-threaded.
WmfBkgnd	0	...	...	Controls background of *.wmf* files:
				0 Background is transparent.
				1 Background is same as AutoCAD's background color.
WmfForegnd	0	...	...	Controls foreground colors of exported *.wmf* images:
				0 Foreground is darker than background.
				1 Foreground is lighter than background.
WorldUcs	1	R/O	...	Matching of WCS with UCS:
				0 Current UCS is not WCS.
				1 UCS is WCS.
WorldView	1	...	DWG	Display during **3DORBIT**, **DVIEW**, and **VPOINT** commands:
				0 Display UCS.
				1 Display WCS.
				2 UCS changes relative to the UCS specified by the UCSBASE system variable.
WriteStat	1	R/O	...	Indicates whether drawing file is read-only:
				0 Drawing cannot be written to.
				1 Drawing can be written to.

X

Variable	Default	R/O	Loc	Meaning
XClipFrame	0	...	DWG	Visibility of xref clipping boundary:
				0 Not visible.
				1 Visible.
XEdit	0	...	DWG	Toggles whether drawing can be edited in-place when referenced by another drawing:
				0 Cannot in-place refedit.
				1 Can in-place refedit.
XFadeCtl	50	...	REG	Fades objects not being edited in-place:
				0 No fading; minimum value.
				90 90% fading; maximum value.

Variable	Default	R/O	Loc	Meaning
XLoadCtl	1	...	REG	Controls demand loading:
				0 Demand loading turned off; entire drawing is loaded.
				1 Demand loading turned on; xref file opened.
				2 Demand loading turned on; a *copy* of the xref file is opened.
XLoadPath	"d:\temp"	...	REG	Path for loading xref file.
XRefCtl	0	...	REG	Determines creation of *.xlg* xref log files:
				0 File not written.
				1 File written.
XrefNotify *(New in 2004)*	2	...	REG	Determines if alert is displayed:
				0 No alert displayed.
				1 Icon indicates xrefs are attached; yellow alert indicates missing xrefs.
				2 Also displays balloon messages when an xref is modified.

Z

Variable	Default	R/O	Loc	Meaning
ZoomFactor	60	...	REG	Controls the zoom level via mouse wheel; valid values range between 3 and 100.

APPENDIX D

AutoCAD Toolbars and Menus

AUTOCAD 2004 TOOLBARS

The indicates the toolbar is new or changed in AutoCAD 2004.

CAD STANDARDS TOOLBAR

DIMENSION TOOLBAR

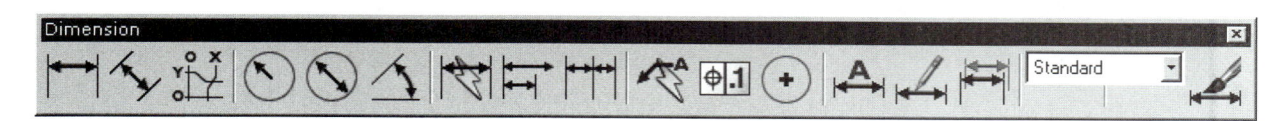

DRAW TOOLBAR

DRAWORDER TOOLBAR

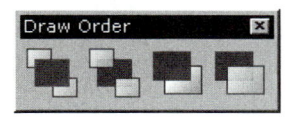

INQUIRY TOOLBAR

INSERT TOOLBAR

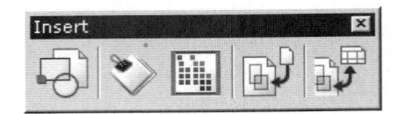

LAYERS TOOLBAR

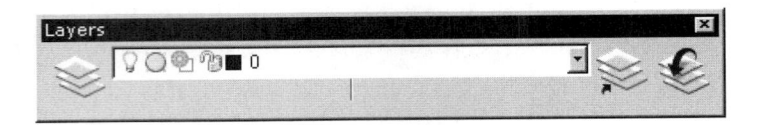

LAYOUTS TOOLBAR

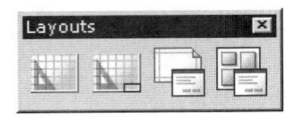

MODIFY TOOLBAR

MODIFY II TOOLBAR

OBJECT SNAP TOOLBAR

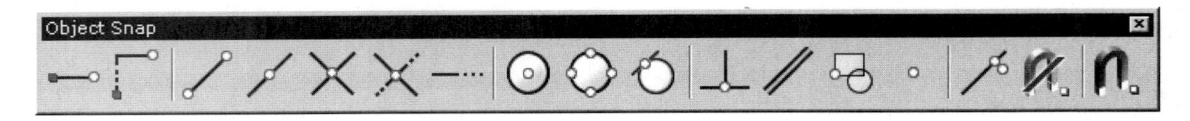

PROPERTIES TOOLBAR

REFEDIT TOOLBAR

REFERENCE TOOLBAR

RENDER TOOLBAR

SHADE TOOLBAR

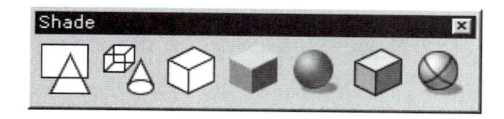

SOLIDS TOOLBAR

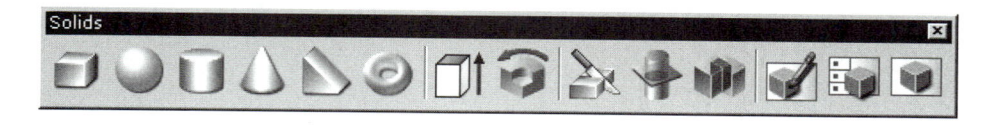

SOLIDS EDITING TOOLBAR

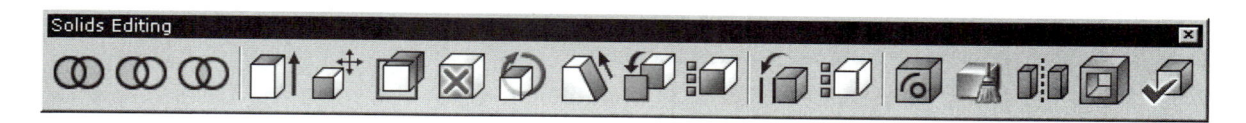

STANDARD TOOLBAR

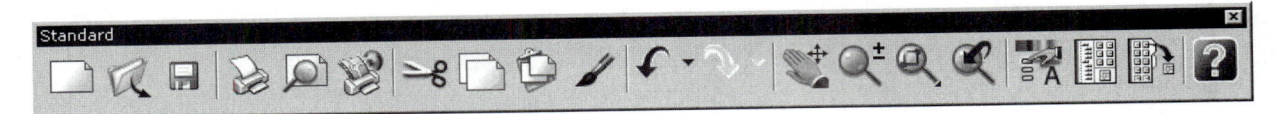

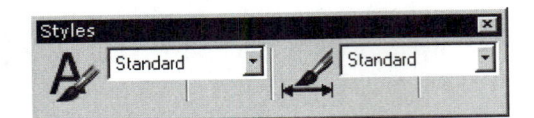

STYLES TOOLBAR

SURFACES TOOLBAR

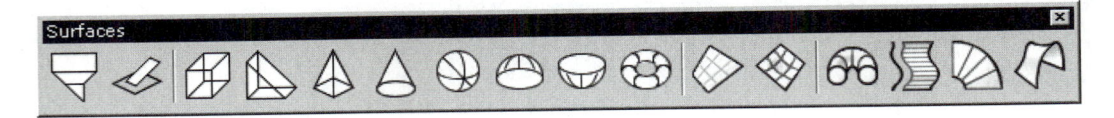

TEXT TOOLBAR

UCS TOOLBAR

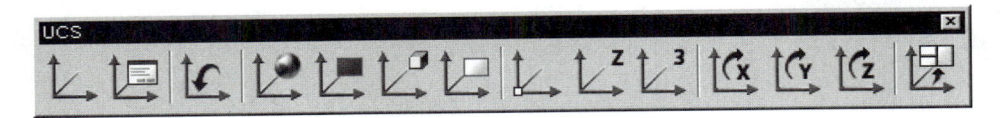

UCS II TOOLBAR

VIEW TOOLBAR

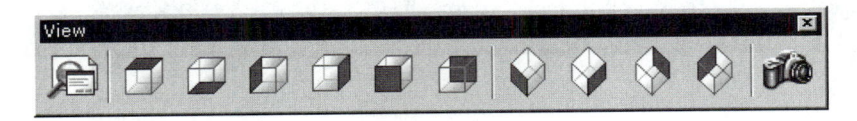

Appendix D: AutoCAD Toolbars and Menus **845**

VIEWPORTS TOOLBAR

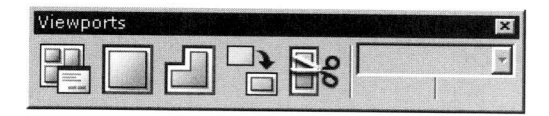

WEB TOOLBAR

ZOOM TOOLBAR

3D ORBIT TOOLBAR

846

AUTOCAD 2004 MENUS

![2004] The icon indicates the menu is new to AutoCAD 2004.

FILE MENU

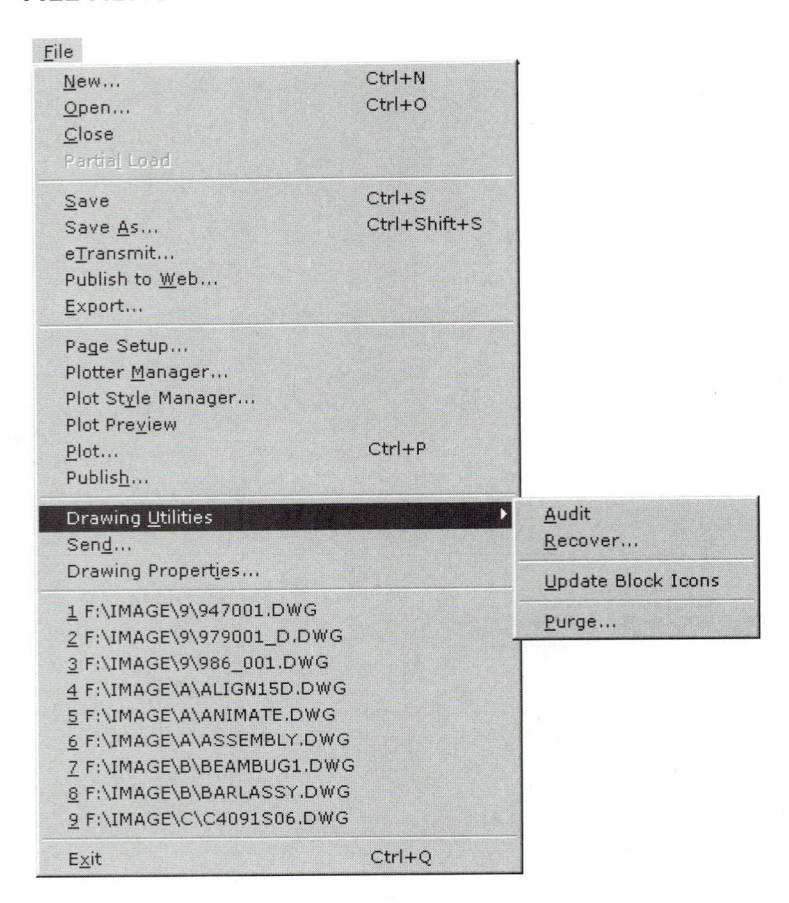

File	
New...	Ctrl+N
Open...	Ctrl+O
Close	
Partial Load	
Save	Ctrl+S
Save As...	Ctrl+Shift+S
eTransmit...	
Publish to Web...	
Export...	
Page Setup...	
Plotter Manager...	
Plot Style Manager...	
Plot Preview	
Plot...	Ctrl+P
Publish...	
Drawing Utilities ▶	
Send...	
Drawing Properties...	
1 F:\IMAGE\9\947001.DWG	
2 F:\IMAGE\9\979001_D.DWG	
3 F:\IMAGE\9\986_001.DWG	
4 F:\IMAGE\A\ALIGN15D.DWG	
5 F:\IMAGE\A\ANIMATE.DWG	
6 F:\IMAGE\A\ASSEMBLY.DWG	
7 F:\IMAGE\B\BEAMBUG1.DWG	
8 F:\IMAGE\B\BARLASSY.DWG	
9 F:\IMAGE\C\C4091S06.DWG	
Exit	Ctrl+Q

Drawing Utilities submenu:
- Audit
- Recover...
- Update Block Icons
- Purge...

EDIT MENU

Edit	
Undo Copy to clipboard	Ctrl+Z
Redo Group of commands	Ctrl+Y
Cut	Ctrl+X
Copy	Ctrl+C
Copy with Base Point	Ctrl+Shift+C
Copy Link	
Paste	Ctrl+V
Paste as Block	Ctrl+Shift+V
Paste as Hyperlink	
Paste to Original Coordinates	
Paste Special...	
Clear	Del
Select All	Ctrl+A
OLE Links...	
Find...	

VIEW MENU

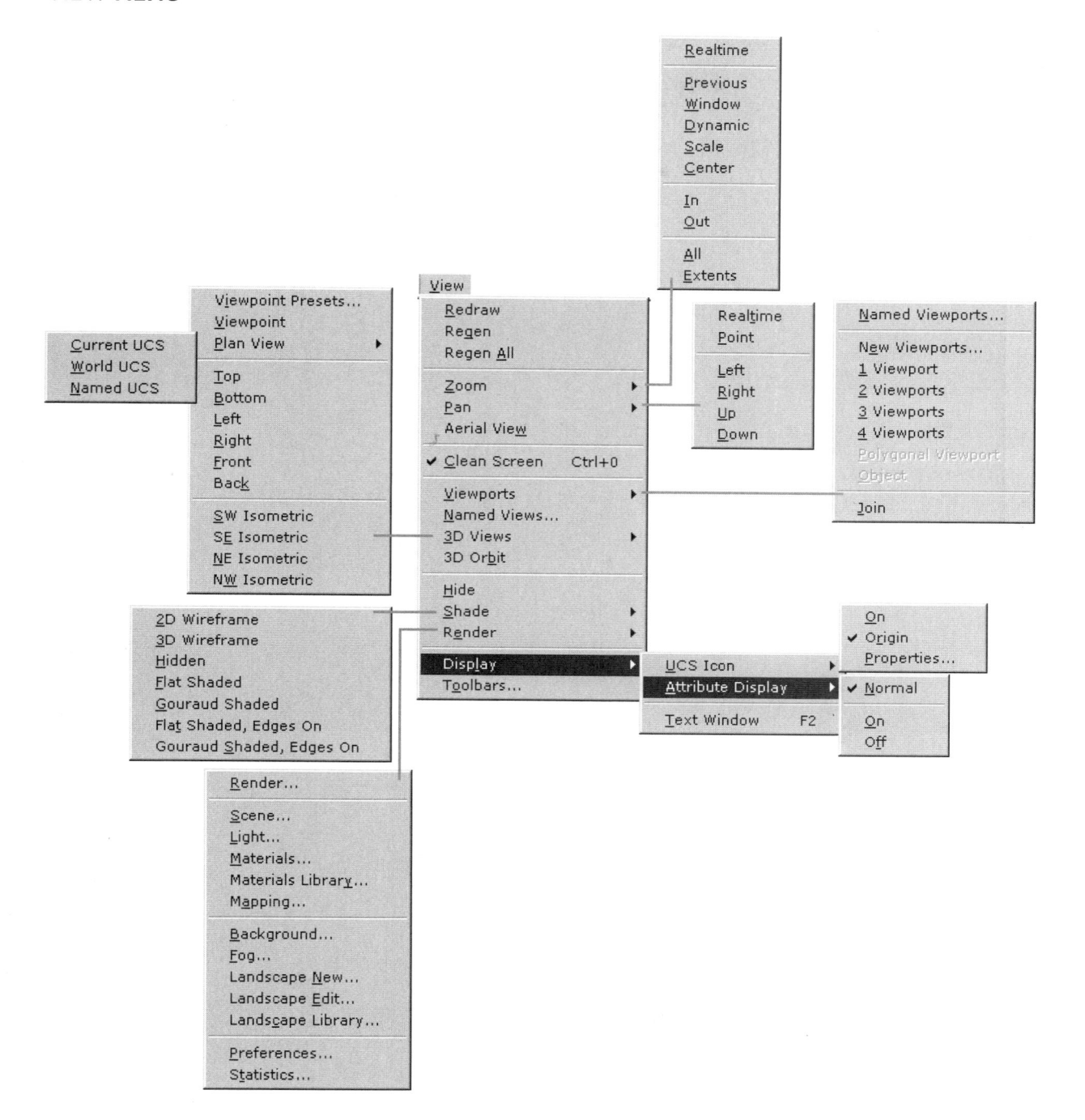

Zoom submenu:
- Realtime
- Previous
- Window
- Dynamic
- Scale
- Center
- In
- Out
- All
- Extents

Pan submenu:
- Realtime
- Point
- Left
- Right
- Up
- Down

Viewports submenu:
- Named Viewports...
- New Viewports...
- 1 Viewport
- 2 Viewports
- 3 Viewports
- 4 Viewports
- Polygonal Viewport
- Object
- Join

Plan View submenu:
- Current UCS
- World UCS
- Named UCS

3D Views submenu:
- Viewpoint Presets...
- Viewpoint
- Plan View ▶
- Top
- Bottom
- Left
- Right
- Front
- Back
- SW Isometric
- SE Isometric
- NE Isometric
- NW Isometric

View menu:
- Redraw
- Regen
- Regen All
- Zoom ▶
- Pan ▶
- Aerial View
- ✓ Clean Screen   Ctrl+0
- Viewports ▶
- Named Views...
- 3D Views ▶
- 3D Orbit
- Hide
- Shade ▶
- Render ▶
- Display ▶
- Toolbars...

Shade submenu:
- 2D Wireframe
- 3D Wireframe
- Hidden
- Flat Shaded
- Gouraud Shaded
- Flat Shaded, Edges On
- Gouraud Shaded, Edges On

Render submenu:
- Render...
- Scene...
- Light...
- Materials...
- Materials Library...
- Mapping...
- Background...
- Fog...
- Landscape New...
- Landscape Edit...
- Landscape Library...
- Preferences...
- Statistics...

Display submenu:
- UCS Icon ▶
- Attribute Display ▶
- Text Window   F2

UCS Icon submenu:
- On
- ✓ Origin
- Properties...

Attribute Display submenu:
- ✓ Normal
- On
- Off

INSERT MENU

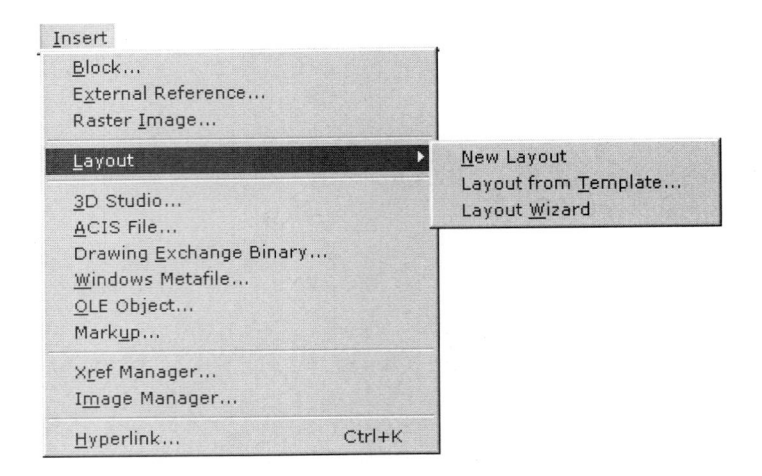

Insert	
Block...	
External Reference...	
Raster Image...	
Layout ▶	New Layout
	Layout from Template...
	Layout Wizard
3D Studio...	
ACIS File...	
Drawing Exchange Binary...	
Windows Metafile...	
OLE Object...	
Markup...	
Xref Manager...	
Image Manager...	
Hyperlink... Ctrl+K	

FORMAT MENU

Format
- Layer...
- Color...
- Linetype...
- Lineweight...
- Text Style...
- Dimension Style...
- Plot Style...
- Point Style...
- Multiline Style...
- Units...
- Thickness
- Drawing Limits
- Rename...

TOOLS MENU

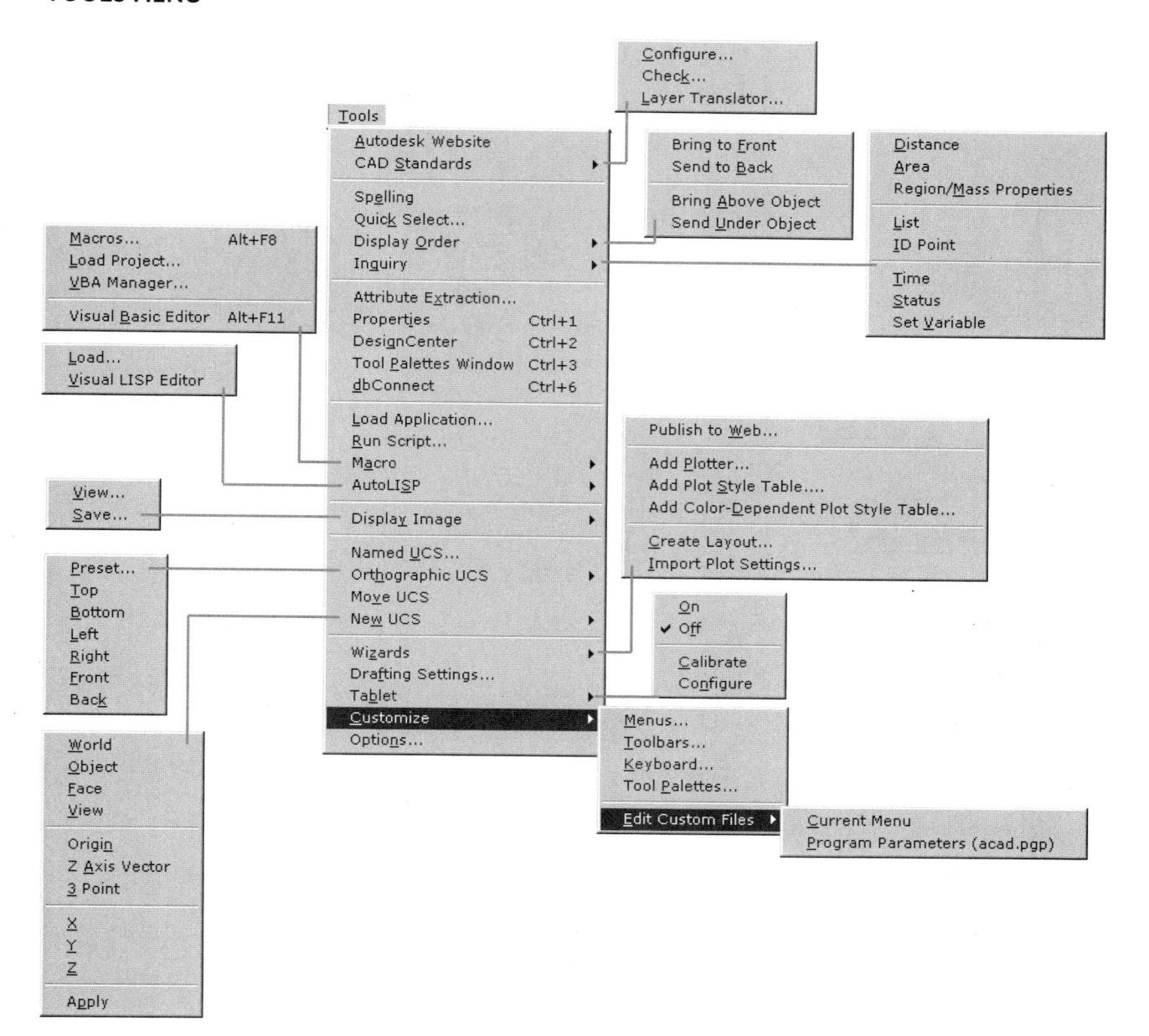

Tools
- Autodesk Website
- CAD Standards ▶
  - Configure...
  - Check...
  - Layer Translator...
- Spelling
- Quick Select...
- Display Order ▶
  - Bring to Front
  - Send to Back
  - Bring Above Object
  - Send Under Object
- Inquiry ▶
  - Distance
  - Area
  - Region/Mass Properties
  - List
  - ID Point
  - Time
  - Status
  - Set Variable
- Attribute Extraction...
- Properties   Ctrl+1
- DesignCenter   Ctrl+2
- Tool Palettes Window Ctrl+3
- dbConnect   Ctrl+6
- Load Application...
- Run Script...
- Macro ▶
  - Macros...   Alt+F8
  - Load Project...
  - VBA Manager...
  - Visual Basic Editor Alt+F11
- AutoLISP ▶
  - Load...
  - Visual LISP Editor
- Display Image ▶
  - View...
  - Save...
- Named UCS...
- Orthographic UCS ▶
  - Top
  - Bottom
  - Left
  - Right
  - Front
  - Back
- Move UCS
- New UCS ▶
  - World
  - Object
  - Face
  - View
  - Origin
  - Z Axis Vector
  - 3 Point
  - X
  - Y
  - Z
- Wizards ▶
  - Publish to Web...
  - Add Plotter...
  - Add Plot Style Table...
  - Add Color-Dependent Plot Style Table...
  - Create Layout...
  - Import Plot Settings...
- Drafting Settings...
- Tablet ▶
  - On
  - ✓ Off
  - Calibrate
  - Configure
- Customize ▶
  - Menus...
  - Toolbars...
  - Keyboard...
  - Tool Palettes...
  - Edit Custom Files ▶
    - Current Menu
    - Program Parameters (acad.pgp)
- Options...

UCS Preset dialog:
- Preset...
- Top
- Bottom
- Left
- Right
- Front
- Back

Apply

DRAW MENU

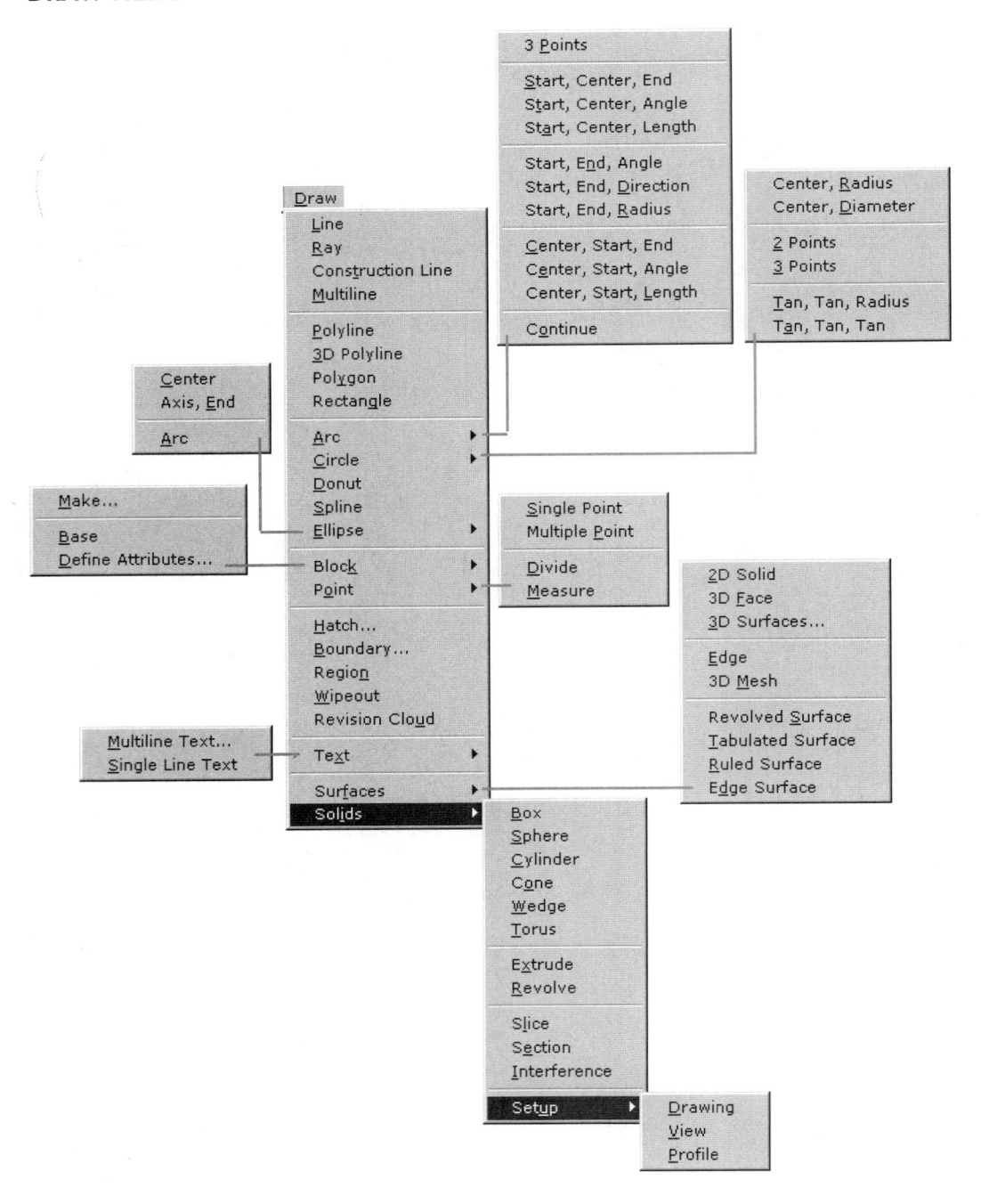

Draw
- Line
- Ray
- Construction Line
- Multiline
- Polyline
- 3D Polyline
- Polygon
- Rectangle
- Arc ▶
- Circle ▶
- Donut
- Spline
- Ellipse ▶
- Block ▶
- Point ▶
- Hatch...
- Boundary...
- Region
- Wipeout
- Revision Cloud
- Text ▶
- Surfaces ▶
- **Solids** ▶

Arc
- 3 Points
- Start, Center, End
- Start, Center, Angle
- Start, Center, Length
- Start, End, Angle
- Start, End, Direction
- Start, End, Radius
- Center, Start, End
- Center, Start, Angle
- Center, Start, Length
- Continue

Circle
- Center, Radius
- Center, Diameter
- 2 Points
- 3 Points
- Tan, Tan, Radius
- Tan, Tan, Tan

Ellipse
- Center
- Axis, End
- Arc

Block
- Make...
- Base
- Define Attributes...

Point
- Single Point
- Multiple Point
- Divide
- Measure

Text
- Multiline Text...
- Single Line Text

Surfaces
- 2D Solid
- 3D Face
- 3D Surfaces...
- Edge
- 3D Mesh
- Revolved Surface
- Tabulated Surface
- Ruled Surface
- Edge Surface

Solids
- Box
- Sphere
- Cylinder
- Cone
- Wedge
- Torus
- Extrude
- Revolve
- Slice
- Section
- Interference
- Setup ▶

Setup
- Drawing
- View
- Profile

Appendix D: AutoCAD Toolbars and Menus

DIMENSION MENU

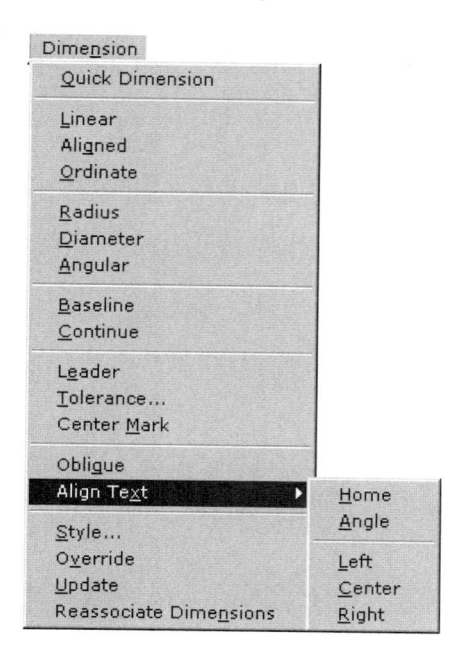

Dimension
- Quick Dimension
- Linear
- Aligned
- Ordinate
- Radius
- Diameter
- Angular
- Baseline
- Continue
- Leader
- Tolerance...
- Center Mark
- Oblique
- Align Text ▶
  - Home
  - Angle
  - Left
  - Center
  - Right
- Style...
- Override
- Update
- Reassociate Dimensions

MODIFY MENU

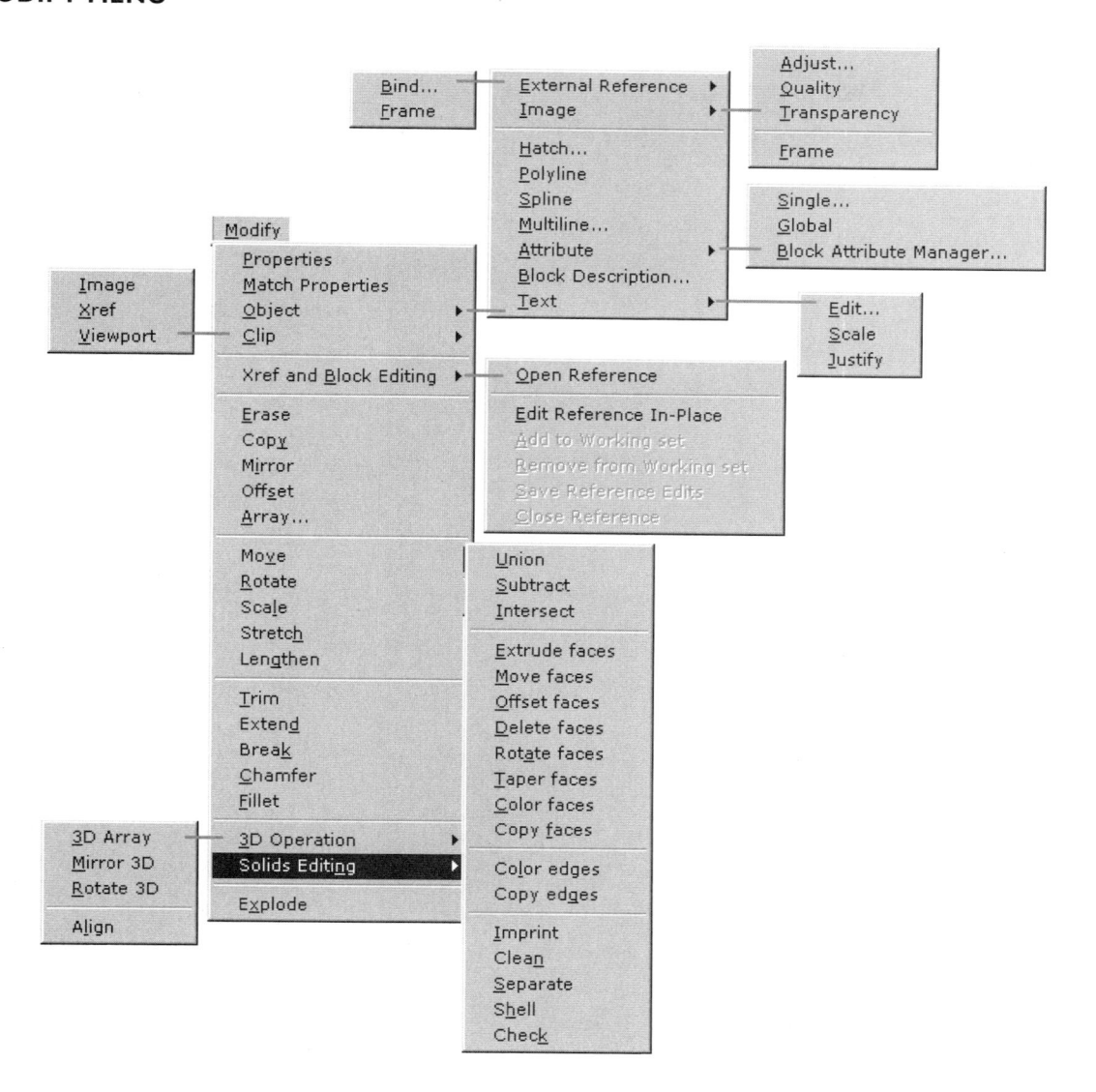

Modify
- Properties
- Match Properties
- Object ▶
  - External Reference ▶
    - Bind...
    - Frame
  - Image ▶
    - Adjust...
    - Quality
    - Transparency
    - Frame
  - Hatch...
  - Polyline
  - Spline
  - Multiline...
  - Attribute ▶
    - Single...
    - Global
    - Block Attribute Manager...
  - Block Description...
  - Text ▶
    - Edit...
    - Scale
    - Justify
- Clip ▶
  - Image
  - Xref
  - Viewport
- Xref and Block Editing ▶
  - Open Reference
  - Edit Reference In-Place
  - Add to Working set
  - Remove from Working set
  - Save Reference Edits
  - Close Reference
- Erase
- Copy
- Mirror
- Offset
- Array...
- Move
- Rotate
- Scale
- Stretch
- Lengthen
- Trim
- Extend
- Break
- Chamfer
- Fillet
- 3D Operation ▶
  - 3D Array
  - Mirror 3D
  - Rotate 3D
  - Align
- Solids Editing ▶
  - Union
  - Subtract
  - Intersect
  - Extrude faces
  - Move faces
  - Offset faces
  - Delete faces
  - Rotate faces
  - Taper faces
  - Color faces
  - Copy faces
  - Color edges
  - Copy edges
  - Imprint
  - Clean
  - Separate
  - Shell
  - Check
- Explode

EXPRESS MENU
New in 2004

This menu is available only after being installed separately from the distribution CD.

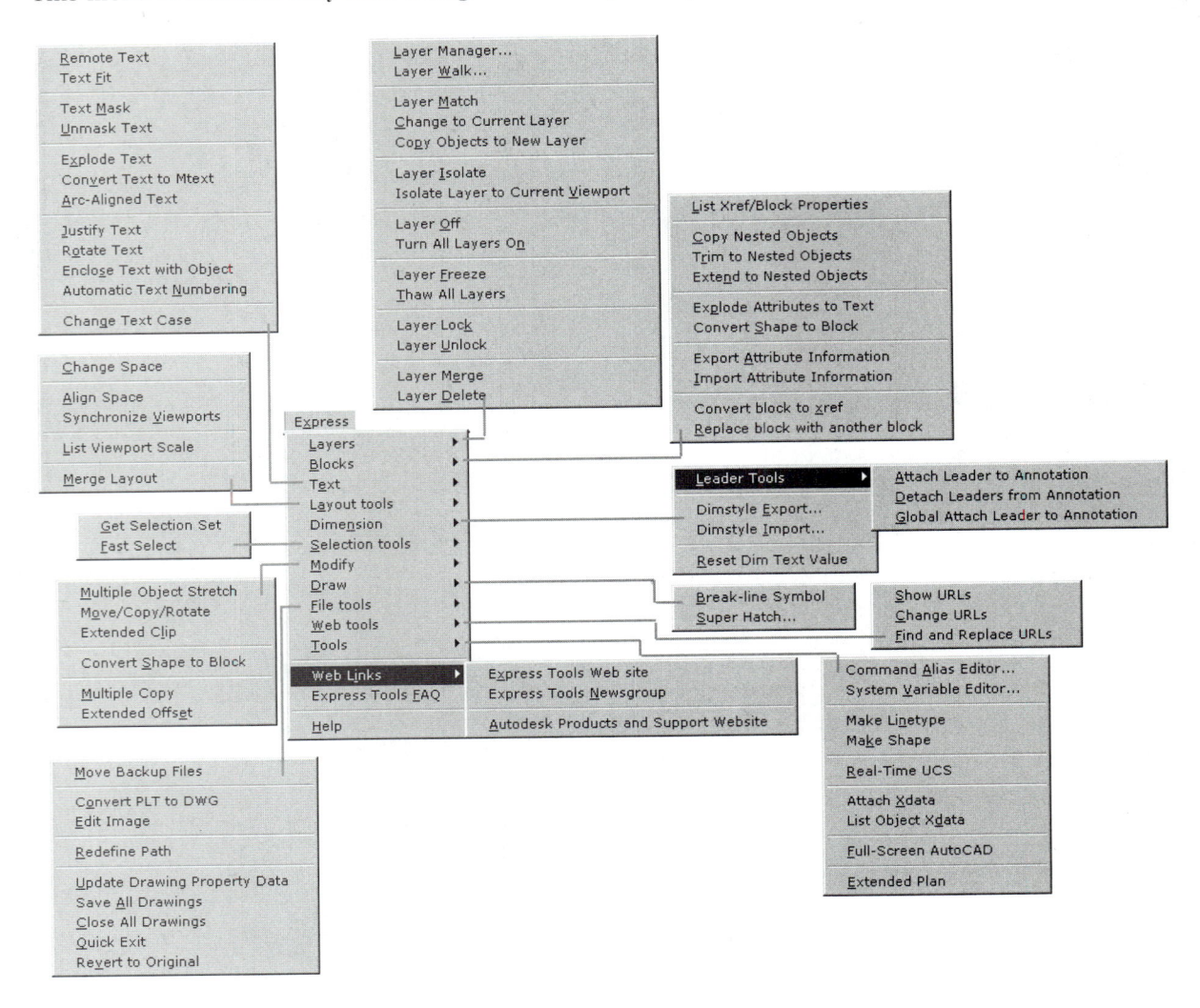

Express
- Layers ▶
- Blocks ▶
- Text ▶
- Layout tools ▶
- Dimension ▶
- Selection tools ▶
- Modify ▶
- Draw ▶
- File tools ▶
- Web tools ▶
- Tools ▶
- Web Links ▶
- Express Tools FAQ
- Help

Text:
- Remote Text
- Text Fit
- Text Mask
- Unmask Text
- Explode Text
- Convert Text to Mtext
- Arc-Aligned Text
- Justify Text
- Rotate Text
- Enclose Text with Object
- Automatic Text Numbering
- Change Text Case

Layers:
- Layer Manager...
- Layer Walk...
- Layer Match
- Change to Current Layer
- Copy Objects to New Layer
- Layer Isolate
- Isolate Layer to Current Viewport
- Layer Off
- Turn All Layers On
- Layer Freeze
- Thaw All Layers
- Layer Lock
- Layer Unlock
- Layer Merge
- Layer Delete

Blocks:
- List Xref/Block Properties
- Copy Nested Objects
- Trim to Nested Objects
- Extend to Nested Objects
- Explode Attributes to Text
- Convert Shape to Block
- Export Attribute Information
- Import Attribute Information
- Convert block to xref
- Replace block with another block

Layout tools:
- Change Space
- Align Space
- Synchronize Viewports
- List Viewport Scale
- Merge Layout

Selection tools:
- Get Selection Set
- Fast Select

Modify:
- Multiple Object Stretch
- Move/Copy/Rotate
- Extended Clip
- Convert Shape to Block
- Multiple Copy
- Extended Offset

Dimension:
- Leader Tools ▶
- Dimstyle Export...
- Dimstyle Import...
- Reset Dim Text Value

Leader Tools:
- Attach Leader to Annotation
- Detach Leaders from Annotation
- Global Attach Leader to Annotation

Draw:
- Break-line Symbol
- Super Hatch...

Web tools:
- Show URLs
- Change URLs
- Find and Replace URLs

Web Links:
- Express Tools Web site
- Express Tools Newsgroup
- Autodesk Products and Support Website

Tools:
- Command Alias Editor...
- System Variable Editor...
- Make Linetype
- Make Shape
- Real-Time UCS
- Attach Xdata
- List Object Xdata
- Full-Screen AutoCAD
- Extended Plan

File tools:
- Move Backup Files
- Convert PLT to DWG
- Edit Image
- Redefine Path
- Update Drawing Property Data
- Save All Drawings
- Close All Drawings
- Quick Exit
- Revert to Original

WINDOW MENU

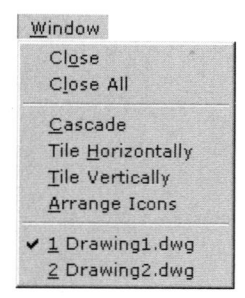

- Window
  - Close
  - Close All
  - Cascade
  - Tile Horizontally
  - Tile Vertically
  - Arrange Icons
  - ✓ 1 Drawing1.dwg
  - 2 Drawing2.dwg

HELP MENU

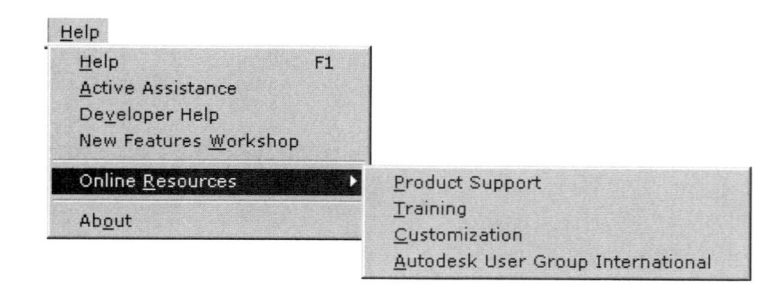

- Help
  - Help   F1
  - Active Assistance
  - Developer Help
  - New Features Workshop
  - Online Resources ▶
    - Product Support
    - Training
    - Customization
    - Autodesk User Group International
  - About

WINDOW CONTROL MENU

- Restore
- Move
- Size
- Minimize
- Maximize
- ✕ Close   Ctrl+F4
- Next   Ctrl+F6

APPENDIX E

AutoCAD Fonts

The fonts shown below are some of thoese provided by Autodesk with AutoCAD 2004 in SHX and TrueType format.

Complex
ABCDEFGHIJKLMNOPQRSTUVWZXYZ
abcdefghijklmnopqrstuvwzxyz
0123456789 ° ± ø
!@#$%^&*()_+-=,./<>?;':"[]{}\|'~

Country Blueprint
ABCDEFGHIJKLMNOPQRSTUVWZXYZ
abcdefghijklmnopqrstuvwzxyz
0123456789 ° ±Ø
!@#$%^&*()_+-=,./<>?;':"[]{}\|'~

Euroroman
ABCDEFGHIJKLMNOPQRSTUVWZXYZ
abcdefghijklmnopqrstuvwzxyz
0123456789 ◻ ± ÿ
!@#$%^&*()_+-=,./<>?;':"[]{}\|'~

GDT
ABCDEFGHIJKLMNOPQRSTUVWZXYZ
∠⌴⌒⌓○//⌰⌖=⌀∩ⓁⓂ⌀Ⓟt Ⓞ Ⓢ ⌲ — ⌵⌴∇▷▷
0123456789 ° ± ø
!@#$%⌒&*()_+-=,./<>?;':"[]{}\|±°

GothicE (English)
𝔄𝔅ℭ𝔇𝔈𝔉𝔊ℌ𝔍𝔍𝔎𝔏𝔐𝔑𝔒𝔓𝔔ℜ𝔖𝔗𝔘𝔙𝔚𝔛𝔜ℨ
abcdefghijklmnopqrstuvwxyz
0123456789 ° ± ø
!@#$%^&*()_+-=,./<>?;':"[]{}\|`~

GothicG (German)
𝔄𝔅ℭ𝔇𝔈𝔉𝔊ℌ𝔍𝔎𝔏𝔐𝔑𝔒𝔓𝔔ℜ𝔖𝔗𝔘𝔙𝔚𝔛𝔜ℨ
abcdefghijklmnopqrstuvwxyz
0123456789 ° ± ø
!@#$%^&*()_+-=,./<>?;':"[]{}\|`~

GothicI (Italian)
ABCDEFGHIJKLMNOPQRSTUVWXYZ
abcdefghijklmnopqrstuvwxyz
0123456789 ° ± ø
!@#$%^&*()_+-=,./<>?;':"[]{}\|`~

GreekC (Complex)
ΑΒΧΔΕΦΓΗΙϑΚΛΜΝΟΠΘΡΣΤΥΩΖΞΨZ
αβχδεφγηιδκλμνοπϑρστυ∈ωξψζ
0123456789 ° ± ø
!@#$%~&*()_+-=,./<>?;':"[]{}\|`~

GreekS (Simplex)
ΑΒΧΔΕΦΓΗΙϑΚΛΜΝΟΠΘΡΣΤΥΩΖΞΨZ
αβχδεφγηιδκλμνοπϑρστυ∈ωξψζ
0123456789 ° ± ø
!@#$%~&*()_+-=,./<>?;':"[]{}\|`~

ISOCP
ABCDEFGHIJKLMNOPQRSTUVWXYZ
abcdefghijklmnopqrstuvwxyz
0123456789 ° ± ø
!@#$%^&×()_+-=,./<>?;':"[]{}\|`~

ISOCP2
ABCDEFGHIJKLMNOPQRSTUVWXYZ
abcdefghijklmnopqrstuvwxyz
0123456789 ° ± ø
!@#$%^&×()_+-=,./<>?;':"[]{}\|`~

ISOCP3
ABCDEFGHIJKLMNOPQRSTUVWXYZ
abcdefghijklmnopqrstuvwxyz
0123456789 ° ± ø
!@#$%^&×()_+-=,./<>?;':"[]{}\|`~

ISOEUR
ABCDEFGHIJKLMNOPQRSTUVWXYZ
abcdefghijklmnopqrstuvwxyz
0123456789 ° ± ø
!@#$%^&*()_+-=,./<>?;':"[]{}\|`~

ISOCT
ABCDEFGHIJKLMNOPQRSTUVWXYZ
abcdefghijklmnopqrstuvwxyz
0123456789 ° ± ø
!@#$%^&×()_+-=,./<>?;':"[]{}\|`~

ISOCT2
ABCDEFGHIJKLMNOPQRSTUVWXYZ
abcdefghijklmnopqrstuvwxyz
0123456789 ° ± ø
!@#$%^&×()_+-=,./<>?;':"[]{}\|`~

ISOCT3
ABCDEFGHIJKLMNOPQRSTUVWXYZ
abcdefghijklmnopqrstuvwxyz
0123456789 ° ± ø
!@#$%^&×()_+-=,./<>?;':"[]{}\|`~

ISOCTEUR
ABCDEFGHIJKLMNOPQRSTUVWXYZ
abcdefghijklmnopqrstuvwxyz
0123456789 ° ± ø
!@#$%^&*()_+-=,./<>?;':"[]{}\|`~

Italic
ABCDEFGHIJKLMNOPQRSTUVWXYZ
abcdefghijklmnopqrstuvwxyz
0123456789 ° ± ø
!@#$%^&()_+-=,./<>?;':"[]{}\|`~*

ItalicC (Complex)
ABCDEFGHIJKLMNOPQRSTUVWXYZ
abcdefghijklmnopqrstuvwxyz
0123456789 ° ± ø
!@#$%^&()_+-=,./<>?;':"[]{}\|`~*

ItalicT (Triplex)
ABCDEFGHIJKLMNOPQRSTUVWXYZ
abcdefghijklmnopqrstuvwxyz
0123456789 ° ± ø
!@#$%^&*()_+-=,./<>?;':"[]{}\|`~

Appendix E: AutoCAD Fonts

Monotxt
ABCDEFGHIJKLMNOPQRSTUVWZXYZ
abcdefghijklmnopqrstuvwzxyz
0123456789 ° ± ø
!@#$%^&*()_+-=,./<>?;':"[]{}\|`~

RomanC (Complex)
ABCDEFGHIJKLMNOPQRSTUVWZXYZ
abcdefghijklmnopqrstuvwzxyz
0123456789 ° ± ø
!@#$%^&*()_+-=,./<>?;':"[]{}\|`~

RomanD (Duplex)
ABCDEFGHIJKLMNOPQRSTUVWZXYZ
abcdefghijklmnopqrstuvwzxyz
0123456789 ° ± ø
!@#$%^&*()_+-=,./<>?;':"[]{}\|`~

RomanS (Simplex)
ABCDEFGHIJKLMNOPQRSTUVWZXYZ
abcdefghijklmnopqrstuvwzxyz
0123456789 ° ± ø
!@#$%^&*()_+-=,./<>?;':"[]{}\|`~

RomandT (Triplex)
ABCDEFGHIJKLMNOPQRSTUVWZXYZ
abcdefghijklmnopqrstuvwzxyz
0123456789 ° ± ø
!@#$%^&*()_+-=,./<>?;':"[]{}\|`~

Romantic
ABCDEFGHIJKLMNOPQRSTUVWZXYZ
abcdefghijklmnopqrstuvwzxyz
0123456789 ° ± Ø
!@#$%^&*()_+-=,./<>?;':"[]{}\|`~

ScriptC (Complex)
ABCDEFGHIJKLMNOPQRSTUVWXYZ
abcdefghijklmnopqrstuvwxyz
0123456789 ° ± ø
!@#$%^&*()_+-=,./<>?;':"[]{}\|`~

ScriptS (Simplex)
ABCDEFGHIJKLMNOPQRSTUVWXYZ
abcdefghijklmnopqrstuvwxyz
0123456789 ° ± ø
!@#$%^&*()_+-=,./<>?;':"[]{}\|`~

Simplex
ABCDEFGHIJKLMNOPQRSTUVWZXYZ
abcdefghijklmnopqrstuvwzxyz
0123456789 ° ± ø
!@#$%^&*()_+-=,./<>?;':"[]{}\|`~

SyAstro (Astronomical Symbols)
☉☿♀⊕♂♃♄♅♆Ψ♇☽☼☀⚹♈♉♊♋♌♍♎♏♐♑♒♓
0123456789 ° ± ø
!@#$%^&*()_+-=,./<>?;':"[]{}\|`~

SyMap (Mapping Symbols)

0123456789 ° ± ø
!@#$%^&*()_+-=,./<>?;':"[]{}\|`~

SyMath (Mathematical Symbols)
ℵ'‖|±∓×·÷=≠≡<>≤≥∝~√∪∩⊃∈→↑
←↓∂∇√∫∮∞§†‡∃∏∑()[]{}⟨⟩√≅∫≈≌
0123456789 ° ± ø
!@#$%^&*()_+-=,./<>?;':"[]{}\|`~

SyMeteo (Weather Symbols)

0123456789 ° ± ø
!@#$%^&*()_+-=,./<>?;':"[]{}\|`~

SyMusic (Music Symbols)

0123456789 ° ± ø
!@#$%^&*()_+-=,./<>?;':"[]{}\|`~

TXT
ABCDEFGHIJKLMNOPQRSTUVWZXYZ
abcdefghijklmnopqrstuvwzxyz
0123456789 ° ± ø
!@#$%^&*()_+-=,./<>?;':"[]{}\|`~

APPENDIX F

AutoCAD Linetypes

The linetypes shown on the next page are provided by Autodesk with AutoCAD 2004 in the *acad.lin* and *acadiso.lin* files.

Linetype	Description
ACAD_ISO02W100	ISO dash __ __ __ __ __ __ __ __ __ __ __
ACAD_ISO03W100	ISO dash space __ __ __ __ __ __
ACAD_ISO04W100	ISO long-dash dot ____ . ____ . ____ . ____ . _
ACAD_ISO05W100	ISO long-dash double-dot ____ .. ____ .. ____ .
ACAD_ISO06W100	ISO long-dash triple-dot ____ ... ____ ... ____
ACAD_ISO07W100	ISO dot . . . .
ACAD_ISO08W100	ISO long-dash short-dash ____ __ ____ __ ____ _
ACAD_ISO09W100	ISO long-dash double-short-dash ____ __ __ ____
ACAD_ISO10W100	ISO dash dot __ . __ . __ . __ . __ . __ . __ .
ACAD_ISO11W100	ISO double-dash dot __ __ . __ __ . __ __ . __
ACAD_ISO12W100	ISO dash double-dot __ . . __ . . __ . .
ACAD_ISO13W100	ISO double-dash double-dot __ __ . . __ __ . .
ACAD_ISO14W100	ISO dash triple-dot __ . . . __ . . . __ . . .
ACAD_ISO15W100	ISO double-dash triple-dot __ __ . . . __ __ .
BATTING	Batting SSSSSSSSSSSSSSSSSSSSSSSSSSSSSSSSSS
BORDER	Border __ __ . __ __ . __ __ . __ __ . __ __ .
BORDER2	Border (.5x) __ __ . __ __ . __ __ . __ __ .
BORDERX2	Border (2x) ____ ____ . ____ ____ . ____
CENTER	Center ____ _ ____ _ ____ _ ____ _ ____
CENTER2	Center (.5x) ____ _ ____ _ ____ _ ____ _ ____
CENTERX2	Center (2x) _____ __ _____ __ _____
DASHDOT	Dash dot __ . __ . __ . __ . __ . __ . __
DASHDOT2	Dash dot (.5x) _._._._._._._._._._._._._.
DASHDOTX2	Dash dot (2x) ____ . ____ . ____ . ____
DASHED	Dashed __ __ __ __ __ __ __ __ __ __ __
DASHED2	Dashed (.5x) _ _ _ _ _ _ _ _ _ _ _ _ _ _ _
DASHEDX2	Dashed (2x) ____ ____ ____ ____ ____
DIVIDE	Divide ____ . . ____ . . ____ . . ____
DIVIDE2	Divide (.5x) __ . . __ . . __ . . __ . . __
DIVIDEX2	Divide (2x) _____ . . _____ . . _
DOT	Dot . . . .
DOT2	Dot (.5x) ....
DOTX2	Dot (2x) . . . .
FENCELINE1	Fenceline circle ----0----0----0----0----0---
FENCELINE2	Fenceline square ----[]----[]----[]----[]----
GAS_LINE	Gas line ----GAS----GAS----GAS----GAS----GAS---
HIDDEN	Hidden __ __ __ __ __ __ __ __ __ __ __ __
HIDDEN2	Hidden (.5x) _ _ _ _ _ _ _ _ _ _ _ _ _ _ _
HIDDENX2	Hidden (2x) ____ ____ ____ ____ ____ ____
HOT_WATER_SUPPLY	Hot water supply ---- HW ---- HW ---- HW ----
PHANTOM	Phantom _____ __ __ _____ __ __
PHANTOM2	Phantom (.5x) ___ _ _ ___ _ _ ___ _ _ ___
PHANTOMX2	Phantom (2x) _____ ____ ____ _
TRACKS	Tracks -I-I-I-I-I-I-I-I-I-I-I-I-I-I-I-I-I-I
ZIGZAG	Zig zag /\/\/\/\/\/\/\/\/\/\/\/\/\/\/\/\

APPENDIX G

AutoCAD Hatch Patterns

The hatch patterns shown below are provided by Autodesk with AutoCAD 2004 in the *acad.pat* and *acadiso.pat* files, as well as generated as solid-color and gradient fills.

The icon indicates that gradient fills are new to AutoCAD 2004

ANSI PATTERNS

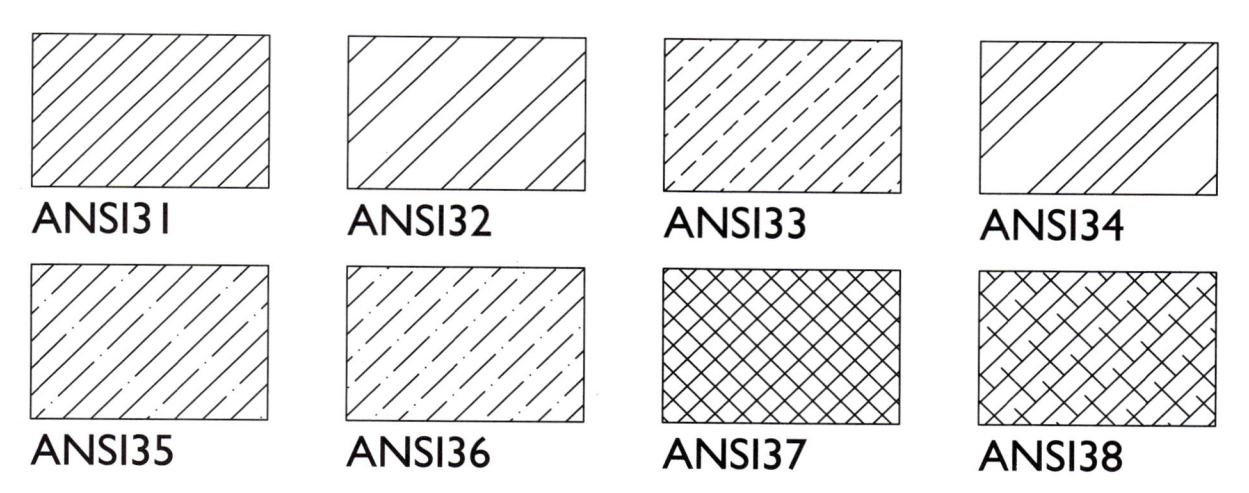

ANSI31   ANSI32   ANSI33   ANSI34

ANSI35   ANSI36   ANSI37   ANSI38

861

ISO PATTERNS

ISO02W100	ISO03W100	ISO04W100	ISO05W100
ISO06W100	ISO06W100	ISO08W100	ISO09W100
ISO10W100	ISO11W100	ISO12W100	ISO13W100
ISO14W100	ISO15W100		

Appendix G: AutoCAD Hatch Patterns **863**

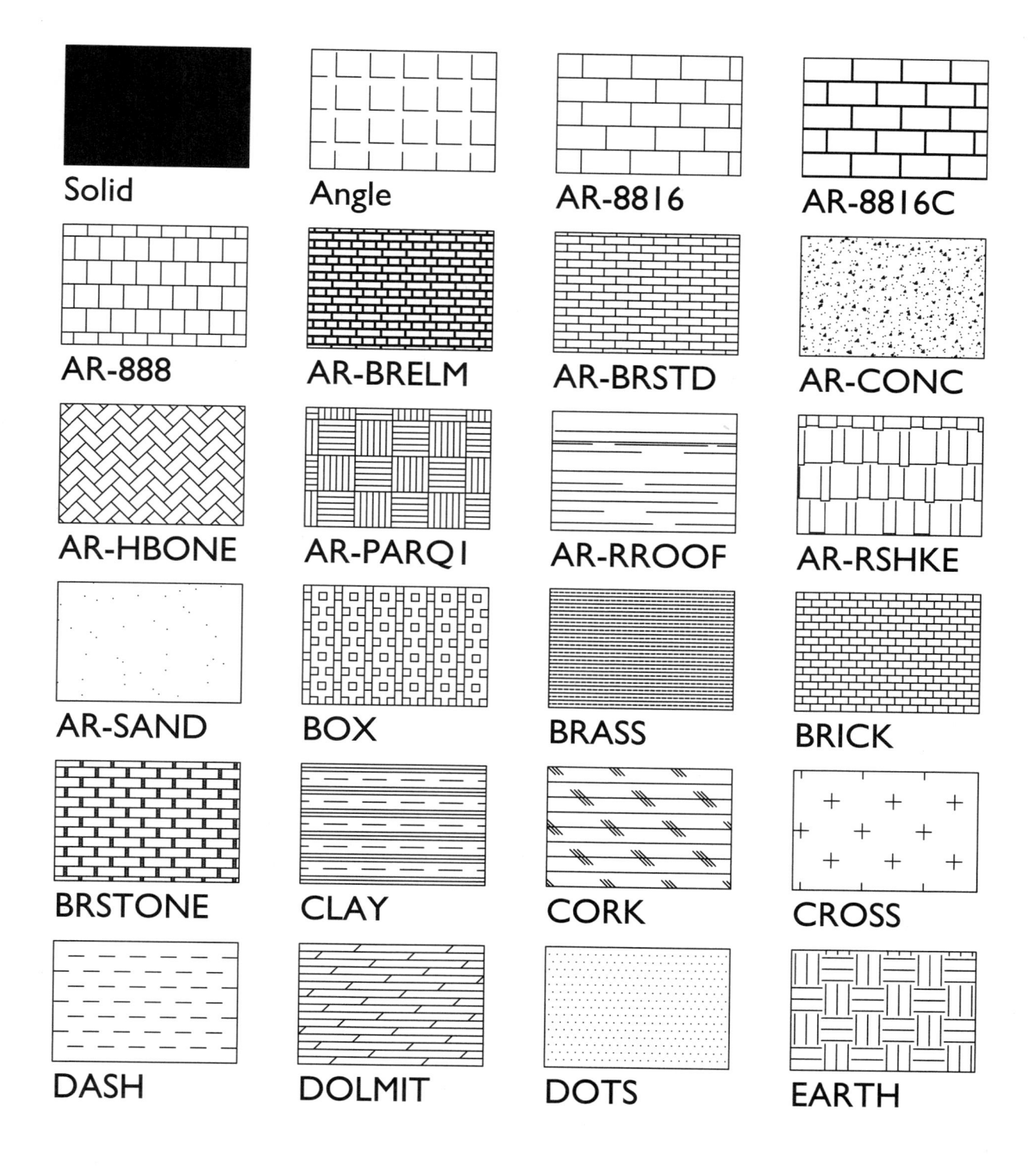

OTHER PATTERNS

Solid	Angle	AR-8816	AR-8816C
AR-888	AR-BRELM	AR-BRSTD	AR-CONC
AR-HBONE	AR-PARQ1	AR-RROOF	AR-RSHKE
AR-SAND	BOX	BRASS	BRICK
BRSTONE	CLAY	CORK	CROSS
DASH	DOLMIT	DOTS	EARTH

OTHER PATTERNS

ESCHER	FLEX	GRASS	GRATE
GRAVEL	HEX	HONEY	HOUND
INSUL	LINE	MUDST	NET
NET3	PLAST	PALSTI	SACNCR
SQUARE	STARS	STEEL	SWAMP
TRANS	TRIANG	ZIGZAG	

Appendix G: AutoCAD Hatch Patterns **865**

GRADIENT PATTERNS

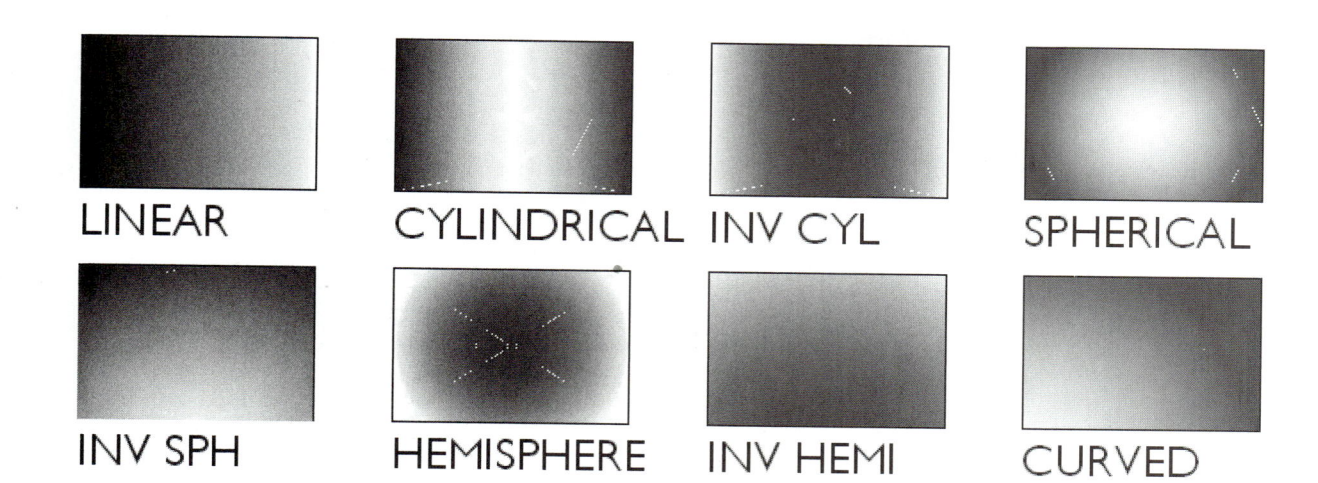

LINEAR   CYLINDRICAL   INV CYL   SPHERICAL

INV SPH   HEMISPHERE   INV HEMI   CURVED

APPENDIX H

AutoCAD Lineweights

The lineweights shown below and on the next page are "hardwired" into AutoCAD 2004. They are measured in inches or millimetres.

Millimeters	Sample
0.00 mm	
0.05 mm	
0.09 mm	
0.13 mm	
0.15 mm	
0.18 mm	
0.20 mm	
0.25 mm	
0.30 mm	
0.35 mm	
0.40 mm	
0.50 mm	
0.53 mm	
0.60 mm	
0.70 mm	
0.80 mm	
0.90 mm	
1.00 mm	
1.06 mm	
1.20 mm	
1.40 mm	
1.58 mm	
2.00 mm	
2.11 mm	

Millimeters	Inches	Points	Pen size	ISO	DIN	JIS	ANSI
0.05	0.002						
0.09	0.003	1/4 pt.					
0.13	0.005				*		
0.15	0.006						
0.18	0.007	1/2 pt.	0000	*	*	*	
0.20	0.008						
0.25	0.010	3/4 pt.	000	*	*	*	
0.30	0.012		00				2H or H
0.35	0.014	1 pt.	0	*	*	*	
0.40	0.016						
0.50	0.020		1	*	*	*	
0.53	0.021	1-1/2 pt.					
0.60	0.024		2				H, F, or B
0.70	0.028	2-1/4 pt.	2-1/2	*	*	*	
0.80	0.031		3				
0.90	0.035						
1.00	0.039		3-1/2	*	*	*	
1.06	0.042	3 pt.					
1.20	0.047		4				
1.40	0.056			*	*	*	
1.58	0.062	4-1/4 pt.					
2.0	0.078			*	*		
2.11	0.083	6 pt.					

APPENDIX I

DesignCenter Symbols

The symbols (blocks) shown on the following pages are provided by Autodesk with AutoCAD 2004 in the *autocad 2004\\sample\\designcenter* folders.

AEC
- Home - Space Planner
- House Designer
- Kitchens
- Landscaping

MECHANICAL
- Fasteners - Metric
- Fasteners - US
- HVAC - Heating Ventilation Air Conditioning
- Hydraulic - Pneumatic
- Pipe Fittings
- Plant Process
- Welding

ELECTRONICS
- Analog Integrated Circuits
- Basic Electronics
- CMOS Integrated Circuits
- Electrical Power

AEC

HOME - SPACE PLANNER

HOUSE DESIGNER

Appendix I: DesignCenter Symbols **871**

KITCHENS

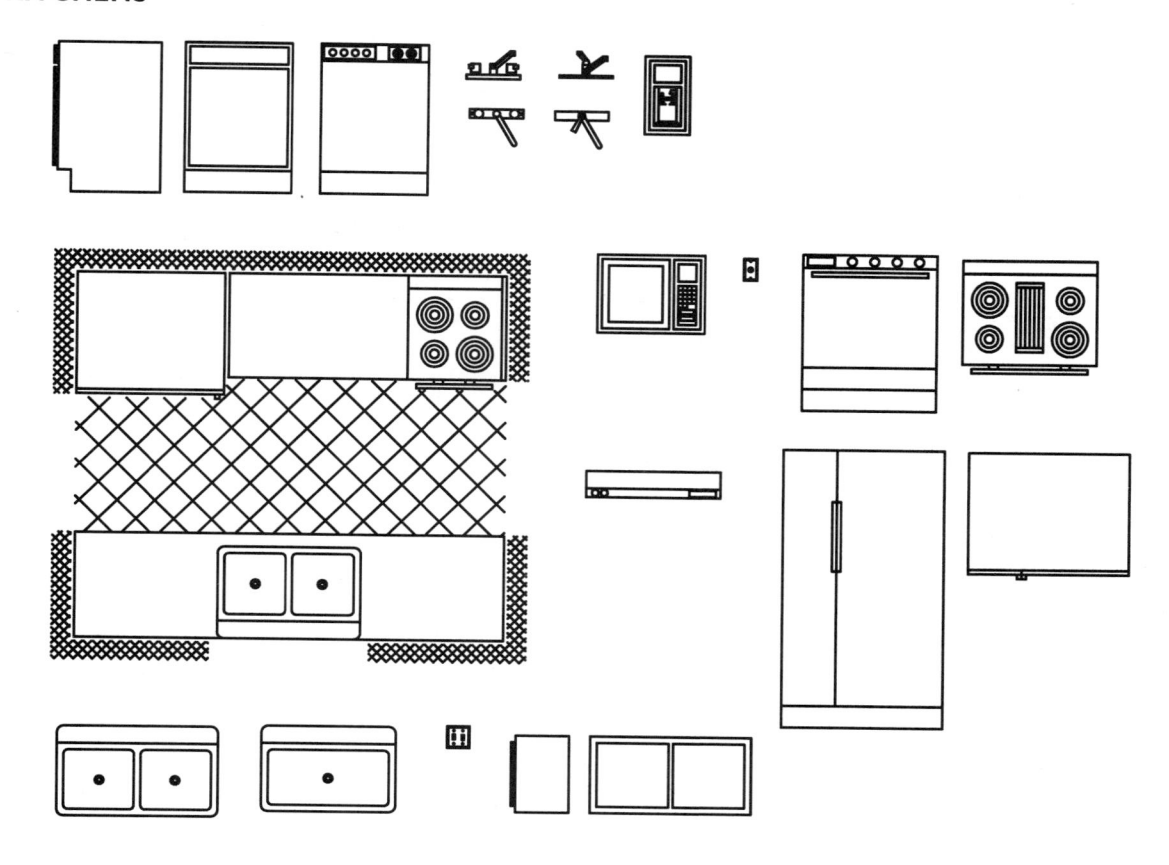

LANDSCAPING

MECHANICAL

FASTENERS - METRIC

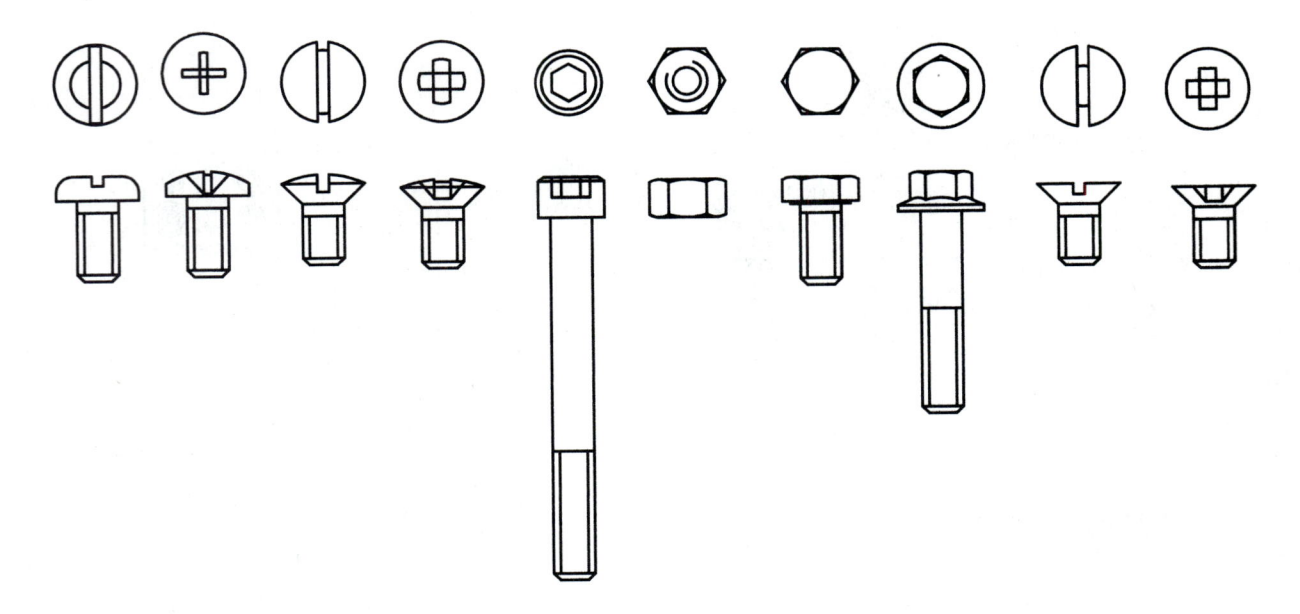

FASTENERS - US

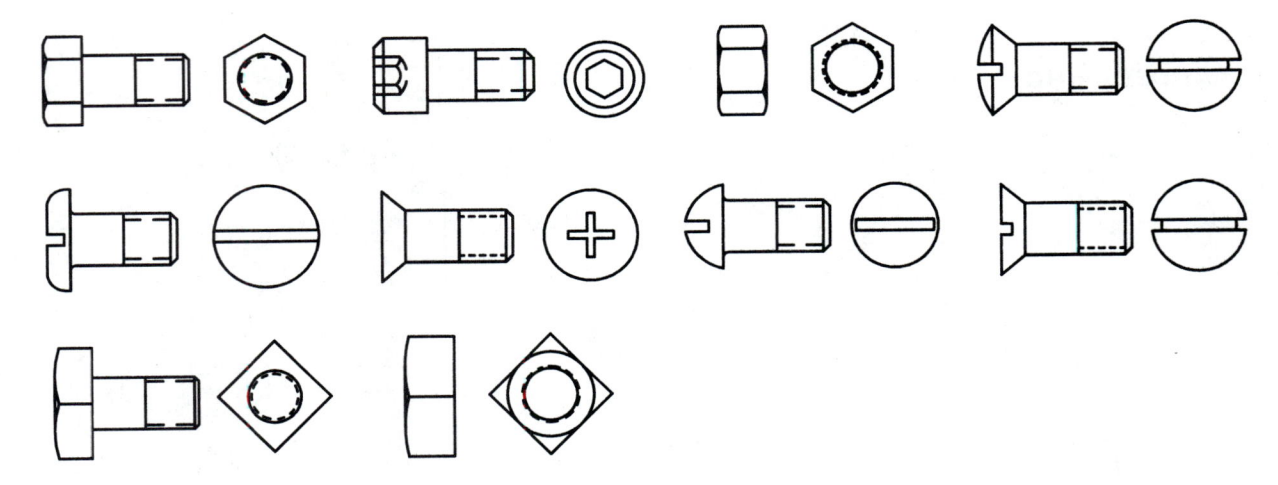

Appendix I: DesignCenter Symbols

HVAC - HEATING VENTILATION AIR CONDITIONING

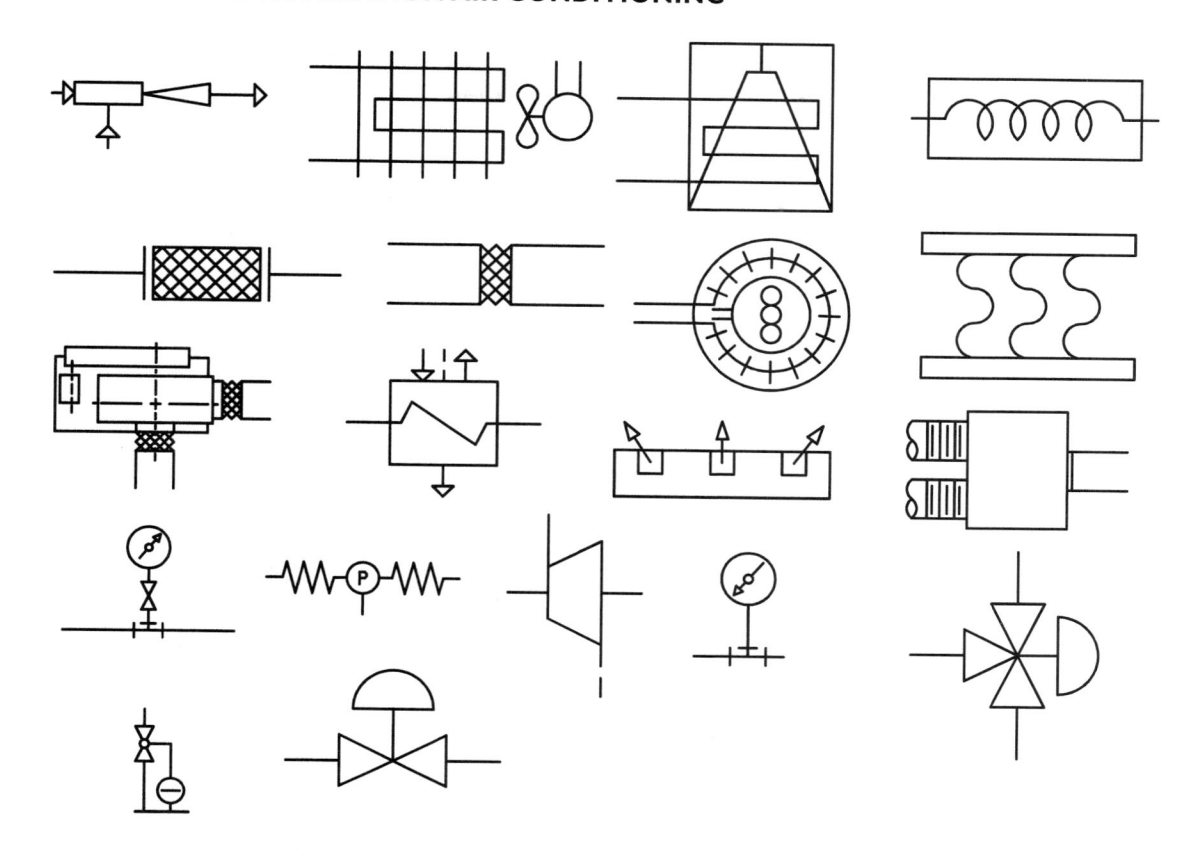

HYDRAULIC - PNEUMATIC

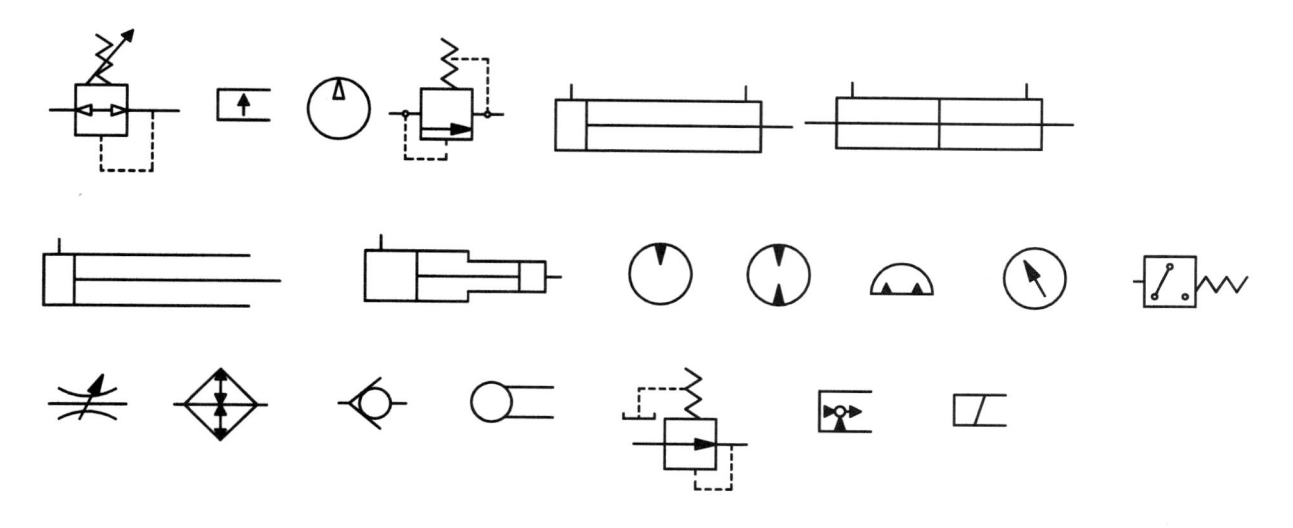

PIPE FITTINGS

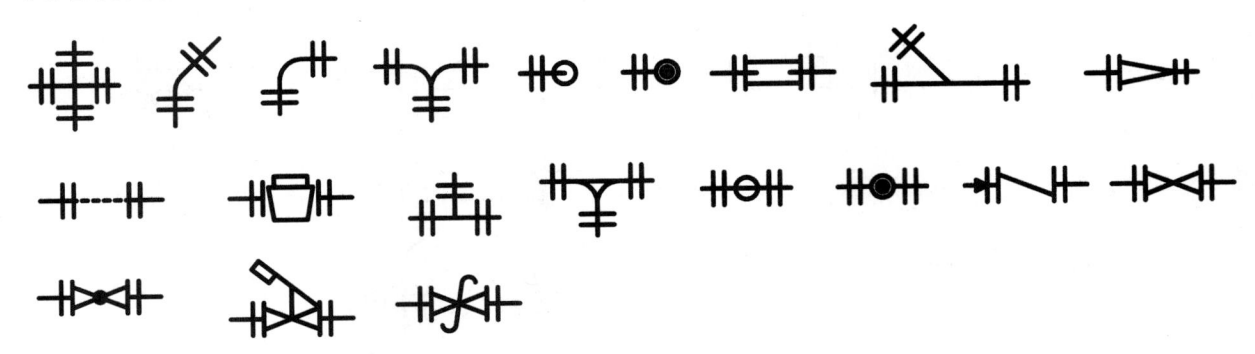

PLANT PROCESS

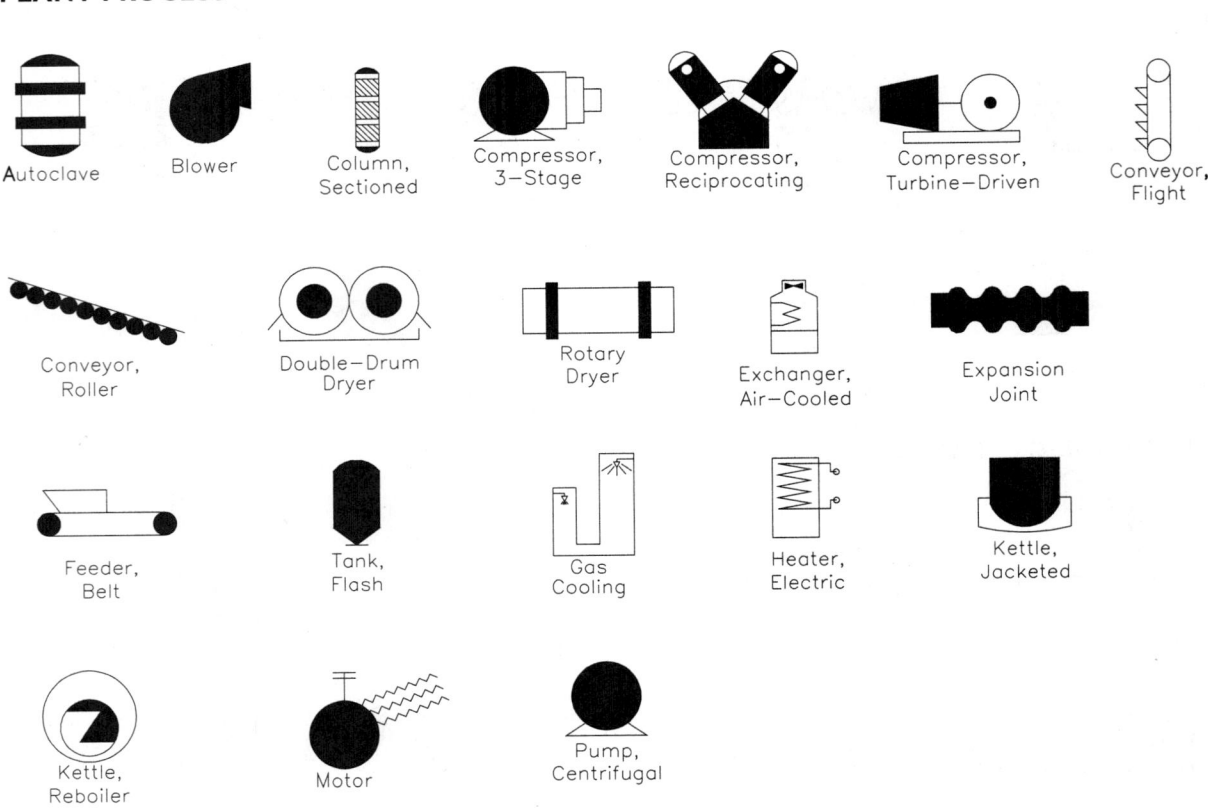

Autoclave | Blower | Column, Sectioned | Compressor, 3-Stage | Compressor, Reciprocating | Compressor, Turbine-Driven | Conveyor, Flight

Conveyor, Roller | Double-Drum Dryer | Rotary Dryer | Exchanger, Air-Cooled | Expansion Joint

Feeder, Belt | Tank, Flash | Gas Cooling | Heater, Electric | Kettle, Jacketed

Kettle, Reboiler | Motor | Pump, Centrifugal

WELDING

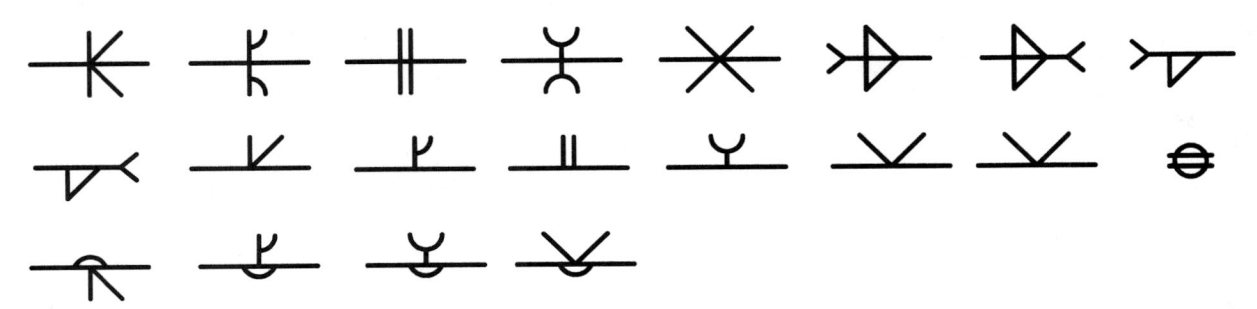

Appendix I: DesignCenter Symbols **875**

ELECTRONICS

ANALOG INTEGRATED CIRCUITS

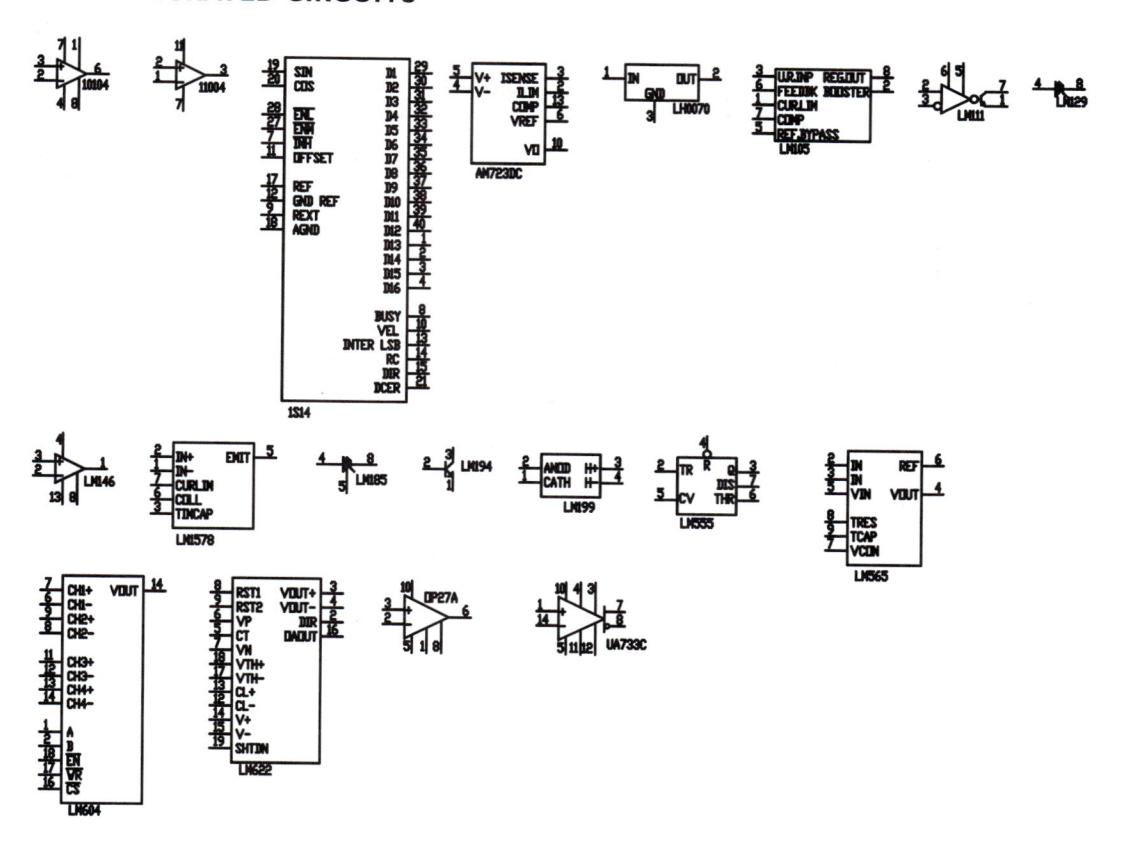

BASIC ELECTRONICS

CMOS INTEGRATED CIRCUITS

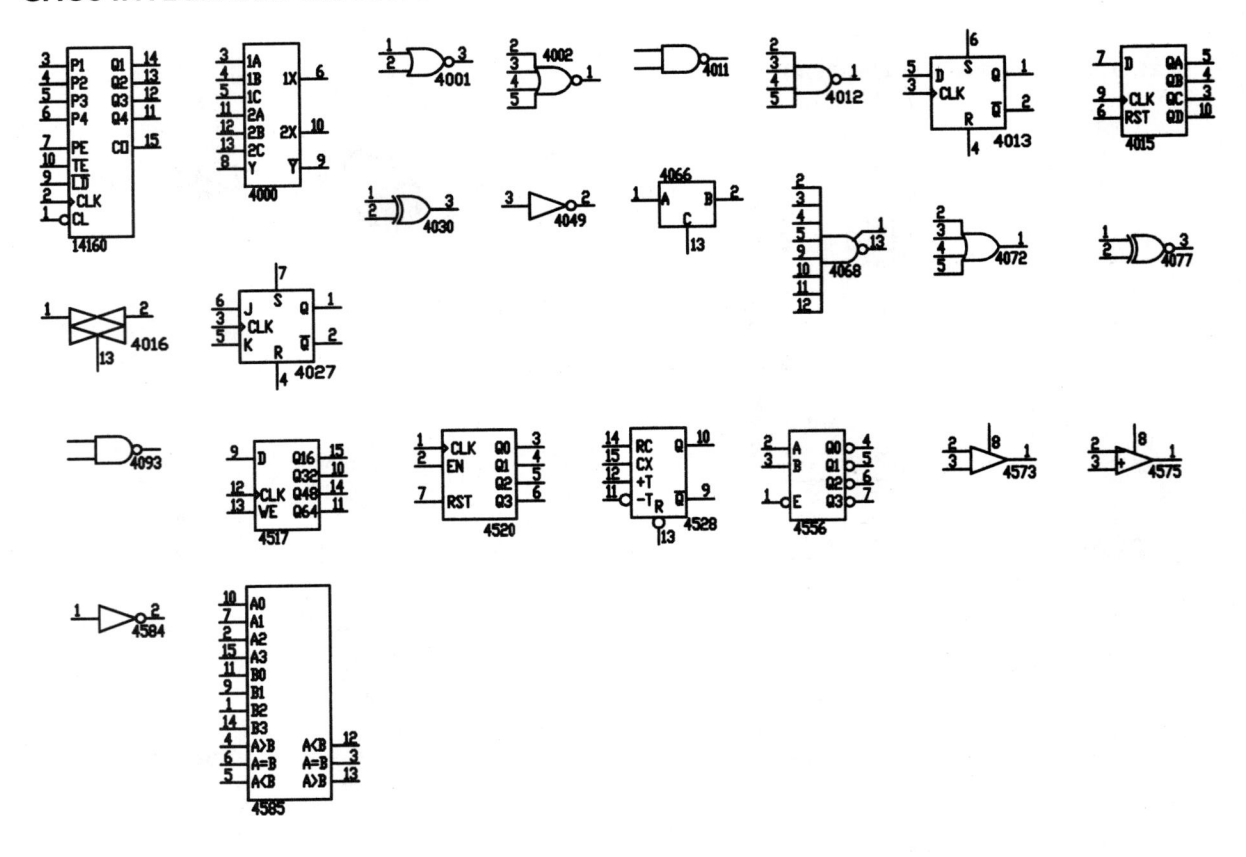

ELECTRICAL POWER

APPENDIX J

AutoCAD Template Drawings

The template drawings shown on the following pages are provided by Autodesk with AutoCAD 2004 in three formats:

- Drawings *.dwg* files.
- Color-dependent plot style *.dwt* templates.
- Named plot style *.dwg* templates.

GENERIC TEMPLATES
- Acad
- AcadIso
- Architectural
- Generic 24in x 36in

AMERICAN NATIONAL STANDARDS INSTITUTE TEMPLATES
- ANSI A (portrait)
- ANSI A
- ANSI B
- ANSI C
- ANSI D
- ANSI E

CHINESE STANDARD TEMPLATES
- Gb_a0
- Gb_a1
- Gb_a2
- Gb_a3
- Gb_a4
- Gb

GERMAN INDUSTRIAL STANDARDS TEMPLATES
- DIN A0
- DIN A1
- DIN A2
- DIN A3
- DIN A4

INTERNATIONAL ORGANIZATION OF STANDARDS TEMPLATES
- ISO A0
- ISO A1
- ISO A2
- ISO A3
- ISO A4 (portrait)

JAPANESE INDUSTRIAL STANDARDS TEMPLATES
- JIS A0
- JIS A1
- JIS A2
- JIS A3
- JIS A4 (landscape)
- JIS A4 (portrait)

GENERIC TEMPLATES

ARCHITECTURAL

GENERIC 24IN X 36IN

Appendix J: AutoCAD Template Drawings **879**

AMERICAN NATIONAL STANDARDS INSTITUTE TEMPLATES

ANSI A (PORTRAIT)

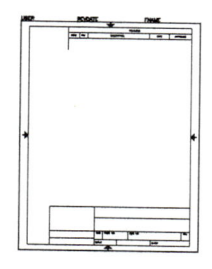

ANSI A

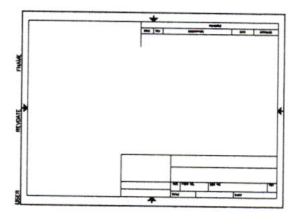

ANSI B

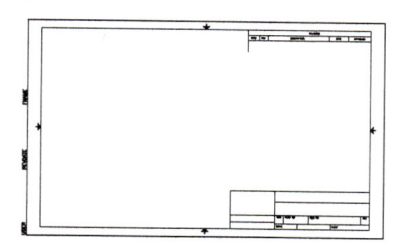

ANSI C

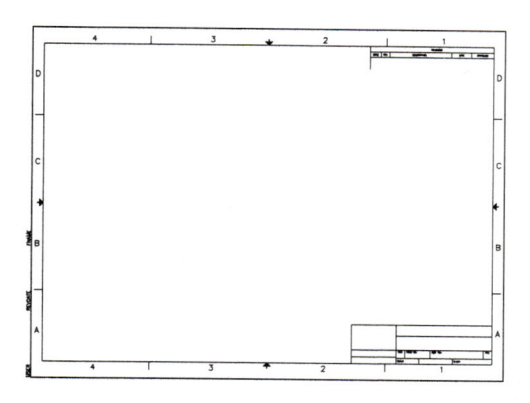

ANSI D

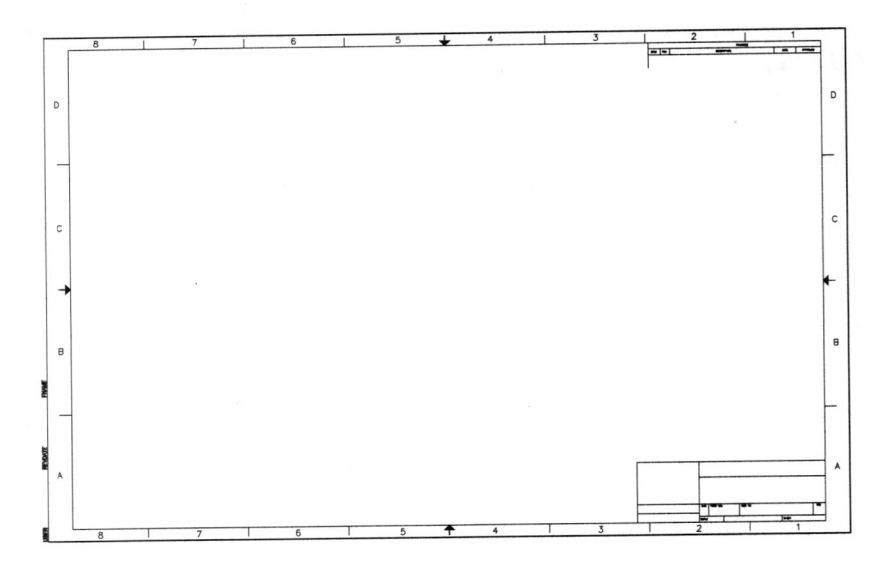

ANSI E

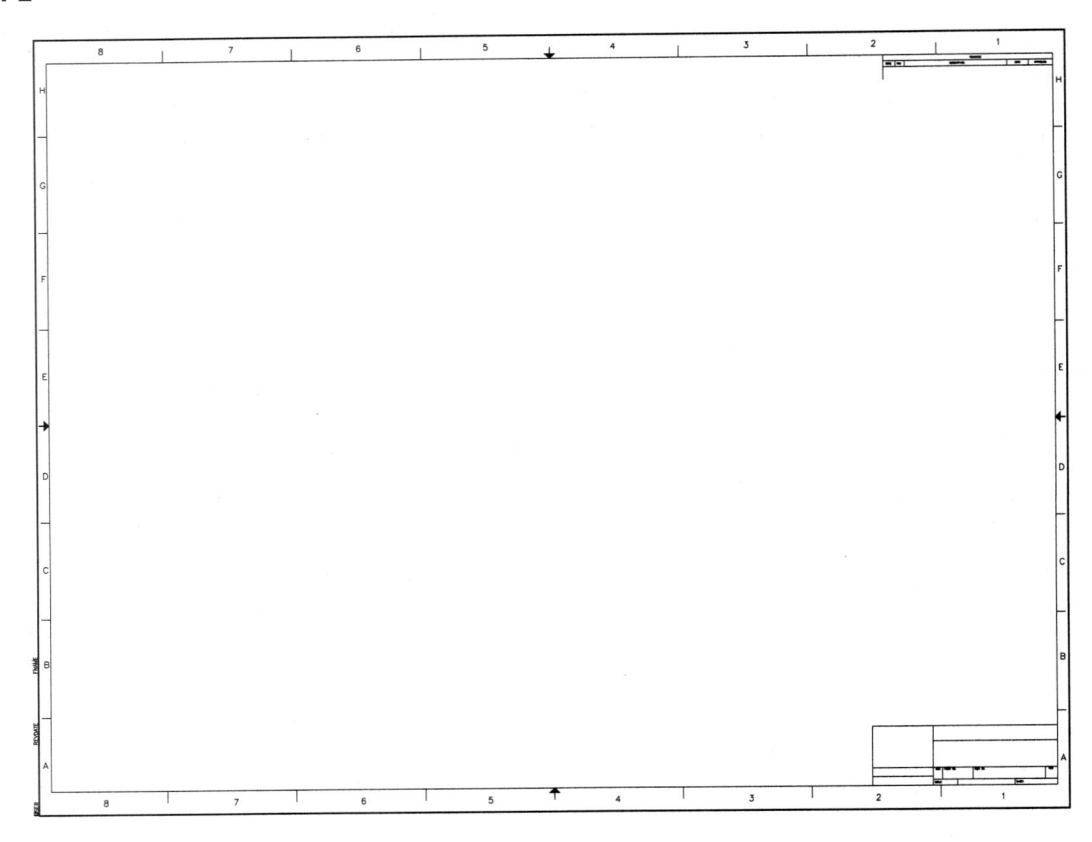

CHINESE STANDARD TEMPLATES

GB_A0

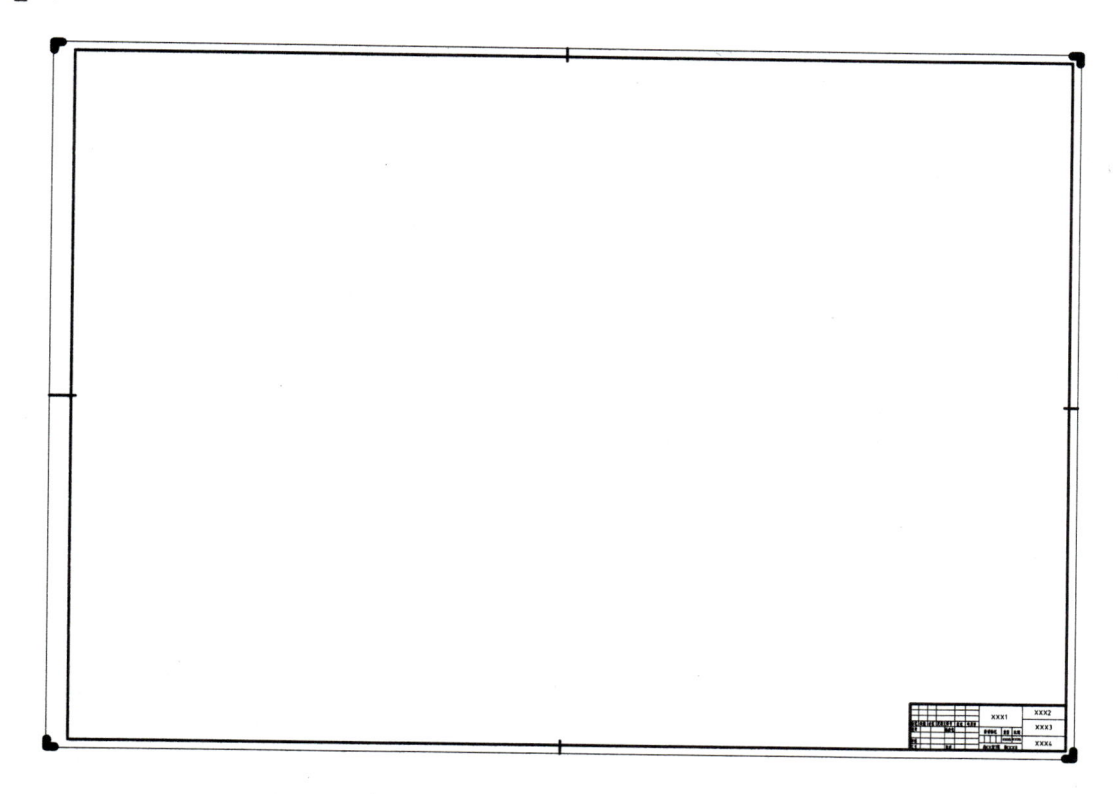

GB_A1

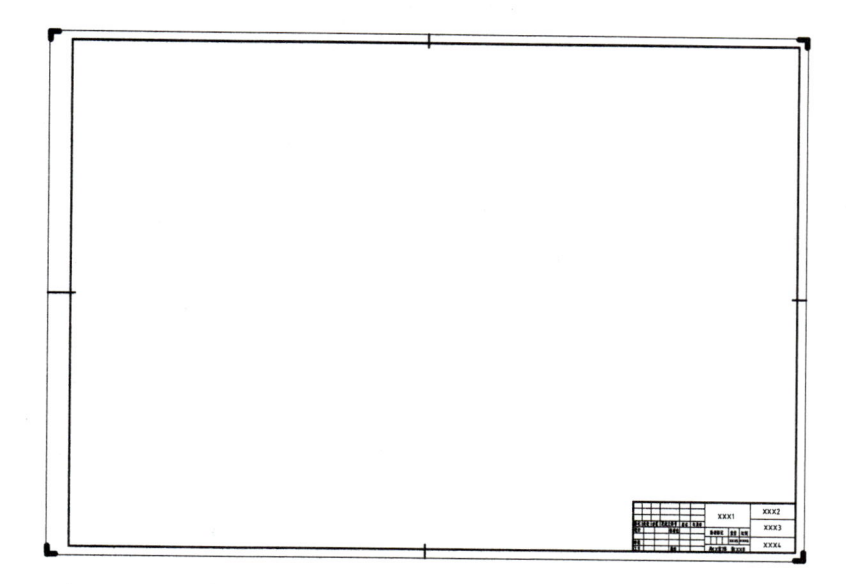

GB_A2

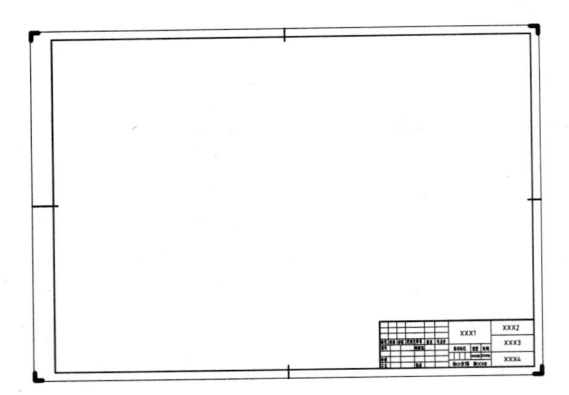

GB_A3

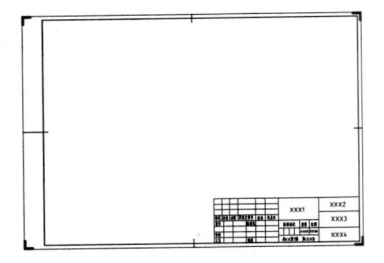

GB_A4

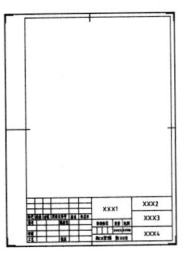

GB

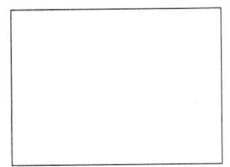

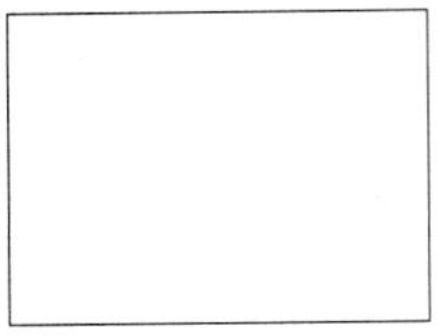

GERMAN INDUSTRIAL STANDARDS TEMPLATES

DIN A0

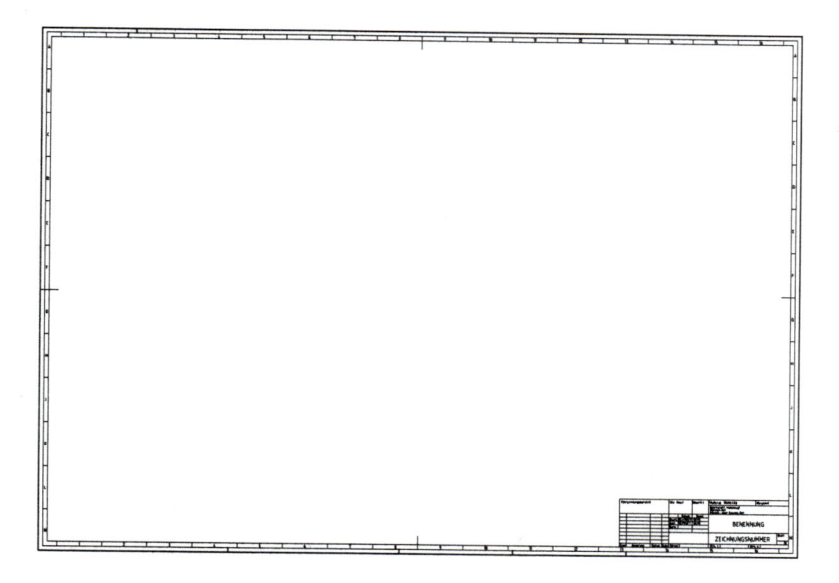

DIN A1

DIN A2

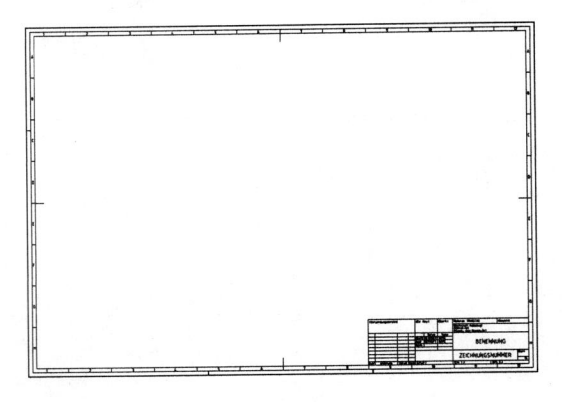

DIN A3

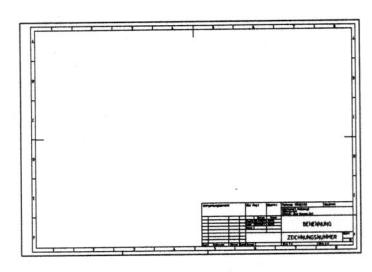

DIN A4

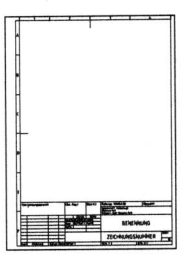

INTERNATIONAL ORGANIZATION OF STANDARDS TEMPLATES

ISO A0

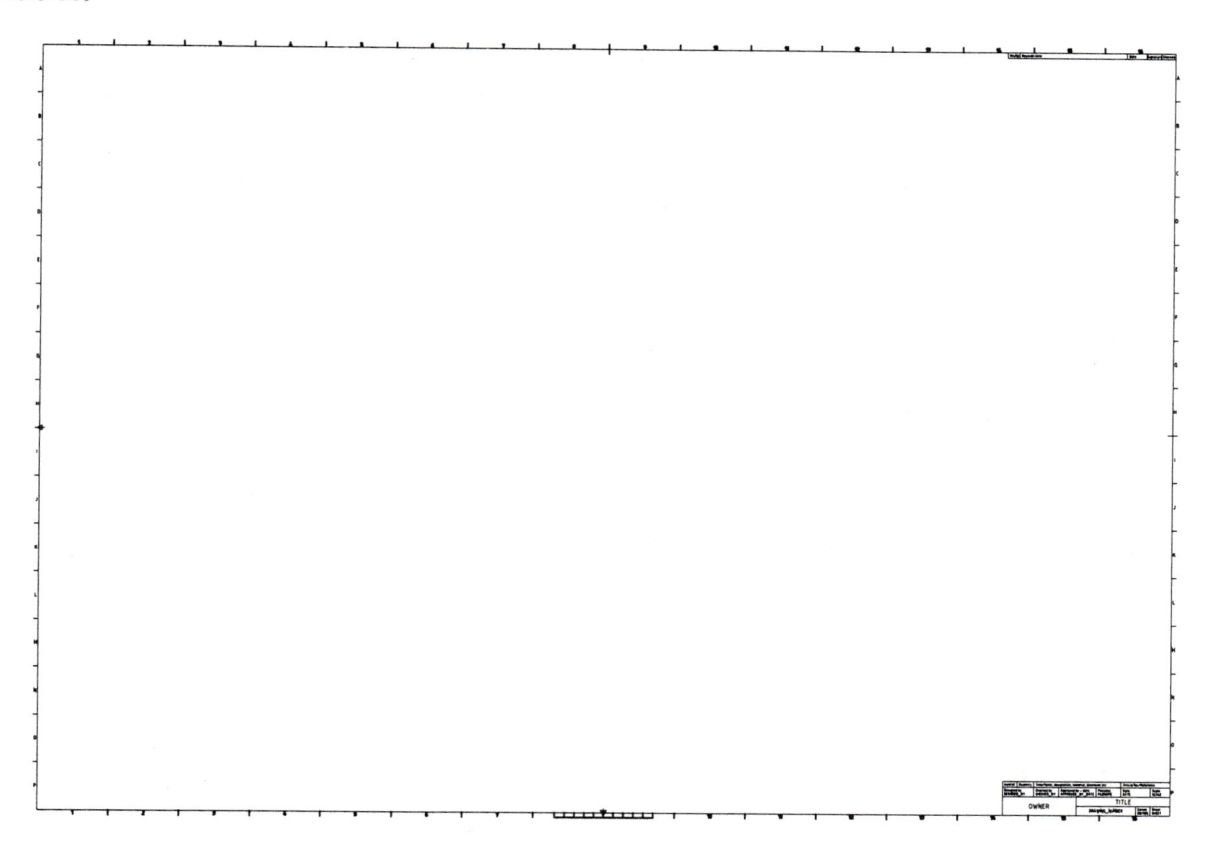

ISO A1

ISO A2

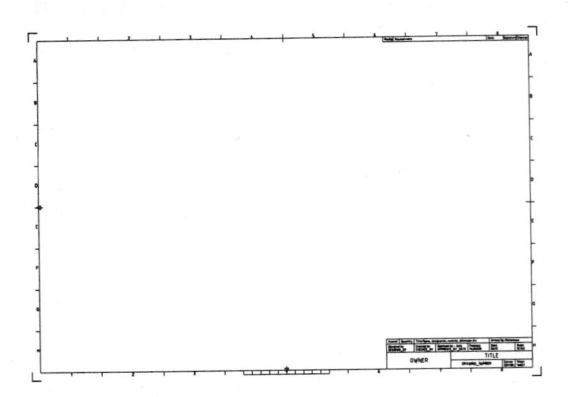

ISO A3

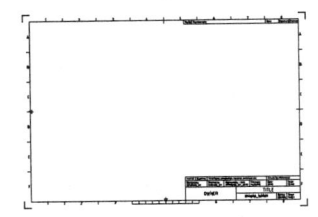

ISO A4 (PORTRAIT)

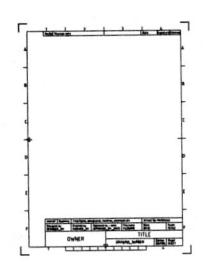

JAPANESE INDUSTRIAL STANDARDS TEMPLATES

JIS A0

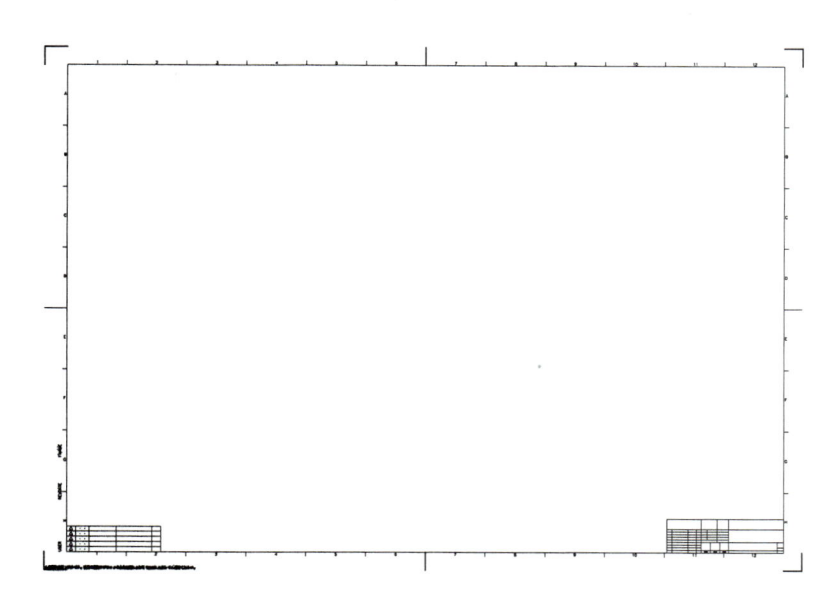

JIS A1

JIS A2

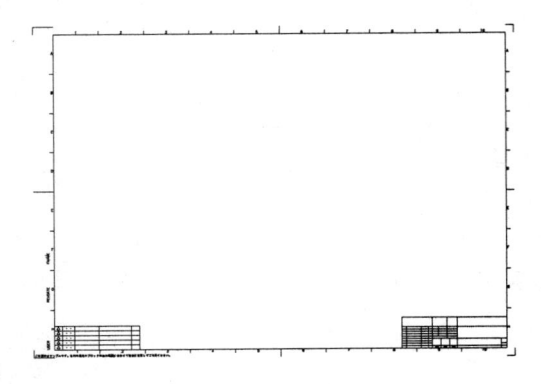

JIS A3

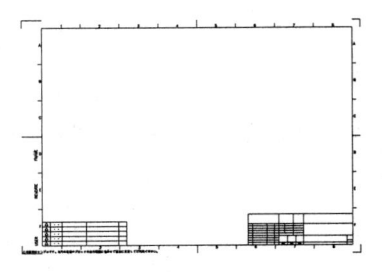

JIS A4 (LANDSCAPE)

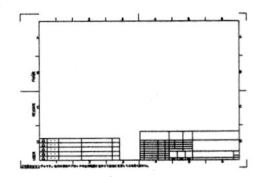

JIS A4 (PORTRAIT)

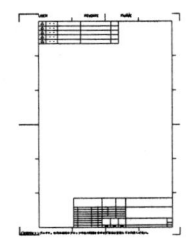

INDEX

A

ADCenter 232
Add Objects to the Selection Set 286
Advanced Wizard 83
Aerial View window 215
Ai_Dim_TextAbove 600
Ai_Dim_TextCenter 600
Ai_Dim_TextHome 600
AIDIM and **AI_DIM** 597
AiDimPrec 598
AiDimTextMove 599
Align 488
Aligned dimensions 552
ALL (Select All Objects) 281
Alternate
    Commands 50
    Keys 804
    Text 608
    Tolerance 608
    Units 540, 625
AngBase 451
AngDir 451
Angle, DimOrdinate 642
Angular dimensions 563, 624
Angularity 650
ANSI Hatch Patterns 861
APParent Intersection 188
Arc 135
Area 756
Areas and perimeters 756
Array 407
-Array 411
Arrowhead 510, 539, 605, 618
ASCII Characters 491
AU (Automatic Option) 284
Audit 363
AutoCAD
    Commands 789
    Colors 314
    Ccoordinates 59
    Program Files 785
    User Interface 34
AutoCAPS 499
Automatic backups 107

AutoSpool 710

B

Backup Files 109
Base 253
Base point 383
Baseline dimensions 553
BatchPlt 742
Beginning a Drawing Session 34
BHatch 263
Blipmode 754
Block 241
-Block 245
Block
    DesignCenter and ToolPalettes 250
    Explode 249
    Measure and Divide 406
    Rotation Angle 249
    Scale 248
BlockIcon 255
Boundary 276
Boundary Set 273
BOX (Select with a Box) 283
Break 438
    First 439
Buttons in Dialog Boxes 51

C

Canceling the Selection Process 286
Cartesian Coordinates 61
CENter 188
Center marks 561, 606, 620
Central Processing Unit 774
Chamfer 413
    Angle 419
    Dimensioning 544
    Method 420
    mUltiple 420
    Points 418
    Polyline 419
    Trim 420
Change 453
Change
    Attribute Definitions 455
    Blocks 455

Case 499
Circles 455
Current Layer 338
Dimension display precision 598
Layers' Status 338
Lines 454
Screen color 96
Selected Items 285
Text 455
Check Boxes 51
Checking spelling 520
Circle 131
Diameter 132
2P 133
3P 133
Tan, Tan, Tan 134
Ttr (tan tan radius) 134
Circular Runout 651
Circularity 651
CleanScreenOn 214
Clearing Local Overrides 630
Close 364
Drawing and Remain in AutoCAD 12
Color 36, 313
Color Books 316
Format 606
Lineweight 605
Combine Paragraphs 499
Command
Aliases 799
Line Keystrokes 803
Prompt Area 4, 39
Compile 526
Components of Computers 774
Concentricity 651
Cones and Pyramids Dimensioning 542
Continuing dimensions 554
Control Codes 491
Control Keys 43, 802
ConvertCtb and **ConvertPstyles** 733
Coordinate systems 59
Copy 383, 397
Displacement 398
Multiple 398
Copying And Moving Files 783
Corrupt drawings 363
CP (Select with a Crossing Polygon) 284
Create
Tool Palette 240
New Dimstyles 616
Crosshair Cursor 4
Crossing Window 282
Custom
Content 235
Properties 731
Cylindrical Coordinates 63

Cylindricity 651

D

Date Modified 93
DbList 759
DC Center 236
DdEdit 512
DdPModes 151
DdPType 406
DdSelect 290
Default Plot Styles 734
Deleting Layers 336
Description 240
DesignCenter 232
Symbols 870
Dialog Boxes 48
Differences Between Local and Network Printers 725
Digitizers 778
DimAligned 552
DimAngular 562
DimBaseline 553
DimCenter 561
DimContinue 553, 642
DimDiameter 559
DimEdit 594
Dimension Line 538, 604, 617, 537
Angle 548
Extension 547
Modes 545
MText 547
Objects 541
Standards 545
Styles 38, 614
Text 539
Text 548
Variables and Styles 545, 601
DimHorizontal 546
DimLinear 550
DimOrdinate 639
DimRadius 559
DimRotated 549
DimStyle 614
DimTEdit 594
DimVertical 549
Direct Distance Entry 68, 387
Direct selection and Grips Editing 286
Discard the Drawing, and Exit AutoCAD 11
Disk
Care 776
Drives 775
Displaying
Coordinates 70
Files 780
Dist 754
Divide 404

Divide, Block   406
Donut 141
DOS Session   781
Drawing
   Area   4, 38
   Circles   132
   Ellipses   147
   Files   785
   Other objects   152
   Points   150
   Polylines   143
Drawing sets   717
DrawOrder   297
Drives and files   779
Drop Lists   52
DSettings   174
DsViewer   215
Duplex Printing   728
DwgProps   761

E

Editing
   dimension text   597
   dimensioned objects   585
   ordinate dimensions   647
   plotter configurations 726
   text   512
Ellipse 145
   Arc   148
   Center   147
   Rotation   148
ENDpoint   188
ePlots   58
Equilateral triangles   152
Erase   437
eXit   383
Exiting Drawings   11
Explode   456
Extend   443

F

F (Select with a Fence)   283
Favorites   237
File
   Dialog Boxes   781
   Extensions   785
   Manager   780
   Names And Extensions   780
   Spec   88
Files List 88
Files of Type   104
Fillet   413
Filter   292
Find   521
Find and Replace   498
Finding Files   93

First Line Indent   496
Fit   488, 622
Flatness   651
Floating DesignCenter   233
Flyouts   9
FontAlt   519
Fonts   855
Form Symbols   651
Freezing and Thawing Layers   332
From   192
FTP   103
Full Preview   708
Function Keys   43, 803

G

G (Select the Group)   285
Geometric
   Characteristic Symbols   650
   Dimensioning and Tolerances (GD&T)   649
Global and Object Linetype Scale   328
Gradient
   Fills 268
   Patterns   865
Graphics 729
Grid   173
Grips   378
   Color and Size   287, 380
   Editing Modes   286, 382, 585
   Editing options   381
Group   294

H

Hardware of an AutoCAD System   773
Hatch Boundary   269
HatchEdit   263
Height   489
Hide Paperspace Objects   714
History   236
Holes Dimensioning   543
HSL   316
Hyphen Prefix   48

I

ID   753
Image Tiles   53
Import-Export Files   786
Import Text   500
Importing Older Configuration Files   726
Indents and Tabs   497
Initialization Strings   731
Inkjet Plotters   776
Input Devices   777
Insert   241
-Insert   252
Inserting
   Blocks   246

Blocks with DesignCenter 250
Blocks with Toolpalettes 251
Exploded Blocks 252
INSertion 189
International Standards 602
INTersection 189
ISO Hatch Patterns 862
Isosceles triangles 153

J

Jointype 465
Justification 498, 607
Justify 487
JustifyText 525
Keyboard 41
    Coordinates 67
    Shortcuts 42, 802

L

L (Select the Last Object) 285
Laser Printers 777
Layer 35, 330
Layout 681
Layout
    Tabs 4, 39
    Plot 705
    Layouts 663
LayoutWizard 676
Leader 511
    Line 509
Least Material Condition (LMC) 652
Lengthen 444
Limits 175, 540, 608
Line 119
Linear dimensions 550, 623
Lines
    and Arrows 617
    and Polyline Segments Dimensioning 541
Linetype 36, 324
Linetypes 859
Lineweight 36, 321, 867
LISP and ObjectARX Programming Files 786
List 759
List Boxes 52
Local and Network Printer Connections 724
Location Symbols 651
Locking and Unlocking Layers 332
Look In 93
LWeight 319

M

M (Select Through the Multiple Option) 285
Manipulating
    Files 782
    Toolbar 10
Marking Points 754

MassProp 758
MatchProp 344
Material Characteristics Symbols 652
Maximum Material Condition (MMC) 652
Measure 404
Measure, Block 406
Measurement Scale 624
Media 728
    Destination 729
    Size 706
Memory 775
Menu Bar 4
Menus 47, 846
Metric 80
MIDpoint 190
MInsert 254
Mirror 399
Mirror Mode 385
MirrText 400
Mode Indicators 40
Monitors 777
Mouse 5, 778
    Buttons 44
    Digitizer Buttons 804
Move 385, 448
Moving dimension text 599
MRedo 356
MText 492
    Control Codes 501, 806
    DimOrdinate 641
    Editor shortcut keys 805
    Shortcut Menu 498
    Style 504
-**MText** 504
MTextEd 504, 513
Multiple 150
MvSetup 94
My Network Places 102

N

NEArest 190
New 79
NODe 190
NONe 192
Nontraditional Printers 725
Notes on Designing Blocks 244
Noun/Verb Selection 291

O

Object Grouping 292
    Snaps 69
    Tracking 69
Obliquing extension lines 595
OFF 193
Offset 401
    Through 403

OffsetDist 403
OffsetGapType 403
Oops 437
Open 86
Options 105
Ordinate
    base point 643
    dimensions 640
Orientation 706
    Symbols 650
Ortho 180
OSnap 186
-OSnap 195
OTrack 196

P

P (Select the Previous Selection Set) 285
PageSetup 701
Pan 212
-Pan 213
Paper Mode 668
Paragraph Indent 496
PARallel 190
Parallelism 651
Partial Preview 708
Parts of dimensions 537
Pattern
    Alignment 274
    Name 266
PEdit 459
Pen Plotters 777
Peripheral Hardware 776
PERpendicular 191
Perpendicularity 651
Physical Pen Configuration 729
Pick (Select Single Object) 280
Pickbox 289
    Size 291
Placement 625
PLine 142
Plot 703
Plot Styles 37, 735
Plot to File 710
PlotStamp 714
Plotter
    Configuration Editor Options 728
    Configurations 720
    Printer 705
PlotterManager 719
Plotters 776
Plotting
    drawings 704
    support Files 786
Point 150
    coordinates 753
    Filters 67

Polar 181
    Coordinates 62
    Semicircular arrays 410
    Snap and Ortho Mode 68
PolarAng 184
PolarDist 184
PolarMode 184
Polygon 127
Polygonal viewports 684
Predefined views 216
Preparing for plotting 701
Press and Drag 291
Preview 702
Primary
    Tolerance 608
    Units 623
Printers 724
Profile
    of a Line 650
    of a Surface 650
    Symbols 650
Projected Tolerance Zone 652
Properties 340, 499, 760
PsLtScale 687
Publish 717
Purge 358
-Purge 361

Q

QDim 555
QLeader 505
QNew 81
QSave 107
QSelect 288
QText 523
QUAdrant 191
QUIck 193
    continuous dimensions 556
Quit 110
Quit 364

R

R (Remove Objects from the Selection Set) 286
Radial dimensions 560
Radio Buttons 52
RAY 197
Recover 362
    Erased objects 437
Rectang 122
Rectangular
    arrays 408
    viewports 682
Redefining Blocks 252
Redlining 420
Redo 356
Redraw 205

Reference 383
Regardless of Feature Size (RFS) 652
Regen 205
RegenAuto 205
Regions 758
Rejustifying text 526
Relative
    Coordinates 65
    Coordinates and Angles 122
Remain in AutoCAD, and Start a New Drawing 12
Remove
    Formatting 499
    Islands 271
    Temporary Files 787
Rename 357
-Rename 358
Renaming
    and deleting dimstyles 615
    and deleting named views 218
Repeating Commands 44
Replacing Blocks with Drawings 252
Repositioning dimension text 596
Resizing text 525
RevCloud 420
RGB 315
Right Mouse Button 45
Right triangles 1572
RMLin 423
Rotate 449
    Mode 385
Running and temporary osnaps 187
Runout Symbols 651

S

SpaceTrans 689
Save and **SaveAs** 101
    Exit AutoCAD 12
    Remain in AutoCAD 12
SaveFilePath 108
Scale 451
    Dimension Features 623
    Mode 385
    Reference 452
Scalene triangles 156
ScaleText 525
Scanner 778
Screen
    Picks 66
    Pointing 41
Scroll Bars 4, 52, 212
Search 237
Select 280
    All 499
    Objects During Commands 286
Set Mtext Width 497
Setting scale factors 99

SetVar 766
Shaded Viewport 712
Shortcut menus 371
Show Details 334
Showing Points by Window Corners 41
SI (Select a Single Object) 285
Single-line Text Editor 512
Single Points Dimensioning 543
Size and Fit 605
Snap 67, 176
    Spacing 68
Special Selection Modes 285
Spell 520
Spherical Coordinates 64
SplFrame 465
SplineSegs 467
SplineType 466
Stack 499
Stamping plots 714
Standard Folders Sidebar 90
Start From Scratch 80
Start, Center, End & Center, Start, End 137
Starting AutoCAD 2
    with a named view 219
Startup Dialog box 79
Status 764
Status Bar 4, 40
Storing and recalling named views 217
Straightness 651
Stretch 383, 447
    The Rules 447
Style 489, 515
    Toolbar 518
-Style 519
StylesManager 732
Summary of object snap modes 187
Support Files 785
Suppression and Position 605
Symbol 500
    Usage 652
Symmetry 651
System
    Board 774
    Printers 725
System Variables 49, 766, 807

T

Tablet Menus 46
Tabs 496
TANgent 191
Template
    and Blocks 55
    Drawings 878
Text 484, 606
    DimOrdinate 641
    Edit Boxes 53

Index

Formatting Toolbar 494
Placement 623
Rotation Angle 490
Styles 37, 515
-Text 492
TextFill 519
TextQlty 519
The Mouse Wheel 46
Through 403
Tilemode 663
Time 763
Time-Sensitive Right-Click 377
Title Bar 35, 240
Tolerance 649
Tolerance 540, 626
Diameter 651
Format 626
Symbols 651
Text 608
Toolbar 4, 35, 46, 91, 236, 841
ToolPalettes 256
Total Runout 651
Tracking 68, 68, 192
Transparent
Panning 212
Redraws 205
Zoom 211
Tray Settings 40
Tree View 240
Trim 440
True Position 651
TT 192
Types 93

U

U 285, 353, 437
UCS
Icon 4, 38
origin and 3point 643
Undo 353, 383
Units 97
Scale, and Precision 607
User-defined
Coordinate Systems 66
Paper Sizes and Calibration 731
Patterns 267
Using
Dialog Boxes 50
Menus 5, 47
Toolbar Buttons 8

V

Vertices Dimensioning 543
View 216
Viewing Files 781
Viewport Border 670

Viewports 682
Viewports from objects 685
ViewRes 220
Volo View 423
VpClip 684
-VPorts 684

W

W (Select Within a Window) 281
WBlock 253
WBlock 361
Wedges and Cylinders Dimensioning 542
Wild-Card Characters 780
Windows Explorer 780
WipeOut 522
Wizards 82
Working with Blocks and Xrefs 239
WP (Select with a Windowed Polygon) 284
Write-Protecting Data 776

X

X, Y, Z Coordinates 40
Xdatum 641
XLINE 197
Xplode 458

Y

Ydatum 641

Z

Zero Suppression 624
ZOOM 206

3

3DArray 412

LICENSE AGREEMENT FOR AUTODESK PRESS

A Thomson Learning Company

Educational Software/Data

You the customer, and Autodesk Press incur certain benefits, rights, and obligations to each other when you open this package and use the software/data it contains. BE SURE YOU READ THE LICENSE AGREEMENT CAREFULLY, SINCE BY USING THE SOFTWARE/DATA YOU INDICATE YOU HAVE READ, UNDERSTOOD, AND ACCEPTED THE TERMS OF THIS AGREEMENT.

Your rights:

1. You enjoy a non-exclusive license to use the enclosed software/data on a single microcomputer that is not part of a network or multi-machine system in consideration for payment of the required license fee, (which may be included in the purchase price of an accompanying print component), or receipt of this software/data, and your acceptance of the terms and conditions of this agreement.

2. You own the media on which the software/data is recorded, but you acknowledge that you do not own the software/data recorded on them. You also acknowledge that the software/data is furnished "as is," and contains copyrighted and/or proprietary and confidential information of Autodesk Press or its licensors.

3. If you do not accept the terms of this license agreement you may return the media within 30 days. However, you may not use the software during this period.

There are limitations on your rights:

1. You may not copy or print the software/data for any reason whatsoever, except to install it on a hard drive on a single microcomputer and to make one archival copy, unless copying or printing is expressly permitted in writing or statements recorded on the diskette(s).

2. You may not revise, translate, convert, disassemble or otherwise reverse engineer the software/data except that you may add to or rearrange any data recorded on the media as part of the normal use of the software/data. 3. You may not sell, license, lease, rent, loan, or otherwise distribute or network the software/data except that you may give the software/data to a student or and instructor for use at school or, temporarily at home. Should you fail to abide by the Copyright Law of the United States as it applies to this software/data your license to use it will become invalid. You agree to erase or otherwise destroy the software/data immediately after receiving note of Autodesk Press' termination of this agreement for violation of its provisions. Autodesk Press gives you a LIMITED WARRANTY covering the enclosed software/data. The LIMITED WARRANTY can be found in this product and/or the instructor's manual that accompanies it. This license is the entire agreement between you and Autodesk Press interpreted and enforced under New York law.

Limited Warranty

Autodesk Press warrants to the original licensee/ purchaser of this copy of microcomputer software/ data and the media on which it is recorded that the media will be free from defects in material and workmanship for ninety (90) days from the date of original purchase. All implied warranties are limited in duration to this ninety (90) day period. THEREAFTER, ANY IMPLIED WARRANTIES, INCLUDING IMPLIED WARRANTIES OF MERCHANTABILITY AND FITNESS FOR A PARTICULAR PURPOSE ARE EXCLUDED. THIS WARRANTY IS IN LIEU OF ALL OTHER WARRANTIES, WHETHER ORAL OR WRITTEN, EXPRESSED OR IMPLIED.

If you believe the media is defective, please return it during the ninety day period to the address shown below. A defective diskette will be replaced without charge provided that it has not been subjected to misuse or damage. This warranty does not extend to the software or information recorded on the media. The software and information are provided "AS IS." Any statements made about the utility of the software or information are not to be considered as express or implied warranties. Delmar will not be liable for incidental or consequential damages of any kind incurred by you, the consumer, or any other user. Some states do not allow the exclusion or limitation of incidental or consequential damages, or limitations on the duration of implied warranties, so the above limitation or exclusion may not apply to you. This warranty gives you specific legal rights, and you may also have other rights which vary from state to state. Address all correspondence to:

AutodeskPress
Executive Woods
5 Maxwell Drive
Clifton Park, NY 12065-2919